TABLE OF RELATIVE ATOMIC MASSES 1987[a]

Scaled to the relative atomic mass, $A_r(^{12}C) = 12$

Values given here apply to elements as they exist naturally on Earth. The uncertainties in the values are indicated by the figures given in parentheses, which is applicable to the last digit.

Atomic Number	Name	Symbol	Relative Atomic Mass	Atomic Number	Name	Symbol	Relative Atomic Mass
1	Hydrogen	H	1.007 94(7)	55	Cesium	Cs	132.905 43(5)
2	Helium	He	4.002 602(2)	56	Barium	Ba	137.327(7)
3	Lithium	Li	6.941(2)	57	Lanthanum	La	138.905 5(2)
4	Beryllium	Be	9.012 182(3)	58	Cerium	Ce	140.115(4)
5	Boron	B	10.811(5)	59	Praseodymium	Pr	140.907 65(3)
6	Carbon	C	12.011(1)	60	Neodymium	Nd	144.24(3)
7	Nitrogen	N	14.006 74(7)	61	Promethium	Pm	
8	Oxygen	O	15.999 4(3)	62	Samarium	Sm	150.36(3)
9	Fluorine	F	18.998 403 2(9)	63	Europium	Eu	151.965(9)
10	Neon	Ne	20.179 7(6)	64	Gadolinium	Gd	157.25(3)
11	Sodium	Na	22.989 768(6)	65	Terbium	Tb	158.925 34(3)
12	Magnesium	Mg	24.305 0(6)	66	Dysprosium	Dy	162.50(3)
13	Aluminum	Al	26.981 539(5)	67	Holmium	Ho	164.930 32(3)
14	Silicon	Si	28.085 5(3)	68	Erbium	Er	167.26(3)
15	Phosphorus	P	30.973 762(4)	69	Thulium	Tm	168.934 21(3)
16	Sulfur	S	32.066(6)	70	Ytterbium	Yb	173.04(3)
17	Chlorine	Cl	35.452 7(9)	71	Lutetium	Lu	174.967(1)
18	Argon	Ar	39.948(1)	72	Hafnium	Hf	178.49(2)
19	Potassium	K	39.098 3(1)	73	Tantalum	Ta	180.947 9(1)
20	Calcium	Ca	40.078(4)	74	Tungsten	W	183.85(3)
21	Scandium	Sc	44.955 910(9)	75	Rhenium	Re	186.207(1)
22	Titanium	Ti	47.88(3)	76	Osmium	Os	190.2(1)
23	Vanadium	V	50.941 5(1)	77	Iridium	Ir	192.22(3)
24	Chromium	Cr	51.996 1(6)	78	Platinum	Pt	195.08(3)
25	Manganese	Mn	54.938 05(1)	79	Gold	Au	196.966 54(3)
26	Iron	Fe	55.847(3)	80	Mercury	Hg	200.59(3)
27	Cobalt	Co	58.933 20(1)	81	Thallium	Tl	204.383 3(2)
28	Nickel	Ni	58.69(1)	82	Lead	Pb	207.2(1)
29	Copper	Cu	63.546(3)	83	Bismuth	Bi	208.980 37(3)
30	Zinc	Zn	65.39(2)	84	Polonium	Po	
31	Gallium	Ga	69.723(1)	85	Astatine	At	
32	Germanium	Ge	72.61(2)	86	Radon	Rn	
33	Arsenic	As	74.921 59(2)	87	Francium	Fr	
34	Selenium	Se	78.96(3)	88	Radium	Ra	
35	Bromine	Br	79.904(1)	89	Actinium	Ac	
36	Krypton	Kr	83.80(1)	90	Thorium	Th	232.038 1(1)
37	Rubidium	Rb	85.467 8(3)	91	Protactinium	Pa	
38	Strontium	Sr	87.62(1)	92	Uranium	U	238.028 9(1)
39	Yttrium	Y	88.905 85(2)	93	Neptunium	Np	
40	Zirconium	Zr	91.224(2)	94	Plutonium	Pu	
41	Niobium	Nb	92.906 38(2)	95	Americium	Am	
42	Molybdenum	Mo	95.94(1)	96	Curium	Cm	
43	Technetium	Tc		97	Berkelium	Bk	
44	Ruthenium	Ru	101.07(2)	98	Californium	Cf	
45	Rhodium	Rh	102.905 50(3)	99	Einsteinium	Es	
46	Palladium	Pd	106.42(1)	100	Fermium	Fm	
47	Silver	Ag	107.868 2(2)	101	Mendelevium	Md	
48	Cadmium	Cd	112.411(8)	102	Nobelium	No	
49	Indium	In	114.82(1)	103	Lawrencium	Lr	
50	Tin	Sn	118.710(7)	104	Unnilquadium	Unq	
51	Antimony (Stibium)	Sb	121.75(3)	105	Unnilpentium	Unp	
52	Tellurium	Te	127.60(3)	106	Unnilhexium	Unh	
53	Iodine	I	126.904 47(3)	107	Unnilseptium	Uns	
54	Xenon	Xe	131.29(2)				

[a] IUPAC Commission on Atomic Weights, *J. Phys. Chem. Ref. Data*, **17**:1791 (1988).

PHYSICAL CHEMISTRY

PHYSICAL CHEMISTRY
First Edition

Robert A. Alberty
Professor of Chemistry
Massachusetts Institute of Technology

Robert J. Silbey
Professor of Chemistry
Massachusetts Institute of Technology

John Wiley & Sons, Inc.
New York   Chichester   Brisbane   Toronto   Singapore

ACQUISITIONS EDITOR / Nedah Rose
PRODUCTION MANAGER / Linda Muriello
DESIGNER / Kevin Murphy
PRODUCTION SUPERVISOR / Micheline A. Frederick
MANUFACTURING MANAGER / Lorraine Fumoso
SUPERVISING COPYEDITOR / Elizabeth Swain
COVER PHOTO / Reginald Wickham

Recognizing the importance of preserving what has been written, it is a policy of John Wiley & Sons, Inc. to have books of enduring value published in the United States printed on acid-free paper, and we exert our best efforts to that end.

Library of Congress Cataloging in Publication Data:

Alberty, Robert A.
   Physical chemistry / Robert A. Alberty, Robert J. Silbey.— 1st
ed.
     p.   cm.
   Includes indexes.
   ISBN 0-471-62181-1
   1. Chemistry, Physical and theoretical. I. Silbey, Robert J.
II. Title.
QD453.2.D36 1992   91-13148
541.3—dc20   CIP

Printed in the United States of America

Printed and bound by the Hamilton Printing Company.

10 9 8 7 6 5 4

Preface

The objective of this book is to make the concepts and methods of physical chemistry clear and interesting to students who have had a year of calculus and a year of physics. The underlying theory of chemical phenomena is complicated, and so it is a challenge to make the most important concepts and methods understandable to undergraduate students. However, these basic ideas are accessible to students, and they will find them useful whether they are chemistry majors, biologists, engineers, or earth scientists. The basic theory of chemistry is presented from the viewpoint of academic physical chemists, but many applications of physical chemistry to practical problems are described.

A course in physical chemistry provides many opportunities for the immediate application of ideas and equations in solving problems. In this edition, the number of examples has been increased to 244, which is 34 percent greater than the previous, seventh, edition. In addition, many calculations are carried out in the body of the text. The number of figures has been increased to 356, (30% greater than the previous edition). The number of problems is essentially the same as in the seventh edition, but 26 percent of the old problems have been replaced. There are also two parallel sets of problems at the end of each chapter. The answers for the first set are given in the Appendix. The *Solutions Manual for Physical Chemistry* contains worked-out solutions for all of the problems in the first set and answers for the second set.

As a new feature in this edition, we have discussed one or two Special Topics at the end of each chapter.

Since the number of credits in physical chemistry courses, and therefore the need for more advanced material, varies at different universities and colleges, more topics have been included in this edition than can be covered in some courses.

The book is still divided into four parts, but they have been reorganized since the seventh edition. The book starts with thermodynamics to show clearly the power and limits of this treatment of equilibrium. In general, the treatment of thermodynamics has been upgraded. There is a more systematic treatment of the second law and more emphasis on the entropy. There is also emphasis on the solution of problems in chemical equilibrium, and the material on electrochemical equilibria has been revised extensively to make it clearer. In addition, the thermodynamic tables in the Appendix have been brought up to date. Finally, the difference between extensive thermodynamic quantities and molar thermodynamic quantities has been clarified by using overbars on the molar thermodynamic quantities (see the Note to Students).

Part II, on quantum mechanics and spectroscopy, describes the behavior of atoms and molecules. The level of the treatment of quantum mechanics has been raised again because of its central importance for understanding modern physical chemistry. Linear operators, eigenvectors, and eigenvalues are introduced early and used throughout. Emphasis is placed on the description of molecular properties and spectroscopy using quantum mechanical ideas.

In Part III, Statistical Mechanics and Kinetics, the chapter on Statistical Mechanics has been completely rewritten to make this subject more understandable and usable to students. Theoretical gas kinetics has been combined with photochemistry to form a new chapter, Chemical Dynamics and Photochemistry.

Part IV, Macroscopic and Microscopic Structures, contains two new chapters on Macromolecules and Surface Dynamics, respectively, that provide increased emphasis on these two important subject areas.

Throughout, we have emphasized the use of SI (Système International d'Unités) units, although some other units are used when they are especially convenient or widely used. The advantage of SI is that it is coherent: when SI units are used in a calculation, the result comes out in SI units without any numerical factors. This feature of SI is particularly important in physical chemistry, which uses such a wide range of physical quantities. The underlying unity of science is emphasized by the use of seven base units to represent all physical quantities. The Appendix contains a list of Symbols for Physical Quantities, their SI units, and the numbers of the Sections where they are introduced. The fundamental physical constants of this edition are from the 1986 report of the CODATA Task Group on Fundamental Constants.

History

Outlines of Theoretical Chemistry, as it was then entitled, was written in 1913 by Frederick Getman, who carried it through 1927 in four editions. The next four editions were written by Farrington Daniels. In 1955, Robert Alberty joined Farrington Daniels, the name of the book was changed to *Physical Chemistry*, and the numbering of the editions was started over. Their collaboration ended in 1972 when Farrington Daniels died. After seven editions, the current senior author welcomes Robert J. Silbey as coauthor. Robert J. Silbey has a well-deserved reputation as an excellent teacher, and he brings a broad experience in physical chemistry to this book, which again is a first edition. It is remarkable that this textbook of *Physical Chemistry* traces its origins back 78 years.

Over the years this book has profited tremendously from the advice of physical chemists all over the world. Many physical chemists who care how their subject is presented have written to us with their comments, and we hope that will continue. We are especially indebted to our colleagues at MIT who have reviewed sections of this editions and have given us the benefit of their advice. These include Sylvia T. Ceyer, Carl W. Garland, Irwin Oppenheim, and Mark S. Wrighton. Parts of the book were reviewed by members of IUPAC Commission I.1 on Physiochemical Symbols, Terminology, and Units, including I. Mills, T. Cvitas, N. Kallay, R. Cohen, and B. Holmstrom. Important advice was also provided by A. J. Bard, M. Chase, R. Parsons, P. Rieger, and R. Sauer.

R. D. Allendoerfer (SUNY University at Buffalo), D. W. Beistel (University of Missouri-Rolla), T. M. Dunn (University of Michigan), R. F. Firestone (Ohio State University), N. Ganapathisubramaninian (Wake Forest), D. K. Gosser (The City College of CUNY), R. C. Millikan (University of California-Santa Barbara), M. O'Keeffe (Arizona State University), J. W. Reed (Kent State University), J. H. Reeves (Wilmington, North Carolina), S. L. Seager (Weber

State College), D. F. Tuan (Kent State University), and S. Widmer (Adelphi University) read parts of this edition and made many useful suggestions.

We are indebted to John Wiley for its support, and especially to our editors, Dennis Sawicki and Nedah Rose, for their advice.

We are also indebted to Lillian Alberty and Vera Spanos for the difficult job of typing the manuscript.

Cambridge, Massachusetts   **Robert A. Alberty**
January, 1991   **Robert J. Silbey**

Note to Students

In this book we will discuss over 300 physical quantities. All are listed in Appendix F, with the section numbers where they are defined. The value of a physical quantity is equal to the product of a numerical factor and a unit:

$$\text{physical quantity} = \text{numerical value} \times \text{unit}$$

The values of all physical quantities can be expressed in terms of SI base units (see Appendix A), but some physical quantities are dimensionless, and so the symbol for the SI unit is taken as 1 because this is what you get when units cancel.

Certain combinations of base units are used so frequently that they are given special symbols. For example, the base unit $kg \, m^2 \, s^{-2}$ for energy is represented by J, the joule, named in honor of Joule.

Note that in print, physical quantities are represented by italic type and units are represented by roman type. Physical quantities that are vectors or matrices are represented by boldface italic type.

The SI system of units has the advantage that it is *coherent*. This means that in using an equation that gives a relationship between physical quantities, if all quantities in a calculation are expressed in SI base units, the result will be expressed in SI base units without including any numerical factors. This is very helpful in physical chemistry where physical quantities from quite different areas of science come together in the same equation. However, for reasons of convenience and history, some non-SI units are used in physical chemistry. They can each be expressed in terms of SI units, and a list of them is given in the Appendix A.

At equilibrium, a system has certain extensive properties, including volume (V), internal energy (U), enthalpy (H), entropy (S), Helmholtz energy (A), and Gibbs energy (G). These physical quantities are extensive because their values depend on the size of the system. If the volume of a system is divided by the amount of substance n, we obtain the molar volume $\overline{V} = V/n$. The molar volume is an intensive quantity because its value does not depend on the size of the system (Section 1.1). The molar volume has the SI units $m^3 \, mol^{-1}$. In a similar way we can define intensive quantities $\overline{U}, \overline{H}, \overline{S}, \overline{A}$, and $\overline{G}$, known as the molar internal energy, and so on. The SI unit of a molar quantity always includes mol^{-1}.

A subscript i on a molar quantity or a chemical formula in parenthesis is used to identify the substance, as in $\overline{V}_i$ or $\overline{V}(i)$. These symbols are also used to represent partial molar quantities (Section 4.12). When it is important to emphasize that we are dealing with a pure substance, a superscript * is used as in $\overline{V}_i^*$ for the molar volume of a pure substance or P_i^* for the vapor pressure of a pure substance.

When a system is in its standard state, the thermodynamic property has a superscript degree sign, as in $H°$, $S°$, or $G°$. The molar property of a substance in its standard state is represented, for example, by $\overline{H}_i^\circ$, $\overline{S}_i^\circ$, and $\overline{G}_i^\circ$. Actually, it is not possible to determine the absolute values of the internal energy, enthalpy, Helmholtz energy, and Gibbs energy, and so the values of these properties are determined relative to those of the elements. In particular, we will make a good deal of use of the standard molar enthalpy of formation $\Delta_f H_i^\circ$ and the standard molar Gibbs energy of formation $\Delta_f G_i^\circ$. The overbar is not used on these quantities because the Δ_f indicates the formation of one mole of i from the elements contained in it. The formation quantities necessarily have mol^{-1} in their units.

Other Δ quantities that are frequently used are $\Delta_{mix}S$, $\Delta_{mix}V$, $\Delta_{fus}H$, $\Delta_{vap}H$, and $\Delta_r G^\circ$. These symbols are used to represent molar quantities and, again, the overbar is not used. The units for these quantities necessarily include mol^{-1}.

Recommendations of the International Union of Pure and Applied Chemistry on these matters are given in I. Mills, *Quantitative, Units, and Symbols in Physical Chemistry*, Blackwell Scientific Publications, Oxford, 1988.

A problem book containing worked-out solutions for the first set of problems and answers for the second set of problems is available as a companion to this text. Please ask for *Solutions Manual for Physical Chemistry*, by Robert A. Alberty and Robert J. Silbey.

Contents

APPENDIX /845

A
Physical Quantities and Units /845

B
Values of Physical Constants /849

C
Thermodynamic Tables /850

D
Mathematical Relations /859

E
Greek Alphabet /863

F
Symbols for Physical Quantities and Their SI Units /864

G
Answers to the First Set of Problems /874

INDEX /889

PHYSICAL CHEMISTRY

PART ONE
THERMODYNAMICS

Thermodynamics deals with the interconversion of various kinds of energy and the changes in physical properties that are involved. Thermodynamics is concerned with equilibrium states of matter and has nothing to do with time. Even so, it is one of the most powerful tools of physical chemistry; because of its importance, the first part of this book is devoted to it. The first law of thermodynamics deals with the amount of work that can be done by a chemical or physical process and the amount of heat that is absorbed or evolved. On the basis of the first law it is possible to build up tables of enthalpies of formation that may be used to calculate enthalpy changes for reactions that have not yet been studied. With information on heat capacities of reactants and products also available, it is possible to calculate the heat of a reaction at a temperature where it has not previously been studied.

The second law of thermodynamics deals with the natural direction of processes and the question of whether a given chemical reaction can occur by itself. The second law was formulated initially in terms of the efficiencies of heat engines, but it also leads to the definition of entropy, which is important in determining the direction of chemical change. The second law provides the basis for the definition of the equilibrium constant for a chemical reaction. It provides an answer to the question, "To what extent will this particular reaction go before equilibrium is reached?" It also provides the basis for reliable predictions of the effects of temperature, pressure, and concentration on chemical and physical equilibrium. The third law provides the basis for calculating equilibrium constants from calorimetric measurements only. This is an illustration of the way in which thermodynamics interrelates apparently unrelated measurements on systems at equilibrium.

After discussing the laws of thermodynamics and the various physical quantities it involves, our first applications will be to the quantitative treatment of chemical equilibrium. These methods are then applied to equilibrium between different phases. This provides the basis for the quantitative treatment of distillation and for the interpretation of the phase changes in mixtures of solids. Then thermodynamics is applied to electrochemical cells and biochemical reactions.

1
Zeroth Law of Thermodynamics and Equations of State

This chapter opens with a discussion of the thermodynamic concept of temperature. The principle involved in defining temperature was not recognized until the establishment of the first and second laws of thermodynamics, and so it is referred to as the zeroth law. This leads to a discussion of the thermodynamic properties of gases and liquids. After discussing an ideal gas we consider the behavior of real gases. In doing that we will see the importance of the concept of the state of a system. The thermodynamic properties of a gas or liquid are represented by an equation of state, such as the virial equation or the van der Waals equation. Equations of state are used extensively in thermodynamics. To represent the properties of a gas or liquid correctly over a wide range of conditions, an equation of state must be able to describe the critical region; we discuss this in general, and then for the van der Waals equation.

1.1 State of a System

A thermodynamic system is that part of the physical universe that is under consideration. A system is separated from the rest of the universe by a real or

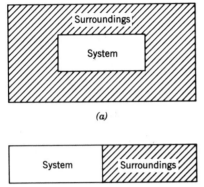

Figure 1.1   (*a*) A system is separated from its surroundings by a boundary, real or imaginary. (*b*) As a simplification we can imagine the system to be separated from the surroundings by a single wall that may be an insulator or a heat conductor. Later we will consider semipermeable boundaries so that the system is open to the transfer of matter.

imaginary **boundary.** The part of the universe outside the boundary is referred to as **surroundings,** as illustrated by Fig. 1.1. Even though a system is enclosed by a boundary, heat may be transferred between system and surroundings, and the surroundings may do work on the system, or vice versa. If the boundary around a system prevents interaction of the system with its surroundings, the system is called an **isolated** system.

If matter can be transferred from the surroundings to the system, or vice versa, the system is referred to as an **open** system; otherwise it is a **closed** system.

When a system is under discussion it must be described precisely. A system is **homogeneous** if its properties are uniform throughout; such a system consists of a single phase. If a system contains more than one phase, it is **heterogeneous.** A simple example of a two-phase system is liquid water in equilibrium with ice. Water can also exist as a three-phase system: liquid, ice, and vapor, all in equilibrium. As we will see in Section 1.12, a gas in the earth's gravitational field is actually not homogeneous. However, we will ordinarily ignore the effects of external fields in chemical thermodynamics.

To describe a homogeneous system, or a phase in a heterogeneous system, it is necessary to specify its composition and other macroscopic properties. Such properties are volume, pressure, temperature, density, and mass. Some of these properties are extensive and some are intensive. Volume and mass are **extensive properties** since they depend on the size of the system. That is, if a system is duplicated exactly and is added to the original system, the volume and the mass are doubled. The pressure, temperature, and density are **intensive properties** since they remain the same when the system is duplicated and added to the original system.

When an extensive property of a homogeneous system is divided by the mass of the system, an intensive property is obtained; for example, the volume per unit mass (the reciprocal of the density) is an intensive property. It is often useful to divide a thermodynamic quantity for a system by the amount of substance n in the system. **The amount of substance n is the number of entities (atoms, molecules, ions, electrons, or specified groups of such particles) expressed in terms of moles.** If a system contains N molecules, the amount of substance $n = N/N_A$, where N_A is the Avogadro constant (6.022×10^{23} mol^{-1}). The ratio of the volume V to the amount of substance is referred to as the molar volume: $\overline{V} = V/n$. The volume V is expressed in SI units of m^3, and the molar volume $\overline{V}$ is expressed in SI units of m^3 mol^{-1}. We will use the overbar regularly to indicate molar thermodynamic quantities.

When a system is at **equilibrium** under a given set of conditions, it is said to be in a definite **state;** that is, each of its properties has a definite value, which is independent of time. For a system to be in a definite thermodynamic state, it is also important that there not be any fluxes through it. For example, heat should not be flowing through the system.

Experience has shown that the intensive state of a system at equilibrium can be specified by the values of a small number of macroscopic variables. This is fortunate, considering that the specification of the microscopic state would require a huge number of variables. To describe the microscopic state of a system using classical mechanics, we would have to give three coordinates and three components of momentum of each molecule, plus information about its vibrational and rotational motion. The macroscopic properties of a system are called **state variables** or **thermodynamic coordinates;** examples are pressure, volume, mass, and density. Thermodynamics leads to the definition of additional variables that can be used to describe the state of a system; examples of additional variables are temperature, internal energy, and entropy.

When a thermodynamic system is at equilibrium its state is defined entirely by the state variables, **and not by the history of the system.** By history of the system, we mean the previous conditions under which it has existed.

Since the state of a system at equilibrium can be specified by a small number of state variables, it should be possible to express the value of a variable that has not been specified as a function of the values of other variables that have been specified. The simplest example of this is the ideal gas law. We will find that for pure gases in general, it is necessary to specify only two intensive variables in order to specify the intensive state of the system. The term **intensive state of a system** refers to its intensive properties. To give a complete description of the state of a system, we must give at least one extensive property so that the size of the system is specified. If a mole of helium has a certain pressure and a certain density, its state can be represented by 1 He(P, ρ), where P and ρ represent the pressure and density, respectively. For some systems, more than two intensive variables must be stated to specify the state of the system. If there is more than one component, the composition has to be given. If a liquid system is in the form of small droplets, the surface area has to be given. If the system is in an electric or magnetic field, this may have an effect on its properties, and then the electric field strength and magnetic field strength become state variables. We will generally ignore the effect of the earth's gravitational field on a system, although this can be important, as we will see in the special topic at the end of this chapter.

The pressure of the atmosphere is measured with a barometer, as shown in Fig. 1.2a, and the pressure of a gaseous system is measured with a closed-end manometer, as shown in Fig. 1.2b.

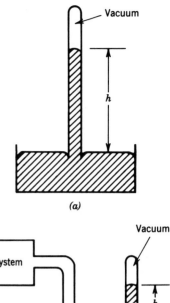

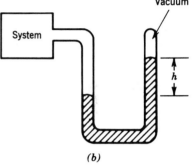

Figure 1.2 (a) The pressure exerted by the atmosphere on the surface of mercury in a cup is given by $P = h\rho g$ (see Example 1.1). (b) The pressure of a system is given by the same equation when a closed-end manometer is used.

1.2 The Zeroth Law of Thermodynamics

Although we all have a commonplace notion of what temperature is, we must define it very carefully so that it is a useful concept in thermodynamics. If two closed systems with fixed volumes are brought together so that they are in thermal contact, changes may take place in the properties of both. Eventually a state is reached in which there is no further change, and this is the state of **thermal equilibrium.** In this state, the two systems have the same temperature. Thus, we can readily determine whether two systems are at the same temperature by bringing them into thermal contact and seeing whether observable changes take place in the properties of either system. If no change occurs, they are at the same temperature.

Now let us consider three systems, A, B, and C, as shown in Fig. 1.3. It is an experimental fact that if system A is in thermal equilibrium with system C, and system B is also in thermal equilibrium with system C, then A and B are in thermal equilibrium with each other. It is not obvious that this should be true, and so this empirical fact is referred to as the **zeroth law of thermodynamics.**

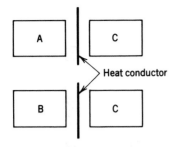

If A and C are in thermal equilibrium, and B and C are in thermal equilibrium then

will be found to be in thermal equilibrium when connected by a heat conductor.

Figure 1.3 The zeroth law of thermodynamics is concerned with thermal equilibrium between three bodies.

To see how the zeroth law leads to the definition of a temperature scale, we need to consider thermal equilibrium between systems A, B, and C in more detail. Assume that A, B, and C each consist of a certain mass of a different fluid. We use the word **fluid** to mean either a gas or a compressible liquid. Our experience is that if the volume of one of these systems is held constant, its pressure may vary over a range of values, and if the pressure is held constant, its volume may vary over a range of values. Thus, the pressure and the volume are independent thermodynamic coordinates. Furthermore, suppose that the experience with these systems is that their intensive states are specified completely when the pressure and volume are specified. That is, when one of the systems reaches equilibrium at a certain pressure and volume, all of its macroscopic properties have certain characteristic values. It is quite remarkable and fortunate that the macroscopic state of a given mass of fluid of a given composition can be fixed by specifying only the pressure and the volume.*

If there are further constraints on the system, there will be a smaller number of independent variables. An example of an additional constraint is thermal equilibrium with another system. Experience shows that if a fluid is in thermal equilibrium with another system, it has only one independent variable. In other words, if we set the pressure of system A at a particular value P_A, we find that there is thermal equilibrium with system C, in a specified state, only at a particular value of V_A. Thus, system A in thermal equilibrium with system C is characterized by a **single** independent variable, pressure or volume; one or the other can be set arbitrarily, but not both. The plot of all the values of P_A and V_A for which there is equilibrium with system C is called an **isotherm**. Figure 1.4 gives this isotherm, which we label Θ_1. Since system A is in thermal equilibrium with system C at any P_A, V_A on the isotherm, we can say that each of the pairs P_A, V_A on this isotherm correspond with the same temperature Θ_1.

When heat is added to system C and the experiment is repeated, a different isotherm is obtained for system A. In Fig. 1.4, the isotherm for the second experiment is labeled Θ_2. If still more heat is added to system C and the experiment is repeated again, the isotherm labeled Θ_3 is obtained.

The zeroth law can be used to show that for any fluid there is a function of pressure and volume that is equal to the temperature Θ:

$$\Phi(P, V) = \Theta \tag{1.1}$$

Such an equation is called an **equation of state.** This is discussed as a Special Topic in Section 1.11. Although this conclusion applies to fluids in general, it is only for ideal gases that a very simple equation of state is obtained.

Temperature is an intensive property because it is independent of the size of the system. The temperature is measured by relating it to some physical property, such as the volume of a fixed mass of a fluid or the electrical resistance of a conductor. The simplest assumption is that the temperature is a linear function of some physical property. The physical property that is most useful for this purpose is the pressure–volume product of a fixed amount of a gas at relatively low pressures. The apparatus for measuring the PV product is called a **gas thermometer.** According to **Boyle's law** the PV product of a fixed amount of gas is a function only of the temperature Θ:

Figure 1.4 Isotherms for fluid A. This plot, which is for a hypothetical fluid A, might look quite different for some other fluid.

---

* This is not true for water in the neighborhood of 4 °C, but the state is specified by giving the temperature and the volume or the temperature and the pressure. See Section 6.2.

$$PV = f(\Theta) \qquad (1.2)$$

but Boyle's law does not specify the function $f(\Theta)$. **Gay-Lussac's law** states that the ratio of PV at temperature Θ_2 to PV at temperature Θ_1 depends only on the two temperatures:

$$\frac{(PV)_2}{(PV)_1} = \phi(\Theta_2, \Theta_1) \qquad (1.3)$$

but Gay-Lussac's law does not specify the function ϕ. The simplest thing to do is to take the ratio of the PV products to be equal to the ratio of the temperatures, thus defining a temperature scale:

$$\frac{(PV)_2}{(PV)_1} = \frac{T_2}{T_1} \quad \text{or} \quad \frac{(PV)_2}{T_2} = \frac{(PV)_1}{T_1} \qquad (1.4)$$

Here we have introduced a new symbol T for the temperature because we have made a specific assumption about the function ϕ.

Equations 1.2 to 1.4 are only exact in the limit of zero pressure, and so T is referred to as the ideal gas temperature.

Since, according to equation 1.4, PV/T is a constant for a fixed mass of gas and since V is an extensive property,

$$PV/T = nR \qquad (1.5)$$

where n is the amount of gas and R is referred to as the **gas constant.**

1.3 The Ideal Gas Temperature Scale

The ideal gas temperature scale is defined by taking the temperature T to be proportional to $P\bar{V}$. In this expression, $\bar{V}$ is the molar volume (V/n, where n is the amount of substance). Since different gases give slightly different scales when the pressure is about one bar (10^5 pascal $= 10^5$ Pa $= 10^5$ N m^{-2}), it is necessary to use the limit of the $P\bar{V}$ product as the pressure approaches zero. When this is done, all gases yield the same temperature scale. We speak of gases under this limiting condition as **ideal.** Thus, the **ideal gas temperature** T is defined by

$$T = \lim_{P \to 0} (P\bar{V}/R) \qquad (1.6)$$

The proportionality constant is called the gas constant R. The unit of thermodynamic temperature, the kelvin, represented by K, is defined as the fraction 1/273.16 of the temperature of the triple point of water. Thus, the temperature of an equilibrium system consisting of liquid water, ice, and water vapor is 273.16 K. When $P\bar{V}$ is measured at the temperature of the triple point of water*

---

* The triple point of water is the temperature and pressure at which ice, liquid, and vapor are in equilibrium with each other in the absence of air. The pressure at the triple point is 611 Pa. The freezing point in the presence of air at 1 atm is 0.0100 °C lower because (1) the solubility of air in liquid water at 1 atm (101 325 Pa) is sufficient to lower the freezing point 0.0024 °C (Section 7.9), and (2) the increase of pressure from 611 Pa to 101 325 Pa lowers the freezing point 0.0075 °C, as shown in Example 6.2. Thus, the ice point is at 273.15 K, and this is frequently taken as a standard temperature in thermodynamics.

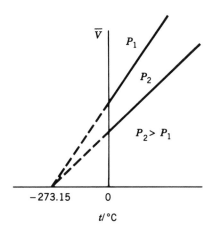

Figure 1.5 Plots of $\overline{V}$ versus temperature for a given amount of a real gas at two low pressures P_1 and P_2, as given by Gay–Lussac's law.

and at a low gas pressure, $P\overline{V} = R(273.16 \text{ K})$. The temperature 0 K is called absolute zero. According to the current best measurements, the freezing point of water is 273.15 K and the boiling point at 1 atmosphere (101 325 Pa, see below) is 373.12 K, but these are experimental values and may be determined more accurately in the future. The Celsius scale t is formally defined by

$$t/°C = T/K - 273.15 \qquad (1.7)$$

The reason for writing the equation in this way is that temperature T on the Kelvin scale has the unit K, and temperature t on the Celsius scale has the unit °C, which need to be divided out before temperatures on the two scales are compared. In Fig. 1.5, the molar volume of an ideal gas is plotted versus the Celsius temperature t at two pressures.

We will find later that the ideal gas temperature scale is identical with one based on the second law of thermodynamics, which is independent of the properties of any particular substance (see Section 3.4). In Chapter 17 the ideal gas temperature scale will be identified with that which arises in statistical mechanics.

The gas constant R can be expressed in various units, but we will emphasize the use of SI units. The SI unit of pressure is the pascal, Pa, which is the pressure produced by a force of 1 N on an area of 1 m². In addition to using the prefixes listed in the front cover of the book to express larger and smaller pressures, it is convenient to have a unit that is approximately equal to the atmospheric pressure. This unit is the bar, which is 10^5 Pa. Earlier the atmosphere, which is defined as 101 325 Pa, had been used as a unit of pressure.

Example 1.1
Calculate the pressure of the earth's atmosphere at a point where the barometer reads 76 cm of mercury at 0 °C and the acceleration of gravity g is 9.806 65 m s^{-2}. The density of mercury at 0 °C is 13.5951 g cm^{-3}, or 13.5951 × 10^3 kg m^{-3}.

Pressure P is force f divided by area A:

$$P = f/A$$

The force exerted by a column of air over an area A is equal to the mass m of the air in a vertical column with a cross section A times the acceleration of gravity g:

$$f = mg$$

The mass of the column of air is equal to the mass of mercury $m = \rho A h$ raised above the flat surface in Fig. 1.2a:

$$f = \rho A h g$$

Thus, the pressure of the atmosphere is

$$P = h\rho g$$

If h, ρ, and g are expressed in SI units, the pressure P is expressed in pascals. Thus, the pressure of a standard atmosphere may be expressed in SI units as follows:

$$1 \text{ atm} = (0.76 \text{ m})(13.5951 \times 10^3 \text{ kg m}^{-3})(9.806 \ 65 \text{ m s}^{-2})$$
$$= 101 \ 325 \text{ N m}^{-2} = 101 \ 325 \text{ Pa} = 1.013 \ 25 \text{ bar}$$

The conversion factor to remember is $1.013 \ 25$ bar atm^{-1}.

To determine the value of the gas constant we also need the definition of a mole. A **mole** is the amount of substance that has as many atoms or molecules as 0.012 kg (exactly) of ^{12}C. The **molar mass** M of a substance is the mass divided by the amount of substance n, and so its SI unit is kg mol^{-1}. Molar masses can also be expressed in g mol^{-1}, but it is important to remember that in making calculations in which all other quantities are expressed in SI units, the molar mass must be expressed in kg mol^{-1}. The molar mass M is related to the molecular mass m by $M = N_A m$, where N_A is the Avogadro constant and m is the mass of a single molecule.

Until 1986 the recommended value of the gas constant was based on measurements of the molar volumes of oxygen and nitrogen at low pressures. The accuracy of such measurements is limited by problems of sorption of gas on the walls of the glass vessels used. In 1986 the recommended value* of the gas constant

$$R = 8.314 \ 51 \text{ J K}^{-1} \text{ mol}^{-1} \tag{1.8}$$

was based on measurements of the speed of sound in argon. The equation used is discussed in Section 18.4.

For an ideal gas

$$P\overline{V} = RT \tag{1.9}$$

where $\overline{V}$ is the molar volume V/n. The ideal gas law can also be written as

$$PV = nRT \tag{1.10}$$

where V is the volume of the gas. Since pressure is force per unit area, the product of pressure and volume has the dimensions of force times distance, which is work or energy. Thus, the gas constant is obtained in joules if pressure and volume are expressed in pascals and cubic meters; note that $1 \text{ J} = 1 \text{ Pa m}^3$. For an ideal gas the intensive state can be described by giving any two of P, $\overline{V}$, or T, and the extensive state can be described by giving any three of P, V, T, and n.

Example 1.2
Calculate the value of R in cal K^{-1} mol^{-1}, L bar K^{-1} mol^{-1}, and L atm K^{-1} mol^{-1}.
Since the calorie is defined as 4.184 J,

$$R = 8.314 \ 51 \text{ J K}^{-1} \text{ mol}^{-1}/4.184 \text{ J cal}^{-1}$$
$$= 1.987 \ 22 \text{ cal K}^{-1} \text{ mol}^{-1}$$

---

* E. R. Cohen and B. N. Taylor, The 1986 Adjustment of the Fundamental Physical Constants, *CODATA Bull.* **63**:1 (1986); *J. Phys. Chem. Ref. Data* **17**:1795 (1988).

Since the liter is 10^{-3} m^3 and the bar is 10^5 Pa,

$$R = (8.314\ 51\ \text{Pa m}^3\ \text{K}^{-1}\ \text{mol}^{-1})(10^3\ \text{L m}^{-3})(10^{-5}\ \text{bar Pa}^{-1})$$
$$= 0.083\ 145\ 1\ \text{L bar K}^{-1}\ \text{mol}^{-1}$$

Since 1 atm is 1.013 25 bar,

$$R = (0.083\ 145\ 1\ \text{L bar K}^{-1}\ \text{mol}^{-1})/(1.013\ 25\ \text{bar atm}^{-1})$$
$$= 0.082\ 066\ 7\ \text{L atm K}^{-1}\ \text{mol}^{-1}$$

1.4   Ideal Gas Mixtures and Dalton's Law

Equation 1.10 applies to a mixture of ideal gases as well as a pure gas, when n is the total amount of gas. Since $n = n_1 + n_2 + \cdots$, then

$$P = (n_1 + n_2 + \cdots)RT/V$$
$$= n_1RT/V + n_2RT/V + \cdots$$
$$= P_1 + P_2 + \cdots = \sum_i P_i \qquad (1.11)$$

where P_1 is the partial pressure of species 1. Thus, the total pressure of an ideal gas mixture is equal to the sum of the partial pressures of the individual gases; this is **Dalton's law.** The partial pressure of a gas in an ideal gas mixture is the pressure that it would exert alone in the total volume at the temperature of the mixture:

$$P_i = n_iRT/V \qquad (1.12)$$

A useful form of this equation is obtained by replacing RT/V by P/n:

$$P_i = n_iP/n = y_iP \qquad (1.13)$$

The dimensionless quantity y_i is the mole fraction of species i in the mixture, and it is defined by n_i/n. Substituting equation 1.13 in 1.11 yields

$$1 = y_1 + y_2 + \cdots = \sum_i y_i \qquad (1.14)$$

so that the sum of the mole fractions in a mixture is unity.

Figure 1.6 shows the partial pressures P_1 and P_2 of two components of a binary mixture of ideal gases at various molar fractions and at constant total pressure for the second component. The various mixtures are considered at the same total pressure P.

The behavior of real gases is more complicated than the behavior of an ideal gas, as we will see in the next section.

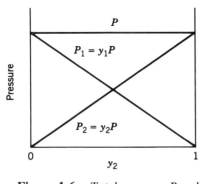

Figure 1.6   Total pressure P and partial pressures P_1 and P_2 of components of binary mixtures of gases as a function of the mole fraction y_2 of the second component at constant total pressure. *Note*: $y_1 = 1 - y_2$.

Example 1.3

A mixture of 1 mol of methane and 3 mol of ethane is held at a pressure of 10 bar. What are the mole fractions and partial pressures of the two gases?

$$y_m = 1 \text{ mol}/4 \text{ mol} = 0.25$$

$$P_m = y_m P = (0.25)(10 \text{ bar}) = 2.5 \text{ bar}$$

$$y_e = 3 \text{ mol}/4 \text{ mol} = 0.75$$

$$P_e = y_e P = (0.75)(10 \text{ bar}) = 7.5 \text{ bar}$$

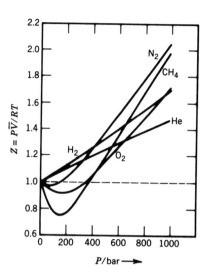

Example 1.4

The maximum partial pressure of water vapor in air at equilibrium at a given temperature is the vapor pressure of water at that temperature. The **actual** partial pressure of water vapor in air is a percentage of the maximum, and that percentage is called the relative humidity. Suppose the relative humidity of air is 50% at a temperature of 20 °C. If the atmospheric pressure is 1 bar, what is the mole fraction of water in the air? The vapor pressure of water at 20 °C is 2330 Pa. Assume the gas mixture behaves as an ideal gas.

$$y_{H_2O} = P_i/P = (0.5)(2330 \text{ Pa})/10^5 \text{ Pa} = 0.0117$$

Figure 1.7 Influence of high pressure on the compressibility factor, $P\overline{V}/RT$, for various gases at 298 K.

1.5 Real Gases: Compressibility Factor

Real gases behave like ideal gases in the limits of low pressures and high temperatures, but they deviate significantly at high pressures and low temperatures. The **compressibility factor** $Z = P\overline{V}/RT$ is a convenient measure of the deviation from ideal gas behavior. Figure 1.7 shows the compressibility factors of a variety of gases as a function of pressure at 0 °C. Ideal gas behavior, indicated by the dashed line, is included for comparison. As the pressure is reduced to zero, the compressibility factor approaches unity, as expected for an ideal gas. At very high pressures the compressibility factor is always greater than unity. This can be understood in terms of the finite size of molecules. At very high pressures the molecules of the gas are pushed closer together, and the volume of the gas is larger than expected for an ideal gas because a significant fraction of the volume is occupied by the molecules themselves. At low pressure a gas may have a smaller compressibility factor than an ideal gas. This is due to intermolecular attractions. The effect of intermolecular attractions disappears in the limit of zero pressure because the distance between molecules approaches infinity.

Figure 1.8 shows how the compressibility factors of nitrogen depend on temperature, as well as pressure. As the temperature is reduced, the effect of intermolecular attraction at pressures of the magnitude of 100 bar increases because the molar volume is smaller at lower temperatures and the molecules are closer together. All gases show a minimum in the plot of compressibility factor versus pressure if temperature is low enough. Hydrogen and helium, which have very low boiling points, exhibit this minimum only at temperatures much below 0 °C.

A number of equations have been developed to represent P-V-T data for real gases. Such an equation is called an **equation of state** because it relates state properties for a substance at equilibrium. Equation 1.9 is the equation of state

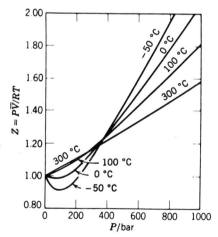

Figure 1.8 Influence of pressure on the compressibility factor, $P\overline{V}/RT$, for nitrogen at different temperatures.

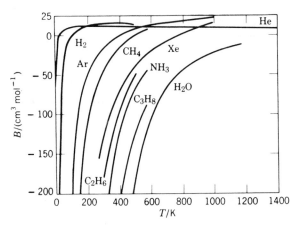

Figure 1.9 Second virial coefficient B. (From K. E. Bett, J. S. Rowlinson, and G. Saville, *Thermodynamics for Chemical Engineers*. Cambridge, MA: MIT Press, 1975. Reproduced by permission of The Athlone Press.)

for an ideal gas. The first equation of state for real gases that we will discuss is closely related to the plots in Figs. 1.7 and 1.8, and is called the virial equation.

1.6 The Virial Equation

In 1901 H. Kamerlingh-Onnes proposed an equation of state for real gases, which expresses the compressibility factor Z as a power series in $1/\overline{V}$ for a pure gas:

$$Z = \frac{P\overline{V}}{RT} = 1 + \frac{B}{\overline{V}} + \frac{C}{\overline{V}^2} + \cdots \tag{1.15}$$

The coefficients B and C are referred to as the second and third virial coefficients, respectively.* For a particular gas these coefficients depend only on the temperature and not on the pressure. The word virial is derived from the Latin word for force.

The second virial coefficient B is shown as a function of temperature for several gases in Fig. 1.9, and some values are given in Table 1.1.

For many purposes it is more convenient to use P as an independent variable. When this is done the pressure–volume product for a mole of a pure gas is generally written as

$$P\overline{V} = RT + BP + C'P^2 + \cdots \tag{1.16}$$

or

---

* Statistical mechanics shows that the term $B/\overline{V}$ arises from interactions involving two molecules, the $C/\overline{V}^2$ term arises from interaction involving three molecules, etc. (Section 17.14).

Table 1.1   Second Virial Coefficient B in $cm^3\ mol^{-1}$

| | T/K | | | |
Gas	200	300	400	500
Ar	−47.4	−15.5	−1.0	+7.0
N_2	−35.2	−4.2	+9.0	+16.9
CO_2	—	−122.7	−60.5	−29.8
CH_4	−105	−42	−15	−0.5
C_2H_6	−410	−182	−96	−52
C_3H_8	—	−382	−208	−124

Source: J. H. Dymond and E. B. Smith, *The Virial Coefficients of Pure Gases and Mixtures*. Oxford: Oxford University Press, 1980.

$$Z = \frac{P\overline{V}}{RT} = 1 + \frac{BP}{RT} + \frac{C'P^2}{RT} + \cdots \tag{1.17}$$

This is the same B as in equation 1.15. The relation between C and C' can be derived as follows: Equation 1.15 is solved for P and the expressions are written for P and P^2:

$$P = \frac{RT}{\overline{V}} + \frac{BRT}{\overline{V}^2} + \frac{CRT}{\overline{V}^3} + \cdots \tag{1.18}$$

$$P^2 = \left(\frac{RT}{\overline{V}}\right)^2 + \frac{2B(RT)^2}{\overline{V}^3} + \cdots \tag{1.19}$$

where terms higher than $\overline{V}^{-3}$ have been omitted. Substitution of these expressions into 1.17 yields

$$Z = 1 + \frac{B}{RT}\left(\frac{RT}{\overline{V}} + \frac{BRT}{\overline{V}^2} + \frac{CRT}{\overline{V}^3} + \cdots\right) \tag{1.20}$$

$$+ \frac{C'}{RT}\left[\left(\frac{RT}{\overline{V}}\right)^2 + \frac{2B(RT)^2}{\overline{V}^3} + \cdots\right]$$

$$= 1 + \frac{B}{\overline{V}} + \frac{1}{\overline{V}^2}(B^2 + RTC') + \cdots \tag{1.21}$$

Comparison with equation 1.15 shows that

$$C = B^2 + RTC' \tag{1.22}$$

so that

$$C' = \frac{C - B^2}{RT} \tag{1.23}$$

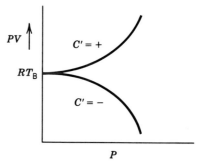

Figure 1.10   At the Boyle temperature ($B = 0$), a gas behaves nearly ideally over a range of pressures.

The second virial coefficient B for nitrogen is zero at 54 °C, which is consistent with Fig. 1.7. A real gas may behave like an ideal gas over an extended range in pressure when the second virial coefficient is zero, as shown in Fig. 1.10. The temperature at which this occurs is called the **Boyle temperature** T_B. The Boyle temperatures of a number of gases are given later in Table 1.2.

Example 1.5

What is the molar volume of $N_2(g)$ at 500 K and 600 bar according to (*a*) the ideal gas law and (*b*) the virial equation? The virial coefficient B of $N_2(g)$ at 500 K is 0.0169 L mol^{-1}.

$$(a) \quad \overline{V} = \frac{RT}{P} = \frac{(8.314 \times 10^{-2} \text{ L bar K}^{-1} \text{ mol}^{-1})(500 \text{ K})}{600 \text{ bar}}$$

$$= 6.93 \times 10^{-2} \text{ L mol}^{-1}$$

$$(b) \quad Z = 1 + \frac{BP}{RT}$$

$$= 1 + \frac{(0.0169 \text{ L mol}^{-1})(600 \text{ bar})}{(8.314 \times 10^{-2} \text{ L bar K}^{-1} \text{ mol}^{-1})(500 \text{ K})}$$

$$= 1.244$$

$$\overline{V} = \frac{ZRT}{P} = (1.244)(6.93 \times 10^{-2} \text{ L mol}^{-1})$$

$$= 8.62 \times 10^{-2} \text{ L mol}^{-1}$$

We can give this result a molecular interpretation by saying that $\overline{V}$ is greater for the real gas because of the finite volume of the molecules.

1.7 P-$\overline{V}$-T Surface for a One-Component System

To discuss more general equations of state, we will now look at the possible values of P, $\overline{V}$, and T for a pure substance. The state of a pure substance is

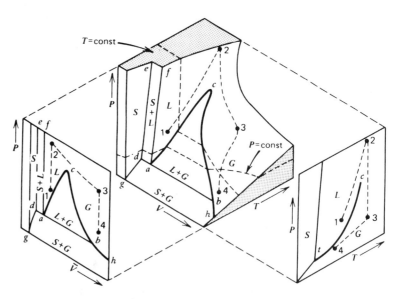

Figure 1.11 P-$\overline{V}$-T surface for a one-component system that contracts on freezing. (From K. E. Bett, J. S. Rowlinson, and G. Saville, *Thermodynamics for Chemical Engineers*. Cambridge, MA: MIT Press, 1975. Reproduced by permission of The Athlone Press.)

represented by a point in a Cartesian coordinate system with P, $\overline{V}$, and T plotted along the three axes. Each point on the surface of the three-dimensional model in Fig. 1.11 describes the state of a one-component system that contracts on freezing. We will not be concerned here with the solid state, but will consider that part of the surface later (Section 6.4). Projections of this surface on the P-$\overline{V}$ and P-T planes are shown. There are three two-phase regions on the surface: S + G, L + G, and S + L. These are ruled surfaces, that is, they may be thought of as being generated by a moving straight line, in this case one perpendicular to the P-T plane. These three surfaces intersect at the **triple point** t where vapor, liquid, and solid are in equilibrium.

The projection of the three-dimensional surface on the P-T plane is shown to the right of the main diagram in Fig. 1.11. The vapor pressure curve goes from the triple point t to the **critical point** c. The sublimation pressure curve goes from the triple point t to absolute zero. The melting curve rises from the triple point. Most substances contract on freezing, and for them the slope dP/dT for the melting line is positive.

At high temperatures the substance is in the gas state, and as the temperature is raised and the pressure is lowered the surface is more and more closely represented by the ideal gas equation of state $P\overline{V} = RT$. However, much more complicated equations are required to represent the rest of the surface that represents gas and liquid. Before discussing equations that can represent this part of the surface, we will consider the unusual phenomena that occur near the critical point. Any realistic equation of state must be able to reproduce this behavior at least qualitatively.

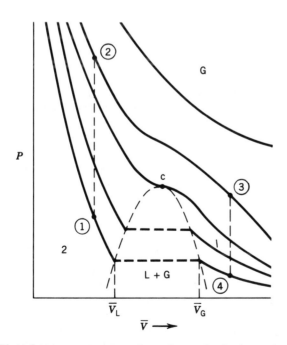

Figure 1.12   Plot of pressure versus molar-volume plot in the region of the critical point. The dashed horizontal lines in the two-phase region are called tie lines. The dashed line 1-2-3-4 shows how a liquid can be converted to a gas without the appearance of a meniscus.

Table 1.2 Critical Constants and Boyle Temperatures

Gas	T_c/K	P_c/bar	$\overline{V}_c/L\ mol^{-1}$	Z_c	T_B/K
Helium 4	5.2	2.27	0.0573	0.301	22.64
Hydrogen	33.2	13.0	0.0650	0.306	110.04
Nitrogen	126.2	34.0	0.0895	0.290	327.22
Oxygen	154.6	50.5	0.0734	0.288	405.88
Chlorine	417	77.0	0.124	0.275	
Bromine	584	103.0	0.127	0.269	
Carbon dioxide	304.2	73.8	0.094	0.274	714.81
Water	647.1	220.5	0.056	0.230	
Ammonia	405.6	113.0	0.0725	0.252	995
Methane	190.6	46.0	0.099	0.287	509.66
Ethane	305.4	48.9	0.148	0.285	
Propane	369.8	42.5	0.203	0.281	
n-Butane	425.2	38.0	0.255	0.274	
Isobutane	408.1	36.5	0.263	0.283	
Ethylene	282.4	50.4	0.129	0.277	624
Propylene	365.0	46.3	0.181	0.276	
Benzene	562.1	49.0	0.259	0.272	
Cyclohexane	553.4	40.7	0.308	0.272	

1.8 Critical Phenomena

For a pure substance there is a critical point (P_c, T_c) at the end of the liquid–gas coexistence curve where the properties of the gas and liquid phases become so nearly alike that they can no longer be distinguished as separate phases. Thus, T_c is the highest temperature at which condensation of a gas is possible, and P_c is the highest pressure at which a liquid will boil when heated.

The critical pressures P_c, volumes $\overline{V}_c$, and temperatures T_c of a number of substances are given in Table 1.2, along with the compressibility factor at the critical point $Z_c = P_c\overline{V}_c/RT_c$, and the Boyle temperature T_B.

Critical phenomena are most easily discussed using the projection of the three-dimensional surface in Fig. 1.11 on the P-$\overline{V}$ plane. Figure 1.12 shows only the parts of the P-$\overline{V}$ plot labeled L, G, and L + G. When the state of the system is represented by a point in the L + G region of this plot, the system contains two phases, one liquid and one gas, in equilibrium with each other. The molar volumes of the liquid and gas can be obtained by drawing a horizontal line parallel to the $\overline{V}$ axis through the point representing the state of the system and noting the intersections with the boundary line for the L + G region. Such a line, which connects the state of one phase with the state of another phase with which it is in equilibrium, is called a **tie line**. Two tie lines are shown in Fig. 1.12. The pressure in this case is the equilibrium vapor pressure of the liquid. As the temperature is raised, the tie line becomes shorter, and the molar volumes of the liquid and gas approach each other. At the critical point c the tie line vanishes and the distinction between liquid and gas is lost. At temperatures above the critical temperature, there is a single fluid phase.

Figures 1.11 and 1.12 also show how a liquid at point 1 can be converted to a gas at point 4 without the appearance of an interface between two phases. To do this, liquid at point 1 is heated at constant volume to point 2, then expanded at constant temperature to point 3, and finally cooled at constant volume to point 4 where it is a gas. Thus, liquid and vapor phases are really the same in terms of molecular organization, and so when the density of these two phases for a substance become equal, they cannot be distinguished and there is a critical point. On the other hand, a solid and a liquid have different molecular organizations, and the two phases do not become identical even if their densities are equal. Therefore, solid–liquid, solid–gas, and solid–solid equilibrium lines do not have critical points, as do gas–liquid lines.

At the critical point the isothermal compressibility [$\kappa = -\overline{V}^{-1}(\partial \overline{V}/\partial P)_T$, equation 1.43] becomes infinite because $(\partial P/\partial \overline{V})_{T_c} = 0$. If the isothermal compressibility is very large, as it is in the neighborhood of the critical point, very little work is required to compress the fluid. Therefore, gravity sets up large differences in density between the top and bottom of the container, as large as 10% in a column of fluid only a few centimeters high. This makes it difficult to determine $P\overline{V}$ isotherms near the critical point. The high isothermal compressibility also permits **spontaneous fluctuations** in the density that extend over macroscopic distances. The distance may be as large as the wavelength of visible light or larger. Since fluctuations in density are accompanied by fluctuations in refractive index, light is strongly scattered and this is called **critical opalescence.**

1.9 The Van der Waals Equation

Although the virial equation is very useful, it is important to have approximate equations of state with only a few parameters. We turn now to such an equation. To modify the ideal gas law to obtain an equation that can represent the P-V-T behavior of real gases, it is necessary to take account of the finite size of real gas molecules and intermolecular attractions. In 1879 van der Waals pointed out that the volume occupied by gas molecules can be taken into account by subtracting an excluded volume b from the actual molar volume $\overline{V}$ in the ideal gas law:

$$P(\overline{V} - b) = RT \qquad (1.24)$$

which corresponds to equation 1.16 with $B = b$ and C' and higher constants equal to zero.

This equation can represent compressibility factors greater than unity, but cannot yield compressibility factors less than unity. In addition, van der Waals recognized that gas molecules attract each other and that real gases are therefore more compressible than ideal gases. The forces that lead to condensation are still referred to as van der Waals forces, and their origin is discussed in Section 12.11. Van der Waals provided for intermolecular attraction by adding to the observed pressure P in the equation of state a term $a/\overline{V}^2$, where a is a constant whose value depends on the gas.

Table 1.3   Van der Waals Constants

Gas	a/L^2 bar mol^{-2}	b/L mol^{-1}	Gas	a/L^2 bar mol^{-2}	b/L mol^{-1}
H$_2$	0.247 6	0.026 61	CH$_4$	2.283	0.042 78
He	0.034 57	0.023 70	C$_2$H$_6$	5.562	0.063 80
N$_2$	1.408	0.039 13	C$_3$H$_8$	8.779	0.084 45
O$_2$	1.378	0.031 83	C$_4$H$_{10}$(n)	14.66	0.122 6
Cl$_2$	6.579	0.056 22	C$_4$H$_{10}$(iso)	13.04	0.114 2
NO	1.358	0.027 89	C$_5$H$_{12}$(n)	19.26	0.146 0
NO$_2$	5.354	0.044 24	CO	1.505	0.039 85
H$_2$O	5.536	0.030 49	CO$_2$	3.640	0.042 67

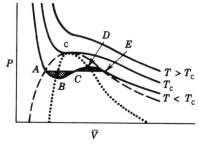

Figure 1.13   Isotherms calculated from the van der Waals equation. The dashed line is the boundary of the L + G region.

Van der Waals' equation is*

$$(P + a/\overline{V}^2)(\overline{V} - b) = RT \tag{1.25}$$

When the molar volume $\overline{V}$ is large, b becomes negligible in comparison with $\overline{V}$, $a/\overline{V}^2$ becomes negligible with respect to P, and van der Waals' equation reduces to the ideal gas law, $P\overline{V} = RT$.

The van der Waals constants for a few gases are listed in Table 1.3. They can be calculated from experimental measurements of P, $\overline{V}$, and T or from the critical constants, as shown later in equations 1.35 and 1.36.

Figure 1.13 shows three isotherms calculated using the van der Waals equation. At the critical temperature the isotherm has an inflection point at the critical point. At temperatures below the critical temperature each isotherm passes through a minimum and a maximum. The locus of these points shown by the dotted line has been obtained from $(\partial P/\partial \overline{V})_T = 0$. The states within the dotted line have $(\partial P/\partial \overline{V})_T > 0$, that is, the volume increases when the pressure increases. These states are therefore mechanically unstable and do not exist. Maxwell showed that states corresponding to the points between A and B and those between D and E are metastable, that is, not true equilibrium states. The dashed line is the boundary of the two-phase region; the part of the isotherm to the left of A represents the liquid and that to the right of E gas. The analysis shows that liquid at A is in equilibrium with gas at E. Any state along the line ACE will separate into liquid at A and gas at E. We see that the van der Waals equation with the Maxwell construction can represent the behavior of real substances.

The van der Waals equation is readily used when the volume and temperature are specified, but is more difficult to use when the pressure and temperature are specified. Multiplying out the terms in van der Waals' equation 1.25 and rearranging in descending powers of $\overline{V}$, we have

$$\overline{V}^3 - \overline{V}^2 \left(b + \frac{RT}{P} \right) + \overline{V}\frac{a}{P} - \frac{ab}{P} = 0 \tag{1.26}$$

---

* The van der Waals equation can also be written in the form
$$(P + an^2/V^2)(V - nb) = nRT$$

At temperatures below the critical temperature this cubic equation has three real solutions, each value of P giving three values of $\overline{V}$, as we have seen in Fig. 1.13.

The compressibility factor for a van der Waals gas is given by

$$Z = \frac{P\overline{V}}{RT} = \frac{\overline{V}}{\overline{V} - b} - \frac{a}{RT\overline{V}}$$

$$= \frac{1}{1 - b/\overline{V}} - \frac{a}{RT\overline{V}} \tag{1.27}$$

At low pressures, $b/\overline{V} \ll 1$ so that we can expand the first term using $(1 - x)^{-1} = 1 + x + x^2 + \cdots$. This yields the virial equation in terms of volume:

$$Z = 1 + \left(b - \frac{a}{RT}\right)\frac{1}{\overline{V}} + \left(\frac{b}{\overline{V}}\right)^2 + \cdots \tag{1.28}$$

From this equation we can see that the value of a is relatively more important at low temperatures, and the value of b is relatively more important at high temperatures. To obtain the virial equation in terms of pressure, we can replace V in the second term by the ideal gas value to obtain, to first order in P,

$$Z = 1 + \frac{1}{RT}\left(b - \frac{a}{RT}\right)P + \cdots \tag{1.29}$$

but this approximation is not good enough to give the correct coefficient for the P^2 term. At the Boyle temperature the second virial coefficient is zero, and so for a van der Waals gas

$$T_B = \frac{a}{bR} \tag{1.30}$$

The values of the van der Waals constants may be calculated from the critical constants for a gas. As may be seen in Fig. 1.12, there is a horizontal inflection point in the P versus $\overline{V}$ curve at the critical point so that $(\partial P/\partial \overline{V})_{T_c} = 0$ and $(\partial^2 P/\partial \overline{V}^2)_{T_c} = 0$. The van der Waals equation may be written

$$P = \frac{RT}{\overline{V} - b} - \frac{a}{\overline{V}^2} \tag{1.31}$$

Differentiating with respect to molar volume and evaluating these equations at the critical point yields

$$\left(\frac{\partial P}{\partial \overline{V}}\right)_{T_c} = \frac{-RT_c}{(\overline{V}_c - b)^2} + \frac{2a}{\overline{V}_c^3} = 0 \tag{1.32}$$

$$\left(\frac{\partial^2 P}{\partial \overline{V}^2}\right)_{T_c} = \frac{2RT_c}{(\overline{V}_c - b)^3} - \frac{6a}{\overline{V}_c^4} = 0 \tag{1.33}$$

A third simultaneous equation is obtained by writing equation 1.31 for the critical point:

$$P_c = \frac{RT_c}{\overline{V}_c - b} - \frac{a}{\overline{V}_c^2} \tag{1.34}$$

These three simultaneous equations may be combined to obtain expressions for a and b in terms of T_c and P_c or T_c and $\overline{V}_c$:

$$a = \frac{27R^2T_c^2}{64P_c} = \frac{9}{8} RT_c \overline{V}_c \tag{1.35}$$

$$b = \frac{RT_c}{8P_c} = \frac{\overline{V}_c}{3} \tag{1.36}$$

The compressibility factor at the critical point is given by

$$Z_c = \frac{P_c \overline{V}_c}{RT_c} = \frac{3}{8} \tag{1.37}$$

The experimental values in Table 1.1 are smaller than this. Equations 1.35 and 1.36 can be used to calculate the van der Waals parameters for a gas if the critical temperature and pressure are known. When these equations are substituted into the van der Waals equation, it can be arranged in the form

$$\left[\frac{P}{P_c} + 3 \left(\frac{\overline{V}_c}{\overline{V}} \right)^2 \right] \left[3 \left(\frac{\overline{V}}{\overline{V}_c} \right) - 1 \right] = 8 \frac{T}{T_c} \tag{1.38}$$

This suggests the definition of a reduced pressure P_r, reduced volume V_r, and a reduced temperature T_r defined by

$$P_r \equiv \frac{P}{P_c} \qquad \overline{V}_r \equiv \frac{\overline{V}}{\overline{V}_c} \qquad T_r \equiv \frac{T}{T_c} \tag{1.39}$$

so that the van der Waals equation becomes

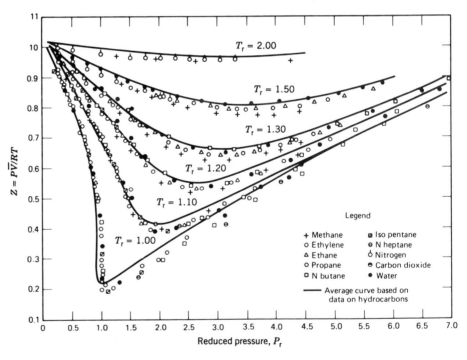

Figure 1.14   Compressibility factors for gases as a function of reduced pressure and temperature. (Reprinted with permission from G. J. Su, *Ind. Eng. Chem.* **38**:803 (1946). Copyright © 1946 American Chemical Society.)

$$\left(P_r + \frac{3}{\overline{V}_r^2}\right)(3\overline{V}_r - 1) = 8T_r \qquad (1.40)$$

This is referred to as a **reduced equation of state.** If the van der Waals equation were exact, all gases would follow this same equation of state because it does not contain any constants associated with the particular gas. This leads to the principle of corresponding states that is more general than equation 1.40. According to the **principle of corresponding states,** all gases obeying an equation of state with two parameters should follow the same equation of state provided their reduced pressures, volumes, and temperatures are used. This principle is exact, but, unfortunately, real gases cannot be accurately represented by two-parameter equations.

Figure 1.14 shows how well the principle of corresponding states applies to a variety of gases. The compressibility factor is plotted versus reduced pressure for a series of reduced temperatures. The principle of corresponding states works pretty well when the molecules are nearly spherical. If they are elongated, their properties differ from those of nearly spherical molecules. By introduction of an acentric factor it is possible to extend the usefulness of the principle of corresponding states.

The principle of corresponding states is very useful for the development of correlation and estimation methods for other properties.*

Example 1.6
What is the molar volume of ethane at 350 K and 70 bar according to (*a*) the ideal gas law, (*b*) Fig. 1.14 and (*c*) the van der Waals equation?

(*a*) $\overline{V} = RT/P = (0.083\ 14\text{ L bar K}^{-1}\text{ mol}^{-1})(350\text{ K})/(70\text{ bar})$

   $= 0.416\text{ L mol}^{-1}$

(*b*)   The critical constants are given in Table 1.2.

$$T_r = T/T_c = 350/305.4 = 1.146$$
$$P_r = P/P_c = 70/48.9 = 1.432$$

From Fig. 1.14, $PV/RT = 0.62$.

$$\overline{V} = \frac{(0.62)(0.083\ 14)(350)}{70} = 0.26\text{ L mol}^{-1}$$

(*c*)   The van der Waals constants are given in Table 1.3.

$$P = \frac{RT}{\overline{V} - b} - \frac{a}{\overline{V}^2}$$

$$70 = \frac{(0.083\ 14)(350)}{\overline{V} - 0.0638} - \frac{5.562}{\overline{V}^2}$$

---

* R. C. Reid, J. M. Prausnitz, and B. E. Poling, *The Properties of Gases and Liquids.* New York: McGraw-Hill, 1987.

This is a cubic equation, but we know it has a single solution because the temperature is above the critical temperature. The most practical way to solve it is to use successive approximations. This yields $\overline{V} = 0.23$ L mol^{-1}.

Example 1.7

A 1-L cylinder of molecular hydrogen is weighed and found to contain 5 mol. What is the pressure inside the cylinder at 25 °C according to the ideal gas law and according to van der Waals equation? Assuming that the van der Waals equation gives the correct answer, what is the percent error in using the ideal gas law?

According to the ideal gas law

$$P = \frac{nRT}{V}$$

$$= \frac{(5 \text{ mol})(0.08314 \text{ L bar K}^{-1} \text{ mol}^{-1})(298 \text{ K})}{1 \text{ L}}$$

$$= 123.9 \text{ bar}$$

According to van der Waals' equation

$$P = \frac{nRT}{V - nb} - \frac{an^2}{V^2}$$

$$= \frac{5(0.08314)(298)}{1 - (5)(0.02661)} - \frac{(0.2476)(5)^2}{1^2}$$

$$= 136.7 \text{ bar}$$

$$\text{Error} = \frac{12.8}{136.7} \times 100 = 9.4\%$$

We will see later that equations of state are very important in the calculation of various thermodynamic properties of gases. Therefore, a variety of them have been developed. To represent the P-V-T properties of a one-component system over a wide range of conditions it is necessary to use an equation with many more parameters. As more parameters are used they lose any simple physical interpretation. The van der Waals equation does not fit the properties of any gas exactly, but it is very useful because it does have a simple interpretation and the qualitatively correct behavior.

1.10   The Chain Rule and the Cyclic Rule

This chapter has been full of functions of two variables, and that is a common situation in thermodynamics. We used several partial derivatives, and so now is a good time to think about how many there are, how they are related, and how many are independent. There are six partial derivatives involving V, P, and T:

$$\left(\frac{\partial V}{\partial T}\right)_P, \left(\frac{\partial V}{\partial P}\right)_T, \left(\frac{\partial P}{\partial V}\right)_T, \left(\frac{\partial P}{\partial T}\right)_V, \left(\frac{\partial T}{\partial V}\right)_P, \left(\frac{\partial T}{\partial P}\right)_V \quad (1.41)$$

Two of these derivatives are used regularly and have names:

$$\alpha = \text{cubic expansion coefficient} = \frac{1}{V}\left(\frac{\partial V}{\partial T}\right)_P \tag{1.42}$$

$$\kappa = \text{isothermal compressibility} = -\frac{1}{V}\left(\frac{\partial V}{\partial P}\right)_T \tag{1.43}$$

The isothermal compressibility κ is always positive because an increase in pressure at fixed T corresponds with a decrease in volume. Two of the remaining derivatives, $(\partial T/\partial V)_P$ and $(\partial P/\partial V)_T$, are simply the reciprocals of the derivatives involved in α and κ. The last two derivatives, $(\partial P/\partial T)_V$, and $(\partial T/\partial P)_V$, are simply reciprocals of each other and may be calculated using the cyclic rule.

If $z = f(x_1, x_2)$ and $x_1 = x_1(y_1, y_2)$, $x_2 = x_2(y_1, y_2)$, then $z = f[x_1(y_1, y_2), x_2(y_1, y_2)]$ is a function of y_1 and y_2. Thus, we can consider z to be a function of either coordinates x_1, x_2 or coordinates y_1, y_2. The following equation can be used to calculate the partial derivative of f with respect to x_1:

$$\left(\frac{\partial f}{\partial x_1}\right)_{x_2} = \left(\frac{\partial f}{\partial y_1}\right)_{y_2}\left(\frac{\partial y_1}{\partial x_1}\right)_{x_2} + \left(\frac{\partial f}{\partial y_2}\right)_{y_1}\left(\frac{\partial y_2}{\partial x_1}\right)_{x_2} \tag{1.44}$$

This is known as the **chain rule** for partial differentiation. Taking $x_1 = T$, $x_2 = P$ and $y_1 = T$, $y_2 = V$ gives

$$\left(\frac{\partial f}{\partial T}\right)_P = \left(\frac{\partial f}{\partial T}\right)_V + \left(\frac{\partial f}{\partial V}\right)_T\left(\frac{\partial V}{\partial T}\right)_P \tag{1.45}$$

Later we will find the chain rule useful for other coordinate transformations. If we take $f = P$, $(\partial f/\partial T)_P = 0$ because P is constant and f is P, and so

$$0 = \left(\frac{\partial P}{\partial T}\right)_V + \left(\frac{\partial P}{\partial V}\right)_T\left(\frac{\partial V}{\partial T}\right)_P \tag{1.46}$$

so that

$$\left(\frac{\partial P}{\partial V}\right)_T\left(\frac{\partial V}{\partial T}\right)_P\left(\frac{\partial T}{\partial P}\right)_V = -1 \tag{1.47}$$

This equation, which is referred to as the **cyclic rule,** can be rearranged to

$$\left(\frac{\partial P}{\partial T}\right)_V = -\left(\frac{\partial P}{\partial V}\right)_T\left(\frac{\partial V}{\partial T}\right)_P = -\frac{(\partial V/\partial T)_P}{(\partial V/\partial P)_T} \tag{1.48}$$

Using equations 1.42 and 1.43 yields

$$\left(\frac{\partial P}{\partial T}\right)_V = \frac{\alpha}{\kappa} \tag{1.49}$$

For an ideal gas, $\alpha = 1/T$ and $\kappa = 1/P$.

1.11 Special Topic: Use of the Zeroth Law to Establish a Temperature Scale

The following mathematical treatment of the experiment described in Fig. 1.3 leads us to the concept of an equation of state.

Let us use C, in a specified state, as a reference state, and bring systems A and B into equilibrium with system C separately.* If we fix the pressure of system A at P_A, the volume must have the value V_A for system A to be in equilibrium with system C. The fact that V_A has this value at equilibrium means that there is a relationship of the form

$$F_1(P_A, V_A, P_C, V_C) = 0 \tag{1.50}$$

Since system B is also in equilibrium with system C, there is an additional equilibrium relationship:

$$F_2(P_B, V_B, P_C, V_C) = 0 \tag{1.51}$$

These equations can each be solved for P_C:

$$P_C = f_1(P_A, V_A, V_C) \tag{1.52}$$

$$P_C = f_2(P_B, V_B, V_C) \tag{1.53}$$

Thus,

$$f_1(P_A, V_A, V_C) = f_2(P_B, V_B, V_C) \tag{1.54}$$

This equation can be solved for P_A to obtain

$$P_A = g(V_A, P_B, V_B, V_C) \tag{1.55}$$

Now, according to the zeroth law, if systems A and B are separately in equilibrium with C, they are also in equilibrium with each other, and thus,

$$F_3(P_A, V_A, P_B, V_B) = 0 \tag{1.56}$$

When this is solved for P_A we obtain

$$P_A = f_3(V_A, P_B, V_B) \tag{1.57}$$

Since P_A is determined by just these three variables, V_C must cancel out of equation 1.55. Since equation 1.55 is just another form of equation 1.54, V_C must also cancel out of equation 1.54, and so equation 1.54 can be written

$$\Phi_A(P_A, V_A) = \Phi_B(P_B, V_B) \tag{1.58}$$

With this equation we have learned something very important about thermal equilibrium between two or more systems. When systems are in thermal equilibrium, there is, for each one, a function of its independent variables that has a common value for all of the systems. We call this common value the **temperature.** If a temperature Θ has been assigned to the reference system (thermometer), in this case C, the **equation of state** for any system at this temperature is

$$\Phi(P, V) = \Theta \tag{1.59}$$

An equation relates the various thermodynamic properties of a substance at equilibrium.

Now we need a method for assigning numerical values to the temperatures of the reference system. If the reference system has a property that varies with the temperature (such as volume), the simplest way to proceed is to take the

---

* The following derivation has been adapted from C. J. Adkins, *Equilibrium Thermodynamics.* Cambridge, UK: Cambridge University Press, 1983.

temperature to be proportional to this property x and calculate the constant a in

$$\Theta = ax \qquad (1.60)$$

by **assigning** the value of Θ at one reference point. As described in Section 1.3, Θ is assigned the value of 273.16 K at the triple point of water.

1.12 Special Topic: Barometric Formula

In applying thermodynamics we generally ignore the effect of the gravitational field, but it is important to realize that if there is a difference in height there is a difference in gravitational potential. For example, consider a vertical column of a gas with a unit cross section and a uniform temperature T, as shown in Fig. 1.15. The pressure at any height h is simply equal to the mass of a gas above that height per unit area times the gravitational acceleration g. The standard acceleration due to gravity is defined as $9.806\ 65$ m s^{-2}. The difference in pressure dP between h and $h + dh$ is equal to the mass of the gas between these two levels times g and divided by the area. Thus,

$$dP = -\rho g\ dh \qquad (1.61)$$

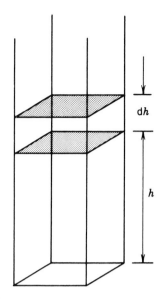

Figure 1.15   Column of an ideal gas of uniform temperature and unit cross section.

where ρ is the density of the gas. If the gas is an ideal gas, then $\rho = PM/RT$, where M is the molar mass, so that

$$dP = -\frac{PMg}{RT}\ dh \qquad (1.62)$$

Separating variables and integrating from $h = 0$, where the pressure is P_0, to h, where the pressure is P, yields

$$\int_{P_0}^{P} \frac{dP}{P} = -\int_{0}^{h} \frac{gM}{RT}\ dh \qquad (1.63)$$

$$\ln \frac{P}{P_0} = -\frac{gMh}{RT} \qquad (1.64)$$

$$P = P_0 e^{-gMh/RT} \qquad (1.65)$$

This relation is known as the **barometric formula.**

Example 1.8
Assuming that air is 20% O_2 and 80% N_2 at sea level and that the pressure is 1 bar, what are the composition and pressure at a height of 10 km, if the atmosphere has a temperature of 0 °C independent of altitude?

$$P = P_0 \exp\left(-\frac{gMh}{RT}\right)$$

For O_2,

$$P_{O_2} = (0.20\ \text{bar}) \exp\left[-\frac{(9.8\ \text{m s}^{-2})(32 \times 10^{-3}\ \text{kg mol}^{-1})(10^4\ \text{m})}{(8.314\ \text{J K}^{-1}\ \text{mol}^{-1})(273\ \text{K})}\right]$$

$$= 0.0503\ \text{bar}$$

For N_2,

$$P_{N_2} = (0.80) \exp\left(-\frac{9.8 \times 28 \times 10^{-3} \times 10^4}{8.314 \times 273}\right)$$

$$= 0.239 \text{ bar}$$

The total pressure is 0.289 bar, and $y_{O_2} = 0.173$ and $y_{N_2} = 0.827$.

References

C. J. Adkins, *Equilibrium Thermodynamics*. Cambridge, UK: Cambridge University Press, 1983.

K. E. Bett, J. S. Rowlinson, and G. Saville, *Thermodynamics for Chemical Engineers*. Cambridge, MA: MIT Press, 1975.

K. G. Denbigh, *The Principles of Chemical Equilibrium*. Cambridge, UK: Cambridge University Press, 1971.

J. H. Dymond and E. B. Smith, *The Virial Coefficients of Pure Gases and Mixtures*. Oxford, UK: Oxford University Press, 1980.

G. N. Lewis, M. Randall, revised by K. S. Pitzer, and L. Brewer, *Thermodynamics*. New York: McGraw-Hill, 1961.

J. M. Smith and H. C. Van Ness, *Introduction to Chemical Engineering Thermodynamics*. New York: McGraw-Hill, 1987.

Problems

1.1   The ideal gas law also represents the behavior of mixtures of gases at low pressures. The molar volume of the mixture is the volume divided by the amount of the mixture. The partial pressure of gas i in a mixture is defined as $y_i P$ for an ideal gas mixture, where y_i is its mole fraction, and P is the total pressure. Ten grams of N_2 is mixed with 5 g of O_2 and held at 25 °C at 0.750 bar. (*a*) What are the mole fractions of N_2 and O_2? (*b*) What are the partial pressures of N_2 and O_2? (*c*) What is the volume of the ideal mixture?

1.2   A mixture of methane and ethane is contained in a glass bulb of 500-cm³ capacity at 25 °C. The pressure is 1.25 bar, and the mass of gas in the bulb is 0.530 g. What is the average molar mass, and what is the mole fraction of methane?

1.3   Nitrogen tetroxide is partially dissociated in the gas phase according to the reaction

$$N_2O_4(g) = 2NO_2(g)$$

A mass of 1.588 g of N_2O_4 is placed in a 500-cm³ glass vessel at 1.0133 bar at 298 K and dissociates to an equilibrium mixture. (*a*) What are the mole fractions of N_2O_4 and NO_2? (*b*) What percentage of the N_2O_4 has dissociated? Assume that the gases are ideal.

1.4   Calculate the second virial coefficient for hydrogen at 0 °C from the fact that the molar volumes at 50.7, 101.3, 202.6, and 303.9 bar are 0.4634, 0.2386, 0.1271, and 0.090 04 L mol⁻¹, respectively.

1.5   The critical temperature of carbon tetrachloride is 283.1 °C. The densities in g/cm³ of the liquid ρ_l and vapor ρ_v at different temperatures are as follows:

$t/°C$	100	150	200	250	270	280
ρ_l	1.4343	1.3215	1.1888	0.9980	0.8666	0.7634
ρ_v	0.0103	0.0304	0.0742	0.1754	0.2710	0.3597

What is the critical molar volume of CCl_4? It is found that the mean of the densities of the liquid and vapor does not vary rapidly with temperature and can be represented by

$$\frac{\rho_l + \rho_v}{2} = AT + B$$

where A and B are constants. The extrapolated value of the average density at the critical temperature is the critical density. The molar volume $\overline{V}_c$ at the critical point is equal to the molar mass divided by the critical density.

1.6 Show that for a gas of rigid spherical molecules, b in the van der Waals equation is four times the molecular volume times Avogadro's constant. If the molecular diameter of Ne is 0.258 nm (Table 18.4), approximately what value of b is expected?

1.7 What is the molar volume of n-hexane at 660 K and 91 bar according to (a) the ideal gas law and (b) the van der Waals equation?

$$T_c = 507.7 \text{ K} \qquad P_c = 30.3 \text{ bar}$$

1.8 Derive the expressions for van der Waals constants a and b in terms of the critical temperature and pressure; that is, derive equations 1.35 and 1.36 from 1.32–1.34.

1.9 Calculate the second virial coefficient of methane at 300 K and 400 K from its van der Waals constants, and compare these results with Fig. 1.9.

1.10 You want to calculate the molar volume of O_2 at 298.15 K and 50 bar using the van der Waals equation, but you don't want to solve a cubic equation. Use the first two terms of equation 1.29. The van der Waals constants of O_2 are $a = 0.138$ Pa m^6 mol^{-1} and $b = 31.8 \times 10^{-6}$ m^3 mol^{-1}. What is the molar volume in L mol^{-1}?

1.11 Show that the equation for the dotted line in Fig. 1.13 is

$$P = \frac{a}{V^2}\left(1 - \frac{2b}{V}\right)$$

This equation gives the boundary of the physically unrealizable region for a van der Waals gas. What is the maximum pressure satisfying this equation?

1.12 Calculate the Boyle temperature for hydrogen from its van der Waals constants.

1.13 What is the equation of state for a liquid for which the coefficient of cubic expansion α and the isothermal compressibility κ are constant?

1.14 The cubic expansion coefficient α is defined by

$$\alpha = \frac{1}{V}\left(\frac{\partial V}{\partial T}\right)_P$$

and the isothermal compressibility κ is defined by

$$\kappa = -\frac{1}{V}\left(\frac{\partial V}{\partial P}\right)_T$$

Calculate these quantities for an ideal gas.

1.15 For a liquid the cubic expansion coefficient α is nearly constant over a narrow range of temperature. Derive the expression for the volume as a function of temperature and the limiting form for temperatures close to T_0.

1.16 Assuming that the atmosphere is isothermal at 0 °C and that the average molar mass of air is 29 g mol^{-1}, calculate the atmospheric pressure at 20 000 ft above sea level.

1.17 Calculate the pressure and composition of air on the top of Mt. Everest, assuming that the atmosphere has a temperature of 0 °C independent of altitude ($h = 29$ 141 ft). Assume that air at sea level is 20% O_2 and 80% N_2.

1.18 A mole of air (80% nitrogen and 20% oxygen by volume) at 298.15 K is brought into contact with liquid water, which has a vapor pressure of 3168 Pa at this temperature. (a) What is the volume of the dry air if the pressure is 1 bar? (b) What is the final volume of the air saturated with water vapor if the total pressure is maintained at 1 bar? (c) What are the mole fractions of N_2, O_2, and H_2O in the moist air? Assume the gases are ideal.

1.19 Using Fig. 1.9 calculate the compressibility factor Z for $NH_3(g)$ at 400 K and 50 bar.

1.20 In this chapter we have considered only pure gases, but it is important to make calculations on mixtures as well. This requires information in addition to that for pure gases. Statistical mechanics shows that the second virial coefficient for an N-component gaseous mixture is given by

$$B = \sum_{i=1}^{N}\sum_{j=1}^{N} y_i y_j B_{ij}$$

where y is mole fraction and i and j identify components. Both indices run over all components of the mixture. The bimolecular interactions between i and j are characterized by B_{ij}, and so $B_{ij} = B_{ji}$. Use this expression to derive the expression for B for a binary mixture in terms of y_1, y_2, B_{11}, B_{12}, and B_{22}.

1.21 The densities of liquid and vapor methyl ether in g cm^{-3} at various temperatures are as follows:

t/°C	30	50	70	100	120
ρ_l	0.6455	0.6116	0.5735	0.4950	0.4040
ρ_v	0.0142	0.0241	0.0385	0.0810	0.1465

The critical temperature of methyl ether is 299 °C. What is the critical molar volume? (See problem 1.5.)

1.22   The critical temperature is generally about 1.6 times greater than the normal boiling point, and the critical volume is generally about 2.7 times greater than the molar volume at the normal boiling point. The compressibility at the critical point is generally about 0.28 for hydrocarbons. On the basis of these observations estimate the critical properties of *n*-butane from its normal boiling point, -0.5 °C, and its molar volume at the normal boiling point, 96.7 cm^3 mol^{-1}.

1.23   Assuming that CH_4 is a spherical molecule, calculate the molecular diameter from the van der Waals constant b. The molecular diameter obtained from viscosity measurements is 0.414 nm. (See problem 1.6.)

1.24   Use the van der Waals constants for CH_4 in Table 1.3 to calculate the initial slopes of the plots of the compressibility factor Z versus P at 300 and 600 K.

1.25   Using the van der Waals equation, calculate the pressure exerted by 1 mol of carbon dioxide at 0 °C in a volume of (*a*) 1.00 L and (*b*) 0.05 L. (*c*) Repeat the calculations at 100 °C and 0.05 L.

1.26   A mole of *n*-hexane is confined in a volume of 0.500 L at 600 K. What will be the pressure according to (*a*) the ideal gas law and (*b*) the van der Waals equation? (See problem 1.7.)

1.27   Gases are in corresponding states when they have the same reduced temperatures and pressures. Under what conditions are hydrogen, carbon dioxide, and water in states corresponding to nitrogen at 400 K and 1 bar?

1.28   A mole of ethane is contained in a 200-mL cylinder at 373 K. What is the pressure according to (*a*) the ideal gas law and (*b*) the van der Waals equation? The van der Waals constants are given in Table 1.3.

1.29   Use the van der Waals constants for H_2 and O_2 in Table 1.3 to calculate the initial slopes of the plots of the compressibility factor Z versus P at 0 °C.

1.30   When pressure is applied to a liquid, its volume decreases. Assuming that the isothermal compressibility

$$\kappa = -\frac{1}{V}\left(\frac{\partial V}{\partial P}\right)_T$$

is independent of pressure, derive an expression for the volume as a function of pressure.

1.31   Calculate α and κ for a gas for which

$$P(\overline{V} - b) = RT$$

1.32   What is the mean atmospheric pressure in Denver, Colorado, which is a mile high, assuming an isothermal atmosphere at 25 °C? Air may be taken to be 20% O_2 and 80% N_2.

1.33   Calculate the pressure and composition of air 100 miles above the surface of the earth assuming that the atmosphere has a temperature of 0 °C independent of altitude.

1.34   The density $\rho = m/V$ of a mixture of ideal gases A and B is determined and is used to calculate the average molar mass M of the mixture; $M = \rho RT/P$. How is the average molar mass determined in this way related to the molar masses of A and B?

2
First Law of Thermodynamics

In this chapter we begin to emphasize processes that take a chemical system from one state to another. The first law of thermodynamics, which is often referred to as the law of conservation of energy, leads to the definition of a new thermodynamic state function, the internal energy U. An additional state function, the enthalpy H, is defined in terms of U, P, and V for reasons of convenience.

Thermochemistry, which deals with the heat produced by chemical reactions and solution processes, is based on the first law. If heat capacities of reactants and products are known, the heat of a reaction may be calculated at other temperatures once it is known at one temperature.

2.1 Work

Force is a vector quantity; that is, it has direction as well as magnitude. Other examples of vector quantities are displacement, velocity, acceleration, and electric field strength. In this book vector quantities are represented by boldface italic type. The magnitude of the vector is represented with lightface italic type. **Force** is defined by

$$f = ma \tag{2.1}$$

where f is the force is the force that will give a mass m an acceleration a.

Work (w) is a scalar quantity defined by

$$w = f \cdot l \tag{2.2}$$

where f is the vector force, l is the vector length of path, and the dot indicates a scalar product (i.e., the product is taken of the magnitude of one vector by the projection of the second vector along the direction of the first). If the force vector of magnitude f and the vector length of magnitude l are separated by the angle θ, the work is given by $fl \cos \theta$.* In SI units, the unit of work is the joule J; 1 J = 1 N m. The differential quantity of work dw done by a force f operating over a distance dl in the direction of the force is $f\, dl$.

Since pressure P is force per unit area, the force on a piston is PA, where A is the surface area perpendicular to the direction of the motion of the piston. Thus, the differential quantity of work done by an expanding gas that causes the piston to move distance dl is $PA\, dl$. But $A\, dl = dV$, the increase in gas volume, and so the differential quantity of pressure–volume work is $P\, dV$.

Work w can be positive or negative since work may be done on a system or a system may do work on its surroundings, as shown in Fig. 2.1. **The IUPAC convention on w is that it is positive when work is done on the system of interest and negative when the system does work on the surroundings.** (As we will see later, a similar convention is applied to heat q; q is positive when heat is transferred from the surroundings to a system, and q is negative when heat is transferred from the system to the surroundings.) Thus, the differential of the PV work done on a system is given by

$$\mathrm{d}w = -P_{\text{ext}}\, dV \tag{2.3}†$$

where P_{ext} is the external or applied pressure.

Work is often conveniently measured by the lifting or falling of masses. The work required to lift a mass m in the earth's gravitational field, which has an acceleration g, is mgh, where h is the height through which the mass is lifted.

The work w required to lift a kilogram 0.1 m is

$$w = mgh = (1 \text{ kg})(9.807 \text{ m s}^{-2})(0.1 \text{ m}) = 0.9807 \text{ J}$$

The total work w on a system when there is a finite change in volume is obtained by summing the infinitesimal amounts of work given by equation 2.3:

---

* Multiplication of vectors is described in Appendix D.

† We use $\mathrm{d}w$ rather than dw as a reminder that work is not an exact differential (Section 2.5), and so the value of its integral depends on the path.

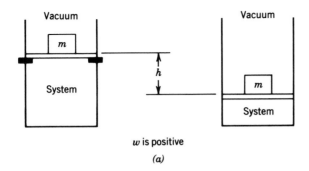

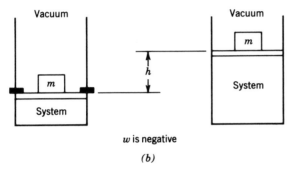

Figure 2.1 (a) Work is done on a system by the surroundings. In this case the stops are pulled out, and the system is compressed to a new equilibrium state. (b) Work is done on the surroundings by the system. When the stops are pulled out, the system expands to a new equilibrium state.

$$w = -\int_1^2 P_{\text{ext}}\,dV \qquad (2.4a)$$

To make this calculation of the work for a finite change in state, P must have a definite value at each instant.

The integral in equation 2.4a is called a **line integral** because its value depends on the path, even when the process is reversible. In this case $P_{\text{ext}} = P$ and the pressure is a function of temperature and volume, and so equation 2.4a should be written

$$w = -\int_1^2 P(T,\ V)\,dV \qquad (2.4b)$$

In an ordinary definite integral, the integrand is a function of one variable. Later in Section 2.8, we will replace P with nRT/V and integrate equation 2.4b at constant temperature, but now we want to take a more general point of view and consider the two processes in Fig. 2.2. The state of a mole of gas can be changed from $(2P_0,\ V_0)$ to $(P_0,\ 2V_0)$ by an infinite number of reversible paths, but we will consider only the two reversible paths shown. In the upper path, the pressure is held constant at $2P_0$ and the gas is heated until it reaches $2V_0$. Then the volume is held constant while the gas is cooled until the pressure

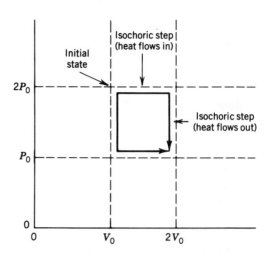

Figure 2.2 For the change in state of a mole of gas from $(2P_0, V_0)$ to $(P_0, 2V_0)$, the work done on the gas depends on the path. By the upper path, $w = -2P_0V_0$; by the lower path, $w = -P_0V_0$. For a clockwise cyclic process, $w = -2P_0V_0 + P_0V_0 = -P_0V_0$. In the cyclic process, the gas is returned to its initial state, and so $\Delta U = 0 = q - P_0V_0$. Thus, $q = P_0V_0$ and heat is absorbed by the system in the cyclic process.

reaches P_0. For this path, $w = -2P_0V_0$. In the lower path, the volume is held constant at V_0 and the gas is cooled until it reaches P_0. Then the gas is heated at constant pressure until it reaches $2V_0$. For this path, $w = -P_0V_0$. Note that in both cases the work is the negative of the area under the path.

2.2 Joule's Experiments

The first law of thermodynamics is often called the law of conservation of energy. This concept first appeared in mechanics and was later extended to include electrostatics and electrodynamics. **Joule performed experiments in 1843–1878 that showed how heat could also be included in conservation of energy.** He showed that under adiabatic conditions a given amount of work heats the water in a calorimeter a certain number of degrees, independent of whether the work is used to turn a paddle wheel (Fig. 2.3) or is dissipated by an electrical current flowing through a resistance or by the friction of rubbing two objects together. An **adiabatic process** is one in which the system under investigation is thermally isolated so that there is no exchange of heat with the surroundings. Since a given change in state of the water in the calorimeter can be accomplished in different ways involving the same amount of work, or by different sequences of steps, the change in state is independent of the path and is dependent only on the total amount of work. This makes it possible to express the change in state of a system in an adiabatic process in terms of the work required, without stating the type of work or the sequence of steps used. The property of the system whose change is calculated in this way is called the **internal energy** U. Since the internal energy U of a system may be increased

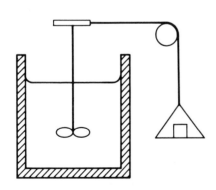

Figure 2.3 Joule heated water by performing work on it, in this case by rotating a paddle wheel, and found that the temperature rise depends only on the amount of work done on the system.

by doing work on it, we may calculate the increase in internal energy from the work w done on a system to change it from one state to another in an adiabatic process:

$$\Delta U = w \qquad \text{(in an adiabatic process)} \qquad (2.5)$$

In words, **the work done on a closed system in an adiabatic process is equal to the increase in internal energy of the system.** The symbol Δ indicates the value of the quantity in the final state minus the value of the quantity in the initial state; $\Delta U = U_2 - U_1$ where U_1 is the internal energy in the initial state and U_2 is the internal energy in the final state. **If the system does work on its surroundings w is negative and, furthermore, ΔU is negative (i.e., the internal energy of the system decreases) if the process is adiabatic.** When equation 2.5 is applied to a system of arbitrary size, the internal energy is an extensive quantity, but in working problems we will often deal with molar quantities and express the change in molar internal energy $\Delta \overline{U}$ in J mol^{-1}.

2.3 Heat

A given change in state of a system can be accomplished in ways other than by the performance of work under adiabatic conditions. A change equivalent to that in the Joule experiment described in the preceding section may be obtained by immersing a hot object in the water. We should not say, however, that the water now has more heat any more than we would say it has more work after it has been heated with moving paddle wheels. In other words, heat and work are forms of energy crossing a boundary. After the experiment, the temperature of the water is higher, and it has a greater internal energy U. Heat is transferred when there is a gradient in temperature, as shown in Fig. 2.4.

Since the same change in state (as determined by measuring properties such as temperature, pressure, and volume) may be produced by doing work on the system or by allowing heat to flow in, the amount of heat q may be expressed in mechanical units. When Joule was doing his experiments the unit of heat was the calorie, which is the heat required to raise the temperature of a gram of water 1 °C, from 14.5 to 15.5 °C. He was able to determine the mechanical equivalent of heat, which is now known to be 1 calorie = 4.184 kg m^2 s^{-2} = 4.184 J. Now we find it more convenient to express heat in joules and to **define the calorie as 4.184 J.** A joule of heat is the amount of heat that produces the same change in a system as a joule of work.

Since heat is an algebraic quantity, it is important to adopt a sign convention. **We will take a positive value of q to indicate that heat is absorbed by the system from its surroundings. A negative value of q means that the system gives up heat to its surroundings.** The change in internal energy U produced by the transfer of heat q to a system when no work is done is given by

$$\Delta U = q \qquad \text{(no work done)} \qquad (2.6)$$

In words, **the heat absorbed by a closed system in a process in which no work is done is equal to the increase in internal energy of the system.** Or, put another way, if no work is done, the heat evolved is equal to the decrease in the internal energy of the system.

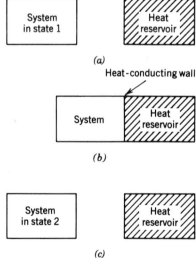

(a)

Heat-conducting wall

(b)

(c)

Figure 2.4 (a) A system in state 1 is insulated from the heat reservoir. (b) The system is brought into contact with the heat reservoir through a heat-conducting wall. (c) The system is then insulated from the heat reservoir and is found to be in state 2.

2.4   The First Law of Thermodynamics and Internal Energy

Since the internal energy of a system can be changed a given amount by either heat or work, these quantities are in this sense equivalent. They are both usually expressed in joules. If both heat and work are added to a system,

$$\Delta U = q + w \tag{2.7}$$

For an infinitesimal change in state

$$dU = đq + đw \tag{2.8}$$

The đ indicates that q and w are not exact differentials, as discussed in the next section.

Equations 2.7 and 2.8 are statements of the **first law of thermodynamics.** This law is the postulate that there exists a property U, referred to as the internal energy, (1) which is a function of the state variables for the system and (2) for which the change ΔU for a process in a closed system may be calculated using equation 2.7.

This mathematical form of the first law seems obvious to us now, but prior to 1850 it was not obvious at all. Before 1850 the principle of conservation of energy in mechanical systems was understood, but the role of heat in this principle was not clear until Joule's experiments led to equation 2.7. Note that the first law provides a means for determining changes in internal energy, but not the absolute value of the internal energy.

If ΔU is negative we may say that the system loses energy in heat that is evolved and work that is done by the system. The first law has nothing to say about how much heat is evolved and how much work is done except that equation 2.7 is obeyed. In other words, the entire decrease in internal energy could show up as work ($q = 0$). Another possibility is that even more than this amount of work would be done and heat would be absorbed ($q > 0$), so that equation 2.7 is obeyed. Although the first law has nothing to say about the relative amounts of heat and work, the second law does, as we will see in Chapter 3. Since the internal energy is a function of the state of a system, there is no change in internal energy when a system is taken through a series of changes that return it to its **initial** state. This is expressed by setting the **cyclic integral** equal to zero:

$$\oint dU = 0 \tag{2.9}$$

The circle indicates integration around a cycle. The cyclic integrals of q and w are not generally equal to zero, and their values depend on the path followed.

The first law is frequently stated in the form that energy may be transferred in one form or another, but it cannot be created or destroyed. Thus, the total energy of an isolated system is constant.*

---

* It is sometimes incorrectly stated that mass can be converted into energy according to Einstein's relation $E = mc^2$, where c is the speed of light. The mass m in this equation is the relativistic mass that is related to the rest mass m_0 by

$$m = m_0 \left(1 - \frac{v^2}{c^2} \right)^{-1/2}$$

where v is the speed of the object. The correct interpretation of $E = mc^2$ is that energy E and mass m are related through the necessarily positive proportionality constant c^2. A given mass m

The internal energy U of a system is an extensive property (Section 1.2); thus, if we double a system the internal energy is doubled. However, the molar internal energy is an intensive property. We will use U for the extensive property and $\overline{U}$ for the intensive property.

2.5 Exact and Inexact Differentials

The internal energy U is a state function, like V, because it depends only on the state of the system. The integral of the differential of a state function along any arbitrary path is simply the difference between values of the function at two limits. For example, if a system goes from state a to state b, we can write

$$\int_a^b dU = U_b - U_a = \Delta U \qquad (2.10)$$

Since the integral is path independent, the differential of a state function is called an **exact differential.** The quantities q and w are not state functions. The integrals of their differentials in going from state a to state b **depend on the path chosen.** Therefore, their differentials are called **inexact differentials.** We will use đ instead of d to indicate inexact differentials. In going from state a to state b the work w done is represented by

$$\int_a^b đw = w \qquad (2.11)$$

Note that the result of the integration is not written $w_b - w_a$, because the amount of work done depends on the particular path that is followed between state a and state b. For example, when a gas is allowed to expand, the amount of work obtained may vary from zero (if the gas is allowed to expand into a vacuum) to a maximum value that is obtained if the expansion is carried out reversibly, as described in Section 2.7.

If an infinitesimal quantity of heat $đq$ is absorbed by a system, and an infinitesimal amount of work $đw$ is done on the system, the infinitesimal change in the internal energy is given by

$$dU = đq + đw \qquad (2.12)$$

where the d is used with U since dU is an **exact** differential and đ is used with q and w because they are **inexact** differentials. In other words, U is a function of the state of the system, and q and w for a process depend on the path.

It is interesting to note that the sum of two inexact differentials can be an exact differential. To illustrate this point further, we consider the path from a to b in Fig. 2.5. We may define the path by a curve $y = y(x)$ connecting a and b.

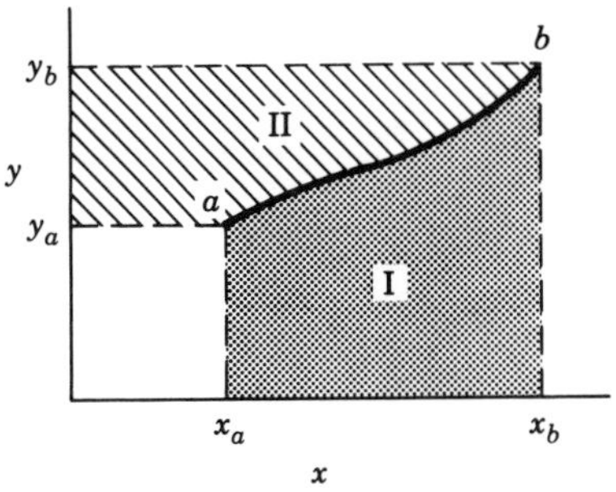

Figure 2.5 Path of a system in going from state a to state b.

---

is therefore equivalent to a certain energy. Since, according to the first law of thermodynamics, energy is conserved, Einstein's relation requires that mass be included in this conservation principle. R. P. Bauman (*J. Chem. Educ.* **43**:366 (1966)) gives a clear illustration of the meaning of relativistic mass and its relation to energy. If a ball of mass m_0 is kicked, its energy and mass are increased at the expense of the energy and mass of the kicker. As the ball bounces to a rest, it gives up energy and mass to the earth. In nuclear fission the rest mass of the fragments is less than the rest mass of the original atom, but the mass of the surroundings is increased through collision with the fission fragments.

The differential $đz = y\,dx$ is not an exact differential

$$\int_a^b đz = z = \int_a^b y\,dx = \text{area I} \tag{2.13}$$

because this area depends on the path between a and b, as may be seen from the figure.

The differential $dz = y\,dx + x\,dy$ is an exact differential. Since $dz = d(xy)$,

$$\int_a^b dz = \Delta z = \int_a^b d(xy) = x_b y_b - x_a y_a \tag{2.14}$$

The reason $dz = y\,dx + x\,dy$ is an exact differential may be seen from Fig. 2.5. The integral of dz from state a to state b may be written

$$\int_a^b dz = \Delta z = \int_a^b y\,dx + \int_a^b x\,dy = \text{area I} + \text{area II} \tag{2.15}$$

The sum of these areas is independent of the shape of the curve (path) between a and b. If $\int dz$ does not depend on the path taken between the points, then dz is said to be an exact differential. Thermodynamic quantities like U, H, S, and G form exact differentials, since their values are dependent on the state variables, and not on the path by which the system got there. There is a simple test to see whether a differential is exact.

2.6 The Test for Exactness

For a system with just two independent degrees of freedom, the total differential dz of a quantity z may be determined by the differentials dx and dy in two other quantities x and y. In general,

$$dz = M(x, y)dx + N(x, y)dy \tag{2.16}$$

where M and N are functions of the independent variables x and y.

To show the test for exactness we now consider a function z that has an exact differential. If z has a definite value at each point in the xy plane, then it must be a function of x and y. If $z = f(x, y)$, then

$$dz = \left(\frac{\partial z}{\partial x}\right)_y dx + \left(\frac{\partial z}{\partial y}\right)_x dy \tag{2.17}$$

Comparing equations 2.16 and 2.17, we find

$$M(x, y) = \left(\frac{\partial z}{\partial x}\right)_y \tag{2.18}$$

$$N(x, y) = \left(\frac{\partial z}{\partial y}\right)_x \tag{2.19}$$

Since the mixed partial derivatives are equal,

$$\left[\frac{\partial}{\partial y}\left(\frac{\partial z}{\partial x}\right)_y\right]_x = \left[\frac{\partial}{\partial x}\left(\frac{\partial z}{\partial y}\right)_x\right]_y \tag{2.20}$$

then

$$\left(\frac{\partial M}{\partial y}\right)_x = \left(\frac{\partial N}{\partial x}\right)_y \qquad (2.21)$$

This equation must be satisfied if dz is an exact differential. **It is Euler's criterion for exactness.** This relation is also very useful for obtaining relations between the derivatives of thermodynamic functions.

To illustrate the use of equation 2.21 let us reconsider the differential dz = y dx. Since M = y and N = 0, $(\partial M/\partial y)_x$ = 1 and $(\partial N/\partial x)_y$ = 0, so that equation 2.21 is not satisfied. Therefore, dz = y dx is not an exact differential. On the other hand, dz = y dx + x dy is an exact differential: M = y, N = x so that $(\partial M/\partial y)_x$ = 1, $(\partial N/\partial x)_y$ = 1, satisfying equation 2.21.

Example 2.1

Show that the differential dV of the volume of an ideal gas is an exact differential.

$$V = \frac{RT}{P}$$

$$dV = -\frac{RT}{P^2}\,dP + \frac{R}{P}\,dT$$

$$\frac{\partial}{\partial T}\left(-\frac{RT}{P^2}\right) = -\frac{R}{P^2} = \frac{\partial}{\partial P}\left(\frac{R}{P}\right)$$

Since the mixed partial derivatives are equal, V is a function of the state of the system. This will always be true for a differentiable function, but the question is whether V is a function of the state of the system when dV = $M(P, T)$ dP + $N(P, T)$ dT.

2.7 Work of Compression and Expansion of a Gas at Constant Temperature

Since work done in compressing a gas is positive we start by considering the compression of a gas **at constant temperature** using the idealized apparatus shown in Fig. 2.6. The gas is contained in a rigid cylinder by a frictionless and

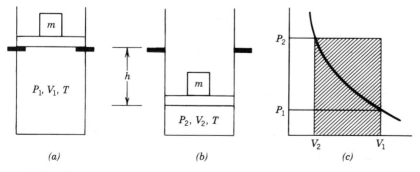

(a)   (b)   (c)

Figure 2.6 Compression of a gas from P_1, V_1, T to P_2, V_2, T in a single step.

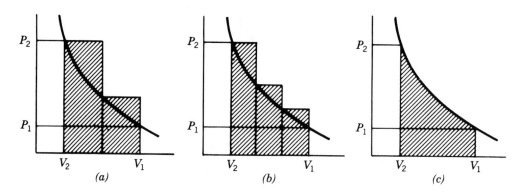

Figure 2.7   Compression of a gas from P_1, V_1, T to P_2, V_2, T in two, three, and an infinite number of steps.

a weightless piston. The cylinder is immersed in a thermostat at temperature T and the space above the cylinder is evacuated so that the final pressure P_2 is due only to the mass m. The gas is initially confined to volume V_1 because the piston is held up by stops. When the stops are pulled out, the piston falls to the equilibrium position, and the gas is compressed to volume V_2. The pressure of the gas at the end of the process is given by

$$P_2 = \frac{mg}{A} \tag{2.22}$$

where g is the acceleration due to gravity, and A is the area of the piston. The amount of work lost in the surroundings is mgh (equation 2.4), where h is the difference in height, and so the work done on the gas is

$$w = mgh = -P_2(V_2 - V_1) \tag{2.23}$$

Since $V_2 < V_1$, the work done on the gas is positive. This is the smallest amount of work that can be used to compress the gas from V_1 to V_2 in a single step at constant temperature. The work done is given by the cross-hatched area in the P–V plot of Fig. 2.6c. Notice that the pressure used in calculating the work is not the pressure of the gas but the external pressure determined by the mass m, cross-sectional area, and acceleration of gravity g.

However, we can carry out the compression with less work if we do it in two or more steps, as shown in Fig. 2.7. We can compress the gas in two steps

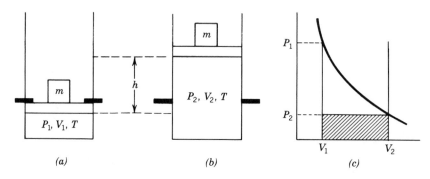

Figure 2.8   Expansion of a gas from P_1, V_1, T to P_2, V_2, T in a single step.

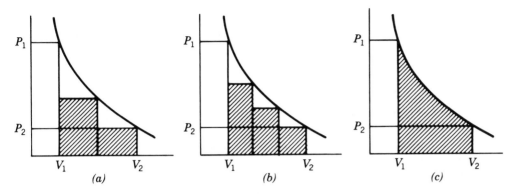

Figure 2.9   Expansion of a gas from P_1, V_1, T to P_2, V_2, T in two, three, and an infinite number of steps.

by first using a mass m' just large enough to compress the gas to volume $(V_1 + V_2)/2$ in the first step, and then using the larger mass m for the second step. The work lost in the surroundings is given by the cross-hatched area in Fig. 2.7a. It is clear that by using more and more steps we arrive eventually at the diagram in Fig. 2.7c which shows that the minimum amount of work is required in the limit of an infinite number of steps. In this case the pressure is changed by an infinitesimal amount for each infinitesimal step, and the work is given by the integral of equation 2.3 at constant temperature.

$$w = \int dw = -\int_{V_1}^{V_2} P \, dV \qquad (2.24)$$

In case c it is not necessary to distinguish between the external pressure and the gas pressure because they differ at most by an infinitesimal amount.

The work w done on a gas in an expansion can be determined using the idealized frictionless piston arrangement shown in Fig. 2.8. The gas is initially confined to volume V_1 because the piston is held by stops. When the stops are pulled out, the gas expands to volume V_2. The mass is chosen so that $P_2 = mg/A$; in other words, this is the maximum mass that the gas will raise to this height. The work gained in the surroundings is mgh, and so the work done on the gas is

$$w = -mgh = -P_2(V_2 - V_1) \qquad (2.25)$$

This is the negative of the largest amount of work that can be obtained in the surroundings by the expansion of the gas from V_1 to V_2 at constant temperature in a single step. The work done on the gas is given by the negative of the cross-hatched area in Fig. 2.8c.

More work can be obtained in the surroundings by using two, three, or an infinite number of steps, as shown in Fig. 2.9. The largest amount of work in the surroundings and the largest negative work on the gas are obtained in the limit of an infinite number of steps. The work done on the gas in the limiting case is given by equation 2.24.

The work obtained in the surroundings in the single-step expansion is clearly not great enough to compress the gas back to its initial state in a single-step compression; this is evident from the cross-hatched areas in Figs. 2.8c and 2.6c. (The subscripts are different in the two figures because we have followed the usual convention of labeling the initial state with a 1 and the final state with a 2.) However, the work obtained in the surroundings in the infinite-step expansion is exactly the amount required to compress the gas back to its initial state by an infinite-step compression; this is evident from the cross-hatched areas in Fig. 2.9c and Fig. 2.7c.

The infinite-step compression described by Fig. 2.7c and the infinite-step expansion described by Fig. 2.9c are referred to as **reversible processes**. These idealized processes at constant temperature are reversible because the energy accumulated in the surroundings in the expansion is exactly the amount required to compress the gas back to the initial state. This can also be seen by applying equation 2.24 to the gas for a complete cycle from P_1, V_1, T to P_2, V_2, T, and back again. (Note that here we are using the same subscripts for the expansion and the compression.)

$$w_{\text{cycle}} = -\int_{V_1}^{V_2} P \, dV - \int_{V_2}^{V_1} P \, dV$$

$$= -\int_{V_1}^{V_2} P \, dV + \int_{V_1}^{V_2} P \, dV = 0 \qquad (2.26)$$

This conclusion applies to any fluid when the path is completely described.

Another important point about reversible processes is that they can be reversed at any point in the process by making an infinitesimal change, in this case in the pressure. Thus, a reversible expansion or compression requires an absence of friction, a balancing of internal and external pressures, and time to reestablish equilibrium after each infinitesimal step. When these conditions are not met the process is **irreversible,** and the system and its surroundings cannot both be restored to their initial conditions. Remember that these are isothermal processes and there is heat flow that we have not talked about.

All real processes are irreversible, yet it is possible to approach reversibility closely in some real processes. Heat may be transferred nearly reversibly if the temperature gradient across which it is transferred is made very small. Electrical charge may be transferred nearly reversibly from a battery if a potentiometer is used so that the difference in electrical potential is very small. A liquid may be vaporized nearly reversibly if the pressure of the vapor is made only very slightly less than that of the equilibrium vapor.

The concept of a reversible process is important because certain thermodynamic calculations can be made only for reversible processes. For processes in the chemical industry, the greater the irreversibility, the greater is the loss in capacity to do work; thus, literally every irreversibility has its cost.

Example 2.2

Two moles of gas at 1 bar and 298 K are compressed at constant temperature by use of a constant pressure of 5 bar. How much work is done on the gas? If the compression is driven by a 100-kg mass, how far will it fall in the earth's gravitational field?

$$w = -P_2(V_2 - V_1)$$

$$= -P_2 \left(\frac{nRT}{P_2} - \frac{nRT}{P_1} \right)$$

$$= -nRT \left(1 - \frac{P_2}{P_1} \right)$$

$$= - (2 \text{ mol})(8.314 \text{ J K}^{-1} \text{ mol}^{-1})(298 \text{ K})(1 - 5)$$

$$= 19\,820 \text{ J}$$

$$h = - \frac{w}{mg}$$

$$= (19\,820 \text{ J})/(100 \text{ kg})(9.8 \text{ m s}^{-2})$$

$$= - 20.22 \text{ m}$$

2.8 Reversible Isothermal Expansion of a Gas

When a gas is expanded in a series of infinitesimal steps with $P_{\text{ext}} = P$ at each step, the pressure throughout the gas is constant (within an infinitesimal amount) during each step. Such a reversible process is often spoken of as one that consists of a series of successive equilibria. Since the gas is at its equilibrium pressure (within an infinitesimal amount) at each step in the expansion, we may substitute the pressure given by an equation of state into equation 2.24 and integrate it. If the gas were allowed to expand rapidly, the pressure and temperature would not be uniform throughout the volume of the gas, and so such a substitution could not be made. If the expansion is carried out reversibly at constant temperature for an ideal gas, the external pressure is always given by $P = nRT/V$. Substituting in equation 2.24, we obtain

$$w_{\text{rev}} = - \int_{V_1}^{V_2} P \, dV = - \int_{V_1}^{V_2} \frac{nRT}{V} \, dV = - nRT \ln \frac{V_2}{V_1} \qquad (2.27)^*$$

since the temperature is constant.

In integration the lower limit always refers to the initial state and the upper limit to the final state. If the gas is compressed, the final volume is smaller and w_{rev} is positive. The positive value means that work is done on the gas. The isothermal expansion of one mole of an ideal gas by a factor of ten yields $w_{\text{rev}} = -(1 \text{ mol})RT \ln 10 = -5229 \text{ J}$ at 273.15 K.

Since for an ideal gas at constant temperature $P_1V_1 = P_2V_2$, the reversible work is also given by

$$w_{\text{rev}} = nRT \ln \frac{P_2}{P_1} \qquad (2.28)$$

The equation for the maximum work of isothermal expansion of a van der Waals gas is obtained by using equation 1.31:

$$w_{\text{rev}} = - \int_{V_1}^{V_2} \left(\frac{nRT}{V - nb} - \frac{an^2}{V^2} \right) dV$$

$$= - nRT \ln \frac{V_2 - nb}{V_1 - nb} + an^2 \left(\frac{1}{V_1} - \frac{1}{V_2} \right) \qquad (2.29)$$

---

* ln represents the natural logarithm and log represents the base 10 logarithm; $\ln x = 2.303 \log x$.

Example 2.3

One mole of an ideal gas expands from 5 to 1 bar at 298 K. Calculate w (a) for a reversible expansion and (b) for an expansion against a constant external pressure of 1 bar?

(a) $\quad w_{\text{rev}} = nRT \ln \dfrac{P_2}{P_1}$

$\quad\quad\quad\quad\quad = (1 \text{ mol})(8.314 \text{ J K}^{-1} \text{ mol}^{-1})(298 \text{ K}) \ln \dfrac{1 \text{ bar}}{5 \text{ bar}}$

$\quad\quad\quad\quad\quad = -3988 \text{ J}$

(b) $\quad w_{\text{irrev}} = -P_2(V_2 - V_1) = -P_2 \left(\dfrac{nRT}{P_2} - \dfrac{nRT}{P_1} \right)$

$\quad\quad\quad\quad\quad = -(1 \text{ bar})(1 \text{ mol})(8.314 \text{ J K}^{-1} \text{ mol}^{-1})(298 \text{ K}) \left(\dfrac{1}{1 \text{ bar}} - \dfrac{1}{5 \text{ bar}} \right)$

$\quad\quad\quad\quad\quad = -1982 \text{ J}$

More work is done on the surroundings when the expansion is carried out reversibly.

2.9   Various Kinds of Work

Work can also be done on a system by extending the area of its surface, stretching it (if it is a solid), and moving the electric charge from a lower to a higher electric potential.

Let us consider the work required to increase the area of a surface. The force f required to increase the area of a liquid film, as illustrated in Fig. 2.10, is

$$f = 2l\gamma \tag{2.30}$$

where γ is the **surface tension** of the liquid and l is the length of the movable bar. The factor 2 is involved because there are two liquid–gas interfaces in this experiment. The surface tension is force per unit length and is usually expressed in N m^{-1}. It is a temperature-dependent quantity in general. The surface tension of water at 25 °C is 71.97×10^{-3} N m^{-1} or 71.97 mN m^{-1}.* The surface tensions of liquid metals and molten salts are large in comparison with those of other liquids, as shown in Table 2.1. The work required to move the bar in Fig. 2.10 to the left by a distance Δx is

$$w = f \, \Delta x = 2l \, \Delta x \gamma = \gamma \, \Delta\sigma \tag{2.31}$$

where $\Delta\sigma$ is the change in surface area $(2l \, \Delta x)$. This is the amount of work done on the liquid system. According to this equation, surface tension is equal to the ratio of work to change in area, so that it can also be expressed in J m^{-2}. The differential of surface work is given by

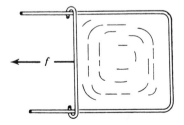

Figure 2.10   Idealized experiment for the determination of the surface tension of a liquid.

---

* In older literature surface tensions are usually expressed in dynes cm^{-1}. In these units the surface tension of water at 25 °C is $(71.97 \times 10^{-3}$ N m$^{-1})(10^5$ dynes N$^{-1})(10^{-2}$ m cm$^{-1}) = 71.97$ dynes cm^{-1}.

$$đw = \gamma \, d\sigma \qquad (2.32)$$

Surface tension arises from the fact that the molecules in the surface of a liquid are attracted into the body of the liquid by the molecules in the body. This inward attraction causes the surface to contract if it can and gives rise to a force in the plane of the surface. Surface tension is responsible for the formation of spherical droplets, the rise of water in a capillary, and the movement of a liquid through a porous solid. Solids also have surface tensions, but it is harder to measure them.

The surface tension of a liquid may be measured by a variety of methods.* For example, since the equilibrium shape of liquid surfaces is determined by a balance of surface tension and gravitational forces, analysis of drop or bubble shape may be used to determine surface tension. The rise of liquid in a capillary or the pull on a thin vertical plate partially immersed in the liquid may be determined and used to calculate the surface tension quite accurately. Less accurate values of the surface tension may be obtained from measurements on moving liquid surfaces. These methods include studies of liquid jets, ripples, drop weight, and the force required to rupture a surface.

Two other forms of work are more familiar, and so we do not need to discuss them in detail. When a piece of rubber is stretched the differential work done on the rubber is given by

$$đw = f \, dl \qquad (2.33)$$

where f is the force and dl is the differential increase in length. When a small charge dQ is moved through an electric potential difference ϕ, the work done on the charge is given by

$$đw = \phi \, dQ \qquad (2.34)$$

These differential work terms become a part of the first law if surface, elongational, and electrical work are involved:

$$dU = đq - P \, dV + \gamma \, d\sigma + f \, dl + \phi \, dQ \qquad (2.35)$$

This is an important equation because it shows how surface tension, surface area, force, elongation, electric potential, and electric charge come into thermodynamics.

The variables involved in work form conjugate pairs of intensive and extensive variables, as shown in Table 2.2. If both the intensive variable and the

Table 2.1 Surface Tensions of Some Liquids

Substance	$t/°C$	γ/N m^{-1}
Copper	1404	1.1
Lead	350	0.453
Mercury	20	0.472
Benzene	20	0.028 8
Ethanol	20	0.022 8
Water	20	0.072 8
	25	0.072 0
	100	0.058 0
Oxygen	−183	0.013 1
Nitrogen	−183	0.006 2

Table 2.2 Some Conjugate Pairs of Thermodynamic Variables

Type of Work	Intensive Variable	Extensive Variable	Differential Work
Hydrostatic	Pressure, P	Volume, V	$-P \, dV$
Surface	Surface tension, γ	Area, σ	$\gamma \, d\sigma$
Elongation	Force, f	Length, l	$f \, dl$
Electrical	Potential difference, ϕ	Electric charge, Q	$\phi \, dQ$

* A. W. Adamson, *Physical Chemistry of Surfaces*, 4th ed. New York: Wiley-Interscience, 1982.

extensive variable are expressed in SI units (see Symbols in the Appendix), the work is expressed in joules.

In this section and the preceding one we have considered processes in thermostats without saying anything about the quantity of heat flowing into or out of the gas. Now it is time to talk about the flow of heat that accompanies such processes.

Example 2.4

(a) A piece of stretched rubber exerts a force of 1 N. How much work has to be done on the rubber to stretch it one centimeter? (b) The surface tension of water is 0.072 N/m at 25 °C. How much work has to be done to increase the water surface by one square meter? (c) A mole of electrons is transported across a potential difference of 1 V from the positive electrode to the negative electrode. How much work is required?

(a) $w = f \, \Delta l = (1 \text{ N})(0.01 \text{ m}) = 0.01 \text{ J}$

(b) $w = \gamma \, \Delta \sigma = (0.072 \text{ N m}^{-1})(1 \text{ m}^2) = 0.072 \text{ J}$

(c) $w = \phi \, \Delta Q = (1 \text{ V})(96\,500 \text{ coulombs}) = 96\,500 \text{ J}$

2.10   Enthalpy

Constant-pressure processes are more common in chemistry than constant-volume processes because many operations are carried out in open vessels. If only pressure–volume work is done and the pressure is constant and equal to the applied pressure, the work done on the system w equals $-P \, \Delta V$, so that equation 2.7 may be written

$$\Delta U = q_P - P \, \Delta V \qquad (2.36)$$

where q_P is the heat for the isobaric (constant pressure) process. If the initial state is designated by 1 and the final state by 2, then

$$U_2 - U_1 = q_P - P(V_2 - V_1) \qquad (2.37)$$

so that the heat absorbed is given by

$$q_P = (U_2 + PV_2) - (U_1 + PV_1) \qquad (2.38)$$

Since the heat absorbed is given by the difference of two quantities that are functions of the state of the system, it is convenient to introduce a new state function, the **enthalpy** H, which is defined by

$$H = U + PV \qquad (2.39)$$

Since the enthalpy is defined in terms of other thermodynamic quantities, the only gain is in terms of convenience. Equation 2.38 may be written

$$q_P = H_2 - H_1 \qquad (2.40)$$

In words, **the heat absorbed in a process at constant pressure is equal to the change in enthalpy if the only work done is reversible pressure–volume work.** For an infinitesimal change

$$đq_P = dH \qquad (2.41)$$

where dH is an exact differential since the enthalpy is a function of the state of the system.

When pressure–volume work is the only kind of work (electrical and other kinds being excluded), it is easy to visualize ΔU and ΔH; in a constant-volume calorimeter (Section 2.20) the evolution of heat is a measure of the decrease in internal energy U, and in a constant-pressure calorimeter the evolution of heat is a measure of the decrease in enthalpy H.

The enthalpy H is an extensive property (Section 1.2). Sometimes, we will be concerned with the molar enthalpy $\overline{H}$, which is an intensive property.

2.11 Change in State at Constant Volume

When a system changes from one state to another, the change in internal energy U or enthalpy H may be calculated from the heat q evolved and the work w done on the system by the surroundings. For a chemically inert system of fixed mass the internal energy U may be taken to be a function of any two of T, V, and P. It is most convenient to take it as a function of T and V. Since U is a state function, the differential dU is given by

$$dU = \left(\frac{\partial U}{\partial T}\right)_V dT + \left(\frac{\partial U}{\partial V}\right)_T dV \qquad (2.42)$$

The first term is the change in internal energy due to the temperature change alone, and the second term is the change in internal energy due to the volume change alone. Since the differential of the internal energy is given by $đq - P_{ext}\, dV$, if only pressure–volume work is involved, and the external pressure is equal to P, then

$$đq = \left(\frac{\partial U}{\partial T}\right)_V dT + \left[P + \left(\frac{\partial U}{\partial V}\right)_T\right] dV \qquad (2.43)$$

If the change in state of system X takes place at constant volume, it may be represented by

$$X(V_1, T_1) \rightarrow X(V_1, T_2)$$

In this case equation 2.43 reduces to

$$đq_V = \left(\frac{\partial U}{\partial T}\right)_V dT \qquad (2.44)$$

Since the change in temperature and the heat transferred are readily measured, it is convenient to define the **heat capacity C_V at constant volume** as

$$C_V \equiv \frac{đq_V}{dT} = \left(\frac{\partial U}{\partial T}\right)_V \qquad (2.45)$$

This equation may be applied to a system of any size, but frequently we will be concerned with the intensive thermodynamic quantity $\overline{C}_V$, which has the SI units $J\ K^{-1}\ mol^{-1}$. Since the heat capacity at constant volume is readily

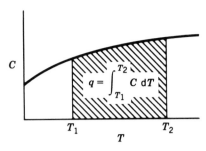

Figure 2.11 The heat q absorbed by a substance when it is heated is equal to the integral of $C\, dT$ from the initial temperature T_1 to the final temperature T_2. If an amount n is heated at constant volume, then $C = nC_V$; and if an amount n is heated at constant pressure, then $C = nC_P$.

measured, equation 2.45 may be integrated to obtain the change in internal energy for a finite change in temperature at constant volume:

$$\Delta U = \int_{T_1}^{T_2} C_V\, dT = q_V \tag{2.46}$$

This is illustrated in Fig. 2.11. Over a small temperature range C_V may be nearly constant so that

$$\Delta U = C_V(T_2 - T_1) = C_V\, \Delta T \tag{2.47}$$

Example 2.5

The molar heat capacity at constant volume of $O_2(g)$ is given by

$$\overline{C}_V = \alpha + \beta T + \gamma T^2$$

where $\alpha = 17.23$ J K^{-1} mol^{-1}, $\beta = 13.61 \times 10^{-3}$ J K^{-2} mol^{-1}, and $\gamma = 42.55 \times 10^{-7}$ J K^{-3} mol^{-1}. What is the change in molar internal energy when oxygen is heated from 298 to 500 K?

$$\Delta \overline{U} = \int_{298}^{500} \overline{C}_V\, dT = 17.23(500 - 298) + (13.61 \times 10^{-3}/2)(500^2 - 298^2)$$
$$- (42.55 \times 10^{-7}/3)(500^3 - 298^3)$$
$$= 4437 \text{ J mol}^{-1}$$

Note this is also the heat absorbed in the process of heating at constant volume.

In principle, the quantity $(\partial U/\partial V)_T$ may be measured in an experiment devised by Joule. Imagine two gas bottles connected with a valve and enclosed in a thermally isolated container, as shown in Fig. 2.12. The two bottles constitute the system under consideration. The first bottle is filled with a gas under pressure, and the second is evacuated. When the valve is opened, gas rushes from the first bottle into the second. Joule found that there was no discernible change in the temperature once thermal equilibrium had been established, and so đq = 0. No work is done in this expansion, and so đw = 0 and dU = đq + đw = 0. Since the temperature is constant, equation 2.42 becomes

$$dU = \left(\frac{\partial U}{\partial V}\right)_T dV = 0 \tag{2.48}$$

Since $dV \neq 0$,

$$\left(\frac{\partial U}{\partial V}\right)_T = 0 \tag{2.49}$$

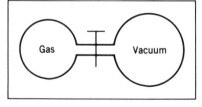

Figure 2.12 Joule's experiment in which a gas expands into a vacuum. Joule found there was no discernable change in temperature and concluded that $(\partial U/\partial V)_T = 0$. We now know that this applies to ideal gases, but not to real gases.

Thus, Joule concluded that the internal energy of the gas is independent of the volume. However, this method is not very sensitive because of the large heat capacity of the gas bottles relative to the gas. Equation 2.49 does apply to an ideal gas. The molecular interpretation of this relation is that there is no interaction between the molecules of an ideal gas, and so the internal energy does not change with the distance between molecules. On the other hand, the internal energy of a real gas depends on the volume at constant temperature, but the second law of thermodynamics is needed to derive equation 4.48 which can be used to obtain $(\partial U/\partial V)_T$ experimentally.*

Example 2.6

Calculate the heat absorbed and the changes in internal energy and enthalpy for the two expansions of an ideal gas described in Example 2.3.

(a)   According to Joule's experiment, $\Delta U = 0$ for the isothermal expansion of an ideal gas. Therefore,

$$q_{rev} = \Delta U - w_{rev}$$
$$= 0 - (-3988 \text{ J})$$
$$= 3988 \text{ J}$$
$$\Delta H = \Delta U + \Delta(PV) = 0$$

(b) $$q_{irrev} = \Delta U - w_{irrev}$$
$$= 0 - (-1982 \text{ J})$$
$$= 1982 \text{ J}$$
$$\Delta H = \Delta U + \Delta(PV) = 0$$

Thus, more heat is absorbed by the gas in the reversible isothermal expansion.

2.12   Change in State at Constant Pressure

These changes in state are of special interest in the laboratory where processes generally take place at constant pressure. For a chemically inert system of fixed mass, it is most convenient to take enthalpy H as a function of temperature and pressure. Since H is a state function, the differential dH is given by

$$dH = \left(\frac{\partial H}{\partial T}\right)_P dT + \left(\frac{\partial H}{\partial P}\right)_T dP \qquad (2.50)$$

If the change in state of a mole of X takes place reversibly at constant pressure, it may be represented by

$$X(P_1, T_1) \rightarrow X(P_1, T_2) \qquad (2.51)$$

For such a change equations 2.41 and 2.50 may be combined to obtain

$$đq_P = \left(\frac{\partial H}{\partial T}\right)_P dT \qquad (2.52)$$

---

* In Example 4.2 we will find that for a van der Waals gas

$$\left(\frac{\partial U}{\partial V}\right)_T = \frac{a}{\bar{V}^2}$$

This should not be surprising because van der Waals added $a/\bar{V}^2$ to the pressure to provide for intermolecular attractions.

Since the change in temperature and the heat transferred are readily measured, it is convenient to define the **heat capacity at constant pressure** C_P as

$$C_P \equiv \frac{\mathrm{d}q_P}{\mathrm{d}T} = \left(\frac{\partial H}{\partial T}\right)_P \qquad (2.53)$$

Since the heat capacity at constant pressure is readily measured, this equation may be integrated to obtain the change in enthalpy for a finite change in temperature:

$$\Delta H = \int_{T_1}^{T_2} C_P \,\mathrm{d}T \qquad (2.54)$$

2.13 Joule–Thomson Expansion

In a Joule–Thomson expansion, a gas flows along a pipe through a porous plate or a constriction to a lower pressure, as illustrated in Fig. 2.13. The pressure is maintained at a lower value on the right side of the porous plate ($P_2 < P_1$) by withdrawing the piston on the right. This process is of interest because under certain conditions it leads to cooling ($T_2 < T_1$). The pipe is insulated and so the gas does not absorb or evolve heat ($q = 0$). To push one mole of gas through the porous plate, work amounting to $P_1\overline{V}_1$ has to be done on the gas by the piston on the left. Work amounting to $P_2\overline{V}_2$ is done on the surroundings by the piston on the right, and so the net work on the gas is

$$w = P_1\overline{V}_1 - P_2\overline{V}_2 \qquad (2.55)$$

Since the pipe is insulated, $q = 0$, and

$$\overline{U}_2 - \overline{U}_1 = P_1\overline{V}_1 - P_2\overline{V}_2 \qquad (2.56)$$

or

$$\overline{U}_2 + P_2\overline{V}_2 = \overline{U}_1 + P_1\overline{V}_1 \qquad (2.57)$$

$$\overline{H}_2 = \overline{H}_1 \qquad (2.58)$$

Thus, we see that there is no change in the enthalpy of the gas in a Joule–Thomson expansion.

The **Joule–Thomson coefficient** μ_{JT} is defined as the derivative of the temperature with respect to pressure in this process:

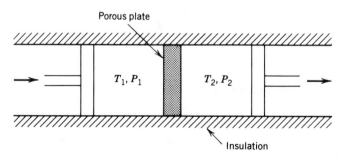

Figure 2.13   Joule–Thompson expansion.

$$\mu_{JT} = \lim_{\Delta P \to 0} \frac{T_2 - T_1}{P_2 - P_1} = \left(\frac{\partial T}{\partial P}\right)_H \tag{2.59}$$

Although the process is irreversible, we can use the fact that the change in state functions is independent of the process to obtain an expression for the Joule–Thomson coefficient in terms of state functions.

Since $dH = 0$ in equation 2.50,

$$(\partial T/\partial P)_H = -\frac{(\partial H/\partial P)_T}{(\partial H/\partial T)_P} \tag{2.60}$$

Since $C_P = (\partial H/\partial T)_P$, the Joule–Thomson coefficient is given by

$$\mu_{JT} = (\partial T/\partial P)_H = -\frac{(\partial H/\partial P)_T}{C_P} \tag{2.61}$$

Thus, the Joule–Thomson coefficient makes it possible to evaluate $(\partial H/\partial P)_T$; in Section 2.11 we noted that Joule was unsuccessful in measuring $(\partial U/\partial V)_T$ directly by allowing a gas to expand into a vacuum. After we have the second law we will see in equation 4.50 that $(\partial H/\partial P)_T$ can be expressed in terms of P-V-T data so that

$$\mu_{JT} = \frac{T(\partial V/\partial T)_P - V}{C_P} \tag{2.62}$$

Thus, the values of Joule–Thomson coefficients can be calculated from C_P and P-V-T data, as well as measured directly.

Example 2.7
Show that the Joule–Thomson coefficient for an ideal gas is zero.

$$(\partial V/\partial T)_P = nR/P$$

$$\mu_{JT} = \frac{nRT/P - V}{C_P} = 0$$

If P_1 and T_1 are chosen arbitrarily, and various values of P_2 are used, plots of T_2 versus P_2 such as those shown in Fig. 2.14 are obtained. These are called isenthalpic curves or isenthalps because the enthalpy is constant along each line. The dashed line is called the **inversion curve** because the sign of $(\partial T/\partial P)_H$ changes from positive (cooling) to negative (heating) along this line. The inversion temperature is the temperature at which μ_{JT} changes a sign. A gas can be used as a refrigerant only at temperatures below the maximum inversion temperature, which is 607 K for nitrogen, 204 K for hydrogen, and 43 K for helium.

Example 2.8
Show that the Joule–Thomson coefficient for a real gas is not zero in the limit of zero pressure.

For a real gas the compressibility factor at low pressures can always be represented by the virial equation:

$$P\overline{V} = RT(1 + BP + C'P^2 + \cdots)$$

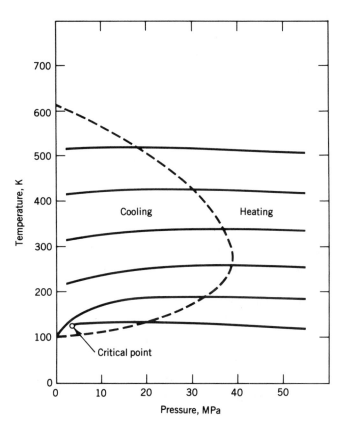

Figure 2.14 Isenthalps and inversion curve for nitrogen. (From M. W. Zemansky and R. H. Dittman, *Heat and Thermodynamics*. New York: McGraw-Hill, 1981. Reprinted with permission.)

where B and C' are functions of temperature only. Thus,

$$P\left(\frac{\partial \overline{V}}{\partial T}\right)_P = R(1 + BP + C'P^2 + \cdots) + RT\left(P\frac{dB}{dT} + P^2\frac{dC'}{dT} + \cdots\right)$$

The Joule–Thomson coefficient is therefore given by

$$\mu_{JT} = RT^2 \frac{(dB/dT) + P(dC'/dT) + \cdots}{\overline{C}_P}$$

$$\lim_{P \to 0} \mu_{JT} = RT^2 \frac{dB/dT}{\overline{C}_P}$$

Thus, not all the properties of a real gas approach those of an ideal gas when the pressure of the real gas is reduced to zero.

2.14 The Relation Between C_P and C_V for an Ideal Gas

We again consider a change in a system such that only pressure–volume work is involved. According to equation 2.43, if the pressure is constant

$$\mathrm{d}q_P = C_V\,\mathrm{d}T + \left[P + \left(\frac{\partial U}{\partial V}\right)_T\right]\mathrm{d}V \qquad (2.63)$$

Dividing by $\mathrm{d}T$ and setting $\mathrm{d}q_P/\mathrm{d}T = C_P$, we obtain

$$C_P - C_V = \left[P + \left(\frac{\partial U}{\partial V}\right)_T\right]\left(\frac{\partial V}{\partial T}\right)_P \qquad (2.64)$$

The quantity on the right side is positive so that $C_P > C_V$. The two terms on the right side may be interpreted as follows: $P(\partial V/\partial T)_P$ is the work produced per unit increase in temperature at constant pressure, and

$$\left(\frac{\partial U}{\partial V}\right)_T \left(\frac{\partial V}{\partial T}\right)_P$$

is energy per unit temperature required to separate the molecules against intermolecular attraction.

Equation 2.64 takes on a particularly simple form for an ideal gas because $(\partial U/\partial V)_T = 0$ and $(\partial V/\partial T)_P = nR/P$. Thus,

$$C_P - C_V = nR \quad \text{or} \quad \overline{C}_P - \overline{C}_V = R \qquad (2.65)$$

This relationship may be visualized as follows. When an ideal gas is heated at constant pressure, the work done in pushing back the piston is $P\,\Delta V = nR\,\Delta T$. For a 1 K change in temperature, the work done is $nR(1\,\text{K})$, and this is just the extra energy required to heat an ideal gas 1 K at constant pressure over that required at constant volume.

We will see later in equation 4.54 that by use of the second law the difference between $\overline{C}_P$ and $\overline{C}_V$ for any material may be expressed in terms of the cubic expansion coefficients α and the isothermal compressibility κ (equations 1.42 and 1.43). The values of $\overline{C}_P$ and $\overline{C}_V$ for liquids and solids are nearly the same.

2.15 Heat Capacities

Values of $\overline{C}_P$ at 25 °C for about 200 substances are given in Appendix C.1, and values for 298 to 3000 K are given for a smaller number of substances in Appendix C.2. The dependence of $\overline{C}_P$ on temperature is shown for a number of gases in Fig. 2.15. In general, the more complex the molecule, the greater its molar heat capacity, and the greater the increase with increasing temperature.

Power series in temperature may be used to represent $\overline{C}_P$ as a function of temperature:

$$\overline{C}_P = \alpha + \beta T + \gamma T^2 \qquad (2.66)$$

Parameters for several gases are given in Table 2.3. The change in enthalpy with temperature at constant pressure is then given by

$$\overline{H}_2 - \overline{H}_1 = \int_{\overline{H}_1}^{\overline{H}_2} \mathrm{d}\overline{H} = \int_{T_1}^{T_2} \overline{C}_P\,\mathrm{d}T \qquad (2.67)$$

$$\overline{H}_2 - \overline{H}_1 = \alpha(T_2 - T_1) + \frac{\beta}{2}(T_2^2 - T_1^2) + \frac{\gamma}{3}(T_2^3 - T_1^3) \qquad (2.68)$$

JANAF Thermochemical Tables and Stull, Westrum, and Sinke, *The Chemical Thermodynamics of Organic Compounds* give values of $\overline{H}_T^\circ - \overline{H}_{298}^\circ$ for various

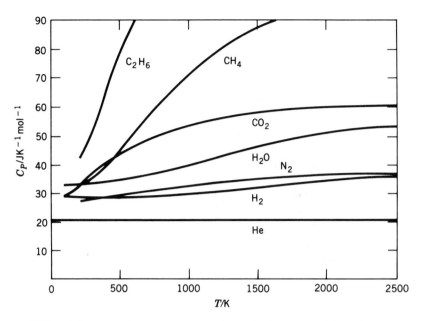

Figure 2.15 Influence of temperature on the molar heat capacities of gases at constant pressure.

Table 2.3 Molar Heat Capacity at Constant Pressure as a Function of Temperature from 300 to 1500 K: $\overline{C}_P = \alpha + \beta T + \gamma T^2$

	$\dfrac{\alpha}{\text{J K}^{-1}\,\text{mol}^{-1}}$	$\dfrac{\beta}{10^{-3}\,\text{J K}^{-2}\,\text{mol}^{-1}}$	$\dfrac{\gamma}{10^{-7}\,\text{J K}^{-3}\,\text{mol}^{-1}}$
$H_2(g)$	29.066	−0.836	20.113
$O_2(g)$	25.503	13.612	−42.553
$Cl_2(g)$	31.696	10.144	−40.375
$N_2(g)$	26.984	5.910	−3.376
$HCl(g)$	28.166	1.809	15.465
$H_2O(g)$	30.206	9.936	11.14
$CO_2(g)$	26.648	42.262	−142.4
$CH_4(g)$	14.143	75.495	−179.64
$C_2H_6(g)$	9.404	159.836	−462.28

temperatures so that $\overline{H}_2^\circ - \overline{H}_1^\circ$ is readily calculated for the substances listed. The superscript zero indicates that the substance is in its standard state; standard states are discussed in Section 2.17. In the absence of data on the heat capacity of a gas, an estimate may be made using the group additivity method as described in Section 2.22.

Example 2.9

Using data in Appendix C.2, calculate the change in the molar enthalpy of methane in going from 500 to 1000 K.

$$\overline{H}^{\circ}_{1000} - \overline{H}^{\circ}_{500} = (\overline{H}^{\circ}_{1000} - \overline{H}^{\circ}_{298}) - (\overline{H}^{\circ}_{500} - \overline{H}^{\circ}_{298})$$

$$= (38.179 - 8.201) \text{ kJ mol}^{-1}$$

$$= 29.978 \text{ kJ mol}^{-1}$$

(Note that this is not the change in the enthalpy of formation of methane; see problem 2.26.)

Thermodynamics does not deal with molecular models, and it is unnecessary to even discuss molecules in connection with thermodynamics. This is one of the strengths of thermodynamics, but it is also a weakness because thermodynamics, by itself, does not provide the means for predicting the numerical values of thermodynamic properties of particular substances. We will see later that kinetic theory and statistical mechanics do lead to quantitative predictions of thermodynamic properties.

Kinetic theory (Chapter 18) shows that the molar translational energy of a monatomic ideal gas is $\frac{3}{2} RT$. The translational energy $\overline{U}_t$ is independent of pressure or molar mass for a monatomic ideal gas so that

$$\overline{U}_t = \tfrac{3}{2} RT \tag{2.69}$$

According to equation 2.39, the molar enthalpy of a monatomic ideal gas is larger than the internal energy by $P\overline{V}$(or RT), so that

$$\overline{H}_t = \tfrac{3}{2} RT + RT = \tfrac{5}{2} RT \tag{2.70}$$

Thus, the translational contribution to the molar heat capacities of monatomic ideal gases are expected to be

$$\overline{C}_V = \left(\frac{\partial \overline{U}}{\partial T}\right)_V = \tfrac{3}{2} R = 12.472 \text{ J K}^{-1} \text{ mol}^{-1} \tag{2.71}$$

$$\overline{C}_P = \left(\frac{\partial \overline{H}}{\partial T}\right)_P = \tfrac{5}{2} R = 20.786 \text{ J K}^{-1} \text{ mol}^{-1} \tag{2.72}$$

Tables C.1 and C.2 show that values of $\overline{C}_P$ for nonatomic gases are constant at 20.876 J K^{-1} mol^{-1} independent of temperature, except for cases where electrons in the atom can be excited to low-lying levels (see especially O(g)).

At room temperature and above, the heat capacities of solid elements heavier than potassium are about 25 J K^{-1} mol^{-1} ($3R$). This relationship, which was first pointed out by Dulong and Petit in 1819, played an important role in the early determination of relative atomic masses because it could be used to select the multiple of the known equivalent weight (obtained from quantitative analysis) that is required to give the relative mass.

However, solid elements with lower atomic numbers have lower heat capacities at room temperature, and the heat capacities of all solids approach zero as the temperature is reduced toward absolute zero. According to the Debye theory (Section 17.15), the heat capacity of an atomic crystal at constant volume should be a universal function of T/Θ_D, where Θ_D is the Debye temperature. This is illustrated in Fig. 2.16. The Debye theory shows that the heat capacity should be proportional to T^3 at low temperatures. This is useful in extrapolating heat capacity data to the neighborhood of absolute zero, where it is difficult to obtain experimental values.

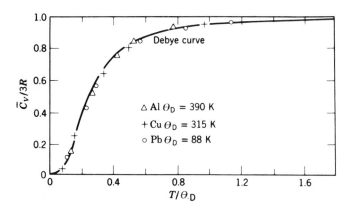

Figure 2.16   Heat capacity data for several solid elements plotted versus T/Θ_D, where Θ_D is the Debye temperature (Section 17.15). (Data from *A Compendium of the Properties of Materials at Low Temperatures*, Part II, *Properties of Solids.* Wadd Technical Report 60-56, Part II, 1960. Reproduced with permission from *Statistical Physics* by Mandl. Copyright © 1971, John Wiley & Sons Ltd.)

2.16   Adiabatic Processes with Gases

In Section 2.7 we discussed the work of compression and expansion of gases in contact with a heat reservoir. Now we consider the compression and expansion of gases in isolated systems. No heat is gained or lost by the gas, so the process is adiabatic and the first law becomes simply $dU = đw$. If only pressure–volume work is involved, $dU = -P_{ext}\, dV$. If the system expands adiabatically, dV is positive and dU is negative; thus, if the expansion is opposed by an external pressure P_{ext}, work is done on the surroundings at the expense of the external energy. The relation $dU = đw = -P_{ext}\, dV$ applies to any adiabatic process, reversible or irreversible, if PV work is the only kind of work involved. If the external pressure is zero (adiabatic expansion into a vacuum), no work is done and there is no change in the internal energy for all gases. If the expansion is opposed by an external pressure, work is done on the surroundings and the temperature drops as internal energy is converted to work. Integrating yields

$$\int_{U_1}^{U_2} dU = -\int_{V_1}^{V_2} P_{ext}\, dV$$

$$\Delta U = U_2 - U_1 = w \tag{2.73}$$

where w is the work done on the gas.

For an ideal gas, the internal energy is a function only of temperature and so $dU = C_V dT$ (equation 2.42). Thus, when an ideal gas expands adiabatically against an external pressure, the temperature drop is simply related to the change in internal energy. If C_V is independent of temperature for an ideal gas at the temperature of interest, then

$$\int_{U_1}^{U_2} dU = C_V \int_{T_1}^{T_2} dT$$

$$\Delta U = U_2 - U_1 = C_V(T_2 - T_1) \tag{2.74}$$

Since $q = 0$, then $\Delta U = w$ and

$$w = \int_{T_1}^{T_2} C_V \, dT = C_V(T_2 - T_1) \tag{2.75}$$

where the second form applies when C_V is independent of temperature. This relation applies to the adiabatic expansion of an ideal gas with C_V independent of temperature whether the process is reversible or irreversible. If the gas expands, the final temperature T_2 will be lower than the initial temperature T_1 and the work done on the gas is negative. If the gas is compressed adiabatically it will heat up.

Example 2.10
An ideal monatomic gas at 1 bar and 273.15 K is allowed to expand adiabatically against a constant pressure of 0.315 bar until it doubles its volume. (*a*) What is the change in molar volume? (*b*) How much work is done on the gas in this process? (*c*) What is the final temperature? (*d*) What is the change in the molar internal energy of the gas? The value of $\overline{C}_V$ is $\frac{3}{2} R$.

(*a*) $\overline{V}_1 = RT/P_1 = (0.083\ 145 \text{ L bar K}^{-1} \text{ mol}^{-1})(273.15 \text{ K})/(1 \text{ bar})$

$\qquad = 22.71 \text{ L mol}^{-1}$

$\qquad \Delta \overline{V} = 2\overline{V}_1 - \overline{V}_1 = 22.71 \text{ L mol}^{-1}$

(*b*) $w = -P_{\text{ext}} \Delta \overline{V} = -(0.315 \times 10^5 \text{ Pa})(22.711 \times 10^{-3} \text{ m}^3 \text{ mol}^{-1})$

$\qquad = -715.4 \text{ J mol}^{-1}$

(*c*) $w = \overline{C}_V \Delta T$

$\qquad \Delta T = (-715.4 \text{ J mol}^{-1})/\frac{3}{2}(8.3145 \text{ J K}^{-1} \text{ mol}^{-1})$

$\qquad = -57.4 \text{ K}$

$\qquad T_2 = T_1 + \Delta T = 273.15 \text{ K} - 57.4 \text{ K} = 215.8 \text{ K}$

(*d*) $\Delta \overline{U} = \overline{C}_V \Delta T = \frac{3}{2}(8.3145 \text{ J K}^{-1} \text{ mol}^{-1})(-57.4 \text{ K})$

$\qquad = -715.4 \text{ J mol}^{-1}$

When an adiabatic expansion is carried out reversibly, the equilibrium pressure is substituted for the external pressure, and so for an ideal gas

$$\overline{C}_V \, dT = -P \, d\overline{V} = -\frac{RT}{\overline{V}} \, d\overline{V}$$

$$\overline{C}_V \frac{dT}{T} = -R \frac{d\overline{V}}{\overline{V}} \tag{2.76}$$

If the heat capacity is independent of temperature, then

$$\overline{C}_V \int_{T_1}^{T_2} \frac{dT}{T} = -R \int_{\overline{V}_1}^{\overline{V}_2} \frac{d\overline{V}}{\overline{V}}$$

$$\overline{C}_V \ln \frac{T_2}{T_1} = R \ln \frac{\overline{V}_1}{\overline{V}_2} \tag{2.77}$$

This equation is a good approximation only if the temperature range is small enough so that $\overline{C}_V$ does not change very much.

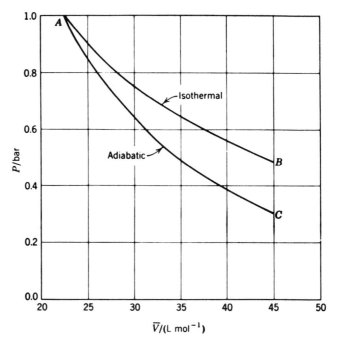

Figure 2.17   Isothermal and reversible adiabatic expansions of one mole of an ideal monatomic gas.

Since $\overline{C}_P - \overline{C}_V = R$, equation 2.77 may be written

$$\frac{T_2}{T_1} = \left(\frac{\overline{V}_1}{\overline{V}_2}\right)^{\gamma - 1} \tag{2.78}$$

where $\gamma = \overline{C}_P/\overline{C}_V$. By use of the ideal gas law we can obtain the following alternative forms of this equation:

$$\frac{T_2}{T_1} = \left(\frac{P_2}{P_1}\right)^{(\gamma - 1)/\gamma} \tag{2.79}$$

$$P_1 \overline{V}_1{}^\gamma = P_2 \overline{V}_2{}^\gamma \tag{2.80}$$

Thus, when a gas expands adiabatically to a larger volume and a lower pressure, the volume is smaller than it would be after an isothermal expansion to the same final pressure. Plots of pressure versus volume for adiabatic and isothermal expansions are shown in Fig. 2.17.

Example 2.11

Figure 2.17 shows that when one mole of an ideal monatomic gas is allowed to expand adiabatically and reversibly from 22.7 L mol^{-1} at 1 bar and 0 °C (at point A on the graph) to a volume of 45.4 L mol^{-1} (at point C), the pressure drops to 0.315 bar. Confirm this pressure and calculate the temperature at C. How much work is done in the adiabatic expansion?

$$\gamma = \frac{\frac{5}{2}R}{\frac{3}{2}R} = \frac{5}{3}$$

$$P_2 = P_1\left(\frac{\overline{V}_1}{\overline{V}_2}\right)^{\gamma} = (1 \text{ bar})\left(\frac{22.7 \text{ L mol}^{-1}}{45.4 \text{ L mol}^{-1}}\right)^{5/3} = 0.315 \text{ bar}$$

$$T_2 = T_1\left(\frac{\overline{V}_1}{\overline{V}_2}\right)^{\gamma-1} = (273.15 \text{ K})\left(\frac{22.7 \text{ L mol}^{-1}}{45.4 \text{ L mol}^{-1}}\right)^{2/3} = 172.07 \text{ K or } -101.08 \text{ °C}$$

$$w = \int_{T_1}^{T_2} \overline{C}_V \, dT = \tfrac{3}{2}R(172.07 \text{ K} - 273.15 \text{ K}) = -1261 \text{ J mol}^{-1}$$

2.17 Thermochemistry

Thermochemistry deals with the heat absorbed or evolved in a chemical re-action or in a phase change. If the temperature rises when a reaction occurs in an isolated system, heat must flow to the surroundings to restore the system to its initial temperature. Such a reaction is said to be **exothermic**, and the heat q is negative. If the temperature falls when a reaction occurs in an isolated system, heat must flow from the surroundings to the system to restore the system to its initial temperature. Such a reaction is said to be **endothermic**, and the heat q is positive.

To connect heat absorbed or evolved with a chemical reaction it is of course necessary to know what the chemical change is and have a measure of its amount. To discuss the thermodynamics of chemical reactions, we will find it convenient to represent chemical reactions in general by

$$0 = \sum_{i=1}^{N} \nu_i A_i \tag{2.81}$$

where the ν_i are the stoichiometric numbers and the A_i are the molecular for-mulas for the N substances involved in the reaction. The stoichiometric num-bers, which are dimensionless, are positive for products and negative for reac-tants. Thus, according to this way of writing a reaction, the reaction $H_2 + \tfrac{1}{2}O_2 = H_2O$ would be written

$$0 = -1H_2 - \tfrac{1}{2}O_2 + 1H_2O \tag{2.82}$$

The reason for using this convention is that it makes it easier to write ther-modynamic equations for chemical reactions.

The amount of reaction that has occurred up to some time is expressed by the **extent of reaction** ξ which is defined by

$$n_i = n_{i0} + \nu_i \xi \tag{2.83}$$

Here n_{i0} is the amount of substance i present initially, and n_i is the amount at some later time. Since n is expressed in moles and ν_i is dimensionless, we see the extent of reaction is expressed in moles. The concept of extent of reaction is important because it provides a connection between the amount of reaction and a particular balanced chemical equation. We will also use the extent of reaction later in calculating equilibrium compositions.

Consider a system containing reactants and products of a reaction. At con-stant temperature and pressure the enthalpy of this system is a function of the amounts of n_i of the various substances present. Therefore, the differential of the extensive enthalpy is given by

$$dH = \sum_{i=1}^{N} \overline{H}_i \, dn_i \tag{2.84}$$

where $\overline{H}_i$ is the molar enthalpy of i and dn_i is the differential of the amount of i. When a reaction takes place at constant T and P, there are increases in the amounts of products and decreases in the amounts of reactants. The differential change in enthalpy is obtained by substituting $dn_i = \nu_i \, d\xi$:

$$dH = \sum_{i=1}^{N} \overline{H}_i \nu_i \, d\xi = d\xi \sum_{i=1}^{N} \nu_i \overline{H}_i \tag{2.85}$$

The **reaction enthalpy** $\Delta_r H$ is defined as $(dH/d\xi)_{T,P}$:

$$\left(\frac{dH}{d\xi} \right)_{T,P} = \Delta_r H = \sum_{i=1}^{N} \nu_i \overline{H}_i \tag{2.86}$$

The reaction enthalpy is the derivative of the enthalpy of the system with respect to extent of reaction. This is perhaps easiest to visualize for a very large system for which we can write $\Delta H / \Delta \xi = \Delta_r H$. If one mole of reaction occurs, $\Delta \xi = 1$ mol and $\Delta H = (1 \text{ mol})\Delta_r H$. To know what a mole of reaction is we must have a balanced chemical equation, since the way an equation is written is arbitrary with respect to direction and with respect to multiplying or dividing by an integer. Thus, the enthalpy of reaction for $2H_2 + O_2 = 2H_2O$ is twice that of the reaction $H_2 + \frac{1}{2}O_2 = H_2O$. To distinguish between the extensive property $\Delta H°$ and the change in standard enthalpy for a specified chemical reaction, we will write $\Delta_r H°$ for a reaction, following the IUPAC recommendation. It is evident from equation 2.86 that the reaction enthalpy has the SI units J mol^{-1}. Here the mol^{-1} refers to a mole of reaction for the **reaction as written**. An overbar is not used on $\Delta_r H°$ because the subscript r indicates that the mol^{-1} unit is involved.

As a further specification of the states of the reactants and products we will usually consider reactions in which the reactants in their standard states are converted to the products in their standard states. When substances are in their standard states, thermodynamic quantities are labeled with superscript zeros (actually degree signs). Thus, if reactants and products are in their standard states, equation 2.86 becomes

$$\Delta_r H° = \sum_{i=1}^{N} \nu_i \overline{H}_i° \tag{2.87}$$

The standard states that are used in chemical thermodynamics are defined as follows:

1. The standard state of a pure gaseous substance, denoted by g, at a given temperature, is the (hypothetical) ideal gas at one bar pressure.
2. The standard state of a pure liquid substance, denoted by l, at a given temperature is the pure liquid at one bar pressure.

3. The standard state of the pure crystalline substance at a given temperature is the pure crystalline substance, denoted by s, at one bar pressure.

4. The standard state of a substance in solution is the hypothetical one of the substance in ideal solution of standard state molality (1 mol kg^{-1}) at one bar pressure, at each temperature. To indicate the standard state of an electrolyte, The NBS Tables of Chemical Thermodynamic Properties (1982), uses two symbols. The thermodynamic properties of completely dissociated electrolyte in water are designated by ai. The thermodynamic properties of undissociated molecules in water are designated by ao. The thermodynamic properties of ions in water are also designated ao to indicate no further ionization occurs.

Lavoisier and Laplace recognized in 1780 that the heat absorbed in decomposing a compound must be equal to the heat evolved in its formation under the same conditions. Thus, if the reverse of a chemical reaction is written, the sign of ΔH is changed. Hess pointed out in 1840 that the overall heat of a chemical reaction at constant pressure is the same, regardless of the intermediate steps involved. These principles are both corollaries of the first law of thermodynamics and are a consequence of the fact that the enthalpy is a state function. This makes it possible to calculate the enthalpy changes for reactions that cannot be studied directly. For example, it is not practical to measure the heat evolved when carbon burns to carbon monoxide in a limited amount of oxygen, because the product will be an uncertain mixture of carbon monoxide and carbon dioxide. However, carbon may be burned completely to carbon dioxide in an excess of oxygen and the heat of reaction measured. Thus, for graphite at 25 °C,

$$C(graphite) + O_2(g) = CO_2(g) \qquad \Delta_r H° = -393.509 \text{ kJ mol}^{-1}$$

The heat evolved when carbon monoxide burns to carbon dioxide can be readily measured also:

$$CO(g) + \tfrac{1}{2}O_2(g) = CO_2(g) \qquad \Delta_r H° = -282.984 \text{ kJ mol}^{-1}$$

Writing these equations in such a way as to obtain the desired reaction, adding, and canceling, we have

$$
\begin{array}{ll}
C(graphite) + O_2(g) = CO_2(g) & \Delta_r H° = -393.509 \text{ kJ mol}^{-1} \\
\underline{ CO_2(g) = CO(g) + \tfrac{1}{2}O_2(g)} & \underline{\Delta_r H° = 282.984 \text{ kJ mol}^{-1}} \\
C(graphite) + \tfrac{1}{2}O_2(g) = CO(g) & \Delta_r H° = -110.525 \text{ kJ mol}^{-1}
\end{array}
$$

In this way an accurate value can be obtained for the heat evolved when graphite burns to CO.

These data may be represented in the form of an enthalpy level diagram, as shown in Fig. 2.18. In addition, this diagram shows the enthalpy changes that are involved in vaporizing graphite to atoms and dissociating oxygen into atoms at 25 °C:

$$C(graphite) = C(g) \qquad \Delta_r H° = 716.682 \text{ kJ mol}^{-1}$$

$$O_2(g) = 2O(g) \qquad \Delta_r H° = 498.340 \text{ kJ mol}^{-1}$$

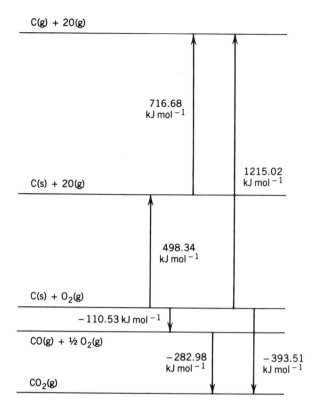

Figure 2.18   Enthalpy level diagram for the system C(s) + O_2(g). The differences in level are standard enthalpy changes at 25 °C and 1 bar.

2.18   Enthalpy of Formation

Since absolute enthalpies are not known, enthalpies relative to a defined **reference state** are used instead. The defined reference state for each substance is made up of the stoichiometric amounts of the elements in the substance, each in its standard state and at the temperature under consideration. These "relative" enthalpies of substances are called **enthalpies of formation** and are represented by $\Delta_f H°$. Since the same reference state is used for reactants and products, the same $\Delta_r H°$ is obtained as would be obtained with equation 2.87 and absolute enthalpies. Thus, standard enthalpy changes for reactions may be calculated using enthalpies of formation as follows:

$$\Delta_r H° = \sum_{i=1}^{N} \nu_i \Delta_f H_i°$$ (2.88)

Note that the enthalpy of formation does not have an overbar because the subscript f, for formation, indicates that the mol^{-1} unit is involved.

   The enthalpy of formation of a substance at a given temperature is the change in enthalpy for the reaction in which one mole of the substance in its standard state at the given temperature is formed from its elements, each in its standard state at that temperature. If there is more than one solid form of an element, one must be selected as a reference. For thermodynamic tables at 25 °C the reference form is usually the most stable form of the element at 25 °C, 1 bar pressure. Thus, the reference form of hydrogen is $H_2(g)$ instead of $H(g)$, the reference form of carbon is graphite, and the reference form of sulfur is rhombic sulfur. For thermodynamic tables that cover a wide range of temperature, different reference states may be used in various temperature ranges. In any case **the enthalpy of formation of an element in its standard state is zero at every temperature.**

From reactions given previously we can see how the following enthalpies of formation at 25 °C are obtained:

	$\Delta_f H°/\text{kJ mol}^{-1}$
$CO_2(g)$	-393.509
$CO(g)$	-110.525
$C(g)$	716.682
$O(g)$	249.170

These enthalpies of formation should be identified in Fig. 2.18. Enthalpies of formation $\Delta_f H°$ at 25 °C for some 200 substances are given in Appendix C.1. These values are from The NBS Tables of Chemical Thermodynamic Properties (1982). Appendix C.2 gives enthalpies of formation from 0 to 3000 K for a smaller group of substances from the JANAF Thermochemical Tables (1985).

The values of enthalpies of formation given in these tables comes from four sources: (1) calorimetrically measured enthalpies of reaction, fusion, vaporization, sublimation, transition, solution, and dilution; (2) temperature variation of equilibrium constants (see Section 5.9); (3) spectroscopically determined dissociation energies (see Section 12.2); (4) calculation from Gibbs energies and entropies (see the third-law method in Section 3.14).

The values of standard enthalpies of formation in Appendixes C.1 and C.2 are for a standard-state pressure of 1 bar, but they are not significantly different for a standard-state pressure of 1 atm. The enthalpies of solids and liquids are not affected significantly by the small increase (1.3%) in pressure in going from 1 bar to 1 atm. The standard enthalpies of formation of gases are also the same whether the standard-state pressure is 1 bar or 1 atm because their standard state is the ideal gas state, and we have seen in Section 2.12 that the enthalpy of an ideal gas is independent of pressure.

You may calculate $\Delta_r H°$ for any reaction for which the reactants and products are listed in tables, but the reaction will not necessarily occur spontaneously in the direction written. The question as to whether or not the reaction can occur is answered by calculations based on the second law of thermodynamics.

Example 2.12

What are the standard enthalpy changes at 298.15 K and 2000 K for the following reaction?

$$CO_2(g) + C(\text{graphite}) = 2CO(g)$$

Using Table C.2, at 298.15 K,

$$\Delta_r H° = 2\Delta_f H°(CO) - \Delta_f H°(CO_2)$$

$$= 2(-110.529 \text{ kJ mol}^{-1}) - (-393.522 \text{ kJ mol}^{-1})$$

$$= 172.464 \text{ kJ mol}^{-1}$$

At 2000 K,

$$\Delta_r H° = 2(-118.708 \text{ kJ mol}^{-1}) - (-396.639 \text{ kJ mol}^{-1})$$

$$= 159.223 \text{ kJ mol}^{-1}$$

2.19 Dependence of Reaction Enthalpy on Temperature

So far we have mainly talked about the reaction enthalpy at 298.15 K. To calculate the standard enthalpy change at some other temperature, given the value at 298.15 K, it is necessary to have heat capacity data on the reactants and the products. Since the enthalpy is a state function we can use the paths indicated below to calculate the standard enthalpy change at any desired temperature.

$$\text{Reactants} \xrightarrow{\Delta_r H_T} \text{Products}$$

$$\int_T^{298} C_{P,\text{react}} \, dT \downarrow \qquad \uparrow \int_{298}^T C_{P,\text{prod}} \, dT \qquad (2.89)$$

$$\text{Reactants} \xrightarrow{\Delta_r H_{298}} \text{Products}$$

$$\Delta_r H_T° = \int_T^{298} C_{P,\text{react}} \, dT + \Delta_r H_{298}° + \int_{298}^T C_{P,\text{prod}} \, dT \qquad (2.90)$$

$$\Delta_r H_T° = \Delta_r H_{298}° + \int_{298}^T (C_{P,\text{prod}}° - C_{P,\text{react}}°) \, dT$$

$$= \Delta_r H_{298}° + \int_{298}^T \Delta_r C_P° \, dT \qquad (2.91)$$

where

$$\Delta_r C_P° = \sum_i \nu_i \overline{C}_{P,i}° \qquad (2.92)$$

The **reaction heat capacity** $\Delta_r C_P°$ does not have an overbar because the subscript r indicates that the unit mol^{-1} is involved.

If data are available on the heat capacities of reactants and products in the form of power series in T (see Table 2.2), $\Delta_r H_T°$ may be expressed as a function of T as follows:

$$\Delta_r C_P° = \Delta_r \alpha + (\Delta_r \beta)T + (\Delta_r \gamma)T^2 \qquad (2.93)$$

where $\Delta_r \alpha = \Sigma \nu_i \alpha_i$, and so on.

$$\Delta_r H_T° = \Delta_r H_{298}° + \int_{298}^T [\Delta_r \alpha + (\Delta_r \beta)T + (\Delta_r \gamma)T^2] \, dT$$

$$= \Delta_r H_{298}° + \Delta_r \alpha(T - 298) + \frac{\Delta_r \beta}{2}(T^2 - 298^2) + \frac{\Delta_r \gamma}{3}(T^3 - 298^3) \qquad (2.94)$$

$$= \Delta_r H_0° + (\Delta_r \alpha)T + (\Delta_r \beta/2)T^2 + (\Delta_r \gamma/3)T^3$$

In the last form of this equation the constant terms have been added together to obtain a hypothetical enthalpy of reaction at 0 K, hypothetical because the power-series representations of C_P are for a limited temperature range. Within this temperature range equation 2.94 does represent the standard enthalpy of formation as a function of temperature.

The JANAF Tables give standard enthalpies of formation at a series of temperatures, and so these values may be used directly to calculate enthalpies of reaction. Some values from the JANAF Tables are given in Appendix C.2.

Example 2.13

The reaction of heated coal (approximated here by graphite) with superheated steam absorbs heat. This heat is usually provided by burning some of the coal. Calculate $\Delta_r H°$ (500 K) for both reactions.

(a) $C(\text{graphite}) + H_2O(g) = CO(g) + H_2(g)$

(b) $C(\text{graphite}) + O_2(g) = CO_2(g)$

How many moles of carbon are required to produce 1 mol of hydrogen if all the heat is produced by reaction (b)? Using Appendix C.2 for reaction (a), we obtain

$$\Delta_r H° \text{ (500 K)} = -110.022 \text{ kJ mol}^{-1} - (-243.831 \text{ kJ mol}^{-1}) = 133.809 \text{ kJ mol}^{-1}$$

Using Appendix C.2 for reaction (b), we find

$$\Delta_r H° \text{ (500 K)} = -393.677 \text{ kJ mol}^{-1}$$

Thus, $133.809/393.677 = 0.340$ mol of graphite has to be burned to provide heat for the first reaction. And 1.340 mol of graphite is required to produce one mole of hydrogen at a constant temperature of 500 K.

Some thermodynamic tables give values of $\overline{H}_T° - \overline{H}_{298}°$ to assist in the calculation of $\Delta_r H_T°$ for a chemical reaction or phase transition:

$$\overline{H}_T° - \overline{H}_{298}° = \int_{298 \text{ K}}^{T} \overline{C}_P° \, dT \qquad (2.95)$$

Depending on the table, $\Delta_r H°$ for phase transitions in the intervening temperature range may be added to the right side of the equation.

Example 2.14

What is the value of $\Delta_r H°$ at 0 K for the following reaction?

$$H_2(g) = 2H(g)$$

The calculation using $\Delta_r H°(298 \text{ K})$ illustrates the use of $\overline{H}° - \overline{H}_{298}°$ from Appendix C.2:

$$\Delta_r H°(298 \text{ K}) = 2(217.999 \text{ kJ mol}^{-1}) = 435.998 \text{ kJ mol}^{-1}$$

$$H_2(g) \xrightarrow{\text{298 K}} 2H(g) \qquad \Delta H°_{298} = 435.998 \text{ kJ mol}^{-1}$$

$$\overline{H}°_{298} - \overline{H}°_0 = 8.468 \text{ kJ mol}^{-1} \uparrow \qquad \downarrow \overline{H}°_0 - \overline{H}°_{298} = -(2)(6.197 \text{ kJ mol}^{-1})$$

$$H_2(g) \xrightarrow{\text{0 K}} 2H(g)$$

$$\Delta_r H°(0 \text{ K}) = (8.468 + 435.998 - 12.394) \text{ kJ mol}^{-1}$$

$$= 432.072 \text{ kJ mol}^{-1}$$

Alternatively, this value may be calculated from the enthalpy of formation of H(g) at 0 K in Appendix C.2:

$$\Delta_r H°(0 \text{ K}) = 2(216.037 \text{ kJ mol}^{-1})$$

$$= 432.074 \text{ kJ mol}^{-1}$$

This value is often referred to as the H–H bond energy. In Chapter 15 we will see how this value can be calculated theoretically; there this energy is referred to as the dissociation energy D_0.

2.20   Calorimetry

Heats of reaction are determined using **adiabatic calorimeters**; that is, the reaction or solution process occurs in a container, which is immersed in a weighed quantity of water and is surrounded by insulation or an adiabatic shield that is kept at the same temperature as the calorimeter so that no heat is gained or lost. A simple adiabatic calorimeter operated at constant pressure is illustrated in Fig. 2.19. Thus, ΔH_A for this adiabatic process is zero. When a certain amount of reactants R are converted completely to products P in a constant pressure calorimeter, the changes in state involved may be represented as follows:

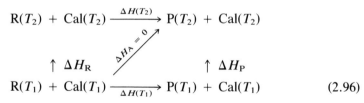

$$\text{R}(T_2) + \text{Cal}(T_2) \xrightarrow{\Delta H(T_2)} \text{P}(T_2) + \text{Cal}(T_2)$$

$$\uparrow \Delta H_R \qquad \qquad \uparrow \Delta H_P$$

$$\text{R}(T_1) + \text{Cal}(T_1) \xrightarrow{\Delta H(T_1)} \text{P}(T_1) + \text{Cal}(T_1) \qquad (2.96)$$

The adiabatic container, thermometer, stirrer, and weighed quantity of water are represented by Cal. Since the enthalpy is a state function, the enthalpy change for the actual process may be written two ways:

$$\Delta H_A = \Delta H(T_1) + \Delta H_P = 0 \qquad (2.97)$$

$$\Delta H_A = \Delta H_R + \Delta H(T_2) = 0 \qquad (2.98)$$

Since the heat capacities of the reactants, products, and calorimeter may be assumed constant over the range T_1 to T_2, these equations become

$$\Delta H(T_1) = -\Delta H_P = -[C_P(\text{P}) + C_P(\text{Cal})](T_2 - T_1) \qquad (2.99)$$

$$\Delta H(T_2) = -\Delta H_R = -[C_P(\text{R}) + C_P(\text{Cal})](T_2 - T_1) \qquad (2.100)$$

where these C_P's are extensive properties. Thus, the results of the calorimetric

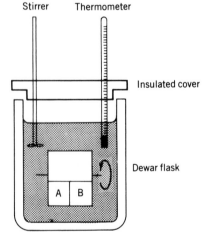

Stirrer   Thermometer

Insulated cover

Dewar flask

A   B

Figure 2.19   Adiabatic calorimeter operated at constant pressure. A reaction between solutions A and B is initiated by rotating the reaction vessel around the axis indicated.

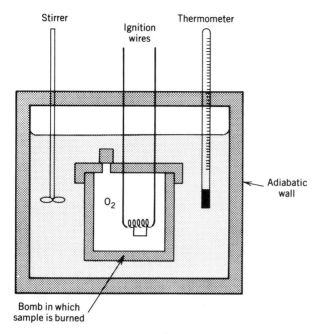

Figure 2.20 Adiabatic bomb calorimeter for carrying out combustions at constant volume.

experiments can be interpreted to obtain ΔH for the conversion of a certain amount of R to P either at T_1 or T_2. The heat capacity term in the first equation can be determined by using a calibrated electric heater coil and measuring I^2Rt with only products present, and the heat capacity term in the second equation can be determined with only reactants present. In this expression I is a constant current that flows through resistor R for time t.

Once ΔH has been determined in a calorimetric experiment, $\Delta_r H$ for a balanced chemical reaction can be calculated using $\Delta_r H = \Delta H / \Delta \xi$.

When a reaction is carried out in a sealed bomb (see Fig. 2.20), no P-V work is done, and the first law may be written $\Delta U = q_V$. Thus, the change in internal energy for the reaction is obtained. When the reaction is carried out at constant pressure, the first law may be written $\Delta H = q_P$. Chemists are usually more interested in ΔH than ΔU because chemical reactions are generally carried out at constant pressure. If ΔU is determined in a bomb calorimeter, the value of ΔH may be calculated using equation 2.39:

$$\Delta_r H = \Delta_r U + RT\sum \nu_g \qquad (2.101)$$

where $\sum \nu_g$ is the sum of stoichiometric numbers of gaseous products and gaseous reactants. In writing equation 2.101 in this way we are ignoring the volume change due to solid and liquid reactants, because this is negligible in comparison with the change in gas volume. We are also assuming that the gases are ideal.

Heats of combustion are useful in calculating other thermochemical data. Also, they have practical as well as theoretical importance. The purchaser of coal is interested in its heat of combustion per ton. The dietician must know, among other factors, the number of calories obtainable from the combustion of various foods. In nutrition the term "calorie" refers to the kilocalorie.

Example 2.15

The combustion of $C_2H_5OH(l)$ in a constant-volume calorimeter produces 1364.34 kJ mol^{-1} at 25 °C. What is the value of $\Delta_r H°$ for the following combustion reaction?

$$C_2H_5OH(l) + 3O_2(g) = 2CO_2(g) + 3H_2O(l)$$

$$\Delta_r H° = \Delta_r U° + RT\sum \nu_g$$
$$= -1364.34 \text{ kJ mol}^{-1} + (8.314 \times 10^{-3} \text{ kJ K}^{-1} \text{ mol}^{-1})(298.15 \text{ K})(-1)$$
$$= -1366.82 \text{ kJ mol}^{-1}$$

This is the quantity of heat that would be evolved at 25 °C and a constant pressure of 1 bar.

Example 2.16

In an adiabatic bomb calorimeter, the combustion of 0.5173 g of ethanol (C_2H_5OH) causes the temperature to rise from 25.0 to 29.289 °C. The heat capacity of the bomb, the reactants, and the other contents of the calorimeter is 3576 J K^{-1}. What is the molar internal energy of combustion of ethanol at 25.0 °C?

The change in the state can be written

$$\left\{\begin{matrix} C_2H_5OH + 3O_2 \\ + \text{ other contents} \end{matrix}\right\} [T = 25 \text{ °C, V}] = \left\{\begin{matrix} 2CO_2 + 3H_2O \\ + \text{ other contents} \end{matrix}\right\} [T = 29.289 \text{ °C, V}]$$

for which $q = 0$ (adiabatic) and $w = 0$ (constant volume) so that $\Delta U = 0$. This change in state can be written as the sum of

$$\left\{\begin{matrix} C_2H_5OH + 3O_2 \\ + \text{ other contents} \end{matrix}\right\} [T = 25 \text{ °C, V}] = \left\{\begin{matrix} 2CO_2 + 3H_2O \\ + \text{ other contents} \end{matrix}\right\} [T = 25 \text{ °C, V}] \quad \Delta U_1$$

$$\left\{\begin{matrix} 3CO_2 + 3H_2O \\ + \text{ other contents} \end{matrix}\right\} [T = 25 \text{ °C, V}] = \left\{\begin{matrix} 2CO_2 + 3H_2O \\ + \text{ other contents} \end{matrix}\right\} [T = 29.289 \text{ °C, V}] \Delta U_2$$

where $\Delta U_1 + \Delta U_2 = \Delta U = 0$, so that

$$\Delta U_1 = -\Delta U_2$$

or

$$\Delta_r U = -\frac{(3.576 \text{ kJ K}^{-1})(4.289 \text{ K})(46.0 \text{ g mol}^{-1})}{0.5173 \text{ g}}$$
$$= -1364 \text{ kJ mol}^{-1}$$

2.21 Enthalpy Changes of Solution Reactions

When a solute is dissolved in a solvent, heat may be absorbed or evolved; in general, the heat of solution depends on the concentration of the final solution. The **integral heat of solution** is the enthalpy change for the solution of 1 mol of solute in n mol of solvent. The solution process may be represented by a chemical equation such as

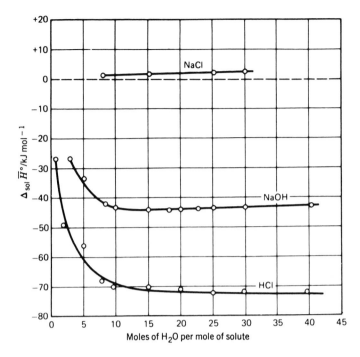

Figure 2.21 Integral heats of solution at 25 °C.

$$HCl(g) + 5H_2O(l) = HCl \text{ in } 5H_2O \qquad \Delta_{sol}\overline{H}°(298 \text{ K}) = -63.467 \text{ kJ mol}^{-1}$$

$$(2.102)$$

where "HCl in 5H₂O" represents a solution of 1 mol of HCl in 5 mol of H₂O. The integral heats of solution of HCl, NaOH, and NaCl are plotted versus the amount of water per mole of solute in Fig. 2.21. As the amount of water is increased, the integral heats of solution approach asymptotic values.

The formation of a solution of HCl in water in which the HCl, assumed to be completely dissociated, is in its standard state is represented by

$$HCl(g) = HCl(ai) \qquad \Delta_{sol}H°(298 \text{ K}) = -74.852 \text{ kJ mol}^{-1} \qquad (2.103)$$

where ai indicates the product is completely dissociated and at unit activity (Section 8.9). This reaction may also be represented by

$$HCl(g) = H^+(ao) + Cl^-(ao) \qquad \Delta_{sol}H°(298 \text{ K}) = -74.852 \text{ kJ mol}^{-1} \quad (2.104)$$

where the ao indicates that the ions do not further dissociate.

The solution of liquid acetic acid in water to form an aqueous solution in which undissociated acetic acid is in its standard state is represented by

$$CH_3CO_2H(l) = CH_3CO_2H(ao) \qquad \Delta_{sol}H°(298 \text{ K}) = -1.3 \text{ kJ mol}^{-1} \quad (2.105)$$

When a solute is dissolved in a solvent that is chemically quite similar to it and there are no complications of ionization or solvation, the heat of solution may be nearly equal to the heat of fusion of the solute. It might be expected that heat would always be absorbed in overcoming the attraction between the molecules or ions of the solid solute when the solute is dissolved. Another process that commonly occurs, however, is a strong interaction with the solvent, referred to as solvation, which evolves heat. In the case of water the solvation is called hydration.

The importance of this attraction of the solvent for the solute in the process of solution is illustrated by the dissolving of sodium chloride in water. In the crystal lattice of sodium chloride, positive sodium ions and negative chloride ions attract each other strongly. The energy required to separate them is so great that nonpolar solvents like benzene and carbon tetrachloride do not dissolve sodium chloride; but a solvent like water, which has a high dielectric constant and a large dipole moment, has a strong attraction for the sodium and chloride ions and solvates them with a large decrease in the energy of the system. When the energy required to separate the ions from the crystal is about the same as the solvation energy, as it is for dissolving NaCl in water, ΔH for the net process is close to zero. When NaCl is dissolved in water at 25 °C, there is only a small cooling effect; q is positive. When Na_2SO_4 is dissolved in water at 25 °C, there is an evolution of heat because the energy of hydration of the ions is greater than the energy required to separate the ions from the crystal.

The integral heat of dilution between two molalities m_1 and m_2 is the heat accompanying the dilution of an amount of solution of concentration m_1 containing 1 mol of solute with pure solvent to make a solution of concentration m_2.

Integral heats of solution, heats of dilution, and heats of reaction in solution may be calculated from tabulated values of heats of formation in solution. The enthalpy of formation of water is neglected in calculations if there is the same number of moles of water on both sides of the balanced chemical equation. Also, the enthalpy of formation of pure water is used for water in aqueous solution. This is done arbitrarily, much as the enthalpies of formation of the elements are taken equal to zero.

Example 2.17

Calculate $\Delta_r H°(298 \text{ K})$ for the following reaction using Appendix C.1.

$$\text{HCl in } 100H_2O + \text{NaOH in } 100H_2O = \text{NaCl in } 200H_2O + H_2O(l)$$

$$\Delta_r H°(298 \text{ K}) = -406.923 - 285.830 + 165.925 + 469.646$$

$$= -57.182 \text{ kJ mol}^{-1}$$

For dilute solutions it is found that the heat of reaction of strong bases, like NaOH and KOH, with strong acids, like HCl and HNO_3, is independent of the nature of the acid or base. This constancy of the heat of neutralization is a result of the complete ionization of strong acids and bases and the salts formed by neutralization. Thus, when a dilute solution of a strong acid is added to a dilute solution of a strong base, the only chemical reaction is

$$OH^-(ao) + H^+(ao) = H_2O(l) \qquad \Delta_r H°(298 \text{ K}) = -55.835 \text{ kJ mol}^{-1}$$

When a dilute solution of a weak acid or base is neutralized, the heat of neutralization is somewhat less because of the absorption of heat in the dissociation of the weak acid or base.

Since for strong electrolytes in dilute solution the thermal properties of the ions are essentially independent of the accompanying ions, it is convenient to use enthalpies of formation of individual ions. The sum of the enthalpies of formation of H^+ and OH^- ions may be calculated from

$$
\begin{array}{lll}
H_2O(l) = H^+(ao) + OH^-(ao) & \Delta_rH° = & 55.835 \text{ kJ mol}^{-1} \\
\underline{H_2(g) + \tfrac{1}{2}O_2(g) = H_2O(l)} & \underline{\Delta_rH° = } & \underline{-285.830 \text{ kJ mol}^{-1}} \\
H_2(g) + \tfrac{1}{2}O_2(g) = H^+(ao) + OH^-(ao) & \Delta_rH° = & -229.995 \text{ kJ mol}^{-1}
\end{array}
$$

The separate enthalpies of formation of H^+ and OH^- cannot be calculated and so, to construct a table of enthalpies of formation of individual ions, it is necessary to adopt an arbitrary convention. Enthalpies of formation of aqueous ions in Appendix C.1 are based on the convention that $\Delta_fH° = 0$ for $H^+(ao)$. In other words, by convention,

$$\tfrac{1}{2}H_2(g) = H^+(ao) + e^- \qquad \Delta_fH° = 0$$

where e^- is the electron, which is understood to reside in a chemically inert electrode. Therefore, the enthalpy of formation of OH^- is given by

$$\tfrac{1}{2}H_2(g) + \tfrac{1}{2}O_2(g) + e^- = OH^-(ao) \qquad \Delta_fH° = -229.995 \text{ kJ mol}^{-1}$$

On the basis of these values for the enthalpies of formation of H^+ and OH^-, the enthalpies of formation of other ions of strong electrolytes may be calculated.

From the enthalpy of formation of HCl(ai) it is possible to calculate the enthalpy of formation of $Cl^-(ao)$.

$$
\begin{array}{ll}
\tfrac{1}{2}H_2(g) + \tfrac{1}{2}Cl_2(g) = H^+(ao) + Cl^-(ao) & \Delta_rH° = -167.159 \text{ kJ mol}^{-1} \\
\tfrac{1}{2}Cl_2(g) + e^- = Cl^-(ao) & \Delta_fH° = -167.159 \text{ kJ mol}^{-1}
\end{array}
$$

2.22 Special Topic: Estimation of Thermodynamic Properties

Thermodynamic properties of molecules in the perfect gas state can be considered to be made up approximately of additive contributions from individual atoms or groups in the molecule. The reason for this approximate additivity is that the forces between atoms in a molecule are short range; therefore, individual groups contribute nearly constant amounts to thermodynamic properties. Benson and Buss* have discussed a hierarchical system of such addivity laws in which the zeroth-order approximation is the additivity of atom properties. This law applies to the molar mass, but does not give useful results for thermodynamic properties. The next higher or first-order approximation is the additivity of bond properties. Tables of bond properties may be developed empirically to be used to obtain $\Delta_fH°$, but these calculations do not distinguish between isomers such as n-butane and isobutane and may be in error by ± 8 kJ mol^{-1}. The second-order approximation to additivity is to treat a molecular property as the sum of contributions due to groups. A group is defined as an atom in a molecule, with attached ligands, that is bound to at least two other

---

* S. W. Benson and J. H. Buss, *J. Chem. Phys.* **29:** 546 (1958).

atoms. The designation of the group identifies first the polyvalent atoms and then its ligands. The Benson groups* required for the alkanes and their contributions to C_P and $\Delta_f H°$ are

	$\overline{C}_P^°$/J K^{-1} mol^{-1}	$\Delta_f H°$/kJ mol^{-1}
C(H)$_3$(C)	25.90	−42.68
C(H)$_2$(C)$_2$	23.01	−20.63
C(H)(C)$_3$	19.00	−7.95
C(C)$_4$	18.28	2.09

The symbol C(H)$_3$(C) represents a carbon atom bonded to three H atoms and another C.

Further corrections (*gauche* and 1,5-H repulsions) are required to give an accuracy of ± 4 kJ mol^{-1} for alkanes of higher carbon number. Group values are available for groups involving many elements.

Example 2.18

Estimate the standard enthalpies of formation at 298.15 K for the three pentane isomers in the ideal gas phase. The experimental values are *n*-pentane, −146.4 kJ mol^{-1}; isopentane, −153.1 kJ mol^{-1}; and neopentane, −168.4 kJ mol^{-1}.

For *n*-pentane,

$$\Delta_f H° = 2(-42.68) + 3(-20.63)$$
$$= -147.3 \text{ kJ mol}^{-1}$$

For isopentane,

$$\Delta_f H° = 3(-42.68) - 20.63 - 7.95$$
$$= -156.6 \text{ kJ mol}^{-1}$$

For neopentane (2, 2-dimethylpropane),

$$\Delta_f H° = 4(-42.68) + 2.09$$
$$= -168.6 \text{ kJ mol}^{-1}$$

References

K. E. Bett, J. S. Rowlinson, and G. Saville, *Thermodynamics for Chemical Engineers.* Cambridge, MA: MIT Press, 1975.

H. B. Callen, *Thermodynamics and an Introduction to Thermostatics,* 2d ed. New York: Wiley, 1985.

J. D. Cox and G. Pilcher, *Thermochemistry of Organic and Organometallic Compounds.* New York: Academic, 1970.

---

* S. W. Benson, *Thermochemical Kinetics,* 2d ed. New York: Wiley, 1976.

K. G. Denbigh, *The Principles of Chemical Equilibrium*. Cambridge, UK: Cambridge University Press, 1981.

N. A. Gokcen, *Thermodynamics*. Hawthorne, CA: Techscience Inc., 1975.

J. G. Kirkwood and I. Oppenheim, *Chemical Thermodynamics*. New York: McGraw-Hill, 1961.

G. N. Lewis, M. Randall, revised by K. S. Pitzer, and L. Brewer, *Thermodynamics*. New York: McGraw-Hill, 1961.

M. Model and R. C. Reid, *Thermodynamics and Its Applications*. Englewood Cliffs, NJ. Prentice-Hall, 1983.

J. M. Smith and H. C. Van Ness, *Introduction to Chemical Engineering Thermodynamics*. New York: McGraw-Hill, 1987.

D. R. Stull, E. F. Westrum, and G. C. Sinke, *The Chemical Thermodynamics of Organic Compounds*. New York: Wiley-Interscience, 1969.

Problems

2.1 How high can a person (assume a weight of 70 kg) climb on one ounce of chocolate, if the heat of combustion (628 kJ) can be converted completely into work of vertical displacement?

2.2 A mole of sodium metal is added to water. How much work is done on the atmosphere by the subsequent reaction if the temperature is 25 °C?

2.3 You want to heat 1 kg of water 10 °C, and you have the following four methods under consideration. The heat capacity of water is 4.184 J K^{-1} g^{-1}.

(a) You can heat it with a mechanical eggbeater that is powered by a 1-kg mass on a rope over a pulley. How far does the mass have to descend in the earth's gravitational field to supply enough work?

(b) You can send 1 A through a 100Ω resistor. How long will it take?

(c) You can send the water through a solar collector that has an area of 1 m^2. How long will it take if the sun's intensity on the collector is 4 J cm^{-2} min^{-1}?

(d) You can make a charcoal fire. The heat of combustion of graphite is -393 kJ mol^{-1}. That is, 12 g of graphite will produce 393 kJ of heat when it is burned to $CO_2(g)$ at constant pressure. How much charcoal will have to burn?

2.4 Show that the differential df is inexact.

$$df = dx - \frac{x}{y} dy$$

Thus, the integral $\int df$ depends on the path. However, we can define a new function g by

$$dg = \frac{1}{y} df$$

which has the property that dg is exact. Show that dg is exact, so that $\oint dg = 0$.

2.5 Over narrow ranges of temperature and pressure, the differential expression for the volume of a fluid as a function of temperature and pressure can be integrated to obtain

$$V = Ke^{\alpha T} e^{-\kappa P}$$

(α and κ are defined in Section 1.10). Show that V is a state function.

2.6 One mole of nitrogen at 25 °C and 1 bar is expanded reversibly and isothermally to a pressure of 0.132 bar. (a) What is the value of w? (b) What is the value of w if the nitrogen is expanded against a constant pressure of 0.132 bar?

2.7 (a) Derive the equation for the work of reversible isothermal expansion of a van der Waals gas from V_1 to V_2. (b) A mole of CH_4 expands reversibly from 1 to 50 L at 25 °C. Calculate the work in joules assuming (1) the gas is ideal, and (2) the gas obeys the van der Waals' equation. For CH_4 (g), $a = 2.283$ L^2 bar mol^{-2} and $b = 0.042\,78$ L mol^{-1}.

2.8 Liquid water is vaporized at 100 °C and 1.013 bar. The heat of vaporization is 40.69 kJ mol^{-1}. What are the values of (a) w_{rev} per mole, (b) q per mole, (c) $\Delta \overline{U}$, and (d) $\Delta \overline{H}$?

2.9 An ideal gas expands reversibly and isothermally from 10 bar to 1 bar. What are the values of (a) w per mole, (b) q per mole, (c) $\Delta \overline{U}$, and (d) $\Delta \overline{H}$?

2.10 For a gas following

$$\overline{V} = \frac{RT}{P} + \left(b - \frac{a}{RT} \right)$$

what is the inversion temperature for the Joule–Thompson effect? What is the sign of ΔT below the inversion temperature?

2.11 Calculate $\overline{H}°(2000 \text{ K}) - \overline{H}°(0 \text{ K})$ for H(g).

2.12 The heat capacities of a gas may be represented by

$$\overline{C}_P = \alpha + \beta T + \gamma T^2$$

For N_2, $\alpha = 26.984 \text{ J K}^{-1} \text{ mol}^{-1}$, $\beta = 5.910 \times 10^{-3} \text{ J K}^{-2} \text{ mol}^{-1}$, and $\gamma = -3.377 \times 10^{-7} \text{ J K}^{-3} \text{ mol}^{-1}$. How much heat is required to heat a mole of N_2 from 300 to 1000 K?

2.13 Methane (considered to be an ideal gas) initially at 25 °C and 1 bar pressure is heated at constant pressure until the volume has doubled. The variation of the molar heat capacity with absolute temperature is given by

$$\overline{C}_P = 22.34 + 48.1 \times 10^{-3} T$$

where $\overline{C}_P$ is in $\text{J K}^{-1} \text{ mol}^{-1}$. Calculate (a) $\Delta \overline{H}$, and (b) $\Delta \overline{U}$.

2.14 Calculate the temperature increase and final pressure of helium if a mole is compressed adiabatically and reversibly from 44.8 L at 0 °C to 22.4 L.

2.15 A mole of argon is allowed to expand adiabatically and reversibly from a pressure of 10 bar and 298.15 K to 1 bar. What is the final temperature and how much work is done on the argon?

2.16 A tank contains 20 L of compressed nitrogen at 10 bar and 25 °C. Calculate w when the gas is allowed to expand reversibly to 1 bar pressure (a) isothermally and (b) adiabatically.

2.17 An ideal monatomic gas at 298.15 K and 1 bar is expanded in a reversible adiabatic process to a final pressure of 1/2 bar. Calculate q per mole, w per mole, and $\Delta \overline{U}$.

2.18 An ideal monatomic gas at 1 bar and 300 K is expanded adiabatically against a constant pressure of 1/2 bar until the final pressure is 1/2 bar. What are the values of q per mole, w per mole, $\Delta \overline{U}$, and $\Delta \overline{H}$? Given: $\overline{C}_V = \frac{3}{2}R$.

2.19 Calculate $\Delta_r H°_{298}$ for

$$H_2(g) + F_2(g) = 2HF(g)$$
$$H_2(g) + Cl_2(g) = 2HCl(g)$$
$$H_2(g) + Br_2(g) = 2HBr(g)$$
$$H_2(g) + I_2(g) = 2HI(g)$$

2.20 The following reactions might be used to power rockets:

(1) $H_2(g) + \frac{1}{2}O_2(g) = H_2O(g)$

(2) $CH_3OH(l) + 1\frac{1}{2}O_2(g) = CO_2(g) + 2H_2O(g)$

(3) $H_2(g) + F_2(g) = 2HF(g)$

(a) Calculate the enthalpy changes at 25 °C for each of these reactions per kilogram of reactants. (b) Since the thrust is greater when the molar mass of the exhaust gas is lower, divide the heat per kilogram by the molar mass of the product (or the average molar mass in the case of reaction 2) and arrange the above reactions in order of effectiveness on the basis of thrust.

2.21 Calculate the enthalpy of formation of $PCl_5(s)$, given the heats of the following reactions at 25 °C:

$$2P(s) + 3Cl_2(g) = 2PCl_3(l) \qquad \Delta_r H° = -635.13 \text{ kJ mol}^{-1}$$
$$PCl_3(l) + Cl_2(g) = PCl_5(s) \qquad \Delta_r H° = -137.28 \text{ kJ mol}^{-1}$$

2.22 Calculate $\Delta_r H$ for the dissociation

$$O_2(g) = 2O(g)$$

at 0, 298, and 3000 K. In Section 12.2 the enthalpy change for dissociation at 0 K will be found to be equal to the spectroscopic dissociation energy D_0.

2.23 Methane may be produced from coal in a process represented by the following steps, where coal is approximated by graphite:

$$2C(s) + 2H_2O(g) = 2CO(g) + 2H_2(g)$$
$$CO(g) + H_2O(g) = CO_2(g) + H_2(g)$$
$$CO(g) + 3H_2(g) = CH_4(g) + H_2O(g)$$

The sum of these three reactions is

$$2C(s) + 2H_2O(g) = CH_4(g) + CO_2(g)$$

What is $\Delta_r H°$ at 500 K for each of these reactions? Check that the sum of the $\Delta_r H°$ values of the first three reactions is equal to $\Delta_r H°$ for the fourth reaction. From the stand-point of heat balance would it be better to develop a process to carry out the overall reactions in three separate reactors, or in a single reactor?

2.24 What is the heat evolved in freezing water at -10 °C given that

$$H_2O(l) = H_2O(cr) \qquad \Delta H°(273 \text{ K}) = -6004 \text{ J mol}^{-1}$$
$$\overline{C}_P(H_2O, l) = 75.3 \text{ J K}^{-1} \text{ mol}^{-1}$$
$$\overline{C}_P(H_2O, s) = 36.8 \text{ J K}^{-1} \text{ mol}^{-1}$$

2.25 What is the enthalpy change for the vaporization of water at 0 °C? This value may be estimated from Table C.1 by assuming that the heat capacities of $H_2O(l)$ and $H_2O(g)$ are independent of temperature from 0 to 25 °C.

2.26 Calculate the standard enthalpy of formation of methane at 1000 K from the value at 298.15 K using the $\overline{H}° - \overline{H}°_{298}$ data in Appendix C.2.

2.27 For a diatomic molecule the bond energy is equal to the change in internal energy for the reaction

$$X_2(g) = 2X(g)$$

at 0 K. Of course, the change in internal energy and the change in enthalpy are the same at 0 K. Calculate the enthalpy of dissociation of $O_2(g)$ at 0 K. The enthalpy of formation of $O(g)$ at 298.15 K is 249.173 kJ mol^{-1}. In the range of 0–298 K the average value of the heat of capacity of $O_2(g)$ is 29.1 J K^{-1} mol^{-1} and the average value of the heat capacity of $O(g)$ is 22.7 J K^{-1} mol^{-1}. What is the value of the bond energy in electronvolts? (When the changes in heat capacities in the range of 0–298 K are taken into account, the enthalpy of dissociation at 0 K is 493.58 kJ mol^{-1}.)

2.28 One gram of liquid benzene is burned in a bomb calorimeter. The temperature before ignition was 20.826 °C and the temperature after the combustion was 25.000 °C. This was an adiabatic calorimeter. The heat capacity of the bomb, the water around it, and the contents of the bomb before the combustion was 10 000 J K^{-1}. Calculate $\Delta_f H°$ for $C_6H_6(l)$ at 298.15 K from these data. Assume that the water produced in the combustion is in the liquid state and the carbon dioxide produced in the combustion is in the gas state.

2.29 An aqueous solution of unoxygenated hemoglobin containing 5 g of protein ($M = 64\ 000$ g mol^{-1}) in 100 cm^3 of solution is placed in an insulated vessel. When enough molecular oxygen is added to the solution to completely saturate the hemoglobin, the temperature rises 0.031 °C. Each mole of hemoglobin binds 4 mol of oxygen. What is the enthalpy of reaction per mole of oxygen bound? The heat capacity of the solution may be assumed to be 4.18 J K^{-1} cm^{-3}.

2.30 Calculate the heat of hydration of $Na_2SO_4(s)$ from the integral heats of solution of $Na_2SO_4(s)$ and $Na_2SO_4 \cdot 10H_2O(s)$ in infinite amounts of H_2O, which are -2.34 and 78.87 kJ mol^{-1}, respectively. Enthalpies of hydration cannot be measured directly because of the slowness of the phase transition.

2.31 We want to determine the enthalpy of hydration of $CaCl_2$ to form $CaCl_2 \cdot 6H_2O$:

$$CaCl_2(s) + 6H_2O(l) = CaCl_2 \cdot 6H_2O(s)$$

We cannot do this directly for a couple of reasons: (1) reactions in the solid state are slow, and (2) there is a series of hydrates and so a mixture of different hydrates would probably be obtained. We can, however, determine the heats of solution of $CaCl_2(s)$ and $CaCl_2 \cdot 6H_2O(s)$ in water at 298 K and take the difference. The experimental heats of solution are as follows:

$$CaCl_2(s) + Aq = CaCl_2(ai) \qquad \Delta_r H = -81.33 \text{ kJ mol}^{-1}$$
$$CaCl_2 \cdot 6H_2O(s) + Aq = CaCl_2(ai) + 6H_2O(l)$$
$$\Delta_r H = 15.79 \text{ kJ mol}^{-1}$$

What is the enthalpy of hydration?

2.32 Calculate the integral heat of solution of one mole of $HCl(g)$ in $200H_2O(l)$:

$$HCl(g) + 200H_2O(l) = HCl \text{ in } 200H_2O$$

2.33 Calculate the enthalpies of reaction at 25 °C for the following reactions in dilute aqueous solutions:

(a) $HCl(ai) + NaBr(ai) = HBr(ai) + NaCl(ai)$
(b) $CaCl_2(ai) + Na_2CO_3(ai) = CaCO_3(s) + 2NaCl(ai)$

2.34 Estimate the molar heat capacity at constant pressure at 298 K for ethane using the Benson method. Compare this result with the value calculated using values in Table 2.3.

2.35 Estimate the enthalpies of formation of n-butane and isobutane at 298.15 K using the Benson method.

2.36 How much work is done when a person weighing 75 kg (165 lb) climbs the Washington monument, 555 ft high. How many kilojoules must be supplied to do this muscular work, assuming that 25% of the energy produced by the oxidation of food in the body can be converted into muscular mechanical work?

2.37 The average person generates about 2500 kcal of heat a day. How many kilowatt-hours of energy is this? If walking briskly dissipates energy at 500 W, what fraction of the day's energy does walking one hour represent? How many kilograms of water would have to be evaporated if this was the only means of heat loss? (The heat of vaporization of water at 35 °C is 2400 J g^{-1}.)

2.38 The surface tension of water is 71.97×10^{-3} N m^{-1} or 71.97×10^{-3} J m^{-2} at 25 °C. Calculate the surface energy in joules of 1 mol of water dispersed as a mist containing droplets 1 μm (10^{-4} cm) in radius. The density of water may be taken as 1.00 g cm^{-3}.

2.39 Are the following expressions exact differentials?

(a) $xy^2\ dx - x^2y\ dy$ (b) $\dfrac{dx}{y} - \dfrac{x}{y^2}\ dy$

2.40   Show that

$$dq = dU + P\,dV = C_V\,dT + RT\,d\ln V$$

is not an exact differential, but

$$\frac{dq}{T} = C_V\,d\ln T + R\,d\ln V$$

is an exact differential.

2.41   Calculate w for a reversible isothermal (298.15 K) expansion of a mole of N_2 from 1 to 10 L assuming it is (a) an ideal gas and (b) a van der Waals gas (see Table 1.3).

2.42   An ideal gas at 25 °C and 100 bar is allowed to expand reversibly and isothermally to 5 bar. Calculate (a) w per mole, (b) the heat absorbed per mole, (c) $\Delta\overline{U}$, and (d) $\Delta\overline{H}$.

2.43   Ammonia gas is condensed at its boiling point at 1.013 25 bar at -33.4 °C by the application of a pressure infinitesimally greater than 1 bar. To evaporate ammonia at its boiling point requires the absorption of 23.30 kJ mol^{-1}. Calculate (a) w_{rev} per mole, (b) q per mole, (c) $\Delta\overline{U}$, and (d) $\Delta\overline{H}$.

2.44   What is w per mole for a reversible isothermal expansion of ethane from 5 to 10 L mol^{-1} at 298 K assuming (a) ethane is an ideal gas and (b) it follows the van der Waals equation (van der Waals constants are in Table 1.3).

2.45   For a gas following

$$P = \frac{RT}{\overline{V} - b}$$

derive the expression for the Joule–Thompson coefficient μ_{JT}. When the pressure drops in a Joule–Thompson expansion for such a gas, what is the sign of ΔT?

2.46   According to Appendix C.2, how much heat is required to raise the temperature of a mole of oxygen from 298 to 3000 K at constant pressure?

2.47   From the following data calculate the value of $(\overline{H}_{298} - \overline{H}_0^\circ)$ for $Al_2O_3(s)$.

T/K	$\overline{C}_P^\circ/J\ K^{-1}$ mol^{-1}	T	$\overline{C}_P^\circ$	T	$\overline{C}_P^\circ$
10	0.009	90	9.69	180	43.79	270	72.37
20	0.076	100	12.84	190	47.53	280	74.84
30	0.263	110	16.32	200	51.14	290	77.19
40	0.691	120	20.06	210	54.60	298.16	79.01
50	1.492	130	23.96	220	57.92	273.16	73.16
60	2.779	140	27.96	230	61.10		
70	4.582	150	31.98	240	64.13		
80	6.895	160	35.99	250	67.01		
		170	39.94	260	69.76		

2.48   One mole of hydrogen at 25 °C and 1 bar is compressed adiabatically and reversibly into a volme of 5 L. Assuming ideal gas behavior, calculate (a) the final temperature, (b) the final pressure, and (c) the work done on the gas.

2.49   One mole of argon at 25 °C and 1 bar pressure is allowed to expand reversibly to a volume of 50 L (a) isothermally and (b) adiabatically. Assuming ideal gas behavior, calculate the final pressure in each case and the work done on the gas.

2.50   A tank contains 20 L of compressed nitrogen at 10 bar and 25 °C. Calculate the maximum work that can be obtained when the gas is allowed to expand reversibly to 1 bar pressure (a) isothermally and (b) adiabatically. The heat capacity of nitrogen at constant volume can be taken to be 20.8 J K^{-1} mol^{-1} independent of temperature.

2.51   Compare the enthalpies of combustion of $CH_4(g)$ to $CO_2(g)$ and $H_2O(g)$ at 298 and 2000 K.

$$CH_4(g) + 2O_2(g) = CO_2(g) + 2H_2O(g)$$

2.52   Compare the enthalpy of combustion of $CH_4(g)$ to $CO_2(g)$ and $H_2O(l)$ at 298 K with the sum of the enthalpies of combustion of graphite and $2H_2(g)$, from which $CH_4(g)$ can, in principle, be produced.

2.53   The enthalpy change for the combustion of toluene to $H_2O(l)$ and $CO_2(g)$ is -3910.0 kJ mol^{-1} at 25 °C. Calculate the enthalpy of formation of toluene.

2.54   Calculate ΔH (298 K) per gram of fuel (exclude oxygen) for

(a) $H_2(g) + \tfrac{1}{2}O_2(g) = H_2O(g)$

(b) $CH_4(g) + 2O_2(g) = CO_2(g) + 2H_2O(g)$

(c) $CH_3OH(l) + \tfrac{3}{2}O_2(g) = CO_2(g) + 2H_2O(g)$

(d) $C_6H_{14}(g) + 9\tfrac{1}{2}O_2(g) = 6CO_2(g) + 7H_2O(g)$

2.55   A 1:3 mixture of CO and H_2 is passed through a catalyst to produce methane at 500 K.

$$CO(g) + 3H_2(g) = CH_4(g) + H_2O(g)$$

How much heat is liberated in producing a mole of methane? How does this compare with the heat obtained from burning a mole of methane at this temperature? How does the heat of combustion of CH_4 compare with the heat of combustion of $CO + 3H_2$?

2.56   Calculate $\Delta_r H^\circ$ for the dissociation

$$H_2(g) = 2H(g)$$

at 0, 298 and 3000 K. The value of 0 K is equal to the spectroscopic dissociation energy D_0.

2.57   In principle, methanol can be produced from methane in two steps or one:

I. $CH_4(g) + H_2O(g) = CO(g) + 3H_2(g)$

$CO(g) + 2H_2(g) = CH_3OH(g)$

II. $CH_4(g) + H_2O(g) = CH_3OH(g) + H_2(g)$

What is $\Delta_r H°$ at 500 K for each of these reactions? From the standpoint of heat balance would it be better to develop a process to carry out the overall reaction in two separate reactors, or in a single reactor?

2.58 Calculate the heat of vaporization of water at 25 °C. The specific heat of water may be taken as $4.18 \text{ J K}^{-1} \text{g}^{-1}$. The heat capacity of water vapor at constant pressure in this temperature range is $33.5 \text{ J K}^{-1} \text{mol}^{-1}$, and the heat of vaporization of water at 100 °C is 2258 J g^{-1}.

2.59 Calculate the enthalpy of dissociation of $H_2(g)$ at 3000 K using $\Delta_f H°[H(g), 298.15 \text{ K}] = 217.999 \text{ kJ mol}^{-1}$ and $\overline{H}° - \overline{H}°_{298}$ values in Appendix C.2.

2.60 Calculate the dissociation energy of CH_4 into atoms at 298.15 K using its $\Delta H°$ for the dissociation reaction at 0 K and $\overline{H}° - \overline{H}°_{298}$ values from Appendix C.2.

2.61 The enthalpy of formation of $CH_4(g)$ at 298.15 K is $-74.873 \text{ kJ mol}^{-1}$. Calculate its enthalpy of formation at 1000 K using $\overline{H}° - \overline{H}°_{298}$ values from Appendix C.2.

2.62 Ammonia is to be oxidized to $NO_2(g)$ to make nitric acid. What temperature will be reached if the only reaction is

$$NH_3(g) + \tfrac{7}{4}O_2(g) = NO_2(g) + \tfrac{3}{2}H_2O(g)$$

and a stoichiometric amount of oxygen is used?

2.63 In an adiabatic bomb calorimeter, oxidation of 0.4362 g of naphthalene $(C_{10}H_8)$ caused a temperature rise of 1.707 °C. The final temperature was 298 K. The heat capacity of the calorimeter and water was $10\,290 \text{ J K}^{-1}$, and the heat capacity of the products can be neglected. If the corrections for the oxidation of the wire and residual nitrogen are neglected, what is the molar internal energy of combustion of naphthalene? What is its enthalpy of formation?

2.64 The combustion of oxalic acid in a bomb calorimeter yields 2816 J g^{-1} at 25 °C. Calculate (a) $\Delta_C U°$ and (b) $\Delta_C H°$ for the combustion of 1 mol of oxalic acid ($M = 90.0$ g mol^{-1}).

2.65 Calculate $\Delta H°(298.15 \text{ K})$ for the solution process

$$CaCl_2 \cdot 6H_2O(s) = CaCl_2 \text{ in } 400H_2O + 6H_2O(l)$$

For $CaCl_2 \cdot 6H_2O(s)$,

$$\Delta_f H° = -2607.9 \text{ kJ mol}^{-1}$$

and for $CaCl_2$ in $400H_2O$,

$$\Delta_f H° = -874.982 \text{ kJ mol}^{-1}.$$

2.66 A mole of liquid sulfuric acid (98 g) is added to a certain quantity of water at 25 °C, and it is found that the temperature is 100 °C! The NBS Tables of Chemical Thermodynamic Properties yields the following information:

	$\Delta_f H°/\text{kJ mol}^{-1}$
$H_2SO_4(l)$	-813.99
$H_2SO_4(ai)$	-909.27

The ai indicates this value is for sulfuric acid that is completely ionized in water. What is the enthalpy of solution of sulfuric acid in enough water to completely ionize it? Let's make some approximations and estimate the mass of the "certain quantity of water" in the first line. Assume that the solution has the same heat capacity as water ($\overline{C}_P = 75.3$ J $\text{K}^{-1} \text{mol}^{-1}$), independent of temperature, and ignore the mass of the sulfuric acid added. How much water was used?

2.67 (a) What is $\Delta_r H°$ for the dilution process NaOH in $100H_2O$ [+ $100H_2O$] = NaOH in $200H_2O$

(b) What is $\Delta_r H°$ for the ion-pair formation $Fe^{3+}(ao) + F^-(ao) = FeF^{2+}(ao)$

The standard enthalpies of formation of $FeF^{2+}(ao)$ and $F^-(ao)$ are -371.1 and -332.63 kJ mol^{-1}, respectively.

2.68 Estimate the standard enthalpies of formation of ethane and propane at 298.15 K using the Benson group additivity method, and compare these values with the values in Appendix C.1.

2.69 When nitroglycerine explodes, the chemical reaction that occurs can be assumed to be

$$C_3H_5N_3O_9(l) = 3CO_2(g) + \tfrac{5}{2}H_2O(g) + \tfrac{3}{2}N_2(g) + \tfrac{1}{4}O_2(g)$$

(a) Calculate $\Delta_r H°$ and $\Delta_r U°$ for this reaction at 298 K, given that the enthalpy of formation of liquid nitroglycerine is -372.4 kJ mol^{-1}.

(b) Consider 0.20 mol of nitroglycerine at 25 °C completely filling a constant volume cell of 0.030 L. Calculate the *maximum* temperature and pressure that would be generated by the explosion of the nitroglycerine if the constant volume cell did not burst (or vaporize). You may assume that (1) the explosion occurs so rapidly that the conditions are adiabatic, (2) the pressure cell comes to immediate thermal equilibrium with the products, (3) the products are ideal gases, and (4) the total constant volume heat capacity of products plus cell has the temperature independent value of 100 J K^{-1}.

3
Second and Third Laws
of Thermodynamics

The first law of thermodynamics states that when one form of energy is converted into another, the total energy is conserved. It does not indicate any other restriction on this process. However, we know that many processes have a natural direction, and it is with the question of direction that the second law is concerned. For example, a gas expands into a vacuum, but, although it would not violate the first law, the reverse never occurs. For a metal bar at uniform temperature to become hot at one end and cold at the other would not be a violation of the first law, yet we know this never occurs spontaneously.

The second law of thermodynamics is one of the most important generalizations in science. It is important in chemistry because it can tell us whether a process or reaction will occur in the forward or backward direction. The quantity that tells us whether a chemical reaction or a physical change can occur spontaneously in an isolated system is the entropy S. Entropy is a function of the state of the system, as is the internal energy U. Thermodynamics does not deal with the rate of approach to equilibrium, but only with the equilibrium state. Some time is required even for a gas to expand into another container, and for some chemical reactions the rate of approach to equilibrium is very slow.

The third law of thermodynamics allows us to obtain the absolute value of the entropy of a substance.

3.1 Spontaneous and Nonspontaneous Changes

We are familiar with the fact that many changes occur spontaneously, that is, when systems are simply left to themselves. For example, heat flows from a hotter body to a colder body, and chemical reactions proceed to equilibrium. It is a matter of experience that spontaneous changes do not reverse themselves; heat never flows spontaneously from a cold body to a hot body, and a chemical reaction that has reached equilibrium never spontaneously produces the original reactants. These latter changes are referred to as **nonspontaneous changes** or **unnatural changes.**

When we see a movie run backward, we often laugh because we know the event could not happen that way. The first law of thermodynamics does not tell us the direction in which a process can occur spontaneously. The laws of classical mechanics and quantum mechanics also do not tell us the direction in which time increases. When a movie of a collision between two particles is run backwards we don't laugh, because it looks as reasonable as the movie run in the forward direction. Indeed, the mechanical equations of motion are invariant under time reversal. It is useful to be able to predict whether a physical change or a chemical reaction will go spontaneously in the forward or backward direction, and so the second law is very important. It was not possible to answer this sort of question until Carnot investigated heat engines, and so we must start there and work our way toward chemical applications.

3.2 Carnot Heat Engine

A heat engine is an engine that uses heat to generate mechanical work by carrying a "working substance" through a cyclic process. The arrangements for a Carnot heat engine are shown in Fig. 3.1. The arrows indicate that in one

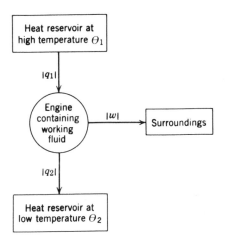

Figure 3.1   Carnot heat engine.

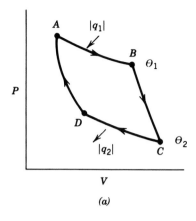

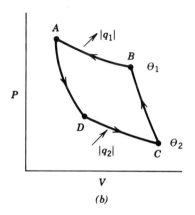

Figure 3.2   The Carnot cycle. (*a*) Plot of *P* versus *V* for a Carnot heat engine. (*b*) Plot of *P* versus *V* for a Carnot refrigerator or Carnot heat pump, which is just a heat engine operating in a reversed cycle.

cycle this engine receives heat $|q_1|$ from the high-temperature reservoir, rejects heat $|q_2|$ to the low-temperature reservoir, and does work $|w|$ on its surroundings. The absolute value signs are used because the signs of these algebraic quantities are set by conventions that require in this case that q_1 is positive, q_2 is negative, and w is negative. For the Carnot refrigerator or heat pumps, these signs are reversed. In the operation of the Carnot heat engine a working fluid, which we will refer to as a gas, is taken through a sequence of four steps that return it to its initial state. The engine itself consists of an idealized cylinder with a piston that can slide without friction and can do work on the surroundings or have the surroundings do work on the gas. The temperatures of the two reservoirs are represented by Θ_1 and Θ_2 because the second law brings in a definition of temperature, which we will discuss in Section 3.3.

The cycle for the Carnot heat engine consists of the following four steps, which are represented in Fig. 3.2:

1. **Reversible isothermal expansion of the gas from state A to state B.** During this expansion step, the piston does work $|w_1|$ on the surroundings, and heat $|q_1|$ is absorbed by the gas from the high-temperature reservoir. Note that according to the convention that w is work done on the gas, w_1 is negative. To assure that the heat transfer is reversible, the temperature of the high-temperature reservoir is only infinitesimally higher than the temperature Θ_1 of the gas.

2. **Reversible adiabatic expansion of the gas from state B to state C.** For this step we assume that the piston and cylinder are thermally insulated so that no heat is gained or lost. The expansion continues until the temperature of the gas has dropped to Θ_2. During this expansion step, the piston does work $|w_2|$ on the surroundings.

3. **Reversible isothermal compression of the gas from state C to state D.** During this compression step the surroundings do work $|w_3|$ on the gas, and heat $|q_2|$ flows out of the gas to the low-temperature heat reservoir. Note that according to the convention that q is heat absorbed by the gas, q_2 is negative. The temperature of the low-temperature reservoir is only infinitesimally lower than the temperature Θ_2 of the gas so that the heat transfer is reversible.

4. **Reversible adiabatic compression of the gas from state D to state A.** This step completes the cycle by bringing the gas back to its initial state at temperature Θ_1. During this compression step, the surroundings do work w_4 on the gas, but no heat is gained or lost.

A first law analysis of a Carnot cycle can be summarized as follows:

Step 1.   Isothermal expansion $\Delta U_1 = q_1 + w_1$
Step 2.   Adiabatic expansion $\Delta U_2 = w_2$
Step 3.   Isothermal compression $\Delta U_3 = q_2 + w_3$
Step 4.   Adiabatic compression $\Delta U_4 = w_4$

The change in the internal energy of the gas for the cycle ΔU_{cy} is the sum of the changes in the four steps, and it is equal to zero because the gas is returned to its initial state. Thus,

$$\Delta U_{cy} = 0 = (q_1 + q_2) + (w_1 + w_2 + w_3 + w_4)$$

$$= q_{cy} + w_{cy} \tag{3.1}$$

Thus, the work $|w_{cy}|$ done by the engine in a cycle is equal to the difference between the "heat in" $|q_1|$ and the "heat out" $|q_2|$:

$$-w_{cy} = q_{cy} = q_1 + q_2 \tag{3.2}$$

$$|w_{cy}| = |q_1| - |q_2| \tag{3.3}$$

The engine absorbs heat q_1 from the high-temperature reservoir and does work on the surroundings so that w is negative, but the first law does not tell us the relative amounts of work and of heat rejected at Θ_2.

In practice, the low-temperature reservoir of a heat engine is often the atmosphere so that the economic cost of producing work in the surroundings is mainly that of supplying q_1. The **efficiency** ϵ of a heat engine is defined as the ratio of the work done on the surroundings to the heat input at the higher temperature. Thus, the efficiency is defined by

$$\epsilon = \frac{-w_{cy}}{q_1} = \frac{q_1 + q_2}{q_1} \tag{3.4}$$

where the negative sign is required by the fact that w_{cy} is negative and the efficiency is, of course, positive. It is convenient to express the efficiency in terms of the magnitudes of the heats

$$\epsilon = 1 - \frac{|q_2|}{|q_1|} \tag{3.5}$$

because this reminds us that $0 < \epsilon < 1$.

From equation 3.5 it is clear that to improve the efficiency of a Carnot heat engine one would like to reduce the magnitude of the heat rejected to the cold reservoir, $|q_2|$. But it has been found impossible to reduce $|q_2|$ to zero. Similarly, to improve the efficiency of a Carnot refrigerator, which is obtained by running the Carnot cycle backwards, as shown in Fig. 3.2b, one would like to reduce the work required for a cycle removing a certain amount of heat $|q_2|$ from the low-temperature reservoir. But it has been found impossible to reduce this to zero. Kelvin and Clausius summarized a great deal of experience into statements of the second law of thermodynamics as follows:

Kelvin's statement (1851): It is impossible to produce work in the surroundings using a cyclic process connected to a single heat reservoir.

Clausius' statement (1854): It is impossible to carry out a cyclic process using an engine connected to two heat reservoirs that will have as its only effect the transfer of a quantity of heat from the low-temperature reservoir to the high-temperature reservoir.

These statements are consistent in the sense that it can be shown that if one is true the other is true. On the other hand, they are both postulates. To get to the chemical applications, we are going to have to go through several steps. These steps are based on Carnot's theorem and a corollary.

Carnot's theorem: No engine operating between two heat reservoirs can be more efficient than a Carnot engine operating between the same two reservoirs.

A corollary: All reversible engines operating between the same heat reservoirs are equally efficient.

These theorems can be derived from the Clausius or Kelvin statement. That is, it can be shown that if these statements are not true, then processes can be invented that violate Kelvin's statement and Clausius' statement of the second law. However, we will not discuss these derivations. Before we can obtain a statement of the second law that is directly applicable to chemistry, we must discuss the temperature scale based on the second law.

3.3 Thermodynamic Temperature

Throughout this derivation absolute values of heats will be used so that the notation will be simplified. The objective here is to show that the ratio of the heat q_c rejected to the cold reservoir to the heat q_h absorbed from the hot reservoir of a reversible Carnot engine is equal to the ratio of a function of the temperature of the cold reservoir to the same function of the temperature of the hot reservoir. This is based on the corollary of Carnot's theorem that all reversible engines operating between the same heat reservoirs are equally efficient. Since the efficiency of a reversible Carnot engine depends on the ratio of the heat q_c rejected to the cold reservoir to the heat q_h absorbed from the hot reservoir, this ratio must be some function of the temperatures of the two reservoirs. This is expressed by

$$\frac{q_h}{q_c} = f(\Theta_c, \Theta_h) \tag{3.6}$$

We have used Θ to represent the empirical temperature scale used in deriving the efficiency of a reversible Carnot engine.

The derivation is based on the two Carnot cycles shown in Fig. 3.3. The upper Carnot cycle operates between reservoirs at Θ_1 and Θ_2, and the lower operates between Θ_2 and Θ_3. The heat q_2 rejected by the upper Carnot cycle is absorbed at the higher temperature of the lower Carnot cycle. For the upper cycle, equation 3.6 becomes

$$\frac{q_1}{q_2} = f(\Theta_1, \Theta_2) \tag{3.7}$$

and for the lower cycle

$$\frac{q_2}{q_3} = f(\Theta_2, \Theta_3) \tag{3.8}$$

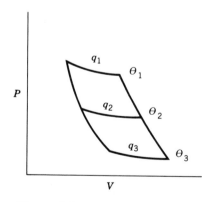

Figure 3.3 Two reversible Carnot cycles with a step in common ($\Theta_1 > \Theta_2 > \Theta_3$).

A heat engine can also be operated between Θ_1 and Θ_3, as can be seen from the figure, and so,

$$\frac{q_1}{q_3} = f(\Theta_1, \Theta_3) \tag{3.9}$$

Substituting these equations into the mathematical identity

$$\frac{q_1}{q_3} = \frac{q_1 q_2}{q_2 q_3} \tag{3.10}$$

yields

$$\frac{q_1}{q_3} = f(\Theta_1, \Theta_3) = f(\Theta_1, \Theta_2) f(\Theta_2, \Theta_3) \tag{3.11}$$

Since the first function does not depend on Θ_2, this empirical temperature has to cancel on the right side. Thus, the function of the two temperatures, $f(\Theta_1, \Theta_3)$, must be of the form $T(\Theta_1)/T(\Theta_3)$, where T is a function that depends on a single empirical temperature Θ. Therefore, a new temperature scale can be defined with the property that

$$\frac{q_h}{q_c} = \frac{T_h}{T_c} \tag{3.12}$$

This equation is the basis for Kelvin's suggestion that a thermodynamic temperature scale be set up with temperatures proportional to the magnitudes of the heat exchanged with heat reservoirs of a reversible Carnot engine. By taking $T = 273.16$ K at the triple point of water, this scale becomes identical with the ideal gas scale. **The thermodynamic temperature scale defined in terms of the second law is more basic than that based on ideal gases because any fluid can be used.** However, the ideal gas scale is more practical for laboratory use and is the one that is actually used in establishing secondary standards for temperature.

Note that when equation 3.12 is used in equation 3.5, we find that the efficiency ϵ of a heat engine is given by

$$\epsilon = 1 - \frac{T_2}{T_1} \tag{3.13}$$

Thus, high efficiencies are obtained if the ratio T_2/T_1 is small; in practice T_2 is generally close to room temperature and so T_1 is made as high as possible.

Lord Kelvin was the first to define temperature in this way and so the thermodynamic scale is called the Kelvin scale in his honor.

3.4   Relation Between the Thermodynamic Temperature and the Ideal Gas Temperature

Equation 3.13 for the efficiency of a Carnot engine can be obtained for the special case that the working fluid is an ideal gas by using equations derived in the preceding chapter. The reversible work for the isothermal steps is given in equation 2.27. The reversible work for the adiabatic steps is given by equation 2.75. Thus, the net work for one cycle of a Carnot heat engine with n moles of an ideal gas is

$$w = -nRT_1 \ln \frac{V_2}{V_1} + \int_{T_1}^{T_2} C_V \, dT - nRT_2 \ln \frac{V_4}{V_3} + \int_{T_2}^{T_1} C_V \, dT$$

$$= -nRT_1 \ln \frac{V_2}{V_1} + nRT_2 \ln \frac{V_3}{V_4} \tag{3.14}$$

This equation may be simplified by use of the equations for the two adiabatic steps. From equation 2.78 for an ideal gas with C_P independent of temperature,

$$T_1 V_2^{\gamma-1} = T_2 V_3^{\gamma-1} \tag{3.15a}$$

$$T_1 V_1^{\gamma-1} = T_2 V_4^{\gamma-1} \tag{3.15b}$$

where $\gamma = C_P/C_V$. Dividing equation 3.15a by equation 3.15b, we obtain

$$\frac{V_2}{V_1} = \frac{V_3}{V_4} \tag{3.16}$$

Substituting this relation into equation 3.14, we obtain

$$w = -nR(T_1 - T_2) \ln \frac{V_2}{V_1} \tag{3.17}$$

Since the first step of the cycle is an isothermal expansion, equation 2.27 yields

$$q_1 = -w = nRT_1 \ln \frac{V_2}{V_1} \tag{3.18}$$

Substituting equations 3.17 and 3.18 into equation 3.2 for the efficiency ϵ of a heat engine yields equation 3.13, which was obtained for any working fluid. The fact that we have obtained the same relationship as that derived from the second law shows that the temperature scales defined by an ideal gas and by the second law are proportional to each other. By choosing 273.16 K for the triple point of water, we have in effect made the size of the degree the same on both scales.

3.5   Entropy

We can express the efficiency of a reversible Carnot engine in two ways:

$$\epsilon = 1 + \frac{q_2}{q_1} = 1 - \frac{T_2}{T_1} \tag{3.19}$$

By subtracting the second form from the first we obtain

$$\frac{q_2}{q_1} + \frac{T_2}{T_1} = 0 \tag{3.20}$$

Multiplying by q_1/T_2, we obtain

$$\frac{q_2}{T_2} + \frac{q_1}{T_1} = 0 \tag{3.21}$$

This is a remarkable result because it suggests that there is a state function S defined by

$$dS \equiv \frac{dq_{rev}}{T} \tag{3.22}$$

where the subscript is required because we are considering a reversible Carnot engine. The change in a state function around any cycle is zero, and this is true for S. This new state function is called the **entropy.** For the reversible Carnot cycle

$$\Delta S_{cy} = \Delta S_1 + \Delta S_2 + \Delta S_3 + \Delta S_4$$

$$= \frac{q_1}{T_1} + 0 + \frac{q_2}{T_2} + 0 = 0 \qquad (3.23)$$

There is no change in entropy in the reversible adiabatic steps. Equation 3.21 can be written as follows for a general cyclic process.

$$\oint \frac{đq_{rev}}{T} = 0 \qquad (3.24)$$

We have seen that $đq$ is an inexact differential, but $1/T$ is an integrating factor for $đq_{rev}$ to generate an exact differential $đq_{rev}/T$. No integrating factor exists for $đ_{irrev}$. Thus, there is a thermodynamic temperature T, defined by equation 3.12, that yields an integrating factor $1/T$ for $đq_{rev}$ and one obtains a new state function.

We have not yet described the useful properties of the entropy, but it is worth noting that we started in Chapter 1 with the properties P and V, introduced the concept of a temperature and its measurement through the zeroth law, U through the first law, and now S and the thermodynamic temperature T through the second law. Since the third, and last, law of thermodynamics deals with a property of the entropy, entropy is the last really new thermodynamic property for systems involving only P-V work. We will introduce some additional and very useful thermodynamic state functions, but they, like the enthalpy H, will be defined in terms of P, V, T, U, and S.

3.6 The Clausius Theorem

Now we come to the second part of the second law, the properties of the entropy.

The derivation of the Clausius theorem starts with Carnot's theorem and takes us one step toward a criterion for spontaneous change that is crucial for thermodynamics. According to Carnot's theorem no engine operating between two heat reservoirs can be more efficient than a reversible Carnot engine operating between the same two reservoirs. In other words, if we compare the efficiency ϵ_{rev} of a reversible Carnot engine with the efficiency ϵ of any arbitrary engine operating between the same two temperatures that is not reversible, then

$$\epsilon \le \epsilon_{rev} \qquad (3.25)$$

Substituting the usual expressions for the efficiencies yields

$$1 + \frac{q_2}{q_1} \le 1 + \frac{q_{2rev}}{q_{1rev}} \qquad (3.26)$$

Remember that q_2 and q_{2rev} are negative. Thus,

$$\frac{-q_2}{q_1} \ge \frac{-q_{2rev}}{q_{1rev}} = \frac{T_2}{T_1} \qquad (3.27)$$

where T is the thermodynamic temperature. Thus,

$$\frac{-q_2}{q_1} \geq \frac{T_2}{T_1} \tag{3.28}$$

or

$$\frac{-q_2}{T_2} \geq \frac{q_1}{T_1} \tag{3.29}$$

or

$$0 \geq \frac{q_1}{T_1} + \frac{q_2}{T_2} \tag{3.30}$$

Earlier we derived the expression with the equal sign for the reversible Carnot engine. If we sum q_i/T_i for the four steps of the cycle, we obtain

$$\sum \frac{q_i}{T_i} \leq 0 \tag{3.31}$$

The derivation of equation 3.31 can be extended to any engine operating in a cycle with many heat reservoirs. A more general way to state the final conclusion is

$$\oint \frac{đq}{T} \leq 0 \tag{3.32}$$

This is the famous **Clausius theorem**. It is to be understood in the following way: (*a*) If any part of the cyclic process is irreversible (natural), the inequality applies. (*b*) If a cyclic process is reversible, the equality applies, as we have seen for the reversible Carnot cycle. (*c*) It is impossible for the cyclic integral to be greater than zero.

3.7   The Second Law

To obtain the second law in the form for practical applications, we apply the Clausius theorem to an irreversible cycle.

$$\text{State 1} \xrightarrow[\text{irreversible}]{} \text{State 2} \xrightarrow[\text{reversible}]{} \text{State 1} \tag{3.33}$$

According to equation 3.32,

$$\int_1^2 \frac{đq}{T} + \int_2^1 \frac{đq_{\text{rev}}}{T} < 0 \tag{3.34}$$

Since the second step is reversible, $đq_{\text{rev}}/T$ can be replaced by dS, and the limits can be interchanged, with a change in sign.

$$\int_1^2 \frac{đq}{T} - \int_1^2 dS < 0 \tag{3.35}$$

Thus,

$$\Delta S = S_2 - S_1 = \int_1^2 dS > \int_1^2 \frac{đq}{T} \tag{3.36}$$

Thus, for an irreversible process the change in entropy is greater than the integral of the heat divided by the temperature. We can also write

$$dS > \frac{dq}{T} \qquad (3.37)$$

This equation can be combined with the definition of the entropy, equation 3.22, to obtain **the second law in the form we will find most useful:**

$$dS \geq \frac{dq}{T} \qquad (3.38)$$

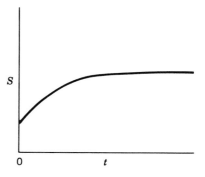

S

0 t

Figure 3.4 Change in the entropy of an isolated system with time. The entropy of the system increases spontaneously until equilibrium is reached. Thermodynamics does not deal with the question of how long it will take to reach equilibrium.

The various forms of the second law are essentially all interconvertible by arguments such as we have made here. In this form the second law provides a criterion for a spontaneous process, that is, one that can occur and can be reversed only by work from outside the system.

Equation 3.38 describes three types of processes:

$$
\begin{aligned}
dS &> dq/T & &\text{spontaneous and irreversible process} \\
dS &= dq/T & &\text{reversible process} \\
dS &< dq/T & &\text{impossible process}
\end{aligned}
\qquad (3.39)
$$

The simplest place to apply equation 3.38 is to an isolated system, because $dq = 0$. For a finite change the three possibilities are

$$
\begin{aligned}
\Delta S &> 0 & &\text{spontaneous and irreversible process} \\
\Delta S &= 0 & &\text{reversible process} \\
\Delta S &< 0 & &\text{impossible process}
\end{aligned}
\qquad (3.40)
$$

We can see that in a spontaneous process in an isolated system the entropy increases, as shown in Fig. 3.4. When a process occurs reversibly in an isolated system the entropy does not change. It is impossible for a process **in an isolated system** to reduce the entropy. When a natural process occurs in an isolated system there is an increase in entropy. The entropy increases to a maximum at equilibrium. Thus, the increase in entropy indicates the time sequence of natural events. This is in contrast with the equations of classical mechanics and quantum mechanics, which are reversible in time.

To use the second law to determine whether or not a process is spontaneous, we need to know the change in entropy. The change in entropy can be calculated using its definition, equation 3.22, if a reversible path can be found.

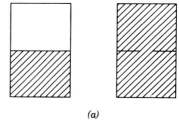

(a)

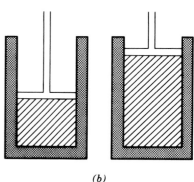

(b)

Figure 3.5 (a) Irreversible expansion of an ideal gas at 298 K into twice the initial volume with no heat or work. (b) Reversible isothermal expansion of an ideal gas to twice the initial volume at 298 K. In (b) the piston must be surrounded by a heat reservoir.

Example 3.1

Is the expansion of an ideal gas into a larger volume thermodynamically spontaneous in an isolated system? More specifically, consider the expansion of an isolated ideal gas initially at 298 K into a volume that is twice as large as its initial volume as shown in Fig. 3.5a.

Remember that Joule found that there is no change in temperature when a dilute gas is allowed to expand in an isolated system. To determine the change in entropy between the initial state and the final state, the reversible isothermal expansion described in Fig. 3.5b) can be used. As we have seen earlier, the work per mole done on the gas is $-RT \ln 2$, so that the heat absorbed by the gas is

$$q_{\text{rev}} = RT \ln 2$$

since $\Delta U = 0$. Thus, the change in entropy of the gas is

$$\Delta \bar{S}_b = R \ln 2 = 5.76 \text{ J K}^{-1} \text{ mol}^{-1}$$

Since the change in state of the gas for the process described in Fig. 3.5a is the same as that for Fig. 3.5b, $\Delta \bar{S}_a > 0$. Thus, we can conclude that the expansion is spontaneous, as we knew all along. The reverse of the process described in Fig. 3.5a, that is, the gas flowing back spontaneously into the initial volume, is impossible since $\Delta \bar{S} < 0$. It is important to note that this problem has nothing to do with minimizing the energy, which is constant.

3.8 The Fundamental Equation for a Closed System

The first and second laws of thermodynamics can be combined for **any** process in a closed system of constant composition. Since U is a state function, the differential dU in any process can be obtained from a reversible process that has the same change in state. But for the reversible process $dq = T\,dS$ and $dw = -P\,dV$ for the pressure–volume work so that

$$dU = T\,dS - P\,dV \tag{3.41}$$

if only expansion work is involved. This equation has so many uses that it is frequently called the **fundamental equation** for a closed system. Note that it involves only state functions. Thus, it applies to any change in state, whether it is accomplished reversibly or irreversibly. If other kinds of work are involved, as discussed in connection with equation 2.35, the fundamental equation can be written

$$dU = T\,dS - P\,dV + f\,dl + \gamma\,d\sigma + \phi\,dQ \tag{3.42}$$

Note that each term consists of an intensive variable times the differential of an extensive variable. We will have to add more terms to equation 3.41 or 3.42 later for open systems and chemical reactions.

Returning to the form for pressure–volume work only, the fundamental equation can be written

$$dS = \frac{1}{T}\,dU + \frac{P}{T}\,dV \tag{3.43}$$

This indicates that the entropy can be considered to be a function of the internal energy and volume. If the volume is constant, an increase in internal energy means there is an increase in entropy. If the internal energy is constant, an increase in volume means there is an increase in entropy.

3.9 Entropy Changes in Reversible Processes

We will now consider some simple processes for which entropy changes are readily calculated. It is especially easy to calculate entropy changes for reversible isothermal processes.

The transfer of heat from one body to another at an infinitesimally lower temperature is a reversible change, since the direction of heat flow can be reversed by an infinitesimal change in the temperature of one of the bodies. For example, consider the vaporization of a pure liquid into its vapor at the equilibrium vapor pressure P:

$$\text{Liquid } (T,P) \rightarrow \text{Vapor } (T,P) \tag{3.44}$$

Since T is constant, the integration of equation 3.22 yields

$$S_2 - S_1 = \Delta S = \frac{q_{rev}}{T} \tag{3.45a}$$

where q_{rev} represents the heat absorbed in the reversible change. Since the pressure is constant, the reversible heat is equal to the change in enthalpy ΔH, so that

$$\Delta S = \frac{\Delta H}{T} \tag{3.45b}$$

This equation may also be used to calculate the entropy of sublimation, the entropy of melting, and the entropy change for a transition between two forms of a solid.

Example 3.2

n-Hexane boils at 68.7 °C at 1.013 25 bar, and the molar enthalpy of vaporization is 28 850 J mol^{-1} at this temperature. If *n*-hexane is vaporized into the saturated vapor at this temperature, the process is reversible and the molar entropy change is given by

$$\Delta \overline{S} = \frac{\Delta \overline{H}}{T} = \frac{28\ 850 \text{ J mol}^{-1}}{341.8 \text{ K}} = 84.41 \text{ J K}^{-1} \text{ mol}^{-1}$$

The overbars indicate that we are dealing with molar changes. (See Trouton's rule in Section 6.5).

The molar entropy of a vapor is always greater than that of the liquid with which it is in equilibrium, and the molar entropy of the liquid is always greater than that of the solid at the melting point. According to the statistical interpretation of entropy to be discussed later in Section 3.12, in which the entropy is a measure of the disorder of the system, the molecules of the gas are more disordered than those of the liquid, and the molecules of the liquid are more disordered than those of the solid.

Now let us apply the second law to the vaporization of a liquid into its saturated vapor. To apply equations 3.40, we must consider an isolated system. In this case the liquid and vapor (the system) and the heat reservoir at T (the surroundings) form an isolated overall system. The total entropy change is given by

$$\Delta S = \Delta S_{syst} + \Delta S_{surr} \tag{3.45c}$$

Since the heat gained by the system is equal to that lost by the surroundings, the entropy change for the surroundings is the negative of the entropy change

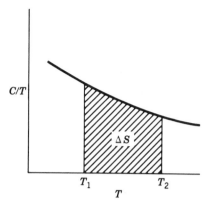

Figure 3.6   When a substance is heated from T_1 to T_2, without a phase change, the increase in entropy is given by the indicated area. If the volume is constant, then $n\overline{C}_V$ is used for C; and if the pressure is constant, then $n\overline{C}_P$ is used.

for the system if the vaporization is carried out reversibly; for both the system and surroundings taken together, the total change of entropy ΔS is zero if the transfer of heat is carried out reversibly. This is in agreement with the second form in equation 3.40.

Heating and cooling a substance is another example of a process that can be carried out reversibly. The change in entropy when a substance is heated or cooled can be calculated using

$$\mathrm{d}S = \frac{\mathrm{d}q_{\text{rev}}}{T} = \frac{C\,\mathrm{d}T}{T} \tag{3.46}$$

where C is C_P for a process at constant pressure and C_V for a process at constant volume. If the heat capacity is independent of temperature, and the temperature is changed from T_1 to T_2, then

$$\Delta S = \int_{T_1}^{T_2} \frac{C}{T}\,\mathrm{d}T = C\ln\frac{T_2}{T_1} \tag{3.47}$$

If the heat capacity is a function of temperature, this function can be substituted into the integral form of equation 3.47, or the entropy change can be obtained from a numerical integration, as shown in Fig. 3.6.

Example 3.3

Oxygen is heated from 300 to 500 K at a constant pressure of 1 bar. What is the increase in molar entropy? The coefficients in an empirical equation for the heat capacity at constant pressure as a function of temperature are given in Table 2.2. Using equation 3.47, we find

$$\Delta S = \int_{T_1}^{T_2} \frac{\overline{C}_P}{T}\,\mathrm{d}T = \int_{T_1}^{T_2} \left(\frac{\alpha}{T} + \beta + \gamma T\right)\mathrm{d}T$$

$$= \alpha\ln\frac{T_2}{T_1} + \beta(T_2 - T_1) + \frac{\gamma}{2}(T_2^2 - T_1^2)$$

$$= 25.503\ln\frac{500}{300} + (13.612 \times 10^{-3})(200) - (42.553 \times 10^{-7})(500^2 - 300^2)$$

$$= 15.07 \text{ J K}^{-1}\text{mol}^{-1}$$

The units have been omitted in this calculation, but you should check that they cancel properly.

The entropy change is readily calculated for a reversible isothermal expansion of an ideal gas. Since the internal energy of an ideal gas is independent of volume at constant temperature (Section 2.11), $\mathrm{d}q = -\mathrm{d}w = P\,\mathrm{d}V$. For a reversible isothermal expansion from V_1 to V_2,

$$\text{Ideal gas } (T,V_1,P_1) \rightarrow \text{Ideal gas } (T,V_2,P_2) \tag{3.48a}$$

$$\Delta S = \int_{V_1}^{V_2} \frac{P}{T}\,\mathrm{d}V = nR\int_{V_1}^{V_2} \frac{1}{V}\,\mathrm{d}V$$

$$= nR\ln\frac{V_2}{V_1}$$

$$= -nR\ln\frac{P_2}{P_1} \tag{3.48b}$$

where the final form simply comes from the fact that the volume is inversely proportional to pressure at constant temperature.

Equations 3.47 and 3.48b may be derived from a more general point of view by starting with the fundamental equation for a closed system (equation 3.43). Since the internal energy is a function of temperature and volume,

$$dU = \left(\frac{\partial U}{\partial T}\right)_V dT + \left(\frac{\partial U}{\partial V}\right)_T dV = C_V dT + \left(\frac{\partial U}{\partial V}\right)_T dT \qquad (3.49)$$

Substituting in equation 3.43 yields

$$dS = \frac{C_V}{T} dT + \frac{1}{T}\left[P + \left(\frac{\partial U}{\partial V}\right)_T\right] dV \qquad (3.50)$$

In Section 2.11 we saw that $(\partial U/\partial V)_T = 0$ for an ideal gas. We will find out how to handle this term for any system in Section 4.6, but here we will consider only ideal gases. For an ideal gas with a heat capacity independent of temperature, equation 3.50 can be integrated to obtain

$$S_{\text{ideal}} = C_V \ln T + nR \ln V + \text{const} \qquad (3.51)$$

or

$$\Delta S_{\text{ideal}} = C_V \ln \frac{T_2}{T_1} + nR \ln \frac{V_2}{V_1} \qquad (3.52)$$

where $\Delta S_{\text{ideal}} = S(T_2, V_2) - S(T_1, V_1)$.

Example 3.4

Show that the differential of the entropy can also be written

$$dS = \frac{1}{T} dH - \frac{V}{T} dP$$

Further show that for an ideal gas

$$dS_{\text{ideal}} = \frac{C_P}{T} dT - \frac{nR}{P} dP$$

and give the indefinite and definite integrated forms when C_P is independent of temperature.

Since $dH = dU + d(PV) = dU + P\,dV + V\,dP$, equation 3.41 can be written as

$$dH = T\,dS + V\,dP$$

or

$$dS = \frac{dH}{T} - \frac{V}{T} dP$$

For any gas

$$dS = C_P \frac{dT}{T} - \frac{V}{T} dP$$

and for an ideal gas

$$dS_{\text{ideal}} = C_P \frac{dT}{T} - \frac{nR}{P} \, dP$$

which can be integrated to obtain

$$S_{\text{ideal}} = C_P \ln \frac{T}{\text{K}} - nR \ln \frac{P}{P^\circ} + \text{const} \qquad (3.53a)$$

and

$$\Delta S_{\text{ideal}} = C_P \ln \frac{T_2}{T_1} - nR \ln \frac{P_2}{P_1} \qquad (3.53b)$$

In the chapter on statistical mechanics we will see that for an ideal monatomic gas the constant term in equation 3.53a for a mole of gas is equal to $-1.151\ 693R + \frac{3}{2}R \ln A_r$ when $P^\circ = 1$ bar, where A_r is the relative atomic mass of the gas. This equation, which will be derived from statistical mechanics in Section 17.7, is referred to as the **Sackur–Tetrode equation.**

Equation 3.53b corresponds to carrying out the change in state for an ideal gas from P_1, T_1 to P_2, T_2 in two reversible steps; in the first the temperature is changed at constant pressure, and in the second the pressure is changed at constant temperature. In general, to compute the change in entropy in any change in state, one must invent a series of reversible changes to accomplish the overall change in state.

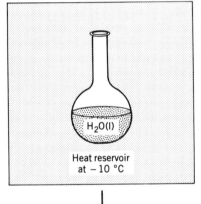

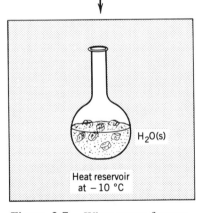

Figure 3.7 When water freezes at $-10\ ^\circ$C, its entropy decreases, but the entropy of the water and the heat reservoir increases, as expected for an irreversible process in an isolated system.

Example 3.5
What is the change in molar entropy of helium in the following process?

$$1 \text{ He}(298 \text{ K}, 1 \text{ bar}) \rightarrow 1 \text{ He}(100 \text{ K}, 10 \text{ bar})$$

For an ideal monatomic gas, $\overline{C}_P = \frac{5}{2}R$. Using equation 3.53b,

$$\Delta \overline{S} = \frac{5}{2}R \ln \frac{100}{298} - R \ln \frac{10}{1}$$
$$= -41.84 \text{ J K}^{-1} \text{ mol}^{-1}$$

3.10 Entropy Changes in Irreversible Processes

To obtain the change in entropy in an irreversible process we have to calculate ΔS along a reversible path between the initial state and the final state. We have already illustrated this in Section 3.7 by the expansion of an isolated ideal gas into a vacuum.

As a second illustration we will consider the spontaneous (irreversible) freezing of water below its freezing point (see Fig. 3.7). The freezing of a mole of supercooled water at $-10\ ^\circ$C is an irreversible change, but it can be carried out reversibly by means of the three steps for which the entropy changes are indicated:

$$H_2O(l,\ 0\ ^\circ C) \xrightarrow{\Delta\overline{S}=\Delta\overline{H}/T} H_2O(cr,\ 0\ ^\circ C)$$

$$\Delta\overline{S} = \int_{263}^{273} \frac{\overline{C}_{liq}}{T}\,dT \uparrow \qquad\qquad \downarrow \Delta\overline{S} = \int_{273}^{263} \frac{\overline{C}_{ice}}{T}\,dT \qquad (3.54)$$

$$H_2O(l,\ -10\ ^\circ C) \longrightarrow H_2O(cr,\ -10\ ^\circ C)$$

For the crystallization of liquid water at 0 °C, $\Delta\overline{H} = -6004$ J mol^{-1}. The heat capacity of water may be taken to be 75.3 J K^{-1} mol^{-1}, and that of ice may be taken to be 36.8 J K^{-1} mol^{-1} over this range. Then the total entropy change of the water when 1 mol of liquid water at -10 °C changes to ice at -10 °C is simply the sum of the three entropy changes:

$$\Delta\overline{S} = (75.3\ \text{J K}^{-1}\ \text{mol}^{-1})\ln\frac{273}{263} + \frac{(-6004\ \text{J mol}^{-1})}{273\text{K}}$$

$$+ (36.8\ \text{J K}^{-1}\ \text{mol}^{-1})\ln\frac{263}{273} \qquad (3.55)$$

$$= -20.54\ \text{J K}^{-1}\ \text{mol}^{-1}$$

The decrease in entropy corresponds to the increase in structural order when water freezes.

The statement that the entropy of an isolated system increases in a spontaneous process may be illustrated by considering supercooled water at -10 °C in contact with a large heat reservoir at this temperature. The entropy change for the isolated system upon freezing includes the entropy change of the reservoir as well as the entropy change of the water. Since the heat reservoir is large, the heat evolved by the water upon freezing is absorbed by the reservoir with only an infinitesimal change in temperature. The transfer of heat to a reservoir at the same temperature is a reversible process. Since the heat of fusion of water at -10 °C is $\Delta H(263\ \text{K}) = (75.3\ \text{J K}^{-1}\ \text{mol}^{-1})(10\ \text{K}) - 6004$ J mol^{-1} $- (36.8\ \text{J K}^{-1}\ \text{mol}^{-1})(10\ \text{K}) = -5619$ J mol^{-1}, the entropy change of the reservoir in the transfer of heat is

$$\Delta\overline{S} = \frac{(5619\ \text{J mol}^{-1})}{263\ \text{K}} = 21.37\ \text{J K}^{-1}\ \text{mol}^{-1} \qquad (3.56)$$

The entropy change of the water is -20.54 J K^{-1} mol^{-1}, and the total entropy change of the system water plus reservoir is

$$\Delta\overline{S} = (21.37 - 20.54)\ \text{J K}^{-1}\ \text{mol}^{-1} = 0.83\ \text{J K}^{-1}\ \text{mol}^{-1} \qquad (3.57)$$

Thus, the total entropy of the isolated system, including water and reservoir, increases, as required by inequality 3.40.

Example 3.6

Earlier in Example 2.7 we found that when an ideal monatomic gas at 1 bar and 273.15 K is allowed to expand adiabatically against a constant pressure of 0.315 bar until it doubles its volume, $T_2 = 215.8$ K and $\Delta\overline{U} = -715.4$ J mol^{-1}. What is the value of $\Delta\overline{S}$ for this process? Given: $\overline{C}_V = \frac{3}{2}R$ and, therefore, $\overline{C}_P = \frac{5}{2}R$.

To calculate $\Delta\overline{S}$, we have to find a reversible path for the change in state, which is

$$1X(273.15\ \text{K, 1 bar}) \rightarrow 1X(215.8\ \text{K, 0.315 bar})$$

This same change in state can be accomplished in two reversible steps:

Reversible isothermal expansion: 1X(273.15 K, 1 bar) → 1X(273.15 K, 0.315 bar)

$$\Delta \overline{S}_1 = -R \ln \frac{P_2}{P_1} = (8.314 \text{ J K}^{-1} \text{ mol}^{-1}) \ln \left(\frac{1 \text{ bar}}{0.315 \text{ bar}} \right)$$

$$= 9.60 \text{ J K}^{-1} \text{ mol}^{-1}$$

Cooling process: 1X(273.15 K, 0.315 bar) → 1X(215.8 K, 0.315 bar)

$$\Delta \overline{S}_2 = \overline{C}_P \ln \frac{T_2}{T_1}$$

$$= \frac{5}{2} (8.314 \text{ J K}^{-1} \text{ mol}^{-1}) \ln \frac{215.8 \text{ K}}{273.15 \text{ K}}$$

$$= -4.90 \text{ J K}^{-1} \text{ mol}^{-1}$$

The entropy change for the reversible two-step process is

$$\Delta \overline{S} = \Delta \overline{S}_1 + \Delta \overline{S}_2 = 9.60 - 4.90$$

$$= 4.70 \text{ J K}^{-1} \text{ mol}^{-1}$$

Note that the entropy of the system increases in this irreversible expansion, even though $q = 0$.

3.11   Entropy of Mixing Ideal Gases

Figure 3.8 shows that amount n_1 of gas 1 at P and T is separated from amount n_2 of gas 2 at the same pressure and temperature. When the partition is withdrawn the gases will diffuse into each other at constant temperature and pressure if they are ideal gases. To calculate the change in entropy in this irreversible process we need to find a way to carry it out reversibly. This can be done in two steps. First, we expand each gas isothermally and reversibly to the final volume $V = V_1 + V_2$. These volumes are not molar volumes, but are actual volumes. When extensive volumes are used, the ideal gas law is written $PV = nRT$.

Using equation 3.48b, the entropy changes for the two gases are

$$\Delta S_1 = -n_1 R \ln \frac{V_1}{V} = -n_1 R \ln \frac{n_1}{n_1 + n_2} = -n_1 R \ln y_1 \qquad (3.58)$$

$$\Delta S_2 = -n_2 R \ln \frac{V_2}{V} = -n_2 R \ln \frac{n_2}{n_1 + n_2} = -n_2 R \ln y_2 \qquad (3.59)$$

where y_i is the mole fraction. The total entropy change for the first step is the sum of the entropy changes for the two gases:

$$\Delta S = -n_1 R \ln y_1 - n_2 R \ln y_2 \qquad (3.60)$$

In the second step the expanded gases are mixed reversibly at constant pressure. To see how this might be done we have to imagine two semipermeable membranes, one of these (represented by dashes) is permeable only to gas 1, and the other, arranged as shown in Fig. 3.9 (represented by dots), is permeable

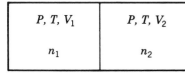

P, T, V_1 n_1	P, T, V_2 n_2

Initial state

$P, T, V_1 + V_2$
$n_1 + n_2$

Final state

Figure 3.8   Mixing of ideal gases. The partition between n_1 mol of gas 1 at P and T and n_2 mol of gas at P and T is withdrawn so that the gases can mix.

only to gas 2. The membrane that is permeable by gas 1 and an impermeable membrane, which is separated from it by volume V, are moved at the same infinitesimally slow rate to the left. As shown by the diagram for the intermediate stage in the reversible mixing process, the $\vdots$ | membrane combination has to be moved to the left against a pressure of P_1 (to the left of $\vdots$ |) plus P_2 (to the left of |). But the pressure on the right sides of this membrane combination is also $P_1 + P_2$. Therefore, no work is required to move the membrane in this frictionless device. The internal energies of the two ideal gases are functions only of T (Section 2.10), and so according to the first law no heat is absorbed by the gas in this step. Consequently, there is no entropy change associated with the second step, and so equation 3.60 gives the total entropy change for the isothermal mixing of the two perfect gases.

Equation 3.60 may be written for the mixing of a total of 1 mol of gas. Dividing both sides of this equation by $n_1 + n_2$ yields the **entropy of mixing** $\Delta_{mix}S$ for 1 mol of mixture

$$\Delta_{mix}S = -R(y_1 \ln y_1 + y_2 \ln y_2) \tag{3.61}$$

This equation may be generalized to the mixing of N ideal gases:

$$\Delta_{mix}S = -R \sum_{i=1}^{N} y_i \ln y_i \tag{3.62}$$

Since $y_i < 1$, $\ln y_i < 0$, and $\Delta_{mix}S$ is always positive.

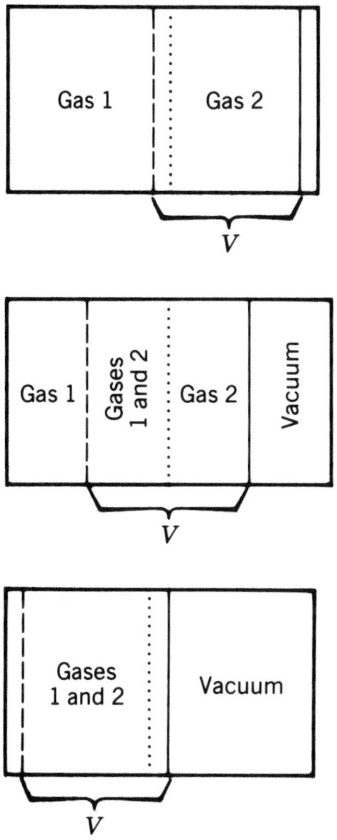

Example 3.7

What is the entropy of mixing of $\frac{1}{2}$ mol of oxygen with $\frac{1}{2}$ mol of nitrogen at 25 °C, assuming that they are ideal gases?

$$\Delta_{mix}S = -R(y_1 \ln y_1 + y_2 \ln y_2)$$
$$= -(8.314 \text{ J K}^{-1} \text{ mol}^{-1})(\tfrac{1}{2} \ln \tfrac{1}{2} + \tfrac{1}{2} \ln \tfrac{1}{2})$$
$$= 5.763 \text{ J K}^{-1} \text{ mol}^{-1}$$

You might try $\frac{1}{4}$, $\frac{3}{4}$, and $\frac{1}{8}$, $\frac{7}{8}$ to see what happens.

Figure 3.9 Reversible isothermal mixing of ideal gases 1 and 2 using a membrane permeable only by gas 1 (– – –), a membrane permeable by gas 2 (···), and an impermeable membrane (—).

3.12 Entropy and Statistical Probability

Using the macroscopic approach of thermodynamics, we have found that the equilibrium state of an isolated system is that state in which the entropy has its maximum value. From a microscopic point of view, we might expect that the equilibrium state of an isolated system would be the state with the maximum statistical probability. For simple systems we can use the molecular point of view to calculate the statistical probabilities of different final states. For example, assume that one mole of an ideal gas is in a container that is connected to a container of equal volume through a stopcock; this expansion has already been discussed in Example 3.1. Actually, it is a little easier to think about the

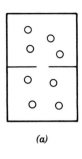

(a)

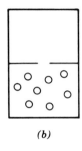

(b)

Figure 3.10 (*a*) Equilibrium gaseous system after a hole is punched in the diaphragm. (*b*) Highly improbable state of the gaseous system after the hole has been punched.

reverse process, and so we will do that. As shown in Fig. 3.10, we will start with an opening between the two chambers, and ask what is the statistical probability that all of the molecules will be in the original chamber? The probability that a particular molecule will be in the original chamber is 1/2. The probability that two particular molecules will be in the original chamber is $(1/2)^2$, and the probability that all the molecules will be in the original chamber is $(1/2)^N$, where N is the number of molecules in the system. If the system contains a mole of gas molecules, the statistical probability that all the molecules will be in the original chamber is

$$(1/2)^{6.022 \times 10^{23}} = e^{-4.174 \times 10^{23}} \tag{3.63}$$

Boltzmann postulated that

$$S = k \ln \Omega \tag{3.64}$$

where k is **Boltzmann's constant** (R/N_A) and Ω is the number of equally probable arrangements for the system. This relation can be used to calculate ΔS for the transformation from an initial state with entropy S and Ω equally probable arrangements to a final state with entropy S' and Ω' equally probable arrangements:

$$\Delta S = S' - S = k \ln \Omega'/\Omega \tag{3.65}$$

It is often difficult to count the number of equally probable arrangements in the final state and in the initial state, but the ratio Ω'/Ω in this case is equal to the ratio of the probability that all the molecules are in one chamber to the probability that they are all in one chamber or the other. The probability that all the molecules are in the original chamber or the other chamber is, of course, unity. Thus, for the system in the preceding paragraph, we have already calculated the ratio of the probabilities that is equal to Ω'/Ω. The change in entropy in going from the state with the gas distributed between the two chambers to the state with all the molecules in the original chamber is

$$\begin{aligned} \Delta S &= k \ln e^{-4.174 \times 10^{23}} \\ &= (1.381 \times 10^{-23} \text{ J K}^{-1})(-4.174 \times 10^{23}) \\ &= -5.76 \text{ J K}^{-1} \end{aligned} \tag{3.66}$$

Since we are considering one mole of an ideal gas, the change in entropy for the expansion process in Example 3.1 is $\Delta \overline{S} = 5.76$ J K^{-1} mol^{-1}, as expected. This confirms that the Boltzmann constant is indeed given by $k = R/N_A$.

If the gas molecules were all to be found in one chamber after having been distributed between the two chambers, we would say that the second law had been violated. We have just seen that the probability that such a thing might happen is not zero. It is, however, so small that we could never expect to be able to observe all the molecules in one chamber, even for systems containing much, much less than one mole of gas. If, however, we considered a system of only two molecules, then we could find both molecules in one chamber with reasonable probability. This shows that the laws of thermodynamics are based on the fact that macroscopic systems contain very large numbers of molecules.

The equation $S = k \ln \Omega$ embodies an important concept, but it is not used very often because it is difficult to calculate Ω. In Chapter 17 on statistical mechanics we will use other equations to calculate the entropy.

The collision of two gas molecules is reversible in the sense that the reverse process can also happen. If, after the molecules are moving away from each other, we could simply reverse the direction of the velocity vectors, the molecules would move along the same trajectories in the reverse direction. In short, the movie of the reverse process is just as reasonable as the movie of the forward process. This is true for both classical mechanics and quantum mechanics. If molecular collisions are reversible, then why is the expansion of a gas into a vacuum, or the mixing of two gases, irreversible? If we could take a movie of the expansion of a gas into a vacuum that would show the locations of all the molecules, we could tell whether the movie was being run forward or backward. However, if we were to look at each molecular collision we would find that each followed the laws of mechanics, and was reversible. If we were to look at the movie being run backward, we would feel that it was depicting something that could not happen. But why couldn't it happen? As a matter of fact, it could, but only if we could give all the molecules the positions they have at the end of the movie but then reverse their velocity vectors. If we then looked at the individual collisions, we would find that they would take all the molecules to the region from which they had expanded. The reason this does not happen in real life is that it takes an extraordinarily special set of molecular coordinates and velocities. This set of coordinates and velocity vectors is so unlikely that thermodynamics says that it can never happen.

3.13   Calorimetric Determination of Entropies

The entropy of a substance at any desired temperature relative to its entropy at absolute zero may be obtained by integrating $đq_{rev}/T$ from absolute zero to the desired temperature. This requires heat capacity measurements down to the neighborhood of 0 K; and enthalpy of transition measurements for all transitions in this temperature range. Since measurements of C_P cannot be carried to 0 K, the Debye function (equation 17.121) is used to represent C_P below the temperature of the lowest measurements.

If data on the enthalpy of fusion at the melting point T_m and the enthalpy of vaporization at the boiling point T_b are available, the entropy at a temperature T above the boiling point T_b relative to that at 0 K may be calculated from

$$\overline{S}_T^\circ - \overline{S}_0^\circ = \int_0^{T_m} \frac{\overline{C}_P^\circ(s)}{T} \, dT + \frac{\Delta_{fus}H^\circ}{T_m} + \int_{T_m}^{T_b} \frac{\overline{C}_P^\circ(l)}{T} \, dT$$
$$+ \frac{\Delta_{vap}H^\circ}{T_b} + \int_{T_b}^{T} \frac{\overline{C}_P^\circ(g)}{T} \, dT \quad (3.67)$$

If there are various solid forms with enthalpies of transition between the forms, the corresponding entropies of transition would have to be included in this sum.

Heat-capacity measurements down to these very low temperatures are made with special calorimeters in which the substance is heated electrically in a carefully insulated system and the input of electrical energy and the temperature are measured accurately.

The attainment of very low temperatures in the laboratory involves successive application of different methods. Vaporization of liquid helium (b.p.

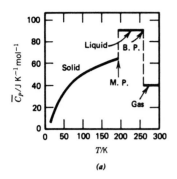

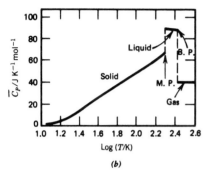

Figure 3.11   Heat capacity of sulfur dioxide at a constant pressure of 1 bar at different temperatures.

4.2 K at 1 bar) at low pressures produces temperatures down to about 0.3 K. Lower temperatures may be reached by use of adiabatic demagnetization. A paramagnetic (Section 16.1) salt such as gadolinium sulfate is cooled with liquid helium in the presence of a strong magnetic field. The salt is thermally isolated from its surroundings and the magnetic field is removed. The salt undergoes a reversible adiabatic process in which the atomic spins become disordered. Since the energy must come from the crystal lattice, the salt is cooled. Temperatures of about 0.001 K may be reached in this way. Adiabatic demagnetization of nuclear spins can then be used to obtain temperatures of the order of 10^{-6} K.

As an illustration of the determination of the entropy of a substance relative to its entropy at 0 K, the measured heat capacities for SO_2 are shown as a function of T and of log T in Figs. 3.11a* and b. Solid SO_2 melts at 197.64 K, and the heat of fusion is 7402 J mol^{-1}. Liquid SO_2 vaporizes at 263.08 K at 1.01325 bar and the heat of vaporization is 24 937 J mol^{-1}. The calculation is summarized in Table 3.1.

Table 3.1   The Entropy of Sulfur Dioxide

T/K	Method of Calculation	$\Delta \overline{S}°$/J K^{-1} mol^{-1}
0–15	Debye function ($\overline{C}_P$ = constant T^3)	1.26
15–197.64	Graphical, solid	84.18
197.64	Fusion, 7402/197.64	37.45
197.64–263.08	Graphical, liquid	24.94
263.08	Vaporization, 24 937/263.08	94.79
263.08–298.1	From $\overline{C}_P$ of gas	5.23
	$\overline{S}°$ (298 K) $- \overline{S}°$(0 K) =	247.85

3.14   The Third Law of Thermodynamics

In the early years of this century T. W. Richards and W. Nernst independently studied the entropy changes of certain isothermal chemical reactions and found that the change in entropy approached zero as the temperature was reduced. The entropy change for a chemical reaction cannot be determined calorimetrically using $\Delta S = q_{rev}/T$ because reactions do not occur reversibly. We will see in Chapter 5 how the entropy change for a chemical reaction may be determined. In the meantime we can discuss the measurement of the change in entropy for a phase change for a single substance. As an example we will consider the phase change

$$S(\text{rhombic}) = S(\text{monoclinic}) \qquad (3.68)$$

A phase change is similar to a chemical reaction in that it is simply a change in state. As for chemical reactions, the change in entropy for a phase change also approaches zero as the temperature is reduced to absolute zero.

---

* W. F. Giauque and C. C. Stephenson, *J. Am. Chem. Soc.* **60:** 1389 (1938).

Figure 3.12 shows the molar entropies of monoclinic and rhombic sulfur down to absolute zero as determined from

$$\overline{S}_T - \overline{S}_0 = \int_0^T \frac{\overline{C}_P}{T} \, dT \qquad (3.69)$$

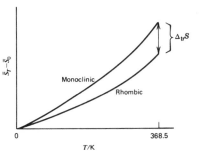

Figure 3.12 Molar entropies of monoclinic and rhombic sulfur from absolute zero to the transition point (368.5 K).

Rhombic sulfur is the stable form below the transition temperature of 368.5 K. However, monoclinic sulfur may be supercooled below this temperature, and its heat capacity may be measured down to the neighborhood of absolute zero. The entropy change for the phase change (equation 3.68) at the transition temperature calculated on the assumption that $\overline{S}_0$ is zero for both forms is 1.09 J K^{-1} mol^{-1}. This is in agreement with the entropy change at 368.5 K calculated from the difference in enthalpy of 401 J mol^{-1} between the two forms at 368.5 K:

$$\Delta_{tr}S = \frac{401 \text{ J mol}^{-1}}{368.5 \text{ K}} = 1.09 \text{ J K}^{-1} \text{ mol}^{-1}$$

where $\Delta_{tr}S$ is the entropy change for the transition per mole. The difference in enthalpy between the two forms is simply equal to the difference between their heats of combustion at 368.5 K.

It can be seen from Fig. 3.12 that as $T \to 0$, $\Delta_{tr}S \to 0$. It has been found that $\Delta_{tr}S \to 0$ as $T \to 0$ for many other isothermal phase transitions and chemical reactions. In 1905 these observations led Nernst to the conclusion that as temperature approaches 0 K, $\Delta_r S$ for all reactions approaches zero.

$$\lim_{T \to 0} \Delta_r S = 0 \qquad (3.70)$$

In 1913 Max Planck took this idea one step further, and we will take his statement as the third law of thermodynamics:

The entropy of each pure element or substance in a perfect crystalline form is zero at absolute zero.

We will see later (Chapter 17) that statistical mechanics gives a reason for picking this value. As the derivation in Section 3.12 suggests, this corresponds with a single quantum state ($\Omega = 1$) for a perfect crystal at absolute zero. Thus, according to the third law, $\overline{S}_0^\circ$ of the preceding section can be taken as zero if the substance has a perfect crystalline form in the neighborhood of absolute zero. Heat-capacity measurements down to temperatures of nearly 0 K are therefore often said to yield "third law entropies." Thus, the calculation of the preceding section yielded the third law entropy of $SO_2(g)$ at 298.15 K.

Third law entropies can be tested against what we would expect from two other types of measurements, measurements of equilibrium constants and of spectroscopic data. As we will see in Chapter 5, if the equilibrium constant for a reaction is measured over a range of temperatures, then both ΔH° and ΔS° can be calculated. This ΔS° can be compared with the value expected from the third law if the heat capacities of all reactants and products have been measured down to the neighborhood of absolute zero.

In Chapter 17 on statistical mechanics we will see that the entropies of relatively simple gases at any desired temperature may be calculated from molar masses and certain spectroscopic information. The molar entropy of a gas at a certain temperature calculated using statistical mechanics may be compared with the molar entropy obtained from calorimetric measurements, assuming that the entropy of the pure crystalline substance is zero at absolute zero.

In general the tests of the third law described in the preceding two paragraphs confirm the third law, but there are some apparent violations. For example, the entropy of $N_2O(g)$ at 298.15 K determined from heat-capacity measurements is $5.8 \text{ J K}^{-1} \text{ mol}^{-1}$ smaller than that calculated from spectroscopic data. This indicates an entropy of $5.8 \text{ J K}^{-1} \text{ mol}^{-1}$ at absolute zero for crystals of N_2O. This is an asymmetric linear molecule, NNO. The residual entropy of the crystal is due to disorder in the arrangement of N_2O molecules. In solid N_2O, the molecules are arranged with random head–tail alignments (such as NNO, ONN, NNO, NNO, ONN) instead of being perfectly ordered (NNO, NNO, NNO, NNO, NNO). If the orientation were perfectly random, the crystal might be regarded as a mixed crystal with equal mole fractions of NNO and ONN. The entropy of the mixed crystal would then be the entropy change of mixing. Using equation 3.61, which applies to ideal crystals as well as ideal gases, we see that

$$\Delta_{mix}S = -R(\tfrac{1}{2} \ln \tfrac{1}{2} + \tfrac{1}{2} \ln \tfrac{1}{2}) = 5.76 \text{ J K}^{-1} \text{ mol}^{-1} \tag{3.71}$$

Therefore, the statistical mechanical value is taken as the correct entropy, and this is the value that will be found in tables. Thus, the apparent violation of the third law is understood and can be calculated for ideal crystals.

Another example of an imperfect crystal from the standpoint of the third law is H_2O. Crystals of H_2O have a residual entropy of $3.35 \text{ J K}^{-1} \text{ mol}^{-1}$ at 0 K. In ice, the hydrogen atoms are arranged around each oxygen atom in a tetrahedral manner. Two of the four atoms are covalently bonded to that oxygen atom, and two are hydrogen bonded (Section 14.9) to that oxygen atom. Since the arrangements of the two types of hydrogen atoms around the oxygen atoms in the crystal are random, it may be shown that the entropy of the crystal should approach $R \ln \tfrac{3}{2} = 3.37 \text{ J K}^{-1} \text{ mol}^{-1}$ at absolute zero.

There are two types of randomness in crystals at absolute zero that are not considered in calculating entropies at absolute zero if these entropies are to be used only for chemical purpose:

1. Most crystals are made up of a mixture of isotopic species, but the entropy of mixing isotopes is ignored because the reactants and products in a reaction or phase change contain the same mixtures of isotopes.

2. There is a nuclear spin degeneracy (Section 17.2) at absolute zero that is ignored because it exists in both reactants and products.

There is a corollary to the third law that is very much in the spirit of the Clausius statement of the second law, since it states an impossibility. According to this corollary, it is impossible to reduce the temperature of a system to 0 K in a finite number of steps. This conclusion that absolute zero is unattainable may be derived from the third law.*

---

* E. A. Guggenheim, *Thermodynamics*. Amsterdam: North-Holland, 1967, p. 157.

The third law is important because it makes possible the calculation of the equilibrium constant for a chemical reaction purely from calorimetric measurements on the reactants and products. The entropies of a number of substances at 298.15 K are given in Appendix C.1. The entropies of a smaller number of substances are given in Appendix C.2 for 0, 298, 500, 1000, 2000, and 3000 K. These values and others in the much larger tables from which they have been taken come from four sources:

1. Heat capacities and enthalpies as a function of temperature
2. Statistical mechanical calculations using molecular structure and energy levels
3. Temperature variation of equilibrium constants (Section 5.9)
4. Calculations from enthalpies and Gibbs energies from other sources such as electromotive force measurements (Section 8.7)

Where the entropies have been determined calorimetrically, corrections have been made for imperfections of crystals encountered in the neighborhood of absolute zero in the few cases where this occurs and for gas nonideality at 1 bar. The standard states are, of course, the same as discussed in Section 2.16. The entropy of H^+ (ao) is arbitrarily assigned the value of zero, and this makes it possible to calculate entropies of other aqueous ions.

In comparing standard entropy values from various sources it is important to be aware of the standard pressure used. The adjustment of standard entropies from a standard state pressure of 1 atm to 1 bar is discussed in problem 3.35.

3.15 Entropy Changes for Chemical Reactions

The **standard reaction entropy** $\Delta_r S°$ for a reaction $\sum \nu_i A_i = 0$ (see Section 2.16) is equal to the change in entropy when the separated reactants, each in its standard state, are completely converted to separated products, each in its standard state, at the specified temperature. The standard reaction entropy $\Delta_r S°$ at a particular temperature is given by

$$\Delta_r S° = \sum_{i=1}^{N} \nu_i \overline{S_i^°} \tag{3.72}$$

where N is the number of substances involved in the reaction. The following equation can be used to calculate the standard entropy of reaction at another temperature:

$$\Delta_r S°(T) = \Delta_r S°(298.15 \text{ K}) + \int_{298.15}^{T} \frac{\Delta_r C_P^°}{T} \, dT \tag{3.73}$$

where $\Delta_r C_P^°$ is $\sum \nu_i \overline{C_{Pi}^°}$ for the reaction. If $\Delta_r C_P^°$ is independent of temperature, the integral can be replaced with $\Delta_r C_P^° \ln (T/298.15)$.

Example 3.8

Calculate the standard reaction entropies for the following reactions at 298 K:

$$(a) \quad H_2(g) + \tfrac{1}{2} O_2(g) = H_2O(l)$$

$$(b) \quad N_2(g) + 3H_2(g) = 2NH_3(g)$$

$$(c) \quad CaCO_3(s, \text{ calcite}) = CaO(s) + O_2(g)$$

$$(d) \quad N_2O_4(g) = 2NO_2(g)$$

The following reaction entropies are calculated using Table C.1:

$$(a) \quad \Delta_r S^\circ = 69.91 - 130.68 - \tfrac{1}{2}(205.14)$$
$$= -163.34 \text{ J K}^{-1} \text{ mol}^{-1}$$

$$(b) \quad \Delta_r S^\circ = 2(194.45) - 191.61 - 3(130.68)$$
$$= -194.75 \text{ J K}^{-1} \text{ mol}^{-1}$$

$$(c) \quad \Delta_r S^\circ = 39.75 + 205.14 - 92.9$$
$$= 151.99 \text{ J K}^{-1} \text{ mol}^{-1}$$

$$(d) \quad \Delta_r S^\circ = 2(240.06) - 304.29$$
$$= 175.83 \text{ J K}^{-1} \text{ mol}^{-1}$$

Note that since the molar entropies of gases are greater than those of liquids and solids, the entropy always increases when the reaction produces more moles of gaseous products than reactant.

Example 3.9

Use Appendix C.1 to calculate the standard reaction entropy for the dissociation of molecular hydrogen into atoms at 298.15 and 1000 K.

$$H_2(g) = 2H(g)$$

$$\Delta_r S^\circ(298.15 \text{ K}) = [2(114.713) - 130.683]J \text{ K}^{-1} \text{ mol}^{-1}$$
$$= 98.743 \text{ J K}^{-1} \text{ mol}^{-1}$$

$$\Delta_r S^\circ(1000 \text{ K}) = 98.743 + \int_{\ln 298.15}^{\ln 1000} [2(20.784) - 28.824]d \ln T$$
$$= 98.743 + 12.744 \ln(1000/298.15)$$
$$= 114.165 \text{ J K}^{-1} \text{ mol}^{-1}$$

where we have neglected the temperature dependence of $\Delta_r C_P^\circ$. A more accurate value at 1000 K can be obtained by using Appendix C.2:

$$\Delta_r S^\circ(1000 \text{ K}) = 2(139.867) - 166.222$$
$$= 113.512 \text{ J K}^{-1} \text{ mol}^{-1}$$

Reactions with more molecules of gaseous products than reactants generally have positive reaction entropies of this magnitude.

3.16   Special Topic: Estimation of Entropies

In Section 17.11 we will see that standard entropies of gases can be calculated quite accurately using statistical mechanics if molecular dimensions, vibration frequencies, and energies of low-lying excited electronic states, if any, are known. In the absence of the third law measurements and of spectroscopic information, entropies of gases may be estimated by use of group additivity. Section 2.21 gave Benson group contributions* to $\Delta_f H°(298 \text{ K})$ for alkanes, and the following table gives the corresponding group contributions to the internal entropy $\overline{S}_{int}°(298 \text{ K})$.

	$\overline{S}_{int}°/\text{J K}^{-1} \text{ mol}^{-1}$
$C(H)_3(C)$	127.24
$C(H)_2(C)_2$	39.41
$C(H)(C)_3$	-50.50
$C(C)_4$	-146.86

The term internal entropy is used because in estimating the entropy the symmetry of the molecule has to be taken into account. The symmetry of molecules is discussed in Chapter 13, and we will see how the symmetry number comes into the calculation of the entropy of small molecules in Section 17.9. The **symmetry number** σ is the number of independent permutations of identical atoms (or groups) that can be arrived at by simple rigid rotations of the entire molecule. We will not discuss the determination of the symmetry number here, but will give some examples. All homonuclear diatomic molecules (H_2, O_2, Cl_2, etc.) have a symmetry number of 2 because they can be rotated end-for-end without changing their appearance. Heteronuclear diatomic molecules have a symmetry number of unity. A water molecule has a symmetry number of 2 because it can be rotated into an equivalent position about a twofold axis through the oxygen atom. The symmetry numbers of other simple molecules are as follows: $NH_3(3)$; $CH_4(12)$; ethylene, $C_2H_4(4)$; cyclopropane, $C_3H_6(6)$; $C_6H_6(12)$. If groups in a molecule can rotate with respect to the rest of the molecule this also contributes to the symmetry number. For example, in organic molecules methyl groups can rotate with respect to the rest of the molecule and each contribute a factor of 3 to the symmetry number. The standard entropy of 298 K calculated by the group method is

$$\overline{S}° = \sum \overline{S}_{int}° - R \ln \sigma \qquad (3.74)$$

where $\overline{S}_{int}°$ are the Benson group values and σ is the symmetry number (Section 17.9).

Example 3.10

Estimate the standard entropies at 298.15 for the three pentane isomers in the ideal gas phase. The experimental values for *n*-pentane, isopentane, and neopentane are 349.56, 343.74, and 306.00 J K^{-1} mol^{-1}, respectively.

---

* S. W. Benson, *Thermodynamical Kinetics*. New York: Wiley, 1976.

The symmetry number for n-pentane is 9 because there are two methyl groups (3^2). Thus, the estimated entropy is

$$\overline{S}^\circ = [2(127.24) + 3(39.41) - 8.314 \ln 9] \text{ J K}^{-1} \text{ mol}^{-1}$$
$$= 354.4 \text{ J K}^{-1} \text{ mol}^{-1}$$

The symmetry number for isopentane (2-methylbutane) is 3^3 because there are three methyl groups, so that

$$\overline{S}^\circ = [3(127.24) + 39.41 - 50.50 - 8.314 \ln 27] \text{ J K}^{-1} \text{ mol}^{-1}$$
$$= 343.2 \text{ J K}^{-1} \text{ mol}^{-1}$$

The symmetry number for neopentane (2,2-dimethylpropane) is 3^4, for the four methyl groups, times 12, since the whole molecule has the symmetry number of CH_4, if the structure of the methyl groups is ignored. Thus,

$$\overline{S}^\circ = [4(127.24) - 146.86 - 8.314 \ln (3^4 \times 12)] \text{ J K}^{-1} \text{ mol}^{-1}$$
$$= 304.9 \text{ J K}^{-1} \text{ mol}^{-1}$$

References

K. E. Bett, J. S. Rowlinson, and G. Saville, *Thermodynamics for Chemical Engineers*. Cambridge, MA: MIT Press, 1975.

H. B. Callen, *Thermodynamics and an Introduction to Thermodynamics*, 2nd ed. New York: Wiley, 1985.

K. Denbigh, *The Principles of Chemical Equilibrium*. Cambridge, UK: Cambridge University Press, 1981.

J. W. Gibbs, *The Collected Works of J. Willard Gibbs*. New Haven, CT: Yale University Press, 1948.

N. A. Gokcen, *Thermodynamics*. Hawthorne, CA: Techscience Inc., 1975.

E. A. Guggenheim, *Thermodynamics*. Amsterdam: North-Holland, 1967.

J. G. Kirkwood and I. Oppenheim, *Chemical Thermodynamics*. New York: McGraw-Hill, 1961.

G. N. Lewis and M. Randall, revised by K. S. Pitzer and L. Brewer, *Thermodynamics*. New York: McGraw-Hill, 1961.

M. Model and R. C. Reid, *Thermodynamics and Its Applications*. Englewood Cliffs, NJ: Prentice-Hall, 1983.

J. M. Smith and H. C. Van Ness, *Introduction to Chemical Engineering Thermodynamics*. New York: McGraw-Hill, 1987.

D. R. Stull, E. F. Westrum, and G. C. Sinke, *The Chemical Thermodynamics of Organic Compounds*. New York: Wiley-Interscience, 1969.

Problems

3.1 Theoretically, how high could a gallon of gasoline lift an automobile weighing 2800 lb against the force of gravity, if it is assumed that the cylinder temperature is 2200 K and the exit temperature 1200 K? (Density of gasoline = 0.80 g cm^{-3}; 1 lb = 453.6 g; 1 ft = 30.48 cm; 1 L = 0.2642 gal. Heat of combustion of gasoline = 46.9 kJ g^{-1}.)

3.2 (a) What is the maximum work that can be obtained from 1000 J of heat supplied to a steam engine with a high-temperature reservoir at 100 °C if the condenser is at 20 °C? (b) If the boiler temperature is raised to 150 °C by the use of superheated steam under pressure, how much more work can be obtained?

3.3 A heat pump is to be used to maintain the temperature of a building at 18 °C when the outside temperature is at −5 °C. For a frictionless heat pump how much work must be expended to obtain a joule of heat?

3.4 Water is vaporized reversibly at 100 °C and 1.013 25 bar. The heat of vaporization is 40.69 kJ mol^{-1}. (a) What is the value of ΔS for the water? (b) What is the value of ΔS for the water plus the heat reservoir at 100 °C?

3.5 Assuming that CO_2 is an ideal gas, calculate $\Delta H°$ and $\Delta S°$ for the following process:

1 CO_2(g, 298.15 K, 1 bar) → 1 CO_2(g, 1000 K, 1 bar)

Given: $\bar{C}_P$ = 26.648 + 42.262 × 10^{-3}T − 142.4 × 10$^{-7}T^2$ in J K^{-1} mol^{-1}

3.6 Calculate the increase in molar entropy of silver that is heated at constant pressure from 0 to 30 °C if the value of $\bar{C}_P$ in this temperature range is considered to be constant at 25.48 J K^{-1} mol^{-1}.

3.7 The temperature of an ideal monatomic gas is increased from 300 to 500 K. What is the change in molar entropy of the gas (a) if the volume is held constant and (b) if the pressure is held constant?

3.8 Ammonia (considered to be an ideal gas) initially at 25°C and 1 bar pressure is heated at constant pressure until the volume has trebled. Calculate (a) q per mole, (b) w per mole, (c) $\Delta \bar{H}$, (d) $\Delta \bar{U}$, and (e) $\Delta \bar{S}$. Given: $\bar{C}_P$ = 25.895 + 32.999 × 10^{-3}T − 30.46 × 10$^{-7}T^2$ in J K^{-1} mol^{-1}.

3.9 Two blocks of the same metal are of the same size but are at different temperatures, T_1 and T_2. These blocks of metal are brought together and allowed to come to the same temperature. Show that the entropy change is given by

$$\Delta S = C_P \ln \frac{(T_1 + T_2)^2}{4T_1 T_2}$$

if C_P is constant. How does this equation show that the change is spontaneous?

3.10 In the reversible isothermal expansion of an ideal gas at 300 K from 1 to 10 L, where the gas has an initial pressure of 20.27 bar, calculate (a) ΔS for the gas and (b) ΔS for all systems involved in the expansion.

3.11 A mole of an ideal gas is expanded isothermally and reversibly from 30 to 100 L at 300 K. (a) What are the values of $\Delta \bar{U}$, $\Delta \bar{S}$, w per mole, and q per mole? (b) If the expansion is carried out irreversibly by allowing the gas to expand into an evacuated container, what are the values of $\Delta \bar{U}$, $\Delta \bar{S}$, w per mole, and q per mole?

3.12 (a) A system consists of a mole of ideal gas that undergoes the following change in state:

1X(g, 298 K, 10 bar) → 1X(g, 298 K 1 bar)

What is the value of $\Delta \bar{S}$ if the expansion is reversible? What is the value of $\Delta \bar{S}$ if the gas expands into a larger evacuated container so that the final pressure is 1 bar? (b) The same change in state takes place, but we now consider the gas plus the heat reservoir at 298 K to be our system. What is the value of $\Delta \bar{S}$ if the expansion is reversible? What is the value of $\Delta \bar{S}$ if the gas expands into a larger container so that the final pressure is 1 bar?

3.13 An ideal gas at 298 K expands isothermally from a pressure of 10 to 1 bar. What are the values of w per mole, q per mole, $\Delta \bar{U}$, $\Delta \bar{H}$, and $\Delta \bar{S}$ in the following cases? (a) The expansion is reversible. (b) The expansion is free. (c) The gas and its surroundings form an isolated system, and the expansion is reversible. (d) The gas and its surroundings form an isolated system, and the expansion is free.

3.14 An ideal monatomic gas is heated from 300 to 1000 K and the pressure is allowed to rise from 1 to 2 bar. What is the change in molar entropy?

3.15 The purest acetic acid is often called glacial acetic acid because it is purified by fractional freezing at its melting point of 16.6 °C. A flask containing several moles of acetic acid at 16.6 °C is lowered into an ice-water bath briefly. When it is removed it is found that exactly 1 mol of acetic acid has frozen. Given: $\Delta_{fus}H(CH_3CO_2H)$ = 11.45 kJ mol^{-1} and $\Delta_{fus}H(H_2O)$ = 5.98 kJ mol^{-1}. (a) What is the change in entropy of the acetic acid? (b) What is the change in entropy of the water bath? (c) Now consider the water bath and acetic acid are in the same system. What is the entropy change for the combined system? Is the process reversible or irreversible? Why?

3.16 In problem 2.18 one mole of ideal monatomic gas at 1 bar and 300 K was expanded adiabatically against a constant pressure of ½ bar until the final pressure was ½ bar; a temperature of 240 K was reached. What is the value of $\Delta \bar{S}$ for this process?

3.17 Ten moles of H_2 and two moles of D_2 are mixed at 25 °C and 1 bar. What is the value of $\Delta S°$? Assume ideal gases.

3.18 Use the microscopic point of view of Section 3.12 to show that for the expansion of amount n of an ideal gas by a factor of two, $\Delta S = nR \ln 2$. In this expression S is an extensive property.

3.19 Calculate the change in molar entropy of aluminum that is heated from 600 to 700 °C. The melting point of aluminum is 660 °C, the heat of fusion is 393 J g^{-1}, and the heat capacities of the solid and the liquid may be taken as 31.8 and 34.4 J K^{-1} mol^{-1}, respectively.

3.20 Steam is condensed at 100 °C and the water is cooled to 0 °C and frozen to ice. What is the molar entropy change of the water? Consider that the average specific heat of liquid water is 4.2 J K^{-1} g^{-1}. The enthalpy of vaporization at the boiling point and the enthalpy of fusion at the freezing point are 2258.1 and 333.5 J g^{-1}, respectively.

3.21 Calculate the molar entropy of carbon disulfide at 25 °C from the following heat capacity data and the heat of fusion, 4389 J mol^{-1}, at the melting point (161.11 K):

T/K	15.05	20.15	29.76	42.22	57.52	75.54	89.37	99.00	108.93	119.91	131.54	156.83	161–298
$\overline{C}_P$/J K^{-1} mol^{-1}	6.90	12.01	20.75	29.16	35.56	40.04	43.14	45.94	48.49	50.50	52.63	56.62	75.48

3.22 What is the standard change in entropy for the dissociation of molecular oxygen at 298.15 and 1000 K? Use Appendix C.2. How do the results compare with those for molecular hydrogen in Example 3.9?

3.23 Using molar entropies from Appendix C.1, calculate $\Delta_r S°$ for the following reactions at 25 °C:

(a) $H_2(g) + \frac{1}{2}O_2(g) = H_2O(l)$

(b) $H_2(g) + Cl_2(g) = 2HCl(g)$

(c) Methane(g) $+ \frac{1}{2}O_2(g) =$ Methanol(l)

3.24 What is $\Delta_r S°$ (298 K) for

$$H_2O(l) = H^+(ao) + OH^-(ao)$$

Why is this change negative and not positive?

3.25 Estimate the standard entropy of n-butane and isobutane at 298.15 K using the Benson method.

3.26 In running a Carnot cycle backward to produce refrigeration, the objective is to remove as much heat $|q_2|$ from the cold reservoir as possible for a given amount of work, and so the coefficient of performance β is defined as

$$\beta = \frac{|q_2|}{|w|} = \frac{|q_2|}{|q_1| - |q_2|} = \frac{T_2}{T_1 - T_2}$$

A household refrigerator operates between 35 and -10 °C. How many joules of heat can in principle be removed per kilowatt-hour of work?

3.27 A heat pump is used to heat a home in the winter when the temperature in the ground is 0 °C and the temperature of the radiator in the house is 35 °C. When a Carnot cycle is run backward for this purpose, the objective is to obtain as much heat in the radiator as possible for a given amount of electrical work. The coefficient of performance β' for a heat pump is defined by

$$\beta' = \frac{|q_1|}{|w|} = \frac{|q_1|}{|q_1| - |q_2|} = \frac{T_1}{T_1 - T_2}$$

What is the minimum amount of electrical work needed to produce a kilowatt-hour of heat?

3.28 When an ideal gas is allowed to expand isothermally in a piston, $\Delta U = q + w = 0$. Thus, the work done by the system on the surroundings is equal to the heat transferred from the reservoir to the gas, and the efficiency of turning heat into work is 100%. Explain why this is not a violation of the second law.

3.29 What is the molar entropy change for the freezing of water at 0 °C? The heat of fusion is 333.5 J g^{-1}.

3.30 Compare the entropy difference between 1 mol of liquid water at 25 °C and 1 mol of vapor at 100 °C and 1.013 25 bar. The average specific heat of liquid water may be taken as 4.2 J K^{-1} g^{-1}, and the heat of vaporization is 2259 J g^{-1}.

3.31 Calculate the increase in the molar entropy of nitrogen when it is heated from 25 to 1000 °C (a) at constant pressure, and (b) at constant volume. Given: $\overline{C}_P = 26.9835 + 5.9622 \times 10^{-3}T - 3.377 \times 10^{-1}T^2$ in J K^{-1} mol^{-1}.

3.32 Assuming that the heat capacity of water is independent of temperature, calculate the net change in entropy when 1 mol of water at 0 °C is mixed with 1 mol of water at 100 °C. Assume that the heat capacity is (4.184 J K^{-1} g^{-1})(18 g mol^{-1}) = 75.3 J K^{-1} mol^{-1}, and that the heat capacity of the calorimeter is negligible.

3.33 One mole of propane gas is allowed to expand isothermally from 2 to 10 L. What is the change in molar entropy of the propane, assuming it is an ideal gas?

3.34 In Section 3.9 we derived the expression

$$S_2 - S_1 = nR \ln \frac{V_2}{V_1}$$

for the reversible isothermal expansion of an ideal gas. We ought to derive this same expression by another reversible path. Do this by imagining that the gas is first expanded adiabatically and reversibly from V_1 to V_2. Since the temperature falls from T to T', heat must be added reversibly and at constant volume in a second step to restore the temperature to T.

3.35 Show that the standard molar entropy $S°(T)$ for an ideal gas at 1 bar can be calculated from the standard molar entropy $S*(T)$ at 1 atm using

$$\bar{S}°(T) - \bar{S}*(T) = R \ln \frac{P*}{P°}$$

$$= 0.109 \text{ J K}^{-1} \text{ mol}^{-1}$$

The conversion for liquids and solids is negligible. Why?

3.36 Argon undergoes the following change in state:

$$P_1 = 20 \text{ bar}, \ T_1 = 300 \text{ K} \rightarrow P_2 = 1 \text{ bar}, \ T_2 = 200 \text{ K}$$

What is the change in molar entropy, assuming that argon is an ideal gas?

3.37 A flask containing several moles of liquid benzene at its freezing temperature (5.5 °C) is placed in thermal contact with an ice-water bath. When the flask is removed from the ice-water bath, it is found that 1 mol of benzene has frozen. Given: $\Delta_{fus}H(C_6H_6) = 9.87$ kJ mol^{-1}, $\Delta_{fus}H(H_2O) = 5.98$ kJ mol^{-1}. (a) What is the change in entropy of benzene? (b) What is the change in entropy of the water bath? (c) Considering the flask of benzene and the water bath as an isolated system, what is the change in entropy of the isolated system? Is the process reversible or irreversible?

3.38 One mole each of $H_2(g)$, $N_2(g)$, and $O_2(g)$ are mixed at 25 °C. What is $\Delta_{mix}S$?

3.39 According to the Debye equation (Section 17.15) the heat capacity of a solid is proportional to the temperature cubed at low temperatures:

$$C_P = (\text{const})T^3$$

Show that the entropy at a temperature T' is given by

$$S(T') = \frac{C_P(T')}{3}$$

when the Debye equation holds.

3.40 Calculate the molar entropy of liquid chlorine at its melting point, 172.12 K, from the following data obtained by W. F. Giauque and T. M. Powell:

T/K	15	20	25	30	35	40	50	60	70	90	110	130	150	170	172.12
$\bar{C}_P$/J K^{-1} mol^{-1}	3.72	7.74	12.09	16.69	20.79	23.97	29.25	33.47	36.32	40.63	43.81	47.24	51.04	55.10	M.P.

The heat of fusion is 6406 J mol^{-1}. Below 15 K it may be assumed that $\bar{C}_P$ is proportional to T^3.

3.41 Calculate $\Delta_r S°$ for the reaction

$$CH_4(g) + 2O_2(g) = CO_2(g) + 2H_2O(g)$$

at 298 and 1000 K.

3.42 What is $\Delta_r S°$ for $H_2(g) = 2H(g)$ at 298, 1000, and 3000 K?

3.43 Calculate the entropy changes for the following reactions at 25 °C:

(a) $Ag^+(ao) + Cl^-(ao) = AgCl(s)$

(b) $HS^-(ao) = H^+(ao) + S^{2-}(ao)$

3.44 From electromotive force measurements it has been found that $\Delta_r S°$ for the reaction

$$\tfrac{1}{2}H_2(g) + AgCl(s) = HCl(aq) + Ag(s)$$

is -62.4 J K^{-1} mol^{-1} at 298.15 K. What is the value of $S°[Cl^-(aq)]$?

3.45 Estimate the standard entropy of n-hexane and 2,2-dimethylbutane at 298.15 K using the Benson method.

4
Gibbs Energy and Helmholtz Energy

With the definitions of T, U, and S we have completed the set of necessary thermodynamic properties. Although the entropy provides a criterion of whether a change in an isolated system is spontaneous, it does not provide a convenient criterion at constant T and V or constant T and P, the usual conditions in the laboratory. We need two additional thermodynamic properties (functions) to make calculations at constant T and V or constant T and P more convenient than they are with entropy. Fortunately, there is a general way to do this using Legendre transforms.

So far we have considered systems containing a fixed amount of material. In this chapter we will also consider open systems. These are systems that can exchange matter with their surroundings. The consideration of open systems leads to the definition of the chemical potential μ. The chemical potential receives its name from the fact that, like the electric potential and the gravitational potential, its gradient is a driving force. The chemical potential makes it possible to discuss chemical equilibrium quantitatively.

4.1 Criteria of Equilibrium for Closed Systems with Only Pressure–Volume Work

In the preceding chapter we saw that $dS > đq/T$ is a criterion for a spontaneous process and $dS = đq/T$ is a criterion for a process at equilibrium. If we consider a closed system with only pressure–volume work, $đq = dU + P\,dV$ so that at constant volume $đq_v = dU$. If the system is isolated, $dU = 0$ and the criterion can be written

$$(dS)_{U,V} \geq 0 \qquad (4.1)$$

This is the basis for saying that **in a spontaneous process in an isolated system the entropy increases to a maximum.** The variables U and V are called the natural variables of S; the significance of the term natural variables is explained in Section 4.4.

The internal energy provides a criterion for a spontaneous process at constant entropy and volume if only reversible pressure–volume work is done. This can be seen by writing the Clausius inequality in the form

$$đq - T\,dS \leq 0 \qquad (4.2)$$

Replacing $đq$ with $dU + P\,dV$ yields

$$dU + P\,dV - T\,dS \leq 0 \qquad (4.3)$$

so that

$$(dU)_{V,S} \leq 0 \qquad (4.4)$$

Thus, if an infinitesimal change takes place in a system of constant volume and entropy, dU is negative if the change is spontaneous and zero if the change is reversible. This is the familiar condition for a conservative mechanical system that the stable state is the one of lowest energy; that is, when the system is in the lowest energy state at constant volume and entropy, any change would increase the energy and such changes cannot occur spontaneously.

Similarly, we can show that if only reversible pressure–volume work is done, then

$$(dH)_{P,S} \leq 0 \qquad (4.5)$$

Note that in the inequalities 4.1, 4.4, and 4.5, the natural variables of S, U, and H are held constant.

None of these three criteria for spontaneous change is immediately useful in the chemistry lab where processes are carried out at constant temperature and volume or, more frequently, at constant temperature and pressure. Fortunately, new thermodynamic potentials can be introduced that have the desired properties. This term has not been used up to now, but U, H, and S are often referred to as thermodynamic potentials in analogy to mechanical potential energy or electrical potential, which in their own areas determine whether a process can happen.

We can get some insight into the procedure for introducing new thermodynamic potentials with the desired properties by reviewing the introduction of the enthalpy. Enthalpy was defined by $H = U + PV$. When the differential of the enthalpy was introduced into the fundamental equation ($dU = T\,dS - P\,dV$) we obtained a new fundamental equation.

$$dH = T\,dS + V\,dP \qquad (4.6)$$

Notice what happened here. Since we defined a new thermodynamic potential by adding PV to U, the new fundamental equation contains $V\,dP$, rather than $-P\,dV$. The natural variables of H are S and P. The definition of H interchanged an intensive variable P with the extensive variable V. In short, adding PV to U interchanged members of the conjugate pair P,V in the fundamental equation and changed a sign. This mathematical operation is called a Legendre transform.

4.2 Legendre Transforms

If a function $Z\,(z_1,z_2,\ldots)$ has the natural variables $z_1, z_2, \ldots$, its differential is of the form

$$dZ = Z_1\,dz_1 + Z_2\,dz_2 + \cdots = \left(\frac{\partial Z}{\partial z_1}\right)_{z_2} dz_1 + \left(\frac{\partial Z}{\partial z_2}\right)_{z_1} dz_2 + \cdots \quad (4.7)$$

where Z_1 is the intensive variable that is conjugate to the extensive variable z_1. We can see that the conjugate variable Z_1 is the derivative of Z with respect to z_i holding all other z_i's fixed. To develop a Legendre transform Y of Z that has Z_1 as an independent variable, we define Y as

$$Y = Z - z_1 Z_1 \qquad (4.8)$$

The differential of the extensive variable Y is

$$dY = dZ - z_1 dZ_1 - Z_1 dz_1 \qquad (4.9)$$

Substituting dZ from equation 4.7 we obtain

$$dY = -z_1 dZ_1 + Z_2\,dz_2 + \cdots \qquad (4.10)$$

The new thermodynamic potential Y is a function of $Z_1, z_2, \ldots$. Comparison with equation 4.7 shows that the intensive variable Z_1 and the extensive variable z_1 have been interchanged in the first term and the sign has been changed. Thus, the enthalpy H is the Legendre transform of the internal energy that is defined by $H \equiv U + PV$.

Since it is difficult to hold S constant in the laboratory and this causes a problem in utilizing the fundamental equation for U, let us interchange it with T, its conjugate variable.

4.3 Helmholtz Energy and Gibbs Energy

The **Helmholtz energy** A is the Legendre transform of the internal energy that is defined by

$$A \equiv U - TS \qquad (4.11)$$

The expression for the total differential of A is obtained by differentiating the definition and substituting the fundamental equation for U:

$$dA = dU - T\,dS - S\,dT$$
$$= -S\,dT - P\,dV \qquad (4.12)$$

Since T and V are the natural variables, this yields the following criterion for spontaneous change and equilibrium in a closed system with only pressure–volume work:

$$(dA)_{T,V} \le 0 \qquad (4.13)$$

Thus, for any irreversible process at constant T and V, the Helmholtz energy A decreases, and for a reversible process it is constant.

The **Gibbs energy*** G is the Legendre transform of the enthalpy that is defined by

$$G \equiv H - TS \qquad (4.14)$$

The expression for the total differential of G is obtained by differentiating the definition and substituting the fundamental equation for H:

$$dG = dH - T\,dS - S\,dT$$
$$= -S\,dT + V\,dP \qquad (4.15)$$

Since T and P are the natural variables, this yields the following criterion for spontaneous change and equilibrium in a closed system with only reversible pressure–volume work:

$$(dG)_{T,P} \le 0 \qquad (4.16)$$

Thus, for any irreversible process at constant T and P in which only pressure–volume work is done, the Gibbs energy decreases, and for a reversible process it is constant (see Fig. 4.1).

The definition of the Gibbs energy completes the definition of the thermodynamic potentials for a closed system that can do only pressure–volume work. The reason there are only four thermodynamic potentials is that the internal energy is determined by two conjugate pairs, (T,S) and (P,V). The thermodynamic potentials U, H, A, and G correspond to all possible combinations of independent variables when one is taken from each pair. If a system can do other types of work (Section 2.9), additional thermodynamic potentials can be defined. The conditions for irreversibility and reversibility for processes involving only pressure–volume work are summarized in Table 4.1. Note that the variables that are held constant are the natural variables in each case. Since the Gibbs energy decreases in an irreversible process at constant T and P, it becomes a minimum in the final equilibrium state. We can, however, *imagine* a reversible process occurring at equilibrium; for example, we may imagine the evaporation of an infinitesimal amount of water from the liquid into the vapor phase that is saturated with water vapor at constant temperature and pressure. For such a process, $dG = 0$.

These same relations may be applied to finite changes as well as infinitesimal changes, replacing the d's by Δ's. It must be remembered, however, that spon-

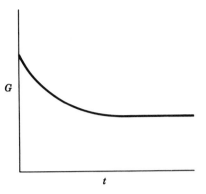

Figure 4.1 When a system undergoes spontaneous change at constant T and P, the Gibbs energy decreases until equilibrium is reached.

Table 4.1 Criteria for Irreversibility and Reversibility for Processes Involving No Work or Only Pressure–Volume Work

For Irreversible Processes	For Reversible Processes
$(dS)_{V,U} > 0$	$(dS)_{V,U} = 0$
$(dU)_{V,S} < 0$	$(dU)_{V,S} = 0$
$(dH)_{P,S} < 0$	$(dH)_{P,S} = 0$
$(dA)_{T,V} < 0$	$(dA)_{T,V} = 0$
$(dG)_{T,P} < 0$	$(dG)_{T,P} = 0$

---

* This is sometimes referred to as the free energy or the Gibbs free energy. The Gibbs energy G is named in honor of J. Willard Gibbs of Yale University, whose many important generalizations in thermodynamics have given him a position as one of the great geniuses of science. The Helmholtz energy H is named in honor of Hermann L. F. von Helmholtz.

taneous changes always go to the minimum (as in the case of the Gibbs energy at constant T and P) or to the maximum (as in the case of the entropy of an isolated system) and not to some other condition, even though the change to some other condition satisfies the required inequality expressed using a Δ.

The Gibbs energy has been defined so that we will have a criterion for spontaneous change at constant T and P that will not require us to specifically consider what is happening in the surroundings, as we did in the derivation in Section 3.9. For an isolated system consisting of the system of interest and the surroundings at constant T and P, the entropy criterion is that $\Delta S_{syst} + \Delta S_{surr}$ must increase for a spontaneous process. Since the temperature is constant, $\Delta S_{surr} = -\Delta H_{syst}/T$, so that $\Delta S_{syst} - \Delta H_{syst}/T = -\Delta G_{syst}/T$ must increase in a spontaneous process at constant T and P, or ΔG_{syst} must decrease. Thus, the Gibbs energy G simply provides a more convenient thermodynamic property than the entropy for the application of the second law at constant T and P.

Although these criteria show whether a certain change is irreversible, it does not necessarily follow that the change will take place with an appreciable speed. Thus, a mixture of 1 mol of carbon and 1 mol of oxygen at 1 bar pressure and 25 °C has a Gibbs energy greater than that of 1 mol of carbon dioxide at 1 bar and 25 °C, and so it is possible for the carbon and the oxygen to combine to form carbon dioxide at this constant temperature and pressure. Although carbon may exist for a very long time in contact with oxygen, the reaction is theoretically possible. The reverse of a thermodynamically spontaneous change is, of course, a nonspontaneous change. Thus, the decomposition of carbon dioxide to carbon and oxygen at room temperature, which involves an increase in Gibbs energy, is nonspontaneous. It can occur only with the aid of an outside agency.

The above discussion applies to systems involving only pressure–volume work. If various kinds of work are involved, the first law is

$$dU = đq + đw \tag{4.17}$$

so that equation 3.38 can be written

$$-dU + T\,dS \geq -đw \tag{4.18}$$

At constant temperature, this becomes

$$-d(U - TS) \geq -đw \tag{4.19}$$

or

$$(dA)_{T,V} \leq đw \tag{4.20}$$

where A is the Helmholtz energy. The symbol A actually comes from *arbeit*, the German word for work. Thus, in a reversible process at constant temperature and volume, the work done on the system is equal to the increase in the Helmholtz energy. In general, equation 4.19 shows that the decrease in A is an *upper bound* on the total work done on the surroundings. When the system does work on the surroundings in a real process, the work done on the surroundings (remember that it is negative) is less than the decrease in Helmholtz energy.

The Gibbs energy is especially useful when non-PV work is involved. In this case the first law can be written

$$dU = đq - P_{ext}dV + đw_{nonpv} \qquad (4.21)$$

so that the inequality $T\,dS \geq đq$ can be written

$$-dU - P_{ext}dV + T\,dS \geq -đw_{nonpv} \qquad (4.22)$$

The external pressure is represented by P_{ext}. At constant temperature and pressure, this can be written

$$-d(U + PV - TS) \geq -đw_{nonpv} \qquad (4.23)$$

$$(dG)_{T,P} \leq đw_{nonpv} \qquad (4.24)$$

For a reversible process at constant T and P the change in Gibbs energy is equal to the non-PV work done on the system by the surroundings. Thus, when work is done on the system, the Gibbs energy increases, and when the system does work on the surroundings, the Gibbs energy decreases. In general, equation 4.23 shows that the decrease in G is an *upper bound* on the non-PV work done on the surroundings. When the system does work on the surroundings, the work done (remember that it is negative) is less than the decrease in Gibbs energy.

Let us consider inequality 4.24 in more detail by applying it to the charging and discharging of an electrochemical cell at constant temperature and pressure. The electrochemical cell is charged by an electrical generator, and we will imagine a perfect direct-current generator that consumes a known amount of mechanical work. When the electrochemical cell is charged, the increase dG in the Gibbs energy of the cell is less than the electrical work done by the generator on the system in a real process and is equal to the electrical work in the theoretical limit of a reversible process.

When the electrochemical cell is discharged by operating an idealized electrical motor that does mechanical work, the Gibbs energy of the cell decreases and the work done is negative. According to inequality 4.24 the work done by the electrical motor is more positive than the decrease in G in a real process. Thus, the amount of work done in the surroundings is less than the decrease in the Gibbs energy of the electrochemical cell, except in the theoretical limit of a reversible process. This provides a simple interpretation for the change in Gibbs energy for a system. For a reversible process at constant temperature and pressure, the change in the Gibbs energy for the system is equal to the non-PV work done on the system by the surroundings. When the system does work on its surroundings, the decrease in Gibbs energy is equal to the loss in the capacity of the system to do work on its surroundings.

4.4 Thermodynamic Derivatives for Closed Systems

In the preceding section we have shown that the fundamental equation 3.42 can also be expressed in terms of H, A, and G. These four fundamental equations are so important that we need to discuss them together. For closed systems involving only pressure–volume work, the fundamental equations are

$$dU = T\,dS - P\,dV \qquad (4.25)$$

$$dH = T\,dS + V\,dP \qquad (4.26)$$

$$dA = -S\,dT - P\,dV \qquad (4.27)$$

$$dG = -S\,dT + V\,dP \qquad (4.28)$$

The differentials on the right sides of these equations indicate the **natural** variables of the thermodynamic potentials on the left sides. The natural variables are the variables that are held constant in writing the criteria of spontaneity and equilibrium in terms of U, H, A, and G; see equations 4.4, 4.5, 4.13, and 4.16. **The natural variables are also important because when a thermodynamic potential of a system is known as a function of its natural variables, all the thermodynamic properties of the system can be calculated,** as we will see by the end of this section. Equations 4.25–4.28 involve only state functions, but integration requires the specification of a reversible path.

Equations 4.25–4.28 are well worth remembering because they are the source of useful derivatives. Since the internal energy U is a state function and may be expressed as a function of S and V, its differential is exact and is given by

$$dU = \left(\frac{\partial U}{\partial S}\right)_V dS + \left(\frac{\partial U}{\partial V}\right)_S dV \qquad (4.29)$$

Comparing equations 4.25 and 4.29, we see that

$$\left(\frac{\partial U}{\partial S}\right)_V = T \qquad (4.30)$$

and

$$\left(\frac{\partial U}{\partial V}\right)_S = -P \qquad (4.31)$$

These equations illustrate the fact that the derivative of an extensive property with respect to another extensive property is an intensive property. Equations 4.30 and 4.31 are quite remarkable in that they indicate new ways of obtaining T and P if only these derivatives could be measured.

Since dH, dA, and dG are exact differentials, equations 4.26–4.28 are convenient sources of the following derivatives:

$$\left(\frac{\partial H}{\partial S}\right)_P = T \qquad (4.32)$$

$$\left(\frac{\partial H}{\partial P}\right)_S = V \qquad (4.33)$$

$$\left(\frac{\partial A}{\partial T}\right)_V = -S \qquad (4.34)$$

$$\left(\frac{\partial A}{\partial V}\right)_T = -P \qquad (4.35)$$

$$\left(\frac{\partial G}{\partial T}\right)_P = -S \qquad (4.36)$$

$$\left(\frac{\partial G}{\partial P}\right)_T = V \qquad (4.37)$$

As an illustration of the usefulness of these derivatives, let us consider the last two. Since the entropy of a system is always positive, G decreases with increasing temperature at constant pressure. Since S is greater for a gas than for the corresponding solid, the temperature coefficient of the Gibbs energy is much more negative for a gas than for the corresponding solid. Since the volume of a system is always positive, G increases with increasing P at constant T. Since V is greater for a gas than for the corresponding solid, the pressure coefficient of G is much larger for a gas than for the corresponding solid.

Now we are in a position to illustrate the fact that if a thermodynamic potential is known as a function of its natural variables, we can calculate all of the thermodynamic properties of the system. Suppose that G for a system has been determined as a function of temperature and pressure. The entropy and volume of the system can be calculated from

$$S = -\left(\frac{\partial G}{\partial T}\right)_P \quad \text{and} \quad V = \left(\frac{\partial G}{\partial P}\right)_T \tag{4.38}$$

Then U, H, and A can be calculated using the equations

$$U = G - PV + TS = G - P\left(\frac{\partial G}{\partial P}\right)_T - T\left(\frac{\partial G}{\partial T}\right)_P \tag{4.39}$$

$$H = G - T\left(\frac{\partial G}{\partial T}\right)_P \tag{4.40}$$

$$A = G - P\left(\frac{\partial G}{\partial P}\right)_T \tag{4.41}$$

This is not possible when G is known as a function of V and T or P and V.

4.5 Maxwell Relations

Since dU, dH, dA, and dG in equations 4.25, 4.26, 4.27, and 4.28 are exact differentials, the mixed second derivatives of the coefficients of the two terms on the right are equal, as shown in Section 2.6. Application of equation 2.21 to equations 4.25, 4.26, 4.27, and 4.28 yields the Maxwell relations:

$$\left(\frac{\partial T}{\partial V}\right)_S = -\left(\frac{\partial P}{\partial S}\right)_V \tag{4.42}$$

$$\left(\frac{\partial T}{\partial P}\right)_S = \left(\frac{\partial V}{\partial S}\right)_P \tag{4.43}$$

$$\left(\frac{\partial S}{\partial V}\right)_T = \left(\frac{\partial P}{\partial T}\right)_V \tag{4.44}$$

$$-\left(\frac{\partial S}{\partial P}\right)_T = \left(\frac{\partial V}{\partial T}\right)_P \tag{4.45}$$

The Maxwell relations are important because they are of considerable help in reaching our objective of expressing any thermodynamic property of a system in terms of easily measured physical quantities. The last two equations are particularly important because they provide the means for obtaining the effects of volume and pressure on the entropy at constant temperature from measurements of the derivatives on the right.

Example 4.1

Derive the first Maxwell equation (4.42).

Since

$$dU = T \, dS - P \, dV$$

then

$$\left(\frac{\partial U}{\partial S}\right)_V = T$$

$$\left(\frac{\partial U}{\partial V}\right)_S = -P$$

Using equation 2.20,

$$\left[\frac{\partial}{\partial V}\left(\frac{\partial U}{\partial S}\right)_V\right]_S = \left(\frac{\partial T}{\partial V}\right)_S$$

$$\left[\frac{\partial}{\partial S}\left(\frac{\partial U}{\partial V}\right)_S\right]_V = -\left(\frac{\partial P}{\partial S}\right)_V$$

Since the mixed partial derivatives are equal, then

$$\left(\frac{\partial T}{\partial V}\right)_S = -\left(\frac{\partial P}{\partial S}\right)_V$$

Example 4.2

Derive the equation for the molar entropy of isothermal expansion of a van der Waals gas.

$$P = \frac{RT}{\overline{V} - b} - \frac{a}{\overline{V}^2}$$

$$\left(\frac{\partial \overline{S}}{\partial \overline{V}}\right)_T = \left(\frac{\partial P}{\partial T}\right)_{\overline{V}} = \frac{R}{\overline{V} - b}$$

$$\int_{\overline{S}_1}^{\overline{S}_2} d\overline{S} = R \int_{\overline{V}_1}^{\overline{V}_2} \frac{1}{\overline{V} - b} \, d\overline{V}$$

$$\Delta \overline{S} = R \ln \frac{\overline{V}_2 - b}{\overline{V}_1 - b}$$

4.6   Thermodynamic Equations of State

A thermodynamic equation of state relates the thermodynamic variables that define the state of a substance. One of these equations may be obtained from equation 4.25 by imposing constant temperature and dividing by dV to obtain

$$\left(\frac{\partial U}{\partial V}\right)_T = T\left(\frac{\partial S}{\partial V}\right)_T - P \tag{4.46}$$

Substituting equation 4.44 yields

$$\left(\frac{\partial U}{\partial V}\right)_T = T\left(\frac{\partial P}{\partial T}\right)_V - P \tag{4.47}$$

This equation makes it possible to calculate the change in internal energy for a given change in volume at constant temperature. It is convenient to write this equation in terms of the cubic expansion coefficient α and the isothermal compressibility κ (Section 1.10):

$$\left(\frac{\partial U}{\partial V}\right)_T = \frac{\alpha T - \kappa P}{\kappa} \tag{4.48}$$

The quantity $(\partial U/\partial V)_T$ is often called the internal pressure. It has the dimensions of pressure and is due to intermolecular attractions and repulsions. The internal pressure changes with the volume because as the volume is increased, the average intermolecular distances increase and the average intermolecular potential energy changes.

Example 4.3
(a) How does the internal pressure of a van der Waals gas depend on the molar volume?
(b) Propane gas is allowed to expand isothermally from 10 to 30 L. What is the change in molar internal energy?

(a) Differentiation of equation 1.31 with respect to T and constant $\overline{V}$ yields $(\partial P/\partial T)_{\overline{V}} = R/(\overline{V} - b)$. Substituting this relation in equation 4.47 yields

$$\left(\frac{\partial \overline{U}}{\partial \overline{V}}\right)_T = \frac{RT}{\overline{V} - b} - P$$

$$= \frac{RT}{\overline{V} - b} - \left(\frac{RT}{\overline{V} - b} - \frac{a}{\overline{V}^2}\right)$$

$$= \frac{a}{\overline{V}^2}$$

Thus, for a van der Waals gas the internal pressure is inversely proportional to the square of the molar volume.

(b) The change in internal energy for a given change in volume at constant temperature is given by

$$\int_{\overline{U}_1}^{\overline{U}_2} d\overline{U} = \int_{\overline{V}_1}^{\overline{V}_2} \frac{a}{\overline{V}^2} \, d\overline{V} = a\left(-\frac{1}{\overline{V}}\right)_{\overline{V}_1}^{\overline{V}_2}$$

$$\Delta \overline{U} = a\left(\frac{1}{\overline{V}_1} - \frac{1}{\overline{V}_2}\right)$$

According to Table 1.2, $a = 8.779$ L^2 bar mol^{-2}, but we need to convert this to SI base units to calculate $\Delta \overline{U}$ in J mol^{-1}:

$$a = (8.779 \text{ L}^2 \text{ bar mol}^{-2})(10^5 \text{ Pa bar}^{-1})(10^{-3} \text{ m}^3 \text{ L}^{-1})^2$$

$$= 0.8779 \text{ Pa m}^6 \text{ mol}^{-2}$$

$$\Delta \overline{U} = a\left(\frac{1}{\overline{V}_1} - \frac{1}{\overline{V}_2}\right)$$

$$= (0.8779 \text{ Pa m}^6 \text{ mol}^{-2})\left(\frac{1}{10 \times 10^{-3} \text{ m}^3 \text{ mol}^{-1}} - \frac{1}{30 \times 10^{-3} \text{ m}^3 \text{ mol}^{-1}}\right)$$

$$= 58.5 \text{ J mol}^{-1}$$

Another equation of state may be obtained from equation 4.26 by imposing constant temperature and dividing by dP to obtain

$$\left(\frac{\partial H}{\partial P}\right)_T = T\left(\frac{\partial S}{\partial P}\right)_T + V \tag{4.49}$$

Substituting in equation 4.45 yields

$$\left(\frac{\partial H}{\partial P}\right)_T = V - T\left(\frac{\partial V}{\partial T}\right)_P$$

$$= V(1 - \alpha T) \tag{4.50}$$

Example 4.4
The pressure on liquid benzene at 25 °C is raised from 1 to 11 bar. What is the change in molar enthalpy? For liquid benzene under these conditions $\alpha = 1.237 \times 10^{-3} \text{ K}^{-1}$ and the density is 0.879 g cm^{-3}.

$$\left(\frac{\partial \overline{H}}{\partial P}\right)_T = \overline{V}(1 - \alpha T)$$

$$= \frac{(0.07811 \text{ kg mol}^{-1})[1 - (1.237 \times 10^{-3} \text{ K}^{-1})(298.15 \text{ K})]}{(0.879 \text{ g cm}^{-3})(10^{-3} \text{ kg g}^{-1})(10^2 \text{ cm m}^{-1})^3}$$

$$= 5.61 \times 10^{-5} \text{ m}^3 \text{ mol}^{-1}$$

$$\Delta \overline{H} = (5.61 \times 10^{-5} \text{ m}^3 \text{ mol}^{-1})\Delta P$$

$$= (5.61 \times 10^{-5} \text{ m}^3 \text{ mol}^{-1})(10 \text{ bar})(10^5 \text{ Pa bar}^{-1})$$

$$= 56.0 \text{ J mol}^{-1}$$

For an ideal gas, $\alpha = 1/T$ and $\kappa = 1/P$ so that

$$\left(\frac{\partial U}{\partial V}\right)_T = 0 \tag{4.51}$$

$$\left(\frac{\partial H}{\partial P}\right)_T = 0 \tag{4.52}$$

Thus, the internal energy of an ideal gas is independent of the volume and the enthalpy is independent of the pressure.

The internal energy of an ideal gas is also independent of the pressure at constant temperature, and the enthalpy is independent of the volume and pressure at constant temperature. It is important to realize that these relations are simply the consequence of the ideal gas law and the second law of thermodynamics. In molecular terms these results show that there are no interactions between molecules of an ideal gas.

Alternatively, we can use equations 4.47 and 4.51 to derive the ideal gas law and thereby show that the temperature introduced in the definition of entropy is the same as that in the ideal gas law.

We found in Section 2.13 that

$$C_P - C_V = \left[P + \left(\frac{\partial U}{\partial V} \right)_T \right] \left(\frac{\partial V}{\partial T} \right)_P \qquad (4.53)$$

Inserting $(\partial V/\partial T)_P = V\alpha$ and equation 4.48 we obtain

$$C_P - C_V = \frac{TV\alpha^2}{\kappa} \qquad (4.54)$$

Since it is difficult to measure C_V, its value is generally calculated from measurements of C_P, the molar volume V, the cubic expansion coefficient α, and the isothermal compressibility κ.

4.7 Effect of Temperature on the Gibbs Energy

Since $(\partial G/\partial T)_P = -S$ and S is a positive quantity, the Gibbs energy G necessarily decreases as the temperature increases at constant pressure. A useful relation is obtained by using this equation to eliminate S from $G = H - TS$:

$$G = H + T \left(\frac{\partial G}{\partial T} \right)_P \qquad (4.55)$$

Since this equation involves both the Gibbs energy and the temperature derivative of the Gibbs energy, it is more convenient to transform it so that only a temperature derivative appears. This may be accomplished by first differentiating G/T with respect to temperature at constant pressure:

$$\left[\frac{\partial (G/T)}{\partial T} \right]_P = -\frac{G}{T^2} + \frac{1}{T} \left(\frac{\partial G}{\partial T} \right)_P \qquad (4.56)$$

Eliminating G from the right side by use of equation 4.55, we have

$$\left[\frac{\partial (G/T)}{\partial T} \right]_P = -\frac{H}{T^2} \qquad (4.57)$$

Since $\partial(1/T)/\partial T = -T^{-2}$,

$$\left[\frac{\partial (G/T)}{\partial (1/T)} \right]_P = \left[\frac{\partial (G/T)}{\partial T} \right]_P \frac{\partial T}{\partial (1/T)} = H \qquad (4.58)$$

This equation may also be written in terms of ΔG and ΔH to obtain

$$\left[\frac{\partial (\Delta G/T)}{\partial (1/T)} \right]_P = \Delta H \qquad (4.59)$$

Thus, ΔH for a reaction may be obtained from a plot of $\Delta G/T$ versus $1/T$ as well as from calorimetric measurements. Since ΔH changes with temperature, the slope is taken at a specific temperature. This equation, which is referred to as the **Gibbs–Helmholtz** equation is important for the calculation of ΔG at another temperature if it is known at one temperature and ΔH is known (see Section 5.9).

4.8   Effect of Pressure on the Gibbs Energy

The equation $(\partial G/\partial P)_T = V$ (equation 4.37) may be integrated to obtain the value of G at another pressure, provided that its value is known at one pressure and V is known as a function of pressure at constant temperature:

$$\int_{G_1}^{G_2} dG = \int_{P_1}^{P_2} V\, dP \tag{4.60}$$

$$G_2 = G_1 + \int_{P_1}^{P_2} V\, dP \tag{4.61}$$

This equation always applies, and, as we have noted, the Gibbs energy of a single substance always increases with the pressure. There are two special cases where equation 4.61 leads to simple relationships. If the volume is nearly independent of pressure, as it is for a liquid or solid, then

$$G_2 = G_1 + V(P_2 - P_1) \tag{4.62}$$

Alternatively, this equation may be written

$$G = G° + V(P - P°) \tag{4.63}$$

where $G°$ is the Gibbs energy when the pressure is equal to the standard-state pressure $P°$.

For gases the Gibbs energy is very much more dependent on pressure. For an ideal gas the dependence of Gibbs energy on pressure is obtained by substituting $V = nRT/P$ in equation 4.60.

$$\int_{G°}^{G} dG = nRT \int_{P°}^{P} d\ln P \tag{4.64}$$

$$G = G° + nRT \ln \frac{P}{P°} \tag{4.65}$$

The standard Gibbs energy $G°$ has different values at different temperatures. The logarithmic dependence of the molar Gibbs energy of an ideal gas on the pressure of the gas is illustrated in Fig. 4.2.

Alternatively, equation 4.64 may be integrated between any two pressures to obtain

$$\Delta G = G_2 - G_1 = nRT \ln \frac{P_2}{P_1} \tag{4.66}$$

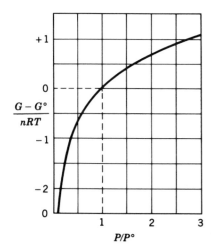

Figure 4.2   Dependence of the molar Gibbs energy of an ideal gas on the pressure of the gas.

Example 4.5

Since the molar Gibbs energy of an ideal gas is given by $\overline{G} = \overline{G}° + RT \ln(P/P°)$, derive the corresponding expressions for $\overline{V}$, $\overline{U}$, $\overline{H}$, $\overline{S}$, and $\overline{A}$. Using equations 4.36, 4.37, 4.39, 4.40, and 4.41, respectively,

$$\overline{V} = \frac{RT}{P}$$

$$\overline{U} = \overline{U}^\circ = \overline{H}^\circ - RT$$

$$\overline{H} = \overline{H}^\circ = \overline{G}^\circ + T\overline{S}^\circ$$

$$\overline{S} = \overline{S}^\circ - R \ln \frac{P}{P^\circ}$$

$$\overline{A} = \overline{A}^\circ + RT \ln \frac{P}{P^\circ}$$

where $\overline{S}^\circ = -(\partial \overline{G}^\circ/\partial T)_P$ and $\overline{U}^\circ = \overline{G}^\circ + T\overline{S}^\circ - RT$. As discussed in Section 4.6, the internal energy U and enthalpy H of an ideal gas are independent of pressure and volume.

Example 4.6

An ideal gas at 27 °C expands isothermally and reversibly from 10 to 1 bar against a pressure that is gradually reduced. Calculate q per mole and w per mole and each of the thermodynamic quantities $\Delta \overline{U}$, $\Delta \overline{H}$, $\Delta \overline{G}$, $\Delta \overline{A}$, and $\Delta \overline{S}$. Since the process is carried out isothermally and reversibly,

$$w_{\text{max}} = -RT \ln \frac{\overline{V}_2}{\overline{V}_1} = -RT \ln \frac{P_1}{P_2} = -(8.314 \text{ J K}^{-1} \text{ mol}^{-1})(300.15 \text{ K}) \ln \frac{10}{1}$$

$$= -5746 \text{ J mol}^{-1}$$

$$\Delta \overline{A} = w_{\text{max}} = -5746 \text{ J mol}^{-1}$$

Since the internal energy of an ideal gas is not affected by a change in volume,

$$\Delta \overline{U} = 0$$

$$q = \Delta \overline{U} - w = 0 + 5746 = 5746 \text{ J mol}^{-1}$$

$$\Delta \overline{H} = \Delta \overline{U} + \Delta(P\overline{V}) = 0 + 0 = 0$$

and since $P\overline{V}$ is constant for an ideal gas at constant temperature,

$$\Delta \overline{G} = \int_{10}^{1} \overline{V} \, dP = RT \ln \tfrac{1}{10} = (8.314 \text{ J K}^{-1} \text{ mol}^{-1})(300.15 \text{ K})(-2.3026)$$

$$= -5746 \text{ J mol}^{-1}$$

$$\Delta \overline{S} = \frac{q_{\text{rev}}}{T} = \frac{5746 \text{ J mol}^{-1}}{300.15 \text{ K}} = 19.14 \text{ J K}^{-1} \text{ mol}^{-1}$$

Also,

$$\Delta \overline{S} = \frac{\Delta \overline{H} - \Delta \overline{G}}{T} = \frac{0 + (5746 \text{ J mol}^{-1})}{300.15 \text{ K}} = 19.14 \text{ J K}^{-1} \text{ mol}^{-1}$$

Example 4.7

An ideal gas expands isothermally at 27 °C into an evacuated vessel so that the pressure drops from 10 to 1 bar; that is, it expands from a vessel of 2.463 L into a connecting vessel such that the total volume is 24.63 L. Calculate the change in thermodynamic quantities.

This process is isothermal, but is not reversible.

$w = 0$ because the system as a whole is closed and no external work can be done.

$\Delta \overline{U} = 0$ because the gas is an ideal gas.

$$q = \Delta \overline{U} - w = 0 + 0 = 0$$

$\Delta \overline{U}$, $\Delta \overline{H}$, $\Delta \overline{G}$, $\Delta \overline{A}$, and $\Delta \overline{S}$ are the same as in Example 4.6 because the initial and final states are the same.

4.9 Fugacity

The Gibbs energy of a real gas is not given by equation 4.65, which we derived for an ideal gas. However, G. N. Lewis recognized that it would be convenient to keep the same form of equation for a real gas. He accomplished this by defining the **fugacity** f using

$$\overline{G} = \overline{G}^{\circ} + RT \ln \frac{f}{P^{\circ}} \tag{4.67}$$

$$\lim_{P \to 0} \frac{f}{P} = 1 \tag{4.68}$$

The fugacity has the units of pressure. As the pressure approaches zero, the gas approaches ideal behavior and the fugacity approaches the pressure. The fugacity is simply a measure of the molar Gibbs energy of a real gas, but it has the following advantage over the molar Gibbs energy: f goes from 0 to ∞, while $\overline{G}$ goes from $-\infty$ to $+\infty$ (see Fig. 4.2).

The fugacity of a real gas at a particular temperature and pressure can be calculated if the equation of state (Sections 1.5 and 1.6) of the gas is known. As we will see in the following derivation, it is convenient to use the virial equation of state written in terms of pressure (equation 1.17). Since $(\partial \overline{G}/\partial P)_T = \overline{V}$, the differential Gibbs energy at constant temperature is $d\overline{G} = \overline{V} \, dP$ for a real gas and $d\overline{G}^{\text{id}} = \overline{V}^{\text{id}} dP$ for an ideal gas. The difference in Gibbs energy between a real gas and an ideal gas can be integrated from some low pressure P^* to the pressure at which we would like to know the fugacity:

$$\int_{P*}^{P} d(\overline{G} - \overline{G}^{\text{id}}) = \int_{P*}^{P} (\overline{V} - \overline{V}^{\text{id}}) \, dP \tag{4.69}$$

$$(\overline{G} - \overline{G}^{\text{id}})_P - (\overline{G}^* - \overline{G}^{*\text{id}})_{P*} = \int_{P*}^{P} (\overline{V} - \overline{V}^{\text{id}}) \, dP \tag{4.70}$$

Now, if we let $P^* \to 0$, then $\overline{G}^* \to \overline{G}^{*\text{id}}$, and

$$(\overline{G} - \overline{G}^{\text{id}})_P = \int_{0}^{P} (\overline{V} - \overline{V}^{\text{id}}) \, dP \tag{4.71}$$

Introducing equation 4.67 for $\overline{G}$ and equation 4.65 for $\overline{G}^{\text{id}}$ yields

$$\ln\left(\frac{f}{P}\right) = \frac{1}{RT} \int_{0}^{P} (\overline{V} - \overline{V}^{\text{id}}) \, dP \tag{4.72}$$

or

$$\frac{f}{P} = \exp\left[\frac{1}{RT}\int_0^P (\overline{V} - \overline{V}^{\text{id}})\, dP\right] \tag{4.73}$$

The ratio of the fugacity to the pressure is called the **fugacity coefficient** ϕ; $\phi = f/P$. The fugacity coefficient is frequently used as a measure of the non-ideality of a gas in connection with its phase equilibrium or chemical equilibrium properties. When P-V-T data are available on a gas a plot may be prepared of the difference between its molar volume and the molar volume of an ideal gas versus pressure at the temperature of interest. The integral of this plot up to the pressure of interest is then used in equation 4.73 to calculate $\phi = f/P$. If the coefficients in the virial equation in terms of pressure are known for the gas, equation 4.73 may be written in terms of compressibility factor Z. Since $\overline{V} = RTZ/P$,

$$\frac{f}{P} = \exp\left[\frac{1}{RT}\int_0^P \left(\frac{RTZ}{P} - \frac{RT}{P}\right) dP\right]$$

$$= \exp\left[\int_0^P \frac{Z-1}{P}\, dP\right] \tag{4.74}$$

Thus, the fugacity of a gas is readily calculated at some pressure P if Z is known as a function of pressure up to that particular pressure. If this information is not available, estimates may be made using the principle of corresponding states (see Section 1.9).

Example 4.8

Given the expression for the compressibility factor Z as a power series in P (equation 1.17), what is the expression for the fugacity in terms of the virial coefficients?

Using equation 4.74, we find

$$\ln \frac{f}{P} = \int_0^P \left(\frac{B}{RT} + \frac{C'P}{RT} + \cdots\right) dP = \frac{BP}{RT} + \frac{C'P^2}{2RT} + \cdots$$

Example 4.9

Using the expression for the compressibility factor Z of a van der Waals gas given in equation 1.29, what is the expression for fugacity of a van der Waals gas?

$$Z = 1 + \left[b - \left(\frac{a}{RT}\right)\right]\frac{P}{RT}$$

$$\ln \frac{f}{P} = \int_0^P \left(\frac{Z-1}{P}\right) dP$$

$$= \int_0^P \left[b - \left(\frac{a}{RT}\right)\right]\frac{1}{RT}\, dP$$

$$= \left[b - \left(\frac{a}{RT}\right)\right]\frac{P}{RT}$$

$$f = P \exp\left[\left(b - \frac{a}{RT}\right)\frac{P}{RT}\right]$$

Example 4.10

Given that the van der Waals constants of nitrogen are $a = 1.408$ L^2 bar mol^{-2} and $b = 0.03913$ L mol^{-1}, estimate the fugacity of nitrogen gas at 50 bar and 298 K.

$$f = P \exp\left[\left(b - \frac{a}{RT}\right)\frac{P}{RT}\right]$$

$$= (50 \text{ bar}) \exp\left\{\left[0.03913 - \frac{1.408}{(0.08314)(298)}\right]\frac{50}{(0.08314)(298)}\right\}$$

$$= 48.2 \text{ bar}$$

When P-V-T data are not available, the fugacity of a pure substance may be estimated if its critical pressure and temperature are known or may be estimated. Figure 4.3 gives the fugacity coefficient ($\phi = f/P$) as a function of reduced pressure and reduced temperature.

Now we are in a better position to understand the standard state of a gas that is used for thermodynamic tables, such as Appendixes C.1 and C.2. The standard state is the pure substance at a pressure of 1 bar in a hypothetical state in which it exhibits ideal gas behavior as shown in Fig. 4.4. The solid line gives the behavior of a real gas. As the pressure is reduced, the real gas approaches ideal behavior. This ideal gas is then compressed to 1 bar along the dashed line as a hypothetical ideal gas.

4.10   Fundamental Equations for Open Systems

Our previous applications of thermodynamics have been restricted to closed systems. We now turn our attention to open systems and systems in which chemical reactions can occur. An open system is one that can exchange matter with its surroundings. We will consider homogeneous open systems because they are simpler than open systems with more than one phase. Since we are considering open systems we need to consider the thermodynamic properties of systems with variable numbers of moles.

If a homogeneous system contains N different species, its internal energy may be considered to be a function of S, V, n_1, n_2, . . . , n_N where n_i is the amount of species i in the system.* The total differential of a function of a set $N + 2$ independent variables is the sum of its $N + 2$ partial differentials each multiplied by the corresponding differential:

$$dU = \left(\frac{\partial U}{\partial S}\right)_{V,n_i} dS + \left(\frac{\partial U}{\partial V}\right)_{S,n_i} dV + \sum_{i=1}^{N}\left(\frac{\partial U}{\partial n_i}\right)_{S,V,n_j} dn_i \qquad (4.75)$$

where the subscript n_i means that the amounts of all N species are held constant, and the subscript n_j means that the amounts of all species are held constant except for the one being varied (i.e., $j \neq i$). The first two derivatives in equation 4.75 are given by equations 4.30 and 4.31, and so

---

* The amount of substance n is the number of entities (atoms, molecules, ions, electrons, or specified groups of such particles) expressed in terms of moles.

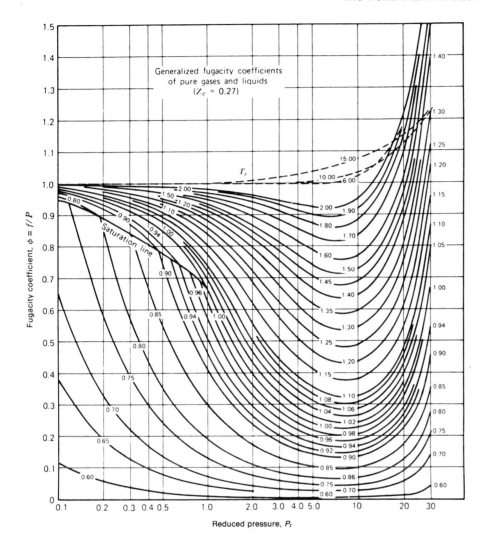

Figure 4.3   Fugacity coefficient ϕ as a function of reduced pressure P_r and reduced temperature T_r. (From O. A. Hougen, K. M. Watson, and R. A. Ragatz, *Chemical Process Principles Charts*, 2d ed. New York: Wiley, 1960.)

$$dU = T\,dS - P\,dV + \sum_{i=1}^{N} \mu_i dn_i \qquad (4.76)$$

where

$$\mu_i = \left(\frac{\partial U}{\partial n_i}\right)_{S,V,n_j} \qquad (4.77)$$

is referred to as the **chemical potential** of species i. The definition shows that the chemical potential of a component of a homogeneous mixture is equal to the rate of change of the internal energy with respect to the amount of that substance in the mixture at constant entropy and volume. Since the chemical potential is an intensive property, its value is independent of the size of the system.

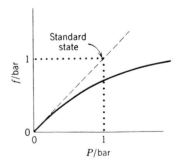

Figure 4.4   Plot of fugacity versus pressure for a real gas. The dashed line is for an ideal gas. The standard state is the pure substance at a pressure of 1 bar in a hypothetical state in which it exhibits ideal gas behavior.

If we make the usual Legendre transforms of equation 4.77 we obtain

$$dH = T\,dS + V\,dP + \sum_{i=1}^{N} \mu_i dn_i \tag{4.78}$$

$$dA = -S\,dT - P\,dV + \sum_{i=1}^{N} \mu_i dn_i \tag{4.79}$$

$$dG = -S\,dT + V\,dP + \sum_{i=1}^{N} \mu_i dn_i \tag{4.80}$$

Thus, the chemical potential μ_i also is given by

$$\mu_i = \left(\frac{\partial H}{\partial n_i}\right)_{S,P,n_j} = \left(\frac{\partial A}{\partial n_i}\right)_{T,V,n_j} = \left(\frac{\partial G}{\partial n_i}\right)_{T,P,n_j} \tag{4.81}$$

It is the last of these relations that we will actually use. It is the chemical potential that determines whether a species will undergo a chemical reaction or diffuse from one part of the system to another.

The above fundamental equations apply to open systems when only PV work is involved. When other kinds of work are involved, there are additional work terms in the fundamental equation for a closed system, as discussed in Section 2.9. Thus, when surface work ($\gamma\,d\sigma$), elongation work ($f\,dl$), and electrical work ($\phi\,dQ$) are involved, equation 4.80 becomes

$$dG = -S\,dT + V\,dP + \sum_{i=1}^{N} \mu_i dn_i + \gamma\,d\sigma + f\,dl + \phi\,dQ \tag{4.82}$$

Example 4.11

When a liquid rises in a capillary at constant temperature, the Gibbs energy is useful for examining the relation between surface tension and pressure for the liquid. Use the fundamental equation to obtain the following Maxwell relation at constant temperature.

$$\left(\frac{\partial V}{\partial \sigma}\right)_{P,T} = \left(\frac{\partial \gamma}{\partial P}\right)_{\sigma,T}$$

4.11   The Additivity Relation for the Gibbs Energy

Equation 4.80 may be expressed in integrated form as follows. Let us suppose that the system being considered is increased in size with the temperature, pressure, and the relative proportions of the components being held constant. Since the relative proportions of the species are unchanged, μ_i values are constant and

$$G = \sum_{i=1}^{N} n_i\,\mu_i \tag{4.83}*$$

This equation shows that the Gibbs energy of a system is the sum of the contributions of the various components. This relation allows us to calculate the Gibbs energy and other thermodynamic properties of a mixture. It is used in the next section to derive ΔG, ΔS, ΔH, and ΔV of mixing of ideal gases.

At equilibrium at constant T and P, G attains its *minimum* value. In Section 5.1 we will use equation 4.80 to obtain a simple expression for chemical equilibrium in terms of the chemical potentials of the components. In Section 5.14 we will use equation 4.83 to obtain the equilibrium composition for a system with many reactions.

If a system contains a single substance, $G = n\mu$. Thus, for a pure substance, the chemical potential is equal to the Gibbs energy per mole. In view of the importance of the chemical potential we will use it a good deal in discussing chemical equilibrium.

Equations 4.76, 4.78, 4.79 may be used to show that

$$U = TS - PV + \sum_{i=1}^{N} n_i \mu_i \tag{4.84}$$

$$H = TS + \sum_{i=1}^{N} n_i \mu_i \tag{4.85}$$

$$A = -PV + \sum_{i=1}^{N} n_i \mu_i \tag{4.86}$$

4.12 Partial Molar Quantities

Since conditions of constant temperature and pressure are so convenient experimentally, we will find it useful to consider all extensive thermodynamic properties as functions of the amounts n_i of the components that make up the system at constant temperature and pressure. For example, at constant temperature and pressure, the volume of a homogeneous mixture is given by

$$V = f(n_1, n_2, \ldots, n_N) \tag{4.87}$$

Thus, the differential of the volume is given by

$$dV = \sum_{i=1}^{N} \left(\frac{\partial V}{\partial n_i}\right)_{T,P,n_j} dn_i$$

$$= \sum_{i=1}^{N} \overline{V}_i \, dn_i \tag{4.88}$$

---

* Alternatively this equation may be derived by using Euler's theorem. A function $f(x_1, x_2, \ldots, x_N)$ is said to be homogeneous of degree n if

$$f(kx_1, kx_2, \ldots, kx_N) = k^n f(x_1, x_2, \ldots, x_N)$$

For such a function Euler's theorem states that

$$nf(x_1, x_2, \ldots, x_N) = \sum_{i=1}^{N} x_i \frac{\partial f}{\partial x_i}$$

Extensive thermodynamic properties are homogeneous of degree 1 in the amounts n_i.

where

$$\overline{V}_i = \left(\frac{\partial V}{\partial n_i}\right)_{T,P,n_j} \tag{4.89}$$

is the **partial molar volume**. The subscript n_j means that the amount of each component is held constant except for component i. This definition may be stated in words by saying that $\overline{V}_i$ is the change in V per mole of i added, when an infinitesimal amount of this component is added to the solution at constant temperature and pressure. Alternatively, it may be said that $\overline{V}_i$ is the change in V when 1 mol of i is added to an infinitely large amount of the solution at constant temperature and pressure. The partial molar volume of a component depends on its concentration and the temperature.

As we have seen in the preceding section, equation 4.88 can be integrated to obtain

$$V = \sum_{i=1}^{N} n_i \overline{V}_i \tag{4.90}$$

The volume V of the solution may be calculated for a given concentration, using this equation, if the partial molar volumes at the given concentration are known.

To discuss the determination of partial molar volumes of liquid solutions, it is convenient to write the molar volume $\overline{V}$ of a binary solution as a function of the partial molar volumes of the two components and their mole fractions:

$$\overline{V} = x_1\overline{V}_1 + x_2\overline{V}_2 = \overline{V}_1 + (\overline{V}_2 - \overline{V}_1)x_2 \tag{4.91}$$

The molar volume of a solution can be calculated from its density and composition. The molar volume of a binary solution at constant T and P is plotted against the mole fraction of the second component in Fig. 4.5. The method of intercepts provides a graphic way of visualizing partial molar quantities. To determine the partial molar volumes of components 1 and 2 at $x_2 = a$, the tangent line is drawn as shown in Fig. 4.5. The partial molar volume of component 1 at $x_2 = a$ is obtained from the intercept on the left, and the partial molar volume of component 2 for this solution is obtained from the intercept on the right. The molar volumes of the pure components $\overline{V}_1^*$ and $\overline{V}_2^*$ are also partial molar volumes, but we use the star here and elsewhere when there is danger of confusing the partial molar volume $\overline{V}_i$ in a solution with the molar volume $\overline{V}_i^*$ of the pure substance. Note that for this particular solution, partial molar volumes of both components are smaller than the molar volumes of the pure components. The reason this construction works is that the equation for the dashed tangent line at this mole fraction is

$$\Phi = \overline{V}_1(a) + [\overline{V}_2(a) - \overline{V}_1(a)]x_2 \tag{4.92}$$

where Φ is the apparent volume. By putting $x_2 = 0$ in this equation, $\Phi(0) = \overline{V}_1(a)$ so that the intercept at $x_2 = 0$ is the partial molar volume of component 1 when $x_2 = a$. By putting $x_2 = 1$ in this equation, $\Phi(1) = \overline{V}_2(a)$ so that the intercept at $x_2 = 1$ is the partial molar volume of component 2 when $x_2 = a$. The method for determining partial molar volumes illustrated in Fig. 4.5 is not very accurate because of the inherent difficulty in obtaining the slope of a curve; however, this method of visualizing the partial molar volumes will be found useful later.

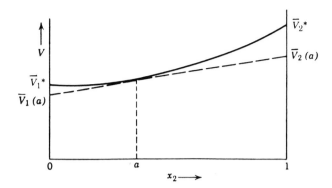

Figure 4.5 Molar volume of a binary solution versus the mole fraction of one of the components. The intercepts of the dashed tangent line are the partial volumes of the two components at $x_2 = a$. The molar volumes of the pure components 1 and 2 are represented by $\overline{V}_1^*$ and $\overline{V}_2^*$.

Equation 4.90 is an important result because it shows how to calculate a thermodynamic property of a solution by simply adding up contributions from the individual components. We can do exactly the same thing with U, H, S, and A. We have already done it with the Gibbs energy (equation 4.83); the partial molar Gibbs energy is the chemical potential. The corresponding expressions for the other extensive thermodynamic properties are

$$V = \sum n_i \overline{V}_i \qquad U = \sum n_i \overline{U}_i \qquad H = \sum n_i \overline{H}_i$$
$$S = \sum n_i \overline{S}_i \qquad G = \sum n_i \overline{G}_i = \sum n_i \mu_i \qquad A = \sum n_i \overline{A}_i \qquad (4.93)$$

where

$$\overline{U}_i = \left(\frac{\partial U}{\partial n_i}\right)_{P,T,n_j} \qquad (4.94)$$

$$\overline{H}_i = \left(\frac{\partial H}{\partial n_i}\right)_{P,T,n_j} \qquad (4.95)$$

$$\overline{S}_i = \left(\frac{\partial S}{\partial n_i}\right)_{P,T,n_j} \qquad (4.96)$$

$$\overline{G}_i = \left(\frac{\partial G}{\partial n_i}\right)_{T,P,n_j} \qquad (4.97)$$

$$\overline{A}_i = \left(\frac{\partial A}{\partial n_i}\right)_{P,T,n_j} \qquad (4.98)$$

Note that μ_i, which is defined by equation 4.81, is equal to the partial molar Gibbs energy $\overline{G}_i$, but is not equal to the partial molar Helmholtz energy $\overline{A}_i$.

The relations between partial molar quantities can be obtained by differentiating the thermodynamic relations between the extensive properties of a homogeneous system. For example, let us differentiate $G = H - TS$ with respect to n_i at constant temperature, pressure, and the amounts of all other species:

$$\left(\frac{\partial G}{\partial n_i}\right)_{T,P,n_j} = \left(\frac{\partial H}{\partial n_i}\right)_{T,P,n_j} - T\left(\frac{\partial S}{\partial n_i}\right)_{T,P,n_j}$$

$$\mu_i \equiv \overline{G}_i = \overline{H}_i - T\overline{S}_i \qquad (4.99)$$

Thus, the relation between partial molar quantities is of the same form as for the extensive quantities for the entire system.

The expressions for derivatives of partial molar quantities can also be obtained by differentiating the thermodynamic relations for extensive properties of a homogeneous system. For example, let us differentiate

$$-S = \left(\frac{\partial G}{\partial T}\right)_P \qquad (4.100)$$

with respect to n_i at constant temperature, pressure, and amounts of all other species:

$$-\left(\frac{\partial S}{\partial n_i}\right)_{T,P,n_j} = \left[\frac{\partial}{\partial n_i}\left(\frac{\partial G}{\partial T}\right)_{P,n}\right]_{T,P,n_j} = \left[\frac{\partial}{\partial T}\left(\frac{\partial G}{\partial n_i}\right)_{T,P,n_j}\right]_{P,n}$$

$$-\overline{S}_i = \left(\frac{\partial \overline{G}_i}{\partial T}\right)_{P,n} = \left(\frac{\partial \mu_i}{\partial T}\right)_{P,n} \qquad (4.101)$$

Similarly, we can show that

$$\overline{V}_i = \left(\frac{\partial \overline{G}_i}{\partial P}\right)_{T,n} = \left(\frac{\partial \mu_i}{\partial P}\right)_{T,n} \qquad (4.102)$$

These relations are actually Maxwell relations obtained from equation 4.80. Equation 4.101 shows that when the temperature is changed at constant pressure and composition, the differential change in μ_i is proportional to the negative of the partial molar entropy of component i. Equation 4.102 shows that when the pressure is changed at constant temperature and composition, the differential change in μ_i is proportional to the partial molar volume of the component. Although the molar volume and entropy of a substance are positive, the partial molar volume and partial molar entropy may be negative.

Example 4.12

The partial molar volumes for a carbon tetrachloride (1)–benzene (2) solution at 25 °C are plotted in Fig. 4.6 and some values are given in the following table:

x_1	0.0	0.3	0.5	0.7	1.0
$\overline{V}_1$/L mol^{-1}	0.179 3	0.112 2	0.100 1	0.098 30	0.097 19
$\overline{V}_2$/L mol^{-1}	0.089 27	0.098 44	0.106 4	0.109 2	0.112 3

(a) What is the molar volume of an equimolar solution?
(b) What is the volume change on mixing 1 and 2 to form one mole of an equimolar solution?

$$(a)\ \overline{V} = 0.5(0.100\ 1) + 0.5(0.106\ 4) = 0.103\ 3\ \text{L mol}^{-1}$$

$$(b)\ \Delta\overline{V}_{\text{mix}} = 0.103\ 3 - 0.5(0.089\ 27) - 0.5(0.097\ 19)$$

$$= 0.010\ 07\ \text{L mol}^{-1}$$

Thus, the volume of a mole of solution is 0.010 07 L greater than that of the two components.

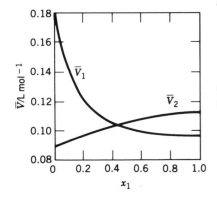

Figure 4.6 Partial molar volumes for CCl$_4$(1)–benzene(2) solutions at 25 °C.

4.13 Gibbs–Duhem Equation

When the composition of a system is changed at constant temperature and pressure, the chemical potentials of the components do not change independently. This can be seen by comparing the differential of G with the fundamental equation. The differential of G derived from $G = \sum n_i \mu_i$ (equation 4.83) is

$$dG = \sum_{i=1}^{N} (n_i \, d\mu_i + \mu_i \, dn_i) \qquad (4.103)$$

Subtracting this from equation 4.80 yields

$$\sum_{i=1}^{N} n_i \, d\mu_i = -S \, dT + V \, dP \qquad (4.104)$$

This equation, which is referred to as the Gibbs–Duhem equation, shows that the possible variations of the intensive variables, T, P, and $\mu_1, \ldots, \mu_N$ for a system are restricted. For a two-component system at constant temperature and pressure that contains 1 mol of material,

$$x_1 \, d\mu_1 + x_2 \, d\mu_2 = 0 \qquad (4.105a)$$

$$x_1 \, d\mu_1 + (1 - x_1)d\mu_2 = 0 \qquad (4.105b)$$

where x_1 is the mole fraction of component 1 and $(1 - x_1)$ is the mole fraction of component 2. Thus, when the composition is changed at constant T and P, the change in the chemical potential of component 2 is not independent of the change in the chemical potential of component 1. Later in Section 7.6 we will use this form of the Gibbs–Duhem equation to show that if Henry's law holds for the solute (component 2), Raoult's law holds for the solvent (component 1).

Example 4.13
There are Gibbs–Duhem equations for other thermodynamic properties as well. Show how the slopes of the plots of the partial molar volumes in Example 4.12 are related to each other.
    The total differential of $\overline{V} = x_1\overline{V}_1 + x_2\overline{V}_2$ is

$$d\overline{V} = x_1 \, d\overline{V}_1 + \overline{V}_1 \, dx_1 + x_2 d\overline{V}_2 + \overline{V}_2 dx_2$$

but according to equation 4.88

$$d\overline{V} = \overline{V}_1 \, dx_1 + \overline{V}_2 \, dx_2$$

The difference between these equations is

$$0 = x_1 \, d\overline{V}_1 + x_2 \, d\overline{V}_2$$

Dividing by dx_1 to obtain slopes in Fig. 4.6 and replacing x_2 with $1 - x_1$ yields

$$0 = x_1 \frac{d\overline{V}_1}{dx_1} + (1 - x_1) \frac{d\overline{V}_2}{dx_1}$$

When $x_1 = 0.5$, the slopes are equal but opposite in sign.

4.14 Thermodynamics of a Mixture of Ideal Gases

Thermodynamics alone cannot lead to the conclusion that a mixture of ideal gases will behave as an ideal gas. Nevertheless, it is found that at low pressures mixtures of real gases do behave as ideal gases; that is, they behave as if there are no interactions between the molecules of the gas. Such mixtures are referred to as **ideal** mixtures to indicate the additional nonthermodynamic assumption involved.

Since the components of an ideal mixture behave independently, equation 4.65 may be used to calculate the Gibbs energy of a component. Now that we have introduced the chemical potential, equation 4.65 can be written for component i of an ideal mixture as

$$\mu_i = \mu_i^\circ + RT \ln \frac{P_i}{P^\circ} \qquad (4.106)$$

where P_i is the partial pressure, P° is the standard pressure, and μ_i° is the standard chemical potential of pure gas i at a pressure of P°. See Fig. 4.7.

The partial pressure P_i of any species i in an ideal gas mixture is defined by

$$P_i \equiv y_i P \qquad (4.107)$$

where y_i is the mole fraction and P is the total pressure. We will use y to represent the mole fraction in the gas phase, and x to represent the mole fraction in the liquid phase. The sum of the partial pressures of all the species in an ideal gas mixture is equal to the total pressure.

$$P_1 + P_2 + P_3 + \cdots = (y_1 + y_2 + y_3 + \cdots)P = P \qquad (4.108)$$

since $\sum_i y_i = 1$.

Thus, the equation for the chemical potential of component i in an ideal gas mixture may be written

$$\mu_i = \mu_i^\circ + RT \ln \frac{y_i P}{P^\circ} \qquad (4.109)$$

To consider changes of composition, but not of total pressure, it is convenient to replace the standard chemical potential by that of the pure component i at the prevailing pressure P, which we represent by μ_i^*. From equation 4.109, with $y_i = 1$,

$$\mu_i^* = \mu_i^\circ + RT \ln \frac{P}{P^\circ} \qquad (4.110)$$

Subtracting this from equation 4.109 yields

$$\mu_i = \mu_i^* + RT \ln y_i \qquad (4.111)$$

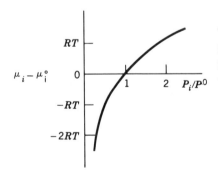

Figure 4.7 Chemical potential μ of an ideal gas as a function of pressure relative to the chemical potential μ° of the gas at the standard pressure of 1 bar. This plot is closely related to Fig. 4.2, but it is given because of the importance of the chemical potential.

This equation defines an ideal mixture. (We will use this equation to define ideal liquid and solid mixtures as well.) The concept of an ideal mixture is useful because it provides a standard with which real mixtures may be compared.

As in the case of a pure component (see Section 4.4), we can obtain thermodynamic properties of an ideal gaseous mixture by finding the expression for the Gibbs energy and calculating the other thermodynamic properties from it. Combining equations 4.83 and 4.111, at constant temperature and pressure, we find

$$G = \sum_{i=1}^{N} n_i \mu_i = \sum_{i=1}^{N} n_i \mu_i^* + RT \sum_{i=1}^{N} n_i \ln y_i \qquad (4.112)$$

$$S = -\left(\frac{\partial G}{\partial T}\right)_{P,n_i} = \sum_{i=1}^{N} n_i \overline{S}_i^* - R \sum_{i=1}^{N} n_i \ln y_i \qquad (4.113)$$

$$H = G + TS = \sum_{i=1}^{N} n_i \overline{H}_i^* \qquad (4.114)$$

$$V = \left(\frac{\partial G}{\partial P}\right)_{T,n_i} = \sum_{i=1}^{N} n_i \left(\frac{\partial \mu_i^*}{\partial P}\right)_{T,n_i} = \sum_{i=1}^{N} n_i \overline{V}_i^* = \frac{RT}{P} \sum_{i=1}^{N} n_i \qquad (4.115)$$

As in the case of equation 4.111, the starred quantities are the molar properties of the pure substances. The enthalpy, volume, and heat capacities of ideal mixtures are simply the sums of those of the amounts of the pure components at the pressure and temperature of the mixture. On the other hand, the Gibbs energies and entropies of ideal mixtures contain terms that are not linear in the amounts of the components.

The partial molar volume $\overline{V}_i$ of ideal gas i in an ideal mixture is equal to $\overline{V}_i^*$, the volume that 1 mol of pure gas i would occupy at a pressure equal to that of the mixture. All ideal gases in an ideal mixture have the same partial molar volume. The total volume of an ideal mixture follows the ideal gas law, with the amount being the sum of the amounts of components.

These equations may be readily applied to calculate the change in thermodynamic properties on mixing. The mixing of two ideal gases, each initially at 1 bar pressure and the same temperature, to yield an ideal mixture at 1 bar pressure is described in Fig. 3.9. Each thermodynamic property for gases before mixing and at the pressure of the final mixture is given by constant terms in equations 4.112 to 4.115. Thus,

$$\Delta_{\text{mix}}G = RT \sum_i n_i \ln y_i = nRT \sum_i y_i \ln y_i \qquad (4.116)$$

$$\Delta_{\text{mix}}S = -R \sum_i n_i \ln y_i = -nR \sum_i y_i \ln y_i \qquad (4.117)*$$

$$\Delta_{\text{mix}}H = 0 \qquad (4.118)$$

$$\Delta_{\text{mix}}V = 0 \qquad (4.119)$$

These changes in thermodynamic properties are extensive, but we will generally apply these equations to intensive properties by setting $n = 1$ mol. Since mole fractions are less than unity, the logarithmic terms are negative, and $\Delta_{\text{mix}}G < 0$. This corresponds with the fact that the mixing of gases is a spontaneous process at constant temperature and pressure. In other words, if two gases at the same pressure and temperature are brought into contact, they will spontaneously diffuse into each other until the gas phase is macroscopically homogeneous.

The Gibbs energy change for mixing two ideal gases is plotted versus the mole fraction of one of the gases in Fig. 4.8a. The greatest Gibbs energy change on mixing is obtained for $y_1 = y_2 = \frac{1}{2}$. The dependence on the entropy of mixing on the mole fraction of one of the components is shown in Fig. 4.8b.

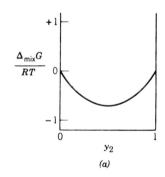

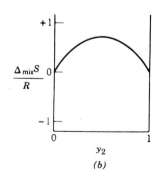

Figure 4.8 Thermodynamic quantities for the mixing of two ideal gases to form 1 mol of ideal mixture.

---

* This equation was derived earlier, as equation 3.62, by a different method.

When ideal gases are mixed at constant temperature and pressure, no heat is produced or consumed. This corresponds with the fact that molecules of ideal gases do not attract or repel each other. Thus, from an energy standpoint, it makes no difference whether the gases are separated or mixed. The driving force for mixing arises exclusively from the change in entropy. From the viewpoint of statistical mechanics (Chapter 17), the mixed state is found at equilibrium because it is more probable, as discussed in Section 3.12 for the expansion of a gas.

Example 4.14

Calculate the changes in the thermodynamic quantities G, S, H, and V for the mixing of $\frac{1}{2}$ mol of oxygen with $\frac{1}{2}$ mol of nitrogen at 25 °C, assuming that they are ideal gases.

$$\Delta_{mix}G = RT(y_1 \ln y_1 + y_2 \ln y_2)$$
$$= (8.314 \text{ J K}^{-1} \text{ mol}^{-1})(298.15 \text{ K})(0.5 \ln 0.5 + 0.5 \ln 0.5)$$
$$= -1718 \text{ J mol}^{-1}$$

$$\Delta_{mix}S = -\frac{\Delta_{mix}G}{T} = 5.763 \text{ J K}^{-1} \text{ mol}^{-1}$$

$$\Delta_{mix}H = 0$$

$$\Delta_{mix}V = 0$$

Example 4.15

At the beginning of this section we assumed that the chemical potential of a component of an ideal gas mixture is given by the same form of equation as the pure gas. Derive equation 4.106 by starting with equation 4.99, which gives the chemical potential of a component of an ideal gas mixture

$$\mu_i = \overline{H}_i - T\overline{S}_i \qquad (a)$$

Since the enthalpy of mixing ideal gases is zero, the molar enthalpy of a component of a mixture of ideal gases is equal to the molar enthalpy of the pure gas:

$$\overline{H}_i = \overline{H}_i^\circ \qquad (b)$$

According to equation 4.93, the entropy of a mixture is equal to the sum of the contributions of the individual components: $S = \sum n_i \overline{S}_i$. Equation 4.113 gives the entropy for a mixture of ideal gases:

$$S = \sum n_i (\overline{S}_i^* - R \ln y_i) \qquad (c)$$

Thus,

$$\overline{S}_i = \overline{S}_i^* - R \ln y_i \qquad (d)$$

According to equation 3.53a for a pure ideal gas

$$\overline{S}_i^* = \overline{S}_i^\circ - R \ln \frac{P}{P^\circ} \qquad (e)$$

So that

$$\overline{S}_i = \overline{S}_i^\circ - R \ln \frac{P}{P^\circ} - R \ln y_i \qquad (f)$$

Substituting equations b and f in equation a yields

$$\mu_i = \overline{H}_i^\circ - T\overline{S}_i^\circ + RT \ln \frac{y_i P}{P^\circ}$$

$$= \mu_i^\circ + RT \ln \frac{y_i P}{P^\circ}$$

$$= \mu_i^\circ + RT \ln \frac{P_i}{P^\circ} \qquad (g)$$

4.15 The Activity

G. N. Lewis introduced the **activity** as a means of dealing with real substances in the gas, liquid, and solid state. In analogy to equation 4.67, the activity of a pure substance, or its activity in a mixture, is defined by

$$\mu_i = \mu_i^\circ + RT \ln a_i \qquad (4.120)$$

Thus, the activity a_i is simply a means for expressing the chemical potential of a species in a mixture. The activity is dimensionless, and $a_i = 1$ in the reference state for which $\mu_i = \mu_i^\circ$. For a real gas $a_i = f_i/P^\circ$, where f_i is the fugacity. For an ideal gas, $a_i = P_i/P^\circ$. We will see later in dealing with solutions that it is convenient to write the activity a_i as the product of an activity coefficient γ_i and a concentration.

The activity of a pure solid or liquid can be taken as unity if the pressure is close enough to the standard state pressure so that the effect of pressure on the chemical potential is negligible. If the effect of pressure is not negligible, the activity of solid or liquid can be readily calculated because the molar volume $\overline{V}$ can be assumed to be constant at all reasonable pressures. For a pure solid or liquid, equation 4.63 can be written

$$\mu(T,P) = \mu^\circ(T) + \overline{V}(P - P^\circ) \qquad (4.121)$$

Comparison with equation 4.120 shows that $RT \ln a = \overline{V}(P - P^\circ)$ or

$$a = e^{\overline{V}(P - P\circ)/RT} \qquad (4.122)$$

Small changes in pressure do not have a significant effect on the activity of a solid or liquid because of the smallness of the exponent.

Example 4.16
What is the activity of liquid water at 1, 10, and 100 bar at 25 °C, assuming that $\overline{V}$ is constant.

At $P = 1$ bar,

$$a = 1$$

At $P = 10$ bar,

$$a = \exp \frac{\overline{V}(P - P°)}{RT}$$

$$= \exp \frac{(0.018 \text{ kg mol}^{-1})(9 \text{ bar})}{(0.08314 \text{ L bar K}^{-1} \text{ mol}^{-1})(298 \text{ K})}$$

$$= 1.007$$

At $P = 100$ bar,

$$a = 1.075$$

4.16 Special Topic: Pressure Drop Across a Curved Surface

The surface tension γ has been introduced in Section 2.9 in connection with surface work. A curved surface of a liquid, or a curved interface between phases, exerts a pressure so that the pressure is higher in the phase on the concave side of the interface. This is analogous to the fact that the pressure inside of a rubber balloon is higher than the atmospheric pressure because of the pressure exerted by the tension of the rubber. However, the difference between a sheet of rubber and the surface of a pure substance is that the tension of the sheet of rubber is roughly proportional to the distance stretched, but the surface tension of a pure substance is independent of area.

To derive the relation between the radius of curvature of a surface, the pressure difference, and the surface tension we will consider a spherical vapor bubble of radius r in a liquid at pressure P, as illustrated in Fig. 4.9. The surface Gibbs energy of the bubble is $4\pi r^2 \gamma$. The surface tension is expected to be a function of radius at sufficiently small radii of curvature, but the nature of the dependence is not known. We will therefore assume that the surface tension is independent of radius of curvature in the calculations made here. If the radius increases by dr, the surface area increases by $8\pi r \, dr$, and the surface energy of the bubble increases by $8\pi r \gamma \, dr$. For this to happen the pressure difference $\Delta P = P_{in} - P_{out}$ across the surface has to increase so that the work against this pressure difference is equal to the increase in surface energy. The PV work of expansion is $4\pi r^2 \, dr \, \Delta P$ so that

$$4\pi r^2 \, dr \, \Delta P = 8\pi r \gamma \, dr \tag{4.123}$$

Thus,

$$\Delta P = \frac{2\gamma}{r} \tag{4.124}$$

Since the radius of curvature in this case is positive, the pressure is greater on the concave side of the interface. As the radius increases to infinity the difference in pressure approaches zero as it must for a planar surface. The pressure drop across a curved surface is responsible for the rise, or depression, of a liquid surface in a capillary.

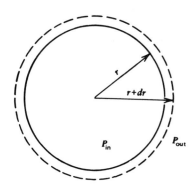

Figure 4.9 Spherical gas bubble in a liquid.

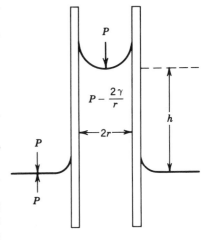

Figure 4.10   Liquid rising in a capillary.

Example 4.17
What pressure in bars is required to prevent water from rising in a 10^{-4} cm diameter capillary a 25 °C?

$$\Delta P = \frac{2\gamma}{r} = \frac{2(71.97 \times 10^{-3} \text{ N m}^{-1})}{(0.5 \times 10^{-6} \text{ m})(10^{5} \text{ N m}^{-2} \text{ bar}^{-1})} = 2.88 \text{ bar}$$

Example 4.18
Assuming that the meniscus in a small capillary has a constant radius of curvature, derive the equation for the height h of capillary rise.

As indicated in Fig. 4.10, the pressure in the liquid under the curved meniscus in the capillary is $P - 2\gamma/r$, where P is the atmospheric pressure. Since the pressure in the liquid just below the flat surface is P, the atmospheric pressure on the flat surface pushes the liquid up the tube until hydrostatic equilibrium is reached. If we ignore the mass of liquid in the capillary above height h, equilibrium is reached when the pressure exerted by the column of liquid of density ρ is equal to the pressure difference across the curved surface:

$$\frac{\rho \pi r^{2} h g}{\pi r^{2}} = \frac{2\gamma}{r}$$

or

$$h = \frac{2\gamma}{g\rho r}$$

where g is the acceleration due to gravity.

4.17   Special Topic: Thermodynamics of Rubberlike Elasticity

Natural rubber and rubberlike synthetic polymers are unusual in that they may be stretched up to 5 to 10 times their unstretched lengths without rupture, and with nearly complete reversibility. Figure 4.11 shows a plot of force per unit cross-sectional area of polymer versus elongation for four temperatures. Note that in contrast with other materials, stretched rubber contracts when its temperature is raised at constant force of elongation.

The small degree of irreversibility in these measurements will be ignored in this discussion. In calculating thermodynamic quantities for rubberlike elasticity it is conventional to calculate all elongations on the basis of the unstrained length at the temperature of the experiment.

Rubberlike materials are made up of long chainlike molecules with enough cross links to prevent flow and to return the material to its original dimensions when the force of extension is removed. Between cross links, which are provided by vulcanization in the case of rubber, a molecular chain may have many conformations, as we will see in Section 22.4.

When a solid is stretched, the differential of the work done by the applied stress is $f \, dl$, where f is the force per unit area and l is the length of the specimen:

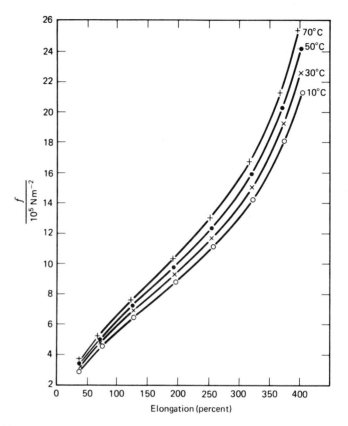

Figure 4.11   Force per unit area versus elongation for a sample of rubber at four temperatures. (From R. L. Anthony, R. H. Caston, and E. Guth, *J. Phys. Chem.* **46:** 826 (1942). Copyright © The Williams & Wilkins Co., Baltimore.)

$$dw = f \, dl \qquad (4.125)$$

We do not usually talk about the work and force per unit area, but in discussing the elasticity of rubber it is convenient to do that since the force per unit area is found to be independent of area. The total work done in a stretching process is, therefore,

$$dw = f \, dl - P \, dV \qquad (4.126)$$

However, when rubbers are stretched the $P \, dV$ term is smaller than the $f \, dl$ term by a factor of 10^3 to 10^4. Therefore, we will ignore the $P \, dV$ term and use equation 4.125 for the differential work. Thus, for a rubberlike material the combined first and second laws may be written

$$dU = T \, dS + f \, dl \qquad (4.127)$$

Since the differential of the Helmholtz energy A is equal to the reversible work at constant temperature (Section 4.3), we have $dA = f \, dl$ under these conditions. Thus,

$$f = \left(\frac{\partial A}{\partial l}\right)_T \qquad (4.128)$$

The force per unit area exerted by a piece of rubber may be expressed as the sum of two terms by substituting equation 4.11 in equation 4.128:

$$f = \left(\frac{\partial A}{\partial l}\right)_T = \left(\frac{\partial U}{\partial l}\right)_T - T\left(\frac{\partial S}{\partial l}\right)_T \tag{4.129}$$

The first contribution to the force is the rate of increase of the internal energy with length, and the second contribution is proportional to the temperature and to the rate of increase of entropy with length. Experimental data show that the second term is by far the more important for rubberlike elasticity. To evaluate the magnitude of this term we need to find a way to measure $(\partial S/\partial l)_T$.

An alternative expression for $(\partial S/\partial l)_T$ may be obtained by starting with the definition of Helmholtz energy, expressed in differential form:

$$dA = dU - T\,dS - S\,dT \tag{4.130}$$

Inserting equation 4.127 yields

$$dA = f\,dl - S\,dT \tag{4.131}$$

Thus,

$$\left(\frac{\partial A}{\partial l}\right)_T = f \quad \text{and} \quad \left(\frac{\partial A}{\partial T}\right)_l = -S \tag{4.132}$$

Taking the cross derivatives, we find

$$\left[\frac{\partial}{\partial l}\left(\frac{\partial A}{\partial T}\right)_l\right]_T = \left[\frac{\partial}{\partial T}\left(\frac{\partial A}{\partial l}\right)_T\right]_l \tag{4.133}$$

$$-\left(\frac{\partial S}{\partial l}\right)_T = \left(\frac{\partial f}{\partial T}\right)_l \tag{4.134}$$

Thus, $(\partial S/\partial l)_T$ may be obtained from measurements of force versus temperature at constant length. Substitution of equation 4.134 into equation 4.129 yields

$$\left(\frac{\partial U}{\partial l}\right)_T = f - T\left(\frac{\partial f}{\partial T}\right)_l \tag{4.135}$$

Rubberlike elasticity provides an interesting example of a situation in which the changes in internal energy, entropy, and Helmholtz energy with length can be obtained without making any measurements of heat. Note that $(\partial U/\partial l)_T$ has the units N m^{-2} so that we are dealing here with an internal energy in J m^{-2}. Since $(\partial S/\partial l)_T$ has the units of N K^{-1} m^{-2}, we are dealing with an entropy with units of J K^{-1} m^{-1}.

Figure 4.12 shows the force per unit area as a function of elongation at a constant temperature of 10 °C for the same sample used in the preceding figure. Using the data from the preceding figure, the values of $-T\,(\partial S/\partial l)_T$ and $(\partial U/\partial l)_T$ were calculated and are shown by dashed lines. It is evident that the force exerted by the rubber is almost entirely entropic in origin. The pressure exerted by an ideal gas is entirely entropic in origin.

The force exerted by a piece of rubber at a fixed percentage elongation is close to being directly proportional to the temperature. This relationship is obtained from equation 4.135 by setting $(\partial U/\partial l)_T$ equal to zero, in accord with the experimental results of Fig. 4.12. Complications arise at extensions over 300 to 400%.

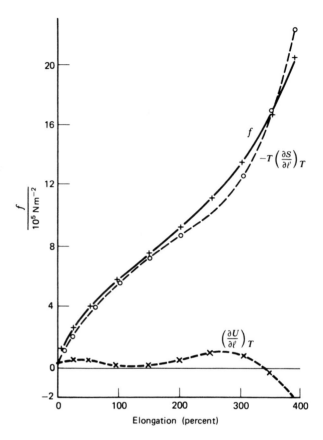

Figure 4.12 Force per unit area versus percentage elongation at a constant temperature of 10 °C for the sample of rubber used in Fig. 4.9. (From. R. L. Anthony, R. H. Caston, and E. Guth, *J. Phys. Chem.* **46:** 826 (1942). Copyright © The Williams & Wilkins Co., Baltimore.)

Qualitatively we can say that the force exerted by an elongated piece of rubber arises from the fact that the molecular chains can adopt many conformations that differ very little in energy. As the rubber is stretched the number of possible conformations for a given chain is decreased. Thus, the entropy decreases as the sample is stretched. Statistical mechanical treatment of a simple model of rubber shows that the entropy is given as a function of relative extension λ ($= l/l_0$) by

$$S = \text{constant} - kK\left(\lambda^2 + \frac{2}{\lambda}\right) \qquad (4.136)$$

Substitution of this relation in equation 4.134 written in terms of relative extension yields after integration

$$f = 2KkT\left(\lambda - \frac{1}{\lambda^2}\right) \qquad (4.137)$$

This equation may be considered to be an equation of state for rubberlike elasticity. According to this simple model, the force per unit area required to maintain a given relative extension λ is proportional to temperature. The model also correctly predicts that if rubber is heated at a constant force f, the sample will contract. Equation 4.137 fits experimental data pretty well for elongation less than 300 to 400%. The quantity K may be interpreted as the number of chain segments per unit volume. The quantity k is the Boltzmann constant, which is R/N_A.

References

J. A. Beattie and I. Oppenheim, *Principles of Thermodynamics*. New York: Elsevier Scientific, 1979.

K. E. Bett, J. S. Rowlinson, and G. Saville, *Thermodynamics for Chemical Engineers*. Cambridge, MA: MIT Press, 1975.

H. B. Callen, *Thermodynamics and an Introduction to Thermostatics*. New York: Wiley, 1985.

K. Denbigh, *The Principles of Chemical Equilibrium*. Cambridge, UK: Cambridge University Press, 1981.

J. W. Gibbs, *The Collected Works of J. Willard Gibbs*. New Haven, CT: Yale University Press, 1948.

N. A. Gokcen, *Thermodynamics*. Hawthorne, CA: Techscience Inc., 1975.

E. A. Guggenheim, *Thermodynamics*. Amsterdam: North-Holland, 1967.

J. G. Kirkwood and I. Oppenheim, *Chemical Thermodynamics*. New York: McGraw-Hill, 1961.

G. N. Lewis and M. Randall, revised by K. S. Pitzer and L. Brewer, *Thermodynamics*. New York: McGraw-Hill, 1961.

S. I. Sandler, *Chemical and Engineering Thermodynamics*. New York: Wiley, 1989.

J. M. Smith and H. C. Van Ness, *Introduction to Chemical Engineering Thermodynamics*. New York: McGraw-Hill, 1987.

Problems

4.1 One mole of nitrogen gas is allowed to expand from 0.5 to 10 L. Calculate the change in molar entropy using (*a*) the ideal gas law and (*b*) the van der Waals equation.

4.2 Derive the relation for $\overline{C}_P - \overline{C}_V$ for a gas that follows van der Waals' equation.

4.3 Earlier we derived the expression for the entropy of an ideal gas as a function of T and P. Now that we have the Maxwell relations, derive the expression for dS for any fluid.

4.4 What is the change in molar entropy of liquid benzene at 25 °C when the pressure is raised to 1000 bar? The coefficient of thermal expansion α is 1.237×10^{-3} K^{-1}, the density is 0.879 g cm^{-3}, and the molar mass is 78.11 g mol^{-1}.

4.5 Derive the expression for $\overline{C}_P - \overline{C}_V$ for a gas with the following equation of state:

$$(P + a/\overline{V}^2)\overline{V} = RT$$

4.6 What is the difference between the molar heat capacity of iron at constant pressure and constant volume at 25 °C? Given: $\alpha = 35.1 \times 10^{-6}$ K^{-1}, $\kappa = 0.52 \times 10^{-6}$ bar^{-1}, and the density is 7.86 g cm^{-3}.

4.7 In equation 1.29 we saw that the compressibility factor of a van der Waals gas can be written as

$$Z = 1 + \frac{1}{RT}\left(b - \frac{a}{RT}\right)P + \cdots$$

(a) To this degree of approximation, derive the expression for $(\partial\overline{H}/\partial P)_T$ for a van der Waals gas. (b) Calculate $(\partial\overline{H}/\partial P)_T$ for $CO_2(g)$ in J bar^{-1} mol^{-1} at 298 K. Given: $a = 3.640$ L^2 bar mol^{-1} and $b = 0.042\,67$ L mol^{-1}.

4.8   (a) Integrate the Gibbs–Helmholtz equation to obtain an expression for ΔG_2 at temperature T_2 in terms of ΔG_1 at T_1, assuming that ΔH is independent of temperature. (b) Obtain an expression for ΔG_2 using the more accurate approximation that $\Delta H = \Delta H_1 + (T - T_1)\Delta C_P$ where T_1 is an arbitrary reference temperature.

4.9   When a liquid is compressed its Gibbs energy is increased. To a first approximation the increase in molar Gibbs energy can be calculated using $(\partial\overline{G}/\partial P)_T = \overline{V}$, assuming a constant molar volume. What is the change in molar Gibbs energy for liquid water when it is compressed to 1000 bar?

4.10   Show that when a liquid is compressed, the change in Gibbs energy is given by

$$\Delta G = V\,\Delta P - \tfrac{1}{2}\kappa V(\Delta P)^2 \qquad \text{if } \kappa\,\Delta P \ll 1.$$

The isothermal compressibility is represented by κ.

4.11   Given the isothermal compressibility κ of water is 45.0×10^{-6} bar^{-1}, calculate a more accurate value for $\Delta\overline{G}$ for the compression to 1000 bar than that calculated in problem 4.9. (See problem 4.10.)

4.12   An ideal gas is allowed to expand reversibly and isothermally (25 °C) from a pressure of 1 bar to a pressure of 0.1 bar: (a) What is the change in molar Gibbs energy? (b) What would be the change in molar Gibbs energy if the process occurred irreversibly?

4.13   The standard entropy of $O_2(g)$ at 1 bar is listed in Appendix C.2 as 205.137 J K^{-1} mol^{-1}, and the standard Gibbs energy of formation is listed as 0 kJ mol^{-1}. Assuming that O_2 is an ideal gas, what will be the molar entropy and molar Gibbs energy of formation at 100 bar?

4.14   Helium is compressed isothermally and reversibly at 100 °C from a pressure of 2 to 10 bar. Calculate (a) q per mole, (b) w per mole, (c) $\Delta\overline{G}$, (d) $\Delta\overline{A}$, (e) $\Delta\overline{H}$, (f) $\Delta\overline{U}$, and (g) $\Delta\overline{S}$, assuming that helium is an ideal gas.

4.15   Toluene is vaporized at its boiling point, 111 °C. The heat of vaporization at this temperature is 361.9 J g^{-1}. For the vaporization of toluene, calculate (a) w per mole, (b) q per mole, (c) $\Delta\overline{H}$, (d) $\Delta\overline{U}$, (e) $\Delta\overline{G}$, and (f) $\Delta\overline{S}$.

4.16   If the Gibbs energy varies with temperature according to

$$G/T = a + b/T + c/T^2$$

How will the enthalpy and entropy vary with temperature? Check that these three equations are consistent.

4.17   Calculate the change in molar Gibbs energy $\overline{G}$ when supercooled water at -3 °C freezes at constant T and P. The density of ice at -3 °C is 0.917×10^3 kg m^{-3}, and its vapor pressure is 475 Pa. The density of supercooled water at -3 °C is 0.9996 kg m^{-3} and its vapor pressure is 489 Pa.

4.18   Calculate the molar Gibbs energy G of fusion when supercooled water at -3 °C freezes at constant T and P. The enthalpy of fusion of ice is 6000 J mol^{-1} at 0 °C. The heat capacities of water and ice in the vicinity of the freezing point are 75.3 and 38 J K^{-1} mol^{-1}, respectively.

4.19   As shown in Example 4.9, the fugacity of a van der Waals gas is given by a fairly simple expression if only the second virial coefficient is used. To this degree of approximation derive the expressions $\overline{G}$, $\overline{S}$, $\overline{A}$, $\overline{U}$, $\overline{H}$, and $\overline{V}$.

4.20   Using the relation derived in Example 4.9, calculate the fugacity of $H_2(g)$ at 100 bar at 298 K.

4.21   Show that if the compressibility factor is given by $Z = 1 + BP/RT$ the fugacity is given by $f = Pe^{Z-1}$. If Z is not very different from unity, $e^{Z-1} = 1 + (Z - 1) + \cdots \approx Z$ so that $f = PZ$. Using this approximation, what is the fugacity of $H_2(g)$ at 50 bar and 298 K using its van der Waals constants?

4.22   Calculate the partial molar volume of zinc chloride in 1-molal $ZnCl_2$ solution using the following data:

% by weight of $ZnCl_2$	2	6	10	14	18
Density/g cm^{-3}	1.0167	1.0532	1.0891	1.1275	1.1665

4.23   Calculate $\Delta_{mix}G$ anf $\Delta_{mix}S$ for the formation of a quantity of air containing 1 mol of gas by mixing nitrogen and oxygen at 298.15 K. Air may be taken to be 80% nitrogen and 20% oxygen.

4.24   A mole of gas A is mixed with a mole of gas B at 1 bar and 298 K. How much work is required to separate these gases to produce a container of each at 1 bar and 298 K?

4.25   The surface tension of toluene at 20 °C is 0.0284 N m^{-1}, and its density at this temperature is 0.866 g cm^{-3}. What is the radius of the largest capillary that will permit the liquid to rise 2 cm?

4.26   Mercury does not wet a glass surface. Calculate the capillary depression if the diameter of the capillary is (a) 0.1 mm and (b) 2 mm. The density of mercury is 13.5 g cm^{-3}. The surface tension of mercury at 25 °C is 0.520 N m^{-1}.

4.27   If the surface tension of a soap solution is 0.050 N m^{-1}, what is the difference in pressure across the film for (a) a soap bubble 2 mm in diameter and (b) a bubble 2 cm in diameter?

4.28   Using experimental data from Fig. 4.9, calculate $(\partial S/\partial l)_T$ and $(\partial U/\partial l)_T$ using only the data at 10 °C and 70 °C, assuming that these quantities are independent of temperature.

These thermodynamic quantities do depend on the elongation, so calculate them for 200% and 400% elongation. (For f and T simply use averages for this temperature range.)

4.29 Show that

$$\left(\frac{\partial U}{\partial S}\right)_V = \left(\frac{\partial H}{\partial S}\right)_P \qquad \left(\frac{\partial H}{\partial P}\right)_S = \left(\frac{\partial G}{\partial P}\right)_T$$

4.30 Earlier we derived the expression for the entropy of an ideal gas as a function of T and V. Now that we have the Maxwell relations, derive the expression for dS for any fluid.

4.31 The coefficient of thermal expansion α of Fe(s) at 25 °C is 355×10^{-7} K^{-1}. What is the change in molar entropy of iron when the pressure is raised to 1000 bar? (The density of iron at 25 °C is 7.86 g cm^{-3}.)

4.32 What is the effect of pressure on the entropy, enthalpy, and internal energy of an incompressible fluid? For an incompressible fluid, α and κ are equal to zero.

4.33 Show that C_P and C_V for an ideal gas are independent of volume and pressure.

4.34 Derive the thermodynamic equation of state

$$\left(\frac{\partial U}{\partial P}\right)_T = V(\kappa P - \alpha T)$$

4.35 Derive the expression for $(\partial \overline{H}/\partial P)_T$ for a gas following the virial equation

$$P\overline{V} = RT + B(T)P$$

4.36 Assuming that the density of water is independent of pressure in the range 1 to 50 bar, what is the change in molar Gibbs energy of water when the pressure is raised this amount?

4.37 An ideal gas is compressed isothermally from 1 to 5 bar at 100 °C. (a) What is the molar Gibbs energy change? (b) What would have been the change in molar Gibbs energy if the compression had been carried out at 0 °C?

4.38 At 298 K, for H$_2$(g), $S° = 130.684$ J K^{-1}, $\Delta_f H° = 0$ kJ mol^{-1}, and $\Delta_f G° = 0$ kJ mol^{-1}. What are the values of the molar entropy, enthalpy of formation, and Gibbs energy of formation at 10^{-2} bar, assuming that H$_2$ is an ideal gas?

4.39 The heat of vaporization of liquid oxygen at 1.013 25 bar is 6820 J mol^{-1} at its boiling point, -183 °C, at that pressure. For the reversible vaporization of liquid oxygen, calculate (a) q per mole, (b) $\Delta \overline{U}$, (c) $\Delta \overline{S}$, and (d) $\Delta \overline{G}$.

4.40 An ideal gas is expanded isothermally and reversibly at 0 °C from 1 to $\frac{1}{10}$ bar. Calculate (a) w per mole, (b) q per mole, (c) $\Delta \overline{H}$, (d) $\Delta \overline{G}$, and (e) $\Delta \overline{S}$ for the gas. One mole of an ideal gas in 22.71 L is allowed to expand irreversibly into

an evacuated vessel such that the final total volume is 227.1 L. Calculate (f) w per mole, (g) q per mole, (h) $\Delta \overline{H}$, (i) $\Delta \overline{G}$, and (j) $\Delta \overline{S}$ for the gas. Calculate (k) $\Delta \overline{S}$ for the system and its surroundings involved in the reversible isothermal expansion and calculate (l) $\Delta \overline{S}$ for the system and its surroundings involved in an irreversible isothermal expansion in which the gas expands into an evacuated vessel such that the final total volume is 227.1 L.

4.41 Steam is compressed reversibly to liquid water at the boiling point 100 °C. The heat of vaporization of water at 100 °C and 1.013 25 bar is 2258 J g^{-1}. Calculate w per mole and q per mole and each of the thermodynamic quantities $\Delta \overline{H}$, $\Delta \overline{U}$, $\Delta \overline{G}$, $\Delta \overline{A}$, and $\Delta \overline{S}$.

4.42 An ideal gas at 300 K has an initial pressure of 15 bar and is allowed to expand isothermally to a pressure of 1 bar. Calculate (a) the maximum work that can be obtained from the expansion, (b) $\Delta \overline{U}$, (c) $\Delta \overline{H}$, (d) $\Delta \overline{G}$, and (e) $\Delta \overline{S}$.

4.43 Calculate the molar entropy changes for the gas plus reservoir in Examples 4.6 and 4.7.

4.44 An ideal gas is compressed reversibly at 100 °C from 2 bar to 10 bar. Calculate $\Delta \overline{H}$, $\Delta \overline{S}$, and $\Delta \overline{G}$. Show that you can get the same value of $\Delta \overline{G}$ in two ways.

4.45 Toluene is vaporized at its boiling point (111 °C). The heat of vaporization at this temperature is 361.9 J g^{-1}. Calculate $\Delta \overline{H}$, $\Delta \overline{S}$, and $\Delta \overline{G}$.

4.46 What is the change in molar Gibbs energy for the freezing of water at -10 °C? The vapor pressure of H$_2$O(l, -10 °C) is 286.5 Pa, and the vapor pressure of H$_2$O(s, -10 °C) is 260 Pa. At -10 °C the molar volume of supercooled water is 1.80×10^{-5} m^3 mol^{-1} and the molar volume of ice is 2.00×10^{-5} m^3 mol^{-1}.

4.47 At low pressures the compressibility factor for a van der Waals gas is given by

$$Z = \frac{P\overline{V}}{RT} = 1 + \left(\frac{b - a}{RT}\right)\frac{P}{RT}$$

Derive the expression for $\Delta \overline{G}$ for a change in pressure from P_1 to P_2.

4.48 For a gas that follows the equation of state $P(\overline{V} - b) = RT$, show that the fugacity is given by

$$f = Pe^{bP/RT}$$

4.49 The apparent specific volume v of a solute (that is, volume contributed to the solution by one kilogram of solute) is equal to the volume V of the solution minus the volume of pure solvent it contains divided by the mass of solute:

$$v = \frac{V - (V\rho_s - m)/\rho_0}{m}$$

where m is the mass of solute, ρ_s is the density of the solution, and ρ_0 is the density of the solvent. The density of a solution of serum albumin containing 15.4 g of protein per liter is 1.0004×10^3 kg m^{-3} at 25 °C ($\rho_0 = 0.977\ 07 \times 10^3$ kg m^{-3}). Calculate the apparent specific volume.

4.50   A solution of magnesium chloride, $MgCl_2$, in water containing 41.24 g L^{-1} has a density of 1.0311×10^3 kg m^{-3} at 20 °C. The density of water at this temperature is 0.998×10^3 kg m^{-3}. Calculate (a) the apparent specific volume (see the equation in problem 4.49), and (b) the apparent molar volume of $MgCl_2$ in this solution.

4.51   Calculate ΔG and ΔS for mixing 2 mol of H_2 with 1 mol of O_2 at 25 °C under conditions where no chemical reaction occurs.

4.52   Acetone has a density of 0.790 g cm^{-3} at 20 °C and rises to a height of 2.56 cm in a capillary having a radius of 0.0235 cm. What is the surface tension of acetone at this temperature?

4.53   If a liquid has a contact angle of Θ with the surface of the capillary, the surface tension can be calculated from the capillary rise h using

$$\gamma = \frac{\rho g r h}{2 \cos \Theta}$$

Derive this equation by assuming that it is the vertical component of the force due to surface tension that causes the capillary rise. For mercury on glass, $\Theta = 180°$, and so there is capillary depression.

4.54   What are the pressures inside bubbles in water at 25 °C if the pressure in the water is 1 bar and the bubbles are (a) 1 mm and (b) 0.1 mm in diameter? The surface tension of water at 25 °C is 7.197×10^{-2} N m^{-1}.

4.55   What is the value of K in equation 4.90 for the sample of rubber described by Fig. 4.4 at 10 and 70 °C?

4.56   Liquid water can be superheated to 120 °C at 1.01325 bar. Calculate the changes in entropy, enthalpy, and Gibbs energy for the process of superheated water at 120 °C and 1.01325 bar changing to steam at the same temperature and pressure. the enthalpy of vaporization is 40.58 kJ mol^{-1} at 100 °C and 1.01325 bar. Given: $\overline{C}_P(H_2O,l) = 75.3$ J K^{-1} mol^{-1} and $\overline{C}_P(H_2O,g) = (36 + 0.013T)$ J K^{-1} mol^{-1}, where T is in kelvins.

5
Chemical Equilibrium

The emphasis in this chapter is on single reactions in the gas phase. However, there are brief discussions of multireaction equilibria and of gas-solid reactions. Chemical equilibrium in the liquid phase is discussed in later chapters.

The idea of the reversibility of chemical reactions was first stated clearly in 1799 by C. Berthollet, while he was acting as scientific adviser to Napoleon in Egypt. He noted the deposits of sodium carbonate in certain salt lakes and concluded that they were produced by the high concentration of sodium chloride and dissolved calcium carbonate, the reverse of the laboratory experiment in which sodium carbonate reacts with calcium chloride to precipitate calcium carbonate. In 1863 the influence that the concentrations of ethyl alcohol and acetic acid have on the concentration of ethyl acetate was reported by M. Berthelot and Saint-Gilles.

In 1864 Guldberg and Waage showed experimentally that in chemical reactions an equilibrium is reached that can be approached from either direction. They were apparently the first to realize that there is a mathematical relation between the concentrations of reactants and products at equilibrium. In 1877 van't Hoff suggested that in the equilibrium expression for the hydrolysis of ethyl acetate, the concentration of each reactant should appear to the first power, corresponding with the stoichiometric numbers in the balanced chemical equation.

5.1 Derivation of the General Equilibrium Expression

As we saw in Section 2.16, a generalized chemical reaction can be represented by

$$0 = \sum_{i=1}^{N} \nu_i A_i \tag{5.1}$$

where the A_i are the formulas of reactant and product species, and where the stoichiometric numbers ν_i are positive for products and negative for reactants. Since chemical equilibrium is generally studied at constant temperature and pressure, the Gibbs energy is the criterion of equilibrium. Let us consider a system in which a single reaction occurs in a single phase. According to equation 4.80 the total differential of the Gibbs energy for a system at constant temperature and pressure is given by

$$(dG)_{T,P} = \sum_{i=1}^{N} \mu_i \, dn_i \tag{5.2}$$

where N is the number of species involved in the reaction, and dn_i is the differential change in the amount of the ith species. We saw earlier in Section 2.16 that when a chemical reaction occurs, the changes in the amounts of species involved in the reaction are proportional to their stoichiometric numbers ν_i in the chemical equation 5.1. If a reaction system initially contains amount n_{i0} of i, the amount n_i of i at a later time may be written

$$n_i = n_{i0} + \nu_i \xi \tag{5.3}$$

where ξ is the extent of reaction. The extent of reaction is an extensive property, expressed in moles, which is zero for the initial state.

Thus, from equation 5.3, dn_i is given by

$$dn_i = \nu_i \, d\xi \tag{5.4}$$

Substituting this into equation 5.2 yields

$$\left(\frac{\partial G}{\partial \xi} \right)_{T,P} = \sum_{i=1}^{N} \nu_i \mu_i \tag{5.5}$$

Since $(\partial G/\partial \xi)_{T,P}$ is a complicated symbol, it is replaced by $\Delta_r G$ and is referred to as the **reaction Gibbs energy.**

$$\Delta_r G = \sum_{i=1}^{N} \nu_i \mu_i \tag{5.6}$$

The reaction Gibbs energy is the change in Gibbs energy when the extent of reaction ξ changes by 1 mol, as specified by a balanced chemical equation, with the reactants and products at specified chemical potentials. Since the stoichiometric numbers are dimensionless, the reaction Gibbs energy has units of joules per mole.

When a reaction takes place at constant temperature and pressure, the Gibbs energy decreases, and the reaction continues until the Gibbs energy has reached its minimum value, as shown by Fig. 5.1. At equilibrium at constant temperature and pressure, the Gibbs energy of the system has its minimum value and so $(\partial G/\partial \xi)_{T,P}$ (or $\Delta_r G$) is equal to zero, and

$$\sum_{i=1}^{N} \nu_i \mu_{i,\text{eq}} = 0 \qquad (5.7)$$

where the chemical potentials in this equation are the values at equilibrium. **This general condition applies to all chemical equilibria, whether they involve gases, liquids, solids, or solutions.** It is of interest to note that the thermodynamic criterion for equilibrium has the same form as the reaction to which it applies (equation 5.1), except that molecular formulas are replaced by the corresponding chemical potentials of the reactants and products in the equilibrium reaction mixture.

The most general form for the reaction Gibbs energy expression is obtained by substituting $\mu_i = \mu_i^\circ + RT \ln a_i$ (equation 4.120) in equation 5.6. This yields

$$\Delta_r G = \sum_{i=1}^{N} \nu_i(\mu_i^\circ + RT \ln a_i)$$

$$= \sum_{i=1}^{N} \nu_i \mu_i^\circ + RT \sum_{i=1}^{N} \ln a_i^{\nu_i}$$

$$= \Delta_r G^\circ + RT \ln \prod_{i=1}^{N} a_i^{\nu_i} \qquad (5.8)$$

where $\Delta_r G^\circ$ is the standard reaction Gibbs energy ($\sum \nu_i \mu_i^\circ$). Equation 5.8 gives the change in Gibbs energy for a specified chemical reaction when the reactants and products have activities a_i. When each reactant and product has an activity of unity or $\prod a_i^{\nu_i} = 1$, $\Delta_r G = \Delta_r G^\circ$.

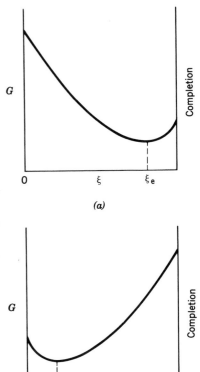

Figure 5.1 (*a*) Gibbs energy as a function of extent of reaction at constant T and P for a reaction in a single phase that goes nearly to completion. (*b*) Gibbs energy as a function of extent of reaction that does not go very far toward completion.

Example 5.1

Consider the formation of a mole of $NH_3(g)$:

$$\tfrac{1}{2}N_2(g, a_{N_2}) + \tfrac{3}{2}H_2(g, a_{H_2}) = NH_3(g, a_{NH_3})$$

at constant temperature and pressure in such a large reactor that the indicated activities do not change appreciably when $\xi = 1$ mol. The change in Gibbs energy in this reaction is

$$\Delta_r G = \Delta_r G^\circ + RT \ln \frac{a_{NH_3}}{a_{N_2}^{1/2} a_{H_2}^{3/2}}$$

Depending on the activities of the reactants and products, $\Delta_r G$ may be positive, in which case the forward reaction is not spontaneous, or negative, in which case the forward reaction is spontaneous. If $\Delta_r G = 0$, the reactants and products are at equilibrium at the indicated activities.

It is convenient to write equation 5.8 as

$$\Delta_r G = \Delta_r G° + RT \ln Q \tag{5.9}$$

where the **quotient of activities** is represented by Q:

$$Q = \prod_{i=1}^{N} a_i^{\nu_i} \tag{5.10}$$

If the reactants and products are in equilibrium, the reaction Gibbs energy is zero, and equation 5.8 becomes

$$0 = \Delta_r G° + RT \ln K \tag{5.11}$$

or

$$\Delta_r G° = -RT \ln K \tag{5.12}$$

where the **equilibrium constant** is given by

$$K = \prod_{i=1}^{N} a_{i,eq}^{\nu_i} \tag{5.13}$$

Notice that this is the definition of the equilibrium constant K, which is a function only of temperature and is dimensionless.* The magnitude of the equilibrium constant depends on the way the chemical equation is written because of the stoichiometric numbers. Therefore, the value for an equilibrium constant should always be accompanied by a balanced chemical equation.

Example 5.2

What is the most general equilibrium constant expression for the following reaction?

$$3C(graphite) + 2H_2O(g) = CH_4(g) + 2CO(g)$$

$$K = \left(\frac{a_{CH_4} a_{CO}^2}{a_C^3 a_{H_2O}} \right)_{eq}$$

If the pressure is not too high, the graphite can be considered to be in its standard state so that $a_C = 1$. The activities of the gases can be replaced by $f_i/P°$, or, if the pressure is low enough, by $P_i/P°$.

5.2 Equilibrium Constant Expressions for Gas Reactions

For real gases the activity is given by $a_i = f_i/P°$, where f_i is the fugacity of the ith species and $P°$ is the standard state pressure.

In Section 4.9 we saw that the partial molar Gibbs energy, which we can now refer to as the chemical potential, of a gas is given by

---

* Later, in Section 17.12, when we calculate K from statistical mechanics, we will also find that we obtain a pure number. Nevertheless, the numerical value obtained depends on the standard states for reactants and products used in the calculation of K.

$$\mu_i = \mu_i^\circ + RT \ln \frac{f_i}{P^\circ} \tag{5.14}$$

This relation can be substituted into equation 5.7 and the same operations carried out to obtain

$$K = \prod_{i=1}^{N} \left(\frac{f_{i,\text{eq}}}{P^\circ}\right)^{\nu_i} \tag{5.15}$$

This equation is not used very often because of the difficulty in evaluating f_i in a mixture of gases (see Section 5.12), but it is the most general expression for the equilibrium constant of a reaction involving real gases.

For ideal gases, we have seen earlier that

$$\mu_i = \mu_i^\circ + RT \ln \frac{P_i}{P^\circ} \tag{5.16}$$

Substituting this relation in equation 5.7 and carrying out the operations in Section 5.1 yields

$$K = \prod_{i=1}^{N} \left(\frac{P_{i,\text{eq}}}{P^\circ}\right)^{\nu_i} \tag{5.17}$$

This equilibrium constant is also only a function of temperature. For real gases the equilibrium constant calculated using this equation will depend on pressure since in general for real gases $f_i \neq P_i$. The term thermodynamic equilibrium constant is often used for the equilibrium constant obtained by using equation 5.17 at low pressure or by use of equation 5.15. Calculations of K using tables of Gibbs energies of formation yield thermodynamic equilibrium constants.

The value of an equilibrium constant cannot be interpreted unless it is accompanied by a balanced chemical equation and a specification of the standard state of each reactant and product. The values of the stoichiometric numbers are arbitrary to the extent that a chemical equation may be multiplied or divided by a positive or negative number. In this section we are considering gas reactions, and so the standard state of each reactant and product is the pure gas at 1 bar in the ideal gas state (Section 4.9). Later in Section 8.5 and 8.9 we will discuss the standard states of substances in liquid solution in more detail. In using equations 5.13, 5.15, and 5.17 and similar equations, we will omit the subscript eq.

In equation 5.8 $\Delta_r G^\circ$ is the change in standard Gibbs energy when the indicated amounts of the *unmixed* reactants in their standard states at temperature T and 1 bar pressure react to form *unmixed* products in their standard states at the same temperature and pressure. The Gibbs energy change for the reaction from the mixed reactants to the equilibrium mixture is not equal to $\Delta_r G^\circ$.

The equilibrium extent of a chemical reaction in ideal gas mixtures depends on just three independent variables: (1) pressure, (2) initial composition, and (3) temperature. We will examine the effects of each of these variables and also the effect of adding inert gases to the reaction mixture.

Example 5.3

(a) A mixture of CO(g), H_2(g), and CH_3OH(g) at 500 K with $P_{CO} = 10$ bar, $P_{H_2} = 1$ bar, and $P_{CH_3OH} = 0.1$ bar is passed over a catalyst. Can more methanol be formed? Given: $\Delta_r G^\circ = 21.21$ kJ mol^{-1}.

$$CO(g, 10 \text{ bar}) + 2H_2(g, 1 \text{ bar}) = CH_3OH(g, 0.1 \text{ bar})$$

$$\Delta_r G = \Delta_r G^\circ + RT \ln Q$$

$$= 21.21 \text{ kJ mol}^{-1} + (0.008\ 314\ 5 \text{ kJ mol}^{-1})(500\text{K}) \ln \frac{(0.1)}{(10)(1)^2}$$

$$= 2.07 \text{ kJ mol}^{-1}$$

Thus, the reaction as written is not spontaneous. (*b*) Can the following conversion occur at 500 K?

$$CO(g, 1 \text{ bar}) + 2H_2(g, 10 \text{ bar}) = CH_3OH(g, 0.1 \text{ bar})$$

$$\Delta_r G = 21.21 \text{ kJ mol}^{-1} + (0.008\ 314\ 5 \text{ kJ K}^{-1} \text{ mol}^{-1})(500 \text{ K}) \ln \left[\frac{(0.1)}{(1)(10)^2} \right]$$

$$= -7.51 \text{ kJ mol}^{-1}$$

Under these conditions the reaction is thermodynamically spontaneous.

5.3 Thermodynamics of a Simple Gas Reaction

To illustrate why gas reactions never go to completion, let us consider a simple isomerization of ideal gas A to ideal gas B at constant pressure:

$$A(g) = B(g) \tag{5.18}$$

According to equation 4.83, the Gibbs energy of the reaction mixture at any extent of reaction is

$$G = n_A \mu_A + n_B \mu_B \tag{5.19}$$

where n_A is the amount of A and n_B is the amount of B. If the reaction is started with 1 mol of A, the amounts of A and B at a later time are given in terms of extent of reaction ξ by

$$n_A = 1 - \xi \tag{5.20}$$

$$n_B = \xi \tag{5.21}$$

Thus,

$$G = (1 - \xi)\mu_A + \xi\mu_B \tag{5.22}$$

From equation 5.16, the chemical potentials of A and B in the mixture of ideal gases are given by

$$\mu_A = \mu_A^\circ + RT \ln (P_A/P^\circ)$$

$$= \mu_A^\circ + RT \ln y_A + RT \ln (P/P^\circ)$$

$$= \mu_A^\circ + RT \ln (1 - \xi) + RT \ln (P/P^\circ)$$

$$\mu_B = \mu_B^\circ + RT \ln y_B + RT \ln (P/P^\circ)$$

$$= \mu_B^\circ + RT \ln \xi + RT \ln (P/P^\circ) \tag{5.23}$$

where P is the total pressure at equilibrium. Substituting these equations into equation 5.22,

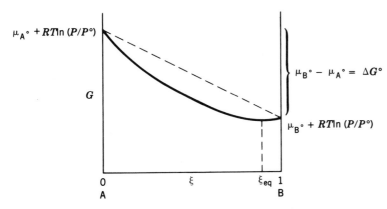

Figure 5.2 Gibbs energy of the reaction system A(g) = B(g) versus the extent of reaction ξ at constant temperature.

$$G = (1 - \xi)\mu_A^\circ + \xi\mu_B^\circ + RT \ln (P/P^\circ)$$
$$+ RT[(1 - \xi) \ln (1 - \xi) + \xi \ln \xi]$$
$$= \mu_A^\circ - (\mu_A^\circ - \mu_B^\circ)\xi$$
$$+ RT \ln (P/P^\circ) + \Delta_{mix}G^\circ \qquad (5.24)$$

where $\Delta_{mix}G^\circ$ is the Gibbs energy of mixing $(1 - \xi)$ mol of A with ξ mol of B. Figure 5.2 gives a plot of G versus ξ. The first three terms in equation 5.24 give the linear function represented by the dashed line. It is the mixing term that causes the minimum in the plot of G versus extent of reaction ξ. At constant temperature and pressure, the criterion of equilibrium is that the Gibbs energy is a minimum (see Section 4.3). Thus, starting with A, the Gibbs energy can decrease along the curve until ξ_{eq} mol of B have been formed. Starting with B the Gibbs energy can decrease until $(1 - \xi_{eq})$ mol of A have formed.

Even though B has the lower value of the standard molar Gibbs energy $(\mu_A^\circ > \mu_B^\circ)$, the system can achieve a lower Gibbs energy by having some A present at equilibrium with the resulting Gibbs energy of mixing. Generalizing from this example we can say that no chemical reaction of gases goes to completion; nevertheless, it may be very difficult to detect reactants at equilibrium if the products have a very much lower Gibbs energy.

5.4 Determination of Equilibrium Constants

If the initial concentrations of the reactants are known and only one reaction occurs, it is necessary to determine the concentration of only one reactant or product at equilibrium to be able to calculate the concentrations or pressures of the others by means of the balanced chemical equation. Chemical methods based on chemical reaction with one of the reactants or products can be used for such analyses only when the reaction being studied can be stopped at equilibrium, as by a very sudden chilling to a temperature where the rate of further chemical change is negligible, or by destruction of a catalyst. Otherwise, the concentrations will shift during the chemical analysis.

Measurements of physical properties, such as density, pressure, light absorption, refractive index, electromotive force, and electrical conductivity, are especially useful for determination of the concentrations of reactants at equilibrium since, for these methods, it is unnecessary to "stop" the reaction.

It is essential to know that equilibrium has been reached before the analysis of the mixture can be used for calculating the equilibrium constant. The following criteria for the attainment of equilibrium at constant temperature are useful:

1. The same value of the equilibrium constant should be obtained when the equilibrium is approached from either side.
2. The same value of the equilibrium constant should be obtained when the concentrations of reacting material are varied over a wide range.

The determination of the density of a partially dissociated gas provides one of the simplest methods for measuring the extent to which the gas is dissociated. When a gas dissociates, more molecules are produced, and at constant temperature and pressure the volume increases.

If the equilibrium extent of reaction ξ has been determined, we need the expression for the equilibrium constant in terms of the equilibrium extent of reaction and the pressure. We will assume that the gases are ideal. The first step is to express the mole fractions of reactants and products, in terms of the extent of reaction ξ.

For example, consider the dissociation of an initial amount $n_0(N_2O_4)$ of N_2O_4 to NO_2 at a given temperature and pressure. If the equilibrium extent of reaction is ξ, the equilibrium amount of N_2O_4 is $n_0(N_2O_4) - \xi$ and the equilibrium amount of NO_2 is 2ξ. However, since the equilibrium composition at a given temperature and pressure is independent of the size of the system, we might as well divide these expressions by $n_0(N_2O_4)$ and represent the equilibrium amounts by the dimensionless quantities $1 - \xi'$ for N_2O_4 and $2\xi'$ for NO_2. The quantity ξ' is a **dimensionless extent** of reaction defined by $\xi' = \xi/n_0(N_2O_4)$. In working equilibrium problems we will leave off the prime to simplify the notation and write $1 - \xi$ rather than $1 \text{ mol} - \xi$. The following format is recommended:

$$N_2O_4(g) = 2NO_2(g)$$

Initial amounts   1   0

Equilibrium amounts $1 - \xi$ 2ξ Total amount $= 1 + \xi$ (5.25)

Equilibrium mole fractions $\dfrac{1 - \xi}{1 + \xi}$ $\dfrac{2\xi}{1 + \xi}$

$$
\begin{aligned}
K &= \frac{(P_{NO_2}/P^\circ)^2}{P_{N_2O_4}/P^\circ} \\[2mm]
&= \frac{\{[2\xi/(1 + \xi)](P/P^\circ)\}^2}{[(1 - \xi)/(1 + \xi)](P/P^\circ)} \\[2mm]
&= \frac{4\xi^2 P/P^\circ}{1 - \xi^2}
\end{aligned}
$$

(5.26)

At equilibrium the amount of $N_2O_4(g)$ is $1 - \xi$ and the amount of $NO_2(g)$ is 2ξ, so that the total amount is $1 + \xi$. The partial pressures of the reactants at equilibrium are obtained by multiplying their equilibrium mole fractions by the total pressure P. Equation 5.26 gives the relation between the equilibrium constant K, the equilibrium extent of reaction ξ, and the total pressure P. This may be solved for the equilibrium extent of reaction to obtain

$$\xi = \frac{1}{[1 + (4/K)(P/P^\circ)]^{1/2}} \qquad (5.27)$$

This is the dimensionless extent of reaction obtained by dividing the extent of reaction ξ by $n_0(N_2O_4)$. Equations 5.26 and 5.27 apply to any dissociation reaction of the type $A(g) = 2B(g)$ (see Fig. 5.3), but not to dissociation reactions of the type $A(g) = B(g) + C(g)$. For reactions of the latter type, the factor 4 in equation 5.27 is replaced by 1.

The equilibrium extent of reaction 5.25 is readily determined by measuring the density of the partially dissociated gas. For example, assume that we start with a mass m of $N_2O_4(g)$. The initial volume is $V_1 = mRT/M_1P$, where M_1 is the molar mass of N_2O_4 (92.01 g mol^{-1}). If the gas is held at a constant pressure and temperature, the equilibrium volume is given by $V_2 = mRT/M_2P$, where M_2 is the average molar mass of the partially dissociated gas, which is defined by $M_2 = M_{N_2O_4}y_{N_2O_4} + M_{NO_2}y_{NO_2}$. Thus, $V_1/V_2 = M_2/M_1$. This ratio is equal to $1/(1 + \xi)$, and so

$$\xi = \frac{M_1 - M_2}{M_2} \qquad (5.28)$$

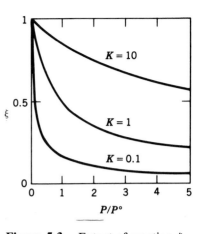

Figure 5.3 Extent of reaction ξ for the reaction $A(g) = 2B(g)$ as a function of P/P°, for various values of K, using equation 5.27.

Example 5.4
If 1.588 g of nitrogen tetroxide gives a total pressure of 1.0133 bar when partially dissociated in a 500-cm^3 glass vessel at 25 °C, what is the extent of reaction? What is the value of K? What is the extent of reaction at a total pressure of 0.5 bar?

$$M_2 = \frac{RT}{P}\frac{m}{V} = \frac{(0.083145 \text{ L bar K}^{-1}\text{ mol}^{-1})(298.1 \text{ K})(1.588 \times 10^{-3} \text{ kg})}{(1.0133 \text{ bar})(0.5 \text{ L})}$$

$$= 77.70 \text{ g mol}^{-1}$$

$$\xi = \frac{92.01 - 77.70}{77.70} = 0.1842$$

$$K = \frac{4\xi^2(P/P^\circ)}{1 - \xi^2} = \frac{(4)(0.1842)^2(1.0133)}{1 - (0.1842)^2} = 0.143$$

The extent of reaction at 0.5 bar is calculated using equation 5.27:

$$\xi = \left[\frac{0.143}{0.143 + 4(0.5)}\right]^{1/2} = 0.258$$

As an example of a more complicated gas reaction, consider that ammonia is produced by holding an initial mixture containing equal amounts of nitrogen and hydrogen at constant temperature and high pressure in contact with a catalyst. Note that reactants are not always added in stoichiometric proportions. In the actual production of ammonia, hydrogen is, of course, the more expensive reactant. Again we use the dimensionless extent of reaction:

$$N_2(g) + 3H_2(g) = 2NH_3(g)$$

Initial amounts   1   1   0

Equilibrium amounts $1 - \xi$ $1 - 3\xi$ 2ξ Total amount $= 2 - 2\xi$

Equilibrium mole fractions $\dfrac{1 - \xi}{2 - 2\xi}$ $\dfrac{1 - 3\xi}{2 - 2\xi}$ $\dfrac{2\xi}{1 - 2\xi}$ (5.29)

Note that when ξ moles of N_2 have been used up, 3ξ moles of H_2 will have reacted, and 2ξ moles of NH_3 will have been formed. Thus, the total amount of reactants and products at equilibrium is $2 - 2\xi$. The equilibrium constant is given by

$$K = \frac{4\xi^2(2 - 2\xi)^2}{(1 - \xi)(1 - 3\xi)^3(P/P°)^2} \tag{5.30}$$

If the balanced chemical equation is divided by 2, the equilibrium constant for it will be the square root of K as expressed by equation 5.30. If the balanced chemical equation is reversed, the equilibrium constant for it will be the reciprocal of K in equation 5.30.

Example 5.5

What total pressure must be used to obtain a 10% conversion of hydrogen to ammonia at 400 °C, assuming an initially equimolar mixture of nitrogen and hydrogen and ideal gas behavior? The equilibrium constant for the formation of $NH_3(g)$ according to reaction 5.29 with a standard state pressure of 1 bar is 1.60×10^{-4} at 400 °C.

We use equation 5.30 to obtain

$$1.60 \times 10^{-4} = \frac{4(0.1)^2(1.8)^2}{(0.9)(0.7)^3(P/P°)^2}$$

$$\frac{P}{P°} = 51.2$$

$$P = 51.2 \text{ bar}$$

Example 5.6

(a) What is the value of the standard Gibbs energy for reaction 5.29 at 400 °C? (b) What is the value of the equilibrium constant and the standard reaction Gibbs energy when reaction 5.29 is divided by two? (c) What is the value of the equilibrium constant and the standard reaction Gibbs energy when reaction 5.29 is reversed?

(a) $\Delta_r G° = -RT \ln K$

$= -(8.315 \text{ J K}^{-1} \text{ mol}^{-1})(673 \text{ K}) \ln (1.60 \times 10^{-4})$

$= 48.91 \text{ kJ mol}^{-1}$

(b) $K = (1.60 \times 10^{-4})^{1/2} = 0.01265$

   $\Delta_r G° = -RT \ln 0.01265$

        $= 24.46 \text{ kJ mol}^{-1}$

(c) $K = \dfrac{1}{(1.60 \times 10^{-4})} = 6250$

   $\Delta_r G° = -RT \ln 6250$

        $= -48.91 \text{ kJ mol}^{-1}$

5.5 Use of Standard Gibbs Energies of Formation to Calculate Equilibrium Constants

There are three ways that $\Delta_r G°$ for a reaction may be obtained: (1) $\Delta_r G°$ may be calculated from a measured equilibrium constant using equation 5.12, (2) $\Delta_r G°$ may be calculated from

$$\Delta_r G° = \Delta_r H° - T \Delta_r S° \tag{5.31}$$

using $\Delta_r H°$ obtained calorimetrically and $\Delta_r S°$ obtained from the third law entropies, and (3) for gas reactions $\Delta_r G°$ may be calculated using statistical mechanics (Chapter 17) and certain information about molecules obtained from spectroscopic data. Methods 2 and 3 make it possible to calculate equilibrium constants of reactions that have never been studied in the laboratory. In method 2 the necessary data are obtained solely from thermal measurements, including heat capacity measurements down to the neighborhood of absolute zero. The calculation of equilibrium constants using statistical mechanics is even more remarkable in that only properties of the individual molecules are used to calculate equilibrium constants for reactions of ideal gases. For the simplest reactions $\Delta_r H°$ may be calculated from spectroscopic data; but for more complicated reactions, the calculation of $\Delta_r H°$ requires calorimetric data.

   Rather than tabulating values of equilibrium constants of reactions, or of $\Delta_r G°$ values calculated using equation 5.12, it is more convenient to tabulate values of **standard Gibbs energies of formation** $\Delta_f G°$, which are the standard Gibbs energies for the formation of a mole of i from its elements. The standard Gibbs energy of formation of i is related to the standard enthalpy of formation of i and the standard entropy of i by

$$\Delta_f G_i° = \Delta_f H_i° - T\left(\overline{S}_i° - \sum v_e \overline{S}_e°\right) \tag{5.32}$$

where $\sum v_e \overline{S}_e°$ is the sum of the standard entropies of the elements in the formation reaction for species i. The change in standard Gibbs energy for a reaction is calculated from the standard Gibbs energies of formation $\Delta_f G_i°$ using

$$\Delta_r G° = \sum_{i=1}^{N} v_i \Delta_f G_i° \tag{5.33}$$

where v_i's are the stoichiometric numbers from the balanced chemical equation. The standard Gibbs energies of formation of elements in their reference states are zero at all temperatures.

Standard Gibbs energies of formation at 298.15 K and 1 bar are provided for about 15 000 species in The NBS Tables of Chemical Thermodynamic Properties.* Some values for this table are given in Appendix C.1. Standard Gibbs energies of formation at temperatures up to 6000 K are given in the JANAF Tables.* Some values from this table are given in Appendix C.2. Gibbs energies of formation of several hundred organic compounds up to 1000 K are given in Stull et al.† The values in this latter table need to be converted to joules and to a standard state pressure of 1 bar, before being used with values in the other tables.‡

Example 5.7

Given the following calorimetric information, calculate the standard Gibbs energy of formation of $H_2O(g)$ at 298.15 K.

	$\Delta_f H°/kJ\ mol^{-1}$	$\overline{S}°/J\ K^{-1}\ mol^{-1}$
$H_2O(g)$	-241.818	188.825
$H_2(g)$	0	130.684
$O_2(g)$	0	205.138

The standard Gibbs energy of formation is the standard reaction Gibbs energy for the following reaction.

$$H_2(g) + \tfrac{1}{2}O_2(g) = H_2O(g)$$

$$\Delta_f G°(H_2O, g) = \Delta_f H°(H_2O, g) - T\,\Delta_f S°(H_2O, g)$$

$$= -241,818 - (298.15)[188.825 - 130.684 - (0.5)(205.138)]$$

$$= -228.582\ kJ\ mol^{-1}$$

Example 5.8

Calculate the equilibrium constants for the following reactions at the indicated temperature.

(a) $3O_2(g) = 2O_3(g)$ at 25 °C

(b) $CO(g) + 2H_2(g) = CH_3OH(g)$ at 500 K

(a)   Use Appendix C.1 to obtain

---

* D. D. Wagman et al., The NBS tables of chemical thermodynamic properties. *J. Phys. Chem. Ref. Data* **11** (Suppl. 2) (1982).

* M. W. Chase et al., JANAF thermochemical tables. *J. Phys. Chem. Ref. Data* **14** (Suppl. 1) (1985).

† D. R. Stull, E. F. Westrum, and G. C. Sinke, *The Chemical Thermodynamics of Organic Compounds.* New York: Wiley, 1969.

‡ R. D. Freeman, *J. Chem. Educ.* **62**:681 (1985).

$$\Delta_r G° = 2\Delta_f G°(O_3, g) - 3\Delta_f G°(O_2, g)$$
$$= 2(163.2) = 326.4 \text{ kJ mol}^{-1}$$
$$K = \exp\left(-\frac{\Delta_r G°}{RT}\right)$$
$$= \exp\left[-\frac{326\,400}{(8.3145)(298.15)}\right]$$
$$= 6.62 \times 10^{-58}$$

(*b*) Use Appendix C.2 to obtain

$$\Delta_r G° = -134.27 - (-155.44) = 21.17 \text{ kJ mol}^{-1}$$
$$K = \exp\left[-\frac{21\,170}{(8.3145)(500)}\right]$$
$$= 6.14 \times 10^{-3}$$

Example 5.9
What is the equilibrium partial pressure of NO in air at 1000 K?

$$\tfrac{1}{2}N_2(g) + \tfrac{1}{2}O_2(g) = NO(g)$$

The value of K is calculated using Appendix C.2. $\Delta_r G° = 77.772$ kJ mol^{-1}.

$$K = \exp\left(-\frac{77\,772}{8.3145 \times 1000}\right)$$
$$= 8.663 \times 10^{-5}$$
$$= \frac{(P_{NO}/P°)}{(P_{N_2}/P°)^{1/2}(P_{O_2}/P°)^{1/2}} = \frac{(P_{NO}/P°)}{(0.80)^{1/2}(0.20)^{1/2}}$$
$$P_{NO} = 3.465 \times 10^{-5} \text{ bar}$$

Example 5.10
By referring to Example 2.14 and Example 3.9, estimate the standard Gibbs energies of formation of *n*-pentane, isopentane, and neopentane at 298.15 K. What are the equilibrium mole fractions within the isomer group at this temperature? The standard entropies of graphite and molecular hydrogen at this temperature are 5.74 and 130.68 J K^{-1} mol^{-1}, respectively.
For *n*-pentane,

$$\Delta_f G° = -147.25 - \frac{(298.15)(354.4 - 5 \times 5.74 - 6 \times 130.68)}{1000}$$
$$= -10.6 \text{ kJ mol}^{-1} \quad (\text{experimental}, -8.33 \text{ kJ mol}^{-1})$$

For isopentane,

$$\Delta_f G° = -156.62 - \frac{(298.15)(343.2 - 5 \times 5.74 - 6 \times 130.68)}{1000}$$
$$= -16.6 \text{ kJ mol}^{-1} \quad (\text{experimental}, -13.27 \text{ kJ mol}^{-1})$$

For neopentane,

$$\Delta_f G^\circ = -168.6 - \frac{(298.15)(304.9 - 5 \times 5.74 - 6 \times 130.68)}{1000}$$

$$= -17.2 \text{ kJ mol}^{-1} \quad (\text{experimental}, -17.37 \text{ kJ mol}^{-1})$$

The equilibrium mole fraction within the group of isomers is equal to the equilibrium pressure fraction. The equilibrium constants for the formation reactions of the pentanes are each given by $K_i = P_i/P_{H_2}^6 = \exp(-\Delta_f G^\circ/RT)$ so that using the experimental values we obtain

$$y_n = \frac{P_n}{P_n + P_{\text{iso}} + P_{\text{neo}}} = \frac{\exp(-\Delta_f G_n^\circ/RT)}{\Sigma_i \exp(-\Delta_f G_i^\circ/RT)}$$

$$= \frac{28.8}{1345} = 0.021$$

$$y_{\text{iso}} = 0.157$$

$$y_{\text{neo}} = 0.822$$

5.6   Effect of Pressure on Gas Reactions

For an ideal mixture of ideal gases the equilibrium partial pressures of the reactants and products can be expressed in terms of their equilibrium mole fractions y_i and the total pressure P of reactants and products:

$$K = \prod_i \left(\frac{y_i P}{P^\circ}\right)^{\nu_i} = \prod_i y_i^{\nu_i} \prod_i \left(\frac{P}{P^\circ}\right)^{\nu_i} = \left(\frac{P}{P^\circ}\right)^{\nu} K_y \qquad (5.34)$$

In this equation $\nu = \sum_i \nu_i$, and K_y is the equilibrium constant written in terms of mole fractions at a particular total pressure.

The value of K_y is a function of temperature only at a constant total pressure P:

$$K_y = \prod_{i=1}^{N} y_i^{\nu_i} = \left(\frac{P}{P^\circ}\right)^{-\nu} K \qquad (5.35)$$

The equilibrium constant written in terms of mole fractions depends on the pressure as well as the temperature, but, as we will see, it is very useful in calculating the equilibrium extent of reaction because it is written in terms of amounts. If the amount of gaseous products is equal to the amounts of gaseous reactants, then $\nu = \sum \nu_i = 0$, $K_y = K$, and changing the total pressure of reactants does not affect the equilibrium mole fractions of the reactants and products. If a reaction causes an increase in the number of molecules, then $\nu > 0$, and K_y decreases as the pressure is increased at constant temperature. Thus, raising the pressure decreases the equilibrium mole fractions of the products and increases the equilibrium mole fractions of the reactants; in short, raising the pressure pushes the reaction backward.

Le Châtelier's principle provides a quick way to check conclusions like this about the effects of changes in independent variables on chemical equilibrium.

According to Le Châtelier, when an independent variable of a system at equilibrium is changed, the equilibrium shifts in the direction that tends to reduce the effect of the change. When pressure is increased, the equilibrium shifts in the direction to reduce the number of molecules.

If a reaction involves only solids and liquids, the effect of pressure on the equilibrium is small. If pure solids or liquids are involved in a reaction along with gases, their stoichiometric numbers do not appear in the expression for the equilibrium constant since their activities are taken as unity.

Example 5.11

In example 5.5, we saw that when an initially equimolar mixture of nitrogen and hydrogen is placed in contact with an ammonia catalyst at 400 °C and 51.2 bar, there is a 10% conversion to ammonia at equilibrium. What pressure is required to obtain a 15% conversion?

Using equation 5.30 we obtain

$$1.60 \times 10^{-4} = \frac{4(0.15)^2(1.7)^2}{(0.85)(0.55)^3(P/P^\circ)^2}$$

$$\frac{P}{P^\circ} = 107.2$$

$$P = 107.2 \text{ bar}$$

According to Le Châteliers principle, raising the total pressure will cause the reaction to shift in the direction of the product NH_3 because there is a decrease in moles of gas in the forward reaction.

5.7 Effect of Initial Composition

To discuss the effect of initial composition on the equilibrium composition for a reaction we will use the equilibrium constant expressed in terms of mole fractions. At any time during a reaction the amounts of each reactant and product may be expressed in terms of its initial amount n_{i0} and the extent of reaction ξ:

$$K_y = \prod_i \left(\frac{n_{i0} + \nu_i \xi}{n_0 + \nu \xi} \right)^{\nu_i} = \left(\frac{1}{n_0 + \nu \xi} \right)^{\nu} \prod_i (n_{i0} + \nu_i \xi)^{\nu_i} \qquad (5.36)$$

In this equilibrium expression the amount of gas is represented by

$$\sum n_i = \sum (n_{i0} + \nu_i \xi) = \sum n_{i0} + \xi \sum \nu_i = n_0 + \xi \nu \qquad (5.37)$$

where the initial amount of gaseous reactants and products is represented by n_0 and $\nu = \sum \nu_i$. Thus, the calculation of the amounts of reactants and products at equilibrium from the initial composition and the value of K_y simply comes down to the solution of a polynomial in ξ. The polynomials arising here always have one positive real root. Quadratic equations are readily solved, and higher-order polynomials can be solved by iterative methods.

5.8 Effect of Inert Gases

If an inert gas is added to an equilibrium mixture of gases at constant temperature and volume, there is no effect on the equilibrium. But adding an inert gas at constant temperature and pressure has the same effect as lowering the pressure. When inert gases are present, equation 5.36 has to be modified to include the number of moles of inert gas in the denominator of each mole fraction; thus $n_0 + \nu\xi$ becomes $n_0 + \nu\xi + n_{\text{inerts}}$. Substituting the modified form of equation 5.36 in equation 5.34 yields

$$K = \left(\frac{P/P^\circ}{n_0 + \nu\xi + n_{\text{inerts}}}\right)^\nu \prod_{i=1}^{N} (n_{i0} + \nu_i\xi)^{\nu_i} \qquad (5.38)$$

If $\nu < 0$, the addition of an inert gas at constant pressure reduces the sum of the partial pressures of the reactants and products, and the reaction shifts to the left to compensate for this. Equation 5.38 applies generally to ideal mixtures of ideal gases, but if the partial pressure of the inert gases is known, this partial pressure may be subtracted from the total pressure, and equation 5.34 may be used with P equal to the sum of the partial pressures of the reactants and products.

Example 5.12

As an illustration of the effect of initial composition, pressure, and the addition of an inert gas, consider the equilibrium for the production of methanol from CO and H_2:

$$CO(g) + 2H_2(g) = CH_3OH(g)$$

The value of K at 500 K is 6.23×10^{-3}. (a) A gas stream containing equimolar amounts of CO and H_2 is passed over a catalyst at 1 bar. What is the extent of reaction at equilibrium?

	CO	H_2	CH_3OH	
Initial amounts	1	1	0	
Equilibrium amounts	$1 - \xi$	$1 - 2\xi$	ξ	Total amount $= 2(1 - \xi)$
Mole fraction	$\dfrac{1}{2}$	$\dfrac{1 - 2\xi}{2(1 - \xi)}$	$\dfrac{\xi}{2(1 - \xi)}$	

$$K = \frac{4\xi(1 - \xi)}{(1 - 2\xi)^2} = 6.23 \times 10^{-3}$$

Solving this quadratic indicates that $\xi = 0.001\,55$ so that $y_{CO} = 0.5000$, $y_{H_2} = 0.4992$, and $y_{CH_3OH} = 0.0008$.

(b) To attain more complete reaction the pressure is raised to 100 bar and 2 mol of hydrogen is used per mole of CO. What is the equilibrium extent of reaction?

	CO	H_2	CH_3OH	
Initial amounts	1	2	0	
Equilibrium amounts	$1 - \xi$	$2 - \xi$	ξ	Total amount $= 3 - 2\xi$
Mole fraction	$\dfrac{1 - \xi}{3 - 2\xi}$	$\dfrac{2 - 2\xi}{3 - 2\xi}$	$\dfrac{\xi}{3 - 2\xi}$	

$$K = \frac{\xi(3 - 2\xi)^2}{(1 - \xi)(2 - 2\xi)^2(100)^2} = 6.23 \times 10^{-3}$$

Solution of this equation for the extent of reaction by successive approximation yields $\xi = 0.817$ so that $y_{CO} = 0.134$, $y_{H_2} = 0.268$, and $y_{CH_3OH} = 0.598$. (You might check that this gives the right value for the equilibrium constant.)

(c) If the reactant gases contain a mole of nitrogen in addition to 1 mol of CO and 2 mol of hydrogen, what is the equilibrium extent of reaction at 100 bar? The first two lines of the table in (b) are unchanged, except that the total number of moles is now $4 - 2\xi$.

$$K = \frac{\xi(4 - 2\xi)^2}{(1 - \xi)(2 - 2\xi)^2(100)^2} = 6.23 \times 10^{-3}$$

Solution of this equation yields $\xi = 0.735$ so that $y_{CO} = 0.105$, $y_{H_2} = 0.210$, $y_{CH_3OH} = 0.291$, and $y_{N_2} = 0.395$. Here the presence of an inert gas reduces the equilibrium conversion to product, but for a reaction for which $\nu = \sum \nu_i$ is positive, addition of an inert gas will cause the reaction to go further to the right.

5.9 Effect of Temperature on the Equilibrium Constant

The effect of temperature on chemical equilibrium is determined by $\Delta_r H°$, as shown by the Gibbs–Helmholtz equation 4.59. If $\Delta_r G° = -RT \ln K$ is substituted into this equation, we obtain

$$\left[\frac{\partial \ln K}{\partial (1/T)} \right]_P = -\frac{\Delta_r H°}{R} \tag{5.39}$$

or

$$\left(\frac{\partial \ln K}{\partial T} \right)_P = \frac{\Delta_r H°}{RT^2} \tag{5.40}$$

Thus, for an endothermic reaction the equilibrium constant increases as the temperature is increased, and for an exothermic reaction the equilibrium constant decreases as the temperature is increased.

According to Le Châtelier's principle, when an equilibrium system is perturbed, the equilibrium will always be displaced in such a way as to oppose the applied change. When the temperature of an equilibrium system is raised, this change cannot be prevented by the system, but what happens is that the equilibrium shifts in such a way that more heat is required to heat the reaction mixture to the higher temperature than would have been required if the mixture were inert. In other words, when the temperature is raised, the equilibrium shifts in the direction that causes an absorption of heat.

If $\Delta_r H°$ is independent of temperature, the integral of equation 5.40 from T_1 to T_2 yields

$$\ln \frac{K_2}{K_1} = \frac{\Delta_r H°(T_2 - T_1)}{RT_1 T_2} \tag{5.41}$$

If $\Delta_r H°$ is independent of temperature, then $\Delta_r C_P°$ is zero. If $\Delta_r C_P°$ is zero, then equation 3.73 shows that $\Delta_r S°$ is also independent of temperature. Thus,

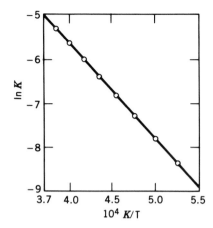

Figure 5.4   A plot of $\ln K$ against reciprocal absolute temperature for the reaction $N_2(g) + O_2(g) = 2NO(g)$. The standard enthalpy of reaction is calculated from the slope of the straight line.

when $\Delta_r C_P^\circ = 0$, the temperature dependence of the equilibrium constant is given by $\Delta_r G^\circ = -RT \ln K = \Delta_r H^\circ - T \Delta_r S^\circ$ or

$$\ln K = -\frac{\Delta_r H^\circ}{RT} + \frac{\Delta_r S^\circ}{R} \tag{5.42}$$

According to this equation a plot of $\ln K$ versus $1/T$ is linear over a temperature range in which $\Delta_r H^\circ$ and $\Delta_r S^\circ$ for the reaction are constant.

Example 5.13
Calculate $\Delta_r H^\circ$ and $\Delta_r S^\circ$ for the reaction

$$N_2(g) + O_2(g) = 2NO(g)$$

from the following values of K:

T/K	1900	2000	2100	2200	2300	2400	2500	2600
$K/10^{-4}$	2.31	4.08	6.86	11.0	16.9	25.1	36.0	50.3

These data are plotted in Fig. 5.4. Since the plot is linear we can use equation 5.42. The slope of the plot is -2.19×10^4 K and so

$$\Delta_r H^\circ = -\text{slope} \times R = -(-2.19 \times 10^4)(8.3145 \text{ J K}^{-1} \text{ mol}^{-1})$$
$$= 182 \text{ kJ mol}^{-1}$$

The intercept of Fig. 5.2 at $1/T = 0$ can be calculated from the experimental value of K at some temperature and the slope. The intercept may be used to calculate the standard entropy change $\Delta_r S^\circ$ according to equation 5.42:

$$\frac{\Delta_r S^\circ}{R} = 3.13$$

$$\Delta_r S^\circ = (3.13)(8.3145 \text{ J K}^{-1} \text{ mol}^{-1}) = 26.0 \text{ J K}^{-1} \text{ mol}^{-1}$$

If $\Delta_r H^\circ$ and $\Delta_r S^\circ$ depend on the temperature, the simplest approximation is to assume that $\Delta_r C_P^\circ$ is constant. According to equations 2.91 and 3.47

$$\Delta_r H_T^\circ = \Delta_r H_{298}^\circ + \Delta_r C_P^\circ (T - 298.15 \text{ K}) \tag{5.43}$$

$$\Delta_r S_T^\circ = \Delta_r S_{298}^\circ + \Delta_r C_P^\circ \ln \frac{T}{298.15 \text{ K}} \tag{5.44}$$

Substituting these relations in $-RT \ln K = \Delta_r H_T^\circ - T \Delta_r S_T^\circ$ yields

$$\ln K = -\frac{\Delta_r H_{298}^\circ}{RT} + \frac{\Delta_r S_{298}^\circ}{R} - \frac{\Delta_r C_P^\circ}{R}\left(1 - \frac{298.15 \text{ K}}{T} - \ln \frac{T}{298.15 \text{ K}}\right) \tag{5.45}$$

If $\Delta_r C_P^\circ$ is not independent of temperature, then its dependence on temperature can often be represented by the first several terms in a power series in temperature, as shown in equation 2.93. Then, according to equation 2.94 and Example 3.3,

$$\Delta_r H_T^\circ = \Delta_r H_0^\circ + \Delta_r \alpha T + \frac{\Delta_r \beta}{2} T^2 + \frac{\Delta_r \gamma}{3} T^3 \tag{5.46}$$

$$\Delta_r S_T^\circ = \Delta_r S_0^\circ + \Delta_r \alpha \ln T + \Delta_r \beta T + \frac{\Delta_r \gamma}{2} T^2 \tag{5.47}$$

Substituting these relations in $-RT \ln K = \Delta_r H_T^\circ - T \Delta_r S_T^\circ$ yields

$$\ln K = -\frac{\Delta_r H_0^\circ}{RT} + \frac{(\Delta_r S_0^\circ - \Delta_r \alpha)}{R} + \frac{\Delta_r \beta T}{2R} + \frac{\Delta_r \gamma T^2}{6R} + \frac{\Delta_r \alpha}{R} \ln T \tag{5.48}$$

where $\Delta_r \alpha = \sum \nu_i \alpha_i$, $\Delta_r \beta = \sum \nu_i \beta_i$, and $\Delta_r \gamma = \sum \nu_i \gamma_i$. If $\Delta_r \alpha$, $\Delta_r \beta$, and $\Delta_r \gamma$ are known, then $\Delta_r H_0^\circ$ can be calculated using equation 5.46 and $\Delta_r S_0^\circ$ can be calculated using equation 5.48 if the equilibrium constant is known at one temperature.

The effect of temperature on an endothermic reaction A(g) = 2B(g) is shown in Fig. 5.5 for three total pressures. As the temperature is increased, the equilibrium extent of reaction increases. At a given temperature the extent of reaction is greater at a lower pressure.

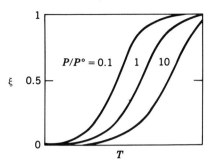

Figure 5.5 Extent of reaction ξ versus temperature for an endothermic reaction A(g) = 2B(g) at three total pressures.

5.10 Equilibrium Constants for Gas Reactions Written in Terms of Concentrations

Since thermodynamic tables for gases are based on a standard state pressure of 1 bar for the ideal gas, the $\Delta_f G_i^\circ$ values lead directly to equilibrium constants in terms of pressure (or fugacities). However, in connection with chemical kinetics it is useful to express equilibrium constants of gas reactions in terms of concentrations, since rate equations are written in terms of concentrations (Section 19.2). These two types of equilibrium constants will be represented by K_P and K_c. To obtain a general expression for the equilibrium constant K_c in terms of concentrations for ideal gases we replace P_i in equation 5.17 with $P_i = n_i RT/V = c_i RT$:

$$K_P = \prod_{i=1}^{N} \left(\frac{P_i}{P^\circ}\right)^{\nu_i} = \prod_i \left(\frac{c_i RT}{P^\circ}\right)^{\nu_i} \tag{5.49}$$

To define a dimensionless equilibrium constant in terms of concentration, we introduce the standard concentration c°, which represents one mole per liter. Introducing this standard concentration into each term of equation 5.49 yields

$$K_P = \prod_{i=1}^{N} \left[\left(\frac{c_i}{c^\circ}\right)\left(\frac{c^\circ RT}{P^\circ}\right)\right]^{\nu_i} = \left(\frac{c^\circ RT}{P^\circ}\right)^{\Sigma \nu_i} \prod_i \left(\frac{c_i}{c^\circ}\right)^{\nu_i}$$

$$= \left(\frac{c^\circ RT}{P^\circ}\right)^{\Sigma \nu_i} K_c \tag{5.50}$$

where the **equilibrium constant expressed in terms of concentration**

$$K_c = \prod_i \left(\frac{c_i}{c^\circ}\right)^{\nu_i} \tag{5.51}$$

is a function only of temperature for a mixture of ideal gases. If $c^\circ = 1$ mol L^{-1} and $P^\circ = 1$ bar, then $c^\circ RT/P^\circ = 24.79$ at 298.15 K.

Example 5.14

What is the value of the equilibrium constant K_c for the dissociation of ethane into methyl radicals at 1000 K?

$$C_2H_6(g) = 2CH_3(g)$$

$$\Delta G° = 2 \Delta_f G°(CH_3) - \Delta_f G°(C_2H_6)$$

$$= 2(159.82) - 109.84 = 209.80 \text{ kJ mol}^{-1}$$

$$K_P = \exp\left(-\frac{\Delta_r G°}{RT}\right)$$

$$= \exp \frac{(-209.80)}{(8.3145 \times 10^{-3})(1000)}$$

$$= 1.102 \times 10^{-11}$$

$$K_c = \frac{([CH_3]/c°)^2}{[C_2H_6]/c°} = K_P \frac{P°}{c°RT}$$

$$= (1.102 \times 10^{-11}) \frac{(1 \text{ bar})}{(1 \text{ mol L}^{-1})(0.083145)(1000 \text{ K})}$$

$$= 1.325 \times 10^{-13}$$

Thus, at equilibrium $[CH_3]^2/[C_2H_6] = 1.325 \times 10^{-13}$ mol L^{-1}, where the brackets indicate concentrations in moles per liter.

5.11   Heterogeneous Chemical Reactions

Chemical reactions may involve reactants in different phases. The following are examples of reactions that involve only pure solids and the gas phase:

$$C(graphite) + H_2O(g) = CO(g) + H_2(g) \tag{5.52}$$

$$CaCO_3(s) = CaO(s) + CO_2(g) \tag{5.53}$$

Here we will discuss only heterogeneous reactions, like the above, which involve pure solid phases. Equilibrium constants for reactions like 5.52 and 5.53 **can be written without terms for the pure solid phases, provided they are present.** The reason for doing this is that the activities of the pure solid phases are very nearly equal to unity for moderate pressures (see Section 5.1). If the gases are ideal then the equilibrium constant expressions for reactions 5.52 and 5.53 are

$$K = \frac{(P_{CO}/P°)(P_{H_2}/P°)}{P_{H_2O}/P°} \tag{5.54}$$

$$K = \frac{P_{CO_2}}{P°} \tag{5.55}$$

There is chemical equilibrium only if all substances in the chemical equation are present in the phases shown. The equilibrium constants of such reactions are independent of the amount of pure solid (or liquid) phase. In contrast with reactions in a single phase, such reactions may go to completion if a solid phase is used up. As long as carbon is present, the first reaction behaves like a gas

Table 5.1 Pressures of $CO_2(g)$ in Equilibrium with $CaCO_3(s)$ and $CaO(s)$

$t/°C$	500	600	700	800
$P_{CO_2}/P°$	9.2×10^{-5}	2.39×10^{-3}	2.88×10^{-2}	0.2217
$t/°C$	897	1000	1100	1200
$P_{CO_2}/P°$	0.987	3.820	11.35	28.31

reaction; no matter what the initial conditions are, the reaction does not go to completion because of the entropy of mixing in the gas phase. The second reaction goes to the right or to the left until the pressure of $CO_2(g)$ is precisely equal to K. Table 5.1 gives the pressure of $CO_2(g)$ in equilibrium with $CaO(s)$ and $CaCO_3(s)$ at a series of temperatures. The natural logarithm of the pressures are plotted versus $1/T$ in Fig. 5.6. The three phases are at equilibrium only along the line. If P_{CO_2}, T are in the area below the line, then $CaCO_3(s)$ dissociates completely to $CaO(s)$ and $CO_2(g)$. If P_{CO_2}, T are in the area above the line, then $CaO(s)$ reacts completely to form $CaCO_3(s)$ and the system becomes $CaCO_3(s)$ and $CO_2(g)$.

If solid or liquid solutions are formed (e.g., if CaO and $CaCO_3$ were somewhat mutually soluble), the position of the equilibrium would depend on the concentration or, more precisely, the activities of the components in the equilibrium solid solution.

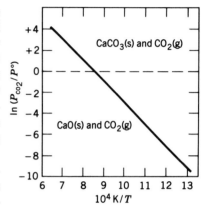

Figure 5.6 Dissociation pressures of $CaCO_3(s)$. All three phases are present only at P_{CO_2}, T along the line.

Example 5.15

Calculate $\Delta_r G°$, $\Delta_r H°$, and $\Delta_r S°$ for reaction 5.53 at 1000 K using data in (*a*) Table 5.1 and (*b*) Appendix C.1 with the assumption that $\Delta_r C_P°$ is independent of temperature.

(*a*) $\Delta_r G° = -RT \ln K = -(8.3145 \text{ J K}^{-1} \text{ mol}^{-1})(1000 \text{ K})(-3.00) = 24.9 \text{ kJ mol}^{-1}$

  $\Delta_r H° = -R(\text{slope}) = -(8.3145 \text{ J K}^{-1} \text{ mol}^{-1})(-2.055 \times 10^4 \text{ K})$

  $= 171 \text{ kJ mol}^{-1}$

  $\Delta_r S° = \dfrac{\Delta H° - \Delta G°}{T}$

  $= \dfrac{(171 - 24.9) \times 10^3 \text{ J mol}^{-1}}{1000 \text{ K}}$

  $= 146 \text{ J K}^{-1} \text{ mol}^{-1}$

(*b*) According to Appendix C.1 the values of 298 K are $\Delta_r G° = 130.40 \text{ kJ mol}^{-1}$, $\Delta_r H° = 178.32 \text{ kJ mol}^{-1}$, and $\Delta_r S° = 160.59 \text{ J K}^{-1} \text{ mol}^{-1}$ when calcite is the reactant.

  $\Delta_r C_P = \overline{C}_P°(\text{CaO}) + \overline{C}_P°(CO_2) - \overline{C}_P°(CaCO_3)$

  $= 42.80 + 37.11 - 81.88 = -1.97 \text{ J K}^{-1} \text{ mol}^{-1}$

Using equations 5.43–5.45,

  $\Delta_r H°_{1000} = 178.32 \text{ kJ mol}^{-1} - (1.97 \times 10^{-3} \text{ kJ K}^{-1} \text{ mol}^{-1})(701.85 \text{ K})$

  $= 176.94 \text{ kJ mol}^{-1}$

$$\Delta_r S^{\circ}_{1000} = 160.59 \text{ J K}^{-1} \text{ mol}^{-1} - (1.97 \text{ J K}^{-1} \text{ mol}^{-1}) \ln (1000/298.15)$$

$$= 158.21 \text{ J K}^{-1} \text{ mol}^{-1}$$

$$\Delta_r G^{\circ}_{1000} = \Delta H^{\circ}_{1000} - T\Delta S^{\circ}_{1000}$$

$$= 176.94 \text{ kJ mol}^{-1} - (1000 \text{ K})(158.21 \times 10^{-3} \text{ kJ K}^{-1} \text{ mol}^{-1})$$

$$= 18.73 \text{ kJ mol}^{-1}$$

Example 5.16
At 1393 K

$$Fe_2O_3(s) + 3CO(g) = 2Fe(s) + 3CO_2(g) \qquad K = 0.0467$$

$$2CO_2(g) = 2CO(g) + O_2(g) \qquad K = 1.4 \times 10^{-12}$$

What is the equilibrium pressure of $O_2(g)$ in a vessel containing $Fe_2O_3(s)$ and $Fe(s)$ at equilibrium at 1393 K? (The standard state pressure $P^{\circ} = 1$ bar.)
   Before adding these reactions, we multiply the first by two (and square the equilibrium constant) and multiply the second by three (and cube the equilibrium constant) to remove CO and CO_2 from the resulting equation:

$$2Fe_2O_3(s) + 6CO(g) = 4Fe(s) + 6CO_2(g) \qquad K = 2.19 \times 10^{-3}$$

$$6CO_2(g) = 6CO(g) + 3O_2(g) \qquad K = 2.74 \times 10^{-36}$$

Adding yields

$$2Fe_2O_2(s) = 4Fe(s) + 3O_2(g) \qquad K = (2.19 \times 10^{-3})(2.74 \times 10^{-36})$$

$$= 6.01 \times 10^{-39}$$

$$= \left(\frac{P_{O_2}}{P^{\circ}}\right)^3$$

$$P_{O_2} = 1.82 \times 10^{-13} \text{ bar}$$

Note that when the reactions are added, their equilibrium constants are multiplied.

5.12 Chemical Equilibrium in Nonideal Gas Mixtures

Certain industrial processes, such as the production of methanol and ammonia, are carried out at sufficiently high pressures that equilibrium calculations assuming ideal gases are in serious error. For nonideal gases the equilibrium constant is expressed by

$$K = \prod_i \left(\frac{f_i}{P^{\circ}}\right)^{\nu_i} \tag{5.56}$$

where f_i is the fugacity of component i in the mixture at equilibrium and P° is the standard-state pressure. The fugacity of i may be expressed by

$$f_i = \phi_i y_i P \tag{5.57}$$

where ϕ_i is the fugacity coefficient of i in the mixture and P is the total pressure. Substituting this into equation 5.56 yields

$$K = \left(\frac{P}{P^{\circ}}\right)^{\Sigma \nu_i} \prod_i (\phi_i y_i)^{\nu_i} \qquad (5.58)$$

$$= \left(\frac{P}{P^{\circ}}\right)^{\Sigma \nu_i} \prod_i \phi_i^{\nu_i} \prod_i y_i^{\nu_i} \qquad (5.59)$$

In these equations, K depends only on temperature, but the fugacity coefficients ϕ_i in the mixture depend on temperature, pressure, and composition.

Lewis and Randall suggested that in some cases it may be a good approximation to assume that the fugacity coefficient of i is equal to the fugacity of pure i at the pressure of the equilibrium; in other words it is assumed that the real gases form an ideal mixture. The fugacity coefficient can be calculated using equation 4.74 and used in equation 5.59 to obtain an approximate value of K.

5.13 Special Topic: Independent Reactions and Chemical Stoichiometry

So far, we have talked about systems in which there is a single chemical reaction. However, in many practical applications of chemical thermodynamics a number of chemical reactions occur simultaneously. The equilibrium composition can be calculated by using a set of independent chemical reactions that will represent all possible chemical changes in the system. A set of reactions is independent if no member of the set may be obtained by adding and subtracting other members of the set. Different sets of reactions may be chosen to describe a given system, but the *number* of independent reactions is always the same. For a simple system a set of independent reactions may be found by inspection, but for more complex systems, a systematic procedure is needed. The procedure* is based on the fact that the atoms of each element have to be conserved in any chemical or physical change in the system.

Let us consider a closed chemical system containing the five gases: $1 = CO$, $2 = H_2$, $3 = CO_2$, $4 = H_2O$, and $5 = CH_4$. This system is important because it is the first step toward representing the Fischer–Tropsch process in which a variety of organic substances are produced from CO and H_2. In the Fischer–Tropsch process, a mixture of carbon monoxide and molecular hydrogen is passed over a catalyst to obtain hydrocarbons. To conserve the elements involved, the amounts n_i of the five species must at all times satisfy the following set of equations.

$$1n_1 + 0n_2 + 1n_3 + 0n_4 + 1n_5 = b_C \qquad (5.60)$$

$$0n_1 + 2n_2 + 0n_3 + 2n_4 + 4n_5 = b_H \qquad (5.61)$$

$$1n_1 + 0n_2 + 2n_3 + 1n_4 + 0n_5 = b_O \qquad (5.62)$$

---

* W. R. Smith and R. W. Missen, *Chemical Reaction Equilibrium Analysis*. New York: Wiley, 1982.

where b_C, b_H, and b_O are the total amounts of the indicated elements in the system. If the system changes from an initial state with n_1^*, n_2^*, n_3^*, n_4^*, and n_5^* to a final state with n_1, n_2, n_3, n_4 and n_5, the difference between the element-balance equations for the final and initial states is given by

$$1 \, \Delta n_1 + 0 \, \Delta n_2 + 1 \, \Delta n_3 + 0 \, \Delta n_4 + 1 \, \Delta n_5 = 0 \qquad (5.63)$$

$$0 \, \Delta n_1 + 2 \, \Delta n_2 + 0 \, \Delta n_3 + 2 \, \Delta n_4 + 4 \, \Delta n_5 = 0 \qquad (5.64)$$

$$1 \, \Delta n_1 + 0 \, \Delta n_2 + 2 \, \Delta n_3 + 1 \, \Delta n_4 + 0 \, \Delta n_5 = 0 \qquad (5.65)$$

where $\Delta n_i = n_i - n_i^*$.

Three simultaneous linear equations do not provide us with enough information to calculate the values of the five unknowns. However, they do restrict the possible sets of values of Δn_i for any change of state. To identify possible sets of values of the five unknowns that are solutions, it is useful to rearrange equations 5.63–5.65. These manipulations can be made more easily if equations 5.63–5.65 are represented by a matrix:

$$\begin{bmatrix} 1 & 0 & 1 & 0 & 1 \\ 0 & 2 & 0 & 2 & 4 \\ 1 & 0 & 2 & 1 & 0 \end{bmatrix} \qquad (5.66)$$

There are three elementary row operations that can be applied to a matrix without altering the solutions of the underlying set of linear equations. These are (1) multiplications of a row by a constant different from zero, (2) adding a constant multiple of any row to another row, and (3) interchange of two rows.

Matrix 5.66 can be changed to

$$\begin{bmatrix} 1 & 0 & 0 & -1 & 2 \\ 0 & 1 & 0 & 1 & 2 \\ 0 & 0 & 1 & 1 & -1 \end{bmatrix} \qquad (5.67)$$

by use of a Gaussian elimination, which uses the elementary row operations. One way to do this is to (1) multiply the first row by -1 and add it to the third row, (2) multiply the second row by $\frac{1}{2}$, and (3) multiply the third row by -1 and add it to the first row. The three equations that are represented by matrix 5.67 can be written

$$\Delta n_1 = \Delta n_4 - 2\Delta n_5$$

$$\Delta n_2 = -\Delta n_4 - 2\Delta n_5$$

$$\Delta n_3 = -\Delta n_4 + \Delta n_5 \qquad (5.68)$$

The most general vector of Δn_i's satisfying these equations is

$$[\Delta n_4 - 2 \, \Delta n_5, \; -\Delta n_4 - 2 \, \Delta n_5, \; -\Delta n_4 + \Delta n_5, \; \Delta n_4, \; \Delta n_5] \qquad (5.69)$$

this general solution is evidently the sum of two vectors:

$$[\Delta n_4, \; -\Delta n_4, \; -\Delta n_4, \; \Delta n_4, \; 0] + [-2 \, \Delta n_5, \; -2 \, \Delta n_5, \; \Delta_5, \; 0, \; \Delta n_5] \qquad (5.70)$$

This sum can be written more conveniently as

$$\Delta n_4[1, \; -1, \; -1, \; 1, \; 0] + \Delta n_5[-2, \; -2, \; 1, \; 0, \; 1] \qquad (5.71)$$

If $\Delta n_4 = 0$, the vector $-2, -2, 1, 0, 1$ must be a solution of equations 5.63–5.65, and if $\Delta n_5 = 0$, the vector $1, -1, -1, 1, 0$ must be a solution. Of course, multiples of these solutions are also solutions. Other solutions are given by equation 5.71 with arbitrary values of Δn_4 and Δn_5. We can test these solutions by substituting in equations 5.63–5.65, but it is more convenient to write equations 5.63–5.65 in matrix form, substitute the solutions, and perform the matrix multiplication.

Equations 5.63–5.65 can be written in the form

$$Av = 0 \qquad (5.72)$$

where A is the **system formula matrix**, v is the **stoichiometric number matrix** of solutions, and 0 is a **matrix of zeros** of the appropriate size. The system formula matrix A is $M \times N$, where M is the number of elements and N is the number of species. The stoichiometric number matrix v is $N \times R$, where R is the number of independent reactions, and 0 is, therefore, $M \times R$. For the system under consideration, equation 5.72 is

$$\begin{bmatrix} 1 & 0 & 1 & 0 & 1 \\ 0 & 2 & 0 & 2 & 4 \\ 1 & 0 & 2 & 1 & 0 \end{bmatrix} \begin{bmatrix} 1 & -2 \\ -1 & -2 \\ -1 & 1 \\ 1 & 0 \\ 0 & 1 \end{bmatrix} = \begin{bmatrix} 0 & 0 \\ 0 & 0 \\ 0 & 0 \end{bmatrix} \qquad (5.73)$$

You should multiply this out to check it. Notice that the v matrix is given by the last two columns of equation 5.67 (with signs reversed) with a unit matrix put below it. This matrix equation is equivalent to two balanced chemical equations. To see this, first consider the matrix multiplication

$$\begin{bmatrix} 1 & 0 & 1 & 0 & 1 \\ 0 & 2 & 0 & 2 & 4 \\ 1 & 0 & 2 & 1 & 0 \end{bmatrix} \begin{bmatrix} 1 \\ -1 \\ -1 \\ 1 \\ 0 \end{bmatrix} = \begin{bmatrix} 0 \\ 0 \\ 0 \end{bmatrix} \qquad (5.74)$$

which can be written

$$1\begin{bmatrix} 1 \\ 0 \\ 1 \end{bmatrix} - 1\begin{bmatrix} 0 \\ 2 \\ 0 \end{bmatrix} - 1\begin{bmatrix} 1 \\ 0 \\ 2 \end{bmatrix} + 1\begin{bmatrix} 0 \\ 2 \\ 1 \end{bmatrix} + 0\begin{bmatrix} 1 \\ 4 \\ 0 \end{bmatrix} = \begin{bmatrix} 0 \\ 0 \\ 0 \end{bmatrix} \qquad (5.75)$$

The column vectors in equation 5.75 each represent one of the reactants. Because of the ordering of the elements in equations 5.60–5.62, the top number represents the number of carbon atoms, the middle number represents the number of hydrogen atoms and the bottom number represents the number of oxygen atoms. In chemical notation this equation is written as

$$1CO - 1H_2 - 1CO_2 + 1H_2O = 0 \qquad (5.76)$$

or

$$H_2 + CO_2 = H_2O + CO \qquad (5.77)$$

Similarly, we can show that the second column vector in $\boldsymbol{v}$ yields

$$2CO + 2H_2 = CH_4 + CO_2 \qquad (5.78)$$

These two equations are independent and can represent all possible changes in composition in the system. Since there are two chemical equations, there are two extents of reaction.

These calculations provide a method for determining the number of independent reactions R required to represent all possible chemical changes between listed species. When a Gaussian elimination is performed on the system formula matrix A, the number of columns to the right of the unit matrix is equal to R. When a matrix has the form of matrix 5.67, the rank of the matrix is equal to the number of rows. If one of the rows in this matrix had turned out to have all zeros, the rank would have been two. The **number of independent reactions** R is equal to the number of species N (that is, the number of columns) minus the rank of the A matrix:

$$R = N - \text{rank } A \qquad (5.79)$$

Usually rank A is equal to the number of elements. But if two elements are in the same proportion in all species, this combination of elements is a pseudo-element, and the rank is reduced by one. If the system is made up of only isomers, the rank is one and there is a single component. If the species had been arranged in a different order in the system formula matrix, a different set of independent reactions would have been obtained, but the value of R is independent of the order of species in the system formula matrix.

Example 5.17

A system contains $CaO(s)$, $CO_2(g)$, and $CaCO_3(s)$. How many independent reactions are there?

The system formula matrix is

$$A = \begin{bmatrix} 1 & 0 & 1 \\ 2 & 1 & 3 \\ 0 & 1 & 1 \end{bmatrix}$$

if the elements (rows) are in the order C, O, Ca and the species (columns) are in the order CO_2, CaO, $CaCO_3$. Gaussian elimination yields

$$A = \begin{bmatrix} 1 & 0 & 1 \\ 0 & 1 & 1 \\ 0 & 0 & 0 \end{bmatrix}$$

This corresponds to the stoichiometric number matrix

$$\boldsymbol{v} = \begin{bmatrix} -1 \\ -1 \\ 1 \end{bmatrix}$$

which corresponds with the equation

$$CaO(s) + CO_2(g) = CaCO_3(s)$$

The number of independent reactions is given by

$$R = N - \text{rank } A = 3 - 2 = 1$$

5.14   Special Topic: Multireaction Equilibria

Many practical calculations of chemical equilibrium involve systems in which a number of chemical reactions occur simultaneously. When reactions occur simultaneously, equilibrium constant expressions must be written in terms of the equilibrium extents of the individual independent reactions. However, these equations become quite complicated, and so other methods are used. Since the Gibbs energy of a system (that is, the extensive property) is the sum of the contributions of the various species (equation 4.83), the equilibrium composition can be calculated by minimizing the extensive Gibbs energy for the system:

$$G = \sum_i n_i \mu_i \qquad (5.80)$$

subject to the constraint that the total number of atoms of each element is constant. **Element conservation** is expressed by

$$\sum_{i=1}^{N} a_{ki} n_i = b_k \qquad k = 1, 2, \ldots, M \qquad (5.81)$$

where b_k is the amount of element k, and M is the number of elements. The number of atoms of element k in species i is represented by a_{ki}, and N is the number of species. This can be done by use of the method of **Lagrange's undetermined multipliers.** In this method the atom balance equations (equations 5.60–5.62) are each multiplied by a Lagrange multiplier λ_k, one for each element, and added to equation 5.80. The sum of the element balance equations, each multiplied by a Lagrange multiplier, is

$$\sum_k \lambda_k \left(\sum_i a_{ki} n_i - b_k \right) = 0 \qquad k = 1, 2, \ldots, M \qquad (5.82)$$

Since this quantity is necessarily zero, it may be added to equation 5.80 to form the Lagrangian function

$$\mathscr{L}(n_i, \lambda_k) = \sum_i n_i \mu_i + \sum_i \lambda_k \left(\sum_i a_{ki} n_i - b_k \right) \qquad (5.83)$$

To find the saddle point of the Lagrangian function the derivatives of $\mathscr{L}$ with respect to the variables are set equal to zero:

$$\frac{\partial \mathscr{L}}{\partial n_i} = \mu_i + \sum_k a_{ki} \lambda_k = 0 \qquad i = 1, 2, \ldots, N \qquad (5.84)$$

$$\frac{\partial \mathscr{L}}{\partial \lambda_k} = \sum_{i=1}^{N} a_{ki} n_i - b_k = 0 \qquad k = 1, 2, \ldots, M \qquad (5.85)$$

These $M + N$ equations are then solved simultaneously to obtain the numbers of moles and Lagrange multipliers. A computer is used because such problems have to be solved by an iterative method.

Example 5.18

A mixture of 2 mol of CH_4 and 3 mol of H_2O reacts to equilibrium at 1000 K and 1 bar. What is the equilibrium composition assuming the only species present at equilibrium are CH_4, H_2O, CO, CO_2, and H_2? (This example is from Smith and Van Ness, Introduction to Chemical Engineering Thermodynamics, New York: McGraw-Hill, 1987.)

The numbers of various types of atoms in each of the molecular species are

	CH_4	H_2O	CO	CO_2	H_2
C	1	0	1	1	0
H	4	2	0	0	2
O	0	1	1	2	0

These are the values of the various a_{ki}, where k is the number of the row and i is the number of the column.

Substituting $\mu_i = \Delta_f G_i^\circ + RT \ln(n_i/\sum n_i)$ in equation 5.84 and dividing through by RT yields

$$\frac{\Delta_f G_i^\circ}{RT} + \ln \frac{n_i}{\sum n_i} + \sum_k^M a_{ki}\lambda_k = 0 \qquad i = 1, 2, \ldots, N$$

Since $N = 5$, this yields five equations:

$$CH_4: \quad \frac{19\,460}{RT} + \ln \frac{n_{CH_4}}{\sum n_i} + \frac{\lambda_C}{RT} + \frac{4\lambda_H}{RT} = 0$$

$$H_2O: \quad \frac{-192\,579}{RT} + \ln \frac{n_{H_2O}}{\sum n_i} + \frac{2\lambda_H}{RT} + \frac{\lambda_O}{RT} = 0$$

$$CO: \quad \frac{-200\,294}{RT} + \ln \frac{n_{CO}}{\sum n_i} + \frac{\lambda_C}{RT} + \frac{\lambda_O}{RT} = 0$$

$$CO_2: \quad \frac{-395\,924}{RT} + \ln \frac{n_{CO_2}}{\sum n_i} + \frac{\lambda_C}{RT} + \frac{2\lambda_O}{RT} = 0$$

$$H_2: \quad \ln \frac{n_{H_2}}{\sum n_i} + \frac{2\lambda_H}{RT} = 0$$

Since there are three elements, equation 5.85 yields three element-balance equations:

$$C: \quad n_{CH_4} + n_{CO} + n_{CO_2} = 2 \text{ mol}$$

$$H: \quad 4n_{CH_4} + 2n_{H_2O} + 2n_{H_2} = 14 \text{ mol}$$

$$O: \quad n_{H_2O} + n_{CO} + 2n_{CO_2} = 3 \text{ mol}$$

When these eight simultaneous nonlinear equations are solved with a computer the following results are obtained by Smith and Van Ness: $y_{CH_4} = 0.0199$, $y_{H_2O} = 0.0995$, $y_{CO} = 0.1753$, $y_{CO_2} = 0.0359$, $y_{H_2} = 0.6694$, $\lambda_C = 0.797RT$, $\lambda_O = 25.1RT$, and $\lambda_H = 0.210RT$. The values of the Lagrange multipliers are not a part of the solution of the physical problem, but they were introduced to make it possible to get a solution in this orderly way which can be converted into a computer program.

5.15   Special Topic: Geothermmometers and Geobarometers

In geology, reactions such as

$$3CaAl_2Si_2O_8(s) = Ca_3Al_2Si_3O_{12}(s) + 2AlSiO_5(s) + SiO_2(s)$$
$$\text{plagioclase} \qquad \text{garnet} \qquad \text{sillamnite} \quad \text{quartz}$$
$$(5.86)$$

can be used as a thermometer or a barometer. At a given depth in the earth, these four substances can be deposited together at equilibrium only at a certain temperature that depends on the pressure. When these solid phases are found together, it is assumed that they are in equilibrium at the ambient temperature and pressure and that less stable phases have been consumed in solid-state reactions. The chemical potentials of solids are not affected very much by moderate changes in pressure, but in the crust of the earth the pressures may be very large and can have a significant effect on the chemical potential of a solid.

The chemical potential of a solid at T, P is given by

$$\mu_i(T, P) = \mu_i^\circ(T) + RT \ln a_i \qquad (5.87)$$

where a_i is activity. Since

$$d\mu_i = d\overline{G}_i = -\overline{S}_i \, dT + \overline{V}_i \, dP \qquad (5.88)$$

then

$$\left(\frac{\partial \mu_i}{\partial P}\right)_T = \overline{V}_i \qquad (5.89)$$

When $\overline{V}_i$ is assumed to be independent of pressure, integration yields

$$\mu_i(T, P) - \mu_i^\circ(T) = \int_{P^\circ}^{P} \overline{V}_i \, dP = \overline{V}_i(P - 1 \text{ bar}) \qquad (5.90)$$

Substituting this in equation 5.87 yields

$$a_i = \exp \frac{\overline{V}_i(P - 1 \text{ bar})}{RT} \qquad (5.91)$$

The activity of a pure solid is unity at 1 bar at any temperature and increases exponentially with the pressure.

To drive the relationship between the equilibrium pressure and temperature for a reaction such as 1, we go back to the basic relationship

$$\Delta_r G(T, P) = \Delta_r G^\circ(T, P^\circ) + RT \ln \prod_{i=1}^{N} a_i^{\nu_i} \qquad (5.92)$$

Since the activities of the pure solids are all equal to unity at 1 bar, $\Delta_r G^\circ(T, P^\circ)$ tells us whether the reaction is spontaneous under these conditions. At 900–1000 K and 1 bar, these four substances are not in equilibrium with each

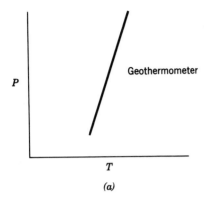

(a)

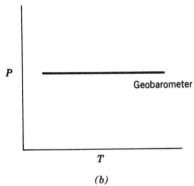

(b)

Figure 5.7 (a) Equilibrium P-T plot for a reaction between solids that is useful as a geothermometer. (b) Equilibrium P-T plot for a reaction between solids that is useful as a geobarometer.

other because $\Delta_r G$ is positive, so that the reaction goes to completion to the left. However, $\Delta_r V$ is negative, and so at a sufficiently high pressure, the reaction is spontaneous.

Substituting equation 5.91 in equation 5.92 and writing the standard reaction Gibbs energy in terms of the reaction enthalpy and reaction entropy yields

$$\Delta_r G(T, P) = \Delta_r H^\circ - T \Delta_r S^\circ + \Delta_r V(P - 1 \text{ bar}) \quad (5.93)$$

At equilibrium at a particular P and T, $\Delta_r G(T, P) = 0$, and so this equation provides the relationship between the equilibrium pressure and the equilibrium temperature for a reaction in which all of the reactants and products are pure solids.

For reaction 5.86, equation 5.93 at equilibrium is

$$0 = -43\,095 + 133.18T - 5.458(P - 1 \text{ bar}) \quad (5.94)$$

where the energies are in J mol^{-1}. At 900 K, $P = 14\,066$ bar, and at 1000 K, $P = 16\,505$ bar. Thus, the crystals in the rock provide evidence of the temperature and pressure combinations at which it could have been formed.

If the plot of P versus T for a reaction is very nearly vertical, as shown in Fig. 5.7, the reaction is a very good geothermometer because an uncertainty in the pressure causes only a small uncertainty in the temperature. If the plot of P versus T for a reaction is very nearly horizontal, the reaction is a very good geobarometer because an uncertainty in the temperature causes a small uncertainty in the pressure.

References

K. E. Bett, J. S. Rowlinson, and G. Saville, *Thermodynamics for Chemical Engineers*. Cambridge, MA: MIT Press, 1975.

K. G. Denbigh, *The Principles of Chemical Equilibrium*. Cambridge, UK: Cambridge University Press, 1981.

N. A. Gokcen, *Thermodynamics*. Hawthorne, CA: Techscience, Inc., 1975.

R. Holub and P. Vónka. *The Chemical Equilibrium of Gaseous Systems*. Boston: Reidel, 1976.

G. N. Lewis, and M. Randall, revised by K. S. Pitzer and L. Brewer, *Thermodynamics*. New York: McGraw-Hill, 1961.

M. Modell and R. C. Reid, *Thermodynamics and Its Applications*. Englewood Cliffs, NJ: Prentice-Hall, 1983.

S. I. Sandler, *Chemical and Engineering Thermodynamics*. New York: Wiley, 1989.

J. M. Smith and H. C. Van Ness, *Introduction to Chemical Engineering Thermodynamics*. New York: McGraw-Hill, 1987.

W. R. Smith and R. W. Missen, *Chemical Reaction Equilibrium Analysis*. New York: Wiley, 1982.

F. van Zeggeren and S. H. Storey, *The Computation of Chemical Equilibria*. New York: Cambridge University Press, 1970.

Problems

5.1 For the reaction $N_2(g) + 3H_2(g) = 2NH_3(g)$, $K = 1.60 \times 10^{-4}$ at 400 °C. Calculate (a) $\Delta_r G°$ and (b) $\Delta_r G$ when the pressures of N_2 and H_2 are maintained at 10 and 30 bar, respectively, and NH_3 is removed at a partial pressure of 3 bar. (c) Is the reaction spontaneous under the latter conditions?

5.2 A 1:3 mixture of nitrogen and hydrogen was passed over a catalyst at 450 °C. It was found that 2.04% by volume of ammonia gas was formed when the total pressure was maintained at 10.13 bar (A. T. Larson and R. L. Dodge, *J. Am. Chem. Soc.* **45**:2918 (1923)). Calculate the value of K for $\frac{3}{2}H_2(g) + \frac{1}{2}N_2(g) = NH_3(g)$ at this temperature.

5.3 At 55 °C and 1 bar the average molar mass of partially dissociated N_2O_4 is 61.2 g mol^{-1}. Calculate (a) ξ and (b) K for the reaction $N_2O_4(g) = 2NO_2(g)$. (c) Calculate ξ at 55 °C if the total pressure is reduced to 0.1 bar.

5.4 A liter reaction vessel containing 0.233 mol of N_2 and 0.341 mol of PCl_5 is heated to 250 °C. The total pressure at equilibrium is 29.33 bar. Assuming that all gases are ideal, calculate K for the only reaction that occurs:

$$PCl_5(g) = PCl_3(g) + Cl_2(g)$$

5.5 An evacuated tube containing 5.96×10^{-3} mol L^{-1} of solid iodine is heated to 973 K. The experimentally determined pressure is 0.496 (M. L. Perlman and G. K. Rollefson, *J. Chem. Phys.* **9**:362 (1941)). Assuming ideal gas behavior, calculate K for $I_2(g) = 2I(g)$.

5.6 Nitrogen trioxide dissociates according to the reaction

$$N_2O_3(g) = NO_2(g) + NO(g)$$

When one mole of $N_2O_3(g)$ is held at 25 °C and 1 bar total pressure until equilibrium is reached, the extent of reaction is 0.30. What is $\Delta_r G°$ for this reaction at 25 °C?

5.7 For the reaction

$$2HI(g) = H_2(g) + I_2(g)$$

at 698.6 K, $K = 1.83 \times 10^{-2}$. (a) How many grams of hydrogen iodide will be formed when 10 g of iodine and 0.2 g of hydrogen are heated to this temperature in 3-L vessel? (b) What will be the partial pressures of H_2, I_2, and HI?

5.8 Express K for the reaction

$$CO(g) + 3H_2(g) = CH_4(g) + H_2O(g)$$

in terms of the equilibrium extent of reaction ξ when one mole of CO is mixed with one mole of hydrogen.

5.9 What are the percentage dissociations of $H_2(g)$, $O_2(g)$, and $I_2(g)$ at 2000 K and a total pressure of 1 bar?

5.10 To produce more hydrogen from "synthesis gas" (CO + H_2) the water gas shift reaction is used.

$$CO(g) + H_2O(g) = CO_2(g) + H_2(g)$$

Calculate K at 1000 K and the equilibrium extent of reaction starting wth an equimolar mixture of CO and H_2O.

5.11 Calculate the extent of reaction ξ of 1 mol of $H_2O(g)$ to form $H_2(g)$ and $O_2(g)$ at 2000 K and 1 bar. (Since the extent of reaction is small, the calculation may be simplified by assuming that $P_{H_2O} = 1$ bar.)

5.12 At 500 K CH_3OH, CH_4 and other hydrocarbons can be formed from CO and H_2. Until recently the main source of the CO mixture for the synthesis of CH_3OH was methane:

$$CH_4(g) + H_2O(g) = CO(g) + 3H_2(g)$$

When coal is used as the source, the "synthesis gas" has a different composition:

$$C(graphite) + H_2O(g) = CO(g) + H_2(g)$$

Suppose we have a catalyst that catalyzes only the formation of CH_3OH. (a) What pressure is required to convert 25% of the CO to CH_3OH at 500 K if the synthesis gas comes from CH_4? (b) If the synthesis gas comes from coal?

5.13 Many equilibrium constants in the literature were calculated with a standard state pressure of 1 atm (1.013 25 bar). Show that the corresponding equilibrium constant with a standard pressure of 1 bar can be calculated using

$$K(\text{bar}) = K(\text{atm})(1.013\ 25)^{\Sigma \nu_i}$$

where the ν_i are the stoichiometric numbers of the gaseous reactants.

5.14 Older tables of chemical thermodynamic properties are based on a standard state pressure of 1 atm. Show that the corresponding $\Delta_f G_i°$ with a standard state pressure of 1 bar can be calculated using

$$\Delta_f G_j°(\text{bar}) = \Delta_f G_j°(\text{atm}) - (0.109 \times 10^{-3}\,\text{kJ}\,\text{K}^{-1}\text{mol}^{-1})T\sum \nu_i$$

where the ν_i are the stoichiometric numbers of the gaseous reactants and products in the formation reaction.

5.15 Show that the equilibrium mole fractions of n-butane and isobutane are given by

$$y_n = \frac{e^{-\Delta_f G_n°/RT}}{(e^{-\Delta_f G_n°/RT} + e^{-\Delta_f G_{iso}°/RT})}$$

$$y_{iso} = \frac{e^{-\Delta_f G_{iso}°/RT}}{(e^{-\Delta_f G_n°/RT} + e^{-\Delta_f G_{iso}°/RT})}$$

5.16   Calculate the molar Gibbs energy of butane isomers for extents of reaction of 0.2, 0.4, 0.6, and 0.8 for the reaction

$$n\text{-butane} = \text{isobutane}$$

at 1000 K and 1 bar. At 1000 K

$$\Delta_f G°(n\text{-butane}) = 270 \text{ kJ mol}^{-1}$$

$$\Delta_f G°(\text{isobutane}) = 276.6 \text{ kJ mol}^{-1}$$

Make a plot and show that the minimum corresponds with the equilibrium extent of reaction.

5.17   In the synthesis of methanol by $CO(g) + 2H_2(g) = CH_3OH(g)$ at 500 K, calculate the total pressure required for a 90% conversion to methanol if CO and H_2 are initially in a 1:2 ratio. Given: $K = 6.09 \times 10^{-3}$.

5.18   At 1273 K and at a total pressure of 30.4 bar the equilibrium in the reaction $CO_2(g) + C(s) = 2CO(g)$ is such that 17 mol% of the gas is CO_2. (a) What percentage would be CO_2 if the total pressure were 20.3 bar? (b) What would be the effect on the equilibrium of adding N_2 to the reaction mixture in a closed vessel until the partial pressure of N_2 is 10 bar? (c) At what pressure of the reactants will 25% of the gas be CO_2?

5.19   When alkanes are heated up, they lose hydrogen and alkenes are produced. For example,

$$C_2H_6(g) = C_2H_4(g) + H_2(g)$$
$$K = 0.36 \text{ at } 1000 \text{ K}$$

If this is the only reaction that occurs when ethane is heated to 1000 K, at what total pressure will ethane be (a) 10% dissociated and (b) 90% dissociated to ethylene and hydrogen?

5.20   At 2000 °C water is 2% dissociated into oxygen and hydrogen at a total pressure of 1 bar. (a) Calculate K for $H_2O(g) = H_2(g) + \frac{1}{2}O_2(g)$. (b) Will the extent of reaction increase or decrease if the pressure is reduced? (c) Will the extent of reaction increase or decrease if argon gas is added, when the total pressure is held equal to 1 bar? (d) Will the extent of reaction change if the pressure is raised by addition of argon at constant volume to the closed system containing partially dissociated water vapor? (e) Will the extent of reaction increase or decrease if oxygen gas is added while holding the total pressure constant at 1 bar?

5.21   At 250 °C, PCl_5 is 80% dissociated at a pressure of 1.013 bar, and so $K = 1.80$. What is the extent of reaction at equilibrium after sufficient nitrogen has been added at constant pressure to produce a nitrogen partial pressure of 0.9 bar? The total pressure is maintained at 1 bar.

5.22   The following exothermic reaction is at equilibrium at 500 K and 10 bar:

$$CO(g) + 2H_2(g) = CH_3OH(g)$$

Assuming that the gases are ideal, what will happen to the amount of methanol at equilibrium when (a) the temperature is raised, (b) the pressure is increased, (c) an inert gas is pumped in at constant volume, (d) an inert gas is pumped in at constant pressure, and (e) hydrogen gas is added at constant pressure?

5.23   The following reaction is nonspontaneous at room temperature and endothermic:

$$3C(\text{graphite}) + 2H_2O(g) = CH_4(g) + 2CO(g)$$

As the temperature is raised, the equilibrium constant will become equal to unity at some point. Estimate this temperature using data from Appendix C.2.

5.24   The measured density of an equilibrium mixture of N_2O_4 and NO_2 at 15 °C and 1.103 bar is 3.62 g L^{-1}, and the density at 75 °C and 1.013 bar is 1.84 g L^{-1}. What is the enthalpy change of the reaction $N_2O_4(g) = 2NO_2(g)$?

5.25   Calculate K_c for the reaction in problem 5.19 at 1000 K and describe what it is equal to.

5.26   The equilibrium constant for the reaction

$$N_2(g) + 3H_2(g) = 2NH_3(g)$$

is 35.0 at 400 K when partial pressures are expressed in bars. Assume that the gases are ideal. (a) What is the equilibrium composition and equilibrium volume when 0.25 mol N_2 is mixed with 0.75 mol H_2 at a temperature of 400 K and a pressure of 1 bar? (b) What is the equilibrium composition and equilibrium pressure if this mixture is held at a constant volume of 33.26 L at 400 K?

5.27   Show that to a first approximation the equation of state of a gas that dimerizes to a small extent is given by

$$\frac{P\overline{V}}{RT} = 1 - \frac{K_c}{\overline{V}}$$

5.28   Water vapor is passed over coal (assumed to be pure graphite in this problem) at 1000 K. Assuming that the only reaction occurring is the water gas reaction

$$C(\text{graphite}) + H_2O(g) = CO(g) + H_2(g) \qquad K = 2.52$$

calculate the equilibrium pressures of H_2O, CO, and H_2 at a total pressure of 1 bar. (Actually the water gas shift reaction

$$CO(g) + H_2O(g) = CO_2(g) + H(g)$$

also occurs, but it is considerably more complicated to take this additional reaction into account.)

5.29 Mercuric oxide dissociates according to the reaction $2HgO(s) = 2Hg(g) + O_2(g)$. At 420 °C the dissociation pressure is 5.16×10^4 Pa, and at 450 °C it is 10.8×10^4 Pa. Calculate (a) the equilibrium constants, and (b) the enthalpy of dissociation per mole of HgO.

5.30 The decomposition of silver oxide is represented by

$$2Ag_2O(s) = 4Ag(s) + O_2(g)$$

Using data from Appendix C.1. and assuming $\Delta_r C_P^\circ = 0$ calculate the temperature at which the equilibrium pressure of O_2 is 0.2 bar. This temperature is of interest because Ag_2O will decompose to yield Ag at temperatures above this value if it is in contact with air.

5.31 The dissociation of ammonium carbamate takes place according to the reaction

$$(NH_2)CO(ONH_4)(s) = 2NH_3(g) + CO_2(g)$$

When an excess of ammonium carbamate is placed in a previously evacuated vessel, the partial pressure generated by NH_3 is twice the partial pressure of the CO_2, and the partial pressure of $(NH_2)CO(ONH_4)$ is negligible in comparison. Show that

$$K = \left(\frac{P_{NH_3}}{P^\circ}\right)^2 \frac{P_{CO_2}}{P^\circ} = \frac{4}{27}\left(\frac{P}{P^\circ}\right)^3$$

where P is the total pressure.

5.32 At 1000 K methane at 1 bar is in the presence of hydrogen. In the presence of a sufficiently high partial pressure of hydrogen, methane does not decompose to form graphite and hydrogen. What is this partial pressure?

5.33 For the reaction

$$Fe_2O_3(s) + 3CO(g) = 2Fe(s) + 3CO_2(g)$$

the following values of K are known.

$t/°C$	250	1000
K	100	0.0721

At 1120 °C for the reaction $2CO(g) = 2CO(g) + O_2(g)$, $K = 1.4 \times 10^{-12}$ bar. What equilibrium partial pressure of O_2 would have to be supplied to a vessel at 1120 °C containing solid Fe_2O_3 just to prevent the formation of Fe?

5.34 Calculate the partial pressure of $CO_2(g)$ over $CaCO_3(calcite)–CaO(s)$ at 500 °C using equation 5.45 and data in Appendix C.1.

5.35 The NBS Tables contain the following data at 298 K:

	$\Delta_f H°/kJ\ mol^{-1}$	$\Delta_f G°/kJ\ mol^{-1}$
$CuSO_4(s)$	−771.36	−661.8
$CuSO_4·H_2O(s)$	−1085.83	−918.11
$CuSO_4·3H_2O(s)$	−1684.31	−1399.96
$H_2O(g)$	−241.818	−228.572

(a) What is the equilibrium partial pressure of H_2O over a mixture of $CuSO_4(s)$ and $CuSO_4·H_2O(s)$ at 25 °C? (b) What is the equilibrium partial pressure of H_2O over a mixture of $CuSO_4·H_2O(s)$ and $CuSO_4·3H_2O(s)$ at 25 °C? (c) What are the answers to (a) and (b) if the temperature is 100 °C and ΔC_P° is assumed to be zero?

5.36 One micromole of CuO(s) and 0.1 μmol of Cu(s) are placed in a 1-L container at 1000 K. Determine the identity and quantity of each phase present at equilibrium if $\Delta_f G°$ of CuO(s) is −66.66 kJ mol^{-1} and of CuO$_2$(s) is −77.94 kJ mol^{-1} at 1000 K. (From H. F. Franzen, *J. Chem. Ed.* **65**:146 (1988).)

5.37 Acetic acid and ethanol react to form ethyl acetate and water according to

$$CH_3CO_2H(l) + C_2H_5OH(l) = CH_3CO_2C_2H_5(l) + H_2O(l)$$

Analytical data obtained by titrating the acetic acid remaining at equilibrium at 298 K indicates that $K = 40$ when mole fractions are used, and the solution is assumed to be ideal. The standard Gibbs energies of formation of ethanol, acetic acid, and water are given in Appendix C.1. Calculate the standard Gibbs energy of formation of ethyl acetate.

5.38 Amylene, C_5H_{10}, and acetic acid react to give the ester according to the reaction

$$C_5H_{10}(l) + CH_3COOH(l) = CH_3COOC_5H_{11}(l)$$

What is the value of K_c if 0.006 45 mol of amylene and 0.001 mol of acetic acid, dissolved in a certain inert solvent to give a total volume of 0.845 L, react to give 0.000 784 mol of ester? Use the molar concentration scale.

5.39 (a) A system contains CO(g), CO_2(g), H_2(g), and H_2O(g). How many chemical reactions are required to describe chemical changes in this system? Give an example. (b) If solid carbon is present in the system in addition, how many independent chemical reactions are there? Give a suitable set.

5.40 For a closed system containing C_2H_2, H_2, C_6H_6, and $C_{10}H_8$, use a Gaussian elimination to obtain a set of independent chemical reactions. Perform the matrix multiplication to verify $Av = 0$.

5.41 In determining equilibrium constants for reactions with large or small constants, the analytical method usually puts a limit on the magnitude of the constant that can be experimentally determined. For a reaction of the type A = B, it is found that there is less than 1 part per 1000 of B at equilibrium at 25 °C. Calculate the *minimum* value for $\Delta_r G^\circ_{298}$ for this reaction.

5.42   At 500 °C, $K = 5.5$ for the reaction

$$CO(g) + H_2O(g) = CO_2(g) + H_2(g)$$

If a mixture of 1 mol of CO and 5 mol of H_2O is passed over a catalyst at this temperature, what will be the equilibrium mole fraction of H_2O?

5.43   At 400 °C, $K = 79.1$ for the reaction

$$NH_3(g) = \tfrac{1}{2}N_2(g) + \tfrac{3}{2}H_2(g)$$

Show that the fraction α of NH_3 dissociated at a total pressure P is given by

$$\xi = \frac{1}{\sqrt{1 + kP}}$$

and calculate the value of k in this equation.

5.44   For the reaction $N_2O_4(g) = 2NO_2(g)$, K at 25 °C is 0.143. What pressure would be expected if 1 g of liquid N_2O_4 were allowed to evaporate into a liter vessel at this temperature? Assume that N_2O_4 and NO_2 are ideal gases.

5.45   The dissociation of N_2O_4 is represented by $N_2O_4(g) = 2NO_2(g)$. If the density of the equilibrium gas mixture is 3.174 g L^{-1} at a total pressure of 1.013 bar at 24 °C, what minimum pressure would be required to keep the degree of dissociation of N_2O_4 below 0.1 at this temperature?

5.46   At 250 °C, 1 L of partially dissociated phosphorus pentachloride gas at 1.013 bar weighs 2.690 g. Calculate the extent of reaction ξ and the equilibrium constant.

5.47   Derive the analog of equation 5.30 for the reaction $N_2(g) + 3H_2(g) = 2NH_3(g)$ when stoichiometric amounts of nitrogen and hydrogen are used.

5.48   Hydrogen is produced on a large scale from methane. Calculate the equilibrium constant K for the production of H_2 from CH_4 at 1000 K using reaction $CH_4(g) + H_2O(g) = CO(g) + 3H_2(g)$.

5.49   (*a*) What is the equilibrium constant for the formation of CH_4 from CO and H_2 at 500 and 1000 K?

$$CO(g) + 3H_2(g) = CH_4(g) + H_2O(g)$$

(*b*) What is the equilibrium constant for the formation of CH_4 from graphite and H_2O at 500 and 1000 K?

$$2C(graphite) + 2H_2O(g) = CH_4(g) + CO_2(g)$$

5.50   (*a*) Calculate the extent of dissociation of $H_2(g)$ at 3000 K and 1 bar. A value of 0.072 was obtained experimentally by Langmuir. (*b*) Calculate the extent of dissociation of $O_2(g)$ at 3000 K at 1 bar.

5.51   What is K for

$$I_2(g) = 2I(g)$$

at 1000 K and what is the degree of dissociation at 1 bar? At 0.1 bar?

5.52   The standard Gibbs energies of the isomers of the pentanes at 100 K are as follows:

	$\Delta_f G^\circ$/kJ mol^{-1}
n-Pentane	−179.9
2-Methylbutane	−185.4
2,2-Dimethylpropane	−195.0

What are the equilibrium mole fractions of the three isomers? If the gases are ideal, will changing the pressure change this distribution?

5.53   Show that in going from a standard-state pressure of 1 atm to 1 bar, thermodynamic quantities for reactions are corrected as follows:

$$\Delta C_P^\circ(bar) = \Delta C_P^\circ(atm)$$
$$\Delta H^\circ(bar) = \Delta H^\circ(atm)$$
$$\Delta S^\circ(bar) = \Delta S(atm) + (0.109 \text{ J K}^{-1} \text{ mol}^{-1}) \sum \nu_i$$
$$\Delta G^\circ(bar) = \Delta G^\circ(atm)$$
$$\qquad - (0.109 \times 10^{-3} \text{ kJ K}^{-1} \text{ mol}^{-1})T \sum \nu_i$$
$$K(bar) = K(atm)(1.013\ 25)^{\sum \nu_i}$$

where $\sum \nu_i$ is the difference between the stoichiometric numbers of gaseous products and gaseous reactants in the balanced chemical equation.

5.54   The reaction

$$2NOCl(g) = 2NO(g) + Cl_2(g)$$

comes to equilibrium at 1 bar total pressure and 227 °C when the partial pressure of the nitrosyl chloride, NOCl, is 0.64 bar. Only NOCl was present initially. (*a*) Calculate $\Delta_r G^\circ$ for this reaction. (*b*) At what total pressure will the partial pressure of Cl_2 be 0.1 bar?

5.55   Acetic acid is produced on a large scale by the carbonylation of methanol at about 500 K and 25 bar using a rhodium catalyst. What is K_y under these conditions? ($\Delta_f G^\circ$ for acetic acid gas at 500 K is −335.28 kJ mol^{-1}.)

5.56   Calculate the total pressure that must be applied to a mixture of three parts of hydrogen and one part nitrogen to give a mixture containing 10% ammonia at equilibrium at 400 °C. At 400 °C, $K = 1.60 \times 10^{-4}$ for the reaction $N_2(g) + 3H_2(g) = 2NH_3(g)$.

5.57   A mixture of one mole of nitrogen and one mole of hydrogen is equilibrated over a catalyst for the ammonia reaction at 500 K and 1 bar. (*a*) What are the equilibrium mole fractions of N_2, H_2, and NH_3? (*b*) The experiment is repeated with 1.2 mol of nitrogen and 1 mol of hydrogen. What is the equilibrium partial pressure of ammonia? (*c*) How do you explain this result in terms of Le Chatelier's principle?

5.58   Assume that the following reaction is in equilibrium at 1000 K:

$$3C(graphite) + 2H_2O(g) = CH_4(g) + 2CO(g)$$

$\Delta_r H°(1000\ K) = 182\ kJ\ mol^{-1}$. (*a*) What will be the effect on the equilibrium composition of raising the temperature at a total pressure of 1 bar? (*b*) What will be the effect of raising the pressure to 5 bar? (*c*) What will be the effect of adding nitrogen at a constant pressure of 1 bar?

5.59   When N_2O_4 is allowed to dissociate from NO_2 at 25 °C at a total pressure of 1 bar, it is 18.5% dissociated at equilibrium, and so $K = 0.141$. (*a*) If N_2 is added to the system at constant volume, will the equilibrium shift? (*b*) If the system is allowed to expand as N_2 is added at a constant total pressure of 1 bar, what will be the equilibrium degree of dissociation when the N_2 partial pressure is 0.6 bar.

5.60   For the formation of nitric oxide

$$N_2(g) + O_2(g) = 2NO(g)$$

K at 2126.9 °C is 2.5×10^{-3}. (*a*) In an equilibrium mixture containing 0.1 bar partial pressure of N_2 and 0.1 bar partial pressure of O_2, what is the partial pressure of NO? (*b*) In an equilibrium mixture of N_2, O_2, NO, CO_2, and other inert gases at 2126.9 °C and 1 bar total pressure, 80% by volume of the gas is N_2 and 16% O_2. What is the percentage by volume of NO? (*c*) What is the total partial pressure of inert gases?

5.61   Prove that for $2C_2 = C_4$ in the presence of other gases in the gas phase, the equilibrium constant expression

$$K = \frac{y_4}{y_2^2(P/P°)}$$

may be interpreted in two ways:

(*i*)   The mole fractions are expressed only in terms of C_2 and C_4 and $P = P_2 + P_4$

(*ii*)   The mole fractions are expressed in terms of all gases present and P is the total pressure.

5.62   The following data apply to the reaction $Br_2(g) = 2Br(g)$:

T/K	1123	1172	1223	1273
$K/10^{-3}$	0.408	1.42	3.32	7.2

Determine by graphical means the reaction enthalpy at 1200 K.

5.63   The average molar mass M of equilibrium mixtures of NO_2 and N_2O_4 at 1.013 bar total pressure are given in the following table at three temperatures:

$t/°C$	25	45	65
$M/g\ mol^{-1}$	77.64	66.80	56.51

(*a*) Calculate the degree of dissociation of N_2O_4 and the equilibrium constant at each of these temperatures. (*b*) Plot log K against $1/T$ and calculate $\Delta_r H°$ for the dissociation of N_2O_4. (*c*) Calculate the equilibrium constant at 35 °C. (*d*) Calculate the degree of dissociation for NO_2 at 35 °C when the total pressure is 0.5 bar.

5.64   One mole of carbon monoxide is mixed with one mole of hydrogen and passed over a catalyst for the following reaction:

$$CO(g) + 2H_2(g) = CH_3OH(g)$$

At 500 K and a total pressure of 100 bar, 0.40 mol of CH_3OH is found at equilibrium. What is the value of the equilibrium constant expressed in terms of $P/P°$, where $P°$ is the reference pressure of 1 bar? What is the equilibrium constant expressed in terms of $c/c°$, where $c°$ is the reference concentration of 1 mol L^{-1}? Assume ideal gases.

5.65   The solubility of hydrogen in a molten iron alloy is found to be proportional to the square root of the partial pressure of hydrogen. How can this be explained?

5.66   Is magnetite (Fe_3O_4) or hematite (Fe_2O_3) the more stable ore thermodynamically at 25 °C in contact with air?

5.67   An equimolar mixture of CO(g), H_2(g), and H_2O(g) at 1000 K is compressed. At what total pressure will solid carbon start to precipitate out if there is chemical equilibrium? (See Problem 5.28.)

5.68   At 50 °C the partial pressure of H_2O(g) over a mixture of $CuSO_4·3H_2O$(s) and $CuSO_4·H_2O$(s) is 4.0×10^3 Pa and over a mixture of $CuSO_4·3H_2O$(s) and $CuSO_4·5H_2O$(s) is 6.3×10^3 Pa. Calculate the change in Gibbs energy for the reaction

$$CuSO_4·5H_2O(s) = CuSO_4·H_2O(s) + 4H_2O(g)$$

5.69   Calculate (*a*) K and (*b*) $\Delta_r G°$ for the following reaction at 20 °C:

$$CuSO_4 \cdot 4NH_3(s) = CuSO_4 \cdot 2NH_3(s) + 2NH_3(g)$$

The equilibrium pressure of NH_3 is 8.26 kPa.

5.70 The vapor pressure of water above mixtures of $CuCl_2 \cdot H_2O(s)$ and $CuCl_2 \cdot 2H_2O(s)$ is given as a function temperature in the following table:

$t/°C$	17.9	39.8	60.0	80.0
P/bar	0.0049	0.0250	0.122	0.327

(*a*) Calculate $\Delta_r H°$ for the reaction

$$CuCl_2 \cdot 2H_2O(s) = CuCl_2 \cdot H_2O(s) + H_2O(g)$$

(*b*) Calculate $\Delta_r G°$ for the reaction at 60 °C. (c) Calculate $\Delta_r S°$ for the reaction at 60 °C.

5.71 The equilibrium constant for the association of benzoic acid to a dimer in dilute benzene solutions is as follows at 43.9 °C:

$$2C_6H_5COOH = (C_6H_5COOH)_2 \qquad K_c = 2.7 \times 10^2$$

Molar concentrations are used in expressing the equilibrium constant. Calculate $\Delta_r G°$, and state its meaning.

5.72 Superheated steam is passed over coal, represented in these calculations by graphite, at 1000 K to produce CO, CO_2, H_2, and CH_4. Since there are 6 species and 3 element balance equations, there are 3 independent chemical reactions, which may be written as follows:

$$C(graphite) + H_2O(g) = CO(g) + H_2(g)$$
$$3C(graphite) + 2H_2O(g) = CH_4(g) + 2CO(g)$$
$$2C(graphite) + 2H_2O(g) = CH_4(g) + CO_2(g)$$

What are the equilibrium constants for these reactions? If the total pressure is raised to 100 bar, one of these reactions will predominate. Neglecting the other two reactions, calculate the equilibrium mole fractions of the gases present.

5.73 The following molecular species are in equilibrium in the gas phase: CH_4, C_2H_6, C_2H_4, and C_3H_6. How many independent chemical reactions are required to represent all possible changes in this system? Derive a set of independent reactions.

5.74 Derive equation 5.27 for calculating the extent of reaction for a gas dissociation reaction of the type

$$A(g) = 2B(g)$$

at constant temperature and pressure.

6

Phase Equilibrium: One-Component Systems

A **phase** is a part of a system, uniform throughout in chemical composition and physical properties, that is separated from other homogeneous parts of the system by boundary surfaces. The phase behavior exhibited by pure substances is quite varied and complicated, but there are powerful generalizations from thermodynamics that help us to understand these phenomena. The Clapeyron equation expresses dP/dT for a two-phase system containing one component at equilibrium in terms of other thermodynamic quantities, and the phase rule deals with the number of phases that can be present at equilibrium.

6.1 Criteria of Equilibrium in Terms of Intensive Properties

So far we have usually considered systems consisting of a single phase. If a system contains several phases that are isolated from each other, the internal energy or entropy of the system is simply the sum of the energies or entropies of the separate phases. However, if heat or matter can be exchanged between phases, or if the volume of one can be increased at the expense of another, there are relations between the thermodynamic properties of different phases at equilibrium. We say that equilibrium introduces **constraints,** and we will be concerned with these constraints in this chapter. We begin by considering three imaginary processes that can be carried out with two phases, α and β (see Fig. 6.1), which are initially at equilibrium, and ask what conditions must be met, if these phases are to remain in equilibrium. These imaginary processes might be carried out with ice water in a Dewar flask.

First, imagine that an infinitesimal quantity of heat dq is transferred from phase α to phase β in an isolated system. If the system is to be at equilibrium after the transfer, there must not be a change in entropy (Section 3.7). Thus,

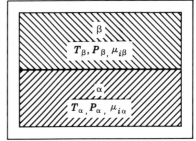

Figure 6.1 When two phases are in equilibrium, the following conditions have to be satisfied: $T_\alpha = T_\beta$, $P_\alpha = P_\beta$, and $\mu_{i\alpha} = \mu_{i\beta}$.

$dS = 0$, and the decrease in entropy of phase α is balanced by the increase in entropy of phase β, if the transfer is made at equilibrium:

$$dS_\alpha + dS_\beta = 0 \qquad (6.1)$$

Since the system is isolated, $dU = 0$ and $dU_\alpha + dU_\beta = 0$. If the volumes of the phases do not change, $dV_\alpha = dV_\beta = 0$. Under these conditions, $dU = T\,dS$ so that

$$T_\alpha dS_\alpha + T_\beta dS_\beta = 0 \qquad (6.2)$$

If we substitute equation 6.1 into equation 6.2, it is evident that **the temperatures of the two phases must be equal at equilibrium:**

$$T_\alpha = T_\beta \qquad (6.3)$$

Second, imagine that the volume of phase α is decreased by an infinitesimal amount dV and the volume of phase β is increased by the same amount. We have seen in Section 4.3, that if the total volume and temperature are fixed, and the system is to remain at equilibrium, there must not be a change in the Helmholtz energy. Thus, for the process we are considering, $dA = 0$, or

$$dA_\alpha + dA_\alpha = 0 \qquad (6.4)$$

Now we substitute expressions from the fundamental equation for dA_α and dA_β:

$$-P_\alpha dV + P_\beta dV = 0 \qquad (6.5)$$

It is evident that **the pressures of the two phases must be equal at equilibrium:**

$$P_\alpha = P_\beta \qquad (6.6)$$

Third, imagine that one phase contains a solute i, that is also present in the other phase, and that the distribution of the solute between the two phases is in equilibrium at constant temperature and pressure. Now we consider the reversible transfer of an infinitesimal amount of i, dn_i, from the phase α to phase β. If the system is to remain at equilibrium at constant temperature and pressure, the second law requires that the Gibbs energy constant be constant. Thus, $dG = 0$, or

$$dG_\alpha + dG_\beta = 0 \qquad (6.7)$$

According to the fundamental equation, since $dn_\alpha = -dn_i$ and $dn_\beta = dn_i$,

$$-\mu_{i\alpha} dn_i + \mu_{i\beta} dn_i = 0 \qquad (6.8)$$

Thus, it is evident that **the chemical potential of a species must be the same in two phases that are in equilibrium:**

$$\mu_{i\alpha} = \mu_{i\beta} \qquad (6.9)$$

This relation applies to all species. Of course, equations 6.3, 6.6, and 6.9 can be applied to any number of phases.

If phases α and β are not in equilibrium, and a small quantity dn_i of species i is transferred from phase α to phase β in the direction approaching equilibrium at constant temperature and pressure, we have

$$dG_\alpha + dG_\beta < 0 \quad \text{or} \quad -\mu_{i\alpha} dn_i + \mu_{i\beta} dn_i < 0 \qquad (6.10)$$

if dn_i is positive, then

$$\mu_{i\alpha} > \mu_{i\beta} \qquad (6.11)$$

Thus, **a substance will pass spontaneously from the phase where it has the higher chemical potential to the phase where it has the lower chemical potential.** In this respect the chemical potential is like other kinds of potential, such as electrical and gravitational, in that spontaneous change is always in the direction from high to low potential.

6.2 Phase Diagrams of One-Component Systems

The conditions for the existence of various phases in a one-component system are conveniently shown in a plot of pressure versus temperature, and several of these phase diagrams are shown in Fig. 6.2. The phase diagram for a one-

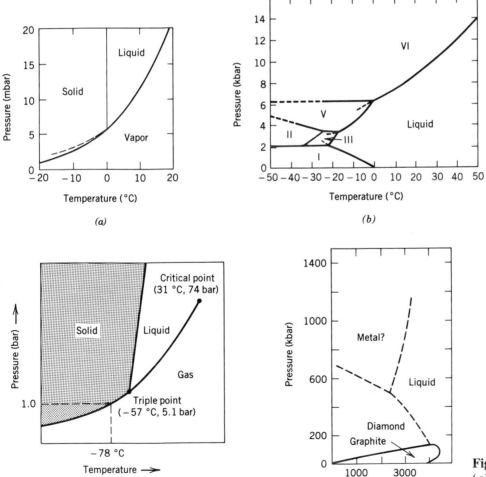

Figure 6.2 Phase diagrams for (*a*) and (*b*) water, (*c*) carbon dioxide, and (*d*) carbon.

component system that contracts on freezing was shown in Fig. 1.5. The phase diagram for water at low pressures is given in Fig. 6.2a, and the phase diagram at high pressures is given in Fig. 6.2b. In Fig. 6.2a, the vapor pressure of water is given as a function of temperature by the line from the **triple point** (at 0.01 °C and 611 Pa) to the **critical point** (Section 1.8). The curve for the sublimation pressure of ice goes down to zero at 0 K. At higher pressures, four other crystal forms of ice are formed. Starting at the triple point, the freezing point of ice *I* is lowered to -22 °C when the pressure is raised to 2000 bar, but higher pressures lead to other crystal forms of ice for which dP/dT is positive, as shown in Fig. 6.1b.

The phase diagram for carbon dioxide in Fig. 6.2c shows that there is equilibrium between solid and gas at 1 bar at -78 °C. Liquid carbon dioxide is produced only above 5.1 bar.

The phase diagram for carbon in Fig. 6.2d shows that graphite and diamond are in equilibrium at room temperature only at pressures above 10 000 bar. Diamonds for industrial use are produced at high pressures and temperatures using catalysts. The details of this phase diagram are not well known because of the difficulty in obtaining equilibrium.

In Section 1.1 we discussed the fact that generally it is necessary to specify only two intensive variables to specify the thermodynamic state of a pure gas. This is true for other phases of a pure substance as well, and it is generally most convenient to use temperature and pressure. However, temperature and density might be used. The independent variables that are chosen are referred to as **degrees of freedom,** and it is the number of degrees of freedom that is important, not the particular intensive variables chosen.

When we want to describe a one-component system with two phases present at equilibrium, however, there is just one degree of freedom. The lines in Fig. 6.2 give the conditions where two phases are in equilibrium. If we specify the temperature, the constraint that there must be equilibrium between two phases means that the pressure cannot be set arbitrarily; it must correspond with the temperature as shown by the line in the phase diagram. If we specify the pressure, the equilibrium constraint means that there will be two phases only at a particular temperature. At a triple point there are zero degrees of freedom because three phases are in equilibrium only at a particular temperature and pressure. Thus, the number of degrees of freedom F for a one-component system is given by $F = 3 - p$, where p is the number of phases.*

6.3   Existence of Phases in a One-Component System

To understand the change from solid to liquid to gas phases when a solid is heated at constant pressure, we may consider a plot of chemical potential versus

---

* According to the phase rule (see Section 6.6), the intensive state of liquid water is characterized by two intensive variables. It is all right if we pick temperature and pressure. There is a problem if we pick pressure and molar volume since the temperature is not uniquely determined by these variables. In the case of water, the molar volume has a minimum in the neighborhood of 4 °C. The lesson from this is that we should pick one conjugate variable from each pair, and not two conjugate variables from the same pair.

temperature at constant pressure for the various phases, as shown in Fig. 6.3*a*. As we have already seen (Section 6.1), the stable phase is that with the lowest value of the chemical potential.

If two or three phases of a single component have the same chemical potential at a certain temperature and pressure, they will coexist at equilibrium as at the melting point T_m, boiling point T_b, or triple point. Below the melting point T_m the solid has the lowest chemical potential and is therefore the stable phase. Between T_m and T_b the liquid is the stable phase. It may be seen from this figure that the phase transitions are sudden and there are no indications of drastic change in the properties of the system as the temperature approaches the transition point.

For a single phase of a pure substance, the chemical potential is a function of temperature and pressure. Thus, the chemical potential can be represented as a surface in μ-P-T space. There are chemical potential surfaces of this type for each phase of a substance: gas, liquid, and one or more solid phases. According to thermodynamics the stable form of a substance at a given temperature and pressure will be that with the lowest chemical potential. The surfaces representing any two phases will intersect along a line, and three surfaces will intersect at a point, called the triple point. The phase diagram for a one-component system is the projection of these intersections on the P-T plane.

Rather than looking at surfaces in three dimensions, it is more convenient to consider the chemical potential as a function of temperature at constant pressure, as in Fig. 6.3*b*.

The slopes of the lines giving the chemical potentials of solid, liquid, and gas in Fig. 6.3*a* are given by (see equation 4.36)

$$\left(\frac{\partial \mu}{\partial T}\right)_P = -\overline{S} \qquad (6.12)$$

Since the entropy is positive the slopes are negative, and since $S_g > S_l > S_s$, the slope is more negative for the gas than the liquid and more negative for the liquid than for the solid.

At a lower pressure the plots of μ versus T are displaced, as shown in Fig. 6.3*b*. The effect of pressure on the chemical potential of a pure substance at constant temperature is given by (see equation 4.37)

$$\left(\frac{\partial \mu}{\partial P}\right)_T = \overline{V} \qquad (6.13)$$

Since the molar volume is always positive, the chemical potential μ decreases as the pressure is decreased at constant temperature. Since $V_g \gg V_l$, V_s, this effect is much larger for a gas than for a liquid or solid. As shown in Fig. 6.3*b*, reducing the pressure lowers the boiling point and normally lowers the melting point. The effect on the boiling point is much larger because of the large difference in the molar volumes of gas and liquid. As a result, the range of temperature over which the liquid is the stable phase has been reduced. It is evident that at a sufficiently low pressure the curve for the chemical potential of the gas will intercept the solid curve below the temperature where the solid and liquid have the same chemical potential. At this low pressure the solid will sublime instead of melt; that is, it passes directly into the vapor without going through the liquid state, as illustrated by dry ice in Fig. 6.2*c*.

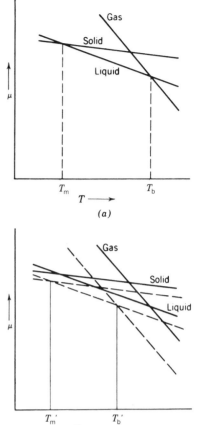

Figure 6.3 Dependence of the chemical potential of solid, liquid, and gas phases on temperature at constant pressure. The dashed lines in (*b*) are for a lower pressure. The plots should be slightly concave downward, since entropy increases with increasing temperature, but they have been drawn as straight lines here for simplicity.

At some particular pressure the solid, liquid, and vapor curves will intersect at a point; the temperature and pressure at which these three phases coexist is referred to as the triple point. If a substance can exist in more than one solid phase, the phase diagram will have more than one triple point, as illustrated in Fig. 6.2b and d.

The transitions we have been discussing in the preceding section are referred to as **first-order phase transitions** because there is a discontinuity in the first derivatives of the chemical potential. Since $(\partial\mu/\partial T)_P$ is different on the two sides of the transition temperature, the two phases have different entropies, and thus different enthalpies. Since $(\partial\mu/\partial P)_T$ is different on the two sides, the two phases have different volumes. The behavior of C_P is more complicated in that dq_P/dT is infinite when two phases coexist at the transition temperature.

In a **second-order phase transition** there is no discontinuity in the first derivatives of μ, but discontinuities occur in the second derivatives (curvature) of the chemical potential. Since $(\partial\mu/\partial T)_P$ is the same on two sides of the transition, there is no discontinuity in the entropy at the transition temperature. Therefore, there is no heat of transition. There is a discontinuity in C_P because $\overline{C}_P = -T(\partial\mu^2/\partial T^2)_P$, as may be derived readily from equation 4.40. However, C_P does not become infinite at the critical point as in the case of first-order transitions. The sudden appearance of superconductivity in certain metals when they are cooled to low temperatures is an example of such a second-order transition.

6.4   The Clapeyron Equation

Consider a one-component system with two phases (α and β) in equilibrium. As shown in Section 6.1, the pressure, temperature, and chemical potential must be the same in the two phases. For the chemical potentials,

$$\mu_\alpha = \mu_\beta \tag{6.14}$$

When the temperature is changed at constant pressure, or the pressure is changed at constant temperature, one of the phases will disappear. However, if the temperature and pressure are both changed in such a way as to keep the two chemical potentials equal to each other, the two phases will continue to coexist. The plot of pressure versus temperature along which the two phases coexist is referred to as the **coexistence curve** (see Fig. 6.4). The necessary relation for dP/dT was derived by Clapeyron.

For a change of pressure and temperature along the coexistence curve,

$$d\mu_\alpha = d\mu_\beta \tag{6.15}$$

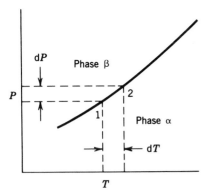

Figure 6.4   Coexistence curve for a one-component system. Along this line $\mu_\alpha = \mu_\beta$. If the temperature is changed dT, the pressure has to be changed dP as indicated to maintain equilibrium. At point 1, $\mu_{1\alpha} = \mu_{1\beta}$. At point 2, $\mu_{1\alpha} + d\mu_\alpha = \mu_{1\beta} + d\mu_\beta$, so that $d\mu_\alpha = d\mu_\beta$.

since the chemical potential is equal to the molar Gibbs energy for a one-component system, $d\mu = d\overline{G} = \overline{V}\,dP - \overline{S}\,dT$ according to equation 4.28. Thus,

$$\overline{V}_\alpha dP - \overline{S}_\alpha dT = \overline{V}_\beta dP - \overline{S}_\beta dT \tag{6.16}$$

or

$$\frac{dP}{dT} = \frac{\overline{S}_\beta - \overline{S}_\alpha}{\overline{V}_\beta - \overline{V}_\alpha} = \frac{\Delta\overline{S}}{\Delta\overline{V}} = \frac{\Delta\overline{H}}{T\,\Delta\overline{V}} \tag{6.17}$$

Although we have considered molar quantities, the quantity unit cancels in this equation. Thus, ΔS and ΔV per unit mass can also be used. This equation is referred to as the Clapeyron equation, and it may be applied to vaporization, sublimation, fusion, or the transition between two solid phases of a pure substance. Note that dP/dT is a total derivative, not a partial derivative; but there is a constraint: the two phases remain in equilibrium so that $\Delta G = 0$. The molar enthalpies of sublimation, fusion, and vaporization at the triple point are related by

$$\Delta_{sub}H = \Delta_{fus}H + \Delta_{vap}H \qquad (6.18)$$

since the heat required to vaporize a given amount of the solid is the same whether this process is carried out directly or by first melting the solid and then vaporizing the liquid.

Water is unusual in that it expands on freezing so that ΔV and thus dP/dT for melting is negative.

Example 6.1

What is the change in the boiling point of water at 100 °C per Pa change in atmospheric pressure?

The molar enthalpy of vaporization is 40.69 kJ mol^{-1}, the molar volume of liquid water is 0.019×10^{-3} m^3 mol^{-1}, and the molar volume of steam is 30.199×10^{-3} m^3 mol^{-1}, all at 100 °C and 1.01325 bar:

$$\frac{dP}{dT} = \frac{\Delta_{vap}H}{T(\overline{V}_g - \overline{V}_l)} = \frac{(40{,}690 \text{ J mol}^{-1})}{(373.15 \text{ K})(30.180 \times 10^{-3} \text{ m}^3 \text{ mol}^{-1})}$$

$$= 3613 \text{ Pa K}^{-1}$$

Thus, $dT/dP = 2.768 \times 10^{-4}$ K Pa^{-1}.

Example 6.2

Calculate the change in pressure required to change the freezing point of water 1 °C. At 0 °C the heat of the fusion of ice is 333.5 J g^{-1}, the density of water is 0.9998 g cm^{-3}, and the density of ice is 0.9168 g cm^{-3}. The reciprocals of the densities, 1.0002 and 1.0908, are the volumes in cubic centimeters of 1 g. The volume change upon freezing $(\overline{V}_l - \overline{V}_s)$ is therefore -9.06×10^{-8} m^3 g^{-1}. For small changes ΔH_{fus}, T, and $(\overline{V}_l - \overline{V}_s)$ are virtually constant, so that

$$\frac{\Delta P}{\Delta T} = \frac{\Delta_{fus}H}{T(\overline{V}_l - \overline{V}_s)} = \frac{33.5 \text{ J g}^{-1}}{(273.15 \text{ K})(-9.06 \times 10^{-8} \text{ m}^3 \text{ g}^{-1})}$$

$$= -1.348 \times 10^7 \text{ Pa K}^{-1}$$

The change in the freezing point of water per bar pressure is

$$\frac{\Delta T}{\Delta P} = \frac{10^5 \text{ Pa bar}^{-1}}{-1.348 \times 10^7 \text{ Pa K}^{-1}} = -0.0075 \text{ K bar}^{-1}$$

This shows that an increase in pressure of 1 bar lowers the freezing point 0.0075 K. The negative sign indicates that an increase in pressure causes a decrease in temperature. The change in pressure required to change the freezing point of water 1 °C is

$$\frac{\Delta P}{\Delta T} = -\frac{1}{0.0075} \text{ K bar}^{-1} = -133 \text{ bar K}^{-1}$$

Example 6.3

Using the data in Appendix C.1, determine the vapor pressures of $H_2O(l)$ and $Br_2(l)$ at 298.15 K. These vaporization processes may be handled as if they were chemical reactions.

$$H_2O(l) = H_2O(g)$$

$$\Delta_{vap}G^\circ_{298} = -RT \ln \frac{P}{P^\circ}$$

$$= -228.572 - (-237.129) = 8.557 \text{ kJ mol}^{-1}$$

$$P = 0.03169 \text{ bar}$$

$$Br_2(l) = Br_2(g)$$

$$\Delta_{vap}G^\circ_{298} = -RT \ln \frac{P}{P^\circ}$$

$$= 3.110 \text{ kJ mol}^{-1}$$

$$P = 0.2852 \text{ bar}$$

Example 6.4

Calculate the equilibrium pressure for the conversion of graphite to diamond at 25 °C. The densities of graphite and diamond may be taken to be 2.25 and 3.51 g cm^{-3}, respectively, independent of pressure, in calculating the change of ΔG with pressure.

$$C(\text{graphite}) = C(\text{diamond})$$

From Appendix C.1,

$$\Delta G^\circ = 2900 \text{ J mol}^{-1}$$

$$\Delta V = 12\left(\frac{1}{3.51} - \frac{1}{2.25}\right) \times 10^{-6} \text{ m}^3 \text{ mol}^{-1} = -1.91 \times 10^{-6} \text{ m}^3 \text{ mol}^{-1}$$

Since $(\partial \Delta G/\partial P)_T = \Delta V$.

$$\int_1^2 d\,\Delta G = \int_{P_1}^{P_2} \Delta V\, dP = \Delta G_2 - \Delta G_2 = \Delta V(P_2 - P_1)$$

$$P_2 = \frac{\Delta G_2 - \Delta G_1}{\Delta V} + P_1$$

$$= \frac{0 - 2900 \text{ J mol}^{-1}}{-1.91 \times 10^{-6} \text{ m}^3 \text{ mol}^{-1}} + 10^5 \text{ Pa}$$

$$= 1.52 \times 10^9 \text{ Pa or } 1.52 \times 10^4 \text{ bar}$$

6.5 The Clausius–Clapeyron Equation

For vaporization and sublimation Clausius showed how the Clapeyron equation may be simplified by assuming that the vapor obeys the ideal gas law and by neglecting the molar volume of the liquid V_l in comparison with the molar volume of the gas V_g. Substituting RT/P for V_g, we have

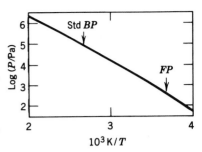

$$\frac{dP}{dT} = \frac{\Delta_{vap}H}{TV_g} = \frac{P\Delta_{vap}H}{RT^2} \tag{6.19}$$

On rearrangement equation 6.19 becomes

$$\frac{dP}{P} = d\ln\frac{P}{P°} = \frac{\Delta_{vap}H}{RT^2}dT \tag{6.20}$$

Figure 6.5 Vapor pressure of water.

where $P°$ is the standard pressure used. Integrating on the assumption that ΔH_{vap} is independent of temperature and pressure yields

$$\int d\ln\frac{P}{P°} = \frac{\Delta_{vap}H}{R}\int T^{-2}\,dT \tag{6.21}$$

$$\ln\frac{P}{P°} = -\frac{\Delta_{vap}H}{RT} + C \tag{6.22}$$

where C is the integration constant. This suggests that a plot of $\ln(P/P°)$ versus $1/T$ should be linear, and this is borne out by data on both vaporization and sublimation, as shown in Fig. 6.5.

Frequently, it is more convenient to use the equation obtained by integrating between limits, P_2 at T_2 and P_1 at T_1, as follows:

$$\int_{P_1/P°}^{P_2/P°} d\ln\frac{P}{P°} = \frac{\Delta_{vap}H}{R}\int_{T_1}^{T_2} T^{-2}\,dT \tag{6.23}$$

$$\ln\frac{P_2}{P_1} = \frac{\Delta_{vap}H}{R}\left[-\frac{1}{T_2} - \left(-\frac{1}{T_1}\right)\right] \tag{6.24}$$

$$\ln\frac{P_2}{P_1} = \frac{\Delta_{vap}H(T_2 - T_1)}{RT_1T_2} \tag{6.25}$$

To represent the vapor pressure as a function of temperature over a wide range of temperature, it is necessary to take the temperature dependence of $\Delta_{vap}H°$ into account. Another deficiency in this simple equation is that the vapor has been assumed to be an ideal gas.

Over narrow ranges of temperature, the enthalpy of vaporization can be taken to be a linear function of temperature; see equation 5.43. However, in calculating vapor pressures over a wider range of temperature, we have to recognize that the enthalpy of vaporization approaches zero as the temperature approaches the critical temperature, as shown in Fig. 6.6.

If the enthalpy of vaporization of a liquid is not known, an approximate value may be estimated using **Trouton's rule** that the molar entropy of vaporization at the standard boiling point (the boiling point at 1 bar pressure) is a constant, about 88 J K^{-1} mol^{-1}:

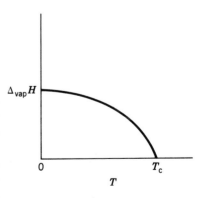

$$\Delta_{vap}S = \frac{\Delta_{vap}H}{T_b} \cong 88 \text{ J K}^{-1}\text{ mol}^{-1} \tag{6.26}$$

Figure 6.6 The enthalpy of vaporization approaches zero as the temperature approaches the critical temperature.

Table 6.1 Enthalpies and Entropies of Fusion at the Melting Point and Vaporization at the Boiling Point[a]

Substance	$\dfrac{T_{\text{fus}}}{K}$	$\dfrac{\Delta_{\text{fus}}H}{\text{kJ mol}^{-1}}$	$\dfrac{\Delta_{\text{fus}}S}{\text{J K}^{-1}\,\text{mol}^{-1}}$	$\dfrac{T_{\text{b}}}{K}$	$\dfrac{\Delta_{\text{vap}}H}{\text{kJ mol}^{-1}}$	$\dfrac{\Delta_{\text{vap}}S}{\text{J K}^{-1}\,\text{mol}^{-1}}$
N_2	63.3	0.720	11.37	77.4	5.577	72.1
C_2H_6	89.9	2.86	31.81	184.52	14.71	79.7
NH_3	195.4	5.653	28.93	239.72	23.33	97.3
CCl_4	250.3	2.5	10.00	349.9	30.0	85.7
H_2O	273.2	6.01	22.00	373.15	40.66	109.0

[a] The boiling points are at 1 atm.

The relative constancy of the entropy of vaporization of many liquids is readily understood in terms of the Boltzmann hypothesis relating entropy to disorder. The change from liquid to vapor leads to increased disorder. The entropy of vaporization is zero at the critical temperature because the liquid and gas are indistinguishable and the enthalpy of vaporization is zero. Most liquids behave alike not only at their critical temperatures but also at equal fractions of their critical temperatures. Hence, different liquids should have about the same entropy of vaporization at their boiling point, provided that there is no association or dissociation upon vaporization. For substances like water and alcohols, which form hydrogen bonds (Section 14.9), the entropy of vaporization is greater than $88\ \text{J K}^{-1}\,\text{mol}^{-1}$. Hydrogen and helium, which boil at only a little above absolute zero, might well be expected to show large departures from this rule. Acetic acid and carboxylic acids, in general, have abnormally low heats of vaporization, since the vapor consists of double molecules and still more energy would be required to break them up into single molecules.

A number of enthalpies and entropies of fusion and vaporization are given in Table 6.1.

Example 6.5

The boiling point of benzene is 80.1 °C at one atm. Estimate the vapor pressure of benzene at 25 °C using Trouton's rule.

The vapor pressure at 353.3 K is 1.013 bar, and the estimated heat of vaporization is $(88\ \text{J K}^{-1}\,\text{mol}^{-1})(353.3\ \text{K}) = 31.1\ \text{kJ mol}^{-1}$:

$$\ln \frac{P_2}{P_1} = \frac{\Delta_{\text{vap}}H(T_2 - T_1)}{RT_1T_2}$$

$$\ln \frac{1.013\ \text{bar}}{P_1} = \frac{(31.1 \times 10^3)(55.1)}{(8.314)(298.2)(353.3)}$$

$$P_1 = 0.143\ \text{bar}$$

6.6 Phase Rule

In 1876, Gibbs* derived a general expression for the number of degrees of freedom for a multiphase system at equilibrium. In Section 6.1 we found that the number of degrees of freedom for a one-component system is given by $F = 3 - p$, where p is the number of phases. Now we want to consider a system with a larger number of components. It is important to remember that we are only interested in describing the intensive state of the system; thus, we are not concerned with the relative amounts of the various phases.

The **number of components** c in a system is defined as the smallest number of species in terms of which the compositions of each of the phases in the system can be described separately. The number of components may be smaller than the number of species s that may be added to form the system, because there may be relationships between the concentrations of various species at equilibrium that make it unnecessary to specify the concentrations of all s species in describing the system. There are two types of relations between concentrations of species: chemical equilibrium expressions and initial conditions. For each independent chemical equilibrium expression, the number of independent mole fractions is reduced by one. The determination of the number of independent chemical reactions in a complex system is described in Section 5.13. The number of independent mole fractions is also reduced by one for each independent relation between mole fractions determined by initial conditions. For example, if NH_3 is added to a reaction system and this is the only source of N_2 and H_2, we know that $3y_{N_2} = y_{H_2}$.

If there are s species, R independent equilibrium constant expressions, and m relations between concentrations due to initial conditions, the number of components c is given by

$$c = s - R - m \, (-1) \tag{6.27}$$

where the 1 represents the electroneutrality condition which is a constraint when ions are present. There are often several possible and equally satisfactory choices of components. The choices are arbitrary, but the **number of components** is an important characteristic of a system.

The **number of degrees of freedom** F of a system is the smallest number of intensive variables (pressure, temperature, and concentrations of components in the various phases) that must be specified to describe completely the state of the system. As we have seen before, to describe the state of a fixed amount of a substance that can exist as a gas or liquid, it is necessary to specify only two variables, T and P, or P and V, or V and T, because the third variable can be calculated from the equation of state. Thus, a system consisting of a pure gas or a pure liquid has two degrees of freedom ($F = 2$).

Consider a system in equilibrium that consists of p phases. When we refer to the number of phases, we mean the number of different kinds of phases; for example, a system containing liquid water and many pieces of ice, but no gas phase, has only two phases. If a phase contains c components, its composition may be specified by stating $(c - 1)$ mole fractions—one less than the number

---

* J. W. Gibbs, *Trans. Conn. Acad. Arts Sci., 1876–1878: The Collected Works of J. Willard Gibbs*, Vol. 1. New Haven, CT: Yale University Press, reprinted 1948.

of components because the mole fraction of one component can be obtained from $\sum x_i = 1$, where x_i represents the mole fraction of component i. Thus, the total number of concentrations to be specified for the whole system is ($c - 1$) for each of the p phases or $p(c - 1)$ concentrations. In general, there are two more variables that have to be considered, temperature and pressure. Thus,

$$\text{Number of variables} = p(c - 1) + 2 \qquad (6.28)$$

Next we consider the number of relationships that must be satisfied for phase equilibrium. The chemical potential μ for each component is the same in each phase α, β, γ, and so on, and so $\mu_{i,\alpha} = \mu_{i,\beta} = \mu_{i,\gamma} = \cdots$ for component i. There are p phases but only ($p - 1$) equilibrium relationships of the type $\mu_{i\alpha} = \mu_{i\beta}$ for each component. For example, if there are two phases, there is only one equilibrium relationship for each component that gives its distribution between the two phases. Altogether there are c components, each one of which can be involved in an equilibrium between phases. Thus,

$$\text{Number of independent equations} = c(p - 1) \qquad (6.29)$$

The difference between the number of variables and the number of independent equations is the number of independent variables, which is referred to as the number of degrees of freedom

$$F = [p(c - 1) + 2] - c(p - 1)$$

or

$$F = c - p + 2 \qquad (6.30)$$

This is the important phase rule of Gibbs. The number of degrees of freedom F is equal to the number of intensive variables that can be set arbitrarily. For example, for a one-component system $F = 3 - p$. Under conditions where a single phase is present, $F = 2$ and the pressure and temperature can both be set arbitrarily. Under conditions where two phases are in equilibrium, $F = 1$ and either the temperature or pressure may be set arbitrarily. Under conditions where three phases are in equilibrium, $F = 0$ and the pressure and temperature are fixed by the equilibrium. Such a system is said to be **invariant** and is represented by a point in a plot of pressure versus temperature.

If the number of intensive variables that is altered is greater than the number of degrees of freedom F, the practical consequence is that one or more phases will be destroyed. For example, consider water at the triple point. If the temperature or pressure is changed, two phases will disappear because the point representing the system in the P-T phase diagram will lie in the solid, liquid, or gas region.

It can be seen from this equation that the greater the number of components in a system, the greater the number of degrees of freedom. On the other hand, the greater the number of phases, the smaller is the number of variables such as temperature, pressure, and mole fraction that must be specified to describe the system completely.

For a given number of components the number of phases is a maximum when $F = 0$. If $F = 0$, phases will be at equilibrium at one set of conditions, and the experimenter cannot change any of these conditions without destroying a phase. For a one-component system, the maximum number of phases that can be present at equilibrium is $p = c + 2 = 3$. For a two-component system, the maximum number of phases is 4, and so on.

Equations 6.28 and 6.30 are based on the assumption that pressure and temperature are both variables. If the pressure, for example, is fixed, the phase rule becomes $F = c - p + 1$. On the other hand, if the system were affected by both temperature and pressure and another variable, such as magnetic field strength, the phase rule becomes $F = c - p + 3$.

Example 6.6

In the preceding chapter we discussed the $CaCO_3(s) = CaO(s) + CO_2(g)$ equilibrium; see Fig. 5.3. (*a*) How many degrees of freedom are there when all three phases are present at equilibrium? (*b*) How many degrees of freedom are there when $CaCO_3(s)$ and $CO_2(g)$ are present?

(*a*) $c = s - R = 3 - 1 = 2$

   $F = c - p + 2 = 2 - 3 + 2 = 1$

Therefore, only the temperature or pressure may be varied independently.

(*b*) $F = 2 - 2 + 2 = 2$

Therefore, both the temperature and pressure may be varied without destroying a phase.

6.7   Special Topic: Effect of Pressure and Surface Curvature on the Vapor Pressure of a Substance

First, we consider the vapor pressure p of a flat surface of a liquid that is under pressure $P = p + p_{inert}$, where p_{inert} is the pressure of an inert gas that is insoluble in the liquid. The system is represented in Fig. 6.7. At equilibrium, the chemical potential of the vapor at pressure p is equal to the chemical potential of the liquid at pressure P:

$$\mu(gas,T,p) = \mu(liquid,T,P) \tag{6.31}$$

The derivative of this equation with respect to total pressure P at constant T is given by

$$\left[\frac{\partial\mu(gas)}{\partial p}\right]_T \left(\frac{\partial p}{\partial P}\right)_T = \left[\frac{\partial\mu(liq)}{\partial P}\right]_T \tag{6.32}$$

We have seen that, in general, $(\partial\mu/\partial P)_T = \overline{V}$ (equation 6.13), so that this equation can be written

$$\overline{V}_g \left(\frac{\partial p}{\partial P}\right)_T = \overline{V}_l$$

or

$$\left(\frac{\partial p}{\partial P}\right)_T = \overline{V}_l/\overline{V}_g \tag{6.33}$$

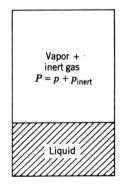

Figure 6.7   Vapor pressure of a liquid under pressure; $P = p + p_{inert}$, where p is the vapor pressure.

Since the molar volume of the liquid is very much smaller than that of the gas, the vapor pressure of the liquid rises very slowly with the total pressure P. If the vapor behaves ideally so that $\overline{V}_g = RT/p$, this equation can readily be integrated from the pressure p_0 of the liquid in contact with vapor only to the pressure with the inert gas present:

$$RT \int_{p_0}^{p} \frac{dp}{p} = \overline{V}_l \int_{p_0}^{P} dP$$

$$RT \ln \frac{p}{p_0} = \overline{V}_l (P - p_0) \qquad (6.34)$$

Thus, the vapor pressure p of the liquid at a higher pressure P can be calculated if the molar volume of the liquid is known.

There is another way that pressure can be exerted on a liquid, and that is by surface tension. When the liquid is in a small droplet, the pressure from surface tension raises the pressure inside the droplet, and this can significantly increase the vapor pressure. In Section 4.16 we have seen that the difference in pressure across a curved surface is given by

$$\Delta P = \frac{2\gamma}{r} \qquad (6.35)$$

where γ is the surface tension (Section 2.9) and r is the radius of curvature (a signed quantity). The vapor pressure of a small droplet exceeds that of a plane surface of the liquid, and the vapor pressure of a concave surface of a liquid (as in a bubble within a liquid) is less than that of a plane surface.

The effect of surface tension on the vapor pressure of a droplet can be obtained by substituting equation 6.35 in equation 6.34 to obtain

$$RT \ln \frac{p}{p_0} = \frac{2\gamma \overline{V}_l}{r} \qquad (6.36)$$

This equation, which is referred to as the **Kelvin equation,** gives the vapor pressure p of a droplet of radius r. The Kelvin equation is based on the assumption that the surface tension is independent of the radius of curvature; therefore, it is not accurate when the radius of curvature becomes quite small. Nevertheless, the Kelvin equation is helpful in understanding the nucleation of condensation and boiling. Figure 6.8 shows the magnitude of the effect on the vapor pressure of water at 25 °C.

We can think of small droplets as having higher vapor pressures than liquids with planar interfaces because surface molecules are not drawn into the interior by so many near neighbors. For a vapor to condense in the absence of foreign surfaces it is necessary for small clusters of molecules to form and to grow and finally coalesce to form the bulk phase. This does not happen if the pressure of the vapor is only slightly higher than the equilibrium vapor pressure, because the very small droplets that are formed first have a higher vapor pressure. However, when the pressure has been increased sufficiently over the equilibrium value, general condensation of droplets occurs.*

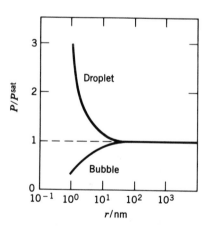

Figure 6.8   Effect of radius of curvature of a surface on the vapor pressure of water at 25 °C. (From R. Aveyard and A. Haydon, *An Introduction to the Principles of Surface Chemistry.* Cambridge, UK: Cambridge University Press, 1973.)

---

* Similar phenomena are involved in the freezing of liquids. Water without impurities or foreign surfaces may be cooled to −40 °C before nucleation begins spontaneously.

Example 6.7

Water vapor is rapidly cooled to 25 °C to find the degree of supersaturation required to nucleate water droplets spontaneously. It is found that the vapor pressure of water must be four times its equilibrium vapor pressure at 25 °C. (*a*) Calculate the radius of a stable water droplet formed at this degree of supersaturation. (*b*) How many water molecules are there in the droplet?

$$(a) \quad r = \frac{2\overline{V}_1\gamma}{RT \ln (P/P^{\text{sat}})} = \frac{2(18 \times 10^{-6} \text{ m}^3 \text{ mol}^{-1})(0.071\,97 \text{ N m}^{-1})}{(8.314 \text{ J K}^{-1} \text{ mol}^{-1})(298 \text{ K})\ln 4} = 0.75 \text{ nm}$$

$$(b) \quad N = \frac{\frac{4}{3}\pi r^3 \rho}{M/N_{\text{A}}} = \frac{\frac{4}{3}\pi(0.75 \times 10^{-9} \text{ m})^3(1 \times 10^3 \text{ kg m}^{-3})}{(18 \times 10^{-3} \text{ kg mol}^{-1})/(6.022 \times 10^{23} \text{ mol}^{-1})} = 59$$

In view of the assumption that the surface tension is independent of the radius of curvature, these values must be considered as approximations.

If the sign of one side of this equation is changed, it yields the vapor pressure of the concave surface of a liquid in a capillary or a small bubble. We can think of the surface of the liquid in a capillary that it wets or in a small bubble as having a lower vapor pressure than the bulk liquid because molecules in the surface are drawn into the interior by more near neighbors than in a flat surface. The Kelvin equation helps us to understand why liquids have a tendency to superheat at their boiling points. If a small bubble starts to form at the boiling point, equation 6.36 (with a sign change) is not satisfied, and the bubble will be squeezed out of existence by the force of surface tension. At a temperature above the boiling point, the vapor pressure will be enough higher that a bubble of a certain radius will be thermodynamically stable.

References

K. E. Bett, J. S. Rowlinson, and G. Saville, *Thermodynamics for Chemical Engineers*. Cambridge, MA: MIT Press, 1975.

R. S. Bradley and D. C. Munro, *High Pressure Chemistry*. New York: Pergamon, 1965.

A. Findlay, A. N. Campbell, and N. O. Smith, *The Phase Rule and Its Applications*. New York: Dover, 1951.

A. Reisman, *Phase Equilibria*. New York: Academic, 1970.

S. I. Sandler, *Chemical and Engineering Thermodynamics*. New York: Wiley, 1989.

J. M. Smith and H. C. Van Ness, *Introduction to Chemical Engineering Thermodynamics*. New York: McGraw-Hill, 1987.

Problems

6.1   Derive

$$\overline{C}_P = -T\left(\frac{\partial^2 \mu}{\partial T^2}\right)_P$$

from equation 4.40:

$$H = G - T\left(\frac{\partial G}{\partial T}\right)_P$$

6.2   The boiling point of hexane at 1 atm is 68.7 °C. What is the boiling point at 1 bar? Given: The vapor pressure of hexane at 49.6 °C is 53.32 kPa.

6.3   What is the boiling point of water 2 miles above sea level? Assume that the atmosphere follows the barometric formula (equation 1.65) with $M = 0.0289$ kg mol^{-1} and $T = 300$ K. Assume the enthalpy of vaporization of water is 44.0 kJ mol^{-1} independent of temperature.

6.4   Liquid mercury has a density of 13.690 g cm^{-3}, and solid mercury has a density of 14.193 g cm^{-3}, both being measured at the melting point, -38.87 °C, at 1 bar pressure. The heat of fusion is 9.75 J g^{-1}. Calculate the melting points of mercury under a pressure of (a) 10 bar, and (b) 3540 bar. The observed melting point under 3540 bar is -19.9 °C.

6.5   From the $\Delta_f G°$ of Br$_2$(g) at 25 °C, calculate the vapor pressure of Br$_2$(l). The pure liquid at 1 bar and 25 °C is taken as the standard state.

6.6   Calculate $\Delta G°$ for the vaporization of water at 0 °C using data in Appendix C.1 and assuming that $\Delta H°$ for the vaporization is independent of temperature. Use $\Delta G°$ to calculate the vapor pressure of water at 0 °C.

6.7   The change in Gibbs energy for the conversion of aragonite to calcite at 25 °C is -1046 J mol^{-1}. The density of aragonite is 2.93 g cm^{-3} at 25 °C and the density of calcite is 2.71 g cm^{-3}. At what pressure at 25 °C would these two forms of CaCO$_3$ be in equilibrium?

6.8 n-Propyl alcohol has the following vapor pressures:

$t/°C$	40	60	80	100
P/kPa	6.69	19.6	50.1	112.3

Plot these data so as to obtain a nearly straight line, and calculate (a) the enthalpy of vaporization, (b) the boiling point at 1 bar, and (c) the boiling point at 1 atm.

6.9   For uranium hexafluoride the vapor pressure (in Pa) for the solid and liquid are given by

$$\ln P_s = 29.411 - 5893.5/T$$

$$\ln P_l = 22.254 - 3479.9/T$$

Calculate the temperature and pressure of the triple point.

6.10   The heats of vaporization and of fusion of water are 2490 J g^{-1} and 333.5 J g^{-1} at 0 °C. The vapor pressure of water at 0 °C is 611 Pa. Calculate the sublimation pressure of ice at -15°C, assuming that the enthalpy changes are independent of temperature.

6.11   The sublimation pressures of solid Cl$_2$ are 352 Pa at -112 °C and 35 Pa at -126.5 °C. The vapor pressures of liquid Cl$_2$ are 1590 Pa at -100 °C and 7830 Pa at -80 °C. Calculate (a) $\Delta_{sub}H$, (b) $\Delta_{vap}H$, (c) $\Delta_{fus}H$, and (d) the triple point.

6.12   The vapor pressure of solid benzene, C$_6$H$_6$, is 299 Pa at -30 °C and 3270 Pa at 0 °C, and the vapor pressure of liquid C$_6$H$_6$ is 6170 Pa at 10 °C and 15,800 Pa at 30 °C. From these data, calculate (a) the triple point of C$_6$H$_6$, and (b) the enthalpy of fusion of C$_6$H$_6$.

6.13   The boiling point of n-hexane at 1 bar is 68.6 °C. Estimate (a) its molar heat of vaporization, and (b) its vapor pressure at 60 °C.

6.14   From tables giving $\Delta_f G°$, $\Delta_f H°$, and $\overline{C}_P°$ for H$_2$O(l) and H$_2$O(g) at 298 K, calculate (a) the vapor pressure of H$_2$O(l) at 25 °C, and (b) the boiling point at 1 atm.

6.15   What is the maximum number of phases that can be in equilibrium in one-, two-, and three-component systems?

6.16   How many degrees of freedom are there for the following systems, and how might they be chosen?

(a)   CuSO$_4$·5H$_2$O(s) in equilibrium with CuSO$_4$(s) and H$_2$O(g).

(b)   N$_2$O$_4$ in equilibrium with NO$_2$ in the gas phase.

(c)   CO$_2$, CO, H$_2$O, and H$_2$ in equilibrium in the gas phase.

(d)   The system described in (c) is made up of specified amount of CO and H$_2$.

6.17   Graphite is in equilibrium with gaseous H$_2$O, CO, CO$_2$, H$_2$, and CH$_4$. How many degrees of freedom are there? What degrees of freedom might be chosen for an equilibrium calculation?

6.18   A gaseous system contains CO, CO$_2$, H$_2$, H$_2$O, and C$_6$H$_6$ in chemical equilibrium. How many components are there? How many independent reactions? How many degrees of freedom are there?

6.19   The vapor pressure of water at 25 °C is 23.756 mm Hg. What is the vapor pressure of water when it is in a container with an air pressure of 100 bar, assuming the dissolved gases do not affect the vapor pressure. The density of water is 0.997 07 g cm^{-3}.

6.20   Ice has the unusual property of a melting point that is lowered by increasing pressure. If this is the reason we can skate on ice, would a 75-kg skater whose skates contact the ice with an area of 0.1 cm² be able to skate at -3 °C?

6.21   Calculate $\Delta G°$ for the vaporization of water at 0 °C using data from Appendix C.1 and assuming that ΔC_P for the vaporization of water is independent of temperature. Use $\Delta G°$ to calculate the vapor pressure of water at 0 °C.

6.22   Calculate the vapor pressure of liquid mercury at 25 °C using data in Appendix C.1.

6.23   Calculate an approximate value for the transition temperature for

$$CaCO_3(\text{calcite}) = CaCO_3(\text{aragonite})$$

using data in Appendix C.1 and assuming $\Delta C_P° = 0$.

6.24   The vapor pressure of Hg(l) is 0.133 bar at 260 °C and 0.533 bar at 330 °C. Assume that $\Delta C_P° = 0$ and that mercury vapor is an ideal gas. What are the values of $\Delta_{vap}H°$, $\Delta_{vap}G°$, and $S°(g)$ at 25 °C? The entropy of Hg(l) is 76.0 J K⁻¹ mol⁻¹ at 25 °C.

6.25   The enthalpy of vaporization of toluene is 38.1 kJ mol⁻¹. Given that the boiling point of toluene at 1 atm is 110.6 °C, what is the boiling point at 1 bar and what is the change in boiling point with this change in pressure?

6.26   What is the boiling point of water on a mountain where the barometer reading is 88 kPa? The heat of vaporization of water may be taken to be 40.67 kJ mol⁻¹.

6.27   The sublimation pressure of solid CO_2 is 133 Pa at -134.3 °C and 2660 Pa at -114.4 °C. Calculate the enthalpy of sublimation.

6.28   Estimate the vapor pressure of ice at the temperature of solid carbon dioxide (-78 °C at 1 bar pressure of CO_2), assuming that the heat of sublimation is constant. The heat of sublimation of ice is 2.83 kJ g⁻¹, and the vapor pressure of ice is 611 Pa at 0 °C.

6.29   Given the thermodynamic information on $H_2O(l)$ and $H_2O(g)$ in Appendix C.1, calculate the vapor pressure of water at 500 °C using equation 5.45 (a) with and (b) without the $\Delta_r C_P°$ term.

6.30   If $\Delta_{vap}C_P = \overline{C}_{P,\text{vap}} - \overline{C}_{P,\text{liq}}$ is independent of temperature, then $\Delta_{vap}H = \Delta_{vap}H_0 + T\,\Delta_{vap}C_P$ where $\Delta_{vap}H_0$ is the hypothetical enthalpy of vaporization at absolute zero. Since $\Delta_{vap}C_P$ is negative, $\Delta_{vap}H$ decreases as the tempera-

ture increases. Show that if the vapor is an ideal gas, the vapor pressure is given as a function of temperature by

$$\ln P = \frac{-\Delta_{vap}H_0}{RT} + \frac{\Delta_{vap}C_P}{R}\ln T + \text{constant}$$

6.31   The vapor pressure of toluene is 8.00 kPa at 40.3 °C and 2.67 kPa at 18.4 °C. Calculate (a) the heat of vaporization, and (b) the vapor pressure at 25 °C.

6.32   At 0 °C ice absorbs 333.5 J g⁻¹ in melting; water absorbs 2490 J g⁻¹ in vaporizing. (a) What is the enthalpy of sublimation of ice at this temperature? (b) At 0 °C the vapor pressure of both ice and water is 611 Pa. What is the rate of change of vapor pressure with temperature dP/dT for ice and liquid water at this temperature? (c) Estimate the vapor pressure of ice and of liquid water at -5 °C.

6.33   According to Trouton's rule, the entropy of vaporization of a liquid at its boiling point is 88 J K⁻¹ mol⁻¹. What is the change in boiling point expected for a liquid with a boiling point of (a) 100 °C and (b) 200 °C at 101 325 Pa in going to a reference state of 1 bar?

6.34   Calculate $\Delta G°$ for

$$H_2O(g, 25 \text{ °C}) = H_2O(l, 25 \text{ °C})$$

The vapor pressure of water at 25 °C is 3168 Pa.

6.35   An aqueous solution contains NaCl, NaBr, KCl, and KBr. How many components are there in the solution?

6.36   How many degrees of freedom do the following systems have? (a) $NH_4Cl(s)$ is allowed to dissociate to $NH_3(g)$ and $HCl(g)$ until equilibrium is reached. (b) A solution of alcohol in equilibrium with its vapor.

6.37   (a) A system consists of H_2, O_2, and H_2O in the gas phase at a sufficiently low temperature that there is no reaction. How many degrees of freedom are there, and what would be a suitable choice of independent variables to define the intensive state of the system? (b) A catalyst is added to the above system and equilibrium is reached. How many degrees of freedom are there now, and what would be a suitable choice of independent intensive variables?

6.38   The vapor pressure of mercury is 133.3 Pa at 126.2 °C. When mercury is enclosed in a container with an air pressure of 500 bar at 126.2 °C, what is the vapor pressure of the mercury? The density of mercury can be taken as 13.6 g cm⁻³.

7
Phase Equilibrium: Two- and Three-Component Systems

Two-component, liquid–liquid systems may show very complicated behavior. The concept of ideal solutions provides a standard with which real solutions may be compared. Ideal solutions follow Raoult's law, and so it is convenient to express the deviations of real solutions from ideality by use of activity coefficients calculated from deviations from Raoult's law. Real solutions of nonelectrolytes follow Henry's law at low concentrations, and activity coefficients can also be calculated from deviations from Henry's law. Ideal solubility, freezing point depression, and osmotic pressure are also discussed in this chapter.

This chapter provides an introduction to phase equilibrium in two-component systems consisting of solid and liquid phases, including the formation of congruently and incongruently melting compounds, and the formation of solid solutions. The chapter ends with discussions of several phase diagrams for systems of three components.

7.1 Vapor–Liquid Equilibria of Binary Liquid Mixtures

When a binary liquid mixture is in equilibrium with its vapor at a constant temperature, the chemical potential of each component is the same in the gas and liquid phases:

$$\mu_i(g) = \mu_i(l) \tag{7.1}$$

Rather than using fugacities for the components in the vapor phase, we will assume the vapor is an ideal gas in this chapter. Thus, the chemical potential of a component in the gas phase is given by

$$\mu_i(g) = \mu_i^\circ(g) + RT \ln \frac{P_i}{P^\circ} \tag{7.2}$$

where P° is the standard state pressure of 1 bar. The most general way to express the chemical potential of a component in a liquid mixture is to use the activity, which was introduced in Section 4.15:

$$\mu_i(l) = \mu_i^\circ(l) + RT \ln a_i \tag{7.3}$$

Thus, equation 7.1 can be written

$$\mu_i^\circ(g) + RT \ln \frac{P_i}{P^\circ} = \mu_i^\circ(l) + RT \ln a_i \tag{7.4}$$

This equation can also be applied to pure liquid i which, of course, has an activity of unity in the liquid phase:

$$\mu_i^\circ(g) + RT \ln \frac{P_i^*}{P^\circ} = \mu_i^\circ(l) \tag{7.5}$$

The equilibrium vapor pressure of pure i at temperature T is P_i^*. Now we subtract this equation from equation 7.4 to eliminate the standard chemical potentials of gas and liquid:

$$RT \ln \frac{P_i}{P_i^*} = RT \ln a_i \tag{7.6}$$

or

$$a_i = \frac{P_i}{P_i^*} \tag{7.7}$$

Thus, if the vapor is an ideal gas, **the activity of a component of a solution is equal to the ratio of its partial pressure above the solution to the vapor pressure of the pure liquid.**

This discussion has been quite general, but so far it has not shown us how to predict the partial pressure P_i of a component of an actual solution. In 1884 Raoult found that for certain solutions, the partial pressure of a component is equal to the mole fraction of that component times the vapor pressure of the pure component:

$$P_i = x_i P_i^* \tag{7.8}$$

This equation is not exact, but it is so useful that it is referred to as **Raoult's law.** It is obeyed most closely when the components are quite similar. Mixtures

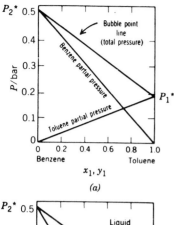

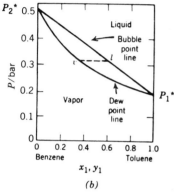

Figure 7.1   Benzene (2)–toluene (1) at 60 °C. (*a*) Partial and total pressures. (*b*) Liquid and vapor compositions.

of benzene and toluene obey Raoult's law, as illustrated by Fig. 7.1. In molecular terms Raoult's law is obeyed by pairs of liquids A and B where A–A, A–B, and B–B interactions are all the same. Since the gas phase has been assumed to be ideal, Raoult's law can also be written

$$y_i P = x_i P_i^* \tag{7.9}$$

where y_i is the mole fraction of i in the gas phase.

When equation 7.8 is substituted in equation 7.7, we see that for solutions with very similar components, the activity of a component is equal to its mole fraction; thus, $a_i = x_i$ for an ideal solution. When this relation is substituted in equation 7.3, we obtain

$$\mu_i(l) = \mu_i^\circ(l) + RT \ln x_i \tag{7.10}$$

We are going to take this equation as the definition of an **ideal solution.**

By use of Raoult's law we can calculate the phase diagram for an ideal liquid mixture. The total vapor pressure of an ideal binary liquid mixture is given by

$$
\begin{aligned}
P &= P_1 + P_2 = x_1 P_1^* + x_2 P_2^* \\
&= P_2^* + (P_1^* - P_2^*) x_1
\end{aligned}
\tag{7.11}
$$

This equation for the **bubble point line** is plotted in Fig. 7.1 and in Fig. 7.2*a*. At points in the phase diagram above the bubble point line, the system is in the liquid state. Suppose the pressure on a specific binary solution of benzene and toluene is reduced from some high value. When the pressure reaches the bubble point line, it is equal to the total vapor pressure of the solution given by equation 7.11 and a further lowering of the pressure will cause bubbles of vapor to form.

The composition of the vapor in equilibrium with a binary solution can readily be calculated using Raoult's law. The mole fraction of component 1 in the vapor is given by

$$y_1 = \frac{P_1}{P_1 + P_2} = \frac{x_1 P_1^*}{x_1 P_1^* + x_2 P_2^*} = \frac{x_1 P_1^*}{P_2^* + (P_1^* - P_2^*) x_1} \tag{7.12}$$

This equation can be solved for x_1 to obtain the following expression for the mole fraction of component 1 in solution that corresponds with a certain mole fraction y_1 in the equilibrium vapor:

$$x_1 = \frac{y_1 P_2^*}{P_1^* + (P_2^* - P_1^*) y_1} \tag{7.13}$$

Now we can use equation 7.9 to calculate the total pressure P that corresponds with a certain mole fraction of component 1 in the vapor phase. Substituting equation 7.13 in equation 7.9 yields

$$P = \frac{P_1^* P_2^*}{P_1^* + (P_2^* - P_1^*) y_1} \tag{7.14}$$

This equation for the **dew point line** is plotted in Fig. 7.1*b* and in Fig. 7.2*b*. At points in the phase diagram below the dew point, the system is in the vapor state. Suppose the pressure on a specific binary vapor of benzene and toluene is raised from some low value. When the pressure reaches the dew point line, it is equal to that given by equation 7.14, and further raising the pressure causes condensation of the first droplets of liquid. In the region of the phase diagram between the dew point line and the bubble point line, two phases are present at equilibrium.

Points on the dew point line and bubble point line at the same pressure represent the compositions of vapor and liquid phases that are in equilibrium. These points are connected by a horizontal line referred to as a **tie line**, one of which is shown in Fig. 7.1*b*. The overall composition of the two-phase system can range from v to l. At v the system is all vapor, and at l the system is all liquid. If the mole fraction of toluene in the system is halfway between v and l, the number of moles of liquid is equal to the number of moles of vapor. If the mole fraction of toluene in the system is x, the ratio of the number of moles of liquid to the number of moles of vapor is equal to $(x - v)/(l - x)$. This rule, which is readily derived from the conservation of moles is referred to as the **lever rule**.

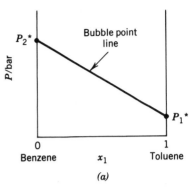

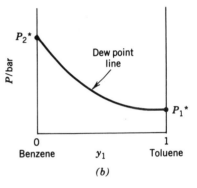

Figure 7.2 (*a*) Plot of the total pressure for the benzene (2)–toluene (1) system versus x_1, as given by equation 7.11. (*b*) Plot of the total pressure for the benzene (2)–toluene (1) system versus y_1, as given by equation 7.14. These plots are superimposed in Fig. 7.1.

Example 7.1
At 60 °C the vapor pressures of pure benzene and toluene are 0.513 and 0.185 bar, respectively. What are the equations of the bubble point line and dew point line? For a solution with 0.60 mole fraction toluene, what are the partial pressures of toluene and benzene, and the mole fraction of toluene in the vapor?

We will consider toluene to be component number 1.

Bubble point line:
$$P = P_2^* + (P_1^* - P_2^*)x_1$$
$$= 0.513 \text{ bar} - (0.328 \text{ bar})x_1$$

Dew point line:
$$P = \frac{P_1^* P_2^*}{P_1^* + (P_2^* - P_1^*)y_1}$$
$$= \frac{0.0949 \text{ bar}^2}{0.185 \text{ bar} - (0.328 \text{ bar})y_1}$$

$$P_1 = x_1 P_1^* = (0.60)(0.185 \text{ bar}) = 0.111 \text{ bar}$$
$$P_2 = x_2 P_2^* = (0.40)(0.513 \text{ bar}) = 0.205 \text{ bar}$$
$$P = 0.513 \text{ bar} - (0.328 \text{ bar})(0.60) = 0.316 \text{ bar}$$
$$y_1 = \frac{x_1 P_1^*}{P_2^* + (P_1^* - P_2^*)x_1}$$
$$= \frac{(0.60)(0.185 \text{ bar})}{0.513 \text{ bar} - (0.328 \text{ bar})(0.60)}$$
$$= 0.351$$

Example 7.2
According to the data in the previous example, what are the activities of toluene (component 1) and benzene (component 2) in a solution containing 0.600 mole fraction toluene according to equation 7.7?

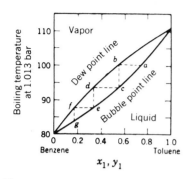

$$a_1 = \frac{P_1}{P_1^*} = \frac{0.111 \text{ bar}}{0.185 \text{ bar}}$$
$$= 0.600$$

$$a_2 = \frac{P_2}{P_2^*} = \frac{0.205 \text{ bar}}{0.513 \text{ bar}}$$
$$= 0.400$$

Since these are ideal solutions, the activities are equal to the mole fractions.

Figure 7.3   Benzene (2)–toluene (1) boiling points: liquid and vapor compositions. The liquid boils at the temperature given by the lower curve.

7.2   Boiling Point Diagrams for Ideal Liquid Mixtures

We have been considering ideal binary liquid mixtures at constant temperature, and now we will consider them at constant pressure because this is of special interest in connection with distillation. Phase diagrams at constant pressure are referred to as **boiling point diagrams.** The boiling point diagram for benzene–toulene solutions at 1.013 bar (1 atm) is shown in Fig. 7.3. At points in the phase diagram above the dew point line, the system is in the vapor state. At points below the bubble point line the system is in the liquid state. Between the two lines, two phases are present and the relative amounts of the two phases is given by the lever rule. Since benzene and toluene form ideal solutions, this diagram may be calculated from the information given in Table 7.1 on the vapor pressures of benzene and toluene at temperatures between their boiling points of 80.1 and 110.6 °C, respectively, at 1.013 bar.

Table 7.1   Vapor Pressures of Toluene (1) and Benzene (2)

	$t/°C$							
	80.1	88	90	94	98	100	104	110.6
P_1^*/bar	—	0.508	0.543	0.616	0.698	0.742	0.836	1.013
P_2^*/bar	1.013	1.285	1.361	1.526	1.705	1.800	2.004	—

Example 7.3
What is the mole fraction x_1 of toluene in the toluene–benzene solution that boils at 100 °C and what is the mole fraction y_1 of toluene in the vapor?
  Equation 7.11 may be solved for x_1:

$$x_1 = \frac{P - P_2^*}{P_1^* - P_2^*} = \frac{1.013 \text{ bar} - 1.800 \text{ bar}}{0.742 \text{ bar} - 1.800 \text{ bar}} = 0.744$$

Now that the mole fraction of toluene in the liquid phase is known, its mole fraction y_1 in the vapor phase is readily calculated using equation 7.9:

$$y_1 = \frac{x_1 P_1^*}{P} = \frac{(0.744)(0.742 \text{ bar})}{1.013 \text{ bar}}$$
$$= 0.545$$

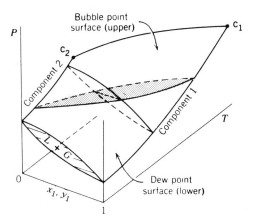

Figure 7.4 Three-dimensional plot for a two-component liquid and vapor system. The plot consists of two surfaces (bubble point surface and dew point surface). (From K. E. Bett, J. S. Rowlinson, and G. Saville, *Thermodynamics for Chemical Engineers*. Cambridge, MA: MIT Press, 1975. Reproduced by permission of The Athlone Press.)

These points are labeled *a* and *b* in Fig. 7.2. For nonideal solutions the points have to be obtained experimentally.

The relationship between the vapor pressure diagram of Fig. 7.1 and the boiling point diagram of Fig. 7.2 is shown in Fig. 7.4. It is possible to plot vapor–liquid data on a two-component system in a three-dimensional diagram because the maximum number of degrees of freedom is $F = c - p + 2 = 2 - 1 + 2 = 3$.

In the P-T-composition diagram the states of pairs of phases in equilibrium with each other define surfaces. These two surfaces come together in the planes that represent the vapor pressures of the two components as a function of temperature. Note that the vapor pressure curve for component 1 ends at the critical point c_1, and the vapor pressure curve for component 2 ends at the critical point c_2. The upper and lower surfaces also come together along a critical locus between c_1 and c_2 made up of the points at which the vapor and liquid phases in equilibrium become identical. In Fig. 7.4 the equilibrium pressures and equilibrium mole fractions of component 1 in the liquid phase x_1 form the upper surface. The lower surface in Fig. 7.4 is made up of points representing the equilibrium pressures and equilibrium mole fractions of component 1 in the vapor phase y_1.

Between the two surfaces two phases are present at equilibrium, saturated vapor and saturated liquid. Since there are two phases, the number of degrees of freedom between the surfaces is $F = 2 - 2 + 2 = 2$. If the temperature and pressure are specified, the compositions of the two phases are fixed, and they are given by the abscissas of the two ends of a horizontal tie line. The relative amounts of the two phases are not fixed by specifying only temperature and pressure, but the phase rule is not concerned with the relative amounts of phases. If the temperature and the composition of the liquid phase are given, the pressure is given by the ordinate and the composition of the vapor phase is given by the other end of the horizontal tie line.

7.3   Fractional Distillation

When a binary solution is partially vaporized, the component that has the higher vapor pressure is concentrated in the vapor phase, thus producing a difference in composition between the liquid and the equilibrium vapor. This vapor may be condensed, and the vapor obtained by partially vaporizing this condensate is still further enriched in the more volatile component. In **fractional distillation** this process of successive vaporization and condensation is carried out in a fractionating column. Figure 7.3 shows that a solution of 0.75 mole fraction toluene and 0.25 mole fraction benzene boils at 100 °C under 1 atm pressure, as indicated by point *a*. The equilibrium vapor is richer in the more volatile compound, benzene, and has the composition *b*. This vapor may be condensed by lowering the temperature along the line *bc*. If a small fraction of this condensed liquid is vaporized, the first vapor formed will have the composition corresponding to *d*. This process of vaporization and condensation may be repeated many times, with the result that a vapor fraction rich in benzene is obtained.

Each vaporization and condensation represented by the line *abcde* corresponds to an idealized process in that only a small fraction of the vapor is condensed and only a small fraction of the condensate is revaporized. It is more practical to effect the separation by means of a distillation column, such as a bubble-cap column illustrated in Fig. 7.5.

Each layer of liquid on the plates of the column is equivalent to the boiling liquid in a distilling flask, and the liquid on the plate above it is equivalent to the condenser. The vapor passes upward through the bubble caps, where it is partially condensed in the liquid and mixed with it. Part of the resulting solution is vaporized in this process and is condensed in the next higher layer, while part of the liquid overflows and runs down the tube to the next lower plate. In this way there is a continuous flow of redistilled vapor coming out the top and a continuous flow of recondensed liquid returning to the boiler at the bottom. To make up for this loss of material from the distilling column, fresh solution is fed into the column, usually at the middle. The column is either well insulated or surrounded by a controlled heating jacket so that there will not be too much condensation on the walls. The whole system reaches a steady state in which the composition of the solution on each plate remains unchanged as long as the composition of the liquid in the distilling pot remains unchanged.

A distillation column may alternatively be packed with material that provides efficient contact between liquid and vapor and occupies only a small volume, so that there is free space to permit a large throughput of vapor. Helices of glass, spirals of screen, and different types of packing are used with varying degrees of efficiency.

The efficiency of a column is expressed in terms of the equivalent number of theoretical plates. The number of **theoretical plates** in a column is equal to the number of successive infinitesimal vaporizations at equilibrium required to give the separation that is actually achieved. The number of theoretical plates

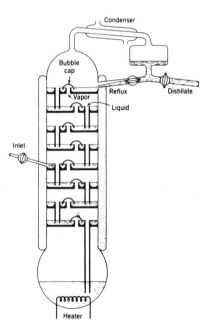

Figure 7.5   Bubble-cap fractionating column.

depends somewhat on the reflux ratio, the ratio of the rate of return of liquid to the top of the column to the rate of distilling liquid off. The number of theoretical plates in a distillation column under actual operating conditions may be obtained by counting the number of equilibrium vaporizations required to achieve the separation actually obtained with the column.

Suppose that in distilling a solution of benzene and toluene with a certain distillation column it is found that distillate of composition g is obtained when the composition of the liquid in the boiler is given by a, in Fig. 7.3. Such a distillation is equivalent to three simple vaporizations and condensations, as indicated by steps abc, cde, and efg. Since the distilling pot itself corresponds to one theoretical plate, the column has two theoretical plates.

7.4   Thermodynamic Properties of Ideal Liquid Mixtures

Before turning to nonideal liquid mixtures, we consider the thermodynamic consequences of equation 7.10, which defines an ideal solution. The molar Gibbs energy of an ideal solution is given by

$$\overline{G} = \sum x_i \mu_i$$
$$= \sum x_i \mu_i^\circ + RT \sum x_i \ln x_i \tag{7.15}$$

Since this equation gives $\overline{G}$ as a function of temperature and pressure, it contains all of the thermodynamic information about an ideal solution. The molar entropy of an ideal solution is obtained from $-(\partial \overline{G}/\partial T)_P$:

$$\overline{S} = \sum x_i \overline{S}_i^\circ - R \sum x_i \ln x_i \tag{7.16}$$

The molar enthalpy of an ideal solution is obtained from $[\partial(\overline{G}/T)/\partial(1/T)]_P$:

$$\overline{H} = \sum x_i \overline{H}_i^\circ \tag{7.17}$$

and the molar volume is obtained from $(\partial \overline{G}/\partial P)_T$:

$$\overline{V} = \sum x_i \overline{V}_i^\circ \tag{7.18}$$

The first terms in these equations give the thermodynamic properties of the amounts of pure liquids required to form the mixture. Therefore, the changes in these thermodynamic properties per mole of solution are given by

$$\Delta_{\mathrm{mix}}G = RT \sum_{i=1}^{N} x_i \ln x_i \tag{7.19}$$

$$\Delta_{\mathrm{mix}}S = -R \sum_{i=1}^{N} x_i \ln x_i \tag{7.20}$$

$$\Delta_{\mathrm{mix}}H = 0 \tag{7.21}$$

$$\Delta_{\mathrm{mix}}V = 0 \tag{7.22}$$

These are the same equations that were obtained for ideal gas mixtures (Section 4.14), except that these equations are written for 1 mol of mixture. **Thus, there is no volume change or heat evolution when liquids are mixed to form ideal solutions at constant temperature and pressure.**

Example 7.4

Derive the equation for the molar enthalpy of an ideal solution (equation 7.17).

$$\left[\frac{\partial(\overline{G}/T)}{\partial T}\right]_P = -\frac{\overline{H}}{T^2}$$

$$\frac{\overline{G}}{T} = \sum \frac{x_i \mu_i^\circ}{T} + R \sum x_i \ln x_i$$

$$\left[\frac{\partial(\overline{G}/T)}{\partial T}\right]_P = \sum x_i \left[\frac{\partial(\mu_i^\circ/T)}{\partial T}\right]_P$$

$$= -\sum \frac{x_i \overline{H}_i^\circ}{T^2}$$

Thus,

$$\overline{H} = \sum x_i \overline{H}_i^\circ$$

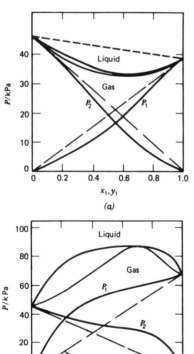

Figure 7.6 (a) Liquid mixture with a negative azeotrope: Chloroform (1)–acetone (2) at 35.17 °C. (b) Liquid mixture with a positive azeotrope: carbon disulfide (1)–acetone (2) at 35.17 °C.

7.5 Vapor Pressure of Nonideal Mixtures

Both negative and positive deviations from Raoult's law are found. Figure 7.6a shows a system with pronounced negative deviations, and Fig. 7.6b shows a system with pronounced positive deviations. Note that in both cases the bubble point line and dew point line are horizontally tangent to each other at the maximum or minimum. Systems with a maximum or minimum are referred to as **azeotropes.** Note that at the azeotropic composition, the vapor has the same composition as the liquid. When a system has a minimum in the vapor pressure plot, it will have a maximum in the boiling point plot, as shown in Fig. 7.7a. When a system has a maximum in the vapor pressure plot, it has a minimum in the boiling point plot, as shown in Fig. 7.7b. When a system forms an azeotrope, its components cannot be separated by simple fractional distillation. For example, in Fig. 7.7 to the left of the maximum, solutions can be separated into component 2 and the azeotrope by fractional distillation, but pure component 1 cannot be obtained. Azeotropes can be "broken" by distilling at another pressure where the system does not form an azeotrope, or by adding a third component.

Now how can we understand systems with a minimum in the vapor pressure plot in molecular terms?

The system acetone–chloroform has a minimum in its vapor pressure curve because of the formation of a weak hydrogen bond between the oxygen of the acetone and the hydrogen of the chloroform:

$$\begin{array}{ccc} \text{Cl} & & \text{CH}_3 \\ | & & | \\ \text{Cl—C—H}\cdots\text{O}{=}\text{C} & & \\ | & & | \\ \text{Cl} & & \text{CH}_3 \end{array}$$

A hydrogen bond is a bond between two molecules, or two parts of one molecule, that results from the sharing of a proton between two atoms, one of which is usually fluorine, oxygen, or nitrogen (Section 12.10). Because of hydrogen bonding, the vapor pressures of both components are less than would be expected of there were no interaction and the mixture obeyed Raoult's law.

Positive deviations from Raoult's law result when A–A and B–B interactions are stronger than A–B interactions. If these deviations are large enough, immiscibility results. When the positive deviations from Raoult's law are larger than in Fig. 7.6*b*, phase separation occurs and the phase diagram takes on the appearance of Fig. 7.8*a*. In addition to the liquid–vapor regions, we now have a region (L_1 and L_2) where two liquid phases are in equilibrium. The corresponding temperature—mole fraction plot is shown in Fig. 7.8*b*. At the lowest temperature shown in this diagram, the solubility of component 1 in component 2 is given by point A. As more component 1 is added, the amount of phase B grows until point B is reached. As the temperature increases above the lowest temperature shown, the solubility of component 1 in component 2 increases and the solubility of component 2 in component 1 increases. Phase separation in binary mixtures is discussed at the end of the chapter as a Special Topic. Phase separation occurs when the Gibbs energy of the two-phase system is lower than that of the homogeneous system.

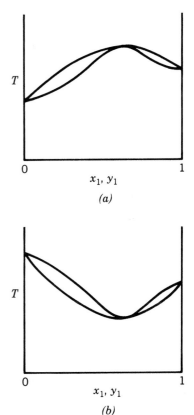

Figure 7.7 (*a*) Boiling point curve for a maximum boiling azeotrope. (*b*) Boiling point curve for a minimum boiling azeotrope.

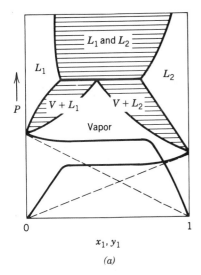

(*a*)

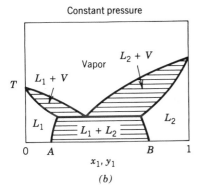

(*b*)

Figure 7.8 (*a*) Vapor pressure diagram for a system showing partial immiscibility. (*b*) Boiling point diagram at constant pressure for a system showing partial immiscibility.

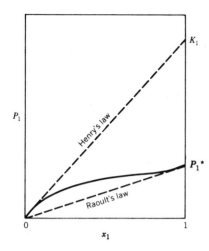

Figure 7.9 Vapor pressure curve for one component of a binary liquid mixture at constant temperature.

7.6 Henry's Law

In all of these phase diagrams Raoult's law is approached for a component as its mole fraction approaches unity. As the mole fraction of a component approaches zero, its partial pressure is given by

$$P_i = y_i P = K_i x_i \qquad (7.23)$$

which is known as **Henry's law.** These two statements are illustrated in Fig. 7.9.

Henry's law results from the circumstance that in dilute enough solutions the environment of the minor component is constant, and its partial pressure is proportional to its mole fraction.

The value of the Henry's law constant K_i (see equation 7.23) is obtained by plotting the ratio P_i/x_i versus x_i and extrapolating to $x_i = 0$. Such a plot is shown later in Fig. 7.11.

It is convenient to express the solubilities of gases in liquids by use of Henry's law constants. A few gas solubilities at 25 °C are summarized in this way in Table 7.2. Up to a pressure of 1 bar Henry's law holds within 1 to 3% for many slightly soluble gases.

Example 7.5

Using the Henry's law constant, calculate the solubility of carbon dioxide in water at 25 °C in moles per liter (represented by using square brackets) at a partial pressure of CO_2 over the solution of 1 bar. Assume that a liter of solution contains practically 1000 g of water.

$$K = \frac{P_i}{x_i} = 0.167 \times 10^9 \text{ Pa} = \frac{101\ 325 \text{ Pa}}{[CO_2]}\left([CO_2] + \frac{1000}{18.02}\right)$$

Since $[CO_2]$ may be considered negligible in comparison with the number of moles of water, 1000/18.02,

$$[CO_2] = \frac{(101\ 325 \text{ Pa})(55.5 \text{ mol L}^{-1})}{0.167 \times 10^9 \text{ Pa}} = 3.37 \times 10^{-2} \text{ mol L}^{-1}$$

Table 7.2 Henry's Law Constants ($K_i/10^9$ Pa) for Gases at 25 °C[a]

Gas	Solvent	
	Water	Benzene
H_2	7.12	0.367
N_2	8.68	0.239
O_2	4.40	
CO	5.79	0.163
CO_2	0.167	0.0114
CH_4	4.19	0.0569
C_2H_2	0.135	
C_2H_4	1.16	
C_2H_6	3.07	

[a] $K_i = P_i/x_i$. The partial pressure of the gas is expressed in pascals.

The solubility of a gas in liquids usually decreases with increasing temperature, since heat is generally evolved in the solution process. There are numerous exceptions, however, especially with the solvents liquid ammonia, molten silver, and many organic liquids. It is common observation that a glass of cold water, when warmed to room temperature, shows the presence of many small air bubbles.

The solubility of an unreactive gas is due to intermolecular attractive forces between gas molecules and solvent molecules. There is a good correlation between the boiling points of gases and their solubilities in solvents at room temperature. Substances with low boiling points (He, H_2, N_2, Ne, etc.) have weak intermolecular attractions and are therefore not very soluble in liquids.

The solubility of gases in water is usually decreased by the addition of other solutes, particularly electrolytes. The extent of this "salting out" varies considerably with different salts, but with a given salt the relative decrease in solubility is nearly the same for different gases. The solubility of liquids in water also shows this salting out phenomenon.

Henry's law gives rise to the concept of **dilute real solutions.** If the vapor pressure of a solute follows Henry's law, then in dilute solutions its chemical potential follows an equation very much like that for an ideal solution (equation 7.10). We can show that as follows. Substituting Henry's law into equation 7.2 yields

$$\mu_i(l) = \mu_i^\circ(g) + RT \ln \frac{K_i x_i}{P^\circ}$$

$$= \mu_i^\circ(g) + RT \ln \frac{K_i}{P^\circ} + RT \ln x_i \tag{7.24}$$

$$= \mu_i^*(l) + RT \ln x_i$$

where

$$\mu_i^*(l) = \mu_i^\circ(g) + RT \ln \frac{K_i}{P^\circ} \tag{7.25}$$

is the chemical potential of solute i in its standard state in the liquid. In this standard state, the solute at $x_i = 1$ has the same properties as in very dilute solutions, where each molecule is surrounded only by molecules of solvent. This standard state is useful, even if it is hypothetical.

For the binary mixture illustrated in Fig. 7.9, the vapor pressure of component 1 in its standard state for a dilute real solution is equal to K_1.

Example 7.6
Show that if Henry's law holds for the solute (component 2), Raoult's law holds for the solvent (component 1).

The Gibbs–Duhem equation (4.104) provides a relationship between the differentials of the chemical potentials of components 1 and 2 at constant temperature and pressure:

$$\text{If } \mu_2 = \mu_2^\circ + RT \ln a_2$$

$$= \mu_2^\circ + RT \ln \frac{K x_2}{P_2^*}$$

$$d\mu_2 = \frac{RT}{x_2} dx_2$$

Using equation 4.104,

$$d\mu_1 = -\frac{x_2}{x_1} d\mu_2 = -\frac{RT}{x_1} dx_2$$

Since $x_1 + x_2 = 1$, $dx_2 = -dx_1$, then

$$d\mu_1 = RT \frac{dx_1}{x_1} = RT \, d \ln x_1$$

$$\mu_1 = RT \ln x_1 + \text{constant}$$

If $x_1 = 1$, then $\mu_1 = \mu_1^\circ$. Therefore,

$$\mu_1 = \mu_1^\circ + RT \ln x_1$$

which can be used to derive Raoult's law. Thus, the range of applicability of Henry's law for the solute is identical with the range of applicability of Raoult's law for the solvent.

7.7 Activity Coefficients Based on Deviations From Raoult's Law

To make quantitative calculations on nonideal solutions, it is convenient to introduce the **activity coefficient** γ_i. Several different activity coefficients can be defined, but we will start with the one based on deviations from Raoult's law. This is the one that is normally used when both components are liquids. One way to look at this is that an activity coefficient is simply inserted into equation 7.10 for the chemical potential of an ideal solution:

$$\mu_i(l) = \mu_i^\circ(l) + RT \ln \gamma_i x_i \qquad (7.26)$$

This equation gives the correct chemical potential for a component of a real solution. Another way of looking at this is to say that we have set the activity of a component equal to $\gamma_i x_i$:

$$a_i = \gamma_i x_i \qquad (7.27)$$

Since the activity of a component always approaches its mole fraction as x_i approaches unity, we can see that

$$\gamma_i \to 1 \quad \text{as} \quad x_i \to 1 \qquad (7.28)$$

If there are positive deviations from Raoult's law, γ_i is greater than unity; and if there are negative deviations from Raoult's law, γ_i is less than unity.

According to equation 7.7, $a_i = P_i/P_i^*$, where P_i^* is the vapor pressure of component i, and according to equation 7.27, $a_i = \gamma_i x_i$, so that we can set these two expressions equal to each other:

$$a_i = \gamma_i x_i = \frac{P_i}{P_i^*} \qquad (7.29)$$

Now we have a way to calculate the activity coefficient of i from experimental data.

$$\gamma_i = \frac{P_i}{x_i P_i^*} \qquad (7.30)$$

Thus, the activity coefficient of component i is equal to the ratio of the partial pressure of i above the solution to the partial pressure of i expected from Raoult's law. Since, for ideal gases, $P_i = y_i P$, this equation can also be written as

$$\gamma_i = \frac{y_i P}{x_i P_i^*} \qquad (7.31)$$

Table 7.3 Activity Coefficients for Acetone–Ether Solutions at 30 °C

Mole Fraction Acetone x_2	Activity Coefficient							
	Rault's Law						Henry's Law	
	Ether			Acetone			Acetone	
	P_1/kPa	$x_1P_1^*$/kPa	γ_1	P_2/kPa	$x_2P_2^*$/kPa	γ_2	K_2x_2/kPa	γ_2
0	86.1	86.1	1.0	0	0	. . .	0	(1.000)
0.2	71.3	68.9	1.04	12.0	7.5	1.60	15.7	0.77
0.4	58.7	51.7	1.14	19.7	15.1	1.31	31.4	0.63
0.5	52.1	43.1	1.21	22.4	18.9	1.19	39.2	0.57
0.6	44.3	34.4	1.28	25.3	22.7	1.12	47.1	0.54
0.8	26.9	17.3	1.56	31.3	30.1	1.04	62.7	0.50
1.0	0	0	. . .	37.7	37.7	1.00	78.4	(0.48)

[a] The activity coefficients for ether are those calculated from Raoult's law.

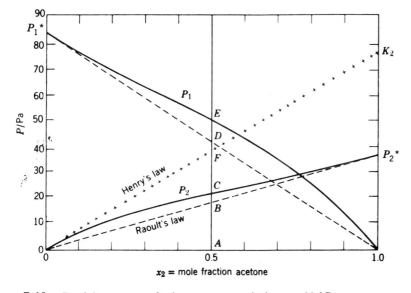

Figure 7.10 Partial pressure of ether–acetone solutions at 30 °C.

Example 7.7

Calculate the activity coefficients for ether (1) and acetone (2) in ether–acetone solutions at 30 °C. The experimental data are given in Table 7.3 and are plotted in Fig. 7.10.

At 0.5 mole fraction acetone, the activity coefficients of the two components are given by

$$\gamma_1 = \frac{P_1}{x_1P_1^*} = \frac{52.1 \text{ kPa}}{(0.5)(86.1 \text{ kPa})} = 1.21$$

$$\gamma_2 = \frac{P_2}{x_2P_2^*} = \frac{22.4 \text{ kPa}}{(0.5)(37.7 \text{ kPa})} = 1.19$$

The activity coefficients of both components, calculated in this way at other mole fractions, are summarized in Table 7.3. It will be noted that as the mole fraction of either component approaches unity, its activity coefficient approaches unity, since the vapor pressure asymptotically approaches that given by Raoult's law.

7.8 Activity Coefficients Based on Deviations from Henry's Law

When one of the components is above its critical temperature, its activity coefficients cannot be calculated from deviations from Raoult's law, and so deviations from Henry's law are used. Activity coefficients of the other component, usually referred to as the solvent, can continue to be based in its deviations from Raoult's law.

When Henry's law is used, the activity coefficient γ_i' is introduced into equation 7.24 for the chemical potential of solute i in a dilute real solution.

$$\mu_i(l) = \mu_i^*(l) + RT \ln \gamma_i' x_i \qquad (7.32)$$

If we replace $\mu_i(l)$ with $\mu_i^\circ(g) + RT \ln(P_i/P^\circ)$ (equation 7.2), and $\mu_i^*(l)$ with $\mu_i^\circ(g) + RT \ln (K_i/P^\circ)$ (equation 7.25), we obtain

$$P_i = \gamma_i' K_i x_i \qquad (7.33)$$

which is the modified form of Henry's law. Thus, the activity coefficient based on Henry's law is calculated from

$$\gamma_i' = \frac{P_i}{x_i K_i} \qquad (7.34)$$

Since $P_i = y_i P$, this can also be written

$$\gamma_i' = \frac{y_i P}{x_i K_i} \qquad (7.35)$$

This activity coefficient is greater than 1 when there are positive deviations from Henry's law, and it is less than 1 when there are negative deviations from Henry's law. Note that

$$\gamma_i' \rightarrow 1 \quad \text{as} \quad x_i \rightarrow 0 \qquad (7.36)$$

because Henry's law is approached as the concentration of the solute goes to zero. In molecular terms this means that the standard state for the solute, in which x_i for the solute is unity, is one in which each molecule of solute has the same interactions that it experiences in very dilute solutions.

To calculate the activity coefficients for acetone based on deviations from Henry's law, we have to calculate the Henry's law constant K_2 from the data in Table 7.3. This is done by extrapolating the apparent Henry's law constant $K_2' = P_2/x_2$ versus x_2 and extrapolating to $x_2 = 0$:

$$K_2' = \frac{P_2}{x_2} \qquad (7.37)$$

The extrapolation of this ratio is illustrated in Fig. 7.11, where values of P_2/x_2 are plotted versus x_2. It is found that the Henry's law constant at infinite dilution (K_2) has a value of 78.3 Pa at 30 °C.

Figure 7.11 Evaluation of Henry's law constant K_2 for acetone in ether–acetone solutions at 30 °C.

If acetone obeyed Henry's law with this value of the constant over the entire concentration range, its vapor pressure for a solution with 0.5 mole fraction would be given by point F in Fig. 7.10. The actual partial pressure is given by point C.

At 0.5 mole fraction acetone, the activity coefficient of acetone based on deviations from Henry's law is

$$\gamma_2 = \frac{P_2}{K_2 x_2} = \frac{22.4 \text{ kPa}}{(0.5)(78.3 \text{ kPa})} = 0.572 \qquad (7.38)$$

The activity coefficients of acetone calculated in this way are summarized in Table 7.3. The activity coefficients for the "solvent" ether remain the same as calculated before on the basis of deviations from Raoult's law.

Since the activity of a component of a real solution is given by P_i/P_i^* (equation 7.7), substituting equation 7.34 yields

$$a_i = \frac{\gamma_i' K_i x_i}{P_i^*} \qquad (7.39)$$

The relation between the two types of activity coefficients can be obtained by substituting $a_i = \gamma_i x_i$,

$$\gamma_i = \frac{\gamma_i' K_i}{P_i^*} \quad \text{or} \quad \gamma_i' = \frac{\gamma_i P_i^*}{K_i} \qquad (7.40)$$

Table 7.4 provides a summary of activity coefficients of nonelectrolytes. **For dilute real solutions, the solvent is usually treated on the basis of deviations from Raoult's law and the solute is usually treated on the basis of deviations from Henry's law.**

For mixtures of two liquids, mole fractions provide a natural way of expressing concentrations, but for other types of solutions, other concentration scales are used. Henry's law may also be written $P_i = K_i m_i$, where m_i is the molal concentration of i (moles per kilogram of solvent), or $P_i = K_i c_i$, where

Table 7.4 Chemical Potential and Activity Coefficients of Nonelectrolytes

Solution	Solvent	Solute
Ideal solutions	$\mu_1(l) = \mu_1^\circ(l) + RT \ln x_1$	$\mu_2(l) = \mu_2^\circ(l) + RT \ln x_2$
		where $\mu_2^\circ(l) = \mu_2^\circ(g) + RT \ln \dfrac{P_2^*}{P^\circ}$
Dilute real solutions	$\mu_1(l) = \mu_1^\circ(l) + RT \ln x_1$	$\mu_2(l) = \mu_2^*(l) + RT \ln x_2$
		where $\mu_2^*(l) = \mu_2^\circ(g) + RT \ln \dfrac{K_2}{P^\circ}$
Real solutions based on deviations from		
Raoult's law	$\mu_1(l) = \mu_1^\circ(l) + RT \ln \gamma_1 x_1$	$\mu_2(l) = \mu_2^\circ(l) + RT \ln \gamma_2 x_2$
Henry's law		$\mu_2(l) = \mu_2^*(l) + RT \ln \gamma_2' x_2$

Note: $\gamma_1 = P_1/x_1 P_1^*$; $\gamma_2 = P_2/x_2 P_2^*$; $\gamma_2' = P_2/x_2 K_2 = \gamma_2 P_2^*/K_2$.

c_i is the molar concentration of i (moles per liter of solution), but we will only use $P_i = K_i x_i$ here.

Although the numerical values of the activity coefficients of acetone depend on which method is employed, the same result is obtained in any thermodynamic calculation using these activity coefficients, independent of method or concentration scale. These thermodynamic calculations involve the comparison of initial and final states, and the standard reference state cancels out.

Example 7.8

What is the change in Gibbs energy for the transfer of 1 mol of acetone from pure liquid acetone to an infinite amount of a solution of an equimolar mixture of acetone and ether at 303 K? Make the calculation with activity coefficients of acetone in the final solution based on deviations from Raoult's law ($\gamma_2 = 1.19$ and $P_2^* = 37.7$ kPa) and based on deviations from Henry's law ($\gamma_2' = 0.572$ and $K_2 = 78.3$ kPa).

Using activity coefficients based on deviations from Raoult's law yields

$$\Delta G = G_{final} - G_{initial} = \mu_2^\circ(l) + RT \ln \gamma_2 x_2 - \mu_2^\circ(l)$$
$$= (8.314 \text{ J K}^{-1} \text{ mol}^{-1})(303 \text{ K})\ln[(1.19)(.5)] = -1311 \text{ J mol}^{-1}$$

Using activity coefficients based on deviations from Henry's law gives

$$\Delta G = G_{final} - G_{initial} = \mu_2^*(l) + RT \ln \gamma_2' x - \mu_2^\circ(l)$$

where $\mu_2^*(l)$ is the chemical potential that acetone would have in the hypothetical standard state in which $x_2 = 1$ if Henry's law ($P_2 = K_2 x_2$) was obeyed over the whole range. This standard state potential is

$$\mu_2^*(l) = \mu_2^\circ(g) + RT \ln \frac{K_2}{P^\circ}$$

We need to eliminate the reference potential of the pure gas in favor of the reference potential of the pure liquid. These are related by

$$\mu_2^\circ(l) = \mu_2^\circ(g) + RT \ln \frac{P_2^*}{P^\circ}$$

where P_2^* is the vapor pressure of pure acetone. Thus,

$$\mu_2^*(l) = \mu_2^\circ(l) + RT \ln \frac{K_2}{P_2^*}$$

Substituting this in the expression for ΔG, we obtain

$$\Delta G = RT \ln \frac{\gamma_2' K_2 x_2}{P_2^*}$$

$$= (8.314)(303) \ln \frac{(0.572)(78.3)(0.5)}{37.7}$$

$$= -1311 \text{ J mol}^{-1}$$

Thus, the same answer is obtained with either convention.

7.9 Determination of Molar Mass by Freezing Point Depression

Now we return to a consideration of ideal solutions and a group of properties that are referred to as **colligative properties.** These properties are freezing point depression, boiling point elevation, osmotic pressure, and the lowering of the vapor pressure by a nonvolatile solute. The latin root of the word colligative means to bind together. The thing that binds these four properties together is that, for ideal solutions, they all depend on the number of particles. Thus, these properties are useful for determining molar masses of solutes. We will consider only two of these properties, freezing point depression and osmotic pressure.

Suppose we want to determine the molar mass of a solute B in a solvent A by the depression of the freezing point of A. Assuming that the solution is ideal and that pure crystalline A freezes out of solution, the equation for equilibrium is

$$\mu_A^{\circ}(s, T) = \mu_A(l, T, x_A) = \mu_A^{\circ}(l, T) + RT \ln x_A \qquad (7.41)$$

Thus, at the T at which the two phases are in equilibrium

$$\ln x_A = \frac{\mu_A^{\circ}(s, T) - \mu_A^{\circ}(l, T)}{RT} = -\frac{\Delta_{fus}G_A^{\circ}(T)}{RT} \qquad (7.42)$$

where $\Delta_{fus}G_A^{\circ}(T)$ is the Gibbs energy of fusion of the solvent at temperature T. Now we make a further assumption that $\Delta_{fus}C_{P,A}^{\circ} = 0$; in other words, we assume that $\Delta_{fus}H_A^{\circ}$ and $\Delta_{fus}S_A^{\circ}$ are independent of temperature in the range near the freezing point. In that case, the Gibbs energy of fusion at temperature T is given by

$$\begin{aligned} \Delta_{fus}G_A^{\circ}(T) &= \Delta_{fus}H_A^{\circ} - T\,\Delta_{fus}S_A^{\circ} \\ &= \Delta_{fus}H_A^{\circ} - T\frac{\Delta_{fus}H_A^{\circ}}{T_{fus,A}} \qquad (7.43) \\ &= \Delta_{fus}H_A^{\circ}\left(1 - \frac{T}{T_{fus,A}}\right) \end{aligned}$$

Substituting this equation into equation 7.42 yields

$$\begin{aligned} \ln x_A &= -\left(\frac{\Delta_{fus}H_A^{\circ}}{R}\right)\left(\frac{1}{T} - \frac{1}{T_{fus,A}}\right) \\ &= -\left(\frac{\Delta_{fus}H_A^{\circ}}{R}\right)\left(\frac{T_{fus,A} - T}{T_{fus,A}T}\right) \end{aligned} \qquad (7.44)$$

If the freezing point depression is small, equation 7.44 can be written as follows:

$$\ln x_A = \ln(1 - x_B) = -\frac{\Delta_{fus}H_A^{\circ}\Delta T_f}{RT_{fus,A}^2} \qquad (7.45)$$

The logarithmic term may be expanded according to

$$\ln(1 - x) = -x - \tfrac{1}{2}x^2 - \tfrac{1}{3}x^3 - \cdots \quad (-1 < x < 1) \qquad (7.46)$$

For dilute solutions the first term is an adequate approximation so that equation 7.45 may be written

$$\Delta T_f = \left(\frac{RT_{fus,A}^2}{\Delta_{fus}H_A^{\circ}}\right)x_B \qquad (7.47)$$

In discussing the depression of the freezing point, the concentration of the solute is generally given in terms of molal concentration m (i.e., moles of solute per kilogram of solvent) rather than as mole fraction. The relation between these concentrations is

$$x_B = \frac{m_B}{1/M_A + m_B} \approx m_B M_A \tag{7.48}$$

where M_A is the molar mass of the solvent, and the approximation applies to dilute solutions. The molal concentration m_B has the units mol kg^{-1}, and $1/M_A$ also has the units mol kg^{-1} when SI units are used.

Substituting equation 7.48 into equation 7.47 yields

$$\Delta T_f = \frac{R T_{fus,A}^2 M_A m_B}{\Delta_{fus} H_A^\circ} = K_f m_B \tag{7.49}$$

where the **freezing point constant** is given by

$$K_f = \frac{R T_{fus,A}^2 M_A}{\Delta_{fus} H_A^\circ} \tag{7.50}$$

The foregoing relations apply only to dilute solutions. Information about activity coefficients may be obtained by studying the freezing points of more concentrated solutions.

Example 7.9
What is the value of the freezing point constant for water? The enthalpy of fusion at 273.15 K is 6.00 kJ mol^{-1}.

$$K_f = \frac{R T_{fus,A}^2 M_A}{\Delta_{fus} H_A^\circ} = \frac{(8.314 \text{ J K}^{-1} \text{ mol}^{-1})(273.1 \text{ K})^2 (18.02 \times 10^{-3} \text{ kg mol}^{-1})}{6000 \text{ J mol}^{-1}}$$

$$= 1.86 \text{ K(mol kg}^{-1})^{-1}$$

According to this value of K_f, 0.1 mol of solute added to 1 kg of water will lower the freezing point 0.186 K, but the relation holds only for dilute solutions. Even a 1-molal solution is too concentrated, and the depression is something less than 1.86 K.

Example 7.10
The addition of a nonvolatile solute to a solvent increases the boiling point above that of the pure solvent. The elevation of the boiling point is given by

$$\Delta T_b = \frac{R T_{b,A}^2 M_A m_B}{\Delta_{mb} H_A^\circ}$$

$$= K_b m_B$$

where $T_{b,A}$ is the boiling point of the pure solvent and M_A is its molar mass. The derivation of this equation parallels that of equation 7.49 very closely, and so it is not given. What is the elevation of the boiling point when 0.1 mol of nonvolatile solute is added to 1 kg of water? The enthalpy of vaporization of water at the boiling point is 40.6 kJ mol^{-1}.

$$K_b = \frac{(8.314 \text{ J K}^{-1} \text{ mol}^{-1})(373.1 \text{ K})^2(0.018 \ 01 \text{ kg mol}^{-1})}{40{,}600 \text{ J mol}^{-1}}$$

$$= 0.513 \text{ K kg mol}^{-1}$$

$$\Delta T_b = (0.513 \text{ K kg mol}^{-1})(0.1 \text{ mol kg}^{-1})$$

$$= 0.0513 \text{ K}$$

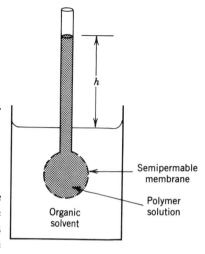

7.10 Osmotic Pressure

When a solution is separated from the solvent by **a semipermeable membrane** that is permeable to solvent but not to solute, the solvent flows through the membrane into the solution, where the chemical potential of the solvent is lower. This process is known as osmosis. This flow of solvent through the membrane can be prevented by applying a sufficiently high pressure to the solution. The **osmotic pressure** Π is the pressure difference across the membrane required to prevent spontaneous flow of solvent in either direction across the membrane. This is illustrated in Fig. 7.12.

Figure 7.12 Determination of the molar mass of a high polymer by use of osmotic pressure.

The phenomenon of osmotic pressure was described by Abbé Nollet in 1748, and Pfeffer, a botanist, made the first direct measurements in 1877. Van't Hoff analyzed Pfeffer's data on the osmotic pressure of sugar solutions and found empirically that an equation quite analogous to the ideal gas law gave approximately the behavior of dilute solutions, namely, $\Pi V = RT$, where V is the volume of solution containing a mole of solute. The origin of the pressure is quite different from that for a gas, however, and the equation of the form of the ideal gas equation is applicable only in the limit of low concentrations.

At equilibrium the chemical potential $\mu_1^\circ(P,T)$ of pure solvent at pressure P is equal to the chemical potential of the solvent in the solution at pressure $P + \Pi$.

$$\mu_1^\circ(P,T) = \mu_1(P + \Pi, T, x_1) \tag{7.51}$$

The osmotic pressure Π that is applied to the solution exactly compensates for the lowering of the chemical potential of the solvent that is caused by the solute. For an ideal solution, equation 7.51 may be written

$$\mu_1^\circ(P, T) = \mu_1^\circ(P + \Pi, T) + RT \ln x_1 \tag{7.52}$$

where $\mu_1^\circ(P + \Pi, T)$ is the chemical potential of the pure solvent at temperature T and pressure $P + \Pi$. According to equation 4.101,

$$d\mu_1 = \overline{V}_1 dP \qquad \text{(constant } T \text{ and composition)} \tag{7.53}$$

where $\overline{V}_1$ is the partial molar volume of the solvent. Thus, the effect on the chemical potential of the solvent of raising the pressure is given by

$$\mu_1^\circ(P + \Pi, T) = \mu_1^\circ(P, T) + \int_P^{P+\Pi} \overline{V}_1^\circ \, dP \tag{7.54}$$

Assuming that $\overline{V}_1^\circ$ is constant and equal to $\overline{V}_1^*$, the molar volume of the solvent, we obtain

$$\mu_1^\circ(P + \Pi, T) = \mu_1^\circ(P, T) + \overline{V}_1^*\Pi \tag{7.55}$$

Substituting equation 7.55 into equation 7.52 yields

$$\bar{V}_1^* \Pi = -RT \ln x_1 = -RT \ln (1 - x_2) \qquad (7.56)$$

This equation is, of course, applicable only to ideal solutions, since equation 7.52 has been used in the derivation.

At sufficiently low mole fraction of the solute, the logarithmic term may be expanded according to equation 7.46. When only the first term in the series is retained, equation 7.56 becomes

$$\bar{V}_1^* \Pi = RTx_2 \qquad (7.57)$$

Since the solution is dilute, $x_2 = n_2/(n_1 + n_2) \simeq n_2/n_1$ and $\bar{V}_1^* = V/n_1$, where V is the volume of the solution and n_2 is the amount of solute, $n_2 = m_2/M_2$, where m_2 represents mass of solute and M_2 is molar mass. Thus, equation 7.57 may be written

$$\Pi V = n_2 RT \qquad (7.58)$$

or

$$\Pi = \frac{m_2}{V} \frac{RT}{M_2} \qquad (7.59)$$

where m_2/V is the concentration in mass per unit volume, and M_2 is the molar mass of the solute. This is the approximate equation that van't Hoff found empirically. It is evident from the approximations introduced why this equation cannot hold for concentrated solutions.

To represent osmotic pressure data on high polymers over a wider range of concentration it is necessary to add terms in higher powers of the concentration, as in the virial equation for gases (Section 1.6):

$$\frac{\Pi}{C} = \frac{RT}{M} + BC + \cdots \qquad (7.60)$$

where C is the concentration (mass per volume). Data on polymer solutions show the analog of the Boyle temperature, that is, the temperature at which the second virial coefficient B is equal to zero. For a polymer solution this temperature is referred to as the **theta temperature** Θ. For a mixed solvent at a given temperature, B may be equal to zero for a particular composition, and so a solvent of this composition is referred to as a theta solvent. It is convenient to use a theta solvent for the determination of the molar mass of a given polymer because equation 7.59 is obeyed over a moderate range of concentrations. Ideal behavior for a polymer solution is obtained when the segment–segment interactions in the polymer are equal to the segment–solvent interactions. At a temperature above the theta temperature ($T > \Theta$), the repulsion between the segments of the polymer is greater and $B > 0$. Since the segment–segment repulsions are greater, the macromolecule assumes an extended form in solution. At a temperature below the theta temperature ($T < \Theta$), the repulsion between segments and solvent is greater and $B < 0$. Since the segments repel solvent and attract each other, the molecules assume a more balled up form.

These changes in the average configuration of a polymer molecule in solution can be confirmed by any method that yields the root-mean-square end-to-end distance (Section 22.4). Thus, the contribution of the polymer to the viscosity of a solution is greater above the theta temperature than below the theta temperature.

In essence, osmotic pressure indicates the number of particles per unit volume. Synthetic high polymers have a distribution of molar masses. In this case equation 7.60 yields the **number-average molar mass** $\overline{M}_n$, which is equal to the mass of the whole sample divided by the amount of substance:

$$\overline{M}_n = \frac{\sum_i n_i M_i}{\sum_i n_i} \tag{7.61}$$

Here n_i is the amount of substance of molar mass M_i per gram of dry polymer. We will see in Section 22.3 that scattering measurements yield the **mass-average molar mass** $\overline{M}_m$. This type of average weights molecules proportionally to their molar masses in the averaging process; that is, the molar mass M_i is multiplied by the mass $n_i M_i$ of substance of that molar mass instead of by the amount. The **mass-average molar mass** $\overline{M}_m$ is defined by

$$\overline{M}_m = \frac{\sum_i w_i M_i}{\sum_i w_i} = \frac{\sum_i n_i M_i^2}{\sum_i n_i M_i} \tag{7.62}$$

In solutions of proteins or other colloidal electrolytes it is necessary to distinguish between the **total osmotic pressure,** which would be obtained with a membrane impermeable to both salt and protein, and the **colloid osmotic pressure,** which is obtained with a membrane permeable to salt ions but not to protein. The latter type of membrane is always used when it is desired to obtain the molar mass of the protein or other colloidal electrolyte.

For colloidal electrolytes in solutions with low concentrations of electrolytes, the measured osmotic pressure is greater than that expected for the colloidal ions alone. This is a result of the fact that, although the salt ions may pass through the membrane, they will not be distributed equally at equilibrium. Donnan showed that because of the high molar mass ion on one side of the membrane, the concentration of the small ion of the same sign as the macroion is lower on that side of the membrane than in the salt solution and that this is compensated by an increased concentration of the small ion of opposite charge. The Donnan effect may be reduced by increasing the salt concentration and, if possible, adjusting the pH to the isoelectric point (Section 9.10) of the colloidal electrolyte.

Example 7.11

A solution of polystyrene in benzene contains 10 g/L. The equilibrium height of the column of solution (density, 0.88 g cm^{-3}) in the osmometer (Fig. 7.12) corrected for capillary rise is 11.6 cm at 25 °C. What is the number average molar mass of the polystyrene, assuming the solution is ideal?

$$P = h\rho g = (0.116 \text{ m})(0.88 \times 10^3 \text{ kg m}^{-3})(9.8 \text{ m s}^{-2})$$

$$= 1000 \text{ Pa}$$

$$\Pi = \frac{mRT}{MV}$$

$$M = \frac{mRT}{\Pi V}$$

$$M = \frac{(10 \text{ g})(8.314 \text{ J K}^{-1} \text{ mol}^{-1})(298 \text{ K})}{(10^3 \text{ Pa})(10^{-3} \text{ m}^3)}$$

$$= 24.8 \times 10^3 \text{ g mol}^{-1}$$

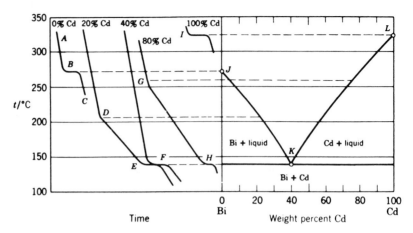

Figure 7.13 Cooling curves and the temperature–concentration phase diagram for the system bismuth–cadmium at constant pressure.

7.11 Two-Component Systems Consisting of Solid and Liquid Phases

First we will consider systems in which the components are completely miscible in the liquid state and completely immiscible in the solid state, so that only the pure solid phases separate out on cooling solutions. Such a phase diagram is illustrated in Fig. 7.13. When molten bismuth or molten cadmium is cooled, the plot of temperature versus time has a nearly constant slope. At the temperature at which the solid crystallizes out, however, the cooling curve becomes horizontal if the cooling is slow enough. The halt in the cooling curve results from the heat evolved when the liquid solidifies. This is shown by the cooling curves for bismuth (labeled 0% Cd) and cadmium in Fig. 7.13 at 273 and 323 °C, respectively.

When a *solution* is cooled, there is a change in slope of the cooling curve at the temperature at which one of the components begins to crystallize out. The change in slope is due to the evolution of heat by the progressive crystallization of the solid as the solution is cooled and to the change in heat capacity. Such changes in slope are evident in the cooling curves for 20% cadmium and 80% cadmium. These curves also show horizontal sections, both at 140 °C. At this temperature both solid cadmium and solid bismuth crystallize out at the same time. The temperature at which this occurs is called a **eutectic temperature.** A solution of cadmium and bismuth containing 40% cadmium shows a single plateau *F* at 140 °C, and so this is the eutectic composition. The eutectic is not a phase; it is a mixture of two solid phases and has a fine grain structure.

The temperatures at which new phases appear, as indicated by the cooling curves, are then transferred to the temperature–composition diagram, as shown at the right in Fig. 7.13. In the area above *JKL* there is one liquid phase. For a two-component system without chemical reaction at constant pressure, the phase rule is $F = 2 - p + 1 = 3 - p$. Therefore, if there is a single phase, there are two degrees of freedom, which can be taken as temperature and one mole fraction. Thus, in the solution region above *JKL*, temperature and composition may be varied without changing the number of phases.

Along *JK*, bismuth freezes out; along *LK*, cadmium freezes out. Thus, in the area under *JK* and down to the eutectic temperature *K* there are two phases: solid bismuth and a solution having a composition that is determined by the temperature. Similarly under *KL* and down to the eutectic temperature *K*, two phases are in equilibrium: solid cadmium and a solution having a composition given by *KL*. In these two regions there is one degree of freedom. Thus, in these regions only the temperature or the composition of the liquid can be specified. If the temperature is specified, the composition of the liquid can be read off of the curved line. If the composition of the liquid is given, the equilibrium temperature can be read of the curved line. Within the two phase regions there are horizontal tie lines (not shown), and for any given point in either region the lever rule (Section 7.1) can be used to calculate the relative amounts of the two phases.

At the eutectic point *K* there are three phases: solid bismuth, solid cadmium, and liquid solution containing 40% cadmium. Then $F = 3 - 3 = 0$, and so this is an invariant point. There is only one temperature and one composition of solution at which these three phases can exist together at equilibrium at a given constant pressure.

The area below the eutectic temperature *K* is a two-phase area in which solid bismuth and solid cadmium are present, and $F = 3 - 2 = 1$. Only the temperature need be specified to describe the system completely at a given constant pressure. The ratio of bismuth to cadmium may change, but there is only a mixture of pure solid bismuth and pure solid cadmium, and there is no need to specify any concentration.

It is evident from Fig. 7.13 that the addition of cadmium lowers the freezing point of bismuth along line *JK*, and that the addition of bismuth lowers the freezing point of cadmium along line *LK*. Alternatively, we may consider that *JK* is the solubility curve for bismuth in liquid cadmium, and *LK* is the solubility curve for cadmium in liquid bismuth. If the solutions are ideal and if the phases that separate are pure solids, solubilities may be calculated.

Earlier in Section 7.9 we discussed the determination of the molar mass of solute B in solvent A in a liquid solution in equilibrium with pure solid A, and derived equation 7.44, which gives the mole fraction of the solvent A as a function of temperature. We can change our point of view and think of A as the solute and B as the solvent. Thus, the solubility x_A of A in solvent B is given by

$$x_A = \exp\left[-\frac{\Delta_{fus}H_A^\circ}{R}\left(\frac{1}{T} - \frac{1}{T_{fus,A}}\right)\right] \qquad (7.63)$$

The remarkable thing about this equation is that it does not contain any parameters for the solvent B. **Thus, the solubility of A is the same in all solvents that form ideal solutions.** It is evident that a high melting point and a large enthalpy of fusion leads to a low solubility. Since the enthalpy of fusion is positive, the ideal solubility increases as the temperature increases.

Example 7.12

Calculate the solubility of bismuth in an ideal solution at 150 and 200 °C and compare the results with Fig. 7.13. The enthalpy of fusion of bismuth at its melting point (273 °C) is 10.5 kJ mol^{-1}. (The enthalpy of fusion is assumed to be independent of temperature.)

$$\ln x_{Bi} = \frac{\Delta_{fus}H_{Bi}\,(T - T_{Bi,fus})}{RTT_{Bi,fus}}$$

At 150 °C

$$\ln x_{Bi} = \frac{(10.5 \times 10^3 \text{ J mol}^{-1})(-123 \text{ K})}{(8.314 \text{ J K}^{-1} \text{ mol}^{-1})(423 \text{ K})(546 \text{ K})}$$

$$x_{Bi} = 0.510$$

At 200 °C

$$\ln x_{Bi} = \frac{(10.5 \times 10^3 \text{ J mol}^{-1})(-73 \text{ K})}{(8.314 \text{ J K}^{-1} \text{ mol}^{-1})(473 \text{ K})(546 \text{ K})}$$

$$x_{Bi} = 0.700$$

The solubility increases with increasing temperature, as we would expect for an endothermic process.

7.12 Compound Formation

The components of a binary system may react to form a solid compound that exists in equilibrium with liquid over a range of composition. If the formation of a compound leads to a maximum in the temperature–composition diagram, as illustrated by Fig. 7.14 for the zinc–magnesium system, we say there is a **congruently melting compound**. The composition that corresponds to the maximum temperature is the composition of the compound. On the mole percent scale such maxima may be achieved at 50%, 33%, 25%, and so on, corresponding to integer ratios of the components of 1:1, 1:2, 1:3, and so on. Figure 7.14 looks very much like two phase diagrams of the type we have discussed placed side by side, but there is a difference. The liquidus curve has a horizontal tangent (zero slope) at the melting point of the congruently melting compound $MgZn_2$, while the slope is not zero at the melting points of the pure components. Thus, additions of small amounts of zinc and magnesium to the compound will not lower the melting or freezing point.

Example 7.13

Six-tenths mole of Mg and 0.40 mol of Zn are heated to 650 °C, represented by point J in Fig. 7.14. Describe what happens when this solution is cooled down to 200 °C, as indicated by the vertical line. (The experiment would have to be done in an inert atmosphere to prevent oxidation by air.) At 470 °C point K is reached and solid $MgZn_2$ separates from solution. The freezing point is gradually lowered as the solution becomes richer in Mg. Finally, at 347 °C, when the liquid is 74 mol % in Mg and 26 mol % in Zn, the whole solution freezes, and solid $MgZn_2$ and solid Mg come out together.

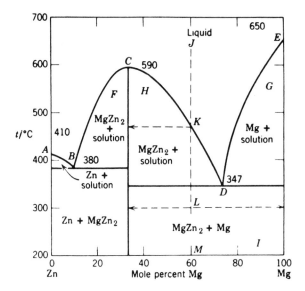

Figure 7.14 Temperature–composition diagram, showing a maximum for the system zinc–magnesium.

From this temperature down there is no further change in the phases. At all temperatures below 347 °C there are pure solids Mg and $MgZn_2$.

Instead of melting, a compound may decompose into another compound and a solution at a definite temperature. This melting point, called an incongruent melting point, is illustrated in Fig. 7.15, which shows part of the phase diagram for the sodium sulfate–water system.

When pure $Na_2SO_4 \cdot 10H_2O$ is heated, it undergoes a transition at 32.38 °C to give anhydrous Na_2SO_4 and solution of composition C. The line BC gives the solubility of $Na_2SO_4 \cdot 10H_2O$ in water, and the line CD gives the solubility

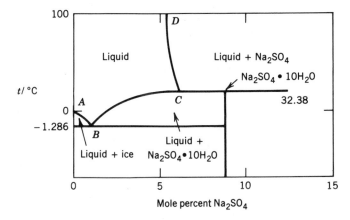

Figure 7.15 Part of the phase diagram for Na_2SO_4–H_2O showing incongruent melting of $Na_2SO_4 \cdot 10H_2O$ to rhombic anhydrous Na_2SO_4.

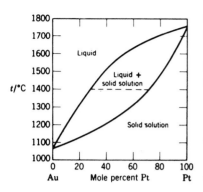

Figure 7.16 Phase diagram for gold–platinum showing solid solutions.

of Na_2SO_4 in water. Figure 7.15 explains the discontinuity in the solubility curve *BCD* for sodium sulfate in water.

When the three phases Na_2SO_4, $Na_2SO_4 \cdot 10H_2O$, and saturated solution are in equilibrium with each other at constant pressure (point *C*), the system is invariant.

7.13 Solid Solutions

Often pure solid freezes out of a solution, but for some systems a solid solution freezes out. A continuous series of solid solutions may be formed, as illustrated in Fig. 7.16 for platinum and gold. The two lines in this diagram give the compositions of the liquid solutions (upper line) and solid solutions (lower line) that are in equilibrium with each other. When these diagrams are studied, it is convenient to remember that the liquid phase is richer in that component of mixture that has the lower melting point.

Above the upper line of Fig. 7.16 the two metals exist in liquid solution; below the lower line the two metals exist in solid solutions. The upper curve is the freezing point curve for the liquid, and the lower one is the melting point curve for the solid. The space between the two curves represents mixtures of the two—one liquid solution and one solid solution in equilibrium. For example, a mixture containing 50 mol % gold and 50 mol % platinum, when brought to equilibrium at 1400 °C, will consist of two phases, a solid solution containing 70 mol % platinum and a liquid solution containing 28 mol % platinum. If the original mixture contained 60 mol % platinum, there would still be the same two liquid and solid solutions at 1400 °C of the same compositions, 70 and 28 mol %, but there would be a relatively greater amount of the solid solution that contains 70 mol % platinum.

The fractional crystallization of solid solutions is seriously complicated by the fact that the attainment of equilibrium is much slower in solid solutions than in liquid solutions. It takes a considerable length of time, particularly at low temperatures, for a change in concentration at the surface to affect the concentration at a point in the interior of the solid solution.

In view of the use of the freezing point as a criterion of purity, it is important to note that when solid solutions are formed the freezing point may be *raised* by the presence of the other component.

Figure 7.16 is analogous to the phase diagram for two miscible liquids and vapor, as shown in Fig. 7.3. For substances forming ideal solid solutions, the phase diagram may be calculated theoretically. Systems exhibiting nonideal solid solution behavior may show maxima or minima in their melting curves (analogous to Fig. 7.7) that have nothing to do with the formation of compounds.

Many properties of alloys, ceramics, and structural materials depend on the presence of solid solutions. The hardening and tempering of steel involve the existence of solid solutions of carbon in different iron–carbon compounds. The solid solution stable at the high temperatures is hard. To retain this hardness, the proper compositions and temperatures are obtained, as indicated by the phase diagrams, and the steel is quenched quickly in oil or water, so that it does not have time to form a different solid solution that is stable at lower temperatures. Reheating the steel to a somewhat lower temperature gives an opportunity for partial conversion to the softer solid solution which is stable at the lower temperature. In this way the steel may be given different degrees of hardening.

Sometimes partial miscibility is encountered in the solid state just as it is in the liquid state. The silver–copper system is an example of partial miscibility of solids. As shown by Fig. 7.17, at 800 °C copper dissolves in solid silver to the extent of 6% by weight, and silver dissolves in copper to the extent of 2% by weight. At the eutectic point, pure copper and silver do not crystallize out, but saturated solid solutions do. The regions α and β represent continuously variable solid solutions, and at constant pressure there are two degrees of freedom F in these regions because there is a single phase.

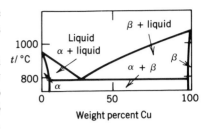

Figure 7.17 Phase diagram for the silver–copper showing partial miscibility of solid solutions.

7.14 Systems of Three Components

For systems of three components the phase rule yields $F = 5 - p$. If there is a single phase, then $F = 4$, and so a complete geometrical representation would require the use of four-dimensional space. If the pressure is constant, a three-dimensional representation may be used. If both the temperature and pressure are constant, then $F = 3 - p$, and so the system may be represented in two dimensions and

$$p = 1 \quad F = 2 \quad \text{bivariant}$$
$$p = 2 \quad F = 1 \quad \text{univariant}$$
$$p = 3 \quad F = 0 \quad \text{invariant}$$

A system of three components has two independent composition variables, say x_2 and x_3. Thus, the composition of a three-component system can be plotted in Cartesian coordinates with x_2 on one axis and x_3 on the other, bounded by the line $x_2 + x_3 = 1$. Since this plot is not symmetrical with respect to the three components, it is more common to plot compositions on an equilateral triangle in which each apex represents a pure component.

In an equilateral triangle, the sum of the distances from any given point to the three sides, along the perpendiculars to the sides, is equal to the height of the triangle. The distance from each apex to the center of the opposite side of the equilateral triangle is divided into 100 parts, corresponding to percentage composition, and the composition corresponding to a given point is readily obtained by measuring the perpendicular distance to the three sides. For example, in Fig. 7.18 point O represents a mixture with a gross composition of 50% by weight acetic acid, 10% by weight vinyl acetate, and 40% by weight water.

Of the many possible kinds of ternary systems, we will consider only certain types formed by three liquids and by a liquid and two solids.

If two pairs of the liquids are completely miscible and one pair is partially miscible, a diagram of the type illustrated in Fig. 7.18 is obtained. This figure represents the system water–acetic acid–vinyl acetate at 25 °C and atmospheric pressure.

When water is added to vinyl acetate along the line BC, the water dissolves at first, forming a homogeneous solution. However, as more water is added,

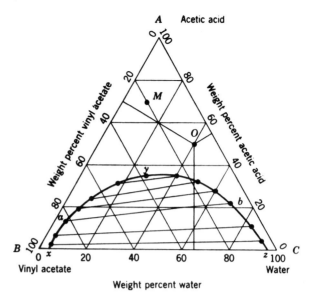

Figure 7.18   Three-component phase diagram at 25 °C and 1 atm showing regions of miscibility.

saturation is reached at composition *x,* and then two liquid phases appear, vinyl acetate saturated with water and water saturated with vinyl acetate, having the composition *z.* As more water is added, the amount of the *z* phase increases and that of the *x* phase decreases, but the composition of each phase remains always the same. Finally, when the percentage of water exceeds that given by *z,* there is only one liquid phase, an unsaturated solution of vinyl acetate in water. At all compositions between *x* and *z* there are two liquid phases with compositions *x* and *z.*

If acetic acid, which is miscible with vinyl acetate and with water in all proportions, is added, it is distributed between the two layers, forming two ternary solutions of vinyl acetate, water, and acetic acid that are in equilibrium with each other, provided that the gross composition of the mixture falls in the region below the *xyz* curve. For example, if the gross composition lies on the line *ab,* the two phases that are in equilibrium are represented by points *a* and *b.*

Other tie lines are shown for other gross compositions; usually tie lines are not parallel to each other or to a side of the triangle. The compositions of the two phases that are in equilibrium with each other, corresponding to the intersection of the tie line with the curves *xy* and *zy,* have to be determined experimentally. As more acid is added, the two phases become more alike, and the tie lines become shorter. Ultimately, when the compositions of the two solutions become identical, the tie line shrinks to a single point *y.* Point *y* is a **critical point,** and further addition of acetic acid will result in the formation of a single homogeneous phase. Any point under the curve represents a ternary mixture that will separate into two liquid phases; any point above the curve represents a single homogeneous liquid phase.

If there are two liquid phases, as in the area below the line *xyz,* the number of degrees of freedom *F* is 1 at fixed *T* and *P,* so that it is necessary to specify the percentage of one component in only one phase to describe the system completely. The percentage of the other components in this phase can be obtained from the intersection of this percentage with the line *xyz,* and the composition of the other phase can be obtained from the intersection of the other end of the tie line with the line *xyz.* For example, if one phase in the two-phase system in Fig. 7.18 contains 5% water, the composition of this phase is given by point *a* and the composition of the other phase by point *b.*

When a substance is added to a two-phase liquid mixture, it is generally distributed with different equilibrium concentrations in the two phases. The distribution of acetic acid between the water-rich and vinyl acetate-rich phases may be calculated from the data of Fig. 7.18. It is apparent from this figure that the ratio of the concentrations of acetic acid in the two phases, as given by the ends of the tie lines, changes with the amount of acetic acid added. However, if the amount of solute added is sufficiently small, it is often found that the distribution coefficient, which is defined as the ratio of the concentrations of the solute in the two phases, is relatively independent of concentration.

Example 7.14

Acetic acid is added to a mixture of 50 g of vinyl acetate and 50 g of water at 25 °C and 1 bar. Describe what happens in terms of the phases present as more and more acetic acid is added, using Fig. 7.18.

The initial system is represented in the middle of the horizontal axis. There are initially two phases: a saturated solution of water in vinyl acetate of composition *x* and a saturated solution of vinyl acetate in water of composition *z*. When acetic acid is added, the gross composition moves up a line from the initial composition to the acetic acid apex. As long as the gross composition is in the two-phase region, the compositions of the two phases that are in equilibrium are given by the two ends of the tie line through the gross composition. When the gross composition is above the two-phase region, there is a single homogeneous phase. As more and more acetic acid is added, the gross composition approaches A.

Substances having slightly different distribution coefficients may be separated by a column operation in which one liquid phase is held by a finely divided solid with a large surface area and the other liquid phase flows through the column. This process is referred to as **partition chromatography** and is closely related to chromatography experiments, depending on adsorption (Section 24.1). In column operation the equivalent of many theoretical plates, each equivalent to a batch extraction, is obtained. The components of a mixture emerge at the bottom of the column one by one at different times if their distribution coefficients are sufficiently different.

Figure 7.19 shows an example of a system involving two solids and a liquid at constant temperature and pressure. Solutions along *RP* are saturated with $Pb(NO_3)_2$, and solutions along *PS* are saturated with $NaNO_3$. At the intersection of these two solubility curves (point *P*), the solution is saturated with respect to both $Pb(NO_3)_2$ and $NaNO_3$; there are three phases in equilibrium, and so this point is invariant if the temperature and pressure are constant. A few tie lines are shown in the two-phase regions.

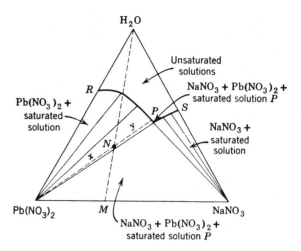

Figure 7.19   Phase diagram for the system lead nitrate–sodium nitrate–water at 25 °C.

A diagram such as Fig. 7.19 is useful in deciding how to obtain the maximum amount of pure substance from a mixture by crystallization from a solvent. For example, if water is added to mixture M, the gross composition moves along the dashed line toward the H_2O apex. If only a small amount of water is added, the phases $Pb(NO_3)_2$, $NaNO_3$, and solution of composition P will be present. If sufficient water is added to reach point N, the only solid phase at equilibrium will be $Pb(NO_3)_2$. If the solution is heated to get all the $NaNO_3$ into solution and then cooled to 25 °C, the solid phase will be pure $Pb(NO_3)_2$. The composition of the mother liquor would be only slightly different from P. The relative amounts of $Pb(NO_3)_2$ and mother liquor are given by y/x. Thus, it is seen that if more water is added the recovery of $Pb(NO_3)_2$ will be reduced.

In a phase diagram like Fig. 7.19, two phase regions are indicated by the tie lines that show the compositions of phases that are in equilibrium. The blank regions may represent either one phase or three phases, and so it is important to understand how to distinguish between these two types of areas. One-phase regions, like the solutions region, have curved boundaries, whereas boundaries of the three-phase regions are straight, so that these regions have a triangular shape.

7.15   Special Topic: Excess Thermodynamic Properties

An excess thermodynamic property M^E is equal to the difference between the actual property M and the property for an ideal solution at the same T, P, and x. Here M represents the extensive properties V, U, H, C_p, S, A, and G. Usually it is convenient to consider a mole of solution, and so we will be concerned with **molar excess properties** that are defined by

$$\overline{M}^E = \overline{M} - \overline{M}^{id} \tag{7.64}$$

where $\overline{M}^{\mathrm{id}}$ is the calculated value for an ideal solution. The property change of mixing $\Delta\overline{M}$ is equal to $\overline{M} - \sum x_i \overline{M}_i^\circ$, and the property change of mixing to form an ideal solution $\Delta\overline{M}^{\mathrm{id}}$ is equal to $\overline{M}^{\mathrm{id}} - \sum x_i \overline{M}_i^\circ$, and so

$$\overline{M}^{\mathrm{E}} = \Delta\overline{M} - \Delta\overline{M}^{\mathrm{id}} \tag{7.65}$$

Thus, an excess thermodynamic property can also be called the excess property change of mixing, but we will use the simpler terminology. Since $\Delta\overline{M}^{\mathrm{id}} = 0$ when $\overline{M}$ is $\overline{V}$, $\overline{U}$, $\overline{H}$, or $\overline{C}_P$, the excess properties $\overline{V}^{\mathrm{E}}$, $\overline{U}^{\mathrm{E}}$, $\overline{H}^{\mathrm{E}}$, and $\overline{C}_P^{\mathrm{E}}$ are simply equal to the corresponding property change of mixing. However, for the entropy and Gibbs energy

$$\overline{S}^{\mathrm{E}} = \Delta\overline{S} + R\sum x_i \ln x_i = S - \sum x_i \overline{S}_i^\circ + R\sum x_i \ln x_i \tag{7.66}$$

$$\overline{G}^{\mathrm{E}} = \Delta\overline{G} - RT\sum x_i \ln x_i = \overline{G} - \sum x_i \overline{G}_i^\circ - RT\sum x_i \ln x_i \tag{7.67}$$

where $\Delta\overline{S}$ is the actual entropy of mixing and $\Delta\overline{G}$ is the actual Gibbs energy of mixing.

The molar Gibbs energy of a solution is given by

$$\overline{G} = \sum x_i \overline{G}_i \tag{7.68}$$

where $\overline{G}_i$ is the partial molar Gibbs energy of component i. Then

$$\overline{G}^{\mathrm{E}} = \sum x_i (\overline{G}_i - \overline{G}_i^\circ) - RT\sum x_i \ln x_i$$
$$= \sum x_i RT \ln a_i - RT\sum x_i \ln x_i \tag{7.69}$$

where equation 4.120 has been used to introduce the activity. Thus,

$$\overline{G}^{\mathrm{E}} = RT\sum x_i \ln \frac{a_i}{x_i} = RT\sum x_i \ln \gamma_i \tag{7.70}$$

where equation 7.27 has been used to introduce the activity coefficient. Since $\overline{G}^{\mathrm{E}}$ is a thermodynamic property that follows the additivity rule with respect to $RT \ln \gamma_i$, then $RT \ln \gamma_i$ can be considered to be a partial molar quantity (Section 4.12).

$$\overline{G}^{\mathrm{E}} = \sum x_i \overline{G}_i^{\mathrm{E}} \tag{7.71}$$

where

$$\overline{G}_i^{\mathrm{E}} = RT \ln \gamma_i \tag{7.72}$$

The excess Gibbs energy $\overline{G}^{\mathrm{E}}$ is very useful because it provides the most compact way of storing information about a real solution. Various mathematical relations have been developed to express $\overline{G}^{\mathrm{E}}$ as a function of T, P, and x. For example, the Margules equation is

$$\overline{G}^{\mathrm{E}} = RT (A_{21}x_1 + A_{12}x_2)x_1 x_2 \tag{7.73}$$

where A_{21} and A_{12} are parameters to be calculated from experimental data or theory. Equations like 7.73 are important because activity coefficients for the two components can be calculated from them:

$$\ln \gamma_1 = x_2^2[A_{12} + 2(A_{21} - A_{12})x_1] \tag{7.74}$$

$$\ln \gamma_2 = x_1^2[A_{21} + 2(A_{12} - A_{21})x_2] \tag{7.75}$$

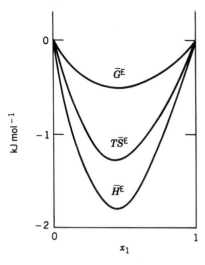

Figure 7.20   Excess thermodynamic properties at 50 °C for acetone (1)–chloroform (2) solutions.

and the other excess properties can be calculated from equations like

$$\overline{H}^{\mathrm{E}} = -RT^2 \left[\frac{\partial(\overline{G}^{\mathrm{E}}/RT)}{\partial T} \right]_{P,x} \tag{7.76}$$

$$\overline{S}^{\mathrm{E}} = -\left(\frac{\partial \overline{G}^{\mathrm{E}}}{\partial T} \right)_{P,x} \tag{7.77}$$

$$\overline{V}^{\mathrm{E}} = \left(\frac{\partial \overline{G}^{\mathrm{E}}}{\partial P} \right)_{T,x} \tag{7.78}$$

It is readily shown that $\overline{G}^{\mathrm{E}} = \overline{H}^{\mathrm{E}} - T\overline{S}^{\mathrm{E}}$.

The values of the excess properties for acetone (1)–chloroform (2) solutions are shown in Fig. 7.20. The formation of these solutions is exothermic, as might be expected from the strong interaction of acetone molecules with chloroform molecules (Section 7.5). However, $\overline{S}^{\mathrm{E}}$ indicates that the solution becomes more "ordered" so that the magnitude of the excess Gibbs energy is smaller than the magnitude of $\overline{H}^{\mathrm{E}}$.

Example 7.15

Calculate the excess Gibbs energies for the acetone–ether solutions in Table 7.3 and determine the parameters A_{12} and A_{21} from a plot of $\overline{G}^{\mathrm{E}}/x_1 x_2 RT$ versus x_2. How well do these parameters represent the activity coefficients based on deviations from Raoult's law for ether and acetone at 30 °C?

The values of $\overline{G}^{\mathrm{E}}/RT$ are calculated using

$$\overline{G}^{\mathrm{E}}/RT = x_1 \ln \gamma_1 + x_2 \ln \gamma_2$$

The values of $\overline{G}^{\mathrm{E}}/x_1 x_2 RT$ were plotted versus x_2 to obtain the values of the parameters in equation 7.73; $A_{12} = 0.68$, $A_{21} = 0.82$. These values were used to calculate γ_1 and γ_2 using equations 7.74 and 7.75.

x_2	$\overline{G}^{\mathrm{E}}/RT$	γ_1 (expt.)	γ_1 (eq. 7.74)	γ_2 (expt.)	γ_2 (eq. 7.75)
0.2	0.1254	1.04	1.04	1.60	1.63
0.4	0.1866	1.14	1.15	1.31	1.29
0.5	0.1823	1.21	1.23	1.19	1.19
0.6	0.1667	1.28	1.33	1.12	1.11
0.8	0.1203	1.56	1.60	1.04	1.02

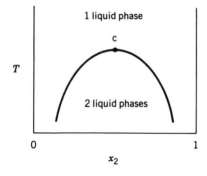

Figure 7.21   Solubility curve for two liquids showing a critical point (c).

7.16   Special Topic: Phase Separation in Binary Mixtures

If positive deviations from Raoult's law are large enough, phase separation results. This means that for a certain range of mole fractions, the mixture will split into two phases, one richer in component 1 and the other richer in component 2. As the temperature is raised, the compositions of the equilibrium phases often approach each other until a critical point is reached, as shown in Fig. 7.21. Sometimes a critical point is reached when the temperature is lowered. It is of interest to see how an expression for the Gibbs energy for the mixture can account for phase separation and what happens at the critical point.

The mixing of x_1 moles of liquid A_1 and x_2 moles of liquid A_2 to form a mole of solution at constant temperature and pressure is represented by

$$x_1 A_1(l) + x_2 A_2(l) = 1 \text{ solution } (x_1, x_2) \tag{7.79}$$

The Gibbs energy of mixing is given by

$$
\begin{aligned}
\Delta_{mix}G &= x_1\mu_1 + x_2\mu_2 - x_1^\circ\mu_1^\circ - x_2^\circ\mu_2^\circ \\
&= x_1\mu_1^\circ + x_1 RT \ln \gamma_1 x_1 + x_2\mu_2^\circ + x_2 RT \ln \gamma_2 x_2 - x_1\mu_1^\circ - x_2\mu_2^\circ \\
&= RT(x_1 \ln x_1 + x_2 \ln x_2) + RT(x_1 \ln \gamma_1 + x_2 \ln \gamma_2) \\
&= \Delta_{mix}G(\text{ideal}) + G^E \tag{7.80}
\end{aligned}
$$

where the first term is the ideal Gibbs energy of mixing (equation 7.19) and the second term is the excess Gibbs energy of mixing. The excess Gibbs energy of mixing is a function of x_1, or x_2, and so we can ask, What is the simplest function that can account for phase separation? We want the effect of the G^E term to vanish at both $x_1 = 0$ and $x_2 = 0$, and so the simplest mathematical form is

$$G^E = b x_1 x_2 = b x_1 (1 - x_1) \tag{7.81}$$

To keep it simple we will assume that b is independent of temperature and pressure. This symmetrical form for the excess Gibbs energy implies that the activity coefficients of the two components are given by

$$\gamma_1 = \exp \frac{b x_2^2}{RT} \quad \text{and} \quad \gamma_2 = \exp \frac{b x_1^2}{RT} \tag{7.82}$$

Thus, as the mole fraction of component 2 goes to zero, the activity coefficient of component 1 goes to unity, as expected, and vice versa.

When equation 7.81 applies, the enthalpy of mixing is given by

$$\Delta_{mix}H = b x_1 x_2 \tag{7.83}$$

as you can readily show by using the Gibbs–Helmholtz equation, and the entropy of mixing is the same as that for an ideal solution (equation 7.20), as can be shown by use of $\Delta_{mix}S = -\partial\Delta_{mix}G/\partial T$. The volume change on mixing is zero as can be seen from $\Delta_{mix}V = \partial_{mix}G^E/\partial P$, since the constant b is independent of temperature and pressure.

Figure 7.22 shows plots of $\Delta_{mix}G/RT$ versus x_2 for various values of b/RT. These plots have been calculated using the following combination of equations 7.80 and 7.81:

$$\frac{\Delta_{mix}G}{RT} = (1 - x_2) \ln(1 - x_2) + x_2 \ln x_2 + \frac{b}{RT} x_2 (1 - x_2) \tag{7.84}$$

The lowest curve is for an ideal solution. As the repulsive interaction between the two components increases, b increases. At large values of b there is a hump in the middle and two minima. When there are two minima, there are two phases, one rich in component 1 and the other rich in component 2, with compositions given by the minima. The system separates into these two phases because it attains a lower Gibbs energy that way.

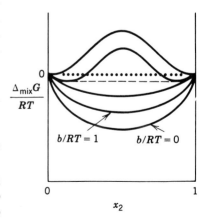

Figure 7.22 Gibbs energy of mixing two liquids. The lowest curve is for an ideal solution.

We can also consider that these plots are for a constant b and variable temperature. At high temperatures there is no phase separation, but when the temperature is lowered, phase separation starts at the critical point. The critical point is given by the curve that has a zero *second* derivative at $x_2 = 0.5$:

$$\frac{d^2(\Delta_{mix}G/RT)}{dx_2^2} = \frac{1}{1 - x_2} + \frac{1}{x_2} - \frac{2b}{RT} = 0 \tag{7.85}$$

Thus, $b = 2RT_c$ or $T_c = b/2R$. As the temperature is lowered further, the immiscibility range broadens. Since we have used a symmetrical function for the excess Gibbs energy, the solubilities of the two components are equal, but real systems are represented by more complicated asymmetric functions.

References

A. Alper, *Phase Diagrams*, Vols. I, II, and III. New York: Academic, 1970.

K. E. Bett, J. S. Rowlinson, and G. Saville, *Thermodynamics for Chemical Engineers*. Cambridge, MA: MIT Press, 1975.

K. G. Denbigh, *The Principles of Chemical Equilibrium*. Cambridge, UK: Cambridge University Press, 1971.

A. Findlay, A. N. Campbell, and N. O. Smith, *The Phase Rule and Its Applications*. New York: Dover, 1951.

R. M. Garrels and C. L. Christ, *Solutions, Minerals, and Equilibria*. New York: Harper & Row, 1965.

J. H. Hildebrand, J. M. Prausnitz, and R. L. Scott, *Regular and Related Solutions*. New York: Van Nostrand-Reinhold, 1970.

J. Prausnitz, T. Anderson, E. Grens, C. Eckert, R. Hsieh, and J. O'Connell, *Computer Calculations for Multicomponent Vapor–Liquid and Liquid–Liquid Equilibria*. Englewood Cliffs, NJ: Prentice Hall, 1980.

A. Reisman, *Phase Equilibria*. New York: Academic, 1970.

J. S. Rowlinson, *Liquids and Liquid Mixtures*. London: Butterworths, 1969.

S. I. Sandler, *Chemical and Engineering Thermodynamics*. New York: Wiley, 1989.

J. M. Smith and H. C. Van Ness, *Introduction to Chemical Engineering Thermodynamics*. New York: McGraw-Hill, 1987.

H. C. Van Ness and M. M. Abbott, *Classical Thermodynamics of Nonelectrolyte Solutions*. New York: McGraw-Hill, 1982.

Problems

7.1 What is the difference in chemical potential of benzene (component 1) in toluene (component 2) at $x_1 = 0.5$ and $x_1 = 0.1$ at 25 °C, assuming ideal solutions?

7.2 A binary liquid mixture of A and B is in equilibrium with its vapor at constant temperature and pressure. Prove that μ_A (g) $= \mu_A$ (l) and μ_B (g) $= \mu_B$ (l) by starting with

$$G = G(g) + G(l)$$

and the fact that $dG = 0$ when infinitesimal amounts of A and B are simultaneously transferred from the liquid to the vapor.

7.3 Ethanol and methanol form very nearly ideal solutions. At 20 °C, the vapor pressure of ethanol is 5.93 kPa, and that of methanol is 11.83 kPa. (*a*) Calculate the mole fractions of methanol and ethanol in a solution obtained by mixing 100 g of each. (*b*) Calculate the partial pressures and the total vapor pressure of the solution. (*c*) Calculate the mole fraction of methanol in the vapor.

7.4 One mole of benzene (component 1) is mixed with two moles of toluene (component 2). At 60 °C the vapor pressures of benzene and toluene are 51.3 and 18.5 kPa, respectively. (*a*) As the pressure is reduced, at what pressure will boiling begin? (*b*) What will be the composition of the first bubble of vapor?

7.5 The vapor pressures of benzene and toluene have the following values in the temperature range between their boiling points at 1 bar:

$t/°C$	79.4	88	94	100	110.0
$P^*_{C_6H_6}/bar$	1.000	1.285	1.526	1.801	
$P^*_{C_7H_8}/bar$		0.508	0.616	0.742	1.000

(*a*) Calculate the compositions of the vapor and liquid phases at each temperature and plot the boiling point diagram. (*b*) If a solution containing 0.5 mole fraction benzene and 0.5 mole fraction toluene is heated, at what temperature will the first bubble of vapor appear and what will be its composition?

7.6 At 1.013 bar pressure propane boils at -42.1 °C and *n*-butane boils at -0.5 °C; the following vapor–pressure data are available:

$t/°C$	-31.2	-16.3
P/kPa (propane)	160.0	298.6
P/kPa (*n*-butane)	26.7	53.3

Assuming that these substances form ideal binary solutions with each other, (*a*) calculate the mole fractions of propane at which the solution will boil at 1.013 bar pressure at -31.2 and -16.3 °C. (*b*) Calculate the mole fractions of propane in the equilibrium vapor at these temperatures. (*c*) Plot the temperature-mole fraction diagram at 1.013 bar, using these data, and label the regions.

7.7 The following table gives mole percent acetic acid in aqueous solutions and in the equilibrium vapor at the boiling point of the solution at 1.013 bar:

B.P., °C	118.1	113.8	107.5	104.4	102.1	100.0
Mol % of acetic acid						
In liquid	100	90.0	70.0	50.0	30.0	0
In vapor	100	83.3	57.5	37.4	18.5	0

Calculate the minimum number of theoretical plates for the column required to produce an initial distillate of 28 mol % acetic acid from a solution of 80 mol % acetic acid.

7.8 If two liquids (1 and 2) are completely immiscible, the mixture will boil when the sum of the two partial pressures exceeds the applied pressure: $P = P^*_1 + P^*_2$. In the vapor phase the ratio of the mole fractions of the two components is equal to the ratio of their vapor pressures:

$$\frac{P^*_1}{P^*_2} = \frac{x_1}{x_2} = \frac{m_1 M_2}{m_2 M_1}$$

where m_1 and m_2 are the masses of components 1 and 2 in the vapor phase, and M_1 and M_2 are their molar masses. The boiling point of the immiscible liquid system naphthalene–water is 98 °C under a pressure of 97.7 kPa. The vapor pressure of water at 98 °C is 94.3 kPa. Calculate the weight percent of naphthalene in the distillate.

7.9 A regular binary solution is defined as one for which

$$\mu_1 = \mu^\circ_1 + RT \ln x_1 + wx^2_2$$
$$\mu_2 = \mu^\circ_2 + RT \ln x_2 + wx^2_1$$

Derive $\Delta_{mix}G$, $\Delta_{mix}S$, $\Delta_{mix}H$, and $\Delta_{mix}V$ for the mixing of x_1 moles of component 1 with x_2 moles of component 2. Assume that the coefficient w is independent of temperature.

7.10 From the data given in the following table construct a complete temperature–composition diagram for the system ethanol–ethyl acetate for 1.013 bar. A solution containing 0.8 mole fraction of ethanol, EtOH, is distilled completely at 1.013 bar. (*a*) What is the composition of the first vapor to come off and (*b*) that of the last drop of liquid to evaporate? (*c*) What would be the values of these quantities if the distillation were carried out in a cylinder provided with a piston so that none of the vapor could escape?

x_{EtOH}	y_{EtOH}	B.P., °C	x_{EtOH}	y_{EtOH}	B.P., °C
0	0	77.15	0.563	0.507	72.0
0.025	0.070	76.7	0.710	0.600	72.8
0.100	0.164	75.0	0.833	0.735	74.2
0.240	0.295	72.6	0.942	0.880	76.4
0.360	0.398	71.8	0.982	0.965	77.7
0.462	0.462	71.6	1.000	1.000	78.3

7.11 The Henry's law constants for oxygen and nitrogen in water at 0 °C are 2.54×10^4 bar and 5.45×10^4 bar, respectively. Calculate the lowering of the freezing point of water by dissolved air with 80% N_2 and 20% O_2 by volume at 1 bar pressure.

7.12 Use the Gibbs–Duhem equation to show that if one component of a binary liquid solution follows Raoult's law, the other component will too.

7.13 The following data on ethanol–chloroform solutions at 35 °C were obtained by G. Scatchard and C. L. Raymond (*J. Am. Chem. Soc.* **60**:1278 (1938)):

$x_{EtOH,liq}$	0	0.2	0.4	0.6	0.8	1.0
$y_{EtOH,vap}$		0.0000	0.1382	0.1864	0.2554	0.4246	1.0000
Total pressure, kPa	39.345	40.559	38.690	34.387	25.357	13.703

Calculate the activity coefficients of ethanol and chloroform based on the deviations from Raoult's law.

7.14 Show that the equations for the bubble point line and dew point line for nonideal solutions are given by

$$x_1 = \frac{P - \gamma_2 P_2^*}{\gamma_1 P_1^* - \gamma_2 P_2^*}$$

$$y_1 = \frac{P\gamma_1 P_1^* - \gamma_1\gamma_2 P_1^* P_2^*}{P\gamma_1 P_1^* - P\gamma_2 P_2^*}$$

7.15 A regular binary solution is defined as one for which

$$\mu_1 = \mu_1^\circ + RT \ln x_1 + wx_2^2$$

$$\mu_2 = \mu_2^\circ + RT \ln x_2 + wx_1^2$$

Derive the expressions for the activity coefficients γ_i and γ_2 in terms of w.

7.16 The expressions for the activity coefficients of the components of a regular binary solution were derived in the preceding problem. Derive the expressions for γ_1 in terms of the experimentally measured total pressure P, the vapor pressures of the two components, and the composition of the solution for the case that the deviations from ideality are small.

7.17 Using the data in problem 7.39, calculate the activity coefficients of water (1) and *n*-propanol (2) at 0.20, 0.40, 0.60, and 0.80 mole fraction of *n*-propanol, based on deviations from Henry's law and considering water to be the solvent.

7.18 If 68.4 g of sucrose ($M = 342$ g mol^{-1}) is dissolved in 1000 g of water: (*a*) What is the vapor pressure at 20 °C? (*b*) What is the freezing point? the vapor pressure of water at 20 °C is 2.3149 kPa.

7.19 The protein human plasma albumin has a molar mass of 69 000 g mol^{-1}. Calculate the osmotic pressure of a so-lution of this protein containing 2 g per 100 cm^3 at 25 °C in (*a*) pascals, and (*b*) millimeters of water. The experiment is carried out using a salt solution for solvent and a membrane permeable to salt as well as water.

7.20 The following osmotic pressures were measured for solutions of a sample of polyisobutylene in benzene at 25 °C.

c/kg m^{-3}	5	10	15	20
Π/Pa	49.5	101	155	211

Calculate the number average molar mass from the value of Π/c extrapolated to zero concentration of the polymer.

7.21 A sample of polymer contains 0.50 mole fraction with molar mass 100 000 g mol^{-1} and 0.50 mole fraction with molar mass 200 000 g mol^{-1}. Calculate (*a*) M_n, and (*b*) M_m.

7.22 Calculate the osmotic pressure of a 1 mol L^{-1} sucrose solution in water from the fact that at 30 °C the vapor pressure of the solution is 4.1606 kPa. The vapor pressure of water at 30 °C is 4.2429 kPa. The density of pure water at this temperature (0.995 64 g cm^{-3}) may be used to estimate V_1 for a dilute solution. To do this problem, Raoult's law is introduced into equation 7.56.

7.23 Calculate the solubility of *p*-dibromobenzene in benzene at 20 and 40 °C assuming that ideal solutions are formed. The enthalpy of fusion of *p*-dibromobenzene is 13.22 kJ mol^{-1} at its melting point (86.9 °C).

7.24 Calculate the solubility of naphthalene at 25 °C in any solvent in which it forms an ideal solution. The melting point of naphthalene is 80 °C, and the enthalpy of fusion is 19.29 kJ mol^{-1}. The measured solubility of naphthalene in benzene is $x_1 = 0.296$.

7.25 The NBS Tables of Chemical Thermodynamic Properties list $\Delta_f G^\circ$ for I_2 in $C_6H_6 : x$ as 7.1 kJ mol^{-1}. The x indicates that the standard state for I_2 in C_6H_6 is on the mole fraction scale. What is the solubility of I_2 in C_6H_6 at 298 K on the mole fraction scale? A chemical handbook lists the solubility as 16.46 g I_2 in 100 cm^3 of C_6H_6. Are these solubilities consistent?

7.26 The following cooling curves have been found for the system antimony–cadmium:

Cd, wt %	0	20	37.5	47.5	50	58	70	93	100
First break in curve, °C	—	550	461	—	419	—	400	—	—
Continuing constant temperature, °C	630	410	410	410	410	439	295	295	321

Construct a phase diagram, assuming that no breaks other than these actually occur in any cooling curve. Label the diagram completely and give the formula of any compound formed. How many degrees of freedom are there for each area and at each eutectic point?

7.27 The phase diagram for magnesium–copper at constant pressure shows that two compounds are formed: $MgCu_2$, which melts at 800 °C, and Mg_2Cu, which melts at 580 °C. Copper melts at 1085 °C, and Mg at 648 °C. The three eutectics are at 9.4% by weight Mg (680 °C), 34% by weight Mg (560 °C), and 65% by weight Mg (380 °C). Construct the phase diagram. How many degrees of freedom are there for each area and at each eutectic point?

7.28 For the ternary system benzene–isobutanol–water at 25 °C and 1 bar the following compositions have been obtained for the two phases in equilibrium:

Water-rich Phase		Benzene-rich Phase	
Isobutanol, wt %	Water, wt %	Isobutanol, wt %	Benzene, wt %
2.33	97.39	3.61	96.20
4.30	95.44	19.87	79.07
5.23	94.59	39.57	57.09
6.04	93.83	59.48	33.98
7.32	92.64	76.51	11.39

Plot these data on a triangular graph, indicating the tie lines. (*a*) Estimate the compositions of the phases that will be produced from a mixture of 20% isobutanol, 55% water, and 25% benzene. (*b*) What will be the composition of the principle phase when the first drop of the second phase separates when water is added to a solution of 80% isobutyl alcohol in benzene?

7.29 The following data are available from the system nickel sulfate–sulfuric acid–water at 25 °C. Sketch the phase diagram on triangular coordinate paper, and draw appropriate tie lines.

Liquid Phase		Solid Phase
$NiSO_4$, wt %	H_2SO_4, wt %	Solid Phase
28.13	0	$NiSO_4 \cdot 7H_2O$
27.34	1.79	$NiSO_4 \cdot 7H_2O$
27.16	3.86	$NiSO_4 \cdot 7H_2O$
26.15	4.92	$NiSO_4 \cdot 6H_2O$
15.64	19.34	$NiSO_4 \cdot 6H_2O$
10.56	44.68	$NiSO_4 \cdot 6H_2O$
9.56	48.46	$NiSO_4 \cdot H_2O$
2.67	63.73	$NiSO_4 \cdot H_2O$
0.12	91.38	$NiSO_4 \cdot H_2O$
0.11	93.74	$NiSO_4$
0.08	96.80	$NiSO_4$

7.30 The Gibbs–Duhem equation in the form

$$\left(\frac{\partial M}{\partial T}\right)_{P,x} dT + \left(\frac{\partial M}{\partial P}\right)_{T,x} dP - \sum (x_i dM_i) = 0$$

applies to any molar thermodynamic property M in a homogeneous phase. If this applied to G^E, it may be shown* that if the vapor is an ideal gas

$$x_1 \frac{d \ln (y_1 P)}{dx_1} + x_2 \frac{d \ln (y_2 P)}{dx_1} = 0 \qquad \text{(constant } T\text{)}$$

Show that this can be rearranged to the coexistence equation

$$\frac{dP}{dy_1} = \frac{P (y_1 - x_1)}{y_1(1 - y_1)}$$

Thus, if P versus y_1 is measured, there is no need for measurements of x_1.

7.31 Use the activity coefficients for acetone–ether solutions at 30 °C from Table 7.3 to calculate G^E/RT at the various mole fractions of acetone.

7.32 Ethylene dibromide and propylene dibromide form very nearly ideal solutions. Plot the partial vapor pressure of ethylene dibromide ($P^* = 22.9$ kPa), the partial vapor pressure of propylene dibromide ($P^* = 16.9$ Pa), and the total vapor pressure of the solution versus the mole fraction of ethylene dibromide at 80 °C. (*a*) What will be the composition of the vapor in equilibrium with a solution containing 0.75 mole fraction of ethylene dibromide? (*b*) What will be the composition of the liquid phase in equilibrium with ethylene dibromide-propylene dibromide vapor containing 0.50 mole fraction of each?

7.33 At 25 °C the vapor pressures of chloroform and carbon tetrachloride are 26.54 and 15.27 kPa, respectively. If the liquids form ideal solutions, (*a*) what is the composition of the vapor in equilibrium with a solution containing 1 mol of each; (*b*) what is the total vapor pressure of the mixture?

7.34 Benzene and toluene form very nearly ideal solutions. At 80 °C, the vapor pressures of benzene and toluene are as follows: benzene, $P^* = 100.4$ kPa; toluene, $P^* = 38.7$ kPa. (*a*) For a solution containing 0.5 mole fraction of benzene and 0.5 mole fraction of toluene, what is the composition of the vapor and the total vapor pressure at 80 °C? (*b*) What is the composition of the liquid phases in equilibrium at 80 °C with benzene–toluene vapor having 0.75 mole fraction benzene?

---

*J. M. Smith and H. C. Van Ness, *Introduction to Chemical Engineering Thermodynamics*. p. 346. New York: McGraw Hill, 1975.

7.35   At 140 °C the vapor pressure of pure C_6H_5Cl is 1.237 bar and that of pure C_6H_5Br is 0.658 bar. These two liquids form ideal solutions to a very high degree of approximation. (*a*) What is the mole fraction of C_6H_5Cl in a $C_6H_5Cl–C_6H_5Br$ solution that just boils at 140 °C at 1 bar? (*b*) What is the mole fraction of C_6H_5Cl in the vapor produced in (*a*)? (*c*) Suppose this vapor is condensed, what is its total vapor pressure at 140 °C?

7.36   At 100 °C benzene has a vapor pressure of 180.9 kPa, and toluene has a vapor pressure of 74.4 kPa. Assuming that these substances form ideal binary solutions with each other, calculate the composition of the solution that will boil at 1 bar at 100 °C and the vapor composition.

7.37   The vapor pressure of the immiscible liquid system diethylaniline–water is 1.013 bar at 99.4 °C. The vapor pressure of water at that temperature is 99.2 kPa. How many grams of steam are necessary to distill 100 g of diethylaniline? (See Problem 7.8.)

7.38   What are the entropy change and Gibbs energy change on mixing to produce a benzene–toluene solution with $\frac{1}{3}$ mole fraction benzene at 25 °C?

7.39   For a solution of *n*-propanol and water, the following partial pressures in kPa are measured at 25 °C. Draw a complete pressure–composition diagram, including the total pressure. What is the composition of the vapor in equilibrium with a solution containing 0.5 mole fraction of *n*-propanol?

$x_{n\text{-propanol}}$	P_{H_2O}	$P_{n\text{-propanol}}$	$x_{n\text{-propanol}}$	P_{H_2O}	$P_{n\text{-propanol}}$
0	3.168	0.00	0.600	2.65	2.07
0.020	3.13	0.67	0.800	1.79	2.37
0.050	3.09	1.44	0.900	1.08	2.59
0.100	3.03	1.76	0.950	0.56	2.77
0.200	2.91	1.81	1.000	0.00	2.901
0.400	2.89	1.89			

7.40   Plot the following boiling point data for benzene–ethanol solutions at 1.013 bar and estimate the azeotropic composition:

B.P., ° C	78	75	70	70	75	80
Mole fraction of benzene						
In liquid	0	0.04	0.21	0.86	0.96	1.00
In vapor	0	0.18	0.42	0.66	0.83	1.00

State the range of mole fractions of benzene for which pure benzene could be obtained by fractional distillation at 1.013 bar.

7.41   The following table gives the mole percent of *n*-propanol ($M = 60.1$ g mol^{-1}) in aqueous solutions and in the vapor at the boiling point of the solution at 1.013 bar pressure:

B.P., ° C	100.0	92.0	89.3	88.1	87.8	88.3	90.5	97.3
Mole % of *n*-propanol								
In liquid	0	2.0	6.0	20.0	43.2	60.0	80.0	100.0
In vapor	0	21.6	35.1	39.2	43.2	49.2	64.1	100.0

With the aid of a graph of these data, calculate the mole fraction of *n*-propanol in the first drop of distillate when the following solutions are distilled with a simple distilling flask that gives one theoretical plate: (*a*) 87 g of *n*-propanol and 211 g of water; (*b*) 50 g of *n*-propanol and 5.02 g of water.

7.42   Using the Henry's law constants in Table 7.1, calculate the percentage (by volume) of oxygen and nitrogen in air dissolved in water at 25 °C. The air in equilibrium with the water at 1 bar pressure may be considered to be 20% oxygen and 80% nitrogen by volume.

7.43   By use of the data of the following table, which gives pressures in kPa at 35.2 °C for carbon disulfide–acetone solutions, calculate the activity coefficients based on deviations from Raoult's law for acetone and carbon disulfide at 32.5 °C for a solution containing 0.6 mole fraction of carbon disulfide:

x_{CS_2}	0	0.2	0.4	0.6	0.8	1.0
P_{CS_2}/kPa	0	37.3	50.4	56.7	61.3	68.3
$P_{acetone}$/kPa	45.9	38.7	34.0	30.7	25.3	0

7.44   Using the data of the following table, which gives vapor pressures at 35.2 °C, calculate the activity coefficients of acetone (2) and chloroform (1) at 35.2 °C, based on the deviations from Raoult's law for mole fractions of chloroform of 0.20, 0.40, 0.60, and 0.80:

x_1	0	0.2	0.4	0.6	0.8	1.00
P_1/kPa	0	4.5	10.9	19.7	30.0	39.1
P_2/kPa	45.9	36.0	24.4	13.6	5.6	0

7.45   The solubility of $I_2(s)$ in CCl_4 is listed in a chemical handbook as 29.1 g in 100 cm^3 of CCl_4 at 25 °C. What is $\Delta_f G°$ for I_2 in CCl_4 on the mole fraction scale?

7.46   (*a*) Use the following data to calculate the Henry's law constant for the solute chloroform in the solvent acetone at 35.2 °C:

x_{CHCl_3}	0	0.0603	0.1853	0.2910
P_{CHCl_3}/kPa	0	1.26	4.25	7.39

(*b*) Using this value of the Henry's law constant and the data in problem 7.4, calculate the activity coefficient of $CHCl_3$ from deviations from Henry's law. The activity coefficient for acetone, considered as the solvent, are the same as in problem 7.44.

7.47 Using the data in problem 7.39, calculate the activity coefficients of water and n-propanol at 0.20, 0.40, 0.60, and 0.80 mole fraction n-propanol, based on deviations from Henry's law and considering n-propanol to be the solvent.

7.48 The logarithms of the activity coefficients for a binary solution may be expressed as power series:

$$\ln \gamma_1 = a_1 x_2 + b_1 x_2^2 + c_1 x_2^3 + \cdots$$
$$\ln \gamma_2 = a_2 x_1 + b_2 x_1^2 + c_2 x_1^3 + \cdots$$

Using the Gibbs–Duhem equation

$$x_1 d \ln \gamma_1 + x_2 d \ln \gamma_2 = 0 \qquad \text{(fixed } P, T\text{)}$$

show that

$$a_2 = a_1 = 0 \qquad b_2 = b_1 + \tfrac{3}{2} c_1 \qquad c_2 = -c_1$$

7.49 The vapor pressure of a solution containing 13 g of a nonvolatile solute in 100 g of water at 28 °C is 3.6492 kPa. Calculate the molar mass of the solute, assuming that the solution is ideal. The vapor pressure of water at this temperature is 3.7417 kPa.

7.50 In acidic aqueous solutions acetic acid exists in the monomer form, but in nonpolar solvents, like benzene, it exists in the form of a dimer. Derive the following expression for the distribution coefficient:

$$K = \frac{x^2_{\text{CH}_3\text{CO}_2\text{H in H}_2\text{O}}}{x_{\text{CH}_3\text{CO}_2\text{H in C}_6\text{H}_6}}$$

7.51 The following osmotic pressures of polyvinyl acetate in dioxane were measured by G. V. Browning and J. D. Ferry at 25 °C:

$c/10^{-2}$ b cm^{-3}	0.292	0.579	0.810	1.140
Π/cm of solvent	0.73	1.76	2.73	4.68

Calculate the number-average molar mass. The density of dioxane is 1.035 g cm^{-3}.

7.52 Human blood plasma contains approximately 40 g of albumin ($M = 69\,000$ g mol^{-1}) and 20 g of globulin ($M = 160\,000$ g mol^{-1}) per liter. Calculate the colloid osmotic pressure at 37 °C, ignoring the Donnan effect.

7.53 Calculate (a) number, and (b) mass-average molar masses for the following mixture of high-polymer fractions: 1 g of $M = 20$ kg mol^{-1}, 2 g of $M = 50$ kg mol^{-1}, and 0.5 g of $M = 100$ kg mol^{-1}.

7.54 Calculate the solubility of anthracene ($M = 178.2$ g mol^{-1}) in toluene ($M = 92.1$ g mol^{-1}) at 100 °C. The enthalpy of fusion of anthracene is 28.9 kJ mol^{-1}, and the melting point of anthracene is 217 °C. The actual solubility is

0.0592 on the mole-fraction scale. How do you explain the difference?

7.55 Calculate the solubility of cadmium in bismuth at 250 °C. The melting point of cadmium is 323 °C and the enthalpy of fusion at the melting point is 6.07 kJ mol^{-1}.

7.56 The following data are obtained by cooling solutions of magnesium and nickel:

Ni, wt %	0	10	28	38	60	83	88	100
Infection in cooling curve, °C	—	608	—	770	1050	—	—	—
Plateau in cooling curve, °C	651	510	510	510	770	1180	1080	1450

It is found that in addition cooling solutions between 28 and 38% Ni deposit Mg_2Ni, whereas solutions containing between 38 and 82% Ni deposit $MgNi_2$. Plot the phase diagram.

7.57 The following are the compositions of the phases in equilibrium with each other in the system methylcyclohexane–aniline–n-heptane at 1 bar and 25 °C. Draw a triangular diagram for the system, including tie lines, and compute the exact composition of the first infinitesimal increment of the new liquid phase that forms when a sufficient quantity of pure aniline is added to a 40% solution of methylcyclohexane in n-heptane to give separation into two phases.

Hydrocarbon Layer		Aniline Layer	
Methyl-cyclohexane, wt %	n-Heptane, wt %	Methyl-cyclohexane, wt %	n-Heptane, wt %
0.0	92.0	0.0	6.2
9.2	83.0	0.8	6.0
18.6	73.4	2.7	5.3
33.8	57.6	4.6	4.5
46.0	45.0	7.4	3.6
59.7	30.7	9.2	2.8
73.6	16.0	13.1	1.4
83.3	4.4	15.6	0.6
88.1	0.0	16.9	0.0

7.58 At 25 °C the solubility of KNO_3 in pure H_2O is 46.2% by weight, the solubility of $NaNO_3$ in pure H_2O is 52.2% by weight, and $NaNO_3$, KNO_3, and saturated solution are in equilibrium when the composition of the solution is H_2O, 31.3%, KNO_3, 28.9%, and $NaNO_3$, 39.8%. No crystalline hydrates or double salts are formed. Sketch this system on a triangular diagram, labeling the areas in which you would expect to find (a) only solution; (b) a mixture of solution and solid KNO_3; (c) a mixture of solution and solid $NaNO_3$; and (d) a mixture of solid KNO_3, solid $NaNO_3$, and solution.

8
Electrochemical Equilibrium

Electrochemical reactions involve free electrons that are transferred from a metal to a component of a solution. The equilibria of electrochemical reactions are important in galvanic cells which produce an electric current and electrolytic cells which consume an electric current.

The measurement of the electromotive force of an electrochemical cell over a range of temperature makes it possible to obtain the thermodynamic quantities for the reaction that occurs in the cell. The activity coefficients of electrolytes may also be calculated from these measurements; this is illustrated in this chapter by the determination of the activity coefficient of hydrochloric acid.

Electrochemical cells are of practical interest in that they offer the means to convert the Gibbs energy change of a chemical reaction into work without the second-law losses of heat engines.

An understanding of the conversion of chemical energy into electrical energy is important for work with batteries, fuel cells, electroplating, corrosion, electrorefining (e.g., the production of aluminum), and electroanalytical techniques. This chapter discusses the thermodynamics of such processes. The kinetics of electrode reactions are discussed briefly in Section 21.12.

8.1 Coulomb's Law, Electric Field, and Electric Potential

Of the four kinds of interactions recognized in physics, strong nuclear interactions, weak interactions, electromagnetic interactions, and gravitation, only
electromagnetic interactions are of importance in chemistry. The most elementary of these is the attractive or repulsive interaction between two charges,
Q_1 and Q_2. Since the direction of the force is along the line connecting the
charges, it is convenient to write **Coulomb's law** in vector notation.

$$f = \frac{1}{4\pi\epsilon_0\epsilon_r}\frac{Q_1Q_2}{r^2}\,r \qquad (8.1)$$

where r is the distance between charges, r is the unit vector in the direction
of the force, ϵ_0 is the **permittivity of vacuum** ($8.854\ 187\ 817 \times 10^{-12}\ C^2\ N^{-1}$
m^{-2}), and ϵ_r is the relative permittivity (dielectric constant). The relative permittivities of a number of gases and liquids are given later in Table 12.3. When
the direction of the force is not being considered Coulomb's law may be written
in the form

$$f = \frac{Q_1Q_2}{4\pi\epsilon_0\epsilon_r r^2} \qquad (8.2)$$

The electric field strength E at a certain point is defined as the electrical
force per unit charge. The electric field strength is a vector because it has
direction as well as magnitude. If a very small test charge Q_1 is used, the electric
field strength is equal to the ratio of the force to the charge:

$$E = \frac{f}{Q_1} \qquad (8.3)$$

From equation 8.2 the magnitude of the electric field strength in a vacuum due
to charge Q_2 is given by

$$E = \frac{Q_2}{4\pi\epsilon_0 r^2} \qquad (8.4)$$

The electric field strength has the SI units of $V\ m^{-1}$. In the remainder of this
chapter we will use E to represent electromotive force, rather than electric
field strength. The electromotive force is the difference in electric potential
between two points and is expressed in volts.

The magnitude of the electric field around a charged particle is the derivative
of a scalar quantity called the electric potential. The difference between the
electric potential at two points is equal to the work per unit charge required
to move a charge from one point to the other. Thus, the unit of potential difference is joules per coulomb; this unit is referred to as a volt. $1\ V = 1\ J\ C^{-1}$.
The choice of zero potential is arbitrary, but it is customary to define the
potential as zero when the particles are at infinite distance. Thus, the **electric
potential** at a point is the work required to bring a unit positive charge from
infinity to the point in question. The electric potential is given by

$$\phi = -\int_{\infty}^{r}\frac{Q_2\mathrm{d}r}{4\pi\epsilon_0\epsilon_r r^2} = \frac{Q_2}{4\pi\epsilon_0\epsilon_r r} \qquad (8.5)$$

Figure 8.1 shows the electric potential as a function of distance from an electric
charge.

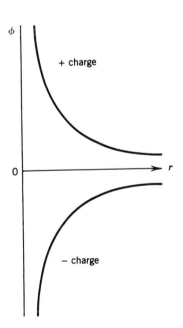

Figure 8.1   Electric potential ϕ
as a function of distance from an
electric charge.

As we begin to treat electrolyte solutions we need to remember that there is an additional constraint on the composition of a phase, and that is the **electroneutrality condition**

$$\sum_i n_i z_i = 0 \qquad (8.6)$$

where n_i is the number of ions of charge $z_i e$ in the phase. The charge number z_i is positive for cations and negative for anions; e is the charge on a proton, 1.6022×10^{-19} C. For a phase to have a nonzero electric potential ϕ there must be a small deviation from the electroneutrality condition, but these deviations are so small that we can neglect them in using equation 8.6.

Example 8.1

What is the electric potential on the surface of a sphere 10 cm in diameter if it contains an excess of 10^{-10} mol of a monovalent cation? The charge on the sphere is 10^{-10} F, where F is the Faraday constant ($F = N_A e = 96\ 485$ C mol^{-1}). The potential on the surface can be calculated by assuming that this charge is located at the center of the sphere. Assuming a relative permittivity of unity,

$$\phi = \frac{(10^{-10}\ \text{mol})(96\ 485\ \text{C mol}^{-1})}{4\pi(8.854\ 187 \times 10^{-12}\ \text{C}^2\ \text{N}^{-1}\ \text{m}^{-2})(0.05\ \text{m})}$$

$$= 1.734 \times 10^6\ \text{V}$$

Since potential differences between phases are much smaller than this, the deviations from equation 8.6 are much smaller than 10^{-10} mol in a volume of $\frac{4}{3}\pi(0.05\ m)^3$. Thus, it is a good approximation to say that equation 8.6 is obeyed, and two phases can have the same chemical composition, but different electric potential.

8.2   Equilibria Involving Potential Differences

In discussing the differential form of the first law of thermodynamics, we saw that when a small charge dQ is moved through an electric potential difference ϕ, the work done on the charge is given by $đw = \phi\ dQ$ (equation 2.34). This term carries over to the equation for the differential of the Gibbs energy (equation 4.82). If ions with charge number z_i are transferred, the differential charge dQ can be expressed in terms of the differential amount dn_i of ion i by

$$dQ = z_i F\ dn_i \qquad (8.7a)$$

where F is the Faraday constant. **The charge number of an ion z_i is the charge in terms of the proton charge with the sign of the ion.** The **Faraday constant** is equal to the product of the Avogadro constant and the proton charge:

$$F = N_A e = (6.022\ 136\ 7 \times 10^{23}\ \text{mol}^{-1})(1.602\ 177\ 33 \times 10^{-19}\ \text{C})$$

$$= 96\ 485.309\ \text{C mol}^{-1} \qquad (8.7b)$$

The charge number z_i of an ion is dimensionless, and so the differential charge

dQ is expressed in coulombs. To write a general expression for the electrical work, we can assume that any of the species in a system may be ions and write

$$\unicode{273}w_{ele} = \sum_{i=1}^{N} z_i F \phi \, dn_i \qquad (8.8)$$

For an open system with transfer of charges, equation 4.82 becomes

$$dG = -S \, dT + V \, dP + \sum_{i=1}^{N} [(\mu_i)_{\phi=0} + z_i F \phi] \, dn_i$$

$$= -S \, dT + V \, dP + \sum_{i=1}^{N} \mu_i \, dn_i \qquad (8.9)$$

Since the chemical potential μ_i is the derivative of G with respect to n_i at constant T, P, and specified amounts of other species, then

$$\mu_i = (\mu_i)_{\phi=0} + z_i F \phi \qquad (8.10)$$

This quantity is sometimes represented by $\tilde{\mu}_i$ and called the electrochemical potential, but it is simply the chemical potential defined in equation 4.81.

The chemical potential of a positive ion in a phase with a more positive potential is higher than in the phase with the less positive potential, if the phases are the same chemically. The chemical potential of a negative ion in a phase with a more positive potential is lower than in the phase with the less positive potential.

When there is a difference in electric potential between two phases that are in equilibrium, the criteria of equilibrium $\mu_{i\alpha} = \mu_{i\beta}$ (equation 6.9) becomes

$$(\mu_{i\alpha})_{\phi=0} + z_i F \phi_\alpha = (\mu_{i\beta})_{\phi=0} + z_i F \phi_\beta \qquad (8.11)$$

At equilibrium for a chemical reaction, equation 5.7 becomes

$$\sum_{i=1}^{N} \nu_i [(\mu_i)_{\phi=0} + z_i F \phi]_{eq} = 0 \qquad (8.12)$$

this relation will be used to derive the fundamental equation for an electrochemical cell.

8.3 Fundamental Equation for an Electrochemical Cell

Electrochemical cells can be classified as **galvanic cells,** in which chemical reactions occur spontaneously, and **electrolytic cells,** in which chemical reactions are caused by an externally applied potential difference. Galvanic cells of commercial importance include the Leclanché Zn/MnO_2 cell and the Zn/Ag_2O_3 cell used, for example, in watches. Fuel cells, such as the H_2–O_2 cell used in space craft, are galvanic cells in which the oxidizable and reduceable fuels are supplied continuously. Electrolytic cells are used in the commercial production of chlorine and aluminum and in the electrorefining of copper. The Pb–PbO_2–H_2SO_4 storage cell is an electrolytic cell when it is being charged and a galvanic cell when it is being used as a battery. To connect what happens in the laboratory to convention, **we must always write the electrode reaction for**

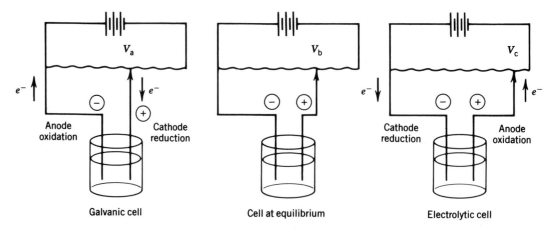

Figure 8.2 Use of a potentiometer to measure the potential difference E for a galvanic cell with no current flowing. (a) The cell operates spontaneously with reduction on the right since $V_a < E$. (b) No current passes through the cell since $V_b = E$. (c) A nonspontaneous reaction in the cell is driven by the battery in the potentiometer since $V_c > E$.

the electrode on the right in a schematic representation of a galvanic cell as a reduction.* The mnemonic for this is "reduction on the right":

$$Ox + ne^- = Red \tag{8.13}$$

The **cathode** is the electrode at which a **reduction reaction** occurs. This electrode is positive in a galvanic cell because electrons are flowing into it and reacting with an oxidized species to form a reduced species, as shown in equation 8.13. The **anode** is the electrode at which an oxidation reaction occurs:

$$Red' = Ox' + ne^- \tag{8.14}$$

The primes indicate that the reactions at the two electrodes are generally different.

The difference in potential between the electrodes of a cell can be measured using a **potentiometer**, as shown in Fig. 8.2. In a potentiometer a steady current from a battery flows through a resistor. A sliding contact is used to apply some fraction of the potential difference across the slide wire to an electrochemical cell. If the applied potential difference is less than the electromotive force of the galvanic cell, the cell discharges spontaneously as shown in Fig. 8.2a.

As the potential difference applied to the cell is increased, a point will be reached at which the current through the cell is zero, as shown in Fig. 8.2b. This is the equilibrium potential difference that we will be primarily concerned with in this chapter. This is a thermodynamic measurement because the direction of the current through the cell can be changed by an infinitesimal change in the applied potential, if the electrode reactions are fast under the conditions in the electrochemical cell.

When the applied potential is further increased, as shown in Fig. 8.2c, it drives the cell reaction in the reverse direction, and the cell is referred to as

---

* Electrochemical Nomenclature, *Pure Appl. Chem.* **37**:501 (1974).

an electrolytic cell. Now oxidation occurs at the right electrode, and so it is called the anode. (Note that our rule "reduction on the right" applies to a galvanic cell, not an electrolytic cell.) The right electrode supplies electrons to the external circuit according to reaction 8.14. The definitions of anode and cathode in terms of oxidation and reduction are independent of whether a cell is a galvanic cell or an electrolytic cell.

A very wide variety of different types of galvanic cells can be constructed. A metal electrode can be immersed in a solution containing its ions. An amalgam (solution of a metal in liquid mercury) electrode can be in contact with a solution containing ions of the metal. A nonmetal can be used by bubbling gas over a platinum electrode. Or a platinum electrode can simply be in contact with a solution containing oxidized and reduced forms of other species (for example, Fe^{3+} and Fe^{2+}).

Galvanic cells can be classified as cells without a liquid junction (see Fig. 8.3a, for example) and those with a liquid junction (see Fig. 8.3b, for example). Cells without a liquid junction can be analyzed exactly by thermodynamics, but an accurate treatment of those with liquid junctions must include consideration of rate processes.

Electrochemical cells are represented schematically by diagrams such as

$$Pt \mid H_2(g) \mid HCl(m) \mid AgCl(s) \mid Ag \qquad (8.15)$$

$$Zn \mid Zn^{2+} \vdots Cu^{2+} \mid Cu \qquad (8.16)$$

$$Zn \mid Zn^{2+} \vdots\vdots Cu^{2+} \mid Cu \qquad (8.17)$$

$$Pt \mid H_2(g), H^+ \vdots\vdots Fe^{3+}, Fe^{2+} \mid Pt \qquad (8.18)$$

$$Zn \mid Zn^{2+} (c_1) \vdots Zn^{2+} (c_2) \mid Zn \qquad (8.19)$$

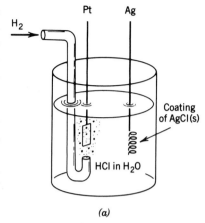

(a)

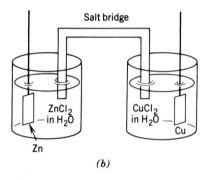

(b)

Figure 8.3 (a) Galvanic cell without liquid junction. (b) galvanic cell with liquid junction.

The single vertical bars represent phase boundaries, the dashed vertical bars represent a junction between two miscible liquids, and the double dashed vertical bars represent a "salt bridge" made up of a concentrated solution of potassium chloride or ammonium nitrate. It is necessary to use a salt bridge when the solutions in contact with the anode and cathode can react with each other. Actually, an electrochemical cell is never in complete equilibrium; however, certain species are kept away from each other pretty effectively by certain barriers in the cell.

The type of cell represented in 8.15 is the most suitable for exact thermodynamic analysis because it does not have any liquid junctions. Liquid junctions, such as those in 8.16 and 8.17, contribute to the electromotive force, and the theory of liquid junction potentials is quite complicated and depends on the mobilities (Section 21.3) of the ions involved. Liquid junction potentials arise from irreversible processes and must be treated by the methods of irreversible thermodynamics. The liquid junction potential arising in a cell like 8.16 can be reduced by using a salt bridge containing a concentrated solution of a salt containing cations and anions that have nearly the same absolute mobilities.

To obtain the relationship between the electromotive force for a cell and the chemical potentials or activities of the reactants and products, we will consider the following cell without a liquid junction:

$$Pt_L \mid H_2(g) \mid HCl(m) \mid AgCl(s) \mid Ag(s) \mid Pt_R \qquad (8.20)$$

where m is the molality. According to the international convention we assume that a reduction reaction occurs in the right electrode and an oxidation occurs in the left electrode. These two electrode reactions are

$$2AgCl(s) + 2e^-(Pt_R) = 2Ag(s) + 2Cl^-(m) \tag{8.21}$$

$$H_2(g) = 2H^+(m) + 2e^-(Pt_L) \tag{8.22}$$

The sum of these electrode reactions is the cell reaction.

$$H_2(g) + 2AgCl(s) + 2e^-(Pt_R) = 2H^+(m) + 2Cl^-(m) + 2Ag(s) + 2e^-(Pt_L) \tag{8.23}$$

At equilibrium, equation 8.12 may be applied to this reaction,

$$0 = 2\mu(H^+) + 2\mu(Cl^-) + 2\mu(Ag) - 2F\phi_L - \mu(H_2) - 2\mu(AgCl) + 2F\phi_R \tag{8.24}$$

where the subscript $\phi = 0$ has been omitted on the chemical potentials to simplify the notation. Note that the chemical potential of an electron in Pt at $\phi = 0$ is the same in the right and left electrodes. Here the reactants are at $\phi = 0$, but the electrons are in the platinum electrodes, which are at ϕ_L and ϕ_R. Thus,

$$-2F(\phi_R - \phi_L) = 2\mu(H^+) + 2\mu(Cl^-) + 2\mu(Ag) - \mu(H_2) - 2\mu(AgCl) \tag{8.25}$$

In considering electrochemical cells, it is customary to represent the difference in potential between the right electrode ϕ_R and the left electrode ϕ_L by E. **In this chapter we will always consider E to be the potential difference when the current is zero.** Thus,

$$-2FE = 2\mu(H^+) + 2\mu(Cl^-) + 2\mu(Ag) - \mu(H_2) - 2\mu(AgCl) \tag{8.26}$$

for the cell reaction

$$H_2(g) + 2AgCl(s) = 2HCl(m) + 2Ag(s) \tag{8.27}$$

which is equation 8.23 written without the electrons.

In general,

$$\Delta_r G = \sum_i \nu_i \mu_i = -nFE \tag{8.28}$$

where $\Delta_r G$ is the reaction Gibbs energy for the cell reaction as written and n is the **charge number of an electrochemical reaction** (the number of electrons transferred) for the cell reaction as written. Note that when the right-hand electrode has a more positive potential than the left-hand electrode, the electromotive force E for the cell is positive. If E is positive, $\Delta_r G$ for the cell reaction is negative (i.e., the cell reaction is spontaneous at constant temperature and pressure). If the right-hand electrode is more positive, its electrode reaction is a reduction reaction when the cell operates spontaneously.

If the schematic representation for a cell is reversed, the electrode reactions and cell reaction are written in the opposite directions, and the signs of E and $\Delta_r G$ are reversed. The cell reaction can be multiplied or divided by a number, and that is the reason equation 8.28 contains the factor n, the charge number. The electromotive force may be expressed in terms of the activities a_i of the reactants and products by use of $\mu_i = \mu_i^\circ + RT \ln a_i$ in equation 8.28 to obtain

$$-nFE = \sum_i \nu_i \mu_i^\circ + RT \sum_i \nu_i \ln a_i$$

$$= -nFE^\circ + RT \ln \prod_i a_i^{\nu_i} \tag{8.29}$$

where $E°$ is the **standard electromotive force of the cell**, the electromotive force when the activities of all reactants and products are equal to unity. Equation 8.29 is the **Nernst equation** and is usually written

$$E = E° - \frac{RT}{nF} \ln \prod_i a_i^{\nu_i} \qquad (8.30)$$

At 25 °C

$$E = E° - \frac{(8.314 \text{ J K}^{-1} \text{ mol}^{-1})(298.15 \text{ K})}{n(96\ 485 \text{ C mol}^{-1})} \ln \prod_i a_i^{\nu_i}$$

$$= E° - \frac{(0.025\ 69 \text{ V})}{n} \ln \prod_i a_i^{\nu_i} \qquad (8.31)$$

Later we will see in detail how $E°$ is determined, but for now we can note that it is the electromotive force of a cell in which all reactants and products are at unit activities.

If the activities of the reactants and products correspond with those of an equilibrium mixture, $E = 0$, and equation 8.30 becomes

$$E° = \frac{RT}{nF} \ln K \quad \text{or} \quad K = e^{nFE°/RT} \qquad (8.32)$$

where K is the equilibrium constant for the cell reaction.

Example 8.2
Three different galvanic cells have standard electromotive forces $E°$ of 0.01, 0.1, and 1.0 V, respectively, at 25 °C. Calculate the equilibrium constants of the reactions that occur in these cells assuming the charge number n for each reaction is unity.

$$K = e^{nFE°/RT}$$

For $E° = 0.01$ V,

$$K = \exp \frac{(96\ 485 \text{ C mol})(0.01 \text{ V})}{(8.314 \text{ J K}^{-1} \text{ mol}^{-1})(298.15 \text{ K})}$$

$$= 1.476$$

For $E° = 0.1$ V, $K = 49.0$, and for $E° = 1$ V, $K = 8.02 \times 10^{16}$.

To determine $E°$ for an arbitrarily chosen cell, we have to know certain things about the activities of electrolytes.

8.4 Activity of Electrolytes

In studying the thermodynamics of nonelectrolyte mixtures, we represented the chemical potential of a component by equation 7.26. Electrolytes have to be treated in a different way because they dissociate, but the ions cannot be studied separately because the condition of electric neutrality applies. In work with electrolyte solutions it is customary to use the molal scale. The molality m_i is equal to the amount of electrolyte per kilogram of solvent, which is given by $n_i/M_1 n_1$, where n_i is the amount of electrolyte in n_1 mol of solvent and M_1 is the molar mass of the solvent. Thus, the molality has the units mol kg^{-1}. The molality has an interesting property, as compared with the mole fraction, that the addition of a second solute does not change the molality of the first. Activity coefficients may be given on the molality scale for nonelectrolytes as well. On the m scale the **activity** of a solute substance i is defined by*

$$a_i = \frac{\gamma_i m_i}{m^\circ} \tag{8.33}$$

where m° is the standard value of the molality (1 mol/kg of solvent), and

$$\lim_{m_i \to 0} \gamma_i = 1 \tag{8.34}$$

When an infinitesimal quantity of an electrolyte is added to a kilogram of solvent, the differential change in the Gibbs energy is given by

$$dG = \mu_+ \, dm_+ + \mu_- \, dm_- \tag{8.35}$$

However, we cannot add cations and anions separately because the solution must be electrically neutral. If the strong electrolyte is $A_{\nu_+} B_{\nu_-}$, where ν_+ is the number of cations and ν_- is the number of anions, electroneutrality requires that

$$m = \frac{m_+}{\nu_+} = \frac{m_-}{\nu_-} \tag{8.36}$$

Equation 8.35 can be written

$$
\begin{aligned}
dG &= (\nu_+ \mu_+ + \nu_- \mu_-) \, dm \\
&= \mu \, dm
\end{aligned} \tag{8.37}
$$

where

$$\mu = \nu_+ \mu_+ + \nu_- \mu_- \tag{8.38}$$

is the chemical potential for the electrolyte, which can be determined experimentally. The chemical potentials of the cation and anion are given by

---

* Activity coefficients may also be given on the molar concentration scale (moles per liter). On the c scale the activity of solute substance i is defined by

$$a_i = \frac{\gamma_i c_i}{c^\circ}$$

where c is the standard value of the molar concentration (1 mol L^{-1}), and

$$\lim_{c_i \to 0} \gamma_i = 1$$

The various activity coefficients have different numerical values, but we will use γ_i for all of them to avoid confusion with other symbols. When there is danger of confusing activity coefficients on various scales, they may be given subscripts as in $\gamma_{x,i}$, $\gamma_{m,i}$, and $\gamma_{c,i}$.

$$\mu_+ = \mu_+^\circ + RT \ln \gamma_+ m_+ \tag{8.39}$$

$$\mu_- = \mu_-^\circ + RT \ln \gamma_- m_- \tag{8.40}$$

where μ_+° and μ_-° are the standard state chemical potentials and γ_+ and γ_- are the activity coefficients of the cation and anion. We will omit the standard values m° of the molality in the remainder of this chapter to simplify the notation. When equations 8.39 and 8.40 are substituted in equation 8.38,

$$\mu = (\nu_+ \mu_+^\circ + \nu_- \mu_-^\circ) + RT \ln \gamma_+^{\nu_+} \gamma_-^{\nu_-} m_+^{\nu_+} m_-^{\nu_-} \tag{8.41}$$

To get a term proportional to the molality m of the electrolyte in the logarithmic term, a **mean ionic molality** $m_\pm$ and a **mean ionic activity coefficient** $\gamma_\pm$ are defined as

$$m_\pm = (m_+^{\nu_+} m_-^{\nu_-})^{1/\nu_\pm} = m(\nu_+^{\nu_+} \nu_-^{\nu_-})^{1/\nu_\pm} \tag{8.42}$$

$$\gamma_\pm = (\gamma_+^{\nu_+} \gamma_-^{\nu_-})^{1/\nu_\pm} \tag{8.43}$$

where

$$\nu_\pm = \nu_+ + \nu_- \tag{8.44}$$

Then equation 8.41 becomes

$$\mu = \mu^\circ + \nu_\pm RT \ln \gamma_\pm m_\pm \tag{8.45}$$

and the **activity of the electrolyte** is given by

$$a_{A_{\nu_+} B_{\nu_-}} = (\gamma_\pm m_\pm)^{\nu_\pm}$$
$$= \gamma_\pm^{\nu_\pm} m^{\nu_\pm}(\nu_+^{\nu_+} \nu_-^{\nu_-}) \tag{8.46}$$

The standard chemical potential μ° of the electrolyte is the chemical potential in a solution of unit activity on the molality scale. It can be determined using deviations from Henrys' law (Section 7.8) and extrapolating to $m = 0$ where $\gamma_\pm = 1$. This extrapolation is guided by the Debye–Hückel limiting law (Section 8.5).

The mean ionic molality $m_\pm$ of a 1-1 electrolyte like NaCl is equal to m, of a 2-1 electrolyte like CaCl$_2$ is equal to $4^{1/3}m$, of a 2-2 electrolyte like CuSO$_4$ is equal to m, and of a 3-1 electrolyte like LaCl$_3$ is equal to $27^{1/4}m$, as may be deduced from equation 8.46. The numbers 1, 2, and 3 refer to the number of charges on the cation and anion.

In equations where the activities of electrolytes occur, these activities may be replaced by expressions involving the molality m and the mean ionic activity coefficient $\gamma_\pm$.

Example 8.3
Write the expressions for the activities of NaCl, CaCl$_2$, CuSO$_4$, and LaCl$_3$ in terms of their molalities and mean ionic activity coefficients.

$$a_{NaCl} = m^2 \gamma_\pm^2 \qquad a_{CaCl_2} = 4m^3 \gamma_\pm^3 \qquad a_{CuSO_4} = m^2 \gamma_\pm^2 \qquad a_{LaCl_3} = 27m^4 \gamma_\pm^4$$

8.5 Debye–Hückel Theory*

Electrolytes containing ions with multiple charges have larger effects on the activity coefficients of ions than electrolytes containing only singly charged ions. To express electrolyte concentrations in a way that takes this into account, G. N. Lewis† introduced the ionic strength I defined by

$$I = \tfrac{1}{2} \sum_i m_i z_i^2 = \tfrac{1}{2}(m_1 z_1^2 + m_2 z_2^2 + \cdots) \tag{8.47}$$

where z_i is the charge (signed) of the ion in units of the charge on a proton. The summation is continued over all the different ionic species in the solution and m is the molal concentration. The greater effectiveness of ions of higher charge in reducing the activity coefficient is provided for by multiplying their concentrations by the square of their charges. According to equation 8.47, the ionic strength of a 1-1 electrolyte is equal to its molality. The ionic strength for a 1-2 electrolyte is $3m$ and for a 2-2 electrolyte is $4m$.

The Coulomb forces between ions are of much longer range than van der Waals forces (Section 12.11), and therefore it is almost impossible to make measurements on electrolyte solutions at sufficiently low concentrations to obtain dilute solution behavior, in the sense of Henry's law. At infinite dilution the distribution of ions in an electrolytic solution can be considered to be completely random because the ions are too far apart to exert any attraction on each other, and the activity coefficient of the electrolyte is unity. At higher concentrations, where the ions are closer together, however, the Coulomb attractive and repulsive forces become important. Because of this interaction of ions the concentration of positive ions is slightly higher in the neighborhood of a negative ion, and the concentration of negative ions is slightly higher in the neighborhood of a positive ion, than in the bulk solution. Because of the attractive forces between an ion and its surrounding ionic atmosphere, the activity coefficient of the electrolyte is reduced. This effect is greater for ions of high charge and is greater in solvents of lower dielectric constant where the electrostatic interactions are stronger.

Debye and Hückel were able to show that in dilute solutions the activity coefficient γ_i of an ion species i with a charge number of z_i is given by

$$\log \gamma_i = -A z_i^2 I^{1/2} \tag{8.48}$$

where I is the ionic strength and

$$A = \frac{1}{2.303} \left(\frac{2\pi N_A m_{\text{solv}}}{V}\right)^{1/2} \left(\frac{e^2}{4\pi\epsilon_0\epsilon_r kT}\right)^{3/2} \tag{8.49}$$

where m_{solv} is the mass of solvent in volume V and ϵ_r is the relative permittivity.

Example 8.4

Calculate the value of the coefficient A in the Debye–Hückel equation for aqueous solutions at 298.15 K. The relative permittivity ϵ_r of water at this temperature is 78.54.

---

* P. Debye and E. Hückel, *Phys. Z.* **24:**185, 305 (1923).

† G. N. Lewis and M. Randall, revised by K. S. Pitzer and L. Brewer, *Thermodynamics*, p. 335. New York: McGraw-Hill, 1961.

$$A = \frac{1}{2.303} \left[\frac{2\pi(6.022 \times 10^{23} \text{ mol}^{-1})(997 \text{ kg})}{1.000 \text{ m}^3} \right]^{1/2}$$

$$\times \left[\frac{(1.602 \times 10^{-19} \text{ C})^2(0.8988 \times 10^{10} \text{ N m}^2 \text{ C}^{-2})}{(78.54)(1.3807 \times 10^{-23} \text{ J K}^{-1})(298.15 \text{ K})} \right]^{3/2}$$

$$= 0.509 \text{ kg}^{1/2} \text{ mol}^{-1/2}$$

Since the ionic strength I has the units mol kg^{-1}, the units of $I^{1/2}$ and A cancel, as they must, to give a logarithm.

Equation 8.48 gives the activity coefficient of a single ion, but the quantity that is accessible to experimental determination is the mean ionic activity coefficient, which for the electrolyte $A_{\nu_+}B_{\nu_-}$ is given by equation 8.43.

Taking the logarithm of equation 8.43, we have

$$\log \gamma_{\pm} = \frac{1}{\nu_+ + \nu_-}(\nu_+ \log \gamma_+ + \nu_- \log \gamma_-) \qquad (8.50)$$

Substituting equation 8.48 for each activity coefficient, we have

$$\log \gamma_{\pm} = -A \left(\frac{\nu_+ z_+^2 + \nu_- z_-^2}{\nu_+ + \nu_-} \right) I^{1/2} \qquad (8.51)$$

Introducing $\nu_+ z_+ = -\nu_- z_-$, we see that

$$\log \gamma_{\pm} = A z_+ z_- I^{1/2} \qquad (8.52)$$

The charge number z has the sign of the ion, and so we see that the effect of the ion atmosphere is to lower the activity coefficient of the electrolyte. The dependence of the mean ionic activity coefficient on ionic strength is shown in Fig. 8.4 for low ionic strengths.

Example 8.5
Use the Debye–Hückel theory to calculate γ_+, γ_-, and $\gamma_{\pm}$, and a_{NaCl} for 0.001 molal sodium chloride in water at 25 °C.

$$\log \gamma_i = -A z_i^2 I^{1/2}$$
$$= -(0.509)(0.001)^{1/2}$$
$$\gamma_+ = \gamma_- = 0.964$$
$$\log \gamma_{\pm} = A z_+ z_- I^{1/2}$$
$$= (0.509)(1)(-1)(0.001)^{1/2}$$
$$\gamma_{\pm} = (\gamma_+ \gamma_-)^{1/2} = 0.964$$

The Debye–Hückel theory has been of great value in interpreting the properties of electrolyte solutions. It is a limiting law at low ionic strengths. At high values of the ionic strength the activity coefficient of an electrolyte usually increases with increasing ionic strength. Equation 8.52 is in excellent agreement with experiment up to an ionic strength of about 0.01, but large deviations are encountered even at this ionic strength if the product of the charge of the highest

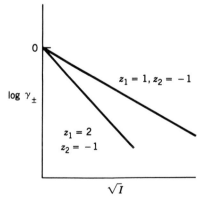

Figure 8.4   Mean ionic activity coefficients $\gamma_{\pm}$ for two electrolytes as a function of ionic strength I in solutions of low ionic strength, according to the Debye–Hückel theory.

charged ion of the salt and the charge of the oppositely charged ion of the electrolyte medium is greater than about 4.

Applications of Debye–Hückel theory are illustrated in the next section and in Sections 9.1 and 9.2

8.6   Determination of the Activity Coefficient of Hydrochloric Acid

The cell discussed in Section 8.3 may be used to determine the activity coefficient of hydrochloric acid. For this purpose, the cell reaction will be written

$$\tfrac{1}{2}H_2(g) + AgCl(s) = HCl(m) + Ag(s) \tag{8.53a}$$

Since the charge number n is equal to unity, the electromotive force, given by equation 8.30, is

$$E = E^{\circ} - \frac{RT}{F} \ln \frac{a_{HCl}}{(P_{H_2}/P^{\circ})^{1/2}} \tag{8.53b}$$

assuming that H_2 is an ideal gas. If the pressure of hydrogen is 1 bar and equation 8.46 is introduced, then

$$E = E^{\circ} - \frac{2.303RT}{F} \log(\gamma_{\pm}^2 m^2)$$

$$= E^{\circ} - 0.059\ 16 \log(\gamma_{\pm}^2 m^2) \tag{8.54}$$

The mean ionic activity coefficient of hydrochloric acid is represented by $\gamma_{\pm}$, and m is the molality.

As it stands, equation 8.54 contains two unknown quantities, E° and $\gamma_{\pm}$. These may be obtained by determining the electromotive force of this cell over a range of hydrochloric acid concentrations, including dilute solutions. Rearranging equation 8.54 and substituting numerical values for 25 °C gives

$$E + 0.1183 \log m = E^{\circ} - 0.1183 \log \gamma_{\pm} \tag{8.55}$$

The exponents in equation 8.54 have been placed in front of the logarithmic term, giving $(2)(0.059\ 16) = 0.1183$. Since at infinite dilution $m = 0$, $\gamma_{\pm} = 1$, and $\log \gamma_{\pm} = 0$, it can be seen that when $E + 0.1183 \log m$ is plotted against m, the extrapolation of $E + 0.1183 \log m$ to $m = 0$ will give E°.

To make a satisfactory extrapolation, use is made of the extended Debye–Hückel theory to furnish a function that will give nearly a straight line. The following expression is a useful empirical extension of equation 8.52 for the mean ionic activity coefficient of a 1-1 electrolyte in dilute aqueous solutions at 25 °C.

$$\log \gamma_{\pm} = -0.509\sqrt{m} + bm$$

where b is an empirical constant.

Substituting into equation 8.55 and rearranging terms, we have

$$E + 0.1183 \log m - 0.0602m^{1/2} = E' = E^{\circ} - (0.1183b)m \tag{8.56}$$

According to this equation, the left side, which we will designate as E', will give a straight line when it is plotted against m, and the intercept at $m = 0$ is $E°$.

In Fig. 8.5, E' is plotted against m. The extrapolated value is $E° = 0.2224$ V when the straight line is drawn through the points at the lower molalities. This is the electromotive force that the cell would deliver with the hydrochloric acid at unit activity.

The value of $E°$ having been determined, the activity coefficient of hydrochloric acid at any other concentration may be calculated from the electromotive force of the cell containing hydrochloric acid at that concentration.

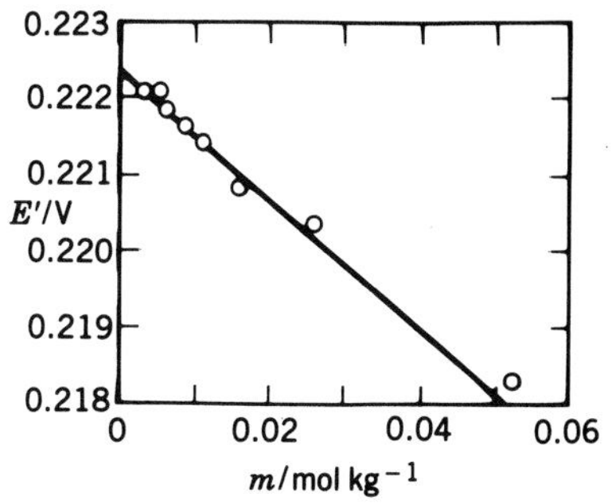

Figure 8.5 Determination of the silver–silver chloride electrode potential by extrapolation of a function of the potential of the cell Pt | H_2(1 bar) | HCl(m) | AgCl | Ag to infinite dilution.

Example 8.6

Calculate the mean ionic activity coefficient of 0.1 mol kg^{-1} hydrochloric acid at 25 °C from the fact that the electromotive force of the cell described in this section is 0.3524 V at 25 °C. Substituting into equation 8.55, we have

$$0.3524 = 0.2224 - 0.1183 \log \gamma_{\pm} - 0.1183 \log 0.1$$

$$\log \gamma_{\pm} = \frac{-0.3524 + 0.2224 + 0.1183}{0.1183} = -0.0989$$

$$\gamma_{\pm} = 0.796$$

In this general manner the activity coefficients of the electrolytes shown in Table 8.1 and Fig. 8.6 have been determined. It should be noted that at high concentrations of electrolytes, activity coefficients may be considerably greater than unity.

The mean ionic activity coefficient for an electrolyte can also be determined by measuring the vapor pressure of the solvent and using the Gibbs–Duhem relation. However, electrochemical cells generally provide much more accurate values of $\gamma_{\pm}$ for electrolytes. The same value of the activity coefficient of an electrolyte is of course obtained whether the equilibrium data come from measurements of vapor pressure, freezing point lowering, boiling point elevation, osmotic pressure, distribution coefficients, equilibrium constants, solubility, or electromotive force.

Table 8.1 Mean Ionic Activity Coefficients $\gamma_{\pm}$ in Water at 25 °C[a] (Concentrations are expressed as molalities)

	m			
Electrolyte	0.1	0.2	0.5	1.0
NaCl	0.778	0.735	0.681	0.657
KCl	0.770	0.718	0.649	0.604
$CaCl_2$	0.518	0.472	0.448	0.500
K_2SO_4	0.441	0.360	0.264	—
$CdSO_4$	0.150	0.102	0.0615	0.0415
$K_3Fe(CN)_6$	0.268	0.212	0.155	0.128

[a] R. C. West, *CRC Handbook of Chemistry and Physics*. Boca Raton, FL: CRC Press, 1988.

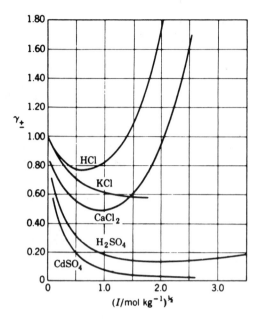

Figure 8.6 Dependence of the mean ionic activity coefficient $\gamma_\pm$ on $I^{1/2}$ for electrolytes at 25 °C.

8.7 Thermodynamics of Electrochemical Cells

Now that we have the standard electromotive force of the galvanic cell described in equation 8.53a, we can calculate the standard reaction Gibbs energy using equation 8.28, which becomes

$$\Delta_r G^\circ = -nFE^\circ \qquad (8.57)$$

and we can calculate the equilibrium constant for the cell reaction by using equation 8.32 with $n = 1$:

$$K = \exp \frac{(96\ 485\ \text{C mol}^{-1})(0.2224\ \text{V})}{(8.3145\ \text{J K}^{-1}\ \text{mol}^{-1})(298.55\ \text{K})}$$

$$= 5745$$

$$= \frac{a_{\text{HCl}}}{(P_{\text{H}_2}/P^\circ)^{1/2}}$$

where hydrogen is assumed to be an ideal gas.

If the standard electromotive force of a cell is measured as a function of temperature, then $\Delta_r S^\circ$, $\Delta_r H^\circ$, and $\Delta_r C_P^\circ$ can be calculated using

$$\Delta_r S^\circ = nF\left(\frac{\partial E^\circ}{\partial T}\right)_P \qquad (8.58)$$

$$\Delta_r H^\circ = -nFE^\circ + nFT\left(\frac{\partial E^\circ}{\partial T}\right)_P \qquad (8.59)$$

$$\Delta_r C_P^\circ = nFT\left(\frac{\partial^2 E^\circ}{\partial T^2}\right)_P \qquad (8.60)$$

The NBS Tables of Chemical Thermodynamic Properties (see Appendix C.1) give the standard thermodynamic properties of many electrolytes. The standard state of an electrolyte or ion is indicated by (ai) if the electrolyte is assumed to be completely ionized (as, for example, NaCl or HCl). For a weak electrolyte (as, for example, acetic acid) properties are given for both the completely ionized standard state (ai) and for the un-ionized (not dissociated) standard state (ao). For an un-ionized solute in aqueous solution, the standard state is the ideal solution at unit molality. The thermodynamic properties of ions are discussed below.

Example 8.7

The standard electromotive force of the cell $Pt \mid H_2(g) \mid HCl(ai) \mid AgCl(s) \mid Ag$ has been determined from 0 to 90 °C by R. G. Bates and V. E. Bower, *J. Res. Nat. Bur. Stand.* **53**:283 (1954). Their data may be represented by

$$\frac{E°}{V} = 0.236\,59 - 4.8564 \times 10^{-4}\left(\frac{t}{°C}\right) - 3.4205$$

$$\times 10^{-6}\left(\frac{t}{°C}\right)^2 + 5.869 \times 10^{-9}\left(\frac{t}{°C}\right)^3$$

What are $\Delta_r G°$, $\Delta_r S°$, $\Delta_r H°$, and $\Delta_r C_P°$ at 25 °C for the reaction

$$\tfrac{1}{2}H_2(g) + AgCl(s) = HCl(ai) + Ag(s)$$

Substituting $t = 25$ °C yields $E° = 0.22240$ V. Thus,

$$\Delta_r G° = -nFE° = -(96\,485 \text{ C mol}^{-1})(0.22240 \text{ V})$$

$$= -21.458 \text{ kJ mol}^{-1}$$

$$\Delta_r S° = nF\left(\frac{\partial E°}{\partial T}\right)_P$$

$$= (96\,485 \text{ C mol}^{-1})\left[-4.8564 \times 10^{-4} - 2(3.4205 \times 10^{-6})\left(\frac{t}{°C}\right)\right.$$

$$\left. + 3(5.869 \times 10^{-9})\left(\frac{t}{°C}\right)^2\right]$$

$$= -62.297 \text{ J K mol}^{-1}$$

$$\Delta_r H° = \Delta G° + T\Delta S° = -21.458 - (298.15)(62.297 \times 10^{-3})$$

$$= -40.032 \text{ kJ mol}^{-1}$$

$$\Delta_r C_P° = nFT\left(\frac{\partial^2 E°}{\partial T^2}\right)_P$$

$$= (96\,485 \text{ C mol}^{-1})(298.15 \text{ K})$$

$$\times \left[-2(3.4205 \times 10^{-6}) + 6(5.869 \times 10^{-9})\left(\frac{t}{°C}\right)\right]$$

$$= -171.4 \text{ J K}^{-1} \text{ mol}^{-1}$$

Since the standard thermodynamic properties of $H_2(g)$, AgCl(s), and Ag(s) are known from other sources, the standard electromotive forces of cell 8.15 over a range of temperature yields the standard thermodynamic properties of aqueous HCl.

Example 8.8

What are the standard thermodynamic properties of HCl($a = 1$) at 25 °C?

The properties of $H_2(g)$, $AgCl(s)$, and $Ag(s)$ at 25 °C are given in Appendix C.1. According to Example 8.7.

$$\Delta_r G° = \Delta_f G°[HCl(ai)] - \Delta_f G°[AgCl(s)]$$

$$\Delta_f G°[HCl(ai)] = -21.458 - 109.805$$

$$= -131.263 \text{ kJ mol}^{-1}$$

$$\Delta_r H° = \Delta_f H°[HCl(ai)] - \Delta_f H°[AgCl(s)]$$

$$\Delta_f H°[HCl(ai)] = -40.032 - 127.068$$

$$= -167.100 \text{ kJ mol}^{-1}$$

$$\Delta_r S° = \overline{S}°[HCl(ai)] + \overline{S}°[Ag(s)] - \overline{S}°[AgCl(s)] - \tfrac{1}{2}\overline{S}°[H_2(g)]$$

$$\overline{S}°[HCl(ai)] = -62.297 - 42.55 + 96.2 + \tfrac{1}{2}(130.574)$$

$$= 56.6 \text{ J K}^{-1} \text{ mol}^{-1}$$

$$\Delta_r C_P° = \overline{C}_P°[HCl(ai)] + \overline{C}_P°[Ag(s)] - \overline{C}_P°[AgCl(s)] - \tfrac{1}{2}\overline{C}_P°[H_2(g)]$$

$$\overline{C}_P°[HCl(ai)] = -171.4 - 25.4 + 50.8 + \tfrac{1}{2}(28.8)$$

$$= -131.6 \text{ J K}^{-1} \text{ mol}^{-1}$$

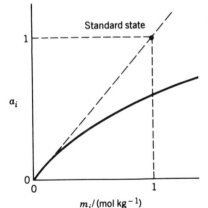

Figure 8.7 Plot of the activity of a solute versus its molality. The dashed line is for an ideal solution. The standard state is a hypothetical $m = 1$ mol kg^{-1} solution in which the solute would have unit activity if it were ideal.

It is important to understand that these thermodynamic properties apply to the hypothetical standard state of HCl in water illustrated in Fig. 8.7. They are the properties that HCl would have in aqueous solution at $m = 1$ mol kg^{-1} if the interactions of the ions with each other and water were the same as at infinite dilution. The activity of a 1 molar solution of HCl is less than 1, as shown by the solid line in Fig. 8.7. Notice that Fig. 8.7 is similar to Fig. 4.4, which was used to explain the concept of the standard state of a gas.

8.8 Electrode Potentials

Rather than simply tabulating the standard electromotive forces of a lot of electrochemical cells, it is more useful to define standard electrode potentials $E°$ and tabulate them. **The standard electrode potential is the potential of a cell in which the hydrogen electrode is on the left and all components of the cell are at unit activity.** The standard electrode potential is a signed quantity, and its value for the cell

$$\tfrac{1}{2}H_2(g) + AgCl(s) = HCl(m) + Ag(s) \tag{8.61a}$$

is 0.2224 V.

Now we adopt the convention that this potential difference is to be attributed entirely to the electrode reaction in the right-hand electrode; that is, the standard electrode potential of the standard hydrogen electrode H^+ ($a = 1$) | H_2 (1 bar) | Pt is arbitrarily assigned the value zero so that its electrode reaction and standard electrode potential are given by

$$H^+(aq) + e^- = \tfrac{1}{2}H_2(g) \qquad E° = 0.0000 \text{ V} \tag{8.61b}$$

Table 8.2 Standard Electrode Potentials at 25 °C[a,e]

Electrode	E°/V	Electrode Reaction
$F^- \mid F_2(g) \mid Pt$	2.87	$\frac{1}{2}F_2(g) + e^- = F^-$
$Au^{3+} \mid Au$	1.50	$\frac{1}{3}Au^{3+} + e^- = \frac{1}{3}Au$
$Pb^{2+} \mid PbO_2 \mid Pb$	1.455	$\frac{1}{2}PbO_2 + 2H^+ + e^- = \frac{1}{2}Pb^{2+} + H_2O$
$Cl^- \mid Cl_2(g) \mid Pt$	1.3604	$\frac{1}{2}Cl_2(g) + e^- = Cl^-$
$H^+ \mid O_2 \mid Pt$	1.2288	$H^+ + \frac{1}{4}O_2 + e^- = \frac{1}{2}H_2O$
$Ag^+ \mid Ag$	0.7992	$Ag^+ + e^- = Ag$
$Fe^{3+}, Fe^{2+} \mid Pt$	0.771	$Fe^{3+} + e^- = Fe^{2+}$
$I^- \mid I_2(s) \mid Pt$	0.5355	$\frac{1}{2}I_2 + e^- = I^-$
$Cu^+ \mid Cu$	0.521	$Cu^+ + e^- = Cu$
$OH^- \mid O_2 \mid Pt^b$	0.4009	$\frac{1}{4}O_2 + \frac{1}{2}H_2O + e^- = OH^-$
$Cu^{2+} \mid Cu$	0.3394	$\frac{1}{2}Cu^2 + e^- = \frac{1}{2}Cu$
$Cl^- \mid Hg_2Cl_2(s) \mid Hg^c$	0.268	$\frac{1}{2}Hg_2Cl_2 + e^- = Hg + Cl^-$
$Cl^- \mid AgCl(s) \mid Ag$	0.2224	$AgCl + e^- = Ag + Cl^-$
$Cu^{2+}, Cu^+ \mid Pt^d$	0.153	$Cu^{2+} + e^- = Cu^+$
$Br^- \mid AgBr(s) \mid Ag$	0.0732	$AgBr + e^- = Ag + Br^-$
$H^+ \mid H_2 \mid Pt$	0.0000	$H^+ + e^- = \frac{1}{2}H_2$
$D^+ \mid D_2 \mid Pt$	−0.0034	$D^+ + e^- = \frac{1}{2}D_2$
$Pb^{2+} \mid Pb$	−0.126	$\frac{1}{2}Pb^{2+} = e^- = \frac{1}{2}Pb$
$Sn^{2+} \mid Sn$	−0.140	$\frac{1}{2}Sn^{2+} + e^- = \frac{1}{2}Sn$
$Ni^{2+} \mid Ni$	−0.250	$\frac{1}{2}Ni^{2+} + e^- = \frac{1}{2}Ni$
$Cd^{2+} \mid Cd$	−0.4022	$\frac{1}{2}Cd^{2+} + e^- = \frac{1}{2}Cd$
$Fe^{2+} \mid Fe$	−0.440	$\frac{1}{2}Fe^{2+} + e^- = \frac{1}{2}Fe$
$Zn^2 \mid Zn$	−0.763	$\frac{1}{2}Zn^{2+} + e^- = \frac{1}{2}Zn$
$OH^- \mid H_2 \mid Pt$	−0.8279	$H_2O + e^- = \frac{1}{2}H_2 + OH^-$
$Mg^{2+} \mid Mg$	−2.37	$\frac{1}{2}Mg^{2+} + e^- = \frac{1}{2}Mg$
$Na^+ \mid Na$	−2.714	$Na^+ + e^- = Na$
$Li^+ \mid Li$	−3.045	$Li^+ + e^- = Li$

[a] All ions are at unit activity (on the molal scale) in water, and all gases are at 1 bar.

[b] See problem 8.49.

[c] The electrode potential of the normal calomel electrode is 0.2802 V and of the calomel electrode containing saturated KCl is 0.2415 V.

[d] The order of writing the ions in the electrolyte solution is immaterial.

[e] Many more standard electrode potentials at 298.15 K are given in S. G. Bratsch, *J. Phys. Chem. Ref. Data* **18**:1(1989).

Thus, the electrode reaction and the standard electrode potential for the $Cl^- \mid AgCl(s) \mid Ag$ electrode are

$$AgCl(s) + e^- = Ag(s) + Cl^-(aq) \qquad E° = 0.2224 \text{ V} \qquad (8.62)$$

Notice that these electrode reactions are both written as reduction reactions, and that standard electrode potentials may be called **reduction potentials.** Thus, electrode potentials are a measure of the tendency of an electrode reaction to occur in the direction of reduction. A brief table of electrode potentials is given in Table 8.2. Notice that the fluorine electrode has the most positive electrode potential; that is, the reaction

$$\tfrac{1}{2}F_2(g) + e^- = F^-(a = 1) \qquad E° = 2.87 \text{ V} \qquad (8.63)$$

has the greatest tendency to go to the right of all of the reactions listed, and the reaction

$$\text{Li}^+(a = 1) + e^- = \text{Li(s)} \qquad E° = -3.045 \text{ V} \qquad (8.64)$$

has the least.

The **standard electromotive force of a cell** can be calculated using

$$E° = E°_R - E°_L \qquad (8.65)$$

Example 8.9

What are the standard electromotive forces $E°$, cell reactions, and standard reaction Gibbs energies for the following cells?

(a)   Pt | Li(s) | Li$^+$ $\vdots$ F$^-$ | F$_2$(g) | Pt

(b)   Pt | F$_2$(g) | F$^-$ $\vdots$ Li$^+$ | Li(s) | Pt

For the first cell

Right electrode	$F_2(g) + 2e^- = 2F^-$	$E°_R = 2.87$ V
Left electrode	$2\text{Li}^+ + 2e^- = 2\text{Li(s)}$	$E°_L = -3.05$ V
Cell reaction	$F_2(g) + 2\text{Li(s)} = 2\text{LiF(aq)}$	$E° = 5.92$ V

$$\begin{aligned}
\Delta_r G° &= -nFE° \\
&= -(2)(96\ 485 \text{ C mol}^{-1})(5.92 \text{ V}) \\
&= -1142 \text{ kJ mol}^{-1}
\end{aligned}$$

For the second cell the reactions are reversed, $E° = -5.92$ V, and $\Delta_r G° = +1142$ kJ mol^{-1}.

In the preceding section we have seen that the cell Pt | H$_2$(g) | HCl ($a = 1$) | AgCl(s) | Ag yields the standard thermodynamic properties of HCl in aqueous solution. A consequence of the assumption that the standard potential for the hydrogen electrode at each temperature is zero is that the thermodynamic properties of H$^+$(aq) are each equal to zero at every temperature.

In the NBS Tables, values of thermodynamic properties of both neutral electrolytes and their constituent ions are given. For the ions the properties are always for the ions in the standard state, and so they are designated by (ao). The properties for ions are based on the convention that $\Delta_f G°$, $\Delta_f H°$, $\bar{S}°$, and $\bar{C}°_P$ for H$^+$(ao) are zero. It follows that a thermodynamic property of a strong electrolyte is equal to the sum of the values for the constituent ions. Although $\bar{S}°$ and $\bar{C}°_P$ for pure gases, liquids, and solids must be positive, these quantities may be negative for a component of a solution. The values tabulated for the properties of the solute are the partial molar properties. The values of $\Delta_f H°$ and $\bar{C}°_P$ for a solute in its standard state are equal to the partial molar enthalpy of formation and the partial molar heat capacity in the infinitely dilute solution.

Example 8.10

What are the values of $\Delta_f G°[Cl^-(ao)]$, $\Delta_f H°[Cl^-(ao)]$, $\overline{S}°[Cl^-(ao)]$, and $\overline{C}_P°[Cl^-(ao)]$?

Since HCl is a strong electrolyte, its thermodynamic properties are the sum of those for $H^+(ao)$ and $Cl^-(ao)$. Therefore, the standard thermodynamic properties of HCl(ai) and $Cl^-(ao)$ are identical. They are the thermodynamic properties for the hypothetical ideal state at a concentration of 1 mol kg^{-1}.

The NBS Tables may also be used to calculate standard electromotive forces at other temperatures because the $\Delta_f H°$ and $\overline{C}_P°$ values may be used to calculate $\Delta_f G°$ at other temperatures. However, a caution to keep in mind is that partial molal enthalpies of formation for ions in aqueous solution may change more rapidly with temperature than their counterparts in the gas phase, so it is important to use $\Delta_r C_P°$.

8.9 Electromotive Forces of Galvanic Cells

The electromotive forces of cells without liquid junctions can be given an exact thermodynamic interpretation, as we have seen for the Pt | H_2(g) | HCl(aq) | AgCl(s) | Ag(s) cell. However, the combination of two arbitrarily chosen electrodes to make a cell generally produces an interface between two different electrolyte solutions. Such an interface contributes a potential difference, referred to as a **junction potential.** In some simple cases junction potentials can be calculated from information about the mobilities of the ions (Section 21.3), but that involves irreversible thermodynamics because diffusion occurs at the interface and the cell is not completely at equilibrium.

In general, it may be stated that the difference of potential resulting from the junction of two solutions is caused by the difference in rates of diffusion of the two ions, **the more dilute solution acquiring a charge with a sign corresponding to that of the faster moving ion.** The difference in potential resulting from the junction of two liquids must be eliminated or corrected to obtain electrode potentials. The most convenient way of minimizing the liquid junction potential is by means of a salt bridge of potassium chloride connecting the two solutions of different electrolytes. Potassium chloride is often used because the mobilities (Section 21.3) of the two ions are about the same. When potassium chloride is not feasible, for example, in a cell containing silver nitrate where AgCl(s) would form, a salt bridge of ammonium nitrate is substituted.

In considering galvanic cells with arbitrarily chosen electrodes, we will represent the salt bridge by $\vdots$ and assume that its contribution to the electromotive force is negligible.

We can either start with a cell and ask for its electromotive force or start with a reaction and ask for the representation of the cell and its electromotive force. Suppose we start with the cell:

$$Pt \mid Ox_L, Red_L \;\vdots\; Ox_R, Red_R \mid Pt \qquad (8.66)$$

where Ox and Red represent the oxidized and reduced species involved in an electrode reaction. To write the electrode reaction for this cell, we have to follow the convention that if the cell reaction is spontaneous, the electromotive force for the cell is positive because of $\Delta_r G = -nFE$ (equation 8.28). For the right-hand electrode to be positive, electrons have to be removed from it by the oxidized form in the right electrode according to the reaction

$$Ox_R + ne^- = Red_R \tag{8.67}$$

The reaction occurring at the left electrode is an oxidation reaction

$$Red_L = Ox_L + ne^- \tag{8.68}$$

Notice that when they are combined to give the net reaction, these reactions must involve the same number n of electrons. The **cell reaction** corresponding with the cell in 8.66 is therefore

$$Ox_R + Red_L = Red_R + Ox_L \tag{8.69}$$

Electrode reactions are always written as reductions with their standard electrode potentials. To obtain the cell reaction, the reaction on the left is subtracted from the reaction on the right, and the standard potential for the cell is obtained by subtracting E_L° from E_R°, according to $E^\circ = E_R^\circ - E_L^\circ$ (equation 8.65).

If we make these calculations with a specific cell reaction, and the standard electromotive force E° is positive, the right-hand electrode is positive and the cell reaction is spontaneous. If the electromotive force is negative, the right-hand electrode is negative, and the cell reaction, as written, is not spontaneous.

Example 8.11

Assume that the following reaction occurs in an electrochemical cell.

$$Cd(s) + Cu^{2+} = Cd^{2+} + Cu(s)$$

(a) What is the representation for the cell for this reaction? (b) What is the standard electromotive force E° for this cell at 25 °C? (c) What is $\Delta_r G^\circ$ for the cell reaction at 25 °C? (d) What is the equilibrium constant for the cell reaction? (e) Which is the positive electrode on open circuit when the activities of Cu^{2+} and Cd^{2+} are each unity?

(a)   Since cupric ion is reduced, it must be in the right electrode.

$$Cd \mid Cd^{2+} \vdots\vdots Cu^{2+} \mid Cu$$

(b)   The standard electrode potentials for the electrode reactions are

$$Cu^{2+} + 2e^- = Cu \qquad E_R^\circ = 0.3394 \text{ V}$$
$$Cd^{2+} + 2e^- = Cd \qquad E_L^\circ = -0.4022 \text{ V}$$

Subtracting, we obtain the cell reaction and its standard elecromotive force:

$$Cd + Cu^{2+} = Cd^{2+} + Cu \qquad E^\circ = 0.7416 \text{ V}$$

(c) $\Delta_r G^\circ = -nFE^\circ = -(2)(96\,485 \text{ C mol}^{-1})(0.7416 \text{ V})$

$$= -143.11 \text{ kJ mol}^{-1}$$

(d)   The equilibrium constant for the cell reaction can also be calculated using equation 8.32:

$$K = \exp \frac{nFE^\circ}{RT}$$

$$= \exp \frac{(2)(96\,485 \text{ C mol}^{-1})(0.7416 \text{ V})}{(8.3145 \text{ J K}^{-1} \text{ mol}^{-1})(298.15 \text{ K})}$$

$$= 1.179 \times 10^{25}$$

(*e*)   The right-hand electrode is positive because the cell reaction is spontaneous and electrons are consumed by the reaction at this electrode.

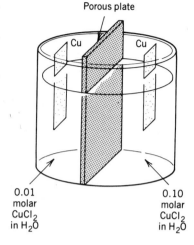

Figure 8.8   Concentration cell.

Example 8.12

Calculate the electromotive force of the concentration cell shown in Fig. 8.8 at 25 °C assuming that activity coefficients are unity.

$$E = -\frac{RT}{nF} \ln \frac{0.01}{0.1}$$

$$= -\frac{0.0591}{2} \log 0.1$$

$$= 0.0296 \text{ V}$$

In interpreting equilibrium constants calculated from electrode potentials it is necessary to know the standard states of all reactants and products. The standard state of an electrolyte in aqueous solution is usually the ideal solution at unit mean molality, as described by Fig. 8.7. If the solvent is also a reactant, it is generally treated on the mole fraction scale, rather than the molal or molar scales used for the other reactants. In this case the equilibrium constant expression is written

$$K = (\gamma_{x,A}x_A)^{\nu_A} \prod_{i \neq A} \left(\frac{\gamma_{m,i}m_i}{m^\circ}\right)^{\nu_i} \tag{8.70}$$

where A is the solvent, $\gamma_{x,A}$ is its activity coefficient on the mole fraction scale, $\gamma_{m,i}$ is the activity coefficient of reactant i on the molal scale, and m° is the standard value of the molality (1 mol kg^{-1}). For dilute solutions $\gamma_{x,A}x_A$ is close to unity, and so the equilibrium constant expression is written

$$K = \prod_{i \neq A} \left(\frac{\gamma_{m,i}m_i}{m^\circ}\right)^{\nu_i} \tag{8.71}$$

which simplies further if the $\gamma_{m,i}$ are close to unity. Although the solvent is left out of equation 8.71, the Gibbs energy of formation of the solvent must be included in calculating $\Delta G°$ for a reaction in liquid solution.

Standard electrode potentials can be used to calculate the solubilities of slightly soluble salts.

Example 8.13

Devise a cell for which the cell reaction is

$$\text{Ag}^+(\text{aq}) + \text{Br}^-(\text{aq}) = \text{AgBr}(\text{s})$$

(*a*)   What is the standard electromotive force of the cell at 25 °C? (*b*)   Calculate $\Delta G°$ for the cell reaction both from $E°$ and from $\Delta_f G°$ values in Appendix C.1. (*c*)   What is the equilibrium constant for the reaction? (*d*)   What is the solubility of AgBr(s) in water?

(a)   Ag | AgBr | Br^- ‖ Ag^+ | Ag

Right electrode	$Ag^+ + e^- = Ag$		$E_R^° = 0.799\ 2$ V
Left electrode	$AgBr + e^- = Br^- + Ag$		$E_L^° = 0.073\ 2$ V
Cell reaction	$Ag^+ + Br^- = AgBr$		$E^° = 0.726\ 0$ V

(b) $\Delta_r G^° = -nFE^° = -(96\ 485\ \text{C mol}^{-1})(0.726\ 0\ \text{V})$

$\qquad\qquad = -70.05\ \text{kJ mol}^{-1}$

$\quad \Delta_r G^° = -96.90 - 77.107 - (-103.96)$

$\qquad\qquad = -70.05\ \text{kJ mol}^{-1}$

(c) $K = \exp \dfrac{-\Delta_r G^°}{RT}$

$\qquad = \exp \dfrac{70\ 050}{8.314\ 5 \times 298.15} = 1.872 \times 10^{12}$

$\qquad = m^{-2}\gamma_{\pm}^{-2}$

This equilibrium constant is equal to $1/K_{sp}$ where $\gamma_{\pm} = 1$. K_{sp} is called the solubility product.

(d) $m = 7.31 \times 10^{-7}\ \text{mol kg}^{-1}$ assuming $\gamma_{\pm} = 1$

If the reactants and products in a galvanic cell are not at unit activity, the electromotive force E of the cell is given by the Nernst equation 8.30. Assuming that the activity coefficients in reaction 8.69 are all unity, the electromotive force is given by

$$E = E^° - \frac{RT}{nF} \ln \frac{[\text{Red}_R][\text{Ox}_L]}{[\text{Red}_L][\text{Ox}_R]} \qquad (8.72)$$

This is a simple example, but it should be obvious how to change this expression if some of the stoichiometric numbers are not unity and the activity coefficients are not unity.

8.10   Calculation of Electrode Potentials from Gibbs Energies of Formation

The standard potential for an electrochemical cell can be calculated from values of $\Delta_f G^°$, if they are available, using $E^° = -\Delta_r G^°/nF$. Since the half-cell reactions are written as reduction reactions, $\Delta_r G^°$ is simply calculated for these reactions, taking $\Delta_f G^° = 0$ for e^-, and is converted to $E^°$ using $-\Delta_r G^°/nF$. This is illustrated by the next example. But in doing this we should realize that the electrode potential reaction is a shorthand notation for a cell with a hydrogen electrode; as we will soon see this is important for the calculation of $\Delta_r S^°$ and $\Delta_r C_P^°$ for an electrode reaction.

Example 8.14

Use Appendix C.1 to calculate the standard electrode potentials for the following three electrodes: $Cd^{2+} \mid Cd$, $Cl^- \mid Cl_2(g) \mid Pt$, $Cl^- \mid AgCl(s) \mid Ag$.

$$Cd^{2+}(ao) + 2e^- = Cd(s)$$

$$\Delta_r G° = -(-77.612) = 77.612 \text{ kJ mol}^{-1}$$

$$E° = -\frac{\Delta G°}{nF} = -\frac{77\ 612 \text{ J mol}^{-1}}{2(96\ 485 \text{ C mol}^{-1})}$$

$$= -0.4022 \text{ V}$$

$$Cl_2(g) + 2e^- = 2Cl^-(ao)$$

$$\Delta_r G° = 2(-131.228 \text{ kJ mol}^{-1})$$

$$E° = \frac{2(131\ 228 \text{ J mol}^{-1})}{2(96\ 485 \text{ C mol}^{-1})}$$

$$= 1.360\ 1 \text{ V}$$

$$AgCl(s) + e^- = Ag(s) + Cl^-(ao)$$

$$\Delta_r G° = -131.228 - (-109.789) = -21.439 \text{ kJ mol}^{-1}$$

$$E° = -\frac{-21\ 439}{96\ 485} = 0.222\ 2 \text{ V}$$

In calculating $\Delta_r G°$ and $\Delta_r H°$ for an electrode reaction we can ignore the fact that we are really talking about a cell with a hydrogen electrode because neither $H^+(ao)$ nor $H_2(g)$ contributes to $\Delta_r G°$. However, $H_2(g)$ does contribute to $\Delta_r S°$ and $\Delta_r C_P°$ for the electrode reaction. The reason for this is that the convention for $H^+(ao)$ is that $\Delta_f G°$, $\Delta_f H°$, $\bar{S}°$, and $\bar{C}_P$ are all zero, but the convention for $H_2(g)$ is that only $\Delta_f G°$ and $\Delta_f H°$ are zero. To calculate $\Delta_r S°$ and $\Delta_r C_P°$ for an electrode reaction we have to remember that the standard electrode potential is the electromotive force of a *cell* in which the standard hydrogen electrode is on the left. Thus, the reaction for the $M^{n+} \mid M$ electrode is actually the cell reaction

$$M^{n+}(ao) + (n/2)H_2(g) = M + nH^+(ao) \tag{8.73}$$

where M is a metal. The reaction for the $X \mid X^{n-}$ electrode is actually the cell reaction

$$X + (n/2)H_2(g) = X^{n-}(ao) + nH^+(ao) \tag{8.74}$$

where X is a nonmetal. $\Delta_f G°$ and $\Delta_f H°$ for these reactions are the same as for

$$M^{n+}(ao) + ne^- = M \tag{8.75}$$

$$X + ne^- = X^{n-} \tag{8.76}$$

but we have to use reactions 8.73 and 8.74 to calculate $\Delta_r S°$ and $\Delta_r C_P°$ for the chemical reaction that corresponds with the electrode potential.

The standard entropy of formation of an ion in aqueous solution is given by

$$\Delta_f S° = \bar{S}_{ion}° - \sum \bar{S}°(\text{elem.}) + (n/2)\bar{S}°(H_2, g) \tag{8.77}$$

where n is the algebraic value of the ionic charge and $\sum \bar{S}°$(elem.) is the sum of the entropies of the elements in the ion, each multiplied by its stoichiometric number in the formation reaction. If $\Delta_f S°$ is calculated this way, then the relation

$$\Delta_f G° = \Delta_f H° - T \Delta_f S° \tag{8.78}$$

is obeyed for the formation of an ion in aqueous solution.

Example 8.15
Appendix C.1 gives $\Delta_f H°$, $\Delta_f G°$, and $\bar{S}°$ for Cd^{2+}(ao). Calculate $\Delta_f S°$ for Cd^{2+}(ao) by two different methods.
  $\Delta_f S°[Cd^{2+}$(ao)] can be calculated by using equation 8.77 and the formation reaction

$$Cd(s) + 2H^+(ao) = Cd^{2+}(ao) + H_2(g)$$
$$\Delta_f S° = \bar{S}°_{ion} - \bar{S}°_{Cd} + \bar{S}°(H_2, g)$$
$$= -73.2 - 51.76 + 130.684$$
$$= 5.7 \text{ J K}^{-1} \text{ mol}^{-1}$$

This quantity can also be calculated using $\Delta_f G° = \Delta_f H° - T \Delta_f S°$:

$$\Delta_f S° = \frac{\Delta_f H° - \Delta_f G°}{T} = \frac{[-75.90 - (-77.612)]1000}{298.15}$$
$$= 5.74 \text{ J K}^{-1} \text{ mol}^{-1}$$

8.11 Determination of pH

The concentrations of hydrogen ions in aqueous solutions range from about 1 mol L^{-1} in 1 mol L^{-1} HCl to about 10^{-14} mol L^{11} in 1 mol L^{-1} NaOH. Because of this wide range of concentrations, Sorenson adopted an exponential notation in 1909. He defined pH as the negative exponent of 10 that gives the hydrogen ion concentration. Now the pH is defined to be as close as possible to the negative base 10 logarithm of the hydrogen ion activity:

$$pH = -\log a_{H^+} \tag{8.79}$$

Strictly speaking, the activity of a single ion cannot be determined, but pH meters are calibrated with buffers for which the pH has been calculated using an extended Debye–Hückel theory.
  The pH may be measured with a hydrogen electrode connected with a calomel electrode through a salt bridge:

$$Pt \mid H_2(P) \mid H^+(a_{H^+}) \vdots Cl^- \mid Hg_2Cl_2 \mid Hg$$

The electromotive force of this cell may be considered to be made up of three contributions

$$E = 0.2802 - 0.0591 \log \left[\frac{a_{H^+}}{(P_{H_2}/P°)^{1/2}} \right] + E_{\text{liquid junction}} \tag{8.80}$$

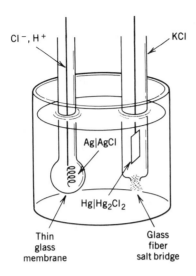

Figure 8.9   Glass electrode and calomel electrode for pH meters.

where the contribution by the normal calomel electrode is 0.2802 V at 25 °C. Although the activities of single ions cannot be determined, equation 8.80 is often used with the assumption that $E_{\text{liquid junction}} = 0$. If $P_{H_2} = 1$ bar, then

$$E - 0.2802 = -0.0591 \log a_{H^+} \qquad (8.81)$$

Hydrogen ion activities obtained in this way are of great practical use, even though they are based on an approximation.

When $pH = -\log a_{H^+}$, equation 8.81 becomes

$$E - 0.2802 = 0.0591 \, pH \qquad (8.82)$$

or

$$pH = \frac{E - 0.2802}{0.0591} \qquad (8.83)$$

Usually the pH is measured with a glass electrode because this avoids the use of hydrogen and the possibility of poisoning the platinized platinum surface.

A glass electrode consists of a reversible electrode, such as a calomel or Ag–AgCl electrode, in a solution of constant pH inside a thin membrane of a special glass. The thin glass bulb of this electrode is immersed in the solution to be studied along with a reference calomel electrode to form the cell indicated by

$$Ag \mid AgCl \mid Cl^-, H^+ \mid \text{glass membrane} \mid \text{solution} \:: \text{calomel electrode}$$

and by Fig. 8.9. It is found experimentally that the potential of such a glass electrode varies with the activity of hydrogen ions in the same way as the hydrogen electrode, that is, 0.0591 V per pH unit at 25 °C. An ordinary potentiometer cannot be used to measure the voltage of such a cell because of the high resistance of the glass membrane, and so an electronic voltmeter must be employed. Electronic devices using the glass electrode have been developed that make it possible to measure pH values to ± 0.01 pH unit with easily portable apparatus. The pH meter, as it is often called, is calibrated by means of a buffer of known pH before it is used to measure the pH of an unknown solution.*

The glass electrode has become the most useful electrode for determining the pH of a solution. It is not affected by oxidizing or reducing agents and is not easily poisoned. It is especially useful in biochemical investigations.

The carbon dioxide electrode is another useful clinical tool. This electrode consists of a glass electrode in contact with a solution of fixed bicarbonate concentration that is separated from the sample solution by a polymer membrane permeable to CO_2. When CO_2 diffuses into the bicarbonate solution, it is hydrated to H_2CO_3 and then rapidly ionizes to $HCO_3^- + H^+$ with a consequent change in pH.

There is a growing number of ion-selective electrodes that use semipermeable membranes to obtain a cell potential that depends on a particular ion.

8.12 Special Topic: Fuel Cells

A fuel cell is a cell that is continuously supplied with an oxidant and a reductant so that it can deliver a current indefinitely. Fuel cells offer the possibility of achieving high thermodynamic efficiency in the conversion of Gibbs energy into mechanical work. Internal combustion engines at best convert only the fraction $(T_2 - T_1)/T_2$ of the heat of combustion into mechanical work. In this relation, which comes from the second law of thermodynamics, T_2 is the temperature of the gas during expansion and T_1 is the temperature of the exhaust (see Section 3.3).

Fuel cells may be classified according to the temperature range in which they operate: low temperature (25 to 100 °C), medium temperature (100 to 500 °C), high temperature (500 to 1000 °C), and very high temperature (above 1000 °C). The advantage of using high temperatures is that catalysts for the various steps in the process are not so necessary. Polarization of a fuel cell reduces the current. Polarization is the result of slow reactions or processes such as diffusion in the cell.

Figure 8.10 indicates the construction of a hydrogen–oxygen fuel cell with a solid electrolyte, which is an ion exchange membrane. The membrane is impermeable to the reactant gases, but is permeable to hydrogen ions, which carry the current between the electrodes. To facilitate the operation of the cell at 40 to 60 °C, the electrodes are covered with finely divided platinum that functions as a catalyst. Water is drained out of the cell during operation. Fuel cells of this general type have been used successfully in the space program and are quite efficient. Their disadvantages for large-scale commercial application are that hydrogen presents storage problems, and platinum is an expensive catalyst. Cheaper catalysts have been found for higher temperature operation of hydrogen–oxygen fuel cells.

---

* Definition of pH scales, standard reference values, measurement of pH and related terminology. *Pure Appl. Chem.* **57**:531 (1985).

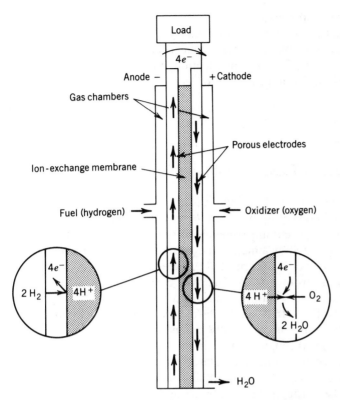

Figure 8.10 Hydrogen–oxygen fuel cell with an ion exchange membrane. (From J. O'M. Bockris and S. Srinivasan, *Fuel Cells, Their Electrochemistry*. New York: Copyright © 1969 by McGraw-Hill, Inc. Used with permission of McGraw-Hill Book Co.)

Fuel cells that use hydrocarbons and air have been developed, but their power per unit weight is too low to make them practical in ordinary automobiles. Better catalysts are needed.

A hydrogen–oxygen fuel cell may have an acidic or alkaline electrolyte. The half-cell reactions are

$$\tfrac{1}{2}O_2(g) + 2H^+ + 2e^- = H_2O(l) \qquad E° = 1.228\ 8\ V$$
$$\underline{\qquad 2H^+ + 2e^- = H_2(g) \qquad\qquad E° = 0 \qquad\qquad}$$
$$H_2(g) + \tfrac{1}{2}O_2(g) = H_2O(l) \qquad E° = 1.228\ 8\ V$$

or

$$\tfrac{1}{2}O_2(g) + H_2O(l) + 2e^- = 2OH^- \qquad\qquad E° = 0.400\ 9\ V$$
$$\underline{\qquad 2H_2O(l) + 2e^- = H_2(g) + 2OH^- \qquad E° = -0.827\ 9\ V}$$
$$H_2(g) + \tfrac{1}{2}O_2(g) = H_2O(l) \qquad\qquad\qquad E° = 1.228\ 8\ V$$

To maximize the power per unit mass of an electrochemical cell, the electronic and electrolytic resistances of the cell must be minimized. Since fused salts have lower electrolytic resistances than aqueous solutions, high-temperature electrochemical cells are of special interest for practical applications.

High temperatures also allow the use of liquid metal electrodes, which make possible higher current densities than solid electrodes.

In this chapter we have been concerned with the thermodynamics of electrochemical cells. Practical applications involve kinetic considerations, which we describe in Section 21.12. Photoelectrochemical cells are discussed in Section 20.14.

8.13   Special Topic: Membrane Potential

If two different electrolyte solutions are separated by a membrane, a potential difference will be set up between the two solutions if the membrane is permeable to some ions and impermeable to others. For example, consider two KCl solutions separated by a membrane permeable to K^+, but impermeable to Cl^-. If solution α is more concentrated than solution β, K^+ ions will diffuse through the membrane from α to β. This will cause solution β to become positively charged relative to solution α and to have a higher electrical potential. Actually the amount of K^+ that has to diffuse through the membrane to produce the potential difference is chemically insignificant, as we have seen in Section 8.1. As K^+ diffuses through the membrane, the electrical potential difference that is set up across the membrane retards the diffusion of more K^+, and eventually an equilibrium is reached. At equilibrium, equation 8.11 applies to each ion that can pass through the membrane and so insertion of $\mu_i = \mu_i^\circ + RT \ln a_i$ yields

$$(\mu_{i\alpha}^\circ)_{\phi=0} + RT \ln a_{i\alpha} + z_i F \phi_\alpha = (\mu_{i\beta}^\circ)_{\phi=0} + RT \ln a_{i\beta} + z_i F \phi_\beta \quad (8.84)$$

where z_i is the charge number of the ion. If the solvent is the same on two sides of the membrane, then

$$(\mu_{i\alpha}^\circ)_{\phi=0} = (\mu_{i\beta}^\circ)_{\phi=0}$$

and

$$\Delta\phi = \phi_\beta - \phi_\alpha = -\frac{RT}{z_i F} \ln \frac{a_{i\beta}}{a_{i\alpha}} \quad (8.85)$$

where $\Delta\phi$ is referred to as the **membrane potential**. This equation may be written in terms of molal concentrations if the activity coefficients of ion i are nearly the same in solutions α and β. The membrane potential $\Delta\phi$ may be measured by placing identical reversible electrodes in solutions α and β.

Example 8.16
The membrane potential for a resting nerve cell is given by $\Delta\phi = \phi_{int} - \phi_{ext} = -70$ mV, where ϕ_{int} is the potential internal to the cell and ϕ_{ext} is the potential external to the cell. Given the fact that the concentration of K^+ inside a resting nerve cell is about 35 times that outside the cell, what membrane potential is expected?

$$\Delta\phi = \phi_{int} - \phi_{ext} = -\frac{(8.314 \text{ J K}^{-1} \text{ mol}^{-1})(298 \text{ K})}{96\ 485 \text{ C mol}^{-1}} \ln 35$$

$$= -91 \text{ mV}$$

This is perhaps as close as equation 8.85 should be expected to come to the observed -70 mV because the resting nerve cell is not actually at equilibrium.

While K^+ is concentrated inside a nerve cell, Na^+ is at about a 10-fold higher concentration outside. The concentration differences for these ions across the cell membrane are maintained by "pumps" utilizing energy from ATP (Section 9.6) and under the control of enzymes.

When a nerve impulse starts at one end of a nerve cell, the membrane potential becomes momentarily positive. When this happens, the membrane's permeability to Na^+ momentarily increases and $\Delta \phi$ moves toward the equilibrium value for Na^+ of about $+60$ mV. This pulse propagates along the nerve cell with a speed of $10-100$ m s^{-1}. After the peak, the permeability to Na^+ decreases and the permeability to K^+ temporarily increases so the $\Delta \phi$ returns to its resting value of -70 mV.

References

A. Bard and L. Faulkner, *Electrochemical Methods*. New York: Wiley, 1980.

A. J. Bard, R. Parsons, and J. Jordan, *Standard Potentials in Aqueous Solutions*. New York: Dekker, 1985.

R. G. Bates, *Determination of pH: Theory and Practice*. New York: Wiley, 1973.

J. O'M. Bockris and D. M. Drazic, *Electro-Chemical Science*. London: Taylor & Francis, 1972.

J. O'M. Bockris and A. K. N. Reddy, *Modern Electrochemistry,* Vols. 1 and 2. New York: Plenum, 1970.

J. Goodisman, *Electrochemistry: Theoretical Foundations*. New York: Wiley, 1987.

A. McDougall, *Fuel Cells*. New York: Wiley, 1976.

M. L. McGlashan, *Chemical Thermodynamics*. New York: Academic, 1979.

J. S. Newman, *Electrochemical Systems*. Englewood Cliffs, NJ: Prentice Hall, 1973.

P. Rieger, *Electrochemistry*. Englewood Cliffs, NJ: Prentice Hall, 1987.

Problems

8.1 How much work is required to bring two protons from an infinite distance of separation to 0.1 nm? Calculate the answer in joules using the protonic charge 1.602×10^{-19} C. What is the work in kJ mol^{-1} for a mole of proton pairs?

8.2 How much work in kJ mol^{-1} can in principle be obtained when an electron is brought to 0.5 nm from a proton?

8.3 A small dry battery of zinc and ammonium chloride weighing 85 g will operate continuously through a 4-Ω resistance for 450 min before its voltage falls below 0.75 V. The initial voltage is 1.60, and the effective voltage over the whole life of the battery is taken to be 1.00. Theoretically, how many kilometers above the earth could this battery be raised by the energy delivered under these conditions?

8.4 (*a*) The mean ionic activity coefficient of 0.1 molar HCl(aq) at 25 °C is 0.796. What is the activity of HCl in this solution? (*b*) The mean ionic activity coefficient of 0.1 molar H_2SO_4 is 0.265. What is the activity of H_2SO_4 in this solution?

8.5 The solubility of Ag_2CrO_4 in water is 8.00×10^{-5} mol kg^{-1} at 25 °C, and its solubility in 0.04 mol kg^{-1} NaNO$_3$ is 8.84×10^{-5} mol kg^{-1}. What is the mean ionic activity coefficient of Ag_2CrO_4 in 0.04 mol kg^{-1} NaNO$_3$?

8.6 A solution of NaCl has an ionic strength of 0.24 mol kg^{-1}. (*a*) What is its molality? (*b*) What molality of Na_2SO_4 would have the same ionic strength? (*c*) What molality of $MgSO_4$?

8.7 Using the limiting law calculate the mean ionic activity coefficients at 25 °C in water of the following electrolytes at 10^{-3} *m*: (*a*) NaCl, (*b*) $CaCl_2$, (*c*) $LaCl_3$.

8.8 The cell Pt | H_2(1 bar) | HBr(*m*) | AgBr | Ag has been studied by H. S. Harned, A. S. Keston, and J. G. Donelson, (*J. Am. Chem. Soc.* **58**: 989 (1936)). The following table gives the electromotive forces obtained at 25 °C:

m/mol kg^{-1}	0.01	0.02	0.05	0.10
E/V		0.3127	0.2786	0.2340	0.2005

Calculate (*a*) $E°$ and (*b*) the activity coefficient for a 0.10-mol kg^{-1} solution of hydrogen bromide.

8.9 Design cells without liquid junction that could be used to determine the activity coefficients of aqueous solutions of (*a*) NaOH and (*b*) H_2SO_4. Give the equations relating electromotive force to the mean ionic activity coefficient at 25 °C.

8.10 The electromotive force of the cell

Pb(s) | $PbSO_4$(s) | $Na_2SO_4 \cdot 10H_2O$(sat) | Hg_2SO_4(s) | Hg(l)

is 0.9647 V at 25 °C. The temperature coefficient is 1.74×10^{-4} V K^{-1}. (*a*) What is the cell reaction? (*b*) What are the values of $\Delta_r G$, $\Delta_r S$, $\Delta_r H$?

8.11 For the galvanic cell

H_2(1 bar) | HCl(ai) | Cl_2(1 bar)

the standard electromotive force at 298.15 K is 1.3604 V and $(\partial E°/\partial T)_P = -1.247 \times 10^{-3}$ V K^{-1}. (*a*) For the cell reaction, what are the values of $\Delta_r G°$, $\Delta_r H°$, and $\Delta_r S°$? (*b*) For Cl$^-$(ao) what are the values of $\Delta_f G°$, $\Delta_f H°$, and $\bar{S}°$?

8.12 In problem 4.8 two equations were derived for calculating $\Delta_r G$ at another temperature if it is known at one. Compare the values of $\Delta_r G°$(323 K) and K_w calculated with these equations for

H_2O(l) = H$^+$(ao) + OH$^-$(ao)

8.13 Calculate $E°$ for the half-cell OH$^-$ | H_2 | Pt at 25 °C using the value of the ion product for water which is 1.006×10^{-14} (Section 9.1).

8.14 What are the values of $\Delta_r G°$ and K for the following reactions 298 K from Appendix C.1?

(*a*) Cu(s) + Zn^{2+}(ao) = Cu^{2+}(ao) + Zn(s)

(*b*) H_2(g) + Cl_2(g) = 2HCl(ai)

(*c*) Ca^{2+}(ao) + CO$_3^{2-}$(ao) = $CaCO_3$(s, calcite)

(*d*) $\frac{1}{2}Cl_2$(g) + Br$^-$(ao) = $\frac{1}{2}Br_2$(g) + Cl$^-$(ao)

(*e*) Ag$^+$(ao) + Fe^{2+}(ao) = Fe^{3+}(ao) + Ag(s)

8.15 For the electrochemical cell

$$Zn(s) \mid Zn^{2+}(aq) \vdots\vdots Cu^{2+}(aq) \mid Cu(s)$$

what are (*a*) the half-cell reactions, (*b*) the cell reaction, (*c*) the standard electromotive force, (*d*) the equilibrium constant for the cell reaction, and (*e*) equilibrium constant expression?

8.16 From standard electrode potentials in Table 8.2 what are the standard Gibbs energies of formation at 25 °C for Cl$^-$(ao), OH$^-$(ao), and Na$^+$(ao)?

8.17 According to Table 8.1 what are the equilibrium constants for the following reactions at 25 °C?

(*a*) H$^+$(ao) + Li(s) = Li$^+$(ao) + $\frac{1}{2}H_2$(g)

(*b*) 2H$^+$(ao) + Pb(s) = Pb^{2+}(ao) + H_2(g)

(*c*) 3H$^+$(ao) + Au(s) = Au^{3+}(ao) + $\frac{3}{2}H_2$(g)

8.18 At 25 °C the standard electrode potential for the Ag$^+$ | Ag electrode is 0.7991 V, and the solubility product for AgI is 8.2×10^{-17}. What is the standard electrode potential for I$^-$ | AgI | Ag?

8.19 Using data from Appendix C.1 calculate the solubility of AgCl(s) in water at 298.15 K. The salt is completely dissociated in the aqueous phase.

8.20 Calculate the standard electrode potentials at 25 °C for the following electrodes using Appendix C.1: (*a*) Li$^+$(ao) | Li(s), (*b*) F$^-$(ao) | F_2(g), and (*c*) Pb^{2+}(ao) | PbO_2(s) | Pb.

8.21 Using Appendix C.1 calculate the values of $\Delta_r G°$, $\Delta_r H°$, $\Delta_r S°$, and $\Delta_r C_P°$ at 25 °C for the electrode reaction for the Na$^+$ | Na electrode.

8.22 The standard electrode potentials $E°$ in the earlier literature are based on a standard state pressure of 1 atm. Show that when the bar is used as the standard state pressure, standard electrode potentials $E°$(atm) need to be corrected to $E°$(bar) using

$$E°(\text{bar}) = E°(\text{atm}) + (0.000\ 169\ \text{V})\ \Delta\nu$$

where $\Delta\nu$ is the increase in the number of gaseous molecules as the cell reaction (including hydrogen) proceeds as written.

8.23 Calculate the standard Gibbs energy of formation of $Na^+(ao)$ from its $\Delta_f H°$ and $\bar{S}°$ using data from Appendix C.1 and equation 8.78.

8.24 When a hydrogen electrode and a normal calomel electrode are immersed in a solution at 25 °C a potential of 0.664 V is obtained. Calculate (a) the pH, and (b) the hydrogen ion activity.

8.25 Calculate the equilibrium constant at 25 °C for the reaction

$$2H^+ + D_2(g) = H_2(g) + 2D^+$$

from the electrode potential for $D^+ \mid D_2 \mid Pt$, which is -3.4 mV at 25 °C.

8.26 A water electrolysis cell operated at 25 °C consumes 25 kWh/lb of hydrogen produced. Calculate the cell efficiency using $\Delta_r G°$ for the decomposition of water.

8.27 (a) When methane is oxidized completely to $CO_2(g)$ and $H_2O(l)$ at 25 °C, how much electrical energy can be produced using a fuel cell, assuming that there are no electrical losses? What is the electromotive force of the fuel cell? (b) When one mole of methane is oxidized completely in a Carnot engine that operates between 500 and 300 K, how much electrical energy can be produced, assuming that the mechanical energy can be converted completely into electrical energy?

8.28 Calculate the electromotive force of

$$Li(l) \mid LiCl(l) \mid Cl_2(g)$$

at 900 K for $P_{Cl_2} = 1$ bar. This high-temperature battery is attractive because of its high electromotive force and low atomic masses. Lithium chloride melts at 883 K and lithium at 453.69 K. (The $\Delta_f G°$ for LiCl(l) at 900 K in JANAF Thermochemical Tables is -336.140 kJ mol^{-1}.)

8.29 A membrane permeable only by Na^+ is used to separate the following two solutions:

α   0.10 mol kg^{-1} NaCl, 0.05 mol kg^{-1} KCl
β   0.05 mol kg^{-1} NaCl, 0.10 mol kg^{-1} KCl

What is the membrane potential at 25 °C and which solution has the highest positive potential?

8.30 What is the electric field strength 0.5 nm from a proton?

8.31 Calculate the energy in kJ mol^{-1} required to separate a positive and negative charge from 0.3 nm to infinity in (a) a vacuum, (b) a solvent of dielectric constant 10, and (c) water at 25 °C, which has a dielectric constant of approximately 80.

8.32 What is the expression for the activity of Na_2SO_4 in terms of the mean ionic activity coefficient and the molality.

8.33 Give the expressions for the mean ionic activity coefficients of LiCl, $AlCl_3$, and $MgSO_4$ in terms of the activity of the electrolyte and its molality.

8.34 What is the ionic strength of each of the following solutions: (a) 0.1 mol kg^{-1} NaCl, (b) 0.1 mol kg^{-1} $Na_2C_2O_4$, (c) 0.1 mol kg^{-1} $CuSO_4$, (d) a solution containing 0.1 mol kg^{-1} Na_2HPO_4 and 0.1 mol kg^{-1} NaH_2PO_4?

8.35 For 0.002 mol kg^{-1} $CaCl_2$ at 25 °C use the Debye–Hückel limiting law to calculate the activity coefficients of Ca^{2+} and Cl^-. What is the mean ionic activity coefficient for the electrolyte?

8.36 According to Appendix C.1 what is the value of the equilibrium constant for the reaction

$$\tfrac{1}{2}H_2(g) + AgCl(s) = Ag(s) + H^+(ao) + Cl^-(ao)$$

at 25 °C and how it is defined?

8.37 Derive the expression for the electromotive force of the cell

$$Pt \mid H_2(g, 1\ bar) \mid KH_2PO_4(m_1),$$
$$Na_2HPO_4(m_2) \vdots NaX(m_3) \mid AgX(s) \mid Ag$$

Substitute the equilibrium expression for the second dissociation of phosphoric acid and describe how the thermodynamic dissociation constant K for that dissociation could be obtained from electromotive force measurements at constant temperature.

8.38 (a) Write the reaction that occurs when the cell

$$Zn \mid ZnCl_2(0.555\ mol\ kg^{-1}) \mid AgCl \mid Ag$$

delivers current and calculate (b) $\Delta_r G$, (c) $\Delta_r S$, and (d) $\Delta_r H$ at 25 °C for this reaction. At 25 °C, $E = 1.015$ V and $(\partial E/\partial T)_P = 4.02 \times 10^{-4}$ V K^{-1}.

8.39 The electromotive force of the cell

$$Cd \mid CdCl_2 \cdot 2\tfrac{1}{2}H_2O,\ sat.\ solution \mid AgCl \mid Ag$$

at 25 °C is 0.675 33 V, and the temperature coefficient is -6.5×10^{-4} V K^{-1}. Calculate the values of $\Delta_r G$, $\Delta_r S$, and $\Delta_r H$ at 25 °C for the reaction

$$Cd(s) + 2AgCl(s) = 2Ag(s) + CdCl_2 \cdot 2\tfrac{1}{2}H_2O(s)$$

8.40 A thallium amalgam of 4.93% Tl in mercury and another amalgam of 10.02% Tl are placed in separate legs of a glass cell and covered with a solution of thallous sulfate to form a concentration cell. The voltage of the cell is 0.029 480 V at 20 °C and 0.029 071 V at 30 °C. (a) Which is the negative electrode? (b) What is the heat of dilution per mole of Tl when Hg is added at 30 °C to change the concentration from 10.02 to 4.93%? (*c*) What is the voltage of the cell at 40 °C?

8.41 What are the values of $\Delta_r G°$, $\Delta_r H°$, and $\Delta_r S°$, and K at 298.15 K for

$$H_2O(l) = H^+(ao) + OH^-(ao)$$

calculated from Appendix C.1? Compare $\Delta_r G°$ with the value obtained in problem 8.45.

8.42 We found in Section 8.6 that $E = 0.2224$ V for

$$Pt \mid H_2 \mid HCl \mid AgCl \mid Ag$$

at 25 °C. Using the value of $\Delta_f G°[Cl^-(ao)]$ obtained in Example 8.6, what is the value of $\Delta_f G°[AgCl(s)]$?

8.43 Calculate the thermodynamic properties of the following strong electrolytes from those of the constituent ions at 25 °C and check that the same values are tabulated under the following entries: HCl(ai), NaCl(ai), and NaOH(ai).

8.44 Calculate the standard electromotive force of the cell

$$Li \mid LiCl(ai) \mid Cl_2(g) \mid Pt$$

at 25 °C using (*a*) electrode potentials and (*b*) standard Gibbs energies of formation.

8.45 Devise a cell for which the cell reaction is

$$H_2O(l) = H^+(ao) + OH^-(ao)$$

Calculate $\Delta_r G°$ at 25 °C from electrode potentials. What is the value of the equilibrium constant at this temperature?

8.46 Devise an electromotive force cell for which the cell reaction is

$$AgBr(s) = Ag^+ + Br^-$$

Calculate the equilibrium constant (usually called the solubility product) for this reaction at 25 °C.

8.47 Using data in Table 8.2 and the Gibbs energies of formation of Ag(ao) + Cl$^-$(ao) calculate the solubility of AgCl(s) in water at 25 °C.

8.48 What are the differences between the standard electrode potentials for a standard-state pressure of 1 bar and 1 atm for the following electrodes at 25 °C:

	$E°$(1 atm)
Cl$^-$ $\mid$ AgCl(s) $\mid$ Ag	0.2224 V
Cl$^-$ $\mid$ Cl$_2$(g) $\mid$ Pt	1.3604 V

See problem 8.22.

8.49 The value of $E°$ for the electrode Pt $\mid$ O$_2$(g) $\mid$ OH$^-$ cannot be measured directly because the electrode is not reversible. Calculate $\Delta_r G°$ and $E°$ at 25 °C for the electrode reaction

$$\tfrac{1}{4}O_2(g) + \tfrac{1}{2}H_2O + e^- = OH^-(ao)$$

from $\Delta_f G°$ values from Appendix C.1.

8.50 Calculate the standard Gibbs energy of formation of NO$_3^-$(ao) from its $\Delta_f H°$ and $\bar{S}°$ at 25 °C using data from Appendix C.1.

8.51 Calculate the standard Gibbs energy of formation of SO$_4^{2-}$(ao) from $\Delta_f H°$ and $\bar{S}°$ values in Appendix C.1.

8.52 A hydrogen–oxygen fuel cell is operated at 25 °C and a total pressure of 5 bar. What is the elecromotive force assuming the gases are ideal?

8.53 A hydrogen electrode and a normal calomel electrode give an electromotive force of 0.435 when placed in a certain solution at 25 °C. (*a*) What is the pH of the solution? (*b*) What is the value of a_{H^+}?

8.54 A mole of H$_2$O(l) is electrolyzed at 298 K and 1 bar. (*a*) How much electrical energy is required if there are no losses due to electrical resistance and overvoltage? (*b*) The hydrogen and oxygen are then burned at constant pressure to produce one mole of H$_2$O(l) at 298 K and 1 bar. How much heat is produced? (*c*) Can this heat be used in a heat engine to produce the amount of electrical energy that was used to electrolyze the water initially? If so what condition has to be met?

8.55 Ammonia may be used as the anodic reactant in a fuel cell. The reactions occurring at the electrodes are

$$NH_3(g) + 3OH^-(ao) = \tfrac{1}{2}N_2(g) + 3H_2O(l) + 3e^-$$
$$O_2(g) + 2H_2O(l) + 4e^- = 4OH^-(ao)$$

What is the electromotive force of this fuel cell at 25 °C?

8.56 For a membrane potential of -70 mV for a resting nerve cell, what ratio of the concentrations of K$^+$ inside and outside would this correspond to if there were equilibrium at 25 °C?

9
Ionic Equilibria and Biochemical Reactions

This chapter provides a number of examples of the application of thermodynamics to solution equilibria. The first two sections are on acid dissociation in aqueous solutions, and the third section begins a transition to equilibria of biochemical interest. We will see that in general biochemical equilibria are handled in a different way from other solution equilibria. In biochemistry it is convenient to express equilibrium constants in terms of total concentrations of reactants and products, including various ionized and complexed species. In this case equilibrium constants and other thermodynamic quantities are functions of pH and the concentrations of metal ions at constant temperature and pressure.

Proteins have remarkable capacities to bind other substances in aqueous solutions. Their binding capacity may be markedly altered by changes in pH and by other substances present in the solution. In contrast with the binding properties of small molecules, the affinity of hemoglobin for oxygen is increased as more oxygen is bound. This is a consequence of changes in the internal structure of the hemoglobin molecule in the binding process.

9.1 Exact Treatment of the Dissociation of Weak Acids and Bases

Some acids, like hydrochloric acid, are believed to be completely dissociated in water, but others, like acetic acid are only partially dissociated and are referred to as **weak acids.** The dissociation of a monoprotic weak acid in water is represented by

$$HA(aq) = H^+(aq) + A^-(aq) \tag{9.1}$$

In water, all ions are hydrated to a greater or lesser extent, and this has a significant effect on their thermodynamic properties. In water the proton may form a hydronium ion, H_3O^+, or other complex species. A hydronium ion H_3O^+ may be hydrated by three water molecules so that $H_9O_4^+$ is formed. Mass spectroscopic studies show that this is a stable species in the gas phase. Since the state of the proton in aqueous solution is not exactly known, the symbol H^+ will be used to represent the hydrated hydrogen ion. Since water is involved in the dissociation reaction, we could also put H_2O on the left side of equation 9.1, but the activity of water does not appear in the equilibrium constant expression because it is the solvent and its activity is taken as unity on the mole fraction scale that is used to treat the solvent (see Section 8.9).

The **acid dissociation constant** K_a is defined by

$$K_a = \frac{a_{H^+} a_{A^-}}{a_{HA}} = \frac{m_{H^+} m_{A^-} \gamma_{\pm}^2}{m^\circ m_{HA} \gamma_{HA}} \tag{9.2}$$

where m is molal concentration (mol kg^{-1} of solvent) and m° is the standard molality (1 mol/kg of solvent). The mean ionic activity coefficient for the dissociated acid is represented by $\gamma_\pm$, and γ_{HA} is the activity coefficient for the undissociated acid. Since the undissociated acid is a nonelectrolyte, its activity coefficient is close to unity in dilute solutions and may be taken as unity as a close approximation. Since acid dissociation constants range over many powers of ten, they are often expressed as pK values, where p$K = -\log K$.

Table 9.1 gives the pK values and other thermodynamic quantities at 25°C for equilibrium constants expressed in terms of activities. Equilibrium constants for these acids at zero ionic strength have been determined by use of extrapolations similar to those described in Section 8.6. From the temperature dependence of K it is possible to calculate $\Delta_r H^\circ$, $\Delta_r S^\circ$, and $\Delta_r C_P^\circ$ for the dissociation.

The weakest acid listed in Table 9.1 is water itself. Its acid dissociation is represented by

$$H_2O(l) = H^+(aq) + OH^-(aq) \tag{9.3}$$

Thus, the acid dissociation constant of water, usually referred to as the **ion product,** is defined by

$$K_w = a_{H^+} a_{OH^-} = \frac{m_{H^+} m_{OH^-} \gamma_\pm^2}{(m^\circ)^2} \tag{9.4}$$

The activity of water, which can be written in the denominator of this expression, is taken as unity since it is the solvent. The value of K_w for water at 25°C is 1.006×10^{-14}. This is listed in Table 9.1 as a pK value. The enthalpy of dissociation of water is the negative of the enthalpy of neutralization of a strong

Table 9.1   Thermodynamic Quantities for Acid Dissociation at 25 °C[a]

	pK	$\Delta_r G°$ kJ mol^{-1}	$\Delta_r H°$ kJ mol^{-1}	$\Delta_r S°$ J K^{-1} mol^{-1}	$\Delta_r C_P°$ J K^{-1} mol^{-1}
Water (K_w)	13.997	79.868	56.563	−78.2	−197
Acetic acid	4.756	27.137	−0.385	−92.5	−155
Chloroacetic acid	2.861	16.322	−4.845	−71.1	−167
Butyric acid	4.82	27.506	−2.900	−102.1	0
Succinic acid, pK_1	4.207	24.016	3.188	−69.9	−134
Succinic acid, pK_2	5.636	31.188	−0.452	−109.2	−218
Carbonic acid, pK_1	6.352	36.259	9.372	−90.4	−377
Carbonic acid, pK_2	10.329	58.961	15.075	−147.3	−272
Phosphoric acid, pK_1	2.148	12.259	−7.648	−66.9	−155
Phosphoric acid, pK_2	7.198	41.099	4.130	−123.8	−226
Glycerol-2-phosphoric acid, pK_1	1.335	7.615	12.103	−66.1	−326
Glycerol-2-phosphoric acid, pK_2	6.650	37.945	−1.724	−133.1	−226
Ammonium ion	9.245	52.777	52.216	−1.7	0
Methylammonium ion	10.615	60.601	54.760	−19.7	33
Dimethylammonium ion	10.765	49.618	49.618	−39.7	96
Trimethylammonium ion	9.791	55.890	36.882	−63.6	184
Tris(hydroxymethyl)aminomethane	8.076	46.099	45.606	−1.3	0
Glycine, pK_1	2.350	13.410	4.837	−28.9	−134
Glycine, pK_2	9.780	55.815	44.141	−39.3	−50
Glycylglycine, pK_1	3.148	17.322	3.607	−54.0	−167
Glycyclglycine, pK_2	8.252	47.112	44.350	−8.4	−42

[a] These values apply at zero ionic strength and are obtained by extrapolation of experimental data at higher ionic strengths. From J. Edsall and J. Wyman, *Biophysical Chemistry*. New York: Academic, 1958.

acid with a strong base (Section 2.21). Weaker acids than H_2O are known. For example, the pK for methanol is 15.53, but special methods, not described here, are required to study such weak acids.

The fact that $\Delta_r S°$ for dissociation for most weak acids is negative is surprising at first. Gas dissociation reactions have positive entropy changes. However, the experimental results for acid dissociations in aqueous solution show that there is a decrease in entropy and, correspondingly, an increase in "order" in the dissociation. This results from a participation of water molecules in the reaction, which is not indicated by the balanced equation 9.1. Water molecules, being dipoles, tend to be oriented in the neighborhood of ions. Thus, the dissociation of an uncharged acid results in the orientation of a number of water molecules about the ions formed, and the consequent decrease in $S°$ overshadows the increase in $S°$ resulting from the formation of two particles from one. The effects that lead to a negative value of $\Delta_r S°$ tend to oppose dissociation. For a number of weak acids in Table 9.1 the value of $\Delta_r H°$ is very small, and so the standard entropy change largely determines the value of the pK according to $2.303RT \, pK = \Delta_r H° - T\Delta_r S°$.

There is almost no entropy change in the dissociation of ammonium ion

$$NH_4^+ = H^+ + NH_3 \qquad (9.5)$$

because there is no change in the number of ions in the reaction, so that the dissociation causes little change in water structure. Water molecules are not so organized around methylammonium ion as around NH_4^+ and, therefore, there is a decrease in entropy upon dissociation of the methylammonium ion. The entropy decrease upon dissociation is even greater for the dimethylammonium and trimethylammonium ions, suggesting decreasing water organization as methyl substitution increases.

The data in Table 9.1 can be used to calculate acid dissociation constants at other temperatures, but it is more important to take $\Delta_r C_P^\circ$ into account in aqueous solutions than in gas reactions. The equations we derived earlier (equations 2.91 and 5.45) can be written in the form

$$\Delta_r G_T^\circ = -RT \ln K = \Delta_r H_\theta^\circ + \frac{T(\Delta_r G_\theta^\circ - \Delta_r H_\theta^\circ)}{\theta} + \Delta_r C_P^\circ \left[T - \theta - T \ln \frac{T}{\theta} \right] \quad (9.6)$$

$$\Delta_r H_T^\circ = \Delta_r H_\theta^\circ + \Delta_r C_P^\circ (T - \theta) \quad (9.7)$$

where θ is the reference temperature, usually 298.15 K.

Example 9.1

What is the ion product of water at 0°C, assuming $\Delta_r C_P^\circ$ is independent of temperature? What is the pH of pure water at 0°C?

$$\Delta_r G^\circ(273.15 \text{ K}) = 55.836 + \frac{(273.15)(79.885 - 55.836)}{298.15}$$

$$+ (-0.2238)\left(273.15 - 298.15 - 273.15 \ln \frac{273.15}{298.15}\right)$$

$$= 78.110 \text{ kJ mol}^{-1}$$

$$K_w = \exp[(-78.110 \times 10^3)/8.3145 \times 273.15]$$

$$= 1.156 \times 10^{-15} = m_{H^+} m_{OH^-}/(m^\circ)^2 = m_{H^+}^2/(m^\circ)^2$$

$$\text{pH} = -\log(K_w^{1/2}) = 7.47$$

If K_a is known, the degree of dissociation of a weak acid like acetic acid at a particular concentration can be calculated using an iterative process. An approximate value of $\gamma_\pm$ is used to calculate m_{H^+} and m_{A^-} and then the ionic strength is calculated, $\gamma_\pm$ is recalculated, and the cycle is repeated. However, for a weaker weak acid, this calculation must take into account the ion product for water, since equation 9.4 must be obeyed and the equilibrium solution must be electrically neutral.

The dissociation of the weak base ammonium hydroxide can be formulated as

$$NH_4OH(aq) = NH_4^+(aq) + OH^-(aq) \quad (9.8)$$

with the base dissociation constant

$$K_b = \frac{m_{NH_4^+} m_{OH^-} \gamma_\pm^2}{m_{NH_4OH} m^\circ} \quad (9.9)$$

where the activity coefficient for NH_4OH has been taken to be unity for this neutral molecule in dilute aqueous solution. Since the degree of hydration of ammonia molecules in water is not known, we take the activities of NH_3 and NH_4OH (or any other hydrated NH_3) to be equal. There is another way of looking at the ionic equilibria of ammonia in aqueous solutions. The acid dissociation of ammonium ion in aqueous solution is represented by

$$NH_4^+(aq) = NH_3(aq) + H^+(aq) \tag{9.10}$$

for which the equilibrium expression is

$$K_a = \frac{m_{NH_3} m_{H^+}}{m_{NH_4^+} m^\circ} \tag{9.11}$$

since the activity coefficients of H^+ and NH_4^+ cancel in dilute solutions. If we take the product of K_a (equation 9.11) and K_b (equation 9.9) for the ammonia dissociation, we obtain (remembering that $a_{NH_3} = a_{NH_4OH}$)

$$K_a K_b = K_w \tag{9.12}$$

Thus, if K_b is obtained experimentally, K_a can be calculated by use of the ion product for water. Because of the availability of pH meters, it is convenient to focus on hydrogen ions and write dissociation reactions as acid dissociations. Thus, we will not use equations 9.8 and 9.9 in discussing ammonia in dilute aqueous solutions.

9.2 Practical Calculations with Weak Acids and Bases

In the preceding section we have discussed precise calculations of acid–base equilibria, but there is another way to formulate these equilibria that starts with the approximation that the pH measures the hydrogen ion activity in dilute aqueous solutions. The calibration of pH meters is accomplished with buffers so that the relation $pH = -\log a_{H^+}$ is accomplished as closely as possible (Section 8.11). The equilibrium expression for the dissociation of HA can be written

$$K = \frac{10^{-pH}[A^-]\gamma_-}{[HA]\gamma_{HA}} \tag{9.13}$$

where the brackets represent molar (mol/L) concentrations and the activity coefficients are on the molar scale. If the ionic strength is held constant it is convenient to use the apparent acid dissociation constant that is expressed in terms of pH and concentrations.

$$K_{app} = \frac{K\gamma_{HA}}{\gamma_-} = \frac{10^{-pH}[A^-]}{[HA]} \tag{9.14}$$

The **apparent acid dissociation constant** K_{app} for a given weak acid depends on the salt concentration. As the electrolyte and acid concentrations approach zero, K_{app} approaches K. In the remainder of this chapter we will use K without the subscript to represent both the limiting value of the dissociation constant and the value at a particular ion strength. In using the dissociation constants of weak acids it is often convenient to write equation 9.14 as

$$pH = pK + \log \frac{[A^-]}{[HA]} \tag{9.15}$$

where $pK = -\log K$. This useful equation is often referred to as the **Henderson–Hasselbalch equation.** It shows that the apparent pK of a weak acid in a particular electrolyte solution can be calculated from the pH of a solution con-

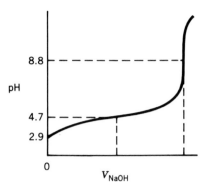

Figure 9.1   Titration of 0.1 molar acetic acid with a concentrated solution of sodium hydroxide.

taining known concentrations of weak acid [HA] and weak base [A$^-$]. For an aqueous solution containing NH$_4$Cl and NH$_3$, the ratio in equation 9.15 would be [NH$_3$]/[NH$_4$Cl]. The Henderson–Hasselbalch equation can be used to calculate the ratio of base and acid forms required to make a buffer of a particular pH. A buffer is most effective in a range of pH values between pK − 1 and pK + 1 because the [base]/[acid] ratio goes from 0.1 to 10 in this range.

We are now in a position to discuss the titration curve of a weak acid with a strong base. As shown in Fig. 9.1, the pH of 0.1 molar acetic acid is about 2.9. As sodium hydroxide is added the pH rises slowly through the buffering region around pH 4.7 (reaching pH 4.7 when one-half of the acid has been neutralized), and then rises most rapidly at the equivalence point at pH 8.8. When sodium hydroxide is added beyond the equivalence point, the pH corresponds with that of a mixture of sodium hydroxide and sodium acetate. The titration curve can be calculated most easily using the approximation that the electrolyte concentration is constant so that an apparent acid dissociation constant for that electrolyte concentration can be used.

The calculation of the titration curve for acetic acid involves the four unknowns [H$^+$], [A$^-$], [HA], and [OH$^-$]. To calculate these unknown concentrations we have two equilibrium expressions.

$$K = \frac{[H^+][A^-]}{[HA]} \qquad (9.16)$$

$$K_w = [H^+][OH^-] \qquad (9.17)$$

Having made the point that equilibrium constants are dimensionless, we will not write the c° in the denominator of equation 9.16 or the (c°)^2 in the denominator of equation 9.17; c° = 1 mol L^{-1}. We also have two conservation equations (conservation of mass and charge)

$$c_A = [HA] + [A^-] \qquad (9.18)$$

$$c_{Na} + [H^+] = [A^-] + [OH^-] \qquad (9.19)$$

where c_A is the total concentration of acetate and c_{Na} is the concentration of sodium ions, which is equal to the concentration of sodium hydroxide added up to that point in the titration. These simultaneous equations yield a cubic equation in [H$^+$], and so we will not go through this calculation in detail. However, it is relatively simple to calculate the initial pH, any pH in the buffering region, and the pH at the equivalence point without solving a cubic equation. At the equivalence point the pH is that of a solution of pure sodium acetate in water. The pH is not 7 because of the hydrolysis reaction

$$A^- + H_2O = HA + OH^- \qquad (9.20)$$

with equilibrium constant

$$K_h = \frac{[HA][OH^-]}{[A^-]} \qquad (9.21)$$

If the numerator and denominator are both multiplied by [H$^+$], we see that

$$K_h = \frac{K_w}{K_a} \qquad (9.22)$$

Thus, the pH of the endpoint of a titration of a weak acid by a strong base can be calculated by use of equations 9.21 and 9.22.

Example 9.2

What is the composition of a 0.1 molar solution of acetic acid in water at 25°C?
As a first approximation we assume that $\gamma_\pm = 1$ so that

$$K_a = 10^{-4.756} = \frac{[H^+][A^-]}{[HA]}$$

If the concentration of OH^- in this acidic solution can be neglected, the concentration of A^- is equal to the concentration of H^+ and

$$K_a = 1.754 \times 10^{-5} = \frac{[H^+]^2}{c_A - [H^+]}$$

where c_A is the amount of acetic acid added to make a liter of solution. The solution of this equation using the quadratic formula yields

$$[H^+] = 1.316 \times 10^{-3} \text{ mol L}^{-1}$$
$$[A^-] = 1.316 \times 10^{-3} \text{ mol L}^{-1}$$
$$[HA] = 0.09868 \text{ mol L}^{-1}$$

The ionic strength of the solution is 1.316×10^{-3}, and so the mean ionic activity coefficient is given by

$$\log \gamma_\pm = 0.509(1)(-1)(1.316 \times 10^{-3})^{1/2}$$
$$\gamma_\pm = 0.9958$$

As a second approximation we take the apparent acid dissociation constant to be $1.754 \times 10^{-5}/(0.9958)^2 = 1.769 \times 10^{-5}$. Use of the quadratic formula again yields

$$[H^+] = 1.321 \times 10^{-3} \text{ mol L}^{-1}$$
$$[A^-] = 1.321 \times 10^{-3} \text{ mol L}^{-1}$$
$$[HA] = 0.09868 \text{ mol L}^{-1}$$

It is not worth making an additional iteration in this case. The concentration of $[OH^-]$ $= K_w/[H^+]$ is 8×10^{-12} mol L^{-1}, which is small enough that we do not need to bring in the equation for electroneutrality (equation 9.19).

Example 9.3

Calculate the hydrogen ion concentration and pH of 0.1 mol L^{-1} sodium acetate at 25 °C. The value of the apparent dissociation constant of acetic acid at this ionic strength is 2.69×10^{-5}.
Since the hydrolysis reaction produces equal numbers of HA molecules and OH^- ions, we can write equation 9.21 as

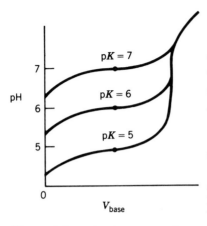

Figure 9.2 Titrations of weak acids with different pK values with a strong base.

$$K_h = \frac{10^{-14}}{2.69 \times 10^{-5}} = \frac{[OH^-]^2}{0.1}$$

Thus $[OH^-] = 6.10 \times 10^{-5}$ mol L^{-1}, $[H^+] = 1.64 \times 10^{-9}$ mol L^{-1}, and pH = 8.79. This is the pH at the equivalence point in a titration of 0.1 mol L^{-1} acetic acid with a concentrated solution of a strong base.

When weaker monoprotic acids are titrated, their buffering regions are found at higher pH values, as shown in Fig. 9.2.

Example 9.4
You want to prepare a liter of an acetate buffer of 0.1 ionic strength and pH 5.0 at 25 °C. How many moles of sodium acetate and acetic acid should you add if the apparent dissociation constant of acetic acid at this ionic strength is 2.69×10^{-5}?

$$pH = pK + \log \frac{[NaA]}{[HA]}$$

$$5.0 = -\log 2.69 \times 10^{-5} + \log \frac{0.1}{[HA]}$$

$$[HA] = 0.0372 \text{ mol } L^{-1}$$

Thus, you would add 0.1 mol of sodium acetate and 0.0372 mol of acetic acid to prepare a liter of buffer.

9.3 Titration of a Polyprotic Acid

In considering the titration of polyprotic acids, it is conventional to shift attention to the average number of protons bound by the base form and use apparent acid dissociation constants; that is, we will assume that the titration is carried out in a solution containing added salt so that the activity coefficients do not change during the titration. For H_3PO_4, the acid dissociation constants are defined by

$$HPO_4^{2-} = H^+ + PO_4^{3-} \qquad K_1 = \frac{[H^+][PO_4^{3-}]}{[HPO_4^{2-}]} \qquad (9.23)$$

$$H_2PO_4^- = H^+ + HPO_4^{2-} \qquad K_2 = \frac{[H^+][HPO_4^{2-}]}{[H_2PO_4^-]} \qquad (9.24)$$

$$H_3PO_3 = H^+ + H_2PO_4^- \qquad K_3 = \frac{[H^+][H_2PO_4^-]}{[H_3PO_4]} \qquad (9.25)$$

(Note that these dissociation constants are usually numbered in the opposite direction in the literature.) The average number of protons bound by phosphate in a particular solution is given by

$$n_H = \frac{[HPO_4^{2-}] + 2[H_2PO_4^-] + 3[H_3PO_4]}{[PO_4^{3-}] + [HPO_4^{2-}] + [H_2PO_4^-] + [H_3PO_4]}$$

$$= \frac{[H^+]/K_1 + 2[H^+]^2/K_1K_2 + 3[H^+]^3/K_1K_2K_3}{1 + [H^+]/K_1 + [H^+]^2/K_1K_2 + [H^+]^3/K_1K_2K_3} \quad (9.26)$$

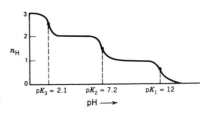

The number of protons bound by inorganic phosphate is shown as a function of pH in Fig. 9.3. This is really a titration curve, but we are focusing on what it tells us, rather than on the volume of base, or acid, added. At 25 °C the successive dissociation constants are $K_1 \cong 10^{-12}$, $K_2 = 6.34 \times 10^{-8}$, and $K_3 = 7.11 \times 10^{-3}$.

Figure 9.3 Proton binding curve for inorganic phosphate at 25 °C.

The way to look at this binding curve is to start at the far right. There phosphate is in the form of PO_4^{3-}, and the hydrogen ion concentration is, let us say, 10^{-14} mol/L. As we increase the concentration of hydrogen ion (decrease the pH), the PO_4^{3-} ion begins to bind protons, and the acid dissociation constant of the HPO_4^{2-} that is formed is about 10^{-12} ($pK_1 = 12$). As the concentration of hydrogen ions is further increased, nearly all of the PO_4^{3-} ions become protonated and we reach the first equivalence point at $n_H = 1$. But then near pH = 7, the HPO_4^{2-} ion begins to bind a second proton. As the concentration of hydrogen ions is further increased, we reach the second equivalence point at $n_H = 2$. In more strongly acidic solutions, $H_2PO_4^-$ begins to bind a third proton to produce H_3PO_4 that has an acid dissociation constant of 7.11×10^{-3} ($pK_3 = 2.1$). The inorganic phosphate is in the form of H_3PO_4 only in strongly acidic solutions. Thus, as PO_4^{3-} binds protons, its proton affinity decreases. For this example, the successive dissociation constants differ by large factors, and so this binding curve is very nearly the sum of three titration curves of monoprotic acids.

9.4 Dissociation Constants of Complex Ions

Because of their negative charges, PO_4^{3-} and HPO_4^{2-} also bind cations such as Mg^{2+}. For example, $MgHPO_4$ is formed, and its dissociation is represented by

$$MgHPO_4 = Mg^{2+} + HPO_4^{2-} \quad (9.27)$$

The **dissociation constant** for this complex ion is defined by

$$K_{MgP} = \frac{[Mg^{2+}][HPO_4^{2-}]}{[MgHPO_4]} \quad (9.28)$$

Protons and Mg^{2+} compete in their binding to HPO_4^{2-}. Since highly charged phosphate ions also tend to bind Na^+ and K^+, acid titrations and binding experiments are often carried out in the presence of cations such as $(CH_3CH_2)_4N^+$, where the bulky substituent groups prevent the close approach of the positive charge to a negatively charged ion. The study of the competition of metal ions and hydrogen ions is often used to determine dissociation constants for complex ions since we do not have metal ion electrodes that can be used to determine the concentration of unbound metal ions. In the neutral pH

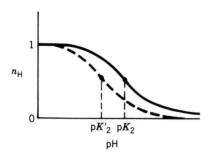

Figure 9.4 Proton binding curve for HPO_4^{2-} on the absence (—) and presence (---) of Mg^{2+}.

region we can neglect the H_3PO_4 concentration and consider a system consisting of Mg^{2+}, H^+, HPO_4^{2-}, $H_2PO_4^-$, and $MgHPO_4$ only. The number of additional protons bound by HPO_4^{2-} per mole of phosphate is given by

$$n_H = \frac{[H_2PO_4^-]}{[HPO_4^{2-}] + [H_2PO_4^-] + [MgHPO_4]} \tag{9.29}*$$

$$= \frac{[H^+]/K_2}{1 + [H^+]/K_2 + [Mg^{2+}]/K_{MgP}}$$

The number of n_{Mg} of magnesium ions bound by HPO_4^{2-} per mole of phosphate is given by

$$n_{Mg} = \frac{[MgHPO_4]}{[HPO_4^{2-}] + [H_2PO_4^-] + [MgHPO_4]} \tag{9.30}$$

$$= \frac{[Mg^{2+}]/K_{MgP}}{1 + [H^+]/K_2 + [Mg^{2+}]/K_{MgP}}$$

Figure 9.4 shows the binding curve for the second proton of phosphate in the presence and absence of Mg^{2+}. The shape of the curve is independent of the concentration of Mg^{2+}, but the curve is displaced to lower pH values as the concentration of Mg^{2+} is increased. This suggests that equation 9.29 might be written in the form

$$n_H = \frac{[H^+]/K_2'}{1 + [H^+]/K_2'} \tag{9.31}$$

where the apparent second acid dissociation constant for phosphate is given by

$$K_2' = K_2 \left(1 + \frac{[Mg^{2+}]}{K_{MgP}} \right) \tag{9.32}$$

This provides a means for obtaining the value of the dissociation constant K_{MgP} from acid titration curves.

Example 9.5
0.01 mol L^{-1} NaH_2PO_4 dissolved in 0.2 mol L^{-1} $(CH_3CH_2)_4NCl$ is titrated with $(CH_3CH_2)_4NOH$ at $25\,°C$. The midpoint of the titration curve is at pH 6.80. An identical solution is made 0.05 mol L^{-1} in $MgCl_2$ and the titration is repeated. This time the midpoint is pH 6.37. What is the dissociation constant of $MgHPO_4$? Using equation 9.32

$$10^{-6.37} = 10^{-6.80} \left(1 + \frac{0.05}{K_{MgP}} \right)$$

$$K_{MgP} = 2.96 \times 10^{-2}$$

---

* Note that this equation gives the number of protons bound by HPO_4^{2-}, rather than the number of protons bound by PO_4^{3-}, which is given by equation 9.26.

9.5 Statistical Effects in Polyprotic Acids

Thermodynamics ordinarily deals with what we call **macroscopic** equilibrium constants. In formulating such equilibrium constant expressions, the existence of isomeric forms that are rapidly interconverted is ignored and the sum of their concentrations is inserted in the equilibrium constant expression. However, in some situations it is important to distinguish between isomeric forms and to define **microscopic** equilibrium constants that involve particular forms. An example of this, the acid dissociation of glycine, is discussed as a special topic in the last section of this chapter. Other examples arise in the acid titration of proteins and of synthetic polymers with acidic groups. Here we will discuss the simplest possible example, the titration of a dicarboxylic acid, $HO_2C(CH_2)_nCO_2H$, **where n is so large that the carboxyl groups are independent of each other.** The macroscopic dissociation constants of this dibasic acid can be written as follows:

$$HA^- = H^+ + A^{2-} \qquad K_1 = \frac{[H^+][A^{2-}]}{[HA^-]} \tag{9.33}$$

$$H_2A = H^+ + HA^- \qquad K_2 = \frac{[H^+][HA^-]}{[H_2A]} \tag{9.34}$$

We can also look at this dissociation from a microscopic point of view and represent the intermediate form by HA if the "right" proton dissociates and by AH if the "left" proton dissociates. Since the acid groups are identical and independent of one another by assumption,

$$K = \frac{[H^+][HA^-]}{[HAH]} = \frac{[H^+][AH^-]}{[HAH]} = \frac{[H^+][A^{2-}]}{[HA^-]} = \frac{[H^+][A^{2-}]}{[AH^-]} \tag{9.35}$$

where K is the microscopic or intrinsic dissociation constant. Thus,

$$K_1 = \frac{[H^+][A^{2-}]}{[HA^-] + [AH^-]} = \frac{1}{1/K + 1/K} = \frac{K}{2} \tag{9.36}$$

$$K_2 = \frac{[H^+]([HA^-] + [AH^-])}{[H_2A]} = 2K \tag{9.37}$$

so that $K_2 = 4K_1$. The second acid dissociation constant is four times larger than the first because of this statistical effect.

One way to look at this is to say that when HAH dissociates, the proton can come off of either end, and so HAH has twice the tendency to dissociate as the corresponding monoprotic acid, $CH_3(CH_2)_nCO_2H$. Conversely, when the base form A^{2-} binds a proton, there are two sites where the proton can be bound.

In contrast with a dibasic acid with different microscopic equilibria, the titration curve for a diprotic acid of the type we are discussing looks exactly like that of a monoprotic acid, except that two moles of base are required to neutralize one mole of acid. The midpoint of this titration curve yields the microscopic dissociation constant K, but we would use $K_1 = K/2$ and $K_2 = 2K$ for the macroscopic constants. The number of protons bound by this diprotic acid is

$$n_H = \frac{[H^+]/K_1 + 2[H^+]^2/K_1K_2}{1 + [H^+]/K_1 + [H^+]^2/K_1K_2} = \frac{2[H^+]/K + 2[H^+]^2/K^2}{1 + 2[H^+]/K + [H^+]^2/K^2} \quad (9.38)$$

$$= \frac{2[H^+]/K(1 + [H^+]/K)}{(1 + [H^+]/K)^2} = \frac{2[H^+]/K}{1 + [H^+]/K}$$

This result is expected because the acidic groups are identical.

It can be shown* that for a polyprotic acid with n independent and equivalent groups

$$K_i = \frac{iK}{n - i + 1} \quad (9.39)$$

These same considerations apply to the binding of metal ions or to the binding of oxygen to a type of hemoglobin with four equivalent binding sites.

Example 9.6
Assuming that the four oxygen binding sites of hemoglobin are identical, express the four successive dissociation constants K_1, K_2, K_3, and K_4 in terms of the microscopic constant for any one of the four binding sites.

$$K_i = \frac{iK}{n - i + 1}$$

$$K_1 = K/4, \quad K_2 = 2K/3, \quad K_3 = 3K/2, \quad \text{and} \quad K_4 = 4K.$$

9.6   Thermodynamics of Biochemical Reactions

Since phosphate and related substances encountered in biology are weak acids and bind metal ions, equilibrium constant expressions in biochemistry are written in terms of sums of species, rather than particular ion species. For example, the hydrolysis of adenosine triphosphate (Fig. 9.5) to adenosine diphosphate and inorganic phosphate is usually represented by

$$ATP + H_2O = ADP + P_i \quad (9.40)$$

where ATP, ADP, and P_i represent the sums of all the molecular species of adenosine triphosphate, adenosine diphosphate, and inorganic phosphate, respectively.

In the next section we will see that such reactions produce or consume acid; when reactions of this type are carried out in a buffer in a calorimeter, the acid produced reacts with the base form of the buffer and causes a heat effect that has to be subtracted from the observed heat in order to calculate $\Delta H°$ for reaction 9.40. However, at a specified pH value the equilibrium expression can be written

$$K' = \frac{[ADP][P_i]}{[ATP]} \quad (9.41)$$

---

* C. Tanford, *Physical Chemistry of Macromolecules*. New York: Wiley, 1961, p. 532.

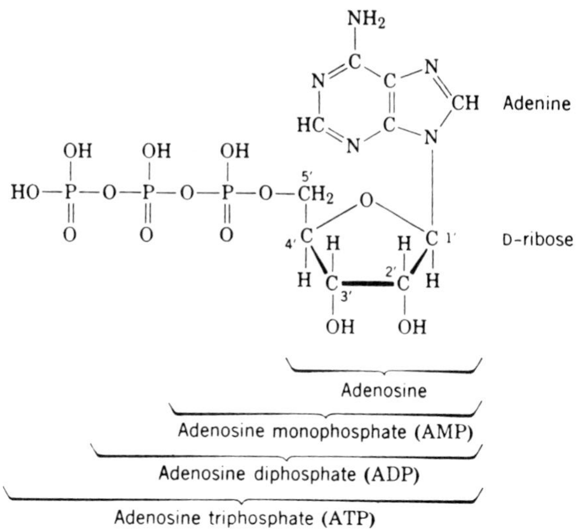

Figure 9.5   Structure of adenosine triphosphate, ATP.

The concentration of water is omitted because it is constant in dilute solutions. This equilibrium constant, which is referred to as the **apparent equilibrium constant**, is constant at a given temperature and pH. Since biological reactions are normally studied in the neighborhood of pH 7, it is customary* to tabulate the value of K' at pH 7. The corresponding **apparent standard reaction Gibbs energy** $\Delta_r G^{\circ\prime}$ is calculated using

$$\Delta_r G^{\circ\prime} = -RT \ln K' \qquad (9.42)$$

The apparent equilibrium constant K' of a reaction involving multiply charged anions also depends on the concentration of magnesium ions and other multiply charged cations. The concentration of magnesium ion is often expressed in terms of pMg, which is defined as $-\log[Mg^{2+}]$.

The $\Delta_r G^{\circ\prime}$ values for a number of hydrolysis reactions at pH 7, pMg 4, and 25 °C are given in Table 9.2. These values can be added and subtracted to obtain apparent equilibrium constants for many important biochemical reactions. Since the equilibrium constant may be calculated for any pair of reactions, such a table summarizes equilibrium data for a very large number of phosphate transfer reactions. Similar tables may be constructed for the transfer of pyrophosphate (pyrophosphoric acid is $H_4P_2O_7$) or any other group. There is an analogy between the transfer of phosphate in these reactions and the transfer of electrons in reactions studied in electrochemistry or protons in acid–base equilibria. These $\Delta_r G^{\circ\prime}$ values are the standard reaction Gibbs energies when the reactant at 1 molar concentration in an aqueous buffer at pH 7 and pMg 4, is converted to the unmixed products, each in 1 molar solution at pH 7 and pMg 4. The ionic strength is held constant so that the unknown activity

---

* Recommendations for measurement and presentation of biochemical equilibrium data, prepared by the Interunion Commission in Biothermodynamics, in *J. Biol. Chem.*, **251**, 6879 (1976).

Table 9.2   Standard Gibbs Energies of Hydrolysis at 25 °C pH 7, pMg 4 at 0.2 mol L^{-1} Ionic Strength (P$_i$ represents inorganic phosphate)

	$\dfrac{\Delta_r G^{\circ\prime}}{\text{kJ mol}^{-1}}$
Phosphoenolpyruvate + H$_2$O = enol pyruvate + P$_i$	−61.9
Creatine phosphate + H$_2$O = creatine + P$_i$	−43.5
Acetyl phosphate + H$_2$O = acetate + P$_i$	−43.1
CoA-S-phosphate + H$_2$O = CoA-SH + P$_i$	−37.7
ATP + H$_2$O = ADP + P$_i$	−39.7
ADP + H$_2$O = AMP + P$_i$	−36.8
Pyrophosphate (PP) + H$_2$O = 2P$_i$	−34.3
Arginine phosphate + H$_2$O = arginine + P$_i$	−29.3
Glucose-6-phosphate + H$_2$O = glucose + P$_i$	−12.6
Fructose-1-phosphate + H$_2$O = fructose + P$_i$	−12.6
AMP + H$_2$O = adenosine + P$_i$	−12.6
Glycerol-1-phosphate + H$_2$O = glycerol + P$_i$	−9.2

coefficients are constant. Apparent equilibrium constants and standard reaction Gibbs energies of the type defined in equation 9.41 and 9.42 are very useful.

Example 9.7

What is the equilibrium constant K' at pH 7 and pMg 4 and 25 °C for

$$\text{creatine phosphate} + \text{ADP} = \text{creatine} + \text{ATP}$$

Since

$$\text{creatine phosphate} + \text{H}_2\text{O} = \text{creatine} + \text{P}_i \qquad \Delta_r G^{\circ\prime} = -43.5 \text{ kJ mol}^{-1}$$

$$\text{ADP} + \text{P}_i = \text{ATP} + \text{H}_2\text{O} \qquad \Delta_r G^{\circ\prime} = 39.8 \text{ kJ mol}^{-1}$$

For the given reaction, $\Delta_r G^{\circ\prime} = -43.5 + 39.8 = -3.7$ kJ mol^{-1}. Using $\Delta_r G^{\circ\prime} = -RT \ln K'$,

$$K' = \frac{[\text{Cr}][\text{ATP}]}{[\text{CrP}][\text{ADP}]} = 4.6$$

As shown in this example, a hydrolysis reaction with a negative Gibbs energy change may be used to drive a phosphorylation reaction with a (smaller) positive Gibbs energy change. Such reactions are said to be **coupled**, and the coupling is provided by the enzyme (Section 21.11) that catalyzes the phosphate transfer reaction. Although we can calculate $\Delta_r G^{\circ\prime}$ by adding two reactions in Example 9.5, this does not mean that the coupled reaction occurs through these steps. For there to be coupling, the mechanism is generally of the following type:

$$\text{creatine phosphate} + \text{E} = \text{EP} + \text{creatine}$$
$$\text{EP} + \text{ADP} = \text{ATP} + \text{E}$$

$$\text{creatine phosphate} + \text{ADP} = \text{creatine} + \text{ATP} \qquad \Delta_r G^{\circ\prime} = -3.7 \text{ kJ mol}^{-1}$$

Table 9.3 Thermodynamic Quantities for Acid Dissociation at 25 °C and 0.2 mol L^{-1} Ionic Strength

Dissociation Reaction	Dissociation Constant	pK	$\Delta_r G°$ kJ mol^{-1}	$\Delta_r H°$ kJ mol^{-1}	$\Delta_r S°$ J K^{-1} mol^{-1}
$HATP^{3-} = H^+ + ATP^{4-}$	K_{1ATP}	6.95	39.66	-7.03	-156.5
$H_2ATP^{2-} = H^+ + HATP^{3-}$	K_{2ATP}	4.06	23.22	0.00	-77.8
$HADP^{2-} = H^+ + ADP^{3-}$	K_{1ADP}	6.88	39.29	-5.73	-151.0
$H_2ADP^{1-} = H^+ + HADP^{2-}$	K_{2ADP}	3.93	22.43	4.18	-61.1
$HAMP^{1-} = H^+ + AMP^{2-}$	K_{1AMP}	6.45	36.82	-3.56	-135.6
$H_2AMP = H^+ + HAMP^-$	K_{2AMP}	3.75	21.34	4.18	-57.7
$H_2PO_4^{1-} = H^+ + HPO_4^{2-}$	K_{2P}	6.78	38.70	3.35	-118.8
$HP_2O_7^{3-} = H^+ + P_2O_7^{4-}$	K_{1PP}	8.95	51.09	1.67	-165.7
$H_2P_2O_7^{2-} = H^+ + HP_2O_7^{3-}$	K_{2PP}	6.12	34.94	0.46	-115.5

where E is enzyme and EP is phosphorylated enzyme. The large change in Gibbs energy for the hydrolysis of creatine phosphate is not lost, but is conserved in the enzyme phosphate compound and then in ATP. This is an absolutely essential process for living things because this is the way a thermodynamically spontaneous reaction can drive a nonspontaneous reaction that synthesizes a needed compound.

9.7   Interpretation of the Apparent Gibbs Energy of Hydrolysis of Adenosine Triphosphate*

If the hydrolysis reaction described by equation 9.40 occurs in an aqueous buffer in the absence of magnesium ion, the effect of pH on the equilibrium in the neutral range can be calculated if the acid dissociation constants of ATP, ADP, and P in the neutral pH range are known.

At pH values well above the neutral range, ATP exists in aqueous solutions as a -4 ion. As the pH is reduced to 7, ATP picks up a proton and becomes a -3 ion. As the pH is reduced to pH 4, ATP picks up another proton and become a -2 ion. Since ATP^{4-} is so highly charged, it has a strong tendency to bind cations from the medium. Significant concentrations of complexes with Na^+ and K^+ are formed even at 0.1 mol L^{-1} concentrations of the cations. Therefore, the acid dissociation constants of ATP are most conveniently determined in media having bulky cations such as $(n\text{-propyl})_4N^+$ that are bound much less strongly. The pK values for ATP, ADP, adenosine-5-monophosphate AMP, and inorganic phosphate determined in 0.2 mol L^{-1} ionic strength (n-propyl)$_4$NCl at 25 °C are summarized in Table 9.3.

The values of $\Delta_r H°$ in Table 9.3 were obtained by determining pK values at two or more temperatures. Since the enthalpy changes are small in all cases, it is the entropy changes that determine the differences in strength of these acids. The magnitude of the standard entropy change correlates with the charge and charge distribution of the base formed. The larger the charge of the base

---

* R. A. Alberty, *J. Biol. Chem.* **244**:3290 (1969).

formed, the more negative $\Delta S°$ is as a result of the greater hydration of the more highly charged ions.

In the neutral pH range and in the absence of cations that are bound, the only ions that have to be considered in the hydrolysis of ATP are ATP^{4-}, $HATP^{3-}$, ADP^{3-}, $HADP^{2-}$, HPO_4^{2-}, and $H_2PO_4^{1-}$. The expression for the apparent equilibrium constant in equation 9.41 can be written

$$K' = \frac{([ADP^{3-}] + [HADP^{2-}])([HPO_4^{2-}] + [H_2PO_4^{1-}])}{([ATP^{4-}] + [HATP^{3-}])} \quad (9.43)$$

Introducing the acid dissociation constants defined in Table 9.3 yields

$$K' = \frac{[ADP^{3-}][HPO_4^{2-}]}{[ATP^{4-}]} \frac{(1 + [H^+]/K_{1ATP})(1 + [H^+]/K_{2P})}{(1 + [H^+]/K_{1ATP})} \quad (9.44)$$

The first factor is closely related to the equilibrium constant K for the hydrolysis reaction written in terms of ionic species:

$$ATP^{4-} + H_2O = ADP^{3-} + HPO_4^{2-} + H^+ \quad (9.45)$$

$$K = \frac{[ADP^{3-}][HPO_4^{2-}][H^+]}{[ATP^{4-}]} \quad (9.46)$$

Reaction 9.45 written in terms of specific species is referred to as a reference reaction, and equation 9.46 gives the usual equilibrium constant expression, assuming ideal solutions. For reaction 9.45, the thermodynamic quantities are $\Delta_r G° = 1.17$ kJ mol^{-1}, $\Delta_r H° = -19.66$ kJ mol^{-1}, and $\Delta_r S° = -69.87$ J K^{-1} mol^{-1}. Substituting equation 9.46 into equation 9.44 yields

$$K' = \frac{K(1 + [H^+]/K_{1ADP})(1 + [H^+]/K_{2P})}{(1 + [H^+]/K_{1ATP})[H^+]} \quad (9.47)$$

Since K is independent of pH, this equation gives us the dependence of K' on pH.

At high pH values, the ATP hydrolysis reaction is represented by equation 9.46, and so we can see that a mole of H^+ is produced for each mole of ATP hydrolyzed. Magnesium and calcium ions are present in many biological systems, and, since they are bound by HPO_4^{2-}, ATP^{4-}, $HATP^{3-}$, and other organic phosphates, these binding equilibria have to be taken into account in more accurate calculations. In addition, enthalpies of reaction and entropies of reaction for biochemical reactions also depend on pH and pMg.

Example 9.8
Given that the reaction Gibbs energy for reaction 9.45 at 25 °C is $\Delta_r G° = 1.17$ kJ mol^{-1}, calculate the apparent equilibrium constant K' at pH 7 and pH 8, using the acid dissociation constants in Table 9.3.

$$K = \exp(-1170/8.314 \times 298) = 0.62$$

At pH 7, equation 9.47 yields

$$K' = \frac{0.62(1 + 10^{-0.12})(1 + 10^{-0.22})}{10^{-7}(1 + 10^{-0.05})}$$

$$= 9.2 \times 10^6$$

At pH 8, equation 9.47 yields

$$K' = \frac{0.62(1 + 10^{-1.12})(1 + 10^{-1.22})}{10^{-8}(1 + 10^{-1.05})}$$

$$= 6.5 \times 10^7$$

Thus, the hydrolysis of ATP can provide a greater driving force for the synthesis of another phosphate ester at pH 8 than at pH 7.

The biochemical equilibria we have been discussing involve small molecules, but many biochemical equilibria involve macromolecules such as proteins and nucleic acids. As an example we will now consider the binding of small molecules by a protein.

9.8 Binding of Oxygen by Myoglobin

Myoglobin is an oxygen storage protein that is found in muscle tissue in many species. Its molar mass is 16 000 g mol^{-1}, and each molecule contains one heme group, one atom of iron, and binds one molecule of oxygen when it is saturated.

The shape of the binding curve for myoblogin is exactly what is expected for the simple reaction

$$MbO_2 = Mb + O_2 \tag{9.48}$$

where Mb represents a molecule of myoglobin. The dissociation constant is defined by

$$K = \frac{[Mb]P_{O_2}}{[MbO_2]} \tag{9.49}$$

where P_{O_2} is the partial pressure of oxygen in the gas phase. Conservation of myoglobin requires that

$$[Mb]_0 = [Mb] + [MbO_2] \tag{9.50}$$

where $[Mb]_0$ is the total molar concentration of myoglobin. Combining equation 9.49 and 9.50 yields

$$Y = \frac{[MbO_2]}{[Mb]_0} = \frac{P_{O_2}/K}{1 + P_{O_2}/K} \tag{9.51}$$

where Y is the fractional saturation. The plot of Y versus P_{O_2} is given in Fig. 9.6.

9.9 Binding of Oxygen by Hemoglobin

Hemoglobin is the oxygen transport protein in many species. It has a molar mass of 64 000 g mol^{-1}, and each molecule contains four heme groups, four atoms of iron, and binds four molecules of oxygen when it is saturated. Hemoglobin may be reversibly dissociated into four molecules of molar mass

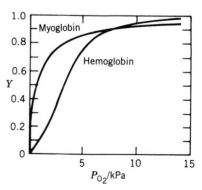

Figure 9.6 Fractional saturation Y of myoglobin and hemoglobin with oxygen at pH 7.4 and 38 °C.

16 000 g mol^{-1}, each of which contains one heme and one atom of iron. These smaller molecules are of two types, represented by α and β, and hemoglobin has the composition $\alpha_2\beta_2$. Many enzymes have a similar subunit structure, and they and hemoglobin have remarkable binding properties, which are a consequence of this subunit structure.

The oxygen binding properties of myoglobin and hemoglobin are distinctively different, as shown by Fig. 9.6. Hemoglobin's S-shaped (sigmoid) binding curve is a great advantage for its physiological function, because the amount of oxygen bound changes rapidly between the partial pressure of oxygen in the lungs (about 13.3 kPa) and in the tissues (about 2.0 kPa).

The equation for the number n of oxygen molecules bound by hemoglobin can be expressed in terms of the partial pressure of oxygen and the successive dissociation constants in the same way that we expressed the number of protons bound by inorganic phosphate in equation 9.26. To compare the binding of hemoglobin with myoglobin it is convenient to use the fractional saturation Y, which in this case is $n/4$ since four oxygen molecules may be bound:

$$Y = \frac{P_{O_2}/K_1 + 2P_{O_2}^2/K_1K_2 + 3P_{O_2}^3/K_1K_2K_3 + 4P_{O_2}^4/K_1K_2K_3K_4}{4[1 + P_{O_2}/K_1 + P_{O_2}^2/K_1K_2 + P_{O_2}^3/K_1K_2K_3 + P_{O_2}^4/K_1K_2K_3K_4]} \quad (9.52)$$

Equation 9.52 can represent the sigmoid binding curve for hemoglobin but, remarkably, only if the successive dissociation constants *decrease* instead of increase. This is contrary to the usual observation that the dissociation constants for successive ligand molecules increase. You might ask, If some of the binding sites have higher affinities for oxygen, shouldn't they be filled first? The answer is that the binding sites with higher affinity do not have this high affinity until some oxygen has been bound. This is called cooperativity.

When binding at the middle of the titration curve increases more rapidly with ligand concentration than can be accounted for by equation 9.51, it is said to be **cooperative**. Cooperativity arises when the binding of the first ligands causes changes in the structure of the protein that increase the affinities of the remaining sites. The origin of this effect in hemoglobin was quite mysterious until the structures of oxygenated and deoxygenated hemoglobin were determined by X-ray diffraction. Myoglobin was the first protein for which the detailed molecular structure was obtained by X-ray diffraction (Kendrew 1959). When the structure of hemoglobin was obtained, it was found that each of its four subunits has a three-dimensional configuration much like that of myoglobin. When oxygen is bound, groups near the heme shift slightly, and these structural changes affect the configurations of the four subunits so that the binding properties of the other heme groups are enhanced.* The cooperativity of the oxygen binding by hemoglobin is treated quantitatively in terms of the Monod–Wymann–Changeux (MWC) concerted mechanism.† According to this mechanism, a small fraction of hemoglobin exists in the quaternary oxy structure that binds oxygen more strongly. When the first oxygen molecule is bound, it is bound preferentially to this structure. As more oxygen molecules are bound, the oxy structure is sufficiently stabilized to be the major structure, and so subsequent binding is strong.

---

* See R. E. Dickerson and I. Geis, *Hemoglobin*. Menlo Park, CA: Benjamin/Cummings, 1983.

† J. Monod, J. Wyman, and J. P. Changeux, *J. Mol. Biol.* **12**:88 (1965); J. Wyman and S. J. Gill, *Binding and Linkage*. Mill Valley, CA: University Science Books, 1990.

In addition to the positive cooperativity of hemoglobin, there is also negative cooperativity. For example, some enzymes have identical subunits, but the binding of the first substrate molecule causes molecular changes that reduce the affinities of the enzymatic sites in neighboring subunits.

Hemoglobin has another remarkable property that makes it even more effective in the oxygen transport system. When hemoglobin is oxygenated at pH 7.4, it produces 0.6 mol of H^+ for each mole of oxygen molecule bound. A corollary of this so-called Bohr effect is that the affinity of hemoglobin for oxygen depends on pH. In the neighborhood of pH 7.4, the partial pressure of oxygen required to half saturate hemoglobin decreases as the pH is increased. The explanation of the Bohr effect is that the binding of oxygen affects the acid dissociation constants of certain acid groups in hemoglobin. This effect is of considerable physiological importance because in the lungs H^+, liberated by hemoglobin upon oxygenation, reacts with HCO_3^- to make H_2CO_3. The carbonic acid then dehydrates and CO_2 diffuses into the air space of the lungs. If hemoglobin did not dissociate H^+, the blood would become alkaline in the lungs as CO_2 was exhaled. In the capillaries this process is reversed; hemoglobin absorbs H^+ as it loses oxygen, and this converts H_2CO_3, produced metabolically, to HCO_3^-.

Example 9.9

Suppose that a particular hemoprotein has two oxygen binding sites that are independent. Express the fractional saturation Y in terms of (a) the macroscopic dissociation constants K_1 and K_2 and (b) the microscopic dissociation constants k_1 and k_2. (c) How are K_1 and K_2 related to k_1 and k_2?

(a) $$Y = \frac{P_{O_2}/K_1 + 2P_{O_2}^2/K_1K_2}{2[1 + P_{O_2}/K_1 + P_{O_2}^2/K_1K_2]}$$

(b) $$Y = \frac{1}{2}\left[\frac{P_{O_2}/k_1}{1 + P_{O_2}/k_1} + \frac{P_{O_2}/k_2}{1 + P_{O_2}/k_2}\right]$$

(c) $$K_1 = \frac{1}{1/k_1 + 1/k_2}$$

$$K_2 = k_1 + k_2$$

9.10 Special Topic: Relation Between Microscopic Dissociation Constants and Macroscopic Dissociation Constants

The acid dissociations of amino acids offer interesting examples of the relations between microscopic dissociation constants and macroscopic dissociation constants. Glycine in strongly acidic solution ($^+H_3NCH_2CO_2H$) may be considered to be a dibasic acid, and two acid dissociation constants K_1 and K_2 (defined below) may be obtained by titration with a base. However, if we want to understand the titration in molecular terms we have to consider the following steps.

$$
\begin{array}{ccccc}
& & {}^+H_3NCH_2CO_2^- & & \\
K_{11} & \nearrow^{-H^+} & & \searrow^{-H^+} & K_{22} \\
{}^+H_3NCH_2CO_2H & _{+H^+}^{} & \Updownarrow K_z & _{+H^+}^{} & H_2NCH_2CO_2^- \\
K_{12} & \searrow_{+H^+} & & \nearrow_{+H^+} & K_{21} \\
& & H_2NCH_2CO_2H & &
\end{array}
\qquad (9.53)
$$

where the microscopic constants are K_{11}, K_{12}, K_{21}, K_{22}, and K_z. These constants cannot ordinarily be determined directly. The macroscopic constants K_1 and K_2 can be determined by titration.

When glycine is titrated, we do not have information on the relative amounts of the dipolar in, ${}^+H_3NCH_2CO_2^-$, and the uncharged molecule, $H_2NCH_2CO_2H$, and so the first acid dissociation constant K_1, a macroscopic dissociation constant, is given by

$$
\begin{aligned}
K_1 &= \frac{[H^+]([{}^+H_3NCH_2CO_2^-] + [H_2NCH_2CO_2H])}{[{}^+H_3NCH_2CO_2H]} \\
&= K_{11} + K_{12} = 10^{-2.35}
\end{aligned}
\qquad (9.54)
$$

where K_{11} and K_{12} are the microscopic dissociation constants for the two possible dissociations of a first proton. The second acid-dissociation constant is given by

$$
\begin{aligned}
K_2 &= \frac{[H^+][H_2NCH_2CO_2^-]}{([{}^+H_3NCH_2CO_2^-] + [H_2NCH_2CO_2H])} \\
&= \frac{1}{1/K_{22} + 1/K_{21}} = 10^{-9.78}
\end{aligned}
$$

$$\qquad (9.55)$$

where K_{21} and K_{22} are the microscopic dissociation constants for the two possible dissociations of a second proton. This same formulation would apply to any dibasic acid. In addition, the principle of detailed balance (Section 19.7) yields a further relationship between the values of the four microscopic constants:

$$
K_{11}K_{22} = K_{12}K_{21}
\qquad (9.56)
$$

Since K_1 and K_2 are known (values are given for 25 °C in Table 9.1), we need a value of only one of the four microscopic constants to calculate values of the other three. Assuming that K_{12} has the same value as the dissociation constant of the methyl ester of glycine,

$$
{}^+H_3NCH_2CO_2CH_3 = H^+ + H_2NCH_2CO_2CH_3
\qquad (9.57)
$$

we can use equation 9.54 to obtain the value of K_{11}. Since the equilibrium constant for reaction 9.57 is $10^{-7.70}$, we obtain $K_{11} = 10^{-2.35}$ from equation 9.54, and

$$
\frac{K_{21}}{K_{22}} = \frac{10^{-2.35}}{10^{-7.70}}
\qquad (9.58)
$$

from equation 9.56. Substituting this ratio in equation 9.55 yields $K_{21} = 10^{-4.43}$, so that $K_{22} = 10^{-9.78}$.

Thus, the ratio K_z of dipolar ion to neutral molecule is

$$K_z = \frac{K_{11}}{K_{12}} = \frac{10^{-2.35}}{10^{-7.70}} = 10^{5.35} \tag{9.59}$$

The dissociation of the most acidic form of glycine accordingly goes almost exclusively by the top path; this is not true for all amino acids.

An amino acid is said to be **isoelectric** at the pH at which there are equal concentrations of the positively and negatively charged forms. Setting

$$[^+H_3NCH_2CO_2H] = [H_2NCH_2CO_2^-]$$

and substituting from equation 9.54 and 9.55 yields

$$[H^+]^2_{\text{isoelectric}} = K_1 K_2$$

which may be written

$$pH_{\text{isoelectric}} = \tfrac{1}{2}(pK_1 + pK_2) \tag{9.60}$$

For glycine the isoelectric point is

$$pH_{\text{isoelectric}} = \tfrac{1}{2}(2.35 + 9.78) = 6.06$$

Equation 9.56 may be written in the form

$$pK_{21} - pK_{11} = pK_{22} - pK_{12} \qquad 4.43 - 2.35 = 9.78 - 7.70 \tag{9.61}$$

This equation says that the strengthening of the carboxyl dissociation by the positive charge on the amino group is equal to the weakening of the dissociation of the NH_3^+ group by the negative charge on the carboxyl group.

References

R. G. Bates, *Determination of pH*. New York: Wiley, 1973.

R. P. Bell, *Acids and Bases*. London: Methuen, 1969.

C. R. Cantor and P. R. Schimmel, *Biophysical Chemistry, Part III, The Behavior of Biological Macromolecules*. San Francisco: Freeman, 1980.

R. E. Dickerson and I. Geis, *Hemoglobin*. Menlo Park, CA: Benjamin/Cummings, 1983.

J. T. Edsall and H. Gutfreund, *Biothermodynamics*. New York: Wiley, 1983.

J. Edsall and J. Wyman, *Biophysical Chemistry*. New York: Academic, 1958.

H. S. Harned and B. B. Owen, *The Physical Chemistry of Electrolytic Solutions*. New York: Reinhold, 1958.

I. Klotz, *Energy Changes in Biochemical Reactions*. New York: Academic, 1967.

A. L. Lehninger, *Bioenergetics*. New York: Benjamin, 1965.

D. Poland, *Cooperative Equilibria in Physical Biochemistry*. Oxford UK: Clarendon, 1978.

H. Rossotti, *The Study of Ionic Equilibrium*. London: Longman, 1978.

Problems

9.1 Show that the slope of the titration curve of a monobasic acid is given by

$$\frac{d\alpha}{dpH} = \frac{2.303K[H^+]}{(K + [H^+])^2}$$

where α is the degree of neutralization.

9.2 Using the Debye–Hückel theory, estimate the apparent pK for acetic acid at 0.1 mol L^{-1} ionic strength. At 25 °C the thermodynamic pK value is 4.756. It is assumed that the activity coefficient for undissociated acetic acid is unity at this value of the ionic strength.

9.3 According to Appendix C.1, what are the values of $\Delta_r G°$, $\Delta_r H°$, and $\Delta_r S°$ at 298 K for

$$H_2O(l) = H^+(ao) + OH^-(ao)$$

Show that the same value of $\Delta_r S°$ is obtained from $\Delta_r G°$ and $\Delta_r H°$ as by using $\Delta_r S° = \sum \nu_i S_i°$. Calculate K_w at 298 K.

9.4 For the acid dissociation of acetic acid, $\Delta_r H°$ is approximately zero at room temperature in H_2O. For the acidic form of aniline, which is approximately as strong an acid as acetic acid, $\Delta_r H°$ is approximately 21 kJ mol^{-1}. Calculate $\Delta_r S°$ for each of the following reactions:

$$CH_3CO_2H = H^+ + CH_3CO_2^- \qquad pK = 4.75$$
$$C_6H_5NH_3^+ = H^+ + C_6H_5NH_2 \qquad pK = 4.63$$

How do you interpret these entropy changes? What compensates for the increase in entropy expected from the increase in number of molecules in the balanced chemical reaction?

9.5 Estimate pK_3 and pK_2 for H_3PO_4 at 25 °C and 0.1 mol L^{-1} ionic strength. The values at zero ionic strength are $pK_3 = 2.148$ and $pK_2 = 7.198$.

9.6 In a strong acid solution, the amino acid histidine binds three protons. The acid dissociation constants numbered from the weakest acid dissociation are 6.92×10^{-10}, 1.00×10^{-6}, and 1.51×10^{-2} at 25°C. Calculate the concentrations of the four forms of histidine (His^-, $HisH$, $HisH^+$, and $HisH_3^{2+}$) in a 0.1 M solution of histidine at pH 7, assuming that these constants apply at the ionic strength of the solution.

9.7 The pK for the dissociation of $CaATP^{2-}$ at 25 °C in 0.2 mol L^{-1} (n-propyl)$_4$NCl is 3.60. The pK for $HATP^{3-} = H^+ + ATP^{4-}$ is 6.95. Calculate the apparent pK of this ATP ionization when ATP is titrated in a solution containing 0.1 mol L^{-1} $CaCl_2$. Assume that the Ca^{2+} concentration is much larger than total ATP concentrations.

9.8 Will 0.01 mol L^{-1} creatine phosphate react with 0.01 mol L^{-1} adenosine diphosphate to produce 0.04 mol L^{-1} creatine and 0.02 mol L^{-1} adenosine triphosphate at 25 °C, pH 7, pMg 4? What concentration of ATP can be formed if the other reactants are maintained at the indicated concentration?

9.9 The cleavage of fructose 1,6-diphosphate (FDP) to dihydroxyacetone phosphate (DHP) and glyceraldehyde 3-phosphate (GAP) is one of a series of reactions most organisms use to obtain energy. At 37 °C and pH 7, $\Delta_r G°'$ for the reaction FDP = DHP + GAP is 23.97 kJ mol^{-1}. What is $\Delta_r G°'$ in an erythrocyte in which [FDP] = 3×10^{-6} mol L^{-1}, [DHP] = 138×10^{-6} mol L^{-1}, and [GAP] = 18.5×10^{-6} mol L^{-1}?

9.10 How many grams of ATP have to be hydrolyzed to ADP to lift 100 lb 100 ft if the available Gibbs energy can be converted into mechanical work with 100% efficiency? It is assumed that [ATP] = [ADP] = [P_i] = 0.01 mol L^{-1} and that $\Delta G°'$ is -39.8 kJ mol^{-1} at 25 °C.

9.11 Biochemistry textbooks give $\Delta_r G°' = -20.1$ kJ mol^{-1} for the hydrolysis of ethyl acetate at pH 7 and 25 °C. Experiments in acid solution show that

$$\frac{[CH_3CH_2OH][CH_3CO_2H]}{[CH_3CO_2CH_2CH_3]} = 14$$

where the equilibrium concentrations are in moles per liter. What is the value of $\Delta_r G°'$ obtained from this equilibrium constant? The pK of acetic acid = 4.60 at 25 °C.

9.12 Fumarase catalyzes the reaction

$$fumarate + H_2O = \text{L-malate}$$

At 25 °C and pH 7

$$K' = 4.4 = \frac{[\text{L-malate}]}{[fumarate]}$$

What is the value of K' at pH 4?
Given: For fumaric acid $K_1 = 10^{-4.18}$
For L-malic acid $K_1 = 10^{-4.73}$

9.13 Given $\Delta_r G° = 49.4$ kJ mol^{-1} for

$$ATP^{4-} + H_2O = AMP^{2-} + P_2O_4^{7-} + 2H^+$$

calculate $\Delta G°'$ at pH 7 and 25 °C and 0.2 mol L^{-1} ionic strength. See Table 9.3.

9.14 Ethyl acetate is hydrolyzed in an aqueous buffer at pH 7 in a calorimeter. The enthalpy of hydrolysis is measured in a calorimeter and does not correspond with what is calculated from the following standard enthalpies of formation from a chemical thermodynamic table.

	$\Delta_f H°$ (298 K)/kJ mol^{-1}
Acetic acid (l)	−484.5
Ethanol (l)	−277.0
Ethyl acetate (l)	−479.0
H_2O (l)	−285.8

Explain why. What additional information would you need to calculate the heat absorbed in this experiment?

9.15 If n molecules of a ligand A combine with a molecule of protein to form PA_n without intermediate steps, derive the relation between the fractional saturation Y and the concentration of A.

9.16 Since it is difficult to determine the values of the four dissociation constants in equation 9.52, the empirical Hill equation

$$Y = \frac{1}{1 + K_h/P_{O_2}^h}$$

is frequently used to characterize binding. Show that the Hill coefficient h may be obtained by plotting $\log[Y/(1 - Y)]$ versus $\log P_{O_2}$.

9.17 The percent saturation of a sample of human hemoglobin was measured at a series of oxygen partial pressures at 20 °C, pH 7.1, 0.3 mol L^{-1} phosphate buffer, and 3×10^{-4} mol L^{-1} heme:

P_{O_2}/Pa	Percent Saturation
393	4.8
787	20
1183	45
2510	78
2990	90

Calculate the values of h and K_h in the Hill equation. (See problem 9.16.)

9.18 What is the ion product of water at 50 °C, assuming that $\Delta_r C_P°$ is independent of temperature? What is the pH of neutrality at this temperature?

9.19 At 298.15 K, H_2O (l) $= H^+$(ao) $+ OH^-$(ao), $K = 1.008 \times 10^{-14}$, and $\Delta_r H° = 55.836$ kJ mol^{-1}. Given that $S°[H_2O(l)] = 69.92$ J K^{-1} mol^{-1}, what is the value of $S°[OH^-$(ao)]?

9.20 Using Appendix C.1, calculate $\Delta_r H°$, $\Delta_r G°$, and $\Delta_r S°$ for

$$CH_3CO_2H(ao) = H^+(ao) + CH_3CO_2^-(ao)$$

and compare the values in Table 9.1 for 298.15 K. Calculate the acid dissociation constant of acetic acid at 298.15 K.

9.21 An aqueous solution of 0.01 mol L^{-1} phosphoric acid is titrated at 25 °C with such a concentrated solution of strong alkali that the solution is not significantly diluted during the titration. Plot $[H_3PO_4]$, $[H_2PO_4^-]$, $[HPO_4^{2-}]$, and $[PO_4^{3-}]$ as a function of pH.

9.22 A protein molecule has two different and independent sites; one binds A and the other binds B. What are the probabilities of P, PA, PB, and PAB?

9.23 At pH 7 and pMg 4 what value of pCa is required to put half the ATP in the form CaATP^{-2}? At 0.2 mol L^{-1} ionic strength and 25 °C the following constants are known:

$$HATP^{3-} = H^+ + ATP^{4-} \qquad pK = 6.95$$
$$MgATP^{2-} = Mg^{2+} + ATP^{4-} \qquad pK = 4.00$$
$$CaATP^{2-} = Ca^{2+} + ATP^{4-} \qquad pK = 3.60$$

9.24 For the weak acid $CH[(CH_2)_5CO_2H]_3$, the carboxyl groups are essentially identical and independent. (a) If the intrinsic acid dissociation constant is K, what are the values of K_1, K_2, and K_3? (b) At $[H^+] = K$, what are the relative proportions of minus three ion, minus two ion, minus one ion, and uncharged molecules?

9.25 In a series of biochemical reactions the product in one reaction is a reactant in the next. This has the effect that spontaneous reactions drive nonspontaneous reactions. For example, reaction 2 follows reaction 1:

1. L-malate $=$ fumarate $+ H_2O$

$$\Delta_r G°' = 2.9 \text{ kJ mol}^{-1}$$

2. fumarate $+$ ammonia $=$ aspartate

$$\Delta_r G°' = -15.6 \text{ kJ mol}^{-1}$$

The $\Delta_r G°'$ values are for pH 7 and 37 °C, and the state of ionization of the reactants is ignored. In reaction 1, the activity of H_2O is to be taken as 1. If the ammonia concentration is 10^{-2} mol L^{-1}, calculate [aspartate]/[L-malate] at equilibrium.

9.26 In the living cell two reactions may be coupled together by having a common intermediate. This is true for the following two reactions that are enzyme catalyzed, creatine + P_i = creatine phosphate + H_2O

$$\Delta_r G^{\circ\prime} = 46 \text{ kJ mol}^{-1}$$

ATP + H_2O = ADP + P_i

$$\Delta_r G^{\circ\prime} = -33 \text{ kJ mol}^{-1}$$

The $\Delta_r G^{\circ\prime}$ values are for pH 7.5 and 25 °C. If in a steady state in a living cell [ATP] = 10^{-3} mol L^{-1} and [ADP] = 10^{-4} mol L^{-1}, calculate the steady state of the ratio [creatine phosphate]/[creatine].

9.27 From the data of Table 9.2 calculate $\Delta_r G^{\circ\prime}$ for

$$\text{ATP} + H_2O = \text{AMP} + \text{PP}$$

9.28 What is the maximum concentration of ATP that can be formed enzymatically from acetyl phosphate and ADP each at 0.01 mol L^{-1} and pH 7 and pMg 4 at 25 °C, assuming that the ambient concentration of acetate is also 0.01 mol L^{-1}? Given

acetylP + H_2O = acetate + P_i

$$\Delta_r G^{\circ\prime} = -43.1 \text{ kJ mol}^{-1}$$

ADP + P_i = ATP + H_2O

$$\Delta_r G^{\circ\prime} = 39.8 \text{ kJ mol}^{-1}$$

9.29 The hydrolysis of adenosine triphosphate ATP to adenosine diphosphate ADP and inorganic phosphate at pH 8 and 25 °C

$$\text{ATP}^{4-} + H_2O = \text{ADP}^{3-} + \text{HPO}_4^{2-} + H^+$$

has a standard enthalpy change of -13 kJ mol^{-1}. The standard enthalpy changes of acid dissociation of HATP^{-3}, HADP^{-2}, and $H_2PO_4^{-1}$ are -8, 0, and $+8$ kJ mol^{-1}, respectively. Calculate the standard enthalpy change for the reaction

$$\text{HATP}^{3-} + H_2O = \text{HADP}^{2-} + H_2PO_4^-$$

9.30 If $\Delta_r G^{\circ\prime}$ for the hydrolysis of acetyl phosphate ($CH_3CO_3PO_3H_2$) is -43.1 kJ mol^{-1} at 25 °C and pH 7, what is the value at pH 4? It may be assumed that the pK of acetyl phosphate in the neighborhood of pH 7 is identical with pK_2 of orthophosphate.

9.31 Given the following system of reactions,

$$
\begin{array}{ccc}
\text{AH} + \text{B} & \overset{K_1}{\rightleftharpoons} & \text{CH} \\
\updownarrow & & \updownarrow \\
\text{A} + \text{B} & \underset{K_2}{\rightleftharpoons} & \text{C}
\end{array}
$$

where the K's represent dissociation constants. (a) Calculate the dependence on hydrogen ion concentration of the apparent equilibrium constant

$$K' = \frac{([C] + [CH])}{([A] + [AH])[B]}$$

(b) What is the relationship between the four dissociation constants?

9.32 Nucleosides associate in aqueous solution to form dimers. For adenosine (A) at 25 °C the equilibrium constant for 2A = A_2 is 4.5 when concentrations are expressed in mol L^{-1}. What is the concentration of dimers in a 0.5 mol L^{-1} solution of adenosine? (This association is due to the stacking of bases.)

9.33 The partial pressure of oxygen required to half saturate hemoglobin at pH 7.4 is 3.7 kPa. If the partial pressure of oxygen in the alveolar spaces of the lungs is 13.3 kPa, and the partial pressure in the capillaries is 5.3 kPa, what percentage of the total oxygen carrying capacity of hemoglobin is being used if h in the Hill equation is 2.7? (See problem 9.16.)

9.34 When myoglobin is in contact with air how many parts per million of CO are required to tie up 10% of the myoglobin? The partial pressure of oxygen required to half saturate myoglobin at 25 °C is 3.7 kPa. The partial pressure of CO required to half saturate myoglobin in the absence of oxygen is 0.009 kPa.

PART TWO
QUANTUM CHEMISTRY

Phenomena on the molecular scale cannot be understood without quantum mechanics. The development of quantum mechanics in this century has led to the possibility of calculating energy levels and other properties of atoms and molecules. We will consider the electronic orbitals of the hydrogen atom in detail and describe how these calculations may be extended to atoms with two or more electrons. The calculations show why properties such as ionization potential, electron affinity, and atom size vary in a complicated periodic way. In fact, quantum mechanics provides the explanation of the periodic table.

The application of quantum mechanics to molecules has made it possible to understand the nature of the chemical bond. The bonding in H_2^+ and H_2 are considered in some detail in Chapter 12. The application of quantum mechanics to molecules (other than H_2^+) involves approximation, but energy levels, bond lengths, and angles may be calculated quite accurately for many small molecules. Approximate calculations for larger molecules are also useful. Applications of molecular quantum mechanics are increasing rapidly because of the increasing power of computers.

In Chapter 13 we consider the symmetry of molecules in their equilibrium configurations. The symmetry of a molecule may greatly simplify quantum mechanical calculations of its energy levels and geometry. Symmetry also determines whether a molecule can be optically active or have a dipole moment.

Quantum mechanics provides the basis for understanding spectroscopy. Spectroscopy is useful for identifying molecules and determining their concentrations, but spectroscopy is especially important in physical chemistry because it yields information about individual molecules. Microwave and far infrared spectra provide information on internuclear distances and bond angles. Infrared and Raman spectroscopy provide information about vibrational frequencies. Visible and ultraviolet spectroscopy provide information on dissociation energies and electronic excited states.

The development of lasers has had a major impact on spectroscopy and its applications because of the monochromaticity and coherence of laser radiation. Nuclear magnetic resonance and electron spin resonance spectroscopy have become so important in the practice of chemistry that they are discussed in a separate chapter. Nuclear magnetic resonance is made possible by nuclei that have spin, and electron spin resonance is made possible by unpaired electrons in molecules, free atoms, or radicals. Fourier transform NMR, with its enhanced sensitivity, has opened up almost the whole periodic table to this form of spectroscopy.

10
Quantum Theory

The early years of the twentieth century saw a great revolution in physics: the birth of quantum mechanics, which replaces classical mechanics as the description of motion on an atomic scale. In 1900, Planck showed that the description of the distribution of energies of electromagnetic radiation in a cavity requires the quantization of energy. This was quickly followed by the application of quantization to atomic and molecular phenomena. Modern chemistry relies on quantum mechanics for the description of most phenomena. In this chapter we consider the basic concepts of quantum mechanics and apply them to a few simple problems such as the harmonic oscillator and rigid rotator.

10.1   The Origins of Quantum Theory

During the latter part of the 19th century, physicists were justly proud of the apparent completeness of their subject. Newton's laws and the electromagnetic wave theory of Maxwell had successfully explained most of the phenomena they were interested in. As experiments were able to go to lower temperatures

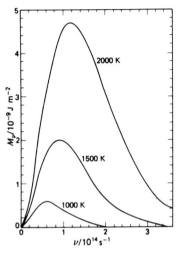

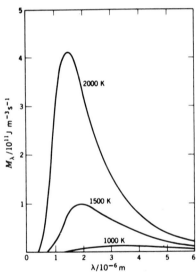

Figure 10.1   (a) Spectral distribution of the radiant exitance M_ν. Note that the frequency of maximum intensity is directly proportional to the temperature. (b) Spectral concentration of the radiant exitance M_λ in terms of wavelength. In both figures the area under a curve is the radiant exitance M in W m^{-2}. The area increases with the fourth power of the temperature.

and to both shorter and longer wavelengths of radiation, problems arose which showed that the classical laws did not explain phenomena on the atomic or molecular level. These problems were the low-temperature heat capacities of solids, the photoelectric effect, the distribution of radiation from a black body, and the spectra of atoms. By 1925, a new mechanics had been invented which successfully explained all these phenomena: quantum mechanics.

The first of the problems to be studied was **black body** or **cavity radiation.** A solid at a high temperature emits radiation (for example, a heated metal bar glows red). Black body or cavity radiation is obtained when a hollow cavity with a pinhole is held at a constant temperature and the distribution of wavelengths of the radiation coming out of the hole is measured. The distribution of wavelengths should have been calculable from the laws of thermodynamics and electromagnetism; however, this attempt failed. Max Planck, a noted thermodynamicist in Berlin where the best experiments were being done, had studied this problem. He and others were able to show that the laws of thermodynamics require that there is a universal distribution of these frequencies ν (or wavelengths $\lambda = c/\nu$, where c is the speed of light) which depends only on the temperature and not on the material out of which the cavity is made. It is most convenient to discuss this distribution by introducing the energy per unit volume (or **energy density**) of the radiation between ν and $\nu + d\nu$ equal to $\rho(\nu)$ $d\nu$. By 1900, Planck showed that the best experiments could be fit by the function

$$\rho(\nu, T) = \frac{8\pi h (\nu/c)^3}{e^{h\nu/kT} - 1} \tag{10.1}$$

where k is the **Boltzmann constant** equal to R/N_A and h is an experimentally determined constant called the Planck constant, the current value of which is

$$h = 6.626\ 075\ 5 \times 10^{-34}\ \text{J s}$$

Note that $\rho(\nu, T)$ has the units J m^{-3} s. The energy per unit volume between ν and $\nu + d\nu$ is $\rho(\nu)$ $d\nu$; this can be converted to a function of wavelength λ by noting that $\nu = c/\lambda$, so $|\,d\nu/d\lambda\,| = c/\lambda^2$. Therefore,

$$\rho(\lambda, T) = \rho\left(\frac{c}{\lambda}\right) \left|\frac{d\nu}{d\lambda}\right| = \frac{8\pi hc}{\lambda^5} \frac{1}{e^{hc/\lambda kT} - 1} \tag{10.2}$$

where $\rho(\lambda)$ has the units J m^{-4}. Planck derived these forms by breaking with classical physics and assuming that the radiation emitted or absorbed by the solid only comes in finite elements (called quanta) of energy $h\nu$ or *integral* multiples of $h\nu$. By combining this assumption with classical thermodynamic arguments he was able to derive the experimentally determined $\rho(\nu, T)$. His bold assumption of quantization of energy was unprecedented—a total break with classical ideas, which allow a continuous range of energies to be absorbed or emitted—and was the foundation of quantum theory. It is an interesting feature of this calculation that the experimental data were good enough to allow Planck to obtain a very good value of k and hence Avogadro's constant N_A, which was not well known at that time, by using the relation $N_A = R/k$.

In Fig. 10.1, the rate at which energy is radiated from a pinhole in a cavity, called the **exitance,** $M_\nu = (c/4)\,\rho(\nu, T)$, is plotted as a function of ν for three temperatures and as a function of wavelength λ for the same temperatures.

Note that there is a maximum in these curves which changes with T. By differentiating $\rho(\lambda,T)$ with respect to T we find

$$\lambda_{max}T = 2.898 \times 10^{-3} \text{ K m} \tag{10.3}$$

which is known as the Wien displacement law, known experimentally since the late 19th century. In Fig. 10.1, both M_ν and M_λ are plotted where

$$M_\nu = \frac{c}{4}\rho(\nu,T) = \frac{2\pi h\nu^3}{c^2}\frac{1}{e^{h\nu/kT} - 1} \tag{10.4}$$

$$M_\lambda = \frac{c}{4}\rho(\lambda,T) = \frac{2\pi hc^2}{\lambda^5}\frac{1}{e^{hc/\lambda kT} - 1} \tag{10.5}$$

If classical laws were obeyed, then $h \to 0$ (since then energy can be absorbed or emitted in any amount), and using l'Hopital's rule*

$$\rho_{cl}(\nu,T) \to \frac{8\pi\nu^2 kT}{c^3} \tag{10.6}$$

which is known as the Rayleigh–Jeans law, derived on the basis of classical ideas only. Notice that the classical expression for $\rho(\nu)$ becomes infinite at large ν while the quantum mechanical expression goes to zero because of the exponential. For small ν, the classical and quantum expressions are very close to one another. Planck's bold assumption that energy was emitted and absorbed in quanta was so successful in describing the experimental results that, in spite of the break with classical theory, it was accepted quickly.

10.2 Line Spectra

Another problem that could not be explained on the basis of classical laws was the spectra of atoms. Excited atoms emit radiation consisting of only certain discrete frequencies or wavelengths. Classical theories of a system of positive and negative charges predict that the radiation should be continuous. In Fig. 10.2, a small region of the hydrogen atom spectrum is shown. In 1885 Balmer discovered that the wavelength λ of the lines in the visible region of the emission spectrum of hydrogen atoms could be expressed by a simple relation that is now written as

$$\frac{1}{\lambda} = \bar\nu = R\left(\frac{1}{2^2} - \frac{1}{n_2^2}\right) \tag{10.7}$$

where n_2 is an integer greater than 2 and R is the **Rydberg constant**, 109 677.5856 cm^{-1}, for hydrogen. Note that we have introduced the wavenumber $\bar\nu$ which is proportional to energy. The value of R may be determined very accurately

---

* If a function of a variable x is indeterminate as $x \to 0$ because the function becomes $0/0$, then the limit can be found by taking the limit of the derivative of the numerator divided by the derivative of the denominator. See, for example, G. B. Thomas and R. Finney, *Calculus and Analytic Geometry*. Reading, MA: Addison-Wesley, 1979.

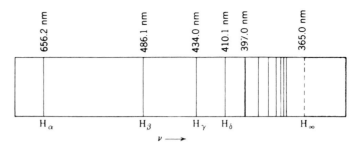

Figure 10.2 Balmer series of lines in the emission spectrum of atomic hydrogen. The wavelengths are given in nanometers (1 nm = 10^{-9} m). The lines are designated with Greek letters, and these letters are used to indicate the corresponding transitions in Fig. 10.3.

because of the high precision with which the wavelengths of spectral lines can be measured.* In spectroscopy it is often more convenient to use frequencies or wavenumbers than wavelengths because frequencies and wavenumbers are proportional to energy, and spectroscopy involves transitions between different energy levels.

In equation 10.7, n_2 must be larger than 2 for the wavelength to be positive and finite. As n_2 becomes larger the value of $\tilde{\nu}$ becomes larger and approaches $\frac{1}{4}R$ (the series limit) as n_2 approaches infinity.

The success of the Balmer formula led to further exploration, and other series of lines were discovered in the atomic hydrogen spectrum that could be represented by the equation

$$\tilde{\nu} = R\left(\frac{1}{n_1^2} - \frac{1}{n_2^2}\right) \tag{10.8}$$

where n_1 is also an integer and $n_2 > n_1$. The series for which $n_1 = 1$ (Lyman series) is in the vacuum ultraviolet; the series for which $n_1 = 3$ (Paschen series), 4 (Brackett series), or 5 (Pfund series) are in the infrared region. It is important to note that every line in the spectrum can be represented as a difference of two terms, R/n_1^2 and R/n_2^2.

The spectra of atoms with more than one electron are more complicated but in all cases it is found that all the frequencies of the line spectra can be represented as differences between a small set of numbers characteristic of the atom. In other words, each atom has a set of energies E_n called energy levels, and all the frequencies in the line spectrum of that atom can be calculated as

---

* According to quantum mechanics the Rydberg constant is given by (see Sec. 11.1)

$$R = \frac{e^4\mu}{8\epsilon_0^2 h^3 c}$$

where μ is the reduced mass (equation 10.113) for the atom and ϵ_0 is the permittivity of a vacuum. The reduced mass of an atom with one electron is given by

$$\mu = \frac{m_{\text{nuc}}m_{\text{e}}}{m_{\text{nuc}} + m_{\text{e}}}$$

where m_{nuc} is the mass of the nucleus and m_{e} is the mass of the electron. For a nucleus with a mass $m_{\text{nuc}} \to \infty$, $\mu = m_{\text{e}}$ and $R_\infty = 109\ 737.315\ 34\ \text{cm}^{-1}$.

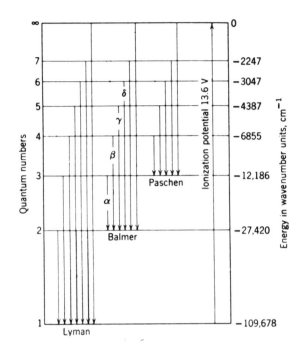

Figure 10.3   Energy levels for the hydrogen atom as calculated from the Bohr theory.

$$\nu = \frac{1}{h}(E_n - E_m) \tag{10.9}$$

where $E_n > E_m$ and h is the Planck constant. The connection of the difference in energy to frequency, $E = h\nu$, is the same as in Planck's theory of blackbody radiation. However, in this case, an additional assumption is made: The electrons in atoms can have only certain specific energy levels and therefore only certain spectral frequencies are observed. For this assumption to be accepted, a theory for such energy levels had to be developed.

A successful theory of the spectrum of the hydrogen atom was developed in 1913 by Bohr. Bohr made a complete break with classical mechanics by assuming that in the hydrogen atom the angular momentum of the orbital electron can only have values that are integral multiples of a quantum of angular momentum of magnitude $\hbar$, referred to as h-bar, which is $h/2\pi$. Bohr assumed that an electron moves in a circular orbit around the positively charged nucleus. We now know that this is not a correct way to describe the orbital motion of electrons, but, nevertheless, Bohr was able to derive a correct expression for the energy levels of hydrogenlike atoms (i.e., atoms with one electron). Bohr's formula for the Rydberg constant was in perfect agreement with the experimental data. However, Bohr's theory could not be extended successfully to helium or other atoms with more than one electron.

The electronic energy levels in the hydrogen atom, as calculated from the Bohr theory, are summarized in Fig. 10.3. The Lyman series of lines is produced by electrons jumping from orbits with quantum numbers 2, 3, 4, . . . into the lowest permitted orbit ($n_1 = 1$). The Balmer series of lines is produced by electrons falling from larger orbits into the second orbit ($n_1 = 2$), and so on. The energies of the various orbits may be expressed in several ways. The energies in wavenumbers given at the right in Fig. 10.3 are the wavenumbers for radiation produced when an electron with no initial kinetic energy falls from an infinite distance into a given orbit, that is, the series limits. The wavenumber $\tilde{\nu}$ of any line in the spectrum may be obtained by subtracting the values at the right for the two energy levels involved. Thus, the second line in the Balmer series is due to an electron falling from the fourth orbit into the second, and its wavenumber is $27\,420 - 6855 = 20\,565$ cm^{-1}.

10.3   Photoelectron Effect and Particlelike Properties of Radiation

Another experiment whose description required quantum theory is the photoelectric effect. It had been observed that light absorbed by a metal surface causes electrons to be ejected from the surface; it was found that the kinetic energy of the fastest ejected electron is independent of the **intensity** of the light, and that there is a cutoff frequency below which no electrons were ejected. Classical physics predicted that the energy absorbed by the surface should be proportional to the intensity of the light, so it should be possible to increase the intensity of the incident light and eject electrons even at low frequencies. In 1905 Einstein pointed out that whereas these effects could not be explained using the classical theory of light, they could be explained if the energy of light is transferred to matter in particlelike bundles that later came to be called photons. He assumed that the energy ϵ of a photon is given by

$$\epsilon = h\nu \qquad (10.10)$$

Planck had assumed that the emission and absorption of light in black body radiation is quantized in units of $h\nu$. Einstein went a step further in suggesting that the energy in the light is quantized, with each photon having energy $h\nu$.

When light is absorbed by a metal, the total energy of a photon $h\nu$ is given to a single electron within the metal. If this quantity of energy is sufficiently large, the electron may penetrate the potential barrier at the surface of the metal (called the work function) and still retain some energy as kinetic energy. The kinetic energy retained by the electron depends on the energy and, therefore, the frequency of the photon that ejected it. The number of electrons ejected depends on the number of incident photons, and therefore to the intensity of light.

The idea that radiation has some particlelike properties (i.e., that it comes in bundles of energy called photons) was shown to be correct in 1923 when Arthur Compton studied the scattering of X-rays by a graphite target. He found that although the incident X-rays were monochromatic, the scattered X-rays contained in addition a component of slightly longer wavelength. These experiments could be explained using the hypothesis that the scattering was due to collisions between an X-ray photon and an individual electron in the target (Fig. 10.4).

These and other experiments showed that electromagnetic radiation has a dual nature: in certain experiments it behaves like a wave and in certain interactions with matter it behaves like particles.

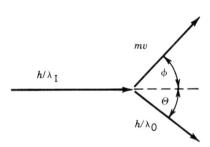

Figure 10.4   The Compton effect: the incident radiation (X-ray) has wavelength λ_I and momentum of magnitude h/λ_I; the electron is initially at rest; the outgoing radiation has wavelength λ_0 and momentum of magnitude h/λ_0; the outgoing electron has momentum mv. By requiring conservation of energy and momentum, the change in wavelength of the photon can be calculated as a function of Θ.

Example 10.1

(a) Calculate the number of photons emitted in 1 s from a 100-W red lamp, assuming for simplicity that all the photons have an average wavelength of 694 nm. (b) Calculate the number of photons emitted from a ruby laser in a 5-ns pulse with 0.1-GW power (λ = 694 nm).

The energy emitted in 1 s from a 100-W lamp is 100 J. Each photon has energy

$$\epsilon = \frac{hc}{\lambda} = \frac{6.626 \times 10^{-34} \text{ J s} \times 3 \times 10^8 \text{ m s}^{-1}}{694 \times 10^{-9} \text{ m}}$$

$$= 2.864 \times 10^{-19} \text{ J}$$

Thus, the number of photons is

$$\frac{100 \text{ J}}{2.864 \times 10^{-19} \text{ J}} = 3.49 \times 10^{20}$$

The energy emitted in a 5-ns pulse of 0.1 GW is 0.5 J. Each photon has $\epsilon = 2.864 \times 10^{-19}$ J; the number of photons emitted in 5 ns is therefore

$$\frac{0.5 \text{ J}}{2.864 \times 10^{-19} \text{ J}} = 1.764 \times 10^{18}$$

10.4   Wavelike Properties of Particles

In classical physics there is a clear distinction between waves and particles in describing phenomena. But, as we have seen in the preceding section, that way of thinking began to change as particlelike properties of electromagnetic radiation were discovered. In 1924, in his doctoral thesis, Louis de Broglie proposed that the wave particle behavior of radiation also applies to matter, that is, that **particles have wavelike properties.**

By use of the special theory of relativity Einstein had shown that the momentum of a photon is given by $p = \epsilon/c$ so that

$$p = \frac{h\nu}{c} = \frac{h}{\lambda} \tag{10.11}$$

It might be thought that since a photon does not have mass it would not have momentum. However, since a photon travels at the speed of light, its relativistic mass has to be considered. de Broglie postulated that this same equation might apply to material particles so that they would have a wavelength given by (for nonrelativistic velocities)

$$\lambda = \frac{h}{p} = \frac{h}{mv} \tag{10.12}$$

For macroscopic particles the de Broglie wavelength is so short that it does not lead to observable phenomena. However, for electrons, neutrons, and other microscopic particles the wavelengths may be of the order of interatomic distances in solids.

It was recognized that de Broglie's hypothesis could be tested by scattering a beam of electrons from a crystalline solid, this being a way of getting a grating of suitable dimensions to match the wavelength of electrons of accessible energies. In 1928 Davisson and Germer obtained a diffraction pattern from electrons impinging on the face of a nickel crystal that confirmed equation 10.12. This suggested that all particles have a wavelike property with a wavelength that is inversely proportional to the momentum. This can be considered to be a basic quantum postulate.

It is important to note that the particle concepts of energy E and momentum p are related to the wave concepts of frequency ν and wavelength λ by the Planck constant h. The particle model and the wave model are used for both matter and radiation, but most experiments emphasize either their particle-like or wave-like nature. One or the other model is generally more useful in a given experiment, but sometimes either may be used. Niels Bohr described this situation with his principle of complementarity. Both models are required to understand matter or radiation.

Example 10.2

What is the de Broglie wavelength λ of an electron that has been accelerated through a potential difference of 100 V?

The energy of an electron of mass m moving with a velocity v well below the velocity of light is given by

$$E = \tfrac{1}{2}mv^2 = \frac{p^2}{2m}$$

Thus, the momentum is given by

$$p = \sqrt{2mE}$$

The energy of the electron is $(1.602 \times 10^{-19}\ \text{C})(100\ \text{V}) = 1.602 \times 10^{-17}\ \text{J}$. Thus, the momentum is

$$p = \sqrt{(2)(9.110 \times 10^{-31}\ \text{kg})(1.602 \times 10^{-17}\ \text{J})}$$
$$= 5.403 \times 10^{-24}\ \text{kg m s}^{-1}$$

and the wavelength is

$$\lambda = \frac{h}{p} = \frac{6.626 \times 10^{-34}\ \text{J s}}{5.403 \times 10^{-24}\ \text{kg m s}^{-1}}$$
$$= 1.226 \times 10^{-10}\ \text{m} = 0.1226\ \text{nm}$$

10.5   The Heisenberg Uncertainty Principle

Classical mechanics does not involve any limitations in the accuracy with which observables may be measured. For example, the position and momentum of a particle may be simultaneously measured to any desired accuracy. This does require an interaction of the observer with the system that can disturb the system, but the disturbance can be made negligible or can be taken into account by suitable calculations. In 1927 Heisenberg formulated his principle that values of particular pairs of observables cannot be determined simultaneously with arbitrarily high precision in quantum mechanics. Examples of pairs of observables that are restricted in this way are momentum and position, and energy and time; such pairs are referred to as **complementary.** The quantitative expressions of the Heisenberg uncertainty principle can be derived by combining the de Broglie relation $p = h/\lambda$ and the Einstein relation $E = h\nu$ with properties of all waves.

The de Broglie wave for a particle is made up of a superposition of an infinitely large number of waves of the form

$$\psi(x,t) = A \sin 2\pi\left(\frac{x}{\lambda} - \nu t\right)$$

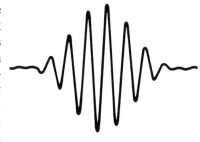

$$= A \sin 2\pi(\kappa x - \nu t) \qquad (10.13)$$

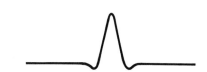

Figure 10.5 Superposition of waves to give (a) a weakly localized and (b) a strongly localized wavepacket.

where A is amplitude and κ is the reciprocal wavelength. The waves that are added together have infinitesimally different wavelengths. This superposition of waves produces a **wave packet** such as that shown in Fig. 10.5. By use of Fourier integral* methods it is possible to show that for wave motion of any type

$$\Delta x \, \Delta\kappa = \Delta x \, \Delta\frac{1}{\lambda} \geq \frac{1}{4\pi} \qquad (10.14)$$

$$\Delta t \, \Delta\nu \geq \frac{1}{4\pi} \qquad (10.15)$$

where Δx is the extent of the wave packet in space, $\Delta\kappa$ is the range in reciprocal wavelength, $\Delta\nu$ is the range in frequency, and Δt is a measure of the time required for the packet to pass a given point. The Δ's are actually standard deviations in these equations.† We can understand equations 10.14 and 10.15 in the following way. If at a given time the pulse extends over a short range of x values, there is a limit to the accuracy with which we can measure the wavelength. If a pulse of radiation is of short duration there is a limit to the accuracy with which we can measure the frequency.

One form of the Heisenberg uncertainty principle may be derived by substituting the de Broglie relation in equation 10.14. Since $1/\lambda = p_x/h$ for motion in the x direction, then

$$\Delta x \, \Delta\frac{p_x}{h} \geq \frac{1}{4\pi} \qquad (10.16)$$

$$\Delta x \, \Delta p_x \geq \frac{\hbar}{2} \qquad (10.17)$$

where $\hbar = h/2\pi$. We can understand this limitation on our ability to determine the simultaneous position and momentum of an electron in this way. To de-

* R. Eisberg and R. Resnick, *Quantum Physics*, 2d ed. New York: Wiley, 1985.

† The calculation of the standard deviation is discussed in Section 10.9.

termine the position of the electron at least one photon would have to strike the electron, and the momentum of the electron would inevitably be altered in the process. This would limit our ability to measure the momentum. If we use a photon of shorter wavelength to determine the position of the electron more accurately, the disturbance of the momentum is greater and Δp_x is greater, according to relation 10.17. Of course, the same uncertainty applies to $\Delta y \, \Delta p_y$ and $\Delta z \, \Delta p_z$.

Another form of the Heisenberg uncertainty principle may be derived by substituting $E = h\nu$ in equation 10.15. This yields

$$\Delta t \, \Delta \frac{E}{h} \geq \frac{1}{4\pi} \qquad (10.18)$$

$$\Delta t \, \Delta E \geq \frac{\hbar}{2} \qquad (10.19)$$

We can understand this limitation on our ability to measure the energy level of an electron in an atom in the following way. Suppose that excited atoms emit electromagnetic radiation in going to a more stable state. If these excited atoms live a long time, the radiation will be nearly monochromatic and the spectral line will be sharp. If the excited atoms have a very short half-life, the electromagnetic radiation will have a wider range in frequencies, in accord with equation 10.19. Thus, spectral lines have natural widths determined by the lifetime of the excited state, and they are further broadened by collisions and the Doppler effect. If the frequency is uncertain by $\Delta \nu$, the energy of the excited atom is uncertain by $\Delta E = h \, \Delta \nu$.

It is important to realize that the uncertainties in equations 10.17 and 10.19 are not experimental errors that are dependent on the quality of the measuring apparatus, but are inherent in quantum mechanics. Because of this uncertainty the results of quantum mechanical calculations are expressed in terms of probabilities. The success of a classical mechanical description for macroscopic systems can be understood from the fact that the de Broglie wavelength of macroscopic systems is extremely small (see the problems at the end of the chapter for examples).

Example 10.3

What is the standard deviation in the velocity of an electron if the uncertainty in its position is 100 pm?

$$\Delta p_x \geq \frac{h}{4\pi \, \Delta x}$$

$$\geq \frac{6.625 \times 10^{-34} \text{ J s}}{4\pi(10^{-10} \text{ m})}$$

$$\geq 5.272 \times 10^{-25} \text{ kg m s}^{-1}$$

$$\Delta v_x = \frac{\Delta p_x}{m} \geq \frac{5.272 \times 10^{-25} \text{ kg m s}^{-1}}{9.110 \times 10^{-31} \text{ kg}}$$

$$\geq 5.79 \times 10^5 \text{ m s}^{-1}$$

10.6 The Schrödinger Equation

After de Broglie's hypothesis that a particle has wavelike properties associated with it, many physicists attempted to derive the correct wave equation for a particle. In 1926, Schrödinger and Heisenberg independently developed theories that looked very different, but were later shown to be equivalent by Dirac, a British scientist. We will consider only the Schrödinger equation. The general form of this equation contains time as a variable, but we will be concerned primarily with the time-independent form because most chemical applications of quantum mechanics need only time-independent states.

The **time-independent Schrödinger** equation for a single particle of mass m moving in the x direction with a potential energy function $V(x)$ is

$$-\frac{\hbar^2}{2m}\frac{d^2\psi(x)}{dx^2} + V(x)\psi(x) = E\psi(x) \tag{10.20}$$

where the wavefunction $\psi(x)$ is a function of the x coordinate, and E is the total energy of the particle. This equation is not derived but is a postulate of quantum mechanics and must be judged on its success in describing experiments Note that the wavefunction ψ, like the amplitude of an electromagnetic wave, can be complex; that is, it may involve $i = \sqrt{-1}$ (see Appendix D). The interpretation of ψ was provided by Born who suggested, based on an analogy to electromagnetic waves, that the **probability** of finding the particle between x and $x + dx$ is given by $\psi^*(x)\psi(x)\,dx$ where ψ^* is the complex conjugate of ψ. (The complex conjugate is found by changing i to $-i$ everywhere in ψ.) This means that $\psi^*(x)\psi(x)$ is a **probability density.** For any ψ, $\psi^*\psi$ is both real and positive, as it must be to have the interpretation of a probability density. For example, if ψ is a complex number, it can be written as $a + ib$, then $\psi^* = a - ib$ and $\psi^*\psi = a^2 + b^2$, which is clearly positive and real. We often write $|\psi|^2$ for $\psi^*\psi$.

With this interpretation of ψ, the probability of finding the particle between x_1 and x_2 is

$$\int_{x_1}^{x_2} \psi^*(x)\psi(x)\,dx \tag{10.21}$$

and, since the probability of finding the particle anywhere on the x axis must be 1,

$$\int_{-\infty}^{+\infty} \psi^*(x)\psi(x)\,dx = 1 \tag{10.22}$$

For this one-dimensional example, the units of ψ are $m^{-1/2}$ to ensure that the probability is a pure number. If we were considering a three-dimensional system, then the integral of $|\psi|^2$ over three dimensions would be the probability of finding the particle anywhere in the space, which is 1. Then the wavefunction would have units $m^{-3/2}$.

Example 10.4

Normalizing a wavefunction. Given that a particle is restricted to the region $-a < x < a$ and has a wavefunction Ψ proportional to $\cos(\pi x/2a)$, normalize the wavefunction.

We must find a constant N such that $\int_{-a}^{a} dx\,|\Psi|^2 = 1$, where $\Psi = N\cos(\pi x/2a)$. Thus,

$$1 = \int_{-a}^{a} dx\, N^2\cos^2\frac{\pi x}{2a} = \frac{1}{2}\int_{-a}^{a} dx\, N^2\left(1 + \cos\frac{\pi x}{a}\right) = N^2 a$$

or $N = (1/a)^{1/2}$.

The wavefunction contains all the information we can have about a particle in quantum mechanics; methods to find this information will be presented in Section 10.7. However, for $|\psi|^2$ to be a probability density, ψ must have certain general properties. First of all, $|\psi|^2$ should be single valued. Second, the probability density should be such that its integral over all space is 1. Finally, the wavefunction ψ must be continuous so that the probability density is continuous and certain other properties (e.g., momentum) are finite (see Section 10.7). All of these limitations allow us to reject many solutions to the Schrödinger equation and keep others. In many cases, this restricts the allowed wavefunctions to only certain values of the energy E. In these cases, we say that the energy is quantized.

We may write the Schrödinger equation in a slightly different way:

$$\left[-\frac{\hbar^2}{2m}\frac{d^2}{dx^2} + V(x)\right]\psi(x) = E\psi(x) \tag{10.23}$$

The quantity in square brackets is called an **operator,** which when acting on the wavefunction $\psi(x)$ yields a constant E multiplied by $\psi(x)$. In general, when an operator $\hat{A}$ operates on a function $\psi_i(x)$ to yield a constant a_i times the function

$$A\psi_i(x) = a_i\psi_i(x) \tag{10.24}$$

we say that ψ_i is an **eigenfunction** of the operator $\hat{A}$ with eigenvalue a_i. Thus, returning to the Schrödinger equation, we call $\psi(x)$ the eigenfunction of the **Hamiltonian operator,**

$$\hat{H} = \left[-\frac{\hbar^2}{2m}\frac{d^2}{dx^2} + V(x)\right] \tag{10.25}$$

with eigenvalue E. In the next section, we discuss operators in a more general way.

10.7 Operators

An operator $\hat{A}$ specifies a mathematical operation to be carried out on a function to obtain a new function. For example, the operator d/dx indicates that a function is to be differentiated with respect to x, and the operator $\hat{x}$ indicates that a function is to be multiplied by x. The operator SQRT indicates that the square root of a function is to be taken; $SQRT(x^2 + y^2)$ is the function $(x^2 + y^2)^{1/2}$.

Quantum mechanics makes use only of linear operators. An operator (identified by a carat over the symbol) is linear if

$$\hat{A}(c_1 f_1 + c_2 f_2) = c_1 \hat{A} f_1 + c_2 \hat{A} f_2 \qquad (10.26)$$

where c_1 and c_2 are constants and f_1 and f_2 are functions. Thus, the operator SQRT is not a linear operator.

Example 10.5

What are the eigenfunctions and eigenvalues of the operator d/dx?

$$\frac{d}{dx} f(x) = kf(x) \qquad \frac{df(x)}{f(x)} = k\,dx \qquad \ln f(x) = kx + c \qquad f(x) = e^c e^{kx} = c' e^{kx}$$

where c and c' are constants. For each different value of k there is an eigenfunction $c' e^{kx}$. Or, to put it another way, the eigenfunction $c' e^{kx}$ has the eigenvalue k.

Operators are extremely important in quantum mechanics because they provide the means for calculating possible measured values of physically observable properties of a system. It is a postulate of quantum mechanics that for every observable in classical mechanics there is a linear quantum mechanical operator $\hat{A}$. It is a further postulate that the possible measured values are the eigenvalues a_i obtained from equation 10.24. Since this process sounds pretty mysterious at this point, let us consider as an example the calculation of the energy of a system, which has already been introduced in the Schrödinger equation.

The operator in the time-independent Schrödinger equation for a particle moving in one dimension is

$$\hat{H} = -\frac{\hbar^2}{2m}\frac{d^2}{dx^2} + V(x) \qquad (10.27)$$

as may be seen from equation 10.23. This is referred to as the Hamiltonian operator for the system. When it operates on an appropriate wavefunction it yields the energy as the eigenvalue. This name comes from Sir William Hamilton who developed an alternative form of classical mechanics involving a function H, called the Hamiltonian function. For a system for which the potential energy is a function of the coordinates only, a so-called conservative system because this ensures that the energy is conserved, the Hamiltonian function is equal to the total energy of the system expressed in terms of coordinates and conjugate momenta. For a Cartesian coordinate system the conjugate momenta are the components of the linear momentum p_x, p_y, and p_z in the x, y, and z directions.

For a particle of mass m moving in one dimension subject to a potential energy $V(x)$ the classical Hamiltonian function is

$$H = \frac{p_x^2}{2m} + V(x) \qquad (10.28)$$

When we compare this equation with equation 10.27 for the quantum mechanical operator for this system, we see that there is a resemblance. The process of converting a function for a classical system into the corresponding operator for the quantum mechanical system is formalized by the following rules:

1. Each Cartesian coordinate in the Hamiltonian function is replaced by the operator multiplication by that coordinate:

$$\hat{q} = q \qquad (10.29)$$

2. Each Cartesian component of linear momentum p_q in the Hamiltonian function is replaced by the operator

$$\hat{p}_q = \frac{\hbar}{i} \frac{\partial}{\partial q} = -i\hbar \frac{\partial}{\partial q} \qquad (10.30)$$

where $i = \sqrt{-1}$. The quantity $1/i$ is equal to $-i$ because $i(-i) = 1$.

In converting the classical Hamiltonian to the quantum mechanical Hamiltonian operator, the potential energy function is not changed because of the first of these postulates. In converting the kinetic energy part of the classical Hamiltonian we have to calculate the operator for p_x^2:

$$\hat{p}_x^2 = \left(\frac{\hbar}{i}\frac{\partial}{\partial x}\right)^2 = \frac{\hbar}{i}\frac{\partial}{\partial x}\frac{\hbar}{i}\frac{\partial}{\partial x} = -\hbar^2\frac{\partial^2}{\partial x^2} \qquad (10.31)$$

Replacing p_x^2 in the classical Hamiltonian with this operator and replacing the potential energy function by "multiply by $V(x)$" yields the Hamiltonian operator given in equation 10.27.

Each classical observable is associated with a quantum mechanical operator. The rules given in equations 10.29 and 10.30 are used to convert a classical observable to the quantum operator. Another postulate of quantum mechanics is that the only possible measured values of an observable are the eigenvalues of the operator representing that observable. If the system has a wavefunction that is an **eigenfunction** of an operator representing an observable, then a measurement of that observable yields that **eigenvalue.** If the system has a wavefunction that is not an eigenfunction, then each measurement of that observable will yield a particular eigenvalue of the operator with a probability that can be calculated from the wavefunction.

The correspondence between a number of quantum mechanical observables and quantum mechanical operators is shown in Table 10.1. We are not ready to consider some of these observables yet, but in Section 10.8 we will calculate the energy, the average value of x, the average value of x^2, the momentum p_x, and the average value of p_x^2 for a particle in a one-dimensional box.

The sum of two operators that operate on a function of x is defined by

$$(\hat{A} + \hat{B})f(x) = \hat{A}f(x) + \hat{B}f(x) \qquad (10.32)$$

The product of two operators is defined by

$$\hat{A}\hat{B}f(x) = \hat{A}[\hat{B}f(x)] \qquad (10.33)$$

First operator $\hat{B}$ is applied to $f(x)$ and then operator $\hat{A}$ is applied to the resulting function. In general, $\hat{A}\hat{B}$ does not necessarily produce the same result as $\hat{B}\hat{A}$, in contrast with ordinary algebra. The commutator $[\hat{A}, \hat{B}]$ of the operators $\hat{A}$ and $\hat{B}$ is defined as

$$[\hat{A},\hat{B}] = \hat{A}\hat{B} - \hat{B}\hat{A} \qquad (10.34)$$

If $\hat{A}\hat{B} = \hat{B}\hat{A}$, the operators are said to **commute,** $[\hat{A}, \hat{B}] = 0$.

If operators $\hat{A}$ and $\hat{B}$ do commute, the observables they represent can in principle be measured simultaneously and precisely. If not, simultaneous mea-

Table 10.1 Classical Mechanical Observables and the Corresponding Quantum Mechanical Operators[a]

Observables		Operators	
Name	Symbol	Symbol	Operation
For one-dimensional systems			
Position	x	$\hat{x}$	Multiply by x
Position squared	x^2	$\hat{x}^2$	Multiply by x^2
Momentum	p_x	$\hat{p}_x$	$\dfrac{\hbar}{i}\dfrac{\partial}{\partial x}$
Momentum squared	p_x^2	$\hat{p}_x^2$	$-\hbar^2\dfrac{\partial^2}{\partial x^2}$
Kinetic energy	$T = \dfrac{p_x^2}{2m}$	$\hat{T}_x$	$-\dfrac{\hbar^2}{2m}\dfrac{\partial^2}{\partial x^2}$
Potential energy	$V(x)$	$\hat{V}(x)$	Multiply by $V(x)$
Total energy	$E = T_x + V(x)$	$\hat{H}$	$-\dfrac{\hbar^2}{2m}\dfrac{\partial^2}{\partial x^2} + V(x)$
For three-dimensional systems			
Position	$\mathbf{r}$	$\hat{\mathbf{r}}$	Multiply by $\mathbf{r}$
Momentum	$\mathbf{p}$	$\hat{\mathbf{p}}$	$-i\hbar\left(\mathbf{i}\dfrac{\partial}{\partial x} + \mathbf{j}\dfrac{\partial}{\partial y} + \mathbf{k}\dfrac{\partial}{\partial z}\right)$
Kinetic energy	T	$\hat{T}$	$-\dfrac{\hbar^2}{2m}\nabla^2$ $= -\dfrac{\hbar^2}{2m}\left(\dfrac{\partial^2}{\partial x^2} + \dfrac{\partial^2}{\partial y^2} + \dfrac{\partial^2}{\partial z^2}\right)$
Potential energy	$V(x, y, z)$	$\hat{V}(x, y, z)$	Multiply by $V(x, y, z)$
Total energy	$E = T + V$	$\hat{H}$	$-\dfrac{\hbar^2}{2m}\nabla^2 + V(x, y, z)$
Angular momentum	$l_x = yp_z - zp_y$	$\hat{L}_x$	$-i\hbar\left(y\dfrac{\partial}{\partial z} - z\dfrac{\partial}{\partial y}\right)$
	$l_y = zp_x - xp_z$	$\hat{L}_y$	$-i\hbar\left(z\dfrac{\partial}{\partial x} - x\dfrac{\partial}{\partial z}\right)$
	$l_z = xp_y - yp_x$	$\hat{L}_z$	$-i\hbar\left(x\dfrac{\partial}{\partial y} - y\dfrac{\partial}{\partial x}\right)$

[a] Adapted from D. A. McQuarrie, *Quantum Chemistry*. Mill Valley, CA: University Science Books, 1983.

surement is impossible beyond a certain level of precision. The failure of $\hat{x}$ and $\hat{p}_x$ to commute leads, for example, to the Heisenberg uncertainty relation $\Delta x \, \Delta p_x \geq \hbar/2$.

Example 10.6
Show that if all the eigenfunctions of two operators $\hat{A}$ and $\hat{B}$ are the same functions, $\hat{A}$ and $\hat{B}$ commute with each other. The eigenvalues of $\hat{A}$ and $\hat{B}$ are represented by a_i and b_i and the eigenfunctions are ψ_i, so that

$$\hat{A}\psi_i = a_i\psi_i \quad \text{and} \quad \hat{B}\psi_i = b_i\psi_i$$

The eigenfunctions of the operator $\hat{A}\hat{B}$ are obtained as follows:

$$\hat{A}\hat{B}\psi_i = \hat{A}(\hat{B}\psi_i) = \hat{A}b_i\psi_i = b_i\hat{A}\psi_i = b_ia_i\psi_i$$

The operator $\hat{B}\hat{A}$ has eigenvalue a_ib_i, as may be seen from

$$\hat{B}\hat{A}\psi_i = \hat{B}(\hat{A}\psi_i) = \hat{B}a_i\psi_i = a_i\hat{B}\psi_i = a_ib_i\psi_i$$

Since $a_ib_i = b_ia_i$, $\hat{A}$ and $\hat{B}$ commute with each other.

Actually, the rules given here relating the operators to classical observables is only one of the many possible ways of constructing these rules. We say that it is a particular representation (here the coordinate representation) of quantum mechanics. There is an equally valid representation (called the momentum representation) in which the operator $\hat{p}_x$ is multiplied by the number p_x and the operator for x is $-(\hbar/i)(\partial/\partial p_x)$. Although there are cases for which these other representations are useful, we will only use the most common one here, the coordinate representation.

10.8   Particle in a One-Dimensional Box

The simplest problem to treat in quantum mechanics is that of a particle of mass m constrained to move in one dimension x of length a. The potential energy $V(x)$ is taken to be 0 for $0 < x < a$ and infinite outside this region (Fig. 10.6). We will see that this leads to quantized energy levels.

In the region between $x = 0$ and $x = a$, the Schrödinger equation 10.23 can be written

$$\frac{\hbar^2}{2m}\frac{d^2\psi}{dx^2} + E\psi = 0 \tag{10.35}$$

or

$$\frac{d^2\psi}{dx^2} = -\frac{2mE}{\hbar^2}\psi \equiv -k^2\psi \tag{10.36}$$

where

$$k = \frac{\sqrt{2mE}}{\hbar^2}$$

The general solution to this equation is

$$\psi(x) = A\cos kx + B\sin kx \tag{10.37}$$

In the region outside $0 < x < a$, the only physical solution to the Schrödinger equation is $\psi = 0$, because only then $V\psi = 0$, even for V infinite. This implies that the probability for finding the electron outside the box is zero (that is, $|\psi|^2 = 0$ outside the box). To avoid a discontinuity in ψ at $x = 0$ and $x = a$, ψ must be zero at those points. To satisfy this condition at $x = 0$, A must

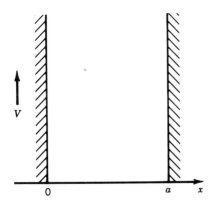

Figure 10.6   Potential for a particle in a one-dimensional box. The potential becomes infinite for $x > a$ and $x < 0$, and is zero for $0 < x < a$.

be equal to zero, since cos 0 = 1. The condition at $x = a$ can be satisfied only if sin $ka = 0$, or

$$ka = n\pi \qquad n = 1, 2, \cdots \qquad (10.38)$$

This forces the quantization

$$E = \frac{h^2 n^2}{8ma^2} \qquad n = 1, 2, \cdots \qquad (10.39)$$

Therefore, a particle constrained to be between $x = 0$ and $x = a$ has **quantized** energy levels, given by equation 10.39. Notice that as a gets larger, the energy levels get closer together. In the limit of a very large box (or a very heavy particle) the energy levels are so close together that the quantization may be unnoticeable. In the limit that a becomes infinite, all energies become allowed, so the perfectly free particle can have any energy.

A particle in a box cannot have zero energy because the lowest energy $h^2/8ma^2$ is given by equation 10.39 for $n = 1$. Although $n = 0$ satisfies the boundary conditions, the corresponding wavefunction is zero everywhere. The **zero-point energy** associated with the state $n = 1$ is found whenever a particle is constrained to a finite region; if this were not so the uncertainty principle would be violated. The next higher energy levels are at 4 times ($n = 2$) and 9 times ($n = 3$) this energy, as shown in Fig. 10.7. The wavefunctions are superimposed on this plot, and we can see that the wavelength is equal to $2a/n$.

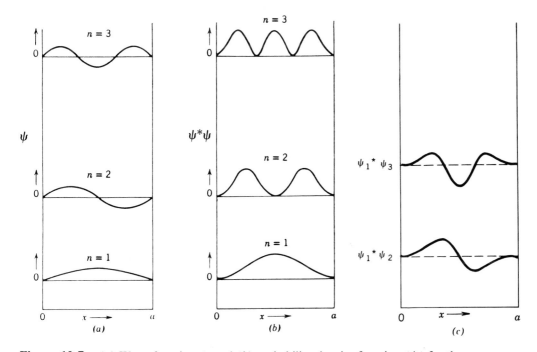

Figure 10.7 (a) Wave function ψ, and (b) probability density function $\psi^*\psi$ for the lowest three energy levels for a particle in a box. The plots are placed at vertical heights that correspond to the energies of the levels. As the number of nodes goes up, the energy goes up. (c) The product of wavefunctions $\psi_1^*\psi_2$ and $\psi_1^*\psi_3$ plotted against x.

Example 10.7

What is the ground-state energy for an electron that is confined to a potential well with a width of 0.2 nm?

$$E = \frac{h^2 n^2}{8ma^2}$$

$$= \frac{(6.626 \times 10^{-34} \text{ J s})^2(1)^2}{8(9.110 \times 10^{-31} \text{ kg})(0.2 \times 10^{-9} \text{ m})^2}$$

$$= 1.506 \times 10^{-18} \text{ J}$$

$$= \frac{(1.506 \times 10^{-18} \text{ J})(6.022 \times 10^{23} \text{ mol}^{-1})}{(10^3 \text{ J kJ}^{-1})}$$

$$= 907 \text{ kJ mol}^{-1}$$

Using equation 10.38 for k in the wavefunction yields

$$\psi_n(x) = B \sin \frac{n\pi x}{a} \qquad (10.40)$$

In order that the square of the wavefunction can be interpreted as a probability, it is necessary to **normalize** it so that the probability that the particle lies between $x = 0$ and $x = a$ is unity (i.e., the wavefunction is normalized):

$$\int_0^a \psi_n^*(x)\psi_n(x) \, dx = 1 \qquad (10.41)$$

$$|B|^2 \int_0^a \sin^2 \frac{n\pi x}{a} \, dx = 1 \qquad (10.42)$$

Since the value of the integral is $a/2$, $B = (2/a)^{1/2}$, the normalized wavefunction for a particle in a one-dimensional box is

$$\psi = \left(\frac{2}{a}\right)^{1/2} \sin \frac{n\pi x}{a} \qquad (10.43)$$

Figure 10.7b gives the probability densities $\psi^*\psi$ for a particle in an infinitely deep box. These are the probabilities per unit distance that the particle will be found at a given position. The most probable position for a particle in the zero-point level ($n = 1$) is in the center of the box. Note that the ψ_n are waves with wavelength $\lambda_n = 2a/n$. This means that ψ_n is zero at values of x equal to an integral number of $\lambda_n/2$. These zeros are called nodes of the wavefunction. In one-dimensional problems, the more nodes in an eigenfunction, the higher its eigenvalue of energy. For this problem, the number of nodes is $n - 1$.

As the value of the quantum number n increases, the probability density oscillates more and more. For very high values of n there are so many oscillations we would not expect to observe anything other than a constant value for the probability density. Classically this is just what we would expect for a particle in a box. This is an example of the **correspondence principle** of Bohr, according to which the quantum mechanical predictions approach the predictions of classical mechanics as the quantum number approaches infinity.

The wavefunctions ψ_i have been normalized so that

$$\int_{-\infty}^{\infty} \psi_i^* \psi_j \, dx = 1 \qquad \text{if } i = j \qquad (10.44)$$

For particle-in-a-box wavefunctions, as for all energy eigenfunctions,

$$\int_{-\infty}^{\infty} \psi_i^* \psi_j \, dx = 0 \qquad \text{if } i \neq j \tag{10.45}$$

Such wavefunctions are said to be orthogonal to each other. We can demonstrate the relation in equation 10.45 if we plot $\psi_i^* \psi_j$ for $i \neq j$ as a function of x (see Fig. 10.7c). We see that the negative contribution to the integral just cancels the positive contribution. Relations 10.44 and 10.45 can be combined by writing

$$\int_{-\infty}^{\infty} \psi_i^* \psi_j \, dx = \delta_{ij} \tag{10.46}$$

where δ_{ij} is called the **Kronecker delta,** which is defined by

$$\delta_{ij} = \begin{cases} 0 & \text{for } i \neq j \\ 1 & \text{for } i = j \end{cases} \tag{10.47}$$

Wavefunctions that satisfy equation 10.46 are said to be **orthonormal.** Solutions of the Schrödinger equation corresponding to different energies for a system are always orthogonal.

Some quantum mechanical systems have **degenerate** eigenfunctions. This means that a series of functions $\phi_1, \phi_2, \ldots, \phi_n$ are all eigenfunctions of operator $\hat{A}$ and correspond to the same eigenvalue a:

$$\hat{A}\phi_i = a\phi_i \qquad i = 1, 2, \ldots, n \tag{10.48}$$

When this is the case, any linear combination of the eigenfunctions is also an eigenfunction of $\hat{A}$ with the same eigenvalue. This can be shown as follows for the linear combination

$$\psi = c_1\phi_1 + c_2\phi_2 + \cdots + c_n\phi_n \tag{10.49}$$

where the c_i's are constants. Operating on this function with $\hat{A}$ yields

$$\hat{A}\psi = \hat{A} \sum_{i=1}^{n} c_i\phi_i = \sum_{i=1}^{n} c_i\hat{A}\phi_i$$

$$= \sum_{i=1}^{n} c_i a\phi_i = a \sum_{i=1}^{n} c_i\phi_i = a\psi \tag{10.50}$$

which is in the form of an eigenvalue equation. This feature of the eigenfunctions of an operator will be used in Section 11.2.

We can see from Fig. 10.7 that the most probable position for the particle is in the middle of the box if the system is in the ground state, but it is more likely to be at $a/4$ and $3a/4$ in the first excited state ($n = 2$). Notice that the observable "position" is not an eigenvalue of the wavefunctions for a particle in a one-dimensional box. The operator $\hat{x}$ does not give an eigenvalue when it operates on the particle-in-a-box energy eigenfunctions. This means that if we measured the position of a particle in a box we would get different answers in different trials. If we performed this measurement many times, we would confirm the probability densities $\psi^*\psi$ shown in Fig. 10.7. However, there is another type of question that does have a definite answer. We can ask for the average position of the particle, or its average x coordinate squared, or its average momentum squared. We will see how to answer these questions in the next section.

10.9 Expectation Values

When the wavefunction ψ is not an eigenfunction of the operator F, a measurement of $\hat{F}$ will give one of a number of possible values, the eigenvalues of $\hat{F}$. To determine the average value of F experimentally we take a number of systems prepared in an identical fashion and measure F for each system. The **average value** of F is then obtained by adding all the measured values and dividing by the number of measurements:

$$\langle F \rangle = \frac{\sum_{i=1}^{n} F_i}{N} \tag{10.51}$$

Alternatively, this expression may be written

$$\langle F \rangle = \frac{\sum n_f F_f}{N} \tag{10.52}$$

where n_f is the number of times that a particular value of F is obtained. This expression may be written as

$$\langle F \rangle = \sum \left(\frac{n_f}{N} \right) F_f = \sum p_f F_f \tag{10.53}$$

where p_f is the probability of observing the value F_f.

These ideas are readily extended to continuous distributions by defining a probability density $f(x)$ such that the probability that x lies between x and $x + dx$ is $f(x)\,dx$ and the probability that x lies between $x = a$ and $x = b$ is

$$\int_a^b f(x)\,dx \tag{10.54}$$

Probability distributions are normalized by requiring that

$$\int_{-\infty}^{\infty} f(x)\,dx = 1 \tag{10.55}$$

The mean and the second moment for a continuous distribution are calculated using

$$\langle x \rangle = \int_{-\infty}^{\infty} x f(x)\,dx \tag{10.56}$$

and

$$\langle x^2 \rangle = \int_{-\infty}^{\infty} x^2 f(x)\,dx \tag{10.57}$$

These values can be readily calculated by substituting $\psi^*\psi$ for $f(x)$, as we will soon see in an example. However, when we come to an observable that is represented by a differential operator a question arises as to whether the differential operator operates on ψ or $\psi^*\psi$. This question is answered by adopting a postulate. To calculate the average or expectation value $\langle a \rangle$ of a quantity represented in quantum mechanics by a differential operator $\hat{A}$ the following expression is used:

$$\langle a \rangle = \int_{-\infty}^{\infty} \psi_n^* \hat{A} \psi_n \, d\tau \tag{10.58}$$

where the wavefunction is normalized and $d\tau$ represents the differential of all coordinates. We can see that this postulate works in the special case that ψ is an eigenfunction of the operator $\hat{A}$. Since $\hat{A}\psi_n = a_n\psi_n$, equation 10.58 becomes

$$\langle a \rangle = \int_{-\infty}^{\infty} \psi_n^* a_n \psi_n \, d\tau = a_n \int_{-\infty}^{\infty} \psi_n^* \psi_n d\tau = a_n \qquad (10.59)$$

Equation 10.58 gives the correct answer in this case because we always obtain the same value.

Example 10.8
What are the values of $\langle x \rangle$ and $\langle x^2 \rangle$ for a particle in any of the eigenstates of a one-dimensional box?

$$\langle x \rangle_n = \frac{2}{a} \int_0^a x \sin^2 \frac{n\pi x}{a} \, dx$$

By use of a table of definite integrals we obtain

$$\langle x \rangle = \frac{a}{2}$$

for all values of the principal quantum number n, which is reasonable.

$$\langle x^2 \rangle_n = \frac{2}{a} \int_0^a x^2 \sin^2 \frac{n\pi x}{a} \, dx$$

$$= \left(\frac{a}{2\pi n} \right)^2 \left(\frac{4\pi^2 n^2}{3} - 2 \right)$$

Example 10.9
What are the values of $\langle p_x \rangle$ and $\langle p_x^2 \rangle$ for a particle in any of the eigenstates of a one-dimensional box? Since the operator for momentum in the x direction is $-i\hbar \, d/dx$,

$$\langle p_x \rangle_n = \int_0^a \left[\left(\frac{2}{a} \right)^{1/2} \sin \frac{n\pi x}{a} \right] \left(-i\hbar \frac{d}{dx} \right) \left[\left(\frac{2}{a} \right)^{1/2} \sin \frac{n\pi x}{a} \right] dx$$

$$= -i\hbar \frac{2\pi n}{a^2} \int_0^a \sin \frac{n\pi x}{a} \cos \frac{n\pi x}{a} \, dx$$

$$= 0$$

where the last result is obtained by using integral tables. Now that we have derived this result we can observe that the average momentum has to be zero because the particle in a box cannot continue to travel in one direction:

$$\langle p_x \rangle_n^2 = \int_0^a \left[\left(\frac{2}{a} \right)^{1/2} \sin \frac{n\pi x}{a} \right] \left(-i\hbar \frac{d}{dx} \right)^2 \left[\left(\frac{2}{a} \right)^{1/2} \sin \frac{n\pi x}{a} \right] dx$$

$$= \frac{n^2 \pi^2 \hbar^2}{a^2}$$

Now that we can calculate the mean value of an observable, there is another question of considerable interest. That is, what is the spread around the mean? The spread around the mean is measured by the **variance** σ_x^2, which is defined by

$$\sigma_x^2 = \langle (x - \langle x \rangle)^2 \rangle = \sum (x_i - \langle x \rangle)^2 f_i \qquad (10.60)$$

where f_i is the probability of x_i. The average deviation from the mean is equal to zero, but by squaring the deviations from the mean we get an inherently positive quantity that is zero if all the values of x are identical, is small if the distribution of x values is narrow, and is large if the distribution of x values is broad. The variance is represented by the symbol σ_x^2 because the square root of the variance is equal to the **standard deviation** σ_x. The standard deviation is especially useful for representing the breadth of a distribution because it has the units of x.

There is a simple way to calculate the variance which is more readily extended to continuous distributions. Equation 10.60 may be rearranged as follows:

$$\begin{aligned}
\sigma_x^2 &= \sum (x_i^2 - 2\langle x \rangle x_i + \langle x \rangle^2) f_i \\
&= \sum x_i^2 f_i - 2\langle x \rangle \sum x_i f_i + \langle x \rangle^2 \sum f_i \\
&= \langle x^2 \rangle - 2\langle x \rangle^2 + \langle x \rangle^2 \\
&= \langle x^2 \rangle - \langle x \rangle^2 \qquad (10.61)
\end{aligned}$$

In the second line $\langle x \rangle$ and $\langle x \rangle^2$ are taken outside the summations because they are simply numbers.

Example 10.10

What is the standard deviation in x for the particle in an eigenstate of a one-dimensional box? What is the standard deviation in p_x? How do these results compare with the Heisenberg uncertainty principle? In this example we will use Δx and Δp_x for these standard deviations as we did in Section 10.5.

Using the results of Example 10.8, we find

$$\begin{aligned}
\Delta x &= (\langle x^2 \rangle - \langle x \rangle^2)^{1/2} \\
&= \left(\frac{a}{2\pi n} \right) \left[\frac{\pi^2 n^2}{3} - 2 \right]^{1/2}
\end{aligned}$$

Using the results of Example 10.9 we find

$$\begin{aligned}
\Delta p_x &= (\langle p_x^2 \rangle - \langle p_x \rangle^2)^{1/2} \\
&= \frac{n\pi\hbar}{a}
\end{aligned}$$

The product of these standard deviations is

$$\Delta x \, \Delta p_x = \left(\frac{a}{2\pi n} \right) \left(\frac{\pi^2 n^2}{3} - 2 \right)^{1/2} \left(\frac{n\pi\hbar}{a} \right) = \frac{\hbar}{2} \left(\frac{\pi^2 n^2}{3} - 2 \right)^{1/2}$$

The product $\Delta x \, \Delta p_x$ for a particle in a box is in agreement with the Heisenberg uncertainty principle since it is necessarily greater than $\hbar/2$. Now we can see what happens to Δx and Δp_x when we change the dimensions of the box. As the length of the box is increased, the standard deviation in x increases, and we say the position of the particle becomes more uncertain. As the length of the box is increased, the standard deviation in p decreases, and we say the momentum becomes more definite.

10.10   Particle in a Three-Dimensional Box

There are several more things we can learn from the particle in a box by extending it to a three-dimensional box. The particle is confined to a rectangular parallelepiped with sides of lengths a, b, and c by having an infinite potential outside the box.

The classical Hamiltonian for a particle that can move in three dimensions is

$$H = \frac{p_x^2}{2m} + \frac{p_y^2}{2m} + \frac{p_z^2}{2m} \tag{10.62}$$

The time-independent Schrödinger equation for a single particle of mass m moving in three dimensions is

$$\hat{H}\psi(x, y, z) = E\psi(x, y, z) \tag{10.63}$$

where the Hamiltonian operator is

$$\hat{H} = -\frac{\hbar}{2m}\nabla^2 + V(x, y, z) \tag{10.64}$$

and

$$\nabla^2 \equiv \frac{\partial^2}{\partial x^2} + \frac{\partial^2}{\partial y^2} + \frac{\partial^2}{\partial z^2} \tag{10.65}$$

is referred to as the **Laplacian operator** or del squared. The wavefunction is normalized so that

$$\int_{-\infty}^{\infty}\int_{-\infty}^{\infty}\int_{-\infty}^{\infty} \psi^*(x, y, z)\psi(x, y, z)\, dx\, dy\, dz = 1 \tag{10.66}$$

If a particle can move in three dimensions, its probability density $p(x, y, z)$ is given by

$$p(x, y, z) = \psi^*(x, y, z)\psi(x, y, z) \tag{10.67}$$

The probability that the x coordinate is between x and $x + dx$, the y coordinate is between y and $y + dy$, and the z coordinate is between z and $z + dz$ is $p(x, y, z)\, dx\, dy\, dz = \psi^*(x, y, z)\psi(x, y, z)\, dx\, dy\, dz$. This last expression is sometimes shortened to $\psi^*\psi\, d\tau$, where $d\tau$ represents the differential element of volume $dx\, dy\, dz$.

Since the potential within the box is zero, we obtain the following partial differential equation for the region inside the box:

$$-\frac{\hbar^2}{2m}\left(\frac{\partial^2}{\partial x^2} + \frac{\partial^2}{\partial y^2} + \frac{\partial^2}{\partial z^2}\right)\psi = E\psi \tag{10.68}$$

Partial differential equations are solved, where possible, by using the technique of separation of variables to obtain a set of ordinary differential equations. In this case we assume that the wavefunction ψ is the product of three functions, each depending on just one coordinate:

$$\psi(x, y, z) = X(x)Y(y)Z(z) \tag{10.69}$$

By substituting this for ψ in equation 10.20 and then dividing by $X(x)Y(y)Z(z)$, we obtain

$$-\frac{\hbar^2}{2m}\left[\frac{1}{X(x)}\frac{d^2X(x)}{dx^2} + \frac{1}{Y(y)}\frac{d^2Y(y)}{dy^2} + \frac{1}{Z(z)}\frac{d^2Z(z)}{dz^2}\right] = E \tag{10.70}$$

Since each term on the left side of the equation is a function of a different independent variable and thus can be varied independently of one another each must equal a constant in order that the sum of the three terms equal a constant for all values of x, y, and z:

$$E_x + E_y + E_z = E \tag{10.71}$$

This converts the partial differential equation 10.70 into three ordinary differential equations that can be easily solved:

$$-\frac{\hbar^2}{2m}\left[\frac{1}{X(x)}\frac{d^2X(x)}{dx^2}\right] = E_x \tag{10.72a}$$

$$-\frac{\hbar^2}{2m}\left[\frac{1}{Y(y)}\frac{d^2Y(y)}{dy^2}\right] = E_y \tag{10.72b}$$

$$-\frac{\hbar^2}{2m}\left[\frac{1}{Z(z)}\frac{d^2Z(z)}{dz^2}\right] = E_z \tag{10.72c}$$

These equations are just like equation 10.35 and may be solved in the same way to obtain

$$X(x) = A_x \sin\frac{n_x \pi x}{a} = A_x \sin\left(\frac{2mE_x}{\hbar^2}\right)^{1/2} x \tag{10.73a}$$

$$Y(y) = A_y \sin\frac{n_y \pi y}{b} = A_y \sin\left(\frac{2mE_y}{\hbar^2}\right)^{1/2} y \tag{10.73b}$$

$$Z(z) = A_z \sin\frac{n_z \pi z}{c} = A_z \sin\left(\frac{2mE_z}{\hbar^2}\right)^{1/2} z \tag{10.73c}$$

where a, b, and c are the lengths of the sides in the x, y, and z directions, respectively, n_x, n_y, and n_z are nonzero integers, called quantum numbers, and $E_x = h^2 n_x^2/8ma^2$ and so on. Thus, there is a quantum number for each coordinate. When the wavefunction is normalized we obtain

$$\psi(x, y, z) = \left(\frac{8}{abc}\right)^{1/2} \sin\frac{n_x \pi x}{a} \sin\frac{n_y \pi y}{b} \sin\frac{n_z \pi z}{c} \tag{10.74}$$

When this eigenfunction is substituted in equation 10.68 we obtain

$$E = \frac{h^2}{8m}\left(\frac{n_x^2}{a^2} + \frac{n_y^2}{b^2} + \frac{n_z^2}{c^2}\right) \tag{10.75}$$

Later in statistical mechanics (Section 17.7) we will use this equation for the translational energy of a molecule in a container.

A new feature arises when the sides of the box are equal. If $a = b = c$, the energy levels are given by

$$E = \frac{h^2}{8ma^2}(n_x^2 + n_y^2 + n_z^2) \tag{10.76}$$

For a one-dimensional box the state of the system could be specified by giving the value of the quantum number or the energy. For a cubical box this is no longer true, because a given energy may be achieved by different combinations of the three quantum numbers n_x, n_y, and n_z. The quantum numbers $n_x = 2$, $n_y = 1$, $n_z = 1$; $n_x = 1$, $n_y = 2$, $n_z = 1$; and $n_x = 1$, $n_y = 1$, $n_z = 2$ describe different states of the system (different wavefunctions), but these states have the same energy. Such an energy level is said to be **degenerate,** and the degeneracy is equal to the number of independent wavefunctions associated with a given energy level. The degeneracies of levels with the following quantum numbers are

$n_x n_y n_z$	111	211	221	311	222	321	322	411	331
Degeneracy	1	3	3	3	1	6	3	3	3

In this list the level with quantum numbers 211 is made up of three states with quantum numbers 211, 121, and 112.

Degeneracies arise in quantum mechanics when there is some element of symmetry. If the symmetry is "broken" by giving the sides of the box different lengths, the degeneracy is "lifted." We will see later that the spherical symmetry of certain atomic states is broken by the application of magnetic or electric fields, and this lifts some of the degeneracies.

The degeneracy of a translational energy level increases rapidly with energy. If we define $n^2 = n_x^2 + n_y^2 + n_z^2$, then $E = (h^2/8ma^2)n^2$. If we think of the allowed values of n_x as points along the x axis, the allowed values of n_y as points along the y axis, and those of n_z as points along the z axis, then n can be thought of as the length of a vector in this three-dimensional space. All such vectors with the same length have the same energy; that is, they represent degenerate states. For very large n, the number of degenerate states is proportional to the surface area of the sphere of radius n in this space; therefore, the degeneracy is proportional to n^2. Actually, since the allowed values of n_x, n_y, and n_z are *integers*, we have counted states as degenerate that are not *exactly* degenerate. However, for large enough n, these states will be so close in energy that for practical (experimental) purposes we can take them to be degenerate.

Example 10.11

We shall see (in Chapter 17) that the most probable translational energy for an atom in a gas at temperature T is equal to $\frac{3}{2}kT$ where $k = R/N_A$ is Boltzmann's constant. Calculate the degeneracy of the most probable energy level for an argon atom at 300 K and 1 bar pressure, assuming that the atom can be treated as a particle in a three-dimensional box.

The volume of a gas at these conditions is approximately 0.022 m³, so that the side of the box a can be taken as $(0.022)^{1/3}$ m. We have

$$E = \frac{h^2}{8ma^2}n^2 = \frac{3}{2}kT$$

or

$$n^2 = \frac{\frac{3}{2}kT8ma^2}{h^2}$$

$$= \frac{\frac{3}{2}(1.38 \times 10^{-23})(300)(8)(40 \times 1.67 \times 10^{-27})(0.022)^{2/3}}{(6.6 \times 10^{-34})^2}$$

$$= 6 \times 10^{20}$$

Thus, the most probable value of n is 2.25×10^{10} and the degeneracy is of the order 6×10^{20} for this level.

The particle in a three-dimensional box illustrates a general point about the separability of the Hamiltonian for a system. The Hamiltonian operator for a particle in a three-dimensional box can be written as the sum of three independent terms:

$$\hat{H} = \hat{H}_x + \hat{H}_y + \hat{H}_z \tag{10.77}$$

When the Hamiltonian is separable in this way, we can factor the wavefunction into a product of three wavefunctions, an eigenfunction for each coordinate. This leads to separation of the eigenvalue problem in three variables into three separate equations, each in one variable. The sum of the three eigenvalues obtained is equal to the eigenvalue for the original problem.

10.11 The Classical Harmonic Oscillator

To understand the vibrations of molecules we need to understand the quantum mechanical treatment of a harmonic oscillator, and as background for that we need to review the classical treatment of a harmonic oscillator. The simplest example of a harmonic oscillator is a mass connected to a wall by means of an idealized spring, in the absence of gravity. As shown in Fig. 10.8a, the displacement of the mass is shown by its x coordinate, and the origin of the coordinate system is taken at the equilibrium position. The mass oscillates about its equilibrium position, and the motion is said to be **harmonic** if the force F due to the spring is directly proportional to the displacement x from its equilibrium position x_{eq}, which we can define as the origin of the x axis:

$$F = -kx \tag{10.78}$$

The negative sign comes from the fact that F is opposite to the displacement x. The proportionality constant k, referred to as a **force constant,** is small for a weak spring and large for a stiff spring. Since force is expressed by mass times acceleration, the equation for motion in the x direction is

$$m \frac{d^2x}{dt^2} + kx = 0 \tag{10.79}$$

The general solution of this differential equation is (compare with equation 10.37)

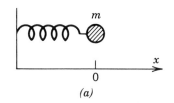

(a)

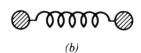

(b)

Figure 10.8 (a) Mass m connected to a wall by a spring in the absence of gravity. The equilibrium position of the mass is at $x = 0$. (b) Two equal masses connected by a spring.

$$x(t) = A \sin \omega t + B \cos \omega t \qquad (10.80)$$

where $\omega = (k/m)^{1/2}$ is the **fundamental vibration frequency** in radians per second. If we initially stretch the spring so that the mass is at position x_0, and its velocity is zero, and then let go, the time course of the motion is represented by

$$x(t) = x_0 \cos \omega t \qquad (10.81)$$

The mass oscillates between x_0 and $-x_0$ with a frequency of ω radians per second or $\nu = \omega/2\pi$ cycles per second.

The energy of a harmonic oscillator is equal to the sum of its potential energy and its kinetic energy. When the mass is at $+x_0$ or $-x_0$, the energy is all potential energy, and as the mass goes through $x = 0$, the energy is all kinetic energy. We can calculate the potential energy V from the fact that the force is the negative derivative of the potential energy

$$F = -\frac{dV}{dx} \qquad (10.82)$$

so that

$$V = -\int F \, dx + \text{constant} \qquad (10.83)$$

Since $F = -kx$, integration of equation 10.83 yields

$$V = \frac{kx^2}{2} \qquad (10.84)$$

if we take the constant to be zero when $x = 0$. Substituting equation 10.81 yields

$$V = \frac{kx_0^2}{2} \cos^2 \omega t \qquad (10.85)$$

Thus, the potential energy of the harmonic oscillator varies between zero and $kx_0^2/2$ during each period of oscillation.

The kinetic energy of the moving mass is

$$T = \tfrac{1}{2}m\left(\frac{dx}{dt}\right)^2 \qquad (10.86)$$

Using equation 10.81 yields

$$T = \tfrac{1}{2}m\omega^2 x_0^2 \sin^2 \omega t$$
$$= \tfrac{1}{2}kx_0^2 \sin^2 \omega t \qquad (10.87)$$

where the second form has been obtained using $\omega = (k/m)^{1/2}$.

The total energy is

$$E = T + V$$
$$= \tfrac{1}{2}kx_0^2 \sin^2 \omega t + \tfrac{1}{2}kx_0^2 \cos^2 \omega t$$
$$= \tfrac{1}{2}kx_0^2 \qquad (10.88)$$

Thus, the total energy is constant. The potential energy and the kinetic energy each oscillate between zero and a maximum value of $\tfrac{1}{2}kx_0^2$, but the total energy is conserved. The harmonic oscillator is a conservative system, and it may be shown that any system is **conservative** for which the force can be expressed as the derivative of a potential energy.

This completes the classical treatment of the harmonic oscillator, but we need to see how these results apply to the model shown in Fig. 10.8*b*, which is closer to a diatomic molecule. It can be shown that for two masses connected by a spring, equation 10.79 is replaced by

$$\mu \frac{d^2x}{dt^2} + kx = 0 \tag{10.89}$$

where μ is the **reduced mass**

$$\frac{1}{\mu} = \frac{1}{m_1} + \frac{1}{m_2} \tag{10.90}$$

and x is the distance between the two masses minus the equilibrium distance. This change in coordinates reduces the two-body problem to a one-body problem. Since we are interested in diatomic molecules, in the rest of the book we will identify the vibrational frequency with

$$\omega = \left(\frac{k}{\mu}\right)^{1/2} \tag{10.91}$$

or

$$\nu = \frac{1}{2\pi}\left(\frac{k}{\mu}\right)^{1/2} \tag{10.92}$$

10.12   The Quantum Mechanical Harmonic Oscillator

To obtain the energy levels for the quantum mechanical harmonic oscillator of mass μ we start with the classical Hamiltonian function found from equations 10.84, 10.86 and (10.91):

$$H = p_x^2/2\mu + \tfrac{1}{2}\omega^2\mu x^2 \tag{10.93}$$

and convert it to the quantum mechanical Hamiltonian operator by replacing x^2 by x^2 and p_x by $\hat{p}_x = -i\hbar\, d/dx$ to obtain

$$\hat{H} = -\frac{\hbar^2}{2\mu}\frac{d^2}{dx^2} + 2\pi^2\nu^2\mu x^2 \tag{10.94}$$

The solution of the Schrödinger equation is too complicated to discuss here in detail, but when it is solved it is found that there are well-behaved solutions only if the harmonic oscillator has energies given by

$$E_v = (v + \tfrac{1}{2})h\nu \qquad v = 0, 1, 2, \ldots \tag{10.95}$$

where v is the **vibrational quantum number** and $\nu = (1/2\pi)(k/\mu)^{1/2}$. The energy levels are equally spaced with a separation of $h\nu$. These levels are shown in Fig. 10.9*b* and *c*. It is especially important to notice that the energy of the ground state is not zero, as it is classically, but $E_0 = h\nu/2$. This **zero point energy** is in accordance with the Heisenberg uncertainty principle, as shown in Example 10.12.

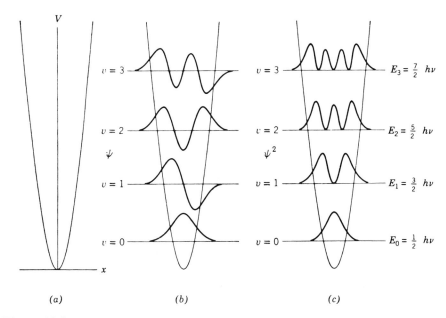

Figure 10.9   (*a*) Potential energy curve for a classical harmonic oscillator. (*b*) Allowed energy levels and wavefunctions for a quantum mechanical harmonic oscillator. (*c*) Probability density functions for a quantum mechanical harmonic oscillator.

The wavefunctions for the first two energy levels are given by

$$\psi_0 = \left(\frac{\alpha}{\pi}\right)^{1/4} e^{-\alpha x^2/2} \tag{10.96}$$

$$\psi_1 = \left(\frac{4\alpha^3}{\pi}\right)^{1/4} x e^{-\alpha x^2/2} \tag{10.97}$$

where $\alpha = (k\mu/\hbar^2)^{1/2}$. The wavefunction for the ground state has the shape of the Gaussian probability function.

In general, the wavefunctions for the harmonic oscillator are given by

$$\psi_v(x) = N_v H_v(\alpha^{1/2}x)e^{-\alpha x^2/2} \tag{10.98}$$

$$N_v = \frac{1}{(2^v v!)^{1/2}} \left(\frac{\alpha}{\pi}\right)^{1/4} \tag{10.99}$$

and the $H_v(\alpha^{1/2}x)$ are polynomials called **Hermite polynomials.** The wavefunctions for the first four levels are plotted in Fig. 10.9*b*, and the probability densities $|\psi_v|^2$ are plotted in Fig. 10.9*c*.

The wavefunctions given above have been normalized. The probability that the x coordinate of the harmonic oscillator is between x and $x + dx$ is given by $\psi^2\, dx$. If a large number of identically prepared systems are examined, the fraction having coordinates between x and $x + dx$ is equal to this probability. The **probability densities** ψ^2 are plotted versus x for the first four energy levels in Fig. 10.9*c*. In the ground state ($v = 0$) the most probable distance occurs at the position of the minimum in the potential well. This is in distinct contrast with that for a classical simple harmonic oscillator, which would spend the longest times at the turning points where the velocity goes to zero. As the quantum number increases, however, the quantum mechanical probability density function approaches that for the classical harmonic oscillator. This is an example of Bohr's "correspondence principle," according to which the quantum mechanical result must approach the classical result in the limit of infinite quantum number.

It is of special interest to note in Fig. 10.9c that the quantum mechanical oscillator has a certain probability of being at a greater distance than is allowed for the classical harmonic oscillator at the same energy. For the classical oscillator to be outside the parabolic potential curve, the kinetic energy would have to be negative (since $V > E$ in that region). In quantum mechanics, a particle can have a nonzero probability of being in the classically forbidden region. The probability of being outside the classical turning points given by the parabolic potential energy curve is 0.16 for the zero-point level.

The Hermite polynomials, which were known before the development of quantum mechanics, for the first four levels of the harmonic oscillator are as follows:

$$H_0(\xi) = 1 \tag{10.100a}$$

$$H_1(\xi) = 2\xi \tag{10.100b}$$

$$H_2(\xi) = 4\xi^2 - 2 \tag{10.100c}$$

$$H_3(\xi) = 8\xi^3 - 12\xi \tag{10.100d}$$

A property of these polynomials is that $H_v(\xi)$ is an even function of ξ if v is even and odd if v is odd. An even function is a function that satisfies

$$f(x) = f(-x) \quad \text{(even)} \tag{10.101}$$

and an odd function is a function that satisfies

$$f(x) = -f(-x) \quad \text{(odd)} \tag{10.102}$$

An even function like x^2 or $\cos x$ is symmetric when reflected across the x axis, and an odd function like x or $\sin x$ changes sign. Since $\exp(-\alpha x^2/2)$ in the wavefunctions for the harmonic oscillator is even, the even–odd character of the wavefunctions is determined by the Hermite polynomials. Thus, the harmonic oscillator wavefunctions are even when v is even and odd when v is odd. This even–odd character makes it easy to evaluate integrals, as shown in the following example.

Example 10.12

Show that the probability density for the zero-point level of the harmonic oscillator is in accord with the Heisenberg uncertainty principle.

To calculate the standard deviation σ_x for the x coordinate using equation 10.61, we need to calculate $\langle x \rangle$ and $\langle x^2 \rangle$:

$$\langle x \rangle = \int_{-\infty}^{\infty} \psi_0 x \psi_0 \, dx = 0$$

Since the wavefunction is real, $\psi_0^* = \psi_0$. Since ψ_0^2 is an even function and x is an odd function, the integrand is odd and so this integral is zero.

$$\langle x^2 \rangle = \int_{-\infty}^{\infty} \psi_0 x^2 \psi_0 \, dx$$

$$= \left(\frac{\alpha}{\pi}\right)^{1/2} \int_{-\infty}^{\infty} x^2 e^{-\alpha x^2} \, dx$$

$$= \left(\frac{\alpha}{\pi}\right)^{1/2} \left[\frac{1}{2\alpha}\left(\frac{\pi}{\alpha}\right)^{1/2}\right]$$

$$= \frac{1}{2\alpha} = \frac{1}{2}\frac{\hbar}{(\mu k)^{1/2}}$$

To calculate the standard deviation σ_p for the momentum using equation 10.61, we need to calculate $\langle p \rangle$ and $\langle p^2 \rangle$:

$$\langle p \rangle = \int_{-\infty}^{\infty} \psi_0 \left(-i\hbar \frac{d}{dx}\right) \psi_0 \, dx = 0$$

Since ψ_0 is an even function, its derivative is odd. Therefore, the integrand is odd, and the integral is zero:

$$\langle p^2 \rangle = \int_{-\infty}^{\infty} \psi_0 \left(-\hbar^2 \frac{d^2}{dx^2}\right) \psi_0 \, dx$$

$$= \left(\frac{\alpha}{\pi}\right)^{1/2} \int_{-\infty}^{\infty} e^{-\alpha x^2/2} \left(-\hbar^2 \frac{d^2}{dx^2}\right) e^{-\alpha x^2/2} \, dx$$

$$= \hbar^2 \left(\frac{\alpha}{\pi}\right)^{1/2} \int_{-\infty}^{\infty} (\alpha - \alpha^2 x^2) e^{-\alpha x^2} \, dx$$

$$= \frac{\hbar^2 \alpha}{2} = \frac{\hbar(\mu k)^{1/2}}{2}$$

Using equation 10.61, $\Delta x = \langle x^2 \rangle^{1/2}$, and $\Delta p_x = \langle p^2 \rangle^{1/2}$, we obtain

$$\Delta x \, \Delta p_x = \frac{\hbar}{2}$$

which is in accord with the Heisenberg uncertainty principle, and shows that in its ground state the harmonic oscillator has the minimum value of $\Delta x \, \Delta p_x$ allowed by the uncertainty principle.

10.13 The Rigid Rotor

A particle rotating around a fixed axis as shown in Fig. 10.10a has angular momentum and rotational kinetic energy. The kinetic energy of the revolving particle is given by $T = \frac{1}{2}mv^2 = p^2/2m$, but it is more convenient to express the kinetic energy in terms of the angular velocity ω. If the particle is rotating about a fixed point at a radius r with a frequency of ν, the velocity of the particle is given by

$$v = 2\pi r\nu = r\omega \tag{10.103}$$

where the angular velocity $\omega = d\theta/dt$ is equal to $2\pi\nu$. The frequency has units of s^{-1} or Hz. The angular velocity is expressed in radians per second.

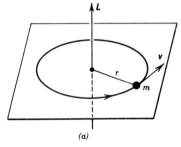

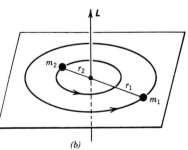

Figure 10.10 (a) Rotation of a mass about a fixed point. (b) Rotation of a diatomic molecule about its center of mass.

The kinetic energy of a particle in circular motion about a fixed point is usually expressed in terms of the angular velocity, or, as we will soon see, in terms of angular momentum:

$$T = \tfrac{1}{2}mv^2 = \tfrac{1}{2}mr^2\omega^2 = \tfrac{1}{2}I\omega^2 \qquad (10.104)$$

The **moment of inertia** I, which is introduced in this equation, is defined by

$$I = mr^2 \qquad (10.105)$$

for the rotation of a classical particle about an axis. In the expression for the rotational kinetic energy the moment of inertia plays the role that mass plays in the expression for the kinetic energy of the linear motion, and the angular velocity ω plays the role of the linear velocity v. This suggests that the angular momentum L should be defined as

$$L = I\omega = mvr = pr \qquad (10.106)$$

so that the kinetic energy of rotational motion can be expressed in terms of the angular momentum:

$$T = \tfrac{1}{2}I\omega^2 = \frac{L^2}{2I} \qquad (10.107)$$

just as translational kinetic energy can be expressed in terms of linear momentum. If no torque is applied, angular momentum is conserved.

Figure 10.10b shows a model of a rigid diatomic molecule that is an example of a rigid rotor. The two masses rotate about their center of mass, satisfying the condition that

$$r_1 m_1 = r_2 m_2 \qquad (10.108)$$

where r_1 is the distance of m_1 from the center of mass and r_2 is the distance of m_2 from the center of mass. The equilibrium distance R between the nuclei is $R = r_1 + r_2$ so that

$$r_1 = \frac{m_2}{m_1 + m_2}R \quad \text{and} \quad r_2 = \frac{m_1}{m_1 + m_2}R \qquad (10.109)$$

The rotational kinetic energy is

$$\begin{aligned}
T &= \tfrac{1}{2}m_1 r_1^2 \omega^2 + \tfrac{1}{2}m_2 r_2^2 \omega^2 \\
&= \tfrac{1}{2}(m_1 r_1^2 + m_2 r_2^2)\omega^2 \\
&= \tfrac{1}{2}I\omega^2 \qquad (10.110)
\end{aligned}$$

where I is the moment of inertia of the diatomic molecule:

$$I = m_1 r_1^2 + m_2 r_2^2 \qquad (10.111)$$

Using equation 10.109 to eliminate r_1 and r_2 we obtain

$$\begin{aligned}
I &= \frac{m_1 m_2}{m_1 + m_2}R^2 \\
&= \mu R^2 \qquad (10.112)
\end{aligned}$$

where μ is the reduced mass:

$$\frac{1}{\mu} = \frac{1}{m_1} + \frac{1}{m_2} = \frac{m_1 + m_2}{m_1 m_2} \qquad (10.113)$$

Since the moment of inertia for a single particle in circular motion is mR^2, we have reduced the two-body problem to a one-body problem by introducing the reduced mass.

The rotational kinetic energy of a diatomic molecule may also be written in terms of the angular momentum L

$$T = \frac{L^2}{2I} = \frac{L^2}{2\mu R^2} \tag{10.114}$$

by using equation 10.112. Because there is no potential energy, the classical Hamiltonian of a rigid rotor is just the kinetic energy. Applying the correspondence between the classical kinetic energy and the quantum mechanical Hamiltonian operator we obtain (see Table 10.1)

$$\hat{H} = -\frac{\hbar^2}{2\mu} \nabla^2 \tag{10.115}$$

for rotational energy, where the Laplacian operator ∇^2 was introduced in equation 10.65. In discussing rotation it is more convenient to use spherical polar coordinates that are defined in Fig. 10.11. In spherical polar coordinates the Laplacian operator is

$$\nabla^2 = \frac{1}{R^2} \frac{\partial}{\partial R}\left(R^2 \frac{\partial}{\partial R}\right) + \frac{1}{R^2 \sin^2 \theta} \frac{\partial^2}{\partial \phi^2} + \frac{1}{R^2 \sin \theta} \frac{\partial}{\partial \theta}\left(\sin \theta \frac{\partial}{\partial \theta}\right) \tag{10.116}$$

Since the two masses of the rigid rotor are at fixed distances from the origin, R is constant and so we can ignore the derivatives with respect to R in ∇^2. Substitution of equation 10.116 into equation 10.115 yields

$$\hat{H} = -\frac{\hbar^2}{2I}\left[\frac{1}{\sin \theta} \frac{\partial}{\partial \theta}\left(\sin \theta \frac{\partial}{\partial \theta}\right) + \frac{1}{\sin^2 \theta} \frac{\partial^2}{\partial \phi^2}\right] \tag{10.117}$$

where $I = \mu R^2$. The rigid rotor wavefunction is a function of the two angles θ and ϕ, and so the eigenvalue problem to be solved is

$$-\frac{\hbar^2}{2I}\left[\frac{1}{\sin \theta} \frac{\partial}{\partial \theta}\left(\sin \theta \frac{\partial}{\partial \theta}\right) + \frac{1}{\sin^2 \theta} \frac{\partial^2}{\partial \phi^2}\right] Y(\theta, \phi) = EY(\theta, \phi) \tag{10.118}$$

This equation is a standard differential equation, whose solutions $Y(\theta, \phi)$ are called **spherical harmonics.** The first several spherical harmonics are given in Table 10.2. Two quantum numbers, l and m, arise in the solution of this eigenvalue equation and so the wavefunctions are represented by $Y_l^m(\theta, \phi)$. It is found that

$$\hat{H}Y_l^m(\theta, \phi) = \frac{l(l + 1)\hbar^2}{2I} Y_l^m(\theta, \phi) \tag{10.119}$$

so that the rigid rotor can have only the energies given by

$$E = \frac{l(l + 1)\hbar^2}{2I} \qquad l = 0, 1, 2, \ldots \tag{10.120}$$

Note there is no zero-point energy for rotation. Unlike the harmonic oscillator the uncertainty principle can be satisfied even when the rotational energy is zero, because that wavefunction (Y_0^0) gives equal probability for all θ and ϕ (i.e., maximum uncertainty in those angles).

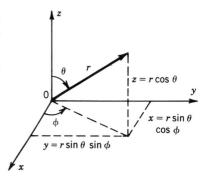

Figure 10.11 Relationship between Cartesian coordinates (x, y, z) and spherical polar coordinates (r, θ, ϕ).

Table 10.2   The First Several Spherical Harmonics

$$Y_0^0 = \frac{1}{(4\pi)^{1/2}}$$

$$Y_2^0 = \left(\frac{5}{16\pi}\right)^{1/2} (3\cos^2\theta - 1)$$

$$Y_1^0 = \left(\frac{3}{4\pi}\right)^{1/2} \cos\theta$$

$$Y_2^1 = \left(\frac{15}{8\pi}\right)^{1/2} \sin\theta \cos\theta\, e^{i\phi}$$

$$Y_1^1 = \left(\frac{3}{8\pi}\right)^{1/2} \sin\theta\, e^{i\phi}$$

$$Y_2^{-1} = \left(\frac{15}{8\pi}\right)^{1/2} \sin\theta \cos\theta\, e^{-i\phi}$$

$$Y_1^{-1} = \left(\frac{3}{8\pi}\right)^{1/2} \sin\theta\, e^{-i\phi}$$

$$Y_2^2 = \left(\frac{15}{32\pi}\right)^{1/2} \sin^2\theta\, e^{2i\phi}$$

$$Y_2^{-2} = \left(\frac{15}{32\pi}\right)^{1/2} \sin^2\theta\, e^{-2i\phi}$$

10.14   Angular Momentum

So far we have neglected the fact that the angular momentum is a vector and has components in the x, y, and z directions. To develop the quantum mechanical operators for the angular momentum in the x, y, and z directions, we need to review the classical expressions for angular momentum in three dimensions. In the preceding section we have seen that the rotational energy of a diatomic molecule may be expressed in terms of its angular momentum. In the next chapter we will find that an electron may have angular momentum. We will also find that electrons and certain nuclei have intrinsic (spin) angular momentum. We will not discuss spin angular momentum here, but will leave that to Section 11.4. Angular momentum is a very important property because it is conserved for many systems we will be interested in.

In classical mechanics the angular momentum of a particle rotating about a fixed point is represented by a vector L in the direction perpendicular to the plane of the circular motion. If a mass m is rotating about a fixed point with linear velocity v, the **angular momentum** L is given by the cross product of the radius r and the linear momentum vector p:

$$L = r \times mv = r \times p \tag{10.121}$$

The cross product of the two vectors r and p is a vector of magnitude $|r||p| \sin\theta$, where θ is the angle between r and p, having a direction that a right-hand screw would travel as r is rotated to p. Thus, the angular momentum vector for the circular motion shown in Fig. 10.10a points up.

The vectors r and p may be expressed in terms of their components and unit vectors i, j, and k pointing along the positive x, y, and z axes:

$$r = xi + yj + zk \tag{10.122}$$
$$p = p_x i + p_y j + p_z k \tag{10.123}$$

The cross product of r and p may conveniently be calculated from a determinant (see Appendix D):

$$L = r \times p = \begin{vmatrix} i & j & k \\ x & y & z \\ p_x & p_y & p_z \end{vmatrix}$$

$$= (yp_z - zp_y)i + (zp_x - xp_z)j + (xp_y - yp_x)k \qquad (10.124)$$

Thus, the three components of the classical angular momentum of a particle rotating about a fixed point are

$$L_x = yp_z - zp_y \qquad (10.125a)$$

$$L_y = zp_x - xp_z \qquad (10.125b)$$

$$L_z = xp_y - yp_x \qquad (10.125c)$$

The square of the angular momentum is given by the scalar product of L with itself:

$$L \cdot L = L^2 = L_x^2 + L_y^2 + L_z^2 \qquad (10.126)$$

The square of the angular momentum is a scalar (see the vector discussion in Appendix D). If no torque acts on a particle, its angular momentum is constant (conserved). In classical mechanics all possible values of L and E are permitted.

The quantum mechanical operators for the angular momentum are obtained by replacing the quantities in equations 10.125a, 10.125b, and 10.125c with their corresponding quantum mechanical operators; specifically $\hat{p}_x = (\hbar/i)$ $(\partial/\partial x)$, and so on.

$$\hat{L}_x = -i\hbar\left(y\frac{\partial}{\partial z} - z\frac{\partial}{\partial y}\right) \qquad (10.127a)$$

$$\hat{L}_y = -i\hbar\left(z\frac{\partial}{\partial x} - x\frac{\partial}{\partial z}\right) \qquad (10.127b)$$

$$\hat{L}_z = -i\hbar\left(x\frac{\partial}{\partial y} - y\frac{\partial}{\partial x}\right) \qquad (10.127c)$$

The operator for the square of the angular momentum is given by

$$\hat{L}^2 = |\hat{L}|^2 = \hat{L} \cdot \hat{L} = \hat{L}_x^2 + \hat{L}_y^2 + \hat{L}_z^2 \qquad (10.128)$$

It is often more convenient to use the angular momentum operators given in equations 10.127a, 10.127b, 10.127c, and 10.128 in spherical polar coordinates r, θ, ϕ, which are defined in Fig. 10.11. In these new coordinates

$$\hat{L}_x = i\hbar\left(\sin\phi\frac{\partial}{\partial\theta} + \cot\theta\cos\phi\frac{\partial}{\partial\phi}\right) \qquad (10.129a)$$

$$\hat{L}_y = i\hbar\left(-\cos\phi\frac{\partial}{\partial\theta} + \cot\theta\sin\phi\frac{\partial}{\partial\phi}\right) \qquad (10.129b)$$

$$\hat{L}_z = -i\hbar\frac{\partial}{\partial\phi} \qquad (10.129c)$$

$$\hat{L}^2 = -\hbar^2\left[\frac{1}{\sin\theta}\frac{\partial}{\partial\theta}\left(\sin\theta\frac{\partial}{\partial\theta}\right) + \frac{1}{\sin^2\theta}\frac{\partial^2}{\partial\phi^2}\right] \qquad (10.130)$$

Note that the terms in ∇^2 that depend on angle (equation 10.116) are equal to $-(1/\hbar^2R^2)\hat{L}^2$.

It is readily shown that $\hat{L}_x$ and $\hat{L}_y$, $\hat{L}_y$ and $\hat{L}_z$, and $\hat{L}_x$ and $\hat{L}_z$ do not commute with each other, but $\hat{L}_x$, $\hat{L}_y$, and $\hat{L}_z$ all commute with $\hat{L}^2$. Therefore, we can measure precisely the square of total angular momentum and one, but only one, of its components. Thus, if the magnitude of the total angular momentum $|L| = \sqrt{L^2} = \sqrt{L_x^2 + L_y^2 + L_z^2})$ is measured and L_z is measured, it is not possible to measure L_x or L_y precisely. That is, the eigenfunction of L^2 is also an eigenfunction of L_z but it is not an eigenfunction of L_x or L_y since neither L_x nor L_y commute with L_z. This is an essential difference between classical and quantum mechanical systems and is in accord with the Heisenberg uncertainty principle.

Since $\hat{L}^2$ and $\hat{L}_z$ commute it is possible to construct a function that is an eigenfunction of both operators. In fact we have already seen these wavefunctions, the spherical harmonics, in the eigenvalue equation (see 10.119) for the energy of the rigid rotor. Since $L^2 = 2IK$ for the classical rigid rotor, the quantum mechanical operator $\hat{L}^2$ is equal to $2I\hat{H}$. Thus, equation 10.119 can be written

$$\hat{L}^2 Y_l^m(\theta, \phi) = l(l + 1)\hbar^2 Y_l^m(\theta, \phi) \qquad l = 0, 1, 2, \ldots \qquad (10.131)$$

According to this eigenvalue equation the total angular momentum for a rigid rotor can only have the values

$$L^2 = l(l + 1)\hbar^2 \qquad l = 0, 1, 2, \ldots \qquad (10.132)$$

Operating on the spherical harmonics with $\hat{L}_z$

$$\hat{L}_z Y_l^m(\theta, \phi) = m\hbar Y_l^m(\theta, \phi) \qquad m = -l, -l + 1, \ldots, l - 1, l \qquad (10.133)$$

yields the following eigenvalues for the z component L_z of the angular momentum:

$$L_z = m\hbar \qquad m = -l, -l + 1, \ldots, l - 1, l \qquad (10.134)$$

where the quantum number l is referred to as the **angular momentum quantum number** (or azimuthal quantum number) and m is referred to as the **magnetic quantum number.** There are two quantum numbers because there are two coordinates, θ and ϕ. Equations 10.133 and 10.134 will recur a number of times in quantum mechanics. They will appear again in connection with the hydrogen atom, and again in several places in spectroscopy. Note that the spherical harmonics are eigenfunctions for two different commuting operators, $\hat{L}^2$ and $\hat{L}_z$.

The possible orientations of the angular momentum vector L with respect to a particular direction are shown in Fig. 10.12 for $l = 1$, and Fig. 10.13 for $l = 2$. In the absence of an external electric or magnetic field, the choice of the z axis is entirely arbitrary, but when such a field is applied (either by the experimenter or by placing the particle in a molecule or a crystal) a unique direction is defined which becomes the axis of quantization. Since the L_x and L_y components are unknown, L can only be described as being in the surface of a cone, as illustrated in the figure. The magnitude of the orbital angular momentum L is $[l(l + 1)]^{1/2}\hbar$, and the *maximum* component L_z in a particular direction is $l\hbar$. Thus, the magnitude of the angular momentum L is greater than its z component so that the angular momentum vector cannot point in the direction of an applied magnetic field or along a unique axis.

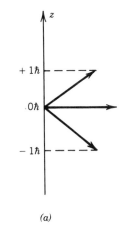

(a)

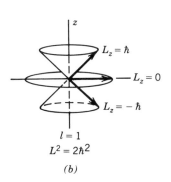

$l = 1$
$L^2 = 2\hbar^2$

(b)

Figure 10.12 (a) Possible orientations for angular momentum vector for $l = 1$. (b) Since the x and y components of angular momentum are indeterminant, the vector can be rotated about the z axis to lie anywhere on the conical surface.

In the absence of an electric or magnetic field, there is a degeneracy of $2l + 1$ since, for an angular momentum l, there are $2l + 1$ values of m.

In considering rotational energy levels of molecules, the rotational quantum number l is denoted by J so that

$$E = \frac{\hbar^2}{2I} J(J + 1) \tag{10.135}$$

The square of the total angular momentum is given by

$$L^2 = J(J + 1)\hbar^2 \qquad J = 0, 1, 2, \ldots \tag{10.136}$$

The magnetic quantum number m is replaced by M so that $L_z = M\hbar$.

The energy depends only on the quantum number J, but the wavefunction depends on J and M (see equations 10.131 and 10.133). Since the values of M range from $-J$ to J, the rotational levels are $(2J + 1)$ -fold degenerate. The degeneracy corresponds to the different possible orientations of the angular momentum vector.

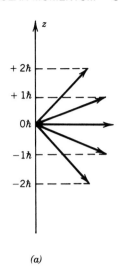

(a)

Example 10.13
What are the reduced mass and moment of inertia of $H^{35}Cl$? The equilibrium internuclear distance R_e is 127.5 pm. What are the values of L, L_z, and E for the state with $J = 1$? Atomic masses of some isotopes are given in the back cover.

$$\mu = \frac{(1.007\ 825 \times 10^{-3}\ \text{kg mol}^{-1})(34.968\ 85 \times 10^{-3}\ \text{kg mol}^{-1})}{[(1.007\ 825 + 34.968\ 85) \times 10^{-3}\ \text{kg mol}^{-1}](6.022\ 367 \times 10^{23}\ \text{mol}^{-1})}$$

$$= 1.626\ 65 \times 10^{-27}\ \text{kg}$$

$$I = \mu R_e^2$$

$$= (1.626 \times 10^{-27}\ \text{kg})(127.5 \times 10^{-12}\ \text{m})^2$$

$$= 2.644 \times 10^{-47}\ \text{kg m}^2$$

$$L = \sqrt{J(J + 1)}\hbar$$

$$= \frac{\sqrt{2}(6.626 \times 10^{-34}\ \text{J s})}{2\pi}$$

$$= 1.491 \times 10^{-34}\ \text{J s}$$

$$L_z = -\frac{\hbar}{2\pi}, 0, \frac{\hbar}{2\pi}$$

$$= -1.054 \times 10^{-34}\ \text{J s}, 0, 1.054 \times 10^{-34}\ \text{J s}$$

$$E = \frac{\hbar^2}{2I} J(J + 1)$$

$$= \frac{(6.626 \times 10^{-34}\ \text{J s})^2(2)}{8\pi^2(2.644 \times 10^{-47}\ \text{kg m}^2)}$$

$$= 4.206 \times 10^{-22}\ \text{J}$$

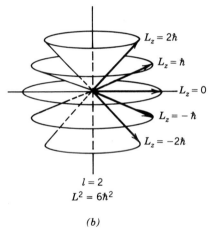

(b)

Figure 10.13 (a) Possible orientations for angular momentum vector for $l = 2$. (b) Since the x and y components of angular momentum are indeterminant, the vector can be rotated about the z axis to lie anywhere on the conical surface.

10.15   Postulates of Quantum Mechanics

A number of postulates of quantum mechanics have been introduced in this chapter as they have been needed. Now is the time to bring them together.

Postulate 1

The state of a quantum mechanical system is completely specified by a wavefunction $\Psi(r,t)$ that is a function of the coordinates of the particles and the time. If time is not a variable its state is completely specified by a time-independent wavefunction $\psi(r)$. These wavefunctions are single-valued, continuous, and square integrable. The wavefunction for a single particle may be interpreted as follows: $\Psi^*(r,t)\Psi(r,t)\,dx\,dy\,dz$ is the probability that the particle is in the volume $dx\,dy\,dz$ located at r at time t.

Postulate 2

For every observable in classical mechanics there is a linear quantum mechanical operator. The operator is obtained from the classical mechanical expression for the observable written in terms of Cartesian coordinates and conjugate momenta by replacing each coordinate q by itself and the conjugate momentum component by $-i\hbar\,\partial/\partial q$.

Postulate 3

The possible measured values of the physical observable A are the eigenvalues a_i of the equation

$$\hat{A}\Psi_i = a_i\Psi_i \tag{10.137}$$

where $\hat{A}$ is the operator corresponding with the observable.

Postulate 4

The average value of the observable corresponding with $\hat{A}$ is given by

$$\langle a \rangle = \int_{-\infty}^{\infty} \Psi^*\hat{A}\Psi\,d\tau \tag{10.138}$$

where Ψ is the normalized wavefunction for the state.

Postulate 5

The wavefunction of a system changes with time according to

$$\hat{H}\Psi(r,\,t) = i\hbar\,\frac{\partial\Psi(r,\,t)}{\partial t} \tag{10.139}$$

where $\hat{H}$ is the Hamiltonian operator for the system.

Postulate 6

The wavefunction of a system of electrons must be antisymmetric to the interchange of any two electrons. We will not discuss this postulate until the next chapter, but it is given here for completeness. This postulate arises in connection with spin and is a more fundamental statement of what is called the Pauli exclusion principle.

It is perhaps worth emphasizing again that we have not derived these postulates, but rather they are to be judged like the laws of classical mechanics or the laws of thermodynamics on the basis of whether or not they are in accord with experimental results.

10.16 Special Topic: The Time-Dependent Schrödinger Equation

To describe the evolution of a one-dimensional quantum mechanical system with time, it is necessary to use the time-dependent Schrödinger equation:

$$\hat{H}\Psi(x,\ t) = -\frac{\hbar}{i}\frac{\partial\Psi}{\partial t} \tag{10.140}$$

This form applies to a one-dimensional system. In general it is difficult to find solutions to the time-dependent Schrödinger equation. However, it is possible to separate variables for conservative systems, that is, systems in which the potential energy is a function of distance but not of time. For a conservative system the time-dependent Schrödinger equation is

$$-\frac{\hbar^2}{2m}\frac{\partial^2\Psi(x,\ t)}{\partial x^2} + V(x)\Psi(x,\ t) = -\frac{\hbar}{i}\frac{\partial\Psi}{\partial t} \tag{10.141}$$

Special solutions may be found by writing the wavefunction as a product of a function of distance and a function of time:

$$\Psi(x,\ t) = \psi(x)f(t) \tag{10.142}$$

Substituting this relation into equation 11.141 and dividing by $\psi(x)f(t)$ yields

$$-\frac{\hbar^2}{2m}\frac{1}{\psi(x)}\frac{d^2\psi(x)}{dx^2} + V(x) = -\frac{\hbar}{i}\frac{1}{f(t)}\frac{df(t)}{dt} \tag{10.143}$$

Since the left side of the equation does not depend on t and the right side of the equation does not depend on x, they must each be equal to a constant that is independent of x and t. If we take this constant to be equal to E, we obtain the following two differential equations:

$$-\frac{\hbar^2}{2m}\frac{1}{\psi(x)}\frac{d^2\psi(x)}{dx_2} + V(x) = E \tag{10.144}$$

$$-\frac{\hbar}{i}\frac{1}{f(t)}\frac{df(t)}{dt} = E \tag{10.145}$$

The first is the time-independent Schrödinger equation that we have been using, and we can see that the constant that was introduced is equal to the energy. The second equation is readily integrated to obtain, with the initial condition $\Psi(x,\ 0) = \psi(x)$,

$$f(t) = e^{-iEt/\hbar} \tag{10.146}$$

which oscillates harmonically in time. Thus, the time-dependent wavefunction for a conservative one-dimensional system in an eigenstate of H is

$$\Psi(x, t) = \psi(x)e^{-iEt/\hbar} \tag{10.147}$$

This wavefunction is complex, but the probability $\Psi^*(x, t)\Psi(x, t)$ is real:

$$\begin{aligned}\Psi^*(x, t)\Psi(x, t) &= [\psi(x)e^{-iEt/\hbar}]^*[\psi(x)e^{-iEt/\hbar}] \\ &= \psi^*(x)\psi(x)\end{aligned} \tag{10.148}$$

Thus, for a conservative system the probability density for an eigenstate of H is independent of time, and we refer to the system as being in a stationary state.

10.17   Special Topic: Tunneling and Reflection

When discussing the harmonic oscillator, we saw that the wavefunction can be nonzero in classically forbidden regions of space. This is a general property associated with wave phenomena and hence with quantum mechanics. An important consequence of this is that there is a finite probability that a particle can pass through a potential energy barrier in quantum mechanics even though a classical particle would be unable to do so. This is called tunneling and can be illustrated by considering the situation in Fig. 10.14 of a particle of energy E moving in one dimension and encountering a barrier whose height is V such that $V > E$.

In the region to the left of the barrier the Schrödinger equation for the particle is

$$-\frac{\hbar^2}{2m}\frac{d^2}{dx^2}\phi_L = E\,\phi_L \tag{10.149}$$

with a general solution

$$\phi_L = A\,e^{+ikx} + B\,e^{-ikx} \qquad k^2 = \frac{2mE}{\hbar^2} \tag{10.150}$$

The term A represents the amplitude of the incoming wave, while the term B represents that of the reflected wave. In the barrier $(0 < x < a)$, the Schrödinger equation is

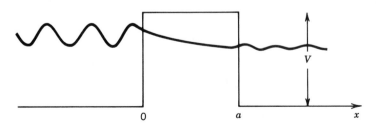

Figure 10.14   Wavefunction for a particle of energy E incident from the left on a potential barrier of height V and width a.

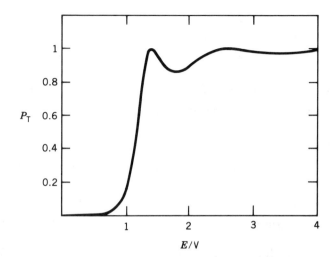

Figure 10.15 Probability of a proton tunneling through a 10-pm-wide potential barrier of height 1 eV.

$$-\frac{\hbar^2}{2m}\frac{d^2}{dx^2}\phi_M + V\phi_M = E\phi_M \tag{10.151}$$

with a general solution

$$\phi_M = A\,e^{+\kappa x} + B\,e^{-\kappa x} \qquad \kappa^2 = \frac{2m(V-E)}{\hbar^2} \tag{10.152}$$

To the right of the barrier the wavefunction is

$$\phi_R = F\,e^{+ikx} \tag{10.153}$$

since the particle is incident from the left so there is no reflected wave here. The amplitude for tunneling through the barrier is then F/A and the probability of tunneling P_T is then $|F/A|^2$. By making the wavefunction and its derivative continuous at the boundaries of the potential step, we can solve for F/A. For barriers such that $\kappa a \gg 1$, P_T is proportional to $e^{-2\kappa a}$, which is small but nonzero. In Fig. 10.15, the probability of a proton tunneling through a barrier of height 1 eV and width 10 pm is shown.

Tunneling processes are important for light particles such as electrons and protons. They are much less important for heavier particles, since κ is proportional to the square root of mass.

Another quantum mechanical phenomenon is the reflection of particles (or waves) from a barrier even when the energy is higher that the barrier. For particles incident from the left on the barrier of Fig. 10.14 the probability of transmission for $E > V$ is also given in Fig. 10.15.

References

P. W. Atkins, *Molecular Quantum Mechanics*. New York: Oxford University Press, 1983.

R. E. Christofferson, *Basic Principles and Techniques of Molecular Quantum Mechanics*. New York: Springer, 1989.

C. Cohen-Tannoudjii, B. Diu, and F. Laloë, *Quantum Mechanics*. New York: Wiley, 1977.

R. Eisberg and R. Resnick, *Quantum Physics*, 2d ed. New York: Wiley, 1985.

H. Haken and H. C. Wolf, *Atomic and Quantum Physics*. New York: Springer, 1987.

M. W. Hanna, *Quantum Mechanics in Chemistry*. New York: Benjamin, 1980.

M. Karplus and R. N. Porter, *Atoms and Molecules*. New York: Benjamin, 1970.

I. N. Levine, *Quantum Chemistry*. Boston: Allyn & Bacon, 1983.

D. A. McQuarrie, *Quantum Chemistry*. Mill Valley, CA: University Science Books, 1983.

L. Pauling and E. B. Wilson, *Introduction to Quantum Mechanics*. New York: McGraw-Hill, 1935.

R. N. Zare, *Angular Momentum*. New York: Wiley, 1988.

Problems

10.1   If the exitance, M_λ (equation 10.5) is integrated over all wavelengths, the total radiant energy emitted per second per square meter, I, is equal to σT^4. The constant σ is called the Stefan–Boltzmann constant and is equal to 5.67×10^{-8} J s^{-1} m^{-2} K^{-4}.

(a)   What is the total radiant energy emitted per second per m^2 at 2000 K?

(b)   If all the radiant energy coming through a 1-cm^2 pinhole from a cavity at 2000 K were to be absorbed by 1 g of water at 373.15 K, how long will it take to vaporize the water?

10.2

(a)   The distribution of wavelengths from a certain star peaks in the visible at $\lambda = 600$ nm. Assuming that the distribution obeys the Planck distribution law, use Wien's displacement law to estimate the temperature of the star.

(b)   A metal bar is heated to "red-heat" so that its radiation peaks at $\lambda = 800$ nm. Estimate the temperature of the bar.

10.3   Calculate the energy per photon and the number of photons emitted per second from

(a)   a 100-W yellow light bulb ($\lambda = 550$ nm)

(b)   A 1-kW microwave source ($\lambda = 1$ cm)

10.4

(a)   Derive the value of the constant in the Wien displacement law (equation 10.2) in terms of h, c, and k.

(b)   If, from experiment, the values of h and c were measured to be 6.6×10^{-34} J s^{-1} and 3.0×10^8 m s^{-1}, and the value of the constant in the Wien displacement law were measured to be 2.9×10^{-3} K m^{-1}, find the value of k from (a). Since R is measured to be 8.3 J K^{-1} mol^{-1}, you can also calculate the Avogadro constant.

10.5   Calculate the wavelengths in mm of the first three lines of the Paschen series for atomic hydrogen.

10.6   Calculate the wavelength of light emitted when an electron falls from the $n = 100$ orbit to the $n = 99$ orbit of the hydrogen atom. Such species are known as high Rydberg atoms. They are detected in astronomy and are more and more studied in the laboratory.

10.7   In the photoelectric effect an electron is emitted from a metal as the result of absorption of a photon of light. Part of the energy of the photon is required to release the electron from the metal; this energy ϕ is called the work function or binding energy. The kinetic energy of the ejected electron is given by

$$\tfrac{1}{2}mv^2 = h\nu - \phi$$

where m and v are the mass and velocity of the electron. For the 100 face of silver metal (see Chapter 23) the velocity of electrons emitted using 200-nm photons is 7.42×10^5 m s^{-1}. Calculate the work function of this face in eV.

10.8 Photoelectron spectroscopy utilizes the photoelectron effect to measure the binding energy of electrons in molecules and solids, by measuring the kinetic energy of the emitted electrons and using the relation in problem 10.7 between kinetic energy, wavelength, and binding energy. One variant of this is photoelectron spectroscopy is X-ray photoelectron spectroscopy (XPS). If the X-ray wavelength is 0.2 nm, calculate the velocity of electrons emitted from molecules in which the binding energies are 10, 100, and 500 eV.

10.9 Electrons are accelerated by a 1000-V potential drop. (*a*) Calculate the de Broglie wavelength. (*b*) Calculate the wavelength of the X-rays that could be produced when these electrons strike a solid.

10.10 An ultraviolet photon ($\lambda = 58.4$ nm) from a helium gas discharge tube is absorbed by a hydrogen molecule which is at rest. Since momentum is conserved, what is the velocity of the hydrogen molecule after absorbing the photon? What is the translational energy of the hydrogen molecule in J mol^{-1}?

10.11 What is the de Broglie wavelength of an oxygen molecule at room temperature? Compare this to the average distance between oxygen molecules in a gas at 1 bar at room temperature.

10.12 What is the de Broglie wavelength of a thermal neutron at 300 K?

10.13 Calculate the de Broglie wavelengths of the following:

(*a*) a 1-g bullet with velocity 300 m s^{-1}

(*b*) a 10^{-6}-g particle with velocity 10^{-6} m s^{-1}

(*c*) a 10^{-10}-g particle with velocity 10^{-10} m s^{-1}

(*d*) a H_2 molecule with energy of $\frac{3}{2} kT$ at $T = 20$ K

10.14 The lifetime of a molecule in a certain electronic state is 10^{-10} s. What is the uncertainty in energy of this state? Give the answer in J and in J mol^{-1}.

10.15 For a particle in a one-dimensional box, the ground-state wavefunction is

$$\phi = \left(\frac{2}{a}\right)^{1/2} \sin \frac{\pi x}{a}$$

(*a*) What is the probability that the particle is in the right half of the box? (*b*) What is the probability that the particle is in the middle third of the box?

10.16 Show that the function $f = 8e^{5x}$ is an eigenfunction of the operator d/dx. What is the eigenvalue?

10.17 What are the results of operating on the following functions with the operator d/dx and d^2/dx^2: (*a*) e^{-ax^2}, (*b*) cos bx, (*c*) e^{ikx}? Which functions are eigenfunctions of these operators? What are the corresponding eigenvalues?

10.18 Show that the operators for the x coordinate and for the momentum in the x direction p_x do not commute. Calculate the operator representing the commutator of x and p_x.

10.19 (*a*) Calculate the energy levels for $n = 1, 2$, and 3 for an electron in a potential well of width 0.25 nm with infinite barriers on either side. The energies should be expressed in kJ mol^{-1}. (*b*) If an electron makes a transition from $n = 2$ to $n = 1$ what will be the wavelength of the emitted radiation?

10.20 For a helium atom in a one-dimensional box calculate the value of the quantum number of the energy level for which the energy is equal to $\frac{3}{2} kT$ at 25 °C (*a*) for a box 1 nm long, (*b*) for a box 10^{-6} m long, and (*c*) for a box 10^{-2} m long.

10.21 Show that the wavefunctions for a particle in a one-dimensional box are orthogonal using

$$\sin \alpha \sin \beta = \tfrac{1}{2} \cos (\alpha + \beta) - \tfrac{1}{2} \cos (\alpha + \beta)$$

10.22 Calculate the degeneracies of the first three levels for a particle in a cubical box.

10.23

(*a*) Later in Table 14.4, we will find that the following molecules have the indicated vibrational frequencies:

$$^{35}Cl_2 \ (560 \ cm^{-1}) \qquad ^{39}K^{35}Cl \ (281 \ cm^{-1})$$
$$^1H_2 \ (4401 \ cm^{-1})$$

What are the force constants for these molecules if we treat them as harmonic oscillators?

(*b*) Assuming that the force constant for $^{37}Cl_2$ is the same as for $^{35}Cl_2$, predict the fundamental vibrational frequency of $^{37}Cl_2$.

10.24 Check the normalization of ψ_0 and ψ_1 for the harmonic oscillator and show they are orthogonal.

10.25 Substitute the $v = 1$ eigenfunction for the harmonic oscillator into the Schrödinger equation for the harmonic oscillator, and obtain the expression for the eigenvalue (energy).

10.26 In the vibrational motion of HI, the iodine atom essentially remains stationary because of its large mass. Assuming that the hydrogen atom undergoes harmonic motion and that the force constant k is 317 N m^{-1}, what is the fundamental vibration frequency ν_0? What is ν_0 if H is replaced by D?

10.27 What are the expectation values for $\langle x \rangle$ and $\langle x^2 \rangle$ for a quantum mechanical harmonic oscillator in the $v = 1$ state? What is the standard deviation Δx?

10.28 What are the reduced mass and moment of inertia of $^{23}Na^{35}Cl$? The equilibrium internuclear distance R_e is 236 pm. What are the values of E for the states with $J = 1$ and $J = 2$.

10.29 The commutator of two operators, $\hat{A}$ and $\hat{B}$, is defined as the operator $\hat{A}\hat{B} - \hat{B}\hat{A}$. Using the definitions of $\hat{L}_x$, $\hat{L}_y$, and $\hat{L}_z$ given in equation 10.127, find the commutator of $\hat{L}_x$ with $\hat{L}_y$ and $\hat{L}_x$ with $\hat{L}_z$.

10.30 The $^{12}C^{16}O$ molecule has an equilibrium bond distance of 112.8 pm. Calculate (a) the reduced mass and (b) the moment of inertia. (c) Calculate the wavelength of the photon emitted when the molecule makes the transition from $l = 1$ to $l = 0$ using equation 10.120 for the energy levels.

10.31 Using the Planck distribution law, equation 10.1, find the *frequency* of maximum emission as a function of temperature.

10.32

(a) Using the result of problem 10.31, find the maximum emission frequency at 3000 K and at 10 000 K.

(b) Using the Wien displacement law, equation 10.3, find the maximum emission wavelength at the same two temperatures.

10.33 Calculate the frequency and the wavelength in nanometers for the line in the Paschen series of the hydrogen spectrum that is due to a transition from the sixth quantum level to the third.

10.34 There is a Brackett series in the hydrogen spectrum where $n_1 = 4$. Calculate the wavelengths in nm of the first two lines of this series.

10.35

(a) Calculate the de Broglie wavelength of an electron accelerated by a potential of 1000 V.

(b) Calculate the de Broglie wavelength of a proton accelerated by a potential of 1000 V.

10.36 Calculate the de Broglie wavelength for thermal neutrons at a temperature of 100 °C.

10.37 What is the momentum of a photon with a wavelength of 500 nm? If this photon is absorbed by a $^{35}Cl_2$ molecule that is at rest, what will be the velocity of the Cl_2 molecule after absorbing the photon?

10.38 An atom makes a transition from an excited state with a lifetime of 10^{-9} s to the ground state and emits a photon with a wavelength 600 nm. What is the uncertainty in the energy of the excited state? What is the percentage uncertainty if the energy is measured from the ground state?

10.39 Since a wavefunction may be complex, it may be represented by

$$\psi = R + iI$$

where $i = \sqrt{-1}$. Show that $\psi^*\psi$ is always real.

10.40 (a) Show that the function $\psi = xe^{-ax^2}$ is an eigenfunction of the operator $d^2/dx^2 - 4a^2x^2$. What is the eigenvalue? (b) Show that the function $\psi = Ke^{r/k}$ is an eigenfunction of the operator d/dr. What is the eigenvalue?

10.41 Derive the expression for the energy of a particle in a one-dimensional box using the de Broglie formula.

10.42 Calculate the first three energy levels, in kJ mol^{-1}, for an electron in a potential well 0.5 nm in width with infinitely high potential outside.

10.43 For a particle in a cubical box calculate $E(8ma^2/h^2)$ for the first 10 states. What is the degeneracy for each energy level?

10.44 Assume the form $\Psi_0 = N_0 e^{-ax^2}$ for the ground-state wavefunction of the harmonic oscillator, and substitute this into the Schrödinger equation. Find the value of a that makes this an eigenfunction.

10.45 Figure 10.9 shows that the quantum mechanical harmonic oscillator can be in regions forbidden to a *classical* harmonic oscillator.

(a) Find an integral expression for the probability that the particle will be in the classically forbidden region for the ground state.

(b) Using tables for Gaussian integrals or error functions (e.g., Abramowitz and Stegun, *Handbook of Mathematical Functions.* New York: Dover, 1964) or a numerical integration program on a microcomputer, compute the value of the integral in (a).

10.46 The expression for the energy levels of a quantum mechanical three-dimensional harmonic oscillator is obtained by expressing the potential energy by

$$V = \tfrac{1}{2}k_x x^2 + \tfrac{1}{2}k_y y^2 + \tfrac{1}{2}k_z z^2$$

where k_x, k_y, and k_z are the force constants in the three directions. What is the expression for the quantum mechanical energy levels and what is the zero-point energy?

10.47 For the three-dimensional *isotropic* harmonic oscillator, $k_x = k_y = k_z$ (see problem 10.46) Write the formula for the energy levels. Give the energies and degeneracies of the first 10 states.

10.48 Using the definitions of L_z and L^2 from equations 10.129c and 10.130 show that these two operators commute.

10.49 Sketch the possible orientation for angular momentum vectors for $l = \frac{3}{2}$.

10.50 Sketch a figure like Fig. 10.13 for $l = 3$.

10.51 What is the reduced mass of $^{14}N^{16}O$? What is its moment of inertia if $R_e = 151$ pm? Using equation 10.120, find the energies of the first three levels of rotational motion.

10.52 Check the normalization and orthogonality of Y_0^0 and Y_l^m, $m = 0, 1, -1$ in Table 10.2.

11
Atomic Structure

This chapter introduces the electronic structure of atoms. The electronic wavefunctions for an atom contain all the information about the electronic properties of the atom. The wavefunctions for the hydrogen atom and the one-electron atoms, such as He^+ and Li^{2+}, can be calculated exactly, but approximate methods have to be used with atoms having two or more electrons. Fortunately, these approximate methods can yield quite precise results, but at the cost of complicated calculations carried out on a computer.

We will find that an additional postulate has to be added to the nonrelativistic quantum theory discussed in the preceding chapter—the Pauli exclusion principle.

We will consider the electronic structure of the hydrogenlike atoms in some detail because of the importance of the concept of orbitals that this introduces. The concept of the atomic orbital and the representation of many-electron systems by products of atomic orbitals will be useful for molecules as well as atoms.

One of the great triumphs of quantum mechanics has been the insight it has provided into the structure of the periodic table and therefore of the periodicity in the physical and chemical properties of the elements.

Only certain transitions between energy levels can occur in the absorption or emission of electromagnetic radiation. The rules governing these transitions are called selection rules, and we will discuss these restrictions on atomic spectra at the end of this chapter.

11.1 The Schrödinger Equation for Hydrogenlike Atoms

A hydrogenlike atom is an atom with atomic number Z and only one electron, such as H, He$^+$, Li^{2+}, Be^{3+}, $\cdots$, U^{91+}. In such an atom, the electron interacts with the nuclear Coulomb potential,

$$V(r) = \frac{-Ze^2}{4\pi\epsilon_0 r} \tag{11.1}$$

where ϵ_0 is the **permittivity of vacuum,** and r is the distance from the nucleus to the electron. Note that the potential energy of interaction between the two oppositely charged particles is negative, since the interaction is zero when they are infinitely far apart and is negative (attractive) for smaller r.

The Hamiltonian for a hydrogenlike atom contains a term for the kinetic energy of the center of mass and a term for the kinetic energy of the relative motion. Since we are not concerned with the motion of the center of mass, we can take the atom to be fixed in space. The relative motion of the two particles can then be represented by the use of the reduced mass, $\mu = m_e m_N/(m_e + m_N)$ (see equation 10.113), in a one-particle Schrödinger equation. Notice that since $m_N \gg m_e$, then $\mu \simeq m_e$ in the H atom. (You should convince yourself that the difference between μ and m_e in the H atom is about 0.05%.) The Hamiltonian can then be written as

$$\hat{H} = -\frac{\hbar^2}{2\mu}\nabla^2 - \frac{Ze^2}{4\pi\epsilon_0 r} \tag{11.2}$$

Since the Coulomb potential has spherical symmetry (i.e., it is independent of angle and depends only on the distance r), it is convenient to express $\hat{H}$ in spherical polar coordinates (Fig. 10.11). The form of ∇^2 was given in equation 10.116 and we have already noted in Section 10.14 that the angle-dependent terms in ∇^2 can be represented as $(-1/\hbar^2 r^2)\hat{L}^2$ (equation 10.116), where $\hat{L}$ is the **angular momentum operator.** Thus,

$$\nabla^2 = \frac{1}{r^2}\frac{\partial}{\partial r}\left(r^2\frac{\partial}{\partial r}\right) - \frac{1}{r^2}\frac{\hat{L}^2}{\hbar^2} \tag{11.3}$$

which, when substituted into the Schrödinger equation and multiplied by $2\mu r^2$, gives

$$-\hbar^2\frac{\partial}{\partial r}\left(r^2\frac{\partial\psi}{\partial r}\right) + \hat{L}^2\psi - 2\mu r^2\left(\frac{Ze^2}{4\pi\epsilon_0 r} + E\right)\psi = 0 \tag{11.4}$$

Notice that the angular dependence of the operator is contained in $\hat{L}^2$, **whose eigenvalues and eigenfunctions we already know** (section 10.14). This immediately suggests that we try to solve equation 11.4 by separation of variables by writing $\Psi(r, \theta, \phi) = R_{nl}(r)\, Y_l^m(\theta, \phi)$, where $R_{nl}(r)$ is called the hydrogenlike radial wavefunction and satisfies the differential equation

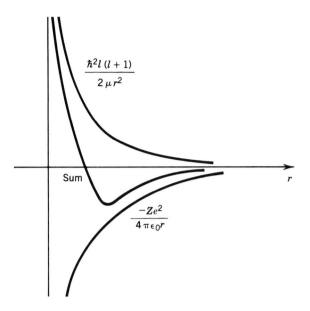

Figure 11.1 Coulomb potential (proportional to $1/r$), centrifugal potential (proportional to $l(l + 1)/r^2$) and the sum or effective potential for a hydrogenlike atom with $l > 0$.

$$-\left(\frac{\hbar^2}{2\mu}\right) \frac{d}{dr}\left(r^2 \frac{d}{dr}R_{nl}\right) + \left[\frac{\hbar^2 l(l + 1)}{2\mu r^2} - \frac{Ze^2}{4\pi\epsilon_0 r} - E\right] R_{nl} = 0 \quad (11.5)$$

A simpler equation can be found by making the substitution $R_{nl}(r) = S_{nl}(r)/r$. Then, after multiplying the entire equation by r, we find

$$-\left(\frac{\hbar^2}{2\mu}\right) \frac{d^2}{dr^2} S_{nl} + \left[\frac{\hbar^2 l(l + 1)}{2\mu r^2} - \frac{Ze^2}{4\pi\epsilon_0 r} - E\right] S_{nl} = 0 \quad (11.6)$$

which looks like the equations we have seen before when studying the harmonic oscillator, for example. Notice that the term proportional to $l(l + 1)/r^2$ (called the centrifugal potential) adds to the Coulomb potential to give an effective potential for $l > 0$, as shown in Fig. 11.1. Solutions to this equation can be found by expanding $S_{nl}(r)$ in a power series in r, and requiring that the wavefunction vanish as r gets large. These functions will represent the **bound** states of the electron in this atom and will have $E < 0$. The mathematical procedure is straightforward but lengthy. It is done carefully in a number of advanced books (see the references at the end of this chapter, for example, Pauling and Wilson). The result of this calculation is that E can have only certain values given by

$$E_n = -\frac{\mu e^4 Z^2}{2(4\pi\epsilon_0)^2 \hbar^2 n^2} \quad (11.7)$$

with the principal quantum number $n = 1,2,3, \ldots, \infty$. Note that E_n does not depend on l; however, in solving the equation, it is found that $n > l$, so that for $n = 1$, l can only be 0, while for $n = 2$, l can be 0 and 1, and so on. The low-lying energy levels are shown in Fig. 11.2.

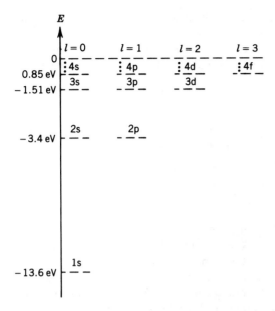

Figure 11.2   Electronic energy levels for the hydrogen atom.

Example 11.1

You may want to get more of a feel for the solutions of equation 11.5. A simple procedure is to guess a solution and see if it works. For example, pick $R = e^{-\gamma r}$ and find the conditions to make this a solution of equation 11.6.

$$\frac{1}{r^2}\frac{d}{dr}\left(r^2 \frac{de^{-\gamma r}}{dr}\right) = \frac{d}{r^2 dr}(-\gamma r^2 e^{-\gamma r}) = \left(\frac{-2\gamma r + \gamma^2 r^2}{r^2}\right)e^{-\gamma r}$$

Thus,

$$-\frac{\hbar^2}{2\mu}\left(\frac{-2\gamma r + \gamma^2 r^2}{r^2}\right)e^{-\gamma r} + \left[\frac{\hbar^2 l(l+1)}{2\mu r^2} - \frac{Ze^2}{4\pi\epsilon_0 r} - E\right]e^{-\gamma r} = 0$$

For the terms multiplying $1/r^2$ and $1/r$ to vanish for all r, we must have $l = 0$ and

$$\frac{\gamma \hbar^2}{\mu} = \frac{Ze^2}{4\pi\epsilon_0 r}$$

and then

$$E = -\frac{\hbar^2 \gamma^2}{2\mu} = -\frac{\mu Z^2 e^4}{2(4\pi\epsilon_0)^2 2\hbar^2}$$

This is the $n = 1$ eigenfunction and energy. Now try it with $R = re^{-\gamma r}$!

From equation 11.7, we can see that atoms with larger Z have more strongly bound electrons (more negative energy for a given n). In addition, we see that the lowest energy state of the atom is the $n = 1$ state and all others are excited states.

As we said above, the reduced mass is almost equal to m_e; it would be exactly m_e if the nucleus had infinite mass. In this case,

$$E_\infty = -\frac{Z^2}{n^2} (13.605\ 80\ \text{eV}) \qquad (11.8)*$$

When a hydrogenlike atom makes a transition from a state with $n = n_2$ to a state with $n = n_1$, light is emitted ($n_2 > n_1$) or absorbed ($n_1 > n_2$). The frequency of the light is given by

$$\nu = \frac{E_{n_2} - E_{n_1}}{h} = RZ^2\, c\left(\frac{1}{n_1^2} - \frac{1}{n_2^2}\right) \qquad (11.9)$$

where R is the Rydberg constant, equal to $1.096\ 775\ 856 \times 10^7\ \text{m}^{-1}$ for hydrogen and $1.097\ 373\ 153\ 4 \times 10^7\ \text{m}^{-1}$ for an infinite nuclear mass. This formula accurately predicts the experimental spectrum of hydrogenlike systems and was one of the great early triumphs of quantum mechanics (Section 10.2). From the formula for the energy levels, we also predict that the energy needed to ionize a hydrogen atom is 13.6 eV because the electron must make a transition from $n = 1$ to $n = \infty$ (unbound). The ionization energy of He$^+$ is 4×13.6 eV since $Z = 2$.

The wavefunctions of hydrogenlike systems have the form $R_{n\ell}(r)\ Y_\ell^m(\theta,\phi)$ indicating that for $n > 1$ there are a number of different wavefunctions with the same energy. For example if $n = 2$, we can have $l = 0$ (and $m = 0$) or $l = 1$ (and $m = +1, 0, -1$). In fact, as we saw in Section 10.14, the allowed values of m for a given l are

$$m = -l, -l + 1, \ldots, l - 1, l \qquad (11.10)$$

The value of l determines the orbital angular momentum, and m determines the Z component of angular momentum (equation 10.134). For historical reasons, we associate letter symbols with the values of l:

$$
\begin{array}{lcccc}
l = & 0 & 1 & 2 & 3 \\
\text{Symbol} = & s & p & d & f
\end{array} \qquad (11.11)
$$

In summary, there are three quantum numbers for each eigenfunction of a hydrogenlike atom which can take on the following values:

$$n = 1, 2, 3, \ldots \qquad (11.12)$$
$$l = 0, 1, 2, \ldots, n - 1 \qquad (11.13)$$
$$m = 0, \pm1, \pm2, \ldots, \pm l \qquad (11.14)$$

Therefore, for a given value of n, there are n^2 degenerate eigenstates. Later, when we discuss electron spin angular momentum, which can have 2 values ($\hbar/2$ or $-\hbar/2$), we will see that the **degeneracy** is actually $2n^2$ for an energy level with principal quantum number n.

It is often better to express energies and distances for atomic problems in units for which these quantities take on numbers close to 1. We therefore

---

* 1 eV is the work done in moving an electron through 1 V and is $1.602\ 177\ 33 \times 10^{-19}$ J.

sometimes use the atomic unit of distance, called the **Bohr radius** a_0, which is the most probable distance of the electron from the nucleus in the ground state of the hydrogen atom. We shall soon see that $a_0 = \hbar^2(4\pi\epsilon_0)/(m_e e^2)$. The atomic unit of energy is the potential energy of two electrons separated by a Bohr radius,

$$E_h = \frac{e^2}{4\pi\epsilon_0 a_0} \tag{11.15}$$

and is called the **hartree**. Both Bohr and Hartree were atomic physicists of the early 20th century.

Example 11.2
Calculate the values of a_0 and E_h.

$$a_0 = \frac{h^2(4\pi\epsilon_0)}{4\pi^2 m_e e^2} = \frac{h^2 \epsilon_0}{\pi^2 m_e e^2} = 52.917\ 724\ 9 \text{ pm}$$

$$E_h = \frac{e^2}{4\pi\epsilon_0 a_0} = \cdots = 4.359\ 748\ 2 \times 10^{-18} \text{ J} = 27.211\ 396\ 1 \text{ eV}$$

(Note that the energy of the H atom in its ground state is $E_h/2$.)

Example 11.3
Show that the bound-state eigenfunctions of a hydrogenlike atom has the factor e^{-Zr/na_0} by examining the Schrödinger equation for large r.

For large r, we neglect all terms in the Schrödinger equation that depend on r^{-1} and r^{-2} so that

$$-\frac{\hbar^2}{2m_e} \frac{d^2 R_{nl}}{dr^2} \sim E_n R_{nl}$$

whose solution is

$$R_{nl} \sim \exp - [-ZE_n m_e/\hbar^2]^{1/2} r$$

Using the form for E_n (equation 11.7) and the form for a_0 given in example 11.2, we find

$$R_{nl} \sim \exp - \left\{ \frac{m_e^2 e^4 Z^2}{(4\pi\epsilon_0)^2 \hbar^4 n^2} \right\}^{1/2} r = \exp\left(-\frac{Zr}{na_0} \right)$$

11.2 Eigenfunctions and Probability Densities for Hydrogenlike Atoms

The eigenfunctions for the hydrogenlike atoms have the form shown just above equation 11.5. A wavefunction for a one-electron system is called an **orbital**. For an atomic system such as H, it is called an atomic orbital. These wave-

Table 11.1   Real Hydrogenlike Wavefunctions[a]

n	l	m	Wavefunction
1	0	0	$\psi_{1s} = \dfrac{1}{\sqrt{\pi}}\left(\dfrac{Z}{a_0}\right)^{3/2} e^{-\sigma}$
2	0	0	$\psi_{2s} = \dfrac{1}{4\sqrt{2\pi}}\left(\dfrac{Z}{a_0}\right)^{3/2}(2 - \sigma)e^{-\sigma/2}$
2	1	0	$\psi_{2p_z} = \dfrac{1}{4\sqrt{2\pi}}\left(\dfrac{Z}{a_0}\right)^{3/2}\sigma e^{-\sigma/2}\cos\theta$
2	1	± 1	$\psi_{2p_x} = \dfrac{1}{4\sqrt{2\pi}}\left(\dfrac{Z}{a_0}\right)^{3/2}\sigma e^{-\sigma/2}\sin\theta\cos\phi$
			$\psi_{2p_y} = \dfrac{1}{4\sqrt{2\pi}}\left(\dfrac{Z}{a_0}\right)^{3/2}\sigma e^{-\sigma/2}\sin\theta\sin\phi$
3	0	0	$\psi_{3s} = \dfrac{1}{81\sqrt{3\pi}}\left(\dfrac{Z}{a_0}\right)^{3/2}(27 - 18\sigma + 2\sigma^2)e^{-\sigma/3}$
3	1	0	$\psi_{3p_z} = \dfrac{\sqrt{2}}{81\sqrt{\pi}}\left(\dfrac{Z}{a_0}\right)^{3/2}(6 - \sigma)\sigma e^{-\sigma/3}\cos\theta$
3	1	± 1	$\psi_{3p_x} = \dfrac{\sqrt{2}}{81\sqrt{\pi}}\left(\dfrac{Z}{a_0}\right)^{3/2}(6 - \sigma)\sigma e^{-\sigma/3}\sin\theta\cos\phi$
			$\psi_{3p_y} = \dfrac{\sqrt{2}}{81\sqrt{\pi}}\left(\dfrac{Z}{a_0}\right)^{3/2}(6 - \sigma)\sigma e^{-\sigma/3}\sin\theta\sin\phi$
3	2	0	$\psi_{3d_{z^2}} = \dfrac{1}{81\sqrt{6\pi}}\left(\dfrac{Z}{a_0}\right)^{3/2}\sigma^2 e^{-\sigma/3}(3\cos^2\theta - 1)$
3	2	± 1	$\psi_{3d_{xz}} = \dfrac{\sqrt{2}}{81\sqrt{\pi}}\left(\dfrac{Z}{a_0}\right)^{3/2}\sigma^2 e^{-\sigma/3}\sin\theta\cos\theta\cos\phi$
			$\psi_{3d_{yz}} = \dfrac{\sqrt{2}}{81\sqrt{\pi}}\left(\dfrac{Z}{a_0}\right)^{3/2}\sigma^2 e^{-\sigma/3}\sin\theta\cos\theta\sin\phi$
3	2	± 2	$\psi_{3d_{x^2-y^2}} = \dfrac{1}{81\sqrt{2\pi}}\left(\dfrac{Z}{a_0}\right)^{3/2}\sigma^2 e^{-\sigma/3}\sin^2\theta\cos 2\phi$
			$\psi_{3d_{xy}} = \dfrac{1}{81\sqrt{2\pi}}\left(\dfrac{Z}{a_0}\right)^{3/2}\sigma^2 e^{-\sigma/3}\sin^2\theta\sin 2\phi$

[a] $\sigma = \dfrac{Z}{a_0}r.$

functions for the hydrogenlike atoms through $n = 3$ are given in Table 11.1.[*]
It is convenient to write these equations in terms of the Bohr radius a_0.

To visualize the nature of these functions it is helpful to consider the radial function and the spherical harmonics separately.

---

[*] The spherical harmonics Y_l^m involve complex functions except for those with $m = 0$. In considering angular dependence it is advantageous to deal with real functions. This may be achieved by taking appropriate linear combinations of spherical harmonics with the same value of l. This has been done in obtaining the wavefunctions in Table 11.1. For example,

$$\psi_{2p_x} = \frac{1}{\sqrt{2}}(Y_1^1 + Y_1^{-1})R_{2,1}$$

$$\psi_{2p_y} = \frac{1}{\sqrt{2}}(Y_1^1 - Y_1^{-1})R_{2,1}$$

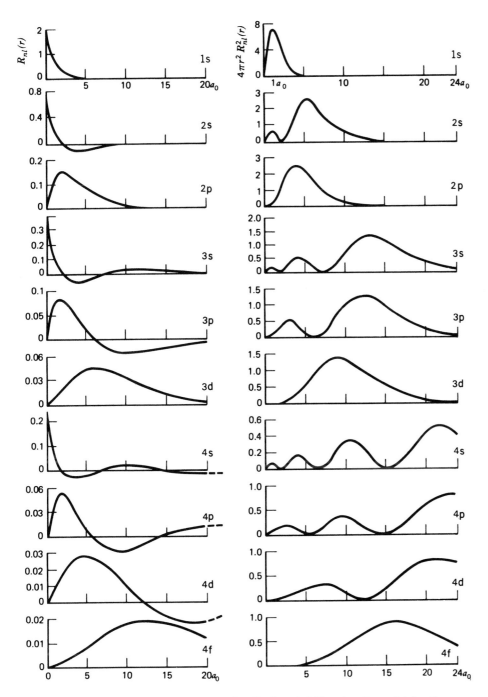

Figure 11.3 Radial function $R_{nl}(r)$ and radical probability density $p_{nl}(r)$ for the hydrogen atom. (From D. A. Davies, *Waves, Atoms and Solids*. London: Longman, 1978.)

The **radial functions** $R_{nl}(r)$ for the hydrogenlike atoms depend on the principal quantum number n, the azimuthal quantum number l, and the atomic number Z. The normalized radial functions for the hydrogen atom are shown in Fig. 11.3. The radial function always contains the factor e^{-Zr/na_0}, where n

is the principal quantum number. As Z is increased, the amplitude of the wave-function falls off more rapidly with increasing r, indicating that the electron is attracted more closely to the higher positively charged nuclei. The radial functions have $n - l - 1$ zero values between $r = 0$ and ∞. These produce spherical nodal surfaces in ψ so that the electron density goes to zero at these surfaces. The existence of nodes is required so that, for example, the 1s and 2s and other orbitals will be orthogonal (Section 10.8); that is

$$\int \psi_{1s}\psi_{2s}\, d\tau = 0 \tag{11.16}$$

where $d\tau$ represents the element of volume.

Probability densities are more useful for visualizing the electronic structure of atoms than are wavefunctions. It is obvious from Fig. 11.3 that $[R_{nl}(r)]^2$ is a maximum at the nucleus for s orbitals. However, if we are interested in the probability that the orbital electron is a certain distance from the nucleus, we need another approach. To calculate this probability density we need to take the product of $[R_{nl}(r)]^2$ and the volume of the spherical shell $4\pi r^2\, dr$ that has a radius of r. This is the **radial probability density** $p_{nl}(r)$ defined by

$$p_{nl}(r) = 4\pi r^2 R_{nl}^2(r) \tag{11.17}$$

The radial probabilities $p(r)$ for a number of orbitals of the hydrogen atom are given in Fig. 11.3. For the hydrogen atom in a 1s orbital, the highest radial probability density occurs at $r = a_0$, the Bohr radius. Thus, a_0 is the most probable radius for a 1s hydrogen atom.

Example 11.4
Find the maximum for the radial probability density $p(r)$ for a hydrogen atom in the 1s state, and thereby show that the radius with the highest radial probability density is a_0.

$$p(r) = 4\pi r^2 \frac{1}{\pi a_0^3} e^{-2r/a_0}$$

$$\frac{dp(r)}{dr} = \frac{8\pi r}{\pi a_0^3} e^{-2r/a_0} - \frac{8\pi r^2}{\pi a_0^4} e^{-2r/a_0} = 0$$

$$1 - \frac{r}{a_0} = 0$$

$$r = a_0$$

As the principal quantum number increases, the electron moves out to greater distances from the nucleus. The average distance for an orbital electron may be expressed by the **expectation value** $\langle r \rangle$:

$$\langle r \rangle_{nl} = \int_0^\infty r p_{nl}(r)\, dr \tag{11.18}$$

$$= \frac{n^2 a_0}{Z}\left\{ 1 + \frac{1}{2}\left[1 - \frac{l(l+1)}{n^2} \right] \right\} \tag{11.19}$$

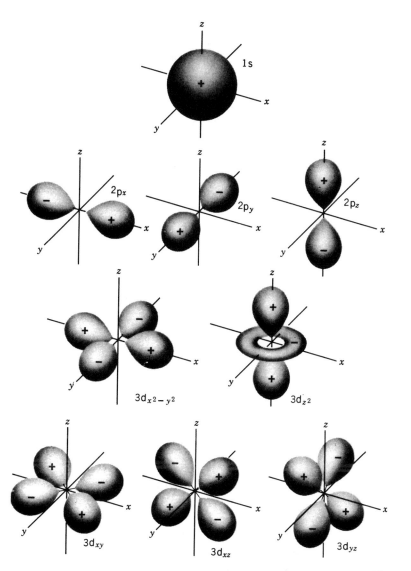

Figure 11.4 Contour surfaces for constant $\Psi^*\Psi$ for one-electron atoms. The indicated signs are those of the wavefunctions. These signs are indicated because they will be of interest later when we discuss molecular orbitals. The probability density is, of course, always positive.

The expectation value for the radius for a hydrogenlike atom in the 1s state is $3a_0/2Z$. The most probable value of r and the expectation value of r are not equal.

The probability density for the electron in a hydrogenlike atom is given by

$$\psi^*\psi = |R(r)\Theta(\theta)\Phi(\phi)|^2 \qquad (11.20)$$

where $Y_l^m(\theta, \phi)$ has been written $\Theta(\theta)\Phi(\phi)$ and where the symbols on the right represent the product of the wavefunction with its complex conjugate. The presentation of electron density as a function of r, θ, and ϕ would require four dimensions. One way to do this is to use the density of dots to represent the probability of finding an electron in a region of space and use stereo plots with a stereo viewer to see probability densities in three dimensions.* The method that is used most frequently is to depict surfaces that enclose some large percentage, say 90%, of the electron density. Diagrams of this type are shown in Fig. 11.4. The probability densities in Fig. 11.4 are all positive because they are squares of the wavefunctions. The signs indicated are those of the wavefunctions themselves before squaring.

The orientations of the p orbitals can be calculated by considering the magnitudes and signs of the trigonometric functions at several angles. In the absence of an electric or magnetic field, electrons in p_x, p_y, and p_z, orbitals all have the same energy; in fact, the energy depends only on the total quantum number n. For the hydrogen atom with quantum number $n = 2$ there are four degenerate states, all having the same energy in the absence of a magnetic or electric field.

The p orbitals do not have to point along the x, y, and z directions. Linear combinations of p_x, p_y, and p_z may be formed to point in any three mutually perpendicular directions. It will be seen in Chapter 12 that the directional character of certain chemical bonds results from the directed orientation of these and other orbitals. Note that opposite lobes of p orbitals have opposite signs.

There are five independent d orbitals. The $3d_{z^2}$ orbital has two large regions of electron density above one axis, by convention the z axis, and a small donut-shaped orbital in the xy plane. The other four d orbitals have four equivalent lobes of electron density with two nodal planes separating them. Note that lobes that are opposite each other in these wavefunctions have the same sign.

One of the deficiencies in the diagrams in Fig. 11.4 for higher wavefunctions is that the nodal surfaces resulting from the radial functions $R(r)$ are not shown. The number of radial nodes for an s orbital is $n - 1$.

11.3 Orbital Angular Momentum of the Hydrogenlike Atom

A hydrogenlike atom may have orbital angular momentum, depending on its angular momentum quantum number l. As we saw in equation 10.131, the square of the magnitude of the angular momentum is obtained by operating on the hydrogenlike wavefunction with the operator $\hat{L}^2$ for the square of the angular momentum. Since the wavefunction is $R(r)Y_l^m(\theta, \phi)$ and $\hat{L}^2$ operates only on θ and ϕ, we obtain equation 10.131, which was discussed earlier in connection with the rigid rotor. Thus, the angular momentum of a hydrogenlike atom can only have the values

$$L = \sqrt{l(l + 1)}\hbar \qquad l = 0, 1, 2,\ldots \qquad (11.21)$$

as given earlier in equation 10.132.

Operating on the spherical harmonics with the operator for the z component of the angular momentum $\hat{L}_z$, as we did in equation 10.133, yields the expression for the z component L_z of the angular momentum in terms of the magnetic quantum number m that we have seen before in equations 10.134 and 11.10:

$$L_z = m\hbar \qquad m = -l, -l + 1, \ldots, l - 1, l \qquad (11.22)$$

---

* D. T. Cromer, *J. Chem. Educ.* **45**:626 (1968).

The possible orientations of the angular momentum vectors of a hydrogenlike atom with $l = 1$ and $l = 2$ are given in Figs. 10.12 and 10.13. In the absence of a magnetic field, and *not including the spin of the electron,* which also is a kind of angular momentum, the energy of the hydrogenlike atom is independent of m. However, in the presence of a magnetic field, the energy depends on m, or the orientation of the angular momentum vector. In fact, the energies of the $2l + 1$ eigenstates with different values of m are all different. We say that the magnetic field has removed the $2l + 1$ degeneracy with respect to m. (Actually, the electron has an intrinsic angular momentum, called spin, in addition to the orbital angular momentum. For the present discussion, we will ignore the spin and treat the hydrogen atom as if the only angular momentum is from orbital motion.) The reason that the degeneracy (see the discussion on degeneracy in Section 10.10) is removed in a magnetic field is that when an atom has angular momentum L, the atom acts like a small magnet. We say that is has a **magnetic dipole moment** μ given by

$$\mu = \gamma_e L \tag{11.23}$$

where γ_e is the **gyromagnetic ratio** of the electron, equal to $-e/2m_e$. The z component of the dipole moment is then given by

$$\mu_z = -\frac{e}{2m_e} L_z \tag{11.24}$$

or in an eigenstate of L_z with eigenvalue $m\hbar$,

$$\mu_z = -\left(\frac{e\hbar}{2m_e}\right)m = -\mu_B m \tag{11.25}$$

where μ_B is called the **Bohr magneton:**

$$\mu_B = e\hbar/2m_e \tag{11.26}$$

which is the natural unit of magnetic dipole moment for electronic states.

When a magnetic dipole is placed in a magnetic field oriented along a given direction, the potential energy is given by

$$E = -\mu \cdot B \tag{11.27}$$

where B is the magnetic flux density. Since we are free to choose the z direction anyway we like, we can pick it to be along B so that $B_z = B$, the magnitude of B. Then,

$$E = -\mu_z B = \frac{eB}{2m_e} L_z \tag{11.28}$$

The Hamiltonian operator for the atom in a magnetic field is then found by adding this potential to $\hat{H}_0$, the Hamiltonian in the absence of the field:

$$\hat{H} = \hat{H}_0 + \frac{eB}{2m_e} \hat{L}_z \tag{11.29}$$

When this is applied to the eigenfunctions of a hydrogenlike atom (that is, the eigenfunctions of $\hat{H}_0$), we find that these functions are eigenfunctions of $\hat{H}$ with eigenvalues

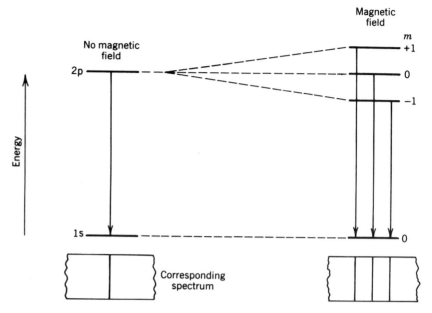

Figure 11.5   Splitting of the 2p level of the hydrogen atom into three closely spaced levels by a magnetic field assuming for illustrative purposes that the electron spin is absent. The spectra for the transition from the 2p level to the 1s are shown in the absence and presence of a magnetic field (From D. A. McQuarrie, *Quantum Chemistry*. Mill Valley, CA: University Science Books, 1983. Reprinted with permission from University Science Books.)

$$E_{nlm} = -\frac{m_e e^4 Z^2}{2(4\pi\epsilon_0)^2 n^2 \hbar^2} + \mu_B m B \qquad (11.30)$$

with $n = 1, 2, \ldots$; $l = 0, 1, \ldots, n - 1$; and $m = l, l - 1, \ldots, -l$. Therefore, in the presence of a magnetic field the energy levels have been split into $2l + 1$ levels (see Fig. 11.5).

The splitting of energy levels in a magnetic field can be studied by measuring the magnetic susceptibility (see Section 16.1), by electron paramagnetic resonance (EPR, Section 16.12), and by optical spectroscopy in a magnetic field. The splitting of spectral lines due to a magnetic field is called the Zeeman effect. When the splitting has a contribution due to the intrinsic (spin) magnetic moment of the electrons, to which we now turn, it is called the anomalous Zeeman effect (for historical reasons).

Example 11.5

For a hydrogenlike atom with a 3d electron, what is the value of the orbital angular momentum? When the atom is placed in a magnetic field of 1 T, what are the relative values of the possible energy levels?

$$L = \sqrt{2(2 + 1)}\hbar = \sqrt{6}\hbar$$

$$L_z = -2\hbar, -\hbar, 0, \hbar, 2\hbar$$

$$E = -\mu_B m B_z$$

$$= -(9.274 \times 10^{-24} \text{ J T}^{-1})(1 \text{ T})m$$

where $m = 0, \pm 1, \pm 2$.

11.4 Electron Spin

The optical spectra of hydrogen atoms and alkali atoms show a fine structure that is not accounted for by the quantum theory that we have described so far. To explain these effects Goudsmit and Uhlenbeck proposed in 1925 that an electron has an **intrinsic angular momentum** S and accompanying **magnetic dipole moment** μ_s. They further proposed that the z component S_z of this spin angular momentum is equal to $m_s\hbar$, where m_s is a fourth quantum number that can have the values $+\frac{1}{2}$ or $-\frac{1}{2}$. In 1928 Dirac showed that spin arises in the relativistic quantum theory for a one-electron system. In nonrelativistic quantum theory for a one-electron system, electron spin must be treated as an additional postulate. The term electron spin is used, but it is not really correct to think of the intrinsic angular momentum of the electron as being due to a spinning motion of the electron mass on its axis.

Since the spin angular momentum of an electron has no analog in classical mechanics, we cannot construct spin angular momentum operators by first writing the classical Hamiltonian. However, it turns out that the treatment of spin angular momentum is closely analogous to the treatment of orbital angular momentum. The magnitudes of the spin angular momentum S and its z component S_z are given by

$$S = \sqrt{s(s + 1)}\hbar = \frac{\sqrt{3}}{2}\hbar \qquad \text{since } s = \frac{1}{2} \qquad (11.31)$$

$$S_z = m_s\hbar \qquad m_s = \pm\frac{1}{2} \qquad (11.32)$$

where the **spin quantum number** s (generally referred to simply as the spin) of the electron has the single value $\frac{1}{2}$, and the quantum number m_s for the z component has two possible eigenvalues, $\pm\frac{1}{2}$. We speak of these two quantum numbers as "spin up" and "spin down."

We have introduced five new quantities in rapid succession. To keep them straight it will help to remember that these quantities and the operators $\hat{S}$ and $\hat{S}_z$ are comparable with quantities introduced for angular momentum. This comparison is shown in Table 11.2.

When the operators $\hat{S}^2$ and $\hat{S}_z$ are applied to spin functions, they yield eigenvalues. Since the spin eigenfunctions do not involve spatial coordinates, the two possible spin functions for an electron are represented by α and β:

$$\hat{S}^2\alpha = \tfrac{1}{2}(\tfrac{1}{2} + 1)\hbar^2\alpha = \tfrac{3}{4}\hbar^2\alpha \qquad (11.33)$$

Table 11.2   Quantities for Representing Orbital Angular Momentum and Spin Angular Momentum for a Single Electron

	Orbital Angular Momentum	Spin Angular Momentum
Angular momentum vector	L	S
Magnitude of above	$L = \sqrt{l(l+1)}\hbar$	$S = \sqrt{s(s+1)}\hbar$
Component of angular momentum vector in z direction	$L_z = m\hbar$	$S_z = m_s\hbar$
Operator for square of angular momentum	$\hat{L}^2$	$\hat{S}^2$
Operator for z component of angular momentum	$\hat{L}_z$	$\hat{S}_z$
Quantum number	$l(=0, 1, 2, \ldots)$	$s(=\frac{1}{2})$
Quantum number for z component	$m(=-l, -l+1, \ldots, 0, \ldots, +l-1, +l)$	$m_s(=\pm\frac{1}{2})$
Magnetic dipole moment vector	$\boldsymbol{\mu}_l$	$\boldsymbol{\mu}_s$

$$\hat{S}^2\beta = \tfrac{1}{2}(\tfrac{1}{2}+1)\hbar^2\beta = \tfrac{3}{4}\hbar^2\beta \tag{11.34}$$

$$\hat{S}_z\alpha = +\tfrac{1}{2}\hbar\alpha \tag{11.35}$$

$$\hat{S}_z\beta = -\tfrac{1}{2}\hbar\beta \tag{11.36}$$

At this level of approximation, the operators $\hat{S}^2$ and $\hat{S}_z$ commute with the Hamiltonian operator $\hat{H}$, $\hat{L}^2$, and $\hat{L}_z$ so that the magnitude of the spin, the z component of the spin, the energy, the magnitude of the orbital angular momentum, and the z component of the orbital angular momentum can all have simultaneous eigenvalues (see Example 10.6).

A *complete* wavefunction for a hydrogenlike atom must indicate the spin state of the electron. Since there are two spin functions, there are twice as many wavefunctions for the hydrogen atom as we indicated earlier: $\psi\alpha$ and $\psi\beta$ for each of the previous ψ's. Thus, a complete state specification for the hydrogen atom requires the four quantum numbers n, l, m, and m_s. This increases the degeneracy of the energy levels from n^2 to $2n^2$.

The two possible orientations for the spin angular momentum vector for an electron in a magnetic field are shown in Fig. 11.6. The vector has magnitude $S = [\tfrac{1}{2}(1+\tfrac{1}{2})]^{1/2}\hbar$. Its component S_z may be $+\tfrac{1}{2}\hbar$ or $-\tfrac{1}{2}\hbar$. The components S_x and S_y cannot be determined simultaneously with S_z, since S_x and S_y do not commute with S_z, just as L_x and L_y do not commute with L_z.

Because of its charge and intrinsic spin angular momentum, an electron has a magnetic dipole moment $\boldsymbol{\mu}_s$. As with orbital angular momentum (equation 11.23), the magnetic moment of an electron is proportional to its spin angular momentum S:

$$\boldsymbol{\mu}_s = -\frac{g_e e}{2m_e}S \tag{11.37}$$

where g_e is the electron g factor, which is 2.002 322.

The component μ_z of the magnetic moment of the electron in the direction of the applied magnetic field is

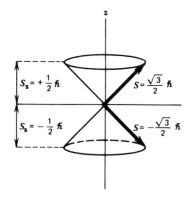

Figure 11.6   Possible orientations for the spin angular momentum S of the electron in a magnetic field. The magnitude S of the spin angular momentum is $[s(s+1)]^{1/2}\hbar$, where s is the spin quantum number. Thus, $S = (\sqrt{3}/2)\hbar$. The z component S_z of the spin angular momentum vector is $m_s\hbar$, where the spin quantum number m_s for the z is $\pm\frac{1}{2}$.

$$\mu_z = -\frac{g_e e}{2m_e} S_z \tag{11.38}$$

where S_z is the component of the spin angular momentum in the direction of the field. Since S_z is given by $m_s \hbar$,

$$\mu_z = -\frac{g_e e \hbar}{2m_e} m_s$$

$$= -g_e \mu_B m_s \tag{11.39}$$

The energy of the spin magnetic moment in a magnetic field B is (see equation 11.28)

$$E = g_e \mu_B m_s B \tag{11.40}$$

The electron spin has two energy states in a magnetic field: $E = +\frac{1}{2} g_e \mu_B B$ and $-\frac{1}{2} g_e \mu_B B$. The transition between these two levels is studied in electron spin resonance.

This energy is the eigenvalue of the operator $\hbar^{-1} g_e \mu_B B \, \hat{S}_z$ or $g_e(eB/2m_e) \, \hat{S}_z$. When this is added to the Hamiltonian of a hydrogenlike atom in a magnetic field, equation 11.29, we find that the total Hamiltonian is

$$\hat{H} = \hat{H}_0 + \frac{eB}{2m_e} \hat{L}_z + \frac{g_e eB}{2m_e} \hat{S}_z$$

$$= \hat{H}_0 + \frac{eB}{2m_e} (\hat{L}_z + g_e \hat{S}_z) \tag{11.41}$$

The eigenvalues of $\hat{H}$ in a magnetic field are now given by

$$E_{nlmm_s} = -\frac{m_e e^4 Z^2}{2(4\pi\epsilon_0)^2 \hbar^2 n^2} + \frac{eB\hbar}{2m_e}(m + g_e m_s) \tag{11.42}$$

In Fig. 11.7, we show the splitting of the 1s and 2p orbital states in a hydrogenlike system and the possible spectral transitions. Since g_e is very close to

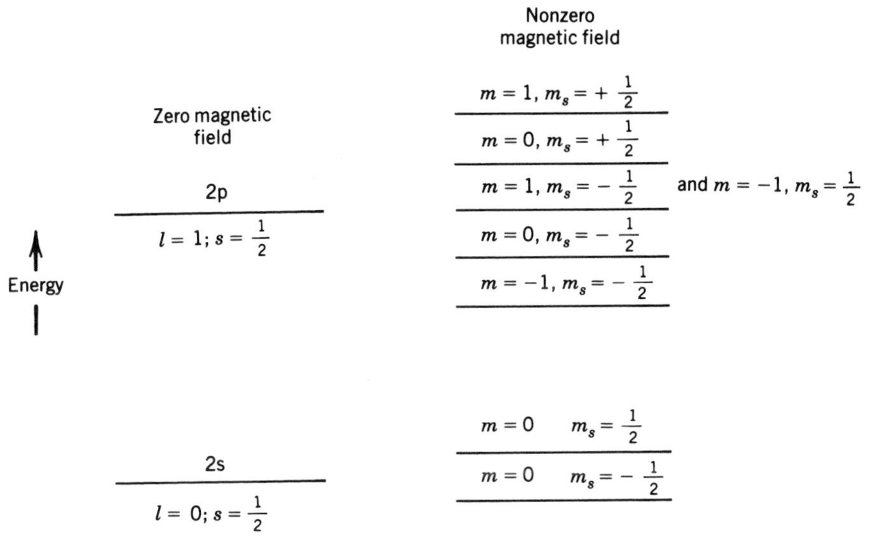

Figure 11.7 The energies of 2s and 2p states of a hydrogenlike atom in a magnetic field including orbital and spin angular momentum.

2, the state with $m = 1$ and $m_s = -\frac{1}{2}$ and the state with $m = -1$ and $m_s = +\frac{1}{2}$ are approximately degenerate.

In general, there is another term in the Hamiltonian which is due to the interaction between the spin and orbital parts of the angular momentum and is called the spin–orbit coupling. When this is important, we must define the total angular momentum of the atomic system, $J = L + S$. We will discuss this briefly at the end of this chapter (Section 11.14).

11.5 Variation Method

Although the Schrödinger equation can be solved for certain simple systems like the particle in a box, the harmonic oscillator, the rigid rotor, and the hydrogenlike atom, it cannot be solved for many electron atoms or molecules. It is therefore necessary to use approximation methods of which the variation method is one of the most important because it allows us to calculate an upper bound for the energy eigenvalue.

The proof of the variation principle depends on the fact that any function that has the same boundary conditions (e.g., vanishing as $r \to \infty$) as the eigenfunctions of the Hamiltonian operator, can be written as a linear combination of the eigenfunctions. Therefore, any function ψ can be written as

$$\psi = \sum c_n \phi_n \tag{11.43}$$

where ϕ_n are the eigenfunctions of H with eigenvalue E_n, and c_n are coefficients. Since ψ is normalized, and the ϕ_n are orthogonal to one another (see Section 10.8); then

$$\sum c_n^* c_n = 1 \tag{11.44}$$

The expectation value of H using ψ is given by

$$\langle H \rangle = \sum c_n^* c_n E_n \tag{11.45}$$

Subtracting $E_0 = (\sum c_n^* c_n) E_0$ from 11.45, where E_0 is the lowest eigenvalue, we find

$$\langle H \rangle - E_0 = \sum c_n^* c_n (E_n - E_0) \tag{11.46}$$

But every term on the right side is ≥ 0, so

$$\langle H \rangle \geq E_0 \tag{11.47}$$

This is the variation theorem: **The expectation value of H using any normalized wavefunction is an upper bound to the lowest eigenvalue.** To obtain an upper bound for an excited-state eigenvalue, ψ must be orthogonal to all the eigenfunctions for lower energy states.

In the variation method, a trial function ψ with one or more variable parameters in it is chosen, and the expectation value of H is calculated. This is then minimized by varying the parameters, to yield the lowest possible expectation value. As more variable parameters are added to ψ, a better approximation to the energy is found, but computation becomes more complex. In principle, as one increases the flexibility of the trial function, one gets closer and closer to the true eigenfunction.

Example 11.6

Use the variation method to obtain an upper bound to the ground-state energy of a particle in a one-dimensional box and compare the result with the true value given in equation 10.39. A normalized trial variation function that satisfies the boundary conditions that $\psi = 0$ at $x = 0$ and $x = l$ is

$$\psi = \frac{\sqrt{30}}{l^{5/2}}x(l - x)$$

$$\hat{H} = -\frac{\hbar^2}{2m}\frac{d^2}{dx^2} \qquad 0 \leqslant x \leqslant l$$

$$\int_0^l \psi^*\hat{H}\psi \, d\tau = -\frac{30\hbar^2}{l^5 2m}\int_0^l (lx - x^2)\frac{d^2}{dx^2}(lx - x^2) \, dx$$

$$= -\frac{30\hbar^2}{l^5 m}\int_0^l (x^2 - lx) \, dx = \frac{5h^2}{4\pi^2 ml^2}$$

$$\frac{5h^2}{4\pi^2 ml^2} \geq E_0$$

The true value is $h^2/8ml^2$.

$$\% \text{ error} = \frac{(5/4\pi^2) - (1/8)}{(1/8)} \times 100 = 1.3\%$$

This is not a true variational calculation because there is no variable parameter, but it does provide an example that an approximate wavefunction always yields a higher energy than the true wavefunction. We will use the variation method in the next section and in Section 12.3.

11.6   Helium Atom

The helium atom has two electrons, and the coordinates used in writing the Hamiltonian are shown in Fig. 11.8. Since we are interested only in the internal motions of the electrons with respect to the nucleus in a heliumlike atom, we may ignore the kinetic energy of the nucleus, and so the Hamiltonian operator may be written

$$\hat{H} = -\frac{\hbar^2}{2m_e}(\nabla_1^2 + \nabla_2^2) - \frac{1}{4\pi\epsilon_0}\left(\frac{Ze^2}{r_1} + \frac{Ze^2}{r_2} - \frac{e^2}{r_{12}}\right) \qquad (11.48)$$

The first term is the kinetic energy operator for the two electrons, the next two terms represent the potential energy of each of the electrons in the field of the nucleus, and the last term represents the interelectronic potential energy.

When this Hamiltonian operator is used in the Schrödinger equation $\hat{H}\psi = E\psi$, we can in principle obtain the eigenfunctions ψ and corresponding eigenvalues E. The wavefunctions ψ are functions of the coordinates of the two electrons $(x_1, y_1, z_1, x_2, y_2, z_2)$. The interelectronic repulsion term $e^2/4\pi\epsilon_0 r_{12}$ makes it impossible to obtain an analytic solution of the Schrödinger equation for a heliumlike atom; thus, it is necessary to use approximation methods to obtain the wavefunctions and energies. Fortunately, in the case of heliumlike

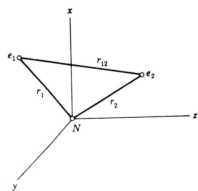

Figure 11.8   Coordinates of the electrons in the helium atom.

atoms these approximation methods yield the wavefunctions and energies to any desired degree of accuracy.

As a first approximation we will ignore the $e^2/4\pi\epsilon_0 r_{12}$ term. This amounts to treating the electrons as if they do not interact with each other. It may be shown* that if a system consists of two particles that do not interact with each other, the Hamiltonian operator for the whole system is the sum of the Hamiltonian operators for the two particles:

$$\hat{H} = \hat{H}_1 + \hat{H}_2 \tag{11.49}$$

Thus, assuming that the two electrons do not interact, the wavefunction for the system can be written as the product of the wavefunctions ψ_1 and ψ_2 for the two independent particles, so that the Schrödinger equation can be divided into two separate equations for the two electrons:

$$\hat{H}_1\psi_1 = E_1\psi_1 \tag{11.50}$$
$$\hat{H}_2\psi_2 = E_2\psi_2 \tag{11.51}$$

The energy of the whole system is given by the sum of the energies of the two electrons:

$$E = E_1 + E_2 \tag{11.52}$$

The Hamiltonian operators for the two electrons are

$$\hat{H}_1 = -\frac{\hbar^2}{2m}\nabla_1^2 - \frac{Ze^2}{4\pi\epsilon_0 r_1} \tag{11.53}$$

$$\hat{H}_2 = -\frac{\hbar}{2m}\nabla_2^2 - \frac{Ze^2}{4\pi\epsilon_0 r_2} \tag{11.54}$$

and the wavefunctions ψ_1 and ψ_2 obtained in this way are hydrogenlike wavefunctions (Table 11.1).

In this first approximation the wavefunction of the ground state of the helium atom is given by the product of the two 1s orbitals:

$$\psi = \frac{1}{\pi^{1/2}}\left(\frac{Z}{a_0}\right)^{3/2}e^{-Zr_1/a_0}\frac{1}{\pi^{1/2}}\left(\frac{Z}{a_0}\right)^{3/2}e^{-Zr_2/a_0} \tag{11.55}$$

This is generally abbreviated $\psi = 1s(1)1s(2)$ where the (1) and (2) refer to electrons 1 and 2.

The trial wavefunction 11.55 can be tested by using it to calculate the approximate energy of the helium atom with respect to the nucleus and two electrons infinitely distant from one another. The exact energy for this is the negative of the experimental energy required to remove the two electrons from the helium atom, which is 79.0 eV. The approximate wavefunction (11.55) is an eigenfunction of $\hat{H}_1 + \hat{H}_2$ with eigenvalue equal to twice the 1s orbital energy of helium ion or $2E_{1s}(Z = 2) = 8(-13.6) = -108.8$ eV. The repulsion energy of the two electrons in the approximate wavefunction (11.55) is given by

$$\int dv_1 \int dv_2\, \psi^*\left(\frac{e^2}{4\pi\epsilon_0 r_{12}}\right)\psi \tag{11.56}$$

*I. N. Levine, *Quantum Chemistry*. Boston: Allyn & Bacon, 1983, p. 104.

When this is calculated it has a value of 37.6 eV, so that the total approximate energy of the helium atom is -71.2 eV compared to the exact value of -79.0 eV. Notice that the approximate value is larger than the true value, as is required by the variation theorem.

A better approximate wavefunction can be found by replacing the nuclear charge Z in the wavefunction (11.55) by an effective nuclear charge Z'. The rationale for this is that we believe that each electron tends to shield the other from the full nuclear charge, so that each "sees" only an effective nuclear charge $Z' < Z$. If we calculate the energy using an approximate wavefunction of type 11.55, but with Z replaced by Z', we find

$$E' = \left[(Z')^2 - 27 \frac{Z'}{8} \right] \frac{e^2}{4\pi\epsilon_0 a_0} \qquad (11.57)$$

We choose Z' to give the lowest energy by minimizing E' with respect to Z'. Differentiating equation 11.57 with respect to Z', we find

$$\frac{dE'}{dZ'} = \left(2Z' - \frac{27}{8} \right) \frac{e^2}{4\pi\epsilon_0 a_0} \qquad (11.58)$$

so that $Z'_{min} = {}^{27}/_{16}$ and $E'_{min} = -77.5$ eV compared to the exact energy of -79.0 eV. We see from this how well we can do using the variation method with only one parameter. By adding further parameters and changing the form of the approximate wavefunction, we can get as accurate an energy as we can measure.

11.7 Pauli Exclusion Principle

The wavefunction that we have just discussed for the ground state of the helium atom is incomplete in that it does not include the spin functions (α or β) for the two electrons. We can write the following four spin functions for the two electrons:

$$\alpha(1)\alpha(2) \qquad \beta(1)\beta(2) \qquad \alpha(1)\beta(2) \qquad \alpha(2)\beta(1) \qquad (11.59)$$

where $\alpha(1)$ indicates electron 1 has the spin function α. However, the last two spin functions cannot be used because they imply that it is possible to distinguish between the two electrons. Electrons are identical to each other, and it is not possible to distinguish between them. The wavefunction for the electrons must reflect this and since the electrons are in identical orbitals, the spin functions for the two electrons must be written in such a way that they do not distinguish between the two electrons. This may be done by writing the spin functions as follows:

$$\alpha(1)\alpha(2) \qquad\qquad (11.60)$$

$$\beta(1)\beta(2) \qquad\qquad (11.61)$$

$$2^{-1/2}[\alpha(1)\beta(2) + \alpha(2)\beta(1)] \qquad\qquad (11.62)$$

$$2^{-1/2}[\alpha(1)\beta(2) - \alpha(2)\beta(1)] \qquad\qquad (11.63)$$

The first three functions are unchanged when the two electrons are interchanged. These three functions are therefore **symmetric** with respect to electron interchange. The fourth function changes sign when the two electrons are interchanged and is therefore **antisymmetric** with respect to electron interchange.

It is found experimentally that wavefunctions (including both spatial and spin functions) of electrons must be antisymmetric with respect to the interchange of any two electrons. This fact was discovered by Pauli in 1926. The Pauli exclusion principle may be stated as follows: **The wavefunction for any system of electrons must be antisymmetric with respect to the interchange of any two electrons.** Another way of stating the Pauli principle in a way that is useful in chemistry is to say that no two electrons in an atom may have the same four quantum numbers n, l, m_l, and m_s. In the nonrelativistic quantum mechanics that we are using, the Pauli principle is an additional postulate.

Since we have described the ground state of the helium atom as 1s(1)1s(2), which is symmetric, we need to multiply it by an antisymmetric spin function so that the complete wavefunction will be antisymmetric. Thus, the complete wavefunction for the ground state of helium can be approximated as

$$\psi = 1s(1)1s(2)2^{-1/2}[\alpha(1)\beta(2) - \alpha(2)\beta(1)] \tag{11.64}$$

In 1929 Slater developed a mathematical method for constructing approximate wavefunctions satisfying the antisymmetry requirement that can be written as determinants. In a **Slater determinant** the elements in a given column involve the same spin orbital, while elements in the same row involve the same electron. The approximate helium wavefunction (equation 11.64) can be written in the form

$$\psi = \frac{1}{\sqrt{2}} \begin{vmatrix} 1s(1)\alpha(1) & 1s(1)\beta(1) \\ 1s(2)\alpha(2) & 1s(2)\beta(2) \end{vmatrix} \tag{11.65}$$

The requirement of the Pauli principle that no two electrons in an atom or molecule have all quantum numbers the same is provided for automatically by the Slater determinant. If two rows or two columns of a determinant are identical, the determinant vanishes. Another useful property of a determinant is that the interchange of two rows or columns changes the sign of the determinant, showing that the Pauli principle is satisfied automatically by a wavefunction expressed by a determinant.

The inclusion of spin does not alter the energy calculations of the preceding section because the Hamiltonian does not include terms of spin. However, spin considerations are decisive for the excited states of helium and for the lithium atom, as we will soon see. This is because of the Pauli principle, not because of explicitly spin-dependent terms occurring in the Hamiltonian.

Particles with half-integral spin ($s = \frac{1}{2}, \frac{3}{2}, \ldots$) all require antisymmetric wavefunctions and are referred to as fermions because they must obey a kind of statistics called Fermi–Dirac statistics. Particles with integral spin ($s = 0$, 1, 2, . . .) all require symmetric wavefunctions and are referred to as bosons because they follow a different statistical law called Bose–Einstein statistics.

11.8 First Excited State of Helium

We have seen that the wavefunction for the ground state of the helium atom may be approximated by the product of two hydrogenlike wavefunctions (orbitals). We might expect that the wavefunction for the first excited state of helium may be approximated by the product of 1s and 2s hydrogenlike wavefunctions. Thus, we will examine $\psi = 1s(1)2s(2)$. This spatial wavefunction suffers from the problem discussed in the preceding section: It implies that it is possible to distinguish between two electrons. The actual wavefunction for the first excited state should be written in one of the two ways that do not distinguish between the two electrons:

$$\psi_a = 2^{-1/2}[1s(1)2s(2) + 1s(2)2s(1)] \qquad (11.66)$$

$$\psi_b = 2^{-1/2}[1s(1)2s(2) - 1s(2)2s(1)] \qquad (11.67)$$

The first function is symmetric and the second is antisymmetric. Note that there is no way to make an antisymmetric function from $1s(1)1s(2)$. However, in either case the probability density ψ^2 is not altered by interchanging the electrons. The two excited states represented by ψ_a and ψ_b have different energies, and ψ_b has the lower energy. The experimental value of the energy for the state approximated by ψ_b is -59.2 eV and for ψ_a is -58.4 eV.

Now we want to incorporate spin into the spatial wavefunction (12.67) for the first excited state of the helium atom. Since this spatial wavefunction is antisymmetric, it must be multiplied by symmetric spin functions according to the Pauli principle. There are three of these. Therefore, the wavefunctions of the first excited state of helium are

$$\psi_1 = 2^{-1/2}[1s(1)2s(2) - 1s(2)2s(1)]\alpha(1)\alpha(2) \qquad (11.68)$$

$$\psi_2 = 2^{-1}[1s(1)2s(2) - 1s(2)2s(1)][\alpha(1)\beta(2) + \alpha(2)\beta(1)] \qquad (11.69)$$

$$\psi_3 = 2^{-1/2}[1s(1)2s(2) - 1s(2)2s(1)]\beta(1)\beta(2) \qquad (11.70)$$

Thus, the first excited state of helium is a **triplet** state because it has a degeneracy of 3 due to the spin. In the presence of a magnetic field the first excited state is split into three energy levels. Since the sum of the spin quantum numbers of the two electrons is 1, this net spin may be oriented in one of three ways in a magnetic field. The z components of the spin angular momentum may be $-\hbar$, 0, and $+\hbar$, and so the energy of the atom in a magnetic field may have three values. In the ground state for helium the electrons are paired, and so the resultant electron spin is zero and the ground state is a singlet.

Since ψ_a is symmetric, it must be multiplied by an antisymmetric spin function to obtain an antisymmetric total wavefunction:

$$\psi = 2^{-1}[1s(1)2s(2) + 1s(2)2s(1)][\alpha(1)\beta(2) - \alpha(2)\beta(1)] \qquad (11.71)$$

The second excited state is a singlet state. Since $l = 0$ and $m = 0$, this state is not split by application of a magnetic field.

11.9 Lithium Atom

The Pauli exclusion principle did not have an effect on our consideration of the hydrogen atom, and it had only a modest effect on our consideration of the helium atom where we found the first excited state to be a triplet. But the Pauli principle has a major effect on the quantum mechanical treatment of the lithium atom. Since by the Pauli principle the 1s orbital can accommodate only two electrons, the third electron must be in a 2s orbital. Following the procedure of Section 11.8, this wavefunction for the ground state of Li can be written as a Slater determinant:

$$\psi = \frac{1}{\sqrt{6}} \begin{vmatrix} 1s(1)\alpha(1) & 1s(1)\beta(1) & 2s(1)\alpha(1) \\ 1s(2)\alpha(2) & 1s(2)\beta(2) & 2s(2)\alpha(2) \\ 1s(3)\alpha(3) & 1s(3)\beta(3) & 2s(3)\alpha(3) \end{vmatrix}$$

$$= \frac{1}{\sqrt{6}} [1s(1)\alpha(1)1s(2)\beta(2)2s(3)\alpha(3) - 1s(1)\alpha(1)1s(3)\beta(3)2s(2)\alpha(2)$$

$$- 1s(1)\beta(1)1s(2)\alpha(2)2s(3)\alpha(3) + 1s(1)\beta(1)1s(3)\alpha(3)2s(2)\alpha(2) \quad (11.72)$$

$$+ 2s(1)\alpha(1)1s(2)\alpha(2)1s(3)\beta(3) - 2s(1)\alpha(1)1s(3)\alpha(3)1s(2)\beta(2)]$$

In contrast with the determinant for the first excited state of He, this wavefunction cannot be written as the product of the spatial function and a spin function. The last column of the determinant could have been written with a β instead of an α; thus, the ground state of the lithium atom is doubly degenerate.

In using equation 11.72 in a variational treatment the nuclear charge Z may be replaced by Z_1 in the 1s function and by Z_2 in the 2s function to allow for the fact that the electrons are partially screened from the nuclear charge of $3+$. The variational treatment yields $Z_1 = 2.69$, $Z_2 = 1.78$. The variational energy is -201.2 eV, compared with the experimental ground-state energy of -203.48 eV.

This treatment is approximate in that the determinantal wavefunction is not an exact solution of the Schrödinger equation containing the interelectronic repulsion terms $(e^2/r_{12} + e^2/r_{13} + e^2/r_{23})/4\pi\epsilon_0$.

11.10   Hartree–Fock Self-Consistent Field Method

For atoms with more electrons the methods for calculating wavefunctions that we have illustrated for helium and lithium rapidly become impractical. In 1928 Hartree introduced the self-consistent field (SCF) method. This method may be used to calculate the ground-state wavefunction and energy for any atom.

If interelectron repulsion terms in the Schrödinger equation are ignored, the Schrödinger equation for an n-electron atom may be separated into n one-electron hydrogenlike equations. The approximate wavefunction obtained in this way is the product of n one-electron functions that are hydrogenlike wavefunctions (orbitals). Hydrogenlike orbitals use the full nuclear charge Z, but we know that the outer electrons of an atom are shielded from the nuclear charge by the inner electrons so that the effective charge is less.

Hartree used a variation function ϕ, which is the product of n orbitals g_i that contain parameters to be evaluated by the variation method (for example, effective nuclear charges):

$$\phi = g_1(r_1, \theta_1, \phi_1,)g_2(r_2, \theta_2, \phi_2) \cdots g_n(r_n, \theta_n, \phi_n) \quad (11.73)$$

Each orbital in this variation function is taken to be the product of radial factor $h_i(r_i)$ and a spherical harmonic $Y_{l_i}^{m_i}(\theta_i, \phi_i)$.

$$g_i = h_i(r_i)Y_{l_i}^{m_i}(\theta_i, \phi_i) \quad (11.74)$$

The spherical harmonics (Section 10.13) are used because the potential in the Schrödinger equation for a many-electron atom is spherically symmetric. Hartree's procedure was to first estimate the form of the orbitals $g_1 \ldots g_n$. Since the wavefunction of the atom is taken as the product of these orbitals, the Schrödinger equation for the atom may be separated into n equations of the type

$$\left[-\frac{\hbar^2}{2m} \nabla_i^2 + V(r_i) \right] g_i = \epsilon_i g_i \qquad (11.75)$$

where ϵ_i is the energy of the orbital for electron i. These equations are solved by successive approximations. The potential energy function $V(r_i)$ for any one electron is obtained on the assumption that the electric charge of all of the other electrons is smeared out to form a spherically symmetric charge cloud. The orbital g_i of the first electron obtained in this way is used to improve the potential energy function $V(r_2)$ for use in the Schrödinger equation for the second electron to obtain an improved orbital g_2 for it. This process is continued for all n electrons, and then the process is started over with electron 1. The calculation of improved orbitals is continued in this way until there is no further change in the orbitals. The product of these orbitals gives the Hartree self-consistent field wavefunction for the atom.

Hartree provided for spin and the Pauli principle by putting no more than two electrons in each orbital, but his wavefunctions did not involve spin and were not made to be antisymmetric with respect to the interchange of electrons. In 1930 Fock (and Slater) pointed out that it is necessary to use spin orbitals and to take linear combinations of antisymmetric products of spin orbitals. A self-consistent field calculation carried out in this way is referred to as a Hartree–Fock calculation.

Figure 11.9 shows the radial electron density of argon orbitals calculated by the Hartree–Fock method. For each value of the principal quantum number n, the probability density is largely concentrated in a narrow range of radii. The set of orbitals having the same principal quantum number n is referred to as a **shell**, and the set of orbitals with the same n value and l value is referred to as a **subshell**. Since the orbitals involve the same angular dependence as the hydrogenlike atomic wavefunctions, the subshells may be referred to as 1s, 2s, 2p, 3s, ... Note that the ns orbitals are more likely to be close to the nucleus than the np, nd, . . .(penetration) so that the ns are less shielded from the nucleus than the np, nd, . . . (see Fig. 11.10).

The theoretically calculated orbital energies for neutral atoms are shown in Fig. 11.11 as a function of atomic number Z. A very approximate value of the energy of a Hartree–Fock orbital may be obtained using an effective nuclear charge Z_{eff} in place of Z. As described earlier in connection with the use of the variation method for the helium atom, the effective charge is less than the atomic number because of the screening by other electrons in the atom.

As the atomic number (nuclear charge) increases, the energies of the inner orbitals become more negative because of the increased attraction between the nucleus and the electrons. The p orbitals have higher (less negative) energies than the s orbitals because p electrons have lower probability densities in the neighborhood of the nucleus and feel the attraction of the nucleus less. At or very near to the nucleus an s orbital is not shielded by inner electrons; therefore, it has a low energy (i.e., a large negative potential).

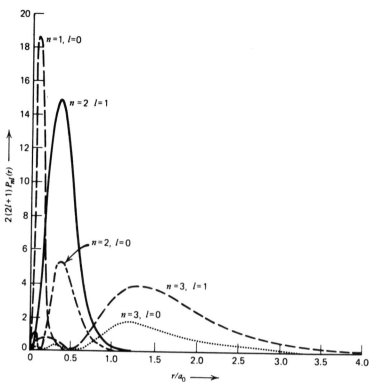

Figure 11.9   Radial probability densities for argon calculated by the Hartree–Fock method. (From R. Eisberg and R. Resnick, *Quantum Physics*. New York: Wiley, 1985.)

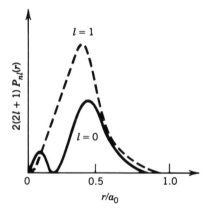

Figure 11.10   Blowup of the n = 2, l = 1, and l = 0 radial probability densities for Ar near the nucleus, showing the greater penetration of the l = 0 (s) orbital.

In Fig. 11.11 $(E/E_h)^{1/2}$ is plotted against Z based on the form of equation 11.8. The energies of ns and np orbitals are closer to each other than those of np and nd orbitals. Note that for some values of Z, the energy of the 4s orbital is less (more negative) than that of the 3d orbital.

The Hartree–Fock calculated energies for atoms usually agree with experimental values to about 1%. The method provides for the interactions between electrons in an average way, but does not provide for their instantaneous interactions. Since electrons tend to stay away from each other, we may speak of correlation in the positions of electrons at any instant. The **correlation energy** is the difference between the exact energy and the Hartree–Fock energy. This energy, which is of the order of an electronvolt or more, is large enough to be a serious problem in the calculation of differences in energies, as, for example, in calculating enthalpy changes for reaction.

The effects of instantaneous electron interaction may be provided for by including excited configurations in a trial wavefunction, using the variation approach. This method is referred to as configuration interaction. By including more configurations in the wavefunction used in the variation method, a representation of the true wavefunction may be approached more closely.

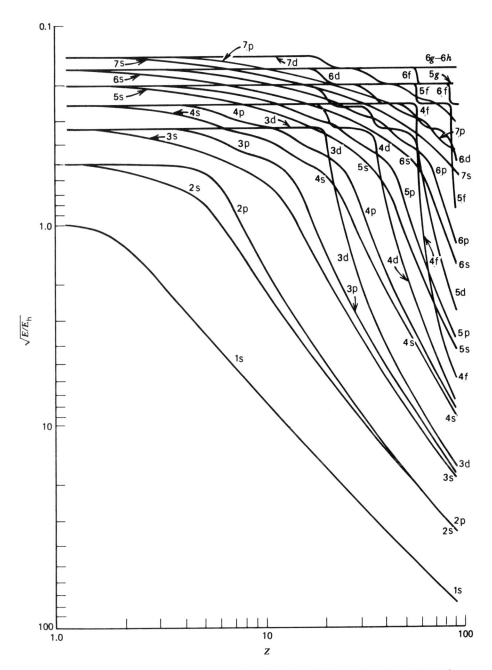

Figure 11.11 Approximate orbital energies in units of E_h as a function of atomic number Z. (From R. Latter, *Phys. Rev.* **99**:510 (1955).)

11.11   The Periodic Table and the Aufbau Principle

The quantum theory of atoms provides an explanation of the structure of the periodic table. As we have seen, the electron subshells in atoms may be des-

Table 11.3 Atomic Properties[a]

Z	Atom	Configuration	Ground-State Term Symbol	First Ionization Energy, eV	Orbital Radius, pm
1	H	1s	$^2S_{1/2}$	13.505	52.9
2	He	$1s^2$	$^1S_0$	24.580	29.1
3	Li	[He]2s	$^2S_{1/2}$	5.390	158.6
4	Be	$[He]2s^2$	$^1S_0$	9.320	104.0
5	B	$[He]2s^22p$	$^2P_{1/2}$	8.296	77.6
6	C	$[He]2s^22p^2$	$^3P_0$	11.264	62.0
7	N	$[He]2s^22p^3$	$^4S_{3/2}$	14.54	52.1
8	O	$[He]2s^22p^4$	$^3P_2$	13.614	45.0
9	F	$[He]2s^22p^5$	$^2P_{3/2}$	17.42	39.6
10	Ne	$[He]2s^22p^6$	$^1S_0$	21.559	35.4
11	Na	[Ne]3s	$^2S_{1/2}$	5.138	171.3
12	Mg	$[Ne]3s^2$	$^1S_0$	7.644	127.9
13	Al	$[Ne]3s^23p$	$^2P_{1/2}$	5.984	131.2
14	Si	$[Ne]3s^23p^2$	$^3P_0$	8.149	106.8
15	P	$[Ne]3s^23p^3$	$^4S_{3/2}$	11.00	91.9
16	S	$[Ne]3s^23p^4$	$^3P_2$	10.357	81.0
17	Cl	$[Ne]3s^23p^5$	$^2P_{3/2}$	13.01	72.5
18	Ar	$[Ne]3s^23p^6$	$^1S_0$	15.755	65.9
19	K	[Ar]1s	$^2S_{1/2}$	4.339	216.2
20	Ca	$[Ar]4s^2$	$^1S_0$	6.111	169.0
21	Sc	$[Ar]4s^23d$	$^2D_{3/2}$	6.56	157.0
22	Ti	$[Ar]4s^23d^2$	$^3F_2$	6.83	147.7
23	V	$[Ar]4s^23d^3$	$^4F_{3/2}$	6.74	140.1
24	Cr	$[Ar]4s3d^5$	$^7S_3$	6.76	145.3
25	Mn	$[Ar]4s^23d^5$	$^6S_{5/2}$	7.432	127.8
26	Fe	$[Ar]4s^23d^6$	$^5D_4$	7.896	122.7
27	Co	$[Ar]4s^23d^7$	$^4F_{9/2}$	7.86	118.1
28	Ni	$[Ar]4s^23d^8$	$^3F_4$	7.633	113.9
29	Cu	$[Ar]4s3d^{10}$	$^2S_{1/2}$	7.723	119.1
30	Zn	$[Ar]4s^23d^{10}$	$^1S_0$	9.391	106.5
31	Ga	$[Ar]4s^23d^{10}4p$	$^2P_{1/2}$	6.00	125.4
32	Ge	$[Ar]4s^23d^{10}4p^2$	$^3P_0$	8.13	109.0
33	As	$[Ar]4s^23d^{10}4p^3$	$^4S_{3/2}$	10.00	100.1
34	Se	$[Ar]4s^23d^{10}4p^4$	$^3P_2$	9.750	91.8
35	Br	$[Ar]4s^23d^{10}4p^5$	$^2P_{3/2}$	11.84	85.1
36	Kr	$[Ar]4s^23d^{10}4p^6$	$^1S_0$	13.996	79.5

[a] From J. T. Waber and D. T. Cromer, *J. Chem. Phys.* **42**:4116 (1965).

ignated 1s, 2s, 2p, 3s, ... According to the Pauli exclusion principle, an s subshell may contain 2 electrons, a p subshell 6, and a d subshell 10. Thus, the subshells each contain $2(2l + 1)$ electrons.

To find the ground-state electron configuration of an atom, we add electrons to the subshells beginning with the lowest energy until we have added the correct number for that atom. This procedure is called the *aufbau* (building-up) principle.

The electron configurations for the first 36 elements found in this way are given in Table 11.3. The symbols [He], [Ne], and [Ar] represent the closed-

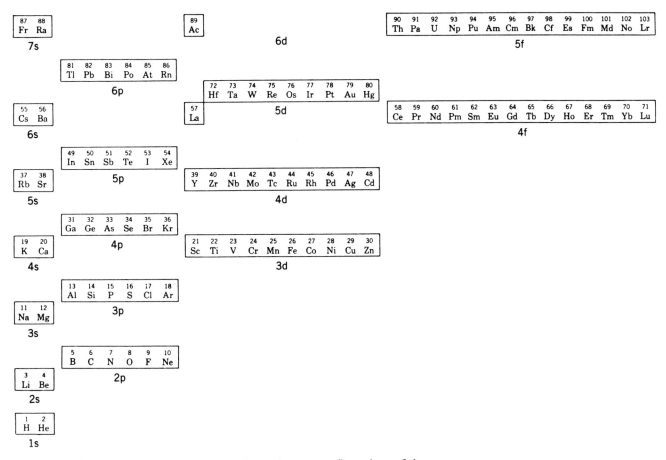

Figure 11.12   Periodic table arranged to show electron configurations of the atoms. The relative energies of the orbitals are shown by their vertical positions. We can think of the electrons being fed successively into higher and higher (i.e., less negative) energy orbitals.

shell electron configurations of these elements. In the section 11.14 we will consider the significance of the atomic term symbols given in the fourth column.

The electron configurations of the elements account for the periodicity of physical and chemical properties of the elements. The periodic table can be arranged to show the electron configurations of the elements, as in Fig. 11.12.* This form of the periodic table makes it clear why the lanthanides (58 to 71) and actinides (90 to 103) have such similar chemical properties. Table 11.3 gives the orbital radii of isolated atoms; this is the radius of the maximum in radial probability density of the outermost orbital. Within a row of the periodic table, the atomic radius tends to decrease with increasing atomic number. Within a column in the periodic table, the atomic radius tends to increase with the atomic number. These general trends in atomic radius can be understood in terms of two factors that primarily determine the radius of the outermost

---

*H. C. Longuet-Higgins, *J. Chem. Educ.* **34**:30 (1957).

orbital. The first factor is that the larger the principal quantum number, the larger the radius of the orbital. The second factor is that increasing the effective nuclear charge reduces the size of the orbital. As we saw in the preceding section, the effective nuclear charge felt by an electron is equal to the nuclear charge less any shielding of this charge by intervening orbital electrons. In a given row of the periodic table the principal quantum number remains constant, but the effective nuclear charge increases so that the radius of the outermost orbital decreases as the atomic number increases. In a column in the periodic table the effective nuclear charge remains nearly constant, but the principal quantum number increases so that the atomic radius increases.

11.12   Ionization Energy

The ionization energy is the energy required to remove an electron completely from a gaseous atom, molecule, or ion. An atom has as many ionization energies as it has electrons. The first ionization energy corresponds with the reaction

$$A = A^+ + e^- \tag{11.76}$$

and the second corresponds with the reaction

$$A^+ = A^{2+} + e^- \tag{11.77}$$

The first ionization energies of the first 36 elements are given in Table 11.3, and the first ionization energies of all of the elements are plotted in Fig. 11.13. The most difficult atom to ionize is helium, and the ionization energies of the other inert gases decrease as the atomic radius increases. The alkali metals are

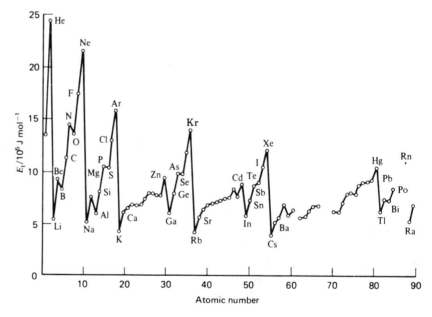

Figure 11.13   First ionization energies of the elements as a function of atomic number.

Table 11.4 Ionization Energies of the First Six Elements[a] in Electronvolts

Atomic Number	Element	Spectrum					
		I	II	III	IV	V	VI
1	H	13.598					
2	He	24.587	54.416				
3	Li	5.392	75.638	122.451			
4	Be	9.322	18.211	153.893	217.713		
5	B	8.298	25.154	37.930	259.368	340.217	
6	C	11.260	24.383	47.887	64.492	392.077	489.981

[a] C. E. Moore, *National Standard Reference Data Series*, Vol. 34. Washington, DC: U.S. Government Printing Office, 1970.

the easiest atoms to ionize because the outer shell is occupied by a single electron. In the series lithium, sodium, potassium, rubidium, and cesium the ionization energy decreases because of the increase in size of the outer orbit containing a single electron.

In contrast, the ionization energies of the halogens are almost as great as those of the inert gases. The electrons in the outer orbits of the halogen atoms are shielded from the nuclear charge mainly by the electrons in inner orbits, since the electrons in the outer orbits are all approximately the same distance from the nucleus. A direct result of this incomplete shielding of the nuclear charge, as far as electrons in the outer orbit are concerned, is the fact that the halogen atoms readily take on an additional electron to form negative ions.

Within a period of the periodic table there is a general increase in ionization energy, but this increase is not regular. The decrease between beryllium and boron is due to the fact that boron has a single 2p electron outside the filled 2s orbital.

Ionization energies may be determined by irradiating atoms with light of very short wavelength (Section 15.10). All of the ionization energies of the first six elements are given in Table 11.4. The successive stages of ionization are indicated in the headings of the columns. Column I gives the ionization energy of the neutral atom. Column II gives the ionization energy for the singly ionized atoms, and so on. After the first electron has been removed, it is more difficult to remove the second, and so on. There is an especially large increase in ionization energy after all the electrons in the outer shell have been removed; this point for each element is shown by the stairstep in Table 11.4.

11.13 Electron Affinity

Since the halogen atoms lack a single electron to complete their outer shell they have a strong affinity for an electron. For example,

$$Cl(g) + e^- = Cl^-(g) \quad \Delta H°(0\ K) = -347\ kJ\ mol^{-1} = \frac{-347\ kJ\ mol^{-1}}{96.485\ kJ\ mol^{-1}\ eV^{-1}}$$

$$= -3.60\ eV \quad (11.78)$$

Table 11.5   Electron Affinities E_{ea} in Electronvolts[a]

H(g)	0.75415		
He(g)	(−0.22)	Ne(g)	(−0.30)
Li(g)	0.602	Na(g)	0.548
Be(g)	(−2.5)	Mg(g)	(−2.4)
B(g)	0.86	Al(g)	0.52
C(g)	1.27	Si(g)	1.24
N(g)	0	P(g)	0.77
O(g)	1.465	S(g)	2.077
F(g)	3.39	Cl(g)	3.614

[a] The values in parentheses are calculated values.

The electron affinity E_{ea} is defined as the energy released in the process of adding an electron, and so the electron affinity of Cl(g) is 3.60 eV. It is possible to calculate electron affinities by using the Hartree–Fock method for both the atom and its negative ion. As shown in Table 11.5 some electron affinities are negative, meaning that the negative ion is unstable with respect to the atom and an electron.

11.14   Atomic Term Symbols

To discuss the orbital angular momentum and spin angular momentum of a many-electron atom, we will use l_i to represent the orbital angular momentum of the ith individual electron in an atom and l_{zi} to represent its component in the z direction. Similarly, we will use s_i to represent the spin angular momentum of a single electron in an atom and s_{zi} to represent its component in the z direction. The symbols used in this section are summarized in Table 11.6, which should be compared with Table 11.2.

The total orbital angular momentum of a light many electron atom is the **vector sum** of the orbital angular momenta l_i for the individual electrons.

Table 11.6   Angular Momentum of a Many-Electron Atom[a]

	Orbital Angular Momentum	
	Electron	Atom
Angular momentum vector	l_i	$L = \Sigma l_i$
Component of angular momentum vector in z direction	l_{zi}	$L_z = \Sigma l_{zi}$
Quantum number of angular momentum	l_i $(0, 1, 2, \ldots)$	$L = l_1 + l_2, l_1 + l_2 - l,$ $\ldots, \lvert l_1 - l_2 \rvert$
Quantum number for z component (magnetic quantum number)	m_i $(-l_i, \ldots, +l_i)$	$M_L = \Sigma m_i$ $(-L, \ldots, +L)$

[a] The four rows in the second half of the table are in the same order as in the first half of the table.

$$L = \sum_i l_i \qquad (11.79)$$

This vector addition is represented schematically in Fig. 11.14. Remember that the orbital angular momentum vectors of the individual electrons and the sum L have only certain orientations with respect to a given direction which we label the z direction. The z component L_z of the total orbital angular momentum of an atom is the **scalar sum** of the z components for the individual electrons, as illustrated in Fig. 11.14.

$$L_z = \sum_i l_{zi} \qquad (11.80)$$

We saw earlier (Section 10.14) that the z component of angular momentum of a single electron is directly proportional to the magnetic quantum number m: $L_z = m\hbar$. When this relation is applied to an atom containing several electrons, the magnetic quantum number for the orbital angular momentum of an atom is capitalized so that $L_z = M_L \hbar$. The z components of the orbital angular momenta of single electrons are represented by $l_{zi} = m_i \hbar$, where m_i is the magnetic quantum number for the ith electron. When these two relations are substituted in equation 11.80 we obtain

$$M_L = \sum_i m_i \qquad (11.81)$$

The magnetic quantum numbers for an atom can range over $-L, \ldots, +L$, depending on the orientation in a magnetic field. Here L is the orbital angular momentum quantum number for the atom. For an atom containing two electrons, the maximum value of L is obtained when the two orbital angular momenta are lined up, and the minimum value is obtained when the two orbital angular momenta are opposed. When the quantum numbers for the orbital angular momenta of the two electrons are represented by l_1 and l_2, the orbital angular momentum quantum numbers for the atom can have the values

$$L = l_1 + l_2, l_1 + l_2 - 1, \ldots, |l_1 - l_2| \qquad (11.82)$$

If the atom contains more than two electrons, this relation can be applied successively.

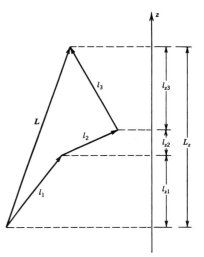

Figure 11.14 Schematic diagram for the addition of angular momentum vectors. (From D. A. McQuarrie, *Quantum Chemistry*. Mill Valley, CA: University Science Books, 1983. Reprinted with permission from University Science Books.)

Table 11.6 Continued

	Spin Angular Momentum		Total Angular Momentum Atom
Electron	Atom		
s_i	$S = \Sigma s_i$		$J = L + S$
s_{zi}	$S_z = \Sigma s_{zi}$		$J_z = L_z + S_z$
s_i ($\frac{1}{2}$)	$S = s_1 + s_2, s_1, + s_2 - 1,$ $\ldots, \mid s_1 - s_2 \mid$		$J = L + S, L + S - 1,$ $\ldots, \mid L - S \mid$
m_{si} ($\pm\frac{1}{2}$)	$M_s = \Sigma m_{si}$ $(-S, \ldots, +S)$		$M_J = M_L + M_S$ $(-J, \ldots, +J)$

[a] The four rows in the second half of the table are in the same order as in the first half of the table.

The total spin angular momentum for a light atom is the vector sum of the spin angular momenta s_i of the individual electrons:

$$S = \sum_i s_i \qquad (11.83)$$

The z component S_z of the total spin angular momentum of an atom is the scalar sum of the z components for the individual electrons:

$$S_z = \sum_i s_{zi} \qquad (11.84)$$

As shown in equation 11.32 the contribution of the ith electron to the total spin angular momentum in the z direction is directly proportional to the quantum number m_{si} for the ith electron: $s_{zi} = m_{si}\hbar$. The total spin angular momentum S_z in the z direction is directly proportional to the spin quantum number M_s for the z component for the whole atom. Thus, equation 11.84 becomes

$$M_S = \sum_i M_{si} \qquad (11.85)$$

The value of M_s can range from $-S$ to $+S$, depending on the orientation of the total spin angular momentum for the atom in a magnetic field. For an atom containing two electrons, the spin quantum number S is a maximum if the spins are parallel and a minimum if the spins are opposed. If the spin quantum numbers of the individual electrons are represented by s_1 and s_2, the possible values of S are

$$S = s_1 + s_2, s_1 + s_2 - 1, \ldots, |s_1 - s_2| \qquad (11.86)$$

For a two-electron atom, $S = 1, 0$.

The total angular momentum J of an atom is the vector sum of all the orbital l_i and spin s_i angular momenta of electrons in it. Like other angular momenta J is quantized and its quantum number J can take on only integer and half-integer values. In principle these can range up to the sum of the orbital angular momentum quantum numbers and spin quantum numbers for the individual electrons in the atom. We are interested in the total angular momentum quantum number J because states with different angular momenta differ in energy. This is partly because the electrons repel each other electrostatically, and the strengths of the repulsions depend on the distribution of electric charge, which we know is connected with the orbital quantum numbers l_i and m_i. This effect is given the shorthand name orbit–orbit, or ll, interaction. In addition, the atomic states must obey the Pauli exclusion principle which, because it dictates which orbital states can be associated with which spin states, indirectly brings in a spin–spin interaction which is determined by s_i and s_{zi}. This effect is called spin–spin, or ss, interaction. In addition to these effects, there is also a direct spin–orbit, or ls, interaction.

The preceding paragraph describes a rather complicated situation, since the relative energies of different atomic states depend on the relative strengths of ll, ss, and ls interactions of the various electrons. In relatively light atoms ($Z < 40$), however, it turns out that the ls interaction is distinctly weaker than the other two. This means that the electron orbits interact to give the total orbital angular momentum described in equation 11.79 by L. At the same time the spins interact to give a resultant spin angular momentum described in equation 11.83 by S. Then the weaker ls interaction can be thought of as coupling L and S to give the total angular momentum J:

$$J = L + S \tag{11.87}$$

The J vector precesses in a magnetic field, and the z component is given by

$$J_z = L_z + S_z \tag{11.88}$$

The quantum number J for the total angular momentum of an atom has values given by

$$J = L + S, L + S - 1, \ldots, |L - S| \tag{11.89}$$

and the total magnetic quantum number M_J is given by

$$M_J = M_L + M_S \tag{11.90}$$

and can range from $-J$ to $+J$. This situation is referred to as LS coupling or Russell–Saunders coupling.

The electron configurations used in describing the various elements in Table 11.3 are incomplete descriptions of the ground states because they do not fully specify how the spin and orbital angular momenta add vectorially. In general, several atomic states with different L, S, and J quantum numbers are represented by a single electron configuration. The energies of these states differ. However, atomic states can be classified according to L, S, and J, and the atomic term symbols used for this purpose have the form

$$^{2S+1}L_J \tag{11.91}$$

In a term symbol the total orbital angular momentum quantum number is not represented by a number, but by a capital letter, just as we represented the angular momentum quantum number for the electron in a hydrogenlike atom by s, p, d, f, . . . for $L = 0, 1, 2, 3, \ldots$:

$$L = 0 \quad 1 \quad 2 \quad 3 \quad 4 \ldots$$
$$\text{Symbol} = S \quad P \quad D \quad F \quad G \ldots$$

The superscript in the term symbol is the spin multiplicity and is referred to as singlet, doublet, triplet, . . . for 1, 2, 3, ...

In heavy atoms this hierarchy breaks down, and spin–orbit coupling eventually becomes much stronger than the other two. In this limit a different coupling scheme (jj coupling) is used in which individual electrons acquire total angular momenta $j_i = l_i + s_i$. The total angular momentum is then obtained using $J = \sum j_i$. In atoms of intermediate Z, neither approximation is valid, and the situation is very complicated, but the eigenfunction of the Hamiltonian is still an eigenfunction of $\hat{J}^2$. Here we will consider only the LS coupling scheme, which is of great utility in unraveling the spectroscopy and chemistry of the atoms in the first two rows of the periodic table.

Example 11.7

What is the atomic term symbol for helium? Since the electron configuration is $1s^2$, the magnetic quantum numbers m_i of both electrons are zero; and since the electrons are paired, the quantum number m_{si} for the z component of the spin is $+\frac{1}{2}$ for one electron and $-\frac{1}{2}$ for the other, as shown in the following summary:

m_1	m_{1s}	m_2	m_{2s}	M_L	M_S	M_J
0	$+\frac{1}{2}$	0	$-\frac{1}{2}$	0	0	0

Since $M_L = 0$ and there is only one possible state, the total orbital angular momentum quantum number L for the helium atom must be equal to zero. Similarly, since $M_s = 0$, the total spin angular momentum quantum number S for the helium atom must be equal to zero. Since $M_J = M_L + M_s$ is equal to zero, the total angular momentum quantum number J for the atom is 0, and the term symbol is $^1S_0$.

The conclusions from the preceding example may be generalized as follows. Although we have not proved it, closed shells do not contribute orbital or spin angular momentum to an atom because the individual angular momenta add vectorially to zero. Thus, it is the valence electrons that determine the term symbol. On the basis of the helium example other atoms with only ns^2 outside of closed shells will also have the atomic term symbol $^1S_0$. This is illustrated by Be, Mg, and Ca in Table 11.3. Since closed shells do not contribute, this is the term symbol for all the inert gases.

Example 11.8

Since lithium and boron each have a single electron outside of closed shells, their possible term symbols are readily identified. Since lithium has a 2s electron with $l = 0$, its magnetic quantum number m must also be zero.

m	m_s	M_L	M_s	M_J
0	$\pm\frac{1}{2}$	0	$\pm\frac{1}{2}$	$\pm\frac{1}{2}$

Thus, $L = 0$, $S = \frac{1}{2}$, and $J = \frac{1}{2}$, so that the term symbol is $^2S_{1/2}$.

Boron has a p electron so that

m	m_s	M_L	M_s	M_J
0	$\pm\frac{1}{2}$	0	$\pm\frac{1}{2}$	$\pm\frac{1}{2}$
± 1	$\pm\frac{1}{2}$	± 1	$\pm\frac{1}{2}$	$\pm\frac{1}{2}, \pm\frac{3}{2}$

Thus, $L = 1$, $S = \frac{1}{2}$, and $J = \frac{3}{2}, \frac{1}{2}$ so that there are two possible term symbols, $^2P_{3/2}$ and $^2P_{1/2}$.

As we have seen from the preceding example, an atom in a given configuration may exist in more than one state. The energies of atoms in levels corresponding with various terms can be calculated, but this is a demanding and time-consuming process. Fortunately, some patterns have emerged that usually make it possible to identify the lowest energy level for the ground-state configuration of an atom. The German spectroscopist Hund summarized these patterns with three empirical rules:

1. The term arising from the ground configuration with the maximum multiplicity $(2S + 1)$ lies lowest in energy.

2. For levels with the same multiplicity, the one with the maximum value of L lies lowest in energy.

3. For levels with the same S and L, the one with the lowest energy depends on the extent to which the subshell is filled.
   a. If the subshell is less than half filled, the state with the smallest value of J is the most stable.

b.   If the subshell is more than half filled, the state with the largest value of J is the most stable.

Thus, according to rule 3a, the ground state of boron discussed in Example 11.8 is $^2P_{1/2}$.

Example 11.9

Suppose an atom with ns^2 valence electrons is excited to a state with valence electrons $nsn's$, where n' is a higher principal quantum number. Now the electron spins do not have to be paired and so there are four possible states, which in general will have different energies:

m_1	m_{1s}	m_2	m_{2s}	M_L	M_s	M_J
0	$+\frac{1}{2}$	0	$+\frac{1}{2}$	0	1	1
0	$+\frac{1}{2}$	0	$-\frac{1}{2}$	0	0	0
0	$-\frac{1}{2}$	0	$+\frac{1}{2}$	0	0	0
0	$-\frac{1}{2}$	0	$-\frac{1}{2}$	0	-1	-1

In the first row the spins are parallel, and so the z component of the spin quantum number M_s for the atom is 1, and the quantum number S for the spin angular momentum is zero. Since the valence electrons are both s electrons, $M = 0$ and $L = 0$. For all four possibilities, the quantum number M for the orbital angular momentum is zero. Therefore, the quantum number L for the total angular momentum is zero. Since the largest value of M_s is unity, the total spin quantum number S for the atom can have values of 1 and 0.

If $S = 1$ and $L = 0$, then $J = 1$ and the term symbol is $^3S_1$. Note that for this case $M_s = 1, 0, -1$, so that we have a triplet state. If $S = 0$ and $L = 0$, then $J = 0$ and the term symbol is $^1S_0$, a singlet state. As indicated by Hund's first rule, the triplet state is the most stable.

As the number of electrons increases, the number of different ways of assigning them increases. For example, there are 15 ways for carbon with two p electrons. Details of these assignments are given in more advanced texts.

Example 11.10

An atom with orbital angular momentum L and spin angular momentum S often has a term in its Hamiltonian of the form $AL \cdot S$ called the spin–orbit coupling term. Since the total angular momentum $J = L + S$ commutes with the Hamiltonian, the states of this atom can be labeled with the eigenvalues of J^2, J_z, L^2, and S^2. Notice that $L \cdot S = \frac{1}{2}(J^2 - L^2 - S^2)$ so that the eigenvalue of $L \cdot S$ is $\frac{1}{2}[J(J + 1) - L(L + 1) - S(S + 1)]$. This can lead to observable energy level splittings. For example the excited states of an alkali atom with the outermost electron excited from the s to the next higher p orbital will have $L = 1$ and $S = \frac{1}{2}$. This leads to two possible J values: $\frac{3}{2}$ and $\frac{1}{2}$. Calculate the energy level splitting due to spin–orbit coupling.

The energy term due to $AL \cdot S$ will be

$$A[J(J + 1) - 1(1 + 1) - \tfrac{1}{2}(\tfrac{1}{2} + 1)] = A[J(J + 1) - \tfrac{11}{4}]$$

For $J = \frac{3}{2}$ the energy term equals A; for $J = \frac{1}{2}$ it equals $-2A$, giving a splitting of $3A$. In Na the observed splitting of the intense yellow flourescent lines is 17 cm^{-1}; thus, $A = 5.7$ cm^{-1} as illustrated in Fig. 11.15.

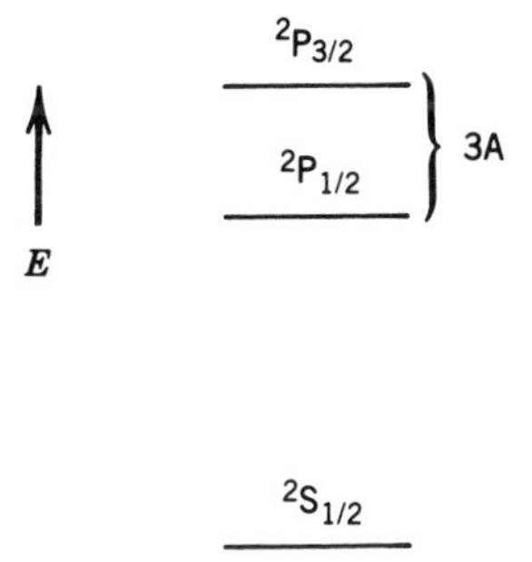

Figure 11.15   Low-lying energy levels in alkali atoms.

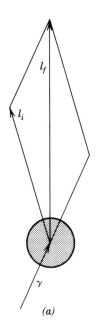

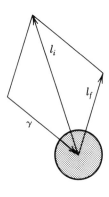

Figure 11.16 Conservation of orbital angular momentum on absorption of a photon (γ): (*a*) the angular momentum of the atom increases in the absorption process; (*b*) the angular momentum of the atom decreases in the absorption process. (From P. W. Atkins, *Molecular Quantum Mechanics*. New York: Oxford University Press, 1983.)

11.15 Special Topic: Atomic Spectra and Selection Rules

We have seen that an atom can exist in a series of states, each of which is identified by a term symbol. Spectroscopic transitions between these states provide information on their energies and quantum numbers. The emission spectra of atoms may be excited in a gas discharge tube or flame. The absorption spectrum may be obtained by passing light through a gas of the atoms.

The energy levels of a hydrogenlike atom depend only on the principal quantum number and are given by equation 11.7, which provides an explanation for the Lyman ($n_1 = 1$), Balmer ($n_1 = 2$), Paschen ($n_1 = 3$), etc., series. However, not all possible transitions can occur in spectroscopic transitions, because photons have an intrinsic angular momentum equal to one unit, and angular momentum is conserved. For left-circularly polarized light $m = 1$, and for right-circularly polarized light $m = -1$. When a photon is absorbed by an atom, its angular momentum is transferred to the electrons of the atom. Since angular momentum has to be conserved in absorption, or emission, we conclude that $\Delta l = \pm 1$ for the atom. These plus and minus signs apply in either absorption or emission because the angular momentum of an electron can increase or decrease during absorption or emission, as illustrated in Fig. 11.16.

We will see in Chapter 14 that the most intense spectroscopic transitions involve the interaction of the electric vector of the radiation with instantaneous electric dipoles in the atom or molecule. These are called **electric dipole transitions.**

The spectroscopic transitions that occur for hydrogen atoms or hydrogenlike atoms are indicated in Fig. 11.17. This diagram, referred to as a Grotrian diagram, shows only lines for which $\Delta l = \pm 1$. The complete selection rule for electric dipole transitions of hydrogenlike atoms is

$$\Delta n, \text{ unrestricted} \qquad \Delta l = \pm 1 \qquad \Delta m = \pm 1, 0 \qquad (11.92)$$

The spectrum of atomic hydrogen has a fine structure that we will not discuss because it is not important for chemistry. These further small splittings are explained by quantum electrodynamics.

Electric dipole transitions are due to the oscillating electric field component of light, and magnetic dipole transitions are due to the oscillating magnetic field component of light. Magnetic dipole transitions are generally about 10^5 weaker (that is, less probable) than electric dipole transitions. Selection rules were initially found experimentally, but they may be derived from the equation for the electric dipole transition moment that is discussed in Chapter 14.

As we have seen the states of many electron atoms depend on the coupling of angular and spin momenta and can be described in terms of L, J, and S. The electric dipole selection rules for atoms in general may be expressed in terms of these quantum numbers:

1. $\Delta L = 0, \pm 1$ except that the transition $L = 0$ to $L = 0$ does not occur. This is an extension of the selection rule $\Delta l = \pm 1$ for the hydrogen atom to transitions involving any number of electrons.

2. For a transition to be permitted there must be a change in **parity.** For an even function

$$\psi(-x) = +\psi(x)$$

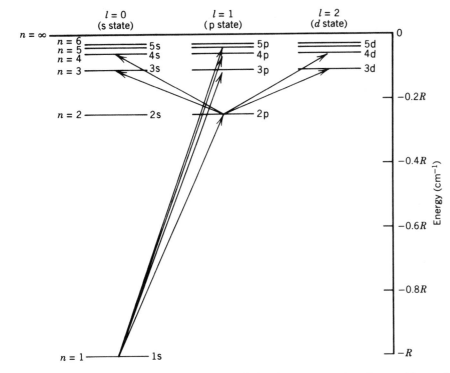

Figure 11.17   Grotrian diagram for some of the shortest wavelength transitions of hydrogenlike atoms, where R is the Rydberg constant.

Such a function is said to have an even parity. For an odd function

$$\psi(-x) = -\psi(x)$$

the parity is odd. The parity of a many electron atom is even if $\sum l_i$ is even and odd if $\sum l_i$ is odd, where the summation extends over all electrons. This selection rule is summarized by even ↔ even, odd ↔ odd, even ↔ odd. This selection rule, which is referred to as the Laporte rule, is consistent with $\Delta l = \pm 1$ when only one electron is promoted from the ground configuration.

3. $\Delta J = 0, \pm 1$ except $J = 0 \leftrightarrow J = 0$

4. $\Delta S = 0$. This selection rule results from the fact that the electric component of the electromagnetic field has no effect on the total spin angular momentum of the electrons in an atom. This is illustrated by the radiative transitions shown by helium, which are shown in Fig. 11.18. There are no transitions between singlet states and triplet states. This selection rule gives rise to the phenomenon of **metastable states.** The lowest triplet state of helium cannot emit a photon and make a transition to the singlet ground state. The transition to the ground state can occur in a collision with another molecule, and so triplet helium has a lifetime that depends on pressure. Metastable atoms play an important rule in photochemistry. This selection rule breaks down for atoms with higher atomic numbers.

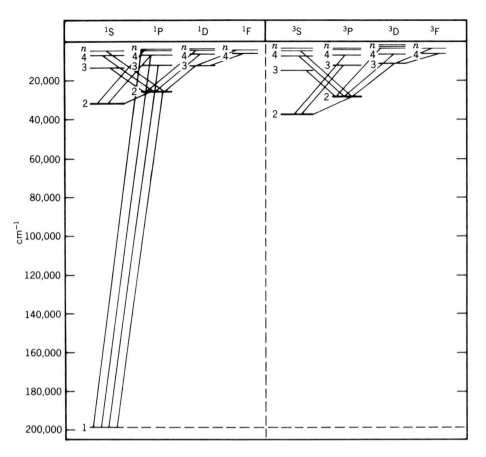

Figure 11.18 Grotrian diagram for the helium atom. (Reproduced from J. M. Hollas, *High Resolution Spectroscopy,* Fig. 6.17, p. 2941. Boston: Butterworth, 1982; by permission of the publishers, Butterworth & Co. (Publishers) Ltd. ©.)

Forbidden transitions (such as $\Delta S \neq 0$) may occur, but they usually occur infrequently and give weak lines. Their occurrence does not mean that quantum mechanics is wrong, but rather results from the approximations used in writing Hamiltonians. Various kinds of higher order interactions are frequently omitted, and when they are included correct (but considerably more complicated) conclusions are reached. When electric or magnetic fields are applied there are changes in various energy levels, and further selection rules are involved. The Zeeman effect is the splitting of lines by the application of a magnetic field. The Stark effect is the splitting of lines by the application of an electric field.

References

P. W. Atkins, *Molecular Quantum Mechanics.* New York: Oxford University Press, 1983.

R. E. Christofferson, *Basic Principles and Techniques of Molecular Quantum Mechanics.* New York: Springer, 1989.

R. Eisberg and R. Resnick, *Quantum Physics*, 2d ed. New York: Wiley, 1985.

C. F. Fischer, *The Hartree–Fock Method for Atoms*. New York: Wiley, 1977.

G. Herzberg, *Atomic Spectra and Atomic Structure*. Englewood Cliffs, NJ: Prentice-Hall, 1937.

M. Karplus and R. N. Porter, *Atoms and Molecules*. New York: Benjamin, 1970.

I. N. Levine, *Quantum Chemistry*. Boston: Allyn & Bacon, 1983.

D. A. McQuarrie, *Quantum Chemistry*. Mill Valley, CA: University Science Books, 1983.

M. A. Morrison, T. L. Estle, and N. F. Lane, *Quantum States of Atoms, Molecules, and Solids*. Englewood Cliffs, NJ: Prentice-Hall, 1976.

L. Pauling and E. B. Wilson, *Introduction to Quantum Mechanics*. New York: McGraw-Hill, 1935.

Problems

11.1 Using data from Table C.2 at 0 K what is the ionization energy of H(g)?

$$H(g) = H^+(g) + e^-$$

11.2 How much energy in eV and kJ mol^{-1} is required to remove electrons from the following orbitals in a H atom: (a) 3d, (b) 4f, (c) 4p, (d) 6s?

11.3 Calculate the ground-state ionization potentials for He$^+$, Li^{2+}, Be^{3+}, and C^{5+}.

11.4 Since H and D have different reduced masses, they also have slightly different electronic energy levels. Calculate the (a) ionization potentials and (b) the wavelengths of the first line in the Balmer series for these atoms.

11.5 A muon is an elementary particle with a negative charge equal to the charge of the electron and a mass approximately 200 times the electron mass. The muonium atom is formed from a proton and a muon. Calculate the reduced mass, the Rydberg constant, and the formula for the energy levels for this atom. What is the most probable radius of the 1s orbital for this atom?

11.6 A hydrogenlike atom has a series of spectal lines at $\lambda = 26.2445$ nm, $\lambda = 19.4404$ nm, $\lambda = 17.3578$ nm, and $\lambda = 16.4028$ nm. What is the nuclear charge on the atom? What is the formula for this spectral series (i.e., n_1 and n_2 in equation 11.9)?

11.7 What are the degeneracies of the following orbitals for hydrogenlike atoms: (a) $n = 1$, (b) $n = 2$, and (c) $n = 3$?

11.8 Show that for a 1s orbital of a hydrogenlike atom the most probable distance from proton to electron is a_0/Z. Find the numerical values for C^{5+} and B^{3+}.

11.9 Find the values of r for which the hydrogenlike atom 2s and 3s wavefunctions are equal to zero (these are the radial nodes). Compare to Fig. 11.1.

11.10 What is the average distance from an orbital electron to the nucleus for a 2s and 2p electron in (a) H and (b) Li^{2+}?

11.11 Use equation 11.18 to derive the expression for the average distance between the electron and proton in a hydrogenlike atom in the 1s level.

11.12 Calculate the expectation value of the distance between the nucleus and the electron of a hydrogenlike atom in the 2p$_z$ state using equation 10.59. Show that the same result is obtained using equation 11.19.

11.13 It can be shown that a linear combination of two eigenfunctions belonging to the same degenerate level is also an eigenfunction of the Hamiltonian with the same energy. In terms of mathematical formulas, if $\hat{H}\psi_1 = E_1\psi_1$ and $\hat{H}\psi_2 = E_1\psi_2$, then $\hat{H}(c_1\psi_1 + c_2\psi_2) = E_1(c_1\psi_1 + c_2\psi_2)$. The wavefunction $c_1\psi_1 + c_2\psi_2$ still needs to be normalized. Equation 11.5 yields the following expressions for the 2p eigenfunctions:

$$\psi_{2p+1} = be^{-Zr/2a}r \sin\theta \, e^{i\phi}$$

$$\psi_{2p-1} = be^{-Zr/2a}r \sin\theta \, e^{-i\phi}$$

$$\psi_{2p0} = ce^{-Zr/2a}r \cos\theta$$

Use this information to find the real functions for the 2p orbitals that are given in Table 11.1. Given:

$$e^{i\phi} = \cos\phi + i\sin\phi$$
$$e^{-i\phi} = \cos\phi - i\sin\phi$$

11.14 For a hydrogenlike atom, what is the magnitude of the orbital angular momentum and what are the possible values of L_z for electrons in the 2p and 3d orbitals?

11.15 What is the magnitude of the angular momentum for electrons in 3s, 3p, and 3d orbitals? How many radial and angular modes are there for each of these orbitals?

11.16 Using equation 11.40, calculate the difference in energy between the two spin angular momentum states of a hydrogen atom in the 1s orbital in a magnetic field of 1 T. What is the wavelength of radiation emitted when the electron spin "flips"? In what region of the electromagnetic spectrum is this?

11.17 For the wavefunction

$$\psi = \begin{vmatrix} \psi_A(1) & \psi_A(2) \\ \psi_B(1) & \psi_B(2) \end{vmatrix}$$

show that (a) the interchange of two columns changes the sign of the wavefunction, (b) the interchange of two rows changes the sign of the wavefunction, and (c) the two electrons cannot have the same spin orbital.

11.18 What are the electron configurations for H^-, Li^+, O^{2-}, F^-, Na^+ and Mg^{2+}?

11.19 How many electrons can enter the following sets of atomic orbitals: 1s, 2s, 2p, 3s, 3p, 3d?

11.20 Since the outer electron for Li is quite a bit further out than the two 1s electrons, this atom is something like a hydrogen atom in the 2s state. The first ionization potential of Li is 5.39 eV. What ionization potential would be expected from this simple model of a Li atom? What effective nuclear charge Z' seen by the outer electron would give the correct first ionization potential?

11.21 The first ionization potentials of Na, K, and Rb are 5.138, 4.341, and 4.166 eV, respectively. Assume that the energy level of the outer electron can be represented by a hydrogenlike formula with an effective nuclear charge Z' and that the relevant orbitals are 3s, 4s, and 5s, respectively. Calculate Z' for these atoms.

11.22 What is the total electronic energy in hartrees of He, Li, and Be with respect to the nuclei and free electrons? Ionization potentials are given in Table 11.4.

11.23 Why is the ionization energy of boron less than beryllium and the ionization energy of oxygen less than nitrogen?

11.24 The enthalpy of formation of $H^-(g)$ at 0 K is given as 143.264 kJ mol^{-1} in Table C.2. What is the electron affinity of H(g)?

11.25 For a carbon atom in the configuration [He]2s^{2}2p^2 the following term symbols are all possible, $^1S_0$, $^3P_0$, $^3P_1$, $^3P_2$, and $^1D_2$. According to Hund's rules, which is the most stable state?

11.26 Which of the following transitions are allowed in the electronic spectrum of a hydrogenlike atom: (a) 2s → 1s, (b) 2p → 1s, (c) 3d → 1s, (d) 3d → 3p?

11.27 If we use Ψ as a trial function for the Hamiltonian $\hat{H}$, where

$$\Psi = \frac{(\phi_0 + a\phi_1)}{(1 + a^2)^{1/2}}$$

and ϕ_0 and ϕ_1 are the two *lowest* eigenfunctions with eigenvalues E_0 and E_1 (with $E_1 > E_0$), show that the variational method yields $a = 0$.

11.28 Calculate the classical energy of an electron and proton separated by 0.0529 nm, relative to the same system at infinite separation. Compare this to the quantum mechanical result for the energy.

11.29 For what value of n do adjacent energy levels in H have a separation of kT at room temperature?

11.30 What are the wavelengths of the first lines in the Balmer series for $^6Li^{2+}$ and $^7Li^{2+}$?

11.31 What is the expectation value for the radius of a hydrogen atom if $n = 50$ and $l = 0$?

11.32 Calculate $\langle r \rangle$ for a 2s electron in a hydrogen atom, given

$$\int_0^\infty e^{-ax}\, dx = \frac{n!}{a^{n+1}} \qquad n > -1, a > 0$$

11.33 Positronium consists of an electron and positron (positive particle with the same mass as an electron). (a) Calculate the wavelength of the radiation emitted when the system goes from the $n = 2$ orbital to the $n = 1$ orbital. (b) Calculate the ionization energy.

11.34 What are the most probable positions for an electron in a 2p$_z$ orbital in hydrogen?

11.35 A hydrogen atom is in a cubical box of 100 Å on a side. For what value of n (for an s state) does the expectation value of the radius equal one-half the box size?

11.36 Show that Ψ_{1s} and Ψ_{2s} are orthogonal for the H atom.

11.37 Using the integrals in problem 11.32, show that the 2s and 3s wavefunctions are normalized.

11.38 What is the most probable distance for a 1s electron in Li^{2+}?

11.39 What is the average distance from the electron to the proton for the 3s, 3p, and 3d states of the hydrogen atom?

11.40 Calculate the splitting in kJ/mol and eV for a H atom in the 2p state in a 10-T magnetic field, neglecting the spin as in Fig. 11.3 and equation 11.30. Compare to kT at room temperature.

11.41 What is the magnitude of the angular momentum for the electrons in 4s, 4p, 4d, and 4f orbitals? How many radial and angular modes are there for each of these orbitals?

11.42 Show that the following wavefunction for the hydrogen atom is antisymmetric to the interchange of the two electrons:

$$\Psi = \begin{vmatrix} 1s\alpha(1) & 1s\beta(1) \\ 1s\alpha(2) & 1s\beta(2) \end{vmatrix}$$

11.43 In a hydrogen atom, the 2s and 2p orbitals have the same energy. However, in a boron atom, the 2s orbital has a lower energy than the 2p. Explain this in terms of the shape of the orbitals.

11.44 Give the electronic configurations for the ground states of the first 18 electrons in the periodic table.

11.45 The first three ionization energies of scandium (Sc, atomic number 21) are

$$
\begin{aligned}
\text{Sc} &= \text{Sc}^+ + e^- & 6.54 \text{ eV} \\
\text{Sc}^+ &= \text{Sc}^{2+} + e^- & 12.8 \text{ eV} \\
\text{Sc}^{2+} &= \text{Sc}^{3+} + e^- & 24.75 \text{ eV}
\end{aligned}
$$

What are the electron configurations of the three species?

12

Molecular Electronic Structure

Quantum mechanics has made it possible to understand the nature of chemical bonding and to predict the structures and properties of simple molecules. Our ideas about covalent bonds go back to 1916, when Lewis described the sharing of electron pairs between atoms. The pairs of electrons held jointly by two atoms were considered to be effective in completing a stable electronic configuration for each atom. This approach provided only a qualitative picture of chemical bonding. The first successful quantum mechanical explanation of a chemical bond, specifically that in molecular hydrogen, was made in 1927 by Heitler and London using the valence bond method. Since then the molecular orbital method has become the method of choice, and so it is emphasized in this chapter. Today the electronic structure, energy levels, bond angles, bond distances, dipole moments, and spectra of simple molecules may be calculated with a high degree of accuracy. For molecules with more than one electron, approximations have to be introduced. However, even approximate calculations are very helpful in understanding molecular structure, chemical properties, and molecular spectra.

12.1 The Born–Oppenheimer Approximation

The Schrödinger equation for a molecule is

$$\hat{H}\psi(r, R) = E\psi(r, R) \qquad (12.1)$$

where the wavefunction depends on the coordinates of all the electrons (represented by r) and the coordinates of all the nuclei (represented by R). The solution of this equation is not known even for the simplest molecules; however, there is an approximation, introduced by Born and Oppenheimer, which gives a very good description of the solutions. The approximation is based on the fact that the nuclei are thousands of times more massive than the electrons, but the forces on each are equal. Thus, to a very good approximation, the nuclei can be treated as stationary when considering the electronic motion, and the nuclei can be considered to move in the average electronic configuration. To implement the approximation for the electronic motion, we simply neglect the nuclear kinetic energy. To simplify the discussion, we consider the hydrogen molecule ion, H_2^+, which has two protons at a separation R and one electron (see Fig. 12.1 for the coordinates). The **electronic Schrödinger equation** is given by

$$\left[-\frac{\hbar^2}{2m}\nabla_{el}^2 - \frac{e^2}{4\pi\epsilon_0}\left(\frac{1}{r_A} + \frac{1}{r_B} - \frac{1}{R}\right) \right]\psi_{el} = E(R)\,\psi_{el} \qquad (12.2)$$

where the electronic Hamiltonian contains the electronic kinetic energy, the electrostatic attraction of the electrons to each nucleus, and the nuclear electrostatic repulsion. Notice that the latter term is a constant since the internuclear distance R is fixed. We can solve this equation for all R, giving us the electronic energy $E(R)$ as a function of R. Actually there will be many solutions of this equation at each value of R, corresponding to different electronic states. The lowest energy electronic state is called the ground state and the higher energy states are called excited states. For the H_2^+ molecule, the ground-state $E(R)$ has the form given in Fig. 12.2a. Note that for other diatomic molecules with more electrons, the electronic Hamiltonian will contain terms similar to those in equation 12.2, but will include, in addition, electron–electron electrostatic repulsion. In spite of this complication, the ground-state electronic energy for *stable* diatomic molecules has the same shape as Fig. 12.2a. At large R, the molecule dissociates into atoms. For bound electronic states the electronic energy is lowered as R decreases until it reaches R_e, the equilibrium internuclear distance. As R gets still smaller, the internuclear repulsions and electron–electron repulsions become very large and the energy rises rapidly.

The **excited electronic states** can be bound (i.e., have the shape of Fig. 12.2a) or repulsive (i.e., have no minimum energy except at $R = \infty$) as in Fig. 12.2b. We examine these states by exciting the ground-state molecule with light or by scattering other molecules or electrons off the ground state. We will discuss these processes in later chapters.

The other aspect of the Born–Oppenheimer approximation is the approximate **Schrödinger equation for the nuclear motion**. Since the nuclei feel the average electronic configuration at each R, the electronic energy $E(R)$ becomes the potential energy for the nuclear motion and the Schrödinger equation becomes

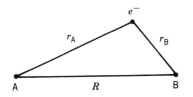

Figure 12.1 Electronic coordinates in H_2^+. The two protons are labeled A and B.

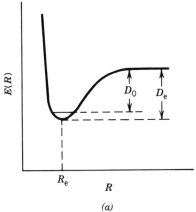

(a)

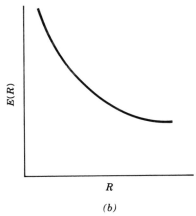

(b)

Figure 12.2 Potential energy $E(R)$ for a diatomic molecule as a function of internuclear distance R. In (a) a bound state exists. In (b), a purely repulsive potential, no bound state exists. D_0 is the dissociation energy from the rovibrational ground state and is often called the spectroscopic dissociation energy. D_e is the equilibrium dissociation energy, which is measured from the minimum of the potential energy.

$$\left[-\frac{\hbar^2}{2\mu} \nabla_R^2 + E(R) \right] \psi_{nu}(R) = E\psi_{nu}(R) \tag{12.3}$$

We will discuss this further in Chapter 14. Since the electronic energy as a function of R, $E(R)$, is the potential energy for nuclear motion, we often call $E(R)$ simply the potential energy curve or potential energy for short. But we should remember that it is the total electronic energy at that internuclear separation.

12.2   Dissociation Energies of Diatomic Molecules

As the atoms of a stable diatomic molecule are slowly brought toward each other from an infinite distance, there is a mutual attraction and a reduction in the electronic energy of the molecule. Then the energy goes through a minimum and rises rather steeply. The equilibrium internuclear distance R_e corresponds with the minimum in the energy curve. At this distance attraction and repulsion are balanced. The depth of the energy minimum with respect to the energy at infinite separation is referred to as the **equilibrium dissociation energy** D_e, or the well depth. This is the dissociation energy that is obtained most directly from the solutions of the Schrödinger equation for a diatomic molecule, but we must be careful to distinguish it from the **spectroscopic dissociation energy** D_0. The spectroscopic dissociation energy D_0 is the energy required to dissociate a molecule in its ground vibrational state into two atoms.

A diatomic molecule can exist in one of a series of vibrational energy levels (see Section 10.12), of which only the ground state is shown in Fig. 12.2. In accordance with the Heisenberg uncertainty principle (Section 10.5), a vibrating diatomic molecule has this zero-point energy, even in the lowest state. If the vibrational motion were that of a harmonic oscillator, the energy would be $\omega_e/2$, the energy of the lowest state of the harmonic oscillator above the potential minimum. Real molecules are not harmonic oscillators, so a correction to this has to be made. In Section 14.8 we will see that the equilibrium dissociation energy D_e for a diatomic molecule may be calculated from the spectroscopic dissociation energy using

Table 12.1   Dissociation Energies for H_2^+ (g) and H_2 (g) and Ionization Potentials (IP) for H_2 (g) and H (g)

	eV	cm^{-1}	kJ mol^{-1}
	H_2^+		
D_0	2.650 79	21 380	255.760
D_e	2.793	22 527	269.481
	H_2		
D_0	4.477 97	36 117	432.055[a]
D_e	4.748 3	38 297	458.135
IP (H_2)	15.425 9	124 417	1488.361
IP (H)	13.598 396	109 677.6	1312.035

[a] This spectroscopic dissociation energy of H_2 is in agreement with $\Delta H_0^\circ = 432.074$ kJ mol^{-1} calculated from Table C.2.

$$D_e = D_0 + \frac{\omega_e}{2} - \frac{\omega_e x_e}{4} + \frac{\omega_e y_e}{8} \qquad (12.4)$$

where ω_e is the harmonic vibration wavenumber, and $\omega_e x_e$ and $\omega_e y_e$ are anharmonicity constants. The constants ω_e, $\omega_e x_e$, and $\omega_e y_e$ are determined from vibrational spectra and are usually expressed in wavenumbers, so one must be careful to express D_e, D_0, and ω_e in wavenumbers when using this equation.

The two types of dissociation energies D_0 and D_e for H_2^+ and H_2 are given in Table 12.1. The values of dissociation energies are given in eV, cm^{-1}, and kJ mol^{-1} because each of these units has certain advantages.* Spectroscopic measurements in the ultraviolet, visible, and infrared regions are usually expressed in terms of wavelengths or wavenumbers. Thus, experimental results are usually expressed in cm^{-1} because conversion to other units involves the use of physical constants whose values may change. The results of theoretical calculations are frequently expressed in electronvolts.

Before discussing the quantum mechanical calculation of the potential energy curves of H_2^+ and H_2, we must discuss the relations between these two molecules and certain aspects of their potential energy curves. The potential energy curves for the ground states of H_2 and H_2^+ are shown in Fig. 12.3. In this figure energies are measured with respect to the ground state of the hydrogen molecule. The energy of two protons and two electrons, infinitely far from each other, is shown. The ionization energy $E_i(H_2)$ for hydrogen is the energy required to remove an electron to an infinite distance from H_2^+, and has been measured accurately:

$$H_2(g) = H_2^+(g) + e^- \qquad E_i(H_2) = 15.4259 \text{ eV} \qquad (12.5)$$

Thus, the zero-point levels of H_2 and H_2^+ are separated by 15.4259 eV, as shown in Fig. 12.3.

The potential energy curves for H_2 and H_2^+ at infinite internuclear distance are separated by the ionization potential of a hydrogen atom in its ground state. This ionization potential was calculated in Example 11.1:

$$H(g) = H^+(g) + e^- \qquad E_i(H) = 13.598\,396 \text{ eV} \qquad (12.6)$$

As can be seen from Fig. 12.3,

$$E_i(H_2) + D_0(H_2^+) = D_0(H_2) + E_i(H) \qquad (12.7)$$

Since it is difficult to measure $D_0(H_2^+)$, this relation has been used to obtain the most accurate value for this quantity.†

---

* The relation between eV and cm^{-1} is given by

$$1 \text{ eV} = \frac{Ee}{hc} = \frac{(1 \text{ V})(1.602\,177\,33 \times 10^{-19} \text{ C})}{(6.626\,075\,5 \times 10^{-34} \text{ J s})(2.997\,924\,58 \times 10^8 \text{ m s}^{-1})(100 \text{ cm m}^{-1})}$$

$$= 8065.541 \text{ cm}^{-1}$$

The relation between eV and kJ mol^{-1} is given by

$$1 \text{ eV} = EeN_A = \frac{(1 \text{ V})(1.602\,177\,33 \times 10^{-19} \text{ C})(6.022\,136\,7 \times 10^{23} \text{ mol}^{-1})}{(10^3 \text{ J kJ}^{-1})}$$

$$= 96.485\,309 \text{ kJ mol}^{-1}$$

† G. Herzberg, *Science* **177**: 123 (1972).

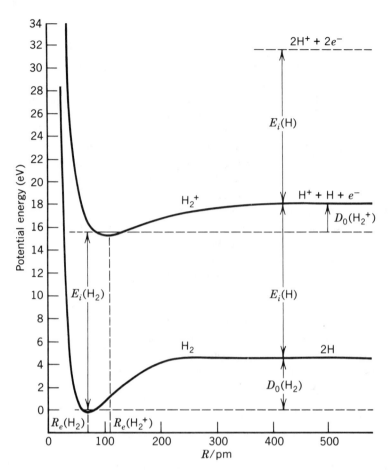

Figure 12.3   Potential energy curves for the ground electronic states of H_2 and H_2^+ with the zero-point vibrational levels shown.

Example 12.1

Calculate the spectroscopic dissociation energy of H_2^+ from the spectroscopic dissociation energy of H_2 and the ionization potential of H and H_2.

$$D_0(H_2^+) = D_0(H_2) + E_i(H) - E_i(H_2)$$

$$= 4.477\ 97 + 13.598\ 396 - 15.4259$$

$$= 2.6505 \text{ eV or } 255.73 \text{ kJ mol}^{-1}.$$

Now that we have placed these potential energy curves in a context, we want to talk about their calculation from quantum mechanics. We want to understand why a stable molecule H_2^+ is formed when two protons and an electron are brought together, and why an even more stable molecule H_2 is formed when two protons and two electrons are brought together.

12.3 The Hydrogen Molecule Ion

The electronic Schrödinger equation for this simplest molecule, which is given by equation 12.3, can be solved exactly. The details of this calculation are not given here, but the results are. At each internuclear distance a series of values of $E_{el}(R)$, and therefore of $E(R)$, are obtained. For the ground electronic state, H_2^+ is dissociated into a proton and a ground-state hydrogen atom at $R = \infty$. The electronic energy $E(R)$ of the hydrogen molecule ion calculated using equation 12.2 is shown in Fig. 12.4.

It is useful to think about the electronic energy $E(R)$ without the internuclear repulsion energy, equal to $1/(4\pi\epsilon_0 R)$. We define

$$E_{el}(R) = E(R) - \frac{e^2}{4\pi\epsilon_0 R} \qquad (12.8)$$

It is also convenient to discuss these energies in units of $E_h = 27.211\ 396\ 1$ eV. We know the value of $E_{el}(R)$ at two values of R. At $R = \infty$, the energy is equal to that of a hydrogen atom and a proton infinitely far from each other, $-E_h/2$. At $R = 0$, $E_{el}(R = 0)$ must be equal to the energy of a helium ion (since at $R = 0$, the two protons have the same attraction to the electron as a helium nucleus), so $E_{el}(0) = -2E_h$. As Fig. 12.4 shows, $E_{el}(R)$ is a smooth function of R. The balance between $E_{el}(R)$ and the internuclear repulsion leads to an equilibrium distance R_e of 106 pm and a dissociation energy D_e of 2.793 eV (269.483 kJ mol^{-1}).

The electronic Schrödinger equation can be solved exactly for H_2^+ and yields a set of orbitals that are molecular in the sense that they extend over the entire molecule. We could use these as a basis to discuss homonuclear diatomic molecules, just as we used the H atom orbitals to discuss the electronic states of many electron atoms. The exact molecular orbitals for H_2^+ are not simple to write down, but we have a very good approximate form for them based on the atomic orbitals of the H atoms. This approach is called the **linear combination of atomic orbitals** (or LCAO) method. In this approach, we begin by noting that if the two protons are far apart, either one could have an electron in the 1s hydrogen atom orbital. Therefore, as a first approximation, the electronic wavefunction of H_2^+ is

$$\psi = c_1 1s_A + c_2 1s_B \qquad (12.9)$$

where $1s_A$ and $1s_B$ are normalized 1s wavefunctions associated with atoms A and B, and the constants c_1 and c_2 must be evaluated by the variation method (Section 11.5).

The variation energy E is given by

$$E = \frac{\int \psi^* \hat{H} \psi \, d\tau}{\int \psi^* \psi \, d\tau} = \frac{\int [c_1 1s_A + c_2 1s_B] \hat{H} [c_1 1s_A + c_2 1s_B] \, d\tau}{\int [c_1 1s_A + c_2 1s_B]^2 \, d\tau}$$

$$= \frac{c_1^2 H_{AA} + 2c_1 c_2 H_{AB} + c_2^2 H_{BB}}{c_1^2 S_{AA} + 2c_1 c_2 S_{AB} + c_2^2 S_{BB}} = \frac{c_1^2 H_{AA} + 2c_1 c_2 H_{AB} + c_2^2 H_{BB}}{c_1^2 + 2c_1 c_2 S + c_2^2}$$

$$(12.10)$$

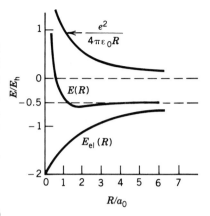

Figure 12.4 Electronic energy of the H_2^+ ground state as a function of the distance between the nuclei. Energy is plotted (in hartrees E_h) with respect to two protons and an electron, all at infinite distance. (From I. N. Levine, *Quantum Chemistry*, 2d ed. Copyright © 1974 by Allyn & Bacon, Inc., Boston. Reprinted with permission of the publisher.)

where the following symbols have been used:

$$H_{AA} = \int 1s_A \hat{H} 1s_A \, d\tau = \int 1s_B \hat{H} 1s_B \, d\tau = H_{BB} \tag{12.11}$$

$$H_{AB} = \int 1s_A \hat{H} 1s_B \, d\tau = \int 1s_B \hat{H} 1s_A \, d\tau \tag{12.12}$$

$$S_{AA} = \int 1s_A 1s_A \, d\tau = \int 1s_B 1s_B \, d\tau = S_{BB} = 1 \tag{12.13}$$

$$S_{AB} = \int 1s_A 1s_B \, d\tau = \int 1s_B 1s_A \, d\tau = S_{BA} = S \tag{12.14}$$

The integrals represented by S are referred to as **overlap integrals.** According to the variation method, the best values of c_1 and c_2 in equation 12.9 are the ones that give the lowest value of E. To find the minimum E, we set the derivatives of E with respect to c_1 and c_2 equal to zero. It is easier to multiply equation 12.10 out and take the derivative of the result to find

$$\frac{\partial E}{\partial c_1} = \frac{2c_1(H_{AA} - E) + 2c_2(H_{AB} - SE)}{c_1^2 + 2c_1 c_2 S + c_2^2} = 0 \tag{12.15}$$

$$\frac{\partial E}{\partial c_2} = \frac{2c_1(H_{AB} - SE) + 2c_2(H_{BB} - E)}{c_1^2 + 2c_1 c_2 S + c_2^2} = 0 \tag{12.16}$$

The values of c_1 and c_2 are then obtained by solving the two simultaneous equations:

$$c_1(H_{AA} - E) + c_2(H_{AB} - SE) = 0 \tag{12.17}$$

$$c_1(H_{AB} - SE) + c_2(H_{BB} - E) = 0 \tag{12.18}$$

There is a nontrivial solution for c_1 and c_2 from these simultaneous equations only if the determinant of the coefficients of c_1 and c_2 is equal to zero:

$$\begin{vmatrix} H_{AA} - E & H_{AB} - SE \\ H_{AB} - SE & H_{BB} - E \end{vmatrix} = 0 \tag{12.19}$$

This equation is referred to as a **secular determinant.** In this case it is a quadratic with two solutions E_g and E_u:

$$E_g = \frac{H_{AA} + H_{AB}}{1 + S} \tag{12.20}$$

$$E_u = \frac{H_{AA} - H_{AB}}{1 - S} \tag{12.21}$$

The integral H_{AA} is called the **Coulomb integral** because the difference between H_{AA} and the energy of a single hydrogen atom is just that of the Coulomb interaction of nucleus B with an electron centered on nucleus A ($H_{AA} = H_{BB}$ in this case). Since this interaction is attractive, its contribution to H_{AA} is negative. The energy of a single hydrogen atom is also negative: thus, H_{AA} is a negative number. The integral H_{AB} is referred to as the **resonance integral.** When equation 12.21 or 12.20 is substituted back into equation 12.17, it is found that

$$c_1 = c_2 \quad \text{or} \quad c_1 = -c_2 \tag{12.22}$$

Substituting this relation into equation 12.9 and normalizing yields

$$\psi_g = \frac{1}{[2(1 + S)]^{1/2}} [1s_A + 1s_B] \tag{12.23}$$

$$\psi_u = \frac{1}{[2(1 - S)]^{1/2}} [1s_A - 1s_B] \tag{12.24}$$

Because of the symmetry of H_2^+, it is not surprising that the orbitals of the two protons are weighted equally. There is no reason for the electron to prefer one nucleus over the other. These functions are good approximations to the exact solutions.

When a molecule has a center of symmetry, the wavefunction may or may not change sign when it is inverted through the center of symmetry. If $\psi(x, y, z) = \psi(-x, -y, -z)$ the wavefunction is said to have **even parity** and is designated with a subscript g for *gerade* (German for even). If $\psi(x, y, z) = -\psi(-x, -y, -z)$ the wavefunction is said to have **odd parity** and is designated with a subscript u for *ungerade* (German for odd). This is the origin of the subscripts on ψ_g and ψ_u.

The electron probability densities along the line of the two nuclei are given for ψ_g and ψ_u in Fig. 12.5. A three-dimensional plot is shown in Fig. 12.6. The squares of the wavefunctions

$$\psi_g^2 = \frac{1}{2(1 + S)} [(1s_A)^2 + 2(1s_A)(1s_B) + (1s_B)^2] \tag{12.25}$$

$$\psi_u^2 = \frac{1}{2(1 - S)} [(1s_A)^2 - 2(1s_A)(1s_B) + (1s_B)^2] \tag{12.26}$$

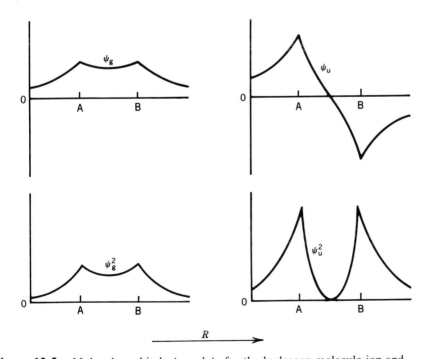

Figure 12.5 Molecular orbitals ψ_g and ψ_u for the hydrogen molecule ion and probability densities ψ_g^2 and ψ_u^2 along the internuclear axis. A and B label the nuclear positions.

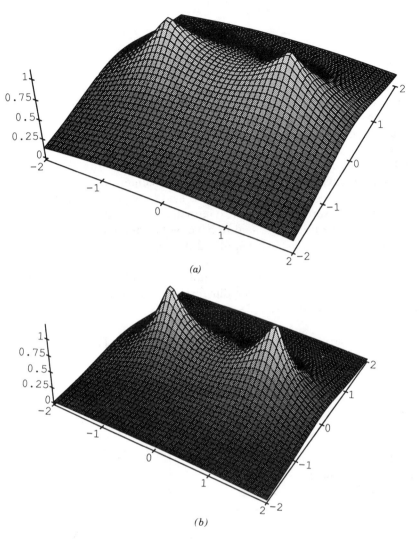

Figure 12.6 (a) ψ_g orbital for H_2^+ for $z = 0$ (i.e., in the molecular plane). The nuclei are at the points $x = +1$ and -1, $y = 0$. (b) ψ_g^2 for H_2^+ for $z = 0$. (c) ψ_u orbital for H_2^+. (d) ψ_u^2 for H_2^+ for $z = 0$.

show that the electron density is increased between the nuclei with ψ_g^2 and decreased with ψ_u^2. In fact, in ψ_u^2 the electron density is zero in a plane bisecting the line between the protons; that is, there is a node midway between the protons. In ψ_g there is **constructive interference** between $1s_A$ and $1s_B$ and in ψ_u there is **destructive interference.** The ground-state wavefunction for H_2^+ is ψ_g. Since this orbital is symmetrical around the internuclear axis, it is referred to as a sigma (σ) orbital, and since it is even (gerade) and is made up of two 1s orbitals, it is designated $\sigma_g 1s$. In addition, because its energy is below that at infinite separation of the atoms, it is called a bonding orbital.

If the electron is placed in the state with the wavefunction ψ_u the energy of the molecule is more than that of the dissociated molecule ($H + H^+$), and so this molecular orbital is referred to as an **antibonding orbital.** The antibonding

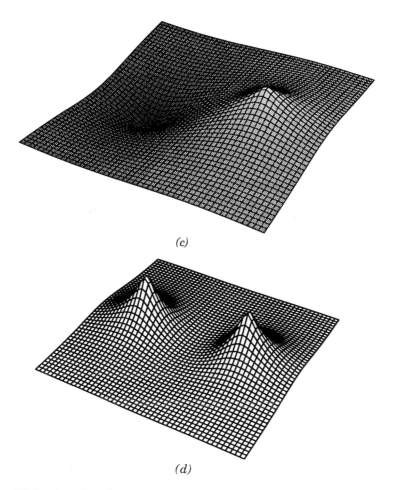

(c)

(d)

Figure 12.6   *(continued)*

orbital is symmetrical about the internuclear axis and is referred to as a $\sigma_u 1s$ orbital. The energy for ψ_u is higher than that for ψ_g for all R.

The value of E_g calculated from equation 12.20 by carrying out the indicated integrations for different values of R does show a minimum, and so this very simple molecular orbital theory does account for the bonding in H_2^+. However, it yields $R_e = 123$ pm (experiment 106 pm) and $D_e = 1.77$ eV (experiment 2.793 eV). The calculations can be improved by adding further terms to the wavefunctions and using the variation method to evaluate further adjustable parameters. However, this simple calculation shows that molecular bonding occurs when electron density accumulates in the region between the nuclei as in the $\sigma_g 1s$ orbital, and does not occur when electron density is pushed away from this region, as in the $\sigma_u 1s$ orbital. A fuller explanation of molecular bond-

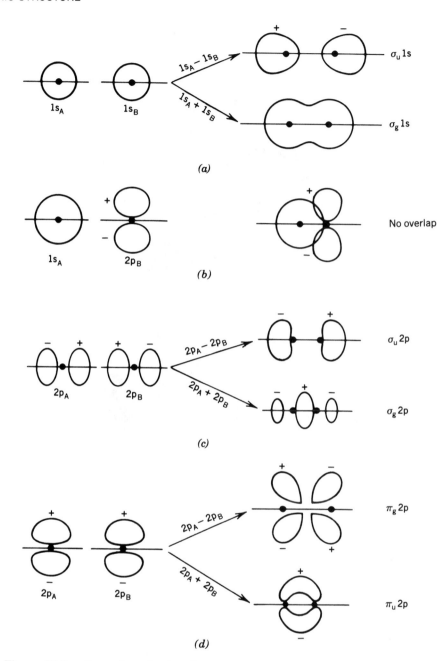

Figure 12.7 Formation of pairs of molecular orbitals from pairs of atomic orbitals. The solid points represent nuclei A and B.

ing requires attention to many details beyond the scope of this book. (See, for example, Hehre et al. in the references at the end of this chapter).

We can use the same LCAO procedure to form other molecular orbitals from the atomic orbitals on the two protons. This is useful only if the molecular orbitals that are found have the bonding characteristics of the $\sigma_g 1s$ or $\sigma_u 1s$ orbitals we have discussed. Two conditions must be met for this to be true:

The two atomic orbitals must have about the same energy and they must have the same symmetry properties with respect to rotations about the internuclear axis. If the latter condition is not satisfied then the integrals corresponding to H_{AB} and to S_{AB} are zero for all R; if the former condition is not satisfied then the integrals corresponding to H_{AA} and H_{BB} will be very different. In either case the coefficients c_1 and c_2 will be very different (one approximately 0 and one approximately 1) so that little or no bonding takes place. It is quite easy to determine whether the overlap integral S_{AB} is zero by examining the positive and negative parts of the orbitals, as indicated in Fig. 12.7.

Figure 12.7a shows the situation when two 1s orbitals (or two ns orbitals) are brought together: The orbitals overlap and bonding can occur. Figure 12.7b shows what happens when a p orbital with axis perpendicular to the internuclear axis is brought up to an s orbital. Here the overlap is zero since the overlap of the positive lobe of the p orbital with the s orbital is exactly canceled by the overlap of the negative lobe of the p orbital and thus no binding can occur. If two parallel p orbitals are brought together, as in Fig. 12.7c and d, bonding can occur. The orbitals formed in the last line of the figure are interesting because they have a nodal plane (the wavefunction is zero everywhere in the plane) perpendicular to the page and containing the internuclear axis. Such an orbital is called a π orbital. The bonding orbital changes sign on inversion and so is ungerade ($\pi_u 2p$), while the antibonding orbital is gerade ($\pi_g 2p$). Note that there is another pair of orbitals, exactly like these, formed from the 2p orbitals perpendicular to the page. Therefore, the π orbitals are doubly degenerate.

This process of combining hydrogenlike orbitals can be continued to higher values of the orbital angular momentum quantum number l. One-electron molecular orbitals are classified according to the quantum number λ for the angular momentum about the internuclear axis. In general, for an electron in a diatomic molecule the axial angular momentum is given by

$$L_z = \pm \lambda \hbar \tag{12.27}$$

where λ corresponds with the absolute value m for an atom. The value of λ for a one-electron orbital for a diatomic molecule is represented by a Greek letter according to

$$\lambda \quad = 0 \quad 1 \quad 2 \quad 3$$
$$\text{Orbital} = \sigma \quad \pi \quad \delta \quad \phi$$

This is, of course, analogous to the classification of atomic orbitals as s, p, d, f, . . . according to $l = 0, 1, 2, 3, ...$

Now we have the problem of arranging these LCAO-MOs in order of increasing energy, so that electrons can be placed in these orbitals, two at a time, to account for the electronic structures of various diatomic molecules. This is the same aufbau process that was used with atoms (Section 11.11) and will be discussed in Section 12.5. The sequence of homonuclear diatomic molecular orbitals is given approximately in Fig. 12.8. We have to say approximately because the order of the energy levels depends on the atomic number of the nuclei and the internuclear distance. As indicated in this diagram, two atomic orbitals combine to form two molecular orbitals, one with lower energy and the other with higher energy than the atomic orbitals from which they were formed. Excited states of H_2^+ are formed by exciting the electron to one of the higher levels, as shown by the potential energy curves in Fig. 12.9.

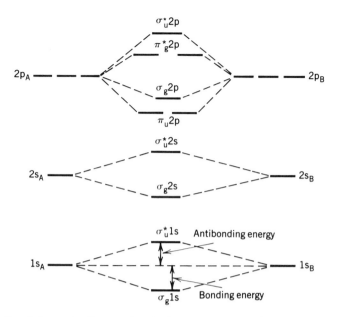

Figure 12.8   Schematic diagram for the lowest energy molecular orbitals of homonuclear diatomic molecules.

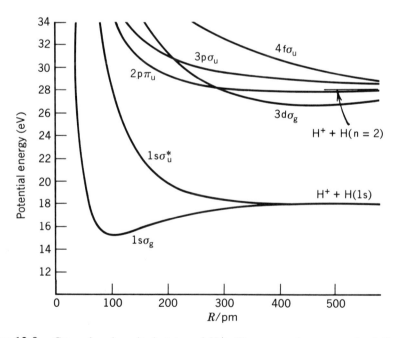

Figure 12.9   Ground and excited states of H_2^+. The energy is measured relative to the ground state of the hydrogen molecule. (From T. Sharp, Potential energy diagram for molecular hydrogen and its ions. In *Atomic Data*, Vol. 2, p. 119. New York: Academic, 1971.)

12.4 The Molecular Orbital Description of the Hydrogen Molecule

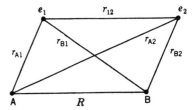

Using the Born–Oppenheimer approximation, the electronic Hamiltonian for the hydrogen molecule may be written

$$\hat{H} = -\frac{\hbar^2}{2m}(\nabla_1^2 + \nabla_2^2) + \frac{e^2}{4\pi\epsilon_0}\left(-\frac{1}{r_{A1}} - \frac{1}{r_{A2}} - \frac{1}{r_{B1}} - \frac{1}{r_{B2}} + \frac{1}{r_{12}}\right) + \frac{e^2}{4\pi\epsilon_0 R}$$

(12.28)

Figure 12.10 Electronic coordinates in the hydrogen molecule. The two protons are represented by A and B.

where the coordinates are defined in Fig. 12.10.

When the electronic Hamiltonian for the hydrogen molecule is used in equation 12.2 an exact solution cannot be obtained because of the $1/r_{12}$ term, just as in the atomic case. This is the reason that the LCAO-MO method is used to obtain an approximate solution. According to the LCAO-MO approach molecular hydrogen is formed by putting two electrons with opposite spin in the $\sigma_g 1s$ orbitals. That is, we assume that each electron can be assigned to an orbital and that the electronic wavefunction for the molecule is the product of the two wavefunctions for the two electrons:

$$\psi_{MO} = \psi_i(1)\psi_j(2)$$

(12.29)

where i and j designate the different orbitals and 1 and 2 designate the two electrons. According to the Pauli principle (Section 11.7), two electrons with opposite spin can be assigned to a given spatial orbital, and so as a first approximation we will assume that in the ground state of the hydrogen molecule the two electrons are placed in the $1\sigma_g$ orbital developed for H_2^+. Thus, the electronic configuration of H_2 will be described as $(1\sigma_g)^2$, just as we described the electronic configuration of He as $(1s)^2$.

The wavefunction for electron 1 in the $1\sigma_g$ molecular orbital given in equation 12.23 is represented by

$$1\sigma_g(1) = \frac{1}{[2(1 + S)]^{1/2}}[1s_A(1) + 1s_B(1)]$$

(12.30)

In discussing the helium atom (Section 11.6) we found that the wavefunction satisfying the antisymmetry requirement is given by a Slater determinant. The same considerations apply here and so the approximate wavefunction for the ground state of the hydrogen molecule is given by the following Slater determinant:

$$\psi_{MO}[(1\sigma_g)^2] = \frac{1}{\sqrt{2}}\begin{vmatrix} 1\sigma_g(1)\alpha(1) & 1\sigma_g(1)\beta(1) \\ 1\sigma_g(2)\alpha(2) & 1\sigma_g(2)\beta(2) \end{vmatrix}$$

(12.31)

This yields the following wavefunction for a hydrogen molecule in its ground state:

$$\psi_{MO}[(1\sigma_g)^2] = [1\sigma_g(1)1\sigma_g(2)\alpha(1)\beta(2) - 1\sigma_g(1)1\sigma_g(2)\beta(1)\alpha(2)](1/2)^{1/2}$$

$$= \frac{[1s_A(1) + 1s_B(1)][1s_A(2) + 1s_B(2)](1/2)^{1/2}[\alpha(1)\beta(2) - \beta(1)\alpha(2)]}{2(1 + S_{AB})}$$

(12.32)

The approximate energy of the hydrogen molecule is obtained by calculating the expectation value of the Hamiltonian (equation 12.28) using this wavefunction,

$$E = \int \psi_{MO}^*[(1\ \sigma_g)^2]\hat{H}_{el}\psi_{MO}[(1\sigma_g)^2]\ d\tau \tag{12.33}$$

This integration leads to a rather complicated equation for E which we will simply write as

$$E = 2E_{1s} + \frac{e^2}{4\pi\epsilon_0 R} - \text{integrals} \tag{12.34}$$

The first term is the electronic energy of two hydrogen atoms at infinite distance. The second term is the energy of electrostatic repulsion of the two nuclei, and the last term is a series of integrals for the interactions of various charge distributions with each other.* These integrals can be evaluated at a series of internuclear distances to obtain the molecular potential energy curve. The minimum is at 84 pm and the calculated dissociation energy is $D_0 = 255$ kJ mol^{-1}. The experimental values are 74.1 pm and 458 kJ mol^{-1}.

Although simple molecular orbital theory does account for a large proportion of the binding energy of the hydrogen molecule, it has to be extended to yield accurate results. A detailed description of the improvements that are possible is beyond the scope of this book, but the general directions can be indicated.

We can see one deficiency in the approximate wavefunction by multiplying out the spatial part of equation 12.32 to obtain

$$1s_A(1)1s_A(2) + 1s_A(1)1s_B(2) + 1s_B(1)1s_A(2) + 1s_B(1)1s_B(2) \tag{12.35}$$

The first term and last term correspond to forms of the hydrogen molecule with ionic bonding, namely $H_A^- H_B^+$ and $H_A^+ H_B^-$, so that this molecular orbital wavefunction describes a state at $R = \infty$ that is 50% H^+ and H^- and 50% $H + H$, which is clearly not correct. This problem can be reduced by introducing variable coefficients c_1 and c_2 in

$$\psi = c_1(R)\psi_{covalent} + c_2(R)\psi_{ionic} \tag{12.36}$$

where

$$\psi_{covalent} = 1s_A(1)1s_B(2) + 1s_A(2)1s_B(1) \tag{12.37}$$

$$\psi_{ionic} = 1s_A(1)s_A(2) + 1s_B(1)1s_B(2) \tag{12.38}$$

Using the variation method the values of c_1 and c_2 can be determined at each value of R. Since equation 12.36 yields $R_e = 74.9$ pm (experimental 74.1 pm) and $D_e = 386$ kJ mol^{-1} (experimental 458 kJ mol^{-1}), the inclusion of one variational parameter leads to considerable improvement.

Further improvements can be obtained by increasing the number of atomic orbitals used, that is, enlarging the basis set, as, for example, by adding 2s and $2p_z$ orbitals, and thereby introducing more variational parameters. The evaluation of parameters can be done in a systematic way by using the Hartree–Fock self-consistent field method (Section 11.10) that is used to obtain atomic orbitals. Equations of the form

$$\hat{H}_{eff}\psi_i = E_i\psi_i \tag{12.39}$$

where $\hat{H}_{eff}$ is the effective one-electron Hamiltonian, ψ_i is the molecular orbital

---

* These integrals are all given in P. W. Atkins, *Molecular Quantum Mechanics*, Appendix 14. New York: Oxford University Press, 1983.

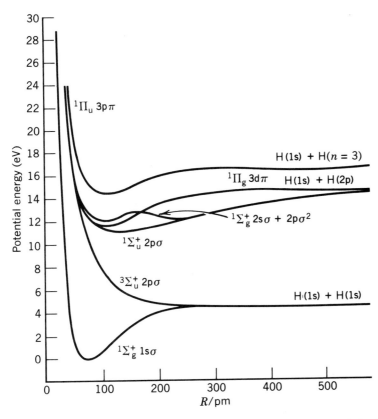

Figure 12.11   Potential energy curves for ground and excited states of H_2. (From T. Sharp, Potential energy diagram for molecular hydrogen and its ions. In *Atomic Data*, Vol. 2, p. 119. New York: Academic, 1971.)

for the ith electron, and E_i is the orbital energy for the ith electron, are solved by the iterative methods described in Section 11.10 for atoms to obtain self-consistent molecular orbitals. These Hartree–Fock wavefunctions do not adequately correlate the motion of electrons with unlike spins, and further improvements can be obtained by using a wavefunction that is a linear combination of functions representing different electronic configurations of the molecule. For example, the doubly excited configuration $(1\sigma_u)^2$ may be added to the molecular orbital because it has the same symmetry. By using 100-term wavefunctions Kolos and Wolniewicz* have obtained $D_0 = 36\ 117.8\ \text{cm}^{-1}$ in comparison with the observed† value of $36\ 117.3 \pm 1.0\ \text{cm}^{-1}$. Their theoretical value of the internuclear distance R_e of H_2 is 74.140 pm, compared with the experimental value from spectroscopic measurements of 74.139 pm.

The hydrogen molecule has a large number of excited electronic states. The potential energy curves for some of the lowest are shown in Fig. 12.11. The states of the H_2 molecule (or any diatomic molecule) are given by molecular

---

* W. Kolos and L. Wolniewicz, *J. Chem. Phys.* **41**: 3663 (1964); **48**: 3672 (1968); **49**: 404 (1968).

† G. Herzberg, *J. Mol. Spectrosc.* **33**: 147 (1970); *Science* **177**: 123 (1972).

term symbols, analogous to the atomic term symbols of Section 11.14. In diatomic molecules the orbital angular momenta of the electrons couple to give a resultant orbital angular momentum L, and the electron spin momenta combine to give a resultant spin angular momentum S. The component of the orbital angular momentum along the axis of the molecule is given by

$$M_L = m_1 + m_2 + \cdots \tag{12.40}$$

where $m_i = 0$ for a σ orbital, $m_i = \pm 1$ for a π orbital, and so on. The quantum number Λ is defined as the absolute value of M_L and is represented by the code letters

$$
\begin{array}{cccc}
\Lambda & = 0 & 1 & 2 & 3 \\
\text{Symbol} & = \Sigma & \Pi & \Delta & \Phi
\end{array}
$$

in analogy to atomic term symbols. The multiplicity of a state of a diatomic molecule is given by $2S + 1$, where S is the sum of the spins of the electrons in the molecule. The term symbol of a molecule is represented by

$$^{2S+1}\Lambda \tag{12.41}$$

For the hydrogen molecule in the ground state, there are two electrons in σ orbitals and so $m_1 = 0$, $m_2 = 0$, and $M_L = 0$. Since $\Lambda = 0$, the molecule is in a Σ state. Since the spins of the electrons are opposed, $S = 0$, and the molecular term symbol is $^1\Sigma$ (a singlet sigma state).

For Σ terms a superscript of plus or minus is added according to the behavior of the wavefunction upon reflection in the plane containing the internuclear axis. A plus sign indicates that the wavefunction is invariant under this operation, and a negative sign indicates the wavefunction changes sign on reflection in this plane. If a diatomic molecule has a center of symmetry a right subscript of g or u is attached to the term symbol to denote the parity of the orbital. As we have seen before (Section 12.3), the parity of an orbital is determined by observing the inversion symmetry. When a point on an orbital is inverted an equal distance through the center of the molecule, the orbital is *gerade* if it has the same sign at the two points. The parity of a multielectron molecule is obtained by noting g or u for every orbital and forming products using g × g = g, g × u = u, and u × u = g.

There are four kinds of angular momenta in a diatomic molecule, of which we have discussed two: the electronic orbital angular momentum L and the electronic spin angular momentum S. The other two are nuclear spin angular momentum I, which we will not consider, and the angular momentum N of the rotational motion of the nuclear framework. These angular momenta are coupled by small terms in the Hamiltonian. For low-mass diatomic molecules, the spin–orbit coupling is weak (called Hund's case (a) in the molecular spectroscopist's jargon), and the result of the coupling is illustrated in Fig. 12.12. The components of L ($\Lambda\hbar$) and S ($\Sigma\hbar$) along the internuclear axis are added to form a total angular momentum component along that axis labeled $\Omega\hbar$. Ω is given by $|\Lambda + \Sigma|$. These quantum numbers will come up again when we discuss selection rules in molecular spectroscopy.

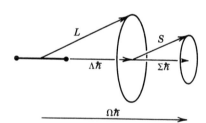

Figure 12.12   Orbital angular momentum L and spin angular momentum S for a diatomic molecule when the spin–orbit coupling is weak, as in Hund's case (a). (Reproduced from J. M. Hollas, *High Resolution Spectroscopy*, Fig. 6.31, p. 316. Boston: Butterworth, 1982; by permission of the publishers, Butterworth & Co. (Publishers) Ltd. ©.)

12.5 Electron Configurations of Homonuclear Diatomic Molecules

As mentioned in Section 12.3, the order in which molecular orbitals of homonuclear diatomic molecules are filled depends upon the nuclear charge and the internuclear distance. The variation in the sequence in energies is similar to that encountered with the relative energies of atomic orbitals of the elements which depend on atomic number (Fig. 11.7). To derive the possible sequences from the lowest energy up, it is useful to introduce a second way of bridging the gap between atomic and molecular systems. In addition to the separated atoms approach, there is the **united atom** approach. In this approach, the molecule is thought of first as the atom obtained from coalescing all the nuclei in the molecule. The united atom for H_2 is ^2He, and the united atom for N_2 is ^{28}Si. The electronic structure of the molecule is obtained by thinking about the changes in orbitals that occur when the nucleus of the united atom is pulled apart to form the nuclei of the molecule at their equilibrium distances. In the **correlation diagram** in Fig. 12.13 the possible energy levels of the united atom for a homonuclear diatomic molecule are given on the left and the sum of the energies of the two separated atoms are given on the right. An important principle for correlation diagrams is the noncrossing rule which states that the lines for the energies of molecular orbitals with the same symmetry cannot cross. Symmetry in this context refers to whether the orbital is $\sigma, \pi, \delta, \ldots$ or whether it is g or u. The orbitals are connected in the order of increasing energy. The molecular orbitals coming out of the united atoms are designated $1s\sigma_g$, $2s\sigma_g$, $2p\sigma_u$, . . ., and the resulting molecular orbitals are designated $1\sigma_g$, $1\sigma_u$, $2\sigma_g$, ... The reason for dropping the s, p, d, . . . in the latter is that different angular momentum quantum numbers may be involved in the united atom and separated atoms approaches; for example, $2p\sigma_u$ correlates with $\sigma_u 1s$.

The electronic structures of successive homonuclear diatomic molecules may be obtained from the correlation diagram by use of the aufbau principle; that is, electrons are added to orbitals in pairs, in order of increasing energy. Notice that the order of orbital energies varies with R. Two electrons may be placed in a σ level and four in a π or δ level (since $L_z = \pm \lambda \hbar$, making these doubly degenerate orbitals). The electron configurations of the ground states of homonuclear diatomic molecules from the first row of the periodic table are given in Table 12.2. In molecules with many electrons the inner electrons tend to be concentrated around the nuclei and take little part in forming bonds. The fact that a configuration is listed does not mean that the molecule is stable. Orbitals that tend to build up electron density between the atoms are referred to as **bonding orbitals.** The bonding orbitals are $1\sigma_g$, $2\sigma_g$, $1\pi_u$, $3\sigma_g$, ... Orbitals with nodal planes between the nuclei are **antibonding orbitals** (Section 12.3). Thus, $1\sigma_u$, $2\sigma_u$, $1\pi_g$, and $3\sigma_u$ are antibonding orbitals. Notice that the orbitals of homonuclear diatomic molecules come in pairs, one bonding and one antibonding. The bonding energy of an electron in a bonding orbital is usually slightly less than the antibonding energy of an electron in an antibonding orbital.

It is useful to define a **bond order** that is roughly proportional to the strength of bonding. The bond order is equal to the number of bonding pairs of electrons minus the number of antibonding pairs. Bond orders of 1, 2, and 3 correspond with what are usually called single, double, and triple bonds.

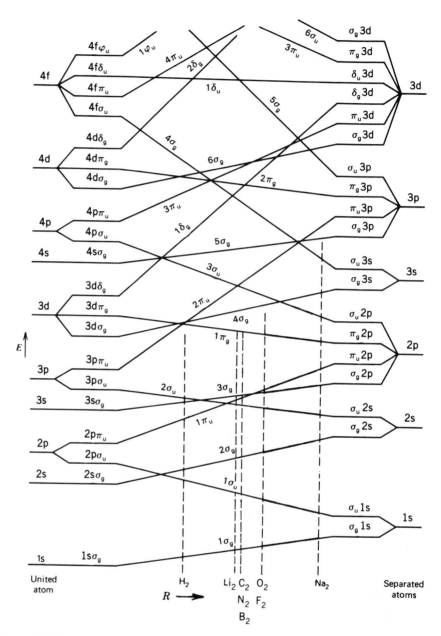

Figure 12.13   Correlation diagram for homonuclear diatomic molecules. The energy scale is schematic, but dashed lines show the correct sequence for filling of orbitals for the molecules indicated. (From R. S. Berry, S. A. Rice, and J. Ross, *Physical Chemistry*. New York: Wiley, 1980.)

We now discuss, one by one, the homonuclear diatomic molecules of the first row elements.

He$_2^+$

The third electron in this molecular ion is in an antibonding orbital, but there

Table 12.2   Ground States of Homonuclear Diatomic Molecules and Ions

Molecule	Number of Electrons	Configuration	Term Symbols	Bond Order	R_e/pm	D_e eV	D_e kJ mol^{-1}
H_2^+	1	$(1\sigma_g)$	$^2\Sigma$	$\frac{1}{2}$	106.0	2.793	269.483
H_2	2	$(1\sigma_g)^2$	$^1\Sigma$	1	74.12	4.7483	458.135
He_2^+	3	$(1\sigma_g)^2(1\sigma_u)$	$^2\Sigma$	$\frac{1}{2}$	108.0	2.5	238
He_2	4	$(1\sigma_g)^2(1\sigma_u)^2$		0	—	—	—
Li_2	6	$[He_2](2\sigma_g)^2$	$^1\Sigma$	1	267.3	1.14	110.0
Be_2	8	$[He_2](2\sigma_g)^2(2\sigma_u)^2$		0	—	—	—
B_2	10	$[Be_2](1\pi_u)^2$	$^3\Sigma$	1	158.9	~3.0	~290
C_2	12	$[Be_2](1\pi_u)^4$	$^1\Sigma$	2	124.2	6.36	613.8
N_2^+	13	$[Be_2](1\pi_u)^4(3\sigma_g)$	$^2\Sigma$	$2\frac{1}{2}$	111.6	8.86	854.8
N_2	14	$[Be_2](1\pi_u)^4(3\sigma_g)^2$	$^1\Sigma$	3	109.4	9.902	955.42
O_2^+	15	$[N_2](1\pi_g)$	$^2\Pi$	$2\frac{1}{2}$	112.27	6.77	653.1
O_2	16	$[N_2](1\pi_g)^2$	$^3\Sigma$	2	120.74	5.213	502.9
F_2	18	$[N_2](1\pi_g)^4$	$^1\Sigma$	1	143.5	1.34	118.8
Ne_2	20	$[N_2](1\pi_g)^4(3\sigma_u)^2$		0	—	—	—

is net bonding because there are two electrons in bonding orbitals. This ion is observed in electric arcs.

He$_2$

According to simple molecular orbital theory, a pair of electrons would occupy the $1\sigma_g$ orbital and a pair would occupy the $1\sigma_u$ orbital. Since both the bonding and antibonding orbitals are filled, there is no decrease in energy as compared to two isolated helium atoms, and a stable He$_2$ molecule is not formed.

Li$_2$

The additional pair of electrons beyond those for He$_2$ go into the $2\sigma_g$ orbital. Thus, there is a single bond between the two nuclei. The vapor of lithium metal is primarily monatomic because the Li$_2$ bond is quite weak and a high temperature is needed to vaporize the metal.

Be$_2$

The additional pair of electrons beyond Li$_2$ goes into the $2\sigma_u$ antibonding orbital. Thus, as in the case of He$_2$, there is no net stabilization as compared with isolated Be atoms.

B$_2$

According to Fig. 12.13 the additional pair of electrons beyond Be$_2$ go into the $1\pi_u$ orbital. A single bond is formed, and the diatomic molecule is reasonably stable. Spectroscopic measurements show that its ground state is a triplet (Section 12.8), so that the outer two electrons are in different $1\pi_u$ orbitals. If there are several orbitals with the same energy, the electrons spread themselves among several orbitals. Since there are two unpaired electrons, the ground state is a triplet.

C_2

The additional two electrons beyond B_2 fill the two half-vacant $1\pi_u$ bonding orbitals. Thus, C_2 has four electrons in bonding orbitals that are not compensated by electrons in corresponding antibonding orbitals. Since C_2 has a bond order of 2, compared with 1 for B_2, we would expect C_2 to have tighter binding and a smaller internuclear distance than B_2.

Although C_2 exists in high-temperature gases rich in carbon, other species $(C_3, C_4, \ldots)$ also exist, and carbon forms networks in the solid state (diamond and graphite) rather than condensing as C_2. Both of these phenomena are due to the fact that the energies of the $1\pi_u$ and 3σ are so close that the molecule C_2 is easily raised into the excited configuration $(1\pi_u)^3(3\sigma_g)$. The unpaired electrons in this molecule can form bonds with further carbon atoms and thereby compensate for the energy of excitation.

N_2

The additional pair of electrons beyond C_2 fills the $3\sigma_g$ bonding orbital. Thus, N_2 has a singlet ground state and a triple bond. Because of this strong bonding to form diatomic molecules, N_2 is a very stable molecule with a short internuclear distance. The first excited state of N_2 is 6.2 eV above the ground state.

Note from Fig. 12.13 that the order of the π_u and $3\sigma_g$ orbitals changes as we go from N_2 to O_2

O_2

According to simple molecular orbital theory, the additional pair of electrons beyond N_2 go into the $1\pi_g$ orbitals. Since there are two degenerate $1\pi_g$ orbitals, the two electrons can go into either the same or different orbitals. If they go into the same orbital a singlet state is formed, and if they go into different orbitals a triplet state is formed. As in the case of atoms where Hund's rule predicts that the triplet state has the lower energy, in O_2 the electrons go into different orbitals and have parallel spins. Therefore, in the ground state O_2 has a spin of one and is paramagnetic (Section 16.1). This prediction was one of the early triumphs of MO theory.

F_2

The additional two electrons beyond O_2 fill the $1\pi_g$ orbitals so that the ground state is a singlet. Since the electron pairs in two $1\pi_g$ antibonding orbitals approximately cancel the bonding due to two electron pairs in the bonding orbitals, the bonding is weaker than in O_2, and the internuclear distance is greater.

Ne_2

The $3\sigma_u$ orbital would be filled, and so the antibonding effects cancel the bonding effects, and no stable molecule is formed.

Correlation diagrams can be constructed for heteronuclear diatomic molecules, but they have to take account of the fact that the energy levels in the two atoms may be quite different. For atomic orbitals in two atoms to be involved in bonding, they must have the same $\sigma, \pi, \ldots$ properties and energies that are not too different. The electrons in heteronuclear diatomic molecules tend to localize more around one nucleus than the other, so that the molecule tends to have a dipole moment (Section 12.12). In the extreme case, ionic molecules (Section 12.9) are formed. The difference in affinity of different atoms in a molecule for electrons is discussed in terms of electronegativity (Section 12.8).

12.6 Electronic Structure of Polyatomic Molecules: Valence Bond Method

The electronic structure of polyatomic molecules can also be described by the LCAO molecular orbital method, but now the electronic energy will depend on many internuclear distances and angles. Therefore, finding the equilibrium geometry by minimizing the energy (calculated with the variational method) with respect to all these coordinates is a difficult and time-consuming process. Such calculations are becoming more routine as more powerful computers become available. The accuracy of the calculation depends on the number of atomic orbitals used. When a very large number of atomic orbitals or basis set is used, the internuclear distances and bond angles of small molecules (less than ~6 atoms) can be calculated as accurately as they can be measured (distances to a few pm and bond angles to a degree or two). However, the energies are usually not calculated accurately enough to be used in thermodynamic calculations.

Since chemical bond energies are the difference between large numbers (of the order of the energy of a 1s orbital in the hydrogen atom or $\sim 1300 \text{ kJ mol}^{-1}$), one must calculate these large numbers to very high accuracy to obtain chemical accuracy in the difference.

It is therefore important to have simple ways of thinking about chemical bonding and molecular structure which, although not accurate, give a qualitative picture with little numerical work. One such method, introduced by Heitler and London for H_2 just after the advent of quantum mechanics, is called the valence bond method. We present a qualitative picture of this method in this section.

The valence bond method is based on the idea that a chemical bond is formed when there is good overlap between the atomic orbitals of the participating atoms. When the idea of **hybrid orbitals** is added to this, an explanation of the bond angles in simple molecules is found. Hybrid orbitals are linear combinations of atomic orbitals on a single atom with definite angular relationships among them. We will illustrate these ideas by examining the molecules BeH_2, BH_3, CH_4, NH_3, and H_2O.

The H–Be–H bond angle is 180°. The ground-state electron configuration of the beryllium atom is $1s^2 2s^2$. To represent the directionality of the BeH_2 bonds and the fact that they are equivalent, two hybrid orbitals of beryllium are formed by taking a linear combination of the 2s orbital and one of the 2p

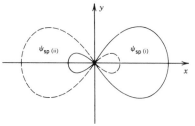

Figure 12.14   The two sp hybrids formed from an s orbital and a p orbital. One hybrid is shown by a solid line and the other is shown by a dashed line.

orbitals. Energy is required to excite a 2s electron to a 2p orbital, but more than enough energy for this is obtained when a stable chemical compound with two bonds is formed. The two sp orbitals formed in this way are

$$\psi_{sp(i)} = 2^{-1/2}(2s + 2p_x) \tag{12.42}$$

$$\psi_{sp(ii)} = 2^{-1/2}(2s - 2p_x) \tag{12.43}$$

where these orbitals have been normalized in the usual way. There are two of these hybrid orbitals because two beryllium orbitals have been used. When an s orbital is hyrbridized with a p_x orbital it must be remembered that the p_x orbital is positive on one side of the nucleus and negative on the other. The node in the radial function for the 2s orbital, inside of which the orbital is negative and outside of which it is positive, may be neglected since the nodal surface is quite close to the nucleus and the contribution of this region to bonding is small. In $\psi_{sp(i)}$ the amplitudes tend to cancel on one side of the nucleus and add on the other as shown in Fig. 12.14. The $\psi_{sp(ii)}$ orbital is equivalent but points in the opposite direction. The lobes of these orbitals extend much further along the x axis than the 2s and $2p_x$, orbitals, and so they provide more overlap with the 1s orbitals of the two hydrogen atoms in BeH_2.

The two Be–H bonds are described by combining the $1s_A$ and $1s_B$ orbitals for the two protons with these hybrid orbitals to obtain the following two bond orbitals:

$$\psi = c_1 1s_A + c_2 \psi_{sp(i)}, \tag{12.44}$$

$$\psi' = c_1' 1s_B + c_2' \psi_{sp(ii)}, \tag{12.45}$$

According to valence bond theory BeH_2 is stabilized by the overlap of the two beryllium sp orbitals and the two hydrogen 1s orbitals. The $1s^2$ electrons of beryllium are not involved.

The three B–H bonds in BH_3 lie in a plane with H–B–H angles of 120°. The boron atom has the configuration $1s^2 2s^2 2p$. The following three hybrid orbitals of boron are constructed to account for three equivalent bonds in BH_3:

$$\psi_{sp^2(i)} = \frac{1}{\sqrt{3}} 2s + \sqrt{\frac{2}{3}} 2p_z \tag{12.46a}$$

$$\psi_{sp^2(ii)} = \frac{1}{\sqrt{3}} 2s - \frac{1}{\sqrt{6}} 2p_z + \frac{1}{\sqrt{2}} 2p_x \tag{12.46b}$$

$$\psi_{sp^2(iii)} = \frac{1}{\sqrt{3}} 2s - \frac{1}{\sqrt{6}} 2p_z - \frac{1}{\sqrt{2}} 2p_x \tag{12.46c}$$

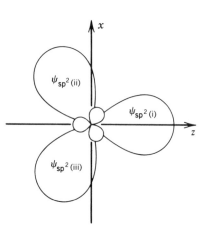

Figure 12.15   The three sp^2 hybrid orbitals formed from an s orbital and two p orbitals.

These wavefunctions have been normalized and are orthogonal. By substituting the expressions for the angular parts of the p_z and p_x orbitals, it is readily shown that the sp^2 orbitals lie in a plane with lobes pointed in directions separated by 120 °, as shown in Fig. 12.15.

Carbon atoms have the electron configuration $1s^2 2s^2 2p^2$, and the outer four valence electrons may be used to form sp^3 hybrid orbitals:

$$\psi_{sp^3(i)} = \frac{1}{\sqrt{4}} (2s + 2p_x + 2p_y + 2p_z) \tag{12.47a}$$

$$\psi_{sp^3(ii)} = \frac{1}{\sqrt{4}} (2s - 2p_x - 2p_y + 2p_z) \tag{12.47b}$$

$$\psi_{sp^3(iii)} = \frac{1}{\sqrt{4}} (2s + 2p_x - 2p_y - 2p_z) \qquad (12.47c)$$

$$\psi_{sp^3(iv)} = \frac{1}{\sqrt{4}} (2s - 2p_x + 2p_y - 2p_z) \qquad (12.47d)$$

These orthonormal orbitals point in the directions shown in Fig. 12.16, in agreement with the tetrahedral structure of methane, and with the geometries of the alkanes.

When we come to NH_3, we must introduce the idea of **lone-pair** electrons. We use the $2s^2 2p_x^1 2p_y^1 2p_z^1$ configuration on N to form four sp^3 hybrids just as in carbon. Now we must put two valence electrons of N in one of these orbitals, leaving three available to bond to the hydrogens. This leads to a tetrahedral structure of NH_3 (bond angle 109°) with one apex of the tetrahedron containing a lone electron pair, as shown in Fig. 12.17a. (The experimentally observed bond angle is 107°.) These lone-pair electrons are available for binding to, for example, a proton, H^+, to form NH_4^+.

The water molecule can be treated in much the same manner, except that we now have two lone pairs, as shown in Fig. 12.17b. This predicts that the bond angle in H_2O is the tetrahedral angle 109° instead of the experimentally observed 104°. The small difference can be accounted for by adding more terms to the wavefunction. The lone pairs of H_2O are available to bond to other atoms. In particular, the interaction between hydrogen atoms on other water molecules and the lone pairs gives rise to the hydrogen bond (see Section 12.10) and the unusual properties of H_2O.

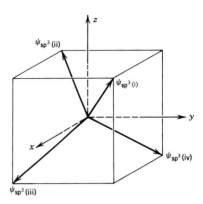

Figure 12.16   Direction of the four sp^3 hybrids formed from an s orbital and three p orbitals.

Example 12.2
We want to construct a hybrid orbital (made up of $2p_x$, $2p_y$, $2p_z$, and 2s orbitals) that points in a particular direction in space. To take a simple example, we will construct two equivalent orbitals in the xy plane, one pointing along a line making an angle χ with the x axis and the other pointing along a line making an angle $-\chi$ with the x axis. Since they are in the xy plane, we need only consider the $2p_x$, $2p_y$, and 2s orbitals. First we form linear combinations of the p orbitals that point in the specified directions:

$$\phi^{(1)} = (\cos \chi)\phi_{2p_x} + (\sin \chi)\phi_{2p_y} \qquad \phi^{(2)} = (\cos \chi)\phi_{2p_x} - (\sin \chi)\phi_{2p_y}$$

Notice that these orbitals are not orthogonal unless $2\chi = 90$:

$$\int dv \phi^{(1)} \phi^{(2)} = \cos^2\chi - \sin^2\chi = \cos 2\chi \qquad (12.48)$$

We can now form linear combinations of $\phi^{(1)}$, $\phi^{(2)}$, and ϕ_{2s} that are orthogonal by using the following relation (note that the coefficients are equal so that the orbitals are equivalent):

$$\Phi_I = c_1 \phi_{2s} + c_2 \phi^{(1)} \qquad \Phi_{II} = c_1 \phi_{2s} + c_2 \phi^{(2)}$$

To find c_1 and c_2, and Φ_I and Φ_{II} are normalized and made orthogonal:

Normalization: $\qquad c_1^2 + c_2^2 = 1 \qquad (12.49)$

Orthogonality: $\quad c_1^2 + c_2^2 \cos 2\chi = 0 \qquad (12.50)$

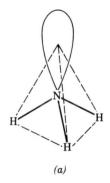

(a)

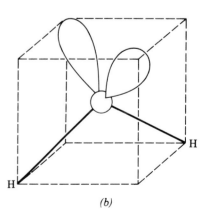

(b)

Figure 12.17   (a) Lone-pair electrons of NH_3. (b) Lone-pair electrons of H_2O.

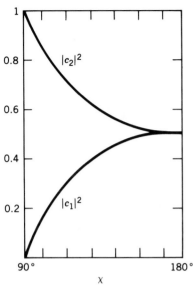

Figure 12.18 The 2s contribution (c_1^2) and the 2p (c_2^2) contribution to the hybrid orbitals as a function of angle χ.

Thus,

$$c_1^2 = \frac{\cos 2\chi}{\cos 2\chi - 1} \tag{12.51a}$$

$$c_2^2 = \frac{1}{1 - \cos 2\chi} \tag{12.51b}$$

In Fig. 12.18, c_1^2 (the 2s contribution) and c_2^2 (the 2p contribution) are plotted versus 2χ. (Note that $2\chi > 90$. Why?) A third orbital orthogonal to these two can be constructed that points along the negative x axis:

$$\Phi_{III} = d_1\phi_{2s} - d_2\phi_{2p_x}$$

with $d_1^2 + d_2^2 = 1$ for normalization and $d_1c_1 = -d_2c_2 \cos \chi$ for orthogonality. After squaring the last expression and using the results for c_1^2 and c_2^2 we find

$$d_1^2 = \frac{1 + \cos 2\chi}{1 - \cos 2\chi} \tag{12.52a}$$

$$d_2^2 = \frac{2 \cos 2\chi}{\cos 2\chi - 1} \tag{12.52b}$$

These are plotted in Fig. 12.19. Note that Φ_{III} is equivalent to Φ_I and Φ_{II} (i.e., $d_1^2 = c_1^2$) only when $2\chi = 120$. For this value of χ, Φ_I, Φ_{II}, and Φ_{III} are the sp^2 hybrid orbitals of equations 12.46.

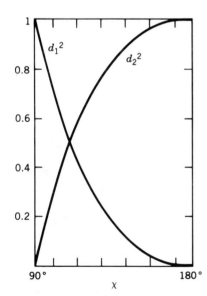

Figure 12.19 The 2s contribution (d_1^2) and the 2p (d_2^2) contribution to the third hybrid orbital as a function of χ.

12.7   Hückel Molecular Orbital Theory

Molecules with extensive π bonding systems, like benzene, are not described very well by valence bond theory. Their electronic structure can be described in a simple, but approximate molecular orbital method developed by Hückel in 1930. The two types of bonds involved in these molecules are illustrated in Fig. 12.20 for ethylene (C_2H_4). Ethylene is a planar molecule, and the bonds between the carbon and hydrogen atoms in the plane are sp^2 hybrid orbitals of carbon and 1s orbitals of hydrogen. This forms the **σ-bond framework** shown in Fig. 12.20a. The 2p$_z$ orbitals of the carbon atoms that are not involved in the σ framework overlap sideways forming the π system, as shown in Fig. 12.20b. This simple picture explains the bond angles in ethylene and its planar structure.

Hückel molecular orbital theory is concerned with the π electron system which is responsible for the special properties of conjugated and aromatic hydrocarbons. The wavefunction of the delocalized π orbital of ethylene is represented by

$$\psi = c_1\phi_1 + c_2\phi_2 \tag{12.53}$$

where ϕ_1 and ϕ_2 are the 2p$_z$ orbitals of the two carbon atoms. The corresponding secular determinant is

$$\begin{vmatrix} H_{11} - ES_{11} & H_{12} - ES_{12} \\ H_{21} - ES_{21} & H_{22} - ES_{22} \end{vmatrix} = 0 \tag{12.54}$$

where $H_{ij} = \int \phi_i^* \hat{H} \phi_j \, dv$ and $S_{ij} = \int \phi_i^* \phi_j \, dv$.

In Hückel molecular orbital theory, the secular equation is simplified by making the following assumptions:

1. Overlap integrals S_{ij} are set equal to zero unless $i = j$, when $S_{ii} = 1$.

2. All the diagonal elements in the secular equation are assumed to be the same; thus, the Coulomb integrals H_{ii} are all set equal to α.

3. The resonance integrals H_{ij} are set equal to zero, except for those on neighboring atoms which are set equal to β.

With these assumptions the secular equation for the π electrons in ethylene becomes

$$\begin{vmatrix} \alpha - E & \beta \\ \beta & \alpha - E \end{vmatrix} = 0 \qquad (12.55)$$

In Hückel theory, the Coulomb integral α and the resonance integral β are regarded as empirical parameters to be evaluated from experimental data on the molecule. Thus, in Hückel theory it is unnecessary to specify the Hamiltonian operator.

Equation 12.55 is readily solved for the energy by using the quadratic formula to obtain $E = \alpha \pm \beta$. Thus, there are bonding and antibonding orbitals, as shown in Fig. 12.21. The resonance integral is negative, and so the energy of the lowest level is $\alpha + \beta$. The bonding orbital is occupied by an electron pair, and so the π electronic energy of ethylene is $2\alpha + 2\beta$.

The wavefunctions for the bonding and antibonding orbitals are obtained by going back to the pair of linear algebraic equations that gave rise to the secular equation:

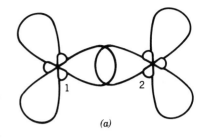

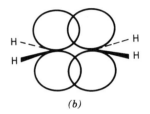

Figure 12.20 Bonding in ethylene. (*a*) In the plane of the nuclei: formation of σ bonds between carbon atoms 1 and 2 using sp² orbitals. (*b*) Perpendicular to the plane of the nuclei: formation of π bonds between p orbitals.

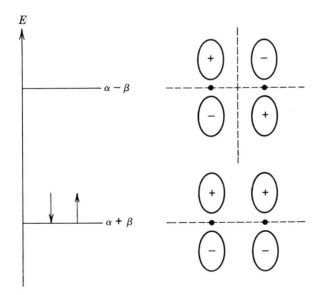

Figure 12.21 Hückel molecular orbitals for ethylene. The carbon nuclei are represented by dots, and the nodal planes for the molecular orbitals are represented by dashed lines.

$$c_1(\alpha - E) + c_2\beta = 0 \qquad (12.56)$$

$$c_1\beta + c_2(\alpha - E) = 0 \qquad (12.57)$$

Substituting $E = \alpha + \beta$ in either equation yields $c_1 = c_2$ so that the wavefunction for the bonding orbital is

$$\psi_1 = 2^{-1/2}(\phi_1 + \phi_2) \qquad (12.58)$$

The bonding orbital is shown in Fig. 12.21 alongside the energy level. Substituting $E = \alpha - \beta$ into equation 12.56 or 12.57 yields $c_1 = -c_2$, so that the wavefunction for the antibonding orbital is

$$\psi_2 = 2^{-1/2}(\phi_1 - \phi_2) \qquad (12.59)$$

The antibonding orbital is also shown in Fig. 12.21. Notice the resemblance to the LCAO orbitals of H_2^+ (Section 12.3).

The π electronic energy of planar 1,3-butadiene ($CH_2{=}CHCH{=}CH_2$) is readily calculated using Hückel theory. The secular determinantal equation is

$$\begin{vmatrix} \alpha - E & \beta & 0 & 0 \\ \beta & \alpha - E & \beta & 0 \\ 0 & \beta & \alpha - E & \beta \\ 0 & 0 & \beta & \alpha - E \end{vmatrix} = 0 \qquad (12.60)$$

If β is factored from each column and $(\alpha - E)/\beta$ is replaced by x we obtain

$$\begin{vmatrix} x & 1 & 0 & 0 \\ 1 & x & 1 & 0 \\ 0 & 1 & x & 1 \\ 0 & 0 & 1 & x \end{vmatrix} = 0 \qquad (12.61)$$

or $x^4 - 3x^2 + 1 = 0$ so that $x = \pm 0.618, \pm 1.618$. Thus, for 1,3-butadiene there are four possible energy levels, two bonding and two antibonding, as shown in Fig. 12.22:

$$E_1 = \alpha + 1.618\beta$$
$$E_2 = \alpha + 0.618\beta$$
$$E_3 = \alpha - 0.618\beta$$
$$E_4 = \alpha - 1.618\beta \qquad (12.62)$$

The four π electrons occupy the two bonding orbitals so that the π electronic energy is

$$E = 2(\alpha + 1.618\beta) + 2(\alpha + 0.618\beta) \qquad (12.63)$$
$$= 4\alpha + 4.472\beta$$

The four Hückel molecular orbitals for 1,3-butadiene are*

---

* The derivation of these wave functions is given in I. N. Levine, *Quantum Chemistry*, 3d ed., p. 469. Boston: Allyn & Bacon, 1983.

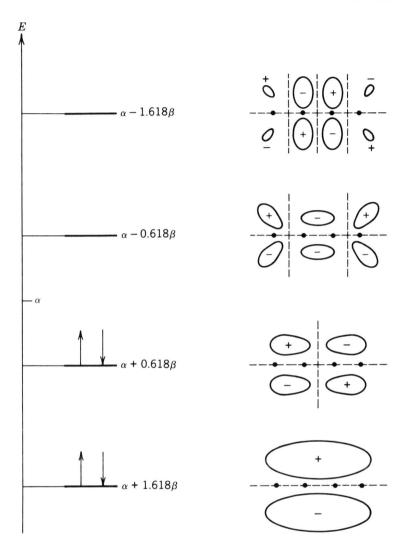

Figure 12.22   Hückel molecular orbitals for 1,3-butadiene. The carbon nuclei are represented by dots, and the nodal planes for the molecular orbitals are represented by dashed lines. (From I. N. Levine, *Quantum Chemistry*, 3d ed. Copyright © 1983 by Allyn & Bacon, Inc., Boston. Reprinted with permission of the publisher.)

$$\psi_1 = 0.372\phi_1 + 0.602\phi_2 + 0.602\phi_3 + 0.372\phi_4$$
$$\psi_2 = 0.602\phi_1 + 0.372\phi_2 - 0.372\phi_3 - 0.602\phi_4$$
$$\psi_3 = 0.602\phi_1 - 0.372\phi_2 - 0.372\phi_3 + 0.602\phi_4$$
$$\psi_4 = 0.372\phi_1 - 0.602\phi_2 + 0.602\phi_3 - 0.372\phi_4 \qquad (12.64)$$

These four molecular orbitals are indicated in Fig. 12.22. Notice that the π orbitals extend the entire length of the molecule.

The Hückel secular determinant for benzene is

$$
\begin{vmatrix}
\alpha - E & \beta & 0 & 0 & 0 & \beta \\
\beta & \alpha - E & \beta & 0 & 0 & 0 \\
0 & \beta & \alpha - E & \beta & 0 & 0 \\
0 & 0 & \beta & \alpha - E & \beta & 0 \\
0 & 0 & 0 & \beta & \alpha - E & \beta \\
\beta & 0 & 0 & 0 & \beta & \alpha - E
\end{vmatrix} = 0 \qquad (12.65)
$$

The six roots are

$$
\begin{aligned}
E_1 &= \alpha + 2\beta \\
E_2 &= E_3 = \alpha + \beta \\
E_4 &= E_5 = \alpha - \beta \\
E_6 &= \alpha - 2\beta
\end{aligned} \qquad (12.66)
$$

Since benzene has six π electrons, pairs of electrons go in the lowest three orbits (those with plus signs in E_i). Thus, the π electronic energy in benzene is

$$
\begin{aligned}
E_\pi &= 2(\alpha + 2\beta) + 4(\alpha + \beta) \\
&= 6\alpha + 8\beta
\end{aligned} \qquad (12.67)
$$

The equations for the six Hückel molecular orbitals for benzene are not given here, but the corresponding electron densities are shown in Fig. 12.23. Note that the π electronic energy in C_6H_6 is more negative than three times the value in C_2H_4, indicating that C_6H_6 is not three double bonds. The difference is called the conjugation energy.

The Hückel theory is an example of a semiempirical molecular orbital method. We have used a simple Hamiltonian (neglecting many terms) to find the orbitals and their energies. We now can use experimental quantities to fit α and β for ethylene. We then use the values to make predictions for butadiene, benzene, and so on. This method does not give quantitative results, but it does provide us with qualitative insights about larger systems for which the more computationally intensive methods are too costly or time-consuming, and it gives us insight into the excited electronic states of conjugated π electron molecules.

Example 12.3
Calculate the Hückel molecular orbital energies of the planar radical (i.e., 1 nonpaired electron) CH_2CHCH_2.

The secular determinantal equation is

$$
\begin{vmatrix}
\alpha - E & \beta & 0 \\
\beta & \alpha - E & \beta \\
0 & \beta & \alpha - E
\end{vmatrix} = 0
$$

or

$$
\begin{vmatrix}
x & 1 & 0 \\
1 & x & 1 \\
0 & 1 & x
\end{vmatrix} = 0 \qquad \text{with } x = \frac{\alpha - E}{\beta}
$$

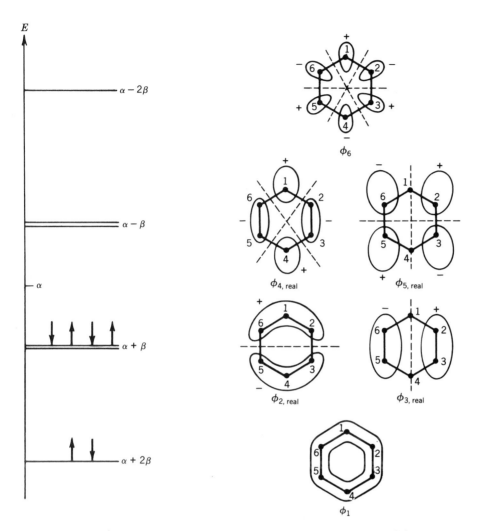

Figure 12.23   Hückel molecular orbitals for benzene. The carbon nuclei are represented by dots, and the nodal planes perpendicular to the molecular plane are represented by dashed lines. (From I. N. Levine, *Quantum Chemistry*, 3d ed. Copyright © 1983 by Allyn & Bacon, Inc., Boston. Reprinted with permission of the publisher.)

This yields

$$x^3 - 2x = 0$$

with roots $x = 0, -\sqrt{2}, +\sqrt{2}$. Therefore,

$$E_1 = \alpha + \sqrt{2}\beta$$
$$E_2 = \alpha$$
$$E_3 = \alpha - \sqrt{2}\beta$$

The orbital corresponding to E_2 is called a nonbonding orbital since its energy is unchanged from the atomic value $\alpha = H_{ii}$, which is the 2p atomic orbital energy in this system.

12.8   Electronegativity

There are several ways of measuring **electronegativity,** which is a quantitative measure of attraction of an atom in a molecule to electrons. For example, in a molecule AB it is measured by the relative stabilities of A^+B^- and A^-B^+. The energy difference between these structures is given by

$$E(A^+B^-) - E(A^-B^+) = (E_{i,A} - E_{ea,B}) - (E_{i,B} - E_{ea,A})$$
$$= (E_{i,A} + E_{ea,A}) - (E_{i,B} + E_{ea,B}) \quad (12.68)$$

where E_i is the ionization energy (Table 11.4) and E_{ea} is electron affinity (Table 11.5). Mulliken defined the electronegativity difference between atoms A and B as $\frac{1}{2}[E(A^+B^-) - E(A^-B^+)]$. Thus, he defined the electronegativity of an atom as $\frac{1}{2}(E_i + E_{ea})$. The electronegativity of an atom depends on its valence state, and so the ionization potential and electron affinity used are not those of the ground state of the atom.

The Pauling electronegativity scale is based on his observation that the A–B bond energy exceeds the mean of the A–A and B–B bond energies if the A–B bond is polar. Pauling defined the electronegativity difference between A and B as $0.050 \, (\Delta_{AB}/\text{kJ mol}^{-1})^{1/2}$, where

$$\Delta_{AB} = E(A-B) - \frac{1}{2}[E(A-A) + E(B-B)] \quad (12.69)$$

The E's are the dissociation energies for the bonds in question. The electronegativity of H was set at 2.1. Values on the Mulliken scale in eV can be converted to the Pauling scale by dividing by 3.17. Exact agreement is not obtained, but the scales are in relatively good agreement. Fluorine is the most electronegative atom (4.0 on Pauling's scale), and cesium is the least electronegative atom (0.7 on Pauling's scale). The electronegativities of a number of elements are given in Fig. 12.24, which shows that the electronegativity depends on the position of the element in the periodic table. As we go down a column of the periodic table, the atoms become less electronegative because of the increasingly effective screening of the charge on the nucleus by inner

						H	
						2.1	
Li	Be	B	C	N	O	F	
1.0	1.5	2.0	2.5	3.0	3.5	4.0	
Na	Mg	Al	Si	P	S	Cl	
0.9	1.2	1.5	1.8	2.1	2.5	3.0	
K	Ca		Ge	As	Se	Br	
0.8	1.0		1.7	2.0	2.4	2.8	
Rb	Sr		Sn	Sb	Te	I	
0.8	1.0		1.7	1.8	2.1	2.4	
Cs	Ba						
0.7	0.9						

Figure 12.24   Electronegativities (Pauling scale) of a number of atoms.

electrons. The alkali-metal atoms have a great tendency to lose their outer electrons and therefore have a low electronegativity.

By use of electronegativities it is possible to predict which bonds will be ionic and which bonds will be covalent. Two elements of very different electronegativity, like a halogen and an alkali metal, form an ionic bond because an electron is almost completely transferred to the atom of higher electronegativity. Two elements with nearly equal electronegativities form covalent bonds. For instance, carbon, which occupies an intermediate position in the electronegativity scale, forms covalent bonds with elements near it in the periodic table. If there is a considerable difference between the electronegativities of the two elements, the bond is polar (i.e., possessed of a high degree of ionic character), as in the case of sodium chloride. In the majority of chemical bonds the sharing of the electron pair is not exactly equal, so that the bond has some ionic character that results in a dipole moment (Section 12.12) for the bond.

12.9 Ionic Bonding

When atoms with nearly the same electronegativity form bonds, the molecular orbitals are spread more or less evenly over the two atoms, and covalent bonds are formed. When atoms have somewhat different electronegativities, the bonding electrons are not evenly shared, but are drawn toward the more electronegative atom. When the electronegativities are quite different an electron moves completely to the more electronegative atom and an **ionic bond** is formed.

The interaction energy $E(R)$ of an ionic bond can be calculated using Coulomb's law (equation 11.1). As two ions approach each other closely because of Coulomb attraction an equilibrium position is reached at some point because of repulsive forces which balance the attraction. The short-range repulsion energy increases very rapidly when the charge clouds of the two ions begin to overlap. The repulsion energy may be represented by an empirical expression such as $b \exp(-aR)$. Thus, the potential energy $E(R)$ of a diatomic molecule is approximately

$$E(R) = \frac{Q_1 Q_2}{4\pi\epsilon_0 R} + be^{-aR} \qquad (12.70)$$

where the energy is measured with respect to the separated ions. Further effects may be taken into account; for example, the electric field of each ion polarizes (Section 11.12) the electron cloud of the other ion and produces a further energy of attraction. However, the short-range repulsion energy and polarization energy are so small close to R_e that the Coulomb attractive term alone gives a good approximation to the dissociation energy of an ionic molecule.

Equation 12.70 applies to dissociation into separated ions. However, the ground state of the dissociated system consists of atoms rather than ions. The dissociation energy D_e into atoms is given by

$$D_e(MX \rightarrow M + X) = D_e(MX \rightarrow M^+ + X^-) - E_i(M) + E_{ea}(X)$$

where $D_e(MX \rightarrow M^+ + M^-)$ is the dissociation energy into ions, $E_i(M)$ is the ionization energy of metal atom, and $E_{ea}(X)$ is the electron affinity of the nonmetal atom. Since the lowest ionization potentials are greater than the highest electron affinities, $D_e(MX \rightarrow M + X)$ is smaller than $D_e(MX \rightarrow M^+ + X^-)$. What happens when MX dissociates to $M + X$ is that at some large internuclear distance the system can decrease its energy by moving the electron from X^- to M^+.

Example 12.4

The dissociation energy of NaCl(g) into atoms is 4.29 eV. What is the dissociation energy into ions and how does this compare with what would be expected from Coulomb's law? The equilibrium internuclear distance in the gaseous NaCl molecule is 0.2361 nm.

$$D_e(\text{NaCl} \rightarrow \text{Na}^+ + \text{Cl}^-) = D_e(\text{NaCl} \rightarrow \text{Na} + \text{Cl}) + E_i(\text{Na}) - E_{ea}(\text{Cl})$$

$$= 4.29 + 5.14 - 3.61 = 5.82 \text{ eV}$$

If only the Coulomb term is used, then

$$D_e(\text{NaCl} \rightarrow \text{Na}^+ + \text{Cl}^-) = \frac{Q_1 Q_2}{4\pi\epsilon_0 R_e}$$

$$= \frac{(1.602 \times 10^{-19} \text{ C})^2}{4\pi(8.854 \times 10^{-12} \text{ C}^2 \text{ N}^{-1}\text{m}^{-2})(0.2361 \times 10^{-9} \text{ m})}$$

$$= \frac{9.770 \times 10^{-19} \text{ J}}{1.602 \times 10^{-19} \text{ J eV}^{-1}}$$

$$= 6.10 \text{ eV}$$

Thus, the Coulomb term accounts for the observed dissociation energy within 0.3 eV.

12.10   Hydrogen Bonds

$[\text{F}{-}\text{H}{-}\text{F}]^-$

Formic acid dimer

Salicylic acid

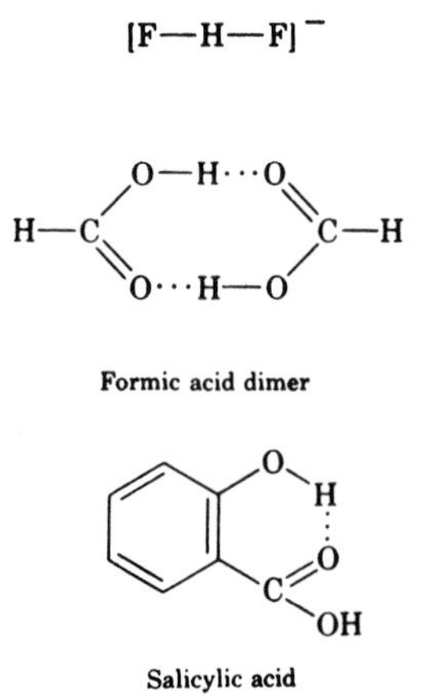

Figure 12.25   Examples of molecules exhibiting hydrogen bonding.

A number of unusual structures such as HF_2^- and formic acid dimer in the gas phase (see Fig. 12.25) are evidence for the formation of hydrogen bonds. The unusually high acid dissociation constant of salicylic acid, as compared with the *meta* and *para* isomers, is also evidence for a hydrogen bond. A hydrogen bond results when a proton may be shared between two electronegative atoms, such as F, O, or N, that are the right distance apart. The proton of the hydrogen bond is attracted by the high concentration of negative charge in the vicinity of these electronegative atoms. Fluorine forms very strong hydrogen bonds; oxygen, weaker ones; and nitrogen still weaker ones. The unusual properties of water are due to a large extent to the formation of hydrogen bonds involving the four lone-pair electrons on oxygen. In ice there is a tetrahedral arrangement with each oxygen atom bonded to four hydrogen atoms. Hydrogen bonds are formed along the axis of each lone pair in ice, and their existence in liquid water is responsible for the high boiling point of water as compared with the boiling points of hydrides of other elements in the same column of the periodic table (H_2S, -62 °C; H_2Se, -42 °C; H_2Te, -4 °C). When water is vaporized, these hydrogen bonds are broken, but in formic and acetic acids the hydrogen bonds are strong enough for dimers of the type illustrated in Fig. 12.25 to exist in the vapor. Hydrogen bonds between N and O are responsible for the stability of the α helix formed by polypeptides and found in protein molecules.

12.11 Intermolecular Forces

When two neutral molecules come close to one another, the various interactions between the electrons and nuclei of one molecule and the electrons and nuclei of the other produce a potential energy of interaction. At very small distances, the molecules repel each other strongly. At certain intermediate distances, the potential energy of interaction is negative; the molecules attract each other, but usually weakly. At very large distances, the potential energy approaches zero. Thus, if we plot the interaction energy versus intermolecular distance, we obtain curves such as Fig. 12.26.

Attractive Forces

The long-range attractive interactions between neutral molecules are called **van der Waals forces** and include a number of terms, all arising from the electrostatic interactions. For example, if the two molecules have permanent dipole moments (Section 12.12), then at fixed angles with respect to one another the interaction between molecules varies as R^{-3}, where R is the intermolecular distance. In the gas or liquid phase, molecules will have random orientations, so the thermal average over the angular part of the interaction causes the first nonzero term in the interaction to be

$$\langle V(R) \rangle_{dd} = -\frac{2}{3kT}\left(\frac{\mu_A \mu_B}{4\pi\epsilon_0}\right)^2 \frac{1}{R^6} \tag{12.71}$$

where μ_A and μ_B are the permanent dipole moments of the two molecules, T is the temperature and dd means dipole–dipole. Notice that the interaction is negative (attractive) since there is a preference in the thermal (statistical mechanical) averaging for the molecules to orient in such a way as to attract each other and lower their mutual energy. As the temperature increases, this tendency is less important, so the attraction decreases.

Another term in the attractive potential energy arises from the **dipole-induced dipole interaction** between molecules. If molecule A has a dipole moment μ_A, this creates an electric field which polarizes the charges on molecule B, creating an induced dipole moment of magnitude $\alpha_B \mu_A$, where α_B is called the polarizability of molecule B. The dipole-induced dipole attractive energy is given by

$$\langle V(R) \rangle_{ind} = -\frac{\alpha_B \mu_A^2 + \alpha_A \mu_B^2}{(4\pi\epsilon_0)^2 R^6} \tag{12.72}$$

Notice that it also varies as R^{-6}.

There is a third term varying as R^{-6} which was explained by London in 1930 and is sometimes called the **London force** or **dispersion force**. It occurs even when the molecules have no permanent dipole moment and can be visualized as the interaction between the fluctuating charge distribution on molecule A with that of molecule B. The rules of quantum mechanics say that even

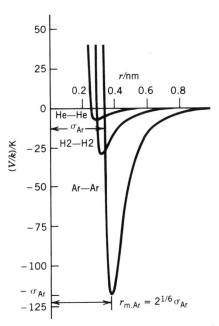

Figure 12.26 Potential energy expressed as V/k for the interaction between two atoms or molecules.

though molecule A has no permanent dipole moment, the average value of the square of the dipole moment operator is not zero. Therefore, we can say that at any instant, the molecule has a dipole moment that polarizes molecule B (and vice versa), leading to a potential varying as R^{-6}. The exact form is complicated, but a good approximation is

$$\langle V(R)\rangle_{disp} = -\frac{3}{2}\left(\frac{E_A E_B}{E_A + E_B}\right)\frac{\alpha_A \alpha_B}{(4\pi\epsilon_0)^2}\frac{1}{R^6} \tag{12.73}$$

where E_A and E_B are approximately equal to the energies of the first electronic transitions of these molecules, α_A and α_B are their polarizabilities, and disp means dispersion.

These three terms add to give the total attractive energy between molecules A and B. In most cases, this energy is on the order of $\sim 2 \times 10^{-3}$ eV at a distance of 0.5 nm. Of course, the larger the dipole moments or the polarizabilities, the stronger is the attraction.

Intermolecular Potential Energy

At small distances, the molecules repel strongly, so when this force is added to the attractive forces we have discussed, we obtain the total intermolecular potential energy. To represent this as a simple function of R, we can take the repulsion to be proportional to R^{-n} or $e^{-\alpha R}$, where n and α have to be fit empirically. One very useful empirical function is named for J. E. Lennard-Jones and is given by

$$V = 4\epsilon\left[\left(\frac{\sigma}{r}\right)^{12} - \left(\frac{\sigma}{r}\right)^{6}\right] \tag{12.74}$$

In this "6–12" potential, the twelfth power dependence for the repulsive energy is chosen for convenience and because it is a reasonable fit to the data. The minimum in this potential is at $r = (2)^{1/6}\sigma$ where $V = -\epsilon$. Note that $V = 0$ at $r = \sigma$, so the energy rises very steeply for small r (see Fig. 12.26). The values of ϵ and σ can be found by fitting experimental data on the second virial coefficient, gas viscosity, and molecular beam scattering cross sections. Typical values for the parameters ϵ and σ for interactions between like atoms or molecules are given in Table 12.3.

Table 12.3   Lennard-Jones Potential Parameters[a]

	$\frac{\epsilon}{k}$/K	σ/pm
Ar	120	341
Xe	221	410
H_2	37	293
O_2	118	358
Cl_2	256	440
CO_2	197	430
CH_4	148	382
C_6H_6	243	860

[a] Note that the depth of the potential well has been expressed as a temperature by dividing the energy ϵ by the Boltzmann constant k.

Example 12.5

By differentiation of the expression for the Lennard-Jones potential, show that the distance r_m at the minimum where $dV/dr = 0$ is $r_m = 2^{1/6}\sigma$. Substituting this relationship into equation 12.74 show that the Lennard-Jones potential may also conveniently be given by

$$V = \epsilon\left[\left(\frac{r_m}{r}\right)^{12} - 2\left(\frac{r_m}{r}\right)^{6}\right]$$

Differentiating equation 12.74 and setting the derivative equal to zero, we obtain

$$\frac{dV}{dr} = 4\epsilon\left[-\frac{12\sigma^{12}}{r^{13}} + \frac{6\sigma^6}{r^7}\right] = 0$$

If we represent the value of r at the minimum by r_m, then

$$r_m = 2^{1/6}\sigma$$

Substituting this in equation 12.74 yields the desired expression for V.

12.12 Special Topic: Dipole Moments of Molecules

Dipole Moments

Other important properties of a molecule that may be understood using quantum mechanics are its dipole moment and its polarizability. These microscopic properties of individual molecules can be obtained from macroscopic measurements of dielectric constants and from spectra.

The dipole moment of a molecule is a measure of the separation of the centers of positive and negative charge within the molecule. If a positive charge $+Q$ is separated from a negative charge $-Q$ by distance r, the dipole moment has the magnitude $\boldsymbol{\mu} = Qr$. The dipole moment $\boldsymbol{\mu}$ is a vector pointing from the negative charge to the positive charge with a magnitude of $|\boldsymbol{\mu}|$. The SI unit for dipole moment is C m.*

The dipole moment for a group of point charges Q_i is

$$\boldsymbol{\mu} = \sum_i r_i Q_i \tag{12.75}$$

where r_i is the vector from the origin to charge Q_i. For a charge density $\rho(x, y, z)$ the dipole moment is defined by

$$\boldsymbol{\mu} = \iiint r(x, y, z)\rho(x, y, z)\, dx\, dy\, dz \tag{12.76}$$

For a distribution of zero total charge the dipole moment is equal to the product of the absolute charge of either the positive or negative distribution times the vector distance r between the centers of the two distributions:

$$\boldsymbol{\mu} = Q^+ r = |Q^-|r \tag{12.77}$$

Dipole moments may be calculated from measurements of the capacitance of an electric capacitor containing the substance. The capacitance C of a parallel plate capacitor is defined as the ratio of the charge on one of the plates to the potential difference between the conductors. Thus, the capacitance has the units coulombs per volt; this unit is referred to as the farad F, that is, 1 F = 1 C/V.

---

* Dipole moments have often been expressed in debye units D after Peter Debye, who made so many contributions to the understanding of polar molecules. The debye unit is 10^{-18} esu cm. Since the charge of a proton is $4.803\ 21 \times 10^{-10}$ esu or $1.602\ 177\ 33 \times 10^{-19}$ C, the debye D may be expressed in SI units as follows:

$$\frac{(10^{-18}\ \text{esu cm})(1.602\ 177\ 33 \times 10^{-19}\ \text{C})(10^{-2}\ \text{m cm}^{-1})}{(4.803\ 21 \times 10^{-10}\ \text{esu})} = 3.335\ 64 \times 10^{-30}\ \text{C m}$$

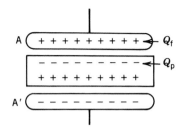

Figure 12.27 Parallel-plate capacitor filled with a dielectric. The plates are labeled A and A'; the dielectric is between the plates.

The capacitance C of a parallel plate capacitor depends on the properties of the insulating material between the plates. A capacitor has the smallest possible capacitance when there is a vacuum between the plates. When an insulating material (dielectric) is placed between the plates of a capacitor, the capacitance increases. To see why the dielectric increases the capacitance, consider that a parallel plate capacitor is charged up and then any electrical conductor between the two plates is disconnected. When a dielectric is inserted between the plates of the capacitor, as shown in Fig. 12.27, the potential difference between the plates is reduced, because the electric field polarizes the dielectric. The dielectric remains electrically neutral inside, but a net surface charge density arises on its surface, as can be seen from Fig. 12.27. Since the sign of the surface charge is opposite to the sign of the charge on the plates, the electric field within the dielectric is reduced below that which would have existed in the capacitor with the same charge on the plates in the absence of the dielectric. Since the potential difference between the plates is reduced by the dielectric and the capacitance is equal to the ratio of the charge on one of the plates to the potential difference between the plates, the capacitance increases.

The relative permittivity (dielectric constant) ϵ_r of a dielectric is defined as the ratio of the capacitance C_d of the capacitor filled with dielectric to the capacitance C when the capacitor is evacuated:

$$\epsilon_r \equiv \frac{C_d}{C} \qquad (12.78)$$

Thus, the relative permittivity of a vacuum is unity. The magnitude of the relative permittivity of a substance depends on the temperature and, if an alternating electric field is used, the frequency. The relative permittivities of a few gases and liquids are given in Table 12.4. These values of the relative permittivity apply at frequencies sufficiently low that equilibrium is maintained as the electric field varies. The electrical permittivity ϵ of a dielectric is given by $\epsilon = \epsilon_r \epsilon_0$, where ϵ_0 is the permittivity of vacuum.

In Fig. 12.27 the free charge on the plates of the capacitor is represented by Q_f and the polarization charge on the surface of the dielectric is represented by Q_p. The surface charge density of the polarization charges is referred to as the electric polarization. The polarization vector P is perpendicular to the surface and points from the negative surface charges on the dielectric to the positive surface charges. The magnitude of the polarization vector has the units

Table 12.4 Relative Permittivity ϵ_r

Gas (1 atm)	At 0 °C	Liquid	At 20 °C
Hydrogen	1.000 272	Hexane	1.874
Argon	1.000 545	Benzene	2.283
Air (CO$_2$ free)	1.000 567	Toluene	2.387
Carbon dioxide	1.000 98	Chlorobenzene	5.94
Hydrogen chloride	1.004 6	Ammonia	15.5
Ammonia	1.007 2	Acetone	21.4
Water (steam at 110 °C)	1.012 6	Methanol	33.1
		Water	80

C m^{-2} or C m m^{-3}, that is, electric dipole moment per unit volume. The magnitude of the polarization is proportional to the applied electric field.

Now we want to obtain the relation between the polarization P, the relative permittivity ϵ_r, and the electric field E. To do that we need to introduce the electric displacement vector D, which is defined by

$$D = \epsilon_0 E + P \qquad (12.79)$$

The electric displacement has the units C m^{-2}. The electric displacement vector is useful because it does not change when a dielectric is introduced into a capacitor and the same free charges remain on the plates of the capacitor. When there is no dielectric in the capacitor, then

$$D = \epsilon_0 E_0 \qquad (12.80)$$

where E_0 is the electric field between the plates in this case. When a dielectric is inserted between the plates, D does not change and so

$$D = \epsilon_0 E_0 = \epsilon_0 E + P \qquad (12.81)$$

Since the polarization of the medium opposes the charge on the capacitor plates, it is responsible for lowering the electric field by a factor equal to the relative permittivity. Thus, $E = E_0/\epsilon_r$ and

$$P = \epsilon_0 E(\epsilon_r - 1) \qquad (12.82)$$

When an electric field is applied to a sample, time is required for the establishment of the equilibrium polarization. Thus, the relative permittivity is a frequency-dependent quantity. We will first consider a static field, or one that is changing so slowly that the polarization has its equilibrium value at all times.

Contributions to the Molecular Polarization

Three contributions to the polarization must be considered: orientation polarization, electronic polarization, and vibrational polarization. Orientation polarization is due to the partial alignment of permanent dipoles. This is illustrated in Fig. 12.28. Molecules that have permanent dipole moments are not completely oriented in an electric field because of the disorienting effect of thermal motion. The extent to which dipoles can be oriented by an applied field was calculated by Debye* using the Boltzmann distribution law (Section 17.1). The electric field acting on the molecule is represented by E_i and referred to as the internal field. The energy of a dipole in the field E_i, is $-\mu \cdot E_i$, where μ is the permanent dipole moment vector of the molecule and the dot refers to the dot product: $-\mu \cdot E_i = -\mu E_i \cos \theta$, where θ is the angle between the two vectors. If the energy of the dipole in the field is small compared with kT, it may be shown that the contribution per molecule in a gas to the mean moment in the direction of the field is given by $\mu^2 E_i/3kT$, where E_i is the magnitude of the internal field. As the temperature is increased, the thermal agitation becomes more vigorous and fewer of the permanent dipoles are oriented in the direction of the field.

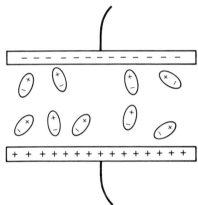

Figure 12.28 Representation of a capacitor containing dipolar molecules. The molecules are shown in their instantaneous positions indicating the partial orientation of the dipoles.

---

* P. Debye, *Polar Molecules*. New York: Chemical Catalog Co., 1929; or New York: Dover.

The second contribution is from the field-induced displacement of the average positions of the electrons relative to the nuclei of the molecule. This electronic contribution $\alpha_e E_i$ is practically independent of the temperature and is proportional to the electric field strength. The proportionality constant α_e is called the **mean molecular electronic polarizability**. The polarizability of a molecule is different in different directions in the molecule. The mean polarizability is the average taken over all possible orientations of the molecule with respect to the field. The mean polarizability is used because thermal motion ensures that all possible orientations are encountered. The polarizability α has the units of dipole moment divided by electric field strength, that is, C m/V m^{-1} = C^2 m^2/J. The ease with which different molecules become polarized by distortion of the electron cloud differs greatly.

The third contribution $\alpha_v E_i$ is from the deformation of the nuclear skeleton of the molecule by the electric field. The proportionality constant α_v is called the **mean molecular vibrational polarizability.** To a first approximation α_v is independent of temperature.

The magnitude P of the polarization, which includes all three contributions, is equal to the mean moment in the direction of the field multiplied by the number of molecules per unit volume:

$$P = \frac{N}{V} \left(\frac{\mu^2}{3kT} + \alpha_e + \alpha_v \right) E_i \tag{12.83}$$

There is a problem in knowing the average internal or local electric field strength E_i. If the concentration of polar molecules is low, it can be shown that $E_i = E(\epsilon_r + 2)/3$, where E_i is the internal field and E is the applied field. Eliminating E_i from equation 12.83 using this relation, eliminating P between this equation and equation 12.82, and introducing $N/V = N_A \rho/M$, we obtain

$$\mathcal{P} = \frac{\epsilon_r - 1}{\epsilon_r + 2} \frac{M}{\rho} = \frac{N_A}{3\epsilon_0} \left(\frac{\mu^2}{3kT} + \alpha_e + \alpha_v \right) \tag{12.84}$$

The quantity $\mathcal{P}$ is called the molar polarization, and ρ is the density. This equation is accurate for dilute gases and holds approximately for slightly polar liquids, but it does not apply to liquids of high dielectric constant because of the approximation of taking $E_i = E(\epsilon_r + 2)/3$. Equation 12.84 neglects short-range molecular interactions; more complicated theories have been developed by Onsager and Kirkwood to take those effects into account.

Measurements of Dipole Moments of Gaseous Molecules

Since the orientation polarization depends on the absolute temperature, it is possible to calculate the polarizability $\alpha_e + \alpha_v$ and the dipole moment μ of gas molecules from measurements of the relative permittivity ϵ_r at a series of temperatures. In the gaseous state the molecules are so far apart that they do not induce dipole moments in each other.

To obtain the dipole moment of a gaseous molecule, the relative permittivities ϵ_r and gas densities are measured at several temperatures, and the corresponding values of the total molar polarization $\mathcal{P}$ are calculated. According to equation 12.84, a plot of $\mathcal{P}$ versus $1/T$ will have an intercept of $N_A(\alpha_e + \alpha_v)/3\epsilon_0$ and a slope of $N_A \mu^2/9k\epsilon_0$. Plots of molar polarization $\mathcal{P}$ versus reciprocal absolute temperature are illustrated in Fig. 12.29 for the chlorine-sub-

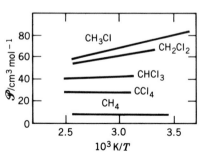

Figure 12.29 Variation of molar polarization with temperature.

Table 12.5 Dipole Moments ($\mu/10^{-30}$ C m) of Gaseous Molecules

$C_6H_5NO_2$	13.21	C_6H_5Cl	5.17	SO_2	5.37
$(CH_3)_2CO$	9.3	C_6H_5OH	5.67	HCl	3.44
H_2O	6.14	C_2H_5OH	5.67	HBr	2.64
CH_4	0	$C_6H_5NH_2$	5.20	HI	1.00
CH_3Cl	6.17	NH_3	4.87	N_2O	0.47
CH_3Br	4.84	H_2S	3.67	CO	0.40
CH_3I	4.50	H_2	0	CS_2	0
CH_2Cl_2	5.30	Cl_2	0	C_2H_4	0
$CHCl_3$	3.84	CO_2	0	C_2H_6	0
CCl_4	0	$C_6H_5CH_3$	1.33	C_6H_6	0

stituted compounds of methane. It is seen that zero slopes are obtained for the symmetrical molecules CH_4 and CCl_4, indicating that these molecules have no permanent dipole moment. Because of the electronegative character of chlorine atoms, CH_3Cl, CH_2Cl_2, and $CHCl_3$ have permanent dipole moments with the chlorine atom at the negative end of the dipole. The dipole moment of CH_3Cl is the largest of the three. Dipole moments of a variety of substances are given in Table 12.5.

The dependence of the relative permittivity of a typical polar substance on frequency is shown in Fig. 12.30. At low frequencies all the terms on the right side of equation 12.84 contribute and the value of ϵ_r is approximately constant, equal to $\epsilon_{r,s}$, the zero-frequency value. As the frequency is increased above the radio-frequency range, the relative permittivity decreases and the orientation polarization eventually becomes negligible because there is insufficient time for molecular orientation to occur before the field is reversed. At these frequencies, ϵ_r reaches a plateau equal to $\epsilon_{r,\infty}$, the so-called high-frequency relative permittivity. In the region where the relative permittivity is labeled n^2, the polarizability is made up of vibrational α_v, and electronic, α_e, terms. The relative permittivity shows dispersion (Section 15.11) in the neighborhood of absorption lines in the infrared and ultraviolet.

Maxwell showed that the relative permittivity measured with a capacitor and the refractive index n of light are related by

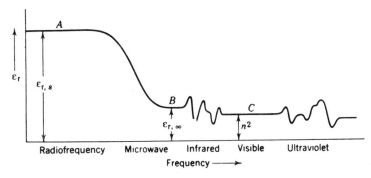

Figure 12.30 Variation of relative permittivity with frequency for a typical polar substance.

$$\epsilon_r = n^2 \qquad (12.85)$$

provided the same frequency is used in both measurements.

The dipole moment of a solute in a nonpolar solvent may be determined from the relative permittivity and density measurements on dilute solutions.* Refractive index measurements are used to evaluate the molecular electronic polarizability α_e.

References

P. W. Atkins, *Molecular Quantum Mechanics*. New York: Oxford University Press, 1983.

R. S. Berry, S. A. Rice, and J. Ross, *Physical Chemistry*. New York: Wiley, 1980.

C. A. Coulson, *The Shape and Structure of Molecules*. Oxford: Clarendon Press, 1973.

W. H. Flygare, *Molecular Structure and Dynamics*, Englewood Cliffs, NJ: Prentice-Hall, 1978.

W. J. Hehre, L. Radom, P. V. R. Schleyer, and J. A. Pople, *Ab Initio Molecular Orbital Theory*. NY: Wiley-Interscience, 1986.

N. E. Hill, W. E. Vaughan, A. H. Price, and M. Davies, *Dielectric Properties and Molecular Behavior*. London: Van Nostrand-Reinhold, 1969.

I. N. Levine, *Quantum Chemistry*. Boston: Allyn & Bacon, 1983.

D. A. McQuarrie, *Quantum Chemistry*. Mill Valley, CA: University Science Books, 1983.

R. S. Mulliken and W. C. Ermler, *Diatomic Molecules*. New York: Academic, 1977.

L. Pauling, *The Nature of the Chemical Bond*. Ithaca, NY: Cornell University Press, 1960.

W. G. Richards and D. L. Cooper, *Ab Initio Molecular Orbital Calculations for Chemists*. Oxford, UK: Clarendon, 1983.

L. Salem, *The Molecular Orbital Theory of Conjugated Molecules*. New York: Benjamin, 1966.

H. F. Schaefer III, *The Electronic Structure of Atoms and Molecules*. Reading, MA: Addison-Wesley, 1972.

---

* D. P. Shoemaker, C. W. Garland, and J. W. Nibler, *Experiments in Physical Chemistry*. New York: McGraw-Hill, 1989.

Problems

12.1 Given that the equilibrium distance in H_2^+ is 106 pm and that of H_2 is 74.1 pm, calculate the internuclear repulsion energy in both cases at R_e. Using $D_e(H_2^+) = 2.79$ eV and $D_e(H_2) = 4.78$ eV, calculate E_{el} at R_e for both.

12.2 Given the equilibrium dissociation energy D_e for N_2 in Table 12.2 and the fundamental vibration frequency 2331 cm^{-1}, calculate the spectroscopic dissociation energy D_0 in kJ mol^{-1}.

12.3 Derive the values of the normalization constants given in equations 12.23 and 12.24.

12.4 Plot ψ_g and ψ_u versus distance along the internuclear axis for H_2^+ in the ground state without worrying about normalization ($R_e = 106$ pm).

12.5 Plot ψ_g^2 and ψ_u^2 versus distance along the internuclear axis for H_2^+ in the ground state without worrying about normalization ($R_e = 106$ pm).

12.6 The overlap integral S for the H_2^+ molecule can be evaluated as

$$S = \left(1 + \frac{R}{a_0} + \frac{R^2}{3a_0^2}\right) e^{-R/a_0}$$

Plot this as a function of R/a_0. At what value of R is it a maximum? At what value of R is it a minimum? What is the value of S at the equilibrium separation, $R = 106$ pm?

12.7 Express the four VB wavefunctions for H_2 as Slater determinants.

12.8 Show that the hybrid orbitals for tetravalent carbon given by equations 12.49 to 12.52 are orthogonal.

12.9 Using the hydrogenlike orbitals of Table 11.1, write out the sp^2 orbitals of equations 12.46–12.48. For $\psi_{sp^2(i)}$, write the wavefunction as a function of r for $\theta = 0$ (along the positive z axis), $\theta = 90°$ (in the xy plane), and $\theta = 180°$ (along the negative z axis).

12.10 Discuss the electronic structure of the methyl radical and the location of the unpaired electron.

12.11 How many electrons are involved in σ and π bonding orbitals in the following molecules: (a) ethylene, (b) ethane, (c) butadiene, (d) benzene?

12.12 The heat of hydrogenation of cyclohexane is -121 kJ mol^{-1} and the heat of hydrogenation of benzene is -209 kJ mol^{-1}. What is the reduction in the energy due to the formation of the π bond system in benzene?

12.13 Consider the Hückel molecular orbitals for butadiene given in equation 12.64. Each ϕ_i is an atomic p$_z$ orbital on carbon atom i, so there is a nodal plane in the xy plane for each molecular orbital. There are other nodes in these orbitals as we move from atom 1 to atom 4 since the orbitals change sign. How many nodes are there for ψ_1, ψ_2, ψ_3, and ψ_4? Where are they? Compare with Fig. 12.22.

12.14 In Hückel theory, the contribution to the electronic energy of a single π bond (see the discussion regarding ethylene in Section 12.7) is 2β. For butadiene, the contribution for two conjugated bonds is 4.472β, while for benzene (three conjugated bonds) it is 8β. What is the extra stabilization in butadiene and benzene due to the conjugation (in terms of β)? Using the data of problem 12.12, calculate the value of β for benzene and predict the value of the extra stabilization for butadiene.

12.15 The Lennard-Jones parameters for nitrogen are $\epsilon/k = 95.1$ K and $\sigma = 0.37$ pm. Plot the potential energy (expressed as V/k in K) for the interaction of two molecules of nitrogen.

12.16 For Ne the parameters of the Lennard-Jones 6-12 potential are $\epsilon/k = 35.6$ K and $\sigma = 275$ pm. Plot V in J mol^{-1} versus r and calculate the distance r_m where $dV/dr = 0$.

12.17 For KF(g) the dissociation constant D_e is 5.18 eV and the dipole moment is 28.7×10^{-30} C m. Estimate these values assuming that the bonding is entirely ionic. The ionization potential of K(g) is 4.34 eV, and the electron affinity of F(g) is 3.40 eV. The equilibrium internuclear distance in KF (g) is 0.217 nm.

12.18 The equilibrium internuclear distance for NaCl(g) is 236.1 pm. What dipole moment is expected? The actual value is 3.003×10^{-29} C m. How do you explain the difference?

12.19 Calculate the dipole moment HCl would have if it consisted of a proton and a chloride ion (considered to be a point charge) separated by 127 pm (the internuclear distance obtained from the infrared spectrum). The experimental value is 3.44×10^{-30} C m. How do you explain the difference?

12.20 Show that the SI units are the same on the two sides of equation 12.84.

12.21 The equilibrium distances in HCl, HBr, and HI are 127, 141 and 161 pm, respectively. Given the dipole moments in Table 12.4, find the effective fractional charges on the H and X ions. Is this in accord with the order of the electronegativities of the halogens?

12.22 The first excited states of He_2 are formed by exciting an electron from the antibonding $1\sigma_u$ molecular orbital to the bonding $2\sigma_g$ orbital. Write the electron configuration. What are the possible spin states? What is the bond order? Are the electronic states g or u?

12.23   Cyclobutadiene, C_4H_4, is a four-carbon atom ring. Write the secular equation similar to equation 12.60 for the π molecular orbitals of this planar molecule. Find the energies of the orbitals. Predict the total π electronic energy of this compound. Is there extra π electron stabilization (as in butadiene or benzene) in this molecule?

12.24   An approximate description of the π electrons in conjugated polyene, $CH_2=CH(CH=CH)CH=CH_2$, is the free electron molecular orbital model. In this model, the π electrons are assumed to be noninteracting and to be in a one-dimensional box of length equal to one less than the number of carbons multiplied by the C–C distance of 150 pm. For butadiene and hexatriene, what are the electron configurations of the ground and first excited states in terms of particle in a box eigenfunctions (see Section 10.8)? What is the excitation energy from ground to first excited state? What is the wavelength of this transition?

12.25   Prove that equations 12.20 and 12.21 are solutions to equation 12.19. Substitute the value of E into equations 12.17 and 12.18 to find the values of c_1 and c_2 given in equation 12.22.

12.26   What is the value of ψ_g^2 (the probability density of finding the electron) at either nucleus in H_2^+ using equation 12.25 for ψ_g? What is the value of ψ_g^2 at the midpoint of the bond? Use $R_{eq} = 106$ pm and $S = 0.86$.

12.27   Using Fig. 12.12, give the electronic configuration and bond orders for the ground states of Li_2^+, Be_2^+, B_2^+, and N_2^+ found by removing an electron from the highest filled molecular orbital.

12.28   In H_2, what percentage is the equilibrium bond energy of the total electron energy?

12.29   In the cyclobutadiene molecule (see problem 12.23), we can allow the bonds to become unequal by making one pair of opposite resonance integrals equal to β_1 and the other pair equal to β_2. This corresponds to unequal bond lengths. Write the secular determinant for this case and solve it as a function of β_2/β_1.

12.30   Consider the Hückel model description of butadiene given in Section 12.7. What is the electronic configuration of the positive ion? Of the negative ion? In electron paramagnetic resonance spectroscopy (EPR), only the unpaired spin densities are probed. For these two ions what are the unpaired π electron densities on each carbon atom (i.e., the square of the coefficients in equation 12.64)?

12.31   Although NaCl is an example of ionic bonding, it dissociates into atoms. The potential curve as a function of R (for R larger than the equilibrium distance) can be approximated by a purely electrostatic attraction, $V = -e^2/(4\pi\epsilon_0 R)$, until at some R, this curve crosses the potential curve for the neutral atoms. Assume that the potential curve for the neutral atoms is independent of R and compute the value of R at the crossing in terms of the ionization potential of Na and the electron affinity of Cl. Given that $IP_{Na} = 5.14$ eV and $E_{ea}(Cl) = 3.60$ eV, compute this R.

12.32   Assume that the R^{-6} part of the Lennard-Jones potential for Ar is given by the dispersion force term, equation 12.73. If the polarizability (α) of Ar is $1.85 \times 10^{-40} J^{-1} C^2 m$, find the approximate energy of the first electronic excitation using the Lennard-Jones parameters in Table 12.5.

12.33   Calculate the spectroscopic dissociation energy D_0 of O_2 from the data in Tables 12.2 and 14.4 and compare it with the dissociation energy for ideal gas O_2 at absolute zero.

12.34   Show that the hybrid orbitals $\psi_{sp(i)}$ and $\psi_{sp(ii)}$ (equation 12.42 and 12.43) are normalized and orthogonal.

12.35   Show that the hybrid orbitals $\psi_{sp^2(i)}$ and $\psi_{sp^2(ii)}$ are orthogonal and normalized.

12.36   Calculate the lattice energy which is the dissociation energy of NaCl(s) into gaseous ions. Given: $\Delta_f H°[Na^+(g), 298 K] = 609.358$ kJ mol^{-1}; $\Delta_f H°[Cl^-(g), 298 K] = -233.13$ kJ mol^{-1}.

12.37   The dipole moments of CH_3Cl, CH_3Br, and CH_3I are given in Table 12.4. Explain the relative magnitudes of these moments in terms of the electronegativity.

12.38   Calculate the dipole moment of fluorobenzene, using a plot of $\mathcal{P}$ versus $1/T$, from the following molar polarizations for gaseous samples measured by K. B. McAlpine and C. P. Smyth (*J. Chem. Phys.* **3**:55 (1935)).

T/K	343.6	371.4	414.1	453.2	507.0
$\mathcal{P}/cm^3\ mol^{-1}$	69.9	66.8	62.5	59.3	55.8

12.39   The Lennard-Jones parameters for argon are $\epsilon/k = 122$ K and $\sigma = 0.34$ nm. Plot the potential energy (expressed as V/k in K) for the interaction of two molecules of argon.

12.40   For methane the parameters for the Lennard-Jones 6-12 potential are $\epsilon/k = 148$ K and $\sigma = 0.382$ nm. Plot V in kJ mol^{-1} versus r and calculate r_m.

13
Symmetry

Ideas about symmetry are of great importance in connection with both theoretical and experimental studies of molecular structure. The basic principles of symmetry are applied in quantum mechanics, spectroscopy, and structural determinations by X-ray, neutron, and electron diffraction. Nature exhibits a great deal of symmetry, and this is especially evident when we examine molecules in their equilibrium configurations. By equilibrium configuration we refer to that with the atoms fixed in their mean positions. When symmetry is present, certain calculations are simplified if the symmetry is taken into account. Aspects of symmetry also determine whether a molecule can be optically active or whether it may have a dipole moment. In this chapter we will see how symmetry may be treated quantitatively using group theory. Modern treatments of rotational, vibrational, and electronic spectroscopy of molecules all make extensive use of group theory.

13.1   Symmetry Operations and Symmetry Elements

A symmetry operation is the transformation of an object in such a way that its final appearance is indistinguishable from its initial appearance. A symmetry element is a point, line, or plane with respect to which a symmetry operation is carried out. The five types of symmetry elements and operations used in describing the symmetry of molecules are introduced in Table 13.1 and discussed in Sections 13.2 to 13.5. In Section 13.7 we will discuss the fact that these operations form a **mathematical group.** The group necessarily includes the identity operation E that leaves the system unchanged.

Table 13.1   Symmetry Elements and Associated Operations

Symbol for Element and Operation	Element	Operation
i	Center of symmetry (or inversion center)	Projection through the center of symmetry to an equal distance on the other side from the center
C_n	Proper rotation axis	Counterclockwise rotation about the C_n axis by $2\pi/n$ (or $360°/n$), where n is an integer
σ	Symmetry plane	Reflection across the plane of symmetry
S_n	Improper rotation axis (also referred to as a rotation-reflection axis or alternating axis)	Counterclockwise rotation about the S_n axis by $2\pi/n$ followed by reflection in a plane perpendicular to the axis (viz., the combined operation of a C_n rotation followed by reflection across a σ_h mirror plane)
E	Identity element	The operation that leaves the system unchanged

13.2   The Inversion Operation and the Center of Symmetry

A molecule has a **center of symmetry** i if there is a point such that a straight line from any nucleus projected through the point encounters an equivalent nucleus equidistant from the center. The center of symmetry (or inversion center) is the symmetry element, and the operation is inversion through the center by which one-half of the molecule can be generated from the other half. The action of the inversion operation is to transform the coordinates (x, y, z) into their respective negatives $(-x, -y, -z)$. The inversion operation may be represented in matrix notation by

$$i \cdot \begin{bmatrix} x \\ y \\ z \end{bmatrix} = \begin{bmatrix} -x \\ -y \\ -z \end{bmatrix} \tag{13.1}$$

If the inversion operation is applied twice we obtain the original configuration:

$$i \cdot i \cdot \begin{bmatrix} x \\ y \\ z \end{bmatrix} = i \cdot \begin{bmatrix} -x \\ -y \\ -z \end{bmatrix} = \begin{bmatrix} x \\ y \\ z \end{bmatrix} \tag{13.2}$$

Thus, the successive application of i an even number of times produces the identity operation E.

Since $i^k = i$ when k is odd, and $i^k = E$ when k is even, the center of symmetry generates only *one distinct* operation other than E. In molecules with a center of symmetry, the nuclei may be thought of as occurring in centrosymmetric

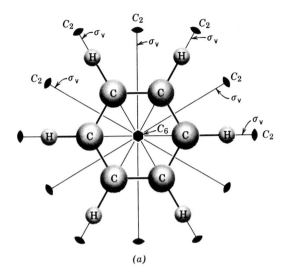

(a)

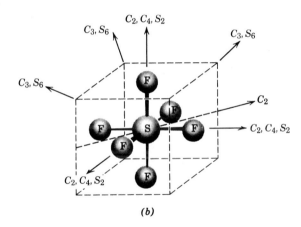

(b)

Figure 13.1   Molecules with a center of symmetry. (a) C_6H_6 molecule with D_{6h} point-group symmetry. (b) SF_6 molecule with O_h point-group symmetry. This centrosymmetric molecule possesses three equivalent C_4 axes at right angles to one another. Three S_4 and three C_2 axes are coincident with the three C_4 axes. Four C_3 axes (and four coincident S_6 axes) are located along the four body diagonals of the cube. A C_2 axis is labeled with ◗ , and a C_6 axis is labeled with ⬢ .

pairs with the exception of the nucleus (if any) lying at the center of symmetry. Centrosymmetric molecules include C_6H_6 (Fig. 13.1a), SF_6 (Fig. 13.1b), the **staggered** conformation of C_2H_6 (Fig. 13.3b), CO_2, and C_2H_4.

13.3   The Rotation Operation and the Symmetry Axis

A symmetry axis is a line about which rotation through an angle $2\pi/n$ radians brings a structure into coincidence with itself. The **rotation operator** is repre-

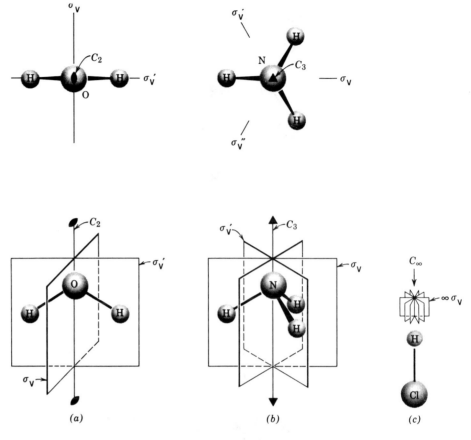

Figure 13.2 (a) H_2O molecule with C_{2v} point-group symmetry. (b) NH_3 molecule with C_{3v} point-group symmetry. (c) HCl molecule with $C_{\infty v}$ point-group symmetry.

sented by C_n, where n is referred to as the order of the rotation, and the rotation is conventionally taken as positive in the counterclockwise direction. If the z axis is a twofold rotation axis, the action of the C_2 rotation operator is to transform the coordinate (x, y, z) to $(-x, -y, z)$. The C_2 rotation operation may be represented by

$$C_2 \cdot \begin{bmatrix} x \\ y \\ z \end{bmatrix} = \begin{bmatrix} -x \\ -y \\ z \end{bmatrix} \tag{13.3}$$

As shown in Fig. 13.2a, a water molecule has a C_2 axis passing through the oxygen and bisecting the angle between the O–H bonds. The ammonia molecule NH_3 (Fig. 13.2b) has a C_3 axis passing through the nitrogen. The benzene molecule (Fig. 13.1a) has a C_6 axis perpendicular to the plane of the ring and six C_2 axes lying in the plane of the ring. Any linear molecule, such as HCl, has a C_∞ axis (Fig. 13.2c), because the appearance of the molecule is not changed by a rotation of any angle (infinity in number) about the internuclear axis.

If the C_2 operation is applied twice in succession, the identity operator is obtained:

$$C_2C_2 = C_2^2 = E$$

If the C_3 operation is applied twice in succession, a 240° rotation is obtained; if it is applied three times, the identity operation is obtained:

$$C_3C_3 = C_3^2 \qquad C_3C_3C_3 = C_3^3 = E$$

The C_3^2 operation is a new operation; thus, the C_3 element of symmetry generates two operations, C_3 and C_3^2.

The operations generated by a fourfold axis are

$$C_4^1 \qquad C_4^2 \qquad C_4^3 \qquad C_4^4$$
$$C_2^1 \qquad\qquad E$$

where equivalent operations are listed below. Since a C_2 axis is always coincident with a C_4 axis, only the operations C_4^1 and C_4^3 are distinct operations of the C_4 axis.

Example 13.1

How many distinct operations are implied by a C_6 axis?

$$C_6^1 \qquad C_6^2 \qquad C_6^3 \qquad C_6^4 \qquad C_6^5 \qquad C_6^6$$
$$C_3^1 \qquad C_2^1 \qquad C_3^2 \qquad\qquad E$$

Thus, two operations (C_6^1 and C_6^5) are characteristic only of a C_6 axis and cannot be represented in any other way.

In discussing symmetry operations, it is convenient to orient the molecules in a right-hand Cartesian coordinate system. The thumb, index, and middle fingers of the right hand are pointed in three mutually perpendicular directions, and these are taken as the x, y, and z directions, respectively. The center of mass of the molecule under consideration is located at the origin of the Cartesian coordinate system, and its principal axis is aligned with the z axis. The **principal axis** is defined as the C_n axis with highest order n; if there are several rotational axes of the same highest symmetry (e.g., three twofold axes at right angles to one another), the z axis is taken along the one passing through the greatest number of atoms.

13.4 The Reflection Operation and the Symmetry Plane

If reflection of all of the nuclei through a plane in a molecule gives a configuration physically indistinguishable from the original one, the molecule is said to have a symmetry plane. The symmetry plane is represented by σ, and the **reflection operation** is represented by σ. Since the operation σ gives a configuration equivalent to the original and since the application of the same σ twice to a molecule produces it original configuration, it follows that a sym-

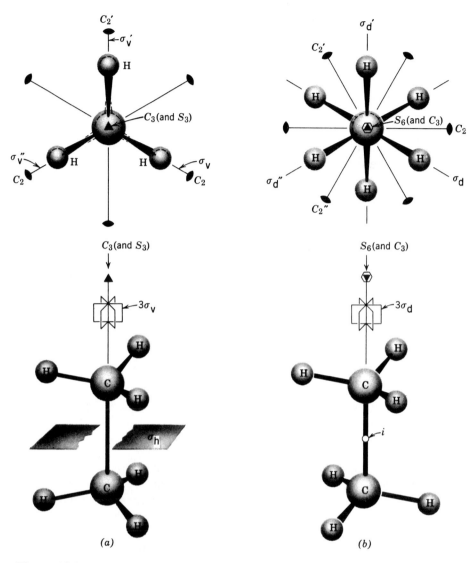

Figure 13.3 (*a*) The eclipsed conformation of ethane, C_2H_6, with D_{3h} point symmetry. The view along the C_3 (and S_3) axis of the C–C bond is shown at the top. (*b*) The staggered conformation of ethane, C_2H_6, with D_{3d} point symmetry. The view along the S_6 (and C_3) axis of the C–C bond is shown at the top.

metry plane generates only one distinct operation in that $\sigma^k = \sigma$ when k is odd, and $\sigma^k = E$ when k is even.

If the xz plane is a symmetry plane the reflection operation σ may be represented by

$$\sigma \cdot \begin{bmatrix} x \\ y \\ z \end{bmatrix} = \begin{bmatrix} x \\ -y \\ z \end{bmatrix} \qquad (13.4)$$

A symmetry plane (and corresponding operation of reflection) perpendicular to the direction of the principal C_n axis (i.e., the normal of the symmetry plane is coincident with the C_n axis of highest order n) is called a **horizontal** symmetry plane and denoted as σ_h. Molecules with a horizontal symmetry plane include C_6H_6 (Fig. 13.1a) (for which σ_h is perpendicular to the C_6 axis and contains all the atoms of this planar molecule and the **eclipsed** conformation of ethane (Fig. 13.3a), which has a σ_h perpendicular to a C_3 principal axis.

Symmetry planes that contain the principal C_n axis are called **vertical** symmetry planes and generally are symbolized as σ_v. **Dihedral** symmetry planes that bisect the angles formed by pairs of horizontal C_2 axes are designated as σ_d. As is evident from Fig. 13.2a, a H_2O molecule has two vertical symmetry planes σ_v and σ_v', which are perpendicular to each other. One of these symmetry planes (σ_v) comprises the plane of the molecule, and the other (σ_v') is perpendicular to it. The twofold axis lies in the intersection of the two symmetry planes. Ammonia has three σ_v containing the C_3 axis, and benzene (Fig. 13.1a) has six σ_v containing the C_6 principal axis. The linear molecule HCl (Fig. 13.2c) has an infinite number of vertical symmetry planes of type σ_v, all of which include the C_∞ rotational axis. Homonuclear diatomic molecules such as H_2 or Cl_2 have a σ_h in addition. Figure 13.3b shows the staggered conformation of ethane to possess three vertical σ_d which contain the principal C_3 axis and which bisect the three horizontal C_2 axes. In the eclipsed conformation of ethane (Fig. 13.3a), however, each of the three vertical symmetry planes contains one of the three horizontal C_2 axes (as well as the C_3 principal axis). Consequently, these symmetry planes are called σ_v.

13.5 The Operation of Improper Rotation and the Improper Axis

The operation of improper rotation consists of a rotation by $2\pi/n$ radians about an axis followed by reflection in a plane perpendicular to the axis. Thus, the **improper rotation** S_n is the product of two operations:

$$S_n = \sigma C_n \tag{13.5}$$

This means that C_n and σ are applied successively. The improper axis is represented by S_n. As we will see, a molecule need not possess the symmetry element σ or C_n to have the symmetry element S_n.

An S_1 element is equivalent to a **plane of symmetry** σ as the operation involving rotation of 360° followed by reflection across the symmetry plane perpendicular to the axis of rotation can be simply represented in terms of reflection across the mirror plane.

An S_2 element is equivalent to a **center of symmetry** i, since the S_2 operation consisting of a counterclockwise rotation about an axis of $2\pi/2$ or 180° followed by reflection across a horizontal symmetry plane perpendicular to this axis yields the same configuration as an inversion through a center of symmetry located at the intersection of the axis of rotation and symmetry plane.

An S_3 element has six operations:

$$
\begin{array}{cccccc}
S_3^1 & S_3^2 & S_3^3 & S_3^4 & S_3^5 & S_3^6 \\
 & C_3^2 & \sigma_h & C_3^1 & & E
\end{array}
$$

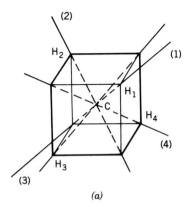

(a)

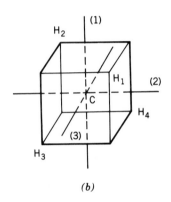

(b)

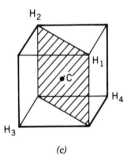

(c)

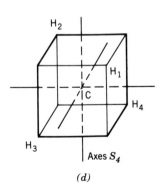

Axes S_4

(d)

Figure 13.4 Elements of the symmetry of methane (T_d). (a) Four C_3 axes, which each have C_3^1 and C_3^2 operations associated with them. (b) Three C_2 axes. (c) Six σ_d planes. (d) Four S_4 axes, which each have S_4^1 and S_4^3 operations associated with them.

The equivalent operations listed below show that an S_3 element introduces only two distinct operations, S_3^1 and S_3^5. Thus, an S_3 axis implies a C_3 axis and horizontal symmetry plane σ_h. The eclipsed conformation of ethane has an S_3 axis, as shown in Fig. 13.3a.

An S_4 axis has four associated operations

$$S_4^1 \quad S_4^2 \quad S_4^3 \quad S_4^4$$
$$C_2^1 \qquad E$$

Thus, an S_4 axis implies a C_2 axis. This is illustrated by methane that has three equivalent S_4 axes at right angles to each other, as shown in Fig. 13.4. Note that $S_n^{2n} = E$ when n is odd, but $S_n^n = E$ when n is even. Since it is a little more difficult to visualize an improper rotation axis than a proper rotation axis, be sure to identify the three S_4 axes of methane in Fig. 13.4. Since S_4^1 and S_4^3 are distinct operations of the S_4 axes, the number of distinct operations is six. Each of the four C_3 axes has two distinct operations (C_3 and C_3^2), so there are 8 in total.

It is always found that $S_n^{n/2} = i$ if n is even and $n/2$ is odd. The staggered configuration of ethane has an S_6 axis, as shown in Fig. 13.3b.

In general, S_n axes with n even contain n operations with neither a σ_h nor a C_n axis (but with a $C_{n/2}$ axis). S_n axes with n odd contain a total of $2n$ operations, including σ_h and the operations generated by C_n.

13.6 Classification of Schoenflies Point Groups

When all the symmetry elements of a molecule including the identity operation are considered, the corresponding symmetry operations form what is called a mathematical group (see Section 13.7). This allows the classification of molecules by symmetry groups. Since all of the symmetry elements of a molecule intersect at some point, the symmetry groups are referred to as point groups. Each point group is designated by a Schoenflies symbol. The Schoenflies symbol designates sufficient symmetry elements to obtain the other symmetry elements and all distinct operations of the point group. The various types of point groups are defined below and are illustrated in Table 13.2. The identity operation is in every group and is therefore not mentioned explicitly. There is, in principle, an infinite number of point groups. The external symmetries of crystals fall into only 32 point groups.

Point Group C_1

Molecules with no symmetry other than the identity are in this class. An example is CHBrClF.

Point Group C_s

This is the group for molecules that only have a reflection plane σ_s. An example is CH_2ClF.

Point Group C_i

Molecules, such as 1,2-dibromo-1,2-dichloroethane, that have only a center of symmetry i belong to this point group.

Point Groups C_n

Molecules possessing only an n-fold axis of rotation belong to a C_n point group.

Point Groups C_{nv}

A molecule with an n-fold axis of rotation and n vertical mirror planes (which are necessarily colinear with the n-fold axis) belong in one of these point groups.

Point Groups C_{nh}

A molecule with an n-fold axis and a plane of symmetry perpendicular to this axis belong to one of these point groups. Such a plane is referred to as a horizontal mirror plane. The C_{2h} point group necessarily involves a center of symmetry as well.

Point Groups D_n

A molecule with a C_n axis and a C_2 axis perpendicular to this axis is in this point group.

Point Groups D_{nd}

A molecule with a C_n axis, a perpendicular C_2 axis, and a dihedral mirror plane is in this point group. The dihedral mirror plane is colinear with the principal axis and bisects the two perpendicular C_2 axes.

Point Groups D_{nh}

As you may have guessed already, molecules in this point group have a horizontal mirror plane, that is, one perpendicular to the principal axis.

Point Groups S_n

To be in one of these point groups a molecule has to have an n-fold improper rotation axis.

Table 13.2   Common Schoenflies Point Groups with Examples

Schoenflies Symbol	Symmetry Elements	Molecular Configuration	Schoenflies Symbol	Symmetry Elements	Molecular Configuration
C_1	E		D_{3h}	E, $2C_3$, $3C_2$, σ_h, $2S_3$, $3\sigma_v$	
C_s	E, σ		D_{4h}	E, $2C_4$, C_2, $2C_2'$, $2C_2''$, i, $2S_4$, σ_h, $2\sigma_v$, $2\sigma_d$	
C_i	E, i		D_{5h}	E, $2C_5$, $2C_5^2$, $5C_2$, σ_h, $2S_5$, $2S_5^2$, $5\sigma_v$	
C_2	E, C_2		D_{6h}	E, $2C_6$, $2C_3$, C_2, $3C_2'$, $3C_2''$, i, $2S_3$, $2S_6$, σ_h, $3\sigma_d$, $3\sigma_v$	
C_{2v}	E, C_2, $\sigma_v(xz)$, $\sigma_v'(yz)$		$D_{\infty h}$	E, $2C_\infty$, $\infty\sigma_v$, i, $2S_\infty$, ∞C_2	H—C≡C—H

Table 13.2   (*continued*)

Schoenflies Symbol	Symmetry Elements	Molecular Configuration	Schoenflies Symbol	Symmetry Elements	Molecular Configuration
C_{3v}	E, $2C_3$, $3\sigma_v$		D_{2d}	E, $2S_4$, C_2, $2C_2'$, $2\sigma_d$	
C_{4v}	E, $2C_4$, C_2, $2\sigma_v$, $2\sigma_d$		D_{3d}	E, $2C_3$, $3C_2$, i, $2S_6$, $3\sigma_d$	
$C_{\infty v}$	E, $2C_\infty$, $\infty\sigma_v$		D_{4d}	E, $2S_8$, $2C_4$, $2S_8^3$, C_2, $4C_2'$, $4\sigma_d$	
C_{2h}	E, C_2, i, σ_h		D_{5d}	E, $2C_5$, $2C_5^2$, $5C_2$, i, $2S_{10}^3$, $2S_{10}$, $5\sigma_d$	
C_{3h}	E, C_3, C_3^2, σ_h, S_3, S_3^5		T_d	E, $8C_3$, $3C_2$, $6S_4$, $6\sigma_d$	
D_{2h}	E, $C_2(z)$, $C_2(y)$, $C_2(x)$, i, $\sigma(xy)$, $\sigma(xz)$, $\sigma(yz)$		O_h	E, $8C_3$, $6C_2$, $6C_4$, $3C_2$, i, $6S_4$, $8S_6$, $3\sigma_h$, $6\sigma_d$	

$C_{\infty h}, D_{\infty h} T_d, O_h, I_h, C_1, C_s, D_{nd}, D_{nh}, D_n, C_{nh}, C_{nh}, C_{nv}, C_n, S_n$

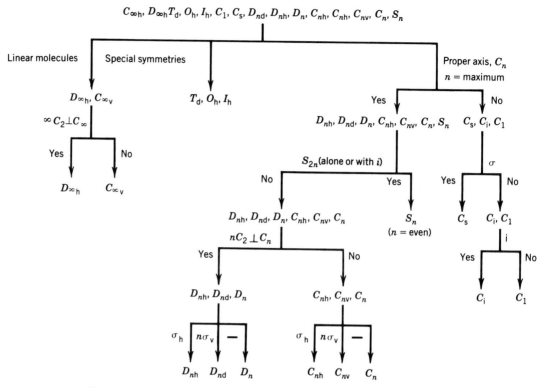

Figure 13.5 Decision tree for assigning a molecule to a point group. (From M. Zeldin, *J. Chem. Ed.* **43**:17(1966) © Copyright American Chemical Society.)

Special Point Groups

Linear molecules are either $C_{\infty v}$ or $D_{\infty h}$. A heteronuclear molecule, such as CO, is $C_{\infty v}$ because the molecular axis is an ∞-fold axis, and they have an infinite number of vertical mirror planes. Homonuclear diatomic molecules or polyatomics such as acetylene are $D_{\infty h}$ because the molecular axis is ∞-fold, and there is an infinite number of perpendicular C_2 axes since the molecule is symmetrical. Tetrahedral molecules are T_d. The T_h point group has all of the symmetry of a cube. Octahedral molecules, such as SF_6, are O_h. Molecules with the symmetry of an icosahedron or dodecahedron are I_h, and atoms with spherical symmetry are K_h.

The classification of a molecule into a point group may be facilitated by the use of a decision tree, such as that in Fig. 13.5.

13.7 Combinations of Symmetry Operations: Point Groups

The set of symmetry operations of a molecule constitute a mathematical group. A set of operations belong to a mathematical group only when they obey the four rules that define a group:

1. If A represents a symmetry operation of the group and B represents another symmetry operation of the same group, the product $AB = F$ is also an operation of the group. The product $AB = F$ means that the operation B followed by the operation A is equivalent to an operation F. This important property (that the product of two operations in the group is an operation in the group), which is illustrated by the multiplication table given in Table 13.3, is referred to as the **group property.** In general, $AB \neq BA$, which means that the operation B followed by operation A is not necessarily equivalent to the operation A followed by operation B. In other words, A does not necessarily commute with B. If $AB = BA$, the multiplication is commutative.

Table 13.3 Multiplication Table for the Group C_{2v}

	Operation B			
	E	C_2^1	σ_v	σ_v'
E	E	C_2^1	σ_v	σ_v'
C_2^1	C_2^1	E	σ_v'	σ_v
Operation A $\quad \sigma_v$	σ_v	σ_v'	E	C_2^1
σ_v'	σ_v'	σ_v	C_2^1	E

The table contains the products AB for the indicated operations. Note that each column and each row has each symmetry operation represented only once.

2. In each group there exists an identity operation E (corresponding to the C_1 rotation of $360°$ about any given axis) such that for any other operation of the group (e.g., A),

$$AE = EA = A \tag{13.6}$$

3. For each operation A there exists in the group an inverse operation A^{-1} such that $A^{-1}A = AA^{-1} = E$. The inverse operation A^{-1} is that which returns the object to its original position. Hence, the inverse of a C_2, σ, or i operation is itself (viz., $C_2 C_2 = E$; $\sigma\sigma = E$; $ii = E$). The inverse of a C_3^1 operation is a C_3^2, while the inverse of an S_4^1 operation is an S_4^3.

4. The associative law of multiplication holds:

$$A(BC) = (AB)C \tag{13.7}$$

Consider H_2O (Fig. 13.2a), which has four symmetry operations: E, C_2^1, σ_v, and σ_v'. The operation of reflection in one vertical symmetry plane (σ_v) followed by the operation of reflection in the other vertical symmetry plane (σ_v') is equivalent to a twofold rotation; that is, $\sigma_v' \, \sigma_v = C_2^1$. Similarly, the successive operations of C_2^1 followed by σ_v yield the same result as the σ_v' operation (viz., $\sigma_v C_2^1 = \sigma_v'$). Each of the four operations is its own inverse (e.g., $\sigma_v \sigma_v = E$). The product operations of H_2O are summarized in Table 13.3 as a group multiplication table, which shows (since no additional operations are generated) that these four symmetry operations form a group and that the operations are commutative. The point group is designated by the Schoenflies symbol C_{2v}. The subscript 2 of this symbol signifies not only that the principal proper rotation axis (C_n) is a C_2, but also that there are two vertical symmetry planes, at right angles to each other, that contain the C_2 axis.

13.8 Applications of Symmetry

The presence of symmetry in a molecule can be used to determine when certain molecular properties will be zero. For example, certain symmetry groups preclude the possibility of a dipole moment or optical activity. We consider these two examples in this section. A dipole moment (Section 12.12) is a vector quantity that is not affected either in direction or in magnitude by any symmetry operation of the molecule. Therefore, the dipole moment vector must be contained in each of the symmetry elements. Consequently, molecules that possess dipole moments belong only to the point groups C_n, C_s, and C_{nv}. The presence or absence of a dipole moment therefore tells something about the symmetry of a molecule. For example, carbon dioxide and water might have structures corresponding to a symmetrical linear molecule, to an unsymmetrical linear molecule, or to a bent molecule. The dipole moments recorded in Table 12.5 show that carbon dioxide has zero moment; therefore, the molecule must be symmetrical and linear. If it were unsymmetrical or bent, there would have been a permanent dipole moment. On the other hand, water has a pronounced dipole moment and cannot have the symmetrical linear structure. A molecule with a center of symmetry cannot have a dipole moment.

If a molecule and its mirror image cannot be superimposed, it is potentially optically active. Since a rotation followed by a reflection always converts a right-handed object to a left-handed object, an S_n axis guarantees that a molecule cannot exist in separate left- and right-handed forms.

All **improper** rotation axes (S_n), including a mirror plane ($\sigma = S_1$) and center of symmetry ($i = S_2$), convert a right-handed object into a left-handed object (i.e., produce a mirror image of the original object), whereas all **proper** rotation axes (C_n) leave a right-handed object unchanged in this respect. Hence, only molecules that have no improper symmetry elements can be optically active.

In a molecule in which internal rotation can take place (e.g., ethane or H_2O_2) it is possible to have optically active conformations, but in a gas or solution these conformers are so rapidly interconverted that optical isomers cannot be resolved.

Example 13.2

To examine the effect of rotation in the xy plane, we rotate about the z axis by an angle θ. Then the x and y coordinates of a point change in the following way:

$$\begin{bmatrix} x \\ y \end{bmatrix} \Rightarrow \begin{bmatrix} x \cos \theta - y \sin \theta \\ x \sin \theta + y \cos \theta \end{bmatrix} = \begin{bmatrix} \cos \theta & -\sin \theta \\ \sin \theta & \cos \theta \end{bmatrix} \begin{bmatrix} x \\ y \end{bmatrix}$$

using matrix multiplication. This can be seen in Fig. 13.6 using some simple trigonometry by the following argument. The initial x,y coordinates can be written as $x = r \cos \phi$, $y = r \sin \phi$. A rotation by θ brings these to

$$X = r \cos(\phi + \theta) = r \cos \theta \cos \phi - r \sin \phi \sin \theta$$
$$Y = r \sin(\phi + \theta) = r \cos \theta \sin \phi + r \sin \theta \cos \phi$$

proving the formula above. What are the matrices representing rotations C_3 and C_3^2?

The C_3 operation implies $\theta = 120°$ and C_3^2 implies $\theta = 240°$. Thus, the matrices for these are

$$C_3 = \begin{bmatrix} \dfrac{-1}{2} & \dfrac{-\sqrt{3}}{2} \\ \dfrac{\sqrt{3}}{2} & \dfrac{-1}{2} \end{bmatrix} \qquad C_3^2 = \begin{bmatrix} \dfrac{-1}{2} & \dfrac{\sqrt{3}}{2} \\ \dfrac{-\sqrt{3}}{2} & \dfrac{-1}{2} \end{bmatrix}$$

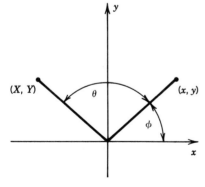

Figure 13.6 The effect of a rotation about the z axis on the coordinates of a point in the xy plane.

Note that when the matrix for C_3 is squared, the matrix for C_3^2 results, as it should. Finally, we note that since the z component of a point does not change for a rotation about the z axis, we can write the matrices in three-dimensional notation; for example,

$$C_3 = \begin{bmatrix} \dfrac{-1}{2} & \dfrac{-\sqrt{3}}{2} & 0 \\ \dfrac{\sqrt{3}}{2} & \dfrac{-1}{2} & 0 \\ 0 & 0 & 1 \end{bmatrix}$$

13.9 Special Topic: Matrix Representations

As we have seen, the application of a symmetry operation transforms the co-ordinates of a point to a set of coordinates for another point that is related to the first by the symmetry operation. For example, a C_2 rotation converts a point with coordinates x_1, y_1, z_1 to point x_2, y_2, z_2, as shown in Fig. 13.7. The new coordinates are given by the equations

$$x_2 = -x_1 + 0y_1 + 0z_1 \tag{13.8}$$

$$y_2 = 0x_1 - y_1 + 0z_1 \tag{13.9}$$

$$z_2 = 0x_1 + 0y_1 + z_1 \tag{13.10}$$

These relations are expressed in matrix notation (see the section on matrices in Appendix D by

$$\begin{bmatrix} x_2 \\ y_2 \\ z_2 \end{bmatrix} = \begin{bmatrix} -1 & 0 & 0 \\ 0 & -1 & 0 \\ 0 & 0 & 1 \end{bmatrix} \begin{bmatrix} x_1 \\ y_1 \\ z_1 \end{bmatrix} \tag{13.11}$$

Thus, the transformation matrix of C_2 is

$$R(C_2) = \begin{bmatrix} -1 & 0 & 0 \\ 0 & -1 & 0 \\ 0 & 0 & 1 \end{bmatrix} \tag{13.12}$$

The other transformation matrices in the C_{2v} group are

$$R(E) = \begin{bmatrix} 1 & 0 & 0 \\ 0 & 1 & 0 \\ 0 & 0 & 1 \end{bmatrix} \tag{13.13}$$

$$R(\sigma_v) = \begin{bmatrix} 1 & 0 & 0 \\ 0 & -1 & 0 \\ 0 & 0 & 1 \end{bmatrix} \tag{13.14}$$

$$R(\sigma_v') = \begin{bmatrix} -1 & 0 & 0 \\ 0 & 1 & 0 \\ 0 & 0 & 1 \end{bmatrix} \tag{13.15}$$

When these matrices are multiplied by each other the results are the same as when the operators are multiplied by each other, as shown in Table 13.3. There-fore, these matrices are referred to as **representatives** of their respective op-erations in the C_{2v} point group. The group multiplication table can be repro-duced by matrix multiplications of the matrix representatives. The set of four matrices is referred to as a **representation** of the C_{2v} point group. The effect of the operations of the C_{2v} point group are shown in Fig. 13.8.

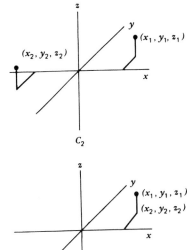

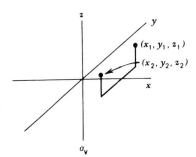

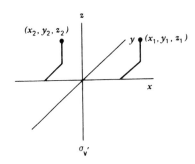

Figure 13.7 Operations in the C_{2v} point group.

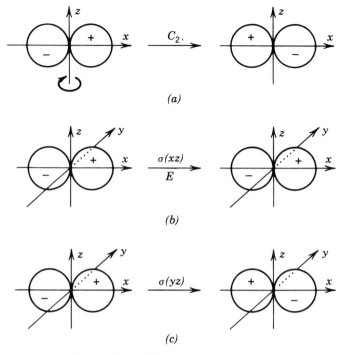

Figure 13.8   Effects of operations of the C_{2v} group on the p_x orbitals.

Example 13.3

Multiply the transformation matrix for C_2 by the transformation matrix for σ_v and identify the operator that corresponds with the product. See Matrices in Appendix D.

$$\begin{bmatrix} 1 & 0 & 0 \\ 0 & -1 & 0 \\ 0 & 0 & 1 \end{bmatrix} \begin{bmatrix} -1 & 0 & 0 \\ 0 & -1 & 0 \\ 0 & 0 & 1 \end{bmatrix} = \begin{bmatrix} -1 & 0 & 0 \\ 0 & 1 & 0 \\ 0 & 0 & 1 \end{bmatrix}$$

Thus, the product yields the transformation matrix for σ_v' so that

$$\sigma_v C_2 = \sigma_v'$$

as shown in Table 13.3.

Although we found this representation by considering the effect of the operations on a point in three-dimensional space, the notion of a representation is more general. Any set of numbers of matrices that have the same multiplication table as the operations in the group form a representation of the group. There are an infinite number of such representations of a group, but there are a finite number that are, in a mathematical sense, more fundamental than the others. These are called **irreducible** representations. The representation we found for the C_{2v} point group (equations 13.12–13.15) is not irreducible. It is, in fact, reducible to three different irreducible representations because the matrices are diagonal. In this case the diagonal elements themselves form an irreducible representation. For example, if we take the xx elements of 13.12–13.15 we have

$$R(E) = 1 \qquad R(C_2) = -1 \qquad R(\sigma_v) = 1 \qquad R(\sigma'_v) = -1 \qquad (13.16)$$

The yy elements give us

$$R(E) = 1 \qquad R(C_2) = -1 \qquad R(\sigma_v) = -1 \qquad R(\sigma'_v) = +1 \qquad (13.17)$$

and the zz elements

$$R(E) = 1 \qquad R(C_2) = +1 \qquad R(\sigma_v) = +1 \qquad R(\sigma'_v) = +1 \qquad (13.18)$$

These are representations because their multiplication table is identical to that of the C_{2v} group operations themselves. Another representation is

$$R(E) = 1 \qquad R(C_2) = +1 \qquad R(\sigma_v) = -1 \qquad R(\sigma'_v) = -1 \qquad (13.19)$$

These turn out to be all the irreducible representations of C_{2v}.

Example 13.4

Show that the representation given in equation 13.16 has the same multiplication table as the operations in the group C_{2v} (Table 13.3).

From the multiplication table of the R's of equation 13.16:

	$R(E)$	$R(C_2)$	$R(\sigma_v)$	$R(\sigma'_v)$
$R(E)$	1	-1	1	-1
$R(C_2)$	-1	1	-1	1
$R(\sigma_v)$	1	-1	1	-1
$R(\sigma'_v)$	-1	1	-1	1

If we compare these numbers to those we would obtain by replacing the operators in Table 13.3 by the R's of equation 13.16, we see they are identical.

Note that the representation in equation 13.18 is almost trivial in the sense that every operation has been represented by the number 1.

In the case of the C_{2v} group, as for all commutative groups, all the irreducible representations are one dimensional (i.e., numbers). Many groups have higher dimensional irreducible representations (e.g., D_{6h}, T_d) and then the matrices in the representation have that dimension.

13.10 Special Topic: Character Tables

Often we do not work with the matrix representations of a group themselves, but with the **character** of the representation, which is defined as the sum of the diagonal elements (trace) of the matrix representation. For one-dimensional representations then, the character and the representation are identical. The

Table 13.4   Character Table for the C_{2v} Group

C_{2v}	E	C_2	$\sigma_v(xz)$	$\sigma_v'(yz)$	
A_1	1	1	1	1	z, z^2, x^2, y^2
A_2	1	1	-1	-1	xy
B_1	1	-1	1	-1	x, xz
B_2	1	-1	-1	1	y, yz

character table for the C_{2v} group is given in Table 13.4. On the left is the label of the irreducible representation (A_1, A_2, B_1, B_2) and on the right are examples of functions of x, y, and z which transform like these representations under the operations of the C_{2v} group. For example, consider the function x. Operating on x with the elements of the C_{2v} group gives

$$E \cdot x = (1)x \tag{13.20}$$
$$C_2 \cdot x = (-1)x \tag{13.21}$$
$$\sigma_v \cdot x = (1)x \tag{13.22}$$
$$\sigma_v' \cdot x = (-1)x \tag{13.23}$$

By comparing this to Table 13.4, we see that x can be labeled B_1.

Example 13.5

Verify the symmetry properties of all the functions on the right side of Table 13.4.

Consider the functions y and z. Under the operation of C_{2v}, they transform in the following way:

$$E \cdot y = (1)y \qquad E \cdot z = (1)z$$
$$C_2 \cdot y = (-1)y \qquad C_2 \cdot z = (1)z$$
$$\sigma_v \cdot y = (-1)y \qquad \sigma_v \cdot z = (1)z$$
$$\sigma_v' \cdot y = (+1)y \qquad \sigma_v' \cdot z = (1)z$$

So that y transforms like B_2 and z like A_1. Composite functions such as yz can be found from the transformations of y and z alone, so that, for example,

$$\sigma_v(yz) = (-1)(+1)yz$$

Symmetry operations may also be applied to wavefunctions and thus wavefunctions can be classified by their symmetry properties. For example, consider the effects of the operators of the C_{2v} group on p orbitals. The effect of the C_2 operation on a p_x orbital is illustrated in Fig. 13.8a. This is summarized by

$$C_2 p_x = -1 p_x \tag{13.24}$$

where -1 indicates the sign reversal. Now what are the effects of the reflection operations $\sigma(xz)$ and $\sigma(yz)$? As shown in Fig. 13.8b, reflection in the xz plane has no effect, and so

$$\sigma(xz) p_x = 1 p_x \tag{13.25}$$

As shown in Fig. 13.8c, reflection in the yz plane causes a sign change so that

$$\sigma(yz)p_x = -1p_x \qquad (13.26)$$

The identity operation E has no effect, and so the number that represents this operation is $+1$. Thus, the p_x orbital can be labeled B_1 in the C_{2v} point group.

Now consider the effects of the operators of the C_{2v} group on a p_y orbital.

$$E p_y = 1 p_y \qquad (13.27)$$

$$C_2 p_y = -1 p_y \qquad (13.28)$$

$$\sigma(xz) p_y = -1 p_y \qquad (13.29)$$

$$\sigma(yz) p_y = +1 p_y \qquad (13.30)$$

and therefore p_y can be labeled B_2 in the C_{2v} group.

The importance of symmetry (or group theory) to chemical problems lies in the fact that if the symmetry of a molecule is that of a given point group, then the wavefunctions must transform like one of the irreducible representations of that group. Thus, the electronic wavefunctions for H_2O (in its C_{2v} equilibrium geometry) can be labeled A_1, A_2, B_1, or B_2 as in Table 13.4.

Furthermore, various operations, such as the Hamiltonian or the dipole moment operation, also transform like particular irreducible representations of the point group. From the transformation properties (i.e., symmetry labels) of the wavefunctions and operations, we can derive rules that tell us when certain integrals involving those operations equal zero. For example, the probability amplitude that a molecule in electronic eigenstate ϕ_i will absorb a photon and end up in state ψ_f depends on the integral

$$\int \psi_f^* \, \boldsymbol{\mu} \, \psi_i \, d\tau \qquad (13.31)$$

(see Section 15.1). The vector operator $\boldsymbol{\mu}$ has three components, μ_x, μ_y, μ_z, which transform like x, y, and z in the molecular symmetry group. The above integral will be zero unless the integrand is unchanged when any of the symmetry operations of the group is applied to it. Suppose the molecule belongs to the C_{2v} group and ψ_i transforms like A_1 (or z) and ψ_f transforms like B_2 (or y). Then the component of the integral with μ_x will vanish because the integrand will then transform as zyx (A_2) and therefore will change sign under σ_v (see Table 13.4). On the other hand, the y component of the integral will not necessarily vanish because the integrand then transforms as zy^2 or A_1, which is unchanged when any operation is applied to it.

Finding out which integrals vanish for symmetry reasons greatly decreases the amount of work we have to do in solving problems, and yields general rules (selection rules) for spectroscopy and other areas of physical chemistry. Such chemical applications are the subject of many of the books listed at the end of the chapter. The programmed introduction to chemical applications by Vincent is especially recommended.

References

P. W. Atkins, *Molecular Quantum Mechanics*. New York: Oxford University Press, 1983.

P. R. Bunker, *Molecular Symmetry and Spectroscopy*. New York: Academic, 1979.

F. A. Cotton, *Chemical Applications of Group Theory*. New York: Wiley, 1971.

B. E. Douglas and C. E. Hollingsworth, *Symmetry in Bonding and Spectra*. New York: Academic, 1985.

R. L. Flurry, *Symmetry Groups*. Englewood Cliffs, NJ: Prentice-Hall, 1980.

D. C. Harris and M. D. Bertolucci, *Symmetry and Spectroscopy*. New York: Oxford University Press, 1978.

S. F. A. Kettle, *Symmetry and Structure*. New York: Wiley, 1985.

A. Vincent, *Molecular Symmetry and Group Theory*. New York: Wiley, 1977.

Problems

In problems 13.1–13.14, list the Schoenflies symbols and symmetry elements for each molecule.

13.1 H_2S

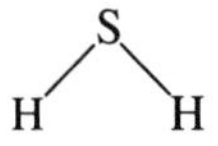

13.2 PCl_3

13.3 *trans*-$[CrBr_2(H_2O)_4]^+$ (ignore the H's)

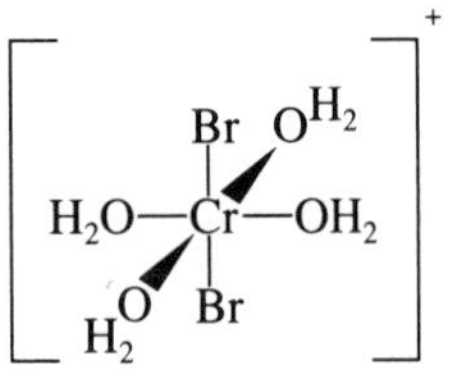

13.4 *gauche*-CH_2ClCH_2Cl

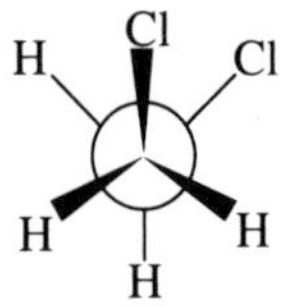

13.5 $C_6H_3Br_3$ (1,3,5-tribromobenzene)

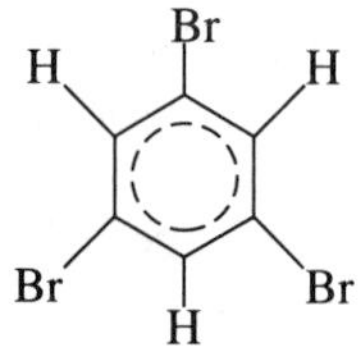

13.6 $CHClBr(CH_3)$

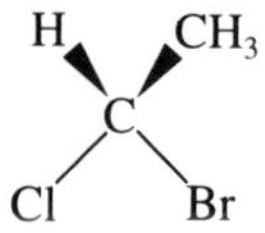

13.7 IF_5

13.8 C_6H_{12} (cyclohexane)

13.9 B_2H_6

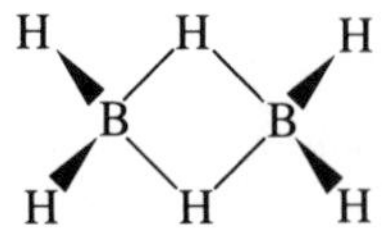

13.10 $C_{10}H_8$ (naphthalene)

(planar)

13.11 C_5H_8 (spiropentane)

(triangles ⊥
to each other)

13.12 C_4H_4S (thiophene)

(planar)

13.13 $C_6H_4Cl_2$ (*p*-dichlorobenzene)

13.14 *trans*-CFClBrCFClBr

13.15 Construct the operator multiplication table for the point group C_{2h}.

13.16 List the operators associated with the S_6 elements and their equivalents, if any. How many distinct operations are produced?

13.17 Consider the three distinct isomers of dichloroethylene, $C_2H_2Cl_2$. To which symmetry group does each belong? Which can have a permanent dipole moment.

13.18 The first excited singlet state of ethylene is twisted so that the two hydrogens and carbon on one side are in a plane perpendicular to the plane containing the other three

atoms. To which symmetry group does it belong? Does it have a dipole moment?

13.19 Which of the molecules in problems 13.1–13.14 can have a permanent dipole moment?

13.20 Which of the molecules in problems 13.1–13.14 can be optically active?

13.21 Consider the three distinct isomers of dichlorobenzene. To which symmetry group does each belong? Which can have a dipole moment?

13.22 There are 10 distinct isomers of dichloronaphthalene, $C_{10}H_6Cl_2$. Two of them do not have a dipole moment. List these two and find the symmetry group to which each belongs.

13.23 Some of the excited electronic states of acetylene are cis-bent and some are trans-bent. What is the symmetry group of these structures? Cis-bent means that the hydrogens bend toward one another, while trans-bent means they bend away from one another.

For the molecules in problems 13.24–13.38, give the Schoenflies symbol and symmetry elements.

13.24 C_8H_8

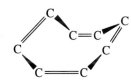

13.25 HCOOH (formic acid)

(planar)

13.26 $UO_2F_5^{3-}$

13.27 $C_{14}H_{10}$ (phenanthrene)

13.28 $Fe(CN)_6^{3-}$

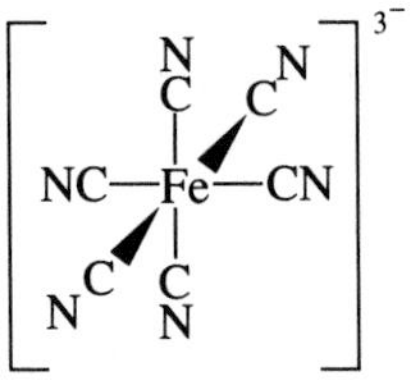

13.29 $[HNBCl]_3$

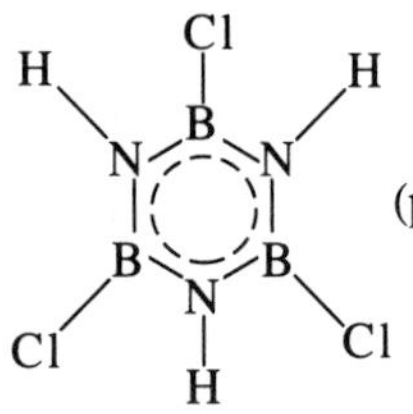

(planar)

13.30 $C_{10}H_{16}$ (adamantane)

13.31 $C_6H_2O_2Cl_2$ (2,5-dichloroquinone)

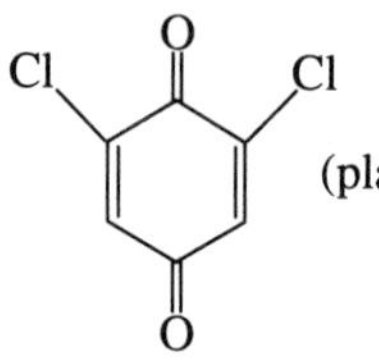

(planar)

13.32 HOCl

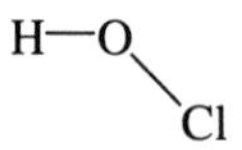

13.33 $(HNBH)_3$

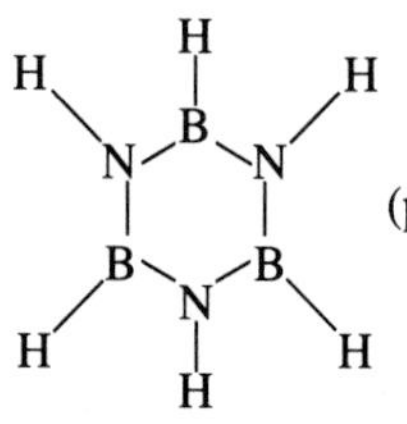

(planar)

13.34 $C_6H_3(C_6H_5)_3$: 1,3,5-triphenylbenzene (nonplanar)

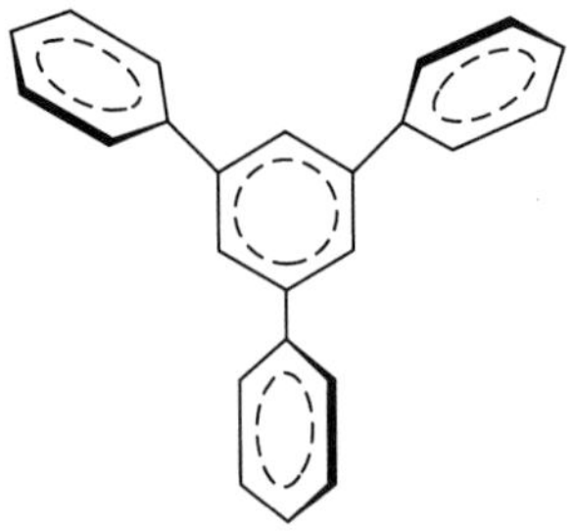

13.35 CH_3Cl (methyl chloride)

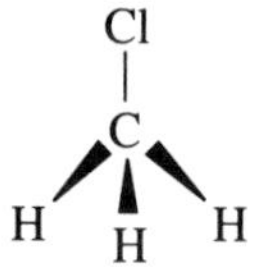

13.36 $Ni(CO)_4$

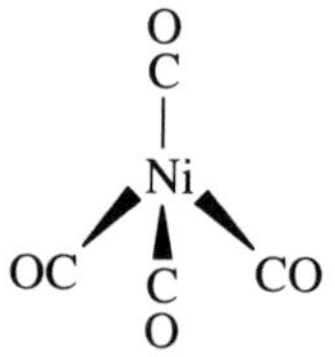

13.37 H_3CCH_2Br

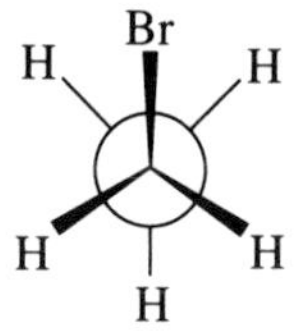

13.38 $Pt(Br)_4(NH_3)_2$ [tetrabromodiamine platinum (IV)] (ignore the H's)

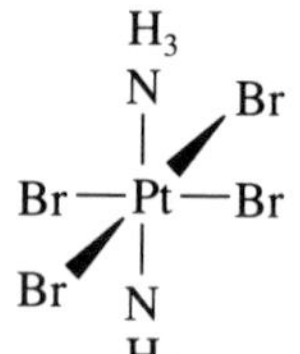

13.39 Construct the operator multiplication table for the point group C_{3v}.

13.40 In some of the excited states of benzene, the molecule is "stretched" so that the hexagon is elongated. What is the symmetry group of the molecule in such a state?

13.41 What is the symmetry group of HD? Can it have a dipole moment?

14
Rotational and Vibrational Spectroscopy

Molecular spectroscopy is a powerful tool for learning about molecular structure and molecular energy levels. The study of rotational spectra gives us information about moments of inertia, interatomic distances, and angles. Vibrational spectra yield fundamental vibrational frequencies and force constants. Electronic spectra yield electronic energy levels and dissociation energies.

The types of spectroscopic transitions that can occur are limited by selection rules. As in the case of atoms, the principal interactions of molecules with electromagnetic radiation are of the electric dipole type, and so we will concentrate on them. Magnetic dipole transitions are about 10^5 times weaker than electric dipole transitions, and electric quadrupole transitions are about 10^8 times weaker. Although the selection rules limit the radiative transitions that can occur, molecular collisions can cause many additional kinds of transitions. Because of molecular collisions the populations of the various molecular energy levels are in thermal equilibrium.

14.1   The Basic Ideas of Spectroscopy

When an isolated molecule undergoes a transition from one quantum eigenstate with energy E_1 to another with energy E_2, energy is conserved by the emission or absorption of a photon. The frequency ν of the photon is related to the difference in energies of the two states by Bohr's relation

$$h\nu \equiv hc\tilde{\nu} = |E_1 - E_2| \tag{14.1}$$

where we have introduced the symbol $\tilde{\nu}$ ($= 1/\lambda$) for the transition energy in **wavenumbers** (SI unit m^{-1}, but usually cm^{-1} is used). The wavenumber $\tilde{\nu}$ is

Table 14.1 Regions of the Electromagnetic Spectrum[a]

	Wavelengths in vacuo, λ_0	Wavenumber in vacuo, $\tilde{\nu}$	Frequency, ν	Photon energy, $h\nu$	Molar energy, $N_A h\nu$
λ-rays	10 pm	10^9 cm^{-1}	30.0 EHz	19.9×10^{-15} J	12.0 GJ/mol
X-rays	10 nm	10^6 cm^{-1}	30.0 PHz	19.9×10^{-18} J	12.0 MJ/mol
Vacuum UV	200 nm	50.0×10^3 cm^{-1}	1.50 PHz	993×10^{-21} J	598 kJ/mol
Near UV	380 nm	26.3×10^3 cm^{-1}	789 THz	523×10^{-21} J	315 kJ/mol
Visible	780 nm	12.8×10^3 cm^{-1}	384 THz	255×10^{-21} J	153 kJ/mol
Near IR	2.5 μm	4.00×10^3 cm^{-1}	120 THz	79.5×10^{-21} J	47.9 kJ/mol
Mid IR	50 μm	200 cm^{-1}	6.00 THz	3.98×10^{-21} J	2.40 kJ/mol
Far IR	1 mm	10 cm^{-1}	300 GHz	199×10^{-24} J	120 J/mol
Microwaves	100 mm	0.1 cm^{-1}	3.00 GHz	1.99×10^{-24} J	12.0 J/mol
Radio waves					

[a] From IUPAC Report "Names, Symbols, Definitions, and Units for Quantities in Optical Spectroscopy," 1984. IR, infrared; UV, ultraviolet. The abbreviations for powers of 10 are given in the front cover of the book.

the number of waves per unit length. If $E_1 > E_2$, the process is photon emission; if $E_1 < E_2$, the process is photon absorption. The frequency range of photons, or the electromagnetic spectrum, is classified into different regions according to custom and experimental methods as outlined in Table 14.1. By measuring the frequency of the photon, we can learn about the eigenstates of the molecule being studied. This is called molecular spectroscopy.

The frequency of the photon in the absorption or emission process tells us the kinds of molecular transitions that are involved. In the radio-frequency region (very low energy), transitions between nuclear spin states can occur (see Chapter 16). In the microwave region, transitions between electron spin states in molecules with unpaired electrons can take place (Chapter 16) and, in addition, transitions between rotational states. In the infrared region, transitions between vibrational states take place (with and without transitions between rotational states). In the visible and ultraviolet regions, the transitions occur between electronic states (accompanied by vibrational and rotational changes). Finally, in the far ultraviolet and X-ray regions, transitions occur that can ionize or dissociate molecules.

Example 14.1

Calculate the energy in joules per quantum, joules per mole, and electronvolts of photons of wavelength 300 nm.

$$h\nu = \frac{hc}{\lambda} = \frac{(6.62 \times 10^{-34} \text{ J s})(3 \times 10^8 \text{ m s}^{-1})}{(300 \times 10^{-9} \text{ m})} = 6.62 \times 10^{-19} \text{ J}$$

$$N_A h\nu = (6.02 \times 10^{23} \text{ mol}^{-1})(6.62 \times 10^{-19} \text{ J}) = 398 \text{ kJ mol}^{-1}$$

$$= \frac{3.98 \times 10^5 \text{ J mol}^{-1}}{96\ 485 \text{ C mol}^{-1}}$$

$$= 4.13 \text{ eV}$$

We shall see that the energy eigenvalues of a molecule can be written as

$$E = E_r + E_v + E_e \tag{14.2}$$

where E_r is the rotational energy, E_v the vibrational energy, and E_e the electronic energy. When the molecule undergoes a transition to another state with the emission or absorption of a single photon of frequency ν, then

$$h\nu = (E_r' - E_r'') + (E_v' - E_v'') + (E_e' - E_e'') \tag{14.3}$$

The primes refer to the state of higher energy and the double primes to the state of lower energy.

The classification of the various regions of the electromagnetic spectrum by the type of transition given above is possible because, in general,

$$E_r' - E_r'' \ll E_v' - E_v'' \ll E_e' - E_e'' \tag{14.4}$$

That is, electronic energy level differences are much greater than vibrational energy level differences, which are much greater than rotational energy level differences.

14.2 Transitional Probabilities and Selection Rules

The spectrum of a molecule consists of a series of lines at the frequencies (or wavelengths) corresponding to all the possible transitions. The strength or intensity of a spectral line depends on the number of molecules per unit volume that were in the initial state (the population density of that state) and the probability that the transition will take place. The population in state n with energy E_n is determined by the temperature T according to the **Boltzmann distribution** (equation 17.18):

$$N_n \propto e^{-E_n/kT} \tag{14.5}$$

so that the higher energy states are less populated than the lower energy states. The transition probability from state n to state m is determined by the square of the **transition dipole moment** $\boldsymbol{\mu}_{nm}$ defined by

$$\boldsymbol{\mu}_{nm} = \int \psi_n^* \hat{\boldsymbol{\mu}} \psi_m \, d\tau \tag{14.6}$$

where $\hat{\boldsymbol{\mu}}$ is the **quantum mechanical dipole moment operator** for the molecule:

$$\hat{\boldsymbol{\mu}} = \sum_i q_i r_i \tag{14.7}$$

where the sum is over all the electrons and nuclei of the molecule, q_i is the charge, and r_i is the position of the ith charged particle. To understand how the transition moment enters, we can think of the molecule interacting with the electric field of the radiation because of a transient or fluctuating dipole moment given by equation 14.6.

From equation 14.6, we see that if the transition dipole moment vanishes (usually because of symmetry), the spectral line has no intensity. The rules governing the nonvanishing of $\boldsymbol{\mu}_{nm}$ are called **selection rules**, and these allow us to make sense out of observed molecular spectra.

If the transition moment from state n to state m is nonzero and there is enough population in the initial state, then the spectral line will be seen in the spectrum. The strength of the line depends on whether the molecule is absorbing a photon from the light field (absorption spectroscopy) or emitting a

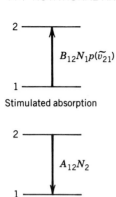

Stimulated absorption

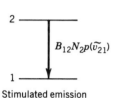

Spontaneous emission

Stimulated emission

Figure 14.1 Definition of Einstein probabilities.

photon to the field (emission spectroscopy). Einstein showed how to relate the absorption probabilities to the emission probabilities by considering a system in which matter (molecules) and radiation (photons) are in equilibrium at a given temperature T. The rate of absorption taking a molecule from state 1 with energy E_1 to state 2 with energy E_2 is proportional to the population density N_1 of molecules in state 1 (equation 14.5) and to the energy density $\rho(\tilde{\nu}_{21})$ per unit wavenumber of radiation satisfying $hc\tilde{\nu}_{21} = E_2 - E_1$. We saw in Chapter 10 that $\rho(\tilde{\nu})\,d\tilde{\nu}$ is the energy per unit volume in the wavenumber interval $\tilde{\nu}$ to $\tilde{\nu} + d\tilde{\nu}$. The rate of absorption from 1 to 2 is then given by

$$\text{Rate}_{2\leftarrow 1} = B_{21}N_1\rho(\tilde{\nu}_{21}) \tag{14.8}$$

where B_{21} is the **Einstein probability of stimulated absorption.** Since the rate has units $m^{-3}\,s^{-1}$, N_1 has the units m^{-3}, ρ has the units $J\,m^{-2}$, and B_{21} has units $s\,kg^{-1}$.

As shown in Fig. 14.1, there are two types of emission, spontaneous and stimulated. The **rate of spontaneous emission** of photons per unit volume is $A_{12}N_2$, where N_2 is the population density in state 2 and A_{12} is called the **Einstein probability of spontaneous emission** with units s^{-1}. Einstein suggested that the radiation field stimulated the transition from state 2 to state 1 (with the emission of a photon) so that the total rate of emission is the sum of $A_{12}N_2$ (spontaneous emission) and $B_{12}N_2\rho(\tilde{\nu}_{21})$, where B_{12} is the **Einstein probability of stimulated emission,** with the same units as B_{21}.

There is a fundamental difference between spontaneous and stimulated emission: In spontaneous emission, the system produces an electromagnetic field with random direction and phase. In the case of stimulated emission, the electromagnetic wave adds in phase and direction (i.e., coherently) to the stimulating wave. This is the characteristic that makes the laser possible (see Section 15.9).

At equilibrium, the total rates of emission and absorption must be equal, so

$$B_{21}N_1\rho(\tilde{\nu}_{21}) = N_2[A_{12} + B_{12}\,\rho(\tilde{\nu}_{21})] \tag{14.9}$$

In addition, at equilibrium, Boltzmann's law (equation 14.5) tells us that

$$N_2 = N_1 e^{-(E_1-E_2)/kT} = N_1 e^{-hc\tilde{\nu}_{21}/kT} \tag{14.10}$$

so that

$$\rho(\tilde{\nu}_{21}) = \frac{A_{12}}{B_{21}e^{-hc\tilde{\nu}_{21}/kT} - B_{12}} \tag{14.11}$$

Einstein then argued that the form of $\rho(\tilde{\nu})$ must be given by Planck's distribution law (equation 10.1), so that

$$B_{21} = B_{12} \tag{14.12}$$

$$A_{12} = 8\pi hc\tilde{\nu}_{21}^3\,B_{21} \tag{14.13}$$

It only remains to determine A_{12} (or B_{12}) in terms of the transition dipole moment, equation 14.6, to complete the discussion. The derivation of the transition probability from quantum mechanics is too advanced for this book*; however, the final result is

---

* See J. Steinfeld, *Molecules and Radiation*. Cambridge, MA: MIT Press, 1985.

$$A_{12} = \frac{16\pi^3 \, \tilde{\nu}_{21}^3 \, |\, \mu_{21}\,|^2}{3h\epsilon_0} \qquad (14.14)$$

Notice that, because of the $\tilde{\nu}_{21}^3$ factor, spontaneous emission is not very important compared to absorption at low frequencies. Thus, for low frequencies (infrared and microwave regions) absorption spectra are generally measured, while at high frequencies (visible and ultraviolet) both emission and absorption spectra are measured.

We can also think of A_{12} as a measure of the lifetime of state 2. Consider molecules in (excited) state 2 with no radiation field present (and so no stimulated emission). The molecules will make a transition to state 1, emitting a photon of frequency $\tilde{\nu}_{21}$, with a probability $A_{12}N_2$. Every time this occurs, N_2 decreases. After a time t, the number of molecules per unit volume in state 2 is given by

$$N_2(t) = N_2(0)\, e^{-A_{12}t} = N_2(0)\, e^{-t/\tau} \qquad (14.15)$$

where we have defined the **lifetime** $\tau = A_{12}^{-1}$. Actually, if a molecule in state 2 can also make transitions to states 3, 4, . . . (with photons of frequency $\tilde{\nu}_{23}, \tilde{\nu}_{24}, \ldots$), then the **total radiative lifetime** is given by

$$\frac{1}{\tau} = \sum_i A_{2i} \qquad (14.16)$$

If other decay processes besides radiative transitions are possible (such as nonradiative transitions) we must add those rates to equation 14.16 to get the total decay rate (inverse lifetime).

Example 14.2
The radiative lifetime of a hydrogen atom in its first excited level (2p) is 1.6×10^{-9} s. What is the magnitude of the electronic transition moment μ_{21} for this transition? The degeneracy g_2 of the 2p level is 3. [$\tilde{\nu} = (2.46 \times 10^{15}\ \text{s}^{-1})/(2.998 \times 10^8\ \text{m s}^{-1}) = 8.21 \times 10^6\ \text{m}^{-1}$.]

$$\mu_{21} = \left[\frac{3h\epsilon_0 g_2}{16\pi^3 \tilde{\nu}_{21}^3 \tau} \right]^{1/2}$$

$$= \left[\frac{(3)(6.626 \times 10^{-34}\ \text{J s})(8.854 \times 10^{-12}\ \text{C}^2\ \text{N}^{-1}\ \text{m}^{-2})(3)}{16\pi^3(8.21 \times 10^6\ \text{m}^{-1})^3(1.6 \times 10^{-9}\ \text{s})} \right]^{1/2}$$

$$= 10.9 \times 10^{-30}\ \text{C m}$$

A dipole moment of this magnitude corresponds to a distance from the proton to the electron of

$$r = \frac{10.9 \times 10^{-30}\ \text{C m}}{1.6 \times 10^{-19}\ \text{C}}$$

$$= 68.1\ \text{pm}$$

This transition dipole moment can be visualized as the movement of an electron 68.1 pm/52.9 pm = 1.29 Bohr radii.

14.3   Schrödinger Equation for Nuclear Motion

We saw in Chapter 12 that the Schrödinger equation for a molecule can be treated in the Born–Oppenheimer approximation so that the **electronic Hamiltonian** is that for fixed nuclei, while the Hamiltonian for nuclear motion contains the kinetic energy operator of the nuclei and the electronic energy (as a function of the nuclear coordinates) as the potential energy operator:

$$\hat{H} = -\frac{\hbar^2}{2\mu}\nabla_R^2 + E(R) \tag{14.17}$$

In the absence of external fields (such as magnetic or electric fields), the potential energy term $E(R)$ can depend only on the relative positions of the nuclei, not on where the molecule is placed or on the orientation of the molecule in space.

The kinetic energy operator consists of the kinetic energy of the center of mass (leading to the translational energy of the molecule), the kinetic energy associated with rotational motion and the kinetic energy of the vibrational motion. Thus, to a very good approximation, we may write

$$H = H_{\text{tr}} + H_{\text{rot}} + H_{\text{vib}} \tag{14.18}$$

where the translational and rotational Hamiltonians contain only kinetic energy terms, while the vibrational Hamiltonian contains $E(R)$, the potential energy depending on the internuclear distances. These internuclear distances are the **vibrational** coordinates of the molecule.

If the Hamiltonian is the sum of three terms, one for each kind of motion, then the wavefunction ψ can be written as a product of wavefunctions:

$$\psi = \psi_{\text{tr}}\,\psi_{\text{rot}}\,\psi_{\text{vib}} \tag{14.19}$$

The Schrödinger equations for the three terms are

$$\hat{H}_{\text{tr}}\psi_{\text{tr}} = E_{\text{tr}}\psi_{\text{tr}} \tag{14.20}$$

$$\hat{H}_{\text{rot}}\psi_{\text{rot}} = E_{\text{rot}}\psi_{\text{rot}} \tag{14.21}$$

$$\hat{H}_{\text{vib}}\psi_{\text{vib}} = E_{\text{vib}}\psi_{\text{vib}} \tag{14.22}$$

The translational wavefunction is that for a free particle (or particle in a very large box) with a mass equal to the mass of the molecule. The translational eigenvalues are very closely spaced and cannot be probed in molecular spectroscopy, so we will neglect them in our discussions.

For a diatomic (or linear) molecule, $\hat{H}_{\text{rot}}$ depends only on two angles, θ and ϕ (see Section 10.13); $\hat{H}_{\text{vib}}$ depends only on R, the internuclear separation. For polyatomic molecules, $\hat{H}_{\text{vib}}$ is more complex, depending on $3N - 6$ coordinates for nonlinear molecules and $3N - 5$ coordinates for linear molecules. We will now turn to a description of the rotational and vibrational eigenstates of both diatomic and polyatomic molecules.

14.4   Rotational Spectra of Diatomic Molecules

To a first approximation the rotational spectrum of a diatomic molecule may be understood in terms of the Schrödinger equation for rotational motion of the rigid rotor (Section 10.13). The wavefunctions are the spherical harmonics $Y_J^M(\theta, \phi)$, and there are two quantum numbers J and M for molecular rotation. The energy eigenvalues are given by

$$E_r = \frac{\hbar^2}{2I} J(J + 1) \qquad J = 0, 1, 2, \ldots$$

$$M = -J, \ldots, 0, \ldots, +J \qquad (14.23)$$

where I is the moment of inertia (Section 10.13). Since the energy does not depend on M, the rotational levels are $(2J + 1)$-fold degenerate.

In spectroscopy it is standard to express the energies of various levels in wavenumbers by dividing E by hc and referring to these values as **term values.** Term values are usually given in cm^{-1}, but the SI unit for a term value is m^{-1}. Rotational term values are represented by $F(J) = E_r/hc$, so that the rotational term values for a diatomic molecule are given by

$$F(J) = \frac{E_r}{hc} = \frac{J(J + 1)h}{8\pi^2 Ic} = J(J + 1)B \qquad (14.24)$$

where the rotational constant is written

$$B = \frac{h}{8\pi^2 Ic} \qquad (14.25)$$

When SI units are used, B comes out in m^{-1}, but this is conveniently converted to cm^{-1} by multiplying by 10^{-2} m cm^{-1}.

The rotational energy levels for a rigid diatomic molecule are given in Fig. 14.2 in terms of the rotational constant.

The selection rules for rotational spectra are derived from equation 14.6. According to the Born–Oppenheimer approximation (Section 12.1), the wavefunction for a molecule in the electronic state ψ_{el}, the vibrational state ψ_v, and having a particular set of rotational quantum numbers JM, can be written as a product $\psi_{el}\psi_v\psi_{JM}$. The transition moment for an electric dipole transition from a rotational state JM to a rotational state $J'M'$ of the same electronic state is therefore given by

$$\iint \psi_{el}^*\psi_v^*\psi_{J'M'}^* \; \hat{\boldsymbol{\mu}}\psi_{el}\psi_v\psi_{JM} \; d\tau_{el} \, d\tau_{rot} \, d\tau_{vib} \qquad (14.26)$$

where $\hat{\boldsymbol{\mu}}$ is the dipole moment operator. The permanent dipole moment $\mu_0^{(e)}$ of a molecule in this electronic state is equal to the expectation value of the operator $\boldsymbol{\mu}$ over the wavefunction for the electronic state.

$$\boldsymbol{\mu}_0^{(e)} = \int \psi_{el}^*\hat{\boldsymbol{\mu}}\psi_{el} \; d\tau_{el} \qquad (14.27)$$

Thus, equation 14.26 becomes

$$\iint \psi_v^*\psi_{J'M'}^* \; \boldsymbol{\mu}_0^{(e)} \; \psi_v\psi_{JM} \; d\tau_{rot} \, d\tau_{vib} \qquad (14.28)$$

The integral over the vibrational coordinate yields the permanent dipole moment in that particular vibrational state. For simplicity, we will write it as $\boldsymbol{\mu}_0$, so that the final result for the integral is

$$\int \psi_{J'M'}^* \; \boldsymbol{\mu}_0\psi_{JM} \; d\tau_{rot} \qquad (14.29)$$

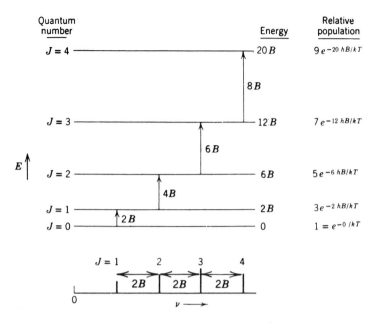

Figure 14.2 Rotational levels for a rigid diatomic molecule and the absorption spectrum that results from $\Delta J = 1$. The energies and relative populations of the levels are indicated on the right. The transitions are labeled by the upper of the two J values involved. Note the degeneracies of the levels have been taken into account in the population and the intensities of the lines depend on the relative populations.

A molecule has a rotational spectrum only if this integral is nonzero. Thus, the gross selection rule for rotational spectra is that a molecule must have a permanent dipole moment to emit or absorb radiation by making a transition between different states of rotation. This is expected from the fact that a rotating dipole produces an oscillating electric field that can interact with the oscillating field of a light wave. A homonuclear diatomic molecule like H_2 or O_2 does not have a dipole moment, and so it does not show a pure rotational spectrum. Heteronuclear diatomic molecules do have dipole moments, and so they do have rotational spectra. Polyatomic molecules are discussed in the next section. To find the specific selection rules we need to find the conditions on the quantum numbers that make the integral in equation 14.29 nonzero. For a linear molecule it can be shown that the transition moment is nonzero for

$$\Delta J = \pm 1 \qquad \Delta M = 0, \pm 1$$

This selection rule may be understood in the same way as that for atoms (Section 11.15). Since a photon has one unit of angular momentum and angular momentum must be conserved in emission or absorption, the angular momentum of a molecule must change by a compensating amount.

For a transition between two adjacent rotational levels the rotational term difference, which is represented by $\Delta_1 F(J + \frac{1}{2})$, is given by

$$\Delta_1 F(J + \tfrac{1}{2}) = F(J + 1) - F(J) = [(J + 1)(J + 2) - J(J + 1)]B$$
$$= 2B(J + 1) \tag{14.30}$$

where J is the rotational quantum number for the lower level. The rotational term difference is expressed in terms of wavenumber units and is the wavenumber of the radiation absorbed or emitted. As shown in Fig. 14.2, the frequencies of the successive lines in the rotational spectrum are given by $2B$, $4B$, $6B$, ... Thus, there is a series of equally spaced lines with separations of $2B$. A separate series of lines is found for each isotopically different species of a given molecule, because the moments of inertia of isotopically substituted molecules are different.

We have been talking about diatomic molecules as if they are rigid rotors, but of course they are not. As the rotational motion increases, the chemical bond stretches due to centrifugal forces, the moment of inertia increases, and, consequently, the rotational energy levels come closer together. This may be taken into account by adding a term to equation 14.23:

$$F(J) = \frac{E_r}{hc} = BJ(J + 1) - DJ^2(J + 1)^2 \qquad (14.31)$$

The quantity D is the centrifugal distortion constant in wavenumbers. With this addition equation 14.30 becomes

$$\Delta_1 F(J + \tfrac{1}{2}) = 2B(J + 1) - 4D(J + 1)^3 \qquad (14.32)$$

The moment of inertia of a diatomic molecule also depends on its vibrational state because of the anharmonicity of vibrational motion. Since molecules are generally in their ground vibrational state at room temperature we do not have to take this into account in considering pure rotational spectra; however, we will have to take it into account by an extension of equation 14.32 in discussing vibration–rotation spectra.

Example 14.3

In early measurements of the pure rotational spectrum of $H^{35}Cl$, Czerny found that the wavenumbers of absorption lines are given by

$$\bar{\nu} = (20.794 \text{ cm}^{-1})(J + 1) - (0.000\ 164 \text{ cm}^{-1})(J + 1)^3$$

where J is the quantum number of the lower state. What is the internuclear distance in $H^{35}Cl$? What is the value of the centrifugal distortion constant?

From equation 14.32, $B = 10.397 \text{ cm}^{-1} = 1039.7 \text{ m}^{-1}$

Since

$$B = \frac{h}{8\pi^2 cI} = \frac{h}{8\pi^2 c\mu R_0^2}$$

$$R_0 = \sqrt{\frac{h}{8\pi^2 c\mu B}}$$

$$= \sqrt{\frac{6.626 \times 10^{-34} \text{ J s}}{8\pi^2(2.998 \times 10^8 \text{ m s}^{-1})(1.626\ 68 \times 10^{-27} \text{ kg})(1039.7 \text{ m}^{-1})}}$$

$$= 129 \text{ pm}$$

(The reduced mass of $H^{35}Cl$ is given in Example 10.10.)

The centrifugal distortion constant is given by

$$D = \tfrac{1}{4}(0.000\ 164 \text{ cm}^{-1}) = 4.1 \times 10^{-5} \text{ cm}^{-1}$$

We have discussed the selection rules that determine the transitions that can give rise to absorption or emission, but there is another factor that determines the observed intensities and that is the population of the initial state. If very few molecules in a sample are in the initial state for a transition, the absorption and emission from this initial state will be weak, other factors being equal. We have seen that molecules can exist in many energy states, but at equilibrium the populations of these states decrease exponentially with the energy of the state. According to the Boltzmann distribution (equation 17.18), the fraction f_i of the molecules in the ith energy state is given by

$$f_i = \frac{e^{-\epsilon_i/kT}}{\sum_i e^{-\epsilon_i/kT}} = \frac{e^{-\epsilon_i/kT}}{q} \tag{14.33}$$

where q is the denominator. If the energy of a state is large compared with kT, the probability of finding a molecule in that state at equilibrium will be small. Because of degeneracy (Section 10.10) many states of a molecule may have the same energy and these degenerate states make up the energy **level**. When energy levels are used the Boltzmann distribution can be written

$$f_i = \frac{g_i e^{-\epsilon_i/kT}}{\sum_i g_i e^{-\epsilon_i/kT}} \tag{14.34}$$

where g_i is the degeneracy (Section 10.10) of the ith level. As discussed earlier, the component of the angular momentum in a particular direction is equal to $M_J \hbar$, where M_J may have values of $J, (J - 1), \ldots, 0, \ldots, -J$, where J is the rotational quantum number. Thus, there are in all $2J + 1$ different possible states with quantum number J. In the absence of an external electric or magnetic field the energies are identical for these various sublevels, and so the Jth energy level is said to have a degeneracy of $2J + 1$. The rotational energy in the absence of an external electric or magnetic field, ignoring D in equation 14.32, is given by $\epsilon_i = hcBJ(J + 1)$ so that, using equation 14.33, the fraction of molecules in the Jth rotational level is given by

$$f_J = \frac{(2J + 1)e^{-[hcJ(J + 1)B]/kT}}{q} \tag{14.35}$$

According to this equation, the number of molecules in level J increases with J at low J values, goes through a maximum and then, because of the exponential term, decreases as J is further increased. The lines in the spectrum at the bottom of Fig. 14.2 have been labeled with the rotational quantum number J of the upper of the two states involved. The intensities of the lines are proportional to the populations in the lower state involved in the transition.

For molecules with larger moments of inertia I the rotational energies are smaller, in fact, small compared with kT. The quantum numbers may become quite large before $e^{-\epsilon_J/kT}$ becomes appreciably different from unity. For small quantum numbers populations are proportional to the degeneracies, since $e^{-\epsilon_J/kT} \approx 1$ for $\epsilon_J \ll kT$.

There is a complication in rotational spectroscopy that we will not be able to discuss. The statistics of nuclear spin affect the number of degenerate states at each J level, and therefore the intensities of the rotational lines. The use of the Boltzmann distribution alone is an oversimplification.*

---

* See one of the references at the end of the chapter, such as Flygare.

Although homonuclear diatomic molecules do not have permanent electric dipole moments and do not exhibit pure rotational spectra, they do show rotational Raman spectra (Section 14.9), and their electronic and vibrational spectra show rotational fine structure.

14.5 Rotational Spectra of Polyatomic Molecules

For the treatment of its pure rotational spectrum we may consider a polyatomic molecule to be a rigid framework with fixed bond lengths and angles equal to their mean values. For a polyatomic molecule the **moment of inertia** about a particular axis that passes through the center of mass of the molecule is simply the sum of the moments due to the various nuclei about that axis:

$$I = \sum_i m_i R_i^2 \tag{14.36}$$

where R_i is the perpendicular distance of the nucleus of mass m_i from the axis. The symmetry axes of a molecule all pass through the center of mass of the molecule.

Molecules may be classified according to their ellipsoid of inertia, constructed as follows. Lines are drawn from the center of mass of the molecule in various directions with length proportional to $1/\sqrt{I_\alpha}$, where I_α is the moment of inertia about that line as an axis. The three mutually perpendicular principal axes of the ellipsoid of inertia are designated a, b, and c. The principal moments of inertia about these axes are labeled so that $I_a \leq I_b \leq I_c$.

The symmetry axes of a molecule help in locating its principal axes. Any symmetry operation of a molecule must apply to its ellipsoid of inertia.

The principal moments of inertia are used to classify molecules, as shown in Table 14.2. If all three principal moments of inertia are equal, the molecule is a **spherical** top. If two principal moments are equal, the molecule is a **symmetric** top. A molecule is a **prolate** top (cigar shaped) if the two larger moments are equal. The molecule is an **oblate** top (discus shaped) if the two smaller moments are equal. The molecule is an **asymmetric** top if all three principal moments are unequal.

The quantum mechanical Hamiltonian operator for the rotational motion of polyatomic molecules is found by first writing the classical mechanical energy in terms of angular momentum operators. Since we know how to convert classical angular momentum to quantum mechanics, we can then find the quantum

Table 14.2 Classification of Polyatomic Molecules According to Their Moments of Inertia

Moments of Inertia	Type of Rotor	Examples
$I_b = I_c$ $I_a = 0$	Linear	HCN
$I_a = I_b = I_c$	Spherical top	CH_4, SF_6, UF_6
$I_a < I_b = I_c$	Prolate symmetric top	CH_3Cl
$I_a = I_b < I_c$	Oblate symmetric top	C_6H_6
$I_a \neq I_b \neq I_c$	Asymmetric top	CH_2Cl_2, H_2O

Hamiltonian, and solve the Schrödinger equation. The last part turns out to be straightforward for all the cases except the asymmetric top. We will not discuss the latter.

In classical mechanics the rotational energy of a rotor with one degree of freedom is

$$E_r = \tfrac{1}{2}I\omega^2 = \frac{(I\omega)^2}{2I} = \frac{L^2}{2I} \tag{14.37}$$

where ω is the angular velocity in radians per second, I is the moment of inertia, and L is the angular momentum. For an object that can rotate in three dimensions the classical expression for the rotational kinetic energy is

$$E_r = \tfrac{1}{2}I_{xx}\omega_x^2 + \tfrac{1}{2}I_{yy}\omega_y^2 + \tfrac{1}{2}I_{zz}\omega_z^2 \tag{14.38}$$

Since we will want to convert this to a quantum mechanical expression, it is more convenient to express it in terms of the angular momentum $L_q = I_{qq}\omega_q$, where q represents a direction.

$$E_r = \frac{L_x^2}{2I_{xx}} + \frac{L_y^2}{2I_{yy}} + \frac{L_z^2}{2I_{zz}} \tag{14.39}$$

in which the components of the total angular momentum about the three principal axes are given by

$$L_x = I_{xx}\omega_x \tag{14.40}$$

$$L_y = I_{yy}\omega_y \tag{14.41}$$

$$L_z = I_{zz}\omega_z \tag{14.42}$$

The total angular momentum is given by

$$L^2 = L_x^2 + L_y^2 + L_z^2 \tag{14.43}$$

The expressions for the energies of spherical tops, linear molecules, and symmetric tops are as follows:

1. *Spherical top.* For a spherical top $I_{xx} = I_{yy} = I_{zz} = I$, the momental ellipsoid is a sphere, and equation 14.37 becomes

$$E_r = \frac{(L_x^2 + L_y^2 + L_z^2)}{2I} = \frac{L^2}{2I} \tag{14.44}$$

where the second form has been obtained by introducing equation 14.43.

The quantum mechanical expression for the rotational energy is obtained by substituting the quantum mechanical expression for the square of the angular momentum, $L^2 = J(J + 1)\hbar^2$:

$$E = \frac{J(J + 1)\hbar^2}{2I} \qquad J = 0, 1, 2, \ldots \tag{14.45}$$

However, spherical top molecules cannot have dipole moments for symmetry reasons. Only molecules belonging to point groups C_n, C_s, and C_{nv} can possess dipole moments. Therefore, spherical top molecules do not have pure rotational spectra. They do, however, have vibrational and electronic spectra with rotational fine structure. The moment of inertia for a symmetrical tetrahedral molecule, such as CH_4, is

$$I = \tfrac{8}{3}mR^2 \tag{14.46}$$

where R is the bond length and m is the mass of each of the four atoms arranged in a tetrahedral manner.

2. *Linear molecule*. For a linear molecule, $I_{xx} = 0$ and $I_{yy} = I_{zz}$. Thus, equation 14.37 becomes

$$E_r = \frac{L_y^2 + L_z^2}{2I_{zz}} = \frac{L^2}{2I_{zz}} \tag{14.47}$$

For a linear polyatomic molecule the equation for the rotational term $F(J)$ is the same as that given earlier for a diatomic molecule.

3. *Symmetric top*. Examples of symmetric top molecules are NH_3, CH_3Cl, and the molecule shown later in Example 14.4. For these molecules $I_{xx} = I_{yy}$, but I_{zz} is different. We will use $I_{\parallel}$ for the moment of inertia parallel to the axis (I_{zz}) and $I_{\perp}$ for the moment perpendicular to the axis (I_{xx} and I_{yy}). Thus, the classical energy of rotation is

$$E_r = \frac{L_x^2 + L_y^2}{2I_{\perp}} + \frac{L_z^2}{2I_{\parallel}} \tag{14.48}$$

This can be written in terms of the magnitude of the angular momentum $L^2 = L_x^2 + L_y^2 + L_z^2$ as follows:

$$E_r = \left(\frac{1}{2I_{\perp}}\right)(L_x^2 + L_y^2 + L_z^2) - \left(\frac{1}{2I_{\perp}}\right)L_z^2 + \left(\frac{1}{2I_{\parallel}}\right)L_z^2$$

$$= \left(\frac{1}{2I_{\perp}}\right)L^2 + \left[\left(\frac{1}{2I_{\parallel}}\right) - \left(\frac{1}{2I_{\perp}}\right)\right]L_z^2 \tag{14.49}$$

The quantum mechanical expression for the energy is obtained by substituting $L^2 = J(J + 1)\hbar^2$ (as we saw in connection with equation 10.132) and $L_z^2 = K^2\hbar^2$ (as we saw in connection with equation 10.134). This latter substitution comes from the fact that in quantum mechanics the component of angular momentum about any axis is restricted to the values $K\hbar$, where $K = 0, \pm 1, \ldots, \pm J$:

$$E_r = \left(\frac{1}{2I_{\perp}}\right)J(J + 1)\hbar^2 + \left[\left(\frac{1}{2I_{\parallel}}\right) - \left(\frac{1}{2I_{\perp}}\right)\right]K^2\hbar^2 \tag{14.50}$$

where $J = 0, 1, 2, \ldots$ and $K = 0, \pm 1, \pm 2, \ldots, \pm J$. This equation is generally used in the form

$$\frac{E_{JK}}{hc} = BJ(J + 1) + (A - B)K^2 \tag{14.51}$$

where

$$B = \frac{\hbar}{4\pi c I_{\perp}} \quad \text{and} \quad A = \frac{\hbar}{4\pi c I_{\parallel}} \tag{14.52}$$

The quantum number K determines the component of the angular momentum along the axis of the symmetric top; this is the angular momentum of rotation about the symmetry axis. When $K = 0$ there is no rotation about the symmetry axis, and the rotation is about the axis perpendicular to the symmetry axis, that is, end-over-end rotation. When K has its max-

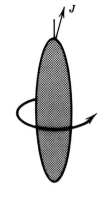

(a) K ≈ J

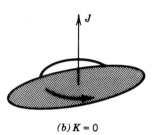

(b) K = 0

Figure 14.3 Meaning of the quantum number K.

imum value ($+J$ or $-J$), most of the molecular rotation is about the symmetry axis (see Fig. 14.3).

The specific selection rules for rotational spectra of symmetric top molecules are $\Delta J = \pm 1$ and $\Delta K = 0$. The reason there cannot be a change in quantum number K is that the dipole vector of the molecule is oriented along the principal axis. The electromagnetic field of radiation can affect the rotation of the dipole, but it cannot affect the rotation of the molecule about its principal axis because there is no dipole moment perpendicular to the principal axis.

Example 14.4

Derive the expressions for the moments of inertia $I_\parallel$ and $I_\perp$ of the octahedral symmetric top molecule AB_2C_4 shown in the diagram

$$I_\parallel = 4m_C R^2$$

$$I_\perp = 2m_C R^2 + 2m_B r^2$$

$$I_\parallel = 4m_C R^2$$

$$I_\perp = 4m_C R^2 + 2m_B r^2$$

The pure rotational spectroscopy of molecules has made possible the most precise evaluations of bond lengths and bond angles. The spectrum of a polyatomic molecule gives at most three principal moments of inertia; since usually more than three bond lengths and angles are involved, isotopically different molecules must be studied, and it must be assumed that isotopically different molecules have the same set of bond lengths and bond angles. In effect, a number of simultaneous equations are solved for the internuclear distances and angles. These spectra are in the microwave region. Microwave radiation is produced by special electronic oscillators called klystrons. Monochromatic radiation is produced, and the frequency may be varied continuously over wide ranges. The usual experimental arrangement is shown in Fig. 14.4. Microwave radiation is transmitted down in a waveguide that contains the gas being studied. The intensity of the radiation at the other end of the waveguide is measured by use of a crystal diode detector and amplifier. The oscillator frequency is swept over a range, and the transmitted intensity is presented on an oscilloscope or a recorder as a function of frequency.

According to the Heisenberg uncertainty principle, the accuracy with which an energy level may be determined is inversely proportional to the time the molecule is in this level. Hence, to obtain sharp rotational lines of a gas, the pressure must be maintained sufficiently low so that the average time between collisions is long compared to the period of a rotation. Usually it is necessary to determine microwave spectra at pressures below 10 Pa to reduce the line-broadening effects of collisions.

Microwave lines at low pressures are very sharp, and their frequencies may often be determined to the order of 1 ppm. Therefore, the moments of inertia of a molecule may be determined with great accuracy.

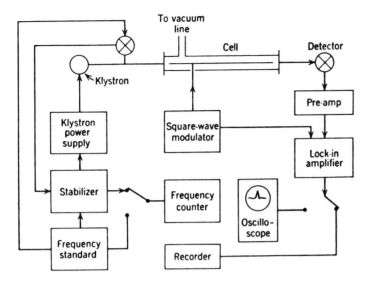

Figure 14.4 Block diagram of a Stark-modulated microwave spectrometer.

The lines in the microwave spectrum are split if the molecules being studied are in an electric field. This so-called Stark effect is due to the interaction of the dipole moment of the gaseous molecule and the electric field. Since the splitting is proportional to the permanent dipole moment, the magnitude of the dipole moment may be derived from the spectrum.

14.6 Vibrational Spectra of Diatomic Molecules

Within the Born–Oppenheimer approximation (Section 12.1), the potential energy for the vibrational motion of a molecule is the electronic energy, $E(R)$. It is difficult to find the solutions of the Schrödinger equation using the potential $E(R)$, so a simple approximation is often made: $E(R)$ is expanded in a **Taylor series** about the equilibrium separation R_e to obtain

$$E(R) = E(R_e) + \left(\frac{dE}{dR}\right)_{R_e} (R - R_e) + \frac{1}{2}\left(\frac{d^2E}{dR^2}\right)_{R_e} (R - R_e)^2$$
$$+ \frac{1}{3!}\left(\frac{d^3E}{dR^3}\right)_{R_e} (R - R_e)^3 + \cdots \qquad (14.53)$$

The first term is simply a constant, the electronic energy at the equilibrium geometry, and the second term is zero since dE/dR is zero at the minimum of the potential energy curve. Therefore, near the bottom of the potential curve, we can approximate $E(R)$ by the first term

$$E(R) = \tfrac{1}{2}k(R - R_e)^2 \qquad (14.54)$$

where k is a **force constant**. In this approximation, the potential is harmonic (i.e., parabolic). In Section 10.12, we have already discussed the solutions of the Schrödinger equation for the simple harmonic oscillator. There we saw that the energy levels are given by

$$E_v = (v + \tfrac{1}{2})h\nu \qquad v = 0, 1, 2, \ldots \tag{14.55}$$

where $\nu = (1/2\pi) (k/\mu)^{1/2}$ and μ is the reduced mass of the diatomic molecule (see Section 10.11). It is standard in spectroscopy to give the energy in terms of wavenumbers, so we divide E_v by hc:

$$G(v) = \frac{E_v}{hc} = \omega(v + \tfrac{1}{2}) \tag{14.56}$$

where ω is ν/c. In this approximation the energy levels are equally spaced. This is not a bad approximation for the lowest vibrational states of a diatomic molecule.

The vibrational frequencies for many diatomics are of the order of 1000 cm^{-1}, with higher values for molecules with hydrogen atoms or strong bonds, and lower values for molecules with heavy atoms or weak bonds (see Table 14.4 in the next section).

Not all diatomic molecules have an infrared (vibrational) absorption spectrum because not all transitions are allowed.

To determine which transitions are possible in a vibrational spectrum, we must use equation 14.26 for the electric dipole transition moment. Since the dipole moment for a diatomic molecule, which is given by equation 14.28, depends on the internuclear distance, we expand this dipole moment in a Taylor series about $R = R_e$:

$$\mu_0^{(e)} = \mu_e + \left(\frac{\partial \mu}{\partial R}\right)_{R_e} (R - R_e) + \frac{1}{2}\left(\frac{\partial^2 \mu}{\partial R^2}\right)_{R_e} (R - R_e)^2 + \ldots \tag{14.57}$$

For a molecule in a given electronic state, the transition dipole moment for a vibrational transition is given by

$$\int \psi_{v''}^* \mu_0 \psi_{v'} \, d\tau = \mu_e \int \psi_{v''}^* \psi_{v'} \, d\tau + \left(\frac{\partial \mu}{\partial R}\right)_{R_e} \int \psi_{v''}^* (R - R_e)\psi_{v'} \, d\tau$$

$$+ \frac{1}{2}\left(\frac{\partial^2 \mu}{\partial R^2}\right)_{R_e} \int \psi_{v''}^* (R - R_e)^2 \psi_{v'} \, d\tau + \cdots \tag{14.58}$$

The first term is equal to zero because the vibrational wavefunctions for different v are orthogonal. The second term is nonzero if the dipole moment depends on the internuclear distance R. Thus, the selection rule for a diatomic molecule is that a molecule will show a vibrational spectrum only if the dipole moment changes with internuclear distance.

Homonuclear diatomic molecules, such as H_2 and N_2, have zero dipole moment for all bond lengths and therefore do not show vibrational spectra. In general, heteronuclear diatomic molecules do have dipole moments that depend on internuclear distance, and so they exhibit vibrational spectra.

The integral in the second term of equation 14.58 vanishes unless $v' = v'' \pm 1$ for harmonic oscillator wavefunctions. According to this specific selection rule a harmonic oscillator would have a single vibrational absorption or emission frequency. In general, we would expect the second and higher derivatives of the dipole moment with respect to internuclear distance to be small; after all, if the dipole moment were due to fixed charges a variable distance apart, then $(\partial^2\mu/\partial R^2)$ and higher derivatives would be equal to zero. Although these higher derivatives are small, they do give rise to overtone transitions with $\Delta v = \pm 2, \pm 3, \ldots$, with rapidly diminishing intensities.

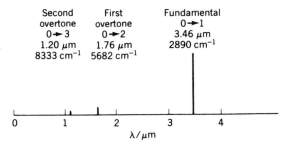

Figure 14.5   "Stick" representation of the vibrational absorption spectrum of HCl. The relative intensities of the lines fall off five times as fast as indicated.

These can be seen in the vibrational absorption spectrum of HCl represented schematically in Fig. 14.5. The strongest absorption band is at 3.46 μm; there is a much weaker band at 1.76 μm and a very much weaker one at 1.198 μm. These are the overtone transitions $v = 0$ to $v = 2$, and $v = 0$ to $v = 3$. The vibrational energy levels of a typical diatomic molecule are shown in Fig. 14.6.

For a harmonic oscillator equation 14.33 indicates that the fraction of the molecules in the ith energy level is given by (note that the levels are nondegenerate)

$$f_i = \frac{e^{-(i+1/2)h\nu/kT}}{\sum\limits_{i=0}^{\infty} e^{-(i+1/2)h\nu/kT}}$$

$$= \frac{e^{-ih\nu/kT}}{\sum\limits_{i=0}^{\infty} e^{-ih\nu/kT}} \qquad (14.59)$$

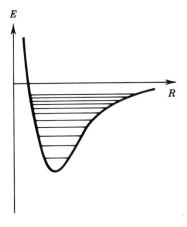

Figure 14.6   Potential energy as a function of internuclear distance and vibrational energy levels of a typical diatomic molecule.

The denominator is a geometric series for which the sum is given by

$$\sum\limits_{i=0}^{\infty} x^i = \frac{1}{1-x} \qquad (14.60)$$

so that

$$\sum\limits_{i=0}^{\infty} e^{-ih\nu/kT} = \frac{1}{1-e^{-h\nu/kT}} \qquad (14.61)$$

Thus, the fraction of the molecules in the ith vibrational state is given by

$$f_i = (1 - e^{-h\nu/kT})e^{-ih\nu/kT} \qquad (14.62)$$

At room temperature this relation predicts that the ratio of the population of $H^{35}Cl$ in $v = 1$ to that in $v = 0$ is 8.9×10^{-7}. Therefore, the molecules with $v = 1$ and higher do not contribute to the spectrum.

Figure 14.4 shows that equation 14.56 is not sufficient to represent the energy levels of a diatomic molecule; if equation 14.56 did apply, the overtones would be at integral multiples of the fundamental. When the Schrödinger equation is solved for equation 14.53 truncated after the cubic term, it is found that the energy levels are given by an equation of the form

$$G(v) = \omega_e(v + \tfrac{1}{2}) - \omega_e x_e(v + \tfrac{1}{2})^2 + \omega_e y_e(v + \tfrac{1}{2})^3 \qquad (14.63)$$

Table 14.3   Band Centers for $H^{35}Cl$ Vibrational Bands[a]

Transition	$G(v)/cm^{-1}$	$\Delta G(v + \frac{1}{2})/cm^{-1}$	$\Delta^2 G(v + 1)/cm^{-1}$
$v = 0 \to 0$	0		
$v = 0 \to 1$	2 885.98	2885.98	
$v = 0 \to 2$	5 667.98	2782.00	−103.98
$v = 0 \to 3$	8 346.78	2678.80	−103.20
$v = 0 \to 4$	10 992.81	2576.03	−102.77
$v = 0 \to 5$	13 396.19	2473.38	−102.65

[a] D. H. Rank, D. P. Eastman, B. S. Rao, and T. A. Wiggins, *J. Opt. Soc. Am.* **52**:1 (1962).

where ω_e is the vibrational wavenumber; x_e and y_e are anharmonicity constants,* and $v = 0, 1, 2 \ldots$.

The coefficients ω_e and $\omega_e x_e$ in the power series for the vibrational energy may be determined in the way indicated in Table 14.3. The second column gives the observed wavenumbers of the successive band centers. The next column gives the differences between the successive lines in wavenumbers. This so-called first difference is equal to the vibrational term difference between states $v + 1$ and v and is represented by $\Delta G(v + \frac{1}{2})$:

$$\Delta G(v + \tfrac{1}{2}) \equiv G(v + 1) - G(v) \quad (14.64)$$

Inserting equation 14.63 and setting y_e to zero since it is usually very small yields

$$\Delta G(v + \tfrac{1}{2}) = \omega_e - 2\omega_e x_e(v + 1) \quad (14.65)$$

where v is the lower of the two levels.

The last column of Table 14.3 gives second differences $\Delta^2 G(v + 1)$, which are defined by

$$\Delta^2 G(v + 1) \equiv \Delta G(v + \tfrac{3}{2}) - \Delta G(v + \tfrac{1}{2}) \quad (14.66)$$

Substituting equation 14.65 yields

$$\Delta^2 G(v + 1) \equiv -2\omega_e x_e \quad (14.67)$$

Note that this is independent of v. In Table 14.3, we see that this is almost exact for $H^{35}Cl$, showing how well this formula works.

The vibrational parameters for a number of diatomic molecules are given in Table 14.4.

Example 14.5
According to the data in Table 14.3 for $H^{35}Cl$, what are the values of the spectroscopic parameters $\omega_e x_e$ and ω_e? What is the zero-point energy?

The second differences in Table 14.3 indicate that the anharmonicity constant $\omega_e x_e$ is equal to $103.0 \ cm^{-1}/2 = 51.5 \ cm^{-1}$. The value of ω_e calculated from the first difference

---

* The anharmonicity constants are tabulated as $\omega_e x_e$ and $\omega_e y_e$ because early in the history of spectroscopy equation 14.63 was written $G(v) = \omega_e[(v + \frac{1}{2}) - x_e(v + \frac{1}{2})^2 + y_e(v + \frac{1}{2})^3]$.

Table 14.4 Constants of Diatomic Molecules[a]

	State	T_e/cm^{-1}	ω_e/cm^{-1}	$\omega_e x_e/cm^{-1}$	B_e/cm^{-1}	α_e/cm^{-1}	R_e/pm	$\dfrac{N_A\mu}{10^{-3}\ kg\ mol^{-1}}$	D_0/eV	IP/eV
$^{79}Br_2$	$^1\Sigma_g^+$	0	325.321	1.077	0.082107	3.187×10^{-4}	228.10	39.459 166	1.9707	10.52
$^{12}C_2$	$^1\Sigma_g^+$	0	1854.71	13.34	1.8198	0.0176	124.25	6.000 000	6.21	12.15
$^{12}C^1H$	$^2\Pi_r$	0	2858.5	63.0	14.457	0.534	111.99	0.929 741	3.46	10.64
$^{35}Cl_2$	$^1\Sigma_g^+$	0	559.7	2.67	0.2439	1.4×10^{-3}	198.8	17.484 427	2.47937	11.50
$^{12}C^{16}O$	$^1\Sigma^+$	0	2169.814	13.288	1.931281	0.017504	112.832	6.856 209	11.09	14.01
$^1H_2$	$^1\Sigma_g^+$	0	4401.21	121.34	60.853	3.062	74.144	0.503 913	4.4781	15.43
	$^1\Sigma_u^+$	91,700	1358.09	20.888	20.015	1.1845	129.28			
$^1H^{81}Br$	$^1\Sigma^+$	0	2648.98	45.218	8.46488	0.23328	141.443	0.995 427	3.758	11.67
$^1H^{35}Cl$	$^1\Sigma^+$	0	2990.95	52.819	10.5934	0.30718	127.455	0.979 593	4.434	12.75
$^1H^{127}I$	$^1\Sigma^+$	0	2309.01	39.644	6.4264	0.1689	160.916	0.999 884	3.054	10.38
$^{127}I_2$	$^1\Sigma_g^+$	0	214.50	0.614	0.03737	1.13×10^{-4}	266.6	63.452 238	1.54238	9.311
$^{39}K^{35}Cl$	$^1\Sigma^+$	0	281	1.30	0.128635	7.89×10^{-4}	266.665	18.429 176	4.34	8.44
$^{14}N_2$	$^1\Sigma_g^+$	0	2358.57	14.324	1.99824	0.017318	109.769	7.001 537	9.759	15.58
	$^3\Pi_g$	59,619	1733.39	14.122	1.6375	0.0179	121.26			
	$^3\Pi_u$	89,136	2047.18	28.445	1.8247	0.0187	114.87			
$^{23}Na^{35}Cl$	$^1\Sigma^+$	0	366	2.0	0.218063	1.62×10^{-3}	236.08	13.870 687	4.23	8.9
$^{14}N^{16}O$	$^2\Pi_r$ $\Omega = \frac{1}{2}$	119.82	1904.04	14.100	1.72	0.0182	150.77	7.466 433	6.496	9.26
	$\Omega = \frac{3}{2}$	0	1904.20	14.075	1.67	0.0171				
$^{16}O_2$	$^3\Sigma_g^-$	0	1580.19	11.98	1.44563	0.0159	120.752	7.997 458	5.115	12.07
	$^1\Delta_g$	7918.1	1483.5	12.9	1.4264	0.0171	121.56			
	$^3\Sigma_u^-$	49793.3	709.31	10.65	0.8190	0.01206	160.43			
$^{16}O^1H$	$^2\Pi_i$	0	3737.76	84.811	18.911	0.7242	96.966	0.948 087	4.392	12.9

[a] K. P. Huber and G. Herzberg, *Molecular Spectra and Molecular Structure IV, Constants of Diatomic Molecules*. New York: Van Nostrand, 1979.

is 2998.2 cm^{-1}. Thus, according to this analysis the zero-point energy for HCl given by equation 14.63 is

$$G(0) = (2998.2\ cm^{-1})(\tfrac{1}{2}) - (51.5\ cm^{-1})(\tfrac{1}{2})^2$$
$$= 1486.2\ cm^{-1}$$

In the preceding chapter we referred to the difference between the equilibrium dissociation energy D_e and the spectroscopic dissociation energy D_0 for H_2^+ and H_2. According to equation 14.63 the energy of the ground state of a diatomic molecule is given by

$$G(0) = \frac{\omega_e}{2} - \frac{\omega_e x_e}{4} + \frac{\omega_e y_e}{8} \qquad (14.68)$$

Thus, the **equilibrium dissociation energy** is given by

$$D_e = D_0 + \frac{\omega_e}{2} - \frac{\omega_e x_e}{4} + \frac{\omega_e y_e}{8} \qquad (14.69)$$

For $^1H^1H$, the values of ω_e, $\omega_e x_e$, and $\omega_e y_e$ are 4401.21, 121.33, and 0.813 cm^{-1}. Therefore, the zero-point energy is $G(0) = 4401.21/2 - 121.33/4 + 0.813/8 = 2170$ cm^{-1}. This is the value used in the preceding chapter. Note, however, that H_2 does not have an infrared spectrum, so these values are determined by other means.

The Taylor series in equation 14.53 only represents the potential energy of a diatomic molecule in the neighborhood of the minimum. What is really needed is a potential energy function for the whole range of R values. In 1929 Morse suggested the following potential function:

$$V(R) = D_e\{1 - \exp[-a(R - R_e)]\}^2 \tag{14.70}$$

When $R \rightarrow \infty$ the potential energy approaches the equilibrium dissociation energy, and the potential energy is zero at $R = R_e$. The Schrödinger equation can be solved for the Morse potential, and the corresponding term value expression (equation 15.71) is

$$G(v) = a\left(\frac{\hbar D_e}{\pi c \mu}\right)^{1/2}\left(v + \frac{1}{2}\right) - \left(\frac{\hbar a^2}{4\pi c \mu}\right)\left(v + \frac{1}{2}\right)^2 \tag{14.71}$$

By comparing this equation with equation 14.63 we find that

$$\omega_e = a\left(\frac{\hbar D_e}{\pi c \mu}\right)^{1/2} \tag{14.72}$$

$$\omega_e x_e = \frac{\hbar a^2}{4\pi c \mu} \tag{14.73}$$

Thus, if $\omega_e x_e$ of a diatomic molecule is known from the vibrational spectrum, a may be eliminated between these equations to obtain the following expression for the equilibrium dissociation energy

$$D_e = \frac{\omega_e}{4x_e} \tag{14.74}$$

This is not a precise method because the Morse equation does not represent the whole potential energy curve accurately over the entire range of R, but it is useful if only the vibrational spectrum of a molecule is known.

14.7   Vibration–Rotation Spectra of Diatomic Molecules

When a molecule in a state with vibrational quantum number v and rotational quantum number J makes a spectral transition to another state, the vibrational quantum number changes to $v \pm 1$ (according to the harmonic oscillator selection rules) and the rotational quantum number can change to $J \pm 1$ or remain the same. The possible transitions are shown in Fig. 14.7. The transitions with $\Delta J = +1$ give rise to lines in the R branch of the spectrum, and the transitions with $\Delta J = -1$ give rise to lines in the P branch of the spectrum. The intensities of the lines in these branches reflect the thermal populations of the initial rotational states. The Q branch, when it occurs, consists of lines corresponding to $\Delta J = 0$. Generally, these transitions are forbidden, except for molecules like NO, which have orbital angular momentum about their axes.

The energies of the levels in Fig. 14.7 are given to the accuracy we need here by the equation

$$E(v, J)/hc = G(v) + F_v(J)$$
$$= \omega_e(v + \tfrac{1}{2}) - \omega_e x_e(v + \tfrac{1}{2})^2 + B_v J(J + 1) \tag{14.75}$$

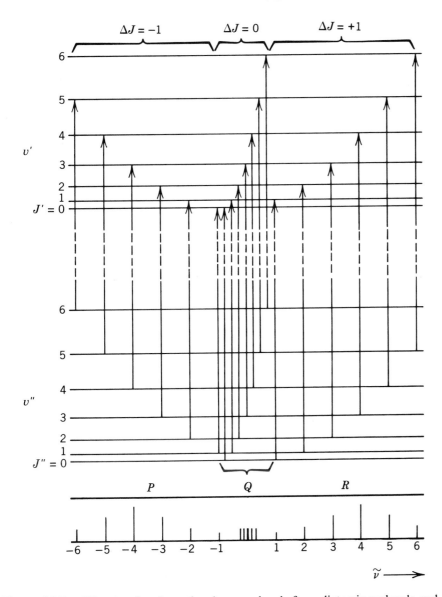

Figure 14.7 Vibrational and rotational energy levels for a diatomic molecule and the transitions observed in the vibration–rotation spectrum when the transition between v' and v'' is allowed. In the spectrum shown at the bottom the lines are indexed with their m values. The relative heights of the spectral lines indicate relative intensities of absorption.

which expresses the energy of a level as the sum of the first term of the rotational term value (equation 14.31) and the first two terms of the vibrational term value (equation 14.63). Now it is necessary to put a subscript v on B_v since the rotational constant depends on the vibrational quantum number v.

The wavenumber of the line resulting from the transition $(v'', J'' \rightarrow v', J')$ is given by

$$\tilde{\nu} = \frac{E' - E''}{hc} = \tilde{\nu}_0 + B'_v J'(J' + 1) - B''_v J''(J'' + 1)$$

$$= \tilde{\nu}_0 + (B'_v + B''_v)m + (B'_v - B''_v)m^2 \qquad (14.76)$$

where $m = \pm 1, \pm 2, \ldots$. The lines in the R branch are obtained from the second form of the equation by letting $m = 1, 2, 3, \ldots$ (i.e., $m = J' = J'' + 1$), and the lines in the P branch are obtained by letting $m = -1, -2, -3, \ldots$ (i.e., $m = -J''$). The missing line at $\tilde{\nu}_0$ is given by

$$\tilde{\nu}_0 = (v' - v'')\omega_e - [(v' + \tfrac{1}{2})^2 - (v'' + \tfrac{1}{2})^2]x_e\omega_e \qquad (14.77)$$

If the rotational constants for the lower B''_v and upper B'_v vibrational levels have the same value, the lines would be equally spaced on the wavenumber scale. However, since the potential energy curve for a diatomic molecule differs from a parabola, the mean internuclear distance increases with the vibrational quantum number. Therefore, $B'_v - B''_v$ is negative, and the separation of successive rotational lines in the R branch decreases as m increases. Correspondingly, in the P branch the spacing increases as m becomes more negative. These features are evident in the fundamental infrared absorption spectrum for HCl shown in Fig. 14.8. The dependence of the rotational constant on the vibrational quantum number is generally represented by

$$B_v = B_e - \alpha_e(v + \tfrac{1}{2}) \qquad (14.78)$$

where α_e is the **vibration–rotation coupling constant.**

By applying equation 14.76 to data for a vibration–rotation band it is possible to obtain values for B'_v, B''_v, and $\tilde{\nu}_0$. Then by use of equation 14.78 the values of B_e and α_e may be obtained. Finally ω_e may be calculated from $\tilde{\nu}_0$ since x_e has been determined in Section 14.6.

Table 14.4 gives the vibrational and rotational constants for a number of diatomic molecules and several of their electronic excited states. The electronic energy relative to the ground state is represented by T_e.

Example 14.6
Calculate the relative populations of the first five rotational levels of the ground vibrational state of $H^{35}Cl$ at 300 K. According to Table 14.4, $B_v = 10.5934$ cm^{-1} $-$ (0.3072 cm^{-1})/2 $= 10.4398$ cm^{-1} for $v = 0$.

$$\frac{N_J}{N_0} = (2J + 1)e^{-hcJ(J+1)B_v/kT}$$

where N_0 is the number of molecules in the $J = 0$ state. First we need to calculate the following factor:

$$\frac{hcB_v}{kT} = \frac{(6.626 \times 10^{-34} \text{ J s})(2.998 \times 10^8 \text{ m s}^{-1})(10.44 \text{ cm}^{-1})(10^2 \text{ cm m}^{-1})}{(1.3806 \times 10^{-23} \text{ J K}^{-1})(300 \text{ K})}$$

$$= 5.007 \times 10^{-2}$$

For $J = 1$

$$\frac{N_1}{N_0} = 3e^{-2(5.007 \times 10^{-2})}$$

$$= 2.71$$

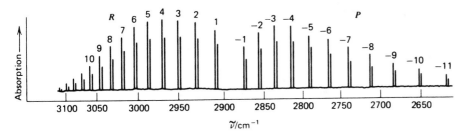

Figure 14.8 Fundamental vibrational band for HCl ($v = 0 \rightarrow 1$). The double peaks are due to the presence of $H^{35}Cl$ (75% abundance) and $H^{37}Cl$ (25% abundance). The integers represent the value of m in equation 14.76. (Reprinted with permission from A. R. H. Cole, *Tables of Wavenumbers for the Calibration of Infrared Spectrometers*, 2d ed. Copyright © 1977, Pergamon Press on behalf of IUPAC.)

The relative populations for $J = 0, 1, 2, 3, 4,$ and 5 are 1.00, 2.71, 3.70, 3.84, 3.31, and 2.45, in excellent agreement with Fig. 14.8.

14.8 Vibrational Spectra of Polyatomic Molecules

We will discuss the vibration of polyatomic molecules first from a classical viewpoint, and then from a quantum mechanical viewpoint. As discussed in Section 14.3, the **number of vibrational degrees of freedom** is $3N - 6$ for nonlinear molecules and $3N - 5$ for linear molecules, where N is the number of nuclei. A diatomic molecule has one vibrational coordinate R. A linear triatomic molecule, like CO_2, has four. A nonlinear triatomic molecule, like H_2O, has three. NH_3, CH_4, and N_2O_4 have 6, 9, and 12 independent vibrational coordinates, respectively.

However, it is easier to start out using all $3N$ coordinates and reduce the numbers to $3N - 6$ or $3N - 5$ later. The kinetic energy T of a polyatomic molecule is given by

$$T = \frac{1}{2} \sum_{k=1}^{N} m_k \left[\left(\frac{dx_k}{dt} \right)^2 + \left(\frac{dy_k}{dt} \right)^2 + \left(\frac{dz_k}{dt} \right)^2 \right] \tag{14.79}$$

This equation can be simplified by introducing mass-weighted Cartesian displacement coordinates $q_1, \ldots, q_{3N}$.

$$q_1 = m_1^{1/2}(x_1 - x_{ie}) \qquad q_2 = m_1^{1/2}(y_1 - y_{ie}) \qquad q_3 = m_1^{1/2}(z_1 - z_{ie})$$
$$q_4 = m_2^{1/2}(x_2 - x_{2e}) \cdots \qquad q_{3N} = m_{3N}^{1/2}(z_{3N} - z_{3Ne}) \tag{14.80}$$

where x_{ie}, and so on are the values of the coordinates at the equilibrium geometry of the molecule. Since these are independent of time, the kinetic energy becomes

$$T = \frac{1}{2} \sum_{i=1}^{3N} \left(\frac{dq_i}{dt} \right)^2 \tag{14.81}$$

Since the potential energy V is a function of the coordinates $x_1, \ldots, z_N$, it is also a function of the mass-weighted coordinates. We will use a Taylor series expansion about the equilibrium position as we did in equation 14.53, this time for a function of several variables:

$$V = V_e + \sum_{i=1}^{3N} \left(\frac{\partial V}{\partial q_i}\right)_e q_i + \frac{1}{2} \sum_{i=1}^{3N} \sum_{k=1}^{3N} \left(\frac{\partial^2 V}{\partial q_i \partial q_k}\right)_e q_i q_k + \cdots \quad (14.82)$$

Since V_e is the potential energy at the equilibrium configuration, it is a constant which we can set equal to zero, and the terms in $(\partial V/\partial q_i)_e$ are all equal to zero because the potential energy is a minimum at the equilibrium configuration by definition. If we neglect terms higher than quadratic, equation 14.82 can be written

$$V = \frac{1}{2} \sum_{i=1}^{3N} \sum_{k=1}^{3N} K_{ik} q_i q_k \quad (14.83)$$

so that the total energy is given by

$$E = T + V = \frac{1}{2} \sum_{i=1}^{3N} \left(\frac{dq_i}{dt}\right)^2 + \frac{1}{2} \sum_{i=1}^{3N} \sum_{k=1}^{3N} K_{ik} q_i q_k \quad (14.84)$$

where K_{ik} is the second derivative of V with respect to q_i and q_k evaluated at the equilibrium configuration. The problem in using this expression is with the cross terms. Fortunately, it is possible to make a linear transformation of the mass-weighted coordinates q to new coordinates Q such that the quadratic term does not contain cross terms:

$$E = \frac{1}{2} \sum_{i=1}^{3N} \left(\frac{dQ_i}{dt}\right)^2 + \frac{1}{2} \sum_{i=1}^{3N-6 \atop \text{or } 3N-5} \kappa_i Q_i^2 \quad (14.85)$$

We have used the fact that translational and rotational motion have only kinetic energy so that there are $3N - 6$ vibrational coordinates for a nonlinear molecule and $3N - 5$ for a linear molecule. These $3N - 6$ or $3N - 5$ coordinates are referred to as **normal coordinates**, and the corresponding $3N - 6$ or $3N - 5$ vibrations are referred to as **normal modes of vibration.**

In a normal mode of vibration, the nuclei move in phase (i.e., the nuclei pass through the extremes of their motion simultaneously). The motions of the nuclei in a normal mode are such that the center of mass does not move, and the molecule as whole does not rotate. This means that different atoms move different distances. Each normal mode has a characteristic vibration frequency. Sometimes several modes have identical vibration frequencies and are referred to as degenerate modes. It can be shown that any vibrational motion of a polyatomic molecule can be expressed as a linear combination of normal modes of vibrations. Although we can think about normal modes in terms of these classical diagrams, we must remember that in quantum mechanics vibrational motion is represented in terms of probability density functions.

Turning now to the quantum mechanical treatment of a molecule, the vibrational Hamiltonian obtained from equation 14.85 is simply a sum of terms, one for each coordinate:

$$\hat{H} = \sum_{i=1}^{3N-6 \atop \text{or } 3N-5} \hat{H}_i \quad (14.86)$$

This indicates that the vibrational wavefunctions for the molecule can be written as the product of harmonic oscillator wavefunctions, one for each coordinate Q_i. We have seen in equation 10.95 that the eigenvalues for the harmonic oscillator are given by $E_i = (v_i + \frac{1}{2})hc\tilde{v}_i$ so that the total vibrational energy of a polyatomic molecule is

$$E = \sum_{i=1}^{\substack{3N-6 \\ \text{or } 3N-5}} (v_i + \frac{1}{2})hc\tilde{v}_i \qquad (14.87)$$

The frequency of a normal mode depends both on the force constant k_Q for the mode and on the reduced mass μ_Q for the mode: $2\pi\nu_i = (k_Q/\mu_Q)^{1/2}$.

The four normal modes of CO_2 are shown in Fig. 14.9a. The first normal mode is a symmetrical stretching vibration in which the carbon atom remains fixed. The third normal mode is an asymmetrical stretching vibration. The other two normal modes are orthogonal bending vibrations. The lower vibration frequency for the bending vibrations indicates that it is generally easier to bend a molecule than to stretch it. Figure 14.9b shows the three normal modes of vibration of H_2O. As indicated in the diagrams, the displacements of various atoms in a normal mode are not equal, but depend on the masses and force constants.

For a polyatomic molecule, some normal modes of vibration are spectroscopically active and some are not. The gross selection rule is still that the displacements of a normal mode must cause a change in dipole moment in order to be spectroscopically active in the infrared.

Of the four normal mode vibrations for CO_2 the symmetric stretch is not active in the infrared, but the other vibrations are. Since CO_2 is linear and symmetrical in its equilibrium state, it does not have a dipole moment, and the symmetric stretching vibration does not create one. The asymmetric stretch and bending vibrations produce a changing dipole moment. The three normal modes of H_2O are all active in the infrared because the magnitude of the dipole moment changes in each type of vibration.

The specific selection rule for vibrational spectroscopy is that $\Delta v = \pm 1$ in the harmonic oscillator approximation. In addition combination bands are formed in which two or more vibrational modes change simultaneously.

The frequencies, in cm^{-1}, of the strongest bands for H_2O vapor are summarized in Table 14.5. The weaker bands in the spectrum are the overtones and combinations shown in the table. As shown in the table, the vibrations are not harmonic, and so the overtones are not exact multiples, and the combinations are not exact sums.

One of the vibrational motions of a polyatomic molecule may be an internal rotation. If there is an appreciable potential energy barrier for an internal rotation about some bond, there will be an oscillation about the mean position. For example, in ethylene, $CH_2{=}CH_2$, there is a large potential barrier for internal rotation so that there are only small oscillations about the C=C bond. In some cases, such as $CH_3{-}CH_3$, the potential energy barrier is small enough that the internal rotation is said to be "free" at room temperature.

The vibrational spectra of polyatomic molecules are very useful for identification and serve also as a criterion of purity. For such practical applications the infrared spectra for a large number of compounds have been cataloged and are used like fingerprints. Groups of atoms within the molecule have quite characteristic absorption bands. The wavelengths at which a certain group

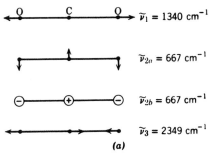

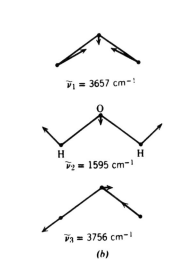

Figure 14.9 (a) Normal modes of vibration of symmetrical linear triatomic molecule CO_2. (b) Normal modes of vibration of nonlinear triatomic molecule H_2O. The vectors representing the magnitudes of the oxygen vibrations have been increased relative to those of hydrogen.

Table 14.5   Infrared Bands of H_2O Vapor

$\tilde{v}/cm^{-1}$	Intensity	Interpretation
1595.0	Very strong	$\tilde{v}_2$
3151.4	Medium	$2\tilde{v}_2$
3651.7	Strong	$\tilde{v}_1$
3755.8	Very strong	$\tilde{v}_3$
5332.0	Medium	$\tilde{v}_2 + \tilde{v}_3$
6874	Weak	$2\tilde{v}_2 + \tilde{v}_3$

absorbs vary slightly, depending on the structure of the rest of the molecule.

The infrared spectrum of a molecule may be considered to be made up of several regions.

1. Hydrogen stretching vibrations, 3700–2500 cm^{-1}. These vibrations occur at high frequencies because of the low mass of the hydrogen atom. If an OH group is not involved in hydrogen bonding (Section 12.10), it usually has a frequency in the vicinity of 3600–3700 cm^{-1}. Hydrogen bonding causes this frequency to drop by 300–1000 cm^{-1} or more. The NH absorption falls in the 3300- to 3400-cm^{-1} range, and the CH absorption falls in the 2850- to 3000-cm^{-1} range. For SiH, PH, and SH it is approximately 2200, 2400, and 2500 cm^{-1}.

2. Triple-bond region, 2500–2000 cm^{-1}. These bonds have high frequencies because of the large force constants. The C≡C group usually causes absorption between 2050 and 2300 cm^{-1}, but this absorption may be weak or absent because of the symmetry of the molecule. The C≡N group absorbs near 2200–2300 cm^{-1}.

3. Double-bond region, 2000–1600 cm^{-1}. Absorption bands of substituted aromatic compounds fall in this range and are a good indicator of the position of the substitution. Carbonyl groups, C=O, of ketones, aldehydes, acids, amides, and carbonates usually show strong absorption in the vicinity of 1700 cm^{-1}. Olefins, C=C, may show absorption in the vicinity of 1650 cm^{-1}. The bending of the C–N–H bond also occurs in this region.

4. Single-bond stretch and bend region, 500–1700 cm^{-1}. This region is not diagnostic for particular functional groups, but it is a useful "fingerprint" region, since it shows differences between similar molecules. Organic compounds usually show peaks in the region between 1300 and 1475 cm^{-1} because of the bending motions of bonds to hydrogen. Out-of-plane bending motions of olefinic and aromatic CH groups usually occur between 700 and 1000 cm^{-1}.

14.9   Raman Spectra

In Raman spectroscopy, light of frequency ν is scattered by molecules and the frequency of the outgoing light can be shifted to ν', a process called **inelastic scattering**. Since this represents a change in the energy of the photons from $h\nu$ to $h\nu'$, the molecule has also changed its energy. If the molecule was originally in quantum state i with energy E_i and ends up in quantum state f with energy E_f, conservation of energy requires that

$$h\nu + E_i = h\nu' + E_f \tag{14.88}$$

or

$$h(\nu' - \nu) = E_i - E_f = h \, \Delta\nu_R = hc \, \Delta\tilde{\nu}_R \tag{14.89}$$

where the shift in frequency is labeled $\Delta\nu_R$ and the shift in wavenumber is labeled $\Delta\tilde{\nu}_R$. Notice that Raman spectroscopy is different from absorption or emission spectroscopy in that the **incident** light need not coincide with a quan-

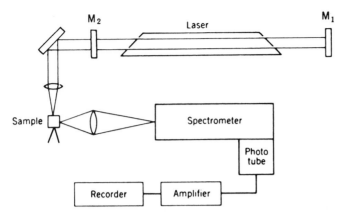

Figure 14.10 Apparatus for obtaining Raman spectra. Mirrors M_1 and M_2 are required to obtain laser action.

tized energy difference in the molecule. Therefore, any frequency of light can be used. Since many final states are possible, of both higher and lower energy than the initial state, many Raman spectral lines can be observed. A typical experimental apparatus is shown in Fig. 14.10 and a Raman spectrum of liquid CCl_4 is shown in Fig. 14.11. The spectral lines with $\nu' < \nu$ are called **Stokes lines** and those with $\nu' > \nu$ are called **anti-Stokes lines.** The Stokes lines are produced when the light loses energy to the molecule and the anti-Stokes lines are produced when the light gains energy from the molecule. Therefore, anti-Stokes lines are caused when the molecule in a higher energy state makes a transition to a lower state. Since the population of molecules in high-energy

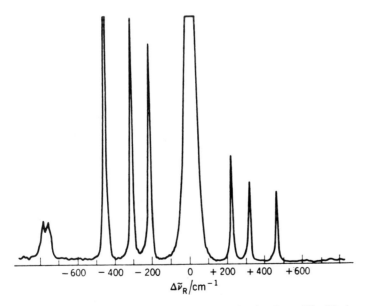

Figure 14.11 Raman spectrum of liquid CCl_4 obtained using a He–Ne laser. The Raman frequencies are measured from the parent line at 632.8 nm. In this spectrum frequency increases to the right so that the lines to the left of the laser line have a lower frequency; they are called Stokes lines.

states is small (because of the Boltzmann factor), anti-Stokes lines are usually weaker than Stokes lines.

Usually less than one part in 10^6 of the incident light is scattered inelastically so that it is necessary to use powerful light sources so that the weak Raman lines can be detected. It is also necessary to use quite monochromatic incident radiation because the scattered frequencies are usually displaced from the parent line by only 10–4000 cm^{-1}. Since lasers are powerful sources of monochromatic radiation, their development has greatly enhanced Raman spectroscopy.

The frequency shifts seen in Raman experiments correspond to vibrational or rotational energy differences, so this kind of spectroscopy gives us information on the vibrational and rotational states of molecules.

The Raman effect arises from the induced polarization of scattering molecules that is caused by the electric vector of the electromagnetic radiation. Some aspects of the Raman effect can be understood classically. First we will consider an isotropic molecule, that is, one that has the properties in all directions (CH_4 is an example). A dipole moment $\boldsymbol{\mu}$ is induced in the molecule by an electric field E:

$$\boldsymbol{\mu} = \alpha \boldsymbol{E} \quad \text{or} \quad \mu = \alpha E \tag{14.90}*$$

For an isotropic molecule the vectors $\boldsymbol{\mu}$ and E point in the same direction, and the **polarizability** α is a scalar. The polarizability α of a molecule that is rotating or vibrating is not constant, but varies with some frequency ν_k (for example, a vibration or rotation frequency) according to

$$\alpha = \alpha_0 + (\Delta\alpha) \cos 2\pi\nu_k t \tag{14.91}$$

where α_0 is the equilibrium polarizability and $\Delta\alpha$ is its maximum variation. Since the electric field of the impinging electromagnetic radiation varies with time according to

$$E = E° \cos 2\pi\nu_0 t \tag{14.92}$$

the induced dipole moment of the molecule is given by

$$\begin{aligned}
\mu &= [\alpha_0 + (\Delta\alpha) \cos 2\pi\nu_k t]E° \cos 2\pi\nu_0 t \\
&= \alpha_0 E° \cos 2\pi\nu_0 t + \tfrac{1}{2}(\Delta\alpha)E°[\cos 2\pi(\nu_0 + \nu_k)t + \cos 2\pi(\nu_0 - \nu_k)t]
\end{aligned}$$

$$\tag{14.93}$$

where the last form has been obtained using the relation $\cos a \cos b = \frac{1}{2}[\cos(a + b) + \cos(a - b)]$. The three terms in this equation provide the classical explanation for Rayleigh scattering (ν_0), anti-Stokes lines ($\nu_0 + \nu_k$), and Stokes lines ($\nu_0 - \nu_k$), respectively. However, the classical treatment incorrectly implies that the Stokes and anti-Stokes lines will occur with equal intensity. The anti-Stokes lines are, of course, weaker because they depend on the populations of excited levels.

In general, a molecule is nonisotropic so that the application of an electric field E in a particular direction induces a moment $\boldsymbol{\mu}$ in a different direction. In this case $\boldsymbol{\alpha}$ is a tensor, and the induced dipole moment is given by

---

* This relation was introduced in Section 12.12.

$$\mu = \alpha E \qquad (14.94)$$

which is expressed by the following matrix equation (see Appendix D on matrices):

$$\begin{bmatrix} \mu_x \\ \mu_y \\ \mu_z \end{bmatrix} = \begin{bmatrix} \alpha_{xx} & \alpha_{xy} & \alpha_{xz} \\ \alpha_{yx} & \alpha_{yy} & \alpha_{yz} \\ \alpha_{zx} & \alpha_{zy} & \alpha_{zz} \end{bmatrix} \begin{bmatrix} E_x \\ E_y \\ E_z \end{bmatrix} \qquad (14.95)$$

This is equivalent to the following set of algebraic equations:

$$\mu_x = \alpha_{xx}E_x + \alpha_{xy}E_y + \alpha_{xz}E_z \qquad (14.96a)$$

$$\mu_y = \alpha_{yx}E_x + \alpha_{yy}E_y + \alpha_{yz}E_z \qquad (14.96b)$$

$$\mu_z = \alpha_{zx}E_x + \alpha_{zy}E_y + \alpha_{zz}E_z \qquad (14.96c)$$

Thus, each component (μ_x, μ_y, μ_z) of the induced dipole moment μ can depend on each component (E_x, E_y, E_z) of the electric field E. Only six of the nine coefficients of the polarizability are independent, since it can be shown that $\alpha_{xy} = \alpha_{yx}$, $\alpha_{xz} = \alpha_{zx}$, and $\alpha_{yz} = \alpha_{zy}$.

The quantum mechanical theory for the selection rules for the Raman effect is more complicated than for pure rotational and vibrational spectra because Raman scattering is a kind of two photon process: The incident photon is absorbed and the leaving photon is emitted by the molecule in a single quantum process.

A molecule scatters light because it is polarizable. Therefore, the gross selection rule for Raman spectroscopy must be concerned with the polarizability of a molecule. For a molecule to show a rotational Raman spectrum the polarizability of the molecule must be anisotropic; that means that the polarizability must depend on the orientation of the molecule with respect to the direction of the electric field. The polarizability of an atom is, of course, isotropic. Similarly, a spherically symmetric molecule (such as CH_4) does not have a rotational Raman spectrum.

The specific selection rules for rotational Raman transitions are as follows for linear and symmetric top molecules:

Linear molecules	$\Delta J = 0, \pm 2$	
Symmetric top	$\Delta J = 0, \pm 2, \Delta K = 0$	when $K = 0$
molecules	$\Delta J = 0, \pm 1, \pm 2, \Delta K = 0$	when $K \neq 0$

where K is the component of the angular momentum J along the principal symmetry axis. The $\Delta J = 0$ applies in vibration–rotation transitions. The fact that $\Delta J = \pm 2$ for linear molecules is a result of the fact that the polarizability of a molecule returns to its initial value twice in a 360° revolution, as shown in Fig. 14.12. The $\Delta K = 0$ is a result of the fact that the dipole of a symmetric top molecule is along the principal axis, and so there cannot be a component of the dipole moment perpendicular to this axis.

The frequencies of the Stokes lines ($\Delta J = 2$) in the rotational Raman spectrum of a linear molecule are given by

$$\Delta \tilde{\nu}_R = BJ'(J' + 1) - BJ''(J'' + 1) \qquad (14.97)$$

where J'' is the initial quantum number. Substituting $J' = J'' + 2$

$$\Delta \tilde{\nu}_R = B[(J'' + 2)(J'' + 3) - J''(J'' + 1)]$$

$$= 2B(2J'' + 3) \qquad (14.98)$$

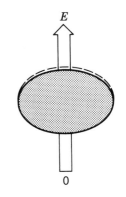

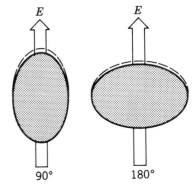

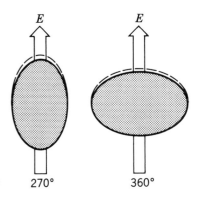

Figure 14.12 The polarizability ellipsoid for a molecule that returns to its initial value twice in a 360° revolution.

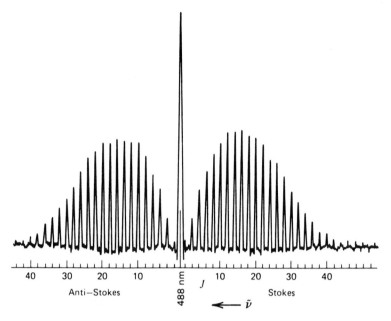

Figure 14.13   Pure rotational Raman spectrum of CO_2. The abscissa gives the quantum numbers for the lower quantum state. The intense peak at 488 nm is due to elastic scattering. (From B. P. Straughan and S. Walker, *Spectroscopy*, Vol. 2. London: Chapman & Hall, 1976.)

These lines appear at lower frequencies than the exciting line and are referred to as the *S* branch. The relative intensities of these lines are determined by the populations of the initial states, as we have discussed for the vibration–rotation spectra in the infrared.

The frequencies of the anti-Stokes lines ($\Delta J = -2$) in the rotational Raman spectrum are given by

$$\Delta \tilde{\nu}_R = -2B(2J'' - 1) \qquad \text{where} \quad J'' \geq 2 \qquad (14.99)$$

The lines appear at higher frequencies, and are referred to as the *O* branch. In addition, there is a *Q* branch for $\Delta J = 0$. The *S*, *Q*, and *O* branches correspond with the *P*, *Q*, and *R* branches of infrared spectroscopy.

The pure rotational Raman spectrum of CO_2 is shown in Fig. 14.13. Notice the large number of initially populated rotational states, since the rotational splitting is small compared to kT.

For a molecule to have a vibrational Raman spectrum it is necessary for the polarizability to change as the molecule vibrates. The polarizabilities of both homonuclear and heteronuclear diatomic molecules change as the molecule vibrates and so both types of molecules have vibrational Raman spectra in contrast to infrared vibrational spectra. The specific selection rule for the vibrational Raman effect is $\Delta v = \pm 1$. The vibrational transitions are accompanied by rotational Raman transitions with the specific selection rules $\Delta J = 0, \pm 2$, as before. The vibrational Raman spectra of homonuclear diatomic molecules are of special interest because they yield force constants and rotational constants that are not available from infrared absorption spectroscopy.

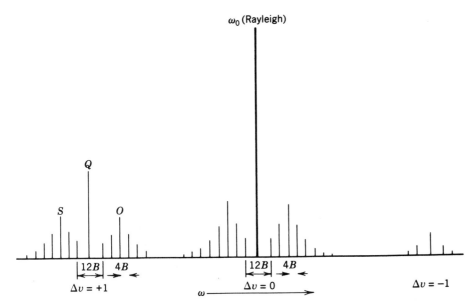

Figure 14.14 Theoretical rotation–vibration Raman spectrum of a linear molecule. The effects of nuclear spin statistics have been omitted from this illustration. (W. A. Guillory, *Introduction to Molecular Structure and Spectroscopy*. Boston: Allyn & Bacon, 1977, by permission.)

Figure 14.14 shows the theoretical rotation–vibration Raman spectrum for $\Delta v = +1, 0, -1$ and $\Delta J = +2, 0, -1$ for a linear molecule. It is assumed that there is a small population in the first excited vibrational state. The center series of lines is the rotational Raman spectrum of the molecule. Since homonuclear diatomic molecules give spectra of this type, Raman spectroscopy provides the possibility, not available in microwave or infrared spectroscopy, of determining their internuclear distances and force constants.

For a polyatomic molecule, some normal modes will be Raman active and some will not, depending on what happens to the polarizability ellipsoid when the generalized coordinate for the normal mode is changed. This is most easily illustrated for a linear symmetric XY_2 molecule, such as CO_2, since the principal axes of the polarizability ellipsoid are coincident with the symmetry axes of the molecule. As shown in Fig. 14.15, only the symmetrical stretching normal mode is Raman active. This particular mode is not active in the infrared (Section 14.8). According to the **mutual exclusion rule** for molecules with a center of symmetry, fundamental transitions that are active in the infared are forbidden in the Raman scattering, and vice versa. Thus, data obtained from Raman and infrared spectra are often complementary. For molecules without a center of symmetry, there are generally some modes that are infrared active only, some which are Raman active only, some which are active in both, and some which are active in neither.

The symmetry group of the molecule can be used to determine whether a particular vibration is Raman active or infrared active. Those normal modes whose symmetry is the same as the functions x, y, or z will be infrared active, while those whose symmetry is the same as x^2, y^2, or z^2 will be Raman active

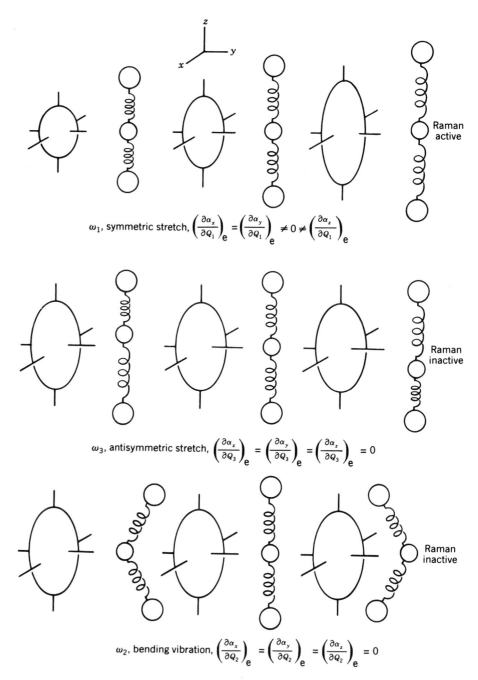

Figure 14.15   Changes in the polarizability ellipsoid in the normal vibrations of CO_2. (R. S. Tobias, *J. Chem. Educ.* **44**:40 (1967).)

(see Chapter 13). When the exciting frequency is within an electronic absorption band for a molecule, the Raman scattering is greatly enhanced. This is referred to as **resonance Raman** spectroscopy. These spectra, which may be determined on rather dilute solutions (10^{-3} to 10^{-6} M) compared with the usual Raman spectra, emphasize vibrations in the region of the molecule that are responsible for the electronic absorption.

References

P. W. Atkins, *Molecular Quantum Mechanics*. New York: Oxford University Press, 1983.

L. J. Bellamy, *Infrared Spectra of Complex Molecules*. London: Chapman & Hall, 1975.

P. R. Bunker, *Molecular Symmetry and Spectroscopy*. New York: Academic, 1979.

H. B. Dunford, *Elements of Diatomic Molecular Spectra*. Reading, MA: Addison-Wesley, 1968.

W. H. Flygare, *Molecular Structure and Dynamics*. Englewood Cliffs, NJ: Prentice-Hall, 1978.

W. A. Guillory, *Introduction to Molecular Structure and Spectroscopy*. Boston: Allyn & Bacon, 1977.

D. C. Harris and M. D. Bertolucci, *Symmetry and Spectroscopy*. Oxford, UK: Oxford University Press, 1978.

G. Herzberg, *Infrared and Raman Spectra of Polyatomic Molecules*. Princeton, NJ: Van Nostrand, 1950.

G. Herzberg, *Spectra of Diatomic Molecules*. Princeton, NJ: Van Nostrand, 1950.

G. Herzberg, *The Spectra and Structures of Simple Free Radicals*. Ithaca, NY: Cornell University Press, 1971.

J. M. Hollas, *High Resolution Spectroscopy*. Boston: Butterworths, 1982.

K. P. Huber and G. Herzberg, *Molecular Spectra and Molecular Structure IV. Constants of Diatomic Molecules*. New York: Van Nostrand-Reinhold, 1979.

I. N. Levine, *Molecular Spectroscopy*. New York: Wiley, 1975.

D. A. Long, *Raman Spectroscopy*. New York: McGraw-Hill, 1977.

J. I. Steinfeld, *Molecules and Radiation*. Cambridge, MA: MIT Press, 1985.

B. P. Straughan and S. Walker, *Spectroscopy,* Vols. 1, 2, 3. New York: Wiley, 1976.

D. H. Whiffen, *Spectroscopy*. London: Longman, 1972.

E. B. Wilson, J. C. Decius, and P. C. Cross, *Molecular Vibrations*. New York: McGraw-Hill, 1955.

Problems

14.1 Since the energy of a molecular quantum state is divided by kT in the Boltzmann distribution, it is of interest to calculate the temperature at which kT is equal to the energy of photons of different wavelength. Calculate the temperature at which kT is equal to the energy of photons of wavelength 10^3 cm, 10^{-1} cm, 10^{-3} cm, and 10^{-5} cm.

14.2 Most chemical reactions require activation energies ranging between 40 and 400 kJ mol^{-1}. What are the equivalents of 40 and 400 kJ mol^{-1} in terms of (a) nm, (b) wavenumbers, and (c) electronvolts?

14.3 (a) What vibrational frequency in wavenumbers corresponds to a thermal energy of kT at 25 °C? (b) What is the wavelength of this radiation?

14.4 Calculate the reduced mass and the moment of inertia of D^{35}Cl, given that $R_e = 127.5$ pm.

14.5 Calculate the frequency in wavenumbers and the wavelength in cm of the first rotational transition ($J = 0 \rightarrow 1$) for D^{35}Cl.

14.6 The pure rotational spectrum of ^{12}C^{16}O has transitions at 3.863 and 7.725 cm^{-1}. Calculate the internuclear distance in ^{12}C^{16}O. Predict the positions, in cm^{-1}, of the next two lines.

14.7 Assume the bond distances in ^{13}C^{16}O, ^{13}C^{17}O, and ^{12}C^{17}O are the same as in ^{12}C^{16}O. Calculate the position, in cm^{-1}, of the first rotational transitions in these four molecules. (Use the information in problem 14.6.)

14.8 The far infared spectrum of HI consists of a series of equally spaced lines with $\Delta\bar{\nu} = 12.8$ cm^{-1}. What is (a) the moment of inertia, and (b) the internuclear distance?

14.9 Using equation 14.35, show that J for the maximally populated level is given by

$$J_{max} = \sqrt{\frac{kT}{2hcB}} - \frac{1}{2}$$

14.10 Using the result of problem 14.9, find the J nearest J_{max} at room temperature for H^{35}Cl and ^{12}C^{16}O. What is the ratio of the population at that J to the population at $J = 0$? What is the energy of that J relative to $J = 0$ in units of kT?

14.11 The moment of inertia of ^{16}O^{12}C^{16}O is 7.167×10^{-46} kg m^2. (a) Calculate the CO bond length, R_{CO}, in CO_2. (b) Assuming that isotopic substitution does not alter R_{CO}, calculate the moments of inertia of

(1) ^{18}O^{12}C^{18}O and (2) ^{16}O^{13}C^{16}O

14.12 Derive the expression for the moment of inertia of a symmetrical tetrahedral molecule like CH$_4$ in terms of the

bond length R and the masses of the four tetrahedral atoms. The easiest way to derive the expression is to consider an axis along one CH bond. Show that the same result is obtained if the axis is taken perpendicular to the plane defined by one group of three atoms HCH.

14.13 What are the values of A and B (from equation 14.51) for the symmetric top NH$_3$ if $I_{\parallel} = 4.41 \times 10^{-47}$ kg m^2 and $I_{\perp} = 2.81 \times 10^{-47}$ kg m^2? What is the wavelength of the $J = 0$ to $J = 1$ transition? What are the wavelengths of the $J = 1$ to $J = 2$ transitions (remember the selection rules, $\Delta J = \pm 1$, $\Delta K = 0$, and find all allowed transitions).

14.14 Consider a linear triatomic molecule, ABC. Find the center of mass (which by symmetry lies on the molecular axis). Show that the moment of inertia is given by

$$I = \frac{1}{M}[R_{AB}^2 m_A m_B + R_{BC}^2 m_B m_C + (R_{AB} + R_{BC})^2 m_A m_C]$$

where R_{AB} is the AB bond distance, R_{BC} is the BC bond distance, m_i are the masses of the atoms, and $M = m_A + m_B + m_C$. Show that if $R_{AB} = R_{BC}$ and $m_A = m_C$, then $I = 2m_A R_{AB}^2$.

14.15 The fundamental vibration frequency of H^{35}Cl is 8.967×10^{13} s^{-1} and of D^{35}Cl is 6.428×10^{13} s^{-1}. What would the separation be between infrared absorption lines of H^{35}Cl and H^{37}Cl on one hand and of D^{35}Cl and D^{37}Cl on the other, if the force constants of the bonds are assumed to be the same in each pair.

14.16 Find the force constants of the halogens ^{127}I$_2$, ^{79}Br$_2$, and ^{35}Cl$_2$ using the data of Table 14.4. Is the order of these the same as the order of the bond energies?

14.17 Given the following fundamental frequencies of vibration, calculate $\Delta H°$ for the reaction

$$H^{35}Cl\,(v = 0) + {}^2D_2\,(v = 0) = {}^2D^{35}Cl\,(v = 0) + H^2D\,(v = 0):$$

$$H^{35}Cl: 2989 \text{ cm}^{-1} \qquad H^2D: 3817 \text{ cm}^{-1}$$
$$^2D^{35}Cl: 2144 \text{ cm}^{-1} \qquad ^2D^2D: 3119 \text{ cm}^{-1}$$

14.18 If the fundamental vibration frequency of ^{1}H$_2$ is 4401.21 cm^{-1}, compute the fundamental vibration frequency of ^{2}D$_2$ and ^{1}H^{2}D assuming the same force constants. If D_0 for ^{1}H$_2$ is 4.4781 eV, what is D_0 for ^{2}D$_2$ and ^{1}H$_2$D? Neglect anharmonicities.

14.19 Using the values for ω_e and $\omega_e x_e$ in Table 14.4 for ^{1}H^{35}Cl, estimate the dissociation energy assuming the Morse potential is applicable.

14.20 (a) What fraction of H$_2$(g) molecules are in the $v = 1$ state at room temperature? (b) What fractions of Br$_2$(g) molecules are in the $v = 1$, 2, and 3 states are room temperatures?

14.21 The first three lines in the R branch of the fundamental vibration–rotation band of $H^{35}Cl$ have the following frequencies in cm^{-1}: 2906.25(0), 2925.78(1), 2944.89(2), where the numbers in parentheses are the J values for the initial level. What are the values of ω_0, B'_v, B''_v, B_e, and α?

14.22 In Table 14.1, D_e for H_2 is given as 4.7483 eV or 458.135 kJ mol^{-1}. Given the vibrational parameters for H_2 in Table 14.4, calculate the value you would expect for $\Delta_f H°$ for H(g) at 0 K.

14.23 Calculate the wavelengths in (a) wavenumbers and (b) micrometers of the center two lines in the vibration spectrum of HBr for the fundamental vibration. The necessary data are to be found in Table 14.4.

14.24 How many normal modes of vibration are there for (a) SO_2 (bent), (b) H_2O_2 (bent), (c) $HC{\equiv}CH$ (linear), and (d) C_6H_6?

14.25 List the numbers of translational, rotational, and vibrational degrees of freedom for (a) Ne, (b) N_2, (c) CO_2, and (d) CH_2O.

14.26 Acetylene is a symmetrical linear molecule. It has seven normal modes of vibration, two of which are doubly degenerate. These normal modes may be represented as follows:

$$\overset{\rightarrow \quad \leftarrow}{\leftarrow H-C{\equiv}C-H\rightarrow} \qquad \overset{\uparrow \qquad \uparrow}{\underset{\downarrow \qquad \downarrow}{H-C{\equiv}C-H}}$$

$$\tilde{\nu}_1 = 3374 \ cm^{-1} \qquad \tilde{\nu}_4 = 612 \ cm$$

$$\overset{\rightarrow \quad \leftarrow}{H{\rightarrow}-C{\equiv}C-{\leftarrow}H} \qquad \overset{\uparrow \qquad \uparrow}{\underset{\downarrow \quad \downarrow}{H-C{\equiv}C-H}}$$

$$\tilde{\nu}_2 = 1974 \ cm^{-1} \qquad \tilde{\nu}_5 = 729 \ cm^{-1}$$

$$\overset{\leftarrow}{H{\rightarrow}{\leftarrow}C{\equiv}C-H\rightarrow}$$

$$\tilde{\nu}_3 = 3287 \ cm^{-1}$$

(a) Which are the doubly degenerate vibrations? (b) Which vibrations are infrared active? (c) Which vibrations are Raman active?

14.27 When CCl_4 is irradiated with the 435.8 nm mercury line, Raman lines are obtained at 439.9, 444.6, and 450.7 nm. Calculate the Raman frequencies of CCl_4 (expressed in wavenumbers). Also calculate the wavelengths (expressed in µm) in the infrared at which absorption might be expected.

14.28 The first several Raman frequencies of $^{14}N_2$ are 19.908, 27.857, 35.812, 43.762, 51.721, and 59.662 cm^{-1}. These lines are due to pure rotational transitions with $J = $ 1, 2, 3, 4, 5, and 6. The spacing between the lines is $4B_e$. What is the internuclear distance?

14.29 What Raman shifts are expected for the first four Stokes lines for CO_2?

14.30 Energies in electronvolts (eV) may be expressed in terms of temperature by use of the relation $e\phi = kT$, where ϕ is the difference in potential in V. What temperature corresponds to 1 V? 100 V? 1000 V? What is the electronvolt equivalent of room temperature?

14.31 The internuclear distance in CO is 0.112 82 nm. Calculate (a) the reduced mass, and (b) the moment of inertia.

14.32 Caculate the frequencies in cm^{-1} and the wavelengths in micrometers for the pure rotational lines in the spectrum of $H^{35}Cl$ corresponding to the following changes in rotational quantum number: $0 \rightarrow 1$, $1 \rightarrow 2$, $2 \rightarrow 3$, and $8 \rightarrow 9$.

14.33 Assuming that the internuclear distance is 0.0742 nm for (a) H_2, (b) HD, (c) HT and (d) D_2, calculate the moments of inertia of these molecules.

14.34 Calculate the energy difference in cm^{-1} and kJ mol^{-1} between the $J = 0$ and $J = 1$ rotational levels of OH, using the data of Table 14.4. Assuming that OD has the same internuclear distance as OH, calculate the energy difference between $J = 0$ and $J = 1$ in OD⟩.

14.35 Consider the molecular radicals ^{12}CH and ^{13}CH. Calculate their moments of inertia using R_e from Table 14.4 and assuming R_e is the same in both. Using the results of problem 14.9, find the value of J closest to J_{max} at room temperature, and compute the difference in energy between this state and the next higher energy state.

14.36 The separation of the pure rotation lines in the spectrum of CO is 3.86 cm^{-1}. Calculate the equilibrium internuclear separation.

14.37 Show that for large J the frequency of radiation absorbed in exciting a rotational transition is approximately equal to the classical frequency of rotation of the molecule in its initial or final state.

14.38 For the rotational Raman effect, what are the displacements of the successive Stokes lines in terms of the rotational constant B? Is the answer the same for the anti-Stokes lines?

14.39 Show that the moments of inertia of a regular hexagonal molecule made up of six identical atoms of mass m are given by

$$I = 6mr^2 \quad \text{and} \quad I_\perp = 3mr^2$$

where r is the bond distance.

14.40 What are the frequencies of the first three lines in the rotational spectrum of $^{16}O^{12}C^{32}S$ given that the O–C distance is 116.47 pm and the C–S distance is 155.76 pm and that the molecule is linear. Atomic masses of isotopes are given in the back covers. The moment of inertia of a linear molecule ABC is given in problem 14.14.

14.41   What are the rotational frequencies for the first three rotational lines in $^{16}O^{12}C^{34}S$, assuming the same bond lengths as in problem 14.40?

14.42   Ammonia is a symmetric top with

$$I_{xx} = I_{yy} = I_{\perp} = 2.8003 \times 10^{-47} \text{ kg m}^2$$

$$I_{zz} = I_{\parallel} = 4.4300 \times 10^{-47} \text{ kg m}^2$$

Calculate the characteristic rotational temperatures Θ_r where

$$\Theta_r = \frac{h^2}{8\pi^2 I k}$$

14.43   Using the Morse potential expression, equation 14.74, estimate D_e for HBr, HCl, and HI from the data in Table 14.4.

14.44   Calculate the values of D_e for HCl, HBr, and HI using the data of Table 14.4 and equation 14.69 (neglect y_e).

14.45   From the data of Table 14.4, calculate the vibrational force constants of HCl, HBr and HI. Are these in the same order as the dissociation energies?

14.46   Using the Boltzmann distribution (equation 14.62), calculate the ratio of the population of the first vibrational excited state to the population of the ground state for $H^{35}Cl$ ($\omega_0 = 2990 \text{ cm}^{-1}$) and $^{127}I_2(\omega_0 = 213 \text{ cm}^{-1})$ at 300 K.

14.47   Use the Morse potential to estimate the equilibrium dissociation energy for $^{79}Br_2$ using ω_e and $\omega_e x_e$ from Table 14.4.

14.48   The wavenumbers of the first several lines in the R branch of the fundamental ($v = 0 \rightarrow 1$) vibrational band for $^2H^{35}Cl$ have the following frequencies in cm^{-1}: 2101.60 (0), 2111.94 (1), 2122.05 (2), where the numbers in parentheses

are the J values for the initial level. What are the values of B'_v, B''_v, B_e, and α? How does the internuclear distance compare with that for $^1H^{35}Cl$?

14.49   Gaseous HBr has an absorption band centered at about 2645 cm^{-1} consisting of a series of lines approximately equally spaced with an interval of 16.9 cm^{-1}. For gaseous DBr estimate the frequency in wavenumbers of the band center and the interval between lines.

14.50   List the numbers of translational, rotational, and vibrational degrees of freedom of Cl_2, H_2O, and C_2H_2.

14.51   List the numbers of translational, rotational, and vibrational degrees of freedom of NNO (a linear molecule) and NH$_3$.

14.52   The rotational Raman spectrum of hydrogen gas is measured using a 488-nm laser. Stokes lines are observed at 355, 588, 815, and 1033 cm^{-1}. Since these transitions are of the type $J \rightarrow J + 2$, it may be shown that the wavenumbers of these lines are given by

$$\bar{\nu} = 4B_e(J + \tfrac{3}{2})$$

where J is the rotational quantum number of the initial state (0, 1, 2, 3, respectively, for the above lines) and B_e is given by equation 15.86. What is R_e? (L. C. Hoskins, *J. Chem. Educ.* **54**:642 (1977).)

14.53   The rotational Raman spectrum of nitrogen gas shows Raman shifts of 19, 27, 34, 53, . . . cm^{-1}, corresponding with rotational quantum numbers of the initial state of $J = 1, 2, 3, 4, \ldots$. Since the spacing is $4B_e$ ignoring centrifugal distortion, what is R_e? (L. C. Hoskins, *J. Chem. Educ.* **52**:568 (1975).)

14.54   Calculate $\Delta H°(298 \text{ K})$ for the reaction

$$H_2 + D_2 = 2HD$$

assuming that the force constant is the same for all three molecules.

15
Electronic Spectroscopy of Molecules

Transitions between electronic levels of molecules lead to absorption and emission in the visible and ultraviolet parts of the spectrum. For some molecules the energy required to change electronic structure is so great that absorption occurs only in the high-energy vacuum ultraviolet part of the spectrum. Electronic spectra contain many lines because electronic excitation is accompanied by change in vibrational and rotational states as well. When the lines can be resolved, electronic spectra are a rich source of information about molecular properties. At higher pressures in the gas phase the lines are so closely spaced that continuous absorption is obtained. The phenomena of fluorescence and phosphorescence involve electronic changes. Electronic spectra of molecules provide a means for learning about excited electronic states that are involved in photochemical reactions.

The development of lasers has revolutionized many areas of spectroscopy because of extraordinary properties of laser radiation.

15.1 Electronic Energy Levels and Selection Rules

In discussing H_2^+ in Chapter 12, we formed one-electron molecular orbitals from atomic orbitals and then used these to describe the electronic states of the H_2 molecule by putting the electrons into these orbitals in various ways. This provided us with an introduction to molecular term symbols that designate the symmetry, multiplicity, and angular momentum properties of the electronic state. **Electronic spectroscopy** is the excitation of electrons from low-energy orbitals to higher energy orbitals by the absorption of light. Since the energies required to do this are generally much larger than vibrational and rotational energies, this form of spectroscopy uses light in the visible, ultraviolet, and even shorter wavelength parts of the spectrum. In addition, vibrational and rotational energy changes almost always occur during an electronic transition, complicating the spectrum. The analysis of such spectra can give us information about dissociation energies, bond lengths, force constants, and potential energy curves.

We will begin by studying the electronic spectra of diatomic molecules which are the most developed and best understood. In the gas phase at low pressure, the vast majority of diatomic molecules are in the ground electronic state and mainly in the lowest vibrational state. However, there is usually a broad distribution of rotational levels (since the rotational energy spacings are small compared to kT). When light is absorbed by these molecules, the electronic energy level is changed, the vibrational level may change, and the rotational level will change. The selection rules for vibrational level changes are not as stringent as in infrared (pure vibration–rotation) spectroscopy; therefore, there are many absorption lines in the electronic spectrum. At low pressures, these are narrow and we can often analyze the spectrum completely. At high pressures, collisions between molecules reduce the lifetimes of the initial and final states, thereby broadening the absorption spectra. In solution, collisions produce a spectrum that appears smooth and continuous, obscuring the very large number of lines making up the spectrum.

To analyze the spectra of diatomic molecules, we need to know the electric dipole selection rules just as we did for atoms (Section 11.15). For atoms, we saw that the levels were labeled by the angular momentum quantum numbers L, S, J. In diatomic molecules, we have to consider in addition the angular momenta due to molecular rotation and due to nuclear spins. The selection rules for diatomic molecules become complicated and depend on the way the angular momenta are coupled in the molecule. The most common situation (especially for diatomics made up of light atoms in the first several rows of the periodic table) is called Hund's case (a). For this case the **selection rules** are as follows:

1. $\Delta \Lambda = 0, \pm 1$. Λ is the component of orbital angular momentum along the z axis; it can have values $0, 1, 2, \ldots, L$. Since electronic states are designated $\Sigma, \Pi, \Delta, \Phi, \Gamma$ corresponding to $\Lambda = 0, 1, 2, 3, 4, \ldots$, we see that Σ–Σ, Π–Σ, Δ–Π transitions are allowed, but Δ–Σ and Φ–Π are not.

2. $\Delta S = 0$. Thus, as in atomic spectra, singlet–singlet and triplet–triplet transitions may occur, but singlet–triplet transitions are forbidden. This rule breaks down in molecules with highly charged nuclei.

3. $\Delta \Sigma = 0$. The quantum number Σ in a molecule is analogous to M_s in an atom and can take the values $S, S - 1, \ldots, -S$. This quantum number, which determines the multiplicity ($2S + 1$) of a state, is reported as a presuperscript in a molecular term symbol. Thus, dipole-allowed transitions can take place only between states of the same multiplicity.

4. $\Delta \Omega = 0, \pm 1$. The total angular momentum $\Omega \hbar$ along the internuclear axis, which is given by $|\Lambda + \Sigma|$, is sometimes given as a postsuperscript in the molecular term symbol.

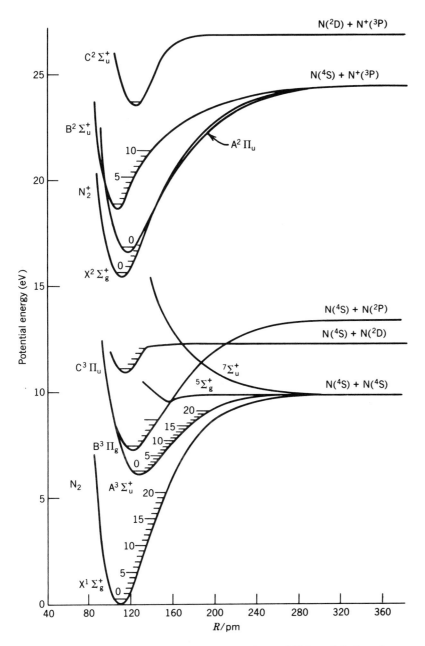

Figure 15.1 Potential energy diagrams for N_2, O_2, and NO and their cations. (From F. R. Gilmore, RAND Corporation Memorandum R-4034-PR, June 1964.)

5. Σ^+–Σ^+ and Σ^-–Σ^- transitions are allowed, but Σ^+–Σ^- are not. These postsuperscripts refer to whether the wavefunction for the electronic state is symmetric $(+)$ or antisymmetric $(-)$ to reflection across any σ_v plane.

6. $g \leftrightarrow u$, $g \nleftrightarrow g$, $u \nleftrightarrow u$. Thus, the only allowed transitions are those involving a change in parity.

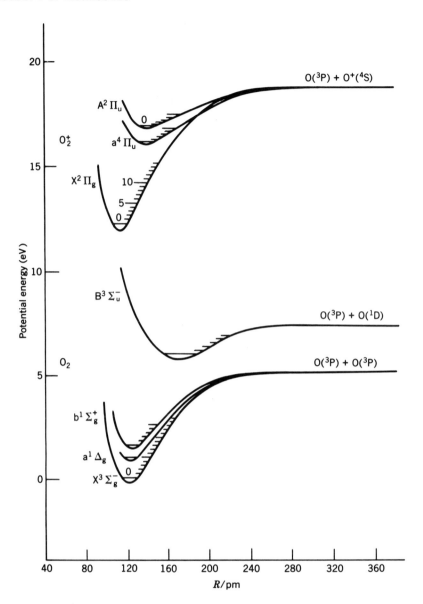

Figure 15.1   (*continued*)

Forbidden transitions may occur, but they generally occur at rates several orders of magnitude slower than permitted transitions. The fact that so-called forbidden transitions do occur is not an indication that quantum mechanics is wrong, but that the approximations in the calculations are not satisfied for the real system. The treatment of a simpler model is often useful, even though the results are approximate.

To illustrate these selection rules we will consider some electronic transitions of the molecules N_2, O_2, and NO. Potential energy diagrams for the ground states and several of the excited electronic states of these molecules and their

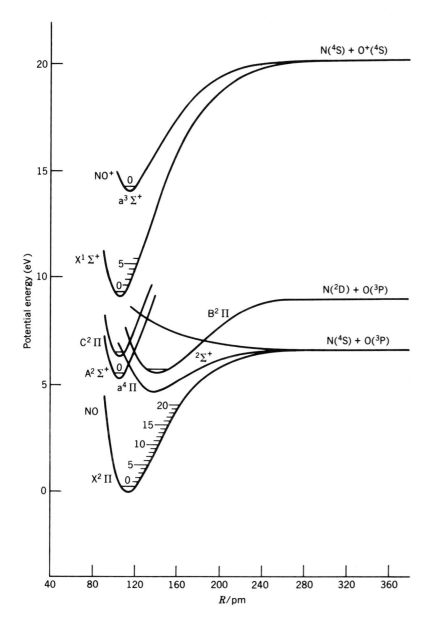

Figure 15.1 (*continued*)

cations are given in Fig. 15.1. The molecular term symbols for these states are given and the atomic term symbols are given for the dissociated atoms.

In addition to the term symbol, electronic states of molecules are also given letter symbols. The ground electronic state is labeled X, and excited states of the same multiplicity are labeled A, B, C, . . . in order of increasing energy. Excited states of different multiplicity are labeled with lowercase letters a, b, c, . . .

Before we can discuss these transitions we have to discuss the manner in which they occur.

Example 15.1

For O_2 and NO, list possible electric dipole transitions from the ground state to excited states of these molecules that are shown.

$$O_2 \quad X^3\Sigma_g^- \rightarrow B^3\Sigma_u^- \quad \text{(Schumann–Runge bands)}$$
$$NO \quad X^2\Pi \rightarrow A^2\Sigma^+$$
$$X^2\Pi \rightarrow B^2\Pi, C^2\Pi$$

Example 15.2

According to Fig. 15.1, what are the changes in internal energy for the reactions

$$(a)\ O_2(X^3\Sigma_g^-) + N(^4S) = NO(X^2\Pi) + O(^3P)$$
$$(b)\ N_2(X^1\Sigma_g^+) + O(^3P) = NO(X^2\Pi) + N(^4S)$$

Reaction (a) is the sum of the reactions

$$O_2(X^3\Sigma_g^-) = O(^3P) + O(^3P) \quad \Delta E = 5.0\ eV$$
$$N(^4S) + O(^3P) = NO(X^2\Pi) \quad \Delta E = -6.5\ eV$$

so that $\Delta E = -1.5\ eV = -145\ kJ\ mol^{-1}$.
Reaction (b) is the sum of the reactions

$$N_2(X^1\Sigma_g^+) = N(^4S) + N(^4S) \quad \Delta E = 9.5\ eV$$
$$O(^3P) + N(^4S) = NO(X^2\Pi) \quad \Delta E = -6.5\ eV$$

so that $\Delta E = 3.0\ eV = 290\ kJ\ mol^{-1}$.

15.2   The Franck–Condon Principle

We will now discuss what happens when a diatomic molecule absorbs a photon and is excited to a higher electronic state. Consider a diatomic molecule in its lowest electronic and vibrational state, in which the most probable internuclear separation is the equilibrium separation. The excited electronic state does *not*, in general, have the same equilibrium internuclear distance. **Franck and Condon recognized that the electronic transition occurs faster than the nuclei can adjust to their new equilibrium position,** so that the most probable position for the nuclei is still the *ground*-state equilibrium position. Thus, an electronic transition can be represented, approximately, by a vertical line, as shown in Fig. 15.2. The vibrational wavefunctions of the upper electronic state will be (approximately) like harmonic oscillator functions (see Section 10.11), so that the largest overlap of probabilities (for the ground- and excited-state vibrations) will occur for vibrational states with quantum numbers greater than 0. This can be seen mathematically from the transition dipole moment. In Chapter 14, we saw that the probability amplitude for a spectroscopic transition is given by

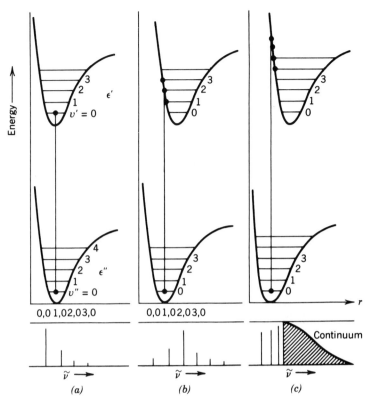

Figure 15.2 Potential energy curves and absorption spectra for three electronic transitions. (From C. N. Banwell, *Fundamentals of Molecular Spectroscopy*. New York: McGraw-Hill. Copyright © 1972. Reproduced with permission.)

$$\boldsymbol{\mu}_{gvJM;ev'J'M'} = \int \psi_g \chi_{gv} \theta_{JM} \boldsymbol{\mu}_e \chi_{ev'} \theta_{J'M'} d\tau_{elec} d\tau_{vib} d\tau_{rot} \qquad (15.1)$$

where ψ_g and ψ_e are the electronic wavefunctions of ground and excited states, respectively, χ_{gv} and $\chi_{ev'}$ are the vibrational wavefunctions of initial and final states, and θ_{JM} and $\theta_{J'M'}$ are the corresponding rotational wavefunctions. We can do the integration over electronic coordinates first, obtaining $\boldsymbol{\mu}_{ge}$. This is nonzero for allowed electronic transitions, and can be taken to be approximately independent of vibrational coordinates (Condon approximation), so that

$$\boldsymbol{\mu}_{gvJM;ev'J'M'} = \left(\int \theta_{JM} \boldsymbol{\mu}_{ge} \theta_{J'M'} d\tau_{rot}\right) \int \chi_{gv} \chi_{ev'} d\tau_{vib} \cdot$$
$$= \left(\int \theta_{JM} \boldsymbol{\mu}_{ge} \theta_{J'M'} d\tau_{rot}\right) S_{vv'} \qquad (15.2)$$

The integral of the vibrational wavefunctions is called a **Franck–Condon overlap integral** $S_{vv'}$. The intensity of a transition is proportional to the square of the transition moment, and therefore to $S_{vv'}^2$. This leads to the fact that the absorption probability is largest into that vibrational state of the excited electronic states whose probability is largest directly above the equilibrium internuclear distance. In the three cases illustrated in Fig. 15.2, the potential energy curve for the excited electronic state lies directly above the ground state in (*a*), and is displaced to larger internuclear separations in (*b*) and (*c*). In Fig. 15.2*a*, the

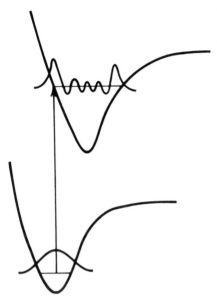

Franck–Condon overlap is largest between $v = 0$ and $v' = 0$. In Fig. 15.2b, the overlap is largest between $v = 0$ and $v' = 2$, so that the line is most intense. Note that this Franck–Condon overlap is greatest because the vibrational wavefunctions for $v > 1$ tend to pile up at the classical turning points (i.e., the potential walls), as shown in Fig. 15.3. There will be intensity in other transitions as well, since the overlap for these is not zero.

In Fig. 15.2c there is a significant probability of excitation to an energy above the dissociation energy of the excited molecule. Since energy levels above this energy are not quantized, continuous absorption occurs. The borderline between absorption lines and continuous absorption is referred to as the **convergence limit.** It, of course, gives the difference in energy between the vibrational ground state of the lower electronic state and the dissociation products of the upper electronic state. Cases b and c are more often encountered than case a because the bonding is generally weaker in the excited electronic state. Since the bonding is weaker, the equilibrium internuclear distance is greater in the excited state.

Figure 15.3 Franck–Condon principle. In an electronic transition, the overlap of the ground vibrational wavefunction in the lower electronic state and the various vibrational wavefunctions in the upper electronic state is greatest for the vibrational level whose classical turning point is at the equilibrium separation in the lower state.

15.3 Determination of Dissociation Energies

The determination of dissociation energies of diatomic molecules from vibrational spectra was referred to in Section 14.8. However, it is virtually impossible to obtain accurate values in this way because of the generally long extrapolation involved.

If the onset between discrete lines and the continuum in Fig. 15.2c is sharp, the dissociation energies D_0'' and D_0' of the ground state and the excited state can be determined quite accurately, as shown in Fig. 15.4. This figure shows that

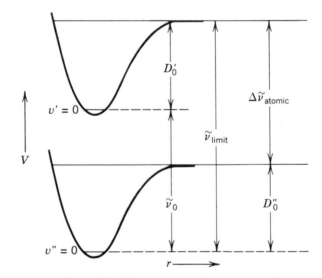

Figure 15.4 Determination of the dissociation energies D_0'' and D_0' of the ground state and the excited state from $\tilde{\nu}_{\text{limit}}$, the wavenumber of the onset of a continuum in a progression.

$$\bar{\nu}_{\text{limit}} = D_0' + \bar{\nu}_0 = D_0'' + \Delta\bar{\nu}_{\text{atomic}} \qquad (15.3)$$

where $\bar{\nu}_{\text{limit}}$ is the wavenumber of the onset of the continuum in the progression and where the dissociation energies are expressed in wavenumbers. If the states of the atoms produced in the dissociation of the ground state and in the dissociation of the excited state are known, $\Delta\bar{\nu}_{\text{atomic}}$ is known, and D_0'' can be calculated from $\bar{\nu}_{\text{limit}}$.

If the wavenumber of the 0–0 band can be obtained from the vibrational spectrum, the dissociation energy D_0' for the excited state can be calculated from $\bar{\nu}_{\text{limit}}$ using equation 15.3. The dissociation energies relative to the minima in the potential energy curves are given by

$$D_e = D_0 + \frac{\bar{\nu}_e}{2} - \frac{\bar{\nu}_e x_e}{4} + \frac{\bar{\nu}_e y_e}{8} + \cdots \qquad (15.4)$$

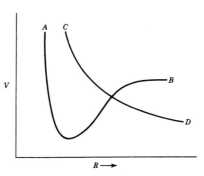

Figure 15.5 Predissociation can result when two potential curves cross in this way.

as may be seen from equation 14.69.

If the potential energy curve for a diatomic molecule is crossed by the curve for a repulsive excited state, as shown in Fig. 15.5, a region of diffuseness is found in the spectrum. When the internuclear distance and energy of the molecule in the ground state are near the crossing point, there is a probability that the molecule will transfer from curve AB to curve CD and dissociate. This has the effect of broadening the vibrational and rotational levels in this region of the spectrum. This effect is referred to as **predissociation** because the molecule with the potential energy curve AB can dissociate at a lower energy when the repulsive curve CD crosses it.

15.4 Spectrophotometers and the Beer–Lambert Law

In considering rotational and vibrational spectra we were concerned primarily with the frequencies of the lines and secondarily with their relative intensities. However, in considering electronic spectra we will be increasingly concerned with the intensity of absorption or emission. The intensity of absorption at a particular wavelength can be determined by passing a monochromatic beam of light through a sample of known thickness and concentration and measuring the intensity I of the transmitted light relative to the intensity I_0 that would be transmitted in the absence of the absorbing substance. **The intensity I of a beam of light is defined as the energy per unit area per unit time.**

The construction of a spectrophotometer is indicated schematically in Fig. 15.6. The principal parts are the source of electromagnetic radiation, the mono-

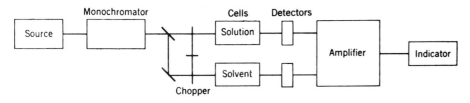

Figure 15.6 Schematic diagram for a simple UV-visible spectrometer. A laser can be used in place of the source and monochromator.

chromator, the cell compartment, the photoelectric detector, and a device for indicating the output from the detector (electric meter, potentiometer, or recording potentiometer). The cell compartment contains an optical absorption cell filled with the solution to be studied and another absorption cell filled with a reference solution, usually pure solvent. The ratio of the intensity I of transmitted light for the solution to the intensity I_0 for the solvent is called the **transmittance.**

The transmittance I/I_0 can be determined at different wavelengths, and the absorption spectrum can be mapped. With some spectrophotometers such a plot is recorded automatically. The positions and intensities of the absorption bands and lines serve for identification and for criteria of purity; the transmittance serves for quantitative analysis of the concentration of material present.

Lambert developed the equation for the attenuation of a light beam as a function of the thickness of a homogeneous medium. Beer developed the equation for the effect of concentration. The Beer–Lambert law may be derived as follows.

The probability that a photon will be absorbed is usually directly proportional to the concentration of absorbing molecules and to the thickness of the sample for a very thin sample. This probability is expressed mathematically by the equation

$$\frac{dI}{I} = -\kappa c \, dx \tag{15.5}$$

where I is the intensity of light of a particular wavelength, that is, energy per unit area per unit time, dI is the change in light intensity produced by absorption in a thin layer of thickness dx and concentration c, and κ is the naperian molar absorption coefficient. Distance x is measured through the cell in the direction of the beam of light that is being absorbed. The concentration c is usually expressed in mol L^{-1}.

The intensity of a beam of light after passing through length l of solution is related to the incident intensity I_0 by equation 15.7, which is obtained by integrating equation 15.5 between the limits I_0 when $x = 0$ and I when $x = l$:

$$\int_{I_0}^{I} \frac{dI}{I} = -\kappa c \int_0^l dx \tag{15.6}$$

$$\ln \frac{I}{I_0} = 2.303 \log \frac{I}{I_0} = -\kappa c l \tag{15.7}$$

Since it is convenient to use logarithms to the base 10, the Beer–Lambert law is used in the form

$$\log \frac{I_0}{I} = A = \epsilon c l \tag{15.8}$$

where the quantity $\log (I_0/I)$ is referred to as the **absorbance** A, and ϵ is referred to as the **molar absorption coefficient** or molar absorptivity. It can be seen from equation 15.8 that the absorbance is directly proportional to the concentration c and to the path length l. The proportionality constant is characteristic of the solute and depends on the wavelength of the light, the solvent, and the temperature. Since the molar absorption coefficient ϵ depends on wavelength, the

Beer–Lambert law is obeyed at each wavelength. If the radiation is not monochromatic the Beer–Lambert law may not be obeyed. The apparent absorption coefficient for a substance that associates or dissociates in solution will change with concentration because of the changing ratio of concentrations of the absorbing species.

Figure 15.7 shows the absorption spectra of benzene and paraxylene.*

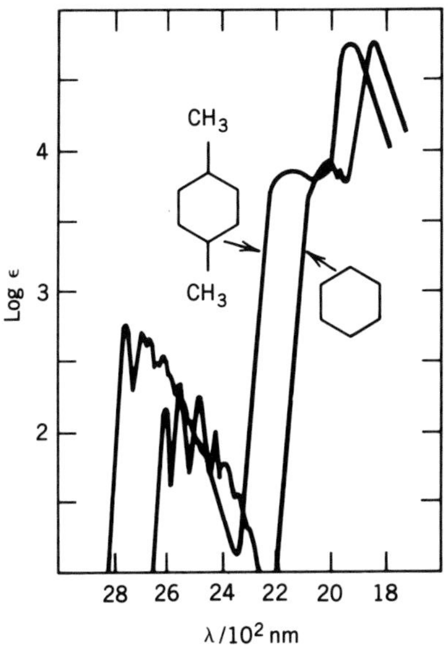

Figure 15.7 Electronic absorption spectra of benzene and paraxylene in solution.

Example 15.3

The percentage transmittance of an aqueous solution of disodium fumarate at 250 nm and 25 °C is 19.2% for a 5×10^{-4} mol L^{-1} solution in a 1-cm cell. Calculate the absorbance A and the molar absorption coefficient ϵ.

$$A = \log \frac{I_0}{I} = \log \frac{100}{19.2} = 0.717$$

$$\epsilon = \frac{A}{lc} = \frac{0.717}{(1\ \text{cm})(5 \times 10^{-4}\ \text{mol L}^{-1})} = 1.43 \times 10^3\ \text{L mol}^{-1}\ \text{cm}^{-1}$$

What will be the percentage transmittance of a 1.75×10^{-5} mol L^{-1} solution in a 10-cm cell?

$$\log \frac{I_0}{I} = (1.43 \times 10^3\ \text{L mol}^{-1}\ \text{cm}^{-1})(10\ \text{cm})(1.75 \times 10^{-5}\ \text{mol L}^{-1}) = 0.251$$

$$\frac{I_0}{I} = 1.782 \quad \text{and} \quad \frac{100I}{I_0} = 56.1\%$$

For mixtures of independently absorbing substances the absorbance is given by

$$\log \frac{I_0}{I} = A = (\epsilon_1 c_1 + \epsilon_2 c_2 + \cdots)l \tag{15.9}$$

where $c_1, c_2, \ldots$ are the concentrations of the substances having molar absorption coefficients of $\epsilon_1, \epsilon_2, \ldots$. A mixture of n components may be analyzed by measuring A at n wavelengths at which the molar absorption coefficients are known for each substance, provided that these coefficients are sufficiently different. The concentrations of the several substances may then be obtained by solving the n simultaneous linear equations.

When a sample is irradiated continuously at low intensity, its absorption coefficient remains constant in the absence of chemical reaction, and this indicates that excited molecules are continuously deactivated so that they do not accumulate. Usually the excitation energy is simply degraded to thermal energy in molecular collisions, but a chemical reaction may occur and change the composition and absorption spectrum of the sample (cf. Chapter 20). An excited molecule may also emit a quantum of radiation. Such emission is referred to as fluorescence or phosphorescence, depending on the difference in excited- and ground-state spin multiplicities (Section 15.8).

---

* P. E. Stevenson, *J. Chem. Educ.* **41**:234 (1964).

The Beer–Lambert law may be written in alternative ways. In a given situation one form may be more convenient to use than another. We will use all of the following forms:

$$I = I_0 10^{-\epsilon c l} \tag{15.10}$$

$$I = I_0 e^{-\kappa c' l'} \tag{15.11}$$

$$I = I_0 e^{-\sigma N x} \tag{15.12}$$

In the first equation, c is expressed in mol L^{-1} and l is expressed in cm, so that ϵ has the units L mol^{-1} cm^{-1}, as we have seen. In the second equation we will express c' in SI base units so that c' is in mol m^{-3}, l' in m, and κ in m^2 mol^{-1}. In the third equation, N is in m^{-3} and x in m. Since σ has the units m^2 it is referred to as the **absorption cross section**. Although σ has the units of area, it is not to be interpreted literally as the area of an absorbing molecule.

Example 15.4

The molar absorbancy index ϵ of a solute is 44 000 L mol^{-1} cm^{-1}. What is the absorption cross section σ? Comparison of equations 15.10 and 15.12 yields

$$\sigma = \frac{2.303\ \epsilon c l}{Nx}$$

Since N is the number of molecules per m^3 and c is the amount per liter,

$$c = \frac{N}{N_A(10^3\ \text{L m}^{-3})}$$

Since x is in meters and l is in centimeters,

$$l = x(10^2\ \text{cm m}^{-1})$$

so that

$$\sigma = \frac{2.303\epsilon(10^2\ \text{cm m}^{-1})}{N_A(10^3\ \text{L m}^{-3})}$$

If ϵ = 44 000 L mol^{-1} cm^{-1}, then

$$\sigma = \frac{2.303(44\ 000\ \text{L mol}^{-1}\ \text{cm}^{-1})(10^2\ \text{cm m}^{-1})}{(6.022 \times 10^{23}\ \text{mol}^{-1})(10^3\ \text{L m}^{-3})}$$

$$= 1.7 \times 10^{-20}\ \text{m}^2$$

Since an electronic absorption band contains many lines that may not be resolved, the intensity is not as accurately measured by the maximum absorption (as represented by the maximum absorbancy index ϵ_{max}) as by the integral over the entire band. The integrated absorption coefficient is defined by

$$\int_{\text{band}} \epsilon\ d\bar{\nu} \tag{15.13}$$

Thus, it has the units L mol^{-1} cm^{-2}.

If the absorption band is Gaussian in shape, then the integrated intensity can be related to the molar absorbancy index ϵ_{max} and the width $\Delta\tilde{\nu}_{1/2}$ at half the maximum absorbancy index:

$$\int_{band} \epsilon \, d\tilde{\nu} = 1.06\epsilon_{max}\Delta\tilde{\nu}_{1/2} \qquad (15.14)$$

For relatively strong absorptions of molecules $\epsilon_{max} = 10^4$ to 10^5 L mol^{-1} cm^{-1}, and $\Delta\tilde{\nu}_{1/2}$ is of the order of 1000 to 5000 cm^{-1}. For weak absorptions $\epsilon_{max} = 10$ L mol^{-1} cm^{-1}, and $\Delta\tilde{\nu}_{1/2}$ is of the order of 100 cm^{-1}. Extremely weak (forbidden) absorptions may have ϵ_{max} of the order of 10^{-3} to 10^{-4} L mol^{-1} cm^{-1}.

Example 15.5
The maximum value of an observed absorbancy index is 44 000 L mol^{-1} cm^{-1} at 30 000 cm^{-1}. If the width of the band at half maximum is 5000 cm^{-1}, what is the value of the integrated absorption coefficient?
    Assuming that the band is Gaussian

$$\int \epsilon \, d\tilde{\nu} = 1.06\epsilon_{max}\Delta\tilde{\nu}_{1/2}$$
$$= 1.06(44\ 000 \text{ L mol}^{-1}\text{ cm}^{-1})(5000 \text{ cm}^{-1})$$
$$= 2.33 \times 10^8 \text{ L mol}^{-1}\text{ cm}^{-2}$$

15.5 Oscillator Strength

The concept of oscillator strength f has been developed to provide a theoretical reference for the intensity of a spectroscopic transition. **The oscillator strength is the ratio of the strength of a transition to the strength of a transition for an electron oscillating harmonically in three dimensions.** The oscillator strength for an actual transition may be calculated from the measured integrated absorption coefficient of the absorption band. Allowed electric dipole transitions yield oscillator strengths of approximately unity. Forbidden transitions have oscillator strengths much less than unity. Singlet–triplet transitions typically have oscillator strengths of the order of 10^{-5}.
    It may be shown that the integrated absorption coefficient is given by

$$\int \kappa \, d\nu = \frac{8\pi^3 N_A \nu}{3hc(4\pi\epsilon_0)} |\, \mu_{21}\,|^2 \qquad (15.15)$$

where μ_{12} is the transition dipole moment given by equation 14.6 and ν is the average frequency for the absorption band. In this equation we are using an integrated absorption coefficient expressed in terms of κ in m^2 mol^{-1} and frequency rather than $\int \epsilon \, d\tilde{\nu}$. It is more convenient to do derivations with quantities in SI base units and convert later, as we will.
    The square of the transition moment for the three-dimensional harmonic oscillator for the transition from the ground state to the first excited state is given by

$$\mid \boldsymbol{\mu}_{12} \mid^2_{osc} = \frac{3he^2}{8\pi^2 m_e \nu} \tag{15.16}$$

We define the oscillator strength f of a transition with transition moment $\boldsymbol{\mu}_{12}$ as

$$f = \frac{\mid \boldsymbol{\mu}_{12} \mid^2}{\mid \boldsymbol{\mu}_{12} \mid^2_{osc}} \tag{15.17}$$

Substituting from equation 15.15 for $\mid \boldsymbol{\mu}_{12} \mid^2$ and from equation 15.16 for $\mid \boldsymbol{\mu}_{12} \mid^2_{osc}$, we find

$$f = \frac{4m_e c\, \epsilon_0}{N_A e^2} \int \kappa\, d\nu \tag{15.18}$$

This is the experimentally measured oscillator strength. To calculate the oscillator strength, which is dimensionless, all the quantities on the right side need to be expressed in SI base units. However, the integrated absorption coefficient is more often given in L mol^{-1} cm^{-2}, as we have seen in Example 15.5, and so some conversion factors must be included in equation 15.18 in order to calculate the oscillator strength from the integrated absorption coefficient expressed in its usual units. If a single electron can undergo more than one transition, the sum of the oscillator strengths for all the transitions arising from any one level to all other levels is unity:

$$\sum_i f_i = 1 \tag{15.19}$$

Example 15.6

Show that the oscillator strength is given by the following equation when the integrated absorption coefficient (equation 15.12) is expressed in L mol^{-1} cm^{-2}:

$$f = (4.32 \times 10^{-9} \text{ L}^{-1} \text{ mol cm}^2) \int \epsilon\, d\bar{\nu}$$

To convert $\int \kappa\, d\nu$, expressed in SI base units, to $\int \epsilon\, d\bar{\nu}$, expressed in SI base units, we need to multiply by $2.303c$, where c is the velocity of light. However, since ϵ is usually expressed in L mol^{-1} cm^{-1} we need to multiply $\int \epsilon\, d\bar{\nu}$ by 10^3 cm^3 L^{-1} to obtain the value in cm mol^{-1}. To complete the conversion to SI base units a further factor 10^{-2} m cm^{-1} is required. Therefore,

$$f = \frac{4m_e c^2 \epsilon_0}{N_A e^2}\, 2.303(10^3 \text{ cm}^3 \text{ L}^{-1})(10^{-2} \text{ m cm}^{-1}) \int \epsilon\, d\bar{\nu}$$

Example 15.7

What is the oscillator strength f of the solute of Example 15.5?

$$f = (4.32 \times 10^{-9} \text{ L}^{-1} \text{ mol cm}^2) \int \epsilon\, d\bar{\nu}$$
$$= (4.32 \times 10^{-9} \text{ L}^{-1} \text{ mol cm}^2)(2.33 \times 10^8 \text{ L mol}^{-1} \text{ cm}^{-2})$$
$$= 1.01$$

There is a relationship between the integrated absorption coefficient and the radiative lifetime τ for a transition. This relationship goes back to the relation between B_{21} and A_{12} (equation 14.13) since $A_{12} = 1/\tau$ for a single transition. Rather than giving the derivation here, we will simply refer to the fact that for the solute described in Example 15.7, the theoretical radiative lifetime is 1.7×10^{-9} s. The radiative lifetime is inversely proportional to the integrated absorption coefficient, and so for the conditions in Example 15.7, a maximum molar absorption coefficient that was tenfold smaller would yield a relaxation time tenfold larger.

15.6 Electronic Spectra of Polyatomic Molecules

The absorption spectra of polyatomic molecules lie mainly in the visible and ultraviolet regions, involving excitation of electrons from the higher energy filled orbitals to the lower energy unfilled orbitals of the molecule. In many unsaturated organic molecules and in inorganic transition metal complexes, the difference in energy between occupied and unoccupied orbitals is small enough so that the absorption of light occurs in the visible. The absorption spectra of saturated organic molecules (e.g., C_2H_6) lie in the vacuum ultraviolet, since it takes a large amount of energy to promote an electron from an occupied orbital (in a single bond) to an unoccupied orbital.

In organic compounds with oxygen, nitrogen, or halogen atoms, there are often filled **nonbonding** molecular orbitals usually associated with the lone pairs. Since these do not participate in the bonding, they may lie relatively high in energy. Excitation of an electron from such an orbital to an unfilled (anti-bonding) π^* orbital leads to absorption in the ultraviolet and visible regions. Thus, groups such as C=O, –N=N–, and –N=O which cause such absorption at wavelengths longer than 180 nm are called **chromophores.** These groups have characteristic absorption wavelengths. For example, C=O groups have a strong absorption band at 120 nm due to $\pi \rightarrow \pi^*$ transitions and a weaker band from 270 to 290 nm due to $n \rightarrow \pi^*$ transitions, where the notation n is used for the nonbonding orbitals. Although the structure of the rest of the molecule affects the details of the spectrum, most compounds with the same chromophore will have the same qualitative ultraviolet spectrum.

Unsaturated molecules containing one C=C bond will have a strong absorption band around 180–190 nm. If there are other unsaturated parts of the molecule, such as other C=C or C=O groups, which are well separated from each other, then the absorption spectrum is the sum of the separate spectra. But, if these other unsaturated groups are conjugated with each other (i.e., separated by only one single bond as in butadiene, $H_2C=CH–CH=CH_2$), then large non-additive effects can occur. In this case, the π orbitals are spread over many atoms reducing the orbital energy and leading to strong absorption in the visible and near ultraviolet. So, single-ring aromatics absorb near 250 nm, naphthalene near 300 nm, and anthracenes and phenanthrenes near 360 nm. Almost all dyes we know contain large conjugated organic systems that absorb strongly in the visible. We will discuss these molecules further in the next section.

The other type of polyatomic molecules that have strong absorptions in the visible are transition metal complexes. These are known for their beautiful colors which arise because of these electronic transitions. Here the low-lying

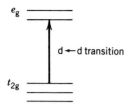

Figure 15.8   d orbital states in an octahedral complex and the meaning of d ← d transitions.

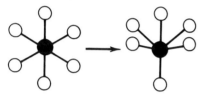

Figure 15.9   Vibrational motion or distortion in an octahedral complex that leads to the destruction of the center of symmetry, thereby making d ← d transitions allowed.

unfilled orbitals are atomic d orbitals from the transition metal. The presence of the ligands splits the five d orbitals into two groups at different energies in octahedral or tetrahedral complexes, as shown in Fig. 15.8. Electrons in orbitals on the ligands can then be excited into an unfilled d orbital on the metal giving a charge transfer transition that is usually very intense. In addition, electrons can be excited from one set of d orbitals to the other, also giving rise to absorption in the visible. These latter transitions are weaker than the charge transfer bands because d → d transitions are forbidden unless some perturbation occurs, such as a distortion of the octahedron (as shown in Fig. 15.9) to a lower symmetry with no inversion center of symmetry.

15.7   Conjugated Molecules: Free-Electron Model

For molecules with conjugated systems of double bonds (i.e., R(CH=CH)$_n$R′) it is found that the electronic absorption bands shift to longer wavelengths as the number of conjugated double bonds is increased.* Approximate quantitative calculations of the absorption frequencies may be made on the basis of the free-electron model for the π electrons of these molecules. The energy for the lowest electronic transition is that required to raise an electron from the highest filled level to the lowest unfilled level. In a system of conjugated double bonds each carbon atom has three σ bonds that lie in a plane, and each σ bond involves one outer electron of that carbon atom. Above and below this plane are the π orbital systems.

Each carbon atom contributes one electron to this π system, but these electrons are free to move the entire length of the series of π orbitals and are not localized at a given carbon atom. In the free-electron model it is assumed that the π system is a region of uniform potential and that the potential energy rises sharply to infinity at the ends of the system (i.e., a square-well potential). Thus, the energy levels E available to the π electrons would be expected to be those calculated for the particle in a one-dimensional box (Section 10.8):

$$E = \frac{n^2 h^2}{8 m_e a^2} \tag{15.20}$$

The length of the box a is usually taken to be the length of the chain between terminal carbon atoms plus a bond length or two.

The π electrons (one for each carbon atom) are assigned to orbitals so that there are two (one with spin $+\frac{1}{2}$ and the other with spin $-\frac{1}{2}$) in each level, starting with the lowest. For a completely conjugated hydrocarbon the number of π electrons is even, and the quantum number of the highest filled level will be $n = N/2$, where N is the number of π electrons (the number of carbon atoms involved). In absorption an electron from the highest filled level is excited to the next higher level with quantum number $n' = N/2 + 1$. The difference in energy of these two levels is

---

* In conjugated molecules double and single bonds alternate.

$$\Delta E = \frac{h^2}{8m_e a^2}(n'^2 - n^2) = \frac{h^2}{8m_e a^2}\left[\left(\frac{N}{2} + 1\right)^2 - \left(\frac{N}{2}\right)^2\right] = \frac{h^2}{8m_e a^2}(N + 1)$$

(15.21)

The absorption frequency in wavenumbers is given by

$$\tilde{\nu} = \frac{\Delta E}{hc} = \frac{h(N + 1)}{8cm_e a^2}$$

(15.22)

For linear molecules, the size of the system a is proportional to N, so that the absorption frequency will vary as $1/N$ for large N.

Example 15.8

Calculate the lowest absorption frequency for octatetraene (C_8H_{10}) that contains a series of four conjugated double bonds. The length of the π bond system is about 0.95 nm.

$$\tilde{\nu} = \frac{h(N + 1)}{8cm_e a^2} = \frac{(6.62 \times 10^{-34} \text{ J s})(9)(10^{-2} \text{ m cm}^{-1})}{8(3 \times 10^8 \text{ m s}^{-1})(9.110 \times 10^{-31} \text{ kg})(0.95 \times 10^{-9} \text{ m})^2}$$

$$= 30\ 200 \text{ cm}^{-1}$$

The observed absorption band is at 33 100 cm^{-1}.

15.8 Fluorescence and Phosphorescence

In **fluorescence,** the radiation is emitted during a transition between electronic states of the same spin or multiplicity, whereas in **phosphorescence** the radiation is emitted in a transition between electronic states of different multiplicities, for example, between a triplet and a singlet state. Since the latter are (approximately) forbidden, the rate is low and therefore the lifetime of the lowest triplet state of a molecule (with singlet ground state) is long (10^{-4} to 100 s). On the other hand, the lifetime of excited singlet states is usually between 10^{-6} and 10^{-9} s.

When a ground-state (S_0) molecule is excited to the first excited singlet state, S_1 (see Fig. 15.10), a number of processes can occur. Since the molecule is generally in an excited vibrational state, collisions with other molecules in the gas or solution can remove vibrational energy from the excited molecule in a process called vibrational relaxation. Thus, the excited molecule ends up in the lowest vibrational state of S_1. Now, when the excited molecule radiates (fluorescence), the frequency is lower than that of the exciting radiation, as shown in Fig. 15.10.

Before a molecule can fluoresce, other processes can occur. The molecule can chemically react, as discussed in Chapter 20, or the molecule can lose its energy in a collision with another molecule, and it may make a transition to another excited electronic state. Such a nonradiative transition from one singlet state to another singlet state, or more generally between states of the same multiplicity, is called **internal conversion,** while nonradiative transitions between states of different multiplicities is called **intersystem crossing.** Since in-

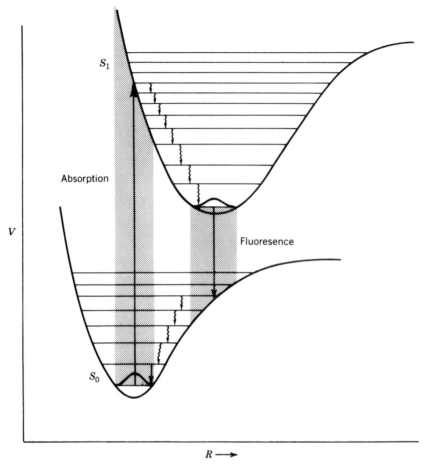

Figure 15.10 Absorption and fluorescence by a diatomic molecule. The shaded areas indicate the transitions to a range of vibrational quantum numbers.

tersystem crossing requires an electron spin flip, it is significantly slower (10^{-2} to 10^{-6}) than internal conversion. Various factors influence the rates of non-radiative transitions, including the difference in energies between the two electronic states involved.

If intersystem crossing occurs, the molecule now can undergo further vibrational relaxation, as shown in Fig. 15.11, finally ending up in the lowest vibrational state of the excited triplet (T_1). The molecule can now undergo collisional loss of energy, or intersystem crossing to the ground state S_0, or can emit a photon in phosphorescence. Triplet-state molecules are especially likely to be involved in a chemical reaction because of their high energy and long lifetime.

The energy of a triplet state is usually lower than the energy of the excited singlet state with the same molecular orbital occupancy because in a triplet state the electrons having the same spin tend to avoid each other (because of the Pauli principle) by staying in different regions. Since the electrons are

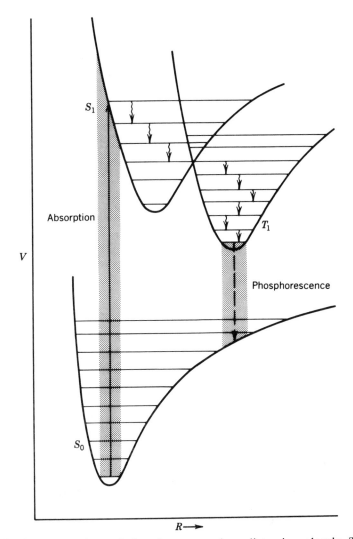

Figure 15.11   Absorption and phosphorescence by a diatomic molecule. The shaded areas indicate transitions to a range of vibrational quantum numbers.

farther apart, there is a decrease in electronic repulsion and the energy of the molecule is lower. Consequently, phosphorescence usually occurs at a lower frequency than fluorescence, as shown in Fig. 15.12. This figure shows that the fluorescence spectrum has an approximate "mirror image" relationship to the absorption spectrum, if the spacings of the vibrational levels in the S_0 and S_1 states are similar.

The transition from the $v = 0$ vibrational level of the upper electronic state to the $v = 0$ vibrational level of the lower electronic state is called the 0–0 band. The small difference in the 0–0 bands for absorption and fluorescence is due to the difference in the solvation of the initial and final states in the two cases. As shown in Fig. 15.13, the excited state does not become equilibrated

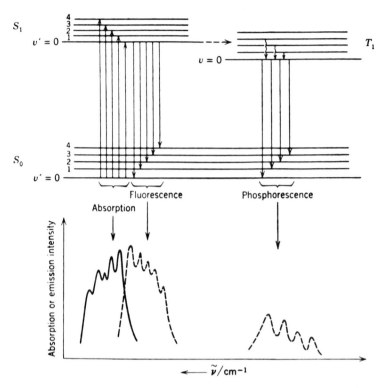

Figure 15.12   Schematic diagram showing absorption, fluorescence, and phosphorescence spectra. (From E. F. H. Brittain, W. O. George, and C. H. J. Wells, *Introduction to Molecular Spectroscopy*. New York: Academic, 1970.)

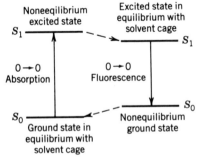

Figure 15.13   Effect of solute–solvent equilibrium in causing a difference in the energy of the 0–0 absorption transition and the 0–0 fluorescence transition. (From C. H. J. Wells, *Introduction to Molecular Photochemistry*. London: Chapman & Hall, 1972. © C. H. J. Wells, 1972.)

with the solvent in the absorption process, and the ground state does not become equilibrated with the solvent in the fluorescence process. As indicated in the diagram, the 0–0 fluorescence transition will be of lower energy than the 0–0 absorption transition.

The above discussion referred to a single excited singlet state and a single excited triplet state, but a molecule will have many singlets and triplets (and higher multiplicities). The higher excited states tend to relax by various processes very quickly (10^{-4}–10^{-12} s) to the lowest excited singlet or triplet, and so can be probed only by using ultrafast spectroscopic techniques.

Example 15.9

Use Fig. 15.1 to calculate the longest wavelength of light that can be used to excite $NO(^2\Pi)$ to an electronically excited state that can fluoresce back to the ground state.

The transition

$$NO(^2\Pi) \rightarrow NO(^2\Sigma)$$

satisfies the selection rules for absorption. The excited state can fluoresce back to the ground state. The energy required for a vertical transition is 6.0 eV, which is equivalent to $(6.0\ \text{eV})(8066\ \text{cm}^{-1}\ \text{eV}^{-1}) = 48 \times 10^3\ \text{cm}^{-1}$ or 2.08×10^{-5} cm or 208 nm.

15.9 Lasers

Normally a beam of light loses intensity as it passes through an absorbing material. However, if molecules are present in an excited state, stimulated emission (Section 14.3) can occur, and the light beam can gain intensity. A laser achieves this condition and produces an intense and coherent beam. By coherent we mean that the light waves are in phase. The name laser comes from **light amplification** by **stimulated emission of radiation.**

When a two-state system (Fig. 14.1) is irradiated, the rate of change in the population of state 2 is equal to the difference between the rate of absorption and the rate of emission (Fig. 14.1):

$$\frac{dN_2}{dt} = BN_1\rho(\bar{\nu}_{21}) - A_{12}N_2 - BN_2\rho(\bar{\nu}_{21})$$

$$= B\rho(\bar{\nu}_{21})(N_1 - N_2) - A_{12}N_2 \qquad (15.23)$$

where the subscripts have been left off of the B's because of equation 14.12. Let us consider the case that spontaneous emission is negligible so that we can ignore the last term in equation 15.23. We can see that as long as $N_1 > N_2$ there will be absorption. However, as irradiation is continued N_2 will approach N_1, and the rate of absorption of radiation will decrease to zero. The system will now be transparent because the net absorption of light of frequency $\bar{\nu}_{21}$ is zero, and the transition is said to be **saturated.** If somehow $N_2 > N_1$, a situation known as population inversion, then dN_2/dt will be negative which means that the number of emitted photons increases, and that the intensity of radiation in the direction of the incident radiation will increase. In other words, there will be amplification. It is this amplification that creates the coherent beam of a laser.

Laser action requires a population inversion, but we have seen that we cannot get a population inversion by irradiating a two-level system. However, population inversion can be obtained in multilevel systems such as those in Fig. 15.14. In Fig. 15.14a the system is raised to level E_3 by the absorption of radiation (pump) and then undergoes a rapid nonradiative transition to level E_2. If the system can be pumped hard enough so that $N_2 > N_1$, then laser action can be obtained on $E_2 \rightarrow E_1$.

Figure 15.14b shows a more suitable system because laser action depends on $N_3 > N_2$. This population inversion is readily achieved because N_2 is essentially zero initially and will remain so if the $E_2 \rightarrow E_1$ nonradiative process is fast. A number of lasers depend on four-level systems such as that in Figs. 15.14c and d. Here the lasing action occurs between two levels, neither of which is involved in the pumping process.

Usually a laser cavity has mirrors at each end to increase the radiation density which stimulates emission. In Fig. 15.15 the two ends of the ruby rod are mirrors. The silver mirror at the back end is nearly 100% reflecting and the silver mirror at the front end has a reflectivity that is less than 100%. The gain of the system is the amount of amplification, usually per round trip through the cavity, for a given population inversion. For a given system some minimum gain is required to overcome the optical losses in the cavity and the energy emission of the laser. Figure 15.15 shows the construction of a ruby laser that is powered by a flash lamp. The output beam is emitted through the partially silvered end of the ruby crystal.

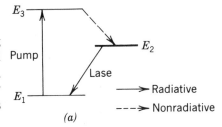

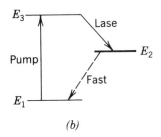

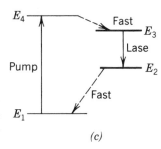

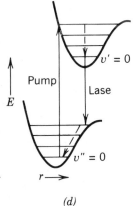

Figure 15.14 Multilevel lasing schemes. (*a*) and (*b*) are three-level systems, and (*c*) and (*d*) are four-level systems represented in different ways. (From W. F. Coleman, *J. Chem. Educ.* **59**:441 (1982).)

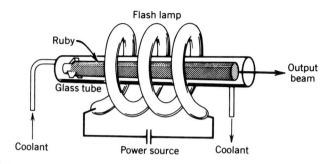

Figure 15.15 Solid-state laser pumped by irradiating a ruby crystal.

The length d of the resonant cavity created by the two mirrors is critical. Standing waves are formed in the cavity so that there is an integral number of waves in the cavity. The electric field of the standing wave in the cavity is zero at the surface of the two mirrors. Since $\lambda = 2d/n$, where n is an integer, the frequency of the standing waves is given by

$$\nu = \frac{nc}{2d} \tag{15.24}$$

where c is the velocity of light.

Lasers may be operated in a **continuous wave** (cw) mode or a **pulse mode.** To operate in cw mode a laser has to be pumped steadily by irradiation with a light source or by electrical discharge at a rate sufficient to supply the emergent laser radiation and overcome any losses. In the pulse mode the laser is excited with a pulsed electric discharge or pulsed lamp and some kind of shutter is used to control the release of energy stored in the laser so that an intense pulse is obtained. The shutter may be mechanical or electro-optical, but the simplest method at visible wavelengths is to use a saturable dye. A cell containing a dye solution with a peak absorption at the laser wavelength is located in the laser cavity. Laser action cannot start until the gain exceeds the loss in the dye plus other losses. However, there is some laser action in the cavity, and if this can build up to a high enough value then the dye begins to bleach because of saturation (equation 15.23), and the radiation density builds up rapidly. The time between pulses is determined by the concentration of the dye.

The initiation of pulses is often called Q switching because the Q factor is a measure of the energy stored to the energy discharged per cycle. Lasers have large Q values. Various types of shutters can be used in a laser cavity so that lasing does not occur while the population inversion is building. Then the shutter is opened, and a large fraction of the energy stored is dumped in a single pulse.

Example 15.10
A widely used solid-state laser is made of $Y_3Al_2O_{15}$ (yttrium aluminum garnet, commonly referred to as YAG) in which some of the Y^{3+} ions are replaced by Nd^{3+}. This laser is generally pumped with a xenon-filled flash lamp. A particular YAG laser yields 10 pulses per second at 1064 nm. Each pulse is 20 ns long and has an energy of 350 mJ. What are the peak and average powers? How many photons are produced per pulse and per minute?

The peak power is 0.35 J/20 $\times$ 10^{-9} s = 17.5 MW. The average power is (0.35 J)(10 s^{-1}) = 3.5 W.

The energy per photon is hc/λ = (6.626 $\times$ 10^{-34} J s)(2.998 $\times$ 10^8 m s^{-1})/(1064 $\times$ 10^{-9} m) = 1.867 $\times$ 10^{-19} J. The number of photons per pulse is (0.35 J)/(1.867 $\times$ 10^{-19} J) = 1.875 $\times$ 10^{18}. The number of photons per minute is (1.875 $\times$ 10^{18})(10)(60 min^{-1}) = 1.125 $\times$ 10^{21} min^{-1}.

Gas lasers are generally pumped by passage of an electric current. The ions and electrons that are produced are accelerated, and the electrons cause excitation by collisions with the molecules of the gas. The first gas laser was the He–Ne laser which can oscillate at three wavelengths, λ_1 = 3.39 μm, λ_2 = 1.15 μm, and λ_3 = 0.633 μm (the most widely used). The helium energy levels are involved in the pumping process, and the laser action occurs between energy levels of Ne.

Some gas lasers use transitions between vibration–rotation levels of a molecule. The CO_2 laser (which also contains N_2 and He) utilizes transitions between two vibrational levels. This laser is one of the most powerful, and it can be operated at 1 MW continuously. It is also one of the most efficient: 15 to 20% of the electrical power put into the electrical discharge is converted to laser radiation.

Figure 15.16 shows the lowest vibrational levels for CO_2 and N_2. Since CO_2 has three modes of vibration the vibrational level may be described by giving the quantum number for each vibrational mode. The level is designated by giving the quantum numbers for (1) symmetric stretching mode, (2) bending mode, and (3) asymmetric stretching mode in that order. Thus, the state is specified by n_1, n_2, n_3. Since the bending mode is doubly degenerate, a bending

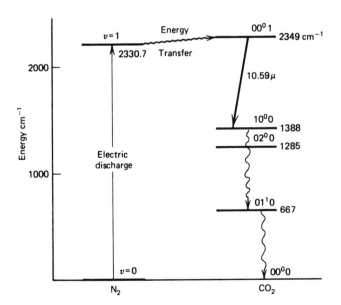

Figure 15.16 Low-lying vibrational energy levels of nitrogen and carbon dioxide molecules. (From J. I. Steinfeld, *Molecules and Radiation*. Cambridge, MA: MIT Press, 1985.)

vibration consists of a combination of the orthogonal bending vibrations. The superscript zero or one on n_2 indicates whether the angular momentum about the axis of the molecule is 0 ($l = 0$) or $\hbar$ ($l = 1$). Thus, 01^10 represents the level with one quantum in the bending vibration with an angular momentum of unity, and 02^00 represents the level with two quanta in the bending vibration with an angular momentum of zero. The 00^01 level is populated by collisions of electrons with ground-state (00^00) molecules and by resonant energy transfer from N_2 molecules. Laser action is produced by the transition between the 00^01 and 10^00 levels ($\lambda \approx 10.6 \ \mu$m). It is also possible to obtain laser action between the 00^01 and 02^00 levels ($\lambda \approx 9.6 \ \mu$m). The transitions shown in the diagram with wavy lines are radiationless transitions that occur rapidly. In this description we have neglected the rotational fine structure.

Some gas lasers may be tuned to several discrete frequencies using a grating or other device that rejects unwanted frequencies from the laser cavity, but dye lasers can provide a continuous range of wavelengths over approximately 40–50 nm. The active medium is a fluorescent organic molecule, and the solution of the dye is pumped with another laser or flashlamp. The optical cavity can be tuned over much of the fluorescence spectrum of the dye. By using a number of dyes it is possible to cover the whole range from 300 to 1000 nm.

In chemical lasers the population inversion is produced by an exothermic chemical reaction. For example, a laser may be produced by mixing hydrogen and fluorine. The following reactions produce HF in higher vibrational levels:

$$F + H_2 \rightarrow HF^* + H \tag{15.25}$$
$$H + F_2 \rightarrow HF^* + F \tag{15.26}$$

The first reaction is exothermic by 132 kJ mol^{-1}. Therefore, HF* may represent molecules in the $v = 3, 2,$ or 1 vibrational states. The second reaction is exothermic by 410 kJ mol^{-1} so that HF* produced may be in a vibrational level as high as $v = 10$. However, the second reaction is not very efficient for pumping the HF laser. Laser action takes place between several vibrational levels. The reaction of hydrogen with fluorine takes place very slowly unless atomic fluorine is provided. Atomic fluorine can be provided by adding SF_6 and dissociating it by electrical means:

$$SF_6 + e^- \rightarrow SF_5 + F + e^- \tag{15.27}$$

Lasers are important in chemistry because they can be used to initiate photochemical reactions, and the short pulses permit the study of very fast reactions. Lasers have also revolutionized Raman spectroscopy because exposures are greatly reduced, and even weak lines may be detected in experiments of short duration. A laser may be used to selectively photodissociate one isotopic species of a molecule, owing to the extreme monochromaticity associated with the laser beam. The dissociated or ionized molecules may be allowed to react with another substance so that the isotopes may be separated by a chemical method.

15.10 Photoelectron Spectroscopy

When a gas M is irradiated with ultraviolet or X-ray photons, the photoionization process can be represented by

$$M + h\nu \rightarrow M^+ + e^- \qquad (15.28)$$

The kinetic energy $E(e^-)$ of the emitted electron is given by

$$E(e^-) = h\nu + E(M) - E(M^+) \qquad (15.29)$$

where $E(M)$ is the energy of M in its initial electronic and vibrational state (usually the ground state) and $E(M^+)$ is the energy of the ion in the electronic and vibrational state in which it is produced. Thus, measurement of the kinetic energy of electrons emitted gives information about the different electronic and vibrational states of M^+.

The most commonly used source of ultraviolet photons is a helium gas discharge tube; the most intense line has a wavelength of 58.4 nm and an energy of 21.22 eV. The kinetic energy of emitted electrons may be measured with a focusing deflection analyzer, utilizing either magnetic or electrostatic fields. An example of an electron energy spectrum obtained by sweeping the appropriate field strength is shown in Fig. 15.17 for hydrogen gas. The most energetic electrons emitted by the sample (line labeled 0) come from the production of hydrogen molecule ions in their ground state:

$$H_2(v = 0) + h\nu = H_2^+(v = 0) + e^- \qquad (15.30)$$

The electrons emitted in this peak have an energy of 5.77 eV so that the energy difference $E(M^+) - E(M)$ for the $v = 0 \rightarrow 0$ transition is $21.22 - 5.77 = 15.45$ eV, which is the adiabatic ionization potential for hydrogen gas (cf. Section 12.2).

The other lines in the spectrum result when H_2^+ is produced in higher vibrational states. Since the minimum of the potential energy curve for H_2^+ is at a somewhat larger internuclear distance than for H_2 (Fig. 12.3), the Franck–Condon principle leads us to expect that the 0–0 transition will not be the most probable. The energy difference for the strongest band (in this case $v = 0 \rightarrow 2$) is referred to as the vertical ionization potential; this ionization potential is $21.22 - 5.24 = 15.98$ eV.

The photoelectron spectra of molecules with more electrons are, of course, considerably more complicated and yield information about the dissociation of inner electrons as well as valence electrons. Photoelectron spectroscopy offers the most direct method for determining the ionization potentials and provides a great deal of information about molecular electronic structure. The photoelectron spectrum from the valence region reflects the higher energy levels in the molecular orbital diagram.

If X-ray photons are used, core electrons may be ejected. The binding energies of core electrons depend on the chemical environment of the atom. These "chemical shifts" offer the possibility of obtaining structural information. For example, the nitrogen 1s spectrum of $Na^+N_3^-$ shows two peaks with relative intensities of 2:1, indicating that the three nitrogen atoms are not chemically equivalent.

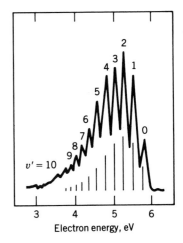

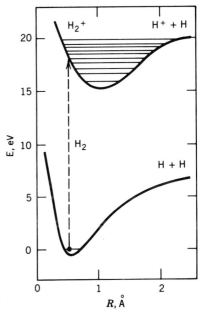

Figure 15.17 (*a*) Photoelectron spectrum of hydrogen gas excited by helium resonance radiation (58.4 nm or 21.22 eV). (From D. W. Turner and D. P. May, *J. Chem. Phys.* **45**:471, 1966.) (*b*) Potential energies of H_2 and H_2^+ with energy levels involved in spectrum.

15.11   Special Topic: Optical Activity* and Optical Rotation

The term optical activity refers to the rotation of the plane of plane-polarized light when it passes through a substance or solution. A closely related phenomenon is the formation of elliptically polarized light from plane-polarized light produced by an optically active medium in the vicinity of its absorption bands. These effects can be understood in terms of the differences in refractive index n and absorbancy index ϵ for left and right circularly polarized light.

Circularly polarized light is light in which the electric vector rotates as the light beam advances. If the electric vector rotates clockwise as observed facing the light source, the light is said to be right-hand circularly polarized light. If it rotates counterclockwise, it is left-hand circularly polarized light. When left and right circularly polarized beams of equal intensity are combined, they yield plane-polarized light. In plane-polarized light the electric vector remains in a plane. A separated beam of circularly polarized light may be obtained by passing plane-polarized light through a quarter wave plate oriented at 45° to the direction of the electric vector of the polarized light. Since the quarter wave plate may be inclined to the right or the left, both right and left circularly polarized light may be obtained in this way.†

Figure 15.18 gives the variation of refractive index (dispersion curve) and the variation of the absorption coefficient (absorption curve) for an optically active material measured with left and right circularly polarized light. The difference in refractive index for the two components is referred to as **circular birefringence,** and the difference in absorption is referred to as **circular dichroism.** The difference curves are shown in the lower part of Fig. 15.18. The plot of Δn versus λ is referred to as the rotatory dispersion curve, and the plot of $\Delta\epsilon$ versus λ is referred to as the circular dichroism spectrum. When an absorption band causes the effects shown in Fig. 15.18, the whole phenomenon is referred to as a Cotton effect. In contrast with ordinary dispersion, a strong absorption band may or may not give a large effect on the rotatory dispersion and a weak absorption band may give a large effect on the rotatory dispersion.

Optically active substances can be divided into two classes: one in which optical activity is found only in the crystal form, for example, quartz, and the other in which it is found in the gaseous, liquid, and certain nonsymmetric crystalline states of the pure substance or in solutions. Optical activity arises in the former group due to the right- or left-hand spiral structure in the crystal and disappears when this structure is melted. Substances in the latter category are optically active because of the asymmetry of the molecule itself. For a molecule whose mirror image is not superimposable on itself, left and right circularly polarized light have different refractive indices and correspondingly different absorption coefficients. This may happen for any molecule having

---

* D. J. Caldwell and H. Eyring, *The Theory of Optical Activity*. New York: Wiley, 1971; C. Djerassi, *Optical Rotatory Dispersion*. New York: McGraw-Hill, 1959; J. G. Foss, *J. Chem. Educ.* **40**:592 (1963); G. Snatzke, *Optical Rotatory Dispersion and Circular Dichroism in Organic Chemistry*. Philadelphia: Sadtler Research Labs., 1967; L. Velluz, M. Legrand, and M. Grosjean, *Optical Circular Dichroism*. New York: Academic, 1965; E. Charney, *The Molecular Basis of Optical Activity*. New York: Wiley, 1979.

† D. Halliday and R. Resnick, *Physics*, Part 2, p. 1081. New York: Wiley, 1978.

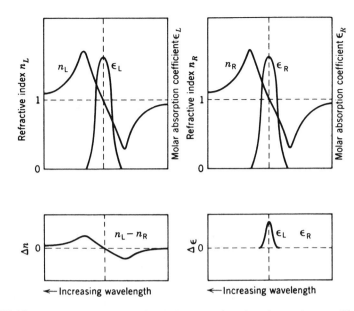

Figure 15.18 Variation of refractive index n and molar absorption coefficient ϵ in the neighborhood of a single absorption line of an optically active substance, as measured with left and right circularly polarized light. The difference curves are referred to as the rotatory dispersion curve (Δn) and the circular dichroism spectrum ($\Delta \epsilon$). (From J. G. Foss, *J. Chem. Educ.* **40**:592, 1963.)

only proper rotation elements of symmetry (Chapter 13). A molecule possessing any improper rotation axis (S_n), a mirror plane, or a center of symmetry, cannot be optically active.

The rotation of plane-polarized light is measured with a polarimeter that consists of a light source, linear polarizer, sample, and analyzer (another linear polarizer). The rotation of the plane of polarization by the sample is measured by rotating the analyzer. If a substance rotates the plane of polarized light to the right, or clockwise, as viewed looking toward the light source, it is said to be dextrorotatory, and the rotation is given a positive sign. If the rotation is counterclockwise, the substance is levorotatory, and the rotation is given a negative sign. the magnitude of the rotation α is directly proportional to the length l of the sample and the concentration c of the optically active molecules, and so it is convenient to calculate a **specific rotation** $[\alpha]$ that is defined by

$$[\alpha] = \frac{\alpha}{cl} \qquad (15.31)^*$$

where l is the path length and c is the concentration in mass per unit volume. For a pure substance, c equals ρ, the density of the pure substance. The specific rotation varies with the wavelength, temperature, and solvent, so these variables must be specified.

---

* In the past it has been customary to express c in g cm^{-3} and l in decimeters. Thus, most values of $[\alpha]$ in the literature are given in deg dm^{-1} cm^3 g^{-1}. In SI it is preferable to express c in kg m^{-3} and l in meters so that $[\alpha]$ is expressed in deg m^2 kg^{-1}.

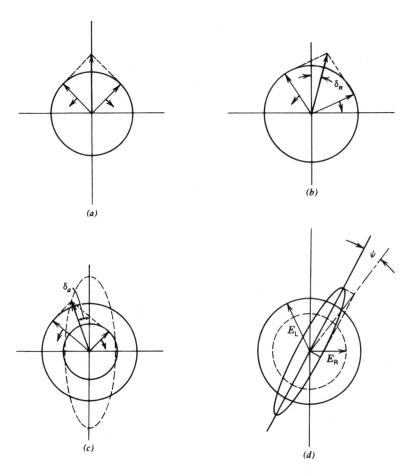

Figure 15.19   Circular birefringence and circular dichroism. (*a*) Representation of plane polarized light as the sum of two components of circularly polarized light rotating in opposite directions. (*b*) Rotation of plane polarized light by an angle δ_n due to different velocities of propagation of left and right circularly polarized light. (*c*) Production of elliptically polarized light from plane polarized light by different absorption coefficients for left and right circularly polarized light. (*d*) Effects on plane polarized light of both differences in refractive indices and absorption coefficients for the two circularly polarized components. In (*c*) and (*d*) the emerging beam is elliptically polarized.

The reason for the rotation of the plane of polarization may be understood by thinking of plane polarized light as being formed by equal contributions of left and right circularly polarized light. When this light enters a medium in which the refractive indices n_L and n_R are different, then the speed of propagation of light in this medium is different for the left and right circularly polarized components. Thus, one moves ahead of the other and they are now out of phase with one another. Since they are equal in amplitude this produces a rotation of the net electric field vector, as shown in Fig. 15.19*a* and *b*. The magnitude of this rotation is linearly related to the path length of the light in the medium.

If the medium has greater absorption for left over right circularly polarized light, then (neglecting the above effect of the rotation of the plane polarization) linearly polarized light will become elliptically polarized in an optically active medium. We can understand this by once again thinking of plane polarized light as being composed of equal components of left and right polarized beams. If the absorption is greater for left polarization, that beam will be diminished in intensity relative to the right polarization. The net result is elliptic polarization of the resultant light, as shown in Fig. 15.19*c*. If both effects are present, the axes of the elliptical polarized light are rotated.

The effects are wavelength dependent, because the closer the light beam is to an absorption, the more pronounced the effect can become. The change in optical rotation with wavelength is called **optical rotatory dispersion** (ORD). The measurements of the rotatory dispersion and circular dichroism can be used to determine the structure, configuration, and conformation of complex optically active molecules (such as proteins, synthetic polypeptides, and steroids).

References

P. W. Atkins, *Molecular Quantum Mechanics*. New York: Oxford University Press, 1983.

P. R. Bunker, *Molecular Symmetry and Spectroscopy*. New York: Academic, 1979.

W. A. Guillory, *Introduction to Molecular Structure and Spectroscopy*. Boston: Allyn & Bacon, 1977.

D. C. Harris and M. D. Bertolucci, *Symmetry and Spectroscopy*. Oxford, UK: Oxford University Press, 1978.

G. Herzberg, *Spectra of Diatomic Molecules*. Princeton, NJ: Van Nostrand, 1950.

J. M. Hollas, *High Resolution Spectroscopy*. Boston: Butterworths, 1982.

J. M. Hollas, *Modern Spectroscopy*. New York: Wiley, 1987.

K. P. Huber and G. Herzberg, *Molecular Spectra and Molecular Structure IV, Constants of Diatomic Molecules*. New York: Van Nostrand-Reinhold, 1975.

H. Lefebre-Brion and R. W. Field, *Perturbations in the Spectra of Diatomic Molecules*. Orlando, FL: Academic, 1986.

I. N. Levine, *Molecular Spectroscopy*. New York: Wiley, 1975.

J. I. Steinfeld, *Molecules and Radiation*. Cambridge, MA: MIT Press, 1985.

W. Struve, *Fundamentals of Molecular Spectroscopy*. New York: Wiley Interscience, 1989.

O. Svelto, *Principles of Lasers*. New York: Plenum, 1976.

Problems

15.1   The spectroscopic dissociation energy of $H_2(g)$ into ground-state hydrogen atoms is 4.4763 eV. What is the spectroscopic dissociation energy of $H_2(g)$ into one ground-state H and one H atom in the 2p state? If H_2 is dissociated with photons of energy 15 eV, what is the velocity of the H atoms coming off in the 1s and 2p states?

15.2   According to the hypothesis of Franck, the molecules of the halogens dissociate into one normal atom and one excited atom. The wavelength of the convergence limit in the spectrum of iodine is 499.5 nm. (*a*) What is the energy of dissociation in kJ mol^{-1} of iodine into one normal and one excited atom? (*b*) The thermochemical value of the heat of dissociation of I_2 into ground-state atoms can be found in Table C.2. Calculate the energy of the excited state of I that is formed from the spectroscopic dissociation in kJ mol^{-1} and eV.

15.3   The ultraviolet absorption of O_2 includes a series of lines (the Schumann–Runge bands) due to transitions from the $^3\Sigma_g^-$ ground state to the excited electronic state $^3\Sigma_u^-$, which are shown in Fig. 15.1. These lines converge to 175.9 nm, which corresponds to dissociation to one O atom in its ground-state ^3P and one O atom in an excited-state ^1D. What is D_0 for O_2? How does this compare with the enthalpy of formation at 0 K? *Given*: The ^1D state of O is 1.970 eV above the ground-state ^3P.

15.4   The spectroscopic dissociation energy of $^{127}I_2$ is 1.542 38 eV according to Table 14.4. What wavelength of light would you use to dissociate ground-state molecules to ground-state atoms if you wanted the atoms to fly away with velocities of 10^3 m s^{-1}?

15.5   A solution of a dye containing 0.1 mol L^{-1} transmits 80% of the light at 435.6 nm in a glass cell 1 cm thick. (*a*) What percentage of light will be absorbed by a solution containing 2 mol L^{-1} in a cell 1 cm thick? (*b*) What concentration will be required to absorb 50% of the light? (*c*) What percentage of the light will be transmitted by a solution of the dye containing 0.1 mol L^{-1} in a cell 5 cm thick? (*d*) What thickness should the cell be to absorb 90% of the light with solution of this concentration?

15.6   Derive equation 15.14 for the integrated intensity of a Gaussian absorption line. A Gaussian line has the form $\epsilon_m e^{-\sigma(\bar{\nu} - \bar{\nu}_m)^2}$, where $\bar{\nu}$ is the frequency at the intensity maximum ϵ_m. [*Hint*: Relate σ to the width of the line at half maximum intensity and use the integral $\int_{-\infty}^{+\infty} e^{-\sigma x^2} \, dx = (\pi/\sigma)^{1/2}$.]

15.7   The following absorption data are obtained for solutions of oxyhemoglobin in pH 7 buffer at 575 nm in a 1-cm cell:

g cm^{-3}	Transmission, %
3×10^{-4}	53.5
5×10^{-4}	35.1
10×10^{-4}	12.3

The molar mass of hemoglobin is 64.0 kg mol^{-1}. (*a*) Is Beer's law obeyed? What is the molar absorption coefficient? (*b*) Calculate the percent transmission for a solution containing 10^{-4} g cm^{-3}.

15.8   The protein metmyoglobin and imidazole form a complex in solution. The molar absorption coefficients in L mol^{-1} cm^{-1} of the metmyoglobin (Mb) and the complex (C) are as follows:

λ	ϵ_{Mb}	ϵ_C
nm	10^3 L mol^{-1} cm^{-1}	10^3 L mol^{-1} cm^{-1}
500	9.42	6.88
630	3.58	1.30

An equilibrium mixture in a cell of 1-cm path length has an absorbance of 0.435 at 500 nm and 0.121 at 630 nm. What are the concentrations of metmyoglobin and complex?

15.9   The absorption spectrum for benzene in Fig. 15.6 shows maxima at about 180, 200, and 250 nm. Estimate the integrated absorption coefficients using ϵ_{max} and $\Delta\bar{\nu}_{1/2}$ and assuming that the width at half maximum is 5000 cm^{-1} in each case. What are the three oscillator strengths? (See Example 15.6.)

15.10   Relatively strong absorption bands have $\epsilon_{max} = 10^4 - 10^5$ L mol^{-1} cm^{-1} and $\Delta\bar{\nu}_{1/2}$ of the order 1000–5000 cm^{-1}, while weak absorption bands have $\epsilon_{max} = 10$ L mol^{-1} cm^{-1} and $\Delta\bar{\nu}_{1/2}$ of the order 100 cm^{-1}. Assuming that the absorption lines are Gaussian, compute the integrated absorption coefficient and the oscillator strengths for these bands.

15.11   The measured oscillator strength of a transition can be used to compute the transition moment, $|\mu_{12}|^2$ by combining equation 15.16 and 15.17*a* to find

$$|\mu_{12}|^2 = \frac{f3he^2}{8\pi^2 m_e \nu}$$

For strong transitions (for which $f \cong 1$), moderately weak transitions ($f \approx 10^{-3}$), and weak transitions ($f \approx 10^{-6}$), calculate $|\mu_{12}|$ and $|R_{12}| = |\mu_{12}|/e$, assuming a transition energy of 25 000 cm^{-1}.

15.12   In Chapter 12, the Hückel molecular orbital model was introduced to describe the electronic states of conjugated molecules. In this chapter, the free electron model (FEMO) was introduced for the same systems. Consider the butadiene molecule in both descriptions. The Hückel model (equation 12.64) gives the energies and wavefunctions for four orbitals, while the FEMO model gives an infinite number of orbitals. Consider the lowest four in the FEMO model. Do they have the same number of nodes as the Hückel orbitals? Is there any way of choosing α and β in the Hückel model or a in the FEMO model to make the predictions for the energies of all four orbitals agree? Suppose we are content to make the lowest electronic absorption energy agree in both models, what is the formula for β in terms of a?

15.13 The lifetimes of vibrationally excited states of molecules of a liquid are limited by the collision rates in the liquid. If one in ten collisions deactivates a vibrationally excited state, what is the broadening of vibrational lines if a molecule undergoes 10^{13} collisions per second?

15.14 The first ionization potentials of Ar, Kr, and Xe are 15.755, 13.966, and 12.130 eV, respectively. Calculate the velocity of the emitted electrons when photons from a He discharge lamp with $\lambda = 58.43$ nm are used to record the photoelectron spectrum of these gases.

15.15 A sample of oxygen gas is irradiated with Mg $K_{\alpha 1 \alpha 2}$ radiation of 0.99 nm (1253.6 eV). A strong emission of electrons with velocities of 1.57×10^7 m s^{-1} is found. What is the binding energy of these electrons?

15.16 The photoelectron spectrum of molecules shows that similar atoms in different chemical environments have slightly different core orbital binding energies. For example, the 1s binding energy of carbon in CH_4 is 290 eV, while it is 293 eV in CH_3F. (a) Explain this shift based on the electronegativity difference between carbon and fluorine. (b) In the molecule $F_3CCOOCH_2CH_3$, predict the order of the carbon 1s binding energies in the 4 carbon atoms.

15.17 When α-D-mannose ($[\alpha]_D^{20} = +29.3°$) is dissolved in water, the optical rotation decreases as β-D-mannose is formed until at equilibrium $[\alpha]_D^{20} = +14.2°$. This process is referred to as mutarotation. As expected, when β-D-mannose ($[\alpha]_D^{20} = -17.0°$) is dissolved in water, the optical rotation increases until ($[\alpha]_D^{20} = +14.2°$ is obtained. Calculate the percentage of α form in the equilibrium mixture.

15.18 The dissociation energies of HCl(g), H_2(g), and Cl_2(g) into normal atoms have been determined spectroscopically and are 4.431, 4.476, and 2.476 eV, respectively. Calculate the enthalpy of formation of HCl(g) at 0 K in kJ mol^{-1} from these data.

15.19 The limit of continuous absorption for Br_2 gas occurs at 19 750 cm^{-1}. The dissociation that occurs is

$$Br_2(\text{ground}) = Br(\text{ground}) + Br(\text{excited})$$

The transition of a ground bromine atom to an excited one corresponds to a wavenumber of 3685 cm^{-1}:

$$Br(\text{ground}) = Br(\text{excited})$$

Calculate the energy increase for the process

$$Br_2(\text{ground}) = 2Br(\text{ground})$$

in (a) cm^{-1}, and (b) electronvolts.

15.20 (a) Calculate the energy levels for $n = 1$ and $n = 2$ for an electron in a potential well of width 0.5 nm with infinite barriers on either side. The energies should be expressed in J and kJ mol^{-1}. (b) If an electron makes a transition from $n = 2$ to $n = 1$, what will be the wavelength of the radiation emitted?

15.21 The Schumann–Runge bands of O_2 are due to absorption from the ground state ($^3\sum_g^-$) to the B $^3\sum_u^-$ excited state. From Fig. 15.1, estimate the longest wavelength for this transition from the lowest vibrational state of ground state O_2.

15.22 When a 1.9-cm absorption cell was used, the transmittance of 436-nm light by bromine in carbon tetrachloride solution was found to be as follows:

c/mol L^{-1}	0.005 46	0.003 50	0.002 10	0.001 25	0.000 66
I/I_0	0.010	0.050	0.160	0.343	0.570

Calculate the molar absorption coefficient. What percentage of the incident light would be transmitted by 2 cm of solution containing 1.55×10^{-3} mol L^{-1} bromine in carbon tetrachloride?

15.23 The absorption band of a certain molecule in solution has a Gaussian shape with maximum molar absorption coefficient of 2×10^4 L mol^{-1} cm^{-1} and a full width at half-maximum of 4000 cm^{-1}. (a) What is the integrated absorption coefficient for this band? (b) What is the oscillator strength f for this transition?

15.24 The absorption coefficient α for a solid is defined by $I = I_0 e^{-\alpha x}$, where x is the thickness of the sample. The absorption coefficients for NaCl and KBr at a wavelength of 28 μm are 14 and 0.25 cm^{-1}. Calculate the percentage of this infrared radiation transmitted by 0.5-cm thicknesses of these crystals.

15.25 Commercial chlorine from electrolysis contains small amounts of chlorinated organic impurities. The concentrations of impurities may be calculated from infrared absorption spectra of liquid Cl_2. Calculate the concentration of $CHCl_3$ in g mL^{-1} in a sample of liquid Cl_2 if the transmittance at $\bar{\nu} = 1216$ cm^{-1} is 45% for a 5-cm cell. At this wavelength liquid Cl_2 does not absorb, and the absorption coefficient for $CHCl_3$ dissolved in liquid Cl_2 is 900 ± 80 $cm^{-1}(g\ cm^{-3})^{-1}$.

15.26   To test the validity of Beer's law in the determination of vitamin A, solutions of known concentrations were prepared and treated by a standard procedure with antimony trichloride in chloroform to produce a blue color. The percent transmission of the incident filtered light for each concentration, expressed in $\mu g\ mL^{-1}$, was as follows:

Concentration, $\mu g\ mL^{-1}$	1.0	2.0	3.0	4.0	5.0
Transmission, %		66.8	44.7	29.2	19.9	13.3

Plot these data so as to test Beer's law. A solution, when treated in the standard manner with antimony trichloride, transmitted 35% of the incident light in the same cell. What was the concentration of vitamin A in the solution?

15.27   The protein metmyoglobin and the azide ion (N_3^-) form a complex. The molar absorption coefficients of the metmyoglobin (Mb) and of the complex (C) in a buffer are as follows:

λ	ϵ_{Mb}	ϵ_C
nm	$10^4\ L\ mol^{-1}\ cm^{-1}$	$10^4\ L\ mol^{-1}\ cm^{-1}$
490	0.850	0.744
540	0.586	1.028

An equilibrium mixture in a 1-cm cell gave an absorbance of 0.656 at 490 nm and 0.716 at 540 nm. (*a*) What are the concentrations of metmyoglobin and complex? (*b*) Since the total azide concentration is 1.048×10^{-4} mol L^{-1}, what is the equilibrium constant for

$$Mb + N_3^- = C$$

15.28   An acid–base indicator is a weak acid (Section 9.1) for which the acidic and basic forms have different absorption spectra and at least one of the forms absorbs strongly. For phenolphthalien, the basic form absorbs strongly and the acidic form absorbs so weakly that this absorption can be neglected. Plot the apparent molar absorbancy index versus pH for an indicator like phenolphthalien, for which the acidic form does not absorb.

15.29   Acetone dissolved in water has a maximum absorption coefficient ϵ of 20 L mol^{-1} cm^{-1} at 38 000 cm^{-1} and the width of the absorption at half-maximum is about 8000 cm^{-1}. What is the value of the integrated absorption coefficient and of the oscillator strength?

15.30   The ionization potential of an atom may be determined by exposing it to high-energy monochromatic radiation and measuring the speed of ejected electrons. When krypton is radiated with 58.4 nm light from a helium discharge lamp, ejected electrons have a velocity of 1.59×10^6 m s^{-1}. What is the ionization potential?

15.31   The most prominent line in the photoelectron spectrum of H_2 is due to the transition

$$H_2(v = 0) + h\nu \rightarrow H_2^+(v = 2) + e^-$$

If helium resonance radiation with an energy of 21.22 eV is used, what will be the electron kinetic energy, assuming that H_2^+ is a harmonic oscillator with a fundamental vibration frequency of 2297 cm^{-1}? The 0–0 ionization potential is 15.45 V.

15.32   When α-D-glucose ($[\alpha]_D^{20} = +112.2°$) is dissolved in water, the optical rotation decreases as β-D-glucose is formed until at equilibrium $[\alpha]_D^{20} = +52.7°$. As expected, when β-D-glucose ($[\alpha]_D^{20} = +18.7°$) is dissolved in water, the optical rotation increases until $[\alpha]_D^{20} = +52.7°$ is obtained. Calculate the percentage of the β form in the equilibrium mixture.

16
Magnetic Resonance Spectroscopy

Magnetic resonance spectroscopy differs from most other kinds of spectroscopy in that a magnetic field is used to provide the energy level separations probed by the radiation. For magnetic fields that can be routinely produced in the laboratory, the transitions between energy levels for nuclear magnetic dipoles occur in the radio-frequency range, and the transitions between energy levels for unpaired electron spins occur in the microwave range. Nuclear magnetic resonance (NMR) and electron paramagnetic resonance (EPR) yield such valuable structural information that they have become indispensable in chemistry.

Different methods for studying nuclear magnetic resonance were developed independently by Purcell and Bloch in 1946. Until ~1980 continuous wave (cw) NMR spectrometers have been used in chemistry. Now Fourier transform spectrometers are increasingly used because of their greater sensitivity. This has opened up almost the whole periodic table to NMR spectroscopy. To relate these magnetic phenomena to the bulk magnetic properties of matter, we begin the chapter with a discussion of the basic ideas of magnetism.

16.1 Basic Facts of Magnetism

The principal measure of the strength of a magnetic field is the **magnetic flux density vector B**. Since magnetic fields are produced by electric currents, it is possible to connect the strength of the field of any magnet, including a single magnetic dipole, with basic mechanical and electrical units. The SI unit of the magnetic flux density is the tesla T*:

$$1 \text{ T} = 1 \text{ N A}^{-1} \text{ m}^{-1} = 1 \text{ kg s}^{-2} \text{ A}^{-1}$$

The magnitude μ of the **magnetic dipole moment vector** $\boldsymbol{\mu}$ of a flat coil enclosing an area $\mathcal{A}$ and carrying a current I is $\mu = I\mathcal{A}$. Thus, the units of the magnetic dipole moment are A m^2. Magnetic dipole moments are more commonly expressed in J T^{-1}, which can be readily shown to be the same as A m^2 by use of the units of T given above. The direction of the $\boldsymbol{\mu}$ vector is perpendicular to the plane of the coil and pointing in the same sense as the linear motion of a right-hand screw turned in the same direction as the current.

In the presence of a magnetic field, magnetic dipoles within a material become partially oriented; the **magnetic dipole moment per unit volume** is referred to as the **magnetization M.** The unit of the magnitude of the magnetization vector M is A m^{-1}. Since the magnetic moments in a diamagnetic material oppose the magnetizing field, M is negative in a diamagnetic material.

It is convenient to introduce another vector quantity, the **magnetic field H,** which is defined by

$$H \equiv \frac{B}{\mu_0} - M \tag{16.1}$$

where μ_0 is the **permeability** of vacuum. The permeability of the vacuum has the value $4\pi \times 10^{-7}$ N A^{-2} exactly. (Note that $\epsilon_0\mu_0 = 1/c^2$, where ϵ_0 is the permittivity of vacuum and c is the speed of light in vacuum.) The magnetic field H is a vector, and its magnitude is expressed in A m^{-1}. The physical significance of H is better seen by writing equation 16.1 as

$$B = \mu_0(H + M) \tag{16.2}$$

The magnetic flux density B, which determines the magnetic force on a moving charged particle, is determined by $\mu_0 H$ arising from the electric currents producing the field and $\mu_0 M$ arising from the magnetic moments induced by the magnetizing field.

For isotropic substances the **magnetic susceptibility** χ is defined by

$$\chi = \frac{M}{H} \tag{16.3}$$

where M and H are the magnitudes of the magnetization and magnetic field vectors. Thus, the magnetic susceptibility is a dimensionless quantity.

In contrast with polarization by an electric field, the magnetic moment may be in the direction of the applied field (χ is positive) or in the opposite direction (χ is negative). For a **diamagnetic** substance χ is negative, small, independent of the magnetic field intensity, and independent of temperature. For a **paramagnetic** substance χ is positive, small, independent of the magnetic field intensity, and decreases with increasing temperature. For a **ferromagnetic** substance χ is positive, large, dependent on the magnetic field and temperature,

---

* The gauss G is the unit for magnetic flux density in the cgs Gaussian system of units; 1 T = 10^4 G.

and dependent on previous history. For an **antiferromagnetic** substance χ is small and positive and is dependent on previous history.

Paramagnetism results from the orientation of permanent magnetic dipoles in a substance. These permanent magnetic dipoles are due to the spin of unpaired electrons or to the angular momentum of electrons in orbitals of atoms or molecules. Electrons in orbitals with $l = 1, 2, 3, \ldots$ have angular momentum and therefore produce a magnetic dipole moment. Nuclei with magnetic moments produce a paramagnetic effect, but this effect is only about a millionth as large as paramagnetism due to orbital moments or unpaired electrons.

In the absence of a magnetic field the magnetic dipoles that produce paramagnetism are not oriented. In the presence of a field the orientation of permanent magnetic dipoles by the field is opposed by the disorganizing effect of thermal motion, as in the alignment of electric dipoles by an electric field (Section 12.12). In most substances the magnetic effects of electron spin and electron orbital motions cancel because electrons are paired in filled shells. Many rare earth and transition metal ions are paramagnetic because they have unpaired electrons. Free radicals have an odd number of electrons and are therefore paramagnetic. The most familiar paramagnetic substance is molecular oxygen. As we have seen earlier (Section 12.5), it has two unpaired electrons. This property of oxygen gas makes it possible to determine its partial pressure in a gas stream with a little torsion balance in the field of a magnet.

The magnetic moment μ of a paramagnetic molecule or ion is commonly measured in units of **Bohr magnetons** μ_B. As we saw in equation 11.26, the Bohr magneton has the value $9.274\,078 \times 10^{-24}$ J T^{-1}. This is equal to the magnetic moment in the direction of the magnetic field associated with the orbital angular momentum of an electron in the 1s orbit of a hydrogen atom.

Diamagnetism results from the induction of microscopic currents in a sample by the external magnetic field. The magnetic dipoles produced in this way are aligned in a direction opposite of that of the external field. Since the induced magnetic dipoles oppose the field, a diamagnetic substance experiences a force in the direction of the weaker part of an inhomogeneous magnetic field. A paramagnetic substance is attracted into the stronger part of an inhomogeneous magnetic field. The diamagnetic effect is produced in all substances, but it is only about a hundredth or a thousandth as strong as the paramagnetic effect when the latter exists.

The metals iron, cobalt, nickel, gadolinium, and dysprosium, and certain of their alloys and compounds are ferromagnetic below a certain critical temperature for each substance. The origin of ferromagnetism was a puzzle that was explained by quantum mechanics. Why do so many electrons in incomplete shells have their spins aligned, and why do they remain aligned even after the applied magnetic field is removed? The explanation is that the lowest energy state for certain solids is one in which spins are parallel instead of opposed, as they are for the two electrons in a hydrogen molecule, for example. The requirements of certain atomic distances and certain radii of the orbitals for d electrons limit this phenomena to just a few elements. Ferromagnetic substances show hysteresis in their magnetic properties. This means that the magnetic moment depends on the magnetic history of the sample; the change in moment with increasing field is not necessarily retraced when the field is reduced. Thus, a ferromagnet will stay aligned in the absence of a magnetic field.

The atoms in an antiferromagnetic substance are arranged so that the magnetic moments of nearest neighbors are opposed.

16.2   Nuclear Magnetism and Nuclear Magnetic Resonance

Both the proton and neutron have a spin angular momentum of $\hbar/2$ just like the electron. And just like many-electron atoms, the nucleus of any atom, which contains many neutrons and protons, has a total spin angular momentum and, therefore, a magnetic dipole moment. All nuclei with an odd mass number have a nonzero spin (an odd integral multiple of $\hbar/2$), and all nuclei with an even mass number but an odd number of protons have a nonzero spin (of an integral multiple of $\hbar/2$). **Nuclei with an even number of both neutrons and protons have zero spin.** Deuterium with one proton and one neutron has a spin of 1. The spins of a number of important nuclei are given in Table 16.1.

For a nucleus, the **total spin angular momentum** is represented by I, the spin quantum number by I, and the **z component of nuclear spin** by I_z. Since I is an angular momentum just like S for electrons, the eigenvalue of I^2 is $I(I + 1)\hbar^2$ and the magnitude of I is given by

$$|I| = [I(I + 1)]^{1/2}\,\hbar \tag{16.4}$$

Similarly, the eigenvalues of I_z are $m_I\hbar$ where

$$m_I = -I, -I + 1, \ldots, I - 1, I \tag{16.5}$$

so there are $2I + 1$ values of m_I, each associated with an eigenstate of I^2 and I_z.

Just as in the case of the electron (see Section 11.4), the magnetic dipole moment of the nucleus μ is proportional to the spin angular momentum,

$$\mu = \frac{g_N e}{2m_p} I \tag{16.6}$$

where g_N is the nuclear g factor and m_p is the mass of the proton. The nuclear g factor can have a negative value, so that μ and I can be parallel or antiparallel. Values of g_N are given in Table 16.1 along with spin I and isotopic abundances

Table 16.1   Nuclear Magnetic Properties

Nucleus	% Abundance	Spin I	g_N	NMR Frequency for 1 T in MHz
^{1}H	99.99	$\frac{1}{2}$	5.585	42.5759
^{2}D	0.01	1	0.857	6.53566
^{7}Li	92.5	$\frac{3}{2}$	2.171	16.546
^{13}C	1.11	$\frac{1}{2}$	1.405	10.7054
^{14}N	99.6	1	0.403	3.0756
^{15}N	0.4	$\frac{1}{2}$	-0.567	4.3142
^{17}O	0.04	$\frac{5}{2}$	-0.757	5.772
^{19}F	100	$\frac{1}{2}$	5.257	40.0541
^{23}Na	100	$\frac{3}{2}$	1.478	11.262

for a few nuclei. Since I has units of $\hbar$, we define the basic unit of nuclear dipole moment, the nuclear magneton, μ_N as

$$\mu_N = \frac{e\hbar}{2m_p} \tag{16.7}$$

Example 16.1

What is the value of the nuclear magneton?

$$\mu_N = \frac{e\hbar}{2m_p} = \frac{(1.602\ 177\ 3 \times 10^{-19}\ \text{C})(6.626\ 076 \times 10^{-34}\ \text{J s})}{4\pi(1.672\ 623\ 1 \times 10^{-27}\ \text{kg})}$$

$$= 5.050\ 787 \times 10^{-27}\ \text{J T}^{-1}$$

The nuclear magneton is $\frac{1}{1836}$ of the Bohr magneton.

Then the z component of the nuclear dipole moment is given as

$$\mu_z = \frac{g_N e}{2m_p} m_I \hbar = g_N \mu_N m_I \tag{16.8}$$

In an external magnetic field B, the energy of a magnetic dipole is given by $E = \mu \cdot B$. Since we can pick the z direction to be along the field, we find

$$E = -\mu_z B = -g_N \mu_N m_I B \tag{16.9}$$

where $m_I = -I, -I + 1, \ldots, I - 1, I$. Therefore, a nucleus of spin I has $2I + 1$ nondegenerate energy states in a magnetic field. Transitions among these levels can be induced by applying electromagnetic radiation with frequency equal to the energy level spacings. This is known as **nuclear magnetic resonance** (NMR). Transitions among the analogous $2S + 1$ levels of a system of electron spin S are called **electron paramagnetic resonance** (EPR) or **electron spin resonance** (ESR). Since these transitions take place between magnetic states, they are induced by the oscillating magnetic field of the radiation rather than the oscillating electric field, and are called magnetic dipole transitions.

The selection rule for such transitions is $\Delta m_I = \pm 1$, leading to a transition at frequency ν, given by

$$\nu = \frac{\Delta E}{h} = \frac{g_N \mu_N B}{h} \tag{16.10}$$

which is called the **Larmor frequency.** Notice that since the energy levels given by equation 16.9 are equally spaced, there is only one Larmor frequency. In Fig. 16.1 we show the energy levels of a single *proton* in a magnetic field. The proton has the lower energy when the proton is parallel to the field ($m_I = \frac{1}{2}$) than when it is antiparallel ($m_I = -\frac{1}{2}$).

Many magnetic resonance spectrometers have a fixed frequency ν of radiation. The magnetic field is varied until the energy level separation becomes resonant with this frequency. At this point the NMR transitions can be observed.

In discussing transitions induced by electromagnetic radiation in Section 14.2, we pointed out that the radiation induces emission from higher to lower

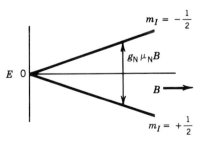

Figure 16.1 Energy levels of a proton in a magnetic field.

levels as well as absorption from lower to higher levels. This occurs in NMR as well. Since spontaneous emission is extremely small and can be neglected for NMR, the processes of absorption and stimulated emission are the only ones that are relevant. The rate of absorption is proportional to the number of nuclei in the lower state, while the rate of stimulated emission is proportional to the population (number) of nuclei in the upper state. Since the proportionality constant is the same for both these processes (see Sections 14.2 and 15.9), we see that the net loss of energy from the electromagnetic field is proportional to the *difference* in populations between the lower and upper levels. Let N_u be the population in the upper state and N_l be the population in the lower state in Fig. 16.1. At thermal equilibrium, these populations obey the Boltzmann equation (see Eq. 14.5)

$$\left(\frac{N_u}{N_l}\right)_{eq} = e^{-\Delta E/kT} = e^{-g_N\mu_N B/kT} \cong 1 - \frac{g_N\mu_N B}{kT} \tag{16.11}$$

where we have expanded the exponential since at temperatures above a few kelvins for any nucleus, the exponent is very small. Then the excess population in the lower state is

$$\frac{N_l - N_u}{N_l + N_u} = \frac{g_N\mu_N B}{2kT} \tag{16.12}$$

The net loss of energy from the radiation is therefore proportional to the right side of equation 16.12 and, in particular, to the value of B. Therefore, magnetic resonance signals increase in intensity as B is raised and T is lowered.

If the intensity of the radiation field is increased, the rate of absorption and stimulated emission may become higher than the rate of **thermal equilibration** so that the populations of the two levels can become equal, leading to saturation, that is, zero net absorption.

Thermal equilibration of the populations is called a **relaxation** process and the rate is represented by T_1^{-1} where T_1 is called the longitudinal or **spin–lattice relaxation time**. The total magnetization M of the sample is the vector sum of the magnetic dipole moments for each of the nuclei. Therefore, the z component of magnetization M_z is given by

$$M_z = -\tfrac{1}{2}(N_u - N_l)\,g_N\mu_N \tag{16.13}$$

for spin $\tfrac{1}{2}$ nuclei. For nuclei with spin larger than $\tfrac{1}{2}$, M_z is related to the sum of the population of a level multiplied by its z component of magnetic dipole moment. As the populations relax to equilibrium, the z component of the magnetization M_z also relaxes to its equilibrium value at the same rate ($1/T_1$). This relaxation is accompanied by a change in energy of the spin system (since the energy levels depend on m_I as given in equation 16.9); therefore, this relaxation process requires an energy reservoir interacting with the spins. Usually, this is the lattice vibrational system; hence, the name spin–lattice relaxation. The equation for $M_z(t)$ is

$$M_z(t) = M_z^{eq}[1 - e^{-t/T_1}] + M_z(0)e^{-t/T_1} \tag{16.14}$$

where M_z^{eq} is the value of M_z at equilibrium ($t = \infty$).

There is another relaxation rate and relaxation time that is related to the x and y (transverse) components of M. At equilibrium, these components are zero, but NMR experiments can be done that produce an initially nonzero M_x

or M_y. These components then relax to equilibrium with a rate T_2^{-1} called the **transverse relaxation time.** Transverse relaxation can be accomplished both with and without energy relaxation. If it is accomplished with energy relaxation, it is due to spin–lattice relaxation; if it is accomplished without energy relaxation, it is due to interactions among the spins (spin–spin relaxation). Normally $T_2 \leqslant T_1$, so that transverse relaxation occurs faster than longitudinal relaxation. The equation for $M_x(t)$ (or $M_y(t)$) is

$$M_x(t) = M_x(0)\, e^{-t/T_2} \tag{16.15}$$

Example 16.2

What is the ratio of the number of proton spins in the lower state to the number in the higher state in a magnetic field of 1 T at room temperature? What is the excess population in the lower state?

$$\frac{N_1}{N_u} = 1 + \frac{g_N \mu_N B}{kT} = 1 + \frac{(5.585)(5.05 \times 10^{-27}\ \text{J T}^{-1})(1\ \text{T})}{(1.38 \times 10^{-23}\ \text{J K}^{-1})(298\ \text{K})}$$

$$= 1 + 6.86 \times 10^{-6}$$

$$\frac{N_1 - N_u}{N_1 + N_u} = 3.43 \times 10^{-6}$$

16.3 High-Resolution NMR Spectrometer

An NMR spectrometer provides the magnetic field to produce the energy levels, the radio frequency to excite transitions, and a radio-frequency receiver to detect emitted radiation. The resolution of an NMR spectrometer depends on the strength and homogeneity of the magnetic field and the constancy of the radio-frequency radiation. High resolution is required for most chemical applications. Since chemical shifts (Section 16.5) are in the range of parts per million, the field needs to be constant to a part in 10^8 or 10^9. The fixed frequency radio-frequency sources that are used are crystal controlled and are stable to a part in 10^9.

Most NMR spectra for chemical purposes are determined for solutions. The NMR absorption lines in liquids are very much narrower than in solids because the dipole–dipole coupling with nuclei in other molecules is averaged to zero by the rapid random rotations of molecules in a liquid. The NMR absorption lines of solids are broad, but they give important information about the distance between nuclei with magnetic moments and about their orientations in the crystal. The nuclei interact by a direct dipole–dipole coupling between their magnetic moments, and this causes very broad absorption lines. More recently, methods have been developed that make it possible to obtain high-resolution spectra on solids.

There are two methods for obtaining high-resolution spectra on liquids: **continuous wave** (cw) and **Fourier transform.** In the continuous wave method, the tube containing the sample is placed in the crossed coils of a radio-frequency transmitter with a fixed frequency and a radio-frequency receiver, as shown

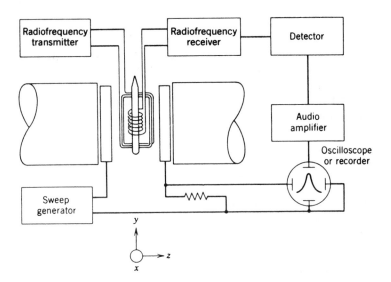

Figure 16.2   Block diagram of a cw high-resolution nuclear magnetic resonance spectrometer. The axis of the transmitter coil is the x axis, and that of the receiver coil is the y axis.

in Fig. 16.2. The transmitter coil is oriented so that it produces a magnetic field that is perpendicular to the B_0 field produced by the electromagnet. The receiver coil is perpendicular to the transmitter coil so that it does not receive a signal directly from the transmitter. The strength of this magnetic field is set at a value just below that required for resonance, and then a secondary field is applied by use of coils around the faces of the pole pieces. The current for these coils is supplied by a sweep generator so that the magnetic field strength at the sample can be swept through the resonance condition. A sweep generator produces a steadily increasing voltage that drops to zero when a certain voltage is reached and then repeats the process. The sweep generator also controls the sweep of the oscilloscope or chart recorder.

The magnetic nuclei in the sample precess about the z direction (axis of the electromagnet). When the magnetic field strength reaches a value where nuclei in the sample are precessing about the direction of the field at the Larmor frequency, transitions from lower to higher levels and vice versa may be excited by radiation from the radio-frequency transmitter. Energy is absorbed from the transmitter into the nuclear spin system only when there is resonance. This energy is radiated by the precessing nuclei and is picked up by the receiver coil. If the signal from the radio-frequency receiver is applied to the vertical amplifier of the oscilloscope, the absorption line is traced out as shown. Very frequently the limiting factor in high-resolution NMR spectroscopy is the inhomogeneity of the magnetic field. An improvement in resolution can be obtained by spinning the sample tube so that the nuclei experience a field that is averaged over the volume of the sample. Even if the field were perfect, an NMR absorption line would have a finite width because of the interaction of a magnetic moment with its surroundings. As we will see in spectra later in the chapter, a sample generally shows a series of resonances and so a series of peaks is obtained. NMR spectra are presented with the magnetic field strength increasing to the right. However, for many purposes it is convenient to think of such a spectrum as being a plot versus frequency at constant magnetic field strength with frequency increasing to the left. Such a spectrum is said to be in the frequency domain.

Magnetic flux densities for available proton NMR spectrometers correspond with frequencies up to 500 MHz. Research is being done on spectrometers with frequencies up to 750 MHz. Higher frequencies lead to both increased sensitivity and more easily interpreted spectra.

Example 16.3

At a certain magnetic flux density the frequency of the electromagnetic radiation that is absorbed or emitted by a sample containing protons is 220 MHz. What is this magnetic flux density?

$$B = \frac{h\nu}{g_N\mu_N} = \frac{(6.6262 \times 10^{-34} \text{ J s})(220 \times 10^6 \text{ s}^{-1})}{(5.585)(5.0508 \times 10^{-27} \text{ J T}^{-1})} = 5.1678 \text{ T}$$

16.4 Fourier Transform Spectroscopy

NMR spectrometers of the cw type are rapidly being replaced by those of the Fourier transform type, which provide much greater sensitivity. The sensitivity of Fourier transform spectrometers is so great that their development has opened up almost the whole periodic table to NMR spectroscopy. In a Fourier transform spectrometer the sample is irradiated with a pulse of radio-frequency radiation that excites all possible NMR resonances. The data obtained directly in this way is referred to as a **free induction decay** (FID), which has to be analyzed by an on-board computer to obtain the usual spectrum in the frequency domain.

When a sample with magnetic moments is in equilibrium with a magnetic field B_0 more nuclei are lined up with the field than are opposed to it, and from Equation 16.13 the sample has a macroscopic magnetization M which is the vector sum of the individual magnetic moments. Although the individual nuclear magnets precess around the direction of the field (the z direction), the macroscopic magnetization M is oriented in the direction of the field, as shown in Fig. 16.3a. In a Fourier transform experiment a pulse of radio-frequency radiation (at the Larmor frequency) is applied to the sample so that it produces a magnetic field B_1 perpendicular to the direction of the main magnetic field B_0. As shown in Fig. 16.3b, the application of B_1 tips M away from the z axis so that it now has a component M_{xy} in the y direction. Note that M remains in the yz plane. This tipping of M occurs only if the frequency of the radio-frequency radiation is close to the frequency of precession in the main magnetic field. Only then is there resonance so that energy can be absorbed from the B_1 field into the spin system. Although a pulse of a certain frequency is used, a Fourier analysis of a pulse shows that it actually contains component frequencies that cover a range that is approximately the reciprocal of the pulse width. Thus, resonances throughout the spectrum are excited simultaneously by a single pulse of radio-frequency radiation.

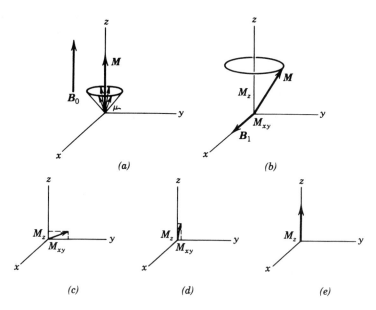

Figure 16.3   The effect of a pulse of radio-frequency radiation on the macroscopic magnetization M of a sample in a magnetic field in the z direction. The pulse B_1 tips the magnetization in the yz plane. After the pulse, the magnetization returns to its initial value with rate constant T_1^{-1}. (From E. D. Becker, *High Resolution NMR.* New York: Academic Press, 1980.)

When a B_1 field is applied in the x direction, M precesses in the yz plane with frequency given by the Larmor equation 16.10. If the radio-frequency pulse lasts t_p seconds, M will "flip" through an angle of

$$\theta = \gamma B_1 t_p \text{ radians}$$

$$= \frac{360}{2\pi}\gamma B_1 t_p \text{ deg} \qquad (16.16)$$

Example 16.4
What pulse length should be used to obtain a 90° flip angle for protons if the B_1 field is 5.8726×10^{-4} T?

$$t_p = \frac{2\pi\theta}{360\gamma B_1}$$

$$= \frac{2\pi 90}{360(2.6748 \times 10^8 \text{ T}^{-1}\text{ s}^{-1})(5.8726 \times 10^{-4}\text{ T})}$$

$$= 10^{-5}\text{ s} = 10\ \mu\text{s}$$

After the pulse of radio-frequency radiation M_{xy} decays back to zero as the nuclei lose energy to their surroundings (spin–lattice relaxation). In this process the magnetization vector in the z direction returns to its initial value, as illustrated in Figs. 16.3c and d. During this decay to the initial situation, the M vector precesses around the B_0 direction (z axis). This induces a signal to the

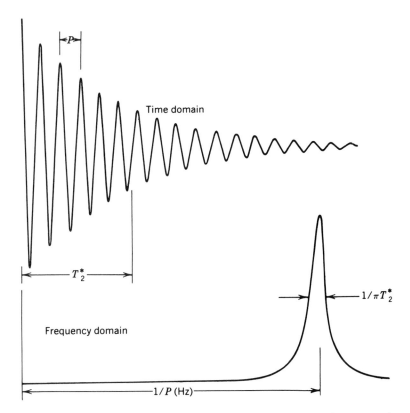

Figure 16.4   The relationship between the time domain and the frequency domain for a single resonance. The period P of the free induction decay gives the chemical shift, and the time constant T_2^* for the decay gives the linewidth. (From J. W. Akitt, *NMR and Chemistry*. New York: Chapman & Hall, 1983.)

coil around the sample. This signal plotted against time, as shown in Fig. 16.4, is referred to as a free induction decay (FID) since the nuclei precess freely. In this FID there is a single resonance frequency, and the radio frequency used in the pulse, and then subsequently as a reference frequency in measuring the magnetic field strength in the y direction, is not exactly the same as the resonance frequency for the nuclei. Therefore, there is interference between the signal from the nuclei and the reference radio frequency. In a perfectly homogeneous magnetic field, the relaxation time for the decay would be the transverse relaxation time T_2. However, the inhomogeneity of the field also contributes to the rate of decay since nuclei in different parts of the field precess at slightly different frequencies and get out of phase with each other. The actual relaxation time T_2^* is given by

$$\frac{1}{T_2^*} = \frac{1}{T_2} + \gamma \, \Delta H_0 \tag{16.17}$$

where ΔH_0 is the field inhomogeneity across the sample volume.

The signal from a free induction decay is a function of time, in contrast with the cw spectrum which exists in the frequency domain. The time and frequency domains are related through the Fourier relationship:

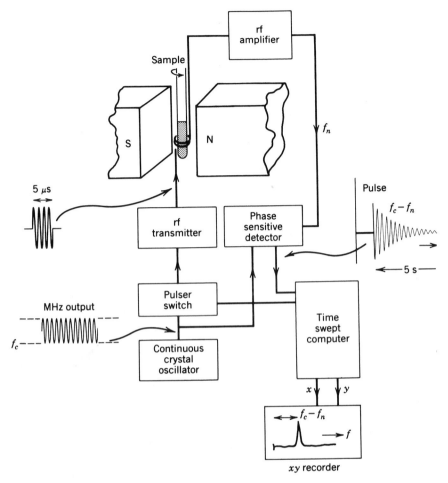

Figure 16.5 Fourier transform NMR spectrometer with a built-in computer to perform a Fourier transform of free induction decay. (From J. W. Akitt, *NMR and Chemistry*. New York: Chapman & Hall, 1983.)

$$F(\omega) = \int_{-\infty}^{\infty} f(t)e^{-i\omega t}dt \qquad (16.18)$$

The cw spectrum is represented by $F(\omega)$, and the free induction decay is represented by $f(t)$. In a Fourier transform spectrometer a built-in computer is used to carry out the integration indicated in equation 16.18. The oscillation time P (time from peak to peak) of the FID is inversely proportional to the chemical shift. The more rapid the oscillations in the FID, the greater the difference in frequency between the radio-frequency oscillator in the spectrometer and the Larmor frequency of the nuclei. The relaxation time T_2^* for the decay of the FID is related to the width of the line in the NMR spectrum. A rapid decay implies strong spin–spin interactions that destroy phase coherence in the precessing nuclei. Thus, according to the Heisenberg uncertainty principle the average lifetime of a state is short and the uncertainty in its energy is large; this leads to a broad line in the spectrum.

A schematic diagram of a Fourier transform NMR spectrometer is given in Fig. 16.5.

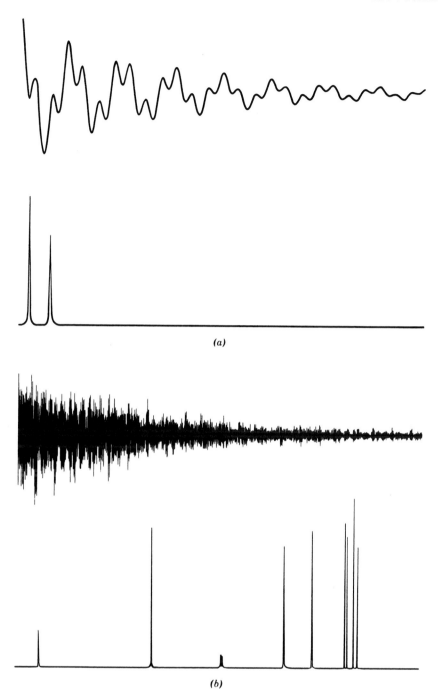

Figure 16.6   Signal versus time for a free induction decay after a pulse of high-frequency radiation, and the plot of intensity versus frequency to which it corresponds for (*a*) two absorption frequencies and (*b*) carbon-13 spectrum of β-pinene obtained on a Bruker WM-300 spectrometer. (Courtesy of J. A. Cooper, Bruker Instruments, Billerica, MA.)

If there are several lines in the NMR spectrum, then several different frequencies are transmitted simultaneously after the radio-frequency pulse. Just as a sound wave with two frequencies has a beat frequency equal to the difference between the two frequencies, the radio-frequency signal after the radio-frequency pulse contains beat frequencies from the spectrum. This is illustrated in Fig. 16.6. For more complicated NMR spectra the free induction signal observed after application of a 90° pulse is too complex to be readily interpreted. However, a Fourier transform yields the NMR spectrum from the free induction decay. In Fig. 16.6a the free induction decay consists of two sine waves so that the plot of absorption versus frequency (i.e., the usual NMR spectrum) has two peaks. The spectrum in Fig. 16.6b consists of a number of sine waves.

The time required to obtain a spectrum by the Fourier transform method is quite short, usually several seconds. Thus, a series of pulses may be used with signal averaging to improve the signal/noise ratio. By the use of Fourier transform methods it is possible to obtain more than an order of magnitude increase in sensitivity over the usual scanning method.

Fourier transform spectroscopy is especially useful in ^{13}C nuclear magnetic resonance spectroscopy because of the low natural abundance of ^{13}C (1.2%). Because of this low natural abundance the ^{13}C spectrum can be regarded as the spectrum of a mixture of molecules, each containing only one atom of ^{13}C. It is not necessary to consider coupling between ^{13}C nuclei, but spectra do show coupling between ^{13}C nuclei and neighboring protons.

16.5 The Chemical Shift

Nuclear magnetic resonance would be of little interest to chemists were it not for the fact that the magnetic field at a nucleus in a molecule is not equal to the magnetic field that is applied to the sample, but is a field that is altered by the screening or shielding of the electrons in the molecule. When a magnetic field is applied to a molecule, the motion of the electrons is affected in such a way as to oppose the applied field. Thus, the magnetic flux density at the nucleus is less than the applied magnetic flux density B_0 by the shielding of the electrons surrounding the nucleus:

$$B(\text{nucleus}) = B_0 - \sigma B_0 = B_0(1 - \sigma) \tag{16.19}$$

The **shielding constant** σ is less than about 10^{-5} for the protons and less than about 10^{-3} for most other nuclei. Since the magnitude of the screening depends on the orientation of the molecule with respect to the applied field, σ is actually a second-rank tensor. However, for a liquid or a gas the directional part of σ averages out so that it may be treated as a scalar.

Because of this screening, identical nuclei in chemically different environments resonate at different values of the **applied** magnetic field. For example, Fig. 16.7 shows a sketch of the proton magnetic resonance spectrum of CH_3CH_2OH under conditions of low resolution. The areas under the three peaks are proportional to the numbers of protons of the three different types. As illustrated in Fig. 16.7, the secondary field opposes the applied field, so that the effective field at a proton is less than the applied field. The more a proton is shielded by the surrounding electrons, the higher the applied field required to produce a resonance at a given frequency. The local field depends on the chemical environment, so the effect is referred to as a **chemical shift.**

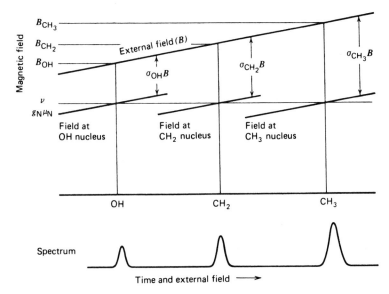

Figure 16.7 Low-resolution proton-magnetic-resonance spectrum of CH_3CH_2OH. As the magnetic field is increased, protons in different chemical environments come into resonance. As indicated the magnetic field strengths at the various protons are different from the strength of the external field. (Adapted from J. C. Davis, *Advanced Physical Chemistry*. New York: Ronald, 1965.)

In ethyl alcohol the protons in CH_3 are shielded to a greater extent than those in CH_2, which are shielded to a greater extent than the proton in OH.

The determination of absolute values of σ would require the study of nuclei without any orbital electrons, and so relative values of σ are determined using a reference compound containing the nucleus being studied. For ^1H NMR tetramethylsilane $Si(CH_3)_4$ (TMS) is used as the reference. For ^{13}C NMR the carbon resonance of TMS is used as the reference. When the TMS is dissolved in the sample it is referred to as an internal reference.

When the shielding constant is included, the Larmor equation for a particular kind of proton in the sample and for protons in the reference substance can be written

$$\nu_S = \frac{\gamma}{2\pi} B_0(1 - \sigma_S) \tag{16.20}$$

$$\nu_R = \frac{\gamma}{2\pi} B_0(1 - \sigma_R) \tag{16.21}$$

where B_0 is the strength of the applied magnetic field. It is advantageous to express chemical shifts on a dimensionless scale defined by

$$\delta \equiv \frac{\nu_S - \nu_R}{\nu_R} \times 10^6 \tag{16.22}$$

The reason for using the δ scale is that the magnitude of the chemical shifts in frequency units is proportional to the magnetic field strength. On the δ scale the chemical shift, which is expressed in parts per million, is independent of

Table 16.2 Chemical Shifts of ^{1}H

Compound	δ/ppm	Compound	δ/ppm
Methyl protons		Olefinic protons	
$(CH_3)_4Si$	0.00	$(CH_3)_2C{=}CH_2$	4.6
$(CH_3)_4C$	0.92	Cyclohexene	5.57
CH_3CH_2OH	1.17		
CH_3COCH_3	2.07	Acetylenic protons	
CH_3OH	3.38	$HOCH_2C{\equiv}CH$	2.33
Methylene protons		Aromatic protons	
Cyclopropane	0.22	Benzene	7.27
Cyclohexane	1.44	Naphthalene	7.73
CH_3CH_2OH	3.59		
		Aldehydic protons	
Methine protons		CH_3CHO	9.72
$(CH_3)_2CHOH$	3.95	C_6H_5CHO	9.96

the magnetic field strength. Substituting equations 16.20 and 16.21 in equation 16.22 yields

$$\delta = \frac{\sigma_R - \sigma_S}{1 - \sigma_R} \times 10^6 \approx (\sigma_R - \sigma_S) \times 10^6 \qquad (16.23)$$

since $\sigma_R \ll 1$. Thus, the chemical shift on the σ scale is equal to the difference between the shielding constant for the reference and the sample, expressed in parts per million.

The protons in TMS are more shielded than most organic protons. Thus, $\sigma_R > \sigma_S$, and chemical shifts for most organic protons are therefore positive numbers. Chemical shifts for protons in various environments, expressed as δ values, are given in Table 16.2.

Chemical shifts for protons are closely related to the electron density around the atoms to which the hydrogen atoms are bonded. If the electron density is high, the protons are shielded from the magnetic field, and so higher magnetic fields have to be applied to achieve resonance at the spectrometer frequency. Protons in methyl groups are highly shielded and therefore have small chemical shifts relative to tetramethylsilane. The protons of benzene are not shielded so much, and therefore they resonate at lower magnetic field strengths than methyl protons. The reason protons of benzene resonate at lower fields can be seen from Fig. 16.8. When a benzene ring is oriented perpendicular to the magnetic field, the circulation of electrons in the π orbitals induces a field that is in the same direction as the applied field at the protons. Therefore, aromatic protons resonate at a lower field than they otherwise would. This effect is reduced by molecular tumbling because, when the benzene ring is oriented parallel to the field, there is no such effect.

Proton chemical shifts in organic compounds can be correlated with the electronegativities of neighboring groups, types of carbon bonding, and hydrogen bonding. In

$$H-\overset{|}{\underset{|}{C}}-X$$

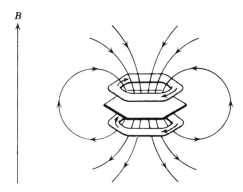

Figure 16.8   Magnetic flux density produced by electron circulation in the π orbitals of the benzene ring. At the protons, the induced field is in the same direction as the applied field.

the chemical shift for the proton depends on the electronegativity of X. The greater the electronegativity (Section 12.8) of X, the more it will draw electrons away from H. The more electrons are drawn away, the lower the magnetic field required for resonance, and the larger the δ value. The electronegativities of the halogens are F > Cl > Br > I, and the proton chemical shifts δ for protons in the methyl halides are $CH_3F > CH_3Cl > CH_3Br > CH_3I$.

The protons of metal hydrides have negative chemical shifts on the δ scale; that is, they are more highly shielded than the protons of TMS and their resonances occur at higher magnetic flux densities.

In recording NMR spectra it is customary to put the low magnetic flux density to the left; this means that the chemical shifts increase to the left as shown in Fig. 16.9. In connection with Fourier transform NMR it is convenient to think about spectra being obtained by increasing the frequency at a constant magnetic flux density. In this case it is useful to think of less screened protons resonating at low frequency and more screened protons resonating at higher frequencies.

The chemical shifts for other nuclei are larger than for protons because they are surrounded by a larger number of electrons. For ^{13}C the chemical shifts, measured with respect to TMS, range from 200 ppm for

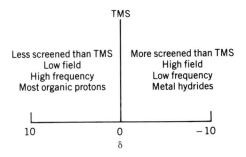

Figure 16.9   Terminology used in discussing proton chemical shifts. (From J. W. Akitt, *NMR and Chemistry*, 2d ed. New York: Chapman & Hall, 1983.)

$$\begin{array}{c} O \\ \| \\ R'\!-\!C\!-\!R \end{array}$$

to approximately zero for methyl groups; the chemical shifts are in much the same order as for protons, except that ^{13}C resonances can be observed for organic groups not including protons. Chemical shifts for ^{14}N and ^{15}N, measured with respect to NH_3 range up to about 800 ppm. Phosphorus chemical shifts, measured with respect to P_4O_6, range from about 100 to -200 ppm.

16.6   Fine Structure and Spin–Spin Coupling

Under high resolution, the NMR spectrum of a molecule in a liquid may split into multiplets (i.e., each line splitting into a number of lines) because of interactions between nuclear spins. There are two kinds of **spin–spin interactions:** one resulting from the direct dipole–dipole coupling and the other from an intermediate coupling to the electrons. The direct coupling depends on the angles between the nuclear spins and the vector joining the nuclei in space. In a magnetic field, this interaction is averaged to zero by rapid tumbling of the molecule, so that for most molecules in solution, the direct interaction does not lead to an observable effect. In solids, this leads to a large broadening of the NMR lines.

The second interaction, however, does lead to a splitting of the spectral lines in solution. This contact interaction comes about in the following way. Consider a molecule with two nuclei A and B each having spin $\frac{1}{2}$. An electron near A will feel the magnetic field of the spin of A and respond by orienting **antiparallel** to A. This electron is coupled to other electrons in the molecule, particularly the one in the same orbital. These two electrons will have antiparallel spins (by the Pauli exclusion principle), so the second electron will tend to have its spin parallel to nuclear spin A. When that second election is near nucleus B, it will tend to make nuclear spin B antiparallel to it. The net result is that the energy of the two nuclear spins, A and B, will be slightly lower if they are antiparallel rather than parallel, because of the interactions with the electrons. Another way of saying this is that there is a term in the expression for the energy which gives antiparallel nuclear spins slightly lower energy than parallel nuclear spins. (Actually, because these interactions are complicated, there are situations in which the reverse is true. What we have described in the usual situation.)

Now consider the energy levels of nuclear spin A in a magnetic field. Besides the effects of direct coupling to the field (Zeeman effect) and the chemical shift, we have to consider the effect of the nearby nuclear spin B. There will now be two energy levels, one for the case B parallel to A and one for the case B antiparallel to A; hence, the nuclear spin resonance lines for A will be split into two lines, with frequency difference proportional to the **spin–spin coupling energy constant** J_{AB} as shown in Fig. 16.10. This constant is independent of magnetic field and is usually expressed in hertz.

As the above argument suggests, the coupling decreases rapidly with the number of bonds between the two nuclei. Typical values of the coupling con-

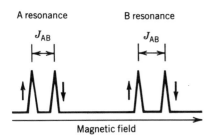

Figure 16.10   The effect of spin–spin coupling J_{AB} on the A and B resonance spectrum. The arrows represent the z component of the other spin coupled to the spin whose resonance is indicated.

Table 16.3 Proton Spin–Spin Coupling Constants

Structure	J/Hz	Structure	J/Hz
(geminal CH_2)	-20 to $+6$	(vinyl cis $C=C$)	12 to 9
(vicinal $-C-C-$)	5.5 to 7.5	(benzene ring)	o 6 to 9 m 0.5 to 4 p 0 to 2.5
(vinyl trans $C=C$)	7 to 10	(cyclohexane chair)	ax, ax 9 to 14 ax, eq 2 to 4 eq, eq 2.5 to 4

stants are given in Table 16.3. Nuclear magnetic resonance spectra with chemical shifts and spin–spin interactions may be interpreted according to first-order analysis under certain conditions. The more general treatment, referred to as second-order analysis, is discussed in the next section. First-order analysis may be made when the following two conditions are satisfied:

1. At magnetic flux densities common to laboratory instruments, the differences in the chemical shifts between nuclei, or groups of nuclei, are large compared with the spin coupling between them. To make this comparison the differences in chemical shifts are multiplied by the frequency ν_0 to obtain the differences in chemical shifts in Hz.

2. The spin–spin couplings involve groups that are magnetically equivalent, not just chemically equivalent. Nuclei in a group are chemically equivalent when they have the same chemical shift. Nuclei in a group are magnetically equivalent when they all have the same chemical shift and when all nuclei in the group are coupled equally with any other single nucleus in the molecule. The protons in the tetrahedral molecule difluoromethane CH_2F_2 are magnetically equivalent because they are equally coupled to the two fluorine atoms (spin $\frac{1}{2}$) by virtue of the symmetry of the molecule. However, the two protons in 1,1-difluoroethylene are not magnetically equivalent because the coupling of the H and F which are cis (on the same side) is different from the coupling of H and F which are trans (on opposite sides). Thus, first-order analysis is applicable to the first spectrum in Fig. 16.11, but not to the second.

When various conformations of a molecule are rapidly interconverted, the equivalence of nuclei should be determined on an average basis, rather than for one of the conformers. If the ethanol molecule CH_2CH_2OH were perfectly rigid the methyl protons would not be equivalent. However, because of rapid rotation about the C–C bond, the electronic environments of the three protons are magnetically equivalent.

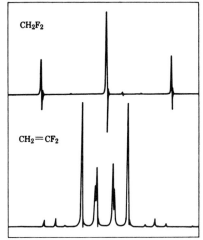

Figure 16.11 Proton-resonance spectra of CH_2F_2 and CH_2=CF_2 at 60 MHz. (From E. D. Becker, *J. Chem. Educ.* **42**:591, 1965.)

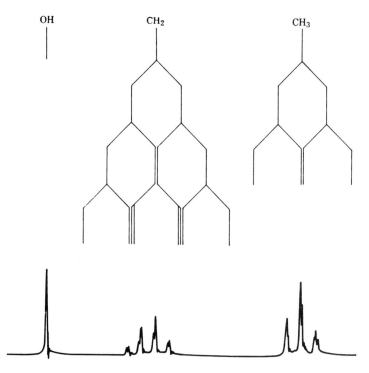

Figure 16.12   Proton resonance spectrum of ethyl alcohol at 40 MHz. The signal from the radio-frequency receiver is plotted vertically, and the magnetic field strength is plotted horizontally. The field increases linearly from left to right.

The proton magnetic resonance spectrum of ethanol shown in Fig. 16.12 may be interpreted by first-order analysis. The absorption line due to methyl (CH_3) protons is split into three components because the neighboring methylene group (CH_2) contains two protons, each with spin $\frac{1}{2}$. We can think of the first methylene proton splitting the methyl proton resonance into a doublet, as shown in the figure, and then the second methylene proton splitting the doublet into a triplet with the center line twice as intense as the other two. The two methylene protons produce the same spin–spin splitting because there is rapid internal rotation around the C–C bond.

The absorption line due to methylene protons is split into four components by the three protons of the neighboring CH_3 group. The reason for the relative intensities of $1:3:3:1$ is evident from the diagram.

In this spectrum the proton in the hydroxyl group does not cause further splitting because it undergoes chemical exchange so rapidly with protons in other molecules that it does not produce a splitting effect. This rapid chemical exchange occurs only when acid catalyzed, but only traces of acid are required to eliminate the splitting.

We can generalize these results by saying that if there are n equivalent protons that interact with the proton being studied, the absorption will be split into $n + 1$ lines, and their relative intensities will be proportional to the coefficients of the binomial expansion of $(1 + x)^n$. These coefficients are given by

Pascal's triangle:

0 proton					1							
1 proton				1		1						
2			1		2		1					
3		1		3		3		1				
4	1		4		6		4		1			
5		1		5		10		10		5		1
6	1		6		15		20		15		6	1

Since the spin–spin splitting is due to neighboring groups, it gives important additional structural information.

For the purpose of discussing different types of spin systems, it is convenient to use letters of the alphabet to represent nuclei. A spin system with two nuclei may be represented by AX if the difference in chemical shifts is large compared with the coupling constant, by AB if they are of the same order of magnitude, or by A_2 if the nuclei are equivalent.

The separation of peaks due to the chemical shift is directly proportional to the field strength, but the separation due to spin–spin splitting J is independent of field strength. Thus, at higher field strengths, the spin–spin splittings do not cause as much complication in the interpretation of the NMR pattern.

16.7 Second-Order Effects

The interpretation of NMR spectra in the preceding section applies when the chemical shifts are large compared with the coupling constant. When the chemical shifts are of the order of the coupling constants or smaller, the interpretation of the spectrum becomes more complicated because second-order effects have to be taken into account. It is the ratio $\nu_0\delta/J$ of the chemical shift separation in frequency units to the coupling constant that is important. We will consider only one example of second-order effects, those encountered in an AB system with two spin $\frac{1}{2}$ nuclei. As shown in Fig. 16.13, the spectrum consists of two doublets if the chemical shift $\nu_0\delta$ is large compared with the coupling constant J. As the chemical shift is reduced, the doublets approach each other, but the question is, How can the two doublets collapse to give a single line when the chemical shift difference is reduced to zero? To answer this question it is necessary to consider the four possible energy states for AB with two spin $\frac{1}{2}$ nuclei. The two possible wavefunctions for each nucleus are represented by α for $I = +\frac{1}{2}$ and β for $I = -\frac{1}{2}$. There are four possible wavefunctions: $\alpha(A)\alpha(B)$, $\alpha(A)\beta(B)$, $\beta(A)\alpha(B)$, and $\beta(A)\beta(B)$. (See Section 11.7.) Spin states $\alpha(A)\beta(B)$ and $\beta(A)\alpha(B)$ have the same component of the spin angular momentum in the direction of the field, and so these states have to be combined to form mixed wavefunctions of the form shown in Table 16.4 with constants C_1 and C_2. The energy levels corresponding with the four wavefunctions* are shown in Table 16.4, where δ is the chemical shift difference.

---

* I. N. Levine, *Molecular Spectroscopy*. New York: Wiley-Interscience, 1975.

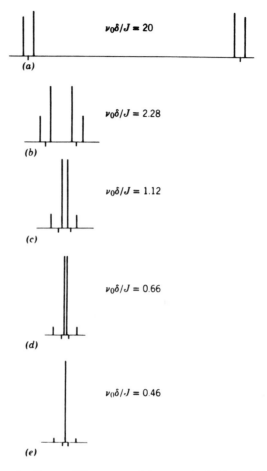

Figure 16.13 Spectra for an AB system with two spin-half nuclei for various ratios of $\nu_0\delta/J$ and constant coupling constant. The A and B chemical shifts are indicated by the small markers under the baseline. (From J. W. Akitt, *NMR and Chemistry* 2d ed. New York: Chapman & Hall, 1983.)

The four possible transitions between the four energy levels are shown in Table 16.5. Only those transitions can take place for which the change in the z component of the spin is ±1. For α the energy eigenvalue is $\frac{1}{2}\hbar$, and for β it is $-\frac{1}{2}\hbar$. The transition energies, which may be calculated from Table 16.4, are

Table 16.4 Wavefunctions and Energy Levels for an AB System with Two Spin $\frac{1}{2}$ Nuclei[a]

Number	Wavefunction	Energy Level (Hz)
1	$\alpha\alpha$	$\frac{1}{2}(\nu_A + \nu_B) + \frac{1}{4}J$
2	$C_1(\alpha\beta) + C_2(\beta\alpha)$	$\frac{1}{2}[(\nu_0\delta)^2 + J^2]^{1/2} - \frac{1}{4}J$
3	$-C_2(\alpha\beta) + C_1(\beta\alpha)$	$-\frac{1}{2}[(\nu_0\delta)^2 + J^2]^{1/2} - \frac{1}{4}J$
4	$\beta\beta$	$-\frac{1}{2}(\nu_A + \nu_B) + \frac{1}{4}J$

[a] From J. W. Akitt, *NMR and Chemistry*. New York: Chapman & Hall, 1983.

Table 16.5 Transition Energies and Line Intensities for an AB System with Two Spin $\frac{1}{2}$ Nuclei[a]

Number		Energy (Hz)	Relative Intensity
a	$3 \rightarrow 1$	$+\frac{1}{2}J + \frac{1}{2}[(\nu_0\delta)^2 + J^2]^{1/2}$	$1 - J/[(\nu_0\delta)^2 + J^2]^{1/2}$
b	$4 \rightarrow 2$	$-\frac{1}{2}J + \frac{1}{2}[(\nu_0\delta)^2 + J^2]^{1/2}$	$1 + J.[(\nu_0\delta)^2 + J^2]^{1/2}$
c	$2 \rightarrow 1$	$+\frac{1}{2}J - \frac{1}{2}[(\nu_0\delta)^2 + J^2]^{1/2}$	$1 + J/[(\nu_0\delta)^2 + J^2]^{1/2}$
d	$4 \rightarrow 3$	$-\frac{1}{2}J - \frac{1}{2}[(\nu_0\delta)^2 + J^2]^{1/2}$	$1 - J/[(\nu_0\delta)^2 + J^2]^{1/2}$

[a] From J. W. Akitt, *NMR and Chemistry*. New York: Chapman & Hall, 1983.

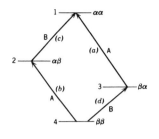

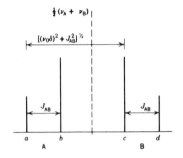

given in Table 16.5 with respect to the mean frequency $(\nu_A + \nu_B)/2$. The transitions between the four energy levels and the corresponding four-line spectrum are shown in Fig. 16.14. The separation of the two lines in each doublet is the spin–spin coupling J_{AB}, but the separation of the two doublets is no longer equal to the chemical shift $\nu_0\delta$. The separation between lines a and c (or b and d or doublet centers) is now given by $[(\nu_0\delta)^2 + J_{AB}^2]^{1/2}$. It is convenient to calculate the chemical shift from the a–d and b–c distances from

$$\nu_0\delta = [(a - d)(b - c)]^{1/2} \qquad (16.24)$$

which may be verified by substituting the energy differences between these pairs of lines.

As shown in Fig. 16.13, the relative intensities of the lines change as the chemical shift difference becomes smaller. In the limit as the chemical shift approaches zero, the spectrum approaches the single-line spectrum characteristic of A_2. In other words, no spin–spin couplings are observed within a group of magnetically equivalent nuclei.

Figure 16.14 Energy level diagram for an AB system with two spin $\frac{1}{2}$ nuclei and the resulting spectrum. (Reprinted by permission of Varian Associates, Palo Alto, CA.)

16.8 Electron Spin Resonance

The basic equations for electron spin resonance (ESR) follow the same pattern as for nuclear magnetic resonance. The magnetic energy of an electron in a magnetic field is

$$E = g_e\mu_B m_s B \qquad (16.25)$$

where g_e is the g factor for the electron (2.002 322 for a free electron), μ_B is the Bohr magneton (Section 11.3), and m_s is the quantum number for the z component of the electron spin. The two energy levels of a single electron in a magnetic field are shown in Fig. 16.15. Because of the negative charge of the electron, the magnetic moment μ_e of an electron is in the direction opposite to its spin angular momentum, and the electron spin quantum number is $-\frac{1}{2}$ in the lower level in contrast with the situation with nuclei. For a transition from $m_s = -\frac{1}{2}$ to $m_s = +\frac{1}{2}$,

$$\Delta E = h\nu = g_e\mu_B B \qquad (16.26)$$

Substances that show ESR spectra include free radicals, odd electron molecules, triplet states of organic molecules, and paramagnetic transition metal ions and their complexes. Any paramagnetic substance can be studied but, as

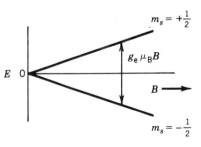

Figure 16.15 Energy levels of an electron in a magnetic field. Compare to Fig. 16.1.

we have seen, most substances are not paramagnetic, because electron spins are usually paired.

An electron resonance experiment is similar to an NMR experiment but, since μ_B for the electron is about 10^3-fold larger than μ_N for nuclei, the frequencies required fall in the microwave range instead of the radio-frequency range when magnetic fields of convenient laboratory strength are used. Usually a frequency of about 10 GHz ($\lambda = 3$ cm) is used with a magnetic field of 0.3 to 0.4 T.

Example 16.5

Calculate the magnetic flux density required to give a precessional frequency of 9500 MHz for a free electron.

$$B = \frac{h\nu}{g_e\mu_B} = \frac{(6.6262 \times 10^{-34} \text{ J s})(9500 \times 10^6 \text{ s}^{-1})}{(2.0023)(9.2741 \times 10^{-24} \text{ J T}^{-1})} = 0.3390 \text{ T}$$

The block diagram for a simple ESR spectrometer is shown in Fig. 16.16. As in the NMR spectrometer, the frequency is held constant, and the magnetic field is swept through resonance. Microwave radiation from a klystron passes down a waveguide to a resonant cavity containing the sample. When there are transitions between electron spin levels in the sample, net energy is absorbed from the microwave radiation, and less microwave energy is received at the crystal detector. By use of field modulation and a phase-sensitive detector the derivative of the absorption or absorption line is recorded on the oscilloscope or strip chart recorder. The shape of the derivative curve is illustrated on the face of the oscilloscope in Fig. 16.16.

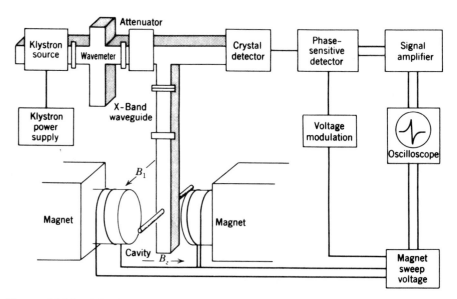

Figure 16.16 Block diagram of simple ESR spectrometer. (W. A. Guillory, *Introduction to Molecular Structure and Spectroscopy*, Copyright © 1977, Allyn & Bacon, Boston. Reprinted with permission of the publisher.)

ESR spectroscopy has been very useful in determining the structure of organic and inorganic free radicals. Free radicals may be produced chemically, photochemically, or by use of high-energy radiation. If a radical has a very short life a flow system or continuous radiation may have to be used to maintain a sufficiently high concentration for detection. Actually, a concentration of only about 10^{-10} mol L^{-1} is required to obtain a spectrum under favorable conditions.

ESR spectra may be determined for unstable radicals trapped in glasses, frozen rare gas matrices, or in crystals. If the radicals are regularly oriented with respect to crystal axes the directional effects of magnetic interactions may be studied.

A molecule in a triplet state has a total electron spin of one ($S = 1$). In this case there are three sublevels that have spin angular momentum S_z about a chosen axis of $+1$, 0, or -1. A triplet molecule has an even number of electrons, two of them unpaired, while a radical with spin $\frac{1}{2}$ has an odd number of electrons. For a molecule to have a triplet state, the unpaired electrons must interact; a molecule with two unpaired electrons a great distance apart is a diradical, not a triplet.

The information about electron densities in molecules that is obtained from ESR spectra has been useful for checking molecular-orbital and valence-bond calculations.

Groups with unpaired electrons (spin labels) may be attached to proteins to provide a very sensitive detector of changes in protein structures.

16.9 Hyperfine Splitting

The splitting of lines in the ESR spectrum provides information of chemical interest. This splitting arises from the fact that unpaired electrons feel the effect of the applied magnetic field and the magnetic field produced by nuclei *within its orbital* that have spin (see Table 16.1). In ESR this splitting of lines is called hyperfine splitting instead of spin–spin splitting as it is in NMR.

The simplest example of hyperfine splitting is the ESR spectrum of atomic hydrogen which consists of two lines; the absorption due to the unpaired electron is split into a doublet by the spin of the proton. The splitting is caused by the magnetic field due to the magnetic moment of the proton. The field experienced by the unpaired electron is the sum of the applied magnetic field and the magnetic field of the proton. Since the proton can take on two orientations with respect to the applied field, the local field at the electron is $B_{\text{loc}} = B_{\text{appl}} \pm a/2$, where a is referred to as the hyperfine splitting constant. The general relation that applies to all nuclei with spin I is

$$B_{\text{loc}} = B_{\text{appl}} + am_I \qquad (16.27)$$

where m_I is the quantum number for the z component of the nuclear angular momentum.

Example 16.6
An electron spin resonance spectrometer operating at 9.302 GHz was used to study the spectrum of atomic hydrogen. Two lines were obtained: one at 357.3 mT and the other at 306.6 mT. What is the hyperfine splitting constant?

The local magnetic flux density is the same for the two lines.

$$B_{\mathrm{loc}} = 357.3 \text{ mT} - a/2$$

$$B_{\mathrm{loc}} = 306.6 \text{ mT} + a/2$$

$$0 = 50.7 \text{ mT} - a$$

$$a = 50.7 \text{ mT}$$

The hyperfine splitting constant may be expressed in terms of frequency by use of the conversion factor calculated in problem 16.17.

$$a = (50.7 \text{ mT})(28 \text{ MHz/mT})$$

$$= 1.42 \text{ GHz}$$

If several magnetic nuclei are present in the same radical, each contributes to the splitting with different hyperfine splitting constants. For two nuclei

$$B_{\mathrm{loc}} = B_{\mathrm{appl}} + a_1 m_{I_1} + a_2 m_{I_2} \tag{16.28}$$

Figure 16.17a shows the effects of the two protons on the possible energy levels for the electron. In the presence of a magnetic field an unpaired electron has two energy levels, $m_s = \frac{1}{2}$ and $m_s = -\frac{1}{2}$. The two protons each split these levels so that the magnetic system has eight energy levels. In electron spin resonance transitions the electron spin flips, but the nuclear spins do not. Thus, in absorbing energy in an ESR transition, the electron goes from an energy level in the lower group ($m_s = -\frac{1}{2}$) to the corresponding level in the upper group ($m_s = +\frac{1}{2}$). As the magnetic field strength is increased, the four possible transitions one-by-one come into resonance and four ESR lines are obtained. Since the four nuclear-spin states ($\alpha_1\alpha_2$, $\alpha_1\beta_2$, $\beta_1\alpha_2$, and $\beta_1\beta_2$) are equally probable, the four lines are of equal intensity. The two hyperfine splittings a_1 and a_2 may be calculated from the spectrum as shown.

If the two protons have the same coupling constant to the unpaired electron, the spectrum has three lines, and the center line is twice as intense because it represents two possible transitions, as shown in Fig. 16.17b. Two or more protons having the same coupling to an unpaired electron are said to be equivalent, and they usually occupy symmetrically equivalent positions in the molecule.

Figure 16.18 shows a simple way of looking at these two spectra. We can describe the effect of two proton spins in the molecule by saying that the first proton splits the original single line due to electron spin into a doublet, and the second proton further splits the two lines into a quartet. If the two hyperfine splittings are different, four absorption lines are obtained. If the two hyperfine splittings are the same (the protons are equivalent), three lines are obtained.

Thus, in general, if n equivalent protons interact with an unpaired electron there will be $n + 1$ lines, and their relative intensities are given by Pascal's

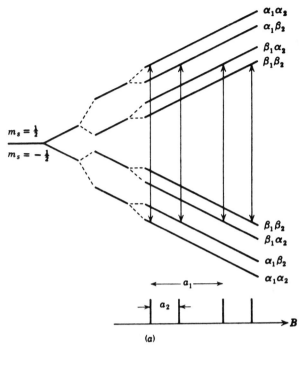

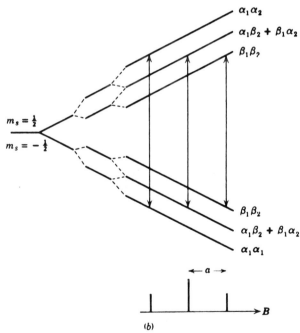

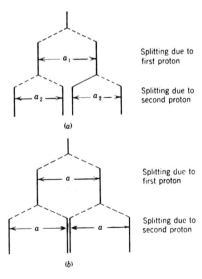

Figure 16.17 (a) Hyperfine energy levels and ESR spectrum for an unpaired electron in the presence of two nonequivalent protons. (b) Hyperfine energy levels and ESR spectrum for an unpaired electron in the presence of two equivalent protons. The spin states of the protons are represented by $\alpha_1\alpha_2$, $\alpha_1\beta_2$, $\beta_1\alpha_2$, and $\beta_1\beta_2$.

Figure 16.18 (a) Hyperfine splitting due to two nonequivalent protons. (b) Hyperfine splitting due to two equivalent protons.

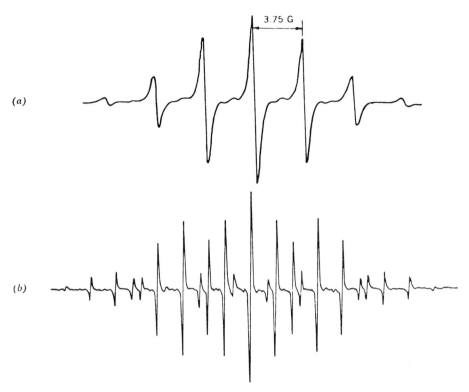

Figure 16.19 (*a*) ESR spectrum of benzene negative ions. (*b*) ESR spectrum of naphthalene negative ions. Note that these spectra are the derivatives of the usual spectra because of the detection method. (From A. Carrington and A. D. McLachlan, *Introduction to Magnetic Resonance*. New York: Harper & Row, 1967.)

triangle (Section 16.6). The ESR spectrum of benzene negative ion ($C_6H_6^-$) in Fig. 16.19*a* illustrates this very nicely. The ESR spectrum of naphthalene negative ion ($C_{10}H_8^-$) shown in Fig. 16.19*b* is more complicated because there are two types of protons, labeled A and B in the following structural formula:

The four A protons alone would give rise to a five-line spectrum. Each of these lines is further split into five lines by the B protons, to give the 25-line spectrum shown in the figure.

The splitting by other magnetic nuclei is similar, except that, of course, nuclei of spin 1, like ^{14}N, cause splitting into triplets of equal intensity, since the quantum number for the ^{14}N nucleus may be $+1$, 0, or -1.

The hyperfine coupling between the proton and electron of the hydrogen atom is 0.0507 T. This contact interaction, mentioned in Section 16.6, is a result of unpaired electron density at a nucleus. This occurs when the unpaired electron occupies an s orbital, but not when it occupies a p or d orbital, since these orbitals have nodes at the nucleus. The coupling between unpaired electrons and protons in organic radicals is much less than this (up to about 3×10^{-3} T) because the unpaired electron occupies orbitals that extend over the whole molecule, and the probability that the unpaired electron is at a particular hydrogen nucleus is quite small.

16.10 Special Topic: Determination of Rate of Exchange by NMR

If two groups of nuclei in a molecule have different chemical shifts at low temperatures, the two NMR lines may coalesce as the temperature is raised. As shown in Fig. 16.20 this occurs when solutions of N,N-dimethylacetamide in carbon tetrachloride are heated from 259 to 368 K. At the lower temperatures two peaks are observed because the skeleton of the molecule tends to be planar, and protons in the two methyl groups have different chemical shifts ν_A and ν_B. This is the situation at 259 K, as shown in Fig. 16.20. However, as the temperature is raised the rate

of internal rotation increases, the protons have a chemical shift that is the average of that for the two positions, and the proton NMR spectrum coalesces into a single line at 368 K. What is important here is the length of time a nucleus spends in a given environment relative to the reciprocal of the chemical shift difference $\nu_0\delta$ between the two lines at low temperatures. Since this time might be 0.1 s, the question is whether the lifetime of a proton in a given environment is long or short compared with 0.1 s.

The coalescence point in the spectrum where the multiplet structure is just lost, at 331 K in Fig. 16.20, provides the most direct information about the rate constant. It may be shown that if the residence times are τ_A and τ_B on the two sites

$$\tau_C = \frac{\sqrt{2}}{\pi(\nu_0\delta)} \qquad \text{where} \qquad \frac{2}{\tau_C} = \frac{1}{\tau_A} + \frac{1}{\tau_B} \qquad (16.29a)$$

and $\nu_0\delta$ is the difference in chemical shift. In the present case the residence times are equal on the two sites and so τ_C is the residence time on either site. The residence time is the reciprocal of a first-order reaction rate constant (Section 19.2), and so the rate constant k for the exchange reaction may be calculated using

$$k = \frac{\pi(\nu_0\delta)}{2^{1/2}} \qquad (16.29b)$$

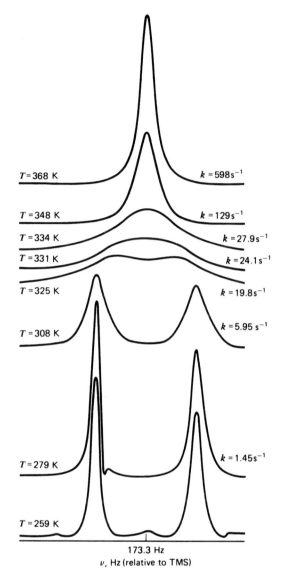

$T = 368$ K $k = 598 s^{-1}$

$T = 348$ K $k = 129 s^{-1}$

$T = 334$ K $k = 27.9 s^{-1}$

$T = 331$ K $k = 24.1 s^{-1}$

$T = 325$ K $k = 19.8 s^{-1}$

$T = 308$ K $k = 5.95 s^{-1}$

$T = 279$ K $k = 1.45 s^{-1}$

$T = 259$ K

173.3 Hz

ν, Hz (relative to TMS)

Figure 16.20 Proton NMR spectrum for the two methyl groups of *N,N*-dimethylacetamide in carbon tetrachloride. The rate constants are for the exchange rate of the two methyl groups. (From F. P. Gasparro and N. H. Kolodny, *J. Chem. Educ.* **54**:258, 1977.)

References

J. W. Akitt, *NMR and Chemistry*. New York: Chapman & Hall, 1983.

E. D. Becker, *High Resolution NMR*. New York: Academic, 1980.

F. A. Bovey, *Nuclear Magnetic Resonance Spectroscopy*. New York: Academic, 1988.

J. W. Cooper, *Spectroscopic Techniques for Organic Chemists*. New York: Wiley, 1980.

T. C. Farrar and E. D. Becker, *Pulse and Fourier Transform NMR*. New York: Academic, 1971.

W. A. Guillory, *Introduction to Molecular Structure and Spectroscopy*. Boston: Allyn & Bacon, 1977.

R. K. Harris, *Nuclear Magnetic Resonance Spectroscopy*. Essex, UK: Longman Scientific & Technical, 1986.

I. N. Levine, *Molecular Spectroscopy*. New York: Wiley-Interscience, 1975.

M. C. R. Symons, *Chemical and Biochemical Aspects of Electro-Spin Resonance*. New York: Wiley, 1978.

Problems

16.1   NMR spectrometers usually have fixed frequencies between 60 and 500 MHz. Calculate the magnetic flux densities needed to give a NMR transition frequency for hydrogen equal to these frequencies.

16.2   For the frequencies of problem 16.1, calculate the corresponding energies in kJ mol^{-1} and compare these with RT at 300 K.

16.3   (*a*) What are the energy levels for a ^{23}Na nucleus in a magnetic field of 2 T? (*b*) What is the absorption frequency?

16.4   The gyromagnetic ratio γ_N for a nucleus is defined by

$$\mu_N = \gamma_N \hbar I$$

What is the value of γ_N for H?

16.5   What is the difference in fractional populations of ^{13}C spins between the upper and lower states in a magnetic field of 2 T at room temperature?

16.6   In a magnetic field of 2 T, what fraction of the protons have their spin lined up with the field at room temperature?

16.7   It is now possible to do NMR experiments at very low temperatures. Calculate the ratio of the number of protons in the upper spin state to that in the lower spin state in a magnetic field of 2 T at 1 and 10 mK.

16.8   Using information from Tables 16.2 and 16.3, sketch the spectrum you would expect for ethyl acetate ($CH_3CO_2CH_2CH_3$).

16.9   Chemical shifts δ are expressed in ppm, but can also be expressed in Hz. What magnetic fields are necessary to produce frequency shifts of 100 and 500 Hz for protons with a δ = 1?

16.10   (*a*) What is the separation of the methyl and methylene proton resonances in ethanol at 60 MHz? (*b*) At 300 MHz? (See Table 16.2.)

16.11   At a magnetic flux density of 1.41 T, the frequency separation between protons in benzene and protons in tetramethylsilane is 436.2 Hz. What is the chemical shift?

16.12   Sketch the proton resonance spectrum of $D_2CHCOCD_3$ (deuteroacetone containing a little hydrogen). Indicate the relative intensities of the lines.

16.13   The proton resonance pattern of 2,3-dibromothiophene shows an AB-type spectrum with lines at 405.22, 410.85, 425.07, and 430.84 Hz measured from tetramethylsilane at 1.41 T [K. F. Kuhlmann and C. L. Braun, *J. Chem. Educ.* **46**:750 (1969)]. (*a*) What is the coupling constant *J*? (*b*) What is the difference in the chemical shifts of the A and B hydrogens? (*c*) At what frequencies would the lines be found at 2 T? Use the results of Section 16.7.

16.14   At room temperature the chemical shift of cyclohexane protons is an average of the chemical shifts of the axial and equatorial protons. Explain.

16.15   The two lines in the proton magnetic resonance spectrum for the two methyl groups connected to nitrogen in *N*,*N*-dimethylacetamide coalesce when the temperature is raised. What is the rate constant for the cis–trans isomerization when the multiplet structure is just lost at 331 K? The difference in chemical shifts between the two peaks is 10.85 Hz.

16.16   Calculate the transition (Larmor) frequency of a free electron in a 3-T field. What energy in cm^{-1} does this correspond to?

16.17   Line separations in ESR may be expressed in G or MHz. Show how the conversion factor 1 T = 2.80 × 10^4 MHz is obtained.

16.18   Sketch the ESR spectrum expected for p-benzosemiquinone radical ion:

The four hydrogens are magnetically equivalent.

16.19   Sketch the ESR spectrum for an unpaired electron in the presence of three protons for the following cases: (a) the protons are not equivalent, (b) the protons are equivalent, and (c) two protons are equivalent and the third is different.

16.20   An unpaired electron in the presence of two protons gives the following four-line ESR spectrum: $\Delta B/10^{-4}$ T = 0, 0.5, 3.5, 4. What are the two coupling constants in T and in MHz?

16.21   Sketch the ESR spectrum of CH_3 radical.

16.22   What is the magnitude of the magnetic moment of the proton?

16.23   Using data from Table 16.1, calculate the magnetic field strengths at which (a) ^{13}C and (b) ^{19}F will have splittings corresponding to 5 MHz.

16.24   With superconducting magnets, it is possible to reach magnetic fields of ~12 T. At this field strength, what is the splitting between spin levels for protons in MHz and cm^{-1}?

16.25   Calculate the magnetic field strength B necessary for resonance at 200 MHz for ^{19}F and ^{13}C.

16.26   A sample containing protons and fluorines (^{19}F) is placed in a magnetic field of 1 T at 25 °C. Calculate the difference in the fraction of spins lined up parallel to the field and those antiparallel for both nuclei.

16.27   Sketch the proton resonance spectrum for a compound, represented by AMX, with three protons with rather different chemical shifts ($\nu_0\delta_A = 100$, $\nu_0\delta_M = 200$, and $\nu_0\delta_X = 700$ Hz). The three protons are coupled with $J_{AM} = 9$, $J_{AX} = 7$, and $J_{MX} = 3$ Hz.

16.28   Describe the proton and deuteron NMR spectra of HD, neglecting the quadrupole magnetic moment of the deuteron.

16.29   On the basis of the spin–spin coupling constants in Table 12.3, describe the spectrum expected for

The protons are labeled to assist with labeling the spectrum.

16.30   Calculate the resonance frequency for electrons at 0.33 T.

16.31   Sketch the ESR spectrum for the ethyl radical with the indicated hyperfine splittings:

$$\cdot CH_2\!-\!CH_3$$
$$2.238 \times 10^{-3} \text{ T} \quad 2.687 \times 10^{-3} \text{ T}$$

16.32   Sketch the ESR spectrum for an unpaired electron in the presence of ^{14}N and ^1H.

PART THREE
STATISTICAL MECHANICS AND KINETICS

Part I dealt with the macroscopic properties of matter at equilibrium, and Part II dealt with the properties of atoms and molecules, again without much consideration of changes with time. In view of the very detailed and accurate information provided by quantum mechanics, it is desirable to use this information to predict thermodynamic properties. Statistical mechanics is the science that connects the properties of individual molecules with the thermodynamic properties of matter in the bulk. For ideal gases it is possible to calculate thermodynamic properties from information obtained from spectroscopy. We will illustrate that by calculating equilibrium constants of reactions of small molecules. The calculation of thermodynamic properties of dense gases, liquids, and solids is more difficult because of intermolecular interactions.

Kinetic theory introduces the calculation of the rates of certain processes by use of a simple model of atoms and molecules in the gas phase. The probabilities of molecular speeds and the values of average speeds depend on the molecular mass and temperature for noninteracting gas molecules. The frequency of collisions and the transport properties (viscosity, diffusion, and heat conduction) for gases of rigid spherical molecules can be calculated. However, the behavior of real gases is more complicated again because of intermolecular interactions.

The prediction of rates of chemical reactions is much more difficult, and so we will first consider the experimental aspects of gas reactions and the use of this information to obtain mechanisms of reactions. Then we turn to chemical dynamics to learn about the role of the transition state and to photochemistry to learn about the various processes that can occur after a molecule has absorbed a photon.

The last chapter in this part of the book deals with the kinetics of reactions in the liquid state. The study of viscosity, diffusion, and electrical transport of ions provides information that is useful in understanding the rates of reactions in liquids. Relaxation methods are useful for studying very fast reactions in the liquid phase, and the theory of diffusion-controlled reactions yields an upper limit for the rate constants of bimolecular reactions. This will help us to better understand acid–base catalysis, enzyme catalysis, and the rates of electrochemical reactions.

17
Statistical Mechanics

The equilibrium properties of matter may be considered from two points of view: the macroscopic and the microscopic. Thermodynamics is a macroscopic view which describes the behavior of large numbers of molecules in terms of pressure, volume, composition, and exchanges of heat and work. The quantitative relationships between various measured properties are not based on any model of the microscopic structure of matter.

On the other hand, quantum mechanics provides a microscopic description for the structure and interactions of molecules. Ideally, we would like to be able to predict the thermodynamic behavior of substances using our knowledge about individual molecules obtained from spectroscopic measurements and from theoretical calculations of wave functions.

Statistical mechanics provides the needed bridge between microscopic mechanics (classical and quantum) and macroscopic thermodynamics. The classical aspects of the science were developed during the latter part of the 19th century by Boltzmann in Austria, Maxwell in England, and Gibbs in the United States. From their work, we can now calculate the thermodynamic properties for ideal gases from information on single molecules. The molecular information required includes vibrational frequencies, moments of inertia, and energies of dissociation. For simple molecules the values of thermodynamic properties so obtained are often more accurate than the ones measured directly. For more complicated systems, especially those involving the strong interactions between molecules, the use of statistical mechanics is much more difficult and is the subject of current research. Statistical mechanics also helps us to understand the properties of real gases, solids, polymers, and biomacromolecules.

Statistical mechanics provides insight into the laws of thermodynamics, and through it we will see heat, work, temperature, irreversible processes, and state functions in a new light.

17.1   The Microscopic State of a System

In thermodynamics the state of a macroscopic system is described by specifying properties such as P, T, V, n, U, H, S, A, G, and μ. Some of these are referred to as **mechanical properties** (P, V, n, U, and H) because their definition does not involve the concept of temperature. We shall see that these properties for ideal gases can be calculated rather directly from statistical mechanics, while the calculation of properties involving temperature (T, S, A, G, and μ), which are referred to as **nonmechanical properties,** requires the use of thermodynamics. Statistical mechanics averages over all the microscopic descriptions of the macroscopic system and shows how the thermodynamic properties depend on the properties of individual atoms or molecules.

In **classical mechanics,** the state of a macroscopic system containing one kind of structureless particle (e.g., atoms) is described in principle by specifying the coordinates and velocity components of all of the individual particles. Thus, $6N$ coordinates are required for a classical description. In **quantum mechanics,** the microstate of a system is described in principle by specifying the total wavefunction Ψ of the N particles. In the case of noninteracting particles, Ψ becomes a product of single-particle wavefunctions, each specified by a set of quantum numbers. If only structureless particles are involved, $3N$ quantum numbers are required, as we have seen in connection with the particle in a three-dimensional box (Section 10.10). If the particles are molecules, more quantum numbers have to be given because the molecules have internal degrees of freedom. Thus, a quantum mechanical system containing N molecules may be described completely at some instant of time by n_1, n_2, . . . , n_N, where n_i represents all the quantum numbers for molecule i. Since the particles are indistinguishable, we can also describe the state by the number of particles with a given set of quantum numbers.

The first thing we must recognize is that a given thermodynamic (macroscopic) state of a system does not correspond to a particular microscopic state. **A system in a given macroscopic state can be in any one of a very large number of microscopic states that all have the same macroscopic properties.** We will represent the number of microstates in a macrostate by Ω. Later we will calculate the number Ω of sets of coordinates and velocities (in classical mechanics) or quantum numbers for the particle in a box (in quantum mechanics) of a mole of ideal gas molecules that correspond to a given temperature, pressure, and volume. The system in a particular macroscopic state could be in any one of these microscopic states.

In discussing the quantum mechanical state of a collection of particles, the concept of symmetry is very important. In Section 11.7 on the Pauli exclusion principle, we found that there are two types of particles with respect to spin. **Fermions** have half-integral spins ($s = \frac{1}{2}, \frac{3}{2}, \ldots$) and antisymmetric wavefunctions. **Bosons** have integral spins ($s = 0, 1, 2, \ldots$) and symmetric wavefunctions. This means that when two particles in a collection of particles are interchanged (see Section 11.7), the wavefunction either is unchanged (symmetric for bosons) or changes sign (antisymmetric for fermions). The number of microscopic states corresponding to a given macroscopic state is different for bosons and fermions, but for now we will be concerned only with collections of particles for which the number N_i of particles in an energy level with energy ϵ_i is much smaller than the degeneracy g_i of that level. When this is true, the number of microscopic states approaches the same value for bosons and fermions, and we obtain what is called **Maxwell–Boltzmann statistics** for both kinds of particles. (Maxwell–Boltzmann statistics had been worked out before the ideas about symmetrical and antisymmetrical wavefunctions had been developed.) Therefore, we do not have to consider the role of symmetry in detail at this point. Maxwell–Boltzmann statistics also involves the assumption that the particles in the system do not interact; later in Section 17.14 we will consider systems in which the particles do interact. In the next section the Maxwell–Boltzmann distribution is derived, but first we must introduce two equations that you may have seen before.

Consider a system containing N independent particles in which each particle can be in one of a series of quantum levels with energies $\epsilon_0, \epsilon_1, \ldots$, where ϵ_0 is the ground-state energy. How many ways W can N **distinguishable** particles be distributed between these molecular energy levels with

N_0 particles in energy level ϵ_0 with degeneracy g_0

N_1 particles in energy level ϵ_1 with degeneracy g_1

$\ldots$

N_i particles in energy level ϵ_i with degeneracy g_i

$\ldots$

At first we ignore the degeneracies and obtain an expression for the **number of ways** W of distributing N distinguishable particles between the various levels with N_0 particles having energy ϵ_0, N_1 particles having energy ϵ_1, and so on. This is the same problem as the calculation of the number of ways that N distinguishable balls can be distributed between boxes with a given number of balls, $N_0, N_1, N_2, \ldots$, in each box:

$$W = \frac{N!}{N_0! N_1! N_2! \cdots N_i! \cdots} = \frac{N!}{\prod_i N_i!} \qquad (17.1)$$

Note that $0! = 1$.

Example 17.1

(*a*) In how many ways W can six distinguishable molecules be placed in three different energy levels with three molecules in the first level, two in the second, and one in the third, without any regard for the energy required?

$$W = \frac{6!}{3!2!1!} = 60$$

(*b*) Write down the different ways of distributing three balls (a, b, and c) between two boxes.

There are two possibilities: three balls in one and none in the other ($W = 2$) or two balls in one and one in the other ($W = 6$). The eight different distributions are abc/–, ab/c, ac/b, bc/a, –/abc, c/ab, b/ac, and a/bc.

Fortunately, $N!$ and $\ln N!$ can be represented in a simple way if N is a large number. According to Stirling's approximation,

$$\ln N! = N \ln N - N \quad \text{or} \quad N! = e^{-N}N^N \quad (17.2)$$

You can try this out with some small numbers, and you will see that the approximation becomes better and better as N increases. We will be dealing with such large numbers (order of 10^{23}) that we will not have to worry about the fact that this is only an approximation.

17.2   The Maxwell–Boltzmann Distribution

This distribution of independent particles between energy levels is calculated for distinguishable particles, but it will turn out later that the Maxwell–Boltzmann distribution also applies to indistinguishable particles. The particles (atoms or molecules) of a given species are indistinguishable. In applying statistical mechanics, we will primarily be interested in systems of indistinguishable particles, but it is easier to make derivations with distinguishable particles (for example, equation 17.1 applies to distinguishable particles) and correct for indistinguishability at the end. We want to calculate the **number Ω of possible microstates** for a system containing a very large number of particles using equation 17.1 for the number W of different ways of distributing particles between energy levels, with a specified number of particles in each level. However, since the molecules are assumed to be distinguishable, we have to take into account the number of different ways that N_i particles can be arranged in an energy level with a degeneracy of g_i. Each of these arrangements constitutes a different microscopic state.

In discussing the eigenstates of simple atomic and molecular systems, we found that the degeneracy of the energy levels arose from the degeneracies of different degrees of freedom: rotational, vibrational, electronic, spin, and translational. The total degeneracy is the product of these individual degeneracies. In Chapter 10, we showed that the **degeneracy*** of the translational levels (of

---

* In calculating this degeneracy, we implicitly lumped into a single level all those states whose translational energies are so slightly different that they are, for all practical purposes, equal.

a particle in a box) at thermal energies is enormous (approximately 10^{20}). Therefore, the N_i distinguishable particles in a level with degeneracy g_i can be arranged in $g_i^{N_i}$ ways. There are g_i possibilities for placing the first particle in any of the g_i states that have energy ϵ_i, g_i for placing the second, and g_i for placing the N_ith. Thus, the total number Ω_{dis} of microstates for the system of distinguishable particles with this set of N_i is $W \prod_i g_i^{N_i}$ or

$$\Omega_{\text{dis}} = \frac{N!}{\prod_i N_i!} \left(\prod_i g_i^{N_i} \right) = N! \prod_i \frac{g_i^{N_i}}{N_i!} \tag{17.3}$$

This is the number of ways that distinguishable particles can be assigned to energy states for a given set of N_i values. However, we have to take into account the fact that there are constraints on the N_i values due to the specified number of particles N and a specified total energy U:

$$N = \sum_i N_i \tag{17.4}$$

$$U = \sum_i N_i \epsilon_i \tag{17.5}$$

There are many distributions $[N_0, N_1, \ldots, N_i, \ldots]$ that satisfy these constraints; they can be represented as follows:

$$[N_0', N_1', \ldots, N_i', \ldots]$$
$$[N_0'', N_1'', \ldots, N_i'', \ldots]$$
$$[N_0''', N_1''', \ldots, N_i''', \ldots]$$

$$\cdots$$

The numbers of ways the N particles can be arranged in each of these distributions are Ω', Ω'', Ω''', ... Thus, a macroscopic system specified by (N, U, V) can be in any one of $\Omega' + \Omega'' + \Omega''' + \cdots$ microstates. **It is a fundamental postulate of statistical mechanics that the system is equally likely to be in any one of these possible microscopic states.** This is often called the principle of **equal a priori probabilities.** The only way to tell whether this assumption is correct is to compare the predictions of statistical mechanics with experimental results. All results indicate that this basic assumption is correct.

Furthermore, for very large N, it can be shown that only the microstate with the largest Ω (Ω_{max}) makes any effective contribution to the sum $\Omega' + \Omega'' + \cdots$. Therefore, we only need to find that distribution that has the maximum Ω, or, equivalently, the maximum $\ln \Omega$. We use the method of Lagrange multipliers (see Section 5.14) to find this distribution subject to the constraints 17.4 and 17.5. Using Stirling's approximation and $N = \sum_i N_i$, equation 17.3 can be written

$$\ln \Omega = \ln N! + \sum_i (\ln g_i^{N_i} - \ln N_i!)$$

$$= N \ln N - N + \sum_i (N_i \ln g_i - N_i \ln N_i + N_i)$$

$$= N \ln N + \sum_i N_i \ln \frac{g_i}{N_i} \tag{17.6}$$

To find the maximum value of $\ln \Omega$, the first step is to express the condition that $\ln \Omega$ is maximized and the two constraints (equations 17.4 and 17.5) in differential form. Since $\ln \Omega$ is a function of the N_i's, we can write the total differential of $\ln \Omega$ and set it equal to zero:

$$\text{d} \ln \Omega = \sum_i \frac{\partial \ln \Omega}{\partial N_i} \, \text{d}N_i = 0 \tag{17.7}$$

This is subject to the conditions that

$$\text{d}N = \sum_i \text{d}N_i = 0 \tag{17.8}$$

$$\text{d}U = \sum_i \epsilon_i \, \text{d}N_i = 0 \tag{17.9}$$

In the method of **Lagrange multipliers,** the constraints are multiplied by undetermined multipliers and added to the differential of the function (equation 17.7) that we want to maximize. This yields

$$\sum_i \left(\frac{\partial \ln \Omega}{\partial N_i} + \alpha - \beta \epsilon_i \right) \text{d}N_i = 0 \tag{17.10}$$

where α and $-\beta$ are the undetermined multipliers. For this sum to be equal to zero, each of the coefficients of $\text{d}N_i$ has to be equal to zero:

$$\frac{\partial \ln \Omega}{\partial N_i} + \alpha - \beta \epsilon_i = 0 \tag{17.11}$$

Now we make use of the fact that $\partial \ln \Omega / \partial N_i$ can be written in a different way by writing out the differential of equation 17.6 (see problem 17.1):

$$\text{d} \ln \Omega = \sum_i \ln \frac{g_i}{N_i} \, \text{d}N_i - \sum_i \text{d}N_i$$

$$= \sum_i \left(\ln \frac{g_i}{N_i} \right) \text{d}N_i \tag{17.12}$$

(Note that $\sum_i \text{d}N_i = 0$.) Since $\ln \Omega = f(N_i)$, $\text{d} \ln \Omega = \sum (\partial \ln \Omega / \partial N_i) \, \text{d}N_i$, and

$$\frac{\partial \ln \Omega}{\partial N_i} = \ln \frac{g_i}{N_i} \tag{17.13}$$

Substituting equation 17.13 into equation 17.11 leads to the Maxwell–Boltzmann distribution:

$$\ln \frac{g_i}{N_i} + \alpha - \beta \epsilon_i = 0 \tag{17.14}$$

or

$$N_i = g_i \exp(\alpha - \beta \epsilon_i) \tag{17.15}$$

We can evaluate α by substituting this relation into $N = \sum_i N_i$:

$$N = (\exp \alpha) \sum_i g_i \exp(-\beta \epsilon_i) = (\exp \alpha)q \tag{17.16}$$

where q is defined as the **molecular partition function:**

$$q \equiv \sum_i g_i \exp(-\beta \epsilon_i) \tag{17.17}$$

Thus, the **Maxwell–Boltzmann distribution** is

$$N_i = \frac{N}{q} g_i \exp(-\beta\epsilon_i) \qquad (17.18)$$

This shows that the most probable number of particles in energy level i is proportional to $\exp(-\beta\epsilon_i)$, where β remains to be evaluated. This equation is very important, and we have already used it in spectroscopy (Section 14.1).

The Maxwell–Boltzmann distribution is the most probable distribution, but there might be many other distributions $[N_0, N_1, \ldots, N_i, \ldots]$ that are nearly as probable. A more complete analysis, which we cannot go into here, shows that, for macroscopic systems, the fluctuations away from the most probable distribution are entirely negligible. Therefore, the most probable distribution is the actual distribution.

The molecular partition function is sometimes referred to as a "sum over states." When it is written in the form of equation 17.17, it is a sum over energy levels. The molecular partition function is important because it is the link between the properties of the molecules in a system and its macroscopic properties.

17.3 Evaluation of β

To evaluate β we compare the expressions for the differential of the internal energy from thermodynamics and from statistical mechanics. For a reversible process involving only PV work, the thermodynamic expression for the differential of the internal energy is

$$dU = dq_{\text{rev}} - P\,dV = T\,dS - P\,dV \qquad (17.19)$$

For a system with independent particles (for example, an ideal gas), the internal energy is given by equation 17.5, and so the differential of the internal energy is given by

$$dU = \sum_i \epsilon_i\,dN_i + \sum_i N_i\,d\epsilon_i \qquad (17.20)$$

The first term in this equation can be identified with $dq_{\text{rev}} = T\,dS$ because the addition of heat to the gas changes the populations of energy levels:

$$T\,dS = \sum_i \epsilon_i\,dN_i \qquad (17.21)$$

The second term in equation 17.20 can be identified with the reversible work term:

$$-P\,dV = \sum_i N_i\,d\epsilon_i \qquad (17.22)$$

The performance of work on the gas changes the volume, which changes the energy levels (see the particle-in-a-box equation 10.76).

From a microscopic point of view, the equilibrium state of a system of particles at specified N, U, V is that with the *maximum* number Ω of microstates. From a thermodynamic point of view, the equilibrium state of an isolated system is that with the *maximum* entropy S. Thus, it is reasonable to expect a simple relation between S and Ω. To establish this relationship consider the entropy of a system composed of two subsystems, one with entropy S_1 having Ω_1 microstates, and the other with entropy S_2 having Ω_2 microstates. The combined system has entropy $S = S_1 + S_2$ and $\Omega = \Omega_1\Omega_2$ microstates, since each state of the first system can be combined with any state of the second. Thus,

$$S = f(\Omega_1\Omega_2) = S_1 + S_2 = f(\Omega_1) + f(\Omega_2) \tag{17.23}$$

The only function $f(\Omega)$ with this property is

$$f(\Omega) = a \ln \Omega + b \tag{17.24}$$

where a and b are constants. Thus, $S = a \ln \Omega + b$. Since the constant b cancels in any calculation of the change in entropy, we can set b equal to zero. We found in Section 3.12 that $a = k$, the **Boltzmann constant,** equal to R/N_A, so that

$$S = k \ln \Omega \tag{17.25}$$

Note that by maximizing Ω at constant N and U (and constant V), we have maximized the entropy. Thermodynamics tells us that this procedure yields the equilibrium state for the system; thus, maximizing the number Ω of microstates is equivalent to maximizing S. **This Boltzmann relation is used as a postulate in statistical mechanics.**

To obtain an expression for β, we compare the expression for the entropy obtained from statistical mechanics with the definition of entropy in thermodynamics, namely,

$$dS = \frac{dq_{rev}}{T} \tag{17.26}$$

where this q is heat. In statistical mechanics the differential of the entropy is given by

$$dS = k \, d \ln \Omega \tag{17.27}$$

Since $\ln \Omega = f(N_1, N_2, \ldots, N_i, \ldots)$, $d \ln \Omega$ can be written in terms of its partial derivatives:

$$dS = k \, d \ln \Omega = k \sum_i \frac{\partial \ln \Omega}{\partial N_i} \, dN_i \tag{17.28}$$

Here again, N_i is used as an independent variable. These partial derivatives are given by equation 17.13, so that

$$dS = k \sum_i \ln \frac{g_i}{N_i} \, dN_i \tag{17.29}$$

The Maxwell–Boltzmann distribution (equation 17.18) can be used to simplify this expression:

$$dS = k \sum_i \left(\ln \frac{q}{N} + \beta\epsilon_i \right) dN_i$$
$$= k\beta \sum_i \epsilon_i \, dN_i \tag{17.30}$$

since $\sum_i dN_i = 0$. Since equation 17.30 must agree with equation 17.26 and $\sum_i \epsilon_i \, dN_i$ corresponds to dq_{rev}, we see that

$$\beta = \frac{1}{kT} \tag{17.31}$$

Thus, we can now write the **molecular partition function** (equation 17.17) as

$$q \equiv \sum_i g_i \exp\left(-\frac{\epsilon_i}{kT}\right) \tag{17.32a}$$

and the Maxwell–Boltzman distribution (equation 17.18) as

$$\frac{N_i}{N} = \frac{g_i \exp(-\epsilon_i/kT)}{q} \tag{17.32b}$$

It is important to notice the effect of temperature on the molecular partition function. As the temperature decreases, the term with the lowest energy becomes the most important relative to the other terms. This is the term for the ground state, which is usually taken to have an energy of zero. If the ground state does have an energy of zero, its contribution to the partition function is g_0 at all temperatures. As $T \to 0$, the higher terms in the partition function approach zero, and so $q \to g_0$. Now consider what happens to the partition function when the temperature is raised to very high values. As the temperature is raised to infinity, the exponential factors in the partition function each approach the value 1, and so the partition function approaches the number of terms in the sum. Therefore, **q is roughly equal to the number of states available to the system at a given temperature.**

Example 17.2
Use the Maxwell–Boltzmann distribution to calculate the fractional populations of the first four vibrational levels of an oxygen molecule of 2000 K. The fundamental vibration frequency ν is $4.737 \times 10^{13} \text{ s}^{-1}$. If the harmonic oscillator approximation applies, then $\epsilon_v = v h\nu$, $v = 0, 1, 2, \ldots$, and $q_v = 1/[1 - \exp(-h\nu/kT)]$, as will be shown in Section 17.8. The vibrational levels are not degenerate.
Equation 17.32b becomes

$$\frac{N_v}{N} = \left[1 - \exp\left(-\frac{h\nu}{kT}\right)\right] \exp\left(-\frac{v h\nu}{kT}\right)$$

Since

$$h\nu/kT = \frac{(6.626 \times 10^{-34} \text{ J s})(4.737 \times 10^{13} \text{ s}^{-1})}{(1.381 \times 10^{-23} \text{ J K}^{-1})(2000 \text{ K})}$$

$$= 1.136$$

and

$$1 - \exp\left(-\frac{h\nu}{kT}\right) = 0.679$$

the fractional populations of the first four levels are given by

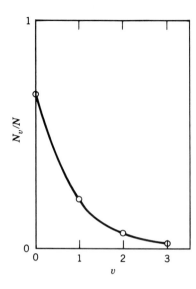

Figure 17.1 Fractional populations at 2000 K for the first four vibrational levels of an oxygen molecule.

$$N_0/N = 0.679 \exp(-0 \times 1.136) = 0.679$$
$$N_1/N = 0.679 \exp(-1 \times 1.136) = 0.218$$
$$N_2/N = 0.679 \exp(-2 \times 1.136) = 0.070$$
$$N_3/N = 0.679 \exp(-3 \times 1.136) = 0.022$$

This distribution at 2000 K is shown in Fig. 17.1. To see the effect of temperature on the distribution, you should calculate the distribution at 3000 K.

Now that we have the Maxwell–Boltzman distribution, we can illustrate the changes that take place in the particle-in-a-box system when the temperature is doubled and when the system is compressed to $2^{-1/2}$ of its initial length. In Fig. 17.2a the black points show the fractional occupations (N_i/N) of the successive energy levels of a particle in a one-dimensional box. The particle is assumed to be structureless; that is, it does not have any internal degrees of freedom. Since $E = h^2 n^2 / 8ma^2$, the energies of these levels are proportional to 1, 4, 9, 16, 25, . . . , and so the ordinate of the plot simply gives these relative energies. When the temperature is doubled, the occupation of the lowest level decreases and the populations of the higher levels increase, as shown by the

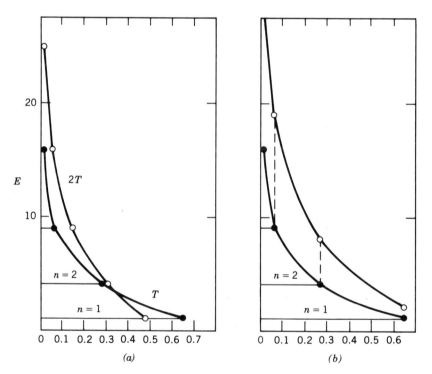

Figure 17.2 Relative energy versus fractional occupation (N_i/N) for a particle in a one-dimensional box. (*a*) The black points give the equilibrium distribution at some temperature T. The open points give the equilibrium distribution at $2T$. (*b*) The black points give the equilibrium distribution for a box of a certain length and the open points give the equilibrium distribution after the system has been compressed adiabatically and reversibly to $2^{-1/2}$ of its original length.

open points. This illustrates the fact, which was referred to above, that when the temperature is raised, there is no effect on the energies of the levels for the system, but the higher levels become more highly populated.

Figure 17.2b illustrates the effect of compressing the one-dimensional box containing the particle. The black points show the populations of the levels before compression. Now assume that the system is compressed adiabatically and reversibly and that enough work is done on the system to reduce the length of the box to $2^{-1/2}$ of its original length. Since $E = h^2n^2/8ma^2$, the energies of the successive energy levels are each doubled by the compression. The compression of the system does not change the occupations of the levels, and so the equilibrium distribution after compression is shown by the open points. The distribution among energy levels does not change because both the energies of the levels and the temperature are doubled by the compression so that there is no effect on the terms in the molecular partition function.

In the next section we will see that the Maxwell–Boltzmann equation is all we need to calculate the mechanical thermodynamic properties of a system of noninteracting particles.

17.4 Calculation of Mechanical Thermodynamic Properties for Independent Particles

The mechanical thermodynamic properties P, U, and H can readily be expressed in terms of the molecular partition function q. The calculation of the nonmechanical properties (S, A, and G) brings in the question as to whether the particles are distinguishable or indistinguishable, and these nonmechanical properties will be discussed in the next section.

The statistical mechanical expression for the pressure can be obtained from equation 17.22. Replacing N_i with the Maxwell–Boltzmann distribution (equation 17.32b) yields

$$-P \, dV = \frac{N}{q} \sum_i g_i \exp\left(-\frac{\epsilon_i}{kT}\right) d\epsilon_i \qquad (17.33)$$

When work is done on the gas, the volume changes and this changes the energy levels. The differential change in the molecular partition function q (equation 17.17) at fixed temperature is

$$dq = \sum_i g_i \exp\left(-\frac{\epsilon_i}{kT}\right)\left(-\frac{1}{kT}\right) d\epsilon_i \qquad (17.34)$$

Eliminating $d\epsilon_i$ between the two preceding equations yields

$$-P \, dV = \frac{N}{q}(-kT \, dq) = -NkT \, d \ln q \qquad (17.35)$$

Thus, the **pressure** of the ideal gas is given by

$$P = NkT \left(\frac{\partial \ln q}{\partial V}\right)_T \qquad (17.36)$$

The internal energy U is given by the first law only within an additive constant, but we will ignore this for the present in order to simplify the notation.

Later we will find it necessary to represent the internal energy calculated from statistical mechanics by $U - U(0)$, where $U(0)$ is the reference value. Substituting the Maxwell–Boltzmann distribution in $U = \sum_i N_i \epsilon_i$ yields

$$U = \frac{N}{q} \sum_i g_i \epsilon_i \exp\left(-\frac{\epsilon_i}{kT}\right) \qquad (17.37)$$

If the volume is constant, the energy levels of the independent particles are constant so that

$$\left(\frac{\partial q}{\partial T}\right)_V = \frac{1}{kT^2} \sum_i g_i \epsilon_i \exp\left(-\frac{\epsilon_i}{kT}\right) = q\left(\frac{\partial \ln q}{\partial T}\right)_V \qquad (17.38)$$

Eliminating the summation between equations 17.37 and 17.38 yields

$$U = NkT^2 \left(\frac{\partial \ln q}{\partial T}\right)_V \qquad (17.39)$$

Thus, the **internal energy** of an ideal gas can be calculated from the derivative of the molecular partition function with respect to temperature at constant volume.

The **enthalpy** is defined by $H = U + PV$, and so it is given by

$$H = NkT \left[T\left(\frac{\partial \ln q}{\partial T}\right)_V + 1 \right] \qquad (17.40)$$

Here we have used $PV = NkT$, which will be derived from statistical mechanics in equation 17.75. Thus, the calculation of the mechanical thermodynamic properties P, U, and H for an ideal gas comes down to the calculation of $(\partial \ln q/\partial V)_T$ and $(\partial \ln q/\partial T)_V$.

17.5   Calculation of Thermodynamic Properties Involving T for Independent Particles

To express the entropy in terms of the molecular partition function, we use the Boltzmann postulate that $S = k \ln \Omega$ (equation 17.25). In contrast with the calculation of the mechanical properties, the calculation of the entropy, and therefore the calculation of the Helmholtz energy and the Gibbs energy, depends upon whether the particles are indistinguishable or distinguishable, as we shall see below. First we will consider that the particles are distinguishable because we have an expression for Ω_{dis} in equation 17.3.

Equation 17.3 may be written

$$\ln \Omega_{\text{dis}} = \ln N! + \sum_i \ln \frac{g_i^{N_i}}{N_i!} \qquad (17.41)$$

Using Stirling's approximation and $\sum_i N_i = N$ yields

$$\ln \Omega_{\text{dis}} = \sum_i \left(N_i \ln \frac{g_i}{N_i} \right) + N \ln N \qquad (17.42)$$

Thus, the entropy of a system containing distinguishable particles is given by

$$S_{dis} = k \left[\sum_i \left(N_i \ln \frac{g_i}{N_i} \right) + N \ln N \right] \qquad (17.43)$$

The ratio g_i/N_i can be expressed in terms of the molecular partition function by use of the Maxwell–Boltzmann distribution (equation 17.32b), and equation 17.5 can be used to obtain the following equation:

$$
\begin{aligned}
S_{dis} &= k \left[\sum_i N_i \left(\ln \frac{q}{N} + \frac{\epsilon_i}{kT} \right) + N \ln N \right] \\
&= k \left(N \ln \frac{q}{N} + \frac{U}{kT} + N \ln N \right) \\
&= Nk \left[T \left(\frac{\partial \ln q}{\partial T} \right)_V + \ln q \right]
\end{aligned} \qquad (17.44)
$$

As we pointed out above, q is proportional to V, the total volume, so q/N is an intensive quantity. Thus, in the formula for S_{dis}, there is a term ($N \ln q$) that is not extensive. Therefore, the form for S_{dis} is not satisfactory for the entropy of a gas because it is not extensive. The problem arises because we have tacitly assumed the particles are distinguishable in deriving our equation.

Since the particles in a gas are actually indistinguishable, we have over-counted by using equation 17.3. We need to divide by $N!$, which is the number of ways N different objects can be permuted. Thus,

$$\ln \Omega_{indis} = \sum_i \ln \frac{g_i^{N_i}}{N_i!} = \sum_i (N_i \ln g_i - \ln N_i!) \qquad (17.45)$$

When Stirling's approximation and $N = \sum_i N_i$ are used, we obtain

$$
\begin{aligned}
\ln \Omega_{indis} &= \sum_i (N_i \ln g_i - N_i \ln N_i + N_i) \\
&= \sum_i N_i \ln \frac{g_i}{N_i} + N
\end{aligned} \qquad (17.46)
$$

Thus, the entropy of an ideal gas is given by

$$S_{indis} = k \left[\sum_i \left(N_i \ln \frac{g_i}{N_i} \right) + N \right] \qquad (17.47)$$

The ratio g_i/N_i can be expressed in terms of the molecular partition function by use of the Maxwell–Boltzmann distribution, and substituting equation 17.32b into equation 17.47 yields

$$
\begin{aligned}
S_{indis} &= k \left[\sum_i N_i \left(\ln \frac{q}{N} + \frac{\epsilon_i}{kT} \right) + N \right] \\
&= k \left(N \ln \frac{q}{N} + \frac{U}{kT} + N \right)
\end{aligned} \qquad (17.48)
$$

where equation 17.5 for the internal energy has been used. Note that all three terms in this equation are proportional to the size of the system, as required for an extensive property. When the expression for U in terms of the partition function (equation 17.39) is introduced, this equation can be written

$$S_{indis} = Nk \left[T \left(\frac{\partial \ln q}{\partial T} \right)_V + \ln \frac{q}{N} + 1 \right] \qquad (17.49)$$

This is the equation we will use later to calculate the **entropy** of an ideal gas consisting of molecules with known properties.

Since the Helmholtz energy is given by $A = U - TS$, we can now derive the expressions for the **Helmholtz energy** of a system of distinguishable particles and a system of indistinguishable particles:

$$A_{dis} = -NkT \ln q \tag{17.50}$$

$$A_{indis} = -NkT \left(\ln \frac{q}{N} + 1 \right) \tag{17.51}$$

Since the **Gibbs energy** is given by $G = H - TS$,

$$G_{dis} = -NkT(\ln q - 1) \tag{17.52}$$

$$G_{indis} = -NkT \ln \frac{q}{N} \tag{17.53}$$

Thus, we have shown that all of the thermodynamic properties of an ideal gas can be calculated if the molecular partition function can be evaluated from molecular properties. The chemical potential μ of an ideal gas can be calculated from $\mu = (\partial A_{indis}/\partial N)_{T,V}$ or from $\mu = (\partial G_{indis}/\partial N)_{T,P}$. In both cases we obtain $-kT \ln(q/N)$. We expect to obtain the same result from the two definitions because of equation 4.81.*

17.6   Molecular Partition Function for an Ideal Gas

An independent particle can have several different types of energy. An atom can have translational energy, electronic energy, and nuclear energy. A molecule can have, in addition, vibrational energy and rotational energy. As a first approximation these various kinds of energy can be considered to be independent. In that case the energy of a particle is given by

$$\epsilon = \epsilon_t + \epsilon_v + \epsilon_r + \epsilon_e + \epsilon_n \tag{17.54}$$

where the subscripts t, v, r, e, and n indicate translational, vibrational, rotational, electronic, and nuclear energy. Since transitions in nuclear energy levels are not involved in chemical reactions, we will simply ignore nuclear energies in this treatment. If the various modes of motion can be treated as independent, the degeneracy of an energy level is equal to the product of the degeneracies for the various modes, and the molecular partition function q can be written as

$$q = \sum_{ijkl} \left[g_{it}g_{jv}g_{kr}g_{le} \exp \left(-\frac{\epsilon_{it} + \epsilon_{jv} + \epsilon_{kr} + \epsilon_{le}}{kT} \right) \right]$$

$$= \sum_{ijkl} \left[g_{it} \exp \left(-\frac{\epsilon_{it}}{kT} \right) \right] \left[g_{jv} \exp \left(-\frac{\epsilon_{jv}}{kT} \right) \right]$$

$$\times \left[g_{kr} \exp \left(-\frac{\epsilon_{kr}}{kT} \right) \right] \left[g_{le} \exp \left(-\frac{\epsilon_{le}}{kT} \right) \right] \tag{17.55}$$

---

* In taking the derivative $(\partial G_{indis}/\partial N)_{T,P}$, we have to express the molecular partition function q as a function of P. This can be done by using $q = q'V = q'NkT/P$, where q' is not a function of P or V.

which is a sum of products. A sum of independent products can always be rewritten as a product of sums. This is illustrated by $(1 + 2)(3 + 4) = 21$ and the sum of all possible products, which is $3 + 4 + 6 + 8 = 21$. Thus, the preceding equation can be rewritten as

$$q = \sum_i \left[g_{it} \exp \left(-\frac{\epsilon_{it}}{kT} \right) \right] \sum_j \left[g_{jv} \exp \left(-\frac{\epsilon_{jv}}{kT} \right) \right]$$

$$\times \sum_k \left[g_{kr} \exp \left(-\frac{\epsilon_{kr}}{kT} \right) \right] \sum_l \left[g_{le} \exp \left(-\frac{\epsilon_{le}}{kT} \right) \right] \quad (17.56)$$

which is a product of sums. For molecules this equation applies only to the ground electronic state because higher electronic states have different vibrational frequencies, different moments of inertia, and, of course, different electronic energies.

Since each of the sums in equation 17.56 has the form of a partition function, the partition function of a molecule in its ground state can be written as

$$q = q_t q_v q_r q_e \quad (17.57)$$

where

$$q_t = \sum_i g_{it} \exp \left(-\frac{\epsilon_{it}}{kT} \right), \quad \text{etc.} \quad (17.58)$$

Since it is the logarithm of q that is involved in calculating thermodynamic properties, we note that

$$\ln q = \ln q_t + \ln q_v + \ln q_r + \ln q_e \quad (17.59)$$

Thus, each thermodynamic property for a system of independent particles can be written as a sum of translational, vibrational, rotational, and electronic contributions. For example, the internal energy calculated using equation 17.39 is given by

$$U = NkT^2 \left(\frac{\partial \ln q_t}{\partial T} \right)_V + NkT^2 \frac{d \ln q_v}{dT}$$

$$+ NkT^2 \frac{d \ln q_r}{dT} + NkT^2 \frac{d \ln q_e}{dT} \quad (17.60)$$

The partial derivative notation has not been used in the last three terms because the vibrational, rotational, and electronic terms are independent of volume, as we will see in the next sections. Thus, when the various modes of motion are independent, **the internal energy can be written as a sum of contributions:**

$$U = U_t + U_v + U_r + U_e \quad (17.61)$$

The translational contribution to the internal energy is given by

$$U_t = NkT^2 \left(\frac{\partial \ln q_t}{\partial T} \right)_V \quad (17.62)$$

and the other contributions are given by the successive terms in equation 17.60. Similar equations can be written out for H, C_P, and C_V.

When we calculate S, A, and G for indistinguishable particles, there is a problem in separating translational, vibrational, rotational, and electronic contributions because $\ln(q/N)$ is involved, rather than $\ln q$. The entropy for indistinguishable particles is given by equation 17.48, which becomes

$$S = Nk \ln \frac{q_t q_v q_r q_e}{N} + \frac{U_t + U_v + U_r + U_e}{T} + kN \qquad (17.63)$$

In breaking this down into contributions from the various kinds of energy, the last term and the N in the logarithm of the first term are associated with the translational contribution because it is the translational motion that makes it impossible to distinguish the particles, and it is the translational partition function that is proportional to the volume, as we shall see. This solves the problem referred to in the first sentence of this paragraph, and yields

$$S_t = Nk \ln \frac{q_t}{N} + \frac{U_t}{T} + kN \qquad (17.64)$$

$$S_v = Nk \ln q_v + \frac{U_v}{T} \qquad (17.65)$$

$$S_r = Nk \ln q_r + \frac{U_r}{T} \qquad (17.66)$$

$$S_e = Nk \ln q_e + \frac{U_e}{T} \qquad (17.67)$$

Example 17.3

Consider a molecule that has only translational energy levels and vibrational energy levels. Assume that the translational partition function is given by

$$q_t = g_{1t}e^{-\epsilon_{1t}/kT} + g_{2t}e^{-\epsilon_{2t}/kT}$$

Assume that the vibrational partition function is given by

$$q_v = g_{1v}e^{-\epsilon_{1v}/kT} + g_{2v}e^{-\epsilon_{2v}/kT}$$

Show that the molecular partition function can be written as a sum of products, that is, in the form of equation 17.55.

Equation 17.57 for the molecular partition function is

$$q = q_t q_v$$

Substituting the expressions given above and multiplying yields

$$q = (g_{1t}e^{-\epsilon_{1t}/kT})(g_{1v}e^{-\epsilon_{1v}/kT}) + (g_{1t}e^{-\epsilon_{1t}/kT})(g_{2v}e^{-\epsilon_{2v}/kT})$$
$$+ (g_{2t}e^{-\epsilon_{2t}/kT})(g_{1v}e^{-\epsilon_{1v}/kT}) + (g_{2t}e^{-\epsilon_{2t}/kT})(g_{2v}e^{-\epsilon_{2v}/kT})$$

This is the sum of products that corresponds to equation 17.55.

17.7   Translational Contributions to Thermodynamic Properties of Ideal Gases

In Chapter 10 we saw that a particle in a cubic box of volume V can have only certain energies specified by the quantum numbers n_x, n_y, and n_z. Using equation 10.76 in equation 17.17 yields the molecular partition function for translational motion:

$$q_t = \sum_{n_x} \sum_{n_y} \sum_{n_z} \exp\left[-\frac{h^2(n_x^2 + n_y^2 + n_z^2)}{8mV^{2/3}kT} \right] \tag{17.68}$$

Replacing the sum of products with a product of sums yields

$$q_t = \sum_{n_x} \exp\left(-\frac{h^2 n_x^2}{8mV^{2/3}kT} \right) \sum_{n_y} \exp\left(-\frac{h^2 n_y^2}{8mV^{2/3}kT} \right) \sum_{n_z} \exp\left(-\frac{h^2 n_z^2}{8mV^{2/3}kT} \right)$$

$$= \left[\sum_{n=1}^{\infty} \exp\left(-\frac{h^2 n^2}{8mV^{2/3}kT} \right) \right]^3 \tag{17.69}$$

To evaluate this sum, we note that for macroscopic containers the exponent is very small, unless T is very small. Since the successive terms in the summation differ by only small amounts, we may replace the summation by integration, considering the quantum number n as a continuous variable that is not restricted to integer values:

$$\int_0^{\infty} e^{-h^2 n^2/8mV^{2/3}kT} \, dn = \frac{(2\pi mkT)^{1/2} V^{1/3}}{h} \tag{17.70a}$$

The value of the definite integral involved here is given in Table 18.1.

We can calculate q_t in a manner that explicitly exhibits the large degeneracy g_t by replacing the triple sum in equation 17.68 by a triple integral:

$$q_t = \iiint dn_x \, dn_y \, dn_z \, e^{-h^2(n_x^2 + n_y^2 + n_z^2)/8mV^{2/3}kT} \tag{17.70b}$$

Defining $n^2 = n_x^2 + n_y^2 + n_z^2$, we can change to spherical coordinates n, Θ, and ϕ (just as in real three-dimensional space):

$$q_t = \iiint d\Theta \cos\Theta \, d\phi n^2 \, dn \, e^{-h^2 n^2/8mV^{2/3}kT} \tag{17.70c}$$

The integrals over angles can be done immediately to find

$$q_t = \int dn \, (4\pi n^2) \, e^{-h^2 n^2/8mV^{2/3}kT} \tag{17.70d}$$

The factor $4\pi n^2$ is the degeneracy q_t.

Thus, the **translational partition function** is given by

$$q_t = \left[\frac{(2\pi mkT)^{1/2} V^{1/3}}{h} \right]^3 = \left(\frac{2\pi mkT}{h^2} \right)^{3/2} V \tag{17.71}$$

Note that the translational partition function for a molecule is proportional to the volume V in which it moves. Thus, the translational partition function is an extensive variable.

The quantity $(h^2/2\pi mkT)^{1/2}$ is called the **thermal de Broglie wavelength** and is represented by Λ. Thus,

$$q_t = \frac{V}{\Lambda^3} \tag{17.72}$$

The condition for the applicability of Maxwell–Boltzmann statistics is that the thermal de Broglie wavelength must be small compared with the mean distance between molecules.

Example 17.4

What is the molecular partition function for translational motion of a hydrogen atom at 3000 K in a volume of 0.2494 m³? (This is the molar volume of an ideal gas at this temperature and 1 bar pressure.) What is the thermal de Broglie wavelength?

Using equation 17.71

$$q_t = \left[\frac{2\pi(1.0080 \times 10^{-3}\,\text{kg mol}^{-1})(1.3806 \times 10^{-23}\,\text{J K}^{-1})(3000\,\text{K})}{(6.022 \times 10^{23}\,\text{mol}^{-1})(6.626 \times 10^{-34}\,\text{J s})^2} \right]^{3/2} \times (0.2494\,\text{m}^3)$$

$$= 7.791 \times 10^{30}$$

Note that the molecular partition function is a dimensionless quantity and may be interpreted approximately as the number of energy levels accessible to the single hydrogen atom in this volume.

$$\Lambda = \left(\frac{V}{q_t} \right)^{1/3} = \left(\frac{0.2494\,\text{m}^3}{7.791 \times 10^{30}} \right)^{1/3} = 3.175 \times 10^{-11}\,\text{m}$$

Note that the thermal wavelength is much smaller than the length of the side of the container, as required by the derivation of equation 17.71.

The translational contribution to the internal energy of an ideal gas can be calculated by substituting equation 17.71 into equation 17.39.

$$U_t = NkT^2 \left(\frac{\partial \ln q_t}{\partial T} \right)_V = \frac{3}{2} NkT \qquad (17.73)$$

Since $Nk = nN_A k = nR$, where n is the amount of gas, $\overline{U}_t = \frac{3}{2}RT$ and $\overline{C}_V = (\partial \overline{U}/\partial T)_V = \frac{3}{2}R$, which is in agreement with the experimental molar heat capacity at constant volume for a monatomic gas. The translational contribution to the enthalpy of an ideal gas can be calculated using equation 17.40:

$$H_t = NkT \left[T \left(\frac{\partial \ln q_t}{\partial T} \right)_V + 1 \right] = \frac{5}{2} NkT \qquad (17.74)$$

Thus, the molar enthalpy is given by $\overline{H}_t = \frac{5}{2}RT$, which is in agreement with the experimental molar heat capacity at constant pressure, $\frac{5}{2}R$, for a monatomic gas.

The pressure of an ideal gas is given by equation 17.36:

$$P = NkT \left(\frac{\partial \ln q_t}{\partial V} \right)_T = \frac{NkT}{V} \qquad (17.75)$$

This is in agreement with $PV = nRT$ or $P\overline{V} = RT$, since $Nk = nR$.

Since the molecules in a gas are indistinguishable, the translational contribution to the entropy can be calculated using equation 17.49:

$$S_t = Nk \left[T \left(\frac{\partial \ln q_t}{\partial T} \right)_V + \ln \frac{q_t}{N} + 1 \right]$$

$$= Nk \left\{ \ln \left[\left(\frac{2\pi mkT}{h^2} \right)^{3/2} \frac{V}{N} \right] + \frac{5}{2} \right\} \qquad (17.76)$$

It is often convenient to calculate the entropy of independent particles as a function of pressure, and this can be done by introducing the ideal gas law:

$$S_t = Nk \left\{ \ln \left[\left(\frac{2\pi m kT}{h^2} \right)^{3/2} \frac{kT}{P} \right] + \frac{5}{2} \right\} \qquad (17.77)$$

If the pressure is equal to the standard state pressure, the standard entropy is obtained. Remember that the pressure must be expressed in pascals in this equation, if the other quantities are in SI units. Since monatomic gases have only translational energy at temperatures where electronic levels are not excited, this is the statistical mechanical basis of the Sackur–Tetrode equation, that was mentioned in Section 3.9.

Example 17.5
Express equation 17.77 in a convenient form for calculating the molar entropy of an ideal monatomic gas, which does not have electronic excitation, at temperature T in kelvins and pressure P in bars. To do this we express the mass of the atom by $A_r m_n$, where A_r is the relative atomic weight and m_n is the atomic mass constant $(m(^{12}C)/12)$. We express the pressure in equation 17.77 by $(P/P^\circ)P^\circ$, where P° is the standard state pressure, and the temperature in equation 17.77 by $(T/K)K$, where T_1 is 1 K. This leads to

$$\bar{S} = R \left\{ \frac{5}{2} + \ln \left[\left(\frac{2\pi m_n kT_1}{h^2} \right)^{3/2} \frac{kT_1}{P^\circ} \right] + \frac{3}{2} \ln A_r - \ln \frac{P}{P^\circ} + \frac{5}{2} \ln \frac{T}{K} \right\}$$

$$= R \left(-1.151\ 693 + \frac{3}{2} \ln A_r - \ln \frac{P}{P^\circ} + \frac{5}{2} \ln \frac{T}{K} \right)$$

This is a convenient form of the **Sackur–Tetrode equation.**

The statistical mechanical expressions for the translational contributions to the Helmholtz energy and the Gibbs energy are given later in Table 17.1.

The thermodynamic properties U_t°, H_t°, A_t°, and G_t° calculated with these equations approach zero as the thermodynamic temperature approaches zero. Thus, we may say that these thermodynamic properties are with respect to a standard state of a hypothetical ideal gas at absolute zero. Since a different standard state is used in chemical thermodynamics we will have to wait until Section 17.11 to compare the results of equations 17.73 and 17.74 with standard thermodynamic tables.

The methods used in deriving these equations becomes inapplicable as $T \to 0$. As $T \to 0$ the de Broglie wavelength becomes so large that the wavefunctions of the particles overlap significantly, and the Boltzmann distribution cannot be used.

Example 17.6
What are the translational contributions to $\bar{C}_P^\circ$, $\bar{S}^\circ$, $\bar{H}^\circ$, and $\bar{G}^\circ$ for O(g) at 298.15 K?

$$m = A_r m_u = (15.9994)(1.660\ 540\ 2 \times 10^{-27} \text{ kg})$$

$$= 2.656\ 77 \times 10^{-26} \text{ kg}$$

$$(\bar{C}_P^\circ)_t = \tfrac{5}{2} R = \tfrac{5}{2}(8.314\ 51 \text{ J K}^{-1} \text{ mol}^{-1}) = 20.786 \text{ J K}^{-1} \text{ mol}^{-1}$$

$$\bar{H}_t^\circ = \tfrac{5}{2} RT = \tfrac{5}{2}(8.314\ 51 \text{ J K}^{-1} \text{ mol}^{-1})(298.15 \text{ K}) = 6.197 \text{ kJ mol}^{-1}$$

$$\left(\frac{2\pi mkT}{h^2}\right)^{3/2} \frac{kT}{P^\circ} = \left[\frac{2\pi(2.656\ 77 \times 10^{-26}\ \text{kg})(1.380\ 658 \times 10^{-23}\ \text{J K}^{-1})(298.15\ \text{K})}{(6.626\ 075\ 5 \times 10^{-34}\ \text{J s})^2}\right]^{3/2}$$

$$\times \frac{(1.380\ 658 \times 10^{-23}\ \text{J K}^{-1})(298.15\ \text{K})}{10^5\ \text{N m}^{-2}} = 2.548\ 789 \times 10^6$$

$$\overline{S}_t^\circ = (8.314\ 51\ \text{J K}^{-1}\ \text{mol}^{-1})(2.5 + \ln 2.548\ 789 \times 10^6)$$

$$= 143.435\ \text{J K}^{-1}\ \text{mol}^{-1}$$

$$\overline{G}_t^\circ = -(8.314\ 51\ \text{J K}^{-1}\ \text{mol}^{-1})(298.15\ \text{K})\ln 2.548\ 789 \times 10^6$$

$$= -36.568\ \text{kJ mol}^{-1}$$

Example 17.7

What are the translational contributions to $\overline{C}_P^\circ$, $\overline{H}^\circ$, $\overline{S}^\circ$, and $\overline{G}^\circ$ for $O_2(g)$ at 298.15 K? The translational contributions to $\overline{C}_P^\circ$ and $\overline{H}^\circ$ are the same as for $O(g)$, or any other ideal monatomic gas. Since the mass of the molecule is twice that of the atom, the translational entropy is larger by $R \ln 2^{3/2} = 8.645\ \text{J K}^{-1}\ \text{mol}^{-1}$ and the translational Gibbs energy is more negative by $RT \ln 2^{3/2} = -2.577\ \text{kJ mol}^{-1}$. Thus, at 298.15 K

$$\overline{C}_P^\circ = 20.786\ \text{J K}^{-1}\ \text{mol}^{-1}$$

$$\overline{S}_t^\circ = 152.080\ \text{J K}^{-1}\ \text{mol}^{-1}$$

$$\overline{H}_t^\circ = 6.197\ \text{kJ mol}^{-1}$$

$$\overline{G}_t^\circ = -39.145\ \text{kJ mol}^{-1}$$

The values of $(\overline{C}_P^\circ)_t$ and $\overline{S}_t^\circ$ are slightly less than $\overline{C}_P^\circ$ and $\overline{S}^\circ$ in Table C.1 because of additional contributions due to rotation, vibration, and electronic excitation. The values of $\overline{H}_t^\circ$ and $\overline{G}_t^\circ$ cannot be compared with $\Delta_f H^\circ$ and $\Delta_f G^\circ$ in Table C.1 because of the difference in reference states, and there are also rotational, vibrational, and electronic contributions.

17.8   Vibrational Contributions to the Thermodynamic Properties of Ideal Gases

The vibrational levels of diatomic molecules have been discussed in Section 14.8 on spectroscopy. If there is excitation to very high vibrational levels, the anharmonicity has to be taken into account in calculating the partition function, but, for many purposes, it is sufficient to use the harmonic oscillator approximation because only lower levels are occupied. In calculating the partition function, vibrational energy is usually measured from the ground state ($v = 0$) rather than from the bottom of the potential energy curve. This simplifies the equations, because the **zero-point energy** ($h\nu/2$, where ν is the vibration frequency) does not appear.

In the harmonic oscillator approximation, the vibrational energy of a molecule with vibrational quantum number v is given by

$$\epsilon_v = v h \nu \qquad v = 0, 1, 2, \ldots \tag{17.78}$$

so that

$$q_v = \sum_{v=0}^{\infty} e^{-vh\nu/kT} = 1 + x + x^2 + x^3 + \cdots \qquad (17.79)$$

where $x = e^{-h\nu/kT}$. When $x < 1$,

$$\frac{1}{1-x} = 1 + x + x^2 + x^3 + \cdots \qquad (17.80)$$

Since $e^{-h\nu/kT} < 1$

$$q_v = \frac{1}{1 - e^{-h\nu/kT}} = \frac{1}{1 - e^{-\Theta_v/T}} \qquad (17.81)$$

where Θ_v is the **characteristic vibrational temperature** defined by $\Theta_v = h\nu/k = hc\bar{\nu}/k$. The characteristic temperature of a vibration Θ_v is the temperature at which the energy of the vibrational quantum is equal to kT. The vibrational partition function for a normal mode is generally not much greater than unity, except at very high temperatures.

When the vibrational partition function in equation 17.81 is substituted into the expressions for the various thermodynamic properties, the vibrational contributions given later in Table 17.1 are obtained. When there are several normal modes, the contributions from the various modes are summed, as indicated. For polyatomic molecules, there are $3N - 5$ normal modes of vibration for linear molecules and $3N - 6$ normal modes for nonlinear molecules, as we have seen in Section 14.8.

Example 17.8
What are the vibrational contributions to $\overline{C}_P^\circ$, $\overline{S}^\circ$, $\overline{H}^\circ$, and $\overline{G}^\circ$ for $O_2(g)$ at 298.15 K? The vibrational frequency is 1580.246 cm^{-1}.

$$\Theta_v = \frac{hc\bar{\nu}}{k} = \frac{(6.626\ 076 \times 10^{-34}\ \text{J s})(2.997\ 925 \times 10^8\ \text{m s}^{-1})(1.580\ 246 \times 10^5\ \text{m}^{-1})}{1.380\ 658 \times 10^{-23}\ \text{J K}^{-1}}$$

$$= 2273.73\ \text{K}$$

$$(\overline{C}_P^\circ)_v = (8.314\ 51\ \text{J K}^{-1}\ \text{mol}^{-1})\left(\frac{2273.73}{298.15}\right)^2 \frac{e^{2273.73/298.15}}{(e^{2273.64/298.15} - 1)^2}$$

$$= 0.236\ \text{J K}^{-1}\ \text{mol}^{-1}$$

$$\overline{S}_v^\circ = (8.314\ 51\ \text{J K}^{-1}\ \text{mol}^{-1})\left[\frac{(2273.73)/(298.15)}{(e^{2273.73/298.15} - 1)} - \ln(1 - e^{-2273.73/298.15})\right]$$

$$= 0.035\ \text{J K}^{-1}\ \text{mol}^{-1}$$

$$\overline{H}_v^\circ = (8.314\ 51\ \text{J K}^{-1}\ \text{mol}^{-1})(298.15\ \text{K})\frac{(2273.73\ \text{K}/298.15\ \text{K})}{e^{2273.73/298.15} - 1}$$

$$= 0.009\ \text{kJ mol}^{-1}$$

$$\overline{G}_v^\circ = (8.314\ 51\ \text{J K}^{-1}\ \text{mol}^{-1})(298.15\ \text{K})\ln(1 - e^{-2273.73/298.15})$$

$$= -0.001\ \text{kJ mol}^{-1}$$

The small value of the vibrational contribution to $\overline{C}_P^\circ$ at 298.15 K indicates that $O_2(g)$ has very little vibrational energy at this temperature. The classical vibrational heat capacity is $R = 8.314\ \text{J K}^{-1}\ \text{mol}^{-1}$ (Section 17.13), and this value of $(\overline{C}_P^\circ)_v$ is approached at 3000 K.

17.9   Rotational Contributions to the Thermodynamic Properties of Ideal Gases

The expression for the rotational energy of a rigid diatomic molecule was derived in Section 14.5:

$$\epsilon_r = \frac{J(J+1)h^2}{8\pi^2 I} \tag{17.82}$$

Here J is the rotational quantum number, and I is the moment of inertia of the diatomic molecule. Although the rotational energy depends only on J, the state of a rigid rotor is specified by the quantum number J and an additional quantum number M, where M can have integral values between $-J$ and $+J$. Thus, there are $2J+1$ values of M for each value of J; in other words, the rotational levels are $(2J+1)$-fold degenerate. The rotational partition function is thus

$$q_r = \sum_{j=0}^{\infty} (2J+1)e^{-J(J+1)h^2/8\pi^2 IkT} \tag{17.83}$$

For molecules with large moments of inertia, the energy levels are so close together that the summation may be replaced by an integration when T is reasonably high (above 10 to 100 K):

$$q_r = \int_0^{\infty} (2J+1)e^{-J(J+1)h^2/8\pi^2 IkT}\, dJ \tag{17.84}$$

Since $(2J+1)dJ$ is the differential of $J(J+1) = J^2 + J$, this equation may be integrated by multiplying and dividing through by $8\pi^2 IkT/h^2$ to obtain

$$q_r = \frac{8\pi^2 IkT}{h^2} \int_0^{\infty} e^{-u}\, du = \frac{8\pi^2 IkT}{h^2} = \frac{T}{\Theta_r} \tag{17.85}$$

where Θ_r is the **characteristic rotational temperature** defined by

$$\Theta_r = \frac{h^2}{8\pi^2 Ik} \tag{17.86}$$

The calculation of the rotational partition function for a homonuclear diatomic molecule (like Cl_2) has to take into account the fact that rotation by 180° interchanges two equivalent nuclei. Since the new orientation is indistinguishable from the first, we have to divide by 2 so that the indistinguishable orientations are counted only once. For a heteronuclear diatomic molecule (like HCl), a 180° rotation produces a distinguishable orientation. The effect of symmetry is taken into account by introducing a **symmetry number** σ that has the

---

* The behavior of H_2, D_2, and T_2 at low temperatures is complicated by the quantum mechanical symmetry requirements applicable to molecules containing identical nuclei. Since the proton has spin $\frac{1}{2}$, the wavefunction for H_2 must be antisymmetric to exchange of the nuclei. The wavefunctions for electronic and vibrational motion are both symmetric. Therefore, either the rotational or nuclear spin wavefunction must be antisymmetric. The rotational wavefunction for a symmetric rotor is symmetric with respect to inversion of the coordinates for even values of the rotational quantum number J, and antisymmetric for odd values of J. Therefore, for even values of J the nuclear wavefunction must be antisymmetric, and these levels have a degeneracy of 1. For odd values of J the nuclear wavefunction must be symmetric, and these levels have a degeneracy of 3. Thus, H_2 is a mixture of two molecular species. Hydrogen with a symmetric nuclear wavefuntion and J

value 2 for a homonuclear molecule and the value 1 for a heteronuclear molecule. In general, then

$$q_r = \frac{T}{\sigma \Theta_r} \qquad (17.87)*$$

Example 17.9
What is the characteristic rotational temperature Θ_r for $H_2(g)$? What is the value of the molecular partition function for rotation at 3000 K? (The moment of inertia is 4.6052×10^{-48} kg m².)

$$\Theta_r = \frac{h^2}{8\pi^2 Ik} = \frac{(6.626\ 08 \times 10^{-34}\ J\ s)^2}{8\pi^2(4.6054 \times 10^{-48}\ kg\ m^2)(1.380\ 66 \times 10^{-23}\ J\ K^{-1})} = 87.544\ K$$

$$q_r = \frac{T}{\sigma \Theta_r} = \frac{3000\ K}{(2)(87.547\ K)} = 17.134$$

The rotational contributions for diatomic molecules obtained by substituting equation 17.87 into the expressions for the various thermodynamic properties are given later in Table 17.1.

Example 17.10
What are the rotational contributions to $\overline{C}_P^\circ$, $\overline{S}^\circ$, $\overline{H}^\circ$, and $\overline{G}^\circ$ for $O_2(g)$ at 298.15 K? (The moment of inertia for $O_2(g)$ is $I = 1.9373 \times 10^{-46}$ kg m².)

$$\Theta_r = \frac{h^2}{8\pi^2 Ik} = \frac{(6.626\ 08 \times 10^{-34}\ J\ s)^2}{8\pi^2(1.9373 \times 10^{-46}\ kg\ m^2)(1.380\ 66 \times 10^{-23}\ J\ K^{-1})} = 2.079\ K$$

$$(\overline{C}_P^\circ)_r = R = 8.314\ J\ K^{-1}\ mol^{-1}$$

$$\overline{S}_r^\circ = R \ln \frac{eT}{\sigma \Theta_r} = (8.314\ 51\ J\ K^{-1}\ mol^{-1}) \ln \frac{(2.7183)(298.15\ K)}{(2)(2.079\ K)}$$

$$= 43.838\ J\ K^{-1}\ mol^{-1}$$

$$\overline{H}_r^\circ = RT = (8.314\ 51\ J\ K^{-1}\ mol^{-1})(298.15\ K) = 2.479\ kJ\ mol^{-1}$$

$$\overline{G}_r^\circ = -RT \ln \frac{T}{\sigma \Theta_r} = -(8.314\ 51\ J\ K^{-1}\ mol^{-1})(298.15\ K) \ln \frac{298.15\ K}{2 \times 2.079\ K}$$

$$= -10.591\ kJ\ mol^{-1}$$

---

odd is called **orthohydrogen.** Hydrogen with an antisymmetric nuclear wavefunction and J even is called **parahydrogen.** The rotational partition function for the equilibrium mixture is

$$q_r = \sum_{J_{even}} (2J + 1) \exp\left(\frac{J(J + 1)\Theta_r}{T}\right) + 3 \sum_{J_{odd}} (2J + 1) \exp\left(\frac{J(J + 1)\Theta_r}{T}\right)$$

If hydrogen is prepared at room temperature or higher where the ortho–para ratio is 3, it will retain this composition at lower temperatures if no catalyst (such as activated charcoal) is present, even though this is a nonequilibrium composition. If a catalyst is present, equilibrium results and a different plot of C_V versus T is obtained at low temperatures, and this plot is in agreement with the equation given above.

The rotational contributions for diatomic molecules can also be used for linear polyatomic molecules such as CO_2, C_2H_2, N_2O, and HCN. The symmetry number is 2 for symmetrical linear molecules such as CO_2 and C_2H_2, and it is 1 for unsymmetrical linear molecules such as N_2O and HCN.

A linear molecule with a small moment of inertia I has a small rotational partition function and a small rotational contribution to $S°$ and $G°$. Molecules with larger moments of inertia have larger rotational contributions to $S°$ and $G°$. The reason for this is that when the moment of inertia is large, the rotational energy levels are close together and more states are populated.

For polyatomic molecules in general the calculation of rotational contributions has to take into account the fact that the molecule may have three different moments of inertia. As shown in Section 14.5, the three moments of inertia for a spherical top molecule, like CH_4, are equal: $I_a = I_b = I_c$. Two moments are equal for a symmetric top: $I_a < I_b = I_c$ for a prolate top like CH_3Cl, and $I_a = I_b < I_c$ for an oblate top like C_6H_6. For an asymmetric top, like H_2O, the three moments of inertia are all different.

The temperatures well above the characteristic rotational temperatures it may be shown* that the rotational partition function for a nonlinear polyatomic molecule is given by

$$q_r = \frac{\pi^{1/2}}{\sigma} \left(\frac{T^3}{\Theta_a \Theta_b \Theta_c} \right)^{1/2} \tag{17.88}$$

where the characteristic rotational temperatures are given by

$$\Theta_a = \frac{h^2}{8\pi^2 I_a k} \tag{17.89}$$

$$\Theta_b = \frac{h^2}{8\pi^2 I_b k} \tag{17.90}$$

$$\Theta_c = \frac{h^2}{8\pi^2 I_c k} \tag{17.91}$$

and σ is the symmetry number. The symmetry number of a molecule is equal to the number of distinct proper rotational operations plus the identity operation (see Chapter 13). For H_2O, $\sigma = 2$. For NH_3, $\sigma = 3$. For ethylene, $\sigma = 4$. For CH_4, $\sigma = 12$. For C_6H_6, $\sigma = 12$. Each freely rotating methyl group in a molecule contributes a factor of 3 to the symmetry number.

The rotational contributions for nonlinear polyatomic molecules are given later in Table 17.1.

Nonlinear polyatomic molecules have three degrees of rotational freedom rather than the two for linear molecules; therefore, the rotational contribution to $\overline{C}_P°$ is $3R/2$, and the enthalpy is $3RT/2$ relative to the hypothetical ideal gas at absolute zero, as suggested by the principle of equipartition.

---

* D. McQuarrie, *Statistical Mechanics*, p. 133. New York: Harper & Row, 1976.

Example 17.11

For $H_2O(g)$, $I_aI_bI_c = 5.7658 \times 10^{-141}$ kg^3 m^6 and $\sigma = 2$. What is the value of the rotational partition function for $H_2O(g)$ and what are the rotational contributions to $\overline{C}_P^\circ$, $\overline{S}^\circ$, $\overline{H}^\circ$, and $\overline{G}^\circ$ at 3000 K?

$$\Theta_a\Theta_b\Theta_c = \left(\frac{h^2}{8\pi^2 k}\right)^3 \frac{1}{I_aI_bI_c} = \left[\frac{(6.626\ 076 \times 10^{-34}\ \text{J s})^2}{8\pi^2(1.380\ 661 \times 10^{-23}\ \text{J K}^{-1})}\right]^3$$

$$\times \frac{1}{5.7658 \times 10^{-141}\ \text{kg}^3\ \text{m}^6} = 1.133\ 05 \times 10^4\ \text{K}^3$$

$$q_r = \frac{\pi^{1/2}}{2}\left(\frac{3000^3}{1.133\ 05 \times 10^4}\right)^{1/2} = 1368.0$$

$$(\overline{C}_P^\circ)_r = \tfrac{3}{2}R = \tfrac{3}{2}(8.314\ 51\ \text{J K}^{-1}\ \text{mol}^{-1}) = 12.472\ \text{J K}^{-1}\ \text{mol}^{-1}$$

$$\overline{S}_r^\circ = (8.314\ 51\ \text{J K}^{-1}\ \text{mol}^{-1})\left\{\ln\left[\left(\frac{1}{2}\right)\left(\frac{3000^3\ \pi}{1.133\ 15 \times 10^4}\right)^{1/2}\right] + \frac{3}{2}\right\}$$

$$= 72.511\ \text{J K}^{-1}\ \text{mol}^{-1}$$

$$\overline{H}_r^\circ = \tfrac{3}{2}RT = \tfrac{3}{2}(8.314\ 51\ \text{J K}^{-1}\ \text{mol}^{-1})(3000\ \text{K}) = 37.415\ \text{kJ mol}^{-1}$$

$$\overline{G}_r^\circ = -(8.314\ 51\ \text{J K}^{-1}\ \text{mol}^{-1})(3000\ \text{K})\ln\left[\frac{\pi^{1/2}}{2}\left(\frac{3000^3}{1.133\ 05 \times 10^4}\right)^{1/2}\right]$$

$$= -180.118\ \text{kJ mol}^{-1}.$$

17.10 Electronic Contributions to the Thermodynamic Properties of Ideal Gases

The electronic state of an atom or of a diatomic molecule is represented by three quantum numbers:

1. An orbital quantum number L for an atom or Λ for a diatomic molecule that can have the values 0, 1, 2, ...

2. A spin quantum number S which determines the multiplicity $(2S + 1)$ of the state.

3. A total angular momentum quantum number J for an atom or Ω for a diatomic molecule.

The characteristic temperature Θ_e for an electronic transition is defined by

$$\Theta_e = \frac{\epsilon_e}{k} \tag{17.92}$$

where ϵ_e is the energy of the excited electronic energy level relative to the ground state. If energies are measured with respect to the electronic ground state, the electronic molecular partition function for an atom or molecule is given by

$$q_e = g_0 + g_1 \exp\left(-\frac{\Theta_{e1}}{T}\right) + g_2 \exp\left(-\frac{\Theta_{e2}}{T}\right) + \cdots \quad (17.93)^*$$

where the degeneracy g_i is given by

$$g_i = 2J_i + 1 \quad \text{or} \quad 2\Omega_i + 1 \quad (17.94)$$

However, for a molecule, only the thermodynamic properties of the ground state can be calculated by combining the first term of equation 17.93 with the partition functions for vibration and rotation **for the ground electronic state.** For the excited electronic levels represented by the second and third terms in equation 17.93, the vibrational frequencies and rotational frequencies are different from those in the ground state. Often excited electronic states of molecules have very much higher energies so that they affect the calculations of thermodynamic properties only at very high temperatures.

Example 17.12

What are the electronic contributions to $\overline{S}^\circ$ and $\overline{G}^\circ$ for O(g) at 298.15 K to the degree of completeness we have used here? The degeneracy in the ground state is 5, the first excited state has an energy of 158.2 cm^{-1} with respect to the ground state and has a degeneracy of 3, and the second excited state has an energy of 226.5 cm^{-1} with no degeneracy.

$$\Theta_{e1} = \frac{hc\bar{\nu}}{k} = \frac{(6.626\,076 \times 10^{-34}\ \text{J s})(2.997\,925 \times 10^8\ \text{m s}^{-1})(1.582 \times 10^4\ \text{m}^{-1})}{1.380\,658 \times 10^{-23}\ \text{J K}^{-1}}$$

$$= 227.6\ \text{K}$$

$$\Theta_{e2} = 325.9\ \text{K}$$

$$q_e = 5 + 3e^{-227.6/298.15} + e^{-325.9/298.15} = 6.7335$$

$$\overline{S}_e^\circ = R \ln q_e = (8.314\,51\ \text{J K}^{-1}\ \text{mol}^{-1}) \ln 6.7335 = 15.853\ \text{J K}^{-1}\ \text{mol}^{-1}$$

$$\overline{G}_e^\circ = -T\overline{S}_e^\circ = -(298.15\ \text{K})(15.853\ \text{J K}^{-1}\ \text{mol}^{-1}) = -4.727\ \text{kJ mol}^{-1}$$

In calculating the electronic partition function for a diatomic molecule it is convenient to measure energy with respect to the atoms in an ideal gas state at absolute zero. When this is done the calculated thermodynamic properties of the molecule are relative to its constituent atoms in their ground states. On

---

* S. J. Stricker (*J. Chem. Educ.* **43**:364 (1966)) discusses the fact that the electronic partition function q_e for the hydrogen atom at 25 °C calculated from

$$q_e = \sum_{n=1}^{\infty} n^2 e^{-\epsilon_n/kT}$$

is infinite. In this equation n^2 is the degeneracy, and ϵ_n is the energy of the orbital with quantum number n which is proportional to n^{-2}. This is contrary to experience, since it would require the population of the ground state to be zero. The problem is that there are an infinite number of terms in the partition function with an energy less than 13.60 eV. The paradox is resolved by considering the size of the orbital of a hydrogen atom for a very large quantum number n. For a finite container the maximum n is finite, and the value of q_e is not detectably different from unity. In fact, the cutoff probably comes from neighboring molecules that limit the radius of a hydrogen atom to about $(V/N)^{1/3}$.

this basis the energy of the electronic ground state of the molecule is $-D_0$, where D_0 is the spectroscopic dissociation energy (Section 14.2). Equation 17.94 may therefore be written as

$$q_e = g_0 e^{D_0/kT} \tag{17.95}$$

when T is low enough that the second and higher terms can be neglected.

The contributions to $(\overline{C}_P^\circ)_e$, $\overline{H}_e^\circ$, $\overline{S}_e^\circ$, and $\overline{G}_e^\circ$ for a molecule are therefore

$$(\overline{C}_P^\circ)_e = 0 \tag{17.96}$$

$$\overline{H}_e^\circ = -D_0 \tag{17.97}$$

$$\overline{S}_e^\circ = R \ln g_0 \tag{17.98}$$

$$\overline{G}_e^\circ = -D_0 - RT \ln g_0 \tag{17.99}$$

Example 17.13

What are the electronic contributions to $\overline{C}_P^\circ$, $\overline{H}^\circ$, $\overline{S}^\circ$, and $\overline{G}^\circ$ for $O_2(g)$ at 298.15 K? The electronic ground state is a triplet, and the spectroscopic dissociation energy D_0 is 491.888 kJ mol^{-1}. The energy of the first excited electronic state is so high that it does not have to be considered.

$(\overline{C}_P^\circ)_e = 0$

$\overline{H}_e^\circ = -D_0 = -491.888$ kJ mol^{-1}

$\overline{S}_e^\circ = (8.314\ 51$ J K^{-1} mol^{-1}) ln 3 $= 9.134$ J K^{-1} mol^{-1}

$\overline{G}_e^\circ = -491.888$ kJ mol^{-1} $-$ (8.314 51 $\times$ 10^{-3} kJ K^{-1} mol^{-1})(298.15 K) ln 3

$\qquad = -494.611$ kJ mol^{-1}

17.11 Thermodynamic Properties of Ideal Gases

The formulas for calculating the molar thermodynamic properties of ideal gases are summarized in Table 17.1.

Now that we have shown how to calculate the translational, rotational, vibrational, and electronic contributions to the thermodynamic properties, we can obtain the properties by summing up the contributions, as shown, for example, in equation 17.61.

Example 17.14

What are the values of $\overline{C}_P^\circ$, $\overline{S}^\circ$, $\overline{H}^\circ$, and $\overline{G}^\circ$ for O(g) at 298.15 K?

Atoms have only translational and electronic contributions, and therefore we sum the contributions calculated in Examples 17.6 and 17.12:

$\overline{C}_P^\circ = (\overline{C}_P^\circ)_t + (\overline{C}_P^\circ)_e = 20.786$ J K^{-1} mol^{-1} since $(\overline{C}_P^\circ)_e$ is taken here as zero

$\overline{S}^\circ = \overline{S}_t^\circ + \overline{S}_e^\circ = (143.435 + 15.853)$ J K^{-1} mol^{-1} $= 159.288$ J K^{-1} mol^{-1}

$\overline{H}^\circ = \overline{H}_t^\circ + \overline{H}_e^\circ = 6.197$ kJ mol^{-1} since $\overline{H}_e^\circ$ is taken here as zero

$\overline{G}^\circ = \overline{G}_t^\circ + \overline{G}_e^\circ = (-36.568 - 4.727)$ kJ mol^{-1} $= -41.295$ kJ mol^{-1}

Table 17.1 Contributions to Molar Thermodynamic Properties of Ideal Gases

	Translation	Vibration	Rotation — Diatomic and Polyatomic Linear	Rotation — Nonlinear	Electronic[a]
$\dfrac{\overline{U}}{R}$	$\dfrac{3}{2}T$	$\displaystyle\sum_i \dfrac{\Theta_{vi}}{\exp(\Theta_{vi}/T) - 1}$	T	$\dfrac{3}{2}T$	0
$\dfrac{\overline{H}}{R}$	$\dfrac{5}{2}T$	$\displaystyle\sum_i \dfrac{\Theta_{vi}}{\exp(\Theta_{vi}/T) - 1}$	T	$\dfrac{3}{2}T$	0
$\dfrac{\overline{C}_V}{R}$	$\dfrac{3}{2}$	$\displaystyle\sum_i \dfrac{(\Theta_{vi}/T)^2 \exp(\Theta_{vi}/T)}{[\exp(\Theta_{vi}/T) - 1]^2}$	1	$\dfrac{3}{2}$	0
$\dfrac{\overline{C}_P}{R}$	$\dfrac{5}{2}$	$\displaystyle\sum_i \dfrac{(\Theta_{vi}/T)^2 \exp(\Theta_{vi}/T)}{[\exp(\Theta_{vi}/T) - 1]^2}$	1	$\dfrac{3}{2}$	0
$\dfrac{\overline{S}}{R}$	$\ln\left[\left(\dfrac{2\pi mkT}{h^2}\right)^{3/2}\dfrac{kT}{P}\right] + \dfrac{5}{2}$	$\displaystyle\sum_i \left\{\dfrac{\Theta_{vi}/T}{[\exp(\Theta_{vi}/T)-1]}\right\} - \ln\left[1 - \exp\left(\dfrac{-\Theta_{vi}}{T}\right)\right]$	$\ln\left(\dfrac{T}{\sigma\Theta_r}\right) + 1$	$\ln\left[\dfrac{(T^3\pi/\Theta_a\Theta_b\Theta_c)^{1/2}}{\sigma}\right] + \dfrac{3}{2}$	$\ln g_0$
$\dfrac{\overline{A}}{R}$	$-T\ln\left[\left(\dfrac{2\pi mkT}{h^2}\right)^{3/2}\dfrac{kT}{P}\right] - T$	$\displaystyle\sum_i T\ln\left[1 - \exp\left(-\dfrac{\Theta_{vi}}{T}\right)\right]$	$-T\ln\dfrac{T}{\sigma\Theta_r}$	$-T\ln\dfrac{(T^3\pi/\Theta_a\Theta_b\Theta_c)^{1/2}}{\sigma}$	$-T\ln g_0$
$\dfrac{\overline{G}}{R} = \dfrac{\mu}{R}$	$-T\ln\left[\left(\dfrac{2\pi mkT}{h^2}\right)^{3/2}\dfrac{kT}{P}\right]$	$\displaystyle\sum_i T\ln\left[1 - \exp\left(-\dfrac{\Theta_{vi}}{T}\right)\right]$	$-T\ln\dfrac{T}{\sigma\Theta_r}$	$-T\ln\dfrac{(T^3\pi/\Theta_a\Theta_b\Theta_c)^{1/2}}{\sigma}$	$-T\ln g_0$

[a] As a simplification, the temperature dependence of the electronic partition function has been neglected.

Example 17.15

What are the values of $\overline{C}_P^\circ$, $\overline{S}^\circ$, $\overline{H}^\circ$, and $\overline{G}^\circ$ for $O_2(g)$ at 298.15 K?

These values are obtained by adding the translational, rotational, vibrational, and electronic contributions calculated in Examples 17.7, 17.8, 17.10, and 17.13.

$$\overline{C}_P^\circ = (\overline{C}_P^\circ)_t + (\overline{C}_P^\circ)_r + (\overline{C}_P^\circ)_v + (\overline{C}_P^\circ)_e = (20.786 + 8.314 + 0.236 + 0)\,\text{J K}^{-1}\,\text{mol}^{-1}$$

$$= 29.336\,\text{J K}^{-1}\,\text{mol}^{-1}$$

$$\overline{S}^\circ = \overline{S}_t^\circ + \overline{S}_r^\circ + \overline{S}_v^\circ + \overline{S}_e^\circ = (152.080 + 43.838 + 0.035 + 9.134)\,\text{J K}^{-1}\,\text{mol}^{-1}$$

$$= 205.088\,\text{J K}^{-1}\,\text{mol}^{-1}$$

$$\overline{H}^\circ = \overline{H}_t^\circ + \overline{H}_r^\circ + \overline{H}_v^\circ + \overline{H}_e^\circ = (6.197 + 2.479 + 0.009 - 491.888)\,\text{kJ mol}^{-1}$$

$$= -483.203\,\text{kJ mol}^{-1}$$

$$\overline{G}^\circ = \overline{G}_t^\circ + \overline{G}_r^\circ + \overline{G}_v^\circ + \overline{G}_e^\circ = (-39.145 - 10.591 - 0.001 - 494.611)\,\text{kJ mol}^{-1}$$

$$= -544.349\,\text{kJ mol}^{-1}$$

Since the value of $\overline{C}_P^\circ$ calculated here for $O_2(g)$ is in good agreement with the value in Table C.2 (29.372 J K^{-1} mol^{-1}), the electronic contribution that we ignored is small at this temperature. The value of $\overline{S}°$ is also in good agreement with the value in Table C.2 (205.033 J K^{-1} mol^{-1}). The values of $\overline{H}°$ and $\overline{G}°$ calculated here cannot be compared with $\Delta_f H°$ and $\Delta_f G°$ because of the difference in reference states, but we will calculate $\Delta_f H°$ and $\Delta_f G°$ for O(g) in Example 17.16.

Molecular parameters of a number of gases are given in Table 17.2. The dissociation energies D_0 are with respect to the constituent atoms in the ground state. For diatomic molecules these dissociation energies have been obtained from spectra. For polyatomic molecules the values of D_0 have been obtained from thermochemical measurements and theoretical calculations of heat capacities of gases. In Table 17.2 values of Θ_e greater than 10 000 K are not given.

The values of $\overline{C}_P°$ and $\overline{S}°$ calculated with the equations given in this chapter may be compared with the values from the JANAF Thermochemical Table (Table C.2), which are more accurate because they take more effects (deviations from rigid rotor harmonic oscillator approximation) into account. For many simple molecules, spectroscopic measurements have yielded the values of the anharmonicity constant x_e, the centrifugal distortion constant D, and the rotation–vibration coupling constant α.

The values of $\overline{H}°$ and $\overline{G}°$ are with respect to ideal gases of the constituent atoms at absolute zero. However, they can be used to calculate $\Delta_r H°$ and $\Delta_r G°$ for chemical reactions because $\overline{H}°$ and $\overline{G}°$ for reactants and products have all been calculated for the same reference state.

Example 17.16

What are the values of $\Delta_r C_P°$, $\Delta_r S°$, $\Delta_r H°$, and $\Delta_r G°$ for $\frac{1}{2}O_2(g) = O(g)$ calculated statisticaly mechanically and what is the equilibrium constant at 298.15 K?

Using values calculated in Example 17.14 and Example 17.15,

$$\Delta_r C_P° = 20.786 - \tfrac{1}{2}(29.336) = 6.118 \text{ J K}^{-1} \text{ mol}^{-1}$$

$$\Delta_r S° = 159.288 - \tfrac{1}{2}(205.088) = 56.744 \text{ J K}^{-1} \text{ mol}^{-1}$$

$$\Delta_r H° = 6.197 - \tfrac{1}{2}(-483.203) = 247.799 \text{ kJ mol}^{-1}$$

$$\Delta_r G° = -41.294 - \tfrac{1}{2}(-544.349) = 230.881 \text{ kJ mol}^{-1}$$

$$K = e^{-\Delta_r G°/RT} = e^{-230\,881/(8.314\,51)(298.15)} = 3.55 \times 10^{-41} = \frac{P_O/P°}{(P_{O_2}/P°)^{1/2}}$$

The values of $\Delta_r H°$ and $\Delta_r G°$ should be equal to $\Delta_f H°$ and $\Delta_f G°$ for O(g). The small differences from Table C.1 are due to the neglect of higher order terms.

Table 17.3 gives $\Delta_r C_P°$, $\Delta_r S°$, $\Delta_r H°$, and $\Delta_r G°$ for two endothermic reactions and two exothermic reactions, calculated with the molecular parameters in Table 17.2.

The reaction for the formation of NH_3 from its elements is an example of a reaction that is thermodynamically spontaneous at room termperature, but does not occur at an appreciable rate. As the temperature is raised to increase the rate of formation of NH_3, the equilibrium constant becomes less favorable. Therefore, it has been important to find catalysts for $\frac{1}{2}N_2(g) + \frac{3}{2}H_2(g) = NH_3(g)$ so that the reaction can be carried out at lower temperatures.

Table 17.2   Molecular Parameters of Gases[a]

Gas	M/g mol^{-1}	D_0/kJ mol^{-1}	σ	Θ_r/K [b]	Θ_v/K [c]	g_0	Θ_e/K [c]
H	1.008 0	0				2	
C	12.001	0				1	23.6(3)
							62.6(5)
N	14.008	0				4	
O	15.994	0				5	228.1(3)
							325.9
Cl	35.453	0				4	1 269.53(2)
I	126.904 5	0				4	10 939.3(2)
H$_2$	2.016	432.073	2	87.547	6338.2	1	
N$_2$	28.013 4	941.4	2	2.875 05	3392.01	1	
O$_2$	31.998 8	491.888	2	2.079	2273.64	3	
Cl$_2$	70.906	239.216	2	0.345 6	807.3	1	
I$_2$	253.82	148.81	2	0.053 76	308.65	1	
HCl	36.465	427.772	1	15.234 4	4301.38	1	
HI	127.918	294.67	1	9.369	3322.24	1	
CO	28.010 55	1070.11	1	2.777 1	3121.48	1	
NO	30.008	627.7	1	2.452 0	2738.87	2	174.2(2)
CO$_2$	44.009 95	1596.23	2	0.561 67	960.10(2)	1	
					1932.09		
					3380.14		
NO$_2$ [d]	46.008	928.3	2	4.243 01	1088.9	2	
					1953.6		
					2396.3		
H$_2$O	18.016	917.773	2	11 331.5	2294.27	1	
					5261.71		
					5403.78		
NH$_3$	17.036 1	1157.77	3	1876.0	1367	1	
					2341(2)		
					4800		
					4955(2)		
CH$_4$	16.043	1640.57	12	435.6	1957(3)	1	
					2207.1(2)		
					4196.2		
					4343.3(3)		
N$_2$O$_4$	92.016	1909.82	4	6.5793×10^{-3}	72	1	
					374		
					554		
					619		
					691		
					971		
					1079		
					1184		
					1814		
					1975		
					2460		
					2515		

[a] These values have all been obtained from M. Chase et al., JANAF Thermochemical Tables. *J. Phys. Chem. Ref. Data* **14** (1985), Supplement 1.

[b] The values of Θ_r for nonlinear polyatomic molecules are values of $\Theta_a\Theta_b\Theta_c$.

[c] The values in parentheses are degeneracies. Otherwise the levels have a degeneracy of unity.

[d] NO$_2$ is a bent molecule with a bond angle of 134° 15′.

Table 17.3 Changes in Standard Thermodynamic Properties for Chemical Reactions of Gases

T/K	$\Delta_r C_P^\circ$ $\mathrm{J\ K^{-1}\ mol^{-1}}$	$\Delta_r S^\circ$ $\mathrm{J\ K^{-1}\ mol^{-1}}$	$\Delta_r H^\circ$ $\mathrm{kJ\ mol^{-1}}$	$\Delta_r G^\circ$ $\mathrm{kJ\ mol^{-1}}$	K
		$\frac{1}{2}O_2(g) = O(g)$			
298.15	6.118	56.744	247.799	230.880	3.55×10^{-41}
500.00	5.305	60.551	248.961	218.685	1.43×10^{-23}
1000.00	3.487	64.321	251.095	186.774	1.75×10^{-10}
2000.00	2.499	66.755	253.950	120.441	7.16×10^{-4}
3000.00	2.272	67.862	256.314	52.728	1.21×10^{-1}
		$\frac{1}{2}N_2(g) + \frac{1}{2}O_2(g) = NO(g)$			
298.15	−0.052	10.638	88.942	85.770	9.41×10^{-16}
500.00	−0.096	11.346	88.922	83.249	2.01×10^{-9}
1000.00	0.136	11.994	88.942	76.948	9.56×10^{-5}
2000.00	0.102	12.426	89.075	64.223	2.10×10^{-2}
3000.00	0.055	12.574	89.151	51.429	1.27×10^{-1}
		$\frac{1}{2}N_2(g) + \frac{3}{2}H_2(g) = NH_3(g)$			
298.15	−22.729	−98.497	−46.195	−16.763	8.65×10^{2}
500.00	−16.977	−108.930	−50.218	4.357	3.50×10^{-1}
1000.00	−5.165	−116.749	−55.516	61.451	6.17×10^{-4}
2000.00	4.347	−116.620	−54.785	178.893	2.13×10^{-5}
3000.00	6.648	−114.328	−49.054	294.584	7.43×10^{-6}
		$CO(g) + 3H_2(g) = CH_4(g) + H_2O(g)$			
298.15	−47.608	−213.781	−206.788	−143.048	1.15×10^{25}
500.00	−36.260	−235.941	−215.340	−97.369	1.49×10^{10}
1000.00	−9.704	−252.178	−226.280	25.899	4.44×10^{-2}
2000.00	9.773	−251.154	−223.691	278.618	5.29×10^{-8}
3000.00	13.941	−246.219	−211.383	527.276	6.59×10^{-10}

The reaction forming methane from carbon monoxide is another reaction that is spontaneous at low temperatures, but has such a low rate that higher temperatures and a catalyst must be used to obtain a reasonable yield.

17.12 Calculation of Equilibrium Constants for Reactions of Ideal Gases

The equilibrium constant can be calculated from thermodynamic properties calculated from spectroscopic properties, but it is also of interest to express the equilibrium constant directly in terms of molecular partition functions. To do that we start with the thermodynamic relation for chemical equilibrium, $\sum_i \nu_i \mu_{i,\mathrm{eq}} = 0$ (equation 5.7). The Gibbs energy of N_i molecules of component i of a mixture of ideal gases is given by equation 17.53, and so the chemical potential μ_i is obtained by multiplying by N_A/N_i to obtain

$$\mu_i = N_A G_i / N_i = -RT \ln \frac{q_i}{N_i} \qquad (17.100)$$

Eliminating N_i from the logarithmic term by use of $P_i V = N_i kT$ yields the chemical potential of i as a function of partial pressure of i:

$$\mu_i = -RT \ln \frac{q_i kT}{P_i V} \qquad (17.101)$$

This equation can be put in the usual form $\mu_i = \mu_i^\circ + RT \ln(P_i/P^\circ)$ for an ideal gas as follows:

$$\mu_i = -RT \ln \frac{q_i kT}{P^\circ V} + RT \ln \frac{P_i}{P^\circ} \qquad (17.102)$$

Substituting this expression into the relation for chemical equilibrium yields

$$-\sum_i \nu_i RT \ln \frac{q_i kT}{P^\circ V} = -RT \sum_i \nu_i \ln \frac{P_i}{P^\circ} = -RT \ln K \qquad (17.103)$$

Thus, the dimensionless **equilibrium constant** K is given by

$$K = \prod_i \left(\frac{q_i kT}{P^\circ V} \right)^{\nu_i} = \left(\frac{kT}{P^\circ} \right)^{\Sigma \nu_i} \prod_i \left(\frac{q_i}{V} \right)^{\nu_i} \qquad (17.104)$$

It is essential that the partition functions in this equation be written with respect to the same zero of energy. This can be done by using the dissociated atoms as the reference state for both the reactants and products. Figure 17.3 further emphasizes the point made at the end of Section 17.10. In considering a single molecule i, the electronic partition function can be written

$$q_{ei} = g_{0i} + g_{1i} \exp \left(-\frac{\epsilon_{1i}}{kT} \right) + \cdots \qquad (17.105)$$

which is simply another way of writing equation 17.93. However, if we want to compare the energy levels of molecule i with molecule j, both partition functions should be calculated by use of energies measured with respect to the

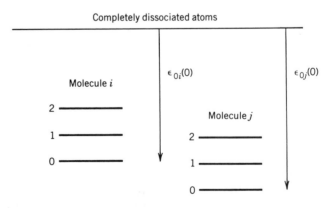

Figure 17.3 Electronic energy levels (labeled 0, 1, 2, . . .) of molecules i and j that contain the same atoms, where $\epsilon_{0i}(0)$ is the negative of the dissociation energy $(-D_{0i})$.

completely dissociated atoms. This can be accomplished with molecule i by multiplying each term by $\exp[-\epsilon_{0i}(0)/kT]$ to obtain

$$q_{ei} = g_{0i} \exp\left[-\frac{\epsilon_{0i}(0)}{kT}\right] + g_{1i} \exp\left[-\frac{\epsilon_{0i}(0) + \epsilon_{1i}}{kT}\right] \qquad (17.106)$$

where $\epsilon_{0i}(0)$ is the negative of the dissociation energy $(-D_{0i})$. If the molecular partition functions are written in terms of the zero-point energies of the molecules involved in a reaction, the expression for K can be written

$$K = \left(\frac{kT}{P^\circ}\right)^{\Sigma \nu_i} \exp\left(-\frac{\sum_i \nu_i \epsilon_{0i}(0)}{kT}\right) \prod_i \left(\frac{q_i}{V}\right)^{\nu_i} \qquad (17.107)$$

where the $\epsilon_{0i}(0)$ are the dissociation energies of the ground-state molecules into ground-state atoms, and the q_i are the usual partition functions with respect to the ground electronic state of the molecule.

The equilibrium constant for a gas reaction can be calculated from spectroscopic information only for diatomic molecules or small polyatomic molecules for which dissociation energies may be determined spectroscopically. For larger polyatomic molecules the statistical mechanical calculation of K requires enthalpy of formation data determined by chemical methods since it is not practical to determine dissociation energies for these larger molecules from spectroscopic measurements.

Example 17.17
Express the equilibrium constant for the dissociation of molecular hydrogen in terms of molecular properties.

$$H_2 = 2H$$

$$K = \frac{(P_H/P^\circ)^2}{P_{H_2}/P^\circ} = \frac{(kT/P^\circ)(q_H/V)^2}{q_{H_2}/V}$$

$$\frac{q_H}{V} = \left(\frac{2\pi m_H kT}{h^2}\right)^{3/2} (2)$$

$$\frac{q_{H_2}}{V} = \left(\frac{2\pi m_{H_2} kT}{h^2}\right)^{3/2} \left(\frac{T}{2\theta_{rH_2}}\right) \left(1 - e^{-\theta_{vH_2}/T}\right)^{-1} e^{D_0/kT}$$

where $D_0 = D_e - \frac{1}{2}h\nu$.

17.13 Equipartition

According to the classical principle of equipartition, each squared term in the classical energy expression for a gas molecule contributes $R/2$ to the molar heat capacity at constant volume. This principle is not exact because the heat capacity decreases with decreasing temperature (quantum mechanics explains why), but the classical calculation does give the limit that the heat capacity approaches at high temperature. Statistical mechanics shows why this is so, and yields the temperature range over which each contribution increases to its classical value.

Table 17.4   Contributions to $\overline{C}_V/R$ for Ideal Gases at High Temperatures[a]

Molecule	Translational	Vibrational	Rotational
Monatomic	$\frac{3}{2}$	0	0
Diatomic	$\frac{3}{2}$	1	1
Linear polyatomic	$\frac{3}{2}$	$3N - 5$	1
Nonlinear polyatomic	$\frac{3}{2}$	$3N - 6$	$\frac{3}{2}$

[a] N is the number of atoms in the molecule.

The full translational contribution is obtained for a gas at very low temperatures. Table 17.4 shows that $\overline{U}_t = \frac{3}{2}RT$ and $\overline{C}_V = \frac{3}{2}R$, independent of the mass or structure of a molecule of an ideal gas. In classical mechanics the translational energy of a gas molecule is given by

$$U_t = \frac{mv_x^2}{2} + \frac{mv_y^2}{2} + \frac{mv_z^2}{2} \qquad (17.108)$$

This suggests that each quadratic term in the classical energy expression contributes $RT/2$ to the $\overline{U}_t$ and $R/2$ to $\overline{C}_V$. This is in agreement with the fact that for monatomic gases, $\overline{C}_V = \frac{3}{2}R$ at pressures where they behave as ideal gases and at temperatures low enough that there is negligible electronic excitation.

In Section 17.8 we have seen that each normal mode of vibration of a polyatomic molecule contributes RT to the internal energy and R to $\overline{C}_V$ in the high-temperature limit. For a one-dimensional harmonic oscillator, classical mechanics yields

$$U_v = \frac{mv_x^2}{2} + \frac{kx^2}{2} \qquad (17.109)$$

Thus, we see again that each squared term contributes $R/2$ to the molar heat capacity in the limit as the temperature is increased.

Well above the characteristic rotational temperature, the contribution of rotation to the molar heat capacity of a diatomic molecule or a linear polyatomic molecule is R. The classical expression for the rotational energy of a diatomic or linear polyatomic molecule is

$$U_r = \frac{I\omega_x^2}{2} + \frac{I\omega_y^2}{2} \qquad (17.110)$$

Since there are two squared energy terms, each contributes $R/2$ to the molar heat capacity at constant volume. There is one more term for a nonlinear polyatomic molecule, and so the contribution of rotation to the molar heat capacity is $\frac{3}{2}R$, as expected.

The high-temperature contributions to the molar heat capacities at constant volume are summarized in Table 17.4.

17.14   Nonideal Gases

In earlier sections of this chapter we have ignored intermolecular interactions and have, therefore, obtained thermodynamic properties of ideal gases. To calculate thermodynamic properties of nonideal gases it is necessary to take into account the potential energy V of intermolecular interactions. This cannot be done by using the molecular partition function. For nonideal gases it is necessary to use the canonical ensemble partition function, which is discussed as a special topic at the end of this chapter. Using the canonical ensemble, it is possible to show that the second virial coefficient $B(T)$ (Section 1.6) is related to the potential energy of interaction $V(R)$ of two molecules (Section 12.11) by

$$B(T) = 2\pi N_A \int_0^\infty (1 - e^{-V(R)/kT})R^2 \, dR \qquad (17.111)$$

The second virial coefficient depends on pairwise interactions, and the third virial coefficient depends on interactions between triplets of molecules. For **an ideal** gas the potential energy $V(R)$ of interaction is zero and so $B(T) = 0$. We will apply equation 17.111 to the calculation of the second virial coefficient of a gas made up of hard spheres and to a gas obeying the Lennard–Jones potential.

Hard-Sphere Potential

The simplest type of potential energy of interaction of two molecules is that due to short-range repulsive interactions. If molecules can be represented by hard spheres, their potential energy of interaction is given by

$$V(R) = \infty \qquad \text{for } 0 \le R \le d \qquad (17.112)$$
$$V(R) = 0 \qquad \text{for } R > d \qquad (17.113)$$

Therefore,

$$B(T) = 2\pi N_A \int_0^d R^2 \, dR + 2\pi N_A \int_d^\infty (1 - 1)R^2 \, dR$$

$$= 2\pi N_A \left(\frac{R^3}{3}\right)_0^d = \frac{2}{3}\pi N_A d^3 = 4N_A b_0 \qquad (17.114)$$

Since the radius of the hard-sphere molecule is half the distance d of closest approach, the volume of a hard-sphere molecule b_0 is equal to $\frac{4}{3}\pi(d/2)^3 = \frac{1}{6}\pi d^3$. Thus, $B(T)$ is equal to four times the volume of a mole of hard-sphere molecules. Alternatively, we can point out that $B(T)$ for a gas of hard spheres is one-half the volume excluded to the center of one hard sphere of diameter d by another of diameter d.

Lennard–Jones Potential

Equation 17.111 for $B(T)$ with the Lennard–Jones potential (Section 12.11) must be evaluated numerically. In using this equation it is convenient to express the results in terms of reduced variables so that it may be readily applied to gases with different Lennard–Jones parameters. In making the numerical integration, distance is expressed as a multiple of the Lennard–Jones σ, and the temperature is expressed in terms of a reduced temperature $T^* = kT/\epsilon$, where ϵ is the depth of the potential well. The results of the numerical integration are

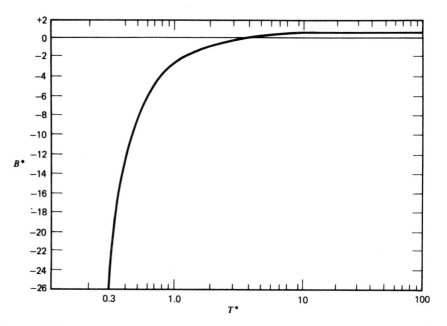

Figure 17.4   Reduced second virial coefficient B^* for the Lennard–Jones potential as a function of reduced temperature T^*. (From J. O. Hirschfelder, C. F. Curtiss, and R. B. Bird, *Molecular Theory of Gases and Liquids*. New York: Wiley, 1954.)

shown in Fig. 17.4, which gives the reduced second virial coefficient $B^*(T^*)$ = $B(T)/b_0$, where b_0 is the hard-sphere second virial coefficient ($b_0 = 2\pi N_A \sigma^3/$ 3). At low temperatures the second virial coefficient reflects the attractive interactions between molecules and at high temperatures it reflects the repulsive interactions.

Example 17.18

Using the Lennard–Jones parameters for argon in Table 12.5 estimate the second virial coefficient B at 300 and 1000 K using Fig. 17.4. Compare these values with values calculated from Fig. 1.4. From Table 12.5, $\sigma = 0.341$ nm and $\epsilon/k = 120$ K.

At 300 K, $T^* = 300$ K/120 K $= 2.50$. From Fig. 17.4, $B^*(T^*) = -0.31$.

$$B = \frac{2\pi}{3}N_A\sigma^3 B^*$$

$$= \frac{2\pi}{3}(6.02 \times 10^{23} \text{ mol}^{-1})(0.341 \times 10^{-9} \text{ m})^3(-0.31)$$

$$= (-1.55 \times 10^{-5} \text{ m}^3 \text{ mol}^{-1})(10^2 \text{ cm m}^{-1})^3$$

$$= -15.5 \text{ cm}^3 \text{ mol}^{-1}$$

Figure 1.4 indicates that $B = -20$ cm^3 mol^{-1}.

At 1000 K, $T^* = 1000$ K/120 K $= 833$. From Fig. 17.4, $B^* = 0.4$.

$$B = \frac{2\pi}{3}N_A\sigma^3 B^*$$

$$= \frac{2\pi}{3}(6.02 \times 10^{23}\ \text{mol}^{-1})(0.341 \times 10^{-9}\ \text{m})^3(0.4)(10^2\ \text{cm m}^{-1})^3$$

$$= 20\ \text{cm}^3\ \text{mol}^{-1}$$

Figure 1.4 indicates $B = 20\ \text{cm}^3\ \text{mol}^{-1}$.

It is important to be able to calculate the equation of state for a nonideal gas from molecular parameters because, once this has been done, other thermodynamic properties may be calculated using the equation of state.

Figure 17.4 is an example of the law of corresponding states. To derive this particular form it was assumed that the intermolecular interactions are spherically symmetric. This is obviously not the case for some molecules, but the law can be extended.

The law of corresponding states provides a means for estimating Lennard–Jones parameters. Since the reduced temperature T_c^* of the critical point calculated from kT_c/ϵ for a number of gases is approximately 1.3, the interaction energy ϵ can be estimated from the critical temperature T_c using $\epsilon/k = T_c/1.3$. Since the reduced molecular volume at the critical point $V_c/N_A\sigma^3$ is approximately 2.7 for a number of gases, the distance parameter σ may be estimated from $\sigma = (V_c/2.7N_A)^{1/3}$.

17.15   Heat Capacities of Solids

To simplify the calculation of the heat capacity of a monatomic solid we can imagine that each atom oscillates about its equilibrium lattice point with a small amplitude. As mentioned in Section 17.13, according to classical theory each mole of atoms would contribute R for each of its three vibrational degrees of freedom, so that the molar heat capacity at constant volume would be $3R = 25\ \text{J K}^{-1}\ \text{mol}^{-1}$. This is observed at high enough temperatures for all atomic solids. However, classical theory was not able to explain the decrease of $\overline{C}_V$ to zero as absolute zero is approached.

To calculate thermodynamic properties at a particular temperature for a crystal made up of atoms, we can regard the crystal as one gigantic molecule with $3N - 6$ internal degrees of freedom. For an atomic crystal these degrees of freedom all correspond to lattice (center-of-mass) vibrations, but for a molecular crystal there are some internal degrees of freedom that correspond to rotations, internal vibrations, or torsions. In principle, we can imagine a normal mode analysis being carried out on any crystal.

In Einstein's first statistical mechanical calculation of the heat capacity of an idealized monatomic crystal in 1907 he assumed that all the vibrational frequencies of the solid have one frequency that we will represent by ν_E. Thus, according to Table 17.1, the molar heat capacity at constant volume is given by $3N$ terms of the type indicated there:

$$\overline{C}_V = \frac{3N_Ak(\Theta_E/T)^2 \exp(\Theta_E/T)}{[\exp(\Theta_E/T) - 1]^2} \tag{17.115}$$

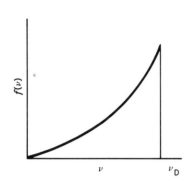

Figure 17.5 Frequency spectrum for an isotropic crystal. The highest frequency permitted is the Debye frequency ν_D.

where $\Theta_E = h\nu_E/k$ is the **characteristic vibrational temperature.** This theoretical result was of great importance because it explained why the heat capacity decreased from the classical result ($3R$) at high temperatures to zero as $T \to 0$. It also helped explain why $\overline{C}_V$ rises more slowly with increasing temperature for diamond and graphite than for silver and copper. High values of ν_E correspond to crystals where the force constant (equation 10.78) is large and/or the reduced mass is small. However, the Einstein theory predicted too rapid a decrease in heat capacity in the neighborhood of absolute zero.

In 1912 Debye introduced the idea of a spectrum of vibrational frequencies for an atomic crystal and was thereby able to derive an expression for the heat capacity that more accurately represented the experimental data at very low temperatures. The theory of sound waves in a continuous medium shows that the probability density (the number of modes between ν and $\nu + d\nu$) $f(\nu)$ for vibrations is proportional to the square of the frequency:

$$f(\nu)\, d\nu = \frac{12\pi\nu^2 V}{\overline{v}^3}\, d\nu \tag{17.116}$$

where $\overline{v}$ is the average speed of the sound waves and V is the volume. The low-frequency vibrations in an atomic crystal have wavelengths that extend over hundreds or thousands of atoms. These vibrations are nearly independent of the atomic-scale structure involved and are characteristic of a continuous medium with given elastic constants. The form of the distribution function used by Debye is shown in Fig. 17.5. He used the probability density given in equation 17.116 up to a frequency ν_D, chosen so that the total number of frequencies would be $3N$:

$$\int_0^{\nu_D} f(\nu)\, d\nu = 3N \tag{17.117}$$

The **Debye wavelength** $\lambda_D = \overline{v}/\nu_D$ corresponds to a minimum wavelength of the order of the interatomic distance in the crystal. According to the theory of elasticity of an isotropic (same properties in each direction) medium, two different velocities of sound have to be taken into account: the longitudinal velocity v_l and the transverse velocity v_t with two polarization directions perpendicular to the direction of propagation. Since these two velocities of sound are different, equation 17.116 becomes

$$f(\nu)\, d\nu = 4\pi\nu^2 V \left(\frac{1}{v_l^3} + \frac{2}{v_t^3} \right) \tag{17.118}$$

When this expression is substituted in equation 17.117 and the integration is carried out, we can use the resulting expression to derive

$$f(\nu) = \frac{9N\nu^2}{\nu_D^3} \tag{17.119}$$

The expression for the energy of the Debye crystal can be derived using this expression, and the resulting expression for the heat capacity of a crystal is

$$C_V = 9Nk \left(\frac{T}{\Theta_D} \right)^3 \int_0^{\Theta_D/T} \frac{x^4 e^x}{(e^x - 1)^2}\, dx \tag{17.120}$$

where the **Debye temperature** $\Theta_D \equiv h\nu_D/kT$ and x is a dimensionless variable equal to $h\nu/kT$. As the temperature approaches absolute zero, this equation reduces to

$$C_V = \frac{12\pi^4}{5} Nk \left(\frac{T}{\Theta_D}\right)^3 \qquad (17.121)$$

which represents the experimental results rather well for many atomic crystals. If a crystal is anisotropic, for example, if it has strong interactions in a plane like graphite does, the low-temperature behavior is more complicated.

The Debye equation is an example of a law of corresponding states (Section 1.9 and 17.14). It indicates that the heat capacities of all atomic solids should lie on the same curve when $\overline{C}_V$ is plotted versus T/Θ_D, as illustrated by Fig. 2.10. Actually, there are various kinds of deviations from the Debye law because the actual frequency spectrum is more complicated than that shown in Fig. 17.5.

17.16   Special Topic: Ensembles

So far, except in the discussion of nonideal gases and crystals, we have discussed only systems in which the particles are independent. A different approach, invented by Gibbs, is required when there are interactions. When there are interactions, the total energy cannot be divided up into contributions from separate molecules. Gibbs showed how to use an ensemble (an imaginary collection of a large number of isolated systems having certain macroscopic properties in common) and average over the ensemble to obtain the thermodynamic properties of a system. The ensemble contains systems with all the possible microstates of the system. The calculations given earlier in this chapter can be described in terms of an ensemble, which is called a microcanonical ensemble. A **microcanonical ensemble** is a collection of systems, having the same number of particles, the same volume, and the same energy (N, V, U), which contains all possible microstates of the system. Averaging over these systems gives the thermodynamic properties that we have calculated in the first part of this chapter. Actually, the most probable microstate predominates, and so we simply take it to represent the system.

A **canonical ensemble** is made up of a very large number η of systems, each of which contains N particles in volume V. The ensemble is prepared by putting systems in contact with a heat reservoir at temperature T, and then removing them to form the ensemble. The systems in the ensemble are isolated from each other, as illustrated in Fig. 17.6. Thus, all the systems in the canonical ensemble are characterized by the same fixed N, T, and V values. The ensemble of systems is also isolated, and contains η systems, which are replicas of the system of interest. The total isolated ensemble contains ηN particles, has a volume of ηV, and a fixed energy U. The energy of a system can have any possible value less than U. The average energy of a system is calculated using $\langle U \rangle = \sum p_i U_i$, where p_i is the probability that a system has energy U_i.

The distribution of systems between the various possible amounts of energy is described in the same way as we described the distribution of molecules between molecular energy levels in Section 17.1:

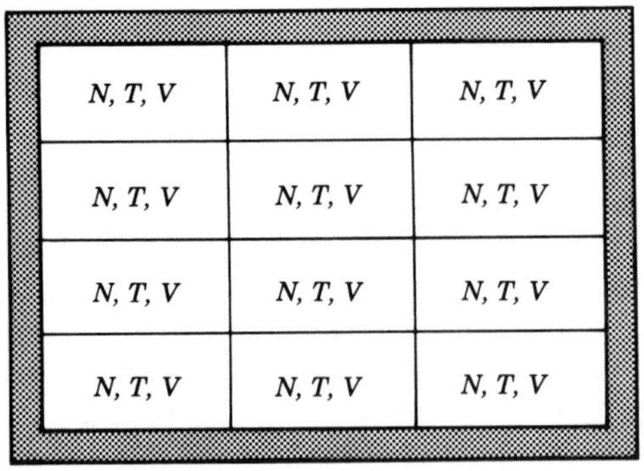

Figure 17.6   Canonical ensemble of systems each having the same N, T, V. The canonical ensemble is an isolated system with energy U_T containing η systems.

η_0 of the systems have energy U_0

η_1 of the systems have energy U_1

$\cdots$

η_i of the systems have energy U_i

$\cdots$

The probability that a system has energy U_i is $p_i = \eta_i/\eta$, where $\eta = \sum \eta_i$. The number of ways of effecting a given distribution of systems is given by

$$\Omega = \frac{\eta!}{\prod_i \eta_i!} \tag{17.122}$$

The only distributions that are permitted are those that satisfy the following two constraints:

$$\eta = \sum_i \eta_i \tag{17.123}$$

$$U = \sum_i \eta_i U_i \tag{17.124}$$

Since the number of systems in the ensemble is very large, it can be shown that there is a most probable distribution that can be effected in Ω_{max} ways. This most probable distribution is by far the most likely. To find this most probable distribution, we will follow the same steps that we used in Section 17.2. The number of systems in the canonical ensemble plays the role of the number of molecules in the earlier example, and the energy of a system in the canonical ensemble plays the role of the energy levels of particles in the earlier example.

For the canonical ensemble, the constraints expressed in terms of differentials are

$$d \ln \Omega = \sum_i \left(\frac{\partial \ln \Omega}{\partial \eta_i} \right) d\eta_i = 0 \tag{17.125}$$

$$\sum_i d\eta_i = 0 \tag{17.126}$$

$$\sum_i U_i \, d\eta_i = 0 \tag{17.127}$$

Lagrange's method of undetermined multipliers leads to

$$\sum_i \left(\frac{\partial \ln \Omega}{\partial \eta_i} + \gamma - \beta U_i \right) = 0 \tag{17.128}$$

so that

$$\frac{\partial \ln \Omega}{\partial \eta_i} + \gamma - \beta U_i = 0 \tag{17.129}$$

for each value of i. As in the earlier derivation (see equation 17.13)

$$\frac{\partial \ln \Omega}{\partial \eta_i} = -\ln \eta_i \tag{17.130}$$

Substituting this in equation 17.129 yields

$$\eta_i = \exp(\gamma - \beta U_i) \tag{17.131}$$

so that the probability that a system is in the ith state is

$$p_i = \frac{\eta_i}{\eta} = \frac{1}{Q} \exp(-\beta U_i) \tag{17.132}$$

where $\beta = 1/kT$ and Q is the **canonical ensemble partition function:**

$$Q(N, T, V) = \sum_{\text{states}} \exp(-\beta U_i)$$

$$= \sum_{\text{levels}} \Omega_i \exp(-\beta U_i) \tag{17.133}$$

where Ω_i is the number of microstates with energy U_i (in other words, the degeneracy of level i).

The average energy of the system is given by

$$\langle U \rangle = \sum_i p_i U_i = \frac{1}{Q} \sum_i U_i \exp(-\beta U_i)$$

$$= kT^2 \left(\frac{\partial \ln Q}{\partial T} \right)_{N,V} \tag{17.134}$$

To calculate the entropy, we use a new and more general form of the fundamental postulate of Boltzmann. This new definition of the entropy is

$$S = -k \sum_i p_i \ln p_i \tag{17.135}$$

If we are considering a microcanonical ensemble in which each microstate is equally probable then

$$p_i = \frac{1}{\Omega} \tag{17.136}$$

where Ω is the total number of microstates. Then $S = -k \sum_i (1/\Omega)(-\ln \Omega) = k \ln \Omega$, which agrees with our earlier definition of entropy. Thus, our new definition contains the old definition as a special case and is the more general form.

The entropy calculated using equation 17.135 for a canonical ensemble is

$$S = -k \sum_i \left[\frac{1}{Q} \exp(-\beta U_i) \right] (-\beta U_i - \ln Q)$$

$$= k\beta \langle U \rangle + k \ln Q \qquad (17.137)$$

This can be compared with the thermodynamic relation

$$S = \frac{U}{T} - \frac{A}{T} \qquad (17.138)$$

Since $\langle U \rangle = U$, we see that

$$A = -kT \ln Q \qquad (17.139)$$

which is the characteristic equation connecting the thermodynamic function $A(N, V, T)$ to the canonical partition function $Q(N, V, T)$. Introducing equation 17.134 for $\langle U \rangle$ and replacing β yields

$$S = k \left[\ln Q + T \left(\frac{\partial \ln Q}{\partial T} \right)_{N,V} \right] \qquad (17.140)$$

This equation applies to a collection of indistinguishable molecules or distinguishable molecules.

The expression for the pressure is obtained from $P = -(\partial A / \partial V)_{N,T}$.

$$P = kT \left(\frac{\partial \ln Q}{\partial V} \right)_{N,T} \qquad (17.141)$$

Since we started with known N, T, and V, we are now in a position to derive expressions for H, G, C_P, and C_V by simply using their definitions in terms of $A(N, V, T)$.

We now have expressions for all the thermodynamic properties in terms of both the molecular partition function q and the canonical ensemble partition function Q. By comparing the two expressions for A we see that for a system of indistinguishable noninteracting particles

$$Q = \frac{q^N}{N!} \qquad (17.142)$$

and for a system of distinguishable noninteracting particles

$$Q = q^N \qquad (17.143)$$

Thus, for a system containing particles that do not interact, the thermodynamic properties can be calculated using either the microcanonical ensemble or the canonical ensemble. If the system is a mixture, or the particles interact, a canonical ensemble must be used. The use of the canonical ensemble partition function for indistinguishable particles that do not interact leads to the relations in Table 17.1.

Other types of ensembles are also useful. The ensemble that yields the Gibbs energy most directly is the isothermal–isobaric ensemble with partition function $\Delta(T, P, N)$. In a grand canonical ensemble, the systems in the ensemble have constant T, V, μ. The systems in the ensemble have the same chemical potential because they are each placed in contact with a reservoir of particles through a semipermeable membrane before being added to the grand canonical ensemble. The number of particles in a system fluctuates, but $\langle N \rangle$ can be calculated.

References

F. C. Andrews, *Equilibrium Statistical Mechanics*. New York: Wiley, 1975.

H. B. Callen, *Thermodynamics and an Introduction to Thermostatics*. New York: Wiley, 1985.

D. Chandler, *Introduction to Modern Statistical Mechanics*. New York: Oxford University Press, 1987.

D. Dessaux, P. Goudmand, and F. Langrand, *Thermodynamique Statistique Chimique*. Paris: Dumond Université, 1982.

H. L. Friedman, *A Course in Statistical Mechanics*. Englewood Cliffs, NJ: Prentice-Hall, 1985.

C. E. Hecht, *Statistical Thermodynamics and Kinetic Theory*. New York: Freeman, 1990.

T. L. Hill, *An Introduction to Statistical Thermodynamics*. New York: Dover, 1960.

B. J. McClelland, *Statistical Thermodynamics*. New York: Wiley, 1973.

D. S. McQuarrie, *Statistical Mechanics*. New York: Harper & Row, 1976.

N. O. Smith, *Elementary Statistical Mechanics*. New York: Plenum, 1982.

Problems

17.1 Show that the differential of ln Ω (equation 17.6) is given by equation 17.12. Start with d ln $\Omega = \sum_i (\partial \ln \Omega / \partial N_i)$ dN_i.

17.2 Using the Boltzmann distribution, calculate the ratio of populations at 25 °C of energy levels separated by (*a*) 1000 cm^{-1}, and (*b*) 10 kJ mol^{-1}.

17.3 Calculate the ratio of populations at 25 °C of energy levels separated by (*a*) 1 eV and (*b*) 10 eV. (*c*) Calculate the ratios at 1000 °C.

17.4 Calculate the fractional occupations (N_i/N) of the energy levels of a particle in a one-dimensional box at a temperature at which the population of the $n = 2$ level is 0.421 of the population of the $n = 1$ level. The energy levels of a particle in a one-dimensional box are nondegenerate. The relative energies of the levels can simply be taken as 1, 4, 9, 16, 25, ... (See Fig. 17.2*a*.)

17.5 The temperature of the system in the preceding problem is doubled by heating it. Calculate the fractional occupations (N_i/N) of the energy levels at the higher temperature. (See Fig. 17.2*a*.)

17.6 Work is done on the system described in problem 17.4 adiabatically and reversibly so that the length of the box is reduced to $2^{1/2}$ of its original length. Calculate the fractional occupations (N_i/N) of the energy levels of the compressed system. (See Fig. 17.2*b*.)

17.7 Starting with the definition of the molecular partition function (equation 17.17) and $U = \sum_i N_i \epsilon_i$, derive equation 17.39 for *U*.

17.8 Show that the same expression is obtained for the chemical potential from A_{indis} (equation 17.51) as from G_{indis} (equation 17.53) for an ideal gas.

17.9 What is the ratio of the thermal de Broglie wavelength to the length of one side of the container for (*a*) a hydrogen atom in a cube 1 nm on a side at 2 K and (*b*) an oxygen molecule in 0.25 m^3 at 300 K?

17.10 Calculate the translational partition function for $H_2(g)$ at 1000 K and 1 bar.

17.11 What are the translational partition functions of hydrogen atoms and hydrogen molecules at 500 K in a volume of 4.157×10^{-2} m^3? (This is the molar volume of an ideal gas at this temperature and a pressure of 1 bar.)

17.12 Calculate the molar entropy of H-atom gas at 1000 K and (*a*) 1 bar and (*b*) 1000 bar.

17.13 Calculate the molar entropy of neon at 25 °C and 1 bar.

17.14 What are the most probable populations of the successive vibrational levels of Cl_2 at 298 and 1000 K? *Given:* $\bar{\nu} = 559.7$ cm^{-1}.

17.15 Derive the expression for the vibrational contribution to the internal energy

$$U_V = \frac{RTx}{e^x - 1}$$

where $x = h\nu/kT$.

17.16 By use of series expansions show the vibrational contribution to $\overline{C}_V$ for a diatomic molecule approaches R as $T \to \infty$.

17.17 What are the rotational contributions to $\overline{C}_P^\circ$, $\overline{S}^\circ$, $\overline{H}^\circ$, and $\overline{G}^\circ$ for $NH_3(g)$ at 25 °C?

17.18 What fraction of HCl molecules is in the state $v = 2$, $J = 7$ at 500 °C? The characteristic vibrational and rotational temperatures are given in Table 17.2.

17.19 Calculate the translational partition functions for H, H_2, and H_3 at 1000 K and 1 bar. What are the rotational partition functions of H_2 and H_3 (linear) at 1000 K? The internuclear distances in H_3 are 94 pm.

17.20 What are the symmetry numbers of the following organic molecules, assuming free rotation of methyl groups: (a) ethane, (b) propane, (c) 2-methylpropane, (d) 2,2-dimethylpropane?

17.21 Calculate the symmetry numbers of methane (CH_4) and ethylene (C_2H_4) by adding up the number of distinct proper rotational operations in Table 13.2 plus the identity operation.

17.22 (a) Calculate the symmetry number for the ethane structures shown in Table 13.2. This is the symmetry number of the rigid structure. (b) Since there is essentially free rotation about the C–C bond, there are three equivalent positions of the second CH_3 group with respect to the first. What is the symmetry number of a freely rotating ethane molecule?

17.23 What are the electronic contributions to $\overline{S}^\circ$ and $\overline{G}^\circ$ for I(g) at 298.15 K and 3000 K?

17.24 Calculate the molar entropies of H(g) and N(g) at 25 °C and 1 bar. The degeneracies of the ground states are 2 and 4, respectively. Compare these values with those in Table C.1.

17.25 A molecule has a ground state and two excited electronic energy levels, all of which are nondegenerate: $\epsilon_0 = 0$, $\epsilon_1 = 1 \times 10^{-20}$ J, and $\epsilon_2 = 3 \times 10^{-20}$ J. What fraction of each level is occupied at 298 and 1000 K?

17.26 The ground state of Cl(g) is fourfold degenerate. The first excited state is 875.4 cm^{-1} higher in energy and is twofold degenerate. What is the value of the electronic partition function at 25 °C? At 1000 K?

17.27 What is the partition function for oxygen atoms at 1000 K according to the data in Table 17.2? What are the relative populations of these levels at equilibrium?

17.28 Derive the expression for the electronic internal energy of an atom or molecule. What is the electronic energy per mole for a chlorine atom at 298 K and 1000 K? (See problem 17.26.)

17.29 Calculate the fraction of hydrogen atoms that at equilibrium at 1000 °C would have $n = 2$.

17.30 A quantum mechanical system has two energy levels, ϵ_1 and ϵ_2. Derive equations for the probability p_1 that the system will be in state 1 and the probability p_2 that the system will be in state 2. What are these probabilities at $T/K = 0$ and ∞? What are the values of p_1 and p_2 at $\Delta\epsilon = kT$?

17.31 Calculate $\overline{C}_P^\circ$ for $NH_3(g)$ at 1000 K. The characteristic vibrational temperatures for the six normal modes are given in Table 17.2.

17.32 Calculate the molar entropy of nitrogen gas at 25 °C and 1 bar pressure. The equilibrium separation of atoms is 109.5 pm and the vibrational wavenumber is 2330.7 cm^{-1}.

17.33 Calculate $\overline{C}_P^\circ$ for CO_2 at 1000 K. Compare the actual contributions to $\overline{C}_P^\circ$ from the various normal modes with the classical expectations.

17.34 Calculate the equilibrium constant for the isotope exchange reaction $D + H_2 = H + DH$ at 25 °C. Assume that the equilibrium distance and force constants of H_2 and DH are the same.

17.35 Calculate the equilibrium constant at 25 °C for the reaction $H_2 + D_2 = 2HD$. It may be assumed that the equilibrium distance and force constant k are the same for all three molecular species, so that the additional vibrational frequencies required may be calculated from $2\pi\nu = (k/\mu)^{1/2}$. Because of the zero-point vibration, $\Delta\epsilon_0$ for this reaction is given by

$$\Delta\epsilon_0 = \tfrac{1}{2}N_A h(2\nu_{HD} - \nu_{H_2} - \nu_{D_2})$$

17.36 Express the equilibrium constant for the reaction $H_2 + I_2 = 2HI$ in terms of molecular properties.

17.37 The classical limits of heat capacities of molecules of ideal gases are readily calculated using the principle of equipartition. Calculate $\overline{C}_V/R$ and $\overline{C}_P/R$ for Ar, O_2, CO_2, and CH_4 and compare $\overline{C}_P/R$ with the values in Table C.2 at 3000 K.

17.38 Considering H_2O to be a rigid nonlinear molecule, what value of $\overline{C}_P^\circ$ for the gas would be expected classically? If vibration is taken into account, what value is expected? Compare these values of $\overline{C}_P^\circ$ with the actual values of 298 and 3000 K in Table C.1.

17.39 Use the law of corresponding states to estimate the Lennard–Jones parameters for methane from $T_c = 191$ K and $\overline{V}_c = 100$ cm^3 mol^{-1}. The values obtained from second virial coefficient data $\sigma = 0.378$ nm and $\epsilon/k = 148.9$ K.

17.40 Estimate the second virial coefficient for methane at 300 and 600 K from its Lennard–Jones parameters in Table 12.5. Compare these values with Fig. 1.4.

17.41 Show how $P = kT(\partial \ln Q/\partial V)_T$ leads to $PV = nRT$.

17.42 The energies of the $n = 2$ and $n = 1$ orbitals of the hydrogen atom are 27 420 and 109 678 cm^{-1}, respectively. What are the relative populations in these levels at (a) 25 °C, and (b) 2000 °C?

17.43 Write out the steps to convert equation 17.41 to equation 17.42 for the number of microstates of a system with distinguishable particles.

17.44 Derive the expression for the chemical potential of an ideal gas of indistinguishable particles as described in the footnote after equation 17.53.

17.45 A helium atom is in a volume of 10^{-9} m^3. What are the values of its translational partition function at 298, 1000, and 5000 K?

17.46 The thermal de Broglie wavelength defined in connection with equation 17.72 is a little different from the de Broglie wavelength. (a) What is the de Broglie wavelength for hydrogen atoms at 3000 K, using the root mean square average momentum as p. (b) How does it compare with the thermal de Broglie wavelength calculated in Example 17.2? (c) How does this thermal de Broglie wavelength compare with the mean distance between hydrogen atoms in a gas of hydrogen atoms at 3000 K and 1 bar?

17.47 Calculate $\overline{S}°$ and $\overline{C}_P°$ for argon ($M = 39.948$ g mol^{-1}) at 25 °C and 1 bar.

17.48 Calculate the molar entropy of helium in the ideal gas state at 25 °C and 1 bar pressure.

17.49 Compare the translational partition function of I(g) at 1000 K and 1 bar with that for H(g) calculated in problem 17.16.

17.50 (a) Calculate the thermal de Broglie wavelength for an O_2 molecule at 1 and 298 K. For Maxwell–Boltzmann statistics to be applicable, the thermal de Broglie wavelength must be small compared with the mean distance between molecules. (b) Calculate the mean distance between gas molecules at 1 bar at these temperatures assuming each molecule is in the center of a cube. (c) Are Maxwell–Boltzmann statistics applicable at both temperatures?

17.51 What are the most probable populations of the first several vibrational levels of O_2(g) at 1000 K? The characteristic vibrational temperature is 2274 K.

17.52 Using

$$\left(\frac{\partial \overline{G}}{\partial T}\right)_P = -\overline{S}$$

and the contribution of vibration to the Gibbs energy for a diatomic molecule in an ideal gas

$$\overline{G} = RT \ln(1 - e^{-x})$$

derive the expression for the corresponding contribution to the entropy.

17.53 What are the characteristic vibrational temperatures for oscillators with frequencies of 10^9 s^{-1} (radio waves), 10^{12} s^{-1} (far infrared), 10^{15} s^{-1} (near ultraviolet), and 10^{18} s^{-1} (X-rays)?

17.54 What are the rotational contributions to $\overline{C}_P°$ and $\overline{S}°$ of CH_4 at 298.15 K?

17.55 What are the symmetry numbers of the following organic compounds, assuming free rotation of methyl groups: (a) ethylene, (b) 1-methylethylene, (c) 1,1-dimethylethylene, (d) 1,1,2-trimethylethylene, (e) 1,1,2,2-tetramethylethylene?

17.56 (a) The water molecule belongs to the C_{2v} point group, which includes the symmetry elements C_2 and E. What is the symmetry number of a water molecule?

17.57 Calculate the ratio of the number of HBr molecules in state $v = 2$, $J = 5$ to the number in state $v = 1$, $J = 2$ at 1000 K. Assume that all of the molecules are in their electronic ground states. ($\Theta_v = 3700$ K, $\Theta_r = 12.1$ K).

17.58 Calculate the temperature at which 10% of the molecules in a system will be in the first excited electronic state if this state is 400 kJ mol^{-1} above the ground state.

17.59 (a) In problem 17.25, what is the electronic energy of the molecule at 298 and 1000 K? (b) Since the internal energy is given in terms of the molecular partition function by

$$U = NkT^2 \left(\frac{\partial \ln q}{\partial T}\right)_V$$

calculate the electronic energy of the molecule at 298 and 1000 K using this equation.

17.60 A molecule exists in singlet and triplet forms with the singlet having the higher energy by 4.11×10^{-21} J per molecule. The singlet level has a degeneracy of 1 and the triplet level has a degeneracy of 3. (a) Ignoring higher levels, what is the electronic partition function? (b) What is the ratio of the concentration of triplets to singlet molecules at 298 K?

17.61 What fraction of hydrogen atoms have $n = 2$ at room temperature according to the Boltzmann distribution? At 3000 K?

17.62 The Sackur–Tetrode equation

$$\overline{S}(V, T) = \overline{S}_0' + R \ln V + \overline{C}_V \ln T$$

seems to predict that $\overline{S} \rightarrow -\infty$ when $T \rightarrow 0$. Explain why this is not in conflict with the third law.

17.63 For actual calculation of the molar entropy of a monatomic gas, the Sackur–Tetrode equation may be written in the form

$$\overline{S} = \overline{S}' + \frac{3}{2} R \ln A_r - R \ln \frac{P}{P°} + \frac{5}{2} R \ln \frac{T}{K}$$

Show that for $P° = 1$ bar, $\overline{S}'/R = -1.151\ 693$.

17.64 Calculate $\overline{C}_P^\circ$ for hydrogen gas at 298.15 and 2000 K. This calculation is discussed in some detail by C. Marzzacco and M. Waldman, *J. Chem. Educ.* **50**:444 (1973).

17.65 Calculate the molar entropy for chlorine gas at 25 °C and 1 bar pressure.

17.66 Calculate the statistical mechanical values of $\overline{C}_P^\circ$, $\overline{S}^\circ$, $\overline{H}^\circ$, and $\overline{G}^\circ$ for H(g) at 3000 K.

17.67 Calculate the statistical mechanical values of $\overline{C}_P^\circ$, $\overline{H}^\circ$, $\overline{S}^\circ$, and $\overline{G}^\circ$ for $H_2(g)$ at 3000 K.

17.68 What are the values of $\Delta_r \overline{C}_P^\circ$, $\Delta_r \overline{H}^\circ$, and $\Delta_r \overline{S}^\circ$, and $\Delta_r \overline{G}^\circ$ for $H_2(g) = 2H(g)$ at 3000 K calculated in the preceding two problems? What is the value of K? What is the degree of dissociation at 1 bar?

17.69 Calculate the values of D_0 in Table 17.2 from data in Table C.2 for $H_2(g)$, $O_2(g)$, $Cl_2(g)$, HCl(g), and CO(g).

17.70 Derive the statistical mechanical expression for the equilibrium constant for the reaction

$$A_2(g) + B_2(g) = 2AB(g)$$

where A and B are isotopes. The contribution of the vibrational partition function may be ignored because it is so close to unity.

17.71 Tabulate the translational, vibrational, and rotational contributions to $\overline{C}_V^\circ/R$ for H, H_2, H_2O, and NH_3 in the ideal gas state that are expected classically. Calculate the classical limits for $\overline{C}_P^\circ$ and compare them with the values in Table C.2 at 3000 K.

17.72 Given the Lennard–Jones parameters for argon in Table 12.5, what values of the second virial coefficient do you expect at 200 and 500 K? Do these values agree with the experimental values in Fig. 1.4?

17.73 Starting with $\langle U \rangle = kT^2 (\partial \ln Q/\partial T)_{N,V}$ show that

$$\langle \epsilon \rangle = \frac{\sum \epsilon_i\, e^{-\epsilon_i/kT}}{\sum e^{-\epsilon_i/kT}}$$

18
Kinetic Theory of Gases

The kinetic theory of gases allows the calculation of the thermodynamic and transport properties of a gas by focusing on the collisions between gas molecules and on the Maxwell–Boltzmann probability distribution of molecular velocities. Combining the Maxwell–Boltzmann distribution and the simplest model for molecular collisions is sufficient to calculate many of the transport properties of gases. Conversely, the theory also allows us to obtain molecular information from experimental data on transport processes (e.g., diffusion, viscosity, and heat conduction).

18.1 The Maxwell–Boltzmann Distribution

In Section 17.2, we found the Maxwell–Boltzmann distribution for a gas of noninteracting molecules. Equation 17.18 gives the number fraction N_i/N of molecules in the gas (at equilibrium at temperature T) in energy level i with energy ϵ_i and degeneracy g_i:

$$\frac{N_i}{N} = \frac{g_i e^{-\beta \epsilon_i}}{q} = \frac{g_i e^{-\beta \epsilon_i}}{\sum g_i e^{-\beta \epsilon_i}} \tag{18.1}$$

where q is the molecular partition function and $\beta = (kT)^{-1}$. If we restrict our attention to a gas of atoms, then the energy is only kinetic energy. For a gas in a rectangular box with sides of different lengths a_1, a_2, a_3, the degeneracy is unity for each energy level and

$$\epsilon_i = \left(\frac{n_1^2}{a_1^2} + \frac{n_2^2}{a_2^2} + \frac{n_3^2}{a_3^2} \right) \left(\frac{\hbar^2 \pi^2}{2m} \right) \tag{18.2}$$

Table 18.1 Definite Integrals Occurring in the Kinetic Theory of Gases

n	0	1	2	3	4	5
$\int_0^\infty x^n \exp(-ax^2)dx$	$\dfrac{1}{2}\left(\dfrac{\pi}{a}\right)^{1/2}$	$\dfrac{1}{2a}$	$\dfrac{1}{4}\left(\dfrac{\pi}{a^3}\right)^{1/2}$	$\dfrac{1}{2a^2}$	$\dfrac{3}{8}\left(\dfrac{\pi}{a^5}\right)^{1/2}$	$\dfrac{1}{a^3}$
$\int_{-\infty}^{+\infty} x^n \exp(-ax^2)\,dx$	$\left(\dfrac{\pi}{a}\right)^{1/2}$	0	$\dfrac{1}{2}\left(\dfrac{\pi}{a^3}\right)^{1/2}$	0	$\dfrac{3}{4}\left(\dfrac{\pi}{a^5}\right)^{1/2}$	0

from equation 10.75, where n_1, n_2, and n_3 are the quantum numbers for x, y, and z motion. We saw in Chapter 10 that these energy levels are very closely spaced for a macroscopic box. Therefore, it is useful to disregard the quantized nature of the energy levels. We do this by considering n_1, n_2, n_3, to be **continuous** variables rather than integers, and replace the sum in q by an integral, as we did in Section 17.7. We can also identify $(\hbar^2\pi^2 n_i^2/a_i^2)$ as p_i^2, the square of the ith component of momentum, so that

$$\frac{N_i}{N} = q^{-1} \exp\left[-\frac{\beta}{2m}(p_x^2 + p_y^2 + p_z^2)\right] \tag{18.3}$$

The last step is to identify the fraction of molecules N_i/N with translational energy ϵ_i as proportional to the **probability distribution function** f_p for finding a molecule with these values of momentum components. Then

$$f_p(p_x, p_y, p_z) = c\, q^{-1} \exp\left[-\frac{\beta}{2m}(p_x^2 + p_y^2 + p_z^2)\right] \tag{18.4}$$

The probability for finding a molecule with *any* momentum must be 1, so the integral of equation 18.4 over all p_x, p_y, and p_z must be 1. Performing the following integral, using the results in Table 18.1, we find

$$\iiint dp_x\, dp_y\, dp_z \exp\left[-\frac{\beta}{2m}(p_x^2 + p_y^2 + p_z^2)\right] = \left(\frac{2m\pi}{\beta}\right)^{3/2} \tag{18.5}$$

so that

$$f_p(p_x, p_y, p_z)\, dp_x\, dp_y\, dp_z$$
$$= (2\pi mkT)^{-3/2} \exp\left[-\frac{\beta}{2m}(p_x^2 + p_y^2 + p_z^2)\right] dp_x\, dp_y\, dp_z \tag{18.6}$$

where we have replaced β by $1/kT$. We can convert this to a probability distribution in v_x, v_y, and v_z by using $p_x = mv_x$, and so on, to find

$$f(v_x, v_y, v_z)\, dv_x\, dv_y\, dv_z$$
$$= \left(\frac{m}{2\pi kT}\right)^{3/2} \exp\left[-\frac{m}{2kT}(v_x^2 + v_y^2 + v_z^2)\right] dv_x\, dv_y\, dv_z \tag{18.7}$$

Note that the function f is a **velocity probability density** so that the probability of finding an atom with velocity components between v_x and $v_x + dv_x$, v_y and $v_y + dv_y$, and v_z and $v_z + dv_z$ is given by $f\, dv_x\, dv_y\, dv_z$.

For a gas of molecules with energy levels that are the sum of vibrational, rotational, electronic, and nuclear energies (see equation 17.54), the partition function is a product of terms, one for each kind of energy. The Boltzmann

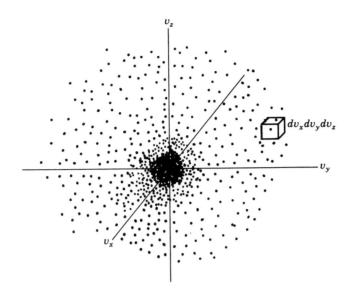

Figure 18.1   Distribution of velocity components for a number of molecules in an ideal gas.

distribution function is therefore also a product of terms, so that the probability distribution function for each kind of energy *separately* is a Boltzmann distribution function. Therefore, for any gas of noninteracting molecules, the velocity distribution function is given by equation 18.7.

In Fig. 18.1, we illustrate equation 18.7 by making the **density** of points in an elemental volume $dv_x\, dv_y\, dv_z$ equal to f at that point in the velocity space. Notice that f is largest near the origin and also that f is symmetrical about the origin.

18.2   Velocity Distribution in One Direction

The Maxwell–Boltzmann distribution law may be used to derive the velocity distribution in a single direction. For example, suppose we are interested in the probable number of molecules that have y components of the velocity in the range v_y to $v_y + dv_y$. This is the probability of finding molecules with velocity components between the two planes in Fig. 18.2. The probability $f(v_y)\, dv_y$ is obtained from

$$f(v_y)\, dv_y = dv_y \int_{-\infty}^{\infty} dv_x \int_{-\infty}^{\infty} dv_z f(v_x, v_y, v_z) \qquad (18.8a)$$

$$f(v_y) = \left(\frac{m}{2\pi kT}\right)^{1/2} \exp\left(-\frac{mv_y^2}{2kT}\right) \qquad (18.8b)$$

There are corresponding relations for the distributions in the x and z directions. This distribution has the form of the Gaussian error curve, as shown in Fig. 18.3. The most probable velocity in any particular direction is zero. The distribution is symmetrical, and so the average velocity in the y direction is zero, as expected for a gas at rest:

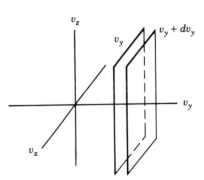

Figure 18.2   Calculation of the probability of velocity components in the range v_y to $v_y + dv_y$.

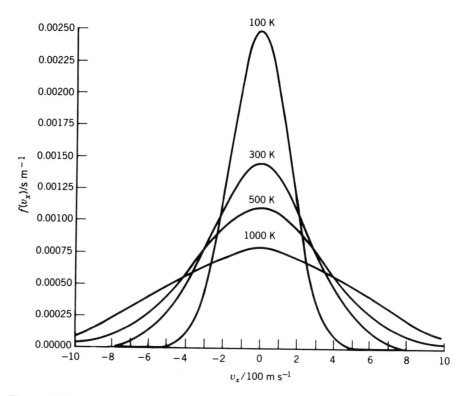

Figure 18.3 Probability density for the velocity of oxygen molecules in an arbitrarily chosen direction at 100, 300, 500, and 1000 K.

$$\langle v_y \rangle = \int_{-\infty}^{\infty} v_y f(v_y) \, dv_y = 0 \qquad (18.9)$$

This integral is readily evaluated by noting that $f(v_y)$ is symmetrical and v_y is an odd function. When the temperature is raised the distribution becomes broader, but the area under the curve remains constant because $f(v_y)$ is normalized.

Equation 18.8 can be used to obtain the following expression for the **average kinetic energy in the x direction** (see Example 18.2):

$$\epsilon_x = \tfrac{1}{2}m\langle v_x^2 \rangle = \tfrac{1}{2}kT \qquad (18.10)$$

Of course, similar expressions apply to the y and z directions. This is an example of the principle of equipartition of energy (Section 17.13).

Example 18.1
Calculate the probability density for v_x of O_2 molecules at 300 K at 0, 300, and 600 m s^{-1}. At 300 m s^{-1}

$$f(v_x) = \left(\frac{M}{2\pi RT}\right)^{1/2} \exp\left(-\frac{Mv_x^2}{2RT}\right)$$

$$= \left[\frac{0.032 \text{ kg mol}^{-1}}{2\pi(8.314 \text{ J K}^{-1}\text{ mol}^{-1})(300 \text{ K})}\right]^{1/2} \exp\left[-\frac{(0.032 \text{ kg mol}^{-1})(300 \text{ m s}^{-1})^2}{2(8.314 \text{ J K}^{-1}\text{ mol}^{-1})(300 \text{ K})}\right]$$

$$= 8.022 \times 10^{-4} \text{ s m}^{-1}$$

The probability densities at 0 and 600 m s^{-1} are 1.429×10^{-3} s m^{-1} and 1.419×10^{-4} s m^{-1}, in agreement with Fig. 18.3.

Example 18.2
Using the distribution function for velocities in the x direction show that

$$\langle v_x^2 \rangle = \frac{kT}{m}$$

The average is the integral of v_x^2 multiplied by the distribution function:

$$\langle v_x^2 \rangle = \int_{-\infty}^{\infty} v_x^2 f(v_x) \, dv_x$$

$$= \left(\frac{m}{2\pi kT}\right)^{1/2} \int_{-\infty}^{\infty} e^{-mv_x^2/2kT} v_x^2 \, dv_x$$

Using the value of the definite integral given in Table 18.1,

$$\langle v_x^2 \rangle = \left(\frac{m}{2\pi kT}\right)^{1/2} \frac{\pi^{1/2}}{2(m/2kT)^{3/2}}$$

$$= \frac{kT}{m}$$

Therefore, equation 18.10 follows.

18.3 Maxwell Distribution of Speeds

Usually we are more interested in the distribution of speeds than in the distribution of component velocities. The **speed** v of a molecule is related to its component velocities by

$$v^2 = v_x^2 + v_y^2 + v_z^2 \tag{18.11}$$

The speed is represented by the distance in velocity space of a point from the origin in Fig. 18.1. Therefore, the probability $F(v) \, dv$ that a molecule has a speed between v and $v + dv$ is given by the probable number of points in a spherical shell of thickness dv, as shown in Fig. 18.4a. The required integration of equation 18.7 is most conveniently carried out by converting to spherical coordinates, which are shown in Fig. 18.4b, using

$$v_x = v \sin\theta \cos\phi \tag{18.12a}$$

$$v_y = v \sin\theta \sin\phi \tag{18.12b}$$

$$v_z = v \cos\theta \tag{18.12c}$$

The differential volume element $dv_x \, dv_y \, dv_z$ can be written in spherical coordinates as

$$dv_x \, dv_y \, dv_z = v^2 \, dv \sin\theta \, d\theta \, d\phi \tag{18.13}$$

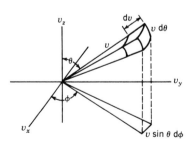

Figure 18.4 (a) Calculation of the probability of velocity components in the range v to $v + dv$. (b) Volume element in spherical coordinates.

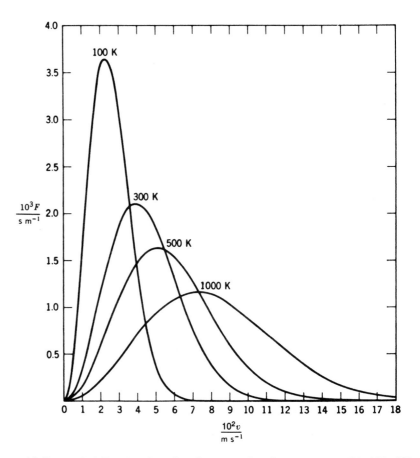

Figure 18.5   Probability density of various speeds v for oxygen at 100, 300, 500, and 1000 K, calculated using equation 18.16.

The probability $F(v)\, dv$ can now be found by integration of $f(v_x, v_y, v_z)\, dv_x\, dv_y\, dv_z$ over the angles θ and ϕ:

$$F(v)\, dv = \int_0^\pi d\theta \int_0^{2\pi} d\phi\; f(v_x, v_y, v_y) \sin\theta\; v^2\, dv \qquad (18.14)$$

Substituting equation 18.7 into this expression and using equation 18.11, we find

$$F(v)\, dv = 4\pi v^2 \left(\frac{m}{2\pi kT}\right)^{3/2} \exp\!\left(-\frac{mv^2}{2kT}\right) dv \qquad (18.15)$$

Thus, the probability density $F(v)$ is

$$F(v) = 4\pi v^2 \left(\frac{m}{2\pi kT}\right)^{3/2} \exp\!\left(-\frac{mv^2}{2kT}\right) \qquad (18.16)$$

As a result the probability density at a speed of 0 is zero. The probability density increases with the speed up to a maximum and then declines.

A plot of $F(v)$ versus the molecular speed v is shown in Fig. 18.5 for oxygen at 100, 300, 500, and 1000 K. The probability that a molecule has a speed between any two values is given by the area under the curve between these two values of the speed. The plot of $F(v)$ versus v is approximately quadratic near the origin. At higher speeds the probability decreases toward zero because the exponential term decreases much more rapidly than v^2 increases. Thus, very few molecules have very high or very low speeds. The fraction of the molecules having speeds greater than 10 times the most probable speed (defined in Section 18.4) is 9×10^{-42} at any temperature. The Avogadro constant times this fraction is so much less than one that we can say that no molecule has a velocity this large.

Notice that the maximum of $F(v)$ moves to higher v as the temperature is raised. The speed at the maximum is the most probable speed for which we will soon derive a formula. Notice also that the width of the curve becomes larger as T increases (and since the area under the curve is always unity, the maximum gets lower). Thus, as T increases, the most probable speed increases and so do the numbers of molecules at high speeds.

18.4 Types of Average Speed

Since there is a distribution of molecular speeds, there are different measures of the average speed. We will discuss the most probable speed v_p, the mean speed $\langle v \rangle$, and the root mean square speed $\langle v^2 \rangle^{1/2}$.

The **most probable speed** v_p is the speed at the maximum of $F(v)$. Setting dF/dv equal to zero, we find

$$\frac{dF(v)}{dv} = \left(\frac{m}{2\pi kT} \right)^{3/2} e^{-mv^2/2kT} \left[8\pi v + 4\pi v^2 \left(-\frac{mv}{kT} \right) \right] = 0 \quad (18.17)$$

or

$$v_p = \left(\frac{2kT}{m} \right)^{1/2} = \left(\frac{2RT}{M} \right)^{1/2} \quad (18.18)$$

The **mean speed** $\langle v \rangle$ is calculated as the average of v using the probability distribution $F(v)$:

$$\langle v \rangle = \int_0^\infty v F(v) \, dv \quad (18.19)$$

Substituting equation 18.16 and performing the integral with the help of Table 18.1, we find

$$\langle v \rangle = 4\pi \left(\frac{m}{2\pi kT} \right)^{3/2} \int_0^\infty \exp\left(-\frac{mv^2}{kT} \right) v^3 \, dv \quad (18.20)$$

$$\langle v \rangle = \left(\frac{8kT}{\pi m} \right)^{1/2} = \left(\frac{8RT}{\pi M} \right)^{1/2} \quad (18.21)$$

The last speed we consider is the **root-mean-square speed** which is defined as the square root of $\langle v^2 \rangle$,

$$\langle v^2 \rangle^{1/2} = \left[\int_0^\infty v^2 F(v) \, dv \right]^{1/2} \quad (18.22)$$

Table 18.2 Various Types of Average Speeds of Gas Molecules at 298 K

Gas	$\langle v^2 \rangle^{1/2}$/m s^{-1}	$\langle v \rangle$/m s^{-1}	v_p/m s^{-1}
H_2	1920	1769	1568
O_2	482	444	394
CO_2	411	379	336
CH_4	681	627	556

Substituting equation 18.16 and using Table 18.1 again, we find

$$\langle v^2 \rangle^{1/2} = \left(\frac{3kT}{m} \right)^{1/2} = \left(\frac{3RT}{M} \right)^{1/2} \qquad (18.23)$$

From these three calculations, we can see that at any temperature

$$\langle v^2 \rangle^{1/2} > \langle v \rangle > v_p \qquad (18.24)$$

Each of these measures of the probability distribution is proportional to $(T/M)^{1/2}$, so that each increases with temperature and decreases with molar mass. Lighter molecules therefore move faster than heavier molecules on average, as shown in Table 18.2.

The **speed of sound** in a gas is, not surprisingly, also about the same magnitude as the average speeds. It can be shown* that sound waves in a gas are longitudinal contractions and rarefactions which are adiabatic and reversible, and which travel at the speed v_s given by the thermodynamic quantity:

$$v_s^2 = - \frac{V}{\rho \left(\dfrac{\partial V}{\partial P} \right)_S} \qquad (18.25)$$

Here V is the volume, S is the entropy, P the pressure and ρ the density of the gas. Since for an ideal gas undergoing a reversible adiabatic expansion or contraction, PV^γ = constant (see equation 2.80), we have

$$\left(\frac{\partial V}{\partial P} \right)_S = - \frac{V}{\gamma P} \qquad (18.26)$$

Substituting this into equation 18.25, and using the ideal gas law, we find

$$v_s^2 = \frac{\gamma P}{\rho} = \frac{\gamma RT}{M} \qquad (18.27)$$

Thus,

$$v_s = \left(\frac{\gamma RT}{M} \right)^{1/2} \qquad (18.28)$$

For monatomic gases $\gamma = \frac{5}{3}$, so that v_s is just smaller than v_p. For real gases, the velocity of sound depends slightly on pressure.

---

* M. Zemansky and R. Dittman, *Heat and Thermodynamics*. New York: McGraw-Hill, 1981.

Example 18.3
Calculate the most probable speed v_p, the mean speed $\langle v \rangle$, and the root-mean-square speed $\langle v^2 \rangle^{1/2}$ for hydrogen molecules at 0 °C.

$$v_p = \left(\frac{2RT}{M}\right)^{1/2} = \left[\frac{(2)(8.314 \text{ J K}^{-1} \text{ mol}^{-1})(273 \text{ K})}{(2.016 \times 10^{-3} \text{ kg mol}^{-1})}\right]^{1/2}$$
$$= 1.50 \times 10^3 \text{ m s}^{-1}$$

$$\langle v \rangle = \left(\frac{8RT}{\pi M}\right)^{1/2} = \left[\frac{(8)(8.314 \text{ J K}^{-1} \text{ mol}^{-1})(273 \text{ K})}{(3.1416)(2.016 \times 10^{-3} \text{ kg mol}^{-1})}\right]^{1/2}$$
$$= 1.69 \times 10^3 \text{ m s}^{-1}$$

$$\langle v^2 \rangle^{1/2} = \left(\frac{3RT}{M}\right)^{1/2} = \left[\frac{3(8.314 \text{ J K}^{-1} \text{ mol}^{-1})(273.15 \text{ K})}{2.016 \times 10^{-3} \text{ kg mol}^{-1}}\right]^{1/2}$$
$$= 1.84 \times 10^3 \text{ m s}^{-1}$$

The root-mean-square speed of a hydrogen molecule at 0 °C is 6620 km h^{-1}, but at ordinary pressures it travels only an exceedingly short distance before colliding with another molecule and changing direction.

18.5 Pressure of an Ideal Gas

The pressure is the average force per unit area that the wall must exert on the molecules to hold them at constant volume. The average force $\overline{F}_x$ in the x direction is equal to the time rate of change in the momentum in the x direction of the molecules that strike the wall:

$$\overline{F}_x = ma_x = m\frac{dv_x}{dt} = \frac{d(mv_x)}{dt} \tag{18.29}$$

Consider the molecules striking area A of the yz plane, as shown in Fig. 18.6. We will assume that when a molecule strikes a flat surface at A it is reflected specularly. Since the momenta in the y and z directions do not change, the change in momentum of the molecule is that in the x direction, namely $-2mv_x$, where $v_x \geq 0$ prior to collision with the wall. In a time dt, a molecule with v_x will hit the surface if that molecule is within a volume $v_x\,dtA$ of the surface ($v_x \geq 0$). Since we assume that the molecules are randomly distributed throughout the volume of the gas, the probable number of molecules having a velocity in the range of v_x to $v_x + dv_x$ and within the necessary distance to hit the surface is

$$Nf(v_x)\,dv_x\left(v_x\,dt\frac{A}{V}\right) \tag{18.30}$$

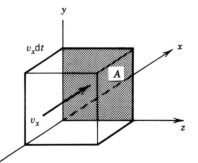

Figure 18.6   Molecule with velocity v_x striking a wall of area A. All molecules in the volume $Av_x\,dt$ with $v_x \geq 0$ will strike this surface in time dt.

The factor in parentheses is the volume in which molecules will hit the surface in a time dt divided by the total volume. This is the probability of finding the molecule in the correct volume to hit the surface for a random spatial distribution. To find the force on the surface exerted by the gas we must multiply the expression 18.30 by the negative of the momentum change of the molecules or $2mv_x$, divide by dt and integrate over all positive v_x. Therefore, the average force in the x direction is

$$\overline{F}_x = N \int_0^\infty dv_x\, (2mv_x)f(v_x)\, v_x\, \frac{A}{V} \tag{18.31}$$

or, by substituting for $f(v_x)$,

$$
\begin{aligned}
\overline{F}_x &= \frac{NA}{V} 2m \left(\frac{m}{2\pi kT}\right)^{1/2} \int_0^\infty dv_x\, v_x^2\, e^{-mv_x^2/2kT} \\
&= \frac{NA}{V} 2m \left(\frac{m}{2\pi kT}\right)^{1/2} \left(\frac{2kT}{m}\right)^{3/2} \int_0^\infty dx\, x^2 e^{-x^2} \\
&= \frac{NA}{V} kT
\end{aligned}
\tag{18.32}
$$

The pressure is the force per unit area; therefore,

$$P = \frac{NkT}{V} \tag{18.33}$$

which is the ideal gas equation of state. We have derived this in Section 17.7 using statistical mechanics. The present derivation shows how the pressure arises from the collisions of molecules with the wall.

18.6   Collisions of Hard-Sphere Molecules

The interactions of molecules in the gas phase are very complicated because of the shape of the intermolecular potential (Section 12.11). In this section we will use a very simple molecular model—the **hard sphere**. This is equivalent to assuming that the intermolecular potential is zero at distances between centers greater than $\frac{1}{2}(d_1 + d_2)$, where d_1 and d_2 are the diameters of the two molecules. The potential energy is infinite at shorter distances, as illustrated in Fig. 18.7. Thus, hard-sphere molecules 1 and 2 do not interact unless the distances between their centers is $\frac{1}{2}(d_1 + d_2)$, and then they bounce like idealized billiard balls.

As shown in Fig. 18.8 hard spherical molecules collide with each other if their centers come within a distance d equal to their diameters if the molecules are alike, or distance $d_{12} = \frac{1}{2}(d_1 + d_2)$ if they are different. The distance d_{12} is called the **collision diameter.**

Let us consider collisions of molecules of type 1 with molecules of type 2. If molecules of type 2 are stationary, a molecule of type 1 will collide in unit time with all molecules of type 2 that have their centers in a cylinder of volume $\pi d_{12}^2 v_1$. According to this simple calculation a molecule of type 1 would undergo $\pi d_{12}^2 v_1 \rho_2$ collisions per unit time where ρ_2 is the number of type 2 molecules per unit volume. However, molecules of type 2 are not stationary and so we need to use the relative speed v_{12} in calculating the rate of collisions $z_{1(2)}$ of a molecule of type 1 with molecules of type 2. Thus,

$$z_{1(2)} = \rho_2 \pi d_{12}^2 \int f(v_1)f(v_2)\, v_{12}\, dv_1\, dv_2 \tag{18.34}$$

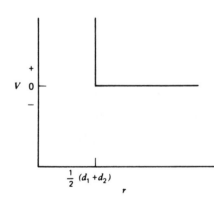

Figure 18.7   Potential energy diagram for two hard sphere molecules with diameters d_1 and d_2.

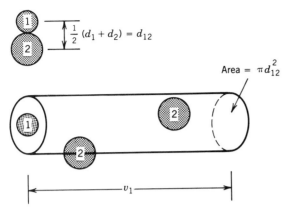

Figure 18.8 Collisions of hard-sphere molecules. If molecules of type 2 are stationary, a molecule of type 1 will collide in unit time with all molecules of type 2 that have their centers in a cylinder of volume $\pi d_{12}^2 v_1$.

where we are averaging over the product of distribution functions for each molecule. Note that the product $f(v_1)f(v_2)$ contains the sum of kinetic energies of particle 1 and particle 2 in the exponential. We have already seen (Section 10.11) that we can convert from the sum of kinetic energies of particle 1 and particle 2 to the kinetic energy of the center of mass plus the relative kinetic energy. In addition, the volume element $dv_1\,dv_2$ becomes the volume element $dv_1\,dv_{CM}$. The integral over the center of mass yields unity; and after the integrations over angles have been done, we have

$$z_{1(2)} = \rho_2 \pi d_{12}^2 \int f(v_{12})\, v_{12}\, dv_{12} \tag{18.35}$$

where

$$f(v_{12}) = 4\pi \left(\frac{\mu}{2\pi kT} \right)^{3/2} v_{12}^2\, e^{-\mu v_{12}^2/2kT} \tag{18.36}$$

and μ is the reduced mass equal to $m_1 m_2/(m_1 + m_2)$. The integration in equation 18.35 can now be done using Table 18.1 to yield

$$z_{1(2)} = \rho_2 \pi d_{12}^2 \left(\frac{8kT}{\pi\mu} \right)^{1/2} = \rho_2 \pi d_{12}^2 \langle v_{12} \rangle \tag{18.37}$$

the **collision frequency** of molecules of type 1 with molecules of type 2. Since the density ρ_2 of molecules of type 2 has the SI unit m^{-3}, the collision diameter d_{12} has the unit m, and the relative mean speed $\langle v_{12} \rangle$ has the unit m s^{-1}, the collision frequency has the unit s^{-1}.

Equation 18.37 introduces a new type of molecular speed, the **mean relative speed** $\langle v_{12} \rangle$:

$$\langle v_{12} \rangle = \left(\frac{8kT}{\pi\mu} \right)^{1/2} \tag{18.38}$$

Let us take a minute to consider why it has the form it does. If we square both sides of equation 18.38 and introduce the definition of the reduced mass μ, we obtain

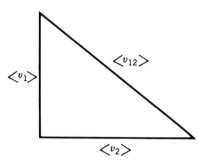

Figure 18.9   The mean relative speed $\langle v_{12} \rangle$ of molecules 1 and 2 can be calculated from this right triangle involving the mean speeds of molecules of types 1 and 2. According to the Pythagorean theorem, $\langle v_{12} \rangle^2 = \langle v_1 \rangle^2 + \langle v_2 \rangle^2$.

$$\langle v_{12} \rangle^2 = \left(\frac{8kT}{\pi} \right) \left(\frac{1}{m_1} + \frac{1}{m_2} \right)$$

$$= \langle v_1 \rangle^2 + \langle v_2 \rangle^2 \qquad (18.39)$$

As shown by Fig. 18.9, we can use the Pythagorean theorem to interpret the mean relative speed in terms of the mean speeds of molecules 1 and 2. Molecules 1 and 2 can collide with each other with any angle between 0° and 180° between their paths, but equation 18.39 shows that the average collision is at 90°. For collisions of indentical particles, $\langle v_{11} \rangle^2 = 2\langle v_1 \rangle^2$, so that $\langle v_{11} \rangle = 2^{1/2}\langle v_1 \rangle = 2^{1/2}\langle v \rangle$.

Example 18.4

What is the mean relative speed of hydrogen molecules with respect to oxygen molcules (or oxygen molecules with respect to hydrogen molecules) at 298 K? The molecular masses are

$$m_1 = \frac{2.016 \times 10^{-3} \text{ kg mol}^{-1}}{6.022 \times 10^{23} \text{ mol}^{-1}}$$

$$= 3.348 \times 10^{-27} \text{ kg}$$

$$m_2 = \frac{32.000 \times 10^{-3} \text{ kg mol}^{-1}}{6.022 \times 10^{23} \text{ mol}^{-1}}$$

$$= 5.314 \times 10^{-26} \text{ kg}$$

$$\mu = [(3.348 \times 10^{-27} \text{ kg})^{-1} + (5.314 \times 10^{-26} \text{ kg})^{-1}]^{-1}$$

$$= 3.150 \times 10^{-27} \text{ kg}$$

$$\langle v_{12} \rangle = \left(\frac{8kT}{\pi\mu} \right)^{1/2}$$

$$= \left[\frac{8(1.381 \times 10^{-23} \text{ J K}^{-1})(298 \text{ K})}{\pi(3.150 \times 10^{-17} \text{ kg})} \right]^{1/2}$$

$$= 1824 \text{ m s}^{-1}$$

Note that the mean relative speed is closer to the mean speed of molecular hydrogen (1920 m s^{-1}) than to that of molecular oxygen (482 m s^{-1}).

If the molecule of type 1 is moving through molecules of type 1, rather than molecules of type 2, equation 18.37 becomes

$$z_{1(1)} = 2^{1/2}\rho\pi d^2\langle v \rangle \qquad (18.40)$$

since $(8kT/\pi\mu)^{1/2}$ becomes $2^{1/2}\langle v \rangle$ because $1/\mu = 1/m + 1/m = 2/m$. The collision frequency $z_{1(1)}$ is the rate of collisions of molecules of type 1 with molecules of type 1.

In connection with chemical kinetics, we will also be interested in the number of collisions per unit time per unit volume. This quantity is referred to as the **collision density,** and it is represented by Z. To calculate the number of collisions of molecules of type 1 with molecules of type 2 per unit time per unit volume of gas Z_{12} we simply multiply $z_{1(2)}$ by the number density ρ_1, so that

$$Z_{12} = \rho_1\rho_2\pi d_{12}^2\langle v_{12} \rangle \qquad (18.41)$$

Table 18.3 Collision Frequencies $z_{1(1)}$ and Collision Densities Z_{11} for Four Gases at 25 °C

| Gas | $z_{1(1)}/s^{-1}$ | | $Z_{11}/\text{mol L}^{-1} \text{s}^{-1}$ | |
	1 bar	10^{-6} bar	1 bar	10^{-6} bar
H_2	14.13×10^9	14.13×10^3	2.85×10^8	2.85×10^{-4}
O_2	6.24×10^9	6.24×10^3	1.26×10^8	1.26×10^{-4}
CO_2	8.81×10^9	8.81×10^3	1.58×10^8	1.58×10^{-4}
CH_4	11.60×10^9	11.60×10^3	2.08×10^8	2.08×10^{-4}

If we are interested in the number of collisions of molecules of type 1 with other molecules of type 1 per unit time per unit volume of gas Z_{11}, equation 18.41 reduces to

$$Z_{11} = \tfrac{1}{2}\rho^2 \pi d^2 \langle v_{11} \rangle$$
$$= 2^{-1/2}\rho^2 \pi d^2 \langle v \rangle \qquad (18.42)$$

where a divisor of 2 has been introduced so that each collision is not counted twice and $\langle v_{11} \rangle$ has been replaced by $2^{1/2}\langle v \rangle$ by means of the reduced mass of like particles. The collision density is readily expressed in mol m^{-3} s^{-1} by simply dividing by the Avogadro constant.

The collision density is of interest because it sets an upper limit on the rate with which two gas molecules can react (see Section 20.1). Actual chemical reaction rates are usually much smaller than the collision rates, indicating that not every collision leads to reaction.

Collision frequencies $z_{1(1)}$ and collision densities Z_{11} for four gases are given in Table 18.3 at 25 °C. The collision densities are expressed in mol L^{-1} s^{-1} because it is easier to think about chemical reactions in these units.

Example 18.5
For molecular oxygen at 25 °C, calculate the collision frequency $z_{1(1)}$ and the collision density Z_{11} at a pressure of 1 bar. The collision diameter of oxygen is 0.361 nm or 3.61×10^{-10} m, as determined in a manner to be described shortly (Section 18.11):

$$\langle v \rangle = \left(\frac{8RT}{\pi M}\right)^{1/2} = \left[\frac{(8)(8.314 \text{ J K}^{-1} \text{ mol}^{-1})(298 \text{ K})}{\pi(32 \times 10^{-3} \text{ kg mol}^{-1})}\right]^{1/2} = 444 \text{ m s}^{-1}$$

The number density is given by

$$\rho = \frac{N}{V} = \frac{PN_A}{RT} = \frac{(1 \text{ bar})(6.022 \times 10^{23} \text{ mol}^{-1})(10^3 \text{ L m}^{-3})}{(0.083145 \text{ L bar K}^{-1} \text{ mol}^{-1})(298 \text{ K})} = 2.43 \times 10^{25} \text{ m}^{-3}$$

The collision frequency is given by

$$z_{1(1)} = \sqrt{2}\rho \pi d^2 \langle v \rangle$$
$$= (1.414)(2.43 \times 10^{25} \text{ m}^{-3})\pi(3.61 \times 10^{-10} \text{ m})^2(444 \text{ m s}^{-1})$$
$$= 6.24 \times 10^9 \text{ s}^{-1}$$

The collision density is given by

$$Z_{11} = \frac{1}{2^{1/2}} \rho^2 \pi d^2 \langle v \rangle$$

$$= (0.707)(2.43 \times 10^{25}\ m^{-3})^2 \pi (3.61 \times 10^{-10}\ m)^2 (444\ m\ s^{-1})$$

$$= 7.58 \times 10^{34}\ m^{-3}\ s^{-1} = \frac{(7.58 \times 10^{34}\ m^{-3}\ s^{-1})(10^{-3}\ m^3\ L^{-1})}{6.022 \times 10^{23}\ mol^{-1}}$$

$$= 1.26 \times 10^8\ mol\ L^{-1}\ s^{-1}$$

18.7 Mean Free Path

The mean free path λ is the average distance traveled between collisions. Although it is not a directly measurable quantity, it is a very useful concept, as we shall see. It can be computed by dividing the average distance traveled per unit time by the collision frequency. For a molecule moving through like molecules

$$\lambda = \frac{\langle v \rangle}{z_{1(1)}} = \frac{1}{2^{1/2}\rho \pi d^2} \tag{18.43}$$

Assuming that the collision diameter d is independent of temperature, the temperature and pressure dependence of the mean free path may be obtained by substituting the ideal gas law in the form $\rho = P/kT$:

$$\lambda = \frac{kT}{2^{1/2}\pi d^2 P} \tag{18.44}$$

Thus, at constant temperature, the mean free path is inversely proportional to the pressure.

Example 18.6
For oxygen at 25 °C the collision diameter is 0.361 nm. What is the mean free path at (a) 1 bar pressure, and (b) 0.1 Pa pressure?

(a) Since from example 18.5, $\rho = 2.43 \times 10^{25}\ m^{-3}$,

$$\lambda = \frac{1}{2^{1/2}\rho \pi d^2}$$

$$\lambda = [(1.414)(2.43 \times 10^{25}\ m^{-3})\pi(3.61 \times 10^{-10}\ m)^2]^{-1} = 7.11 \times 10^{-8}\ m$$

(b) $\rho = \dfrac{PN_A}{RT} = \dfrac{(0.1\ Pa)(6.022 \times 10^{23}\ mol^{-1})}{(8.314\ J\ K^{-1}\ mol^{-1})(298\ K)} = 2.43 \times 10^{19}\ m^{-3}$

$$\lambda = [(1.414)(3.14)(3.61 \times 10^{-10}\ m)^2(2.43 \times 10^{19}\ m^{-3})]^{-1} = 0.071\ m = 7.1\ cm$$

At pressures so low that the mean free path becomes comparable with the dimensions of the containing vessel, the flow properties of the gas become markedly different from those at higher pressures.

18.8 Collisions with a Surface or Escape from an Opening

In studying the reaction of a gas with a solid it is necessary to calculate the number of gas molecules that hit the plane surface per unit time. In addition, it is often necessary to compute the rate at which molecules pass through a small opening into an evacuated vessel (effusion). For small holes the rate is small enough *not* to upset the equilibrium speed distribution in the bulk gas. In addition, the mean free path is assumed to be large compared to the diameter of the hole, so that collisions in the neighborhood of the hole can be neglected.

Actually, we have already done all the necessary calculations in Section 18.5 where we calculated the pressure of an ideal gas. We use Fig. 18.6 again and calculate the number of molecules hitting an area A in unit time to be given by the integral of equation 18.30. Dividing this by A gives the **flux** J_N of particles in the x direction:

$$J_N = \rho \int_0^\infty f(v_x) v_x \, dv_x$$

$$= \rho \left(\frac{kT}{2\pi m} \right)^{1/2} \tag{18.45}$$

The use of the expression for the mean velocity (equation 18.21) leads to a simple equation for the flux:

$$J_N = \frac{\rho \langle v \rangle}{4} \tag{18.46}$$

As previously described, the flux is the number of molecules striking the surface per unit area per unit time or the number of molecules that will escape through a hole into a vacuum per unit area per unit time.

Since we are assuming that the gas is ideal, the number density ρ may be eliminated from equation 18.45 by use of the ideal gas law $P = \rho kT$. Thus,

$$J_N = \frac{P}{(2\pi mkT)^{1/2}} \tag{18.47}$$

For a pure substance, the measurement of the rate of escape J_N through a small hole can be used to calculate the pressure. This is the basis of the Knudsen method for measuring the vapor pressure of a solid or a liquid. The solid or liquid is placed in a container with a small hole. This container is placed in an evacuated chamber and the loss in mass Δw of the container and sample is measured after time t. If the area of the hole is A, the flux is given by

$$J_N = \frac{\Delta w}{mtA} \tag{18.48}$$

A sufficiently large surface area of the solid or liquid must be exposed to maintain the saturation vapor pressure. These simple equations cannot be used if the gaseous sample has molecules with several different masses. For example, the vapor in equilibrium with graphite at high temperatures contains C_1, C_2, C_3, C_4, ... It was not possible to obtain a precise value for $\Delta H°$ for the reaction $C(\text{graphite}) = C(g)$ until the composition of the vapor at a series of temperatures had been obtained by mass spectrometry.

Example 18.7

The vapor pressure of solid beryllium was measured by R. B. Holden, R. Speiser, and H. L. Johnston (*J. Am. Chem. Soc.* **70**:3897 (1948)) using a Knudsen cell. The effusion hole was 0.318 cm in diameter, and they found a mass loss of 9.54 mg in 60.1 min at a temperature of 1457 K. What is the vapor pressure?

$$J_N = \frac{\Delta w}{mtA} = \frac{\Delta w N_A}{MtA}$$

$$= \frac{(9.54 \times 10^{-6}\ \text{kg})(6.022 \times 10^{23}\ \text{mol}^{-1})}{(9.013 \times 10^{-3}\ \text{kg mol}^{-1})(60 \times 60.1\ \text{s})\pi(0.159 \times 10^{-2}\ \text{m})^2}$$

$$= 2.23 \times 10^{22}\ \text{m}^{-2}\ \text{s}^{-1}$$

$$P = J_N (2\pi m k T)^{1/2}$$

$$= (2.23 \times 10^{22}\ \text{m}^{-2}\ \text{s}^{-1})$$

$$\times \left[\frac{2\pi(9.013 \times 10^{-3}\ \text{kg mol}^{-1})(1.381 \times 10^{-23}\ \text{J K}^{-1})(1457\ \text{K})}{(6.022 \times 10^{23}\ \text{mol}^{-1})} \right]^{1/2}$$

$$= 0.968\ \text{Pa} = 0.968 \times 10^{-5}\ \text{bar}$$

Since the flux is inversely proportional to the square root of the mass, diffusion through a porous barrier can be used to separate different isotopic species of gas molecules.

18.9   Effects of Molecular Interactions on Collisions

Collisions between gas molecules are more complicated than indicated in the preceding sections because of intermolecular attractive and repulsive forces. In connection with the discussion of the **Lennard–Jones potential** (Section 12.11), we saw that as molecules approach each other there is first intermolecular attraction and then, at shorter distances, repulsion. This is illustrated by the paths of two colliding molecules that are illustrated in Fig. 18.10. This figure has been drawn so that the center of mass of the two molecules is stationary and the motion is confined to the plane of the paper. The numbers 1, 2, 3, . . . indicate the successive positions of the two molecules. As the molecules approach, they first attract each other so that their paths are drawn together. As the molecules approach each other more closely, they repel each other, and their paths begin to diverge. After the interaction, the paths of the molecules make an angle χ with the directions of the initial paths.

The initial parameters of the collision are the relative kinetic energy ($\frac{1}{2}\mu v_{12}^2$) and the **impact parameter** b. The impact parameter is the minimum distance at which the molecules would pass each other if there were no molecular interactions. If b is large, the angle of deflection χ will be small.

The trajectories for collisions at various values of the impact parameters and for two values of the kinetic energy of approach are shown in Figs. 18.11a and b. In these figures one of the molecules approaches from the top of the figure with various values of the impact parameter b, and the other molecule

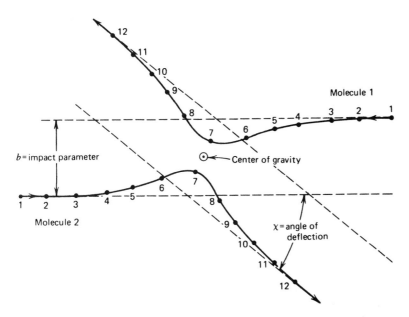

Figure 18.10 Collision of two spherically symmetrical molecules with an impact parameter b. The diagram is drawn so that the center of gravity of the system does not shift during the collision. The angle of deflection is χ. (From W. Kauzmann, *Thermal Properties of Matter*, Vol. 1, *Kinetic Theory of Gases*, © 1966. The Benjamin/Cummings, Menlo Park, CA. Reprinted with permission.)

approaches from below in a symmetrical fasion. In Fig. 18.11*a* the low-energy collisions lead to a very complicated pattern. For large values of the impact parameter the molecules attract each other along the whole trajectory, and the deflection is negative by definition (although it is not possible experimentally to distinguish positive from negative deflections). As the value of the impact parameter decreases, the deflection becomes more and more negative, as shown in the diagram, until the repulsive force begins to be felt. As the impact parameter is further reduced, the repulsive force becomes dominant and there are large positive deflections. For a head-on collision ($b = 0$) the deflection is 180°. In Fig. 18.11*b* the high-energy collisions give results that are close to, but not identical with, what would be expected for collisions of rigid spheres. The scattering angle χ can be calculated from the parameters for the molecular interaction, the impact parameter, and the relative kinetic energy of the two molecules. However, this calculation cannot be made analytically, and so we will not pursue it here.

When collisions of molecules interacting according to a Lennard–Jones potential are considered classically, we encounter the rather unsatisfactory situation that the cross section is infinite; notice in Fig. 18.11 that even "collisions" with large impact parameters b have some deflection. This problem is resolved by quantum mechanics, but we will not be able to go into quantum mechanical scattering theory, which shows that the cross section is similar in size to the hard sphere result, but energy dependent.

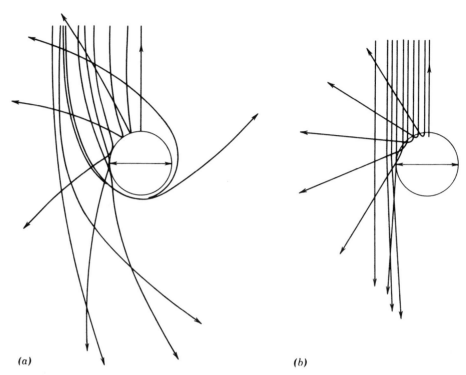

Figure 18.11 Collision trajectories for a pair of molecules interacting by a Lennard–Jones 6–12 potential. The center of gravity is stationary at the center of the circle, which has a diameter equal to the Lennard–Jones constant σ. (a) Trajectories for molecules that approach at energies equal to 0.1 ϵ, where ϵ is the depth of the potential well in the Lennard–Jones potential. (b) Trajectories for molecules that approach at energies equal to 50 ϵ. (From W. Kauzmann, *Thermal Properties of Matter*, Vol. 1; *Kinetic Theory of Gases*, © 1966. The Benjamin/Cummings, Menlo Park, CA. Reprinted with permission.)

18.10 Transport Phenomena in Gases

If a gas is not uniform with respect to composition, temperature, and velocity, transport processes occur until the gas does become uniform. The transport of matter in the absence of bulk flow is referred to as **diffusion.** The transport of heat from regions of high temperature to regions of lower temperature without convection is referred to as **thermal** conduction, and the transfer of momentum from a region of higher velocity to a region of lower velocity gives rise to the phenomenon of **viscous** flow. In each case the rate of flow is proportional to the rate of change of some property with distance, a so-called gradient.

The flux of component i in the z direction due to diffusion is proportional to the concentration gradient dc_i/dz, according to Fick's law:

$$J_{iz} = -D\frac{dc_i}{dz} \tag{18.49}$$

Figure 18.12 Schematic diagrams of apparatus for the measurements of (*a*) the diffusion coefficient *D*, (*b*) the thermal conductivity κ, and (*c*) the viscosity η of gases.

(a)

(b)

(c)

The proportionality constant is the **diffusion coefficient** *D*. The flux J_{iz} is expressed in terms of quantity per unit area per unit time. If SI units are used, J_{iz} has the units mol m^{-2} s^{-1}, dc_i/dz has the units of mol m^{-4}, and *D* has the units of m^2 s^{-1}. The negative sign comes from the fact that if c_i increases in the positive *z* direction, dc_i/dz is positive, but the flux is in the negative *z* direction because the flow is in the direction of lower concentrations.

The diffusion coefficient for the diffusion of one gas into another may be determined by use of a cell such as that shown schematically in Fig. 18.12*a*. The heavier gas is placed in chamber A and the lighter in chamber B. The sliding partition is withdrawn for a definite interval of time. From the average composition of one chamber or the other, after a time interval, *D* may be calculated.

The transport of heat is due to a gradient in temperature. Thus, the flux of energy q_z in the *z* direction due to the temperature gradient in that direction is given by

$$q_z = -\kappa \frac{dT}{dz} \tag{18.50}$$

where the proportionality constant κ is the **thermal conductivity**. When q_z has the units of J m^{-2} s^{-1} and dT/dz has the units of K m^{-1}, κ has the units of J m^{-1} s^{-1} K^{-1}. The negative sign in equation 18.50 indicates that if dT/dz is positive, the flow of heat is in the negative *z* direction, which is the direction toward lower temperature.

The determination of the thermal conductivity by the hot-wire method is illustrated schematically in Fig. 18.12*b*. The outer cyclinder is kept at a constant temperature by a thermostaticaly controlled bath. The tube is filled with the gas under investigation, and the fine wire at the axis of the tube is heated electrically. When a steady state is achieved, the temperature of the wire is measured by determining its electrical resistance. The thermal conductivity is calculated from the temperature of wire and wall, the heat dissipation, and the dimensions of the apparatus.

Thermal diffusion is the flux of material due to a temperature gradient of dT/dz. The fact that the thermal diffusion coefficients depend on mass makes it possible to separate isotopes by use of this effect.

Viscosity is a measure of the resistance that a fluid offers to an applied shearing force. Consider what happens to the fluid between parallel planes, illustrated in Fig. 18.13 when the top plane is moved in the *y* direction at a constant speed relative to the bottom plane while maintaining a constant distance between the planes (coordinate *z*). The planes are considered to be very large, so that edge effects may be ignored. The layer of fluid immediately adjacent to the moving plane moves with the velocity of this plane. The layer next to the stationary plane is stationary; in between the velocity usually changes linearly with distance, as shown. The velocity gradient (i.e., the rate of change of velocity with respect to distance measured *perpendicular* to the

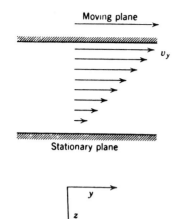

Figure 18.13 Velocity gradient in a fluid due to a shearing action.

direction of flow) is represented by dv_y/dz. The viscosity η is defined by the equation

$$F = -\eta \frac{dv_y}{dz} \qquad (18.51)$$

Here F is the force per unit area required to move one plane relative to the other. The negative sign comes from the fact that if F is in the $+y$ direction, the velocity v_y decreases in successive layers away from the moving plane and dv_y/dz is negative. If F has the units of kg m s^{-2}/m^2 and dv_y/dz has the units of m s^{-1}/m, then the coefficient of viscosity η has the units of kg m^{-1} s^{-1}. The SI unit of viscosity is the Pascal second. Since 1 N = 1 kg m s^{-2}, 1 Pa s = 1 kg m^{-1} s^{-1}. A fluid has a viscosity of 1 Pa s if a force of 1 N is required to move a plane of 1 m^2 at a velocity of 1 m s^{-1} with respect to a plane surface a meter away and parallel with it.*

Although the viscosity is conveniently defined in terms of this hypothetical experiment, it is easier to measure it by determining the rate of flow through a tube, the torque on a disk that is rotated in the fluid, or other experimental arrangement. An experimental arrangement is illustrated in Fig. 18.12c. The outer cylinder is rotated at a constant velocity by an electric motor. The inner coaxial cylinder is suspended on a torsion wire. A torque is transmitted to the inner cyclinder by the fluid, and this torque is calculated from the angular twist of the torsion wire.

18.11 Calculated Transport Coefficients

To calculate the transport coefficients introduced in the last section (D, κ, and η), even for hard-sphere molecules, we would need to consider how the Maxwell–Boltzmann distribution is disturbed by a gradient of concentration, temperature or velocity. This calculation is too advanced for this book, but can be found in some of the references (e.g., Hirschfelder et al.) listed at the end of the chapter.

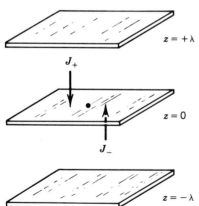

Figure 18.14 Planes constructed at distance $\pm \lambda$ (the mean free path) from the origin. The concentration gradient is in the z direction.

In spite of this, we can get a good qualitative understanding by a highly simplified discussion. Consider the diffusion of molecules in a concentration gradient in the z direction. Imagine that we are at $z = 0$ and we construct planes parallel to the xy plane at $z = \pm \lambda$, where λ is the mean free path (see Fig. 18.14). We choose planes at the mean free path because molecules from more distant points will, on average, have suffered collisions before reaching $z = 0$. Now let us calculate the flux of particles (see Section 18.8) across $z = 0$ due to the molecules above ($z > 0$) and below ($z < 0$). The flux across $z = 0$ from above is

$$J_+ = \left[\rho_0 + \lambda \left(\frac{d\rho}{dz} \right) \right] \frac{\langle v \rangle}{4} \qquad (18.52)$$

where ρ_0 is the number density of particles in the plane at $z = 0$. We have used equation 18.46 and the density of particles at $z = +\lambda$ is given by the term

---

* The cgs unit of viscosity is the poise, that is, 1 g s^{-1} cm^{-1}. 0.1 Pa s = 1 poise.

in brackets. Similarly, the flux across $z = 0$ due to the molecules below $z = 0$ is

$$J_- = \left[\rho_0 - \lambda \left(\frac{d\rho}{dz} \right) \right] \frac{\langle v \rangle}{4} \tag{18.53}$$

The net flux of particles across the plane $z = 0$ is then

$$J = -\frac{1}{2} \langle v \rangle \lambda \frac{d\rho}{dz} \tag{18.54}$$

This equation can be compared with equation 18.49 to obtain

$$D_a = \frac{1}{2} \langle v \rangle \lambda = \left(\frac{kT}{\pi m} \right)^{1/2} \frac{1}{\rho \pi d^2} \tag{18.55}$$

where the subscript a indicates approximate.

The exact theoretical expression for the diffusion coefficient of hard spheres is

$$D = \frac{3\pi}{8} \left(\frac{kT}{\pi m} \right)^{1/2} \frac{1}{\rho \pi d^2} \tag{18.56}$$

so that our highly simplified model yields a good qualitative result.

A similar simplified model for thermal conductivity of hard spheres yields the approximate value

$$\kappa_a = \frac{1}{3} \frac{\overline{C}_V}{N_A} \lambda \langle v \rangle \rho = \frac{2}{3} \frac{\overline{C}_V}{N_A} \left(\frac{kT}{\pi m} \right)^{1/2} \frac{1}{\pi d^2} \tag{18.57}$$

The exact expression for hard spheres is

$$\kappa = \frac{25\pi \overline{C}_V}{32 N_A} \left(\frac{kT}{\pi m} \right)^{1/2} \frac{1}{\pi d^2} \tag{18.58}$$

Finally, the approximate model for the viscosity of hard spheres yields

$$\eta_a = \frac{1}{3} \rho \langle v \rangle m \lambda = \frac{2}{3} \left(\frac{kT}{\pi m} \right)^{1/2} \frac{m}{\pi d^2} \tag{18.59}$$

whereas the exact expression for hard spheres is

$$\eta = \frac{5\pi}{16} \left(\frac{kT}{\pi m} \right)^{1/2} \frac{m}{\pi d^2} \tag{18.60}$$

Note that although the approximate theory yields results that are too low, the dependences on T, m, and ρ agree with the exact theory. The exact expressions can be used to calculate molecular diameters d from experimental transport coefficients. Note that this does not imply that real molecules are hard spheres; in fact, we are forcing a model on the experiment. Nevertheless, the results in Table 18.4 show that a consistent set of molecular diameters result from this analysis of the data.

Example 18.8
Calculate the viscosity of molecular oxygen at 273.2 K and 1 bar. The molecular diameter is 0.360 nm.

Table 18.4 Viscosity and Thermal Conductivity of Gases at 273.2 K and 1 bar and Calculated Molecular Diameters

Gas	η 10^{-5} kg m^{-1} s^{-1}	κ 10^{-2} J K^{-1} m^{-1} s^{-1}	Molecular diameter, d/nm	
			From η	From κ
He	1.85	14.3	0.218	0.218
Ne	2.97	4.60	0.258	0.258
Ar	2.11	1.63	0.364	0.365
H$_2$	0.845	16.7	0.272	0.269
O$_2$	1.92	2.42	0.360	0.358
CO$_2$	1.36	1.48	0.464	0.458
CH$_4$	1.03	3.04	0.414	0.405

Using the exact equation for hard spheres, we find

$$m = \frac{32.00 \times 10^{-3} \text{ kg mol}^{-1}}{6.022 \times 10^{23} \text{ mol}^{-1}}$$

$$= 5.314 \times 10^{-26} \text{ kg}$$

$$\eta = \frac{5\pi}{16} \left(\frac{kT}{\pi m}\right)^{1/2} \frac{m}{\pi d^2}$$

$$= \frac{5\pi}{16} \left[\frac{(1.380 \times 10^{-23} \text{ J K}^{-1})(273.2 \text{ K})}{\pi(5.314 \times 10^{-26} \text{ kg})}\right]^{1/2} \frac{5.314 \times 10^{-26} \text{ kg}}{\pi(0.360 \times 10^{-9} \text{ m})^2}$$

$$= 1.926 \times 10^{-5} \text{ kg m}^{-1} \text{ s}^{-1}$$

References

R. B. Bird, W. E. Stewart, and E. N. Lightfoot, *Transport Phenomena.* New York: Wiley, 1960.

S. Chapman and T. G. Cowling, *The Mathematical Theory of Non-Uniform Gases,* 3d ed. New York, Cambridge University Press, 1970.

C. E. Hecht, *Statistical Thermodynamics and Kinetic Theory.* New York: Freeman, 1990.

J. O. Hirschfelder, C. F. Curtiss, and R. B. Bird, *The Molecular Theory of Gases and Liquids.* New York: Wiley, 1954.

P. C. Jordan, *Chemical Kinetics and Transport.* New York: Plenum, 1979.

W. Kauzmann, *Kinetic Theory of Gases.* Reading, MA: Addison-Wesley, 1966.

D. A. McQuarrie, *Statistical Mechanics.* Chapter 16. New York: Harper & Row, 1976.

R. D. Present, *Kinetic Theory of Gases.* New York: McGraw-Hill, 1958.

Problems

18.1 If the diameter of a gas molecule is 0.4 nm and each is imagined to be in a separate cube, what is the length of the side of the cube in molecular diameters at 0 °C and pressures of (a) 1 bar, and (b) 1 Pa?

18.2 Plot the probability density $f(v)$ of molecular speeds versus speed for oxygen at 25 °C.

18.3 What is the ratio of the probability that gas molecules have two times the mean speed to the probability that they have the mean speed?

18.4 Calculate the mean speed and the root-mean-square speed for the following set of molecules: 10 molecules moving 5×10^2 m s^{-1}, 20 molecules moving 10×10^2 m s^{-1}, and 5 molecules moving 15×10^2 m s^{-1}.

18.5 Calculate the most probable, mean, and root-mean-square speeds for oxygen molecules at 25 °C.

18.6 Derive equations for $\overline{U}$ and $\overline{C}_V$ for any monatomic gas from kinetic theory.

18.7 Calculate the velocity of sound in nitrogen gas at 25 °C. (See Section 18.4.)

18.8 (a) Calculate the collision frequency for a nitrogen molecule in nitrogen at 1 bar pressure and 25 °C. (b) What is the collision density? What is the effect on the collision density (c) of doubling the absolute temperature at constant press, and (d) of doubling the pressure at constant temperature?

18.9 (a) Calculate the mean free path for hydrogen gas ($d = 0.247$ nm) at 1 bar and 0.1 Pa at 25 °C. (b) Repeat the calculation for chlorine gas ($d = 0.496$ nm).

18.10 The pressure in interplanetary space is estimated to be of the order of 10^{-14} Pa. Calculate (a) the average number of molecules per cubic centimeter, (b) the collision frequency and (c) the mean free path in miles. Assume that only hydrogen atoms are present and that the temperature is 1000 K. Assume that $d = 0.2$ nm.

18.11 Calculate the collision frequency $z_{1(1)}$ and the collision density Z_{11} for molecular chlorine at 25 °C and 1 bar. The collision diameter is 0.544×10^{-9} m.

18.12 A gas mixture contains H_2 at 0.666 bar and O_2 at 0.333 bar at 25 °C. (a) What is the collision frequency $z_{1(2)}$ of a hydrogen molecule with an oxygen molecule? (b) What is the collision frequency $z_{2(1)}$ of an oxygen molecule with a hydrogen molecule? (c) What is the collision density Z_{12} between hydrogen molecules and oxygen molecules in mol L^{-1} s^{-1}? The collision diameters of H_2 and O_2 are 0.272 nm and 0.360 nm, respectively.

18.13 For $O_2(g)$, $d = 0.361$ nm, $\Theta_r = 2.079$ K, $\Theta_v = 2273.64$ K, and $M = 31.9988$ g mol^{-1}. At 1 bar and 25 °C what is the average time between collisions? How many vibrational oscillations will have occurred during this time?

18.14 (a) How many molecules of H_2 strike the wall per unit area per unit time at 1 bar at 298 K? 1000 K? (b) How many molecules of O_2 strike the wall per unit area per unit time at 1 bar at 298 K? 1000 K?

18.15 A Knudsen cell containing crystalline benzoic acid ($M = 122$ g mol^{-1}) is carefully weighed and placed in an evacuated chamber thermostated at 70 °C for 1 h. The circular hole through which effusion occurs is 0.60 mm in diameter. Calculate the sublimation pressure of benzoic acid at 70 °C in Pa from the fact that the weight loss is 56.7 mg.

18.16 R. B. Holden, R. Speiser, and H. L. Johnston (*J. Am. Chem. Soc.* **70**:3897 (1948)) found the rate of loss of weight of a Knudsen effusion cell containing finely divided beryllium to be 19.8×10^{-7} g cm^{-2} s^{-1} at 1320 K and 1210×10^{-7} g cm^{-2} s^{-1} at 1537 K. Calculate $\Delta_{sub}H$ for this temperature range.

18.17 A 5-mL container with a hole 10 μm in diameter is filled with hydrogen. This container is placed in an evacuated chamber at 0 °C. How long will it take for 90% of the hydrogen to effuse out?

18.18 The vapor pressure of naphthalene ($M = 128.16$ g mol^{-1}) is 17.7 Pa at 30 °C. Calculate the weight loss in a period of 2 h of a Knudsen cell filled with naphthalene and having a round hole 0.50 mm in diameter.

18.19 The viscosity of helium is 1.88×10^{-5} Pa s at 0 °C. Calculate (a) the collision diameter, and (b) the diffusion coefficient at 1 bar.

18.20 What is the self-diffusion coefficient of radioactive CO_2 in ordinary CO_2 at 1 bar and 25 °C? The collision diameter is 0.40 nm.

18.21 Calculate the mean kinetic energy in electronvolts of a molecule in a gas at 300 K.

18.22 In the text, we derive $U_t = \frac{3}{2}RT$ by using the ideal gas law. Show that the same result may be obtained by averaging over the Maxwell speed distribution to obtain the kinetic energy of an average molecule.

$$U = \int_0^\infty \tfrac{1}{2}mv^2 F(v) \, dv$$

18.23 What is the ratio of the number of molecules having twice the most probable speed to the number having the most probable speed?

18.24 Suppose that a gas contains 10 molecules having an instantaneous speed of 2×10^2 m s^{-1}, 30 molecules with a speed of 4×10^2 m s^{-1}, and 15 molecules with a speed of 6×10^2 m s^{-1}. Calculate $\langle v \rangle$, v_p, and $\langle v^2 \rangle^{1/2}$.

18.25 Calculate the root-mean-square speed of oxygen molecules having a kinetic energy of 10 kJ mol^{-1}. At what temperature would this be the root-mean-square speed?

18.26 What is the root-mean-square speed of a hexane molecule at 0 °C?

18.27 Calculate the velocity of sound in (a) He, and (b) N$_2$ at 25 °C.

18.28 The speed of sound in argon at the triple point of water (273.16 K) has been measured at a series of low pressures and extrapolated to zero pressure to obtain

$$v_s^2 = 94\ 756.75\ \text{m}^2\ \text{s}^{-2}$$

What is the value of the gas constant R if the molar mass of argon is 39.947 753 g mol^{-1} and $\gamma = \frac{5}{3}$?

18.29 What is the average time between collisions of an oxygen molecule in oxygen at 298 K and (a) 1 bar, (b) 10^{-6} bar, and (c) 10^{-12} bar? ($d = 0.36 \times 10^{-9}$ m.)

18.30 What is the mean free path of nitrogen at 1 bar and 25 °C? What is the average time between collisions?

18.31 Oxygen is contained in a vessel at 250 Pa pressure and 25 °C. Calculate (a) the number of collisions between molecules per second per cubic meter, and (b) the mean free path. $d_{O_2} = 0.361$ nm.

18.32 For O$_2$ at 10^{-3} bar at 25 °C, (a) what is the collision frequency $z_{1(1)}$? (b) What is the collision density Z_{11}? (c) What is the average time between collisions of a single molecule?

18.33 An equal number of moles of H$_2$ and Cl$_2$ are mixed and held at 298 K and a total pressure of 1 bar. (a) Calculate the collision frequencies $z_{1(2)}$ and $z_{2(1)}$, where hydrogen is component 1 and chlorine is component 2. (b) Calculate the collision density Z_{12}. *Given:* $d_1 = 0.272$ nm and $d_2 = 0.544$ nm.

18.34 Ultrahigh vacuum is defined as about 10^{-3} Pa. What is the mean free path and average number of collisions per second for a single O$_2$ molecule at this pressure and 25 °C? ($d_{O_2} = 0.361$ nm)

18.35 Calculate the number of collisions per square centimeter per second of oxygen molecules with a wall at a pressure of 1 bar and 25 °C.

18.36. Large vacuum chambers have been built for testing space vehicles at 10^{-6} Pa. Calculate (a) the mean free path of nitrogen at this pressure, and (b) the number of molecular impacts per square meter of wall per second at 25 °C. $d_{N_2} = 0.375$ nm.

18.37 The vapor pressure of water at 25 °C is 3160 Pa (a) If every water molecule that strikes the surface of liquid water sticks, what is the rate of evaporation of molecules from a square centimeter of surface? (b) Using this result, find the rate of evaporation in g cm^{-2} min^{-1} of water into perfectly dry air.

18.38 A substance of $M = 200$ g mol^{-1} has a vapor pressure of 10^{-5} Pa at 25 °C. What mass of the substance will effuse from a Knudsen cell in 2 h through a hole 0.1 cm in diameter?

19
Experimental Kinetics and Gas Reactions

Our consideration of chemical kinetics starts with the quantitative representation of rate data for chemical reactions. The rate of a chemical reaction is often a simple function of the concentrations of the reactants and products; this relationship is referred to as the rate equation. This macroscopic description of the kinetics of a chemical reaction tells us something about how the chemical reaction occurs; that is, it tells us something about the mechanism of the reaction. Other information from thermodynamics, quantum mechanics, spectroscopy, and statistical mechanics is useful in devising a mechanism. Since a mechanism is a hypothesis to explain experimental facts, it is not necessarily unique, even if it accounts for all of the facts; however, some otherwise plausible mechanisms may be definitely excluded using kinetic data.

The specific examples discussed in the chapter are primarily gas reactions, and more detailed consideration of solution reactions is given in Chapter 21.

19.1   Rate of Reaction

The most general expression for the rate of a reaction

$$0 = \sum \nu_i A_i \tag{19.1}$$

is the **rate of conversion** $d\xi/dt$, where ξ is the extent of reaction (Section 2.17). To interpret a rate of conversion, we, of course, need a balanced chemical reaction. Since $dn_i = \nu_i d\xi$, then

$$\frac{dn_i}{dt} = \frac{\nu_i d\xi}{dt} \tag{19.2}$$

Usually we use concentrations in describing kinetics, and so this equation is divided by the volume V and rearranged to define the **rate of reaction** v:

$$v = \frac{1}{V}\frac{d\xi}{dt} = \frac{1}{\nu_i V}\frac{dn_i}{dt} = \frac{1}{\nu_i}\frac{d[A_i]}{dt} \tag{19.3}$$

Therefore, the rate of the reaction

$$A + 2B = X \tag{19.4}$$

is

$$v = \frac{1}{-1}\frac{d[A]}{dt} = \frac{1}{-2}\frac{d[B]}{dt} = \frac{d[X]}{dt} \tag{19.5}$$

Thus, we need a balanced chemical equation to interpret the rate of reaction v. If the reaction goes in the forward direction, the rate of reaction v is positive. If the reaction goes in the backward direction, the rate of reaction is negative. If the reaction is at equilibrium, the rate of reaction v is zero.

We will generally be concerned with reactions occurring at constant volume, but that is not always the case. If the volume changes during the reaction, equation 19.3 can be written

$$v = \frac{1}{\nu_i V}\frac{d([A_i]V)}{dt} = \frac{1}{\nu_i}\frac{d[A_i]}{dt} + \frac{[A_i]}{\nu_i V}\frac{dV}{dt} \tag{19.6}$$

Example 19.1
The reaction $H_2 + Br_2 = 2HBr$ is carried out in a 0.250-L reaction vessel. The change in the amount of Br_2 in 0.01 s is -0.001 mol. (a) What is the rate of conversion $d\xi/dt$? (b) What is the rate of reaction v? (c) What are the values of $d[H_2]/dt$, $d[Br_2]/dt$, and $d[HBr]/dt$?

(a) $\dfrac{d\xi}{dt} = \dfrac{0.001 \text{ mol}}{0.01 \text{ s}} = 0.1 \text{ mol s}^{-1}$

(b) $v = \left(\dfrac{1}{V}\right)\left(\dfrac{d\xi}{dt}\right) = \dfrac{0.1 \text{mol s}^{-1}}{0.25 \text{ L}} = 0.40 \text{ mol L}^{-1} \text{ s}^{-1}$

(c) $\dfrac{d[H_2]}{dt} = -0.40 \text{ mol L}^{-1} \text{ s}^{-1}$

$\dfrac{d[Br_2]}{dt} = -0.40 \text{ mol L}^{-1} \text{ s}^{-1}$

$\dfrac{d[HBr]}{dt} = 0.80 \text{ mol L}^{-1} \text{ s}^{-1}$

The rates of chemical reactions are obtained from measurements of concentration as a function of time. Chemical analytical methods may be used when the reaction can be stopped suddenly. This may be done by rapid cooling for high-temperature reactions, or by catalyst inactivation for a catalyzed reaction. **Physical methods** are especially useful for determining the rate of a chemical reaction because they offer the possibility of continuous measurement of the extent of reaction. A wide variety of physical methods have been used, but spectroscopic methods are the most generally useful. Although the focus in chemical kinetics is on rates, we will often use integrated rate equations so that concentrations measured at various times can be used directly in the quantitative representation of kinetic data, as shown in the next section.

An important characteristic of any measurement method is its response time. The measuring device must obviously respond more rapidly than the concentration is changing. Pulsed lasers have opened up many new opportunities for studying very fast reactions. Reactions occurring in picoseconds (10^{-12} s) may be studied in this way. Special mixing methods have been developed for studying very fast reactions; their use for solution reactions is discussed in Section 21.6.

To study certain gas reactions at high temperatures it is necessary to heat the gas to the higher temperature very quickly because the reaction occurs rapidly. This may be accomplished by means of a **shock tube** in which a shock wave is used to heat the gas suddenly. A tube is divided into two sections separated by a diaphragm that can be ruptured. The gas to be studied is placed on one side of the diaphragm, and a driver gas at higher pressure on the other. When the diaphragm is ruptured, a shock wave passes through the reacting gas, heating it suddenly to a higher temperature. In some reactions the extent of reaction may be determined as a function of time after passage of the shock wave by measuring the absorption of a beam of light passing perpendicularly across the tube.

Photochemical reactions may be initiated rapidly by a light pulse from a flash lamp or a laser. In the **flash photolysis** method, a reaction vessel is exposed to a very high intensity flash of visible or ultraviolet radiation. The flash dissociates molecules in the sample and the concentrations of these species are then determined over a period of time using subsequent flashes at a much lower intensity.

19.2 Order of Reaction

At constant temperature the rate of reaction v depends on the concentrations of reactants and products; it may also depend on catalysts and inhibitors, but we will neglect that for now. For example, if reaction 19.4 goes essentially to completion to the right, the rate of reaction may be given by

$$v = k[A]^\alpha[B]^\beta \tag{19.7}$$

In this **rate equation,** k is the **rate constant** (or rate coefficient) and α and β are independent of concentration and time. Note that α and β are not the stoichiometric numbers in the balanced chemical equation, but have to be obtained from rate experiments. The exponent α is referred to as the **order** of the reaction with respect to reactant A. The order is not necessarily an integer. When the rate law has this general form, the sum of the orders for the reactants is referred to as the **overall order** of the reaction. In this case the overall order is $\alpha + \beta$.

The rate equation for a reaction is frequently more complicated than equation 19.7. For example, the rate of reaction may be affected by products, even if the reaction goes essentially to completion (see Section 20.10). The concentration of a catalyst or inhibitor may also have to be included in the rate law. If a reaction can go by two paths, for example, a catalyzed path and an uncatalyzed path, the rate equation will consist of two additive terms, one for each path (see Section 21.8). The complete rate equation for a reversible reaction contains positive and negative terms and when the rate is set equal to zero, the equilibrium constant expression is obtained.

Now we need to discuss the simplest forms of rate equations that can actually be integrated to obtain the concentration of a reactant as a function of time. This can be done for first-order reactions, second-order reactions, zero-order reactions, fractional-order reactions, and higher order reactions under special conditions. All the equations derived in this section apply only at constant temperature and volume.

The rate equation for a **first-order reaction** A $\rightarrow$ products,

$$-\frac{d[A]}{dt} = k[A] \tag{19.8}$$

may be integrated after it is written in the form

$$-\frac{d[A]}{[A]} = k \, dt \tag{19.9}$$

If the concentration of A is $[A]_1$ at t_1 and $[A]_2$ at t_2, then

$$-\int_{[A]_1}^{[A]_2} \frac{d[A]}{[A]} = k \int_{t_1}^{t_2} dt$$

$$\ln \frac{[A]_1}{[A]_2} = k \, (t_2 - t_1) \tag{19.10}$$

An especially useful form of this equation is obtained if t_1 is taken to be zero, and the initial concentration is represented by $[A]_0$:

$$\ln \frac{[A]_0}{[A]} = kt \tag{19.11}$$

or

$$[A] = [A]_0 e^{-kt} \tag{19.12}$$

$$\ln[A] = \ln[A]_0 - kt \tag{19.13}$$

The last form indicates that the rate constant k may be calculated from a plot of $\ln[A]$ versus t; the slope of the line in such a plot is $-k$. The rate constant for a first-order reaction has the units of reciprocal time.

For a first-order reaction aA $\rightarrow$ products,

$$v = -\frac{1}{a}\frac{d[A]}{dt} = k[A] \tag{19.14}$$

so that

$$-\frac{d[A]}{dt} = ak[A] = k_A[A] \qquad (19.15)$$

where $k_A = ak$.

It is evident from equation 19.10 that to determine the rate constant for a first-order reaction it is only necessary to determine the **ratio** of the concentrations at two times. Physical quantities proportional to concentration may be substituted for concentrations in these equations, since the proportionality constants cancel.

The **half-life** $t_{1/2}$ of a reaction is the time required for half of the reactant to disappear. For the first-order reaction A → products, equation 19.11 leads to

$$k = \frac{1}{t_{1/2}} \ln \frac{1}{1/2} = \frac{0.693}{t_{1/2}} \quad \text{or} \quad t_{1/2} = \frac{0.693}{k} \qquad (19.16)$$

For the first-order reaction $aA \rightarrow$ products, equation 19.15 leads to $t_{1/2} = 0.693/k_A = 0.693/ak$. For a first-order reaction the half-life is independent of the initial concentration. Thus, 50% of the substance remains after one half-life, 25% remains after two half-lives, 12.5% after three, and so on. See Fig. 19.1.

The **relaxation time** τ for a first-order reaction is equal to the reciprocal of the first-order rate constant: $\tau = 1/k$ Thus, equation 19.12 may be written

$$[A] = [A]_0 e^{-t/\tau} \qquad (19.17)$$

A reaction is **second order** if the rate is proportional to the square of the concentration of one reactant or is proportional to the product of the concentrations of two reactants. If the rate is proportional to the square of the concentration of A in the reaction A → products, the rate law

$$-\frac{d[A]}{dt} = k[A]^2 \qquad (19.18)$$

may be integrated after arranging it in the form

$$-\frac{d[A]}{[A]^2} = k \, dt \qquad (19.19)$$

If the concentration is $[A]_0$ at $t = 0$ and $[A]$ at time t, integration yields

$$kt = \frac{1}{[A]} - \frac{1}{[A]_0} \qquad (19.20)$$

Thus, a plot of $1/[A]$ versus t is linear for such a second-order reaction, and the slope is equal to the second-order rate constant. This integrated rate equation also applies if the rate is given by $k[A][B]$, the stoichiometry is represented by A + B = products, and A and B are initially at the same concentration. As may be seen from equation 19.20, the half-life for such a second-order reaction is given by

$$t_{1/2} = \frac{1}{k[A]_0} \qquad (19.21)$$

Thus, the half-life is inversely proportional to the initial concentration. See Fig. 19.2.

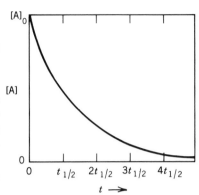

Figure 19.1 For a first-order reaction, half of the reactant disappears in $t_{1/2}$ independent of the initial concentration.

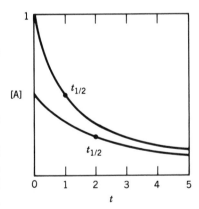

Figure 19.2 For a second-order reaction A + B → products, with $[A]_0 = [B]_0$, the half-life is inversely proportional to $[A]_0$; that is, $t_{1/2} = 1/k[A]_0$. Thus, when $[A]_0$ is reduced by a factor of 2, the half-life doubles.

If the reaction $aA \rightarrow$ products is second order, then

$$v = -\frac{1}{a}\frac{d[A]}{dt} = k[A]^2 \tag{19.22}$$

so that equation 19.20 becomes

$$akt = \frac{1}{[A]} - \frac{1}{[A]_0} \tag{19.23}$$

The half-life for this reaction is given by

$$t_{1/2} = \frac{1}{ak[A]_0} \tag{19.24}$$

A different integrated rate law for a second-order reaction is obtained if the rate is given by $k[A][B]$ and the stoichiometry is given by $aA + bB \rightarrow$ products. The rate constant is defined by

$$v = -\frac{1}{a}\frac{d[A]}{dt} = -\frac{1}{b}\frac{d[B]}{dt} = k[A][B] \tag{19.25}$$

If the initial reactants are not in stoichiometric proportions (i.e., $b[A]_0 \neq a[B]_0$), then the integrated rate equation is

$$kt = \frac{1}{b[A]_0 - a[B]_0}\ln\frac{[A][B]_0}{[A]_0[B]} \tag{19.26}$$

If the stoichiometric numbers of A and B are unity, equation 19.26 becomes

$$kt = \frac{1}{[A]_0 - [B]_0}\ln\frac{[A][B]_0}{[A]_0[B]} \tag{19.27}$$

Thus, for a second-order reaction in which the initial reactants are not in stoichiometric proportions, a plot of $\ln([A]/[B])$ versus t is linear.

The rate constant for a second-order reaction has the units (concentration)$^{-1}$ s^{-1}. If concentrations are expressed in mol L^{-1}, the second-order rate constant has the units L mol^{-1} s^{-1}. If concentrations are expressed in mol m^{-3}, the second-order rate constant has the units m^3 mol^{-1} s^{-1}. If concentrations are expressed in molecules per cubic centimeter, the second-order rate constant has the units cm^3 s^{-1}.

A reaction is **zero order** if the rate is independent of the concentration of the reactant:

$$-\frac{d[A]}{dt} = k \tag{19.28}$$

This can occur if the rate is limited by the concentration of a catalyst; in this case k may be proportional to the concentration of the catalyst. This can also occur in a photochemical reaction if the rate is determined by the light intensity; in this case k may be proportional to the light intensity. Integration of equation 19.28 yields

$$[A]_0 - [A] = kt \tag{19.29}$$

The zero-order rate constant has the units mol L^{-1} s^{-1}, or m^{-3} s^{-1} when concentrations are expressed in molecules per cubic meter.

The rate equation for a reaction A → products may also be integrated if it is of the form $k[A]^n$, where $n = \frac{1}{2}, \frac{3}{2}, 2, 3, \ldots$. All of these rate equations yield

$$\frac{1}{[A]^{n-1}} - \frac{1}{[A]_0^{n-1}} = (n-1)kt \qquad (19.30)$$

In Section 19.15 we will see that the rate of reaction of H_2 with Br_2 is proportional to $[Br_2]^{1/2}$ in the absence of HBr and proportional to $[Br_2]^{3/2}$ if sufficient HBr is added initially. The rate of the formation of phosgene ($COCl_2$) from CO and Cl_2 is given by $k[Cl_2]^{3/2}[CO]$.

19.3 Determination of the Rate Equation

The first objective in studying the kinetics of a chemical reaction is to determine the rate equation. As pointed out in the preceding section the rate equation for a reaction A + 2B = X may be of the form $v = k[A]^\alpha[B]^\beta$, but it may be more complicated (see Sections 19.4 and 19.5), and, in particular, the concentration of the product X may also occur in the rate equation for the forward reaction. To avoid this complication, it is desirable to determine the **initial reaction rate,** that is, the rate at the initial concentrations of the reactants in the absence of product.

Figure 19.3 shows the concentrations of A at various times determined for the reaction

$$A = X \qquad (19.31)$$

by use of a physical method, such as light absorption. To determine the initial rate at each of the four initial concentrations, the rates may be estimated from successive experimental points using $\Delta c/\Delta t$ and extrapolated to $t = 0$. However, it is better to fit the concentration of a product to the power series $[X] = bt + ct^2$ using the method of least squares. If data on only the first 10% of reaction are used, two terms are sufficient for most analytical data.* Linear regression may be used if the equation is written $[X]/t = b + ct$. The parameter b is the initial (forward) velocity v_f of equation 19.31. Then a plot of $\ln v_f$ versus $\ln [A]_0$ may be used to calculate the order with respect to A as shown in Fig. 19.3b. The advantage of this method is that it avoids complications due to products.

In determining initial velocities it is advantageous to use high concentrations of all the reactants but one, so that only one concentration changes significantly during the kinetics experiment. Under these conditions the reaction will have the order for the reactant at the lowest concentration. This is referred to as the **method of isolation.** If the reaction is first order with respect to the substance at low concentration the overall reaction under these conditions is said to be pseudo-first order. The pseudo-first-order rate constant k' is directly proportional to the concentration of the reactant at the higher concentration if the rate is also first order with respect to that substance:

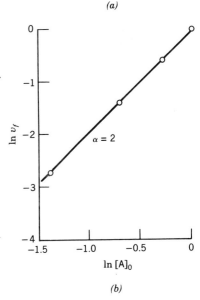

Figure 19.3 (a) Determination of the initial rate of a second-order reaction A → X at four initial concentrations. The initial slopes b determined by least squaring $[X]/t = b + ct$ are equal to the initial velocities v_f. (b) Determination of the order of the reaction with respect to A by plotting $\ln v_f$ verus $\ln [A]_0$.

---

* K. J. Hall, T. I. Quickenden, and D. W. Watts, *J. Chem. Educ.* **53**:493 (1976).

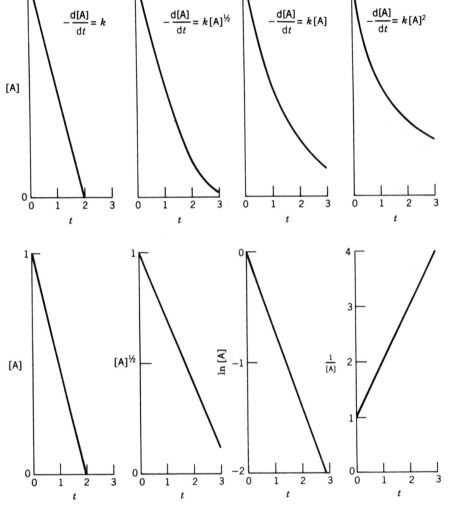

Figure 19.4   (*a*) Plots of [A] versus *t* for zero; half-, first-, and second-order reactions with $[A]_0 = 1$ mol L^{-1}, each having a half-life of 1 min. (*b*) Linear plots for these zero-, half-, first-, and second-order reactions.

$$v = k[A][B]_0 = (k[B]_0)[A] = k'[A] \qquad (19.32)$$

since $[B]_0 \gg [A]_0$.

The concentration of A is plotted versus time for zero-, half-, first-, and second-order reactions in Fig. 19.4*a*. In each case the initial concentration was 1 mol L^{-1}, and the volume was constant. The rate constant was adjusted so that the half-life in each case was 1 min. It is apparent that the change in concentration with time is not very different for these different orders during the first part of the reaction. Thus, if a reaction is followed only during the first half-life it takes very accurate analytical data to determine the order. After longer times the differences are greater, and the plot shown in Fig. 19.3b can be used to distinguish the order and to determine the value of the rate constant.

The study of initial rates may not reveal the full rate law. For example, in Sections 19.10, 19.12, and 19.15, we will see that products may be inhibitory. The effect of a product on the reaction rate may be determined by adding it initially. A reaction is said to be autocatalytic if a product of the reaction causes it to go faster.

So far we have considered reactions that go essentially to completion. Now we will consider first-order reactions that do not go to completion, parallel first-order reactions, and consecutive first-order reactions.

19.4 Reversible First-Order Reactions

When we consider a system in which both the forward and backward reactions are important, the net rate of reaction can generally be expressed as the difference between the rate in the forward direction and that in the backward direction. As a simple example we will consider the reversible reaction

$$A \underset{k_2}{\overset{k_1}{\rightleftharpoons}} B \tag{19.33}$$

The rate law for this reversible* reaction is

$$\frac{d[A]}{dt} = -k_1[A] + k_2[B] \tag{19.34}$$

If initially only A is present, then

$$\frac{d[A]}{dt} = -k_1[A] + k_2([A]_0 - [A]) = k_2[A]_0 - (k_1 + k_2)[A]$$

$$= -(k_1 + k_2)\left([A] - \frac{k_2}{k_1 + k_2}[A]_0\right) = -(k_1 + k_2)([A] - [A]_{eq}) \tag{19.35}$$

where the expression for $[A]_{eq}$ is obtained as follows:

$$\frac{[B]_{eq}}{[A]_{eq}} = \frac{[A]_0 - [A]_{eq}}{[A]_{eq}} = \frac{k_1}{k_2} = K \tag{19.36a}$$

where K is the equilibrium constant for reaction 19.33. This equation can be solved for $[A]_{eq}$.

$$[A]_{eq} = \frac{k_2}{k_1 + k_2}[A]_0 \tag{19.36b}$$

The expression for the equilibrium constant in terms of the forward and backward rate constants can be used to eliminate k_2 from equation 19.34:

---

* Note that "reversible" here means something different than in thermodynamics. In thermodynamics a reversible process goes through equilibrium states. In kinetics this term is used to indicate that the backward reaction is significant.

$$\frac{d[A]}{dt} = -k_1[A] + \frac{k_1}{K}[B]$$

$$= -k_1[A](1 - [B]/[A]K) \tag{19.37}$$

Thus, we can see that the reaction has an initial rate of k_1 [A], and that it slows down as B accumulates. When $[B]/[A] = K$, the reaction is at equilibrium and the rate is zero. Similar equations can be derived for more complicated mechanisms. Integrating equation 19.35 yields

$$-\int_{[A]_0}^{[A]} \frac{d[A]}{[A] - [A]_{eq}} = (k_1 + k_2) \int_0^t dt \tag{19.38}$$

$$\ln \frac{[A]_0 - [A]_{eq}}{[A] - [A]_{eq}} = (k_1 + k_2)t \tag{19.39}$$

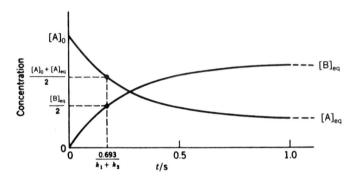

Figure 19.5   Reversible first-order reaction starting with A at concentration $[A]_0$. The values of the rate constants are $k_1 = 3$ s^{-1} and $k_2 = 1$ s^{-1}.

For such a reaction the concentrations of A and B as functions of time are illustrated in Fig. 19.5. The concentrations of A and B will be halfway to their equilibrium values in a time of $0.693/(k_1 + k_2)$.

Thus, a plot of $-\ln([A] - [A]_{eq})$ versus time is linear, and $(k_1 + k_2)$ may be calculated from the slope. It should be especially noted that the rate of approach to equilibrium in this reaction is determined by the sum of the rate constants of the forward and reverse reactions, not by the rate constant for the forward reaction. Since the ratio k_1/k_2 may be calculated from the equilibrium concentrations by use of equation 19.36a, the values of k_1 and k_2 may be obtained.

For some purposes it is more convenient to have equations for [A] and [B] in exponential form. Equation 19.39 may be written as

$$[A] = \frac{k_2[A]_0}{k_1 + k_2}\left(1 + \frac{k_1}{k_2}e^{-(k_1 + k_2)t}\right) \tag{19.40}$$

Since $[B] = [A]_0 - [A]$,

$$[B] = \frac{k_1[A]_0}{k_1 + k_2}(1 - e^{-(k_1 + k_2)t}) \tag{19.41}$$

19.5 Parallel First-Order Reactions

Parallel first-order reactions that compete for the reactant are often encountered because many products may be kinetically accessible and thermodynamically possible.

For the reactions

$$A \xrightarrow{k_1} B$$
$$\searrow{k_2} C$$

(19.42)

the rate equation for A is

$$-\frac{d[A]}{dt} = k_1[A] + k_2[A] = (k_1 + k_2)[A]$$

(19.43)

Thus, the disappearance of A will be first order, and on the basis of the earlier discussion of first-order reactions we can write

$$[A] = [A]_0 e^{-(k_1 + k_2)t}$$

(19.44)

The rate equation for B is

$$\frac{d[B]}{dt} = k_1[A] = k_1[A]_0 e^{-(k_1 + k_2)t}$$

(19.45)

Integration yields

$$[B] = \frac{-k_1[A]_0}{(k_1 + k_2)} e^{-(k_1 + k_2)t} + \text{constant}$$

(19.46)

If $[B] = 0$ at $t = 0$, the constant is $k_1[A]_0/(k_1 + k_2)$ and

$$[B] = \frac{k_1[A]_0}{(k_1 + k_2)} [1 - e^{-(k_1 + k_2)t}]$$

(19.47)

Thus, the fraction of A that is converted to B at infinite time is $k_1/(k_1 + k_2)$. At any time the sum of [A], [B], and [C] must be equal to the total concentration of A at the beginning, $[A]_0$. Consequently, if $[C]_0 = 0$, then

$$[C] = \frac{k_2[A]_0}{(k_1 + k_2)} [1 - e^{-(k_1 + k_2)t}]$$

(19.48a)

It is apparent from equations 19.47 and 19.48a that the ratio of the concentrations of B and C is always given by k_1/k_2, even though they are formed in first-order reactions with different rate constants.

In general, we have to recognize that reactions are reversible, and so the reactions in 19.42 should be rewritten as

$$A \underset{k_{-1}}{\overset{k_1}{\rightleftharpoons}} B$$
$$k_{-2} \updownarrow k_2$$
$$C$$

(19.48b)

This allows us to discuss the concepts of **kinetic control** and **thermodynamic control** of the distribution of products B and C. The equilibrium constants for the formation of B and C from A are given by

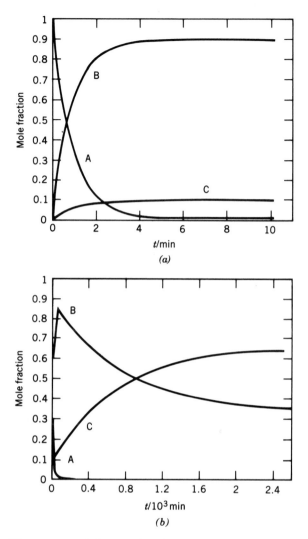

Figure 19.6 Time course of the reaction B $\rightleftharpoons$ A $\rightleftharpoons$ C, starting with A, with rate constants given in the text. (*a*) At short times the reaction is said to be under kinetic control. (*b*) At long times, the reaction is said to be under the thermodynamic control. (From K-C. Lin, *J Chem Educ.* **65**:857 (1988). Copyright © 1988 *Journal of Chemical Education.*)

$$K_1 = \frac{[\text{B}]_{\text{eq}}}{[\text{A}]_{\text{eq}}} = \frac{k_1}{k_{-1}} \quad \text{and} \quad K_2 = \frac{[\text{C}]_{\text{eq}}}{[\text{A}]_{\text{eq}}} = \frac{k_2}{k_{-2}} \tag{19.48c}$$

At chemical equilibrium, these equilibrium constants determine the composition. If the initial concentration of A is $[\text{A}]_0$, the equilibrium composition is given by

$$[\text{A}]_{\text{eq}} = \frac{[\text{A}]_0}{1 + K_1 + K_2} \tag{19.48d}$$

$$[B]_{eq} = \frac{K_1[A]_0}{1 + K_1 + K_2} \qquad (19.48e)$$

$$[C]_{eq} = \frac{K_2[A]_0}{1 + K_1 + K_2} \qquad (19.48f)$$

On the other hand, we have seen above that $[B]/[C] = k_1/k_2$, if the reverse reactions are ignored. This ratio of $[B]$ to $[C]$ will be observed at short times. Figure 19.6 shows calculated mole fractions for $k_1 = 1$, $k_{-1} = 0.01$, $k_2 = 0.1$, and $k_{-2} = 0.000\ 5\ s^{-1}$. The equilibrium constants are $K_1 = 100$ and $K_2 = 200$. Thus, B is formed initially 10 times faster than C, but C is the more stable product. At infinite time, the composition is $[A]/[A]_0 = 3.3 \times 10^{-3}$, $[B]/[A]_0 = 0.332$, and $[C]/[A]_0 = 0.665$. As shown by Fig. 19.6, the reaction goes through two stages. In the first stage the reaction is said to be under kinetic control, and B predominates. However, at much longer times, the reaction is said to be under thermodynamic control, and C predominates. The necessary equations, which are rather complicated, are given by Lin.*

19.6 Consecutive First-Order Reactions

Consecutive reactions occur when the product of a reaction undergoes further reaction. Two consecutive irreversible first-order reactions may be represented by

$$A \xrightarrow{k_1} B \xrightarrow{k_2} C \qquad (19.49)$$

To determine the way in which the concentrations of the substances in such a mechanism depend on time, the rate equations are first written down for each substance. It is then necessary to obtain the solution of these simultaneous differential equations. For the foregoing reactions the rate equations are as follows:

$$\frac{d[A]}{dt} = -k_1[A] \qquad (19.50)$$

$$\frac{d[B]}{dt} = k_1[A] - k_2[B] \qquad (19.51)$$

$$\frac{d[C]}{dt} = k_2[B] \qquad (19.52)$$

It will be assumed that, at $t = 0$, $[A] = [A]_0$, $[B] = 0$, and $[C] = 0$. The rate equation for A is readily integrated to obtain

$$[A] = [A]_0 e^{-k_1 t} \qquad (19.53)$$

Substitution of this expression into equation 19.51 yields

$$\frac{d[B]}{dt} = k_1[A]_0 e^{-k_1 t} - k_2[B] \qquad (19.54)$$

---

* K.-C. Lin, *J. Chem. Educ.* **65**:857 (1988).

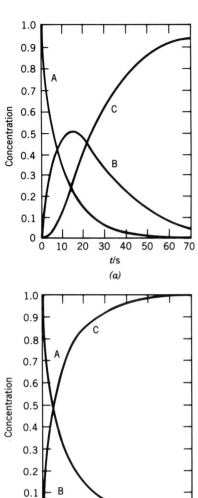

Figure 19.7   Consecutive first-order reaction $A \xrightarrow{k_1} B \xrightarrow{k_2} C$; (a) $k_1 = 0.10 \text{ s}^{-1}$, $k_2 = 0.05 \text{ s}^{-1}$; (b) $k_1 = 0.10 \text{ s}^{-1}$, $k_2 = 1 \text{ s}^{-1}$.

which may be integrated to obtain

$$[B] = \frac{k_1[A]_0}{k_2 - k_1} (e^{-k_1 t} - e^{-k_2 t}) \qquad (19.55)$$

Because of conservation of the number of moles, $[A]_0 = [A] + [B] + [C]$ at any time, and so the concentration of C is given by

$$[C] = [A]_0 - [A] - [B] = [A]_0 \left[1 + \frac{1}{k_1 - k_2}(k_2 e^{-k_1 t} - k_1 e^{-k_2 t}) \right] \qquad (19.56)$$

The concentrations of A, B, and C are shown in Fig. 19.7a for $[A]_0 = 1$ mol L^{-1}, $k_1 = 0.1 \text{ s}^{-1}$, and $k_2 = 0.05 \text{ s}^{-1}$. Note the induction period in the formation of C. The concentrations of A, B, and C are shown in Fig. 19.7b for $[A]_0 = 1$ mol L^{-1}, $k_1 = 0.1 \text{ s}^{-1}$, and $k_2 = 1 \text{ s}^{-1}$. In this case the concentration of B never rises very high. If the calculations were repeated with $k_2 = 20 \text{ s}^{-1}$, the concentration of B would not be visible on this scale. In that case we could think of it as an unstable intermediate in the reaction $A \rightarrow C$. In general, reactions are reversible, and rather different plots are obtained for $A \rightleftharpoons B \rightleftharpoons C$, as shown in Fig. 19.8.

Many examples of first-order reactions are found among the disintegration reactions of the radioactive nuclei. These radioactive decay processes can be expressed with exactness by simple first-order equations.

Complex systems of interconnected first-order reactions may be treated mathematically to find the path the system will take to equilibrium.*

The number of rate equations that can be integrated in closed form is distinctly limited. We will discuss methods for analyzing much more complex kinetic systems in Section 19.17.

19.7   Microscopic Reversibility and Detailed Balance†

The principle of microscopic reversibility is a consequence of the invariance of the equations of classical mechanics when time is reversed. Consider a particle that moves under the action of a force that is a function of position only. The particle moves from position $r(0)$ at $t = 0$ to $r(t_1)$ at time t_1, as shown in Fig. 19.9a. The initial velocity of the particle is $v(0)$, and its velocity at t_1, is $v(t_1)$. Now suppose that at time t_1 we could instantaneously reverse all of the components of the velocity $v(t_1)$ and allow the particle to move for another time period t_1. The particle would retrace its path to its initial position $r(0)$, but the velocity components would be reversed. As shown in Fig. 19.9b, the reversed trajectory can be thought of as beginning at a time $-t_1$, and evolving as time goes forward to $t = 0$. A system is said to be invariant under time reversal if the equation is invariant under the following transformation:

$$t = -t$$
$$r(t) = r(-t)$$
$$v(t) = -v(-t) \qquad (19.57)$$

---

* J. Wei, *Adv. Catal.* **13**:203 (1962); R. Aris, *Elementary Chemical Reactor Analysis*, p. 104. Englewood Cliffs, NJ: Prentice-Hall, 1969.

† L. Onsager, *Phys. Rev.* **37**:405 (1931); B. H. Mahan, *J. Chem. Educ.* **52**:299 (1975).

Quantum mechanics is also time-reversal invariant. The principle of microscopic reversibility illustrated here for a classical particle can be extended to transition probabilities and cross reactions (Section 20.1).

The measurement of a rate constant is quite different from the type of experiment illustrated in Fig. 19.9 because it involves averaging over reactant molecules with many different velocities and angles of approach and product molecules scattered in various directions with various velocities. Starting with the principle of microscopic reversibility, it can be shown that, for an elementary reaction (Section 19.10), the ratio of the foward and backward rate constants is equal to the equilibrium constant obtained from statistical mechanics. For an elementary reaction

$$A + B = C + D \tag{19.58}$$

at equilibrium,

$$\frac{[C]_{eq}[D]_{eq}}{[A]_{eq}[B]_{eq}} = \frac{k_f}{k_b} \tag{19.59}$$

so that

$$k_f[A]_{eq}[B]_{eq} = k_b [C]_{eq}[D]_{eq} \tag{19.60}$$

Thus, for an elementary reaction at equilibrium, the rate of the foward reaction is equal to the rate of the backward reaction. This is an example of what is referred to as the **principle of detailed balance.** In a system with many reactions this principle applies to each reaction individually.

As an application of this principle, consider the suggestion that the isomers A, B, C can be interconverted by the following mechanism:

$$(19.61)$$

Although this mechanism at first sight appears reasonable, no actual process follows this mechanism because it violates the principle of detailed balance. According to this principle, at equilibrium the forward rate of *each* step is equal to the backward rate of that step. The mechanism has to be written as

$$(19.62)$$

with

$$\frac{[B]_{eq}}{[A]_{eq}} = \frac{k_1}{k_2} \tag{19.63a}$$

$$\frac{[C]_{eq}}{[B]_{eq}} = \frac{k_3}{k_4} \tag{19.63b}$$

$$\frac{[A]_{eq}}{[C]_{eq}} = \frac{k_5}{k_6} \tag{19.63c}$$

Multiplying the left sides of equations 19.63a, 19.63b, and 19.63c yields unity, so that

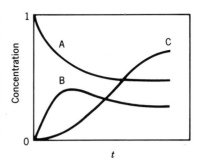

Figure 19.8 Consecutive reactions that do not go to completion: $A \rightleftharpoons B \rightleftharpoons C$.

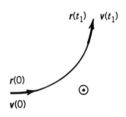

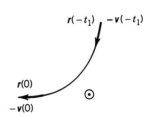

Figure 19.9 (a) A particle with coordinates $r(0)$ and velocity vector $v(0)$ moves under the action of a force to position $r(t_1)$, where it has velocity vector $v(t_1)$. (b) We imagine that the direction of motion is reversed and the clock is set to read $-t_1$. At $t = 0$, the particle will have returned to its original position, but its velocity components are reversed. (From B. H. Mahan, *J. Chem. Educ.* **52**:299 (1975). Copyright © 1975 *Journal of Chemical Education.*)

$$\frac{k_1 k_3 k_5}{k_2 k_4 k_6} = 1 \tag{19.64}$$

and the six rate constants of mechanism 19.62 are not independent.

Example 19.1.
Derive the equilibrium constant expressions 19.63a, 19.63b, and 19.63c from mechanism 19.62 by kinetics.

    This is done by writing down equations for d[A]/dt, d[B]/dt, and d[C]/dt, letting these rates be equal to zero, and solving the three simultaneous equations for [A]$_{eq}$, [B]$_{eq}$, and [C]$_{eq}$ to obtain

$$\frac{[B]_{eq}}{[A]_{eq}} = \frac{k_1 k_4 + k_1 k_5 + k_4 k_6}{k_2 k_4 + k_2 k_5 + k_3 k_5}$$

$$\frac{[C]_{eq}}{[B]_{eq}} = \frac{k_1 k_3 + k_2 k_6 + k_3 k_6}{k_1 k_4 + k_1 k_5 + k_4 k_6}$$

$$\frac{[A]_{eq}}{[C]_{eq}} = \frac{k_2 k_4 + k_2 k_5 + k_3 k_5}{k_1 k_3 + k_2 k_6 + k_3 k_6}$$

By use of equation 19.64, it can be shown that equations 19.63a, 19.63b, and 19.63c are obtained.

19.8   Flow Reactors

We have been discussing batch processes; but when reactions are carried out in the chemical industry, they are generally carried out in flow systems. There are two different kinds of flow systems: **plug flow reactors** and **continuously stirred tank reactors.** In a plug flow reactor, the reactants are mixed and flow through a tube of constant cross section at a constant velocity, in the simplest case. If there is no change in volume in the reaction, various distances from the point of mixing correspond to various reaction times. Thus, the concentration of a reactant at various distances from the point of mixing corresponds to what is expected for a reaction time of $t = V_0/u$ where V_0 is the volume from the point of mixing to that point in the reaction and u is the flow rate in volume per second. See Fig. 19.10. If the reaction is first order, the concentration along the length of the reactor is given by

$$c = c_0 e^{-kV_0/u} \tag{19.65}$$

For plug flow V_0/u may be substituted for t in integrated rate equations for reactions of other orders. Flow methods are used in the laboratory to determine rate constants for fast reactions, both in the gas phase and in liquids (Section 21.5).

    In a continuously stirred tank reactor, the reactor contents are perfectly mixed so that the solution is of uniform composition throughout. In the simplest situation the inlet and outlet flow rates are equal. The rate of reaction in the reactor is constant and is determined by the composition of the **exit stream** from the reactor. Notice that there is a broad distribution in the residence times in the reactor for different molecules. Since the reaction rate is slower than

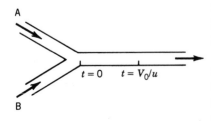

Figure 19.10   Plug flow reactor.

the average rate in a plug flow reactor of the same volume, stirred tank reactors have to have larger volumes. However, stirred tank reactors have certain advantages over plug flow reactors. For example, it is possible to operate the reactor under isothermal conditions even when the heat of reaction is high. If there are multiple reactions, stirred flow reactors may have advantages in accentuating one of the reactions, depending on the form of the rate law.

19.9 Effect of Temperature

The dependence of rate constants on temperature over a limited range can usually be represented by an empirical equation proposed by Arrhenius in 1889:

$$k = Ae^{-E_a/RT} \tag{19.66}$$

where A is the **pre-exponential factor** and E_a is the **activation energy.** The pre-exponential factor A has the same units as the rate constant. Equation 19.66 may be written in logarithmic form:

$$\ln k = \ln A - \frac{E_a}{RT} \tag{19.67}$$

According to this equation, a straight line should be obtained when the logarithm of the rate constant is plotted against the reciprocal of the absolute temperature. This is often called an Arrhenius plot. Differentiating equation 19.67 with respect to temperature yields

$$E_a = RT^2 \frac{d \ln k}{dT} \tag{19.68}$$

This equation may be regarded as the definition of the activation energy. When this equation is integrated we obtain

$$\ln \frac{k_2}{k_1} = \frac{E_a}{R} \left(\frac{T_2 - T_1}{T_1 T_2} \right) \tag{19.69}$$

Example 19.2
The rate constants for the first-order gas reaction $N_2O_5 = 2NO_2 + \frac{1}{2}O_2$ are as follows:

T/K	273	298	308	318	328	338
$k/10^{-5} \text{ s}^{-1}$	0.0787	3.46	13.5	49.8	150	487

What are the values of the activation energy and the pre-exponential factor?
    The plot of $\ln k$ versus $1/T$ is given in Fig. 19.11. The slope of the line in Fig. 19.11 is $-12\,400$ K, and E_a has a value of $-R$ (slope) $= 103$ kJ mol^{-1}. Equation 19.67 becomes

$$\ln k = 31.31 - \frac{103 \times 10^3 \text{ J mol}^{-1}}{(8.314 \text{ J K}^{-1} \text{ mol}^{-1})T}$$

Thus, $A = e^{31.31} = 3.96 \times 10^{13}\text{s}^{-1}$ and equation 19.66 becomes

$$k = (3.96 \times 10^{13} \text{ s}^{-1}) \exp \left[\frac{-103 \times 10^3 \text{ J mol}^{-1}}{(8.314 \text{ J K}^{-1} \text{ mol}^{-1})T} \right]$$

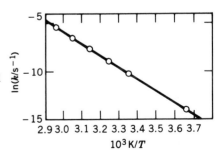

Figure 19.11 Plot of $\ln k$ versus $1/T$ for the decomposition of N_2O_5 from which the Arrhenius activation energy E_a may be calculated.

It should be realized that for any given pre-exponential factor there is a fairly narrow range of activation energies that will give reaction rates in the range measurable by conventional techniques, that is, with half-lives from 1 min to 10 days. For example, if $A = 10^{13} \text{s}^{-1}$ and the temperature is 298 K, reactions with activation energies less than about 80 kJ mol^{-1} will be too fast to study with ordinary methods, and reactions with activation energies greater than about 100 kJ mol^{-1} will be too slow.

The Arrhenius plots for some reactions are curved, and so the activation energy is a function of temperature. As we will see in the next chapter there are theoretical reasons for expecting k for some reactions to vary with temperature according to

$$k = aT^m e^{-E_0/RT} \tag{19.70}$$

When m is known, E_0 may be calculated from the slope of a plot of $\ln(k/T^m)$ versus $1/T$. If m is not provided by a theory, it is difficult to determine its value experimentally because the exponential dependence on $1/T$ is usually much stronger than the temperature dependence of T^m.

Some reactions actually proceed more slowly at higher temperatures. For them, the activation energy is negative, but for such reactions other equations may be more suitable in representing the rate constant as a function of temperature.

Example 19.3

For a reaction that follows equation 19.70, what are the Arrhenius parameters at temperature T'?

The Arrhenius activation energy is defined by equation 19.68, and so it may be obtained as follows:

$$\ln k = \ln a + m \ln T - \frac{E_0}{RT}$$

$$E_a = RT^2 \frac{d \ln k}{dT} = RT^2 \left(\frac{m}{T} + \frac{E_0}{RT^2} \right)$$

$$= E_0 + mRT$$

Thus, E_0 is a hypothetical activation energy at absolute zero. Substitution of this relation in equation 19.70 yields

$$k = aT^m e^m e^{-E_a/RT} \tag{19.71}$$

Thus, the pre-exponential factor A at temperature T' is equal to $a(T')^m e^m$.

When the Arrhenius plot is curved and E_a is calculated at two temperatures, E_0 and m can be calculated. Then the factor a can be calculated from the rate constant at either temperature.

19.10 Mechanisms of Chemical Reactions

Many reactions that follow simple rate laws such as we have been discussing actually occur through a series of steps. These steps are called **elementary** reactions because they cannot be broken down further into simpler chemical reactions. However, when we look at elementary reactions in more detail in Sections 19.12 to 19.14 we will see that they may involve identifiable **physical** steps. The sequence of elementary reactions that add up to give the overall reaction is called the **mechanism** of the reaction. A mechanism is a hypothesis about the elementary steps through which chemical change occurs. Sometimes the elementary steps can be studied in isolation. However, the evidence for a mechanism is often indirect, and there is always the possibility that a different mechanism is also in accord with all the facts about the kinetics of the reaction and is in better accord with other knowledge about the reactants and intermediates involved. A valid mechanism must of course explain the rate law of the backward reaction as well as the forward reaction.

Often it is possible to devise several mechanisms that are consistent with an experimentally determined rate law. Sometimes these mechanisms can be distinguished by use of nonkinetic data. For example, optical and mass spectroscopy may be used to detect intermediates. Isotopically labeled reactants can be used to trace the paths of atoms in a reaction.

In discussing elementary reactions we refer to their **molecularity.** The molecularity is the number of reactant molecules in an elementary step. Thus, elementary reactions are referred to as **unimolecular, bimolecular,** and **trimolecular,** depending on whether one, two, or three molecules are involved as reactants. In contrast to what we said earlier about the lack of relationships between the stoichiometric equation and reaction order, the rate law of an elementary step can be obtained directly from its chemical equation. A unimolecular reaction is first order, a bimolecular reaction is second order, and a trimolecular reaction is third order. Most elementary reactions are unimolecular or bimolecular. Trimolecular reactions are uncommon because three-body collisions are improbable.

The mechanism of a reaction leads directly to a set of differential equations that completely describe the kinetic behavior of that mechanism. A differential equation may be written for each molecular species in the mechanism by writing positive terms for each reaction by which the species is formed and negative terms for each reaction in which the species disappears. However, these differential equations are not all independent. There are conservation equations that must also be satisfied, and so the number of independent rate equations is reduced by this number. The concentrations of various species as functions of time may be obtained by solving the boundary value problem; for a mechanism with a number of steps this can only be done numerically using a computer (Section 19.17). This of course requires numerical values for all the rate constants. However, there are two approximation methods that yield expressions for the rate of the overall reaction in terms of the concentrations of the various reactants and products and the rate constants. These methods are the steady-state method and the rapid equilibrium method.

The steady-state method is based on the fact that the concentrations of intermediates may not change much during a reaction after an initial buildup. Therefore, rate equations are written for intermediates and these rates are set equal to zero. This yields a set of algebraic equations that can be solved for the concentrations of intermediates. These equations can then be used to eliminate these concentrations from the rate equations for the formation of product. As we will see, this provides the relations between the experimentally determined rate parameters and the rate constants for the elementary steps in the mechanism.

A reaction $A \rightleftharpoons B$ may go through an unstable intermediate I; thus, there are two elementary reactions as represented by the following mechanism:

$$A \underset{k_{-1}}{\overset{k_1}{\rightleftharpoons}} I \underset{k_{-2}}{\overset{k_2}{\rightleftharpoons}} B \qquad K = \frac{[B]}{[A]} = \frac{k_1 k_2}{k_{-1} k_{-2}} \tag{19.72}$$

Since $[A] + [I] + [B] = [A]_0$ if we start with A, there are only two independent rate equations, and they can be written as

$$\frac{d[I]}{dt} = k_1[A] + k_{-2}[B] - (k_{-1} + k_2)[I] \tag{19.73}$$

$$\frac{d[B]}{dt} = k_2[I] - k_{-2}[B] \tag{19.74}$$

If I is an unstable intermediate, its concentration rises to a low value which remains rather constant after an initial increase. For this reaction, the steady-state approximation is $d[I]/dt = 0$. By use of this approximation, [I] can be eliminated from equation 19.74 to obtain

$$\frac{d[B]}{dt} = \frac{k_1 k_2 [A] - k_{-1} k_{-2}[B]}{k_{-1} + k_2}$$

$$= k_f[A] - k_b[B] \tag{19.75}$$

Thus, the reaction behaves like a reversible first-order reaction (Section 19.4) with $k_f = k_1 k_2/(k_{-1} + k_2)$ and $k_b = k_{-1} k_{-2}/(k_{-1} + k_2)$. Note that $k_f/k_b = K$. **If the concentration of the intermediate is very small during the entire reaction, it may be very difficult to distinguish mechanism 19.72 from the mechanism A $\rightleftharpoons$ B.** The Arrhenius plots for k_f and k_b will probably be curved since they are each a composite of three rate constants.

If the reaction $A = B$ goes to completion because $k_{-2} = 0$ and, in addition, $k_{-1} \gg k_2$, the unstable intermediate will essentially be in equilibrium with A and

$$\frac{d[B]}{dt} = \frac{k_1 k_2}{k_{-1}}[A]$$

$$= k_2 K_1[A] = k_f[A] \tag{19.76}$$

where $K_1 = k_1/k_{-1}$.

The decomposition of ozone is represented by the reaction $2O_3 = 3O_2$. Thus, the reaction rate is defined by

$$v = -\frac{1}{2}\frac{d[O_3]}{dt} = \frac{1}{3}\frac{d[O_2]}{dt} \tag{19.77a}$$

The rate law for this decomposition in the presence of relatively high concentrations of a chemically inert gas M is

$$v = \frac{k[O_3]^2[M]}{k'[O_2][M] + [O_3]} \tag{19.77b}*$$

---

* S. W. Benson and A. E. Axworthy, Jr., *J. Chem. Phys.* **29**:1718 (1957).

Thus,

$$-\frac{d[O_3]}{dt} = \frac{2k[O_3]^2[M]}{k'[O_2][M] + [O_3]} \qquad (19.77c)$$

This and other information has led to the following mechanism:

$$1. \ O_3 + M \underset{k_{-1}}{\overset{k_1}{\rightleftharpoons}} O_2 + O + M \qquad (19.78a)$$

$$2. \ O + O_3 \overset{k_2}{\longrightarrow} 2O_2 \qquad (19.78b)$$

$$\text{Overall reaction: } 2O_3 = 3O_2 \qquad (19.78c)$$

Two independent rate equations for this mechanism are 19.79 and 19.80; we do not have to write a separate rate equation for O_2 since at constant volume $\frac{1}{3}[O_3] + \frac{1}{2}[O_2] + [O] = \text{constant}$:

$$\frac{d[O]}{dt} = k_1[O_3][M] - k_{-1}[O_2][O][M] - k_2[O][O_3] \qquad (19.79)$$

$$-\frac{d[O_3]}{dt} = k_1[O_3][M] - k_{-1}[O_2][O][M] + k_2[O][O_3] \qquad (19.80)$$

Since oxygen atoms are not produced in the overall reaction and since their concentration is always low, it is a good approximation to set $d[O]/dt = 0$. Thus, in the steady state

$$[O] = \frac{k_1[O_3][M]}{k_{-1}[O_2][M] + k_2[O_3]} \qquad (19.81)$$

This relation may be substituted into equation 19.80 for the decomposition of O_3 to obtain

$$-\frac{d[O_3]}{dt} = \frac{2k_1k_2[O_3]^2[M]}{k_{-1}[O_2][M] + k_2[O_3]} \qquad (19.82)$$

This rate equation is in accord with the empirical rate equation (19.77b). Early in the reaction starting with pure ozone $k_{-1}[O_2][M] \ll k_2[O_3]$, and this rate equation reduces to

$$-\frac{d[O_3]}{dt} = 2k_1[O_3][M] \qquad (19.83)$$

Under these conditions the **first step is rate determining,** and two molecules of O_3 are decomposed each time step 1 occurs because the oxygen atom produced causes the destruction of a second molecule of O_3. On the other hand, if the **second step is rate determining,** that is, $k_{-1}[O_2][M] \gg k_2[O_3]$, then

$$-\frac{d[O_3]}{dt} = \frac{2k_2K_1[O_3]^2}{[O_2]} \qquad (19.84)$$

where $K_1 = k_1/k_{-1}$ is the equilibrium constant for the first step. Oxygen (O_2) inhibits the forward reaction because it reduces the concentration of oxygen atoms in equilibrium with O_3 and O_2 (step 1). (Note that an incorrect rate equation is obtained by starting with the assumption that reaction 1 remains in equilibrium because this step, as well as the second step, destroys O_3 (D. C. Tardy and E. D. Cater, *J. Chem. Educ.* **60**:109 (1983).) This reaction is discussed again in the chapter on photochemistry because ozone is formed in the upper atmosphere by a photochemical reaction. The photochemical formation of ozone and the various reactions by which it can be converted to oxygen are extremely important because ozone in the upper atmosphere shields the surface of the earth from harmful ultraviolet radiation that would otherwise be transmitted through the atmosphere.

The various steps in a mechanism do not necessarily all run at the same rate in a steady state. This is illustrated by the mechanism* for the decomposition of N_2O_5:

$$1. \quad N_2O_5 + M \underset{k_2}{\overset{k_1}{\rightleftharpoons}} NO_2 + NO_3 + M \qquad s_1 = 2 \qquad (19.85a)$$

$$2. \quad NO_2 + NO_3 \overset{k_3}{\longrightarrow} NO + O_2 + NO_2 \qquad s_2 = 1 \qquad (19.85b)$$

$$3. \quad NO + NO_3 \overset{k_4}{\longrightarrow} 2NO_2 \qquad s_3 = 1 \qquad (19.85c)$$

The **stoichiometric** numbers s_1, s_2, and s_3 are the numbers of times the indicated steps have to occur to produce the overall reaction:

$$2N_2O_5 = 4NO_2 + O_2 \qquad (19.86)$$

You should use the stoichiometric numbers of the steps to be sure you can obtain equation 19.86. This mechanism also illustrates the fact that there may be more than one steady-state relation for a mechanism. This mechanism involves five molecular species and two element conservation relations. Thus, there are three independent concentrations at constant volume, and we can choose them to be $[NO_3]$, $[NO]$, and $[N_2O_5]$. In using the steady-state method, the net rates of formation of the intermediates NO_3 and NO are set equal to zero:

$$\frac{d[NO_3]}{dt} = k_1[N_2O_5] - (k_2 + k_3)[NO_2][NO_3] - k_4[NO][NO_3] = 0 \quad (19.87a)$$

$$\frac{d[NO]}{dt} = k_3[NO_2][NO_3] - k_4[NO][NO_3] = 0 \qquad (19.87b)$$

This leads to two simultaneous equations that may be solved to give the following expressions for the steady-state concentrations of NO and NO_3:

$$[NO] = \frac{k_3}{k_4}[NO_2] \qquad (19.88)$$

$$[NO_3] = \frac{k_1[N_2O_5]}{(k_2 + 2k_3)[NO_2]} \qquad (19.89)$$

The rate equation for the decomposition of N_2O_5 is

$$\frac{d[N_2O_5]}{dt} = -k_1[N_2O_5] + k_2[NO_2][NO_3] \qquad (19.90)$$

Substituting equation 19.89 yields the rate equation

$$\frac{d[N_2O_5]}{dt} = -\frac{2k_1k_3}{k_2 + 2k_3}[N_2O_5] \qquad (19.91)$$

---

* H. S. Johnston, *Gas Phase Reaction Rate Theory*, Chapter 1. New York: Ronald, 1966.

so that the overall reaction occurring by this assumed mechanism behaves in a first-order manner.

A more drastic assumption that is sometimes useful is to assume that a reaction has a rate-determining step. If one of the steps is rate determining, all the steps in the mechanism up to that point remain at equilibrium. Thus, the concentrations of all of the substances involved in these prior steps may be calculated from the equilibrium expressions.

19.11 Relation Between the Rate Constants for the Forward and Backward Reactions

In discussing the principle of detailed balance (Section 19.7) we referred to the fact that at equilibrium the forward rate of each step is equal to the reverse rate of that step. For elementary reactions this leads to a relationship between the rate constants for the forward and reverse reactions that is not necessarily obeyed for the overall reaction.

For an elementary reaction there is an exact correspondence between the chemical equation and the rate equation. For the elementary reaction

$$AB \underset{k_b}{\overset{k_f}{\rightleftharpoons}} A + B \tag{19.92}$$

the complete rate equation is

$$-\frac{d[AB]}{dt} = k_f[AB] - k_b[A][B] \tag{19.93}$$

At equilibrium, $d[AB]/dt = 0$ and so

$$\frac{[A][B]}{[AB]} = \frac{k_f}{k_b} = K_c \tag{19.94}$$

where K_c is the equilibrium constant written in terms of concentrations (Section 5.10). Here, as elsewhere in this chapter, we assume that gas mixtures are ideal.

Thus, if the rate constant for the forward reaction is known, the rate constant for the backward reaction can be calculated using the equilibrium constant for the elementary reaction. Since rate equations are generally written in terms of concentrations, it is necessary to use K_c for a gas reaction. The value of K_c can be calculated using

$$K_P = \left(\frac{c^{\circ}RT}{P^{\circ}}\right)^{\Sigma \nu_i} K_c \tag{19.95}$$

which was derived earlier as equation 5.50. Standard Gibbs energies of formation of gases may be used to calculate K_P, and equation 19.95 may then be used to calculate K_c. In using equation 19.94 we need K_c written as

$$K_c = \Pi \, c_i^{\nu_i} \tag{19.96}$$

so that it has units if $\Sigma \nu_i \neq 0$.

Example 19.4

The rate constant for the elementary reaction

$$C_2H_6 \rightarrow 2CH_3$$

is $1.57 \times 10^{-3} \text{ s}^{-1}$ at 1000 K. What is the rate constant for the backward reaction at this temperature? For $CH_3(g)$, $\Delta_f G° = 159.74 \text{ kJ mol}^{-1}$ at 1000 K. Using the value of $\Delta_f G°$ for C_2H_6 at 1000 K from Table C.2 yields $K_P = 1.083 \times 10^{-11}$. Using equation 19.95

$$K_P = 1.083 \times 10^{-11} = \frac{(1 \text{ mol L}^{-1})(0.08314 \text{ L bar K}^{-1} \text{ mol}^{-1})(1000 \text{ K})}{(1 \text{ bar})} K_c$$

$$K_c = 1.302 \times 10^{-13} = \frac{([CH_3]/c°)^2}{[C_2H_6]/c°}$$

Since we need the equilibrium constant written in terms of concentrations, we have

$$1.302 \times 10^{-13} \text{ mol L}^{-1} = \frac{k_f}{k_b} = \frac{1.57 \times 10^{-3} \text{ s}^{-1}}{k_b}$$

so that the rate constant for the reverse reaction at 1000 K is $1.21 \times 10^{10} \text{ L mol}^{-1} \text{ s}^{-1}$.

The principle of detailed balance also makes it possible to calculate the activation energy and pre-exponential factor for the backward reaction, if these quantities are known for the forward reaction. For temperatures in a range where $\Delta_r H°$ and $\Delta_r S°$ are independent of temperature, equation 19.94 can be written

$$\frac{k_f}{k_b} = K_c = K_P \left(\frac{P°}{c°RT}\right)^{\Sigma \nu_i} = e^{\Delta_r S°/R} e^{-\Delta_r H°/RT} \left(\frac{P°}{c°RT}\right)^{\Sigma \nu_i} \quad (19.97)$$

The activation energy is internal energy rather than enthalpy, and so we eliminate $\Delta_r H°$ from the right side of this equation by using the equation

$$\Delta_r H° = \Delta_r U° + RT\Sigma \nu_i \quad (19.98)$$

We then rearrange the right side of equation 19.97 so that it is in the form of a factor times $\exp(-\Delta U°/RT)$.

$$\frac{k_f}{k_b} = K_c = e^{\Delta_r S°/R} \left(\frac{P°}{c°RT}\right)^{\Sigma \nu_i} e^{-\Sigma \nu_i} e^{-\Delta_r U°/RT} \quad (19.99)$$

$$= e^{\Delta_r S°/R} e^{\Sigma \nu_i \ln(P°/c°RT)} e^{-\Sigma \nu_i} e^{-\Delta_r U°/RT} \quad (19.100)$$

$$= e^{1/R\{\Delta_r S° + R\Sigma \nu_i [\ln(P°/c°RT) - 1]\}} e^{-\Delta_r U°/RT} \quad (19.101)$$

The left side of the equation may be written

$$\frac{k_f}{k_b} = \frac{A_f e^{-E_{af}/RT}}{A_b e^{-E_{ab}/RT}} = \frac{A_f}{A_b} e^{-(E_{af} - E_{ab})/RT} \quad (19.102)$$

where E_{af} and E_{ab} are the activation energies for the forward and backward reaction. Since the temperature dependent factors in equations 19.101 and 19.102 may be identified with each other, we obtain

$$E_{af} - E_{ab} = \Delta_r U° \quad (19.103)$$

$$\frac{A_f}{A_b} = \exp\left\{\frac{\Delta S°}{R} + \Sigma\nu_i\left[\ln\left(\frac{P°}{c°RT}\right) - 1\right]\right\} \qquad (19.104)$$

Although there is a term in T in the expression for A_f/A_b, this ratio does not vary much with the temperature compared with $\exp[-(E_{af} - E_{ab})/RT]$. Thus, if we know the activation energy and pre-exponential factor for the forward reaction and $\Delta_r S°$ and $\Delta_r U°$(or $\Delta_r H°$) for the overall reaction, the principle of detailed balancing makes it possible to calculate the activation energy and pre-exponential factor for the reverse reaction.

Example 19.5

The rate constant for the unimolecular dissociation of C_2H_6 into methyl radicals in the range 800–1000 K is given by

$$k_f = (10^{16.0} \text{ s}^{-1}) \, e^{-(360 \text{ kJ mol}^{-1})/RT}$$

What is the Arrhenius equation for the reverse reaction? At 1000 K $S°$ for CH_3 (g) is 251.3 J K^{-1} mol^{-1} and $\Delta_f H°$ is 137.53 kJ mol^{-1}. Using Table C.2 for $C_2H_6(g) = 2CH_3(g)$ yields $\Delta S° = 2(251.3) - 332.3 = 170.3$ J K^{-1} mol^{-1} and $\Delta_r H° = 2(137.53) + 105.77 = 380.8$ kJ mol^{-1}. Thus,

$$\Delta_r U° = 380.8 \text{ kJ mol}^{-1} - (8.314 \times 10^{-3} \text{ kJ K}^{-1} \text{ mol}^{-1})(1000 \text{ K})$$

$$= 372.5 \text{ kJ mol}^{-1}$$

$$= 360 \text{ kJ mol}^{-1} - E_{ab}$$

$$E_{ab} = -12.5 \text{ kJ mol}^{-1}$$

The activation energy for the backward reaction is quite small, as has been found experimentally. The pre-exponential factor for the reverse reaction is calculated from equation 19.104:

$$\frac{10^{16.0}}{A_b} = \exp\left\{\frac{170.3}{8.314} + \ln\left[\frac{(1 \text{ bar})}{(1 \text{ mol L}^{-1})(8.314 \times 10^{-3} \text{ L bar K}^{-1} \text{ mol}^{-1})(1000 \text{ K})}\right] - 1\right\}$$

$$A_b = 2.88 \times 10^8 \text{ L mol}^{-1} \text{ s}^{-1}$$

At 1000 K

$$k_b = 2.88 \times 10^8 e^{12.5 \times 10^3/(8.314)(1000)}$$

$$= 1.30 \times 10^9 \text{ L mol}^{-1} \text{ s}^{-1}$$

The equation $k_f/k_b = K_c$ derived for elementary reactions does not, in general, apply to overall reactions. For overall reactions there is not a unique relationship between the rate law and the balanced chemical equation. However, the equilibrium constant expression does limit the possibilities.* For example, for the reaction

$$A + 2B = C \qquad (19.105)$$

---

* K. Denbigh, *The Principles of Chemical Equilibrium*, 3d ed., p. 442. Cambridge, UK: Cambridge University Press, 1971.

suppose that the rate equation for the forward reaction is

$$-\left(\frac{d[A]}{dt}\right)_f = k_f[A][B] \tag{19.106}$$

Assuming that the rate equation for the backward reaction is of the form

$$\left(\frac{d[A]}{dt}\right)_b = k_b[A]^\alpha[B]^\beta[C]^\gamma \tag{19.107}$$

The net rate is given by

$$-\frac{d[A]}{dt} = -\left(\frac{d[A]}{dt}\right)_f - \left(\frac{d[A]}{dt}\right)_b = k_f[A][B] - k_b[A]^\alpha[B]^\beta[C]^\gamma \tag{19.108}$$

At equilibrium the net rate is zero, and so

$$\frac{[A]_{eq}^\alpha[B]_{eq}^\beta[C]_{eq}^\gamma}{[A]_{eq}[B]_{eq}} = \frac{k_f}{k_b} \tag{19.109}$$

If the equilibrium expression is written in the usual way for reaction 19.105,

$$\frac{[C]_{eq}}{[A]_{eq}[B]_{eq}^2} = K \tag{19.110}$$

we conclude that $\alpha = 0$, $\beta = -1$, and $\gamma = 1$, so that the rate equation for the backward reaction is

$$\left(\frac{d[A]}{dt}\right)_b = \frac{k_b[C]}{[B]} \tag{19.111}$$

However, the balanced chemical equation may be written in other ways, which leads us to admit the possibility of other rate equations for the backward reaction. For example, if the equilibrium expression is written

$$\frac{[C]_{eq}^2}{[A]_{eq}^2[B]_{eq}^4} = K' \tag{19.112}$$

then the rate equation expected for the backward reaction would be

$$\left(\frac{d[A]}{dt}\right)_b = k_b[A]^{-1}[B]^{-3}[C]^2 \tag{19.113}$$

Thus, the form of the rate equation for the forward reaction implies something about the form of the rate equation for the backward reaction, but does not define it uniquely.*

If a rate equation has a sum of terms for the forward reaction, indicating multiple paths, the principle of detailed balancing requires that *each* term for the forward reaction be balanced by a thermodynamically appropriate term for the backward reaction at equilibrium.

Now we need to take a closer look at elementary reactions. Most elementary reactions are unimolecular or bimolecular.

---

* E. L. King, *J. Chem. Educ.* **63**:21 (1986).

Table 19.1 Arrhenius Parameters for k_∞ for Unimolecular Gas Reactions[a]

Reaction	$\log\left(\dfrac{A^\infty}{s^{-1}}\right)$	$\dfrac{E_a}{kJ\ mol^{-1}}$	Temperature Range K
Isomerizations			
$CH_3NC \rightarrow CH_3CN$	13.6	160.5	470–530
cyclo-$C_3H_6 \rightarrow CH_3CH{=}CH_2$	15.45	274	700–800
Dissociation to stable molecules			
$C_2H_5Cl \rightarrow C_2H_4 + HCl$	14.6	254	670–770
cyclo-$C_4H_8 \rightarrow 2C_2H_4$	15.6	262	690–740
Dissociation to free radicals			
$C_2H_6 \rightarrow 2CH_3$	16.0	360	820–1000
$HNO_3 \rightarrow OH + NO_2$	15.3	205	890–1200

[a] From I. W. M. Smith, *Kinetics and Dynamics of Elementary Gas Reactions*. Boston: Butterworths, 1980.

19.12 Unimolecular Reactions

Unimolecular reactions are either isomerizations or dissociations as illustrated by

$$CH_3NC \rightarrow CH_3CN \qquad (19.114)$$

$$C_2H_6 \rightarrow 2CH_3 \qquad (19.115)$$

At a pressure of 1 bar or more these reactions are first order, and Arrhenius parameters in Table 19.1 are for this pressure range. If unimolecular reactions were caused directly by molecular collisions we would expect them to be second order, and so there is a puzzle here. Since unimolecular reactions are first order, we conclude that they are reactions of isolated gas molecules. But why should an isolated gas molecule suddenly isomerize or dissociate? Another puzzling thing about unimolecular reactions is that, although their rate constants are independent of pressure at pressures above about a 1 bar, the rate constant falls off at lower pressures. Figure 19.12 shows this effect for the isomerization of methyl isocyanide to methyl cyanide (reaction 19.114). Unimolecular reactions generally have large activation energies, as shown in Table 19.1; how is the activation energy supplied?

A first approximation of the answers to these questions was provided by Lindemann in 1922. He pointed out that when a molecule is excited to a higher energy state by a bimolecular collision, there is a time lag before decomposition or isomerization. During this time lag the **excited molecule** may lose its extra energy in a second bimolecular collision. Since such reactions are usually studied by diluting the reacting gas A with an excess of inert gas M, the excitation step and deexcitation step usually involve collisions of A and M, as indicated in the following mechanism:

$$A + M \xrightarrow{k_1} A^* + M \qquad (19.116)$$

$$A^* + M \xrightarrow{k_2} A + M \qquad (19.117)$$

$$A^* \xrightarrow{k_3} B + C \qquad (19.118)$$

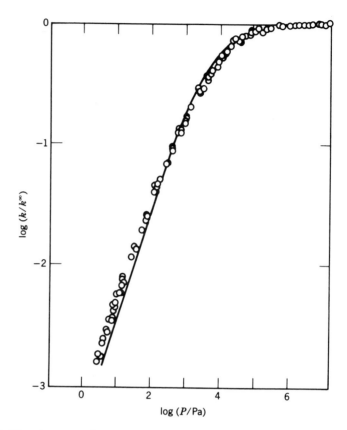

Figure 19.12   Apparent first-order rate constants for the isomerization of CH_3NC at 503.5 K. The curve shows the results of the RRKM theory. (Reprinted with permission from F. W. Schneider and B. S. Rabinovitch, *J. Am. Chem. Soc.* **84**:4215, 1962. Copyright © 1962 American Chemical Society.)

Here A* is an A molecule with enough vibrational energy to isomerize or decompose.* In other words, part of the kinetic energy of a bimolecular collision has been used to raise an A molecule to a higher vibrational level. Although A* has enough energy to react, it does not do so immediately because the energy has to become distributed within the molecule in such a way that the reaction can occur. Any collision can raise A to a higher vibrational level; thus, M in this mechanism might be another A molecule, or product molecule, or a molecule of the added inert gas. If A* collides with another molecule before it undergoes a unimolecular reaction, it will almost certainly lose its high level of vibrational energy. This possibility is represented by the step with rate constant k_2. The other possibility is that A* undergoes a unimolecular reaction with rate constant k_3.

---

* In the next chapter we will discuss A molecules in the transition state, which are represented by $A^{\ddagger}$. The symbol $A^{\ddagger}$ indicates that A is in the middle of the process of chemical change. Excited molecules A* are not in the process of chemical change, but are simply A molecules with additional internal energy.

Only molecules with three or more atoms undergo unimolecular reactions. A diatomic molecule cannot dissociate in this way because it has a single mode of vibrational freedom. If this mode is excited by an amount equal to the dissociation energy, the molecule dissociates in about 10^{-12} s. The kinetics of the dissociation of diatomic molecules is discussed in Section 19.14.

Since A* is never present at a very high concentration we can use the steady-state approximation to obtain an expression for the rate of reaction in terms of [A] and [M]:

$$\frac{d[A^*]}{dt} = k_1[A][M] - (k_2[M] + k_3)[A^*] = 0 \qquad (19.119)$$

Thus,

$$-\frac{d[A]}{dt} = k_3[A]^* = \frac{k_1 k_3[A][M]}{k_2[M] + k_3} = k[A] \qquad (19.120)$$

This rate equation has two limiting forms:

1.   In the high-pressure limit, $k_2[M] \gg k_3$ so that

$$-\frac{d[A]}{dt} = \frac{k_1 k_3}{k_2}[A] \qquad (19.121)$$

Thus, at high pressures the reaction is first order, as observed. The first-order rate constant at high pressures is generally referred to as k^∞, which is equal to $k_1 k_3/k_2$, and the values of the pre-exponential factor A^∞ and E_a are given in Table 19.1.

2.   In the low pressure limit, $k_2[M] \ll k_3$ so that

$$-\frac{d[A]}{dt} = k_1[A][M] \qquad (19.122)$$

Under these conditions all A* isomerize or dissociate, and so the rate of reaction is determined by the number of collisions that are sufficiently energetic. At very low pressures the reaction is found to be second order when the partial pressure of M is changed. However, in a single kinetic experiment, the product can also activate or quench so that [M] is constant and the reaction is pseudo first order. In a unimolecular dissociation, [M] increases during a kinetic experiment, but since the products are less efficient in energizing A, $k_1[M]$ remains approximately constant, and the reaction is again pseudo first order.

At a sufficiently low pressure the rate constant of a unimolecular reaction decreases, and the reaction is said to be in the **falloff region.** According to the Lindemann mechanism the first-order rate constant for a unimolecular reaction is given by

$$k = \frac{k_1 k_3[M]}{k_2[M] + k_3} \qquad (19.123)$$

This equation is correct in the sense that a unimolecular reaction is first order at high pressure and the first-order rate drops off as the pressure is reduced (see Fig. 19.12), but this equation predicts a steeper falloff in k than is actually observed. The RRKM theory (Section 21.6) has been developed to account for this behavior.

Note that mechanism 19.116–19.118 is a little different from other mechanism in this chapter because 19.116 and 19.117 are energy-transfer processes that occur in any system. They are written out here to help us understand the formation and de-energization of reactant molecules. Note the reverse of a unimolecular dissociation reaction is necessarily bimolecular at high pressure and trimolecular at low pressure.

19.13   Bimolecular Reactions

When two uncharged, nonpolar molecules approach each other at the short distances required for chemical reaction, the predominant force is a strong repulsion. This, of course, reflects the Pauli exclusion principle (Section 12.7). If these molecules can react in an exothermic reaction, the activation energy is a measure of the energy required to deform the electron clouds of the reactants so that the reaction can occur. When two molecules with closed electronic shells react in an exothermic reaction the activation energy is generally found to be in the range 80–200 kJ mol^{-1}. As a result, such reactions are rare because there are frequently other reaction paths involving free radicals that are faster, and are therefore the ones by which reaction occurs.

Table 19.2   Activation Energies for Exothermic Bimolecular Reactions[a]

Reaction Type	Electronic Structures	E_a/kJ mol^{-1}
Molecule + Molecule	Two closed shells	80–200
Radical + Molecule	One closed shell/one open shell	0–60
Radical + Radical	Two open shells	~0

[a] S. W. Benson, *Thermochemical Kinetics*. New York: Wiley, 1976.

When one of the reactants is a radical the activation energy is generally in the range 0–60 kJ mol^{-1}, as indicated in Table 19.2. By the term radical we refer to molecules with at least one orbital vacancy in their valence shells and to atoms, except for the rare gases. Since reactions of radicals with closed-shell molecules are faster than reactions of closed-shell molecules with each other, many reactions observed in the gas phase involve radicals. The reactions of radicals with each other are generally even faster because they have activation energies of about zero, as indicated in Table 19.2. In fact, the activation energies for the recombination of radicals may be negative so that the reaction goes more slowly at a higher temperature.

A common type of bimolecular reaction is a metathesis reaction involving an atom or a radical. The enthalpy change for such a reaction may be positive or negative, depending on whether the newly formed bond has a higher or lower dissociation energy than the bond broken. The Arrhenius parameters for some bimolecular reactions are given in Table 19.3. The pre-exponential factors for metathesis reactions involving atoms are all in the range of about $10^{10.5}$–$10^{11.5}$ L mol^{-1} s^{-1}, but the activation energies differ appreciably. The pre-exponential factors for metathesis reactions not involving atoms are smaller, as are those for association reactions of radicals. The activation energies of association reactions of radicals are all about zero. We will see in the next

Table 19.3   Arrhenius Parameters for Bimolecular Reactions

Reaction	$\log\left(\dfrac{A}{\text{L mol}^{-1}\,\text{s}^{-1}}\right)$	$\dfrac{E_a}{\text{kJ mol}^{-1}}$
Metathesis reactions involving atoms		
$Br + H_2 \rightarrow HBr + H$	10.8	76.2
$I + H_2 \rightarrow HI + H$	11.4	143
$Cl + H_2 \rightarrow HCl + H$	10.9	23.0
$O + O_3 \rightarrow 2O_2$	10.5	23.9
$O + NO_2 \rightarrow O_2 + NO$	10.3	4.2
$O + H_2 \rightarrow OH + H$	10.5	42.7
$N + O_2 \rightarrow NO + O$	9.3	26.4
$N + NO \rightarrow N_2 + O$	10.2	0
$O + OH \rightarrow O_2 + H$	10.3	0
Metathesis reactions not involving atoms		
$CH_3 + C_2H_6 \rightarrow CH_4 + C_2H_5$	8.5	45.2
$2C_2H_5 \rightarrow C_2H_4 + C_2H_6$	9.6	0
$C_6H_5 + CH_4 \rightarrow C_6H_6 + CH_3$	8.6	46.4
Association reactions of radicals		
$2CH_3 \rightarrow C_2H_6$	10.5	0
$2C_2H_5 \rightarrow C_4H_{10}$	10.4	0
$CH_3 + NO \rightarrow CH_3NO$	8.8	0

chapter that the pre-exponential factors of such reactions can be estimated quite well using collision theory for rigid spherical molecules.

Some bimolecular reactions are more complex than simple two-body reactions. In some cases the bimolecular rate constant depends on the pressure, and this is an indication that the reaction leads to a **collision complex.** A collision complex is a weakly bound molecule which survives for a time that is longer than the characteristic periods of its vibrations and rotations. If a weakly bound molecule is formed the reaction may be written

$$A + B \rightleftharpoons AB \rightarrow \text{products} \qquad (19.124)$$

When a weakly bound intermediate is formed, the reaction is no longer really a bimolecular reaction, but the dividing line is not sharp. If the pressure is sufficiently low so that the unimolecular dissociation of AB is in the falloff region, the overall reaction will approach third order, and may appear to be a trimolecular reaction. Under these conditions the distinction between bimolecular and trimolecular becomes less meaningful. The reaction

$$ClO + NO \rightleftharpoons ClONO \rightarrow Cl + NO_2 \qquad (19.125)$$

is an example of such a reaction. Its rate constant is given by $(6.2 \times 10^{-12}$ cm^3 s$^{-1})$ exp$(294 \text{ K}/T)$. A further example of the complications that may be encountered when a reaction is studied in detail is the reaction

$$HO_2 + HO_2 \rightarrow H_2O_2 + O_2 \qquad (19.126)$$

This reaction goes by two paths: One is a bimolecular path with a negative activation energy and the other is a trimolecular reaction that also has a negative activation energy.

If the molecule formed in an association reaction of radicals has enough bonds, the energy of the exothermic reaction can be absorbed by various molecular vibrations, without causing dissociation. When this is the situation the forward reaction is bimolecular and the reverse reaction is unimolecular. However, when we go from radicals to atoms the energized product of the association reaction has to be deactivated by collision or it will dissociate.

Example 19.6
A weak complex AB is formed in the reaction

$$A + B \underset{k_{-1}}{\overset{k_1}{\rightleftharpoons}} AB \overset{k_2}{\longrightarrow} C$$

Assuming that $k_{-1} \gg k_2$, derive the rate equation for the formation of C.
Since $k_{-1} \gg k_2$, AB remains in equilibrium with A and B. Therefore,

$$\frac{[AB]}{[A][B]} = \frac{k_1}{k_{-1}} = K$$

$$\frac{d[C]}{dt} = k_2[AB] = \frac{k_1 k_2}{k_{-1}}[A][B]$$

$$= k_2 K[A][B]$$

19.14 Trimolecular Reactions

Two atoms cannot combine in a bimolecular reaction in the gas phase to form a diatomic molecule because the energy release causes the molecule to dissociate. The reaction can occur if a third atom or molecule is involved and carries away the energy produced. Atom recombination reactions are third order in the gas phase, and so the elementary reactions are represented by

$$2A + M \rightarrow A_2 + M \tag{19.127}$$

$$A + B + M \rightarrow AB + M \tag{19.128}$$

The corresponding rate equations are

$$\frac{d[A_2]}{dt} = k[A]^2[M] \tag{19.129}$$

$$\frac{d[AB]}{dt} = k[A][B][M] \tag{19.130}$$

The values of the trimolecular rate constants for the recombination of atoms in Table 19.4 show that they are all about $10^{9.5 \pm 0.05}$ L^2 mol^{-2} s^{-1} at 300 K and about a power of 10 smaller at 3000 K. This corresponds to an activation energy of about -6 kJ mol^{-1}, but the data on trimolecular rate constants are actually better represented by $k = (\text{const})/T$ than by the Arrhenius equation. As indicated in the table, various third bodies M have different efficiencies in trimolecular reactions. As shown by several entries at the bottom of the table, reactions of atoms with diatomic molecules may be trimolecular.

Table 19.4 Rate Constants for Trimolecular Reactions[a]

Reaction	T/K	$\log\left(\dfrac{k}{L^2\ mol^{-2}\ s^{-1}}\right)$
$H + H + M \rightarrow H_2 + M$	300	10.0 (H_2)
	1072	9.5 (H_2)
$O + O + M \rightarrow O_2 + M$	300	8.9 (O_2)
	2000	7.4 (Ar)
$I + I + M \rightarrow I_2 + M$	300	9.3 (Ne)
	300	10.8 ($n\text{-}C_5H_{12}$)
$O + O_2 + M \rightarrow O_3 + M$	380	8.1 (O_2)
$O + NO + M \rightarrow NO_2 + M$	300	10.46 (0_2)
$H + NO + M \rightarrow HNO + M$	300	10.17 (H_2)

[a] From S. W. Benson, *Thermochemical Kinetics*. New York: Wiley, 1976. The third body is the molecule listed in parentheses.

There is an alternative way to looking at trimolecular reactions that avoids nearly simultaneous three-body collisions and also provides an explanation of the negative activation energies. If the third body M is a polyatomic molecule capable of forming a long-lived complex with one of the recombining atoms, the mechanism can be written as

$$A + M \underset{k_{-1}}{\overset{k_1}{\rightleftharpoons}} AM \qquad (19.131)$$

$$AM + A \overset{k_2}{\longrightarrow} A_2 + M \qquad (19.132)$$

where AM is a short-lived intermediate. The steady-state rate equation is

$$\frac{d[A_2]}{dt} = \frac{k_1 k_2 [A]^2 [M]}{k_{-1} + k_2[A]} \qquad (19.133)$$

If the intermediate is very short lived, $k_{-1} \gg k_2[A]$, and equation 19.133 becomes

$$\frac{d[A_2]}{dt} = K_1 k_2 [A]^2 [M] \qquad (19.134)$$

where $K_1 = k_1/k_{-1}$. If k_2 follows the Arrhenius equation with activation energy E_2, the activation energy for the trimolecular reaction is given by

$$E_a = RT^2 \frac{d \ln K_1}{dT} + RT^2 \frac{d \ln k_2}{dT} = \Delta H_1^\circ + E_2 \qquad (19.135)$$

If the first step is exothermic and the activation energy for the second step is not too large, the activation energy for the trimolecular reaction can be negative. If the combination of an atom with another atom or diatomic molecule is trimolecular, the reverse reaction is bimolecular and the Arrhenius equation for the reverse reaction may be calculated using K_c (Section 19.11). If the activation energy for the combination of two atoms is negative, the activation energy for the dissociation of the diatomic molecule is *less* than the dissociation energy for the diatomic molecule. When a dissociation reaction is bimolecular, this indicates that the process of energy transfer (the reverse of k_2 in equation

19.132) is rate limiting. Energy transfer is usually rate limiting at 1 bar above 300 K for three- or four-atom molecules and above 1000 K for five- to seven-atom molecules.

19.15   Unbranched Chain Reactions

As discussed in Section 19.13, reactions of molecules with closed shells generally have high activation energies even when the change in Gibbs energy for the overall reaction is favorable. Faster reaction paths are frequently provided by radical–molecule reactions. As a consequence, many chemical reactions occur through a sequence of elementary reactions involving radicals. These may be nonchain reactions, or they may be chain reactions. Since a radical has an unpaired electron, its reaction with a molecule having paired electrons gives rise to another radical. In this way the reactive center is maintained and can give rise to a **chain of reactions.** We may ask why such a reaction ever stops. Sometimes, as a matter of fact, the chain reaction does not stop until all the material is consumed. At other times, however, the chain is broken when one of the radicals reacts at the wall of the containing vessel or with another radical to form a spin-paired molecule. The length of the chain (i.e., the number of molecules reacting per molecule activated) is determined by the relative rates of the chain-propagating and the chain-breaking reactions.

Mechanisms involving radicals may be **nonchain, straight chain,** and **branched chain.** Branched chain reactions produce explosions if they are highly exothermic; they are discussed in Section 19.16. The rate parameters for several types of elementary radical reactions are given in Table 19.3. Although the rate constants for radical fission reactions may have very large pre-exponential factors, they may be slow, except at very high temperatures, because of high activation energies. The activation energy of a radical fission reaction may be lower when rearrangement to a molecular product compensates for some of the energy required to break a bond.

The reaction of gaseous bromine with hydrogen

$$H_2 + Br_2 = 2HBr \tag{19.136}$$

is an example of an unbranched chain reaction. In the temperature range 500 to 1500 K, Bodenstein found that the rate of this reaction is given by the empirical rate equation

$$v = \frac{1}{2}\frac{d[HBr]}{dt} = \frac{k[H_2][Br_2]^{1/2}}{1 + k'[HBr]/[Br_2]} \tag{19.137}$$

If no HBr is initially present, the initial rate is proportional to $[H_2][Br_2]^{1/2}$. If sufficient HBr is added initially so that $k'[HBr]/[Br_2] \gg 1$, then the initial rate is proportional to $[H_2][Br_2]^{3/2}/[HBr]$, and it is seen that doubling the initial HBr concentration cuts the initial velocity in half. Thus, the presence of the product inhibits the forward reaction. The fact that the rate is proportional to $[Br_2]^{1/2}$ at low [HBr] suggests that Br_2 dissociates and Br atoms play a role in the mechanism; note that if the dissociation is in equilibrium, $[Br] = K^{1/2}[Br_2]^{1/2}$. In 1919 Christiansen, Herzfeld, and Polanyi recognized that Br could react with H_2 to start to chain, and they proposed the following mechanism:

$$\text{Initiation} \quad \text{Br}_2 + \text{M} \underset{k_{-1}}{\overset{k_1}{\rightleftharpoons}} 2\text{Br} + \text{M} \qquad (19.138)$$

$$\text{Propagation} \quad \text{Br} + \text{H}_2 \underset{k_{-2}}{\overset{k_2}{\rightleftharpoons}} \text{HBr} + \text{H} \qquad (19.139)$$

$$\text{H} + \text{Br}_2 \underset{k_{-3}}{\overset{k_3}{\rightleftharpoons}} \text{HBr} + \text{Br} \qquad (19.140)$$

The carrier gas is represented by M. The net reaction is the sum of steps 2 and 3. The reaction is initiated by the dissociation of Br_2, rather than H_2, because the bond in bromine is much weaker. Step 2 is endothermic and has a high activation energy, so that it is slow relative to the other. Step -2 is fast and causes inhibition because it destroys product. The chains are terminated whenever two bromine atoms recombine in the presence of a third body to produce a bromine molecule.

The chain reaction rapidly gets into a steady state in which $d[\text{Br}]/dt$ and $d[\text{H}]/dt$ are both approximately zero. Using the expressions for these derivatives obtained from the mechanism and setting them equal to zero we can obtain a steady-state rate equation. We will do this first for the case that k_{-3} is zero, so that the overall reaction goes to completion to the right. This is the situation that is treated by the rate law in equation 19.137. The steady-state rate law derived from the mechanism under these conditions is

$$v = \frac{1}{2}\frac{d[\text{HBr}]}{dt} = \frac{k_2(k_1/k_{-1})^{1/2}[\text{H}_2][\text{Br}_2]^{1/2}}{1 + k_{-2}[\text{HBr}]/k_3[\text{Br}_2]} \qquad (19.141)$$

Thus, the constant k in equation 19.137 can be identified with $k_2(k_1/k_{-1})^{1/2}$, where k_1/k_{-1} is the dissociation constant for the dissociation of bromine.

If the -3 reaction is included, the steady-state rate equation that is obtained may be written in the following form (see Problem 19.36).

$$v = \frac{1}{2}\frac{d[\text{HBr}]}{dt} = \frac{k_2(k_1/k_{-1})^{1/2}[\text{H}_2][\text{Br}_2]^{1/2}}{1 + k_{-2}[\text{HBr}]/k_3[\text{Br}_2]}\left(1 - \frac{[\text{HBr}]^2}{K[\text{H}_2][\text{Br}_2]}\right) \qquad (19.142)$$

where $K = k_2k_3/k_{-2}k_{-3}$ is the equilibrium constant for the overall reaction. For low extents of reaction, the steady-state rate law reduces to equation 19.141. The rate law for the reverse reaction, obtained under conditions where $[\text{HBr}] \gg (K[\text{H}_2][\text{Br}_2])^{1/2}$, is

$$v = -\frac{1}{2}\frac{d[\text{HBr}]}{dt} = k_{-3}\left(\frac{k_1}{k_{-1}}\right)^{1/2}[\text{HBr}][\text{Br}_2]^{1/2} \qquad (19.143)$$

Thus, the reverse reaction is first order in HBr at constant $[\text{Br}_2]$. The advantage of equation 19.142 over 19.141 is that it gives the steady-state rate law all the way to equilibrium, whether we start with $\text{H}_2 + \text{Br}_2$ or HBr.

At low temperatures the thermal reaction becomes very slow, but the reaction may be carried out photochemically. This is discussed in Section 20.11.

The hydrogen–chlorine thermal reaction is faster than the hydrogen–bromine thermal reaction because the propagation step

$$\text{Cl} + \text{H}_2 \rightarrow \text{HCl} + \text{H} \qquad (19.144)$$

is only endothermic by 5 kJ mol^{-1} and is much faster than reaction 19.139. Since the reverse of this step is slower than for bromine, there is no inhibitory term in the denominator of the rate law. The hydrogen–iodine thermal reaction is very slow below 800 K because the propagation step

$$I + H_2 \rightarrow HI + H \qquad (19.145)$$

has an activation energy of 140 kJ mol^{-1}. At lower temperatures the hydrogen–iodine reaction goes by a nonchain mechanism.*

The pyrolysis of hydrocarbons occurs by radical chain reactions. These reactions are very important industrially; for example, ethylene is produced from ethane and the "cracking" of higher hydrocarbons is used to produce butadiene, butene, and propylene. In industry these reactions are generally carried out heterogeneously, that is, using solid catalysts, but it is important to understand the mechanism of the homogeneous reactions. In fact, these reactions played an important role in the discovery of the mechanisms of radical chain reactions.

As an example, let us consider the pyrolysis of ethane to ethylene:

$$C_2H_6 = C_2H_4 + H_2 \qquad (19.146)$$

At temperatures of 700 to 900 K and pressures above about 0.2 bar, this reaction is first order in its early stages. Later in the reaction methane and propylene are formed, but we will consider only the early stages. There are various kinds of evidence that the reaction has the following mechanism:

Initiation	$C_2H_6 \xrightarrow{k_1} 2CH_3$	(19.147a)
Chain transfer	$CH_3 + C_2H_6 \xrightarrow{k_2} CH_4 + C_2H_5$	(19.147b)
Propagation	$C_2H_5 \xrightarrow{k_3} C_2H_4 + H$	(19.147c)
	$H + C_2H_6 \xrightarrow{k_4} H_2 + C_2H_5$	(19.147d)
Termination	$H + C_2H_5 \xrightarrow{k_5} C_2H_6$	(19.147e)

To the extent that reaction 19.147b occurs, the net chemical change is not given by equation 19.146.

The rate of reaction of ethane is

$$\frac{d[C_2H_6]}{dt} = -(k_1 + k_2[CH_3] + k_4[H])[C_2H_6] \qquad (19.148)$$

The rate of formation of ethane in the last step is ignored because termination occurs only after long chains producing the products. In the steady state the rates of change of the concentrations of the radicals may be taken equal to zero:

$$\frac{d[CH_3]}{dt} = 2k_1[C_2H_6] - k_2[CH_3][C_2H_6] = 0 \qquad (19.149)$$

$$\frac{d[C_2H_5]}{dt} = (k_2[CH_3] + k_4[H])[C_2H_6] - (k_3 + k_5[H])[C_2H_5] = 0 \quad (19.150)$$

$$\frac{d[H]}{dt} = k_3[C_2H_5] - k_4[H][C_2H_6] - k_5[H][C_2H_5] = 0 \qquad (19.151)$$

The steady-state concentrations of the free radicals CH_3, C_2H_5, and H may be obtained from these three simultaneous equations. We find

$$[CH_3] = \frac{2k_1}{k_2} \qquad (19.152)$$

---

* J. Nicholas, *Chemical Kinetics*, p. 120. New York: Wiley, 1976.

$$[C_2H_5] = \frac{2k_1 + k_4[H]}{k_3 + k_5[H]} [C_2H_6] \tag{19.153}$$

$$[H] = \frac{2k_1k_5 \pm \sqrt{(2k_1k_5)^2 + 16k_1k_3k_4k_5}}{-4k_4k_5} \tag{19.154}$$

In general, we can expect that k_1 is small so that

$$[H] = \left(\frac{k_1k_3}{k_4k_5}\right)^{1/2} \tag{19.155}$$

Substituting equation 19.152 and 19.155 in equation 19.148 yields the rate of reaction of ethane

$$\frac{d[C_2H_6]}{dt} = -\left[3k_1 + k_4\left(\frac{k_1k_3}{k_4k_5}\right)^{1/2}\right][C_2H_6] \tag{19.156}$$

so that the reaction is first order in spite of its complicated mechanism.

According to this mechanism the termination reaction involves two different kinds of radicals colliding with each other. Other possibilities exist; for example, H radicals may collide with each other in the presence of a third body (Section 19.14). Depending on the orders of the initiation and termination reactions, the overall order of a pyrolysis reaction may be 0, $\frac{1}{2}$, 1, $\frac{3}{2}$, or 2.[*]

19.16 Branched Chain Reactions

If a propagation step in a chain reaction produces two or more radicals from one, there is a possibility of a rapid increase in rate and, for an exothermic reaction, an explosion. The reaction of hydrogen with oxygen can be explosive above about 700 K. The ranges of temperature and pressure within which there are spontaneous thermal explosions are shown in Fig. 19.13. For example, at about 550 °C stoichiometric hydrogen–oxygen mixtures react very slowly at pressures below 10^{-3} bar. As the pressure is increased, the reaction rate increases slowly, but at a pressure of about 10^{-3} bar, depending on the volume of the vessel, there is a sudden explosion. On the other hand, if the gases are at a considerably higher pressure, the rate is again quite low. Hinshelwood found that if hydrogen at 0.26 bar and oxygen at 0.13 bar are placed in a 300-cm^3 quartz vessel at 550 °C, the rate of reaction is quite slow and becomes slower if the pressure is further reduced to 0.20 bar. If the pressure is reduced to 0.19 bar, however, an explosion occurs. Finally, as the total pressure is increased above the explosion zone, the reaction rate increases until it becomes so fast that the reaction mixture may be said to explode. The fact that the exact limits depend on the vessel surface and the vessel diameter indicates that radical chains may be terminated by reaction at the wall. If the vessel surface is coated with potassium chloride, radicals disappear when they strike the wall; if the vessel surface is coated with boric oxide, radicals are not destroyed so rapidly by collisions with the wall.

---

[*] M. F. R. Mulcahy, *Gas Kinetics*, p. 89. New York: Wiley, 1973.

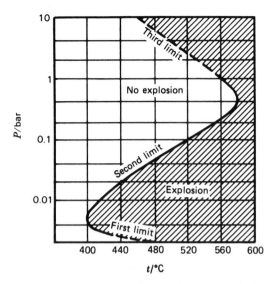

Figure 19.13   Explosion limits of a stoichiometric oxygen–hydrogen mixture.

The first explosion limit can be understood in terms of the following mechanism:

Initiation	$H_2 + O_2 \xrightarrow{\text{wall}} 2OH$	(19.157a)
Propagation	$OH + H_2 \xrightarrow{k_2} H_2O + H$	(19.157b)
Branching	$H + O_2 \xrightarrow{k_3} OH + O$	(19.157c)
	$O + H_2 \xrightarrow{k_4} OH + H$	(19.157d)
Termination	$H + \text{wall} \xrightarrow{k_5}$	(19.157e)

The propagation reaction is exothermic and fast. The third and fourth reactions are called branching reactions because two radicals are formed from one. If the rate of branching is greater than the rate of termination, the number of radicals increases exponentially with time, and an explosion results. Reaction 19.157c is endothermic and slow below 700 K. The conditions of the first explosion limit are governed by the relative rates for branching $2k_3[H][O_2]$ and termination $k_5[H]$. As the concentration of oxygen is increased, the rate of branching becomes greater than the rate of termination and an explosion occurs.

To explain the second limit, above which there is no explosion, it is necessary to invoke a new termination step to prevent the exponential increase in the number of radicals. For a new termination step to become more important as the pressure in increased, the new termination step must be higher order than the branching reaction. Thus, to explain the second limit, the following reaction must be added to the previous reactions:

$$\text{Termination} \qquad H + O_2 + M \xrightarrow{k_6} HO_2 + M \qquad (19.158)$$

In a stoichiometric mixture of oxygen and hydrogen, M may be hydrogen or oxygen, but these two gases have different efficiencies in this reaction. The HO_2 radical is relatively unreactive and does not produce another radical before it is quenched on the wall.

The third explosion limit results from the fact that the following reaction diminishes termination:

$$\text{Propagation} \qquad HO_2 + H_2 \xrightarrow{k_7} H_2O + OH \qquad (19.159)$$

19.17 Analysis of Complex Reaction Systems

In the laboratory we try to study single chemical reactions, but in applying chemical kinetics we generally find a large number of reactions occurring at the same time and sharing reactants and products. Many industrial processes have this characteristic. Another example is the upper atmosphere where solar radiation causes dissociation, and the radicals and ions produced are involved in many reactions. To model the photochemical processes and chemical composition of the stratosphere it is necessary to include 150 chemical reactions and photochemical processes.[*] Another example is a flame, where oxidation and other reactions proceed through many interrelated reactions.[†] These examples are complicated by the fact that compositions vary over distance as well as time. However, calculations can be simplified by first considering regions over which compositions are uniform.

To calculate the concentrations of various species as a function of time, the first step is to write down the mechanism. For example, consider

$$A \underset{k_{-1}}{\overset{k_1}{\rightleftharpoons}} B \qquad (19.160)$$

$$B \underset{k_{-2}}{\overset{k_2}{\rightleftharpoons}} C \qquad (19.161)$$

If we start with n_A°, n_B°, n_C° at $t = 0$, the amounts at a later time are given by

$$n_A = n_A^\circ - \xi_1 \qquad (19.162)$$
$$n_B = n_B^\circ + \xi_1 - \xi_2 \qquad (19.163)$$
$$n_C = n_C^\circ + \xi_2 \qquad (19.164)$$

where ξ_1 and ξ_2 are the extents of reaction of steps 1 and 2. At any time the rates of conversion of A, B, and C are given by

$$\frac{dn_A}{dt} = \frac{-d\xi_1}{dt} \qquad (19.165)$$

$$\frac{dn_B}{dt} = \frac{d\xi_1}{dt} - \frac{d\xi_2}{dt} \qquad (19.166)$$

$$\frac{dn_C}{dt} = \frac{d\xi_2}{dt} \qquad (19.167)$$

These equations can be written in matrix form as follows:

---

[*] W. B. De More et al., *Chemical Kinetics and Photochemical Data for Use in Stratospheric Modelling*, National Aeronautics and Space Administration. Washington, DC: Government Printing Office, 1983.

[†] W. C. Gardiner, Jr. (Ed.), *Combustion Chemistry*. New York: Springer, 1984.

$$\begin{bmatrix} \dfrac{dn_A}{dt} \\[2ex] \dfrac{dn_B}{dt} \\[2ex] \dfrac{dn_C}{dt} \end{bmatrix} = \begin{bmatrix} -1 & 0 \\ 1 & -1 \\ 0 & 1 \end{bmatrix} \begin{bmatrix} \dfrac{d\xi_1}{dt} \\[2ex] \dfrac{d\xi_2}{dt} \end{bmatrix} \tag{19.168}$$

The matrix on the left can be referred to as the amount rate matrix, the first matrix on the right is the stoichiometric number matrix (see Section 5.13), and the second matrix on the right is the rate of conversion matrix. Thus, this equation can be written as

$$\frac{d\boldsymbol{n}}{dt} = \boldsymbol{\nu}\frac{d\boldsymbol{\xi}}{dt} \tag{19.169}$$

This set of rate equations completely defines the time course of the concentration of the various species if the system is initially uniform in composition and the temperature is constant. (If the pressure is constant the volume of the system will in general change with time.)

Now we need to express $d\xi_1/dt$ and $d\xi_2/dt$ in terms of the amounts of A, B, and C. The rate laws for the two steps in this mechanism are

$$\left(\frac{1}{V}\right)\left(\frac{d\xi_1}{dt}\right) = k_1[A] - k_{-1}[B] = \frac{k_1 n_A}{V} - \frac{k_{-1} n_B}{V} \tag{19.170}$$

$$\left(\frac{1}{V}\right)\left(\frac{d\xi_2}{dt}\right) = k_2[B] - k_{-2}[C] = \frac{k_2 n_B}{V} - \frac{k_{-2} n_C}{V} \tag{19.171}$$

After multiplying by V these equations become

$$\frac{d\xi_1}{dt} = k_1 n_A - k_{-1} n_B = r_1 \tag{19.172}$$

$$\frac{d\xi_2}{dt} = k_2 n_B - k_{-2} n_C = r_2 \tag{19.173}$$

where r_i is the rate of conversion for step i. Thus, in matrix notation

$$\frac{d\boldsymbol{n}}{dt} = \boldsymbol{\nu r} \tag{19.174}$$

This **rate equation for a complex reaction system** can be integrated numerically by picking a short time interval h and writing the preceding equation as

$$\frac{\Delta \boldsymbol{n}}{h} = \boldsymbol{\nu r} \tag{19.175}$$

Thus, the species abundance matrix after a short time interval is given by

$$\boldsymbol{n}_{k+1} = \boldsymbol{n}_k + h\boldsymbol{\nu r} \tag{19.176}$$

where $\boldsymbol{r}$ is calculated with $\boldsymbol{n}_k$. Starting with n_A°, n_B°, and n_C° and carrying out many small steps with a computer yields n_A, n_B, and n_C at a later time. The differential equations of chemical kinetics require special computer methods because of the wide ranges in rates with which concentrations can change. When concentrations change rapidly, small steps are required, but when concentrations change slowly, larger steps can be used. Sets of differential equations of this type are said to be "stiff," in analogy with the equations for the vibration of a stiff rod, and so the computer is programmed to take small steps when any concentration is changing rapidly and large steps when all concentrations are changing slowly.

19.18   Special Topic: Solution of Coupled Linear Rate Equations

The simplest example of a chemical system yielding two coupled ordinary (one independent variable, time) differential equations is

$$A \underset{k_2}{\overset{k_1}{\rightleftarrows}} B \tag{19.177}$$

$$\frac{d[A]}{dt} = -k_1[A] + k_2[B] \qquad [A] = [A]_0 \qquad \text{at } t = 0 \tag{19.178}$$

$$\frac{d[B]}{dt} = k_1[A] - k_2[B] \qquad [B] = 0 \qquad \text{at } t = 0 \tag{19.179}$$

Although we have already derived the expressions for [A] and [B] as functions of time in Section 19.4, we want to do this in another way that can be extended to larger systems of equations. This provides another illustration of the usefulness of eigenvalues and eigenfunctions. More information is provided by Strang*, and the more complicated case of $A \rightleftharpoons B \rightleftharpoons C$ is treated by Jordan† and Moore and Pearson.‡

This system can be represented in matrix form:

$$u(t) = \begin{bmatrix} [A] \\ [B] \end{bmatrix} \qquad u_0 = \begin{bmatrix} [A]_0 \\ 0 \end{bmatrix} \qquad A = \begin{bmatrix} -k_1 & k_2 \\ k_1 & -k_2 \end{bmatrix} \tag{19.180}$$

$$\frac{du}{dt} = Au \qquad u = u_0 \quad \text{at } t = 0 \tag{19.181}$$

If this was a scalar differential equation, rather than a vector differential equation,

$$\frac{du}{dt} = au \qquad u = u_0 \quad \text{at } t = 0 \tag{19.182}$$

the solution would be

$$u(t) = e^{at}u_0 \tag{19.183}$$

This suggests that we look for solutions of the form

$$[A] = e^{\lambda t}y \tag{19.184}$$

$$[B] = e^{\lambda t}z \tag{19.185}$$

---

* G. Strang, *Linear Algebra and Its Applications*. New York: Academic, 1980.

† P. C. Jordan, *Chemical Kinetics and Transport*. New York: Plenum, 1979.

‡ J. W. Moore and R. G. Pearson, *Kinetics and Mechanism*. New York: Wiley, 1981.

where y and z are not functions of time. These two equations can be written in vector notation as

$$u(t) = e^{\lambda t}x \tag{19.186}$$

When we substitute equations 19.184 and 19.185 in equations 19.178 and 19.179, we obtain

$$-k_1 y + k_2 z = \lambda y \tag{19.187}$$

$$k_1 y - k_2 z = \lambda z \tag{19.188}$$

since $e^{\lambda t}$ can be canceled out. These linear algebraic equations can be written as

$$(-k_1 - \lambda)y + k_2 z = 0 \tag{19.189}$$

$$k_1 y + (-k_2 - \lambda)z = 0 \tag{19.190}$$

Equations 19.187 and 19.188 can be written in vector notation as

$$Ax = \lambda x \tag{19.191}$$

where λ is the **eigenvalue** and x is the **eigenvector.** Equations 19.189 and 19.190 can be written as

$$(A - \lambda I)x = 0 \tag{19.192}$$

where I is the identity matrix; in this case

$$I = \begin{bmatrix} 1 & 0 \\ 0 & 1 \end{bmatrix} \tag{19.193}$$

Equations 19.189 and 19.190 have the trivial solution $y = z = 0$; there are nontrivial solutions only if the determinant of the coefficients vanishes, that is, when

$$\begin{vmatrix} -k_1 - \lambda & k_2 \\ k_1 & -k_2 - \lambda \end{vmatrix} = 0 \tag{19.194}$$

When this determinant is expanded, we obtain the **characteristic polynomial**

$$\lambda^2 + (k_1 + k_2)\lambda = 0 \tag{19.195}$$

which has the solution $\lambda_1 = 0$ and $\lambda_2 = -(k_1 + k_2)$. These are the eigenvalues for this system.

If we replace λ by $\lambda_1 = 0$ in equation 19.189 and 19.190, we know that the two equations have a nontrivial solution:

$$-k_1 y + k_2 z = 0 \tag{19.196}$$

$$k_1 y - k_2 z = 0 \tag{19.197}$$

If $y = 1$, $z = k_1/k_2$. Thus,

$$x = \begin{bmatrix} 1 \\ \dfrac{k_1}{k_2} \end{bmatrix} \tag{19.198}$$

or any multiple of it, satisfies equations 19.191 and 19.192.

If we replace λ by $\lambda_2 = -(k_1 + k_2)$ in equations 19.189 and 19.190, we know that the two equations will have another nontrivial solution:

$$k_2 y + k_2 z = 0 \tag{19.199}$$

$$k_1 y + k_1 z = 0 \tag{19.200}$$

If $y = 1$, $z = -1$. Thus,

$$x = \begin{bmatrix} 1 \\ -1 \end{bmatrix} \tag{19.201}$$

or any multiple of it, also satisfies equations 19.191 and 19.192.

Now we have two solutions and equation 19.186 can be written as

$$u(t) = e^{0t} \begin{bmatrix} 1 \\ \dfrac{k_1}{k_2} \end{bmatrix} \quad \text{and} \quad u(t) = e^{-(k_1 + k_2)t} \begin{bmatrix} 1 \\ -1 \end{bmatrix} \tag{19.202}$$

The general solution is a superposition of these two particular solutions:

$$u = c_1 e^{0t} \begin{bmatrix} 1 \\ \dfrac{k_1}{k_2} \end{bmatrix} + c_2 e^{-(k_1 + k_2)t} \begin{bmatrix} 1 \\ -1 \end{bmatrix} \tag{19.203}$$

The values of the two parameters c_1 and c_2 are determined by the initial conditions $[A] = [A]_0$ and $[B] = 0$ at $t = 0$. Thus, equation 19.203 becomes

$$u_0 = c_1 \begin{bmatrix} 1 \\ \dfrac{k_1}{k_2} \end{bmatrix} + c_2 \begin{bmatrix} 1 \\ -1 \end{bmatrix} = \begin{bmatrix} [A]_0 \\ 0 \end{bmatrix} \tag{19.204}$$

Solving this pair of equations for c_1 and c_2 yields

$$c_1 = \frac{k_2[A]_0}{k_1 + k_2} \quad \text{and} \quad c_2 = \frac{k_1[A]_0}{k_1 + k_2} \tag{19.205}$$

When these values are substituted in equation 19.203, we obtain

$$[A] = \frac{k_2[A]_0}{k_1 + k_2} \left(1 + \frac{k_1}{k_2} e^{-(k_1 + k_2)t} \right) \tag{19.206}$$

$$[B] = \frac{k_1[A]_0}{k_1 + k_2} \left(1 - e^{-(k_1 + k_2)t} \right) \tag{19.207}$$

which are the same as equations 19.40 and 19.41. Notice that it is the eigenvalues that give us the rate of change with respect to time and the eigenvectors that give us the "normal modes."

References

S. W. Benson, *Thermochemical Kinetics*. New York: Wiley, 1976.

J. H. Espenson, *Chemical Kinetics and Reaction Mechanics*. New York: McGraw-Hill, 1981.

W. C. Gardiner, Jr., *Rates and Mechanisms of Chemical Reactions*. New York: Benjamin, 1969.

W. C. Gardiner, Jr. (Ed.), *Combustion Chemistry*. New York: Springer, 1984.

G. G. Hammes, *Principles of Chemical Kinetics*. New York: Academic, 1978.

H. S. Johnston, *Gas Phase Reaction Rate Theory*. New York: Ronald, 1966.

P. C. Jordan, *Chemical Kinetics and Transport*. New York: Plenum, 1979.

K. J. Laidler, *Chemical Kinetics,* 3d ed. New York: Harper & Row, 1987.

J. W. Moore and R. G. Pearson, *Kinetics and Mechanism*. New York: Wiley, 1981.

J. Nicholas, *Chemical Kinetics*. New York: Wiley, 1976.

M. F. R. Mulcahy, *Gas Kinetics*. New York: Wiley, 1973.

C. N. Satterfield, *Heterogeneous Catalysis in Practice*. New York: McGraw-Hill, 1980.

I. W. M. Smith, *Kinetics and Dynamics of Elementary Gas Reactions*. Boston: Butterworths, 1980.

J. I. Steinfeld, J. S. Francisco, and W. L. Hase, *Chemical Kinetics and Dynamics*. Englewood Cliffs, NJ: Prentice Hall, 1989.

K. Tamaru, *Dynamic Heterogeneous Catalysis*. New York: Academic, 1978.

R. E. Weston, Jr., and H. A. Schwartz, *Chemical Kinetics*. Englewood Cliffs, NJ: Prentice-Hall, 1972.

Problems

19.1   Nitrogen pentoxide (N_2O_5) gas decomposes according to the reaction

$$2N_2O_5 = 4NO_2 + O_2$$

At 328 K, the rate of reaction v under certain conditions is 0.75×10^{-4} mol L^{-1} s^{-1}. Assuming that none of the intermediates have appreciable concentrations, what are the values of $d[N_2O_5]/dt$, $d[N_2]/dt$, and $d[O_2]/dt$?

19.2   In studying the decomposition of ozone

$$2O_3(g) = 3O_2(g)$$

in a 2-L reaction vessel, it is found that $d[O_3]/dt = -1.5 \times 10^{-2}$ mol L^{-1} s^{-1}. (*a*) What is the rate of reaction v? (*b*) What is the rate of conversion $d\xi/dt$? (*c*) What is the value of $d[O_2]/dt$?

19.3   The decomposition of N_2O_5

$$2N_2O_5 = 4NO_2 + O_2$$

is studied by measuring the concentration of oxygen as a function of time, and it is found that

$$\frac{d[O_2]}{dt} = (1.5 \times 10^{-4} \text{ s}^{-1})[N_2O_5]$$

at constant temperature and pressure. Under these conditions the reaction goes to completion to the right. What is the half-life of the reaction under these conditions?

19.4   The following data were obtained on the rate of hydrolysis of 17% sucrose in 0.099 mol L^{-1} HCl aqueous solutions at 35 °C:

t/min	9.82	59.60	93.18	142.9	294.8	589.4
Sucrose remaining, %	96.5	80.3	71.0	59.1	32.8	11.1

What is the order of the reaction with respect to sucrose and the value of the rate constant k?

19.5   Methyl acetate is hydrolyzed in approximately 1 mol L^{-1} HCl at 25 °C. Aliquots of equal volume are removed at intervals and titrated with a solution of NaOH. Calculate the first-order rate constant from the following experimental data:

t/s	339	1242	2745	4546	∞
V/cm³	26.34	27.80	29.70	31.81	39.81

19.6 Prove that in a first-order reaction, where $dn/dt = -kn$, the average life, that is, the average life expectancy of the molecules, is equal to $1/k$.

19.7 The hydrolysis of l-chloro-l-methylcycloundecane in 80% ethanol has been studied by H. C. Brown and M. Borkowski (*J. Chem. Soc.* **74**: 1894 (1952)) at 25 °C. The extent of hydrolysis was measured by titrating the acid formed after measured intervals of time with a solution of NaOH. The data are as follows:

t/h	0	1.0	3.0	5.0	9.0	12	∞
x/cm^3	0.035	0.295	0.715	1.055	1.505	1.725	2.197

(*a*) What is the order of the reaction? (*b*) What is the value of the rate constant? (*c*) What fraction of the 1-chloro-1-methylcycloundecane will be left unhydrolyzed after 8 h?

19.8 The following values of percent transmission are obtained with a spectrophotometer at a series of times during the decomposition of a substance absorbing light at a particular wavelength. Calculate k, $t_{1/2}$, and τ assuming the reaction is first order.

t	Percent Transmission
5 min	14.1
10 min	57.1
∞	100

Beer's law: $\log 100/T = abc$, Where T = percent transmission, a = absorbancy index, b = cell thickness, and c = concentration.

19.9 Since radioactive decay is a first-order process, the decay rate for a particular nuclide is commonly given as the half-life. Given that potassium contains 0.0118% ^{40}K, which has a half-life of 1.27×10^9 years, how many disintegrations per second are there in a gram of KCl?

19.10 The decomposition of HI to $H_2 + I_2$ at 508 °C has a half-life of 135 min when the initial pressure of HI is 0.1 atm and 13.5 min when the pressure is 1 atm. (*a*) Show that this proves that the action is second order. (*b*) What is the value of the rate constant in L mol^{-1} s^{-1}? (*c*) What is the value of the rate constant in bar^{-1} s^{-1}? (*d*) What is the value of the rate constant in cm^3 s^{-1}?

19.11 The reaction between propionaldehyde and hydrocyanic acid has been studied at 25 °C by W. J. Svirbely and J. R. Roth (*J. Am. Chem. Soc.* **75**:3106 (1953)). In a certain aqueous solution at 25 °C the concentrations at various times were as follows:

t/min	2.78	5.33	8.17	15.13	19.80	∞
[HCN]/ mol L^{-1}	0.0990	0.0906	0.0830	0.0706	0.0653	0.0424
[C₃H₇CHO]/ mol L^{-1}	0.0566	0.0482	0.0406	0.0282	0.0229	0.0000

What is the order of the reaction and the value of the rate constant k?

19.12 Hydrogen peroxide reacts with thiosulfate ion in slightly acidic solution as follows:

$$H_2O_2 + 2S_2O_3^{2-} + 2H^+ \rightarrow 2H_2O + S_4O_6^{2-}$$

This reaction rate is independent of the hydrogen ion concentration in the pH range 4 to 6. The following data were obtained at 25 °C and pH 5.0. Initial concentrations: $[H_2O_2] = 0.036$ mol L^{-1}; $[S_2O_3^{2-}] = 0.020\,40$ mol L^{-1}

t/min	16	36	43	52
$[S_2O_3^{2-}]/10^{-3}$ mol L^{-1}	10.30	5.18	4.16	3.13

(*a*) What is the other of the reaction? (*b*) What is the rate constant?

19.13 The reaction A = B is nth order (where $n = \frac{1}{2}, \frac{3}{2}, 2, 3, \ldots$) and goes to completion to the right. Derive the expression for the half-life in terms of k, n, and $[A]_0$.

19.14 A gas reaction 2A = B is second order in A and goes to completion in a reaction vessel of constant volume and temperature with a half-life of 1 h. If the initial pressure of A is 1 bar, what are the partial pressures of A, of B, and the total pressure at 1 h, 2 h, and at equilibrium?

19.15 The rate constant for the reaction

$$I + I + Ar \rightarrow I_2 + Ar$$

is 0.59×10^{16} cm^6 mol^{-2} s^{-1} at 293 K [G. Porter and J. A. Smith, *Proc. R. Soc. London Ser. A* **261**:28 (1961)]. What is the half-life of I if $[I]_0 = 2 \times 10^{-5}$ mol L^{-1} and $[Ar] = 5 \times 10^{-3}$ mol L^{-1}?

19.16 A solution of A is mixed with an equal volume of a solution of B containing the same number of moles, and the reaction A + B = C occurs. At the end of 1 h, A is 75% reacted. How much of A will be left unreacted at the end of 2 h if the reaction is (*a*) first order in A and zero in B; (*b*) first order in both A and B; (*c*) zero order in both A and B?

19.17 Show that for a first-order reaction R → P the concentration of product can be represented as a function of time by

$$[P] = a + bt + ct^2 + \cdots$$

and express a, b, and c in terms of $[R]_0$ and k.

19.18 For a reaction A → X, the following concentrations of A were found in a single kinetics experiment:

[A]/mol L^{-1}	1.000	0.952	0.909	0.870	0.833	0.800
t/h	0	0.05	0.10	0.15	0.20	0.25

What is the rate v of this reaction at $[A] = 1.000$ mol L^{-1}.

19.19 The following table gives kinetic data (Y. T. Chia and R. E. Connick, *J. Phys. Chem.* **63**:1518 (1959)) for the following reaction at 25 °C.

$$OCl^- + I^- = OI^- + Cl^-$$

$\dfrac{[OCl^-]}{\text{mol } L^{-1}}$	$\dfrac{[I^-]}{\text{mol } L^{-1}}$	$\dfrac{[OH^-]}{\text{mol } L^{-1}}$	$\dfrac{d[IO^-]/dt}{10^{-4} \text{ mol } L^{-1} s^{-1}}$
0.0017	0.0017	1.00	1.75
0.0034	0.0017	1.00	3.50
0.0017	0.0034	1.00	3.50
0.0017	0.0017	0.5	3.50

What is the rate law for the reaction and what is the value of the rate constant?

19.20 For a reversible first-order reaction

$$A \underset{k_2}{\overset{k_1}{\rightleftharpoons}} B$$

$k_1 = 10^{-2}$ s^{-1} and $[B]_{eq}/[A]_{eq} = 4$. If $[A]_0 = 0.01$ mol L^{-1} and $[B]_0 = 0$, what will be the concentration of B after 30 s?

19.21 The first three steps in the decay of ^{238}U are

$$^{238}U \xrightarrow[4.5 \times 10^9 y]{\alpha} {}^{234}Th \xrightarrow[24.1 d]{\beta} {}^{234}Pa \xrightarrow[1.14 m]{\beta} {}^{234}U$$

If we start with pure ^{238}U, what fraction will be ^{234}Th after 10, 20, 40, and 80 days?

19.22 The initial rate of the reaction

$$BrO_3^- + 3SO_3^{2-} = Br^- + 3SO_4^{2-}$$

was found by F. S. Williamson and E. L. King (*J. Am. Chem. Soc.* **79**:5397 (1953)) to be given by $k[BrO_3^-][SO_3^{2-}][H^+]$. Give one of the thermodynamically possible rate laws for the backward reaction.

19.23 Suppose the transformation of A to B occurs by both a reversible first-order reaction and a reversible second-order reaction involving hydrogen ion:

$$A \underset{k_2}{\overset{k_1}{\rightleftharpoons}} B \qquad A + H^+ \underset{k_4}{\overset{k_3}{\rightleftharpoons}} B + H^+$$

What is the relationship between these four rate constants?

19.24 The hydrolysis of

$$(CH_2)_6 C \overset{\displaystyle Cl}{\underset{\displaystyle CH_3}{\diagup}}$$

in 80% ethanol follows the first-order rate equation. The values of the specific reaction-rate constants, as determined by H. C. Brown and M. Borkowski (*J. Am. Chem. Soc.* **74**:1896 (1952)), are as follows

$t/°C$	0	25	35	45
k/s^{-1}	1.06×10^{-5}	3.19×10^{-4}	9.86×10^{-4}	2.92×10^{-3}

(*a*) Plot log k against $1/T$; (*b*) calculate the activation energy; (*c*) calculate the pre-exponential factor.

19.25 If a first-order reaction has an activation energy of 104 600 J mol^{-1} and, a pre-exponential factor A of 5×10^{13} s^{-1}, at what temperature will the reaction have a half-life of (*a*) 1 min, and (*b*) 30 days?

19.26 Isopropenyl allyl ether in the vapor state isomerizes to allyl acetone according to a first-order rate equation. The following equation gives the influence of temperature on the rate constant (in s^{-1}):

$$k = 5.4 \times 10^{11} e^{-123\,000/RT}$$

where the activation energy is expressed in J mol^{-1}. At 150 °C, how long will it take to build up a partial pressure of 0.395 bar of allyl acetone, starting with 1 bar of isopropenyl allyl ether (L. Stein and G. W. Murphy, *J. Am. Chem. Soc.* **74**:1041 (1952))?

19.27 The pre-exponential factor for the trimolecular reaction

$$2NO + O_2 \rightarrow 2NO_2$$

is 10^9 cm^6 mol^{-2} s^{-1}. What is the value in L^2 mol^{-2} s^{-1} and cm^6 s^{-1}?

19.28 A reaction $A + B + C \rightarrow D$ follows the mechanism

$$A + B \rightleftharpoons AB \qquad AB + C \rightarrow D$$

in which the first step remains essentially in equilibrium. Show that the dependence of rate on temperature is given by

$$k = Ae^{-(E_a - \Delta H)/RT}$$

where ΔH is the enthalpy change for the first reaction.

19.29 For the mechanism

$$A + B \underset{k_2}{\overset{k_1}{\rightleftharpoons}} C \qquad C \xrightarrow{k_3} D$$

(*a*) Derive the rate law using the steady-state approximation to eliminate the concentration of C. (*b*) Assuming that $k_3 \ll k_2$, express the pre-exponential factor A and E_a for the apparent second-order rate constant in terms of A_1, A_2, and A_3 and E_{a1}, E_{a2}, and E_{a3} for the three steps.

19.30 For the two parallel reaction $A \xrightarrow{k_1} B$ and $A \xrightarrow{k_2} C$, show that the activation energy E' for the disappearance of A is given in terms of activation energies E_1 and E_2 for the two paths by

$$E' = \frac{k_1 E_1 + k_2 E_2}{k_1 + k_2}$$

19.31 Set up the rate expressions for the following mechanism:

$$A \underset{k_2}{\overset{k_1}{\rightleftharpoons}} B \qquad B + C \xrightarrow{k_3} D$$

If the concentration of B is small compared with the concentrations of A, C, and D, the steady-state approximation may be used to derive the rate law. Show that this reaction may follow the first-order equation at high pressures and the second-order equation at low pressures.

19.32 The reaction $NO_2Cl = NO_2 + \frac{1}{2}Cl_2$ is first order and appears to follow the mechanism

$$NO_2Cl \xrightarrow{k_1} NO_2 + Cl \qquad NO_2Cl + Cl \xrightarrow{k_2} NO_2 + Cl_2$$

(*a*) Assuming a steady state for the chlorine atom concentration, show that the empirical first-order rate constant can be identified with $2k_1$. (*b*) The following data were obtained by H. F. Cordes and H. S. Johnston (*J. Am. Chem. Soc.* **76**:4264 (1954)) at 180 °C. In a single experiment the reaction is first order, and the empirical rate constant is represented by k. Show that the reaction is second order at these low gas pressures and calculate the second-order rate constant.

$c/10^{-8}$ mol cm^{-3}	5	10	15	20
$k/10^{-4}$ s^{-1}	1.7	3.4	5.2	6.9

19.33 The reaction

$$2SO_2 + O_2 = 2SO_4$$

is catalyzed by the mechanism

$$2\,NO + O_2 \underset{k_{-1}}{\overset{k_1}{\rightleftharpoons}} 2NO_2$$

$$NO_2 + SO_2 \underset{k_{-2}}{\overset{k_2}{\rightleftharpoons}} NO + SO_3$$

To obtain the overall reaction from this mechanism, the second step has to be taken twice, and so the stoichiometric number s_2 of the second step is said to be 2. The equilibrium constant K_c for an overall reaction is related to the rate constants for the individual steps k_i and k_{-i} by

$$K_c = \prod_{i=1}^{S} \left(\frac{k_i}{k_{-i}} \right)^{s_i}$$

where s_i is the stoichiometric number of the ith step and S is the number of steps. Verify this relation for the above mechanism.

19.34 What is the rate constant for the following reaction at 500 K?

$$H + HCl \rightarrow Cl + H_2$$

The data required are to be found in Tables 19.3 and C.2.

19.35 For the gas reaction

$$O + O_2 + M \underset{k'}{\overset{k}{\rightleftharpoons}} O_3 + M$$

where $M = O_2$, Benson and Axworthy (*J. Chem. Phys.* **26**:1718 (1957)) obtained

$$k = (6.0 \times 10^7 \text{ L}^2 \text{ mol}^{-2} \text{ s}^{-1})e^{2.5/RT}$$

where the activation energy is in kJ mol^{-1}. Calculate the values of the parameters in the Arrhenius equation for the reverse reaction assuming $\Delta H°$ and $\Delta S°$ are independent of temperature.

19.36 For the mechanism

$$H_2 + X_2 \underset{k_{-1}}{\overset{k_1}{\rightleftharpoons}} 2HX \qquad X + H_2 \underset{k_{-2}}{\overset{k_2}{\rightleftharpoons}} HX + H$$

$$X_2 \underset{k_{-3}}{\overset{k_3}{\rightleftharpoons}} 2X \qquad H + X_2 \underset{k_{-4}}{\overset{k_4}{\rightleftharpoons}} HX + X$$

show that the steady-state rate law is

$$\frac{d[HX]}{dt} = 2k_1[H_2][X_2]\left(1 - \frac{[HX]^2}{K[H_2][X_2]}\right)$$

$$\times \left[1 + \frac{(k_3/k_1)\sqrt{2k_2/k_{-2}[X_2]}}{1 + k_{-3}[HX]/k_4[X_2]} \right]$$

19.37 The mechanism of the pyrolysis of acetaldehyde at 520 °C and 0.2 bar is

$$CH_3CHO \xrightarrow{k_1} CH_3 + CHO$$

$$CH_3 + CH_3CHO \xrightarrow{k_2} CH_4 + CH_3CO$$

$$CH_3CO \xrightarrow{k_3} CO + CH_3$$

$$CH_3 + CH_3 \xrightarrow{k_4} C_2H_6$$

What is the rate law for the reaction of acetaldehyde, using the usual assumptions? (As a simplification further reactions of the radical CHO have been omitted and its rate equation may be ignored.)

19.38 For the reaction

$$H_2(g) + Br_2(g) = 2HBr(g)$$

spectroscopic measurements show that $d[Br_2]/dt = -1.2 \times 10^{-3}$ mol L^{-1} s^{-1}. (a) What is the rate of reaction v? (b) What is the value of $d[HBr]/dt$? (c) What is the rate of conversion $d\xi/dt$ if the reaction occurs in a 3-L vessel? (d) What amount of HBr is produced per second in the 3-L vessel under these conditions?

19.39   Under certain conditions, it is found that ammonia is formed from its elements at a rate of 0.10 mol L^{-1} s^{-1}.

$$N_2(g) + 2H_2(g) = 2NH_3(b)$$

(a) What is the rate of reaction v? (b) What is the value of $d[N_2]/dt$? (c) What is the value of $d[H_2]/dt$?

19.40   The rate of the gas reaction $H_2 + Br_2 = 2HBr$ doubles when the concentration of hydrogen is doubled and it increases by a factor of 1.4 when the concentration of bromine is doubled. What is the order with respect to hydrogen, the order with respect to bromine, and the overall order?

19.41   The half-life of a first-order chemical reaction A $\rightarrow$ B is 10 min. What percentage of A remains after 1 h?

19.42   A reaction is carried out with 1-cyclohexenyl allyl malonitrile at 135.7 °C. Calculate the first-order rate constant from the data on the first 5 min and the second 5 min.

t/min	0	5	10
% reaction	19.8	34.2	46.7

19.43   The reaction

$$SO_2Cl_2 = SO_2 + Cl_2$$

is first order with a rate constant of 2.2×10^{-5} s^{-1} at 320 °C. What percentage of SO_2Cl_2 is decomposed after being heated at 320 °C for 2 h?

19.44   The kinetics of the hydrolysis of an ester is studied by titrating the acid produced. A sample is withdrawn and titrated with alkali. The volumes required at various times are

t/min	0	27	60	∞
V/mL	0	18.1	26.0	29.7

(a) Prove that this reaction is first order. (b) Calculate the half-life.

19.45   A gas reaction A = 2B is first order in A and goes to completion in a reaction vessel of constant volume and temperature with a half-life of 10 min. If the initial pressure of A is 1 bar, what are the partial pressures of A and B at 10 min, 20 min, and at equilibrium?

19.46   Modern carbon is radioactive because ^{14}C is produced by cosmic rays by the reaction $^{14}N(n, p)^{14}C$. This nuclide of carbon has a half-life of 5720 years. Carbon recently incorporated into growing plants has a specific activity of 16 disintegrations per minute per gram. (a) What percentage of the carbon in growing plants is ^{14}C? (b) How many grams of modern carbon does it take to provide 0.05 microcuries of ^{14}C? A curie is 3.7×10^{10} nuclear transformations per second.

19.47   Living trees incorporate ^{14}C ($t_{1/2} = 5720$ y) into their wood because there is ^{14}C in CO_2 due to cosmic rays and the nuclear reaction $^{14}N(n, p)^{14}C$. When a tree dies, this radioactivity of the wood slowly disappears. An archeological sample of wood has 42% of the ^{14}C found in living trees. Assuming the level of cosmic rays has been constant, what is the age of the archeological sample?

19.48   The reaction $2NO + O_2 \rightarrow 2NO_2$ is third order and $d[NO_2]/dt = k[NO]^2[O_2]$. The rate constant k has a value of 7.1×10^3 L^2 mol^{-2} s^{-1} at 25 °C. Air blown through a certain hot chamber and cooled quickly at 25 °C and 1 bar contains 1% by volume of nitric oxide, NO, and 20% of oxygen. How long will it take for 90% of this NO to be converted into nitrogen dioxide, NO_2 (or N_2O_4)?

19.49   The second-order rate constant for an alkaline hydrolysis of ethyl formate in 85% ethanol (aqueous) at 29.86 °C is 4.53 L mol^{-1} s^{-1} (H. M. Humphreys and L. P. Hammett, *J. Am. Chem. Soc.* **78**:521 (1956)). (a) If the reactants are both present at 0.001 mol L^{-1}, what will be the half-life of the reaction? (b) If the concentration of one of the reactants is doubled and of the other is cut in half, how long will it take for half the reactant present at the lower concentration to react?

19.50   The reaction

$$CH_3CH_2NO_2 + OH^- \rightarrow H_2O + CH_3CHNO_2^-$$

is of second order, and k at 0 °C is 39.1 L mol^{-1} min^{-1}. An aqueous solution is 0.004 molar in nitroethane and 0.005 molar in NaOH. How long will it take for 90% of the nitroethane to react?

19.51   The second-order rate constant for the reaction of ClO and NO is 6.2×10^{-12} cm^3 s^{-1}. What is its value in L mol^{-1} s^{-1}?

19.52   A solution of ethyl acetate and sodium hydroxide was prepared that contained (at $t = 0$) 5×10^{-3} mol L^{-1} ethyl acetate and 8×10^{-3} mol L^{-1} sodium hydroxide. After 400 s at 25 °C a 25-mL aliquot was found to neutralize 33.3 mL of 5×10^{-3} mol L^{-1} hydrochloric acid. (a) Calculate the rate constant for this second-order reaction. (b) At what time would you expect 20.0 mL of hydrochloric acid to be required?

19.53 Derive the integrated rate equation for a reaction of $\frac{1}{2}$ order. Derive the expression for the half-life of such a reaction.

19.54 The gas-phase formation of phosgene, $CO + Cl_2 \rightarrow COCl_2$, is $\frac{3}{2}$ order with respect to CO. Derive the integrated rate equation for a $\frac{3}{2}$-order reaction. Derive the expression for the half-life.

19.55 Equal molar quantities of A and B are added to a liter of a suitable solvent. At the end of 500 s half of A has reacted according to the reaction A + B = C. How much of A will be reacted at the end of 800 s if the reaction is (a) zero order with respect to both A and B; (b) first order with respect to A and zero order with respect to B; (c) first order with respect to both A and B?

19.56 The reaction between selenious acid and iodide ion in acid solution is

$$H_2SeO_3 + 6I^- + 4H^+ = Se(s) + 2I_3^- + 3H_2O$$

The initial reaction rates were measured at 0 °C at a variety of concentrations, as indicated in the following table, in moles per liter. These initial rates were evaluated from plots of H_2SeO_3 versus time. Determine the form of the rate law. (*Note:* This rate law holds only as long as insignificant quantities of I_3^- are present.)

	$[H_2SeO_3]$ 10^{-4} mol L^{-1}	$[H^+]$ 10^{-2} mol L^{-1}	$[I^-]$ 10^{-2} mol L^{-1}	Initial Rate 10^{-7} mol L^{-1} s^{-1}
1	2.40	2.06	3.0	14.6
2	7.20	2.06	3.0	44.6
3	0.712	2.06	3.0	4.05
4	0.712	2.06	9.0	102
5	0.712	2.06	3.0	4.05
6	0.712	12.5	3.0	173.0

19.57 When an optically active substance is isomerized, the optical rotation decreases from that of the original isomer to zero in a first-order manner. In a given case the half-time for this process is found to be 10 min. Calculate the rate constant for the conversion of one isomer to another.

19.58 The equations for [B] and [C] in Section 19.6 give an indeterminate result if $k_1 = k_2$. Rederive the equations, giving [B] and [C] as functions of time for the special case that

$$A \xrightarrow{k_1} B \xrightarrow{k_1} C$$

19.59 For the reaction 2A = B + C the rate law for the forward reaction is

$$-\frac{d[A]}{dt} = k[A]$$

Give two possible rate laws for the backward reaction.

19.60 The following rate constants were obtained by Wiig for the first-order decomposition of acetone dicarboxylic acid in aqueous solution:

$t/°C$	0	20	40	60
$k/10^{-5}$ s^{-1}	2.46	47.5	576	5480

(a) Calculate the energy of activation. (b) Calculate the pre-exponential factor A. (c) What is the half-life of this reaction at 80 °C?

19.61 Although the thermal decomposition of ethyl bromide is complex, the overall rate is first order and the rate constant is given by the expression $k = (3.8 \times 10^{14} \text{ s}^{-1}) e^{-229\ 000/RT}$, where the activation energy is in J mol^{-1}. Estimate the temperature at which (a) ethyl bromide decomposes at the rate of 1% per second, and (b) the decomposition is 70% complete in 1 h.

19.62 Given that the first-order rate constant for the overall decomposition of N_2O_5 is $k = (4.3 \times 10^{13} \text{ s}^{-1})e^{-103\ 000/RT}$, calculate (a) the half-life at -10 °C; and (b) the time required for 90% reaction at 50 °C. The activation energy is in J mol^{-1}.

19.63 Suppose that a substance X decomposes into A and B in parallel paths with rate constants given by

$$k_A = (10^{15} \text{ s}^{-1})e^{-126\ 000/RT} \qquad k_B = (10^{13} \text{ s}^{-1})e^{-83\ 700/RT}$$

where the activation energies are given in J mol^{-1}. (a) At what temperature will the two products be formed at the same rate? (b) At what temperature will A be formed 0.1 as fast as B? (c) State a generalization concerning the effect of temperature on the relative rates of reactions with different activation energies.

19.64 For the reaction

$$O + NO + M \rightarrow NO_2 + M$$

$k_{300 \text{ K}} = 6 \times 10^9 \text{ L}^2 \text{ mol}^{-2} \text{ s}^{-1}$ and $k_{1000 \text{ K}} = 3 \times 10^{10} \text{ L}^2 \text{ mol}^{-2} \text{ s}^{-1}$. Calculate the parameters in the Arrhenius equation.

19.65 (a) The viscosity of water changes about 2% per degree at room temperature. What is the activation energy for this process? (b) The activation energy for a reaction is 62.8 kJ mol^{-1}. Calculate k_{35}/k_{25}.

19.66 The reaction $2NO + O_2 \rightarrow 2NO_2$ is third order. Assuming that a small amount of NO_3 exists in rapid reversible equilibrium with NO and O_2 and that the rate-determining step is the slow bimolecular reaction $NO_3 + NO \rightarrow 2NO_2$, derive the rate equation for this mechanism.

19.67 The apparent activation energy for the recombination of iodine atoms in argon is -5.9 kJ mol^{-1}. This negative temperature coefficient may result from the following mechanism:

$$I + M = IM \qquad K = \frac{[IM]}{[I][M]} \qquad IM + I \underset{k_{-1}}{\overset{k_1}{\rightleftharpoons}} I_2 + M$$

Assuming that the first step remains at equilibrium, derive the rate equation that includes both the forward and backward reactions. Show that the backward reaction is bimolecular and the equilibrium constant expression for the dissociation of iodine is independent of the concentration of the third body.

19.68 Derive the steady-state rate equation for the following mechanism for a trimolecular reaction:

$$A + A \underset{k_{-1}}{\overset{k_1}{\rightleftharpoons}} A_2^* \qquad A_2^* + M \overset{k_2}{\longrightarrow} A_2 + M$$

19.69 Show that the interconversion of ortho- and para-hydrogen will be $\frac{3}{2}$ order, as obtained experimentally in the range 600 to 750 °C, if the rate-determining step is that between atoms and molecules of hydrogen.

$$H + \text{para-H}_2 \underset{k_2}{\overset{k_1}{\rightleftharpoons}} \text{ortho-H}_2 + H$$
$$\uparrow \quad \downarrow \quad \quad \uparrow \quad \uparrow$$

where the arrows represent the directions of the nuclear spins (cf. Section 16.1).

19.70 What are the Arrhenus parameters for the following elementary reaction?

$$H + HCl \rightarrow Cl + H_2$$

The data required are found in Tables 19.3 and C.2

19.71 The Arrhenius parameters for the reaction

$$Cl + H_2 \rightarrow HCl + H$$

are given in Table 19.3. What is the rate constant for the reverse reaction at 1000 K? The thermodynamic parameters for these substances are given in Table C.2.

19.72 Ozone is decomposed by the catalytic chain

$$NO + O_3 \overset{k_1}{\longrightarrow} NO_2 + O_2 \qquad NO_2 + O \overset{k_2}{\longrightarrow} NO + O_2$$

What is the steady-state rate law for the formation of O_2?

19.73 The formation of phosgene by the reaction

$$CO + Cl_2 = COCl_2$$

appears to follow the mechanism

$$Cl_2 \underset{k_{-1}}{\overset{k_1}{\rightleftharpoons}} 2Cl$$
$$Cl + CO \underset{k_{-2}}{\overset{k_2}{\rightleftharpoons}} COCl$$
$$COCl + Cl_2 \overset{k_3}{\longrightarrow} COCl_2 + Cl$$

Assuming that the intermediate Cl and COCl are in a steady state, what is the rate law for this reaction?

20
Chemical Dynamics and Photochemistry

The preceding chapter was concerned with macroscopic kinetics; this one is concerned with microscopic kinetics, that is, elementary reactions at the molecular level. The calculation of rate constants from properties of individual atoms and molecules is challenging because reactions occur as a result of collisions with a variety of energies, angles of approach, and states of reactants and products. The simplest collision theory of bimolecular reaction is based on consideration of collisions of rigid spherical molecules. To go further and take electronic structure into account it is necessary to use the concept of the potential energy surface for a reaction.

Transition-state theory attempts to simplify the problem by making a "dynamical bottleneck assumption." Transition-state theory is not an exact theory, but is based on a series of approximations. However, it has been useful over a 50-year period. Transition-state theory may also be formulated in terms of changes in thermodynamic properties in the activation process.

Since the absorption of light produces excited states of atoms and molecules, photochemistry is really the study of the chemistry of excited states. As pointed out in Section 14.1 electromagnetic radiation in the visible and ultraviolet is generally required to produce chemical reactions because changes in electronic energy levels are required. More recently it has been found that the absorption of many infrared photons from a high-intensity laser can also cause reaction.

20.1   Simple Collision Theory of Bimolecular Reactions

As we have seen in the preceding chapter, many reactions of atoms or radicals with small molecules in the gas phase have pre-exponential factors in the range $10^{10.5}$–$10^{11.5}$ L mol^{-1} s^{-1}. Since bimolecular reactions of small radicals have zero activation energies, their actual rate constants may be in this range. If the activation energy is zero, we might expect molecules to react on their first collision so that the bimolecular rate constant can be estimated from the collision density Z_{12} between molecules of type 1 and type 2, as calculated for rigid spheres. The collision *density* Z_{12}, given by equation 18.41, is the number of collisions between molecules of type 1 and type 2 per unit volume per unit time. If we divide by the Avogadro constant N_A, we obtain a reaction rate as it is usually defined:

$$\frac{-d[A_1]}{dt} = \frac{Z_{12}}{N_A} = \frac{\pi d_{12}^2 \langle v_{12} \rangle}{N_A} \rho_1 \rho_2$$

$$= N_A \pi d_{12}^2 \left(\frac{8k_B T}{\pi \mu} \right)^{1/2} [A_1][A_2] \tag{20.1}*$$

where ρ_1 and ρ_2 are number densities, d_{12} is the collision diameter $\frac{1}{2}(d_1 + d_2)$, $\langle v_{12} \rangle$ is the relative speed of molecule A_1 with respect to molecule A_2, and μ is the reduced mass $[(1/m_1) + (1/m_2)]^{-1}$. Thus, if a bimolecular reaction occurs on the first collision, the rate constant is given by

$$k = N_A \pi d_{12}^2 \left(\frac{8k_B T}{\pi \mu} \right)^{1/2} \tag{20.2}$$

When SI units are used, k has the units of m^3 mol^{-1} s^{-1} and has to be multiplied by 10^3 L m^{-3} to obtain the value in the usual units L mol^{-1} s^{-1}. If A_1 and A_2 are identical, a factor of 2 must be included on the right side of equation 20.2. Equation 20.2 is only approximate even for bimolecular reactions that occur on the first collision because molecules are not hard spheres that interact only when they touch.

Example 20.1

Calculate the bimolecular rate constant at 298 K for the reaction of two "average" small radicals with a reduced mass μ of 30×10^{-3} kg mol$^{-1}/N_A = 4.98 \times 10^{-26}$ kg and a collision diameter d_{12} of 500 pm.

Using equation 20.2, we find

$$k = (6.022 \times 10^{23} \text{ mol}^{-1})(3.14)(500 \times 10^{-2} \text{ m})^2 \left[\frac{8(1.38 \times 10^{-23} \text{ J K}^{-1})(298 \text{ K})}{(3.14)(4.98 \times 10^{-26} \text{ kg})} \right]^{1/2}$$

$$= 2.17 \times 10^8 \text{ m}^3 \text{ mol}^{-1} \text{ s}^{-1}$$

$$= (2.17 \times 10^8 \text{ m}^3 \text{ mol}^{-1} \text{ s}^{-1})(10^3 \text{ L m}^{-3})$$

$$= 2.17 \times 10^{11} \text{ L mol}^{-1} \text{ s}^{-1}$$

---

* Throughout this chapter we will write the Boltzmann constant as k_B to distinguish it from the rate constant k.

In general, reactions do not occur unless collisions are sufficiently energetic, but the concept of a collision diameter is a useful one. However, in continuing our treatment of bimolecular reactions using collision theory, we will use a quantity S called the **reaction cross section,** which is an area per molecule. To define the reaction cross section we can imagine shooting a beam of A_1 molecules at a volume of A_2 molecules at rest and measuring the number of reactive collisions that result. The number of reactive collisions is proportional to the probability P that a given A_1 molecule will be involved in a reactive collision times the cross-sectional area $\mathcal{A}$ of the beam of A_1 molecules. Since the reactive collisions are due to A_2 molecules *individually* we can also express the number of reactive collisions as the number N_2 of A_2 molecules times the reaction cross section S for the $A_1 + A_2$ reaction. Thus,

$$N_2 S = P\mathcal{A} \quad \text{or} \quad S = \frac{P\mathcal{A}}{N_2} \tag{20.3}$$

The reaction cross section depends on the energy of the molecules in the A_1 beam. More generally the reaction cross section depends on the energy of the collision, since the A_2 molecules are moving too. Rather than assuming that the energy of collision is the sum of the translational energies of spherical molecules A_1 and A_2 at the moment they collide, it is more reasonable to expect that S will be a function of the component of relative kinetic energy along the line of centers at the moment of collision. This is the energy with which the centers of the two molecules are pushed together. In simple collision theory it is assumed that this energy must exceed some value ϵ_0 for a collision to result in reaction. These assumptions lead* to the following relations for the reactive cross section as a function of energy:

$$S = 0 \qquad\qquad \epsilon < \epsilon_0 \tag{20.4}$$

$$S = \pi d_{12}^2 \left(1 - \frac{\epsilon_0}{\epsilon} \right) \qquad \epsilon > \epsilon_0 \tag{20.5}$$

where d_{12} is the collision diameter for rigid spherical molecules. If the collision energy ϵ along the line of centers is less than the minimum value ϵ_0 to cause reaction, the collision cross section is zero; but as the energy increases the reaction cross section S approaches the collision cross section πd_{12}^2 for rigid spherical molecules.

To derive the equation for the rate constant in terms of the reaction cross section, we return to the experiment in which a beam of A_1 molecules is passed through a layer of A_2 molecules at rest. The decrease in intensity I_{A_1} of the A_1 beam due to reactions is given by

$$dI_{A_1} = S I_{A_1} [A_2]\, dx \tag{20.6}$$

where dx is the distance through the layer of A_2 molecules measured in the x direction of the A_1 beam. This equation is analogous to equation 15.5 used to derive the Beer–Lambert law, except that here we must remember that S is a

---

* For a much more complete discussion of bimolecular reactions, see W. C. Gardiner, Jr., *Rates and Mechanisms of Chemical Reactions*. New York: Benjamin, 1969, or I. W. M. Smith, *J. Chem. Educ.* **59**:9 (1982).

function of the velocity v of the A_1 molecules. Since $I_{A_1} = v[A_1]$ and $v = dx/dt$, equation 20.6 can be written

$$-\frac{d[A_1]}{dt} = vS[A_1][A_2] \tag{20.7}$$

Thus, we see that vS is the rate constant for this idealized process; this rate constant has the SI units $m^3 \, s^{-1}$. If gases A_1 and A_2 are reacting in a gas mixture, v in this equation may be interpreted as the magnitude of the *relative* velocity $v = v_1 - v_2$. However, to obtain the rate constant for a mixture of gases we have to integrate vS over the Maxwell distribution of relative speeds $f(v)$†:

$$k = \int_{(2\epsilon_0/\mu)^{1/2}}^{\infty} vSf(v) \, dv \tag{20.8}$$

The lower limit comes from the fact that $\epsilon_0 = \frac{1}{2}\mu v^2$. Since it is more convenient to think of the reaction cross section as a function of energy (equation 20.4 and 20.5), equation 20.8 can be written

$$k = \int_{\epsilon_0}^{\infty} \left(\frac{2\epsilon}{\mu}\right)^{1/2} S(\epsilon)f(\epsilon) \, d\epsilon \tag{20.9}$$

If equation 20.5 is used for $S(\epsilon)$, integration leads to

$$k = N_A \pi d_{12}^2 \left(\frac{8k_B T}{\pi\mu}\right)^{1/2} e^{-\epsilon_0/k_B T} \tag{20.10}$$

where N_A has been inserted to convert to molar units. The pre-exponential factor in this equation was obtained in equation 20.2 in a much simpler way. The exponential factor gives the fraction of the collisions in which the energy along the line of centers exceeds ϵ_0.

The pre-exponential factor in equation 20.10 is proportional to $T^{1/2}$, rather than being independent of temperature as in the Arrhenius equation. However, it is not ordinarily possible to distinguish between $AT^{1/2}e^{-E_a/RT}$ and $Ae^{-E_a/RT}$ in chemical kinetic data. The exponential term tends to dominate the temperature dependence, and experimental errors in the determination of k limit our ability to determine the functional dependence on T.

According to collision theory, a collision that is sufficiently energetic to supply the activation energy may still fail to produce a reaction if the colliding molecules are not oriented in such a way that they can react with each other. This can be corrected by introducing a steric factor into equation 20.10. However, this is really not very useful because there is no simple theory for calculating its value. Since there are also very drastic assumptions underlying equation 20.10, we cannot expect it will yield very accurate results. For example, it ignores the changes in electronic structure that occur in any chemical reaction, and how these changes influence the reactive cross section. To correct

---

† The Maxwell distribution of speeds in a one-component gas is given by equation 18.16, but it may be shown (J. W. Moore and R. G. Pearson, *Kinetics and Mechanism*. New York: Wiley, 1981) that if the mass m is changed to the reduced mass μ of A_1 and A_2 and v is interpreted as the magnitude of the relative speeds, the same equation gives the probability density function for the relative speeds of A_1 and A_2 molecules.

this we must look at chemical reactions from the viewpoint of quantum mechanics.

20.2 Potential Energy Surfaces

Kinetic theory is helpful in that it tells us about molecular collisions, but it does not deal with the changes that take place on a molecular level when reactants are converted to products. When two molecules are very close to each other, they cannot be considered separately because their wavefunctions overlap. Thus, from the time the reactant molecules are close to each other until the products are well separated, the system is a kind of **supermolecule.** This supermolecule is different from an ordinary molecule because it is in the process of change, but it is a molecule in the sense that its energy and electron distribution can be calculated for each nuclear configuration by use of quantum mechanics. According to the Born–Oppenheimer approximation (Section 12.1), the electrons move much more rapidly than the nuclei, and so the molecular electronic energy and wavefunction can be calculated for a given nuclear configuration by use of the electronic Schrödinger equation. This approximation was used earlier to calculate the electronic potential energy function so that the Schrödinger equation could be used to obtain molecular vibrational energy levels.

If a reaction involves N nuclei, there are $3N$ nuclear coordinates, but the group of nuclei has three translational coordinates of the center of mass and two or three rotational coordinates (about the center of mass) that do not affect the potential energy. Thus, the potential energy is a function of $3N - 5$ nuclear coordinates if the nuclei are constrained to a straight line and $3N - 6$ nuclear coordinates in general. For the simplest type of reaction

$$A + BC \rightarrow AB + C \qquad (20.11)$$

where A, B, and C are atoms, three coordinates are required. It is not possible to plot the potential energy as a function of three coordinates, but if the angle θ of approach of A to BC is fixed, the potential energy of the system can be plotted as a function of R_{AB} and R_{BC}, where R is intermolecular distance. Such a plot is shown in Fig. 20.1. If R_{AB} is rather large, as on the left face of this diagram, the potential energy is essentially that of the BC molecule, as in Fig. 12.3. Similarly, the right face gives the potential energy of AB. Thus, Fig. 20.1 omits two monotonous valleys to the left and right that extend indefinite distances. Initially, the R_{AB} distance is very large and as A approaches BC, the lowest energy path is given by the dashed line from reactants R to products P. This dashed line gives the minimum energy path, which is sometimes referred to as the reaction coordinate. We will soon see that the configuration of the system does not actually move along the reaction coordinate in the reaction, but the reaction coordinate does help us visualize the surface. The highest point along the reaction coordinate is a saddle point. At the saddle point, the potential energy is a maximum along the reaction coordinate, but it is a minimum in the direction perpendicular to the reaction coordinate. The reaction system at this point is said to be in the **transition state.** In Fig. 20.1, D is a high plateau giving the potential energy of three atoms well separated from each other.

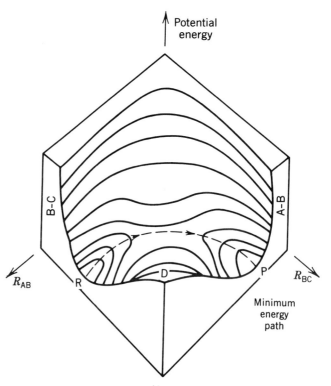

Figure 20.1   Potential energy surface for the reaction A + BC → AB + C with the nuclei constrained to a line.

As a first simple example, consider what happens when A approaches a nonvibrating BC molecule along the internuclear axis. The point representing the configuration of the system moves along the minimum energy path, the dashed line in Fig. 20.1. As R_{AB} decreases, kinetic energy is converted to potential energy as the point representing the system of three nuclei moves up the valley from the left. If there is initially enough kinetic energy for the system to go over the saddle point, AB and C are formed and gain energy as the system goes down the valley to the right. If the kinetic energy is too low, the system returns down the valley to the left, and we would say that the reactants bounced off each other.

Figure 20.1 applies only when the nuclei are constrained to a line, and the potential energy surface will be different if there is a different angle θ of approach. The quantum mechanical calculation of an accurate potential surface for a reaction such as 20.11 is a difficult process, and so surfaces have been calculated only for a few reactions.

A great deal of attention has been focused on the reaction of a hydrogen atom with a hydrogen molecule:

$$H_A + H_B H_C \rightarrow H_A H_B + H_C \tag{20.12}$$

The potential energy surface for this reaction for $\theta = 180°$ is described by means of a contour diagram in Fig. 20.2. This surface has been calculated using ab initio methods with configuration interaction (as discussed in Section 12.4),

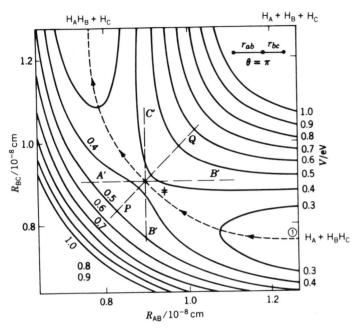

Figure 20.2 Potential energy surface for the reaction $H_A + H_B H_C \rightarrow H_A H_B +$ H_C for a linear approach and departure. (Based on R. N. Porter and M. Karplus, *J. Chem. Phys.* **40**:1105 (1964).)

and the error at any point on the surface is believed to be less than 0.03 eV (2.9 kJ mol^{-1}). As H_A approaches $H_B H_C$, along the minimum energy path, the potential energy of the system increases until the saddle point is reached at ‡. At this point $R_{AB} = R_{BC} = 93$ pm, and the potential energy of the system is 0.43 eV (42 kJ mol^{-1}), the highest along the dashed line. Since the saddle point is 0.43 eV higher than the potential of H_A and H_{BC} at an infinite distance, this energy must be supplied from relative kinetic energy or vibrational energy in order for the reaction to occur. In the upper right corner of Fig. 20.2, there is a high plateau with energy of 432 kJ mol^{-1}. This is the energy of three hydrogen atoms infinitely far apart, with respect to separated reactants or products.

20.3 Theoretical Calculation of a Rate Constant

Once the potential energy curve has been obtained for various approach angles θ, the probability of a reaction for certain initial conditions (relative kinetic energy, vibrational energy, and θ) can, in principle, be calculated using the time-dependent Schrödinger equation. However, this is a very difficult calculation, and so classical mechanics is ordinarily used. **The force on a particular nucleus is given by the gradient in the potential energy.** For example, the component of the force on nucleus i in the x direction is given by (see Section 10.11)

$$F_{x,i} = - \frac{\partial V}{\partial x_i} \qquad (20.13)$$

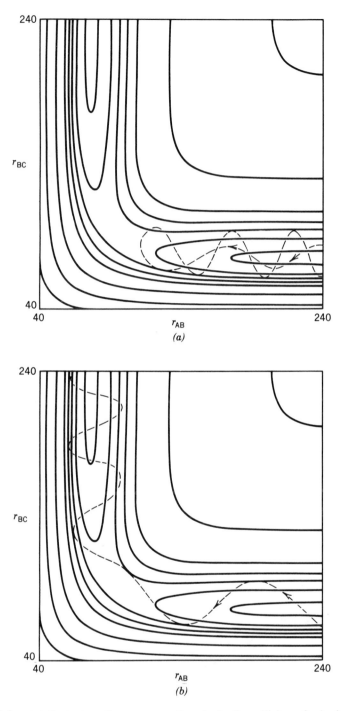

Figure 20.3 (*a*) Trajectory for a nonreactive, inelastic collision of a hydrogen atom and a hydrogen molecule. (*b*) Trajectory for a reactive collision. (From P. Siegbahn and B. Liu, *J. Chem. Phys.* **68**:2457 (1978) and C. J. Horowitz, *J. Chem. Phys.* **68**:2466 (1978).)

where V is the potential energy and x_i is the x coordinate of the nucleus. At each instant the system is represented by a point on the surface, and **Newton's**

law $F = ma$ is integrated numerically to obtain the coordinates of the system as a function of time. Calculations can also be made when $H_B H_C$ initially has vibrational motion, and as an approximation this is also treated classically. Rotational energy is not important in these calculations. These trajectory calculations yield a reaction probability of 0 for certain initial conditions and of 1 for others. Figure 20.3 shows the results of two calculations of this type for colinear collisions. In Fig. 20.3a the $H_B H_C$ molecule is vibrating and H_A approaches with a certain initial velocity, but reaction does not occur. Note that in this nonreactive, inelastic collision, translational energy is converted into vibrational energy in $H_B H_C$. In Fig. 20.3b reaction does occur.

To calculate a rate constant in this way, it is necessary to make a very large number of trajectory calculations with initial states chosen to give a statistically representative sample of possible initial states at the chosen temperature. The initial conditions can be chosen by a Monte Carlo procedure to ensure that the distribution of each initial parameter approaches the correct distribution as the number of calculated trajectories increases. The relative kinetic energies of H_A and H_{BC} are given by the Boltzmann equation. All angles θ of approach have to be included, but most reactive collisions for reaction 20.12 occur at angles near 180°.

To see how a rate constant can be calculated from a series of trajectory calculations, we need to consider the reaction probability $P(b)$, often referred to as the opacity function, and the way it is used to calculate a reaction cross section $S(v)$. The **reaction probability** $P(b)$ is simply the fraction of the total number of trajectories at a selected reactant relative velocity and impact parameter b (Section 18.9) that result in reaction. The reaction probability $P(b)$ for the H + H_2 reaction for a relative velocity of 1.17×10^6 cm s^{-1} is shown in Fig. 20.4 as a function of the impact parameter b. The reaction probability is greatest for an impact parameter of zero, and it decreases to zero at some finite impact parameter.

The contribution to the reaction cross section $S(v)$ of collisions with an impact parameter b is $2\pi b P(b) \, db$, and so the cross section is given by

$$S(v) = \int_0^{b_{max}} 2\pi b P(b) \, db \qquad (20.14)$$

When a chemical reaction occurs in bulk, molecules collide at all possible relative velocities, and so the rate constant $k(T)$ at temperature T is made up of a sum of terms for all possible relative velocities, with each weighted by the fraction f_i of collisions with that relative velocity:

$$k(T) = f_1 k(v_1) + f_2 k(v_2) + \cdots$$
$$= f_1 S(v_1) v_1 + f_2 S(v_2) v_2 + \cdots \qquad (20.15)$$

or

$$k(T) = \int_0^{\infty} f(v,T) v S(v) \, dv \qquad (20.16)$$

where $f(v,T)$ is the Maxwell–Boltzmann distribution for relative velocity v at temperature T. We have already seen a version of this equation in equation 20.8. In 1965 Karplus, Porter, and Sharma* calculated a very large number of

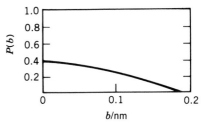

Figure 20.4 Reaction probability $P(b)$ for the reaction H + H_2 for a relative velocity of 1.17×10^6 cm s^{-1} as a function of impact parameter b. (From J. Nicholas, *Chemical Kinetics*. New York: Wiley, 1976. Copyright © 1978 J. Nicholas. Reprinted with permission.)

---

* M. Karplus, R. N. Porter, and R. D. Sharma, *J. Chem. Phys.* **43**:3259 (1965).

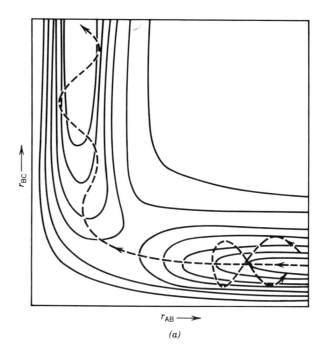

(a)

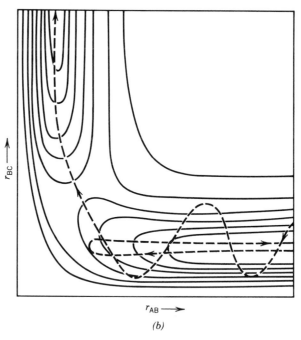

(b)

Figure 20.5   (*a*) Attractive potential energy surface for reaction 20.11.
(*b*) Repulsive potential energy surface for reaction 20.11

trajectories for the H + H₂ reaction and found that their results could be
expressed by the following Arrhenius equation,

$$k = (4.3 \times 10^{13} \text{ mol}^{-1} \text{ cm}^3 \text{ s}^{-1}) \exp\left[-\frac{31\,000 \text{ J mol}^{-1}}{(8.314 \text{ J K}^{-1} \text{ mol}^{-1})T}\right] \quad (20.17)$$

This result is in pretty good agreement with experimental results obtained by studying the reaction

$$D + H_2 \rightarrow DH + H \qquad (20.18)$$

or

$$H + para\text{-}H_2 \rightarrow ortho\text{-}H_2 + H \qquad (20.19)$$

and with the value calculated using transition-state theory, which we will discuss in the next section. The transition-state rate constant over the same range of temperature is given by

$$k = (7.4 \times 10^{13} \text{ mol}^{-1} \text{ cm}^3 \text{ s}^{-1}) \exp \left[- \frac{34\ 440 \text{ J mol}^{-1}}{(8.314 \text{ J K}^{-1} \text{ mol}^{-1})T} \right] \qquad (20.20)$$

Classical calculations have been made on a number of simple reactions, and the general conclusion is that classical calculations can provide an adequate description of the collision dynamics for some purposes. There are quantum mechanical effects in reaction kinetics, and penetration into classically forbidden regions ("tunneling") may be important, especially at lower temperatures.

The biggest difficulty in the quantum mechanical calculation of a reaction rate is calculating the potential energy surface with sufficient accuracy. The shape of this surface is extremely important in determining the role of vibrational energy of a reactant in affecting the likelihood of reaction.

The potential energy diagram is symmetrical for the reaction $H_A + H_B H_C = H_A H_B + H_C$, but this is not true in general. The shape of the potential energy surface determines whether translational motion or vibrational motion will be most effective in causing reaction. This behavior can be understood by consideration of the so-called attractive and repulsive potential energy surfaces shown in Fig. 20.5a and b. The saddle point occurs early in an attractive surface, as illustrated in Fig. 20.5a. If BC has a lot of vibrational energy, the reactants may separate without reaction as shown in Fig. 20.5a. On the other hand, if the system has the same amount of energy, but this energy is mainly relative translational energy, the point representing the system has a higher probability of crossing the barrier. This produces an AB molecule with a significant amount of vibrational energy. Thus, an attractive surface favors reaction when the energy is mainly relative translational motion, and the product molecule is vibrationally excited.

When the potential energy surface is of the repulsive type, collisions with relatively high translational energy and little vibrational energy in the original molecule may not be successful. As shown in Fig. 20.5b, the path of the system is reflected back to products. However, if the original molecule has vibrational energy and the relative kinetic energy is no larger than in the first case, reaction may occur to produce a product with very little vibrational energy. The same surface has been used in discussing both attractive and repulsive situations because a reaction that is attractive in one direction is repulsive in the other.

20.4 Transition-State Theory

Transition-state theory was developed before much was known about potential energy surfaces, and it, in effect, bypasses the problem of the dynamics of a reactive collision. Nevertheless, it is very useful because quantitative calcu-

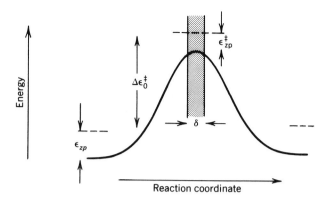

Figure 20.6   Potential energy diagram showing the relationship between the height of the barrier in the potential energy surface and the activation energy $\Delta\epsilon_0^{\ddagger}$. (Reproduced from I. W. M. Smith, *Kinetics and Dynamics of Elementary Gas Reactions*, 1980, Fig. 4.2, p. 116, by permission of the publishers, Butterworth & Co (Publishers) Ltd. ©.)

lations can be made with estimated properties of the transition state for a reaction. The development of transition-state theory goes back to Eyring and Evans and Polanyi* in 1935. The basis of transition-state theory is that it is possible to define a surface in coordinate space and to calculate the flux of trajectories that pass through this surface from the reactant side to the product side *without turning back*. This flux is identified as the reactive flux. If the potential energy barrier is high, there is no difficulty in locating this surface; it is placed at the top of the energy barrier perpendicular to the reaction coordinate. In addition, transition-state theory is based on the Born–Oppenheimer approximation and the assumption that molecules are distributed among their states according to the Boltzmann distribution.

To derive an expression for the rate constant, we will focus our attention on what happens at the top of the potential energy barrier as shown in Fig. 20.6, for the reaction A + BC = AB + C. The dashed lines show the quantum mechanical zero-point energies for reactants, products, and the **transition state** ABC‡. The transition-state species are those systems within a length δ along the reaction coordinate, centered at the top of the potential energy barrier. At chemical equilibrium half of the ABC‡ complexes are moving forward and half are moving backward. It is important to remember that ABC‡ is not an intermediate compound in the reaction, but is a structure that is in the process of falling apart in the direction of the products or in the direction of the reactants. At chemical equilibrium the rate of the reaction in the forward direction can be written either in terms of the macroscopic rate expression for the forward reaction $k_f[A][BC]$ or in terms of the rate that molecules in the transition state are passing over the potential energy barrier to products, $[ABC^{\ddagger}]\langle\vec{v}_x\rangle/2\delta$:

$$k_f[A][BC] = \frac{[ABC^{\ddagger}]\langle\vec{v}_x\rangle}{2\delta} \tag{20.21a}$$

---

* H. Eyring, *J. Chem. Phys.* **3**:107 (1935); M. G. Evans and M. Polanyi, *Trans. Faraday Soc.* **31**:875 (1935).

In this expression $\langle \vec{v}_x \rangle$ is the mean velocity along the reaction coordinate averaged over the 50% of the transition-state species moving in the direction of products. The ratio $\langle \vec{v}_x \rangle / \delta$ has the units of reciprocal time.

Similarly, the rate that species in the transition state are passing over the potential energy barrier from products to reactants can be written in terms of macroscopic kinetics or transition-state theory:

$$k_b[AB][C] = \frac{[ABC^{\ddagger}]\langle \overleftarrow{v}_x \rangle}{2\delta} \qquad (20.21b)$$

It is important to remember that transition-state theory applies to the back reaction as well, and that $k_f/k_b = [AB][C]/[A][BC]$, as discussed in Section 19.11.

We will now concentrate on the calculation of the rate constant for the forward reaction. Using equation 20.21a, the rate constant for the forward reaction can be calculated from

$$k_f = \left(\frac{\langle \vec{v}_x \rangle}{2\delta} \right) \left(\frac{[ABC^{\ddagger}]}{[A][BC]} \right) \qquad (20.22)$$

Since $[ABC^{\ddagger}]/[A][BC]$ looks like an equilibrium constant we will represent it by $K^{\ddagger}$ although $ABC^{\ddagger}$ is not a normal molecular species:

$$k_f = \left(\frac{\langle \vec{v}_x \rangle}{2\delta} \right) K^{\ddagger} \qquad (20.23)$$

The equilibrium constant $K^{\ddagger}$ for the formation of the transition state can be written in terms of the molecular partition functions q_i for the reactants and q^{ts} for the transition state. Following equation 17.107, $K^{\ddagger}$ can be calculated from properties of the reactant molecules and of the transition state using

$$K^{\ddagger} = \frac{(q^{ts}/V)}{\prod_i (q_i/V)} \exp(-\Delta \epsilon_0^{\ddagger}/k_B T) \qquad (20.24)$$

The volume must be included in equation 20.24 because of the use of concentrations in rate laws.

The last assumption of transition-state theory is that in the transition state, motion along the reaction coordinate is separable. In other words, we will assume that the total partition function q^{ts} for the transition state can be written as the product of the partition function q_x for the motion along the reaction coordinate and the partition function $q^{\ddagger}$ for all of the other degrees of freedom of the transition state:

$$q^{ts} = q_x q^{\ddagger} \qquad (20.25)$$

The partition function q_x for one-dimensional motion in the restricted length δ is

$$q_x = \frac{(2\pi \mu_x k_B T)^{1/2} \delta}{h} \qquad (20.26)$$

where μ_x is the appropriate reduced mass. The average velocity across the top of the barrier in the direction of products is a weighted average for all the velocities of the systems crossing the barrier to products:

$$\langle \vec{v}_x \rangle = \frac{\int_0^\infty v_x \exp(-p_x^2/2\mu_x k_B T)\, dp_x}{\int_0^\infty \exp(-p_x^2/2\mu_x k_B T)\, dp_x}$$

$$= \left(\frac{2k_B T}{\pi \mu_x}\right)^{1/2} \tag{20.27}$$

Thus, the factor $(\langle \vec{v}_x \rangle / 2\delta)q_x$ in equation 20.22 reduces to $k_B T/h$ so that we do not need actual values for δ and μ. The resulting equation for transition-state theory is

$$k = \left(\frac{k_B T}{h}\right)\left[\frac{q^\ddagger/V}{\prod_i (q_i/V)}\right] \exp\left(\frac{-\Delta\epsilon_0^\ddagger}{k_B T}\right) \tag{20.28}$$

For a bimolecular reaction this yields a rate constant with units of $m^3\ s^{-1}$.

The partition functions for reactants and for the transition state are separable into contributions due to translational motion and to internal motions:

$$q = q_t q_{int} \tag{20.29}$$

For a unimolecular reaction $q_t^\ddagger/V$ cancels $\prod_i (q_{i,t}/V)$ so that the transition-state expression for the rate constant is

$$k = \left(\frac{k_B T}{h}\right)\left(\frac{q_{int}^\ddagger}{q_{int}}\right) \exp\left(-\frac{\Delta\epsilon_0^\ddagger}{k_B T}\right) \tag{20.30}$$

This yields a rate constant in s^{-1}, as it must. This is the rate constant in the high pressure limit (see Section 19.12), where there are enough collisions to maintain the Boltzmann distribution of A*. However, in the falloff region the rate of formation of vibrationally excited A* molecules is too slow to maintain the Boltzmann distribution. To calculate rate constants in the falloff region it is necessary to calculate the rate constant for the activation process and its reverse. Since these are physical processes they do not depend very strongly on the nature of M.

The classical theory for unimolecular reaction was developed by Rice, Ramsberger, and Kassel. This theory has been improved in several aspects by Marcus, and it is now referred to as the RRKM theory. Since phenomena in the falloff region are quite complicated, we will not be able to discuss them here.

For a bimolecular reaction

$$\frac{q_t^\ddagger/V}{\prod_i (q_{i,t}/V)} = \frac{1}{q_t/V} \tag{20.31}$$

where $q_t/V = (2\pi\mu k_B T)^{3/2}/h^3$ is the partition function per unit volume for the relative motion of the reactants, for which the reduced mass is μ. Thus, the transition-state expression for the rate constant of a bimolecular reaction is

$$k = \left(\frac{k_B T}{h}\right)\left[\frac{q_{int}^\ddagger}{(q_t/V)\prod_i q_{i,int}}\right] \exp\left(-\frac{\Delta\epsilon_0^\ddagger}{k_B T}\right) \tag{20.32}$$

The internal partition functions for the reactants are calculated in the way described in Chapter 17. The calculation of the internal partition function $q_{int}^\ddagger$ for the transition state is a problem because in general we do not know the structure of the transition state. However, various hypotheses about the structure of the transition state may be made, or it can be calculated by ab initio methods.

Table 20.1   Pre-exponential Factors for Some Bimolecular Reactions Calculated by TST Compared with Experimental Values[a]

Reaction	$\dfrac{A_{exp}}{cm^3\ mol^{-1}s^{-1}}$	$\dfrac{A_{calc}}{cm^3\ mol^{-1}s^{-1}}$
$H + H_2 \rightarrow H_2 + H$	5.4×10^{13}	3.5×10^{13}
$Br + H_2 \rightarrow HBr + H$	3×10^{13}	1×10^{14}
$H + CH_4 \rightarrow H_2 + CH_3$	1×10^{13}	2×10^{13}
$H + C_2H_6 \rightarrow H_2 + C_2H_5$	3×10^{12}	1×10^{13}
$CH_3 + H_2 \rightarrow CH_4 + H$	2×10^{12}	1×10^{12}
$CH_3 + CH_3COCH_3 \rightarrow H_4 + CH_3COCH_2$	4×10^{11}	1×10^{11}
$CD_3 + CH_4 \rightarrow CD_3H + CH_3$	1×10^{11}	2×10^{11}
$2ClO \rightarrow Cl_2 + O_2$	6×10^{10}	1×10^{11}

[a] From J. Nicholas, *Chemical Kinetics*. New York: Wiley, 1976.

The pre-exponential factors for a number of bimolecular reactions are given in Table 20.1. The pre-exponential factors may be compared with 10^{13}–10^{15} cm^3 mol^{-1} s^{-1} calculated from simple collision theory (Section 20.1). The last column gives pre-exponential factors calculated from transition-state theory using *estimated* frequencies and geometrical parameters for the activated complex. The activated-complex theory is a considerable advance over simple collision theory.

Example 20.2

Using transition-state theory, calculate the pre-exponential factor for the rate constant for the reaction

$$H + H_2 \rightarrow H_2 + H$$

at 500 K. Assume a linear activated complex with the nuclei each separated by 0.94×10^{-10} m. The vibrational partition functions may be ignored because their values are so close to unity. (The experimental value is 5.4×10^{13} mol^{-1} cm^3 s^{-1}.)

From equation 20.32 the pre-exponential factor is given by

$$A = \frac{k_B T}{h} \frac{q_{int}^{\ddagger}}{(q_t/V) q_{H,int} q_{H_2 int}}$$

The internal partition function $q_{int}^{\ddagger}$ for the transition state is the rotational partition function for a symmetrical and linear arrangement of three hydrogen atoms. Since the central hydrogen atom is on the axis of rotation it does not contribute to the moment of inertia:

$$I = \mu R^2 = \frac{m_1 m_2}{m_1 + m_2} R^2 = \frac{(1.0078 \times 10^{-3}\ kg\ mol^{-1})^2 (1.88 \times 10^{-10}\ m)^2}{(2.0156 \times 10^{-3}\ kg\ mol^{-1})(6.022 \times 10^{23}\ mol^{-1})}$$

$$= 2.96 \times 10^{-47}\ kg\ m^2$$

The internal partition function for the transition state is therefore given by

$$q_{int}^{\ddagger} = \frac{8\pi^2 IkT}{2\ h^2} = \frac{8\pi^2 (2.96 \times 10^{-47}\ kg\ m^2)(1.381 \times 10^{-23}\ J\ K^{-1})(500\ K)}{2(6.63 \times 10^{-34}\ J\ s)^2} = 18.4$$

The transitional partition function for the relative motion of the reactants is given by

$$\frac{q_t}{V} = \frac{(2\pi\mu k_B T)^{3/2}}{h^3}$$

$$= \frac{[2\pi(1.116 \times 10^{-27} \text{ kg})(1.381 \times 10^{-23} \text{ J K}^{-1})(500 \text{ K})]^{3/2}}{(6.626 \times 10^{-34} \text{ J s})^3}$$

$$= 1.157 \times 10^{30} \text{ m}^{-3}$$

where μ is the reduced mass of a hydrogen atom and a hydrogen molecule.

The rotational partition function for H_2 at 3000 K was calculated in Example 17.9 to be 17.13. At 500 K the rotational partition function is

$$q_{rH_2} = 17.13 \frac{500}{3000} = 2.86$$

Thus, the pre-exponential factor is

$$A = \frac{(1.38 \times 10^{-23} \text{ J K}^{-1})(500 \text{ K})(18.4)}{(6.626 \times 10^{-34} \text{ J s})(1.157 \times 10^{30} \text{ m}^{-3})(2.86)}$$

$$= 5.79 \times 10^{-17} \text{ m}^3 \text{ s}^{-1}$$

To compare this result with values in Table 20.1 we need to multiply by Avogadro's number and convert from meters to centimeters:

$$(5.79 \times 10^{-17} \text{ m}^3 \text{ s}^{-1})(6.022 \times 10^{23} \text{ mol}^{-1})(10^2 \text{ cm m}^{-1})^3$$

$$= 3.49 \times 10^{13} \text{ cm}^3 \text{ mol}^{-1} \text{ s}^{-1}$$

20.5   Thermodynamic Formulation of Transition-State Theory

The quantity $[(q^{\ddagger}/V)\prod_i(q_i/V)] \exp(-\Delta\epsilon_0^{\ddagger}/k_B T)$ in equation 20.28 has the same general form as the statistical mechanical expression for an equilibrium constant, and this equation can be written as

$$k = \frac{k_B T}{h} K^{\ddagger}(c^{\circ})^{1-m} \tag{20.33}$$

where c° is the standard-state concentration of 1 mol L^{-1} and m is the order of the reaction. The factor $(c^{\circ})^{1-m}$ has to be included because an equilibrium constant is dimensionless, but the units for a rate constant depend on the order of the reaction. The equilibrium constant $K^{\ddagger}$ may be expressed in terms of an entropy of activation $\Delta S^{\ddagger}$ and an enthalpy of activation $\Delta H^{\ddagger}$ by use of the following familiar relation from thermodynamics:

$$K^{\ddagger} = e^{\Delta S^{\ddagger}/R} e^{-\Delta H^{\ddagger}/RT} \tag{20.34}$$

The quantities $\Delta S^{\ddagger}$ and $\Delta H^{\ddagger}$ are for standard states (in this case unit concentrations) of the reactants and the activated complex, but we will omit the superscript zeros in order to simplify the notation. Substituting equation 20.34 into equation 20.33 yields

$$k = \frac{RT}{N_A h} (c^\circ)^{1-m} e^{\Delta S^\ddagger/R} e^{-\Delta H^\ddagger/RT} \tag{20.35}$$

This equation has approximately the form of the Arrhenius equation. To write equation 20.35 in the Arrhenius form we need to find the relationship between the Arrhenius activation energy E_a and $\Delta H^\ddagger$. The Arrhenius activation energy E_a is defined by

$$E_a = RT^2 \frac{d \ln k}{dT} \tag{20.36}$$

Substituting equation 20.33 and using equation 19.98 yields

$$E_a = RT + RT^2 \frac{d \ln K^\ddagger}{dT} = RT + \Delta U^\ddagger \tag{20.37}$$

Since $\Delta U^\ddagger = \Delta H^\ddagger - P \Delta V^\ddagger$ at constant pressure,

$$E_a = \Delta H^\ddagger + RT - P \Delta V^\ddagger \tag{20.38}$$

For a reaction with ideal gases, $P \Delta V^\ddagger = (1 - m)RT$, where m is the order of the reaction. Thus,

$$E_a = \Delta H^\ddagger + mRT \tag{20.39}$$

Substituting this relation into 20.35 yields

$$k = e^m (c^\circ)^{1-m} \frac{RT}{N_A h} e^{\Delta S^\ddagger/R} e^{-E_a/RT} \tag{20.40}$$

so that the pre-exponential factor A in the Arrhenius equation is given by

$$A = e^m (c^\circ)^{1-m} \frac{RT}{N_A h} e^{\Delta S^\ddagger/R} \tag{20.41}$$

Thus, the standard entropy of activation may be calculated from the pre-exponential factor in the Arrhenius equation. The value of $\Delta S^\ddagger$ depends on the choice of standard state for bimolecular or trimolecular reactions, but not for unimolecular reactions.*

For many unimolecular, bond-breaking gas reactions $\Delta S^\ddagger \cong 0$ because the activated complex is much like the original reactants, and there is very little change in configuration in going from the reactants to the activated complex. In this case $e^{\Delta S^\ddagger/R} \cong 1$ and, at room temperatures,

$$A = \frac{eRT}{N_A h} = \frac{(2.718)(8.31 \text{ J K}^{-1} \text{ mol}^{-1})(300 \text{ K})}{(6.02 \times 10^{23} \text{ mol}^{-1})(6.63 \times 10^{-34} \text{ J s})} = 1.70 \times 10^{13} \text{ s}^{-1} \tag{20.42}$$

for a unimolecular reaction.

If the activation of the molecule involves a rearrangement of atoms or a change in configuration, there will be a change in entropy and $e^{\Delta S^\ddagger/R}$ is not unity. The values of $\Delta S^\ddagger$ are rarely large enough to give a value of more than 10^2 or less than 10^{-2} to the term $e^{\Delta S^\ddagger/R}$, and so frequency factors $e(RT/N_A h)e^{\Delta S^\ddagger/R}$ for unimolecular reactions may range from about 10^{11} to 10^{15}.

---

* P. J. Robinson, *J. Chem. Educ.* **55**:509 (1978).

If there is an increase in rotational and vibrational freedom in the activated complex, $\Delta S^{\ddagger}$ is positive, and the pre-exponential factor is larger than 1.7×10^{13} s^{-1}. If there is a decrease in rotational and vibrational freedom in the activated complex, $\Delta S^{\ddagger}$ is negative, and the pre-exponential factor is smaller than 1.7×10^{13} s^{-1}.

Example 20.3

What is the entropy of activation $\Delta S^{\ddagger}$ for the reaction

$$H + CH_4 = H_2 + CH_3$$

which has a pre-exponential factor of 10^{13} cm^3 mol^{-1} s^{-1} in the neighborhood of 500 K?

$$A = \frac{RTe^2}{N_A hc^{\circ}} e^{\Delta S^{\ddagger}/R}$$

The normal standard state on a concentration basis is $c^{\circ} = 1$ mol L^{-1} = 10^3 mol m^{-3}.

$$\Delta S^{\ddagger} = R \ln \frac{A N_A hc^{\circ}}{RTe^2} = (8.314 \text{ J K}^{-1} \text{ mol}^{-1})$$

$$\times \ln \frac{\begin{array}{c}(10^{13} \text{ cm}^3 \text{ mol}^{-1} \text{ s}^{-1})(10^{-2} \text{ m cm}^{-1})^3(6.02 \times 10^{23} \text{ mol}^{-1}) \\ \times (6.63 \times 10^{-34} \text{ J s})(10^3 \text{ mol m}^{-3})\end{array}}{(8.314 \text{ J K}^{-1} \text{ mol}^{-1})(500 \text{ K})(2.72)^2}$$

$$= -74 \text{ J K}^{-1} \text{ mol}^{-1}$$

The negative value of $\Delta S^{\ddagger}$ indicates that the activated complex has a more organized structure than the separated reactants.

20.6 Molecular Beam Experiments

In gas kinetics experiments reactant molecules approach each other on trajectories with random angles and impact parameters and with relative energies that range around those having the highest probabilities at the reaction temperature. Colliding molecules may also have different amounts of vibrational, rotational, and electronic energies. Reaction rate constants are averages over these various angles of approach, relative energies, and the like.

To get a microscopic view of a bimolecular gas chemical reaction, we would need to carry out the reaction in such a way that reactant molecules are in known quantum states, and the quantum states of product molecules are identified. In the simplest type of apparatus, illustrated schematically in Fig. 20.7, A and B are sources of beams of the two reactants that collide in region C. These collisions occur in a chamber evacuated with a high-speed pump so that the only collisions are between the molecules from A and B. Product molecules and scattered reactant molecules are detected at D. The effect of changing the angle of approach may be studied by moving A or B, and the effect of the relative velocity of the reactants may be studied by use of velocity selectors

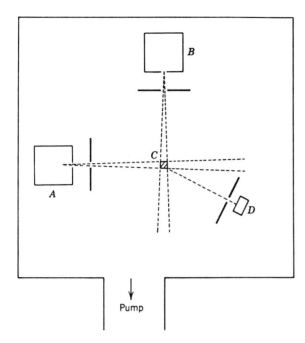

Figure 20.7 Schematic diagram for a molecular beam apparatus for studying the reaction of molecules from source A with molecules from source B. Products are detected at D.

on the beams as they leave *A* and *B*. Sometimes the molecules in molecular beams are put in selected electronic, vibrational, and rotational states so that the effects of these quantum numbers on the observed cross sections can be determined. The term scattering includes three types of phenomena:

1. Elastic scattering in which there are no changes in rotational, vibrational, or electronic quantum numbers in the collision.
2. Inelastic scattering in which there are changes in quantum numbers, but no reaction.
3. Reactive scattering in which new products are formed.

   To study the effects of the quantum numbers of molecules in colliding beams the molecules must be prepared in a certain state and there must not be collisions in the beams that would change these quantum numbers. Molecules can be put into desired vibrational and rotational states by laser excitation. Molecules with dipole moments can be oriented in an inhomogeneous electric field so that the effects of molecular orientation in collision can be studied:

$$\text{Na} + \text{Cl}_2 \rightarrow \text{NaCl} + \text{Cl}$$

This reaction and some similar reactions have quite large cross sections (10^6 pm^2), which indicates that reaction occurs at distances of the order of 500 pm. The sodium atom loses an electron to the chlorine at a distance of this magnitude so that an ionic intermediate is formed:

$$\text{Na} + \text{Cl}_2 \rightarrow (\text{Na}^+ \cdots \text{Cl}^- \cdots \text{Cl}) \rightarrow \text{NaCl} + \text{Cl}$$

This is referred to as a "harpooning" mechanism because the electron from the sodium atom can be thought of as a harpoon, and the sodium and chlorine atoms can be thought of as being drawn together by electrostatic attraction.

20.7   Principles of Photochemistry

In studying the electronic spectroscopy of molecules we have seen how the absorption of a photon by a molecule can raise it to a higher energy level, which may have a minimum in the potential energy curve, or may not. According to the Franck–Condon principle (Section 15.2), the internuclear distance in the molecule is not initially changed by the absorption, and the electric dipole transition moment (Section 14.2) is proportional to the overlap integral of the initial and final vibrational states. In general, selection rules are obeyed, but we must remember that the actual system may be more complicated than the ones for which the selection rules were developed. We will soon see that the selection rule $\Delta S = 0$ (Section 15.2), which prohibits transition between states of different multiplicities (i.e., singlet–triplet), is quite important in photochemistry. When a molecule absorbs a photon it may lose energy through fluorescence or phosphorescence, but now we will be interested in the fact that chemical reactions may ensue. As pointed out in Section 14.1 electromagnetic radiation in the visible and ultraviolet is generally required to produce chemical reactions because changes in electronic energy levels are required. More recently it has been found that the absorption of many infrared photons from a high-intensity laser can also cause reaction.

The **first principle** of photochemistry, which was stated by Grotthus in 1817 and Draper in 1843, is that only light that is absorbed can produce photochemical change. The **second principle,** which was proposed by Stark and Einstein from 1908 to 1912, is that a molecule absorbs a single quantum of light in becoming excited:

$$A + h\nu \rightarrow A^* \tag{20.43}$$

Thus, a mole of photons can excite a mole of molecules. If the electromagnetic radiation is extremely intense, as in a laser beam, two photons may be absorbed essentially simultaneously:

$$A + 2h\nu \rightarrow A^{**} \tag{20.44}$$

In discussing the intensity of light in connection with the Beer–Lambert law (Section 15.4) we used I to represent the intensity of light of a particular wavelength in terms of energy per unit area per unit time. Because of the two principles in the preceding paragraph we will find it convenient in discussing photochemistry to use I_a to represent the **intensity of light absorbed** by a system, expressed as the amount per unit volume per unit time, where, of course, the amount is expressed in moles of photons. A mole of photons is a convenient unit in photochemistry, and it is frequently referred to as an **einstein.** The intensity I_a is calculated from the radiant energy of a specific wavelength absorbed per unit volume per unit time by dividing by $N_A h\nu$.

Example 20.4
Monochromatic radiation at 400 nm, produced by a laser, is completely absorbed by a

reaction mixture with a volume of 0.5 L. If the intensity of the radiation is 50 W, what amount of photons is absorbed in 10 min? What is the value of I_a?

$$\frac{E\lambda}{N_A hc} = \frac{(50 \text{ J s}^{-1})(600 \text{ s})(400 \times 10^{-9} \text{ m})}{(6.022 \times 10^{23} \text{ mol}^{-1})(6.626 \times 10^{-34} \text{ J s})(2.998 \times 10^8 \text{ m s}^{-1})}$$

$$= 0.100 \text{ mol} = 0.100 \text{ einstein}$$

$$I_a = \frac{0.100 \text{ mol}}{(10 \text{ min})(60 \text{ s min}^{-1})(0.5 \text{ L})} = 3.33 \times 10^{-4} \text{ mol L}^{-1} \text{ s}^{-1}$$

For the photochemical reaction

$$Cl_2 + h\nu \underset{k_{-1}}{\overset{I_a}{\rightleftharpoons}} 2Cl \tag{20.45}$$

the rate law is

$$-\frac{d[Cl_2]}{dt} = \frac{1}{2}\frac{d[Cl]}{dt} = I_a - k_{-1}[Cl]^2 \tag{20.46}$$

Note that each term in this equation has the units $\text{mol L}^{-1} \text{ s}^{-1}$.

If the gas or solution strongly absorbs the light, the reaction will occur only near the surface where the light enters. If the gas or solution is weakly absorbing, reaction will occur throughout the volume, but only a fraction of the light will be absorbed (see problem 20.14).

The absorption of light can lead to one or more of several processes: fluorescence, phosphorescence, energy transfer to another molecule, or chemical reaction. It is convenient to express the yield of each of these processes as a quantum yield ϕ. The **quantum yield** ϕ is equal to a ratio of rates. For a chemical reaction, it is equal to the reaction rate v divided by the intensity of light absorbed I_a:

$$\phi = \frac{v}{I_a} \tag{20.47}$$

For fluorescence or phosphorescence it is the rate of emission of radiation divided by the intensity of light absorbed. The quantum yield of chemical reaction is sometimes expressed in terms of a particular reactant or product, but equation 20.47 has the advantage that the quantum yield applies to a specified reaction.

Example 20.5
For the reaction $A + 2B = C$, express $d[A]/dt$, $d[B]/dt$, and $d[C]/dt$ in terms of I_a and the quantum yield ϕ.

Use of equation 20.47 yields

$$\frac{d[A]}{dt} = -\phi I_a$$

$$\frac{d[B]}{dt} = -2\phi I_a$$

$$\frac{d[C]}{dt} = \phi I_a$$

The quantum yield for fluorescence or phosphorescence is necessarily less than unity, and usually much less. The quantum yield for a chemical reaction, however, may be a very large number if the absorption of light produces a radical that starts a chain reaction of a thermodynamically spontaneous reaction. The quantum yield for the first step of a chemical reaction, the so-called primary process, is equal to unity or less.

An electronically excited state of a molecule has a different electron distribution and nuclear configuration than the ground state. An electronically excited state of a molecule may be converted spontaneously into more possible products than the ground state because of the additional energy it has.

Before considering chemical reactions we must consider the physical processes that result from the absorption of electromagnetic radiation. The energy absorbed may produce electronically excited molecules that can react chemically, but often the energy is rapidly dissipated as heat.

Example 20.6

In the photobromination of cinnamic acid, radiation at 435.8 nm with an intensity of 1.4×10^{-3} J s^{-1} was 80.1% absorbed in a liter of solution during an exposure of 1105 s. The concentration of Br_2 decreased by 7.5×10^{-5} mol L^{-1} during this period. What is the quantum yield?

$$E = \frac{N_A hc}{\lambda} = \frac{(6.02 \times 10^{23}\ \text{mol}^{-1})(6.62 \times 10^{-34}\ \text{J s})(3 \times 10^8\ \text{m s}^{-1})}{(435.8 \times 10^{-9}\ \text{m})}$$

$$= 2.74 \times 10^5\ \text{J mol}^{-1}$$

$$I_a = \frac{(1.4 \times 10^{-3}\ \text{J s}^{-1})(0.801)}{(2.74 \times 10^5\ \text{J mol}^{-1})(1\ \text{L})} = 4.09 \times 10^{-9}\ \text{mol L}^{-1}\ \text{s}^{-1}$$

$$v = \frac{7.5 \times 10^{-5}\ \text{mol L}^{-1}}{1105\ \text{s}} = 6.79 \times 10^{-8}\ \text{mol L}^{-1}\ \text{s}^{-1}$$

$$\phi = \frac{v}{I_a} = \frac{6.79 \times 10^{-8}\ \text{mol L}^{-1}\ \text{s}^{-1}}{4.09 \times 10^{-9}\ \text{mol L}^{-1}\ \text{s}^{-1}} = 16.6$$

20.8   Rates of Intramolecular Processes

The electronic excitation of a molecule was discussed in Chapter 15, and in Section 15.8 we saw that the excitation energy may be dissipated by internal conversion, intersystem crossing, fluorescence, or phosophorescence. An important characteristic of each of these processes is the **average lifetime of an excited molecule** before it undergoes the process. In this chapter we will emphasize the rate constant k, that is, the reciprocal of the lifetime τ, because we will be comparing the rates of these physical processes with the rates of chemical reactions. The approximate first-order rate constants for the various intramolecular processes are summarized in Fig. 20.8. Since the absorbing molecule is usually in a singlet ground state S_0, it is excited to a singlet excited state; this may be higher than state S_1, but we will assume that excitation is to S_1 for simplicity.

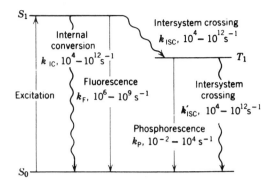

Figure 20.8 First-order rate constants for the various intramolecular processes. The wavy lines represent radiationless transitions. (From C. H. J. Wells, *Introduction to Molecular Photochemistry*. London: Chapman & Hall, 1972. © C. H. J. Wells, 1972.)

When a substance is illuminated with constant intensity, a steady state is reached in which the rates of formation of intermediates are equal to their rates of disappearance. If the absorbed intensity I_a is expressed in moles of photons per unit volume per unit time, the steady-state concentration of S_1 is given by

$$I_a = k_{IC}[S_1] + k_F[S_1] + k_{ISC}[S_1] \tag{20.48}$$

which may be written

$$[S_1] = \frac{I_a}{k_{IC} + k_F + k_{ISC}} \tag{20.49}$$

The steady-state rate equation for T_1 is

$$k_{ISC}[S_1] = k'_{ISC}[T_1] + k_P[T_1] \tag{20.50}$$

Solving for $[T_1]$ yields

$$[T_1] = \frac{k_{ISC}[S_1]}{k'_{ISC} + k_P} \tag{20.51}$$

Substituting equation 20.48 yields

$$[T_1] = \frac{k_{ISC}I_a}{(k'_{ISC} + k_P)(k_{IC} + k_F + k_{ISC})} \tag{20.52}$$

The rate constant k_F for fluorescence is equal to the Einstein probability A_{12} for spontaneous emission (Section 14.2) if there is a single lower state and excitation energy is not lost in various radiationless processes. We saw earlier (Section 15.5) that the radiative lifetime $\tau_0 = 1/k_F$ is equal to $1/A_{12}$, and is approximately inversely proportional to the molar absorbancy index ϵ:

$$\tau_0 \approx \frac{10^{-4}\ \text{L mol}^{-1}\ \text{cm}^{-1}\ \text{s}}{\epsilon} \tag{20.53}$$

In other words, states that are populated readily are also depopulated readily. A strongly absorbing compound with $\epsilon = 10^5\ \text{L mol}^{-1}\ \text{cm}^{-1}$ would be expected to have a natural radiative lifetime of about 10^{-9} s, and a weakly absorbing

Table 20.2   Fluorescence Lifetimes and Quantum Yields of Some Molecules in Solution at Room Temperature[a]

Molecule	Solvent	τ_S/ns	ϕ_F	τ_0/ns
Benzene	Hexane	26	0.070	370
Naphthalene	Hexane	106	0.380	280
Anthracene	Benzene	4	0.240	17
Chlorophyll a	Methanol	6.9	0.280	25
Chlorophyll b	Methanol	5.9	0.080	74
Eosin	Water	4.7	0.150	31

[a] R. B. Cundall and A. Gilbert, *Photochemistry*. London: Nelson, 1970.

compound with $\epsilon = 10^{-2}$ L mol^{-1} cm^{-1} would be expected to have a natural radiative lifetime of about 10^{-2} s.

The observed lifetime τ_S of an excited singlet state is less than the radiative lifetime τ_0 because there are other deactivation processes. According to Fig. 20.8, the rate of decay of S_1 is given by

$$-\frac{d[S_1]}{dt} = (k_{IC} + k_F + k_{ISC})[S_1] \tag{20.54}$$

so that the singlet lifetime is given by

$$\tau_S = \frac{1}{k_{IC} + k_F + k_{ISC}} \tag{20.55}$$

according to equation 19.17.

The singlet lifetime τ_S may be measured in the laboratory by observing the decay of the intensity of the fluorescence after a short ($\ll \tau_S$) pulse of excitation because the intensity of fluorescence is proportional to the excited state concentration. The singlet lifetimes for some organic molecules in solution at room temperature are given in Table 20.2.

The quantum yield for fluorescence ϕ_F is equal to the ratio of fluorescence $k_F[S_1]$ to the total rate of deactivation of the S_1 state, which is given by $(k_{IC} + k_F + k_{ISC})[S_1]$. Thus,

$$\phi_F = \frac{k_F}{k_{IC} + k_F + k_{ISC}} \tag{20.56}$$

This is the quantum yield for fluorescence in the absence of any quenching or chemical reaction. Substituting equations 20.52 and 20.55 yields

$$\tau_S = \tau_0 \phi_F \tag{20.57}$$

Thus, measurement of the singlet lifetime τ_S and the quantum yield for fluorescence makes it possible to calculate the radiative lifetime τ_0. The radiative lifetimes for several molecules calculated in this way are given in Table 20.2. The quantum yield ϕ_{IC} for internal conversion and the quantum yield for intersystem crossing ϕ_{ISC} are given by

$$\phi_{IC} = \frac{k_{IC}}{k_{IC} + k_F + k_{ISC}} \tag{20.58}$$

$$\phi_{ISC} = \frac{k_{ISC}}{k_{IC} + k_F + k_{ISC}} \qquad (20.59)$$

Thus,

$$\phi_F + \phi_{IC} + \phi_{ST} = 1 \qquad (20.60)$$

Returning to Fig. 20.8, we can see that if the rate constant k_{ISC} for the intersystem crossing S_1 to T_1 is fast enough, T_1 will be present at an appreciable concentration. If this happens it is very significant for photochemistry, since triplet state molecules may have long lifetimes compared with singlet state molecules, and therefore have a higher probability of undergoing chemical reaction.

For the T_1 state of Fig. 20.8 the phosphorescence lifetime will be

$$\tau_{T_1} = \frac{1}{k_P + k'_{ISC}} \qquad (20.61)$$

the quantum yield for phosphorescence is given by

$$\phi_P = \frac{\text{rate of phosphorescence emission}}{\text{rate of absorption of radiation}} = \frac{k_P[T_1]}{I_a} \qquad (20.62)$$

$$= \frac{k_P k_{ISC}}{(k'_{ISC} + k_P)(k_{IC} + k_F + k_{ISC})}$$

where the second form has been obtained by substituting the expression for the steady-state value of $[T_1]$ given in equation 20.51 and equation 20.48.

20.9 Quenching

Excited-state molecules may be deactivated rapidly by other molecules in a solution or gas. Such substances are called quenchers, and we will represent them by Q. The quenching of phosphorescence provides a means of determining the rate constant for the reaction of the triplet state molecules and the quencher. The steady-state concentration of triplet molecules in the absence of a quencher is given by equation 20.51. In the presence of a quencher this equation becomes

$$[T_1] = \frac{k_{ISC}[S_1]}{k_P + k'_{ISC} + k_Q[Q]} \qquad (20.63)$$

because of the additional pathway for reaction of the triplet-state molecules with the quencher. The quantum yield for phosphorescence in the absence of a quencher is given by equation 20.62 and will be represented by ϕ_P° in this section. If the steady-state concentration of T_1 in the presence of quencher is given by equation 20.63, the ratio of the quantum yields in the absence and presence of the quencher is given by

$$\frac{\phi_P^\circ}{\phi_P} = \frac{k_P + k'_{ISC} + k_Q[Q]}{k_P + k'_{ISC}} \qquad (20.64)$$

$$= 1 + k_Q \tau_{T_1}[Q]$$

where τ_{T_1}, the phosphorescence lifetime, is defined by equation 20.61. The ratio of the intensity I_P° of phosphorescence in the absence of quencher to the intensity I_P of phosphorescence in the presence of quencher is proportional to the ratio of the quantum yields and so the preceding equation may be written

$$\frac{I_P^\circ}{I_P} = 1 + k_Q \tau_{T_1}[Q] \qquad (20.65)$$

This equation is referred to as the **Stern–Volmer equation.** The quenching of an excited-state molecule may yield an excited electronic state of the quencher as described in the next section.

20.10   Intermolecular Processes

Excited molecules may undergo several different types of reaction. Generally chemical reaction of one excited molecule with another molecule is accompanied by deactivation of the excited molecule, and the excited state is said to be quenched. Quenching an excited molecule (donor D) with a second molecule (acceptor A) may result in the electronic excitation of A with concomitant deactivation of D. **This quenching process is often referred to as electronic energy transfer.** Since minute traces of impurities (quenchers) can rapidly deactivate excited molecules, substances and solvents used in photochemical studies must be carefully purified. For example, molecular oxygen reacts rapidly with excited molecules ($k \approx 10^9 - 10^{10}\,\text{L mol}^{-1}\,\text{s}^{-1}$), and it is therefore often important to deoxygenate the solutions under study.

Electronic energy transfer process may be classified as radiative transfer, short-range transfer, or long-range resonance transfer. In radiative transfer the donor D emits radiation which is absorbed by the acceptor A:

$$D^* \rightarrow D + h\nu \qquad (20.66)$$
$$A + h\nu \rightarrow A^* \qquad (20.67)$$

Short-range energy transfer can occur if the distance between donor and acceptor molecules approaches the collision diameter. A collision is not necessarily required, since the energy transfer can occur at distances slightly greater than the collision diameter. According to the Wigner spin conservation rule, the overall spin angular momentum of the interacting pair of molecules must be unchanged in electronic energy transfer. Like spectroscopic selection rules, this rule is not absolute, but it is an important guide. When only S_0, S_1, and T_1 states are considered, this rule amounts to limiting short-range energy transfer to

$$D^*(S_1) + A(S_0) \rightarrow D(S_0) + A^*(S_1) \qquad (20.68)$$
$$D^*(T_1) + A(S_0) \rightarrow D(S_0) + A^*(T_1) \qquad (20.69)$$

where D represents the donor molecule and A the acceptor molecule. An example of this second type of short-range energy transfer is

$$^3\text{benzaldehyde} \quad + \quad \text{naphthalene}$$
(excited at 366 nm)   (does not absorb at 366 nm)

$$\rightarrow\text{benzaldehyde} + \quad ^3\text{naphthalene}$$
(which phosphoresces)   (20.70)

The efficiencies of these energy transfer processes depend on the relative energies of the states involved. The transfer is relatively efficient if the excited state of D has a greater energy than that of the excited state of A. Energy transfer of this type may be detected by the appearance of fluorescence radiation of $A^*(S_1)$ or phosphorescence radiation of $A^*(T_1)$ from direct excitation of D where A does not absorb.

Energy transfer may occur in a long-lived collision complex. A molecule M* in an electronically excited state may be a very polarizable species and may, as a result, form a collision complex with a ground state molecule Q. The collision complex M*Q generally has a longer lifetime than the corresponding MQ collision complex. The metastable species M*Q is called an **exciplex;** if M and Q are the same the resulting excited complex is called an **excimer.**

The formation of an exciplex can provide a collisional mechanism for the transfer of electronic energy, as illustrated in Fig. 20.9. When the ground-state molecules M and Q approach, their interactions are repulsive. However, if either M or Q is electronically excited, the energy of the collision pair, M*Q or MQ*, will generally be lower than the separated molecules. At the geometry at which the potential energy surfaces intersect, an internal conversion to the lower surface can occur. In Fig. 20.9 the system starts out as M* + Q and ends up as M + Q*, with the consequent electronic energy transfer.

In long-range energy transfer the donor and acceptor molecules are separated by a distance much greater than the collision diameter. The efficiency of the energy transfer depends on the extent of the overlap of the emission spectrum of the donor and the absorption spectrum of the acceptor. The energy transfer

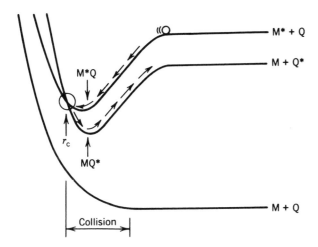

Figure 20.9 Schematic diagram for potential energy surfaces for the formation of exciplexes M*Q and MQ* and internal conversions. (From N. J. Turro, *Modern Molecular Photochemistry*. Menlo Park, CA: Benjamin/Cummings, 1978.)

is not efficient unless the decay process $D^* \rightarrow D$ and the excitation process $A \rightarrow A^*$ are allowed electronic transitions. Thus, the most likely long-range energy transfer processes are those given in equations 20.68 and 20.69. Under favorable circumstances energy transfer can occur over distances of 5 to 10 nm with rate constants of 10^{10}–10^{11} L mol^{-1} s^{-1}.

The rate constant k_{ET} for the energy transfer to an acceptor may be obtained by measuring the lifetime of triplet D in the presence and in the absence of A. If the triplet state of the donor is deactivated solely by phosphorescence, $T_1 \rightsquigarrow S_0$ intersystem crossing, and bimolecular reaction with an acceptor A, the triplet lifetime is given by

$$\tau_{ET} = \frac{1}{k_P + k'_{ISC} + k_{ET}[A]} \tag{20.71}$$

Since the lifetime τ in the absence of the acceptor is given by equation 20.61,

$$\frac{1}{\tau_{ET}} = \frac{1}{\tau} + k_{ET}[A] \tag{20.72}$$

The fact that second-order rate constants k_{ET} obtained using this relation may equal values expected for diffusion-controlled reactions (Section 21.5) indicates that energy transfer may take place as rapidly as donor and acceptor can diffuse together.

20.11   Chemical Reactions and Their Quantum Yields

When an excited species undergoes a chemical reaction, we can add another step to the simplified mechanism that we have been discussing. If the excited species (here assumed to be T_1) reacts with reactant R with a bimolecular rate constant k_R it may be possible to evaluate the rate constant

$$T_1 + R \rightarrow \text{stable products} \qquad \text{rate} = k_R[T_1][R] \tag{20.73}$$

In the steady state

$$[T_1] = \frac{k_{ISC}I_a}{(k_P + k'_{ISC} + k_R[R])(k_{IC} + k_F + k_{ISC})} \tag{20.74}$$

The quantum yield for the production of stable products is

$$\phi = \frac{k_R[T_1][R]}{I_a} \tag{20.75}$$

Substituting equation 20.74 and rearranging yields

$$\frac{1}{\phi} = \frac{k_F + k_{IC} + k_{ISC}}{k_{ISC}} \left(1 + \frac{k_P + k'_{ISC}}{k_R[R]} \right) \tag{20.76}$$

Even for this simplified mechanism there are so many rate constants to be determined before k_R can be obtained that this is not generally practical and so, in further discussions, we will emphasize quantum yields. Note that from equation 20.47 the quantum yield can be determined by measuring the number of quanta absorbed and the number of molecules reacted. This requires no knowledge of rate constants.

Table 20.3 Quantum Yields in Photochemical Reactions at Room Temperature[a]

Reaction	Approximate Wavelength Region, nm	Approximate ϕ
1. $2HI \rightarrow H_2 + I_2$	300–280	2
2. $C_{14}H_{10} \rightleftharpoons \frac{1}{2}(C_{14}H_{10})_2$	<360	0–1
3. $2NO_2 \rightarrow 2NO + O_2$	>435	0
	366	2
4. $CH_3CHO \rightarrow CO + CH_4(+C_2H_6 + H_2)$	310	0.5
	253.7	1
5. $(CH_3)_2CO \rightarrow CO + C_2H_6(+CH_4)$	<330	0.2
6. $NH_3 \rightarrow \frac{1}{2}N_2 + \frac{3}{2}H_2$	210	0.2
7. $H_2C_2O_4(+UO_2^{2+}) \rightarrow CO + CO_2 + H_2O(+UO_2^{2+})$	430–250	0.5–0.6
8. $Cl_2 + H_2 \rightarrow 2HCl$	400	10^5
9. $Co(CN)_6^{3-} + H_2O \rightarrow Co(CN)_5(OH_2)^{2-} + CN^-$	313	0.3^b

[a] W. A. Noyes, Jr., and P. A. Leighton, *Photochemistry of Gases*, Appendixes, pp. 415–465, New York: Reinhold, 1941; F. Daniels, *J. Phys. Chem.* **41**: 713 (1938).

[b] V. Balzani and V. Carassiti, *Photochemistry of Coordination Compounds*. New York: Academic, 1970.

To determine a quantum yield it is necessary to measure the intensity of the light. This may be done by use of a thermopile, which is a series of thermocouples with one set of junctions blackened to absorb all the radiation, which is then converted into heat. The other set of junctions is protected from radiation.

The amount of radiation may also be measured with a chemical **actinometer**, in which the amount of chemical change is determined. The yield of the photochemical reaction in the actinometer was determined originally by use of a thermopile. The quantum yields of a few photochemical reactions are summarized in Table 20.3.

Reaction 1 has the same value of ϕ from 280 to 300 nm, at low pressures and high pressures, in the liquid state or in solution in hexane. The primary process $HI + h\nu = H + I$ is followed by the reactions $H + HI = H_2 + I$ and $I + I = I_2$, thus giving two molecules of HI decomposed for each photon absorbed. Reaction 2, the dimerization of anthracene, has a quantum yield of unity initially, but the reverse thermal reaction can occur at the product accumulates. In reaction 3 at 366 nm the quantum yield is 2 if correction is made for internal screening by the accompanying N_2O_4, which absorbs some light at 366 nm. The reactions are $NO_2 + h\nu = NO_2^*$, $NO_2^* + NO_2 = 2NO + O_2$, where the asterisk indicates an excited molecule. At 435 nm and longer wavelengths no reaction occurs when the radiation is absorbed.

Reaction 4 similarly shows a greater quantum yield at shorter wavelengths. This reaction is interesting because at 300 °C ϕ has a value of more than 300, indicating that the free radicals that are first produced by the absorption of light are able to propagate a chain reaction at the higher temperatures. At room temperature the reactions involved in the chain do not go fast enough to be detected. The products given in parentheses are present also but in small amounts.

The experimental determination of the quantum yield constitutes an excellent method for detecting **chain reactions** (Section 19.15). If several molecules of products are formed for each proton of light absorbed, the reaction is obviously a chain reaction in which the products of the reaction are able to promote reaction of other molecules.

Reaction 5 is an example of the fact that the absorption of light in a particular bond does not necessarily cause the rupture of that bond. Acetone, like other aliphatic ketones, absorbs ultraviolet light at about 280 nm. The C=O bond, which we designate as the chromophore, is very strong and does not break to give atomic oxygen. Instead, the absorption energy leads to the cleavage of an adjacent –C– bond that is weaker; thus,

$$\begin{matrix} CH_3 \\ \diagdown \\ \diagup \\ CH_3 \end{matrix} C{=}O + h\nu \rightarrow CH_3\cdot + CH_3\dot{C}{=}O \qquad (20.77)$$

giving a methyl radical and an acetyl radical. The acetyl radical can then decarbonylate, giving CO and $CH_3\cdot$, or it can react with $CH_3\cdot$ to give back acetone. The methyl radicals can couple to form ethane.

In the photolysis of ammonia, reaction 6, hydrogen atoms are split off, and the low yield is probably due to partial recombination of the fragments. The quantum yield varies with pressure and reaches a maximum at 0.6 to 0.7 Pa.

Reaction 7 illustrates a photosensitized reaction. The photodecomposition of oxalic acid, sensitized by uranyl ion, is so reproducible that it is suitable for use as a chemical actinometer. In the uranyl oxalate actinometer the light is absorbed by the colored uranyl ion, and the energy is transferred to the colorless oxalic acid, which then decomposes. The uranyl ion remains unchanged and can be used indefinitely as a sensitizer. The fact that the molar absorption coefficient of uranyl ion is increased by the addition of colorless oxalic acid indicates the formation of a complex. The formation of a chemical complex is often necessary for photosensitization.

Reaction 8 is the best known example of a chain reaction. About 10^5 molecules react for each quantum absorbed. The molecules of hydrogen chloride formed undergo further reaction with the hydrogen and chlorine atoms produced (Section 19.15). The measurement of the number of molecules per photon gives a measure of the average number of molecules involved in the chain. Initiation of the reaction with a flash of light can result in an explosively fast reaction.

Reaction 9 illustrates the enhanced reactivity of excited state molecules. The inert complex ion $Co(CN)_6^{3-}$ can be recovered unchanged from boiling water as the potassium salt. However, 313-nm irradiation at 25 °C results in aquation to yield $Co(CN)_5(OH_2)^{2-}$ with a quantum yield of 0.30. Given that the excited state has a lifetime of less than 10^{-10} s under the reaction conditions and that a generous estimate of the rate constant for ground-state aquation would be $10^{-6}\,s^{-1}$, the excited state is viewed as being 10^{16} times more reactive than the ground state! Such increases in reactivity are fairly commonly encountered in photochemical work.

The reaction of H_2 and Br_2 to form HBr has been discussed in Section 19.15 as an example of a chain reaction. At temperatures as low as 200 °C the thermal reaction is very slow, but the chain reaction may be initiated photochemically. The empirical rate equation for the photochemical reaction is

$$\frac{d[HBr]}{dt} = \frac{k[H_2]I_a^{1/2}}{P^{1/2}(1 + k'[HBr]/[Br_2])} \tag{20.78}$$

where I_a is the intensity of absorbed light (in mol L^{-1} s^{-1}) and P is the total pressure. This steady-state rate equation for the photochemical reaction may be derived by replacing the forward reaction in equation 19.138 by

$$Br_2 + h\nu \rightarrow 2Br \tag{20.79}$$

with $d[Br]/dt = 2I_a$. In the steady state it can be shown that $2I_a = 2k_{-1}[Br]^2[M]$ so that

$$[Br] = \left(\frac{I_a}{k_{-1}[M]}\right)^{1/2} \tag{20.80}$$

instead of

$$[Br] = \left(\frac{k_1[Br_2]}{k_{-1}}\right)^{1/2} \tag{20.81}$$

in the thermal reaction. Using equation 20.80 in the derivation of the rate equation expected for the photochemical reaction we obtain

$$\frac{d[HBr]}{dt} = \frac{2k_2(I_a/k_{-1})^{1/2}[H_2]}{[M]^{1/2}(1 + k_2[HBr]/k_3[Br_2])} \tag{20.82}$$

which has the same form as equation 20.78 if [M] is proportional to the total pressure P.

The chain length for a photochemical reaction may be defined in various ways, but for this reaction it is conveniently defined as the number of HBr molecules formed per Br atom produced photochemically. At 500 K and 0.1 bar the chain length is about 100. The quantum yield ϕ is about 200 since each photon absorbed produces two Br atoms. The chain length for the $H_2 + Cl_2$ reaction approaches 10^6 since the chain propagation steps are much faster; as a result a steady state is not reached in this photochemical reaction above 200 °C and an explosion occurs.

20.12 The Ozone Layer in the Stratosphere

The formation of ozone in the stratosphere is an example of a photochemical stationary state. The ozone formed in this way is important because it absorbs ultraviolet radiation that would otherwise cause damage to life at the surface of the earth. A simplified mechanism for the formation and destruction of ozone in the stratosphere is the following

$$O_2 \xrightarrow[J_{O_2}]{h\nu} 2O \tag{20.83}$$

$$O + O_2 + M \xrightarrow{k_2} O_3 + M \tag{20.84}$$

$$O_3 \xrightarrow[J_{O_3}]{h\nu} O_2 + O \tag{20.85}$$

$$O_3 + O \xrightarrow{k_4} 2O_2 \tag{20.86}$$

where the J's are photodissociation constants defined below. The photolysis reaction in the first step occurs only at wavelengths less than 242 nm. Ozone is formed in the three-body reaction in step 2. The protective role of ozone is due to the third step in which radiation in the 190- to 300-nm range dissociates ozone. The fourth step is a slow reaction. The recombination of oxygen atoms with a third body to form O_2 is very slow in the stratosphere and can be neglected. If the absorption of solar radiation in the first and third steps is steady, the concentration of ozone rises to a steady level that can be calculated as follows. The rate of change of ozone concentration is given by

$$\frac{d[O_3]}{dt} = k_2[O][O_2][M] - J_{O_3}[O_3] - k_4[O][O_3] \tag{20.87}$$

where J_{O_3} is the **photodissociation coefficient** for ozone. The photodissociation coefficient is the probability of dissociation of a molecule per second by light absorption. This coefficient is calculated using

$$J = \int_0^\infty \phi_{\bar\nu} I_{\bar\nu} \sigma_{\bar\nu} \, d\bar\nu \tag{20.88}$$

where $\phi_{\bar\nu}$ is the quantum yield of dissociation of the molecule at wavenumber $\bar\nu$, $I_{\bar\nu}$ is the intensity of sunlight in quanta per unit area per unit time per wavenumber, and $\sigma_{\bar\nu}$ is the absorption cross section of the molecule at wavenumber $\bar\nu$. Thus, $I_{\bar\nu}$ has SI units of $m^{-1} s^{-1}$, and J has SI units of s^{-1}.

The rate of change in the concentration of oxygen atoms is assumed to be zero in the steady state since they are present at very low concentrations:

$$\frac{d[O]}{dt} = 2J_{O_2}[O_2] - k_2[O][O_2][M] + J_{O_3}[O_3] - k_4[O][O_3] = 0 \tag{20.89}$$

Adding equations 20.87 and 20.89 yields

$$\frac{d[O_3]}{dt} = 2J_{O_2}[O_2] - 2k_4[O][O_3] \tag{20.90}$$

From experimental measurements it is known that $J_{O_3}[O_3] \gg J_{O_2}[O_2]$ and $k_2[O][O_2][M] \gg k_4[O][O_3]$, and so equation 20.89 becomes

$$J_{O_3}[O_3] \cong k_2[O][O_2][M] \tag{20.91}$$

Substituting this for [O] in equation 20.93 yields

$$\frac{d[O_3]}{dt} = 2J_{O_2}[O_2] - \frac{2k_4 J_{O_3}[O_3]^2}{k_2[O_2][M]} \tag{20.92}$$

Therefore, the steady-state concentration of O_3 is given by

$$[O_3]_{ss} = [O_2] \sqrt{k_2 J_{O_2}[M]/k_4 J_{O_3}} \tag{20.93}$$

Since J_{O_2} increases with altitude and $[O_2]$ decreases with altitude, there is a maximum steady-state concentration in the stratosphere, at an altitude of about 20 km. However, equation 20.93 is only approximate because ozone is involved in more reactions, several of which are catalytic cycles with an environmental impact.

In the late 1960s it was found that nitrogen oxides can catalyze reaction 20.86:

$$NO + O_3 \rightarrow NO_2 + O_2$$
$$\underline{NO_2 + O \rightarrow NO + O_2}$$
$$O_3 + O \rightarrow 2O_2 \qquad (20.94)$$

In the mid 1970s it was discovered that chlorine atoms from the photolysis of chlorofluorocarbons ($CFCl_3 + h\nu \rightarrow CFCl_2 + Cl$ and $CF_2Cl_2 + h\nu \rightarrow CF_2Cl + Cl$) at the level of the ozone layer can also catalyze the decomposition of ozone:

$$Cl + O_3 \rightarrow ClO + O_2$$
$$\underline{ClO + O \rightarrow Cl + O_2}$$
$$O_3 + O \rightarrow 2O_2 \qquad (20.95)$$

Quantitative calculations based on this mechanism and the discovery of an "ozone hole" over Antarctica have led to international controls on the manufacture of $CFCl_3$ and CF_2Cl_2.

20.13 Femtosecond Transition-State Spectroscopy*

The development of dye lasers capable of delivering pulses as short as 6×10^{-15} s (6 fs) has made it possible to observe the transition region of a photolysis reaction. In a photolysis reaction the excited molecule ABC* goes through a transition state $[A \cdots BC]^{\ddagger *}$ and dissociates into products. Note that $[A \cdots BC]^{\ddagger *}$ is different from an unstable intermediate in a reaction because it is in the process of flying apart:

$$ABC^* \rightarrow [A \cdots BC]^{\ddagger *} \rightarrow A + BC \qquad (20.96)$$

The potential energy curves for the molecule $ABC(V_0)$, for the first dissociative state (V_1) yielding A + BC, and the second dissociative state (V_2) yielding A + BC* are given in Fig. 20.10. The potential energy curves for A + BC and A + BC* do not have minima, and therefore lead to immediate dissociation. An initial pulse, represented by λ_1, is absorbed at $t = 0$ by the ABC molecule at an internuclear distance close to the potential energy minimum. The absorption of a photon by a molecule involves "instantaneous" vertical transition of the molecule from the ground-state potential energy V_0 to an excited-state potential energy V_1. For there to be appreciable absorption, the wavelength of the initial pulse must satisfy the relation $V_1(R_0) - V_0(R_0) \approx hc/\lambda_1$, where R_0 is the internuclear distance at the potential minimum. Classically the ABC molecules dissociate along V_1, with $R \rightarrow \infty$ as $t \rightarrow \infty$. After a time delay τ, the gas is irradiated with a probe pulse with wavelength λ_2.

In the simplest type of experiment, the wavelength λ_2^{∞} of the probe pulse is tuned to the absorption wavelength of one of the free fragments. If the probe pulse follows the initial pulse closely, the absorption is negligible initially, as shown by the upper plot of optical absorption in Fig. 20.10b. The absorption of the probe pulse λ_2^{∞} becomes substantial only when the fragments achieve large internuclear distance and are no longer interacting. The build up of the

---

* A. H. Zewail, *Science* **242**:1645 (1988).

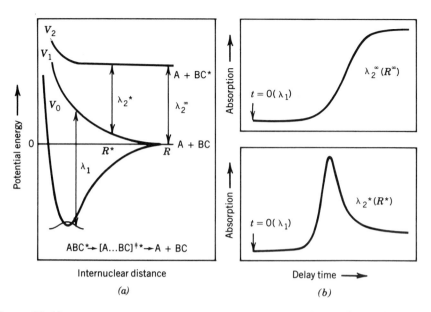

Figure 20.10 (a) Potential versus internuclear distance R for $ABC(V_0)$, $A + BC(V_1)$, and $A + BC^* (V_2)$. (b) Optical absorption as a function of time in femtoseconds after a pulse at λ_1. In the upper plot the absorption is for a probe pulse with wavelength λ_2^∞ that is absorbed by the completely dissociated products ($R = \infty$). In the lower plot the absorption is for a probe pulse λ_2^* that is absorbed by $[A\cdots BC]^{\ddagger*}$ with a particular internuclear distance. (From A. H. Zewail, *Science* **242**:1645 (1988). Reprinted with permission.)

signal from the λ_2^∞ pulse to half of its maximum takes about 200 fs, as shown by Fig. 20.10*b*.

The molecules in transition from the initial excited state can be detected by tuning λ_2 away from λ_2^∞. In this second type of experiment, the probe pulse, which is sent at time τ, will be absorbed significantly only if the transient-state configuration at time τ is such that $V_2(R^*) - V_1(R^*) = hc/\lambda_2^*$, where R^* is the internuclear distance at $t = \tau$. The lower absorption plot in Fig. 20.10*b* shows that the probe pulse λ_2^* is absorbed when the transition state has the internuclear distance R^* and that the absorption of λ_2^* subsequently falls as the internuclear distance increases and separated products are finally formed, after about 300 fs.

The particular reaction studied by Zewail and coworkers was

$$ICN^* \rightarrow [I\cdots CN]^{\ddagger*} \rightarrow I + CN \qquad (20.97)$$

Thus, it is possible to essentially measure absorption spectra of $[I\cdots CN]^{\ddagger*}$ at various times, and distances, as it flies apart. Since the recoil velocity of the fragments is about 1 km s^{-1}, the distance spanned in 100 fs is 100 pm.

This method cannot be used directly for bimolecular reactions such as $A + BC \rightarrow AB + C$ because the bimolecular collisions occur at random times and cannot be controlled. However, this can be done for a class of bimolecular reactions in which a van der Waals "precursor molecule" contains the potential reagent molecules in close proximity. The van der Waals complex $IH\cdots OCO$ was formed in a free jet expansion in an excess of helium carrier gas. A femtosecond pulse was used to dissociate the HI in the van der Waals complex. This caused a hot H atom to be ejected in the direction of the nearest-neighbor O atom in CO_2, thus initiating the reaction

$$H + OCO \rightarrow [HOCO]^{\ddagger} \rightarrow OH + CO \qquad (20.98)$$

By use of a probe pulse suitable for the detection of OH, it was possible to measure the lifetime of the $[HOCO]^{\ddagger}$ collision complex. The lifetime depends on the translational energy, and is shorter at higher collision energies.

Thus, it is now possible, in favorable cases, to view bond breaking or formation in real time and study the dynamics in the region of the transition state.

20.14 Special Topic: Applications of Photochemistry

In **photosynthesis** CO_2 and H_2O are converted to the glucose moiety as starch. The overall process can be represented by

$$CO_2 + H_2O \xrightarrow[\text{photons}]{\text{eight}} (CH_2O) + O_2 \qquad (20.99)$$

where (CH_2O) represents one-sixth of a glucose moiety. This process, which occurs entirely in the chloroplasts of green cells, involves a large number of steps catalyzed by enzymes. The chloroplasts contain chlorophyll a, chlorophyll b, carotenes, electron carriers, and enzymes, and have internal membranes that keep reactants separated. The first step in photosynthesis is the absorption of light by a chlorophyll molecule. Chlorophylls a and b contain networks of alternating single and double bonds, and have strong absorption bands in the visible part of the spectrum with molar absorbancy indices greater than 10^5 L mol^{-1} cm^{-1}. The energy of the absorbed photon is transferred from one chlorophyll molecule to another until it reaches a site called a reaction center.

Example 20.7
Given that the energy change for reaction 20.99 is 477 kJ mol^{-1} and that the energy input between 400 and 700 nm at the earth's surface is equivalent to that at 575 nm, what is the theoretical maximum energy efficiency of photosynthesis by white light?

$$E = N_A h\nu = \frac{(6.02 \times 10^{23} \text{ mol}^{-1})(6.62 \times 10^{-34} \text{ J s})(3 \times 10^8 \text{ m s}^{-1})}{(575 \times 10^{-9} \text{ m})(10^3 \text{ J kJ}^{-1})}$$

$$= 208 \text{ kJ mol}^{-1}$$

$$\text{Efficiency} = \frac{477 \text{ kJ mol}^{-1}}{(8)(208 \text{ kJ mol}^{-1})} = 0.29$$

The efficiency calculated on the basis of total solar radiation at the earth's surface that a plant could actually absorb leads to a maximum efficiency during the maximum growing season of 0.066. Efficiencies of 0.032 have been achieved with corn.

The energy of light can partially be made available as electric current with photocells, but their efficiencies are limited to about 10% for visible light. This

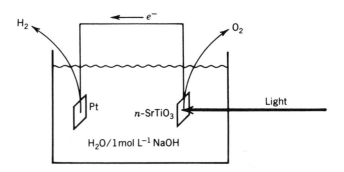

Figure 20.11 Photochemical cell for the conversion of light to chemical energy in the form of H_2 and O_2. (From M. S. Wrighton, *J. Chem. Educ.* **60**:877 (1983).)

has led to research to improve efficiencies and to produce chemical substances with high energy densities that can be stored. A photoelectrochemical cell is an electrochemical cell in which the energy of light (rather than an electric power supply) is utilized to produce electrochemical change. In Section 8.12 the standard electromotive force for a hydrogen–oxygen fuel cell was shown to be 1.229 V at 25 °C. Since this calculation assumed reversibility, we also know that a *minimum* applied potential of 1.229 V is required to electrolyze water at 25 °C. Actually, an overvoltage is required and the electrolysis of water does not occur at this reversible potential (Section 21.12). In a photoelectrochemical cell of the type illustrated in Fig. 20.11, the energy of light absorbed by the anode may provide the energy for the electrolysis of water. Whether or not light energy can be converted wholly or partially into stored chemical energy in the form of $H_2(g) + \frac{1}{2}O_2(g)$ or electrical energy that can be dissipated in the load obviously depends on the properties of the anode. If the anode is the *n*-type semiconductor TiO_2 (see Section 23.15) and ultraviolet light is used, electrolysis of water occurs if a potential of 0.25 V is applied across the electrodes. In this cell the TiO_2 is a **photoassistance agent** in that it serves as a receptor for the light and channels excitation energy into the chemical reaction. The optical energy storage efficiency of such a cell may be expressed as

$$\text{efficiency} = \frac{(\text{energy stored as } H_2) - (\text{energy from power supply})}{(\text{light energy})} \times 100$$

(20.100)

and efficiencies of several percent have been obtained for the TiO_2 electrode. The higher efficiencies obtained with $SrTiO_3$ anodes and ultraviolet light rival the light-to-chemical-energy conversion efficiency of any other man-made photochemical conversion device.

The energy levels in the semiconductor CdTe make it an even more suitable photoassistance anode, since visible light can be used, but this electrode material decomposes unless a stabilizing electrolyte solution is used. When a stabilizing electrolyte solution is used, the energy of the light absorbed may be converted into electrical energy instead of being stored in $H_2(g) + \frac{1}{2}O_2(g)$.

References

M. Baer, *The Theory of Chemical Reaction Dynamics*. Boca Raton, FL: CRC Press, 1985.

S. W. Benson, *Thermochemical Kinetics*. New York: Wiley, 1976.

C. F. Bernasconi (Ed.), *Investigation of Rates and Mechanisms of Reactions*, 4th ed. New York: Wiley, 1985.

R. B. Bernstein, *Chemical Dynamics via Molecular Beam and Laser Techniques*. New York: Oxford University Press, 1982.

G. L. Geoffroy and M. S. Wrighton, *Organometallic Photochemistry*. New York: Academic, 1979.

G. G. Hammes, *Principles of Chemical Kinetics*. New York: Academic, 1978.

H. S. Johnston, *Gas Phase Reaction Rate Theory*. New York: Ronald, 1966.

P. C. Jordan, *Chemical Kinetics and Transport*. New York: Plenum, 1979.

R. D. Levine and R. B. Bernstein, *Molecular Reaction Dynamics and Chemical Reactivity*. New York: Oxford University Press, 1987.

J. W. Moore and R. G. Pearson, *Kinetics and Mechanisms*. New York: Wiley, 1981.

M. F. R. Mulcahy, *Gas Kinetics*. New York: Wiley, 1973.

J. Nicholas, *Chemical Kinetics*. New York: Wiley, 1976.

H. Okabe, *Photochemistry of Small Molecules*. New York: Wiley-Interscience, 1978.

I. W. M. Smith, *Kinetics and Dynamics of Elementary Gas Reactions*. Boston: Butterworths, 1980.

J. I. Steinfeld, J. S. Francisco, and W. L. Hase, *Chemical Kinetics and Dynamics*. Englewood Cliffs, NJ: Prentice Hall, 1989.

B. P. Straughan and S. Walker, *Spectroscopy*, Vol. 3. New York: Wiley, 1976.

D. G. Truhlar, *Potential Energy Surfaces and Dynamics Calculations*. New York: Plenum, 1981.

N. J. Turro, *Modern Molecular Photochemistry*. Menlo Park, CA: Benjamin/Cummings, 1978.

C. H. J. Wells, *Introduction to Molecular Photochemistry*. London: Chapman & Hall, 1972.

Problems

20.1 According to the Boltzmann distribution, the number of molecules $dN(\epsilon)$ in the energy range dE is given by

$$dN(\epsilon) = Ke^{-\epsilon/kT}\,dE$$

if the density of states is independent of ϵ. (a) Derive the expression for the probability that these molecules have energies greater than ϵ_a. (b) What is the probability that the molecules have energies greater than 20 kJ mol^{-1} at 298 K? (c) What is the probability that the molecules have energies greater than 20 kJ mol^{-1} at 1000 K?

20.2 Use equation 19.68 to derive the expression for the activation energy for a reaction following simple collision theory.

20.3   Use the pre-exponential factor $A = 3 \times 10^{13}$ cm^3 mol^{-1} s^{-1} for the reaction Br + H$_2$ → HBr + H to calculate the cross section and collision diameter for this reaction at 400 K.

20.4   (a) Calculate the second-order rate constant for collisions of dimethyl ether molecules with each other at 777 K. It is assumed that the molecules are spherical and have a radius of 0.25 nm. If every collision were effective in producing decomposition, what would be the half-life of the reaction (b) at 1 bar pressure, and (c) at a pressure of 0.13 Pa?

20.5   Show that the transition-state theory yields the simple collision theory result when it is applied to the reaction of two rigid spherical molecules.

20.6   What is the expression for the pre-exponential factor of a second-order gas reaction according to the thermodynamic formulation of transition-state theory? What is the value of the pre-exponential factor at 500 K if the entropy of activation is zero? When this value is compared with values in Table 19.3, what do you conclude about the sign of $\Delta S^{\ddagger}$?

20.7   F. W. Schuler and G. W. Murphy (*J. Am. Chem. Soc.* **72**:3155 (1950)) studied the thermal rearrangement of vinyl allyl ether to allyl acetaldehyde in the range 150 to 200 °C and found that

$$k = 5 \times 10^{11} e^{-128\ 000/RT}$$

where k is in s^{-1} and the activation energy is in J mol^{-1}. Calculate (a) the enthalpy of activation, and (b) the entropy of activation, and (c) give an interpretation of the latter.

20.8   Table 20.1 shows that the pre-exponential factor for the reaction

$$CD_3 + CH_4 \rightarrow CD_3H + CH_3$$

is 1×10^{11} cm^3 mol^{-1} s^{-1}. Calculate $\Delta S^{\ddagger}$ for this elementary bimolecular reaction at 300 K. How do you account for the sign of $\Delta S^{\ddagger}$?

20.9   For the unimolecular dissociation of ethane to methyl radicals

$$C_2H_6(g) \underset{k_b}{\overset{k_f}{\rightleftharpoons}} 2CH_3(g)$$

In the temperature range 800–1000 K and about 1 bar, it has been found that

$$-\frac{d[C_2H_6]}{dt} = k_f[C_2H_6] - k_b[CH_3]^2$$

$$k_f = (10^{16}\ \text{s}^{-1}) \exp\left(-\frac{360\ \text{kJ mol}^{-1}}{RT}\right)$$

$$k_b = 1.19 \times 10^{10}\ \text{L mol}^{-1}\ \text{s}^{-1}$$

At 1000 K, calculate $\Delta S^{\ddagger}$ for (a) the forward reaction and (b) the backward reaction. (c) How do you interpret these values? (d) Calculate $\Delta S°$ for the overall reaction. This is different from the value of $\Delta S°$ that you would calculate from data on $K_P = (P_{CH_3}/P°)^2/(P_{C_2H_6}/P°)$ at a series of temperatures. Why?

20.10   A certain photochemical reaction requires an excitation energy of 126 kJ mol^{-1}. To what values does this correspond in the following units: (a) frequency of light, (b) wavenumber, (c) wavelength in nanometers, and (d) electronvolts?

20.11   How many moles of photons does a laser with an intensity of 0.1 W at 560 nm produce in one hour?

20.12   A sample of gaseous acetone is irradiated with monochromatic light having a wavelength of 313 nm. Light of this wavelength decomposes the acetone according to the equation

$$(CH_3)_2CO \rightarrow C_2H_6 + CO$$

The reaction cell used has a volume of 59 cm^3. The acetone vapor absorbs 91.5% of the incident energy. During the experiment the following data are obtained:

Temperature of reaction = 56.7 °C

Initial pressure = 102.16 kPa

Final pressure = 104.42 kPa

Time of radiation = 7 h

Incident energy = 48.1 × 10^{-4} J s^{-1}

What is the quantum yield?

20.13   A 100-cm^3 vessel containing hydrogen and chlorine was irradiated with light of 400 nm. Measurements with a thermopile showed that 11×10^{-7} J of light energy was absorbed by the chlorine per second. During an irradiation of 1 min the partial pressure of chlorine, as determined by the absorption of light and the application of Beer's law, decreased from 27.3 to 20.8 kPa (corrected to 0 °C). What is the quantum yield?

20.14   Show that if a solute follows the Beer–Lambert law, the intensity of absorbed radiation I_a in moles of photons per unit volume per second is given by

$$I_a = \frac{I_0}{lN_A h\nu}(1 - e^{-\kappa cl})$$

where l is the length of the cell in the direction of the incident monochromatic radiation and I_0 is in energy per unit area per unit time.

20.15   When CH$_3$I molecules in the vapor state absorb 253.7 nm light, they dissociate into methyl radicals and iodine atoms. The energy required to rupture the C–I bond is

209 kJ mol^{-1}. What are the velocities of the iodine atom and the methyl radical, assuming all of the excess energy goes into translational motion?

20.16 The phosphorescence of butyrophenone in acetonitrile is quenched by 1,3-pentadiene (P). The following quantum yields were measured at 25 °C (see N. J. Turro, *Modern Molecular Photochemistry*, p. 248. Menlo Park, CA: Benjamin/Cummings, 1978).

[P]/10^{-3} mol L^{-1}	0	1.0	2.0
ϕ/ϕ_0	1	0.61	0.43

Assuming that the quenching reaction is diffusion controlled and the rate constant has a value of 10^{10} L mol^{-1} s^{-1}, what is the lifetime of the triplet state?

20.17 Biacetyl triplets have a quantum yield of 0.25 for phosphorescence and a measured lifetime of the triplet state of 10^{-3} s. If its phosphorescence is quenched by a compound Q with a diffusion-controlled rate (10^{10} L mol^{-1} s^{-1}), what concentration of Q is required to cut the phosphorescence yield in half?

20.18 For 900 s, light of 426 nm was passed into a carbon tetrachloride solution containing bromine and cinnamic acid. The average power absorbed was 19.2 × 10^{-4} J s^{-1}. Some of the bromine reacted to give cinnamic acid dibromide, and in this experiment the total bromine content decreased by 3.83 × 10^{19} molecules. (*a*) What was the quantum yield? (*b*) State whether or not a chain reaction was involved.

20.19 The following calculations are made on a uranyl oxalate actinometer, on the assumption that the energy of all wavelengths between 254 and 435 nm is completely absorbed. The actinometer contains 20 cm^3 of 0.05 mol L^{-1} oxalic acid, which also is 0.01 mol L^{-1} with respect to uranyl sulfate. After 2 h of exposure to ultraviolet light, the solution required 34 cm^3 of potassium permanganate, KMnO$_4$, solution to titrate the undecomposed oxalic acid. The same volume, 20 cm^3, of unilluminated solution required 40 cm^3 of the KMnO$_4$ solution. If the average energy of the quanta in this range may be taken as corresponding to a wavelength to 350 nm, how many joules were absorbed per second in this experiment? ($\phi = 0.57$).

20.20 A solution of a dye is irradiated with 400 nm light to produce a steady concentration of triplet state molecules. If the triplet state yield is 0.9, and the triplet state lifetime is 20 × 10^{-6} s, what light intensity, expressed in watts, is required to maintain a steady triplet concentration of 5 × 10^{-6} mol L^{-1} in a liter of solution. Assume that all of the light is absorbed.

20.21 The photochemical chlorination of chloroform

$$CHCl_3 + Cl_2 = CCl_4 + HCl$$

is believed to proceed by the following mechanism:

$$Cl_2 + h\nu \xrightarrow{I_a} 2Cl$$

$$Cl + CHCl_3 \xrightarrow{k_1} CCl_3 + HCl$$

$$CCl_3 + Cl_2 \xrightarrow{k_2} CCl_4 + Cl$$

$$2CCl_3 + Cl_2 \xrightarrow{k_3} 2CCl_4$$

Derive the steady-state rate law for the production of carbon tetrachloride.

20.22 Sunlight between 290 and 313 nm can produce sunburn (erythema) in 30 min. The intensity of radiation between these wavelengths in summer and at 45° latitude is about 50 μW cm^{-2}. Assuming that 1 photon produces chemical change in 1 molecule, how many molecules in a square centimeter of human skin must be photochemically affected to produce evidence of sunburn?

20.23 The pre-exponential factor for the reaction

$$H_2 + I_2 = 2HI$$

is 10^{11} L mol^{-1} s^{-1} and the activation energy is 165 kJ mol^{-1} in the range 300 to 500 °C. If the collision diameter is 320 pm, what value of the pre-exponential factor is expected from collision theory at 600 K and what is the value of the steric factor p? (At higher temperatures this reaction goes by the unbranched chain mechanism described in Section 19.15).

20.24 Estimate the pre-exponential factor for the reaction

$$2CH_3 \rightarrow C_2H_6$$

using collision theory. The molecular diameter of CH$_4$ obtained from gas viscosity measurements at 0 °C is 0.414 nm (Table 18.2). The experimental value of A in Table 19.3 is 10$^{10.5}$ L mol^{-1} s^{-1}.

20.25 Estimate the pre-exponential factor in the neighborhood of 500 K for the reaction

$$Cl + H_2 \rightarrow HCl + H$$

assuming the activated complex is linear. The H–H and H–Cl bond distances in the activated complex may be assumed to be 0.092 and 0.145 nm, respectively, and the vibrational partition functions may be taken as unity. The experimental value from Table 19.3 is 10$^{10.9}$ L mol^{-1} s^{-1}.

20.26 The vapor-phase decomposition of di-*tert*-butyl peroxide is first order in the range 110 to 280 °C and follows the equation

$$k = 3.2 \times 10^{16} e^{-163\,600/RT}$$

where k is in s^{-1} and E_a is in J mol^{-1}. Calculate (a) $\Delta H^{\ddagger}$, and (b) $\Delta S^{\ddagger}$.

20.27 The first-order rate constant for the thermal decomposition of $C_2H_5Br(g)$ is given by

$$k = (3.8 \times 10^{14} \text{ s}^{-1})e^{-230\,000/RT}$$

where the activation energy is in J mol^{-1}. Calculate (a) $\Delta H^{\ddagger}$, and (b) $\Delta S^{\ddagger}$ at 500 °C.

20.28 For the first-order gaseous decomposition of propylene oxide $\Delta H^{\ddagger} = 238.1$ kJ mol^{-1} and $\Delta S^{\ddagger} = 25$ J K^{-1} mol^{-1} in the neighborhood of 285 °C. Calculate (a) the frequency factor A, and (b) the first-order rate constant at 285 °C.

20.29 A solution absorbs 300 nm radiation at the rate of 1 W. What does this correspond to in amount of photons absorbed per second?

20.30 What intensities of light in J s^{-1} are required to produce 10^{-6} mol s^{-1} at (a) 700 nm, and (b) 300 nm. (c and d) What are these powers in watts?

20.31 Discuss the economic possibilities of using photochemical reactions to produce valuable products with electricity at 5 cents per kilowatt-hour. Assume that 5% of the electric energy consumed by a quartz–mercury vapor lamp goes into light, and 30% of this is photochemically effective. (a) How much will it cost to produce 1 lb (453.6 g) of an organic compound having a molar mass of 100 g mol^{-1}, if the average effective wavelength is assumed to be 400 nm and the reaction has a quantum yield of 0.8 molecule per photon? (b) How much will it cost if the reaction involves a chain reaction with a quantum yield of 100?

20.32 The quantum yield is 1 for the photolysis of gaseous HI to $H_2 + I_2$ by light of 253.7 nm wavelength. Calculate the amount of HI that will be decomposed if 300 J of light of this wavelength is absorbed.

20.33 In the stratosphere molecular oxygen absorbs solar radiation in the 185–220 nm wavelength region.

$$O_2 + h\nu = O + O$$

If the absorption cross section is 1.1×10^{-23} cm^2, what thickness of a layer of O_2 at 298 K and 1 bar is required to absorb half of the radiation?

20.34 The quantum yield for the photolysis of acetone

$$(CH_3)_2CO = C_2H_6 + CO$$

at 300 nm is 0.2. How many moles per second of CO are formed if the intensity of the 300-nm radiation absorbed is 10^{-2} J s^{-1}?

20.35 The fluorescence quantum yield for benzene at 25 °C is 0.070. The lifetime of the excited state is 26 nm. What is the radiative lifetime τ_0?

20.36 If the lifetime of a triplet is 1 s and the bimolecular rate constant for the quenching of the triplet by O_2 is 10^9 L mol^{-1} s^{-1}, what concentration of O_2 in a solution will reduce the intensity of fluorescence to 10%?

20.37 A cold high-voltage mercury lamp is to be used for a certain photochemical reaction that responds to ultraviolet light of 253.7 nm. The chemical analysis of the product is sensitive to only 10^{-4} mol. The lamp consumed 150 W and converts 5% of the electric energy into radiation of which 80% is at 253.7 nm. The amount of the light that gets into the monochromator and passes out the exit slit is only 5% of the total radiation of the lamp. Fifty percent of this 253.7-nm radiation from the monochromator is absorbed in the reacting system. The quantum yield is 0.4 molecule of product per quantum of light absorbed. How long an exposure must be given in this experiment if it is desired to measure the photochemical change with an accuracy of 1%?

20.38 The quantum yield is unity for the dissociation of acetone vapor using 254 nm radiation at 150 °C. How long will it take to dissociate 10^{-2} mol using a 100-W laser producing 254 nm radiation?

20.39 A uranyl oxalate actinometer is exposed to light of wavelength 390 nm for 1980 s, and it is found that 24.6 cm^3 of 0.004 30 mol L^{-1} potassium permanganate is required to titrate an aliquot of the uranyl oxalate solution after illumination, in comparison with 41.8 cm^3 before illumination. Using the known quantum yield of 0.57, calculate the number of joules absorbed per second. The chemical reaction for the titration is

$$2MnO_4^- + 5H_2C_2O_4 + 6H^+ = 2Mn^{2+} + 10CO_2 + 8H_2O$$

20.40 A photochemical reaction of biological importance is the production of vitamin D, which prevents rickets and brings about the normal deposition of calcium in growing bones. Steenbock found that rickets could be prevented by subjecting the food as well as the patient to ultraviolet light below 310 nm. When ergosterol is irradiated with ultraviolet light below 310 nm, vitamin D is produced. When irradiated ergosterol was included in a diet otherwise devoid of vitamin D, it was found that absorbed radiant energy of about 7.5 $\times 10^{-5}$ J was necessary to prevent rickets in a rat when fed over a period of 2 weeks. The light used has a wavelength of 265 nm. (a) How many quanta are necessary to give 7.5 $\times 10^{-5}$ J? (b) If vitamin D has a molar mass of the same order of magnitude as ergosterol (382 g mol^{-1}), how many grams of vitamin D per day are necessary to prevent rickets in a rat? It is assumed that the quantum yield is unity.

20.41 Given that the intensity of solar radiation is 4.2 J cm^{-2} min^{-1}, how much carbon has to be burned to obtain the same amount of heat as the solar radiation on 1 m^2 in an 8-h day?

20.42 Given that solar radiation at noon at a certain place on the earth's surface is 4.2 J cm^{-2} min^{-1}, what is maximum power output in W m^{-2}?

20.43 If a good agricultural crop yields about 2 tons $acre^{-1}$ of dry organic material per year with a heat of combustion of about 16.7 kJ g^{-1}, what fraction of a year's solar energy is stored in an agricultural crop if the solar energy is about 4184 J min^{-1} ft^{-2} and the sun shines about 500 min day^{-1} on the average? One acre = 43 560 ft^2 and 1 ton = 907 000 g.

21
Kinetics in the Liquid Phase

When chemical reactions occur in solution, the dynamics are different than for gas-phase reactions, for which kinetic theory has been so useful. We can learn about the motion of molecules in a liquid by studying the rate processes, viscosity, diffusion, and electrical conductivity. These processes can be treated by transition-state theory, but here we will emphasize their phenomenological aspects and their relation to the kinetics of reactions in solution. Measurements of viscosity and diffusion provide a means for learning about the rates with which reactants can come together in solution. For ions there is an additional possibility of studying their motion in a solution that is subjected to an electric field. The interpretation of rates in the liquid phase is necessarily more complicated from a molecular viewpoint because of the much greater interaction between molecules. However, bimolecular reactions in solution cannot occur more rapidly than the reactant molecules can diffuse together, and this rate may be calculated from measured diffusion coefficients of the reactants.

Acids and bases catalyze many reactions, and we will discuss two types of mechanisms for acid catalysis. Enzymes catalyze the reactions in living things and provide the mechanisms by which rates of these reactions are controlled. These reactions provide interesting examples of solution kinetics.

The last section of the chapter is about the kinetics of reactions occurring at the surface of an electrode. The rates of surface reactions, which are important in corrosion, in batteries, and in electroanalytical methods, are extremely sensitive to electrical potential differences.

21.1 Viscosity of a Liquid

The viscosity was defined in the discussion of the kinetic theory of gases in Section 18.10. This definition applies to laminar flow, that is, flow in which one layer (lamina) slides smoothly relative to another. When this flow velocity is great enough, turbulence develops. The viscosity of a liquid may be measured by a number of methods which include the determination of the rate of flow through a capillary, the rate of settling of a sphere in a liquid, and the force required to turn one of two concentric cylinders at a certain angular velocity (Fig. 18.11c).

Since the viscosity is the ratio of the force per unit area to the velocity gradient, it has the units kg m^{-1} s^{-1} or Pa s.

When a force F is applied to a particle in solution—as in an electric field, if the particle is charged, or in a centrifugal field—the particle will be accelerated. As the velocity of the particle increases, it experiences an increasing frictional force. For low velocities the frictional force is given by vf, where v is the velocity and f is the **frictional coefficient** of the particle. When the velocity is sufficiently high for the frictional force to be equal to the applied force

$$vf = F \qquad (21.1)$$

and the particle will move with constant velocity.

The frictional coefficient f is of interest because it provides some information about the size and shape of the particle. For spherical particles Stokes showed that for nonturbulent flow,*

$$f = 6\pi\eta r \qquad (21.2)$$

where η is the viscosity and r is the radius of the spherical particle. The frictional coefficients of prolate and oblate ellipsoids and long rods may be expressed in terms of the radius of a sphere of equal volume and a factor depending on the ratio of the major axis to the minor axis.

The **viscosity** η of a liquid may be determined by measuring the rate of settling of a sphere of known density. The force causing the sphere to settle in the fluid is equal to its effective mass times the acceleration of gravity; the effective mass is the mass of the sphere minus the mass of the fluid it displaces. If the sphere has a density ρ and the density of the medium is ρ_0, the force causing motion is $\frac{4}{3}\pi r^3 (\rho - \rho_0)g$, where g is the acceleration of gravity. When the rate of settling of the sphere in the liquid is constant, the retarding force is equal to the force due to gravity, and so

$$\frac{4}{3}\pi r^3(\rho - \rho_0)g = 6\pi\eta r \left(\frac{dx}{dt}\right) \qquad (21.3)$$

$$\frac{dx}{dt} = \frac{2r^2(\rho - \rho_0)g}{9\eta} \qquad (21.4)$$

Thus, by measuring the velocity dx/dt of settling of a sphere of known r and ρ in a liquid of known density ρ_0, the viscosity η may be obtained. This method is especially valuable for solutions of high viscosity, such as concentrated so-

---

* For a derivation see M. A. Lauffer, *J. Chem. Educ.* **58**:250 (1981).

Table 21.1 Viscosities of Liquids in Pa s (kg m^{-1} s^{-1})

t/°C	0	25	50	75
Water	0.001 793	0.000 895	0.000 549	0.000 380
Ethanol	0.001 79	0.001 09	0.000 698	—
Benzene	0.000 90	0.000 61	0.000 44	—
Glycerol	—	0.945	—	—

lutions of high polymers. Conversely, a determination of the rate of settling of colloidal particles of known density in a liquid of known viscosity provides a means for determining the effective particle radius.

The viscosity η may also be determined by passing a liquid through a capillary tube and making use of the **Poiseuille equation:**

$$\eta = \frac{P\pi r^4 t}{8Vl} \tag{21.5}$$

where t is the time required for volume V of liquid to flow through a capillary tube of length l and radius r under an applied pressure P. The viscosities of most liquids decrease with increasing temperature. According to the "hole theory," there are vacancies in a liquid, and molecules are continually moving into these vacancies so that the vacancies move around. This process permits flow, but requires energy because a molecule must surmount an activation barrier to move into a vacancy. The viscosities of several liquids are shown in Table 21.1. The variation of the viscosity with temperature may be represented quite well by

$$\frac{1}{\eta} = Ae^{-E_a/RT} \tag{21.6}$$

where E_a is the activation energy for the fluidity, $1/\eta$.

The viscosity of a liquid increases as the pressure is increased because the number of holes is reduced, and it is therefore more difficult for molecules to move around each other.

In contrast with liquids, the viscosity of a gas increases as the temperature increases. The viscosity of an ideal gas is independent of pressure (Sections 18.10 and 18.11).

Example 21.1
How fast does an air bubble rise in water at 25 °C if its diameter is 1 mm? See Table 21.1.

The rate of settling can be calculated using equation 21.4. The density of air is negligible compared with the density of liquid water, and so

$$\frac{dx}{dt} = \frac{-2r^2\rho_0 g}{9\xi}$$

$$= \frac{(-2)(5 \times 10^{-4}\ \text{m})^2(10^3\ \text{kg m}^{-3})(9.8\ \text{m s}^{-1})}{9(8.95 \times 10^{-4}\ \text{kg m}^{-1}\ \text{s}^{-1})}$$

$$= -0.61\ \text{m s}^{-1}$$

The rate of settling comes out negative because the bubble rises rather than settles.

21.2 Diffusion

Fick's first law of diffusion was introduced in Section 18.10 as an illustration that a flux is proportional to the gradient of something, in this case the concentration. In this section we want to emphasize that diffusion occurs as a result of a gradient in the chemical potential μ. If the chemical potential for a species is uniform in a system, there is no driving force for diffusion.

A force F, equal to the negative gradient of the chemical potential, causes a substance to move from a region where its chemical potential is high to a region where it is low. For an ideal solution $\mu = \mu° + RT \ln c$, and so the force F causing diffusion in the x direction in an ideal solution is given by

$$F = -\frac{d\mu}{dx} = -\frac{RT}{c}\frac{dc}{dx} \qquad (21.7)$$

The force opposing the diffusion of a molecule or ion is the frictional coefficient f times the velocity v. Setting these forces on a molecule or ion equal to each other yields

$$fv = -\frac{RT}{N_A c}\frac{dc}{dx} \qquad (21.8)$$

or

$$vc = -\frac{RT}{N_A f}\frac{dc}{dx} \qquad (21.9)$$

which corresponds with **Fick's first law** (equation 18.49):

$$J = -D\frac{dc}{dx} \qquad (21.10)$$

where J is the flux and D is the **diffusion coefficient.** Thus, the diffusion coefficient for ideal solutions is given by

$$D = \frac{RT}{N_A f} \qquad (21.11)$$

which shows that the diffusion coefficient is inversely proportional to the frictional coefficient. This relation was first derived by Einstein.

In the preceding section we saw that $f = 6\pi\eta r$ for spherical particles, and so

$$D = \frac{RT}{N_A 6\pi\eta r} \qquad (21.12)$$

In the next chapter (Section 22.12) we will see that this relation can be used to determine the molar mass of a protein molecule that is nearly spherical.

In studying transport processes the flux J is seldom measured directly. What is measured is the change in concentration with time at various points. The continuity equation is the expression of the fact that since mass is conserved, the change in concentration in a region is due to the difference between the flows into and out of the region. To relate the flux to the change in concentration, consider the situation illustrated in Fig. 21.1. Flow of solute is occurring in the x direction in a cell of uniform cross section A. We want to calculate the change in concentration in a thin slab of thickness δx. The quantity of

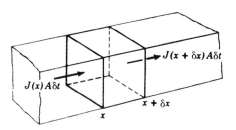

Figure 21.1 Cell of uniform cross section A in which there is transport by diffusion, sedimentation, or electrical migration.

material crossing the plane at x in time δt is $J(x)A\ \delta t$, whereas the quantity leaving through the plane at $x + \delta x$ in the same time is $J(x + \delta x)A\ \delta t$, which may be written

$$\left(J(x) + \frac{\partial J}{\partial x}\delta x \right)A\ \delta t \tag{21.13}$$

The net gain in the quantity of material between these hypothetical planes may be expressed in terms of the change of concentration in the volume $A\ \delta x$ or in terms of the difference between these two quantities of material transported:

$$A\ \delta c\ \delta x = JA\ \delta t - \left(J + \frac{\partial J}{\partial x}\delta x \right)A\ \delta t = -\frac{\partial J}{\partial x}A\ \delta x\ \delta t \tag{21.14}$$

In the limit, as the distances and times are made smaller,

$$\frac{\partial c}{\partial t} = -\frac{\partial J}{\partial x} \tag{21.15}$$

This is referred to as the **equation of continuity.** It relates the change in concentration at a given value of x in the cell to the rate of change of flux with distance.

Fick's first law (equation 21.10) is substituted in the equation of continuity (equation 21.15) to obtain Fick's second law:

$$\frac{\partial c}{\partial t} = \frac{\partial}{\partial x}D\frac{\partial c}{\partial x} \tag{21.16}$$

If the diffusion coefficient D is independent of the concentration and therefore of distance, then

$$\frac{\partial c}{\partial t} = D\frac{\partial^2 c}{\partial x^2} \tag{21.17}$$

which is known as **Fick's second law.**

Diffusion coefficients in solution are usually measured by first forming a sharp boundary between a solution and the solvent, as shown in Fig. 21.2. At a later time the boundary is diffuse.

To derive the expression for concentration as a function of distance and time for the experiment illustrated in Fig. 21.2, equation 21.17 is integrated with the following boundary conditions: When $t = 0$, $c = c_0$ for $x > 0$, and $c = 0$ for $x < 0$; and when $t > 0$, c approaches c_0 as x approaches ∞ and c approaches 0 as x approaches $-\infty$. The result is

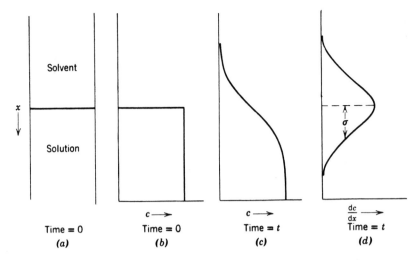

Figure 21.2 Diffusion of an initially sharp boundary in a cell of uniform cross section.

$$c = \frac{c_0}{2}\left(1 + \frac{2}{\sqrt{\pi}}\int_0^{x/2\sqrt{Dt}} e^{-\beta^2}d\beta\right) \qquad (21.18)$$

where the second term in the parentheses is referred to as the Gaussian error function. The equation for the derivative curve in Fig. 21.2d is

$$\frac{\partial c}{\partial x} = \frac{c_0}{2\sqrt{\pi Dt}}e^{-x^2/4Dt} \qquad (21.19)$$

The function $(2\pi)^{-1/2}e^{-y^2/2}$ is referred to as the normal probability function of y. This bell-shaped probability curve is referred to as a **Gaussian curve.**

The square of the standard deviation of the experimental bell-shaped curve is

$$\sigma^2 = \frac{\int_{-\infty}^{\infty} x^2(\partial c/\partial x)dx}{\int_{-\infty}^{\infty} (\partial c/\partial x)dx} \qquad (21.20)$$

Substituting equation 21.19,

$$\sigma^2 = \frac{\int_{-\infty}^{\infty} x^2 e^{-x^2/4Dt}dx}{\int_{-\infty}^{\infty} e^{-x^2/4Dt}dx} = 2Dt \qquad (21.21)$$

where the last form is obtained by using the values of the definite integrals.* Since the **standard deviation** σ of a Gaussian curve is the half-width at the inflection point, and the inflection points are at a height of 0.606 of the maximum ordinate, σ is readily obtained from the experimental curve, and D may be calculated using equation 21.21. The diffusion coefficients of a number of protein molecules are given in Table 22.3.

---

* See Table 18.1.

Example 21.2

(a) The diffusion coefficient of sucrose in water at 25 °C is 5.1×10^{-10} m^2 s^{-1}. What is the standard deviation of an initially sharp boundary (see Fig. 21.2) after 1 h? After 24 h? (b) The diffusion coefficient of hemoglobin in water at 25 °C is 6.9×10^{-11} m^2 s^{-1}. What is the standard deviation of an initially sharp boundary after 1 h? After 24 h?

(a)   After 1 h,

$$\sigma = (2Dt)^{1/2}$$
$$= [2(5.1 \times 10^{-10} \text{ m}^2 \text{ s}^{-1})(1 \text{ h})(60 \text{ min h}^{-1})(60 \text{ s min}^{-1})]^{1/2}$$
$$= 0.00192 \text{ m} = 1.92 \text{ mm}$$

After 24 h,

$$\sigma = (1.92 \text{ mm})24^{1/2} = 11.8 \text{ mm}$$

(b)   After 1 h,

$$\sigma = [2(6.9 \times 10^{-11} \text{ m}^2 \text{ s}^{-1})(1 \text{ h})(60 \text{ min h}^{-1})(60 \text{ s min}^{-1})]^{1/2}$$
$$= 0.705 \text{ mm}$$

After 24 h,

$$\sigma = (0.705 \text{ mm})24^{1/2} = 3.45 \text{ mm}$$

21.3   Mobility of an Ion

A solution containing ions conducts an electric current because the ions move under the influence of an electric field. To measure the electric resistance R of a solution containing ions, the solution is placed in a cell with two electrodes that have a coating of platinum black, and an alternating current is used. When an alternating current is used the electrolysis that occurs when the current passes in one direction is reversed when the current passes in the other direction, and the formation of a nonconducting gas film is prevented.

The electric resistance R of a uniform conductor is directly proportional to its length l and inversely proportional to its cross-sectional area A:

$$R = \frac{rl}{A} = \frac{l}{\kappa A} \tag{21.22}$$

where the proportionality constant r is called the **resistivity** and the proportionality constant $\kappa = 1/r$ is called the **electric conductivity**. In the SI system the electric conductivity κ has the units Ω^{-1} m^{-1}, where the ohm is represented by Ω. Since the potential difference ϕ between two points on a uniform conductor in which a current I is flowing is $\phi = IR$, the unit of electric potential difference ϕ, the volt V, is equal to the product of the unit of electric current, the ampere A, and the unit of resistance, the ohm Ω; 1 V = (1 A)(1 Ω).

Electric conductivities range from 10^8 Ω^{-1} m^{-1} for a metallic conductor at room temperature to 10^{-15} Ω^{-1} m^{-1} for an insulator like SiO_2. The electric

Table 21.2 Electric Mobilities at 25 °C in Water at Infinite Dilution

	$u/10^{-8}$ m^2 V^{-1} s^{-1}		$u/10^{-8}$ m^2 V^{-1} s^{-1}
H^+	36.25	OH^-	20.64
Li^+	4.01	F^-	5.74
Na^+	5.192	Cl^-	7.913
K^+	7.617	NO_3^-	7.406
NH_4^+	7.62	ClO_3^-	6.70
$N(CH_3)_4^+$	4.66	$CH_3CO_2^-$	4.24
Mg^{2+}	5.50	$C_6H_5CO_2^-$	3.36
Ca^{2+}	6.17	SO_4^{2-}	8.29
Pb^{2+}	7.20	CO_3^{2-}	7.18

conductivity κ of an electrolyte solution is made up of contributions from each of the types of ions present.

The **electric mobility** u of an ion is its drift velocity in the direction of the electric field, divided by the electric field strength E:

$$u = \frac{dx/dt}{E} \tag{21.23}$$

(Strictly speaking, the electric field strength is a vector and should be represented by E, but here we will write equations in terms of the magnitude E of the electric field strength, which is a scalar.) The drift velocity of an ion is the average velocity in the direction of the field. Because of Brownian motion, an ion undergoes random displacements so that it does not move in a straight line over macroscopic distances. The electric field strength E is the negative gradient of the electric potential ϕ. When the electric potential varies only in the x direction, then

$$E = -\frac{d\phi}{dx} \tag{21.24}$$

For a uniform conductor the difference in the potential per unit distance may be calculated using Ohm's law. For a conductor of unit cross section the difference in potential between two points is equal to the current density I/A, where I is the current and A the area, multiplied by the resistivity $1/\kappa$:

$$E = \frac{I}{A\kappa} \tag{21.25}$$

Since the electric field strength is expressed in V m^{-1} in the SI system, the electric mobility u has the units m^2 V^{-1} s^{-1}.

The drift velocities of ions can be determined in various ways, and the electric mobilities of a number of small ions at infinite dilution in water at 25 °C are given in Table 21.2.

The remarkably high electric mobility of the hydrogen ion is due to the fact that a proton may be transferred along a series of hydrogen-bonded (Section 12.10) water molecules by rearrangement of the hydrogen bonds. In the following figure (*a*) shows the initial bonding in a group of oriented water molecules and (*b*) shows the final bonding:

$$H—\overset{+}{\underset{\underset{H}{|}}{O}}—H \cdots \underset{\underset{H}{|}}{O}—H \cdots \underset{\underset{H}{|}}{O}—H \cdots \underset{\underset{H}{|}}{O}—H \qquad H—\underset{\underset{H}{|}}{O} \cdots H—\underset{\underset{H}{|}}{O} \cdots H—\underset{\underset{H}{|}}{O} \cdots H—\overset{+}{\underset{\underset{H}{|}}{O}}—H$$

$$(a) \qquad\qquad\qquad\qquad (b)$$

The net effect is rapid long-range proton mobility even though no single proton moves a long distance in a short time. For another hydrogen ion to be transferred to the right through this group of water molecules, molecular rotations must occur to produce again a favorable orientation for charge transfer.

This model for hydrogen ion mobility helps us to understand the remarkable fact that hydrogen ions move about 50 times more rapidly through ice than through liquid water. In ice each oxygen atom is surrounded by four oxygen atoms at a distance of 276 pm in a tetrahedral arrangement. Each hydrogen is near the line through the centers of the oxygen atoms and is about 100 pm from one oxygen and 176 pm from the other. Hydrogen ions may be conducted rapidly through this structure by the above mechanism when the water molecules are oriented correctly. A similar mechanism can be written for the transfer of hydroxyl ions in the opposite direction.

The separation of different macromolecular ions (like protons and nucleic acids) according to their electric mobilities is referred to as **electrophoresis.**

The electric conductivity κ of an electrolyte solution is the sum of the contributions of all the ionic species in the electrolyte. The electric current contributed by an ion depends on its charge number z_i as well as on its electric mobility. The concentration of ion i expressed in faradays of electric charge is $|z_i|c_i$, where c_i is in moles per unit volume. If the ions of type i all move with a velocity of u_i in a field of 1 V/m, the transport of electric charge through a plane perpendicular to the direction of motion is $F|z_i|c_iu_i$. Thus, for the solution as a whole,

$$\kappa = F\sum_i |z_i|c_iu_i = Fc\sum_i |z_i|\nu_iu_i \qquad (21.26)$$

The second form gives the conductivity of a solution of a single electrolyte of concentration c with ν_i ions of type i.

Mobilities increase with the temperature, and the temperature coefficients are very nearly the same for all ions in a given solvent and are approximately equal to the temperature coefficient of the viscosity; for water this is 2% per degree in the neighborhood of 25 °C.

Example 21.3

The electric conductivity κ of pure water is $5.5 \times 10^{-6}\ \Omega^{-1}\ m^{-1}$ at 25 °C. What is the value of the ion product $K_w = [H^+][OH^-]$? The concentrations of hydrogen and hydroxyl ion are, of course, equal and may be calculated using equation 21.26 and the values of the limiting ion mobilities in Table 21.2.

$$\kappa = Fc(u_{H^+} + u_{OH^-})$$

$$c = \frac{\kappa}{F(u_{H^+} + u_{OH^-})} = \frac{5.5 \times 10^{-6}\ \Omega^{-1}\ m^{-1}}{(96\ 485\ C\ mol^{-1})(5.689 \times 10^{-7}\ m^2\ V^{-1}\ s^{-1})}$$

$$= 1.00 \times 10^{-4}\ mol\ m^{-3} = 1.00 \times 10^{-7}\ mol\ L^{-1}$$

Thus, $K_w = (1.00 \times 10^{-7})^2 = 1.00 \times 10^{-14}$ at 25 °C.

Example 21.4
(*a*) What is the electrical conductivity of a solution of 0.01 mol L^{-1} sodium chloride in water at 25 °C, assuming that the electric mobilities of the ions are not significantly different than at infinite dilution? (*b*) Suppose a milliampere is passed through a 1 cm^3 cube of solution between opposite faces. How far will the sodium and chloride ions move in 10 min?

(*a*) Using equation 21.26,

$$\kappa = (96\ 485\ \text{C mol}^{-1})(0.01\ \text{mol L}^{-1})(10^3\ \text{L m}^{-3})[(5.192 + 7.913) \times 10^{-8}\ \text{m}^2\ \text{V}^{-1}\ \text{s}^{-1}]$$

$$= 0.1264\ \text{C s}^{-1}\ \text{m}^{-1}\ \text{V}^{-1} = 0.1264\ \Omega^{-1}\ \text{m}^{-1}$$

(*b*)
$$E = \frac{I}{A\kappa} = \frac{0.001\ \text{A}}{(0.01\ \text{m})^2(0.1264\ \Omega^{-1}\ \text{m}^{-1})}$$

$$= 79.11\ \text{V m}^{-1}$$

For Na$^+$,

$$\Delta x = Eu\ \Delta t = (79.11\ \text{V m}^{-1})(5.192 \times 10^{-8}\ \text{m}^2\ \text{V}^{-1}\ \text{s}^{-1})(10\ \text{min})(60\ \text{s min}^{-1})$$

$$= 2.46\ \text{mm}$$

For Cl$^-$,

$$\Delta x = Eu\ \Delta t = (79.11\ \text{V m}^{-1})(7.913 \times 10^{-8}\ \text{m}^2\ \text{V}^{-1}\ \text{s}^{-1})(10\ \text{min})(60\ \text{s min}^{-1})$$

$$= 3.75\ \text{mm}$$

21.4 Encounter Pairs

We have seen that in a gas, molecules may move many, many molecular diameters between collisions (Section 18.7). But in a liquid a molecule cannot move very far before it collides with a neighbor. This gives rise to the concept that a reactant molecule in a liquid is surrounded by a solvent **cage**. As a result, a given molecule has many collisions with its immediate neighbors before it moves to a new cage. This concept can be applied to two reactant molecules, say A and B. If they do diffuse together, they will be surrounded by a solvent cage that will tend to keep them together until one or the other escapes from the cage. Therefore, collisions between reactant molecules in a liquid will have a very different time sequence in a liquid than in a gas. In a gas the collision frequency is independent of time, but in a liquid the collisions occur in groups, as shown in Fig. 21.3. The groups of collisions are referred to as **encounters.**

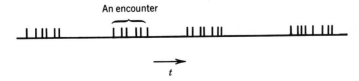

Figure 21.3 Time sequence of collisions between reactants A and B in a liquid. A group of collisions is referred to as an encounter.

At room temperature, an encounter may involve 10 to 10^5 collisions. If there is a significant probability that A and B will react when they collide, it is evident that there is a high probability that they will react during an encounter. When this is true, the rate of the reaction will be controlled by the rate with which A and B can diffuse together to form an encounter pair. Reaction under these conditions is said to be **diffusion controlled.**

We can distinguish between two types of bimolecular reactions in solution by consideration of the following simple mechanism:

$$ A + B \underset{k_{-1}}{\overset{k_1}{\rightleftharpoons}} \{AB\} \xrightarrow{k_2} \text{products} \qquad (21.27) $$

where $\{AB\}$ is the encounter pair. By use of the steady-state approximation (Section 19.10) it is readily shown that

$$ -\frac{d[A]}{dt} = \frac{k_1 k_2}{k_{-1} + k_2} [A][B] \qquad (21.28) $$

If $k_2 \gg k_{-1}$, the reaction rate is determined by the rate k_1 [A][B], which is the rate of reactants diffusing together, and we refer to such a reaction as a **diffusion-controlled reaction.** In a diffusion-controlled reaction, reactant molecules within the same solvent cage collide enough times so that reaction is highly likely before they can diffuse away from each other. For aqueous solutions it has been estimated that the cage lifetime for a pair of noninteracting molecules is of the order 10^{-12} to 10^{-8} s, during which time they may undergo 10 to 10^5 collisions with each other.

If $k_2 \ll k_{-1}$, the reaction rate is

$$ -\frac{d[A]}{dt} = k_2 K_{AB}[A][B] \qquad (21.29) $$

where $K_{AB} = k_1/k_{-1}$ is the equilibrium constant for the formation of the encounter pair. This is an **activation-controlled reaction** because the reaction is largely determined by the activation energy for k_2.

21.5   Diffusion-Controlled Reactions in Liquids

The maximum rate with which reactants can diffuse together in liquids may be calculated using the macroscopic theory of diffusion and experimentally determined diffusion coefficients of the reactants. The elementary theory of diffusion-controlled reactions was developed in 1917 by Smoluchowski in connection with his theoretical study of the coagulation of colloidal gold.

The diffusion coefficient D is defined in terms of Fick's first law, which is given in Sections 18.10 and 21.2. The diffusion coefficients of low molar mass solutes in aqueous solution at 25 °C are of the order of 10^{-9} m^2 s^{-1}. Experimental methods for determining diffusion coefficients in dilute aqueous solutions are discussed in Section 21.2. The diffusion coefficients for ions may be calculated from their ionic mobilities (Section 21.3).

Smoluchowski considered spherical particles with radii R_1 and R_2 that could be considered to react when they diffused within a distance $R_{12} = R_1 + R_2$ of each other.

We may imagine one reactant molecule stationary and serving as a sink. Since the concentration is zero at distance R_{12}, a spherically symmetrical concentration gradient is set up. The flux through this concentration gradient is calculated and is expressed as a **second-order rate constant** k_a for association by

$$k_a = 4\pi N_A(D_1 + D_2)R_{12}f \qquad (21.30)$$

where D_1 and D_2 are the diffusion coefficients of the reactants and f is an electrostatic factor. The electrostatic factor f is different from unity if the reactants are ions. It is larger than unity if the reactants have opposite charges and attract each other, and it is smaller than unity if the reactants have the same charge and repel each other. One ion can be visualized as moving in the electric field created by the other ion.

Because of the **electrostatic factor** f, the effective reaction radius ($R_{eff} = R_{12}f$) is substantially increased for the reaction of oppositely charged ions and substantially decreased for the reaction of two ions with the same sign. If the ionic strength is so low that ion atmospheres may be neglected, f is given by

$$f = \frac{z_1 z_2 e^2}{4\pi\epsilon_0\epsilon_r kTR_{12}}\left[\exp\left(\frac{z_1 z_2 e^2}{4\pi\epsilon_0\epsilon_r kTR_{12}}\right) - 1\right]^{-1} \qquad (21.31)$$

where the charges on the ions are $z_1 e$ and $z_2 e$, ϵ_r is the relative permittivity, and ϵ_0 is the permittivity of free space. More details on the derivation of the equations for diffusion-controlled reactions are given by Hammes.*

The temperature coefficients for diffusion-controlled reactions in water are small because they correspond with the temperature coefficient of the viscosity of liquid water ($E_a = 17.4$ kJ mol^{-1} at 25 °C).

Example 21.5
A typical diffusion coefficient for a small molecule in aqueous solution at 25 °C is 5×10^{-9} m^2 s^{-1}. If the reaction radius is 0.4 nm, what value is expected for the second-order rate constant for a diffusion controlled reaction of neutral molecules?

$k = 4\pi N_A(D_1 + D_2)R_{12} = 4\pi(6.022 \times 10^{23} \text{ mol}^{-1})(10^{-8} \text{ m}^2 \text{ s}^{-1})(0.4 \times 10^{-9} \text{ m})$

$= 3.0 \times 10^7 \text{ m}^3 \text{ mol}^{-1} \text{ s}^{-1} = (3.0 \times 10^7 \text{ m}^3 \text{ mol}^{-1} \text{ s}^{-1})(10^3 \text{ L m}^{-3})$

$= 3.0 \times 10^{10} \text{ L mol}^{-1} \text{ s}^{-1}$

Example 21.6
What is the electrostatic factor f in water at 25 °C if the reaction radius R_{12} is 0.2 nm for opposite unit charges? For like unit charges? The relative permittivity ϵ_r of water is 78.3. For opposite charges,

$$\frac{z_1 z_2 e^2}{4\pi\epsilon_0\kappa kTR_{12}} = \frac{(-1)(1)(1.602 \times 10^{-19} \text{ C})^2(8.988 \times 10^9 \text{ N C}^{-2} \text{ m}^2)}{78.3(1.3807 \times 10^{-23} \text{ J K}^{-1})(298.15 \text{ K})(0.2 \times 10^{-9} \text{ m})} = -3.58$$

$$f = -3.58[e^{-3.58} - 1]^{-1} = 3.68$$

---

* G. G. Hammes, *Principles of Chemical Kinetics*. New York: Academic, 1978.

Thus, a diffusion-controlled reaction is expected to be 3.68 times faster in this case than for uncharged particles.

For ions of the same charge,

$$f = 3.58[e^{3.58} - 1]^{-1} = 0.103$$

Thus, a diffusion-controlled reaction is expected to be 0.103 as fast for single charged particles with the same sign as for neutral particles.

If a reaction occurs so rapidly that appreciable reaction occurs during the process of mixing the reactants conventionally, a flow method may be used. An early example was the study of the reaction of hemoglobin and oxygen.[*] A hemoglobin solution was forced into one arm of a Y-mixer and a solution of oxygen in a buffer into the other. In this way it is possible to mix liquids in about 10^{-3} s. In the stopped-flow method, reagents are forced into the mixer, the flow is brought to a sudden stop, and observations are made of the extent of reaction. In the continuous flow method the solutions are mixed and forced down the tube at a steady rate; the extent of reaction is constant at any given distance down the tube, but increases with distance from the mixing chamber.

Some reactions occur in much less than 10^{-3} s, so their kinetics may not be studied by mixing methods. The time range has been extended down to about 10^{-9} s by the use of **relaxation methods** developed by M. Eigen and coworkers[†] in Göttingen, Germany. A solution in equilibrium is perturbed by rapidly changing one of the independent variables (usually temperature or pressure) on which the equilibrium depends. The change of the system to the new equilibrium is then followed by use of a rapidly responding physical method, for example, light absorption or electrical conductivity.

Equilibria may be shifted by changing the temperature (if $\Delta H \neq 0$) or by changing the pressure (if $\Delta V \neq 0$). A solution may be heated in a microsecond by use of a pulsed laser or by discharging a large electrical capacitor through a special conductivity cell containing the sample. Equilibria may also be shifted by reducing the pressure suddenly by allowing high-pressure gas to escape through a rupture disk. Figure 21.4 is a schematic diagram of a temperature-jump apparatus in which an increase in temperature in a small volume of solution is produced by passing a large current for about 1 μs. If there is a single reaction and the displacement from equilibrium is small, the return to equilibrium at the new higher temperature is presented by

$$\Delta c = \Delta c_0 e^{-t/\tau} \tag{21.32}$$

where τ is the relaxation time (equation 19.17) and Δc_0 is the difference of the concentration of one of the reactants from its equilibrium value at $t = 0$. If several reactions are involved in the return to equilibrium, Δc is expressed by a sum of exponential terms with different relaxation times.

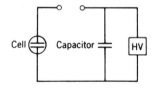

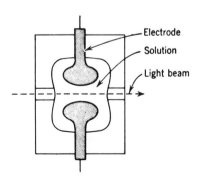

Figure 21.4 Schematic diagram of a temperature-jump apparatus.

---

[*] H. Hartridge and F. J. W. Roughton, *Proc. R. Soc. London Ser. A.* **104**:395 (1923).

[†] M. Eigen and L. De Maeyer, in G. G. Hammes (Ed.), *Investigation of Rates and Mechanisms of Reactions*, 3d ed., p. 63. New York: Wiley, 1974; G. G. Hammes, *Principles of Chemical Kinetics.* New York: Academic, 1978; D. N. Hague, *Fast Reactions.* New York: Wiley-Interscience, 1971.

In another form of relaxation method an independent parameter is varied in a sinusoidal manner and the phase lag in the response of the chemical system is measured. In a sound wave both the pressure and temperature vary and, if the frequency is in the neighborhood of $1/\tau$, there will be enhanced sound absorption.

The linewidth in various types of spectroscopic experiments gives information about the rates of molecular processes. For example, nuclear magnetic resonance (NMR) spectroscopy may be used to determine the rate of a chemical reaction (Section 16.10). Electron spin resonance (ESR) spectroscopy may also be used if the chemical environment of an unpaired electron is affected by a reaction.

21.6 Relaxation Time for a One-Step Reaction

To derive the relationship between the relaxation time τ and the rate constants for a one-step reaction, consider the reaction

$$A + B \underset{k_{-1}}{\overset{k_1}{\rightleftharpoons}} C \tag{21.33}$$

for which the rate equation is

$$\frac{d[C]}{dt} = k_1[A][B] - k_{-1}[C] \tag{21.34}$$

At equilibrium

$$0 = k_1[A]_{eq}[B]_{eq} - k_{-1}[C]_{eq} \tag{21.35}$$

Equation 21.34 may be written in terms of the difference $\Delta[C]$ from the final equilibrium concentrations by introducing

$$[A] = [A]_{eq} - \Delta[C] \tag{21.36}$$

$$[B] = [B]_{eq} - \Delta[C] \tag{21.37}$$

$$[C] = [C]_{eq} + \Delta[C] \tag{21.38}$$

The fact that the differences from equilibrium are the same for the three reactants, except for sign, comes from the stoichiometry of reaction 21.33. Substituting equations 21.36 to 21.38 into equation 21.34 yields

$$\frac{d\,\Delta[C]}{dt} = k_1([A]_{eq} - \Delta[C])([B]_{eq} - \Delta[C]) - k_{-1}([C]_{eq} + \Delta[C]) \tag{21.39}$$

If the displacement from equilibrium $\Delta[C]$ is small,

$$\frac{d\,\Delta[C]}{dt} = -\{k_1([A]_{eq} + [B]_{eq}) + k_{-1}\}\Delta[C] = -\frac{\Delta[C]}{\tau} \tag{21.40}$$

where equation 21.35 has been used and the term in $(\Delta[C])^2$ has been neglected because $\Delta[C]$ is small. Thus, the rate of approach to equilibrium is proportional to the displacement from equilibrium $\Delta[C]$. It is customary to use the relaxation time τ (equation 21.32) to characterize the rate of return to equilibrium. From equation 21.40 we see that for this example

$$\tau = \{k_{-1} + k_1([A]_{eq} + [B]_{eq})\}^{-1} \qquad (21.41)$$

Thus, k_1 and k_{-1} may be obtained as slope and intercept of a plot of τ^{-1} versus $[A]_{eq} + [B]_{eq}$.

Example 21.7
When a sample of pure water in a small conductivity cell is heated suddenly with a pulse of microwave radiation, equilibrium in the water dissociation reaction does not exist at the new higher temperature until additional dissociation occurs. It is found that the relaxation time for the return to equilibrium at 25 °C is 36 μs. Calculate k_1 and k_{-1}.

$$H^+ + OH^- \underset{k_{-1}}{\overset{k_1}{\rightleftharpoons}} H_2O$$

$$\tau = \frac{1}{k_{-1} + k_1([H^+] + [OH^-])}$$

$$K = \frac{[H^+][OH^-]}{[H_2O]} = \frac{k_{-1}}{k_1} = \frac{10^{-14}}{55.5} = 1.8 \times 10^{-16} \text{ mol L}^{-1}$$

Eliminating k_{-1}, we have

$$\tau = \frac{1}{k_1(K + [H^+] + [OH^-])} = \frac{1}{k_1[(1.8 \times 10^{-16}) + (2 \times 10^{-7})]} = 36 \times 10^{-6} \text{ s}$$

$$k_1 = 1.4 \times 10^{11} \text{ L mol}^{-1} \text{ s}^{-1}$$

$$k_{-1} = Kk_1 = (1.8 \times 10^{-16} \text{ mol L}^{-1})(1.4 \times 10^{11} \text{ L mol}^{-1} \text{ s}^{-1}) = 2.5 \times 10^{-5} \text{ s}^{-1}$$

The expressions for the relaxation times of other one-step reactions are readily derived:

$$A \underset{k_{-1}}{\overset{k_1}{\rightleftharpoons}} B \qquad (21.42)$$

$$\tau = (k_{-1} + k_1)^{-1} \qquad (21.43)$$

$$A + B \underset{k_{-1}}{\overset{k_1}{\rightleftharpoons}} C + D \qquad (21.44)$$

$$\tau = \{k_{-1}([C]_{eq} + [D]_{eq}) + k_1([A]_{eq} + [B]_{eq})\}^{-1} \qquad (21.45)$$

If the return to equilibrium involves two steps, there will be two independent rate equations. If the reactions are both near to equilibrium, these equations may be linearized, and the two linear differential equations will yield two relaxation times. The return to equilibrium will then be given by the sum of two exponential terms. In general, the number of exponential terms is equal to the number of independent reactions.*

21.7 Rate Constants for Elementary Reactions in Water

The rate constant of 1.4×10^{11} L mol^{-1} s^{-1} for the combination of hydrogen ions and hydroxyl ions in water at 25 °C is the largest second-order rate constant

---

* G. G. Hammes, *Principles of Chemical Kinetics*, p. 192. New York: Academic, 1978.

Table 21.3 Rate Constants at 25 °C for Elementary Reactions in Dilute Aqueous Solutions (k_a applies to the association reaction and k_d to the dissociation reaction)

Reaction	k_a/L mol^{-1} s^{-1}	k_d/s^{-1}
$H^+ + OH^- \rightleftharpoons H_2O$	1.4×10^{11}	2.5×10^{-5}
$D^+ + OD^- \rightleftharpoons D_2O$	8.4×10^{10}	2.5×10^{-6}
$H^+ + F^- \rightleftharpoons HF$	1.0×10^{11}	7×10^{7}
$H^+ + CH_3CO^- \rightleftharpoons CH_3CO_2H$	4.5×10^{10}	7.8×10^{5}
$H^+ + C_6H_5CO_2^- \rightleftharpoons C_6H_5CO_2H$	3.5×10^{10}	2.2×10
$H^+ + NH_3 \rightleftharpoons NH_4^+$	4.3×10^{10}	24.6
$H^+ + C_3N_2H_4 \rightleftharpoons C_3N_2H_5^+$ (imidazole)	1.8×10^{10}	1.1×10^{3}
$OH^- + NH_4^+ \rightleftharpoons NH_3 + H_2O$	3.4×10^{10}	6×10^{3}
$OH^- + C_3N_2H_5^+ \rightleftharpoons C_3N_2H_4 + H_2O$ (imidazole)	2.5×10^{10}	2.5×10^{3}
$OH^- + {}^+H_3NCH_2CO_2^- \rightleftharpoons H_2NCH_2CO_2^- + H_2O$ (glycine)	1.4×10^{10}	84

known for aqueous solutions. The rate for this reaction would be expected to be large because of the anomalously high mobilities for these ions in aqueous solutions (Section 21.3). However, the reaction is even faster because of an interesting additional effect. Substitution of 1.4×10^{11} L mol^{-1} s^{-1} in equation 21.30, along with the proper electrostatic factor, yields $R_{12} = 0.75$ nm. Since this reaction radius is equal to about three O–H bond distances, the proton apparently "tunnels" to the hydroxyl ion once it is close. **Tunneling** is a quantum mechanical effect in which a particle can penetrate a region even though it does not have the energy to penetrate it classically.

The rate constants k_a for some diffusion-controlled association reactions of H^+ and OH^- are given in Table 21.3, as measured by Eigen and coworkers. Since the association reactions are diffusion controlled, the difference in acid dissociation constants for various weak acids can be attributed primarily to the rate of dissociation of the acid. The reaction of hydroxyl ions with various weak acids is also approximately diffusion controlled because the rate constants are in accord with equation 21.30.

21.8 Acid and Base Catalysis

Acids and bases catalyze many reactions. Suppose the rate of disappearance of a substance S (often called the substrate in a catalytic reaction) is first order in S; $-d[S]/dt = k[S]$. The first-order rate constant k for the reaction in a buffer solution may be a linear function of $[H^+]$, $[OH^-]$, $[HA]$, and $[A^-]$, where HA is the weak acid in the buffer and A^- is the corresponding conjugate base:

$$k = k_0 + k_{H^+}[H^+] + k_{OH^-}[OH^-] + k_{HA}[HA] + k_{A^-}[A^-] \quad (21.46)$$

In this expression k_0 is the first-order constant for the uncatalyzed reaction. The so-called catalytic coefficients k_{H^+}, k_{OH^-}, and k_{A^-} may be evaluated from experiments with different concentrations of these species. If only the term $k_{H^+}[H^+]$ is important, the reaction is said to be subject to **specific hydrogen-**

ion catalysis. If the term k_{HA} [HA] is important, the reaction is said to be subject to **general acid catalysis,** and if the term $k_{A^-}[A^-]$ is important, the reaction is said to be subject to **general base catalysis.**

By considering two types of catalytic mechanisms, we can see how different types of terms arise in equation 21.46. In the first mechanism a proton is transferred from an acid AH^+ to the substrate S, and then the acid form of the substrate reacts with a water molecule to form the product P:

$$S + AH^+ \underset{k_{-1}}{\overset{k_1}{\rightleftharpoons}} SH^+ + A$$

$$SH^+ + H_2O \overset{k_2}{\longrightarrow} P + H_3O^+$$

$$H_3O^+ + A \underset{k_{-3}}{\overset{k_3}{\rightleftharpoons}} AH^+ + H_2O \qquad (21.47)$$

The net reaction is S = P. Assuming that SH^+ is in a steady state, then

$$\frac{d[SH^+]}{dt} = 0 = k_1[S][AH^+] - (k_{-1}[A] + k_2)[SH^+] \qquad (21.48)^*$$

The rate of appearance of product is given by

$$\frac{d[P]}{dt} = k_2[SH^+] = \frac{k_1 k_2[S][AH^+]}{k_{-1}[A] + k_2} \qquad (21.49)$$

where the second form is obtained by solving equation 21.48 for $[SH^+]$. If $k_2 \gg k_{-1}[A]$, then

$$\frac{d[P]}{dt} = k_1[S][AH^+] \qquad (21.50)$$

and the reaction is said to be general acid catalyzed. However, if $k_2 \ll k_{-1}[A]$, then

$$\frac{d[P]}{dt} = \frac{k_1 k_2[S][AH^+]}{k_{-1}[A]} = \frac{k_1 k_2}{k_{-1}K} [S][H^+] \qquad (21.51)$$

Where the second form is obtained by inserting $K = [A][H^+]/[AH^+]$. In this case the reaction is specifically hydrogen-ion catalyzed.

In the second mechanism the acid form of the substrate reacts with a base A instead of a water molecule:

$$S + AH^+ \underset{k_{-1}}{\overset{k_1}{\rightleftharpoons}} SH^+ + A \qquad SH^+ + A \overset{k_2}{\longrightarrow} P + AH^+ \qquad (21.52)$$

The steady-state treatment of this mechanism leads to

$$\frac{d[P]}{dt} = k_2[SH^+][A] = \frac{k_1 k_2[S][AH^+]}{k_{-1} + k_2} \qquad (21.53)$$

which is an example of general acid catalysis.

For mechanisms of the type of 21.47 and 21.52 we might expect a relationship between the rate constant k_a for the acid-catalyzed reaction or k_b for the base-catalyzed reaction to depend on the strength of the acid or base. Indeed, Brønsted found that the rate constant k_a for acid catalysis or k_b for base catalysis

---

* Note that $[H_2O]$ is not written after k_2 because in dilute aqueous solutions $[H_2O]$ cannot be appreciably changed, and so k_2 represents a first-order rate constant.

is proportional to the ionization constant K_a for the acid or K_b for the base raised to some power:

$$k_a = C_A K_a^\alpha \tag{21.54}$$

$$k_b = C_B K_b^\beta \tag{21.55}$$

The exponents α and β are positive and have values between zero and one. The constants C_A, C_B, α, and β apply to a single reaction at a particular temperature catalyzed by different acids and bases. In the Brønsted equation (equations 21.54 and 21.55), low values of α and β indicate a low sensitivity of the catalytic constant to the strength of the catalyzing acid or base.

21.9 Primary Kinetic Salt Effect

Bronsted and Bjerrum investigated the effect of electrolyte concentration on the rate constants of a number of reactions involving ions in aqueous solutions and obtained an unexpected result. We can show that the result they found can be derived from transition state theory and the Debye–Hückel theory. The rate constant k for reactions of the type

$$A^{z_A} + B^{z_B} \rightarrow \text{products} \tag{21.56}$$

is defined by

$$\frac{-d[A]}{dt} = k[A][B] \tag{21.57}$$

Brønsted and Bjerrum found that the rate constant varied with ionic strength I according to

$$k = k° 10^{2A z_A z_B I^{1/2}} \tag{21.58}$$

in the region of low ionic strength where the Debye–Hückel theory is obeyed. In equation 21.58, A is the Debye–Hückel constant ($0.509 \text{ kg}^{1/2} \text{ mol}^{-1/2}$ at 25 °C), $k°$ is the rate constant at zero ionic strength, and z_A and z_B are the charges on ions A and B, with signs. The magnitude of the effect is indicated by Fig. 21.5.

According to **transition-state theory** (equation 20.21), the rate of a reaction is given by

$$\frac{-d[A]}{dt} = \left(\frac{\langle \vec{v}_x \rangle}{2\delta}\right)[X^‡] \tag{21.59}$$

where $X^‡$ represents the transition state for the reaction. In considering electrolyte solutions, nonideality has to be taken into account, and so the expression for the apparent equilibrium constant $K^‡$ has to be written

$$K^‡ = [X^‡]\frac{\gamma_x c°}{[A][B]\gamma_A \gamma_B} \tag{21.60}$$

where $c°$ is the standard state concentration, and the γ's are activity coefficients. Eliminating $[X^‡]$ between these two equations yields

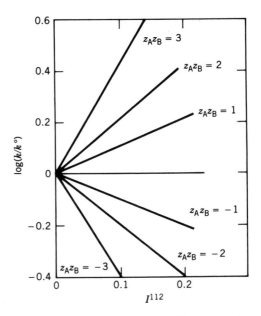

Figure 21.5 Dependence of the rate constants of ionic reactions of various charge types on ionic strength at 25 °C in aqueous solutions. The reactants studied can all be represented by $A^{z_A} + B^{z_B} \rightarrow$ products.

$$\frac{-d[A]}{dt} = \left(\frac{\langle \vec{v}_x \rangle}{2\delta}\right)\left(\frac{K^{\ddagger}\gamma_A\gamma_B}{\gamma_x c^{\circ}}\right)[A][B] \tag{21.61}$$

Comparison with equation 21.57 shows that

$$k = \frac{k^{\circ}\gamma_A\gamma_B}{\gamma_x} \tag{21.62}$$

where $k^{\circ} = (\langle \vec{v}_x \rangle/2\delta)K^{\ddagger}/c^{\circ}$.

At low ionic strengths, the activity coefficients of the ions are given by

$$\log \gamma_i = -Az_i^2 I^{1/2} \tag{21.63}$$

We do not know much about the structure of the transition state $X^{\ddagger}$, but we do know that its charge is $z_A + z_B$. When equation 21.62 is written in logarithmic form, and equation 21.63 is used, we obtain

$$
\begin{aligned}
\log k &= \log k^{\circ} + \log \gamma_A + \log \gamma_B - \log \gamma_x \\
&= \log k^{\circ} - AI^{1/2}[z_A^2 + z_B^2 - (z_A + z_B)^2] \\
&= \log k^{\circ} + 2z_Az_BAI^{1/2}
\end{aligned} \tag{21.64}
$$

If the reacting ions are oppositely charged, raising the ionic strength reduces the effective rate constant because the ions are shielded from each other to a greater extent.

Example 21.8

The rate constant of a reaction

$$A^{+1} + B^{-2} \rightarrow \text{products}$$

is measured at 0.001 ionic strength and at 0.01 ionic strength at 25 °C in water. What is the expected ratio of the rate constants?

To derive a general relation, consider

$$A^{z_A} + B^{z_B} \rightarrow products$$

$$\log k_{0.001} = \log k° + 2z_A z_B (0.509)(0.001)^{1/2}$$

$$\log k_{0.01} = \log k° + 2z_A z_B (0.509)(0.01)^{1/2}$$

Therefore,

$$\log \frac{k_{0.001}}{k_{0.01}} = 2z_A z_B (0.509)(0.001^{1/2} - 0.01^{1/2})$$

$$= 2(1)(-2)(0.509)(0.001^{1/2} - 0.01^{1/2})$$

$$= 0.183$$

$$k_{0.001} = 1.52 k_{0.01}$$

21.10 Oscillating Chemical Reactions

According to everything we have discussed so far, a system at constant temperature and pressure would be expected to approach equilibrium monotonically, without oscillations. But we have also seen that an intermediate can go through a maximum concentration. However, for certain complicated reactions the concentrations of intermediates can oscillate, sometimes in very complicated temporal patterns. This can happen when one substance in a series of reactions is either an activator or inhibitor for a step that occurs earlier in the series. This constitutes a chemical feedback loop. As the overall reaction approaches equilibrium, the oscillations are damped out, and equilibrium is approached asymptotically.

The best known chemical oscillator is the Belousov–Zhabotinskii reaction:

$$2H^+ + 2BrO_3^- + 3CH_2(CO_2H)_2 = 2BrCH(CO_2H)_2 + 3CO_2 + 4H_2O \quad (21.65)$$

When this reaction is catalyzed with Ce^{3+}, the concentration of the intermediate Br^- and the ratio $[Ce^{4+}]/[Ce^{3+}]$ oscillate as shown in Fig. 21.6. The catalyzed reaction involves about 18 steps and 21 different chemical species.

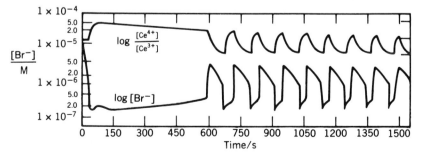

Figure 21.6 Oscillation in $[Br^-]$ and $[Ce^{4+}]/[Ce^{3+}]$ in reaction 21.65 catalyzed by cerium ion. Note that Br^-, Ce^{4+}, and Ce^{3+} are not reactants or products in the net reaction. (Reprinted with permission from R. J. Field, E. Körös, and R. M. Noyes, *J. Am. Chem. Soc.* **94**:8649, 1972. Copyright © 1972 American Chemical Society.)

The underlying theory of oscillating reactions is becoming better understood and is of interest in connection with oscillations observed in biological systems. Chemical systems that produce oscillating reactions can also form spatial structures in initially homogeneous systems.

21.11   Enzyme Catalysis

The most amazing catalysts are the enzymes, which catalyze the multitudinous reactions in living organisms and also provide the means for controlling reaction rates so that the rates of various steps in a series of reactions are compatible. **Enzymes** are proteins, copolymers of aminoacids with specific aminoacid sequences and definite three-dimensional structures. Proteins provide various functional groups at the catalytic site that can interact with a substrate molecule and thereby catalyze a reaction. Some enzymes catalyze a single reaction. An example is fumarase, which catalyzes the hydration of fumarate to L-malate:

$$+ H_2O = \qquad\qquad (21.66)$$

The reactants in an enzyme reaction are generally referred to as substrates. This reaction may be represented by $S = P$, since the concentration of water is constant. Other enzymes catalyze a class of reactions of a given type like ester hydrolysis. Some enzymes require particular metal ions or coenzymes to operate.

Since enzymes are very effective catalysts, they are usually used in laboratory experiments at concentrations much lower than the concentration of the substrate. Generally, the reaction rate that is measured in the laboratory is a steady-state rate. In studies of enzyme kinetics it is advantageous to use initial steady-state rates because the product may be inhibitory.

The initial rate of an enzyme-catalyzed reaction of the type $S = P$ is generally found to be directly proportional to the enzyme concentration. When the substrate concentration is varied, the initial rate is first order with respect to substrate at low [S] and zero order at high [S]. In 1913 **Michaelis and Menten** pointed out that these observations can be explained with the mechanism

$$E + S \underset{k_2}{\overset{k_1}{\rightleftharpoons}} X \overset{k_3}{\longrightarrow} E + P \qquad\qquad (21.67)$$

Since $[E] + [X] = [E]_0$ and $[S] + [P] = [S]_0$, there are two independent rate equations for this mechanism. These can be taken to be

$$\frac{d[X]}{dt} = k_1[E][S] - (k_2 + k_3)[X] \qquad\qquad (21.68)$$

$$\frac{d[P]}{dt} = k_3[X] \qquad\qquad (21.69)$$

These two rate equations cannot be solved to obtain analytic expressions for [E], [S], [X], and [P] as functions of time, but these concentrations may be calculated by use of a computer for specific values of the three rate constants.

Since enzymatic reactions are generally studied with enzyme concentrations (strictly speaking molar concentrations of enzymatic sites) much lower than the concentrations of substrates, it is a good approximation to assume that the enzymatic reaction is in a steady state in which $d[X]/dt = 0$. By introducing the equation for the conservation of enzyme, $[E] = [E]_0 - [X]$, in equation 21.68, we obtain

$$[X] = \frac{k_1[E]_0[S]}{k_1[S] + k_2 + k_3} \tag{21.70}$$

Substituting this expression in equation 21.69 yields

$$\frac{d[P]}{dt} = \frac{k_3[E]_0}{1 + (k_2 + k_3)/k_1[S]} \tag{21.71}$$

This steady-state rate equation for the overall reaction is frequently written

$$v = \frac{k_{cat}[E]_0}{1 + K_M/[S]} \tag{21.72}$$

where k_{cat} is the **turnover number,** in this case k_3, and K_M is the **Michaelis constant,** in this case $(k_2 + k_3)/k_1$.

The initial steady-state velocity v is plotted as a function of the substrate concentration in Fig. 21.7a. At low substrate concentrations the reaction is first order with respect to substrate; but as the substrate concentration is increased, the velocity asymptotically approaches a maximum of $k_{cat}[E]_0$. The quantity k_{cat} is called the turnover number because it is the number of product molecules produced per enzyme molecule (strictly, per catalytic site) per second. The turnover number is about 10^6 s^{-1} for catalase, which catalyzes the decomposition of H_2O_2 to $H_2O + \frac{1}{2}O_2$ and about 100 s^{-1} for chymotrypsin, which catalyzes the hydrolysis of a number of esters and amides. It is evident from equation 21.72 that K_M is equal to the concentration of substrate required to give half the maximum velocity.

To obtain the best values of k_{cat} and K_M, it is useful to be able to plot the kinetic data as a straight line. There are three ways that this can be done, but since it is desirable to show the full accessible range of experiments along both axes, the Eadie–Hofstee plot is best.* This method uses the Michaelis–Menten equation (21.72) written in the following form:

$$\frac{v}{[E]_0[S]} = \frac{k_{cat}}{K_M} - \frac{v}{K_M[E]_0} \tag{21.73}$$

As shown in Fig. 21.7b, the Michaelis constant can be obtained from the slope, and the turnover number can be obtained directly from the intercept on the abscissa.

The reversibility of the overall reaction can be provided for by including the reverse reaction for the second step:

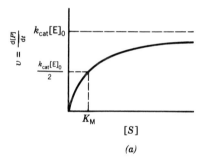

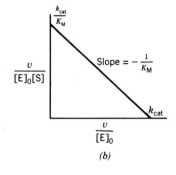

Figure 21.7 (a) Plot of initial steady-state velocity v of appearance of product P at different initial concentrations of substrate S. (b) Eadie–Hostee plot of enzyme kinetic data to obtain the parameters K_M and k_{cat}.

---

* H. B. Dunford, *J. Chem. Educ.* **61**:129 (1984).

$$E + S \underset{k_2}{\overset{k_1}{\rightleftharpoons}} X \underset{k_4}{\overset{k_3}{\rightleftharpoons}} E + P \qquad (21.74)$$

The rate equations for this mechanism are

$$\frac{d[X]}{dt} = k_1[E][S] - (k_2 + k_3)[X] + k_4[E][P] \qquad (21.75)$$

$$\frac{d[P]}{dt} = k_3[X] - k_4[E][P] \qquad (21.76)$$

Substituting $[E]_0 = [E] + [X]$ in equation 21.75 and assuming that $d[X]/dt = 0$ makes it possible to solve equation 21.75 for $[X]$. When this expression and the expression for the conservation of enzymatic sites are substituted in equation 21.76, it is found that the steady-state rate is given by

$$\frac{d[P]}{dt} = v = \frac{k_1 k_3 [S][E]_0 - k_2 k_4 [P][E]_0}{k_2 + k_3 + k_1[S] + k_4[P]} \qquad (21.77)$$

It is convenient to rearrange this rate equation by introducing expressions for the Michaelis constants for the substrate and product:

$$K_S = \frac{k_2 + k_3}{k_1} \qquad K_P = \frac{k_2 + k_3}{k_4}$$

to obtain

$$v = \frac{d[P]}{dt} = \frac{(k_3/K_S)[S] - (k_2/K_P)[P]}{1 + [S]/K_S + [P]/K_P}[E]_0$$

$$= \frac{k_3[E]_0(1 - [P]/[S]K)}{1 + K_S/[S] + K_S[P]/K_P[S]} \qquad (21.78)$$

Thus, the equilibrium constant for the overall reaction is given by

$$K = \frac{[P]_{eq}}{[S]_{eq}} = \frac{k_3 K_P}{k_2 K_S} = \frac{k_1 k_3}{k_2 k_4} \qquad (21.79)$$

Equation 21.77 gives the steady-state velocity for any mixture of substrate and product. It reduces to equation 21.71 if product is not added initially and the reaction goes essentially to completion. The addition of product has two effects: The [P] term in the numerator of equation 21.77 results from the reverse reaction and the [P] term in the denominator is due to product inhibition. It is evident from equation 21.79 that the four rate constants in the mechanism are not independent; they are related through the equilibrium constant.

Example 21.9
At 25 °C and pH 8 the turnover numbers and Michaelis constants for the fumarase reaction fumarate + H_2O = L-malate are

$$k_3 = 0.20 \times 10^3 \text{ s}^{-1} \qquad k_2 = 0.60 \times 10^3 \text{ s}^{-1}$$
$$K_S = 7.0 \times 10^{-6} \text{ mol L}^{-1} \qquad K_P = 100 \times 10^{-6} \text{ mol L}^{-1}$$

What are the values of k_1 and k_4 and the equilibrium constant $[P]_{eq}/[S]_{eq}$?

$$k_1 = \frac{k_2 + k_3}{K_S} = \frac{0.80 \times 10^3 \text{ s}^{-1}}{7.0 \times 10^{-6} \text{ mol L}^{-1}} = 1.14 \times 10^8 \text{ L mol}^{-1} \text{ s}^{-1}$$

$$k_4 = \frac{k_2 + k_3}{K_P} = \frac{0.80 \times 10^3 \text{ s}^{-1}}{100 \times 10^{-6} \text{ mol L}^{-1}} = 8.0 \times 10^6 \text{ L mol}^{-1} \text{ s}^{-1}$$

$$K_{eq} = \frac{[\text{P}]_{eq}}{[\text{S}]_{eq}} = \frac{k_1 k_3}{k_2 k_4}$$

$$= 4.8$$

Compounds that are structurally related to the substrate or product often combine with the catalytic site of the enzyme and cause **inhibition;** that is, the enzyme-catalyzed reaction is slowed down by the inhibitor. Since the substrate and the inhibitor compete for the same site, the effect of the inhibitor may be reduced by raising the substrate concentration. This type of inhibition, called competitive inhibition, may be represented by the mechanism

$$\text{E} + \text{S} \rightleftharpoons \text{ES} \rightarrow \text{E} + \text{P} \qquad \text{E} + \text{I} \rightleftharpoons \text{EI} \qquad K_I = \frac{[\text{E}][\text{I}]}{[\text{EI}]} \qquad (21.80)$$

where I is the inhibitor. The steady-state rate law is

$$v = \frac{V_S}{1 + (K_S/[\text{S}]) (1 + [\text{I}]/K_I)} \qquad (21.81)$$

where K_I is the dissociation constant of EI into E and I and $V_S = k_{cat}[\text{E}]_0$.

Substances that bind to the enzyme, although not at the active site, may not interfere with the binding of the substrate at the active site, but may alter K_S and V_S. Such inhibitors are referred to as noncompetitive. The binding of substances that are not directly involved in an enzymatic reaction can also increase the reaction rate, and such substances are called **activators.**

The way in which V_S and K_S depend on pH, salt concentration, coenzyme concentration, and so on, gives further information about the enzymatic mechanism. In general, an enzymatic reaction has an optimum pH; the maximum velocity V_S decreases as the pH is raised or lowered from the optimum pH. In the neutral pH range the effects are generally reversible, but proteins are irreversibly denatured at extreme pH values. Reversible effects of pH on V_S may be attributable to the ionization of the enzyme–substrate complex. If the enzyme–substrate complex exists in three states with different numbers of protons, and if only the intermediate form breaks down to give product, the expression for the effect of pH on the maximum velocity may be derived from

$$
\begin{array}{c}
\text{ES} \\
K_{bES} \updownarrow \\
\text{HES} \xrightarrow{k} \text{enzyme} + \text{product} \qquad (21.82) \\
K_{aES} \updownarrow \\
\text{H}_2\text{ES}
\end{array}
$$

where K_{aES} and K_{bES} are acid dissociation constants, and k is the rate constant for the rate-determining step. Since

$$[E]_0 = [ES] + [HES] + [H_2ES] \tag{21.83}$$

$$= [HES]\left(1 + \frac{[H^+]}{K_{aES}} + \frac{K_{bES}}{[H^+]}\right)$$

then

$$V_S = k[HES] = \frac{k[E]}{1 + [H^+]/K_{aES} + K_{bES}/[H^+]} \tag{21.84}$$

If this is the mechanism, then a plot of V_S versus pH is a symmetrical bell-shaped curve.

Since a protein has many dissociable acid groups, it is perhaps surprising that the experimental results are sometimes as simple as they are. There is a simple explanation of why two acid groups in the catalytic site might have the total effect on the kinetics that is represented by equation 21.84. If the catalytic function involves both an acidic function and a basic function, we would expect H_2ES to be inactive because the basic site is occupied by a proton and ES to be inactive because it cannot donate a proton to the substrate. Only HES can yield product because it has one group that can donate a proton and another that can accept a proton.

Mechanisms like the ones we have been discussing can give a variety of complicated effects; that is, the rate law may have a very complicated form. But, in general, the steady-state rate is somewhere between zero order and first order in substrate concentration. However, there are some enzymes for which the steady-state rate varies with a higher power of the substrate concentration. In other words, curves analogous to the sigmoid oxygen-binding curve for hemoglobin (Section 9.9) are obtained. This has been found to be especially true of enzymes of importance in the regulation of metabolic pathways. These cooperative effects are encountered with multisite enzymes, not single-site enzymes, because the cooperative effect involves an increased affinity of a second site for a substrate when a first site is occupied. As in the case of hemoglobin, this interaction involves a structural change. According to the Monod–Changeaux–Wyman model*, the multisite enzyme can exist in at least two states. In each of the two states the conformations of all the subunits are assumed to be the same. The binding of substrate shifts the equilibrium toward one or the other of these two states. If the effector drives the equilibrium in the direction that produces an enhanced rate of reaction, the effector is called an activator. If it causes a reduction in rate it is called an inhibitor. As we have seen in the case of hemoglobin, the effect is multiplied by the fact that one effector molecule affects several catalytic sites on the molecule. The fact that enzymatic activities may be affected by various substances present in the cell provides a mechanism for the control of the rates of reactions in living things so that metabolic intermediates do not accumulate.

---

* J. Wyman and S. J. Gill, *Binding and Linkage*. Mill Valley, CA: University Science Books, 1990.

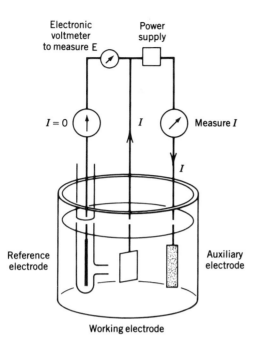

Figure 21.8 Three-electrode cell for the study of electrochemical kinetics. The power supply is adjusted to give the desired current I between the working electrode and the auxiliary electrode. A reference electrode is used to measure the potential E of the working electrode without any current passing through the reference electrode.

21.12 Electrochemical Kinetics*

Electrochemical cells provide a means for studying the rates and mechanisms of reactions that occur at electrodes. Studies of electrochemical kinetics are carried out with cells with three electrodes, rather than the two we considered in Chapter 8. This is done so that the potential of a **working electrode** can be measured with respect to a stable **reference electrode** through which no current is being passed. The three electrodes and the measurements that are made are shown in Fig. 21.8. A direct-current power supply causes a current I to flow between the working electrode and an **auxillary electrode,** usually a calomel electrode. The potential E of the working electrode is measured versus the reference electrode. We will use E_{eq} for the equilibrium potential for the working electrode when no current is flowing from the auxiliary electrode. When a current is being passed through the working electrode, it is said to be **polarized.** An extreme example of this is a surface of mercury in contact with a 1 molar solution of potassium chloride. A current-potential plot for this electrode is illustrated in Fig. 21.9a. No current flows over a range of about 1.5 V. Over

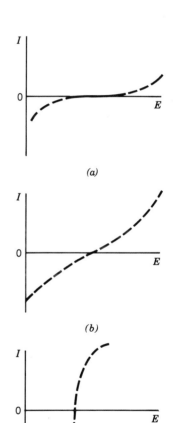

Figure 21.9 Current–potential curves for (a) ideal polarizable electrode, (b) actual electrode, and (c) ideal nonpolarizable electrode. The slope in (c) at low current is due to the resistance of the electrolyte. The dashed lines show the behavior of actual electrodes.

---

* A. J. Bard and L. R. Faulkner, *Electrochemical Methods*. New York: Wiley, 1980; P. H. Rieger, *Electrochemistry*. Englewood Cliffs, NJ: Prentice-Hall, 1987; J. Albery, *Electrode Kinetics*. New York: Oxford, 1975; R. Parsons, *Pure Appl. Chem.* **37**: 499 (1974); **52**:233 (1980); D. R. Crow, *Principles and Applications of Electrochemistry*. London: Chapman & Hall, 1988.

this range of potential, no electrochemical reactions are possible. For reasons that we will see later, no hydrogen is produced because of the kinetic barrier to the production of hydrogen resulting from the large activation energy for the reaction $H_2O + e^- = \frac{1}{2}H_2 + OH^-$. This range of potential over which $I \simeq 0$ is referred to as the ideal polarized range. Figure 21.9*b* shows that the plot for an actual electrode has a finite slope in the neighborhood of the equilibrium potential, which is the potential where $I = 0$.

The other extreme type of behavior is shown by a saturated calomel electrode. If an electrochemical cell consists of two saturated calomel electrodes, the plot of current versus applied potential looks something like Fig. 21.9*c*. The potential of the calomel electrode does not depend very much on the current. This limiting type of electrode is referred to as an **ideal nonpolarizable electrode.**

When a current I flows through an electrochemical cell, the applied potential E is made up of three contributions:

$$E = E_{eq} + IR + \eta \qquad (21.85)$$

The first term E_{eq} is in principle obtained even when no current is flowing. When a current I flows between the electrodes, there is a potential drop IR, where R is the resistance of the electrolyte between the electrodes; this contribution can perhaps be made negligible by putting the electrodes close together. The third contribution η is referred to as the **overpotential.** It is the deviation of the potential of an electrode from its equilibrium value required to cause a given current to flow. The overpotential is a measure of the Gibbs energy difference required to drive an electrode reaction at a given rate. For certain electrodes, like the calomel electrode, the overpotential is very small, but for most electrodes a significant overpotential is required to cause any significant current to flow. The rate of an electrode reaction is controlled by two factors: the kinetics of the reaction itself and the rate of transport of reactants to the electrode interface. The latter process can be speeded up by vigorous stirring and will be assumed here to be fast compared with the rate of electrochemical reaction.

An electrical current I is by convention a current of positive charges, even if the actual carriers are not positive. In discussing currents that flow at an electrode, the convention is that the (positive) current is positive if it flows from the electrode into the electrolyte. When the reaction that predominates at an electrode is an oxidation (Red $\rightarrow$ Ox $+ ne^-$), that electrode is referred to as the **anode** (see Section 8.3). Since the electrons produced pass into the electrode, the positive current flows into the electrolyte. This current is referred to as the **anodic partial current** and is represented by I_a. Note that I_a is positive.

When the reaction that predominates at an electrode is a reduction (Ox $+ ne^- \rightarrow$ Red), that electrode is referred to as the **cathode.** Since electrons flow out of the electrode, the positive current flows into the electrode. This current is referred to as the **cathodic partial current** and is represented by I_c. Note that I_c is negative; it is going in the opposite direction from I_a.

Figure 21.10 shows the partial currents at an electrode where oxidation predominates, at equilibrium, and at an electrode where reduction predominates when a single electrochemical reaction occurs. The total current I is given by

$$I = I_a + I_c \qquad (21.86)$$

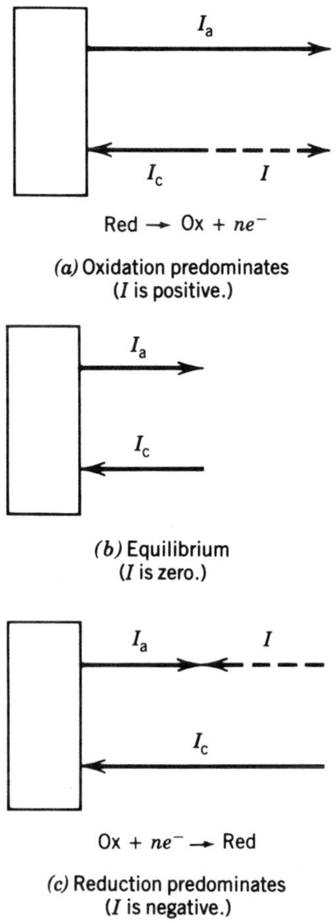

(a) Oxidation predominates
(I is positive.)

(b) Equilibrium
(I is zero.)

(c) Reduction predominates
(I is negative.)

Figure 21.10 Partial anodic and cathodic currents I_a and I_c are shown by solid lines, and the total current I is shown by a dashed line. Arrows to the right indicate that positive current flows from the electrode to the electrolyte. *(a)* The potential of the electrode is sufficiently positive so that oxidation predominates. *(b)* The electrode is at its equilibrium potential so that $I = 0$. *(c)* The potential of the electrode is sufficiently negative so that reduction predominates.

When an electrode at which a single reaction occurs is at equilibrium, $I = 0$ and $I_{a,eq} = -I_{c,eq}$. Thus, equilibrium does not mean the absence of reaction, but rather the equality of rates in the forward and backward directions. The partial currents at equilibrium are referred to as the **exchange current** I_0 of the reaction; thus,

$$I_0 = I_{a,eq} = -I_{c,eq} \tag{21.87}$$

When the potential of an electrode does not equal its equilibrium potential, the current I provides information about the rate constants of the electrode reaction

$$\text{Ox} + ne^- \underset{k_{ox}}{\overset{k_{red}}{\rightleftarrows}} \text{Red} \tag{21.88}$$

To keep the following discussion as simple as possible, we will assume that this is the only reaction that occurs. We will also assume that the solutions are ideal so that the Nernst equation (equation 8.30) for reaction 21.88 can be written

$$E_{eq} = E^\circ - \frac{RT}{nF} \ln \frac{[\text{Red}]}{[\text{Ox}]} \tag{21.89}$$

where E° is the standard potential for the electrode. For an actual reaction, activity coefficients have to be taken into account, but that is often done by keeping the ionic strength constant and incorporating the activity coefficients into E°, which is then referred to as $E^{\circ\prime}$.

If the potential applied to a working electrode is sufficiently positive, an oxidation reaction will occur, and in the limit the current will be equal to the anodic partial current I_a because the cathodic partial current will be negligible. The anodic partial current is proportional to the rate constant for the oxidation reaction and is given by

$$I_a = nFAk_{ox}[\text{Red}] \tag{21.90}$$

where n is the charge number of the electrode reaction (as written), F is the Faraday constant ($96\,485\ \text{C mol}^{-1}$), and A is the area of the electrode. Since n is a number, the electrode reaction rate constant k_{ox} has the SI units of m s^{-1}. In some cases the rate law has a more complicated form, but we will not consider those cases. Strictly speaking, [Red] is the concentration at the surface of the electrode, but we will consider low currents and vigorous stirring so that [Red] can be taken as the bulk concentration.

If the potential applied to the working electrode is sufficiently negative, a reduction reaction will occur and in the limit the current will be equal to the cathodic partial current I_c because the anodic current will be negligible. The cathodic partial current is proportional to the rate constant for the reduction reaction k_{red} and is given by

$$I_c = -nFAk_{red}[\text{Ox}] \tag{21.91}$$

where [Ox] is the bulk concentration.

Substituting equations 21.90 and 21.91 into equation 21.86 yields

$$I = nFA(k_{ox}[\text{Red}] - k_{red}[\text{Ox}]) \tag{21.92}$$

where k_{red} and k_{ox} are exponential functions of the potential E. The extent of this dependence is represented by the value of the transfer coefficient α. The **anodic transfer coefficient** α_a is defined by

$$\alpha_a = \frac{RT}{nF}\left(\frac{\partial \ln k_{ox}}{\partial E}\right)_{T,P} \tag{21.93}$$

and the **cathodic transfer coefficient** α_c is defined by

$$\alpha_c = -\frac{RT}{nF}\left(\frac{\partial \ln k_{red}}{\partial E}\right)_{T,P} \tag{21.94}$$

If the transfer coefficients are independent of potential E, then these equations can be integrated from the standard potential $E°$ to potential E to obtain

$$k_{ox} = k_{ox}° \exp\left[\frac{\alpha_a nF}{RT}(E - E°)\right] \tag{21.95}$$

$$k_{red} = k°_{red} \exp\left[-\frac{\alpha_c nF}{RT}(E - E°)\right] \tag{21.96}$$

where $k°_{red}$ and $k°_{ox}$ are the rate constants at the standard potential $E°$.

Example 21.10

(a) By what factor is k_{ox} changed if E is changed by 1 V at 25 °C? Assume that α_a is 0.5 and $n = 1$. (b) What change in potential E is required to change k_{ox} by a factor of 10?

(a) $k_{ox2}/k_{ox1} = \exp\left[\frac{\alpha_a nF}{RT}(E_2 - E_1))\right]$

$\qquad\qquad = \exp\left[\frac{(0.5)(96\ 485\ \text{C mol}^{-1})(1\ \text{V})}{(8.3145\ \text{J K}^{-1}\text{ mol}^{-1})(298\ \text{K})}\right]$

$\qquad\qquad = 2.83 \times 10^8$

Thus, a change in applied potential of one volt causes a tremendous change in the rate of one electrode reaction.

(b) $10 = \exp[(19.47\ \text{V}^{-1})(E_2 - E_1)]$

$\qquad E_2 - E_1 = 0.118\ \text{V}$

Now that we have expressions for k_{red} and k_{ox}, let us consider an electrode at its equilibrium potential, for which $I = 0$. Equation 21.92 indicates that

$$k_{ox}[\text{Red}] = k_{red}[\text{Ox}] \tag{21.97}$$

Substituting equations 21.95 and 21.96 in equation 21.97 yields

$$(\alpha_a + \alpha_c)(E_{eq} - E°) = \frac{RT}{nF}\ln\frac{k°_{red}}{k°_{ox}} - \frac{RT}{nF}\ln\frac{[\text{Red}]}{[\text{Ox}]} \tag{21.98}$$

Since we have assumed that the electrode reaction is at equilibrium, this equation has to agree with the Nernst equation (equation 21.89). For equation 21.98 to reduce to equation 21.89, we can see that

$$\alpha_a + \alpha_c = 1 \tag{21.99}$$

and

$$\frac{k_{red}^{\circ}}{k_{ox}^{\circ}} = 1 \qquad (21.100)$$

Note that we have assumed that the four parameters are independent of the applied potential E. Thus, we can see that in our kinetic treatment we have used two more constants than we needed to. This situation was encountered earlier in Chapter 19 when rate equations were compared with equilibrium equations; the forward and backward rate constants for a reaction in a single phase are related through the equilibrium constant. Since $k_{red}^{\circ} = k_{ox}^{\circ}$, we will drop the subscripts and refer to them as the **conditional rate constant** k°.

Equations 21.95 and 21.96 for k_{ox} and k_{red} can be written in terms of the conditional rate constant k° and the cathodic transfer coefficient α_c as follows

$$k_{ox} = k^{\circ} \exp\left[(1 - \alpha_c) \frac{nF}{RT} (E - E^{\circ}) \right] \qquad (21.101)$$

$$k_{red} = k^{\circ} \exp\left[-\frac{\alpha_c nF}{RT} (E - E^{\circ}) \right] \qquad (21.102)$$

The total current can be expressed in terms of the applied potential E by substituting these expressions into equation 21.92:

$$I = nFAk^{\circ} \left\{ [\text{Red}] \exp\left[(1 - \alpha_c) \frac{nF}{RT} (E - E^{\circ}) \right] \right.$$
$$\left. - [\text{Ox}] \exp\left[-\frac{\alpha_c nF}{RT} (E - E^{\circ}) \right] \right\} \qquad (21.103)$$

This equation is very important because it gives the total current I as a function of the applied potential E. In principle, it provides a way to evaluate the conditional rate constant and the cationic transfer coefficient. We will soon see that it can be written in a simpler way.

At equilibrium, the anodic and cathodic partial currents are equal in magnitude, but of opposite sign, as we saw in equation 21.87. If we use equations 21.91 and 21.102, the exchange current I_0 is given by

$$I_0 = nFAk^{\circ} [\text{Ox}] \exp\left[-\frac{\alpha_c nF}{RT} (E_{eq} - E^{\circ}) \right] \qquad (21.104)$$

The Nernst equation can be written in exponential form and raised to the $-\alpha_c$ power to obtain

$$\exp\left[-\frac{\alpha_c nF}{RT} (E_{eq} - E^{\circ}) \right] = \left(\frac{[\text{Ox}]}{[\text{Red}]} \right)^{-\alpha_c} \qquad (21.105)$$

Substituting equation 21.105 in equation 21.104 yields

$$I_0 = nFAk^{\circ} [\text{Ox}]^{\alpha_a} [\text{Red}]^{\alpha_c} \qquad (21.106)$$

where the concentrations are bulk concentrations. Thus, the partial currents at equilibrium are directly proportional to the conditional rate constant k°.

Equation 21.103 is written in terms of I, but the corresponding expression for I/I_0 can be obtained by dividing by equation 21.106. Furthermore, the re-

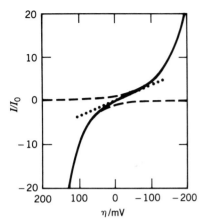

Figure 21.11 Current–potential curve predicted by the Butler–Volmer equation for $\alpha_c = 0.4$ at 25 °C. The dashed lines represent the anodic and cathodic components of the current, and the dotted line represents the linear region near the equilibrium potential. (From P. H. Rieger, *Electrochemistry*, © 1987, p. 274. Reprinted by permission of Prentice-Hall Inc., Englewood Cliffs, NJ.)

sulting equation can be expressed in terms of $E - E_{eq}$, rather than $E - E°$, by use of the Nernst equation. If IR in equation 21.85 is negligible, $E - E_{eq}$ can be replaced by the overpotential η. This yields the following much simpler form of equation 21.103:

$$\frac{I}{I_0} = \exp\left[\frac{(1 - \alpha_c)nF\eta}{RT}\right] - \exp\left[\frac{-\alpha_c nF\eta}{RT}\right] \tag{21.107}$$

which is known as the **Butler–Volmer equation.** A current-potential curve predicted by the Butler–Volmer equation for $\alpha_c = 0.4$ and 25 °C is shown in Fig. 21.11. This figure illustrates the point that at the equilibrium potential E_{eq}, the anodic and cathodic components of the current are equal. It also shows that for small positive and negative overpotentials the current is linear in the overpotential.

Under certain conditions equation 21.107 reduces to simpler forms that are often useful in determining parameters. When the overpotential is sufficiently large, one of the terms will be negligible. For example, when the overpotential is large and negative, the anodic current is negligible, and

$$I = I_0 \exp\left(-\frac{\alpha_c nF\eta}{RT}\right) \tag{21.108}$$

or

$$\ln I = \ln I_0 - \frac{\alpha_c nF\eta}{RT} \tag{21.109}$$

This is known as the Tafel equation. Since it becomes a better approximation as the overpotential increases, there is a danger that the current may become limited by the rate of transport; when this occurs, equation 21.109 cannot be used. Another limiting form of equation 21.107 is obtained when the overpotential is very small. Since $e^x = 1 + x + \cdots$, equation 21.107 reduces to

$$I = \frac{I_0 \eta nF}{RT} \tag{21.110}$$

This provides a means for determining the exchange current I_0.

If the exchange current is large, the slope of the current-potential curve is large near the equilibrium potential. A calomel electrode has a large exchange current. As the exchange current increases, the behavior of an electrode approaches that for an ideal nonpolarizable electrode (see Fig. 21.9c). As I_0 decreases, the plot for an electrode approaches that for an ideal polarizable electrode. In comparing exchange currents for various electrodes, it is convenient to use the exchange current density j_0, which is defined by I_0/A, where A is the area of the working electrode. Hydrogen on zinc has an exchange current density j_0 of about 10^{-5} A/m^2, and the electrode reaction is said to be sluggish. The exchange current density for hydrogen on platinized platinum is about 10 A/m^2, and the electrode reaction is said to be facile.

We have considered only one type of one-step electrode reaction. The rate law may be quite different and much more complicated if there are several steps in the reaction. There are also many different types of electrochemical experiments, which yield other interesting types of information.

21.13   Special Topic: Relation Between Diffusion and Brownian Motion

Diffusion may be considered to result from the random motion of molecules. The random motion of larger particles (but not too large) may be observed directly. This phenomenon is called **Brownian motion** because it was first observed by the English botanist Robert Brown in studying pollen grains with a microscope in 1827.

The Brownian motion of particles that are too small to be seen with a microscope can be studied with an ultramicroscope. In an ultramicroscope a bright beam of light is brought to focus within the colloidal solution, and this region of the solution is viewed with a microscope at right angles to the beam of light. If the particles have a sufficiently different refractive index from the medium, it is possible to see spots of light even though the particles are below the resolving power of the microscope.

The theory of Brownian motion is closely related to the theory of **random flight** discussed in Section 22.4. As a simplification we will consider here only the component of Brownian motion in a single direction; this is referred to as one-dimensional random walk. If the steps are not of equal length, the probability $W(\xi)\,d\xi$ of a displacement ξ along a line is given by equation 22.22. In studying Brownian motion, the positions of a particle are determined after equal intervals of time Δt. From these observations the mean square displacement in time Δt can be determined. Since the steps are defined by the time interval Δt, the number of steps n is equal to $t/\Delta t$ so that equation 22.22 becomes

$$W(x) = \frac{1}{[2\pi t\langle(\Delta x)^2\rangle/\Delta t]^{1/2}}\, e^{-x^2\Delta t/2t\langle(\Delta x)^2\rangle} \tag{21.111}$$

where $\langle(\Delta x)^2\rangle$ is the average value of the square of the step length in the x direction. This equation provides a quantitative description of Brownian motion in one dimension.

Now we want to look at the Brownian motion experiment in terms of the macroscopic observation of diffusion. To do this let us imagine a diffusion experiment in which a sharp band of solute is formed at $x = 0$ and $t = 0$ in a cell of uniform cross section. (This is a different experiment from that in Fig. 21.2 where a solution and solvent have a sharp interface at $x = 0$ and $t = 0$.) The solution of Fick's second law (equation 21.17) for diffusion from a sharp band containing n/A moles per unit area is

$$c(x, t) = \frac{n}{A\sqrt{4\pi Dt}}\, e^{-x^2/4Dt} \tag{21.112}*$$

Since equations 21.111 and 21.112 describe the same experiment, the exponents must be equal and, therefore,

$$D = \frac{\langle(\Delta x)^2\rangle}{2\Delta t} \tag{21.113}$$

---

* C. R. Cantor and P. R. Schimmel, *Biophysical Chemistry*, Part II, p. 579. San Francisco: Freeman, 1980.

Thus, the diffusion coefficient of a particle that can be "seen" with a microscope may be calculated from measurements of values of its x coordinate after successive equal intervals of time. Perrin used this equation with data on spherical particles of known radius to calculate one of the early values of the Avogadro constant using equation 22.86.

Example 21.11
What is the root-mean-square displacement in the x direction of a sucrose molecule due to its Brownian motion in one minute in water at 20 °C? ($D = 4.65 \times 10^{-10}$ m^2 s^{-1})

$$\langle(\Delta x)^2\rangle^{1/2} = \sqrt{2D\,\Delta t}$$
$$= [(2)(4.65 \times 10^{-10} \text{ m}^2 \text{ s}^{-1})(60 \text{ s})]^{1/2}$$
$$= 2.36 \times 10^{-4} \text{ m} = 0.236 \text{ mm}$$

21.14   Special Topic: Kinetics of the Hydration of CO$_2$

In water, carbon dioxide exists largely as dissolved CO_2, not as H_2CO_3. The half-life for the hydration–dehydration reaction depends on the pH, and at neutral pH values is long enough that it may be studied with simple equipment. It is the slowness of this reaction that accounts for the fading of the endpoint when carbonate ion is titrated with acid. The half-time for this uncatalyzed reaction in aqueous buffers is of special interest because this reaction is a necessary step in our exhaling CO_2. In fact, blood passes through our lungs so fast that this reaction would greatly restrict the rate of elimination of CO_2, if it were uncatalyzed. In living things this reaction is catalyzed by carbonic anhydrase, an enzyme, and CO_2 is hydrated and H_2CO_3 dehydrated rapidly enough for the physiological requirements.

To discuss the kinetics of the uncatalyzed hydration of CO_2, it is necessary to consider the following reactions:

$$CO_2 + H_2O \underset{k_{-1} = 13.7 \text{ s}^{-1}}{\overset{k_1 = 0.075 \text{ s}^{-1}}{\rightleftharpoons}} H_2CO_3$$

$$\underset{k_{-2} = 4.7 \times 10^{10} \text{ L mol}^{-1} \text{ s}^{-1}}{\overset{k_2 = 8 \times 10^6 \text{ s}^{-1}}{\rightleftharpoons}} H^+ + HCO_3^- \quad (21.114)$$

$$OH^- + CO_2 \underset{k_{-3} = 1.9 \times 10^{-4} \text{ s}^{-1}}{\overset{k_3 = 8.5 \times 10^5 \text{ L mol}^{-1} \text{ s}^{-1}}{\rightleftharpoons}} HCO_3^-$$

The rate constants are for 25 °C.
The rate equation for CO_2 is

$$\frac{d[CO_2]}{dt} = -k_1[CO_2] + k_{-1}[H_2CO_3] - k_3[OH^-][CO_2] + k_{-3}[HCO_3^-] \quad (21.115)$$

The reaction $H^+ + HCO_3^- \rightleftharpoons H_2CO_3$ occurs so rapidly that it remains in equilibrium in an experiment in which we follow the hydration of CO_2. Since HCO_3^- and H_2CO_3 remain in equilibrium, we may eliminate H_2CO_3 by use of

$$\frac{[\text{H}^+][\text{HCO}_3^-]}{[\text{H}_2\text{CO}_3]} = K_{\text{H}_2\text{CO}_3} = 1.70 \times 10^{-4} \qquad (21.116)$$

to obtain

$$\frac{d[\text{CO}_2]}{dt} = -(k_1 + k_3[\text{OH}^-])[\text{CO}_2] + \left(k_{-3} + \frac{k_{-1}[\text{H}^+]}{K_{\text{H}_2\text{CO}_3}}\right)[\text{HCO}_3^-] \quad (21.117)$$

Since the total concentration of dissolved CO_2 is $[\text{CO}_2] + [\text{H}_2\text{CO}_3] + [\text{HCO}_3^-] \approx [\text{CO}_2] + [\text{HCO}_3^-]$, this equation has the form

$$\frac{d[\text{A}]}{dt} = -k_\text{f}[\text{A}] + k_\text{b}[\text{B}] \qquad (21.118)$$

which corresponds to a reversible first-order reaction. As shown in equation 21.43, the half-life of such a reaction is given by

$$t_{1/2} = \frac{0.693}{k_\text{f} + k_\text{b}} \qquad (21.119)$$

Applying this relation to the hydration of CO_2 yields

$$\begin{aligned} t_{1/2} &= \frac{0.693}{k_1 + k_3 K_\text{w}/[\text{H}^+] + k_{-3} + k_{-1}[\text{H}^+]/K_{\text{H}_2\text{CO}_3}} \\ &= \frac{0.693}{0.0752 + (8.5 \times 10^{-9})/[\text{H}^+] + 8.06 \times 10^4[\text{H}^+]} \end{aligned} \qquad (21.120)$$

Values of the half-life calculated with this equation are given in Table 21.4. The reaction

$$\text{OH}^- + \text{HCO}_3^- \underset{1.3 \times 10^3 \text{ s}^{-1}}{\overset{6 \times 10^9 \text{ L mol}^{-1} \text{ s}^{-1}}{\rightleftharpoons}} \text{CO}_3^{2-} + \text{H}_2\text{O} \qquad (21.121)$$

has to be considered at higher pH values.

Table 21.4 Half-Lives for the Hydration–Dehydration Reaction for CO_2–HCO_3^- at 25 °C

pH	$t_{1/2}$/s
3	0.0086
4	0.085
5	0.79
6	4.2
7	4.1
8	0.75
9	0.081

References

A. J. Bard and L. R. Faulkner, *Electrochemical Methods*. New York: Wiley, 1980.

A. Cornish-Bowden, *Principles of Enzyme Kinetics*. London: Butterworths, 1976.

J. H. Espenson, *Chemical Kinetics and Reaction Mechanisms*. New York: McGraw-Hill, 1981.

A. Fersht, *Enzyme Structure and Mechanism*. San Francisco: Freeman, 1977.

W. C. Gardiner, Jr., *Rates and Mechanisms of Chemical Reactions*. New York: Benjamin, 1969.

H. Gutfreund, *Enzymes: Physical Principles*. New York: Wiley-Interscience, 1972.

G. G. Hammes (Ed.), *Investigation of Rates and Mechanisms of Reactions*, 3d ed. New York: Wiley, 1974.

G. G. Hammes, *Principles of Chemical Kinetics*. New York: Academic, 1978.

P. H. Rieger, *Electrochemistry*. Englewood Cliffs, NJ: Prentice-Hall, 1987.

I. H. Segal, *Enzyme Kinetics*. New York: Wiley-Interscience, 1975.

E. Zeffren and P. L. Hall, *The Study of Enzyme Mechanisms*. New York: Wiley-Interscience, 1973.

Problems

21.1   A steel ball (ρ = 7.86 g cm^{-3}) 0.2 cm in diameter falls 10 cm through a viscous liquid (ρ_0 = 1.50 g cm^{-3}) in 25 s. What is the viscosity at this temperature?

21.2   Estimate the rate of sedimentation of water droplets of 1-μm diameter in air at 20 °C. The viscosity of air at this temperature is 1.808×10^{-5} Pa s.

21.3   The viscosity of mercury is 1.661×10^{-3} Pa s at 0 °C and 1.476×10^{-3} Pa s at 35 °C. What is the activation energy, and what viscosity is expected at 50 °C?

21.4   A sharp boundary is formed between a dilute aqueous solution of sucrose and water at 25 °C. After 5 h the standard deviation of the concentration gradient is 0.434 cm. (*a*) What is the diffusion coefficient for sucrose under these conditions? (*b*) What will be the standard deviation after 10 h?

21.5   (*a*) Calculate the time required for the half-width of a freely diffusing boundary of dilute potassium chloride in water to become 0.5 cm at 25 °C ($D = 1.99 \times 10^{-9}$ m^2 s^{-1}). (*b*) Calculate the corresponding time for serum albumin ($D = 6.15 \times 10^{-11}$ m^2 s^{-1}).

21.6   The standard deviation σ of a freely diffusing boundary between dilute salt solution and water at 25 °C is 3.8 mm after 1 h. What is the diffusion coefficient of the salt in water? What will the standard deviation be after 2 h?

21.7   Using a table of the probability integral, calculate enough points on a plot of *c* versus *x* (like Fig. 21.2*c*) to draw in the smooth curve for diffusion of 0.1 mol L^{-1} sucrose into water at 25 °C after 4 h and 29.83 min ($D = 5.23 \times 10^{-10}$ m^2 s^{-1}).

21.8   Calculate the conductivity of 0.001 mol L^{-1} HCl at 25 °C. The limiting ion mobilities may be used for this problem.

21.9   One hundred grams of sodium chloride is dissolved in 10 000 L of water at 25 °C, giving a solution that may be regarded in these calculations as infinitely dilute. (*a*) What is the conductivity of the solution? (*b*) This dilute solution is placed in a glass tube of 4-cm diameter provided with electrodes filling the tube and placed 20 cm apart. How much current will flow if the potential drop between the electrodes is 80 V?

21.10   It is desired to use a conductance apparatus to measure the concentration of dilute solutions of sodium chloride. If the electrodes in the cell are each 1 cm^2 in area and are 0.2 cm apart, calculate the resistance that will be obtained for 1, 10, and 100 ppm NaCl at 25 °C.

21.11   Show that if A and B can be represented by spheres of the same radius that react when they touch, the second-order rate constant is given by

$$k_a = \frac{8 \times 10^3\, RT}{3\eta} \; \text{L mol}^{-1}\,\text{s}^{-1}$$

where R is in J K^{-1} mol^{-1}. To obtain this result the diffusion coefficient is expressed in terms of the radius of a spherical particle by use of equation 21.12. For water at 25 °C, $\eta = 8.95 \times 10^{-4}$ kg m^{-1} s^{-1}. Calculate k_a at 25 °C.

21.12   What is the reaction radius for the reaction

$$H^+ + OH^- \xrightarrow{\; 1.4\, \times\, 10^{11}\; \text{L mol}^{-1}\,\text{s}^{-1}\;} H_2O$$

at 25 °C, given that the diffusion coefficients of H$^+$ and OH$^-$ at this temperature are 9.1×10^{-9} m^2 s^{-1} and 5.2×10^{-9} m^2 s^{-1}?

21.13   For acetic acid in dilute aqueous solution at 25 °C, $K = 1.73 \times 10^{-5}$ and the relaxation time is 8.5×10^{-9} for a 0.1 M solution. Calculate k_a and k_d in

$$CH_3CO_2H \underset{k_a}{\overset{k_d}{\rightleftharpoons}} CH_3CO_2^- + H^+$$

21.14   Derive the relation between the relaxation time τ and the rate constants for the reaction A + B $\underset{k_2}{\overset{k_1}{\rightleftharpoons}}$ C + D, which is subjected to a small displacement from equilibrium.

21.15   Derive the relation between the relaxation time τ and the rate constants for the mechanism which is subjected to a small displacement from equilibrium. It is assumed that the equilibria, $A \rightleftharpoons A'$, $K_A = [A']/[A]$, and $B \rightleftharpoons B'$, $K_B = [B']/[B]$, are adjusted very rapidly so that these steps remain in equilibrium.

21.16 Calculate the first-order rate constants for the dissociation of the following weak acids: acetic acid, acid form of imidazole $C_3N_2H_5^+$, NH_4^+. The corresponding acid dissociation constants are 1.75×10^{-5}, 1.2×10^{-7}, and 5.71×10^{-10}, respectively. The second-order rate constants for the formation of the acid forms from a proton plus the base are 4.5×10^{10}, 1.5×10^{10}, and 4.3×10^{10} L mol^{-1} s^{-1}, respectively.

21.17 The solution reaction

$$I^- + OCl^- = OI^- + Cl^-$$

is believed to go by the mechanism

$$OCl^- + H_2O \overset{K_1}{\rightleftharpoons} HOCl + OH^- \quad \text{(fast)}$$
$$I^- + HOCl \overset{k}{\longrightarrow} HOI + Cl^- \quad \text{(slow)}$$
$$HOI + OH^- \overset{K_2}{\rightleftharpoons} H_2O + OI^- \quad \text{(fast)}$$

Derive the rate equation for the forward rate of this reaction that shows the effect of the concentration of OH^-.

21.18 The mutarotation of glucose is first order in glucose concentration and is catalyzed by acids (A) and bases (B). The first-order rate constant may be expressed by an equation of the type that is encountered in reactions with parallel paths:

$$k = k_0 + k_{H^+}[H^+] + k_A[A] + k_B[B]$$

where k_0 is the first-order rate constant in the absence of acids and bases other than water. The following data were obtained by J. N. Brønsted and E. A. Guggenheim (*J. Am. Chem. Soc.* **49**:2554 (1927)) at 18 °C in a medium containing 0.02 mol L^{-1} sodium acetate and various concentrations of acetic acid:

$[CH_3CO_2H]$/mol L^{-1}	0.020	0.105	0.199
$k/10^{-4}$ min^{-1}	1.36	1.40	1.46

Calculate k_0 and k_A. The term involving k_{H^+} is negligible under these conditions.

21.19 The rate of a reaction between oppositely charged ions is measured at an ionic strength of 0.01 mol L^{-1}. How will the rate be affected if the ionic strength is raised to 0.05 mol L^{-1} if the reaction is (a) $A^+ + B^- \rightarrow$ or (b) $A^{2+} + B^{2+}$?

21.20 Suppose that an enzyme has a turnover number of 10^4 min^{-1} and a molar mass of 60 000 g mol^{-1}. How many

moles of substrate can be turned over per hour per gram of enzyme if the substrate concentration is twice the Michaelis constant? It is assumed that the substrate concentration is maintained constant by a preceding enzymatic reaction and that products do not accumulate and inhibit the reaction.

21.21 The kinetics of the fumarase reaction

$$\text{fumarate} + H_2O = \text{L-malate}$$

is studied at 25 °C using an 0.01 ionic strength buffer of pH 7. The rate of the reaction is obtained using a recording ultraviolet spectrometer to measure the fumarate concentration. The following rates of the forward reaction are obtained using a fumarase concentration of 5×10^{-10} mol L^{-1}.

$[F]/10^{-6}$ mol L^{-1}	$v_F/10^{-7}$ mol L^{-1} s^{-1}
2	2.2
40	5.9

The following rates of the reverse reaction are obtained using a fumarase concentration of 5×10^{-10} mol L^{-1}:

$[M]/10^{-6}$ mol L^{-1}	$v_M/10^{-7}$ mol L^{-1} s^{-1}
5	1.3
100	3.6

(a) Calculate the Michaelis constants and turnover numbers for the two substrates. In practice many more concentrations would be studied. (b) Calculate the four rate constants in the mechanism

$$E + F \underset{k_{-1}}{\overset{k_1}{\rightleftharpoons}} X \underset{k_{-2}}{\overset{k_2}{\rightleftharpoons}} E + M$$

where E represents the catalytic site. There are four catalytic sites per fumarase molecule. (c) Calculate K_{eq} for the reaction catalyzed. The concentration of H_2O is omitted in the expression for the equilibrium constant because its concentration cannot be varied in dilute aqueous solutions.

21.22 Derive the steady-state rate equation for the mechanism

$$E + S \underset{k_2}{\overset{k_1}{\rightleftharpoons}} X \overset{k_3}{\longrightarrow} E + P \qquad E + I \underset{k_5}{\overset{k_4}{\rightleftharpoons}} EI$$

for the case that $[S] \gg [E]_0$ and $[I] \gg [E]_0$.

21.23 The following initial velocities were determined spectrophotometrically for solutions of sodium succinate to which a constant amount of succinoxidase was added. The velocities are given as the change in absorbancy at 250 nm in 10 s. Calculate V, K_M, and K_I for malonate.

$$\frac{A \times 10^3}{10\ s}$$

[Succinate] 10^{-3} mol L^{-1}	No Inhibitor	15×10^{-6} mol L^{-1} Malonate
10	16.7	14.9
2	14.2	10.0
1	11.3	7.7
0.5	8.8	4.9
0.33	7.1	—

21.24 In the Eadie–Hofstee method for determining k_{cat} and K_M for an enzymatic reaction, $v/[E]_0[S]$ is plotted versus $v/[E]_0$. How are the kinetic parameters obtained from this plot?

21.25 The maximum initial velocities ($v = k_{cat}[E]_0$) for an enzymatic reaction are determined at a series of pH values

pH	6.0	6.4	7.0	7.5	8.0	8.5	9.0
V	11	30	74	129	147	108	53

Calculate the values of the parameters V', K_a and K_b in

$$V = \frac{V'}{1 + [H^+]/K_a + K_b/[H^+]}$$

Hint: A plot of V versus pH may be constructed and the hydrogen ion concentration at the midpoint on the acid side referred to as $[H^+]_a$ and the hydrogen ion concentration at the midpoint on the basic side is referred to as $[H^+]_b$. Then

$$K_a = [H^+]_a[H^+]_b - 4\sqrt{[H^+]_a[H^+]_b} \qquad K_b = \frac{[H^+]_a[H^+]_b}{K_a}$$

21.26 (*a*) If the anodic transfer coefficient α_a is 0.5 for an oxidation reaction, by what factor will k_{ox} be changed if E is changed by 0.01 V at 25 °C if $n = 1$? (*b*) What will the factor be if $\alpha_a = 0.3$?

21.27 What is the root-mean-square displacement in the x direction of a molecule of tobacco mosaic virus due to Brownian motion during one minute in water at 20 °C? (see Table 22.3.)

21.28 Calculate the time necessary for a quartz particle 10 μm in diameter to sediment 50 cm in distilled water at 25 °C. The density of quartz is 2.6 g cm^{-3}. The coefficient of viscosity of water is 8.95×10^{-4} kg m^{-1} s^{-1}.

21.29 How long will it take a spherical air bubble 0.5 mm in diameter to rise 10 cm through water at 25 °C?

21.30 Using data in Table 21.1 and equation 21.6, estimate the activation energy for water molecules to move into a vacancy at 25 °C.

21.31 A sharp boundary is formed between a solution of hemoglobin in a buffer and the buffer solution at 25 °C. After 10 h the half-width of the concentration–gradient curve at the inflection point is 0.226 cm. What is the diffusion coefficient of hemoglobin under these conditions?

21.32 A sharp boundary is formed between a dilute buffered solution of hemoglobin ($D = 6.9 \times 10^{-11}$ m^2 s^{-1}) and the buffer at 20 °C. What is the half-width of the boundary after 1 and 4 h?

21.33 Since σ varies as $\sqrt{2Dt}$, the gradient curve has a certain width after time t, it will be twice as wide after time $4t$, and three times as wide after time $9t$. Sketch $\partial c/\partial x$ versus x for sucrose in water at 25 °C after 1, 4, and 9 h for $c_0 = 0.1$ mol L^{-1}. *Given:* $D = 4.65 \times 10^{-10}$ m^2 s^{-1}.

21.34 For an electrolyte such as HCl it can be shown that the diffusion coefficient in a dilute solution in water is given by

$$D = \frac{2u_1u_2RT}{(u_1 + u_2)F}$$

where u_1 and u_2 are the electric mobilities of the two ions. What is the diffusion coefficient of dilute HCl in water at 25 °C? The electric mobilities are given in Table 21.2.

21.35 Using a table of the normal probability function, calculate enough points on a plot of dc/dx versus x (like Fig. 21.2d) to draw in the smooth curve for diffusion of 0.01 mol L^{-1} sucrose into water at 25 °C after 3 h.

21.36 Calculate the conductivity at 25 °C of a solution containing 0.001 mol L^{-1} hydrochloric acid and 0.005 mol L^{-1} sodium chloride. The limiting ionic mobilities at infinite dilution may be used to obtain a sufficiently good approximation.

21.37 Estimate the conductivity at 25 °C of water that contains 70 ppm by weight of magnesium sulfate.

21.38 It may be shown that the diffusion coefficient at infinite dilution of an electrolyte with two univalent ions is given by

$$D = \frac{2u_1u_2RT}{(u_1 + u_2)F}$$

where u_1 and u_2 are the limiting values of the mobilities of the two ions. What is the diffusion coefficient of potassium chloride in water at 25 °C?

21.39 A study of conductivities at high electric field strengths reveals that the conductivity increases slightly with increasing electric field strength. A microsecond pulse at 10^7 V m^{-1} may be used. Approximately how far will a sodium ion move during such a pulse at room temperature?

21.40 The diffusion coefficient D of an ion is related to its ionic mobility u by

$$D = \frac{uRT}{zF}$$

The ionic mobilities of H$^+$ and OH$^-$ are 3.63×10^{-7} m^2 V^{-1} s^{-1} and 2.05×10^{-7} m^2 V^{-1} s^{-1} at 25 °C. What is the rate constant for the reaction

$$H^+ + OH^- \rightarrow H_2O$$

The reaction radius is 0.75 nm, because once the proton is this close the reaction can proceed very rapidly by quantum mechanical tunneling. The electrostatic factor f is 1.70.

21.41 Pure solutions of the α and β chains of hemoglobin $\alpha_2\beta_2$ can be prepared. Assuming that α and β exist only as monomers in these solutions, and that they react on the first collision, estimate the half-life for the reaction

$$\alpha + \beta \rightarrow \alpha\beta$$

in water at 25 °C. The viscosity of water at this temperature is 8.95×10^{-4} kg m^{-1} s^{-1}. Calculate the half-life if equal volumes of 10^{-6} mol L^{-1} solutions of α and β are mixed. (See problem 21.11.)

21.42 Derive the relation between the relaxation time τ and the rate constants for the reaction

$$A \underset{k_2}{\overset{k_1}{\rightleftharpoons}} B$$

which is subjected to a small displacement from equilibrium.

21.43 An imidazole buffer of pH 7 containing 0.05 mol L^{-1} imidazole has a relaxation time of 2.9×10^{-9} s at 25 °C. What are the values of the rate constants for the reaction

$$C_3N_2H_4 + H^+ \underset{k_{-1}}{\overset{k_1}{\rightleftharpoons}} C_3N_2H_5^+$$

The pK for the imidazole at this temperature is 7.21.

21.44 For the mechanism

$$2A \underset{k_{21}}{\overset{k_{12}}{\rightleftharpoons}} A_2$$

show that

$$\frac{1}{\tau} = 4k_{12}[A]_{eq} + k_{21}$$

21.45 Calculate the first-order rate constants for the following reactions at 25 °C.

H$^+$ Production	OH$^-$ Production
HOAc $\rightarrow$ H$^+$ + OAc$^-$	OAc$^-$ + H$_2$O $\rightarrow$ HOAc + OH$^-$
ImH$^+$ $\rightarrow$ H$^+$ + Im	Im + H$_2$O $\rightarrow$ ImH + OH$^-$
NH$_4^+$ $\rightarrow$ H$^+$ + NH$_3$	NH$_3$ + H$_2$O $\rightarrow$ NH$_4^+$ + OH$^-$

where HOAc is acetic acid and Im is imidazole (C$_3$N$_2$H$_4$). The reverse reactions given above may all be assumed to be diffusion controlled with $k = 10^{10}$ L mol^{-1} s^{-1}. Acid dissociation constants at 25 °C are

HOAc	1.75×10^{-5}
ImH$^+$	1.2×10^{-7}
NH$_4^+$	5.71×10^{-10}

Which conjugate acid–base pair can play both H$^+$ and OH$^-$ production roles about equally effectively?

21.46 The mutarotation of glucose is catalyzed by acids and bases and is first order in the concentration of glucose. When perchloric acid is used as a catalyst, the concentration of hydrogen ions may be taken to be equal to the concentration of perchloric acid, and the catalysis by perchlorate ion may be ignored since it is such a weak base. The following first-order constants were obtained by J. N. Brønsted and E. A. Guggenheim (*J. Am. Chem. Soc.* **49**:2554 (1927)) at 18 °C:

[HClO$_4$]/mol L^{-1}	0.0010	0.0048	0.0099	0.0192	0.0300	0.0400
$k/10^{-4}$ min^{-1}	1.25	1.38	1.53	1.90	2.15	2.59

Calculate the values of the constants in the equation $k = k_0 + k_{H^+}[H^+]$.

21.47 The initial rate v of oxidation of sodium succinate to form sodium fumarate by dissolved oxygen in the presence of the enzyme succinoxidase may be represented by equation 21.72. Calculate k_{cat}, $[E]_0$ and K_M from the following data:

[S]/10^{-3} mol L^{-1}	10	2	1	0.5	0.33
$v/10^{-6}$ mol L^{-1} s^{-1}	1.17	0.99	0.79	0.62	0.50

21.48 For the fumarase reaction

$$\text{fumarate} + H_2O = \text{L-malate}$$

at pH 7, 25 °C, and 0.01 ionic strength, the Michaelis–Menten parameters have the following values:

$$V_F = (1.3 \times 10^3 \text{ s}^{-1})[E]_0 \qquad V_M = (0.8 \times 10^3 \text{ s}^{-1})[E]_0$$
$$K_F = 4 \times 10^{-6} \text{ mol L}^{-1} \qquad K_M = 10 \times 10^{-6} \text{ mol L}^{-1}$$

where $[E]_0$ is the molar concentration of the enzyme, which has four catalytic sites per molecule. Calculate (a) the four rate constants in the mechanism

$$E + F \underset{k_{-1}}{\overset{k_1}{\rightleftharpoons}} EX \underset{k_{-2}}{\overset{k_2}{\rightleftharpoons}} E + M$$

and (b) $\Delta G^{\circ\prime}$ for the overall reaction.

21.49 At 25 °C and pH 7.8, the following values are obtained for the Michaelis constant and maximum initial velocity for the forward reaction catalyzed by fumarase:

$$F + H_2O = M \qquad K = \frac{[M]_{eq}}{[F]_{eq}} = 4.4.$$

$$V_F = (0.8 \times 10^3 \text{ s}^{-1})[E]_0$$

$$K_F = 7 \times 10^{-6} \text{ mol L}^{-1}$$

where the enzyme concentration is in moles of enzyme per liter. The enzyme has four catalytic sites per molecule. In some experiments L-malate was added and was found to be inhibitory with a constant

$$K_M = 100 \times 10^{-6} \text{ mol L}^{-1}$$

Calculate the values of the four rate constants in the mechanism

$$E + F \underset{k_{-1}}{\overset{k_1}{\rightleftharpoons}} X \underset{k_{-2}}{\overset{k_2}{\rightleftharpoons}} E + M$$

where E represents an enzymatic site.

21.50 The Michaelis constant for succinate or succinoxidase is 0.5×10^{-3} mol L^{-1} and the competitive inhibition constant for malonate is 10×10^{-3} mol L^{-1}. In an experiment with 10^{-3} mol L^{-1} succinate and 15×10^{-3} mol L^{-1} malonate, what is the percent inhibition?

21.51 Show how to plot rate data on an enzymatic reaction that is inhibited competitively by an inhibitor I at $[I]_1$ to obtain the value of K_I by using an Eadie–Hofstee plot.

21.52 Derive the steady-state rate equation for the mechanism

$$E + S \underset{k_{-1}}{\overset{k_1}{\rightleftharpoons}} ES \overset{k_2}{\rightarrow} E + P$$
$$\updownarrow K_{EH} \qquad\qquad \updownarrow K_{EHS}$$
$$EH \qquad\qquad EHS$$

Sketch the shape of the plots of V_S and K_S versus pH.

21.53 Calculate the slope of the plot in Fig. 21.11 at an overpotential η of 0 V. Assume $n = 1$.

21.54 What is the root-mean-square displacement in a particular direction of a spherical particle of radius 1 μm in water at 25 °C after 1 min?

21.55 For the reaction

$$CO_2(+H_2O) \underset{k_2}{\overset{k_1}{\rightleftharpoons}} H_2CO_3$$

(The parentheses indicate that H_2O is not included in the equilibrium constant expression or in the rate equation)

$$\Delta H^\circ = 4730 \text{ J mol}^{-1} \qquad \Delta S^\circ = -33.5 \text{ J K}^{-1} \text{ mol}^{-1}$$
At 25 °C $\qquad\qquad k_1 = 0.0375 \text{ s}^{-1}$
At 0 °C $\qquad\qquad k_1 = 0.0021 \text{ s}^{-1}$

Calculate (a) the activation energy for the forward reaction, (b) activation energy for the backward action, (c) k_{-1} at 25 °C, and (d) k_{-1} at 0 °C, assuming that ΔH° is independent of temperature in this range.

PART FOUR
MACROSCOPIC AND MICROSCOPIC STRUCTURES

The second and third parts of the book have been primarily concerned with the properties and dynamics of small molecules. Now we turn our attention to macromolecules, solid-state chemistry, and surface dynamics.

The chapter on macromolecules is concerned with high polymers, proteins, nucleic acids, and other macromolecules that have molar masses ranging from about 10^4 to 10^6 g mol^{-1}, or higher. These substances form viscous solutions, and so measurements of viscosity provide information about the size and shape of these molecules in solution. More information can be obtained from measurements of diffusion, ultracentrifugation, and light scattering.

Structural information about solids is primarily obtained by use of X-ray diffraction. This process is simplified by the recognition of the various types of symmetry that the internal structure of a crystal may have. Since the location of individual atoms is obtained, this is a powerful method for determining interatomic distances and angles. The study of the electronic structure of solids introduces new ideas not encountered with isolated molecules. The application of quantum mechanics to crystals helps us understand why some are excellent conductors of electricity and others are insulators; electrical conductivities of solids range over a factor of 10^{20}. Semiconductors have intermediate properties, and the modification of these properties by the addition of traces of other substances has provided the basis for the semiconductor industry.

The last chapter is concerned with the equilibrium and dynamics of processes that occur at the interface between a solid and a gas. When a molecule strikes a solid surface, it may rebound elastically or inelastically, or it may be adsorbed. An adsorbed molecule may dissociate on the surface or react with other species on the surface, or desorb from the surface. The catalysis of reactions by the surfaces of solids is of tremendous practical importance. The development of a number of "surface sensitive" experimental methods, like low-energy electron diffraction, electron emission from surfaces, and scanning tunneling microscopy, have made it possible to learn about processes at the interface.

22
Macromolecules

Although macromolecules (or polymers), both naturally occurring and man-made, play an immensely important role in our lives, the study of these systems is a relatively new activity. In fact, it was not until the 1920s that scientists became convinced that macromolecules could exist based largely on the work of Staudinger in Germany and Carothers in the United States. Now it is commonplace to say that life is basically the biochemistry of macromolecules, and that the major part of the chemistry industry is devoted to the production of macromolecules.

22.1 Size and Shape of Macromolecules

The term macromolecules covers a very wide range of types from synthetic high polymers, which exist in solution in chains of variable length, to proteins, which are tightly coiled molecules. Nevertheless, there are certain methods and concepts that apply quite generally to macromolecules; these are the principal objectives of this chapter.

Synthetic high polymers consist of long chains of atoms held together by covalent bonds. Such a chain is formed through the process of polymerization in which monomer molecules chemically react to form linear chains, branched chains, or three-dimensional networks. (See Fig. 22.1.) The properties of polymers depend on both their chemical structure and their physical structure. Thus, plastics are usually linear or branched polymers that can be melted at reasonable temperatures and formed into various shapes, while rubbers are lightly cross-linked networks. These cross-links give rubbers their elastic properties.

Linear

Branched

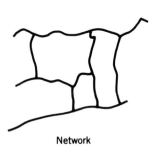

Network

Figure 22.1   Structures of polymers. (From R. J. Young, *Introduction to Polymers,* Chapman and Hall Ltd., London, 1983, p. 3. Reprinted with permission.)

Naturally occurring macromolecules include proteins, nucleic acids, and polysaccharides (e.g., cellulose and starch). Of these, we will discuss only proteins that are made up of some twenty kinds of amino acids joined together by peptide linkages:

$$
\begin{array}{ccc}
\overset{\displaystyle O}{\underset{\displaystyle \parallel}{}} & \overset{\displaystyle H}{\underset{\displaystyle |}{}} & \overset{\displaystyle R}{\underset{\displaystyle |}{}} \\
-C- & N- & C- \\
& & \underset{\displaystyle H}{|}
\end{array}
$$

The term **primary structure** describes the sequence of amino acids in a peptide chain. Peptide chains can form a number of well-ordered local structures such as the alpha helix illustrated in Section 22.5. These are referred to as **secondary structures.** The three-dimensional structure of a protein molecule formed from a single chain is called the **tertiary structure.** Proteins tend to be globular in solution, rather than random chains (as synthetic polymers often are). Some proteins, for example, hemoglobin, consist of noncovalently bound subunits. When a protein is heated or put in extreme conditions, such as a very high or low pH, the folded structure is disrupted, the protein becomes insoluble, and it is said to be **denatured.**

A basic property of a macromolecule is its molar mass or molecular weight. A sample of a polymer almost always contains molecules of different molar mass (polydispersity), so that the average molar mass for a polymer is an important measure. There are two different averages used in the literature, the number average molar mass $\overline{M}_n$ and the mass average molar mass $\overline{M}_m$. The number average is defined as

$$
\overline{M}_n = \frac{\sum N_i M_i}{\sum N_i} \tag{22.1}
$$

where N_i is the number of polymer molecules of molar mass M_i in the sample. The weight or mass average is defined as

$$
\overline{M}_m = \frac{\sum N_i M_i^2}{\sum N_i M_i} \tag{22.2}
$$

If the sample contains molecules of a single mass, these two definitions coincide, but in a sample with dispersity, they are different, so that the ratio $\overline{M}_m/\overline{M}_n$ is a measure of the dispersity of the sample.

The size and dispersity of most synthetic macromolecules make the determination of their structure a very difficult task. In most cases, it is impossible to form a single crystal, so that X-ray diffraction methods are not useful. For certain systems (such as the nucleic acids and the proteins myoglobin and cytochrome c) single crystals are formed and the structure can be found from X-rays, but these are special (if very important) cases. Even if single crystals can be formed, we are often more interested in the structure of the macromolecule in solution than in the crystalline state, and it is always difficult to know whether the structure is affected by crystallinity. For these reasons, a number of methods have been devised to study the size and shape of macromolecules; they give less detailed information than X-ray diffraction, but they are less time-consuming and more easily applicable to most polymers. In the following, we discuss the measurements of osmotic pressure, viscosity, and light scattering to determine the size and shape of macromolecules. In the special topics at the end of the chapter, we discuss sedimentation measurements.

Osmotic pressure and light scattering are used for polymers in solution. The solvent can play an important role in determining the size of a polymer. For example, in a good solvent the polymer–solvent interaction results in a swelling of the polymer. The Gibbs energy is lowered significantly by this. In a poor solvent, the polymer lowers its Gibbs energy by having less contact with the solvent (contracting), and its complete solution may be impossible.

22.2   Osmotic Pressure of Polymer Solutions

In Section 7.10 on colligative properties, we introduced the **osmotic pressure** Π of a solution. The osmotic pressure Π is related to the concentration c (mass per unit volume) of the polymer in solution by

$$\Pi = RT \frac{c}{M} + RT\, b(T) \frac{c^2}{M^2} + \cdots \qquad (22.3)$$

where the first term is the ideal solution result and the succeeding terms are the corrections due to nonideality. This equation is the result for a monodisperse polymer in solution. As we have already indicated, most polymer solutions are polydisperse, so we must make corrections for this. The term c/M in equation 22.3 is proportional to the number of polymer molecules N,

$$\frac{c}{M} = \frac{N}{N_A V} \qquad (22.4)$$

where N_A is Avogadro's constant and V is the volume of the solution. For a polydisperse solution, Π is given by

$$\Pi = RT \frac{N}{N_A V} + RT\, b(T) \left(\frac{N}{N_A V}\right)^2 + \cdots \qquad (22.5)$$

where N is the total number of molecules. The concentration in a polydisperse solution is given by

$$c = \sum_i \frac{N_i M_i}{N_A V} = \overline{M}_n \frac{N}{N_A V} \qquad (22.6)$$

and therefore

$$\Pi = \frac{RTc}{\overline{M}_n} + RT\, b(T) \frac{c^2}{\overline{M}_n^2} + \cdots \qquad (22.7)$$

By taking the limit of Π/c as $c \to 0$, we can find the **number average molar mass** $\overline{M}_n$ of the polymer (see Fig. 22.2). The expansion of Π in powers of concentration is called a virial expansion just as in the equation of state of gases (Section 1.6). The temperature at which $b(T) = 0$ is called the theta point. **Polymer solutions at the theta point act ideally.**

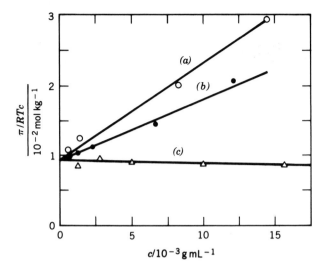

Figure 22.2   Plot of π/RTc versus c for nitrocellulose in (a) acetone, (b) methanol, and (c) nitrobenzene.

Example 22.1

Osmotic pressures were measured for two dilute solutions of a sample of nitrocellulose in methanol at 25 °C. Use the following experimental values of π/RTc to calculate the number average molar mass $\overline{M}_n$ and the second virial coefficient $b(T)$.

c/g cm^{-3}	2.5×10^{-3}	7.0×10^{-3}
(Π/RTc)/mol kg^{-1}	0.0102	0.0145

These concentrations can be expressed in SI base units by multiplying them by $(10^{-3}$ kg g$^{-1})$ $(10^2$ cm m$^{-1})^3$.

The two data points allow us to write equation 22.7 in two ways.

$$0.0102 \text{ mol kg}^{-1} = \frac{1}{\overline{M}_n} + \frac{b(T)\ (2.5 \text{ kg m}^{-3})}{\overline{M}_n^2}$$

$$0.0145 \text{ mol kg}^{-1} = \frac{1}{\overline{M}_n} + \frac{b(T)\ (7.0 \text{ kg m}^{-3})}{\overline{M}_n^2}$$

Solving these equations simultaneously yields

$$\overline{M}_n = 98.1 \text{ kg mol}^{-1}$$
$$b(T) = 9.2 \text{ m}^3 \text{ mol}^{-1}$$

22.3   Light Scattering

Light scattering can also be used to measure the size of polymer molecules in solution, but it is a more general physical chemical technique, so we will describe it in some detail.

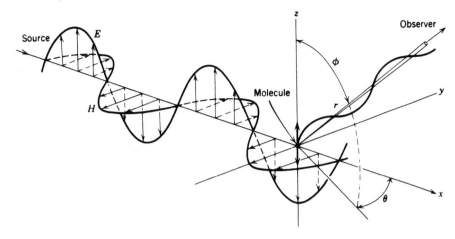

Figure 22.3 Scattering of plane polarized light by a single molecule. For an isotropic molecule the induced dipole moment is along the z axis. Therefore, no light is scattered in the z direction.

When the oscillating electromagnetic field of light impinges on a molecule, the molecule is polarized by the electric field vector and at the frequency of the light. This causes an oscillating dipole to be set up in the molecule, which then acts to radiate a new electromagnetic field. Since the field of a dipole radiates in all directions, we can say that the incident light has been scattered by the molecule. The theory of light scattering was developed by Rayleigh, who then explained the blue color of the sky and the red color at sunset as being due to the preferential scattering of blue light by the molecules and particles of the atmosphere.

If every molecule scatters light, then it might seem as if all liquids and solids would scatter light powerfully. However, if the concentration of molecules is the same everywhere in the system, then the summation of the scattered electromagnetic fields from all the molecules gives zero intensity except in the forward direction. Thus, as Rayleigh showed, it is the **fluctuations** in the concentration that give rise to fluctuations in the refractive index of the system, which then cause net scattering of light. This is why liquids at the critical point scatter light and become cloudy: At this point the concentration fluctuations are very large and thus so are the refractive index fluctuations.

We can use classical electromagnetic theory to calculate the scattering of light from a single molecule. As we said above, the light field induces an oscillating dipole in the molecule. This in turn produces an oscillating electric field at a distance r and angle ϕ with respect to the polarization of the dipole (see Fig. 22.3) given by

$$E_{\mathrm{r}} = -\frac{\alpha E_0 \pi \sin \phi}{\epsilon_0 r \lambda^2} \cos 2\pi\nu\left(t - \frac{r}{c}\right) \qquad (22.8)$$

where α is the **polarizability** (Section 12.12b) of the molecule, E_0 is the magnitude of the electric vector of the incident light wave, ν is the frequency of the light, and c is the velocity of light. The intensity i of the radiation scattered at angle ϕ is proportional to the square of the amplitude (i.e., the square of the

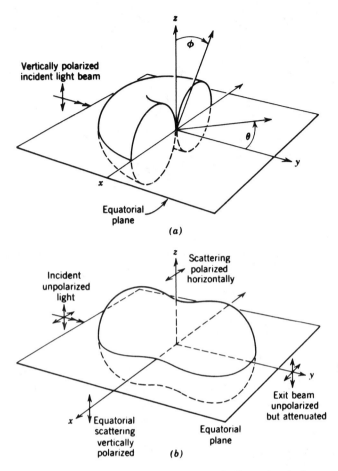

Figure 22.4 (a) Intensity of light scattered by a single molecule from a light wave with its electric vector in the z direction. The length of the radius vector from the origin indicates the scattered amplitude. (The front half of the diagram is omitted for clarity.) (b) Intensity of light scattered by a single molecule from an unpolarized light wave. (From D. F. Eggers et al., *Physical Chemistry*. New York: Wiley, 1964.)

coefficient in equation 22.8). To compare i with the intensity I_0 of the incident light, we divide the square of the amplitude by E_0^2 to obtain

$$\frac{i}{I_0} = \frac{\pi^2 \alpha^2 \sin^2 \phi}{\epsilon_0^2 r^2 \lambda^4} \qquad (22.9)$$

The dependence of the scattering of incident light polarized in the z direction on ϕ is shown in Fig. 22.4a. The scattering is zero in the z direction.

Often unpolarized light is used in scattering experiments. The angular dependence of the intensity of scattering of a small (compared with the wavelength of the light) isotropic molecule may be obtained by adding the scattered intensities from the two perpendicularly polarized light waves. This yields a scattered intensity of

$$\frac{i}{I_0} = \frac{\pi^2 \alpha^2}{2\epsilon_0^2 r^2 \lambda^4}(1 + \cos^2 \theta) \qquad (22.10)$$

where θ is the angle between the incident beam and the direction of observation. This angular dependence is shown in Fig. 22.4b. This shows that the scattered intensities in the forward and reverse directions are equal. The scattered intensity is inversely proportional to the fourth power of the wavelength λ of the light used. Thus, blue light is scattered to a greater extent than red light, and therefore the light transmitted through a suspension of particles is reddish.

As we said above, to calculate the scattering from the polymer molecules in a solution, we must look at the concentration fluctuations in small volumes of the solution smaller than the cube of the wavelength of light, but still large enough to contain a large number of molecules. This procedure can be found in advanced texts on light scattering. Here we outline a simpler version of the calculation which gets the correct answer with less attention to the statistical averaging procedure.

The polarizability of the polymer α can be related to the refractive index n of the solution and the refractive index n_0 of the pure solvent by

$$n^2 - n_0^2 = \frac{N\alpha}{\epsilon_0} \qquad (22.11)$$

where N is the number of polymer particles per unit volume. This equation may be rearranged as follows to express the polarizability in terms of the specific refractive index increment $dn/dc = (n - n_0)/c$, where c is the concentration in mass per unit volume and M is the molar mass of the polymer:

$$\alpha = \frac{(n^2 - n_0^2)\epsilon_0}{N} = (n + n_0)\frac{n - n_0}{c}\frac{c\epsilon_0}{N} = (n + n_0)\frac{dn}{dc}\frac{c\epsilon_0}{N} = 2n_0\frac{dn}{dc}\frac{M\epsilon_0}{N_A} \qquad (22.12)$$

where the last form has been obtained by replacing $n + n_0$ by $2n_0$, since dilute solutions are considered, and c/N is replaced by M/N_A. Substituting this expression into equation 22.10 yields

$$\frac{i}{I_0} = \frac{2\pi^2 n_0^2 (dn/dc)^2 M^2 (1 + \cos^2\theta)}{\lambda^4 r^2 N_A^2} \qquad (22.13)$$

for the scattering of one polymer molecule. Multiplying by $N = cN_A/M$ to obtain the scattering from N particles per unit volume yields

$$\frac{i}{I_0} = \frac{2\pi^2 n_0^2 (1 + \cos^2\theta)(dn/dc)^2 Mc}{N_A \lambda^4 r^2} \qquad (22.14)$$

This equation shows that the excess scattering of the solution over that of the solvent is proportional to the concentration of the solute on a mass per unit volume basis and to the molar mass of the solute.

This equation can be generalized for a solution of polymer molecules of different molar masses M_i. Each species gives rise to a term similar to the above. Since $c_i = M_i N_i/N_A$, where N_i is the number of polymers of mass M_i per unit volume, the result is

$$\frac{i}{I_0} = \frac{2\pi^2 n_0 (1 + \cos^2\theta)(dn/dc)^2}{N_A \lambda^4 r^2}\overline{M}_m c \qquad (22.15)$$

with c the total concentration (mass per unit volume) of polymer molecules. Therefore, the light scattering experiment can be used to measure the mass average molar mass of the solute. The accuracy to which $\overline{M}_m$ can be measured

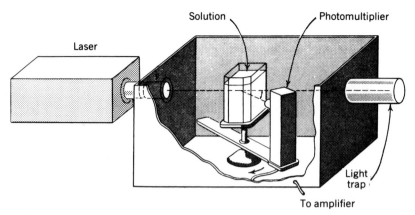

Figure 22.5   Apparatus for measuring light scattered by a solution.

depends on how well n_0 and dn/dc are known; they must be measured in separate experiments.

A straightforward experiment is to measure the transmitted intensity of a light beam through a polymer solution. If there is little absorption of light, the loss of intensity is due to scattering. The total scattered intensity is the integral of i over all angles, so that the transmitted intensity is given by

$$I = I_0 e^{-\tau x} \tag{22.16}$$

where τ is the **turbidity** and x is the thickness of the cell in the direction of the incident beam. This is of the form of Beers' law for light absorption (Section 15.4). The result of this calculation is that

$$\tau = \frac{32\pi^3 n_0^2 (dn/dc)^2 Mc}{3N_A\lambda^4} = HMc \tag{22.17}$$

where

$$H = \frac{32\pi^3 n_0^2 (dn/dc)^2}{3N_A\lambda^4} \tag{22.18}$$

is a constant for a particular solvent, solute, and wavelength. Since the turbidity τ is proportional to molar mass, this method is more useful as the molar mass increases, in contrast to osmotic pressure, which is inversely proportional to molar mass.

Instead of measuring the transmitted intensity and calculating τ using equation 22.17, it is much more sensitive experimentally to determine the scattered intensity using an apparatus such as that shown in Fig. 22.5.

The above theory is for ideal solutions and, in practice, it is necessary to plot Hc/τ versus c and extrapolate to zero concentration to obtain the value of Hc/τ that will yield the correct molar mass from equation 22.18. In fact, it can be shown that for a nonideal polymer solution for which the osmotic pressure is given by

$$\frac{\Pi}{cRT} = \frac{1}{M} + Bc + \cdots \tag{22.19}$$

the turbidity is given by

$$\frac{Hc}{\tau} = \frac{1}{M} + 2Bc + \cdots \qquad (22.20)$$

If one of the dimensions of the polymer is comparable to or greater than the wavelength of light, then there will be interference of the scattered waves from different parts of the same molecule. It is found that large molecules tend to scatter more light in the forward direction, and so the angular dependence of light scattering can also be used to obtain information about polymer shape.

22.4 Spatial Configuration of Polymer Chains

Polymers are differentiated at the molecular level from other substances by their long chains of covalently bonded atoms. The useful mechanical properties of polymers are a consequence of the special attributes of these long chains. Since chains can have many spatial configurations, statistical mechanics is required to obtain a quantitative understanding of these molecular properties. The bond lengths and bond angles in polymer chains are the same as in substances of lower molecular mass within the limits of experimental measurements. But even with these restrictions on bond angles the numbers of possible configurations of a chain several thousand bonds in length is prodigious. In solution the conformation of a particular chain undergoes continual change due to thermal agitation.

In this section we are interested in the average distance between ends of a chain containing a certain number of bonds. To make this calculation we will first consider an idealized model of the simplest possible kind. We will assume that the chain has n links of length l and that the direction in space of any link with respect to the preceding link is entirely random. This **freely jointed chain model** is an oversimplification because (1) angles of successive bonds in real molecules are limited to certain angles and (2) real chains cannot double back on themselves or occupy space filled with another part of the chain. This is known as the excluded volume effect. The mathematical treatment of this simple model involves an idea that is encountered in other areas of physical science, the idea of random flight. The simplest example of random flight is Brownian motion in one dimension; in this case the term random walk is used.

Let us consider a walker who starts at the origin and takes a step of length h along a line to the right or left at successive intervals of time with equal probability ($p = \frac{1}{2}$). If the number n of steps is larger, it is readily shown* that the probability $W(\xi)\,d\xi$ that the walker is at a distance ξ to $\xi + d\xi$ from the origin is given by

$$W(\xi)\,d\xi = \frac{1}{(2\pi n h^2)^{1/2}}\, e^{-\xi^2/2nh^2}\,d\xi \qquad (22.21)$$

where h is the length of each step. If the steps are not the same length, this equation becomes

---

* D. A. McQuarrie, *Statistical Thermodynamics*, Chapter 14. New York: Harper & Row, 1973.

$$W(\xi)\, d\xi = \frac{1}{(2\pi n\overline{h^2})^{1/2}}\, e^{-\xi^2/2n\overline{h^2}}\, d\xi \qquad (22.22)$$

where $\overline{h^2}$ is the average of the square of the step length. This equation may be used to obtain one component of the displacement after n steps in random flight in three dimensions with a constant step length of l. The probability $W(x)\, dx$ that there is a displacement in the range x to $x + dx$ after n steps of length l in three dimensions is obtained by replacing ξ by x and h by $l \cos \theta$, where θ is the angle between the random flight vector and the x axis. The angle θ may take on all values with equal probabilities and hence $\overline{h^2}$ in equation 22.22 is given by $\overline{l^2 \cos^2 \theta} = l^2/3$:

$$W(x)\, dx = \left(\frac{3}{2\pi nl^2}\right)^{1/2} e^{-3x^2/2nl^2}\, dx \qquad (22.23)$$

The probability $W(x, y, z)\, dx\, dy\, dz$ that the coordinates after a random flight of n steps are between x and $x + dx$, y and $y + dy$, and z and $z + dz$ is given by the product of three probabilities of the type given by equation 22.23:

$$W(x, y, z)\, dx\, dy\, dz = \left(\frac{3}{2\pi nl^2}\right)^{3/2} e^{-3(x^2+y^2+z^2)/2nl^2}\, dx\, dy\, dz \qquad (22.24)$$

This equation shows that the most probable coordinates after a random flight process are $x = 0$, $y = 0$, and $z = 0$. However, we are more interested in another question and that is, what is the root-mean-square value of the end-to-end distance after a three-dimensional random flight? Equation 22.24 can be converted to spherical coordinates to obtain

$$W(r)4\pi r^2\, dr = 4\pi \left(\frac{3}{2\pi nl^2}\right)^{3/2} e^{-3r^2/2nl^2}\, r^2\, dr \qquad (22.25)$$

by use of $r^2 = x^2 + y^2 + z^2$. The **mean-square end-to-end distance** is obtained from

$$\overline{r^2} = \int_0^\infty W(r)4\pi r^2\, dr = nl^2 \qquad (22.26)$$

so that the root-mean-square end-to-end distance is

$$(\overline{r^2})^{1/2} = n^{1/2}l \qquad (22.27)$$

Thus, a very simple result has been obtained. For a freely jointed linear polymer with a fully extended length of nl, the root-mean-square end-to-end distance is proportional to $n^{1/2}$ and, therefore, to the square root of the molar mass.

Equation 22.27 was obtained for a freely joined chain, and so now we must consider how it is affected when actual bond angles are taken into account. Figure 22.6 shows three successive bonds of a polymethylene chain. Carbon atoms C_1, C_2, and C_3 define a plane. In the idealized chain we are considering here, atom C_4 may occur any place in the circle which is the base of the cone described by rotation of bond 3. If there is free rotation about each bond, it may be shown* that equation 22.26 becomes

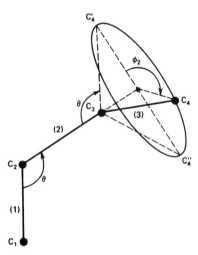

Figure 22.6   Three successive bonds in a singly bonded carbon chain. The first two bonds are in the plane of the page.

---

* P. J. Flory, *Polymer Chemistry*. Ithaca, NY: Cornell University Press, 1953.

$$\overline{r^2} = nl^2\left(\frac{1 - \cos\theta}{1 + \cos\theta}\right) \qquad (22.28)$$

For polymethylene $\theta = 109° 28'$, $\cos\theta = -\frac{1}{3}$, and $\overline{r^2} = 2nl^2$.

Equation 22.28 illustrates the general conclusion that the root-mean-square end-to-end length is proportional to the square root of the number of bonds or links it contains, even if there are restrictions on the bond angles.

Example 22.2

The length of a carbon–carbon bond is 155 pm. What is the root-mean-square end-to-end distance for a freely jointed polymethylene chain with $M = 10^5$ g mol^{-1}? The number n of units in the chain is $(10^5$ g mol$^{-1})/(14$ g mol$^{-1}) = 7.14 \times 10^3$.

$$\begin{aligned}(\overline{r^2})^{1/2} &= (2n)^{1/2}\, l \\ &= [(2)(7.14 \times 10^3)]^{1/2}(155 \text{ pm}) \\ &= 18.5 \text{ nm}\end{aligned}$$

So far in this discussion we have assumed that there is free rotation around each bond; that is, angle ϕ in Fig. 22.6 can have any value. However, this rotation is not free because there is a potential energy associated with such internal rotation.

The simplest example of the restrictions on bond rotations is provided by *n*-butane. The three conformations with the lowest energies are shown in Fig. 22.7, and the corresponding potential energy curve is shown in Fig. 22.8. The eclipsed form ($\phi = 180°$) is not shown because it has a high energy relative to the other three forms. The *gauche* minima lie about 2.1 kJ mol^{-1} above the energy of the *trans* conformation. The *gauche* forms are each about half as probable as the *trans* form; $p = e^{-E/RT}$. When the potential energy of internal rotation in a singly bonded carbon chain is taken into account, equation 22.28 becomes

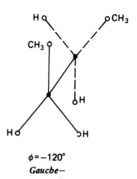

$\phi = -120°$
Gauche−

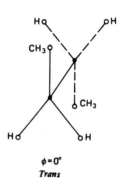

$\phi = 0°$
Trans

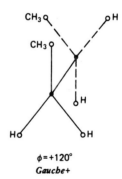

$\phi = +120°$
Gauche+

Figure 22.7 Conformations of *n*-butane.

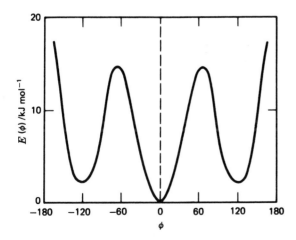

Figure 22.8 Potential energy of rotation about the central (C$_2$–C$_3$) bond of *n*-butane.

$$\overline{r^2} = nl^2 \left(\frac{1 - \cos\theta}{1 + \cos\theta}\right)\left(\frac{1 + \overline{\cos\phi}}{1 - \overline{\cos\phi}}\right) \tag{22.29}$$

where

$$\overline{\cos\phi} = \frac{\int_0^{2\pi} \cos\phi \, e^{-E(\phi)/RT} \, d\phi}{\int_0^{2\pi} e^{-E(\phi)/RT} \, d\phi} \tag{22.30}$$

where $E(\phi)$ is the potential energy of internal rotation given in Fig. 22.8.

Note that in both equations 22.28 and 22.29 $\overline{r^2}$ is still of the form $\beta^2 n$. Thus, the effect of restricting the bond angle and the free rotation about bonds is to increase the "effective bond length" β. Therefore, the form of equation 22.29 is not affected by bringing in these restrictions. These theoretical results are of great importance for an understanding of rubberlike elasticity (Section 4.17) and of hydrodynamic and thermodynamic properties of dilute polymer solutions.

In the model of a polymer we have considered, there are no interactions between different segments of the polymer. In a real polymer in solution, however, there will be interactions of which the most important is the excluded volume interaction, which is due to the finite size of the segments. Even parts of the chain which are far apart *along the chain* can produce this effect (see Fig. 22.9). The real interactions in polymers are very complicated; however, as far as the long length scale properties such as end-to-end distance are concerned, the *details* of the interaction are not important. Flory worked out a simple form of the theory including excluded volume, which gave as the root-mean-square end-to-end distance

$$\langle r^2 \rangle^{1/2} \sim n^{3/5} \tag{22.31}$$

instead of $\langle r^2 \rangle^{1/2} \sim n^{1/2}$. Thus, the excluded volume interactions tend to swell the chain from the ideal random chain. Since the intersegment interactions that give rise to the excluded volume effect are mediated by the solvent in which the polymer is dissolved, the form of $\langle r^2 \rangle$ and the probability distribution function for r^2 depend on solvent. In a good solvent, the polymer is surrounded by solvent molecules and is therefore swollen. In a poor solvent, the polymer prefers its own segment–segment interactions to the segment–solvent interactions, so the polymer tends to be small. In between these limiting cases, there is a point (the theta point) where the interactions are more or less equal, and the excluded volume effects are negligible.

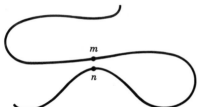

Figure 22.9   Excluded volume interaction between segments m and n.

22.5   Helix–Random Coil Transitions in Polypeptides

Proteins are polymers of amino acids ($H_2NCHRCO_2H$) that have molar masses greater than about 10 000 g mol^{-1}. About 22 different amino acids occur in natural proteins, and they are all of the L configuration. In proteins and polypeptides, amino acid residues are connected through amide bonds:

The nuclei of H, N, C, and O lie in a plane, and the connecting bonds are *trans*. Smaller polymers of amino acids called oligopeptides form random coils in solution, but proteins have a more or less fixed three-dimensional structure held together by hydrogen bonds (Section 12.10), disulfide bonds (–S–S–) between cystine residues, and ionic and van der Waals interactions. In the amide group the bond lengths and angles can be considered to be fixed. Only two angles have to be given per α carbon to specify the configuration of the chain. If certain angles, which are not prohibited by steric hindrance, are repeated indefinitely, a helical structure results. A peptide helix is characterized by the number n of amino acid residues per turn of helix and by the distance d traversed parallel to the helix axis per amino acid residue.

The **α helix** is sufficiently important to show the structures of the right-handed and left-handed varieties in Fig. 22.10. It is primarily the right-handed

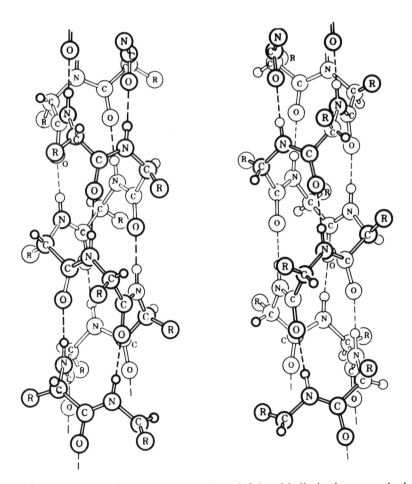

Figure 22.10   Alpha helix of a polypeptide. A left-hand helix is shown on the left and a right-hand helix is shown on the right.

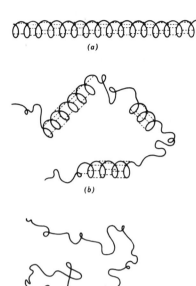

Figure 22.11 Configurational states of a polypeptide: (a) helix; (b) intermediate form; (c) random coil. (Reprinted with permission from L. Peller, *J. Phys. Chem.* **63**:1194 (1959). Copyright © by the American Chemical Society.)

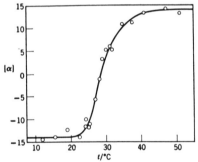

Figure 22.12 Temperature dependence of the optical rotation of poly-γ-benzyl-L-glutamate ($M = 350\,000$ g mol^{-1}) in dichloroacetic acid–dichloroethane (80:20). (Reprinted with permission from P. Doty and J. T. Yang, *J. Am. Chem. Soc.* **78**:499 (1956). Copyright © by the American Chemical Society.)

helix that exists in proteins. The α helix is held together by hydrogen bonds between a carbonyl oxygen and the N–H of the fourth residue along the chain. This produces an especially stable structure with $n = 3.6$ residues per turn. In the direction of the axis of the helix there is a residue every 0.15 nm.

A number of synthetic polypeptides show structural transitions in solution when there is a change in temperature or pH. A single polypeptide chain can exist in three states in solution, as illustrated in Fig. 22.11.

Polymers generally have the helical form at low temperatures and the random coil form at high temperatures; that is, the helix is the low-energy form and the random coil is the high-energy form. However, since the energy of a conformation depends on the difference between unit–unit interactions and unit–solvent interactions, the helix is sometimes the high-energy form. This is the situation with poly-γ-benzene-L-glutamate in a mixture of dichloroacetic acid and dichlolorethane. The transition of this polymer from a random coil at low temperatures to a helix at high temperatures is shown in Fig. 22.12.

The **helix–coil transition** is sometimes referred to as the "melting" of the helix. In contrast to the melting of a crystal, melting of a helix generally takes place over a range of temperature. If the angles of a monomer unit are incompatible with the formation of a helix, we say there is a kink in the polymer chain. If the probability of occurrence of a kink at one position is strongly coupled to the existence of a kink in a neighboring position, the helix–coil transition is said to be **cooperative.** If there is a strong interaction, the transition has a strong all-or-none character; that is, the melting takes place over a narrow range of temperature. On the other hand, if the probability of a kink occurring at a given position is independent of whether there are neighboring kinks, the helix melts gradually and the transition is said to be noncooperative.

22.6 Step-Growth Polymerization

Synthetic high polymers are produced by the reaction of monomer molecules to produce linear or branched chain molecules. Polymerization reactions can be classified according to the mechanism of the reaction. There are basically two types of polymerization reactions: **step-growth polymerization** and **free-radical polymerization.** Their kinetics are quite different. In step-growth polymerization the chains grow in a slow, stepwise manner. In free-radical polymerization individual chains grow rapidly to their final length.

In step-growth polymerization, a linear polymer is produced by the stepwise condensation or addition of the reactive groups in bifunctional monomers. Common functional groups for this type of polymerization are –OH, –CO$_2$H, or –NH$_2$. For example, a polyester can be formed by the reaction of diacids with dialcohols via the polycondensation reaction:

$$x\text{HOOC}\!-\!\text{R}\!-\!\text{COOH} + x\text{HO}\!-\!\text{R}'\!-\!\text{OH} \rightarrow \text{H}\!\!\left[\!\text{O}\!-\!\underset{\underset{\text{O}}{\|}}{\text{C}}\!-\!\text{R}\!-\!\underset{\underset{\text{O}}{\|}}{\text{C}}\!-\!\text{O}\!-\!\text{R}'\!\right]_{\!x}\!\!\text{OH}$$

$$(22.32)$$

$$+ (2x - 1)\text{H}_2\text{O}$$

Alternatively, a polyester can be formed by polymerization of a hydroxy acid:

$$x\text{HO}-\text{R}-\text{COOH} \rightarrow \text{H} \!\!+\!\! \text{O}-\text{R}-\overset{\displaystyle \|}{\underset{\displaystyle \text{O}}{\text{C}}} \!\!+_{\!x}\!\! \text{OH} + (x-1)\text{H}_2\text{O} \qquad (22.33)$$

In the production of polyesters or polyamides, the reactions are reversible so that water must be removed as the reaction progresses. In the production of polyurethanes there is no elimination of a small molecule. Two molecules react to form a dimer. The dimer then reacts with another monomer to form a trimer, and so on. Therefore, the average molar mass of the product increases as the reaction proceeds.

The mathematical treatment of step-growth polymerization would be extremely complicated if the rate constant for the coupling reaction depended on the degree of polymerization. Fortunately, it is found that the rate constant is approximately independent of chain length.

Let us consider the polymerization of a hydroxyacid, HO–R–CO$_2$H or a stoichiometric mixture of a dicarboxylic acid and a glycol. The unreacted carboxylic acid groups can be titrated during a polyesterification reaction to determine the rate. The rate of a polyesterification reaction is acid catalyzed and is proportional to the concentration of –CO$_2$H groups and –OH groups

$$\frac{d[\text{CO}_2\text{H}]}{dt} = k[\text{CO}_2\text{H}][\text{OH}][\text{acid}] \qquad (22.34)$$

The carboxyl groups of the reactant molecules may serve as the acid catalyst (case 1) or an acid catalyst may be added (case 2).

Case 1

If the carboxylic groups of the reactant serve as the acid catalyst, the rate equation becomes

$$\frac{d[\text{CO}_2\text{H}]}{dt} = k[\text{CO}_2\text{H}]^2[\text{OH}] \qquad (22.35)$$

If there are initially equal concentrations of –CO$_2$H and –OH groups their concentrations remain equal during reaction, and the rate equation is

$$-\frac{d[\text{CO}_2\text{H}]}{dt} = k[\text{CO}_2\text{H}]^3 \qquad (22.36)$$

which may be integrated to yield

$$\frac{1}{[\text{CO}_2\text{H}]^2} = \frac{1}{[\text{CO}_2\text{H}]_0^2} + 2kt \qquad (22.37)$$

It is convenient to express the extent of reaction in terms of the fraction p of –CO$_2$H groups reacted.

$$[\text{CO}_2\text{H}] = [\text{CO}_2\text{H}]_0(1 - p) \qquad (22.38)$$

so that

$$\frac{1}{(1-p)^2} = 1 + 2[\text{CO}_2\text{H}]_0^2 kt \qquad (22.39)$$

This rate equation is obeyed very well by step-growth polymerizations after

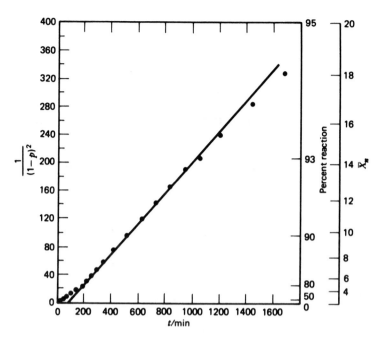

Figure 22.13   Extent of reaction p and degree of polymerization X_n for an equimolar mixture of adipic acid and ethyleneglycol at 166 °C. (From F. A. Bovey and F. H. Winslow, *Macromolecules*. New York: Academic, 1979.)

80% conversion. The early stages of the reaction do not fit this equation very well because the reaction medium is changing from a mixture of the reactants to the polymer product. The kinetics of the last 20% of the reaction is of greatest interest because this is the phase of the reaction in which the high molar masses are attained.

Example 22.3

Adipic acid and ethylene glycol were mixed in equimolar amounts and polymerized at 166 °C (P. J. Flory, *J. Am. Chem. Soc.* **61**:3334 (1939)). The extent of reaction p was determined by titration of the acid in samples that were withdrawn:

t/min	203	398	793	1370
p	0.8161	0.8837	0.9220	0.9405

What is the value of $2k[CO_2H]_0^2$ at this temperature? The foregoing table gives only some of the experimental points; all of the data are plotted in Fig. 22.13. The straight line through the points after 80% reaction does not go through the origin. To calculate the value of $2k[CO_2H]_0^2$ in the linear section, we subtract equation 22.39 for the first point in the above table from equation 22.39 to obtain

$$\frac{1}{(1-p)^2} - \frac{1}{(1-0.8161)^2} = 2[CO_2H]_0^2 k(t - 203)$$

The values of $2[CO_2H]_0^2 k$ at $t = 398$, 793, and 1370 min are 0.228, 0.228, and 0.217 min^{-1}.

Case 2

If an acid catalyst is added to a polyesterification with equal concentrations of $-CO_2H$ and $-OH$, then the rate law is

$$-\frac{d[CO_2H]}{dt} = [CO_2H]^2(k[CO_2H] + k_{cat}[H^+]) \tag{22.40}$$

If $k_{cat}[H^+] \gg k[CO_2H]$ and the concentration of the added acid catalyst does not change during the reaction, then the rate equation becomes

$$-\frac{d[CO_2H]}{dt} = k_2[CO_2H]^2 \tag{22.41}$$

where $k_2 = k_{cat}[H^+]$. Integration and substitution of equation 22.38 yield

$$\frac{1}{1-p} = 1 + k_2[CO_2H]_0 t \tag{22.42}$$

During a step-growth polymerization the average molar mass or degree of polymerization increases steadily. The number-average degree of polymerization $\overline{X}_n$ is equal to the average number of monomer units in the polymer molecules. If the initial number of monomer molecules in the reaction mixture is n_0 and the number present at time t is n, all the rest are in polymer molecules, so that the extent of p of reaction is

$$p = \frac{n_0 - n}{n_0} = 1 - \frac{n}{n_0} \tag{22.43}$$

and we will show in the next section that the degree of polymerization is

$$\overline{X}_n = \frac{n}{n_0} = \frac{1}{1-p} \tag{22.44}$$

where the second form has been obtained by use of equation 22.43. For the average number of monomer units in the polymer molecules to be 100, it is evident that the extent of reaction p will have to be equal to 0.99.

22.7 Molar Mass Distributions of Step-Growth Polymers

We continue to consider the polymerization of a hydroxy acid HO–R–COOH which we represent as AB. The number of monomer units in a given chain i is called the **degree of polymerization**. It is therefore equal to M/M_0 where M is the molar mass of the chain and M_0 the molar mass of the monomer.

To calculate the molar mass distribution in solution, we consider the **probabilities** π_i of forming a chain of size i. This probability must be equal to the probability of having $i - 1$ elementary steps. Since the probability of having any one step is equal to the fraction p of monomers reacted, then π_i must be proportional to p^{i-1}:

$$\pi_i = cp^{i-1} \tag{22.45}$$

To find c, we note that the sum of π_i over all i must equal 1, since the sum of all probabilities is unity:

$$1 = \sum_{i=1}^{\infty} c\, p^{i-1} = c(1 + p + p^2 + \cdots) = \frac{c}{1 - p} \tag{22.46}$$

Therefore, $c = 1 - p$. This makes sense since then the probability of finding a chain with zero bonds, that is, a monomer, is just $1 - p$, the probability of finding an unreacted monomer. We therefore find that the probability of finding a chain of length i (or equivalently the mole fraction of chains of length i) is

$$\pi_i = (1 - p)p^{i-1} \tag{22.47}$$

We can now calculate the number and mass average molar masses. The number average molar mass $\overline{M}_n$ is given by

$$\overline{M}_n = M_0 \sum_{i=1}^{\infty} i\, \pi_i = M_0(1 - p) \sum_{i=1}^{\infty} i\, p^{i-1} \tag{22.48}$$

The sum $1 + 2p + 3p^2 + \cdots = (1 - p)^{-2}$, so that

$$\overline{M}_n = \frac{M_0}{1 - p} = \overline{X}_n M_0 \tag{22.49}$$

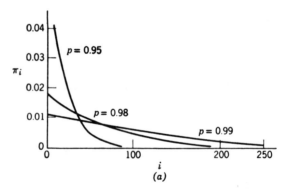

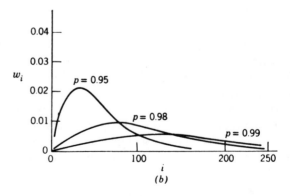

Figure 22.14 (*a*) Mole fraction distribution of condensation polymer for extents of reaction p of 0.95, 0.98, and 0.99. (*b*) Weight fraction distribution of condensation polymer for extents of reaction p of 0.95, 0.98, and 0.99. Note that the extent of reaction must be very close to unity to obtain a high polymer.

We can calculate the **mass average molar mass** by noting that

$$\overline{M}_m = \frac{M_0(\sum_i i^2 \pi_i)}{\sum_i i \, \pi_i} \qquad (22.50)$$

from the definition of $\overline{M}_m$ (see equation 22.2). Equivalently, we can define the probabilities w_i defined on the basis of mass rather than those based on number π_i by

$$w_i = \frac{i\pi_i}{\sum i\pi_i} = i \, \pi_i(1 - p) \qquad (22.51)$$

Then the preceding definition of mass average molar mass follows directly. From this, we find

$$\overline{M}_m = M_0\left(\frac{1 + p}{1 - p}\right) \qquad (22.52)$$

since $\sum_i i^2 p^{i-1} = (1 + p)(1 - p)^{-3}$. $\qquad (22.53)$

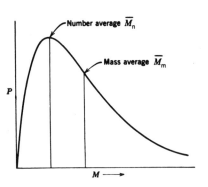

Figure 22.15 Probability density for molar masses.

The number and mass fractions or probabilities for different values of p are plotted in Fig. 22.14. Note that the mass fractions w_i go through a maximum while the number fractions π_i do not. These results have been confirmed by experiments in which polymer samples are fractionated (by solubility methods) into narrow ranges of polymer size and the various fractions analyzed.

If we multiply w_i by M_0 we have the probability of finding a chain of mass $M = iM_0$ in the solution. Therefore, the graph of the probability density for molar mass M, $P(M)$, is the same as the graph for w_i versus i (Fig. 22.14b), and is shown in Fig. 22.15. It can be shown that for p close to 1, the maximum of the curve $P(M)$ occurs for $M \simeq \overline{M}_n$. From the definitions of $\overline{M}_m$ and $\overline{M}_n$, it is clear that $\overline{M}_m \geqslant \overline{M}_n$. The equal sign can occur only if there is only one size polymer in the sample (monodisperse), so that the ratio $\overline{M}_m\overline{M}_n^{-1}$ is a good measure of the polydispersity of the sample. From above, we see that

$$\frac{\overline{M}_m}{\overline{M}_n} = 1 + p \qquad (22.54)$$

We have already seen that p must be close to 1 for the high polymer to be formed, so that for such systems $\overline{M}_m$ is twice $\overline{M}_n$.

Example 22.4
For the condensation polymerization of $HO–CH_2CH_2–CO_2H$, calculate the mole fractions π_i and weight fractions w_i of polymers with 10, 20, 30, and 40 monomer units when the fraction p of monomers reacted is 0.95.

$$\pi_i = (1 - p)p^{i-1}$$
$$= (0.05)0.95^{i-1}$$
$$w_i = i\pi_i(1 - p) = i\pi_i(0.05)$$

i	10	20	30	40
π_i	3.16×10^{-2}	1.89×10^{-2}	1.13×10^{-2}	6.76×10^{-3}
w_i	1.57×10^{-2}	1.89×10^{-2}	1.69×10^{-2}	1.35×10^{-2}

22.8 Branched and Cross-Linked Condensation Polymers

We have been discussing linear condensation polymers that arise from bifunctional monomers. However, if one or more monomers with more than two functional groups are present, the polymer will be branched and under some conditions may become cross linked. For example, if a small amount of glycerol, $HOCH_2CH(OH)CH_2OH$, is added to a hydroxyacid monomer AB, the polymer produced will be branched as shown by

```
A—BA—BA—BA┬AB—AB—AB— A
           A
           B
           |
           A                    A
           B                    |
           |                    B
           A                    A
           B                    B
           |                    A
           A           A—BA—BA─┴AB—AB—A
```

When only a small amount of glycerol is added, there is only one branch point per molecule. The joining of two branched chains is impossible because all growing chains possess A-type functional groups at their ends. However, if B–B molecules are added to the system as well, polymerization can lead to the cross-linked polymer shown by

```
A—BA—BA—BA┬AB—AB—AB—A—B
           A             |
           B             B
           |             A
           A             B
           B             A
           |             |
           A             B
           B             A
           |             |
  A—B—BA—BA—BA─┴AB—AB—A
```

If a trifunctional monomer is added to a mixture of two difunctional monomers A–A and B–B, highly branched polymer molecules are found with both free A groups and free B groups. Such polymer molecules may react with each other with a very large increase in degree of polymerization and the possibility of forming an infinite network. This process is referred to as **gelation,** and the point of transition from a soluble branched polymer to an insoluble polymer is referred to as the **gel point.** Linear polymers are referred to as thermoplastics, and cross-linked polymers, which do not flow when heated, are referred to as thermosetting polymers. Since cross links impart dimensional stability, they are needed for many practical applications.

22.9 Free-Radical Polymerization

Polyethylene, polymethyl methacrylate, polystyrene, polyacrylonitrile, poly-vinyl chloride and many other important polymers are produced by free-radical chain reactions. The monomers are unsaturated organic compounds, and the reaction to form linear high polymers may be represented by

$$n\,C{=}C \rightarrow \left[-C-C- \right]_n \tag{21.55}$$

with H R / H H and H R / H H H groups

These polymerizations are initiated by organic peroxides, azo compounds, or ultraviolet light or X-rays. These agents produce free radicals of the unsaturated organic compounds, and these radicals react with monomer molecules in a **chain reaction** (Section 19.15). The growing chain end bears an unpaired elec-tron, and the unpaired electron is transferred to the new chain end in each addition step.

Chain-reaction polymerizations involve four distinct types of processes:

1   *Chain initiation*. The initiator forms free radicals $R' \cdot$ from itself, and these radicals react with monomer M to form the first link of the chain $R_1 \cdot$, which is also a radical:

$$\text{Initiator} \xrightarrow{k_i} 2R' \cdot \tag{22.56}$$

$$R' \cdot + M \xrightarrow{k'} R_1 \cdot \tag{22.57}$$

2. *Chain propagation*. Further monomer units are added to the active chain

$$R_1 \cdot + M \xrightarrow{k_{1p}} R_2 \cdot \tag{22.58}$$

$$\cdots \cdots$$

$$R_n \cdot + M \xrightarrow{k_{np}} R_{n+1} \cdot \tag{22.59}$$

3. *Chain transfer*. The active site may be transferred to another molecule. The molecule that has lost the active site stops growing, but the molecule that has received the active site starts to grow:

$$R_n \cdot + XY \rightarrow R_nY + Z \cdot \tag{22.60}$$

The molecule represented here by XY may be another polymer molecule and the radical site may be introduced somewhere along the chain. This leads to the formation of a branched polymer.

4. *Chain termination*. The free radical of a growing polymer chain can react with the free radical of another chain to form either one or two molecules of inactive polymer P:

$$R_n \cdot + R_m \cdot \xrightarrow{k_{tc}} P_{n+m} \tag{22.61}$$

$$R_n \cdot + R_m \cdot \xrightarrow{k_{td}} P_n + P_m \tag{22.62}$$

The entire sequence of chain-propagating reactions has no effect on the total concentration of propagating radicals. Thus, it is an excellent approximation to assume that the concentration [R] of radicals is in a steady state and that the rate r_i of initiation of radicals is equal to the rate r_t of termination of radicals:

$$\frac{d[R]}{dt} = r_i - r_t = 0 \tag{22.63}$$

The rate of initiation is given by

$$r_i = 2fk_i[I] \tag{22.64}$$

according to reaction 22.56. Since some free radicals formed in the initiation reaction are lost, f represents the fraction of initiating radicals that actually add to monomer to form R_1.

The rate of termination is a bimolecular process so that

$$r_t = 2k_t[R]^2 \tag{22.65}$$

where the rate constant $2k_t$ is actually $(k_{tc} + k_{td})$.

Inserting equations 22.64 and 22.65 in equation 22.63 yields

$$2k_t[R]^2 = 2fk_i[I] \tag{22.66}$$

This indicates that the steady-state concentration of radicals is

$$[R] = \left(\frac{fk_i[I]}{k_t}\right)^{1/2} \tag{22.67}$$

The overall rate of polymerization r_p is the rate of decrease of monomer concentration. It is a good approximation to assume that the rate constants for successive steps $k_{1p}, k_{2p}, \ldots, k_{np}$ are all the same so that the rate of polymerization is given by

$$r_p = k_p[R][M] \tag{22.68}$$

where [R] is the total concentration of propagating radicals. Since the steady state concentration of radicals is given by equation 22.67,

$$r_p = k_p\left(\frac{fk_i[I]}{k_t}\right)^{1/2}[M] \tag{22.69}$$

Note that the rate of the polymerization reaction is proportional to the square root of the initiator concentration.

The **average kinetic chain length** ν is the average number of monomer molecules polymerized per chain initiated. This quantity is important because the properties of a polymer depend on the average molar mass and so it is of interest to see how this is affected by the kinetic parameters. The average kinetic chain length is equal to the rate of polymerization divided by the rate of initiation.

$$\nu = \frac{r_p}{r_i} = \frac{k_p[M][R]}{2k_t[R]^2} = \frac{k_p[M]}{2k_t[R]} \tag{22.70}$$

Eliminating [R] by use of equation 22.67 yields

$$\nu = \frac{k_p[M]}{2k_t}\left[\frac{k_t}{fk_i[I]}\right]^{1/2} = \left(\frac{k_p^2}{2k_t}\right)^{1/2}\frac{[M]}{r_i^{1/2}} \tag{22.71}$$

Thus, polymer of higher molar mass is obtained with lower rate r_i of initiation.

22.10 Viscosity of Polymer Solutions

In Section 21.1, we discussed the viscosity of a liquid, and pointed out that the viscosity of high polymer solutions is very large. Let us now consider a solution of large particles (like polymer molecules) and ask how the viscosity of the solution depends on the size of the particles.

The ratio of the viscosity η of a suspension of particles to the viscosity of the solvent η_0 may be represented by an equation of the type

$$\frac{\eta}{\eta_0} = 1 + \nu\phi + \kappa\phi^2 + \cdots \tag{22.72}$$

where ϕ is the volume fraction occupied by particles and ν and κ are constants that depend on the shape of the particles. The quantity η/η_0 is referred to as the **relative viscosity,** and it is convenient to use the term **specific viscosity** η_{sp} for $(\eta/\eta_0) - 1$, although neither quantity is, strictly speaking, a viscosity. Thus, equation 22.72 may be written

$$\frac{\eta_{sp}}{\phi} = \nu + \kappa\phi + \cdots \tag{22.73}$$

Einstein showed in 1906 that for spheres $\nu = \frac{5}{2}$. This coefficient increases with axial ratio for prolate and oblate ellipsoids of revolution. It is interesting to note that ν is independent of the size of the particles.

The volume fraction ϕ of a solute may be written cv_2, where c is the concentration of the solute in mass per unit volume of solution and v_2 is the partial specific volume. The partial specific volume v_i is equal to the partial molar volume V_i (Section 4.12) divided by the molar mass. Thus, equation 22.73 may be written

$$\frac{\eta_{sp}}{c} = \nu v_2 + \kappa c v_2^2 + \cdots \tag{22.74}$$

The limit of η_{sp}/c as c approaches zero is called the **intrinsic viscosity** $[\eta]$. Thus, from equation 22.74,

$$\lim_{c \to 0} \frac{\eta_{sp}}{c} = [\eta] = \nu v_2 \tag{22.75}$$

Intrinsic viscosities are usually expressed in $cm^3 \ g^{-1}$.

For a solvated macromolecule the volume occupied by a molecule is greater than that calculated using the partial specific volume. If this effect is taken into account, equation 22.75 becomes

$$[\eta] = \nu(v_2 + \delta v_1) \tag{22.76}$$

where δ is the mass of solvent bound by a gram of solute and v_1 is the partial specific volume of the solvent. If the particle is not spherical, ν and δ are both unknown, but values of ν may be calculated for ellipsoids of revolution. If the particle is spherical or is an ellipsoid of revolution of a known axial ratio, the solvation δ may be calculated from a measurement of the intrinsic viscosity.

Intrinsic viscosities give strong indications of the axial ratios of molecules in solution. For a spherical, unhydrated protein with $v = 0.75 \ cm^3 \ g^{-1}$, we

Table 22.1 Intrinsic Viscosities in Water at 25 °C[a]

Molecule	M/g mol^{-1}	$[\eta]$/cm^3 g^{-1}
Ribonuclease	13 683	2.3
Bovine serum albumin	66 500	3.7
Bushy stunt virus	10 700 000	3.4
Fibrinogen	330 000	27
Tobacco mosaic virus	40 000 000	37
Myosin	493 000	217
DNA	6 000 000	5000

[a] From C. Tanford, *Physical Chemistry of Macromolecules*. New York: Wiley, 1961. Reprinted with permission.

would expect $[\eta] = 1.9$ cm^3 g^{-1}. As shown in Table 22.1, the intrinsic viscosity of ribonuclease is not very much greater than this value. Bushy stunt virus is very nearly spherical, even though it has a molar mass of 1.07×10^7 g mol^{-1}. The elongated molecules of fibrinogen provide the structure for blood clots, and the even more elongated molecules of myosin are the contractile part of muscle. The very high intrinsic viscosity of double-stranded deoxyribonucleic acid (DNA) shows that the molecules are very long and thin. High molar mass DNA may be degraded (reduced in molar mass) by the shear gradients encountered in pipetting solutions. To measure the viscosities of DNA solutions at very low shear it has been necessary to use rotating cylinder, or Couette, viscometers.

In Section 22.4 we found that the root-mean-square end-to-end distance of a linear polymer in a theta solvent is proportional to the square root of its molar mass. On the basis of this result we can derive the form of the relation between the intrinsic viscosity and the molar mass.

The polymer molecule in solution may be visualized as a spherical cloud of segments, with the cloud getting thinner as we go out from the center. The size of this spherical cloud is measured by the root-mean-square end-to-end distance $(\overline{r^2})^{1/2}$. As an approximation of the spherical cloud, we may replace it with a rigid sphere with a radius proportional to $(\overline{r^2})^{1/2}$ and a volume proportional to $(\overline{r^2})^{3/2}$. Since the intrinsic viscosity of a sphere is proportional to the volume per unit mass of the macromolecule, the intrinsic viscosity of a random coil is expected to be proportional to $(\overline{r^2})^{3/2}/M$. Since $(\overline{r^2})$ is proportional to the number of segments in the chain (see equation 22.27), or to M, the intrinsic viscosity of a random coil is expected to be proportional to $M^{1/2}$:

$$[\eta] = KM^{1/2} \tag{22.77}$$

This is exactly what is found for polystyrene in cyclohexane at 34.5 °C. In general the intrinsic viscosity is related to the molar mass of the polymer by the Mark–Houwink equation, which is

$$[\eta] = KM^a \tag{22.78}$$

where a is an empirical constant that usually has a value in the range 0.5 to 0.8. The higher values of a are obtained in good solvents where the polymer chain is more extended than a random coil. The values of K and a for a number of polymer–solvent systems are shown in Table 22.2.

Table 22.2[a]   Parameters in $[\eta] = KM^a$

Polymer	Solvent	$t/°C$	$K/10^{-2}$ cm^3 g^{-1}	a
Natural rubber	Toluene	25	5.0	0.67
Polymethyl methacrylate	Acetone	25	0.75	0.70
	Chloroform	25	0.48	0.80
	Methyl ethyl ketone	25	0.68	0.72
Polystyrene	Benzene	20	1.23	0.72
	Methyl ethyl ketone	20–40	3.82	0.58
	Toluene	20–30	1.05	0.72
Polyvinyl alcohol	Water	25	30.0	0.50

[a] H. R. Allcock and F. W. Lampe, *Contemporary Polymer Chemistry*. Englewood Cliffs, NJ: Prentice-Hall, 1981.

Since K and a have to be determined experimentally, this is a secondary method for the determination of the average molar mass of a high polymer. The constants are determined as follows: A sample of a high polymer is fractionated into a series of fractions, each of which has a narrow range of molar masses. If the average molar masses of each of these fractions are determined by use of an absolute method such as light scattering and the intrinsic viscosity of each fraction is also determined in the same solvent, the empirical relation between intrinsic viscosity and average molar mass can be determined. In the case of a heterogeneous polymer the molar mass determined using viscosity measurements is closer to a mass-average molar mass than a number-average molar mass (Section 22.1).

Example 22.5

Given that the intrinsic viscosity of a sample of DNA ($M = 6 \times 10^6$ g mol^{-1}) is 5.0×10^3 cm^3 g^{-1}, approximately what concentration of DNA in water would have a relative viscosity of 1.1?

$$\frac{(\eta/\eta_0) - 1}{c} = \frac{0.1}{c} = 5.0 \times 10^3 \text{ cm}^3 \text{ g}^{-1}$$

$$c = 2.0 \times 10^{-5} \text{ g cm}^{-3} = 2.0 \times 10^{-2} \text{ g L}^{-1}$$

Since the molar mass of DNA per nm measured along the axis of the helix is 1920 g mol^{-1}, the length of a molecule of DNA with $M = 6 \times 10^6$ g mol^{-1} is 3100 nm or 3.1 μm.

22.11   Special Topic: Sedimentation

Particles suspended in a fluid sediment if their density is higher than that of the suspension medium and float to the top if their density is lower. Their equivalent radius can be calculated from the rate of sedimentation using the expression for the frictional coefficient, $f = 6\pi\eta r$, given in Section 21.1.

Dissolved molecules tend to sediment in the earth's gravitational field or to float upward, depending on their density relative to that of the solvent, but this

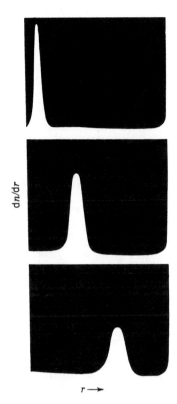

Figure 22.16 Schlieren patterns for an ultracentrifuge experiment with fumarase at 50 400 rpm. The lower photographs were taken 35 and 75 min later than the top photograph. The protein is dissolved in pH 6.8 phosphate buffer.

tendency is counteracted by the random translational motion of the molecules. However, sufficiently powerful **ultracentrifuges** have been built to cause even molecules as small as sucrose to sediment at measureable rates. Svedberg* was the leader in the development of ultracentrifuges, which he defined as centrifuges adapted for quantitative measurements of convection-free and vibration-free sedimentation. There are two distinct types of ultracentrifuge experiments: (1) those where the velocity of sedimentation of a component of the solution is measured (sedimentation velocity), and (2) those where the redistribution of molecules is determined at equilibrium (sedimentation equilibrium).

The acceleration a of a centrifugal field is equal to $\omega^2 r$, where ω is the angular velocity of the centrifuge in radians per second (i.e., 2π times the number of revolutions per second) and r is the distance from the axis of rotation. Ultracentrifuges in which r is about 6 cm are commonly operated at 60 000 rpm or 1000 rps, and so the acceleration is

$$a = \omega^2 r = (2\pi 1000 \text{ s}^{-1})^2 (0.06 \text{ m}) = 2.36 \times 10^6 \text{ m s}^{-2} \quad (22.79)$$

Since the acceleration of the earth's field is 9.80 m s^{-2}, the acceleration is 240 000 times greater than in the earth's field.

A solution to be studied in the velocity ultracentrifuge is placed in a cell with thick quartz windows. The cell has a sector shape when viewed at right angles to the plane of rotation of the centrifuge rotor, since the sedimentation takes place radially. As the high-molar-mass component throughout the solution sediments, a moving boundary is formed behind which there is only solvent. When the solute is a colored substance, its concentration may be determined photometrically as a function of height. One of the most generally useful methods for determining the diffuseness of the boundary depends on measuring the deflection of light by the refractive-index gradient associated with the concentration gradient. Since the bending of a light ray in such a boundary is proportional to the refractive index gradient dn/dx, the curve obtained with such a schlieren optical system has the shape indicated in Fig. 22.16.

Figure 22.16 shows the schlieren patterns for an ultracentrifuge experiment with the enzyme fumarase at 50 400 rpm. The second and third photographs were taken 35 and 70 min later than the top photograph. If additional components with different rates of sedimentation had been present, additional peaks would be evident in the schlieren photograph. Thus, the ultracentrifuge is useful in analyzing complex mixtures such as blood plasma.

When a solution is subjected to a centrifugal field, the solute molecules move away from the axis of rotation if the reciprocal of their partial specific volume v is greater than the density ρ of the solution, and otherwise they move toward the axis of rotation. In other words, the direction of "sedimentation" depends on the sign of the Archimedes factor $(1 - v\rho)$.

The ratio of the velocity to the centrifugal acceleration is called the **sedimentation coefficient** S:

$$S = \frac{1}{\omega^2 r} \frac{dr}{dt} \quad (22.80)$$

---

* T. Svedberg and K. O. Pedersen, *The Ultracentrifuge*. Oxford, UK: Oxford University Press, 1940.

Since ω^2 has the units s^{-2}, the sedimentation coefficient has the unit seconds. The sedimentation coefficients of proteins fall in the range 10^{-13} s to 200×10^{-13} s, and the unit 10^{-13} s is called a **svedberg.** If a boundary is r_1 centimeters from the axis of the centrifuge at time t_1 and r_2 centimeters from the axis at time t_2, the sedimentation coefficient may be calculated from

$$S = \frac{1}{\omega^2(t_2 - t_1)} \ln \frac{r_2}{r_1} \qquad (22.81)$$

which is obtained by integrating equation 22.80.

Example 22.6
In the ultracentrifuge experiment illustrated in Fig. 22.16 the distance from the axis of the ultracentrifuge to the boundary was 5.949 cm in the top photograph and 6.731 cm in the bottom photograph taken 70 min later. Since the speed of the rotor was 50 000 rpm, $\omega^2 = 2.79 \times 10^7 \ s^{-2}$. Using equation 24.81 we see that

$$S = \frac{1}{\omega^2(t_2 - t_1)} \ln \frac{r_2}{r_1} = \frac{\ln(6.731/5.949)}{(2.79 \times 10^7 \ s^{-2})(60 \times 70 \ s)} = 10.5 \times 10^{-13} \ s$$

This is the sedimentation coefficient at 28.2 °C, the temperature of the experiment. Making a correction to 20 °C in water, taking into account the change of viscosity and density, a value of 8.90 svedbergs is obtained.

In a centrifugal field a solute molecule is accelerated until the frictional force resisting its motion is equal to the acceleration of the centrifugal field times the effective mass $m (1 - v\rho)$, where m is the mass of the molecule. The frictional force is the product of the velocity dr/dt and the frictional coefficient f (Section 21.1). Thus, when the steady-state velocity dr/dt is reached,

$$f\frac{dr}{dt} = m(1 - v\rho)\omega^2 r = \frac{M(1 - v\rho)\omega^2 r}{N_A} \qquad (22.82)$$

where M is the molar mass.

Substituting equation 22.82 into equation 22.80 yields

$$S = \frac{M(1 - v\rho)}{N_A f} \qquad (22.83)$$

The sedimentation coefficient by itself cannot be used to determine the molar mass of the sedimenting component unless the molecules are spherical. If the molecules are spherical, $f = 6\pi\eta r$, and equation 22.83 may be used to calculate the molar mass. Since the velocity of sedimentation is so low that there is no appreciable orientation of the molecules, the frictional coefficient involved in sedimentation is taken to be the same as that involved in diffusion. Introduction of equation 21.11 into equation 22.83 yields

$$M = \frac{RTS}{D(1 - v\rho)} \qquad (22.84)$$

To calculate the molar mass from measured values of S and D, it is necessary to correct sedimentation and diffusion coefficients to the same temperature,

Table 22.3 Physical Constants of Proteins at 20 °C in Water

Protein	$S/10^{-13}$ s	$D/10^{-11}$ m^2 s^{-1}	v/cm^3 g^{-1}	M/g mol^{-1}
Beef insulin	1.7	15	0.72	12 000
Lactalbumin	1.9	10.6	0.75	17 400
Myoglobin	2.06	12.4	0.749	16 000
Ovalbumin	3.6	7.8	0.75	44 000
Serum albumin	4.3	6.15	0.735	64 000
Hemoglobin	4.6	6.9	0.749	64 400
Serum globulin	7.1	4.0	0.75	167 000
Urease	18.6	3.4	0.73	490 000
Tobacco mosaic virus	185	0.53	0.72	40 000 000

usually 20 °C, and if S and D depend appreciably on concentration, to zero concentration. Equation 22.84 has probably been the most widely used in the calculation of molar masses of proteins, and the wide range of molar masses that can be obtained by this method is indicated by Table 22.3.

Example 22.7
Using the data of Table 22.3, the molar mass of hemoglobin may be calculated as follows (the density of water at 20 °C is 0.9982 $\times$ 10^3 kg m^{-3}):

$$M = \frac{RTS}{D(1 - v\rho)}$$

$$= \frac{(8.31 \text{ J K}^{-1} \text{ mol}^{-1})(293 \text{ K})(4.6 \times 10^{-13} \text{ s})}{(6.9 \times 10^{-11} \text{ m}^2 \text{ s}^{-1})[1 - (0.749 \times 10^{-3} \text{ m}^3 \text{ kg}^{-1})(0.9982 \times 10^3 \text{ kg m}^{-3})]}$$

$$= 64.4 \text{ kg mol}^{-1} = 64,400 \text{ g mol}^{-1}$$

In measuring the sedimentation coefficient of DNA significant concentration effects are encountered at concentrations as low as 10 mg L^{-1}. Taking advantage of the strong nucleotide absorption around 260 nm, the velocity of the sedimenting boundary may be measured even at this low concentration using an ultraviolet absorption optical system. In another method for studying such dilute solutions a thin layer of DNA solution may be placed on a preformed density gradient of CsCl solution created in an ultracentrifuge cell by the centrifugal field causing the sedimentation. In these experiments the density gradient stabilizes the system from convection.

22.12 Special Topic: Diffusion of Spherical Macromolecules

For spherical macromolecules or colloidal particles, the particle radius r and molar mass M may be calculated from the experimental value of the diffusion coefficient by using equation 21.12. The molar mass of a spherical particle is given by

$$M = \frac{4\pi r^3 N_A}{3v} \tag{22.85}$$

where v is the specific volume (that is, the reciprocal of the density). If equation 22.85 is substituted in equation 21.12, we obtain

$$D = \frac{RT}{N_A 6\pi\eta}\left(\frac{4\pi N_A}{3Mv}\right)^{1/3} \tag{22.86}$$

For spherical particles, then, the diffusion coefficient is inversely proportional to the cube root of the molar mass. Of course, if the particles or molecules are not spherical, the value of the molar mass calculated from equation 22.86 will not be correct. This equation, however, does give the maximum molar mass that is consistent with a given D and v. For a nonspherical particle the frictional coefficient is larger than $6\pi\eta r$ and the molar mass is smaller than that calculated from equation 22.86.

Example 22.8
Using the diffusion coefficient for myoglobin from Table 22.3, calculate the upper limit of the molar mass of this protein. The coefficient of viscosity of water at 20 °C is 0.001 005 Pa s. Rearranging equation 22.86 gives

$$M = \frac{4\pi N_A}{3v}\left(\frac{RT}{DN_A 6\pi\eta}\right)^3 = \frac{4\pi(6.022 \times 10^{23}\ \text{mol}^{-1})}{3(0.749 \times 10^{-3}\ \text{m}^3\ \text{kg}^{-1})}$$

$$\times\left[\frac{(8.314\ \text{J K}^{-1}\ \text{mol}^{-1})(293\ \text{K})}{(12.4 \times 10^{-11}\ \text{m}^2\ \text{s}^{-1})(6.022 \times 10^{23}\ \text{mol}^{-1})6\pi(1.005 \times 10^{-3}\ \text{Pa s})}\right]^3$$

$$= 17.2\ \text{kg mol}^{-1} = 1.72 \times 10^4\ \text{g mol}^{-1}$$

The actual molar mass is 1.60×10^4 g mol^{-1}.

References

H. R. Allcock and F. W. Lampe, *Contemporary Polymer Chemistry*. Englewood Cliffs, NJ: Prentice-Hall, 1981.

B. J. Berne and R. Pecora, *Dynamic Light Scattering*. New York: Wiley, 1976.

F. W. Billmeyer, *Textbook of Polymer Science*. New York: Wiley, 1984.

R. B. Bird, W. E. Stewart, and E. N. Lightfoot, *Transport Phenomena*. New York: Wiley, 1960.

V. A. Bloomfield, D. A. Crothers, and I. Tinoco, Jr., *Physical Chemistry of Nucleic Acids*. New York: Harper & Row, 1974.

F. A. Bovey and F. H. Winslow, *Macromolecules*. New York: Academic, 1979.

C. R. Cantor and P. R. Schimmel, *Biophysical Chemistry*. San Francisco: Freeman, 1980.

P. J. Flory, *Principles of Polymer Chemistry*. Ithaca, NY: Cornell University Press, 1953.

E. G. Richards, *An Introduction to Physical Properties of Large Molecules in Solution*. Cambridge, UK: Cambridge University Press, 1980.

R. S. Seymour and C. E. Carraher, *Polymer Chemistry*. New York: Dekker, 1988.

L. H. Sperling, *Physical Polymer Science*. New York: Wiley-Interscience, 1986.

C. Tanford, *Physical Chemistry of Macromolecules*. New York: Wiley, 1961.

K. E. Van Holde, *Physical Biochemistry*. Englewood Cliffs, NJ: Prentice-Hall, 1971.

J. W. Williams, *Ultracentrifugation of Macromolecules: Modern Topics*. New York: Academic, 1973.

Problems

22.1 A polymer solution contains 250 molecules of molar mass 75 000 g mol^{-1}, 500 molecules of molar mass 100 000 g mol^{-1}, and 250 molecules of molar mass 125 000 g mol^{-1}. Calculate $\overline{M}_n$, $\overline{M}_m$, and the ratio $\overline{M}_m/\overline{M}_n$ (the polydispersity).

22.2 A beam of sodium D light (589 nm) is passed through 100 cm of an aqueous solution of sucrose containing 10 g sucrose per 100 cm^3. Calculate I/I_0, where I_0 is the intensity that would have been obtained with pure water, given that $M = 342.30$ g mol^{-1} and $dn/dc = 0.15$ g^{-1} cm^3 for sucrose. The refractive index of water at 20 °C is 1.333 for the sodium D line.

22.3 Plot the probability density $W(r)$ for random walk in three dimensions after 1000 steps with a step length of unity. Indicate the root-mean-square end-to-end distance on this plot.

22.4 In polyethene H(CH$_2$–CH$_2$)$_n$H the bond length l is 0.15 nm. What is the root-mean-square end-to-end distance for a molecule with universal joints with a molar mass of 10^5 g mol^{-1}? Taking into account the fact that carbon forms tetrahedral bonds, what is $(\overline{r^2})^{1/2}$?

22.5 Adipic acid and 1,10-decamethylene glycol were mixed in equimolar amounts and polymerized at 161 °C and 191 °C (S. D. Hamann, D. H. Solomon, and J. D. Swift, *J. Macromol. Sci. Chem. A* **2**:153 (1968)). The extents of reaction, as determined by acid titration, are as follows:

t/min	0	200	400	600	800
p(161 °C)	0.820	0.900	0.927	0.940	0.947
p(191 °C)	0.820	0.937	0.955	0.963	0.968

Time was measured from the point at which there was 82% esterification. Taking [CO$_2$H]$_0$ = 1.25 mol kg^{-1} at both temperatures, what are the two values of k and the activation energy?

22.6 Adipic acid and diethylene glycol were polymerized at 109 °C using *p*-toluene sulfuric acid as a catalyst (P. J. Flory, *J. Am. Chem. Soc.* **61**:3334 (1939)). The extents of reaction at various times are as follows:

t/min	0	40	80	120
p	0.800	0.909	0.944	0.960

(Note that the time is taken as zero at $p = 0.80$.) At what time will the number average molar mass reach 10 000 g mol^{-1} for this concentration of reactants and catalyst?

22.7 For a condensation polymerization of a hydroxyacid in which 99% of the acid groups are used up, calculate (*a*) the average number of monomer units in the polymer molecules, (*b*) the probability that a given molecule will have the number of residues given by this value, and (*c*) the weight fraction having this particular number of monomer units.

22.8 A hydroxyacid HO–(CH$_2$)$_5$–CO$_2$H is polymerized and it is found that the product has a number-average molar mass of 20 000 g mol^{-1}. (*a*) What is the extent of reaction p? (*b*) What is the degree of polymerization $\overline{X}_n$? (*c*) What is the mass-average molar mass?

22.9 The relative viscosities of a series of solutions of a sample of polystyrene in toluene were determined with an Ostwald viscometer at 25 °C:

c/10^{-2} g cm^{-3}	0.249	0.499	0.999	1.998
η/η_0	1.355	1.782	2.879	6.090

The ratio η_{sp}/c is plotted against c and extrapolated to zero concentration to obtain the intrinsic viscosity. If the constants in equation 22.78 are $K = 3.7 \times 10^{-2}$ and $a = 0.62$ for this polymer, when concentrations are expressed in g/cm^3, calculate the molar mass.

22.10 At 34 °C the intrinsic viscosity of a sample of polystyrene in toluene is 84 cm^3 g^{-1}. The empirical relation between the intrinsic viscosity of polystyrene in toluene and molar mass is

$$[\eta] = 1.15 \times 10^{-2} M^{0.72}$$

What is the molar mass of this sample?

22.11 Given that the intrinsic viscosity of myosin is 217 cm^3 g^{-1}, approximately what concentration of myosin in water would have a relative viscosity of 1.5?

22.12 The sedimentation coefficient of myoglobin at 20 °C is 2.06×10^{-13} s. What molar mass would it have if the molecules were spherical? Given: $v = 0.749 \times 10^{-3}$ m^3 kg^{-1}, $\rho = 0.9982 \times 10^3$ kg m^{-3}, and $\eta = 0.001\ 005$ Pa s.

22.13 The sedimentation and diffusion coefficients for hemoglobin corrected to 20 °C in water are 4.41×10^{-13} s and 6.3×10^{-11} m^2 s^{-1}, respectively. If $v = 0.749$ cm^3 g^{-1} and $\rho_{H_2O} = 0.998$ g cm^{-3} at this temperature, calculate the molar mass of the protein. If there is 1 mol of iron per 17 000 g of protein, how many atoms of iron are there per hemoglobin molecule?

22.14 Given the diffusion coefficient for sucrose at 20 °C in water ($D = 45.4 \times 10^{-11}$ m^2 s^{-1}), calculate its sedimentation coefficient. The partial specific volume v is 0.630 cm^3 g^{-1}.

22.15 A solution of high polymer in benzene has a concentration of 1 g/100 cm^3 and a refractive index for the sodium D line (589 nm) of 1.5021. The refractive index of benzene under these conditions is 1.5011. The turbidity τ is 2×10^{-4} cm^{-1}. What is the molar mass of the polymer? What is I/I_0 for a 10-cm cell?

22.16 In the polymerization of 1,10-decamethylene glycol with adipic acid at 161 °C using toluene sulfonic acid as a catalyst, the following extents of reaction were obtained (S. D. Hamann, D. H. Solomon, and J. D. Swift, *J. Macromol. Sci. Chem. A* **2**:153, (1968)):

t/min	40	80	120
p	0.9677	0.9825	0.9878

Time zero corresponds with 82% esterification. The initial concentration of carboxyl groups is 1.25 mol kg^{-1} of mixture. What is the value of the rate constant?

22.17 In the polymerization of adipic acid [HO$_2$C(CH$_2$)$_4$CO$_2$H] and diethylene glycol [(HOCH$_2$CH$_2$)$_2$O] at what time should the reaction be stopped to obtain a polymer with a number-average molar mass of 10,000 g mol^{-1}? *Given*: 2[CO$_2$H]k = 0.2 min^{-1}.

22.18 For a condensation polymerization of a hydroxyacid in which 95% of the acid groups is used up, calculate (*a*) the average number of monomer units in the polymer molecules, (*b*) the probability that a molecule chosen at random will have this number of residues, and (*c*) the weight fraction having this particular number of monomer units.

22.19 For the polymer described in problem 22.18, calculate the number-average and mass-average molar mass.

22.20 In the condensation polymerization of a hydroxyacid with a residue mass of 200 it is found that 99% of the acid groups are used up. Calculate (*a*) the number-average molar mass, and (*b*) the mass-average molar mass.

22.21 The relation between M and $[\eta]$ for double-stranded linear DNA is $0.665 \ln M = 1.987 + \log([\eta] + 500)$ when $[\eta]$ is expressed in cm^3 g^{-1}. What is the molar mass of DNA that has an intrinsic viscosity of 5000 cm^3 g^{-1}?

22.22 A sample of polystyrene was dissolved in toluene, and the following flow times in an Ostwald viscometer at 25 °C were obtained for different concentrations:

c/10^{-2} g cm^{-3}	0	0.1	0.3	0.6	0.9
t/s	86.0	99.5	132	194	301

If the constants in equation 22.78 are $K = 3.7 \times 10^{-2}$ and $a = 0.62$ for this polymer, calculate the molar mass.

22.23 Calculate the sedimentation coefficient of tobacco mosaic virus from the fact that the boundary moves with a velocity of 0.454 cm h^{-1} in an ultracentrifuge at a speed of 10 000 rpm at a distance of 6.5 cm from the axis of the centrifuge rotor.

22.24 The sedimentation coefficient of gamma globulin at 20 °C is 7.1×10^{-13} s. Calculate how far the protein boundary will sediment in $\frac{1}{2}$ h if the speed of the centrifuge is 60 000 rpm and the initial boundary is 6.50 cm from the axis of rotation.

22.25 Using data from Table 22.4, calculate the molar mass of serum albumin.

22.26 The diffusion coefficient for serum globulin at 20 °C in a dilute aqueous salt solution is 4.0×10^{-11} m^2 s^{-1}. If the molecules are assumed to be spherical, calculate their molar mass. *Given*: $\eta_{H_2O} = 0.001\ 005$ Pa s at 20 °C and $v = 0.75$ cm^3 g^{-1} for the protein.

22.27 The diffusion coefficient of hemoglobin at 20 °C is 6.9×10^{-11} m^2 s^{-1}. Assuming its molecules are spherical, what is the molar mass? *Given*: $v = 0.749$ s 10^{-3} m^3 kg^{-1} and $\eta = 0.001\ 005$ J m^{-3} s.

22.28 The diffusion coefficient of a certain virus having spherical particles is 0.50×10^{-11} m^2 s^{-1} at 0 °C in a solution with a viscosity of 0.001 80 Pa s. Calculate the molar mass of this virus, assuming that the density of the virus is 1 g cm^{-3}.

23
Solid-State Chemistry

In this chapter we first consider crystal geometry, then X-ray diffraction and the structure of specific crystals, and finally semiconductors and synthetic high polymers. The ideas of point-group symmetry developed in Chapter 13 are basic to the understanding of crystals, but the introduction of translations produces new kinds of symmetry and requires space groups (Section 23.3) rather than point groups.

In 1912 Laue suggested that the wavelength of X-rays might be about the same as the distance between atoms in a crystal so that a crystal could serve as a diffraction grating for X-rays. This experiment was carried out by Frederick and Knipping, who observed the expected diffraction. Almost immediately afterward, W. L. Bragg (1913) improved on the Laue experiment, mainly by substituting monochromatic for polychromatic radiation and by providing a more physical interpretation to the Laue theory of the scattering experiment. Bragg also determined the structures of a number of simple crystals, including those of NaCl, CsCl, and ZnS. Since that time, single-crystal X-ray diffraction has developed into the most powerful method known for obtaining the atomic arrangement in the solid state. Since the 1950s, with the advent of high-speed computers capable of handling X-ray data, it has been possible to determine the structures of compounds as complex as proteins.

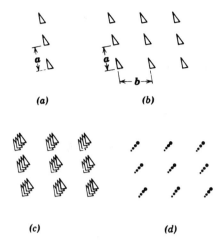

Figure 23.1 (*a*) One-dimensional pattern with vector *a*. (*b*) Two-dimensional pattern with vectors *a* and *b*. (*c*) Three-dimensional pattern with vectors *a*, *b*, and *c*. (*d*) Lattice of pattern *c*.

The chapter ends with a discussion of the relation between structure and properties for several very different types of crystalline solids, specifically metals, semiconductors, and synthetic high polymers.

23.1 Lattices and Unit Cells

A crystal may be described as a three-dimensional pattern in which a structural motif is repeated in such a way that the environment of every motif is the same throughout the crystal. The motif may be an atom or a molecule, or it may be a group of atoms or a group of molecules.

A linear pattern may be described by saying that there is a set of parallel motifs at the end of a vector *a* and its multiples. These vectors are given by

$$T = ua \qquad (23.1)$$

where u is an integer. An example of a linear pattern is shown in Fig. 23.1*a*. The linear pattern formed in this way can be repeated in a second direction represented by the vector *b* to form a two-dimensional pattern in which every structural pattern has the same environment as every other, as shown in Fig. 23.1*b*. Finally, the whole two-dimensional pattern can be repeated in a third direction *c* to produce a translationally ordered pattern in three dimensions, as shown in Fig. 23.1*c*. The three vectors *a*, *b*, and *c* are called primitive vectors of the crystal. The **structural pattern** is repeated at the end of every vector of the form

$$T = ua + vb + wc \qquad \text{where } u, v, w \text{ are integers} \qquad (23.2)$$

For many purposes it is convenient to concentrate on the geometry of the repetition and replace the structural pattern by a point, as shown in Fig. 23.1*d*, to obtain a **lattice.** The lattice may be generated from a single starting point by

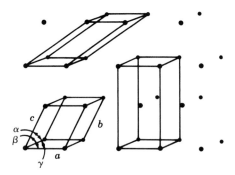

Figure 23.3   Cells in a lattice.

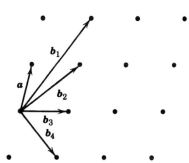

Figure 23.2   Alternative choices for the second translational vector in a two-dimensional lattice.

the infinite repetition of a set of fundamental translations that characterize the lattice. Any three noncoplanar vectors *a, b,* and *c* describe a lattice, but a given lattice can be described by an infinite number of sets of three vectors. This is illustrated in two dimensions in Fig. 23.2. The *a* vector together with any one of the *b* vectors may be chosen to generate the pattern.

The space occupied by a lattice may be divided into **unit cells.** The repetition of a cell (with everything in it) in three dimensions generates the entire pattern of a crystal. A given lattice can be blocked out in cells in different ways, as shown in Fig. 23.3. If the corners of the cells include all of the lattice points in the crystal, the cell is called a **primitive unit cell.** Primitive cells have one lattice point per cell because each of the corner lattice points is shared by eight cells. A lattice can also be blocked out in cells that do not include all lattice points as corners. This is illustrated by one of the cells in Fig. 23.3. Such cells, referred to as **multiple unit cells,** are useful in simplifying the geometry of crystals for which the primitive unit cell is oblique, but the multiple unit cell has two or more edges that are at 90°.

23.2   Classification of Crystal Structures

The classification of crystals is based on their **symmetry,** rather than on the dimensions of their unit cells. It is not satisfactory to use the equality of all edges, for example, because they may become unequal when the temperature changes. The use of symmetry avoids this difficulty because, if two directions in a crystal are equivalent by symmetry, they will necessarily have the same thermal expansion coefficient. The recognition of the types of symmetry we have discussed in Chapter 13 reduces the number of coordinates of atoms that have to be specified to describe the structure of a unit cell because of the relationship between the coordinates of symmetry-related atoms.

We saw in Chapter 13 that a molecule can have various types of rotational symmetry. For a molecule with n-fold symmetry, rotation through an angle of $360°/n$ brings it into an equivalent position, and a molecule may have $n = 1$, 2, 3, . . . , ∞. However, in contrast with individual molecules, crystals and crystal lattices can only have $n = 1, 2, 3, 4,$ or 6. This statement can be proved by use of Fig. 23.4. The lattice shown in this figure has an axis of n-fold symmetry perpendicular to the page. The lattice points A_1, A_2, A_3, and A_4 are each separated by distance a. Because of the assumed symmetry, rotation of the lattice about any lattice point through an angle $\alpha = 2\pi/n$ will produce a lattice

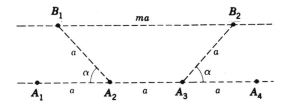

Figure 23.4   Restriction on rotational order in a crystal. There are n-fold rotational axes at lattice points A_1, A_2, A_3, A_4, B_1, and B_2.

indistinguishable from the original. Therefore, clockwise rotation by α about A_3 and counterclockwise rotation by α about A_2 requires that there be lattice points at B_1 and B_2. Since the line B_1B_2 is parallel to A_1A_4, B_1 and B_2 must be separated by an integer multiple of a, represented by ma. Thus,

$$a + 2a \cos \alpha = ma \qquad \cos \alpha = \frac{N}{2}$$

where $N = m - 1$. The only values of α that satisfy this equation are 0°, 60°, 90°, 120°, 180°, 240°, and 360°, which means that **the rotational symmetry of the lattice can be only one-, two-, three-, four-, or sixfold.**

As a result of this restriction on rotational symmetry, all crystals can be classified as belonging to one of only 32 **crystallographic point groups.** The Schoenflies symbols of these 32 crystallographic point groups are shown in Table 23.1, where they are divided into seven crystal systems. The seven crystal systems can also be described in terms of unit cell axes and angles (see Fig. 23.5), but the classification according to symmetry elements is more fundamental.

The symmetries of the 32 crystallographic point groups can also be represented by the **Hermann–Mauguin symbols*** in Table 23.1. Crystallographers prefer the Hermann–Mauguin symbols because they can be extended to include translational symmetry, as we will soon see. Rotation axes are represented by a number; for example, a fourfold rotation axis is represented by 4. Crystallographers use rotation–inversion axes, while spectroscopists use rotary-reflection axes. In a rotary inversion, a rotation by 360°/n is followed by inversion through the center of symmetry. Rotary-inversion axes are represented by a number with an overbar; for example, a fourfold rotary-inversion axis is represented by $\overline{4}$. An inversion center is just a onefold rotary-inversion axis, or $\overline{1}$. A mirror plane is represented by the letter m. Each component of a symbol refers to a different direction; for example, $4/m$ indicates that there is a mirror plane perpendicular to a fourfold rotation axis. If we label axes x, y, z, the symbol $mm2$ indicates that mirror planes are perpendicular to x and y, and a twofold rotation axis is parallel to z. In Chapter 13 a decision tree was given for assigning Schoenflies symbols; a similar decision tree for assigning Hermann–Mauguin symbols to the 32 crystallographic point groups is given by Breneman.*

* N. F. M. Henry and K. Lonsdale (Eds.), *International Tables for X-Ray Crystallography*, Vol. 1, *Symmetry Groups*. Birmingham, UK: Kynoch, 1952.

* G. L. Breneman, *J. Chem. Educ.* **64:**216 (1987).

Table 23.1 Crystallographic Point Groups

Crystal System	Schoenflies Symbol	Hermann–Mauguin Symbol
Triclinic	C_1	1
	C_i	$\bar{1}$
Monoclinic	C_2	2
	C_s	m
	C_{2h}	$2/m$
Orthorhombic	D_2	222
	C_{2v}	$mm2$
	D_{2h}	mmm
Tetragonal	C_4	4
	S_4	$\bar{4}$
	C_{4h}	$4/m$
	D_4	422
	C_{4v}	$4mm$
	D_{2d}	$\bar{4}2m$
	D_{4h}	$4/mmm$
Trigonal	C_3	3
	C_{3i}	$\bar{3}$
	D_3	32
	C_{3v}	$3m$
	D_{3d}	$\bar{3}m$
Hexagonal	C_6	6
	C_{3h}	$\bar{6}$
	C_{6h}	$6/m$
	D_6	622
	C_{6v}	$6mm$
	D_{3h}	$\bar{6}m2$
	D_{6h}	$6/mmm$
Cubic	T	23
	T_h	$m3$
	O	432
	T_d	$\bar{4}3m$
	O_h	$m3m$

In addition to the primitive unit cells we have been talking about, unit cells can also have lattice points that are **body-centered** (I), **face-centered** (F), or **end-centered** (C), and still have the same crystallographic point group symmetry. This leads to the 14 Bravais lattices shown in Fig. 23.5.

To complete the classification of crystal structures, we must consider translational symmetry. The lattice translations discussed in Section 23.1 also satisfy our definition of symmetry operations since the translation of the whole crystal in this way leads to an identical configuration. The Hermann–Mauguin system uses two other types of symmetry elements that result from combining the motions of rotations or reflections with the translational symmetry of the lattice. The operation corresponding to **screw axis**, for which the symbol is n_p where n an p are integers, is a rotation of $360°/n$ followed by a translation of p/n in the direction of the axis. For example, a 3_1 screw axis involves rotation by

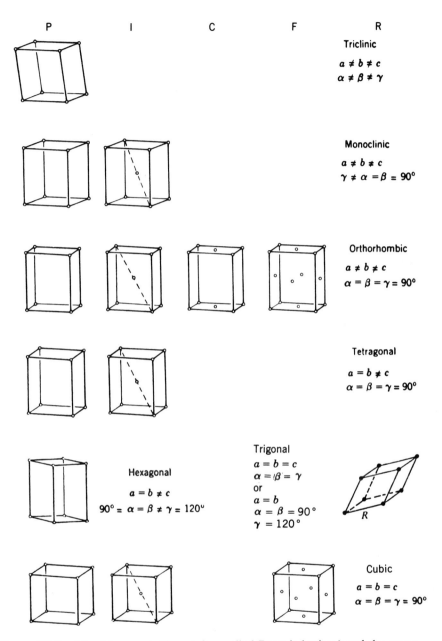

P	I	C	F	R

Triclinic

$a \neq b \neq c$
$\alpha \neq \beta \neq \gamma$

Monoclinic

$a \neq b \neq c$
$\gamma \neq \alpha = \beta = 90°$

Orthorhombic

$a \neq b \neq c$
$\alpha = \beta = \gamma = 90°$

Tetragonal

$a = b \neq c$
$\alpha = \beta = \gamma = 90°$

Hexagonal

$a = b \neq c$
$90° = \alpha = \beta \neq \gamma = 120°$

Trigonal
$a = b = c$
$\alpha = \beta = \gamma$
or
$a = b$
$\alpha = \beta = 90°$
$\gamma = 120°$

R

Cubic

$a = b = c$
$\alpha = \beta = \gamma = 90°$

Figure 23.5   The 14 space lattices (often called Bravais lattices) and the seven crystal systems. P refers to primitive, I to body-centered, C to end-centered, F to face-centered, and R to rhombohedral.

120° followed by translation by one-third of a unit cell parallel to the axis. A 3_2 screw axis implies a rotation of 120° and a translation of two-thirds. The possible screw axes are 2_1, 3_1, 3_2, 4_1, 4_2, 4_3, 6_1, 6_2, 6_3, 6_4, and 6_5.

The operation corresponding to a **glide plane** is a reflection in a plane followed by a translation. If the glide is parallel to the a axis, the symbol for the glide plane is simply a, and the operation is reflection in the plane and translation by $a/2$. There are diagonal and other glides that we will not go into.

Space groups are groups whose elements include both the point symmetry elements and screw axes and glide planes. There are 230 space groups, and all possible crystal structures fall into one of these space groups. The symbol for a space group always starts with P, I, C, F, or R to indicate the type of Bravais lattice (see Fig. 23.5).

The fact that there are 230 ways in which these symmetry operations may be combined in the three-dimensional patterns of crystals was derived independently by three men: Federow, a Russian crystallographer, in 1890; Schoenflies, a German mathematician, in 1891; and Barlow, a British amateur, in 1895. The actual determination of space groups of crystals did not become possible until diffraction techniques made it possible to determine internal symmetry of crystals. Knowledge of the space group of a crystal simplifies the determination of the structure because only the asymmetric portion of the unit cell needs to be studied; the rest of the contents may be obtained from symmetry operations.

23.3 Designation of Crystal Planes

The position of an atom in a unit cell is designated by giving its coordinates as fractions x, y, z of the unit cell edges a, b, c. The point at (xyz) is located by starting at the origin $(0, 0, 0)$ and moving a distance xa along the a axis, then a distance yb parallel to the b axis, and finally a distance zc parallel to the c axis. For example, the fractional coordinates for the lattice points of a face-centered cubic structure are $0, 0, 0; \frac{1}{2}, \frac{1}{2}, 0; \frac{1}{2}, 0, \frac{1}{2}$; and $0, \frac{1}{2}, \frac{1}{2}$. The remaining lattice points may be obtained by adding unity to each of these coordinates. In expressing the locations of atoms in a unit cell the set of coordinates (000) stands for the locations of all eight corners, that is, (100), (111), (101), (110), (001), (011), (010), and (000). In a crystal a lattice point is not necessarily occupied by an atom or molecule, but it does represent a collection of atoms that is repeated in three dimensions.

The distance l between the points x_1, y_1, z_1 and x_2, y_2, z_2 is

$$l = [(x_1 - x_2)^2 a^2 + (y_1 - y_2)^2 b^2 + (z_1 - z_2)^2 c^2$$
$$+ 2(x_1 - x_2)(y_1 - y_2)ab \cos \gamma + 2(y_1 - y_2)(z_1 - z_2)bc \cos \alpha \quad (23.3)$$
$$+ 2(z_1 - z_2)(x_1 - x_2)ca \cos \beta]^{1/2}$$

where the angles are defined in Fig. 23.3. For a cubic crystal this equation simplifies to

$$l = a[(x_2 - x_1)^2 + (y_2 - y_1)^2 + (z_2 - z_1)^2]^{1/2} \quad (23.4)$$

The volume of a unit cell is given by

$$V = abc(1 - \cos^2 \alpha - \cos^2 \beta - \cos^2 \gamma + 2 \cos \alpha \cos \beta \cos \gamma)^{1/2} \quad (23.5)$$

If the unit cell is cubic, orthorhombic, or tetragonal, this equation reduces to $V = abc$.

The planes through the lattice points of crystals are important because they represent possible crystal faces and because they help us understand X-ray

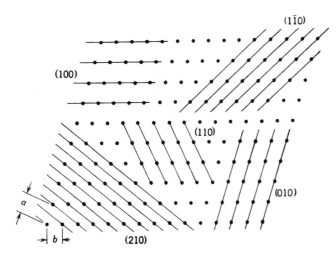

Figure 23.6 Sets of planes through lattice points as seen along the c axis of a crystal.

diffraction phenomena. Figure 23.6 shows the lattice points in one plane of a crystal. The c axis is taken as perpendicular to the page. Various sets of planes that are parallel to the c axis are indicated in the figure. Actually, there is an infinite number of such sets of planes.

The orientation of a set of planes of a crystal lattice may be specified by means of the intercepts of one of the planes on the three axes a, b, and c of the unit cell. Suppose that a plane intercepts the a axis at a/h, measured from the origin, the b axis at b/k, and the c axis at c/l. **This plane is referred to by the indices** ***hkl***.* The indices of a set of planes through a lattice may be obtained by counting the number of planes crossed in moving one lattice space in the ***a***, ***b***, and ***c*** directions, respectively. For the set of planes in the lower left corner of Fig. 23.6, two planes are crossed in going one lattice distance in the direction of the a axis, and one plane is crossed in going one lattice direction horizontally in the direction of the b axis, whereas no plane would be crossed in going one lattice distance into the paper, since the planes are parallel to the c axis. Thus, this set of planes is designated by the indices 210. The indices of the set of planes in the upper right corner are $1\bar{1}0$. The negative sign indicates that, if a particular plane is intercepted by going in the positive direction along a, it is necessary to go in the negative direction along b in order to intercept the same plane. This representation for the exterior faces of a crystal and for the internal planes within the crystal will specify the orientation, but not the spatial position of a plane. The faces of crystals are usually planes with high densities of atoms or molecules, and so they are planes with low indices.

The perpendicular distance d between adjacent planes of a set is given by

---

* The indices are often called "Miller indices" because this designation, invented by Whewell in 1825 and Grossman in 1829, was popularized by Miller's textbook of crystallography in 1829. They showed that faces of crystals could be designated by three integers, although nothing was then known about the internal structures of crystals.

$$d = V[h^2b^2c^2 \sin^2 \alpha + k^2a^2c^2 \sin^2 \beta + l^2a^2b^2 \sin^2 \gamma$$
$$+ 2hlab^2c(\cos \alpha \cos \gamma - \cos \beta) + 2hkabc^2(\cos \alpha \cos \beta - \cos \gamma)$$
$$+ 2kla^2bc(\cos \beta \cos \gamma - \cos \alpha)]^{-1/2} \qquad (23.6)$$

where V is the unit cell volume given by equation 23.5. For the special case that the unit cell axes are mutually perpendicular (i.e., for orthorhombic, tetragonal, and cubic unit cells),

$$\frac{1}{d} = \left(\frac{h^2}{a^2} + \frac{k^2}{b^2} + \frac{l^2}{c^2} \right)^{1/2} \qquad (23.7)$$

23.4 Diffraction Methods

In a crystal it is the electrons that scatter X-rays. Bragg pointed out that it is convenient to consider that the X-rays are "reflected" from a stack of planes in the crystal. For a given stack of planes (hkl) the reflected monochromatic radiation occurs only at certain angles that are determined by the wavelength of the X-rays and the perpendicular distance between adjacent planes. The relationship between these three variables is the **Bragg equation,** which may be derived by referring to Fig. 23.7. The horizontal lines represent planes in the crystal separated by the distance d. The plane ABC is perpendicular to the incident beam of parallel monochromatic X-rays, and the plane LMN is perpendicular to the reflected beam. As the angle of incidence θ is changed, a reflection will be obtained only when the waves are in phase at plane LMN, that is, when the difference in distance between planes ABC and LMN, measured along rays reflected from different planes, is a whole-number multiple of the wavelength. This occurs when

$$FS + SG = n\lambda \qquad (23.8)$$

Since $\sin \theta = FS/d = SG/d$,

$$2d \sin \theta = n\lambda \qquad (23.9)$$

This is the Bragg equation, and θ is called the Bragg angle.

This important equation gives the relationship of the distance between planes

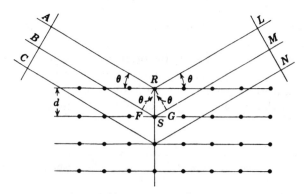

Figure 23.7 Diagram used in proving that $n\lambda = 2d \sin \theta$.

in a crystal and the angle at which the reflected radiation has a maximum intensity for a given wavelength λ; that is, all the scattered X-ray waves are in phase. If λ is longer than $2d$, there is no solution for n and no diffraction. Thus, light waves pass through crystals without being diffracted by the planes of scattering centers. If $\lambda \ll d$, the X-rays are diffracted through inconveniently small angles. The Bragg equation does not indicate the intensities of the various diffracted beams. The intensities depend on the nature and arrangement of the atoms within each unit cell.

The reflection corresponding to $n = 1$ for a given family of planes is called the first-order reflection; the reflection corresponding to $n = 2$ is the second-order reflection; and so on. Each successive order exhibits a larger Bragg angle. In discussing X-ray reflections it is customary to set $n = 1$ in equation 23.9 and consider that the second-order reflection is from a parallel stack of planes separated by half the lattice distance, and so on. Equation 23.9 may be written

$$\lambda = 2\,\frac{d}{n}\,\sin\,\theta = 2d_{nh,nk,nl}\,\sin\,\theta \qquad (23.10)$$

where $d_{nh,nk,nl}$ is the perpendicular distance between adjacent planes having the indices nh,nk,nl. The planes nh,nk,nl are parallel to the hkl planes, and the perpendicular interplanar distance is $d_{nh,nk,nl} = d/n$.

To determine the angles at which X-rays are diffracted, an oriented single crystal may be rotated in an X-ray beam and the intensity of X-rays at the reflection angle determined with a counter. Various types of X-ray cameras have been developed in which the photographic film is moved as the crystal is rotated.

Instead of scattering X-rays from a single large crystal, it is convenient (and sometimes necessary when suitable single crystals are not available) in some types of work to pass a collimated beam of X-rays through a powdered sample containing microcrystals presumably oriented in random directions. The reflections may be recorded on a circular photographic film, as illustrated in Fig. 23.8. If coarse crystals are used, the powder pattern is seen to be made up of

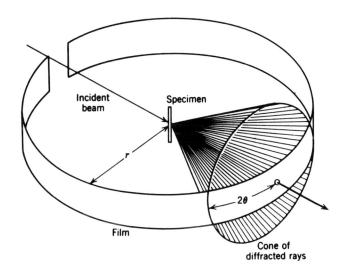

Figure 23.8 X-ray powder camera.

rings of spots, each spot being produced by a suitably oriented small crystal. If the crystals are very fine, a large number of spots of reflected beams are produced by the different crystal planes, and continuous arcs are obtained on the film.

Example 23.1

At what angles θ will X-rays of wavelength 1.542×10^{-10} m be reflected by planes separated by 3.5×10^{-10} m? What is an alternative interpretation of these reflections?

$$\theta = \sin^{-1} \frac{n\lambda}{2d}$$

For $n = 1$,

$$\theta = \sin^{-1} \frac{(1.542 \times 10^{-10} \text{ m})}{2(3.5 \times 10^{-10} \text{ m})}$$

$$= 12.73°$$

For $n = 2$, $\theta = 26.14°$.
For $n = 3$, $\theta = 41.37°$.

These reflections can also be interpreted as being due to first-order reflections from (100), (200), and (300) planes, which have interplanar spacings of d_{hkl} of 3.5×10^{-10} m, 7.0×10^{-10} m, and 10.5×10^{-10} m.

Neutrons can also be diffracted from the planes in a crystal. The **average de Broglie wavelength** (Section 10.4) of thermal neutrons is 252 pm at room temperature. An essentially monochromatic beam may be obtained by diffraction from a crystal monochromator that selects a small band of wavelengths from the beam of **thermal neutrons** from a nuclear reactor. The diffraction of neutrons from a crystal is different from X-rays because neutrons are scattered primarily by the nuclei in the crystal, while the X-rays are scattered by electrons. This means that neutron diffraction, in contrast to X-ray diffraction, is especially useful for accurately locating hydrogen atoms in a crystalline structure. For example, in a compound such as uranium hydride, X-ray diffraction can be utilized to determine the uranium coordinates and neutron diffraction the hydrogen coordinates. Hydrogen atoms scatter electrons weakly because X-rays are scattered by electrons.

Since neutrons possess a magnetic moment by virtue of having a spin of $\frac{1}{2}$, there is an additional scattering if the compound contains paramagnetic atoms or ions with unpaired electrons. Thus, neutron diffraction has been widely utilized to investigate structures of magnetic materials such as MnO and Fe_3O_4 in order to determine the arrangement of the atomic magnetic moments in the solids.

Example 23.2

What is the wavelength of neutrons moving in a particular direction if they are in thermal equilibrium with their surroundings at 25 °C?

The kinetic energy of the neutrons moving in a particular direction can be expressed in terms of their temperature $kT/2$ (because of equipartition) or in terms of their momentum $p^2/2m$, where p is the momentum and m is the mass of a neutron:

$$\frac{kT}{2} = \frac{p^2}{2m}$$

$$p = (mkT)^{1/2}$$

The de Broglie wavelength is given by

$$\lambda = \frac{h}{p} = \frac{h}{(mkT)^{1/2}}$$

Thus, at 25 °C,

$$\lambda = \frac{6.626 \times 10^{-34} \text{ J s}}{[(1.675 \times 10^{-27} \text{ kg})(1.38 \times 10^{-23} \text{ J K}^{-1})(298 \text{ K})]^{1/2}}$$

$$= 252 \text{ pm}$$

23.5   Cubic Lattices

Since the cubic system is the simplest, it is explored in some detail here. There are three-independent Bravais lattices that have all the symmetry of a cube: primitive, body-centered, and face-centered. These are illustrated in Fig. 23.9. Since these Bravais lattices all have the symmetry of a cube, they cannot be distinguished by macroscopic examination of the crystals.

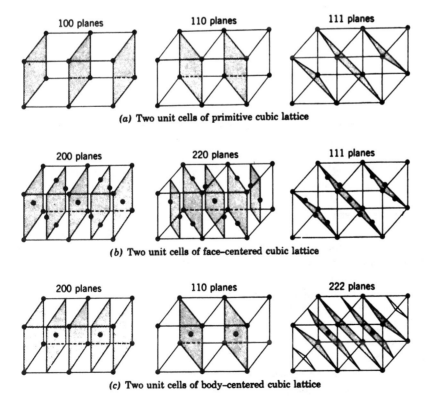

(a) Two unit cells of primitive cubic lattice

(b) Two unit cells of face–centered cubic lattice

(c) Two unit cells of body–centered cubic lattice

Figure 23.9   Planes through cubic lattices.

In the **primitive cubic lattice** in Fig. 23.9*a*, three types of reflecting planes are illustrated: 100, 110, and 111. The reflections from these planes occur at the smallest angles of all the possible planes that can be imagined in a cubic crystal, since planes with higher indices are closer together and θ will be larger, according to equation 23.9.

In a cubic crystal $a = b = c$, and so equation 23.7 for the perpendicular distance d between adjacent planes of a set may be written

$$d_{hkl} = \frac{a}{\sqrt{h^2 + k^2 + l^2}} \qquad (23.11)$$

where a is the length of the side of the unit cell. The perpendicular distances between adjacent planes in a cubic crystal are obtained by substituting 0, 1, 2, 3, . . . for h, k, and l in this equation.

Example 23.3

For a primitive cubic crystal with $a = 3 \times 10^{-10}$ m, what are the smallest diffraction angles θ for (*a*) 100, (*b*) 110, and (*c*) 111 planes for $\lambda = 1.50 \times 10^{-10}$ m?

$$\theta = \sin^{-1} \frac{\lambda}{2d_{hkl}} = \sin^{-1} \frac{\lambda(h^2 + k^2 + l^2)^{1/2}}{2a}$$

$$(a) \quad \theta = \sin^{-1} \frac{1.5 \times 10^{-10} \text{ m}}{6 \times 10^{-10} \text{ m}}$$

$$= 15.48°$$

$$(b) \quad \theta = \sin^{-1} \frac{(1.5 \times 10^{-10} \text{ m})2^{1/2}}{6 \times 10^{-10} \text{ m}}$$

$$= 20.70°$$

$$(c) \quad \theta = \sin^{-1} \frac{(1.5 \times 10^{-10} \text{ m})3^{1/2}}{6 \times 10^{-10} \text{ m}}$$

$$= 25.66°$$

For **primitive cubic crystals** d_{hkl} may have the following values: a, $a/\sqrt{2}$, $a/\sqrt{3}$, $a/\sqrt{4}$, $a/\sqrt{5}$, $a/\sqrt{6}$, $a/\sqrt{8}$, and so on, where $a/\sqrt{7}$ is missing because 7 cannot be obtained from $h^2 + k^2 + l^2$, where h, k, and l are integers.

In the **face-centered cubic lattice** illustrated in Fig. 23.9*b* there are lattice points in the center of each face of the unit cell in addition to the lattice points at the corners. One-half of the face-centered lattice points and one-eighth of the corner lattice points belong to the unit cell, making a total of four per cell. Hence, the equivalent positions are xyz; $\frac{1}{2} + x, \frac{1}{2} + y, z$; $\frac{1}{2} + x, y, \frac{1}{2} + z$; $x, \frac{1}{2} + y, \frac{1}{2} + z$. These four equivalent positions also can be given as $000 + xyz$; $\frac{11}{22}0 + xyz$; $\frac{1}{2}0\frac{1}{2} + xyz$; $0\frac{11}{22} + xyz$. NaCl has a face-centered cubic lattice (Fig. 23.12). Since all of the lattice points can be considered to be occupied by Na^+, there are four Na^+ per unit cell. Since the Cl^- ions on the 12 cell edges (in which each edge is shared by four unit cells) and at the center of the unit cell also are related to one another by face-centering, there are four Cl^- per unit cell. The positions of the four Na^+ and four Cl^- ions are as follows:

Four Na$^+$ 000, $\tfrac{1}{2}\tfrac{1}{2}0$; $\tfrac{1}{2}0\tfrac{1}{2}$; $0\tfrac{1}{2}\tfrac{1}{2}$

Four Cl$^-$ $\tfrac{1}{2}00$; $0\tfrac{1}{2}0$; $00\tfrac{1}{2}$; $\tfrac{1}{2}\tfrac{1}{2}\tfrac{1}{2}$

The diffraction patterns of NaCl and any other face-centered crystal (regardless of the crystal system) show an *absence* of all reflections for which the indices *hkl* are not all even or all odd. Hence, only the reflections 111, 200, 220, 311, 222, 400, 331, 420, etc., are observed. It can be shown mathematically (Section 23.6) that the X-ray scattering from the face-centered, symmetry-related atoms due to the equivalent points at 000, $\tfrac{1}{2}\tfrac{1}{2}0$; $\tfrac{1}{2}0\tfrac{1}{2}$; $0\tfrac{1}{2}\tfrac{1}{2}$ is completely in phase for reflections with indices all even or all odd, but completely out of phase for all other *hkl* reflections. Consequently, the spacings corresponding to the *hkl* reflections that may appear on a powder diagram of a face-centered cubic crystal are (from equation 23.11) $a/\sqrt{3}$, $a/\sqrt{4}$, $a/\sqrt{8}$, $a/\sqrt{11}$, $a/\sqrt{12}$, $a/\sqrt{16}$, $a/\sqrt{19}$, $a/\sqrt{20}$, and so on.

A **body-centered cubic lattice** has two positions per unit cell related by translational symmetry, namely xyz, $\tfrac{1}{2} + x$, $\tfrac{1}{2} + y$, $\tfrac{1}{2} + z$. Hence, there are two lattice points per unit cell at 000 and $\tfrac{1}{2}\tfrac{1}{2}\tfrac{1}{2}$ that have identical environments, as shown in Fig. 23.9c. An examination of the diffraction data for body-centered crystals shows that *hkl* reflections for which the sum $h + k + l$ is odd are not observed. This means that the scattering by each atom in a body-centered unit cell is completely in phase with that of its corresponding body-centered equivalent atom if the sum of the indices is even, but 180° out of phase if the sum is odd. Accordingly, the interplanar spacings found for a body-centered cubic lattice are $a/\sqrt{2}$, $a/\sqrt{4}$, $a/\sqrt{6}$, $a/\sqrt{8}$, and so on, which are the distances between the (110), (200), (211), and (220) planes.

The reflections for the three types of cubic crystals are summarized in Fig. 23.10, in which the presence of a reflection is indicated by a line at the corresponding Bragg angle of incidence. For the purposes of this illustration, the ratio λ/a is arbitrarily taken as 0.500 for primitive cubic, 0.353 for body-centered

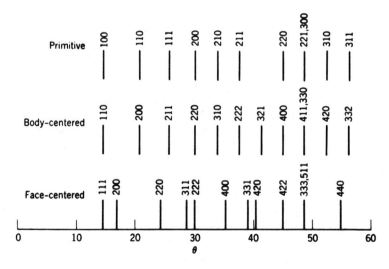

Figure 23.10 Angles of incidence θ and indices for cubic crystals. The values of λ/a have been chosen arbitrarily to cause the first reflection to fall at the same angle for each type of crystal. For primitive cubic, $\lambda/a = 0.500$; for face-centered cubic, $\lambda/a = 0.289$; for body-centered cubic, $\lambda/a = 0.353$.

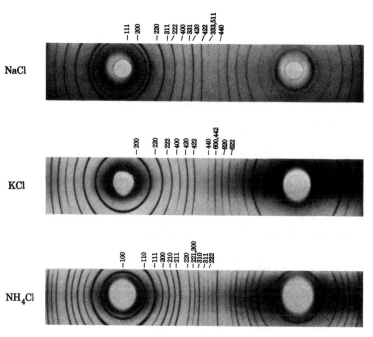

Figure 23.11   X-ray powder patterns for cubic crystals. The X-ray beam enters through the hole at the right and leaves through the hole at the left (see Fig. 23.8). (Courtesy Professor S. Baily of the University of Wisconsin.)

cubic, and 0.289 for face-centered cubic. It may be seen that the various types of cubic crystals may be distinguished by their diffraction patterns, since the patterns are qualitatively different. In the powder pattern for primitive cubic there is a gap after the sixth line. In the powder pattern for body-centered cubic this gap is filled in. In the powder pattern for face-centered cubic the first two lines are relatively close together, the third line is by itself, and the next two lines are close together. Thus, by use of powder patterns, it is possible to determine whether a cubic crystal is primitive, body-centered or face-centered.

The powder patterns of three substances forming cubic crystals are shown in Fig. 23.11. As may be seen by comparison with Fig. 23.10, the reflections for sodium chloride are found to correspond to those expected for a face-centered cubic lattice. The reflection indices have been assigned on this basis. The 100 reflection is missing, and so none of the spacings calculated using the Bragg equation is equal to the length of the side of the unit cell a. However, a may be calculated from the angle of any reflection by use of equations 23.10 and 23.11. The value of a for sodium chloride is 564 pm.

The structural units of the face-centered cubic lattice that have thus been found could be sodium chloride molecules, or there could be a lattice of equal numbers of sodium and chloride ions. If the lattice were made up of sodium chloride molecules, all units of the lattice would be the same, and a more detailed consideration of the theory shows that the intensities of the X-ray reflections would decrease progressively from first- to second- to third-order reflections. This is true for the 200, 400, and 600 reflections, but not for the 111, 222, 333, . . . reflections. It may be noted in Fig. 23.11 that the 111 re-

Figure 23.12   Face-centered lattice of sodium chloride; small spheres, sodium ions; large spheres, chloride ions. Each Na^+ is octahedrally surrounded by six Cl^-, and each Cl^- is octahedrally surrounded by six Na^+.

flection is weak, whereas 222 is strong and 333 is apparently missing. This fact leads to the requirement of a lattice with ions at the lattice points.

The square connecting sodium ions is drawn in Fig. 23.12, showing that there is a face-centered sodium ion. Similar squares could be drawn on any part of the faces. It is evident that the chloride ions are displaced half a cell edge from the sodium ions.

A further examination of Fig. 23.12 shows that the (111) planes that cut diagonally through the sodium chloride crystal include only sodium ions or only chloride ions. Thus, the (222) planes are alternately planes of sodium and chloride ions. It may be remembered that the maxima in X-ray reflection occur when the angle is such that the paths between successive layers of ions are equal to one wavelength of the reflected radiation. If the rays are reflected from these planes at such an angle that the rays from successive planes of chloride ions differ in pathlength by one wavelength, the rays coming from successive sodium ion planes, which are spaced equally between them, will then differ by half a wavelength and cause interference. The interference would be complete except for the fact that the chloride ions have more electrons and scatter X-rays more efficiently than the sodium ions. The reflections from the (222) planes, however, have a difference of a whole wavelength between the reflections from the chloride and from the sodium planes, so that there is no interference, and the 222 reflection is intense. The 333 reflection again corresponds to a difference of one-half wavelength between the two sets of reflecting planes, and this interference, combined with the fact that the third-order spectrum is naturally weaker, leads to a very weak reflection.

The powder pattern for potassium chloride given in Fig. 23.11 is superficially that of a primitive cubic lattice. This seems surprising, since the structure would be expected to be face-centered like that for sodium chloride. The reflections look very much like those from a simple cubic structure because the scattering power of the potassium ion is almost exactly equal to that of the chloride ion, since both ions have the argon electronic structure. A more accurate determination of the intensities of reflection shows that potassium chloride does indeed form a face-centered lattice.

The powder pattern for ammonium chloride given in Fig. 23.11 shows that it has a primitive cubic lattice. If the center of a chloride ion is taken as the corner of the unit cell, the ammonium ion lies at the center of the cell, but the crystal is not body-centered cubic, since the ions are not equivalent.

Although the powder camera technique has been widely utilized for finger-printing compounds, its use to determine the arrangement of atoms is mainly limited to the simple crystal structures comprising the cubic, hexagonal, and tetragonal systems. In general, it is much more convenient to utilize single crystal X-ray techniques to determine the arrangement of atoms. The main advantage of X-ray diffraction over other structural methods is that X-ray diffraction in practically all cases provides a direct, unique solution of the structure.

23.6 Electron Density Function

Since X-rays are scattered by the electrons of a crystal, the objective of X-ray diffraction experiments is to obtain the electron density $\rho(xyz)$ as a function of the coordinates x, y, and z. The electron density $\rho(xyz)$ is defined so that $\rho(xyz)\,dx\,dy\,dz$ is the number of electrons in the volume $dx\,dy\,dz$. Since the electron density is a periodic function, it is convenient to represent it by a **Fourier series.** A periodic function in one direction x may be represented by

$$f(x) = \sum_{n=0}^{\infty} A_n \cos \frac{2\pi nx}{a} + \sum_{n=1}^{\infty} B_n \sin \frac{2\pi nx}{a} \quad (23.12)$$

where A_n and B_n are coefficients chosen to give $f(x)$ the desired shape. A Fourier series may also be represented by

$$f(x) = \sum_{n=-\infty}^{\infty} C_n \exp \frac{-2\pi inx}{a} \quad (23.13)$$

where $i = \sqrt{-1}$. Series 23.12 and 23.13 are equivalent because

$$\exp \frac{-2\pi inx}{a} = \cos \frac{2\pi nx}{a} - i \sin \frac{2\pi nx}{a} \quad (23.14)$$

and the coefficients C_n in general are complex. Since the electron density is a function of three variables, we may write

$$\rho(xyz) = \frac{1}{V} \sum_{h=-\infty}^{\infty} \sum_{k=-\infty}^{\infty} \sum_{l=-\infty}^{\infty} F(hkl) \exp[-2\pi i(hx + ky + lz)] \quad (23.15)$$

where V is the volume of the unit cell, and x, y, and z are the fractional co-ordinates of a point in the unit cell. The **Fourier coefficients** $F(hkl)$ are referred to as **structure factors.** Each structure factor is associated with a particular reflection, the one from the hkl planes. The summations over h, k, l go from $-\infty$ to ∞, but in practice the electron density can be represented quite well with a finite number of terms. Higher resolution is obtained by including more reflections.

The values of the structure factors are determined by the nature of the atoms in the unit cell and their positions. The structure factor for a particular reflection is made up of additive contributions from each of the atoms in the unit cell. The contribution of an atom is equal to the product of an atomic scattering factor f_j and an exponential depending on its fractional coordinates in the unit cell:

$$F(hkl) = \sum_j f_j e^{2\pi i(hx_j + ky_j + lz_j)} \tag{23.16}$$

The atomic scattering factor f_j is a function of $(\sin \theta)/\lambda$. Remember that 2θ is the deviation of the diffracted beam from the direct X-ray beam. At very small values of θ, the atomic scattering factor f_j of a neutral atom is equal to its atomic number since all the electrons scatter in phase. But at larger values of θ, the intensity of scattered radiation is less because there is interference between radiation scattered from various parts of the electron cloud of the atom.

The form of the structure factor expression may be considered to arise as follows. When Bragg's law is satisfied for a given reflection, the amplitude of the wavelet scattered from an atom in one unit cell of the crystal is in phase with the amplitudes of the scattered wavelets from the corresponding atoms in the millions of other unit cells of the crystal. The wavelet scattered by one atom, however, generally will not be in phase with the wavelet scattered by a different atom within the same unit cell, and hence the intensity of the reflection will depend on the extent to which the amplitudes of the different atoms (denoted by the atomic scattering factor f_j of the jth atom) are in phase with one another.

To illustrate the usefulness of the structure factor, let us consider its dependence on the reflection indices hkl for several lattices. In primitive lattices with atoms at the lattice sites, x_j, y_j, and z_j are zero, and so $F(hkl) = \sum f_j$. Thus, the structure factor $F(hkl)$ has the same value for all values of h, k, and l, and there will be reflections for all integer values of h, k, and l.

Example 23.4
Which reflections will be absent in the diffraction pattern for a body-centered cubic unit cell?

The eight corner atoms are each only $\frac{1}{8}$th in the unit cell and their contribution is equivalent to one atom at $x = y = z = 0$. The atom at the center of the unit cell is at $x = y = z = \frac{1}{2}$, and so the structure factor is given by

$$F(hkl) = f e^{2\pi i(0 + 0 + 0)} + f e^{2\pi i(h/2 + k/2 + l/2)}$$
$$= f(1 + e^{i\pi(h + k + l)})$$

where f is the atomic scattering factor. Since $e^{i\pi} = -1$,

$$F_{hkl} = f[1 + (-1)^{h + k + l}]$$

If $h + k + l$ is even, then $F(hkl) = 2f$, but if $h + k + l$ is odd the reflection will be absent. Thus, the reflections (100), (111), (210), (300), . . . will be absent.

Example 23.5

What is the structure-factor expression for a face-centered cubic lattice? Derive a rule for absences in reflections in terms of the values of h, k, l. The four atoms in the unit cell may be assigned the positions (000), $(\frac{1}{2}\frac{1}{2}0)$, $(\frac{1}{2}0\frac{1}{2})$ and $(0\frac{1}{2}\frac{1}{2})$.

$$
\begin{aligned}
F(hkl) &= fe^{2\pi i(h\cdot 0 + k\cdot 0 + l\cdot 0)} + fe^{2\pi i(h/2 + k/2 + 0)} \\
&\quad + fe^{2\pi i(h/2 + 0 + l/2)} + fe^{2\pi i(0 + k/2 + l/2)} \\
&= f[1 + e^{\pi i(h + k)} + e^{\pi i(h + l)} + e^{\pi i(k + l)}] \\
&= f[1 + (-1)^{h + k} + (-1)^{h + l} + (-1)^{k + l}]
\end{aligned}
$$

If h, k, l are all even or all odd, then $F(hkl) = 4f$ and there are reflections. If one is even and the other two are odd, or the reverse, the reflections are absent. This is illustrated by the powder pattern for NaCl in Fig. 23.11.

These examples show that the structure factors contain all the information about all the atoms in a unit cell. For a known structure they may be used to calculate the electron density throughout the unit cell using equation 23.15. But the X-ray diffractionist faces a much more difficult problem. The structure factors $F(hkl)$ are related to the intensities $I(hkl)$ of radiation reflected from the planes (hkl) by

$$I(hkl) \propto |F(hkl)|^2 \tag{23.17}$$

Measurements of the densities of spots on photographic film or of counts recorded by a Geiger counter for each reflection may be subjected to routine corrections to obtain $I(hkl)$ values. Thus, a set of $|F(hkl)|^2$ values may be obtained. Unfortunately, what is needed to calculate $\rho(xyz)$ using equation 25.15 are values of $F(hkl)$ instead of $|F(hkl)|^2$. Since $F(hkl)$ is a complex number, we can write

$$F(hkl) = A(hkl) + iB(hkl) \tag{23.18}$$

so that

$$
\begin{aligned}
|F(hkl)|^2 &= [A(hkl) + iB(hkl)][A(hkl) - iB(hkl)] \\
&= [A(hkl)]^2 + [B(hkl)]^2
\end{aligned}
\tag{23.19}
$$

Since values of $A(hkl)$ and $B(hkl)$ are not obtained directly, indirect methods must be used to obtain these quantities and the electron density $\rho(xyz)$. This is called the **phase problem**. Fortunately, the number of parameters needed to describe a crystal structure is far smaller than the number of reflections, so that the problem is greatly overdetermined. Several methods are used to get around the phase problem. If heavy atoms are present in the unit cell, they may by themselves determine enough phases so that a Fourier map of the electron density may reveal the position of some of the lighter atoms. Another method that has been used is isomorphous replacement. In this method, which has been especially useful in determining the structures of protein crystals, crystals are prepared containing different heavy atoms and information about phases is obtained by comparing the intensities from the different crystals.

X-ray analyses of thousands of crystal structures have led to detailed knowledge of the geometrical properties of different groups of atoms, including well-established values of bond lengths and angles. Modern crystallographic analyses using data-collecting diffractometers and high-speed computers have made it possible to determine the molecular structure of proteins. X-ray diffraction data were used in the determination of the structure of deoxyribonucleic acid, and this led to an understanding of the hydrogen bonding that makes that structure stable.

In certain cases X-ray diffraction may be used to determine the absolute configuration of an optically active substance. In 1951 Bijvoet, Peerdeman, and van Bommel studied sodium rubidium (+)-tartaric acid by X-ray diffraction and found that the absolute configuration was the one arbitrarily chosen from the two possible enantiomorphic structures by Fischer 60 years earlier.

23.7   Ion Radii and Atom Radii

For cubic crystals, the length a of the side of the unit cell can be obtained from equation 23.11. This length must, of course, be consistent with the density ρ of the crystal and the number z of molecules per unit cell. The mass of the contents of a unit cell is zM/N_A, where z is the number of molecules of molar mass M in a unit cell. Thus, the density ρ for a perfect crystal is

$$\rho = \frac{zM}{N_A V} \tag{23.20}$$

where V is the volume of the unit cell.

Example 23.6
The density of sodium chloride at 25 °C is 2.163×10^3 kg m^{-3}. When X-rays from a palladium target having a wavelength of 58.1 pm are used, the 200 reflection of sodium chloride occurs at an angle of 5.91°. How many sodium and chloride ions are there in a unit cell?
  According to Bragg's law,

$$d_{200} = \frac{\lambda}{2 \sin \theta} = \frac{58.1 \text{ pm}}{2 \sin 5.9°} = 282 \text{ pm}$$

Equation 23.11 yields $a = 564$ pm. Using equation 23.20, we find

$$2.163 \times 10^3 \text{ kg m}^{-3} = \frac{z(58.443 \times 10^{-3} \text{ kg mol}^{-1})}{(6.022\ 14 \times 10^{23} \text{ mol}^{-1})(564 \times 10^{-12} \text{ m})^3}$$

$$z = 3.999$$

Thus, as expected, the unit cell contains four sodium ions and four chloride ions.

Example 23.7
Potassium crystallizes with a body-centered cubic lattice and has a density of 0.856×10^3 kg m^{-3}. What is the length of the side of the unit cell a and the distance between

(200), (110), and (222) planes? What is the closest distance between atoms and what is the potassium atom radius r?

$$0.856 \times 10^3 \text{ kg m}^{-3} = \frac{2(39.102 \times 10^{-3} \text{ kg mol}^{-1})}{(6.022\ 136 \times 10^{23} \text{ mol}^{-1})a^3}$$

$$a = 533.3 \times 10^{-12} \text{ m} = 533.3 \text{ pm}$$

Using equation 23.11, we obtain

$$\text{For (200) planes, } d_{200} = 533.3/\sqrt{4} = 266.7 \text{ pm}$$

$$\text{For (110) planes, } d_{110} = 533.3/\sqrt{2} = 377.1 \text{ pm}$$

$$\text{For (222) planes, } d_{222} = 533.3/\sqrt{12} = 154.0 \text{ pm}$$

$$(2r)^2 = \left(\frac{a}{2}\right)^2 + \left(\frac{a}{2}\right)^2 + \left(\frac{a}{2}\right)^2 \qquad 2r = 461.9 \text{ pm} \qquad r = 231.0 \text{ pm}$$

Diamond has a face-centered cubic lattice with atoms at 000 and $\frac{1}{4}\frac{1}{4}\frac{1}{4}$ associated with each lattice point. This structure is represented in two different ways in Fig. 23.13. Since there are two atoms per lattice point, there are eight

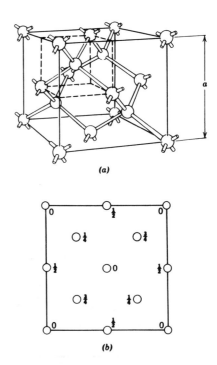

(a)

(b)

Figure 23.13   Two representations of the diamond structure: (*a*) space model showing tetrahedral bonds; (*b*) projection showing fractional coordinates. It is instructive to draw lines showing the bonds in this projection.

atoms per unit cell. The unit cell distance for diamond is 356.7 pm. Silicon, germanium, and gray tin also have this structure with unit cell distances of 543.1, 565.7, and 649.1 pm.

Example 23.8

Calculate the C–C bond distance in diamond and the C–C–C angle. Using equation 25.4 for the points 000 and $\frac{111}{444}$,

$$l = a[(x_2 - x_1)^2 + (y_2 - y_1)^2 + (z_2 - z_1)^2]^{1/2}$$

$$= (356.7 \text{ pm})(\tfrac{3}{16})^{1/2} = 154.5 \text{ pm}$$

which is the C–C distance.

The distance l between a carbon atom at a corner of the unit cell and at a face-centered position is given by

$$l^2 = \left(\frac{356.7 \text{ pm}}{2}\right)^2 + \left(\frac{356.7 \text{ pm}}{2}\right)^2$$

$$l = \frac{\sqrt{2}}{2}(356.7 \text{ pm})$$

$$\sin \theta = \frac{2^{1/2}(356.7 \text{ pm})}{4(356.7 \text{ pm})(\tfrac{3}{16})^{1/2}}$$

$$\theta = 54.736°$$

The C–C–C bond angle is 2θ or 109.471°, which is referred to as the tetrahedral angle.

The value of the Avogadro constant N_A can be calculated from the density of a crystal, the relative atomic mass, and the unit cell length. To be useful for this purpose, a crystal must be free of defects (Section 23.12). Very accurate values of these quantities for silicon have been measured at the National Institute for Standards and Technology*. The value of the Avogadro constant determined in this way $(6.022\ 097\ 6(63) \times 10^{23} \text{ mol}^{-1})$ is tied to measurements of other fundamental constants through a least-squares adjustment to obtain the best values given in Appendix B.

There is a condition on the ratio of ion radii R that must be satisfied for an ionic substance MX to have the NaCl structure. Since the ions are in contact along a cell edge,

$$a = 2(R_+ + R_-) \tag{23.21}$$

In addition, ions cannot overlap along the diagonal of the face of the unit cell. Therefore,

$$(4R_-)^2 \leq 2a^2 \tag{23.22}$$

$$(4R_+)^2 \leq 2a^2 \tag{23.23}$$

---

* R. D. Deslattes, A. Hemins, R. M. Schoonover, C. L. Carroll, and H. A. Bowman, *Phys. Rev. Lett.* **36**:898 (1976).

Thus,

$$a \geq 2\sqrt{2}R_- \quad \text{and} \quad a \geq 2\sqrt{2}R_+ \tag{23.24}$$

Using equation 23.21,

$$2(R_+ + R_-) \geq 2\sqrt{2}R_- \tag{23.25}$$

$$\frac{R_+}{R_-} \geq \sqrt{2} - 1 = 0.414 \tag{23.26}$$

Atom and ion radii for a number of elements are summarized in Table 23.2. It is seen that in each column of the periodic table the ionic radius increases with the principle quantum number of the valence orbital electrons. The radius of an ion is nearly the same in different crystals because the repulsive force increases very sharply as the internuclear distance becomes smaller than a certain value.

Table 23.2   Crystal Structure Data[a]

Atomic Number	Element	Structure[b]	Atom Radius/pm	Ion	Ion Radius/pm[c]
3	Li	b.c.c.	152	Li^+	60
4	Be	c.p.h.	112	Be^{2+}	31
6	C	Cubic (diamond)	77		
8	O			O^{2+}	140
9	F			F^-	136
11	Na	b.c.c.	186	Na^+	95
12	Mg	c.p.h.	161	Mg^{2+}	65
14	Si	Cubic (diamond)	118	Si^{4+}	41
17	Cl			Cl^-	181
19	K	b.c.c.	232	K^+	133
20	Ca	f.c.c.	197	Ca^{2+}	99
26	Fe	b.c.c.	124	Fe^{2+}	80
				Fe^{3+}	64
29	Cu	f.c.c.	128	Cu^+	96
30	Zn	c.p.h.	133	Zn^{2+}	74
32	Ge	Cubic (diamond)	128	Ge^{4+}	53
35	Br			Br^-	195
37	Rb	b.c.c.	245	Rb^+	148
53	I	Orthorhombic	136	I^-	216
55	Cs	b.c.c.	263	Cs^+	169
78	Pt	f.c.c.	139	Pt^{4+}	65

[a] A. Kelly and G. W. Groves, *Crystallography and Crystal Defects*. Reading, MA: Addison-Wesley, 1970.

[b] b.c.c., body-centered cubic; c.p.h., close-packed hexagonal; f.c.c., face-centered cubic.

[c] The radius of an ion depends on the number of neighboring ions of opposite sign, or, in other words, the coordination number. The values in the table apply to an ion with a coordination number of 6. In structures where the coordination number is 4 the radius should be decreased by about 7%, with coordination number 8 the radius should be increased by 3%, and with coordination number 12 by 6%.

It has been found that the distance between two kinds of atoms connected by a covalent bond of a given type (single, double, etc.) is nearly the same in different molecules. The distance between two atoms is taken to be equal to the sum of the bond radii of the two atoms. Since the C–C bond distance is 154 pm in many compounds, the radius for a carbon single bond is taken to be 77 pm. Since the C≡C distance in acetylene is 120 pm, the radius for a carbon triple bond is taken to be 60 pm. By consideration of the bond distances in many compounds it has been possible to build up tables of bond radii, such as Table 23.3, which are useful in predicting the structures of molecules. It must be realized, however, that the effective radius of an atom depends in part also on its environment and on the nature of bonds in the molecule under consideration.

23.8 Binding Forces in Crystals

A number of different types of binding forces are involved in holding crystals together. The physical properties of a crystal are very dependent on the type of bonding.

Ionic crystals are held together by the strong Coulomb attractions of the oppositely charged ions. The lattice energy determined from heat of formation and heat of vaporization measurements agrees with that calculated on the assumption that the units of the crystal are ions held together by electrostatic forces.

In ionic crystals there is no fixed directed force of attraction. Although the ionic crystals are strong, they are likely to be brittle. They have very little elasticity and cannot be easily bent or worked. The melting points of ionic crystals are generally high (NaCl, 800 °C; KCl, 790 °C). In ionic crystals some of the atoms may be held together by covalent bonds to form ions having definite positions and orientations in the crystal lattice. For example, in calcium car-

Table 23.3 Covalent Radii for Atoms[a] (radii in pm)

	H	C	N	O	F
Single-bond radius	30	77.2	70	66	64
Double-bond radius		66.7	60	56	
Triple-bond radius		60.3			
		Si	P	S	Cl
Single-bond radius		117	110	104	99
Double-bond radius		107	100	94	89
Triple-bond radius		100	93	87	
		Ge	As	Se	Br
Single-bond radius		122	121	117	114
Double-bond radius		112	111	107	104
		Sn	Sb	Te	I
Single-bond radius		140	141	137	133
Double-bond radius		130	131	127	123

[a] Taken from L. Pauling, *The Nature of the Chemical Bond*. Ithaca, NY: Cornell University Press, 1960, which should be consulted for details concerning the source and constancy of these radii.

bonate a carbonate ion does not "belong" to a given calcium ion, but three particular oxygen atoms are bonded to a given carbon atom.

Covalent crystals, which are held together by covalent bonds in three dimensions, are strong and hard and have high melting points. Diamond is an example of this type of crystal.

The great difference between graphite and diamond can be understood in terms of the crystal lattice. Graphite has hexagonal networks in sheets like benzene rings. The distance between atoms in the plane is 142 pm, but the distance between these atomic layer planes is 335 pm. In two directions, then, the carbon atoms are tightly held as in the diamond, but in the third direction the force of attraction is much less. As a result, one layer can slip over another. The crystals are flaky, and yet the material is not wholly disintegrated by a shearing action. This planar structure is part of the explanation of the lubricating action of graphite, but this action also depends on absorbed gases, and the coefficient of friction is much higher in a vacuum.

Covalently bonded crystals are insulators because the bonding orbitals are fully occupied. They become conductors only if electrons are excited to unoccupied levels. Thus, they become photoconductors when they are irradiated at a wavelength sufficiently short to raise electrons to excited levels.

Molecular crystals are held together by van der Waals forces (Section 12.11). Examples are provided by crystals of neutral organic compounds and rare gases. Since van der Waals forces are weak, such molecular crystals have low melting points and low cohesive strengths.

Hydrogen-bonded crystals are held together by the sharing of protons between electronegative atoms (Section 12.10). Hydrogen bonds are involved in many organic and inorganic crystals and in the structure of ice and water. They are comparatively weak bonds, but they play an extremely important role in determining the atomic arrangement in hydrogen-bonded substances such as proteins and polynucleotides.

Metallic bonds exist only between large aggregates of atoms. This type of bonding gives metals their characteristic properties: opaque, lustrous, malleable, and good conduction of electricity and heat. Metallic bonding is due to the outer, or valence, electrons. The overlapping of the wavefunctions for the valence electrons in metals results in orbitals that extend over the entire crystal. The electrons pass throughout the volume of the crystal and, for certain purposes, we may consider that there is an electron gas—except that we will see that it is fundamentally different from other gases.

There is a gradual transition between metallic and nonmetallic properties. Atoms with fewer and more loosely held electrons form metals with the most prominent metallic properties. Examples are sodium, copper, and gold. As the number of valence electrons increases and they are held more tightly, there is a transition to covalent properties.

The close-packed structures are often found in metals because the binding energy per unit volume is maximized.

23.9 Close Packing of Spheres

When the bonding of atoms is not highly directional, it is often found that the lowest energy structure is that in which each atom is surrounded by the greatest possible number of neighbors. It is, therefore, of interest to consider the ways in which uniform spheres can be stacked to form close-packed structures. When spheres are packed in a plane, they arrange themselves so that each sphere is surrounded hexagonally by six others. When the second layer is formed by placing spheres in the hollows on top of the first layer, it is evident that all of the hollows in the first layer are not occupied, as may be seen from Fig. 23.14. When a third layer is added, there is a choice as to whether the spheres in this layer are stacked so that they are not above the spheres in the first layer, as in Fig. 23.14a, or are, as in Fig. 25.14b. If the spheres in the third layer are not directly above the spheres in the first layer, as shown in Fig. 23.14a, the structure has cubic symmetry and the cubic unit cell is face-centered. The fact that **cubic close packing** is really face-centered cubic may be seen from Fig. 23.15. Since this is a close-packed structure, it is of interest to calculate the fraction of the volume occupied by spheres. Since the length of the diagonal of the face of a unit cell is $\sqrt{2}a$, the radii of the spheres that just touch are given by $(\sqrt{2}/4)a$. Since there are four spheres per unit cell, the fraction of the volume occupied by spheres is

$$\frac{4(\frac{4}{3}\pi)\left[(\sqrt{2}/4)a\right]^3}{a^3} = 0.7405 \qquad (23.27)$$

In cubic close packing each sphere has twelve nearest neighbors: six within its own layer, three in the layer above, and three in the layer below. Metals and inert gases often form cubic close-packed structures.

If the spheres in the third layer are placed over the spheres in the first layer, as shown in Fig. 23.14b, and the spheres in the fourth layer are placed over those in the second layer, and so on, the unit cell is hexagonal and the packing is referred to as **hexagonal close packing.** As in the case of cubic close packing, each sphere has twelve nearest neighbors, and the fraction of the volume occupied by spheres is again 0.7405. The coordinates of the atoms in the unit cell are (000) and ($\frac{1}{3}\frac{1}{3}\frac{1}{2}$). There are two atoms in the unit cell.

The unit cell dimensions in terms of the radius of a sphere are $a = b = 2r$, $c = 4\sqrt{2}r/\sqrt{3}$, and $c/a = 2\sqrt{2}/\sqrt{3} = 1.633$. A number of metals have hexagonal close-packed structures, but c/a usually deviates a little from the ideal ratio of 1.633. This indicates that the atoms are not exactly spherical in shape.

The layers of hexagonal close packing may be described as ABABAB···. The layers of cubic close packing may be described as ABCABC···.

Hexagonal close packing and cubic close packing are the only two ways of close packing identical spheres so that the environment of each sphere is identical with the environment of all the other spheres, but there are other ways of close packing spheres so that the environment of each sphere is not identical, for example, ABCABABCAB···. In principle there is an infinity of these other ways.

In a number of crystal structures containing two types of atoms, one type of atoms forms a close-packed structure and the other type occupies interstices between the close-packed spheres. There are two types of interstices between close-packed spheres: tetrahedral sites and octahedral sites. When one sphere rests on three others, the centers of the four spheres lie at the apices of a regular tetrahedron, and the space at the center of this tetrahedron is called a tetrahedral site. Since in any close-packed structure each sphere is in contact with three spheres in the layer below it and three spheres in the layer above it, there are two tetrahedral sites per sphere. Thus, if X atoms form a close-

(a)

(b)

Figure 23.14 (a) Cubic close packing (face-centered cubic). (b) Hexagonal close packing.

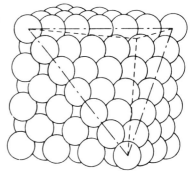

Figure 23.15 Cubic close packing (face-centered cubic). Some atoms have been omitted to show that the close packed planes are (111) planes.

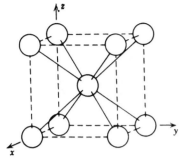

Figure 23.16 Body-centered cubic structure.

packed structure, and Y atoms are small enough to fit in the tetrahedral sites, this might be a convenient structure for a compound XY_2. Half of the sites would be occupied in a compound XY. For the smaller sphere to occupy a tetrahedral site without disturbing the closest-packed lattice, the radius of the smaller spheres should be no greater than 0.225 of that of the larger spheres.

The other type of interstitial site in a close-packed structure is the octahedral site that is surrounded by six spheres whose centers lie at the apices of a regular octahedron. There is one octahedral site for every sphere in a close-packed structure so that a compound of the type XY can be accommodated. For Y atoms to fit into octahedral sites their radii must be less than 0.414 of the radii of the larger spheres.

Although the majority of metallic elements crystallize with hexagonal close packing or cubic close packing, some crystallize with the body-centered cubic arrangement, which is not a close-packed structure. In this structure, which is illustrated in Fig. 23.16, each atom has eight nearest neighbors and six other next nearest neighbors slightly further away at the body-centered positions of neighboring cells. By use of the Pythagorean theorem it is readily shown that the distance from the body-centered point to one of the corners of the cubic unit cell is $(\sqrt{3}/2)a$. If the structure is made up of spheres that touch they must have a radius of $(\sqrt{3}/4)a$. The fraction of the volume of the unit cell (and hence of the entire crystal) occupied by spheres is

$$\frac{2(\tfrac{4}{3}\pi)[(\sqrt{3}/4)a]^3}{a^3} = 0.6802 \tag{23.28}$$

The alkali metals and tungsten crystallize in a body-centered cubic structure.

The characteristics of cubic lattices for spheres are summarized in Table 23.4. It is very rare to find spheres packed in a simple cubic lattice.

Example 23.9
Magnesium forms hexagonal close-packed crystals with $a = 320.9$ pm at 25 °C. What is the density of the metal and the magnesium atom radius?

$$V = a^2c(1 - \cos^2 \gamma)^{1/2} = a^2c \sin \gamma$$

Since $c = 1.633a$,

$$V = 1.633(320.9 \times 10^{-12}\ \text{m})^3 \sin 120° = 4.673 \times 10^{-29}\ \text{m}^3$$

$$\rho = \frac{2(24.305 \times 10^{-3}\ \text{kg mol}^{-1})}{(6.022\ 137 \times 10^{23}\ \text{mol}^{-1})(4.673 \times 10^{-29}\ \text{m}^3)} = 1.727 \times 10^3\ \text{kg m}^{-3}$$

$$r = \tfrac{1}{2}a = \tfrac{1}{2}(320.9\ \text{pm}) = 160.5\ \text{pm}$$

23.10 Structure of Liquids*

In a perfect crystal the atoms, ions, or molecules occur at definite distances from any individual atom, ion, or molecule that is taken as the origin of a

---

* Y. Marcus, *Introduction to Liquid State Chemistry*. New York: Wiley, 1977.

Table 23.4   Characteristics of Cubic Lattices

	Simple	Body-Centered	Face-Centered
Volume of unit cell	a^3	a^3	a^3
Lattice points per cell	1	2	4
Nearest neighbors	6	8	12
Distance to nearest neighbor	a	$(\sqrt{3}/2)a$	$(1/\sqrt{2})a$
Fraction of volume occupied	0.524	0.680	0.740

coordinate system. In a gas the molecules have random positions at a given time. Liquids are intermediate between crystals and gases in that the molecules are not arranged in a definite lattice, but there is some local order. By a detailed analysis of the intensity of the scattered X-rays, it is possible to calculate the distribution of atoms or molecules in a liquid.

When X-rays are scattered by an amorphous phase the intensity $I(\theta)$ of scattered radiation is a function only of the scattering angle θ, which is defined as the angle of deviation from the propagation axis of the incident beam. (Note that this angle is twice the Bragg angle used earlier.) The intensity tends to fall off as θ increases. It may be shown that the intensity is given by

$$I(\theta) \propto \int_0^\infty \mathcal{P}(R) \frac{\sin kR}{kR} \, dR \qquad (23.29)$$

where

$$k = \frac{4\pi}{\lambda} \sin \frac{\theta}{2}$$

and $\mathcal{P}(R) \, dR$ is the probability of finding a particle between R and $R + dR$, R being measured from one particle at the origin. For a uniform medium $\mathcal{P}(R)$ would be proportional to $4\pi R^2$, and so it is convenient to introduce a pair correlation function $g_2(R)$ defined by

$$\mathcal{P}(R) = 4\pi R^2 \rho g_2(R) \qquad (23.30)$$

where ρ is the number density for the medium. If the medium were continuous and homogeneous, the pair correlation function $g_2(R)$ would be constant. As shown in Fig. 23.17, the pair correlation function for a liquid is zero for small values of R, has a maximum at the most probable nearest-neighbor distance, and has diminishing maxima that correspond to second-nearest neighbor distance and so on. The first maximum in Fig. 23.17 is due to the 8 or 12 nearest neighbors that surround each molecule. The pair correlation function $g_2(R)$ is essentially zero at distances less than one molecular diameter because of the strong short-range intermolecular repulsions (Section 12.11). The existence of a shell of nearest neighbors means that there will be a relatively high probability of finding molecules one molecular diameter away, as well as 2, 3, and 4 molecular diameters removed. Since the liquid has no long-range order, $g_2(R)$ is essentially constant after a few molecular diameters. As the temperature of the liquid is raised, the maxima and the minima in the pair correlation function become less pronounced.

The theory of liquids is in a much less satisfactory state than the theories of gases and crystals, but important progress is being made in our understanding of the structure of liquids.

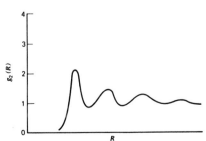

Figure 23.17   Pair correlation function for a liquid.

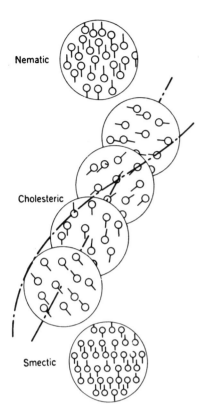

Figure 23.18 Structures of liquid crystals.

23.11 Liquid Crystals

In certain liquids new phases, which are intermediate between liquid and solid phases, appear upon cooling. These phases have a translucent or cloudy appearance and are called liquid crystals.

In a liquid of asymmetric molecules the molecular axes are arranged at random. But in liquid crystals there is some kind of alignment. As shown in Fig. 23.18, there are three types of liquid crystals. In **nematic** liquid crystals the long axes of the molecules are aligned parallel to each other, but the molecules are not arranged in layers. The word nematic was coined from the Greek root for thread to describe the appearance of this particular type of liquid crystal under a microscope. Nematic liquid crystals have a translucent appearance because they scatter light strongly.

In **cholesteric** liquid crystals the molecular axes are aligned, and the molecules are arranged in layers in which the orientation of the axes shifts in a regular way in going from one layer to the next, as shown in Fig. 23.18. The distance measured perpendicular to the layers through which the direction of alignments shifts 360° is of the order of the wavelength of visible light. As a result of the strong Bragg reflection of light, cholesteric liquid crystals have vivid iridescent colors. The pitch of the spiral and the reflected color depends sensitively on the temperature, and so these liquid crystals have been used to measure skin and other surface temperatures. The name cholesteric comes from the fact that many derivatives of cholesterol (but not cholesterol itself) form this type of liquid crystal.

The third type of liquid crystals, **smectic,** are formed by certain molecules with chemically dissimilar parts. The chemically similar parts attract each other, and there is a tendency to form layers as well as to have the molecules aligned in one direction, as illustrated in Fig. 23.18. Smectic phases are soaplike in feel and structure and may have some relationships with cell membranes.

23.12 Defects in Crystals

Real crystals have imperfections or defects. Properties such as electrical conductivity, color, and mechanical properties are strongly dependent on imperfections. These imperfections may be at a point, along a line, or over a surface.

A point imperfection may result from the absence of an atom (vacancy), presence of an impurity atom at a lattice site (substitutional impurity atom), presence of an impurity atom at an interstice (interstitial impurity atom), or displacement of an atom to an interstitial site (self-interstitial), as shown in Fig. 23.19.

If the repeating units in the crystal lattice are electrically neutral, vacancies produce no particular problem with respect to the overall balance of electric charge. But in an ionic crystal vacancies must be balanced so that the crystal as a whole is electrically neutral. In a **Frenkel defect** the vacancies are compensated for by an interstitial atom of the same type. In a **Schottky defect** anion and cation vacancies occur in equal numbers. Schottky defects lower the density of the crystal, but Frenkel defects do not change the density significantly.

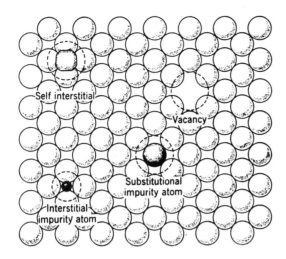

Figure 23.19   Point imperfections in crystals.

At thermal equilibrium an otherwise perfect crystal will have a certain number of lattice vacancies, because the Gibbs energy of a crystal is decreased by the presence of disorder in the structure. The probability that a given lattice site is vacant is proportional to the Boltzmann factor $\exp(-E_v/kT)$, where E_v is the energy required to move an atom from a lattice site in the crystal to a lattice site on the surface. For a crystal with N lattice sites and n vacancies,

$$\frac{n}{N-n} = e^{-E_v/kT} \tag{23.31}$$

Since E_v is of the order of 1 eV, the number of vacancies is very small at ordinary temperatures.

Pure single crystals deform permanently at shear forces orders of magnitude lower than predicted theoretically for perfect crystals. Line defects were postulated to explain the ease with which single crystals deform permanently. The two principal line defects are called **edge dislocations** and **screw dislocations.** An edge dislocation is named for the edge of an atomic plane that terminates within the crystal instead of passing all the way through. The way in which an edge dislocation facilitates shear in a crystal is illustrated in Fig. 23.20. During

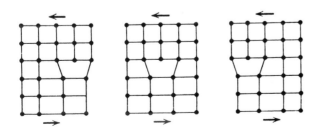

Figure 23.20   Motion of an edge dislocation under shear. (From N. B. Hannay, *Solid-State Chemistry*, Copyright © 1967. Reprinted by permission of Prentice-Hall, Inc., Englewood Cliffs, NJ.)

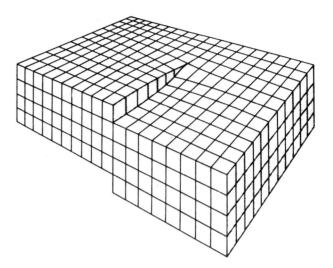

Figure 23.21 Screw dislocation. (From N. B. Hannay, *Solid-State Chemistry*, Copyright © 1967. Reprinted by permission of Prentice-Hall, Inc., Englewood Cliffs, NJ.)

shear the dislocation moves across the crystal with the net effect that the top half of the crystal is displaced one lattice distance with respect to the lower half of the crystal. The number of dislocation lines passing through a unit area within an ordinary crystal is of the order of 10^6 cm^{-2} or more.

Near an edge dislocation, the atoms are pushed together above the edge and pulled apart below the edge. Thus, impurity atoms with larger diameters than the solvent atoms tend to concentrate below the edge, and impurity atoms with smaller diameters tend to concentrate above the edge. This binding of impurity atoms at the dislocation tends to make it more difficult to move a dislocation in an impure material than in a pure material. Therefore, alloys require greater shear forces for permanent deformation than do pure crystals.

A screw dislocation forms a continuous helical ramp of one set of atomic planes about the dislocation line, as shown in Fig. 23.21. A screw dislocation provides for easy crystal growth because atoms can be added at the step. Screw dislocations can move in a crystal subjected to appropriate shear forces, and they give rise to the same type of permanent deformation as an edge dislocation. Real dislocations are often mixtures of edge dislocations and screw dislocations.

Plane defects include the surfaces around the tiny grains that make up crystalline metals. Neighboring grains have unrelated crystallographic orientations, and the boundaries between grains are regions of strain in which impurities tend to concentrate. Grain boundaries may be detected by microscopic examination of a metal surface that has been polished and etched.

Another type of plane defect in connection with the close packing of spheres is the irregular stacking of atomic planes (Section 23.9). For example, in face-centered cubic packing the normal stacking sequence is ABCABC$\cdots$. If a B layer accidentally forms on top of a C layer and the reverse stacking sequence proceeds, the sequence is ABCBACBA$\cdots$. The two parts of the crystal on either side of the plane defect are mirror images and are referred to as twins.

Example 23.10

What fraction (n/N) of the lattice sites are vacant at (a) 298 K, (b) 500 K, and (c) 1000 K for a crystal for which $E_v = 1$ eV.

(a)

$$1 \text{ eV} = 1.602 \times 10^{-19} \text{ J}$$

$$\frac{n}{N} = \exp\left(-\frac{E_v}{kT}\right)$$

$$= \exp\left[-\frac{1.602 \times 10^{-19} \text{ J}}{(1.381 \times 10^{-23} \text{ J K}^{-1})(298 \text{ K})}\right]$$

$$= 1.24 \times 10^{-17}$$

(b) At 500 K, $n/N = 8.40 \times 10^{-11}$.

(c) At 1000 K, $n/N = 9.16 \times 10^{-6}$.

23.13 Free-Electron Theory of Metals

The high electrical conductivity of metals is a result of the ease with which electrons in the metal can move under the influence of a static or low-frequency electric field. In the free-electron model of a metal each valence electron is treated as a particle in a three-dimensional box the size of the metal crystal. In this oversimplified theory the energy levels of the electron are given by equation 10.39 for a particle in a box. For each eigenstate there are actually two possible states of the electron, corresponding to the two values of the spin. It can be shown that the density-of-state function $g(\epsilon)$ is given by

$$g(\epsilon) = C\epsilon^{1/2} \tag{23.32}$$

The number of one-electron-states with energy between ϵ and $\epsilon + d\epsilon$ is given by $g(\epsilon)\, d\epsilon$.

The density of states $g(\epsilon)$ is shown as a function of energy in Fig. 23.22a. As electrons are added at 0 K, the energy levels are filled up to some maximum energy ϵ_F determined by the number of electrons. This is indicated by the cross hatching in the figure. This maximum energy is called the **Fermi energy.** The Fermi energy is the electrochemical potential of the electrons and determines their tendency to move at an interface, just as the chemical potential does for a substance. It is sometimes convenient to present the same information by plotting $g(\epsilon)/\epsilon^{1/2}$ versus ϵ, as shown in Fig. 23.22b. In contrast to the translational energies of gas molecules, where Maxwell–Boltzmann statistics allow any number of particles to have exactly the same energy, electrons follow Fermi–Dirac statistics, which means that only one particle is allowed in each state of the system. At a temperature above absolute zero the number of occupied states in the energy range ϵ to $\epsilon + d\epsilon$ is given by $d(N/V)$:

$$d\left(\frac{N}{V}\right) = f(\epsilon, T)g(\epsilon)\, d\epsilon \tag{23.33}$$

where $f(\epsilon, T)$ is the Fermi–Dirac distribution function,

$$f(\epsilon, T) = \frac{1}{e^{(\epsilon - \mu)/kT} + 1} \tag{23.34}$$

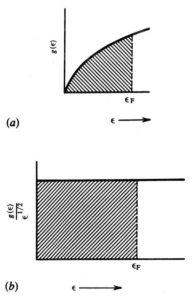

Figure 23.22 (a) Density of states for a free electron gas as a function of energy. (b) $g(\epsilon)/\epsilon^{1/2}$ versus energy for a free electron gas. The cross hatching indicates the levels occupied at 0 K.

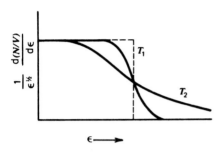

Figure 23.23 Distribution of electron energies in a metal at 0 K (dashed line) and two higher temperatures $T_2 > T_1$.

The quantity μ is a constant at any given temperature. The value of μ can be obtained by integrating 23.34 over all energies since this integration must yield the number of electrons per unit volume. Since $f(\epsilon, T) = \frac{1}{2}$ when $\epsilon = \mu$, the quantity μ is equal to the energy at which $f(\epsilon, T)$ has half its maximum value. At 0 K, $f(\epsilon, 0) = 1$ for energies ϵ less than the Fermi energy ϵ_F, and $f(\epsilon, 0) = 0$ for energies ϵ greater than the Fermi energy ϵ_F. As $T \to 0$, $\mu \to \epsilon_F$. It is a good approximation to take $\mu \approx \epsilon_F$ at other temperatures provided that $\epsilon_F \gg kT$. The approximation that $\mu \approx \epsilon_F$ is a good one because $\epsilon_F \approx 5$ eV for most metals. Combining the preceding two equations and using this approximation yields

$$d\left(\frac{N}{V}\right) = \frac{C\epsilon^{1/2}\,d\epsilon}{e^{(\epsilon - \epsilon_F)/kT} + 1} \qquad (23.35)$$

Figure 23.23 shows the distribution of electron energies at temperatures T_1 and T_2 where $0 < T_1 < T_2$. At room temperature the distribution of electron energies differs only slightly from that at 0 K for $\epsilon_F \approx 5$ eV. A small fraction of the electrons have energies greater than the Fermi energy ϵ_F, and they leave behind holes at $\epsilon < \epsilon_F$. The excited electrons and holes both contribute to the electrical conductivity.

Figure 23.23 provides the explanation for the very small contribution of electrons to the heat capacity for a metal. When the temperature of a metal is raised a small amount, only a small fraction of the electrons have their energies raised. Since the energies of most electrons are not affected by raising the temperature, the electronic heat capacity is negligible compared with the vibrational heat capacity $3R$ predicted by the Einstein and Debye theories (Section 17.15).

The free-electron theory of metals is only an approximation. It would predict an infinite electrical conductivity because the excited electrons could in principle be accelerated to infinite velocities. In a real crystal friction is introduced by the scattering of electrons from vibrating atoms. According to the simplest theory the electrons in a metal have a characteristic electric mobility u. Then the electric conductivity κ is given by

$$\kappa = Nqu \qquad (23.36)$$

where N is the number of current electrons per unit volume and q is the charge carried by an electron.

The theory of the electronic properties of metals is really a special case of a more general theory referred to as the band theory of solids, to which we now turn.

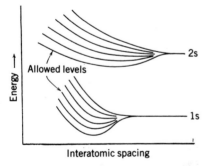

Figure 23.24 Energy levels for a linear array of six hydrogen atoms as a function of internuclear distance.

23.14 Band Theory of Solids

The band theory of solids is an extension of the Hartree–Fock model (Section 11.10) in that it utilizes linear combinations of wavefunctions of atoms. For example, Fig. 23.24 illustrates the splitting that occurs when six hydrogen atoms are brought together in a linear array. The combination of six 1s wavefunctions produces six orbitals, three bonding and three antibonding. As electrons are fed into this energy level scheme, they will first fill the lower energy bonding

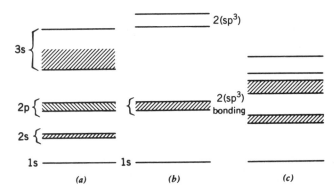

Figure 23.25   Energy levels for electrons in (*a*) a metal (potassium), (*b*) an insulator (diamond), and (*c*) a semiconductor. (From R. S. Berry, S. A. Rice, and J. Ross, *Physical Chemistry*. New York: Wiley, 1980.)

orbitals, two at a time. As the number of interacting atoms is increased the number of energy levels increases, and they become more and more closely spaced within a band, but the width of the band at a given internuclear separation does not increase substantially. We use the term **band** to distinguish the groups of levels arising from different atomic orbitals. In Fig. 23.24 there is the 1s band and the 2s band. Thus, in contrast to molecules, we find in a solid bands of energy levels containing very large numbers of discrete levels, but with relatively large separations in energy between bands. These separations between allowed bands are called energy gaps. The very different properties of conductors, insulators, and semiconductors may be understood in terms of these energy bands and band gaps.

As shown in Fig. 23.25, the bands of an insulator are completely filled. These filled bands are referred to as **valence bands.** The electrons in them are immobile and do not contribute to electrical conductivity. For an electron to be able to move it must be excited to the next band. In an insulator the band gap to the first unfilled band is much larger than kT, and so the population of the excited electrons is negligible.

In a metal the highest occupied band is only partially filled, as shown in Fig. 23.25. This band is referred to as a **conduction band.** Since the energy levels within it are continuous, an electron at the top of the conduction band, where it is immobile, may be excited to a higher level with an infinitesimally higher energy. Since the levels are continuous a conduction electron is essentially free.

23.15   Semiconductors

Semiconductors are covalent solids in which the valence band is completely full and the conduction band is completely empty at 0 K. There is an energy gap between the valence band and the conduction band, as shown in Fig. 23.26. For semiconductors the energy gap is small enough so that even at room temperature a significant number of electrons are thermally excited from the valence band to the conduction band. The excitation of electrons from the valence

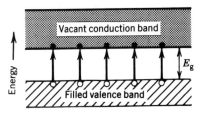

Figure 23.26   Energy bands for intrinsic conductivity in a semiconductor.

Table 23.5   Electric Conductivity and Band Gaps for Elements with the Same Crystal Structure

	$\kappa/\Omega^{-1}\,m^{-1}$	E_g/eV
C (diamond)	$<10^{-4}$	6.0
Si	1.5×10^{-3}	1.14
Ge	2	0.67
Sn (gray)	>100	0.08

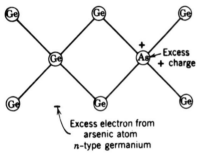

(a)

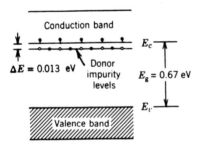

(b)

Figure 23.27   (*a*) Charges resulting from an arsenic impurity in germanium. (*b*) Energy level diagram for a germanium crystal containing arsenic.

band to the conduction band leaves holes in the valence band, as shown in the figure. Both the electrons and the holes contribute to the electrical conductivity. In an applied electric or magnetic field a hole acts as if it had a positive charge $+e$. The conductivity due to electrons in the conduction band and holes in the valence band is referred to as **intrinsic conductivity.** The conductivities of four semiconductors are summarized in Table 23.5. These solids all have the same crystal structure; the atoms are bonded covalently with four adjacent atoms in a tetrahedral arrangement. Cuprous oxide (Cu_2O), silicon carbide (SiC), and gallium arsenide (GaAs) crystals are also semiconductors.

Pure germanium and pure silicon are weakly ionized solids, just as water is a weakly ionized liquid. At room temperature the intrinsic carrier concentrations ($[e^-] = [h^+]$) in Si and Ge are the same order of magnitude as the ion concentration $[(H^+] = [OH^-])$ for water. A hole is represented by h^+.

The resistance of a semiconductor decreases as the temperature is raised because of the thermal excitation of electrons to higher energy levels. The resistance of a metal increases as the temperature increases because the scattering of conduction electrons by atoms in the lattice increases as the atoms acquire more vibrational energy.

Semiconductors exhibit **photoconductivity** when radiated with light of sufficiently short wavelength. The wavelength required depends on the energy gap.

The electrical conductivity of a semiconductor may be increased by adding impurities; this additional conductivity is called **extrinsic conductivity.** The addition of impurities to a semiconductor is referred to as doping. For example, if arsenic is present in crystalline germanium, arsenic atoms replace germanium atoms in the crystal structure. Since an arsenic atom has five electrons in the valence band and germanium has four electrons per atom in the valence band, the fifth electron from arsenic cannot go into the filled valence band. The arsenic atom in the doped germanium has a positive charge, as shown in Fig. 23.27, and the electron is weakly bound in an "orbit" of very large radius around this positive charge. The binding energy of this electron may be calculated to be about -0.01 eV by using equation 11.7 for the electronic energy of a hydrogenlike atom. The experimental value is 0.0127 eV. The binding energy is much smaller than for a hydrogen atom because the relative permittivity of the crystal is 16 and the effective mass of the electron in the crystal lattice is about one-fifth that of a free electron.

Since the fifth electron of arsenic has such a small binding energy to arsenic, it can be ionized (i.e., transferred to the conduction band) at a much lower temperature than an electron at the top of the valence band. Thus, an electron level just below the conduction band is created by the addition of arsenic to germanium. This is illustrated in Fig. 23.27*b*. Arsenic is referred to as a donor impurity because it contributes electrons. The arsenic-doped germanium is called an **n-type** (negative) semiconductor because it has an excess of free electrons.

If boron is added to germanium a **p-type** (positive) semiconductor is produced. Boron is referred to as an **acceptor** impurity because it is deficient in electrons. Boron has three valence electrons per atom, so when a boron atom is tetrahedrally bonded to four germanium atoms, there is an electron deficiency or hole in the valence band of the crystal, as shown in Fig. 23.28. This hole can drift through the crystal as successive electrons fill this hole and create

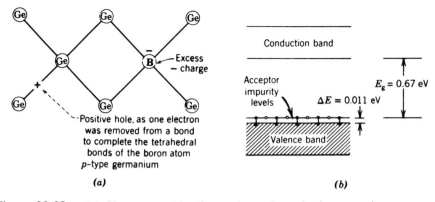

Figure 23.28 (*a*) Charges resulting from a boron impurity in germanium. (*b*) Energy level diagram for a germanium crystal containing boron.

another. Thus, the addition of the acceptor impurity boron produces acceptor impurity levels 0.0108 eV above the valence band of germanium, as shown in Fig. 23.28*b*. Added atoms are not necessarily ionized. If the energy required to produce this ionization is quite small, the ionization is essentially complete at room temperature. However, as the temperature is progressively lowered, charge carriers are progressively "frozen out" on the impurity atoms. Thus, the ionization of added atoms can be treated as a problem in chemical equilibrium. The donor D and acceptor A ionizations are

$$D = D^+ + e^- \tag{23.37}$$

$$A = A^- + h^+ \tag{23.38}$$

The concentrations of the species on the right side increase with increasing temperature. The slope of the plot of the logarithm of the concentration of e^- or h^+ versus $1/T$ gives information about the ionization energy of the donor or acceptor.

When a p-type semiconductor is brought into contact with an n-type semiconductor, a double layer of charge is set up a the junction, as shown in Fig. 23.29. The reason for the double layer of charge can be seen from Fig. 23.29*a*. On the p side of the junction there is an excess of free holes and a very low concentration of electrons. (The relative concentration of electrons is exaggerated in this figure.) On the n side of the junction there is an excess of electrons relative to holes. The holes on the p side have a tendency to diffuse to the n side, and the electrons on the n side have a tendency to diffuse to the p side. This diffusion occurs to a small extent but, when the electrons diffuse into the p-type semiconductor, positive charges do not diffuse with them and a layer of negative charge develops. Similarly, the diffusion of positive holes into the n-type conductor is not accompanied by the diffusion of negative charges and a double layer of charge is formed, as shown in Fig. 23.29*b*. Because of the double layer, the electrostatic potential changes at the junction, and there is a built-in electric field.

These p-n junctions are useful because they can be used as rectifiers, solar cells, and other electronic devices. When an electron and a hole recombine, light may be emitted, and this is the mechanism for luminescent p-n diodes.

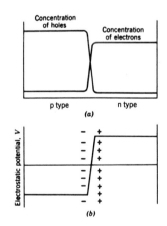

Figure 23.29 (*a*) Variations of concentrations of holes and electrons across a p-n junction. (*b*) Electrostatic potential V from acceptor (−) and donor (+) ions near the junction. (From C. Kittel, *Introduction to Solid State Physics,* Copyright © 1976. Reprinted by permission of John Wiley & Sons, Inc., New York.)

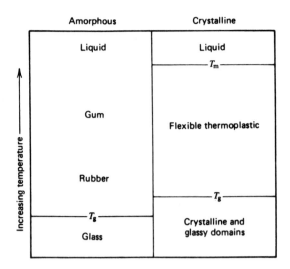

Figure 23.30 Changes in amorphous and crystalline polymers as the temperature is raised.

23.16 Crystallinity in Synthetic Polymers*

In earlier chapters we have considered how synthetic high polymers are formed (Sections 22.7 and 22.9), the approximation of chain conformation by random flight (Section 22.4), and the thermodynamics of rubberlike elasticity (Section 4.17). In this section we consider the mechanical properties of long-chain synthetic polymers that are not extensively cross-linked. Polymeric materials display a very wide range of physical properties: they may be hard or soft, leathery or rubbery, brittle or tough, and meltable or nonmeltable. These properties depend on the molecular structure of the polymer and the temperature.

To discuss the effects of temperature on polymers we need to distinguish between **amorphous polymers** and **crystalline polymers.** In an amorphous polymer the chains are distributed in a completely random manner. In a crystalline polymer some of the chains are incorporated in regions of three-dimensional order, called crystallites. So-called crystalline polymers contain various proportions of ordered and disordered regions. The dependence of physical properties of amorphous and crystalline polymers on temperature is shown in Fig. 23.30.

Both amorphous and crystalline polymers are glasses at low temperatures. Glassy polymers are dimensionally stable and have moderate strength, but they are brittle and fracture like a glass. Poly(methyl methacrylate) at room temperature is an example of a glassy polymer. When a glassy polymer is heated it undergoes a rather sharp change in properties at what is called the glass **transition temperature** T_g. This is not a first-order phase transition (Section 6.3) because it takes place over a range of several degrees, but it is more like a second-order phase transition. There is no change in density at T_g, but there

---

* H. R. Allcock and F. W. Lampe, *Contemporary Polymer Chemistry.* Englewood Cliffs, NJ: Prentice-Hall, 1981; F. A. Bovey and F. W. Winslow, *Macromolecules.* New York: Academic, 1979; J. M. G. Cowie, *Polymers: Chemistry and Physics of Modern Materials.* New York: Intext Educational, 1973.

Table 23.6 Glass Transition Temperatures (T_g) and Crystalline Melting Temperatures (T_m) for Selected Polymers[a]

Polymer	$T_g/°C$	$T_m/°C$
Polystyrene (isotactic)	100	240
Poly(*m*-methylstyrene)(isotactic)	70	215
Poly(methyl methacrylate)(atactic)	114	—
Poly(methyl methacrylate)(isotactic)	48	160
Poly(methyl methacrylate)(syndiotactic)	126	200
Poly(*cis*-1,4-isoprene)	−67	36
Poly(*trans*-1,4-isoprene)	−68	74
Polyethylene	−20	141

[a] A compilation of T_g and T_m values can be found in O. G. Lewis, *Physical Constants of Linear Homopolymers*. New York: Springer, 1968.

is a change in slope of the plot of density versus temperature. The glass transition temperature of a polymer is very important in determining its physical properties and its practical utility.

As shown in Fig. 23.30 amorphous and crystalline polymers behave differently above the glass transition temperature. As the temperature is raised above T_g an amorphous polymer becomes rubbery. The elasticity of a rubbery polymer arises from the fact that the polymer chains can exist in different conformations (Section 22.4).

As an amorphous polymer is heated to high temperatures the rubber is gradually converted to a gum and finally to a viscous liquid. Many polymeric solids have properties that are intermediate between those of ideal solids and ideal liquids. In a perfectly elastic solid the stress is directly proportional to the strain, but independent of the rate of strain. In a perfectly viscous fluid the stress is directly proportional to the rate of strain, but independent of the strain itself. If the stress depends on both the strain and the rate of strain, the material is said to be **viscoelastic**.* Polymers in particular show such behavior because of the complicated way in which such long-chain-like molecules interact with each other. For example, a silicon polymer known as Silly Putty can be bounced like a rubber ball, but over a longer period it flows like a viscous liquid.

A crystalline polymer also has a glass transition temperature below which it is rigid and brittle. Above T_g crystalline polymers become thermoplastic, rather than rubbery. Above T_g they are more flexible, more elastic, and have greater impact resistance than below T_g. In a crystalline polymer the crystallites usually exist only in small domains within the polymer. These materials can be viewed as amorphous polymers in which are imbedded cross-links due to microcrystalline domains. These cross-links give the thermoplastic considerable dimensional stability. Microcrystalline polymers are also usually tougher than amorphous polymers. As the temperature of a crystalline polymer is raised further, the crystals melt rather sharply at the **crystalline melting temperature** T_m. Above this temperature the polymer is a viscous liquid.

Table 23.6 gives the glass transition temperature T_g and the crystalline melting temperature T_m for a few linear polymers. The glass transition temperature

* J. D. Ferry, *Viscoelastic Properties of Polymers*. New York: Wiley, 1980.

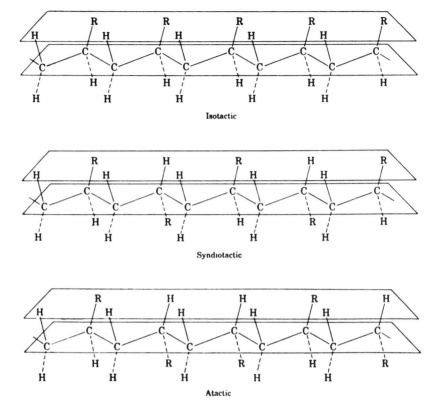

Figure 23.31 Steric configuration of polymer chains.

is closely related to the flexibility of the polymer chain. The glass transition temperature is in general low for a polymer with a highly flexible chain, but the relation between T_g and structure is a complex one involving a number of factors including the stereoregularity of the chains.

Polymers of monomers like styrene have asymmetric carbon atoms. By using certain solid catalysts it is possible to prepare polymers in which all the adjacent asymmetric atoms, at least for long periods of the chain, have the same, or opposite, steric configuration.

There are two types of stereoregular vinyl polymers, **isotactic** and **syndiotactic**, depending on whether successive pseudoasymmetric carbon atoms have the same or opposite enantiomorphic configurations. These two types of structures are illustrated in Fig. 23.31, along with a structure without order. Structures with a lesser degree of order than isotactic or syndiotactic are referred to as **atactic**. Some polymers (stereoblock polymers) have alternating isotactic and syndiotactic sections. K. Ziegler and G. Natta, who first prepared such isotactic polymers, shared the 1963 Nobel prize for chemistry. The regularity in structure of isotactic polymers makes it possible for the polymer chains to fit into a crystal lattice. This increased orderliness increases the melting point, rigidity, and toughness above those of corresponding atactic polymers.

Figure 23.32   Spherulites in isotactic polystyrene. (From F. A. Bovey and F. H. Winslow, *Macromolecules*. New York: Academic, 1979.)

The extent of crystallinity increases when the polymer is stretched because this draws the chains together and reduces the random thermal motions. For example, natural rubber is ordinarily amorphous at room temperature, but becomes oriented and crystalline when stretched. The amount of crystallinity may be investigated by X-ray diffraction and by studying the volume of the polymer as a function of temperature.

Crystallites occur most frequently in polymers having a comparatively simple and symmetrical structure or polymer chains that tend to associate by the formation of hydrogen bonds. There are crystalline regions in nylon, polyethylene, and other polymers in which the chains have sufficient chemical regularity. Large crystals of long-chain polymers may be obtained from solution.

The crystalline regions in polymers are generally in the form of spherulites, which are aggregates of crystals with a radiating fibrillar structure. Figure 23.32 shows spherulites observed under a polarizing microscope between crossed polarizers. These spherulites are grown in two dimensions in a film of molten polymer. In a bulk polymer the spherulites grow in three dimensions. They are generally of the order of 100 μm or less in diameter. The use of microbeam X-ray diffraction shows that the molecular chains are normal to the radial direction in a spherulite.

Polymers that are heavily cross-linked retain their rigidity when heated until the cross-links become thermally broken.

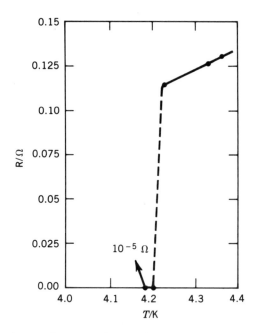

Figure 23.33 Resistance in ohms of a specimen of mercury versus temperature. This is the original plot by Kamerlingh Onnes showing his discovery of superconductivity. (From C. Kittel, *Introduction to Solid State Physics*. New York: Wiley, 1976.)

23.17 Superconductivity

In 1911, Kamerlingh Onnes, a Dutch physicist at the University of Leiden, discovered that the resistivity of mercury suddenly drops to zero as the temperature is lowered below 4.2 K. (see Fig. 23.33.) The same phenomenon occurs (at different characteristic temperatures) for many other metals, alloys, and compounds. We describe this by saying that at a critical temperature T_c the sample undergoes a phase transition from a normal to a superconducting state. In addition to zero resistivity, superconductors have unusual magnetic properties. When a sample in a weak magnetic field is cooled through T_c, the magnetic flux originally in the sample is ejected. This is called the **Meissner effect**. On the other hand, a superconducting state can be destroyed by a strong enough magnetic field (called the critical field H_c). The thermodynamic properties of superconductors are also interesting. For example, the entropy of a superconductor decreases considerably on cooling, indicating that the superconducting state is more ordered than the normal state. Detailed studies suggest that the superconducting state consists of two kinds of electrons: "ordered" pairs and "normal" electrons. As the temperature decreases, the number of ordered pairs increases. The temperature dependence of the populations suggests that the normal electrons are in energy states separated from the energy states of the ordered pairs by a gap Δ, which is on the order of $\sim 3-5 k_B T_c$.

The critical temperature T_c has been found to vary with isotopic substitution, suggesting that the motion of atoms is intimately connected with superconductivity. The first successful theory of superconductivity was given by Bardeen, Cooper, and Schrieffer in 1957, and is called the BCS theory after these

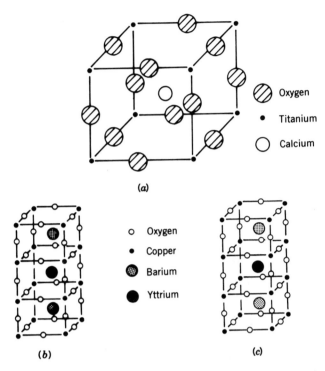

Figure 23.34 (*a*) The perovskite structure of CaTiO$_3$ has a cubic unit cell. The ionic sizes are not drawn to scale. (*b*) Idealized unit cell of the hypothetical YBa$_2$Cu$_3$O$_9$ based on a pervoskite substructure. (*c*) Structure of the 1-2-3 oxide compound YBa$_2$Cu$_3$O$_{7-x}$, obtained from electron diffraction analyses. The ionic sizes are not drawn to scale.

physicists. The theory suggests that the attractive interaction between the electrons and the lattice of metal ions can be large enough so that it overcomes the repulsion between electrons, leading to pairing of electrons (with opposite spins). These pairs then interact very weakly with lattice vibrations and so experience no friction (or scattering) as they move through the sample. The BCS theory explained the energy gap, the thermodynamic properties, and many of the electromagnetic properties, and also gave a formula for T_c in terms of parameters in the Hamiltonian of the system.

The usefulness of superconducting materials for practical purposes is decreased because of the very low temperatures at which most materials become superconducting. Until 1986, the highest known T_c was about 23 K. In that year, Bednorz and Müller, at the IBM lab in Zurich, discovered that an oxide of La, Ba, and Cu became superconducting at about 30 K. The next year another oxide was discovered to be superconducting at ~90 K, above the boiling point of liquid N$_2$ (77 K), which is a plentiful and cheap refrigerant. Suddenly, the possibility of using superconductors at reasonable temperatures became real.

The most studied of the "high-T_c" superconductors is YBa$_2$Cu$_3$O$_{7-x}$ ($x \leq$ 0.1). The nonstoichometry seems to be necessary for superconductivity. Various researchers have tried to substitute other atoms for Cu, but Cu seems to be necessary for its unusual properties. The solid-state structure of this compound is in the pervoskite family (Fig. 23.34). The nonstoichometry means that there are oxygen vacancies in the lattice, which undoubtedly play a role in the properties of the material. Speculation has focused on the sheets and chains of Cu and O atoms in the structure as being necessary for the superconducting state. Through 1990, no agreement has been reached on whether these new materials require a new theoretical model or whether they are examples of a BCS-type theory.

References

H. J. M. Bowen et al. and E. L. Sutton et al. (Suppl. Eds.), *Tables of Inter atomic Distances in Molecules and Ions.* London: Chemical Society, 1965

M. J. Buerger, *Introduction to Crystal Geometry.* New York: McGraw-Hill 1971.

J. P. Glusker and K. N. Trueblood, *Crystal Structure Analysis, A Primer* London: Oxford University Press, 1985.

T. Hahn, *International Tables for Crystallography; Brief Teaching Edition of Volume A, Space-Group Symmetry.* Dordrecht: Reidel, 1985.

A. Kelly and G. W. Groves, *Crystallography and Crystal Defects.* Reading MA: Addison-Wesley, 1970.

C. Kittel, *Introduction to Solid State Physics.* New York: Wiley, 1976.

M. F. C. Ladd and R. A. Palmer, *Structure Determination by X-Ray Crystal lography,* 2d ed. New York: Plenum, 1985.

D. E. Sands, *Introduction of Crystallography.* New York: Benjamin, 1969.

P. Wilkes, *Solid State Theory in Metallurgy.* Cambridge, UK: Cambridge University Press, 1973.

M. M. Woolfson, *An Introduction to X-Ray Crystallography,* Cambridge, UK Cambridge University Press, 1970.

Problems

23.1   What is the equation for the distances between 110 planes for a crystal with mutually perpendicular axes?

23.2   Calculate the angles at which the first-, second-, and third-order reflections are obtained from planes 500 pm apart, using X-rays with a wavelength of 100 pm.

23.3   Calculate the structure factor for a cubic unit cell of AB in which the B atoms occupy the body-centered position. Which reflections will be strong and which weak?

23.4   The crystal unit cell of magnesium oxide is a cube 420 pm on an edge. The structure is interpenetrating face-centered. What is the density of crystalline MgO?

23.5   Platinum forms face-centered cubic crystals. If the radius of a platinum atom is 139 pm, what is the length of the side of the unit cell? What is the density of the crystal?

23.6   Tungsten forms body-centered cubic crystals. From the fact that the density of tungsten is 19.3 g cm^{-3}, calculate (a) the length of the side of this unit cell, and (b) d_{200}, d_{110} and d_{222}.

23.7   Copper forms cubic crystals. When an X-ray powder pattern of crystalline copper is taken using X-rays from a copper target (the wavelength of the K_α line is 154.05 pm) reflections are found at $\theta = 21.65°$, 25.21°, 37.06°, 44.96° 47.58°, and other larger angles. (a) What type of lattice is formed by copper? (b) What is the length of a side of the unit cell at this temperature? (c) What is the density of copper?

23.8   (a) Metallic iron at 20 °C is studied by the Bragg method, in which the crystal is oriented so that a reflection is obtained from the planes parallel to the sides of the cubic crystal, then from planes cutting diagonally through opposite edges, and finally from planes cutting diagonally through opposite corners. Reflections are first obtained at $\theta = 11°\ 36'$, $8°\ 3'$, and $20°\ 26'$, respectively. What type of cubic lattice does iron have at 20 °C? (b) Metallic iron also forms cubic crystals at 1100 °C, but the reflections determined as described in (a) occur at $\theta = 9°\ 8'$, $12°\ 57'$, and $7°\ 55'$, respectively. What type of cubic lattice does iron have at 1100 °C? (c) The density of iron at 20 °C is 7.86 g cm^{-3}. What is the length of the side of the unit cell at 20 °C? (d) What is the wavelength of the X-rays used? (e) What is the density of iron at 1100 °C?

23.9 Cesium chloride, bromide, and iodide form interpenetrating simple cubic crystals instead of interpenetrating face-centered cubic crystals like the other alkali halides. The length of the side of the unit cell of CsCl is 412.1 pm. (a) What is the density? (b) Calculate the ion radius of Cs^{+}, assuming that the ions touch along a diagonal through the unit cell and that the ion radius of Cl^{-} is 181 pm.

23.10 Deslattes et al. (*Phys. Rev. Lett.* **33**:463 (1974)) found the following values for a single crystal of very pure silicon at 25 °C: $\rho = 2.328\ 992$ g cm^{-3}, $a = 543.1066$ pm. Silicon has a face-centered cubic lattice like diamond. The atomic mass is 28.085 41 g mol^{-1}. What value of Avogadro's constant is obtained from these values?

23.11 Insulin forms crystals of the orthorhombic type with unit-cell dimensions of $13.0 \times 7.48 \times 3.09$ nm. If the density of the crystal is 1.315 g cm^{-3} and there are six insulin molecules per unit cell, what is the molar mass of the protein insulin?

23.12 Molybdenum forms body-centered cubic crystals and, at 20 °C, the density is 10.3 g cm^{-3}. Calculate the distance between the centers of the nearest molybdenum atoms.

23.13 Si has a face-centered cubic structure with two atoms per lattice point, just like diamond. At 25 °C, $a = 543.1$ pm. What is the density of silicon?

23.14 The common form of ice has a tetrahedral structure with protons located on the lines between oxygen atoms. A given proton is closer to one oxygen atom than the other and is said to belong to the closer oxygen atom. How many different orientations of a water molecule in space are possible in this lattice?

23.15 The diamond has a face-centered cubic crystal lattice, and there are eight atoms in a unit cell. Its density is 3.51 g cm^{-3}. Calculate the first six angles at which reflections would be obtained using an X-ray beam of wavelength 7.12 pm.

23.16 Calculate the ratio of the radii of small and large spheres for which the small spheres will just fit into octahedral sites in a close-packed structure of the large spheres.

23.17 A close-packed structure of uniform spheres has a cubic unit cell with a side of 800 pm. What is the radius of the spherical molecule?

23.18 Titanium forms hexagonal close-packed crystals. Given the atomic radius of 146 pm, what are the unit cell dimensions and what is the density of crystals?

23.19 What neutron energy in electron volts is required for a wavelength of 100 pm?

23.20 A solution of carbon in face-centered cubic iron has a density of 8.105 g cm^{-3} and a unit cell edge of 358.3 pm. Are the carbon atoms interstitial, or do they substitute for iron atoms in the lattice? What is the weight percent carbon?

23.21 At 550 °C the conductivity of solid NaCl is 2×10^{-4} Ω^{-1} m^{-1}. Since the sodium ions are smaller than the chloride ions (see Table 23.2), they are responsible for most of the electric conductivity. What is the ionic mobility of Na^{+} under these conditions?

23.22 What fraction of the lattice sites of a crystal are vacant at 300 K if the energy required to move an atom from a lattice site in the crystal to a lattice site on the surface is 1 eV? At 100 K?

23.23 The only metal that crystallizes in a primitive cubic lattice is polonium, which has a unit cell side of 334.5 pm. What are the perpendicular distances between planes with indices (110), (111), (210) and (211)?

23.24 Calculate the highest-order diffraction line that can be observed for the 100 planes of NaCl using an X-ray tube with a copper target ($\lambda = 154$ pm).

23.25 For a C lattice (Fig. 23.5), what is the expression for the structure factor? Derive the rule for absent reflections.

23.26 The density of platinum is 21.45 g cm^{-3} at 20 °C. Given the fact that the crystal is face-centered cubic, calculate the length of the side of the unit cell.

23.27 The X-ray powder pattern for molybdenum has reflections at $\theta = 20.25°, 29.30°, 36.82°, 43.81°, 50.69°, 58.00°, 66.30°$, and other larger angles when Cu Kα X-rays are used ($\lambda = 154.05$ pm). (a) What type of cubic crystal is formed by molybdenum? (b) What is the length of a side of the unit cell at this temperature? (c) What is the density of molybdenum?

23.28 A substance forms face-centered cubic crystals. Its density is 1.984 g cm^{-3}, and the length of the edge of the unit is 630 pm. Calculate the molar mass.

23.29 Potassium bromide has a face-centered cubic lattice, and the edge of the unit cell is 654 pm. What is the density of the crystal?

23.30 Calculate the density of a diamond from the fact that it has a face-centered cubic structure with two atoms per lattice point and a unit cell edge of 356.7 pm.

23.31 Tantalum crystallizes with a body-centered cubic lattice. Its density is 17.00 g cm^{-3}. (*a*) How many atoms of tantalum are there in a unit cell? (*b*) What is the length of a unit cell? (*c*) What is the distance between (200) planes? (*d*) What is the distance between (110) planes? (*e*) What is the distance between (222) planes?

23.32 Iron crystallizes in body-centered cubic packing at room temperature. Since the density is 7.88 g cm^{-3} at 25 °C, what is the length of the side of the unit cell at this temperature? What is the radius of an iron atom in this crystalline form?

23.33 Cobalt has a hexagonal close-packed structure with $a = 250.7$ pm. What is its density?

23.34 Aluminum forms face-centered cubic crystals, and the length of the side of the unit cell is 405 pm at 25 °C. Calculate (*a*) the density of aluminum at this temperature, and (*b*) the distances between (200), (220), and (111) planes.

23.35 From the fact that the length of the side of a unit cell for lithium is 351 pm, calculate the atomic radius of Li. Lithium forms body-centered cubic crystals.

23.36 Platinum forms face-centered cubic crystals, and the length of the side of the unit cell is 393 pm. What is the number of atoms per cm^2 of surface on (100), (110), and (111) planes?

23.37 If spherical molecules of 500 pm radius are packed in cubic close packing and in body-centered cubic, what are the lengths of the side of the cubic unit in cells in the two cases?

23.38 What is the de Broglie wavelength of thermal neutrons at 200 °C?

23.39 Metallic sodium forms a body-centered cubic unit cell with $a = 424$ pm. What is the sodium atom radius?

23.40 What concentration of donor atoms in germanium is required to produce an electric conductivity of 10^2 Ω^{-1} m^{-1}? Assume that all donor atoms are ionized and calculate the fraction of the atoms that must be donor atoms. The mobility of electrons in germanium is 0.39 m^2 V^{-1} s^{-1}.

23.41 Silicon is doped with aluminum to the extent of 10^{-8} of the atoms. The conductivity κ is 4.0 Ω^{-1} m^{-1}. What is the mobility of holes? (The density of silicon is 2.4 g cm^{-3}.)

24
Surface Dynamics

Previous chapters have been largely concerned with the properties of matter under conditions where surface effects are negligible. Now we shall consider the equilibrium and dynamics of processes that occur at the interface between a solid and a gas. When a molecule strikes a solid surface, it may rebound elastically or inelastically, undergo a reaction, or be adsorbed. If it is adsorbed, it may diffuse around on the surface, remain fixed, or dissolve in the bulk phase, but we will concentrate on the processes that occur on the surface. An adsorbed molecule may dissociate on the surface or react with another molecule on the surface. If a chemical reaction occurs on the surface, the products may desorb into the gas phase. The use of solid surfaces as catalysts in chemical technology is of tremendous practical importance.

24.1 Physisorption and Chemisorption

In the preceding chapter, we have seen that a crystal can have a number of faces with different Miller indices. Faces with low Miller indices and high densities of atoms or molecules are more likely to be observed because in general they are more stable than surfaces with high Miller indices. The various faces of a crystal have different properties because of their different structures. In a Section 24.9, we will see that the surface of even a pure atomic solid can be more complicated than the flat surfaces illustrated in Chapter 23, where the positions of surface atoms are simple extensions of the bulk structure. Even simple surfaces may have **defects** of the types shown in Fig. 24.1. These defects have different adsorptive and catalytic properties. As a simplification we will largely ignore defects.

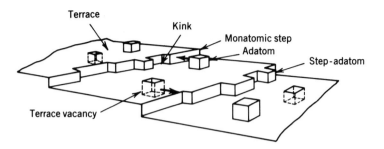

Figure 24.1 Defects in the form of steps, kinks, adatoms, and vacancies on a solid surface. (Reprinted from G. A. Somorjai, *Chemistry in Two Dimensions: Surfaces,* Copyright © 1981 by Cornell University. Used by permission of the publisher, Cornell University Press, Ithaca, NY.)

When a molecule approaches a surface, it encounters a net attractive potential that is similar to the potential between two molecules (Section 12.2) and arises for the same reasons. However, a gas molecule near a surface is attracted not by a single molecule but by several closely spaced surface atoms. The adsorption of a molecule on a solid surface is always an exothermic process. If we represent the gas molecule, the adsorbate, by A and the adsorption site on the surface by S, the process of **adsorption** can be represented as a chemical reaction:

$$A + S = AS \qquad \Delta_{ads}H < 0 \tag{24.1}$$

In Section 24.7 we will see that this process may involve more than one step, but here we simply want to establish the definition of the enthalpy of adsorption $\Delta_{ads}H$.

The surface of an atomic solid has about 10^{15} atoms per square centimeter of surface. If we assume that one molecule adsorbs on each atom of the solid surface, then there are 10^{15} sites on which molecules can adsorb. When one molecule is adsorbed on each site, the surface is said to be covered with a monolayer. Kinetic theory makes it possible to calculate the maximum rate at which surface sites can be occupied by gas molecules. In Section 18.5 we saw that the **flux** J_N of molecules of an ideal gas through a hole is given by

$$J_N = \frac{PN_A}{(2\pi MRT)^{1/2}} \tag{24.2}$$

The same equation applies to collisions with a surface, and, when SI units are used, J_N is the number of molecules striking the surface per square meter per second. If every molecule striking a clean surface is adsorbed and exactly one gas molecule is adsorbed per surface site, this equation provides the means for calculating the time for a surface to become covered with a monolayer when it is exposed to a gas with molar mass M at a specific temperature and pressure.

Example 24.1
How many molecules of oxygen strike 1 cm^2 of surface in 1 s when the pressure is 10^{-6} torr and the temperature is 298 K? (1 torr $= \frac{1}{760}$ atm $= (\frac{1}{760}$ atm) (1.01325 × 10^5 Pa atm^{-1}) = 133.3 Pa.) Using equation 24.2,

$$J_N = \frac{(133.3 \times 10^{-6} \text{ Pa})(6.022 \times 10^{23} \text{ mol}^{-1})}{[2\pi(32 \times 10^{-3} \text{ kg mol}^{-1})(8.314 \text{ J K}^{-1} \text{ mol}^{-1})(298 \text{ K})]^{1/2}}$$

$$= (3.60 \times 10^{18} \text{ m}^{-2} \text{ s}^{-1})(0.01 \text{ m cm}^{-1})^2$$

$$= 3.60 \times 10^{14} \text{ cm}^{-2} \text{ s}^{-1}$$

For a surface with 10^{15} sites per cm^2, the exposure of a clean surface to oxygen at 10^{-6} torr for 1 s is sufficient to form 36% of a monolayer, if every molecule sticks. To express the exposure of a surface to a gas that is adsorbed, surface scientists have developed the unit 10^{-6} torr s, which is called the **langmuir.** Thus, exposure of the surface to 1 langmuir of oxygen results in 36% of a monolayer of adsorbed O_2 at 298 K.

By use of calculations of this type it is readily shown that to keep a surface clean even for a few minutes, it is necessary to evacuate the chamber containing the sample surface to a pressure less than 10^{-8} torr. Thus, surfaces that we encounter daily are always covered with adsorbed molecules. To study surfaces, special vacuum chambers are used; the term ultrahigh vacuum is used to refer to pressures less than 5×10^{-10} torr. By directing a molecular beam (Section 20.6) on crystallographically distinct surfaces, it is possible to determine the rate of filling surface sites as a function of beam energy and angle.

It is convenient to distinguish between **physical adsorption** and **chemisorption.** The forces causing physical adsorption are of the same type as those that cause the condensation of a gas to form a liquid and are generally referred to as van der Waals forces. The heat evolved in a physisorption process is of the order of magnitude of the heat evolved in the process of condensing the gas, and the amount adsorbed may correspond to several monolayers at a high pressure. The extent of physisorption is smaller at higher temperatures.

Chemisorption involves the formation of chemical bonds. However, it is usually not possible to make a sharp distinction between these two kinds of adsorption, except to say that the enthalpy change in chemisorption is much larger than for physical adsorption, lying in the range 40 to 200 kJ mol^{-1}.

Physical adsorption and chemisorption may often be distinguished by the rates with which these processes occur. Equilibrium in physical adsorption is generally achieved rapidly and is readily reversible. Physical adsorption is reversed by lowering the pressure of the gas or raising the temperature of the surface. Chemisorption, on the other hand, may not occur at an appreciable rate at low temperatures if the chemisorption reaction has an activation energy (Section 19.9). In this case, the rate of chemisorption increases rapidly as the temperature is raised. In chemisorption the bonding may be so tight that the original species may not desorb. For example, heating a graphite surface after adsorbing atomic oxygen results in desorption of carbon monoxide.

In the following section, we consider a very simple model for adsorption presented by Langmuir.

24.2 Langmuir Adsorption Isotherm

The simplest equation for adsorption under equilibrium conditions was derived by Langmuir using kinetic theory. He considered a surface with a specific number of binding sites that are identical and can each adsorb one molecule.

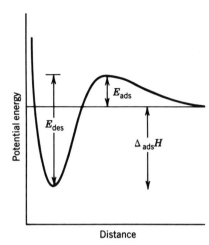

Figure 24.2 Potential energy for the interaction of a molecule with a surface assumed in the derivation of the Langmuir adsorption isotherm.

Thus, the uptake is limited to a **monolayer.** If a uniform surface has n_0 equivalent sites and n are occupied, the **surface coverage** Θ is defined by $\Theta = n/n_0$. Langmuir also assumed that binding at a site has no influence on the properties of neighboring sites; this means that the enthalpy of adsorption is independent of coverage.

To derive the expression for the Langmuir adsorption isotherm, we will set the rate of adsorption equal to the rate of desorption; in other words, the equilibrium is assumed to be dynamic. The rate of adsorption is taken to be equal to the product of the rate J_N of collisions of molecules of molar mass M with the surface, the fraction of the sites that are not covered $(1 - \Theta)$, the **sticking coefficient** s^*, and the fraction $\exp(-E_{ads}/RT)$ of the molecules with the **activation energy for adsorption** E_{ads}. As the molecule approaches the surface it usually encounters a potential barrier with a height E_{ads} that has to be overcome before the molecule can be adsorbed. The sticking coefficient s^* is the fraction of the molecules with energy in excess of the activation energy E_{ads} that actually stick. Thus,

$$\text{rate of adsorption} = \frac{PN_A}{(2\pi MRT)^{1/2}} (1 - \Theta)s^* \exp\left(-\frac{E_{ads}}{RT}\right) \quad (24.3)$$

This equation for the rate of adsorption of molecules on a surface is based on the simple view of the potential for the interaction between a molecule and the surface that is shown in Fig. 24.2; more details are given in Section 24.7.

For an adsorbed molecule to be desorbed, it has to overcome a **potential barrier** E_{des}, as shown in Fig. 24.2. The rate of desorption is taken to be equal to the product of the specific rate constant for desorption k_d, the fraction Θ of the sites that are occupied, and the fraction of the molecules with the activation energy E_{des} for desorption. Thus,

$$\text{rate of desorption} = k_d \Theta \exp\left(-\frac{E_{des}}{RT}\right) \quad (24.4)$$

At equilibrium, the rate of adsorption is equal to the rate of desorption, and so the pressure is given by

$$P = \frac{(2\pi MRT)^{1/2} k_d \Theta}{N_A s^* (1 - \Theta)} \exp\left(\frac{\Delta_{ads}H}{RT}\right) \quad (24.5)$$

where $\Delta_{ads}H$ is the enthalpy of adsorption ($E_{ads} - E_{des}$). The enthalpy of adsorption is negative, and so $E_{ads} < E_{des}$. The Langmuir adsorption isotherm can be written as

$$P = \frac{\Theta}{K(1 - \Theta)} \quad \text{or} \quad \Theta = \frac{KP}{1 + KP} \quad (24.6)$$

where the constant K is given by

$$K = \frac{N_A s^* \exp(-\Delta_{ads}H/RT)}{k_d (2\pi MRT)^{1/2}} \quad (24.7)$$

The enthalpy of adsorption introduced in equation 24.1 and appearing in the Langmuir adsorption isotherm (equation 24.5) can be determined from measurements of equilibrium pressures at two or more temperatures and at the same surface coverage Θ. Adsorption isotherms following the Langmuir equation are shown for two temperatures in Fig. 24.3. If we write equation 24.5 for

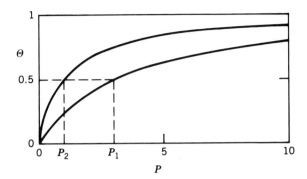

Figure 24.3 Adsorption isotherms at two temperatures for an adsorption following the Langmuir adsorption isotherm.

the equilibrium pressures at T_1 and T_2, and ignore the $T^{1/2}$ dependence in comparison with the exponential dependence involving $\Delta_{ads}H$, we can readily derive

$$\left(\ln \frac{P_1}{P_2}\right)_\Theta = \frac{(\Delta_{ads}H/R)(T_2 - T_1)}{T_1 T_2} \qquad (24.8)$$

Note that this is very nearly the Clausius–Clapeyron equation (Section 6.5), but the enthalpy of adsorption has the opposite sign from the enthalpy of vaporization; $\Delta_{vap}H$ applies to the process liquid $\rightarrow$ vapor, while $\Delta_{ads}H$ applies to the process $A + S \rightarrow AS$ (equation 24.1).

Example 24.2
Use the Langmuir method to derive expressions for the fractions Θ_A and Θ_B of a surface covered by adsorbed molecules A and B, assuming that the molecules compete for the same sites.
In order to discuss the competition between A and B, it is convenient to write equation 24.6 as

$$r_A \Theta_A = k_A(1 - \Theta_A)P_A$$

for pure A and

$$r_B \Theta_B = k_B(1 - \Theta_B)P_B$$

for pure B. When gases A and B are both present, these equations become

$$r_A \Theta_A = k_A(1 - \Theta_A - \Theta_B)P_A$$
$$r_B \Theta_B = k_B(1 - \Theta_A - \Theta_B)P_B$$

Solving these two simultaneous equations for Θ_A and Θ_B yields

$$\Theta_A = \frac{(k_A/r_A)P_A}{1 + (k_A/r_A)P_A + (k_B/r_B)P_B} = \frac{K_A P_A}{1 + K_A P_A + K_B P_B}$$

$$\Theta_B = \frac{(k_A/r_B)P_B}{1 + (k_A/r_A)P_A + (k_B/r_B)P_B} = \frac{K_B P_B}{1 + K_A P_A + K_B P_B}$$

where $K_A = k_A/r_A$ and $K_B = k_B/r_B$.

Example 24.3

If a molecule dissociates on being adsorbed, the process is referred to as dissociative adsorption. Derive the Langmuir adsorption isotherm for dissociative adsorption.

When dissociation occurs on the surface, two sites are required, and so the probability of sticking is proportional to the pressure and to the availability of adjacent sites, $k(1 - \Theta)^2 P$. The probability of desorption is proportional to the probability that adjacent sites are occupied, $r\Theta^2$, since recombination has to take place on the surface prior to desorption. At equilibrium,

$$r\Theta^2 = k(1 - \Theta)^2 P$$

$$\frac{\Theta}{1 - \Theta} = \frac{k^{1/2}P^{1/2}}{r^{1/2}}$$

$$\Theta = \frac{KP^{1/2}}{1 + KP^{1/2}}$$

where $K = (k/r)^{1/2}$.

24.3 Use of Adsorption Measurements to Determine Surface Area

The adsorption of a gas by a solid can be determined by admitting known quantities of a gas into a chamber and measuring the volume and pressure of the gas at equilibrium. If a monolayer is formed, the fractional surface coverage Θ is equal to the ratio of the volume v of the gas adsorbed to the volume v_m required to form a monolayer; that is, $\Theta = v/v_m$, where the volumes are at the same temperature and pressure. Thus, equation 24.6 can be written

$$v = \frac{v_m}{1 + 1/KP} \tag{24.9}$$

If the experimental data follow the Langmuir equation, the parameters v_m and K can be determined from the slope and intercept of the plot of $1/v$ versus $1/P$, which is linear if equation 24.9 is followed. However, in general, there are deviations from the Langmuir equation, especially at higher pressures. If the gas being adsorbed is below its critical point, it is customary to plot the amount adsorbed per gram of adsorbent versus P/P^*, where P^* is the vapor pressure of the bulk liquid adsorbate at the temperature of the experiment. Figure 24.4 shows the shapes of plots of the ratio of the volume v of gas adsorbed (under standard conditions) to the volume v_m required to form a monolayer versus the pressure, specifically P/P^*. These curves have been calculated using the BET theory, which has an adjustable constant c, but, before discussing this theory, we need to reconsider the assumptions of the Langmuir theory.

The derivation of the Langmuir adsorption isotherm involves five implicit assumptions: (1) the gas is ideal, (2) the adsorbed gas is confined to a monolayer, (3) the $\Delta_{ads}H$ of each binding site is the same, (4) there is no lateral interaction between adsorbate molecules, and (5) the adsorbed gas molecules stay at the position where they first collide with the surface. Although the first assumption is good at low pressure, it is not valid if the pressure approaches the critical

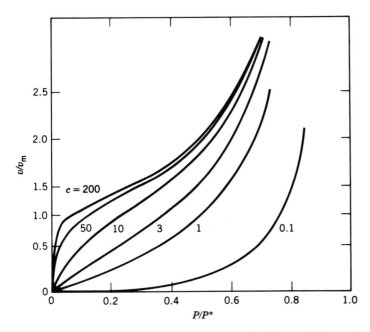

Figure 24.4 Shapes of adsorption isotherms obtained when multilayer adsorption of a gas occurs. These curves were calculated using the BET theory (equation 24.10) with various values of the constant *c*. (From A. W. Adamson, *Physical Chemistry.* New York: Wiley, 1982, with permission.)

pressure. The second assumption fails when there is adsorption on top of the monolayer. The third assumption is poor because $\Delta_{ads}H$ is different on different kinds of binding sites. The incorrectness of the fourth assumption was first shown experimentally when it was found that in certain experimental cases, the enthalpy of adsorption may increase with the amount adsorbed. This effect is caused by lateral attractions between adsorbed molecules. The fifth assumption is incorrect because there is much evidence that adsorbed molecules can be mobile. In spite of these problems, the Langmuir adsorption isotherm is very useful at low pressures where only monolayer adsorption is involved.

We now return to the increase in physical adsorption beyond that required to form a monolayer, which is observed at high pressures. This is referred to as **multilayer adsorption,** and Brunauer, Emmett, and Teller (BET) developed a theory for it. They assumed that the surface possesses uniform, localized sites and that adsorption on one site does not affect adsorption on neighboring sites, just as in the Langmuir theory. It is further assumed that molecules can be adsorbed in second, third, . . . , and *n*th layers with the surface area available for the *n*th layer equal to the coverage of the $(n - 1)$th layer. The energy of adsorption of the first layer is E_1, and the energy of adsorption in succeeding layers is assumed to be E_L, the energy of liquefaction of the gas. By use of these assumptions it is possible to derive the following equation for the ratio of the volume v of gas adsorbed (under standard conditions) to the volume required to form a monolayer:

$$\frac{v}{v_m} = \frac{cx}{(1 - x)[1 + (c - 1)x]} \qquad x = \frac{P}{P^*} \qquad (24.10)$$

This equation can be written in the form

$$\frac{x}{v(1 - x)} = \frac{1}{cv_m} + \frac{(c - 1)x}{cv_m} \tag{24.11}$$

and so it is possible to determine the parameters v_m and c from a plot of $x/v(1 - x)$ versus x. The value of v_m, the volume of gas required to form a monolayer, is of considerable interest because it makes it possible to calculate the surface area of a porous solid. The surface area occupied by a single molecule of adsorbate on the surface can be estimated from the density of the liquified adsorbate. For example, the area occupied by a nitrogen molecule at -195 °C is estimated to be 16.2×10^{-20} m^2 on the assumption that the molecules are spherical and that they are close packed in the liquid. Thus, from the value of v_m obtained from the BET theory, if multilayer adsorption is involved, the surface area of the adsorbant can be calculated. The surface areas of porous solids can be as large as several hundred square meters per gram of solid.

24.4   Low-Energy Electron Diffraction (LEED)

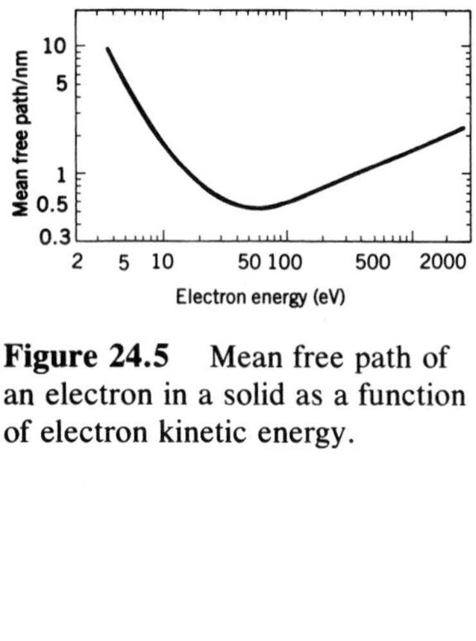

Figure 24.5   Mean free path of an electron in a solid as a function of electron kinetic energy.

Many recent advances in surface science have been made using methods that are surface sensitive; that is, they are sensitive only to the outermost atomic layers of the bulk solid. These methods, which are discussed in this section and the next, largely employ **low-energy electrons.** Electrons with energies in the range of 10 to 250 eV are sensitive to the surface because they have a low mean free path in solids. They penetrate the surface about 0.5 to 2 nm, which is about 1 to 4 atomic layers. Figure 24.5 shows the mean free path of an electron in a solid as a function of its energy. The mean free path for electrons of a given energy is nearly independent of the atomic weight of the element making up the solid, and so this curve is often called the "universal curve." This curve also shows us that in the emission of electrons with energies in the range 10 to 250 eV from solids, the electrons must originate from the top few atomic layers. Therefore, the electron spectroscopies discussed in the next section, which depend on emission from the solid, also provide information on the surface layer, rather than bulk properties.

When a surface is bombarded with electrons, the electrons may be scattered elastically (that is, no energy is lost) or inelastically. Electrons scattered elastically will be diffracted if their de Broglie wavelength is small enough. **Low-energy electron diffraction** (LEED) therefore provides a means for studying the atomic geometry of a surface. In a LEED apparatus the electrons are accelerated in an electron gun, and strike the surface normally, as shown in Fig. 24.6. The surface backscatters a portion of the electrons. The grids in front of the screen are used to remove the inelastically backscattered electrons, while the elastically backscattered electrons are post-accelerated onto the phosphorescent screen for viewing the diffraction pattern. The presence of sharp diffraction spots indicates that the surface is ordered on an atomic scale.

It is evident from the Bragg equation (Section 23.4) that for there to be diffraction, the de Broglie wavelength of the electrons must be less than $2d$, where d is the distance between atomic planes. The **de Broglie wavelenth** λ of electrons accelerated through a potential difference ϕ is given by

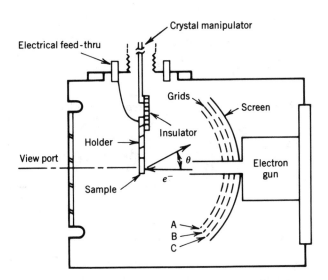

Figure 24.6 Schematic diagram of a LEED apparatus. (From D. G. Castner and G. A. Somorjai, *Chem. Rev.* **79**:233 (1979). Copyright © 1979 by the American Chemical Society)

$$\lambda/\text{nm} = \left(\frac{1.504 \text{ V}}{\phi}\right)^{1/2} \qquad (24.12)$$

As in X-ray diffraction from crystals, the angles at which X-rays are scattered gives information about the symmetry and type of surface unit cell, but intensity measurements are required to obtain atomic coordinates. For some crystals the surface structure is quite different from that of the bulk crystal (Section 24.9). LEED can also be used to determine the structure and order of an adsorbed layer. For example, the structural model for CO adsorbed on palladium (100), as determined by LEED, is shown in Fig. 24.7. The CO is bound at an interstitial site with the carbon atom down; this is referred to as a bridge-bonded site.

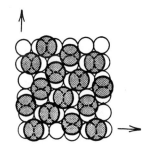

Example 24.4
Derive equation 24.12 and calculate the minimum accelerating potential required to obtain electron diffraction from a crystal with an interplanar spacing of 0.1227 nm. The de Broglie wavelength must, therefore, be less than 0.2454 nm.

From Example 11.2, the momentum p of an electron with energy E is given by

$$p = (2mE)^{1/2} = \frac{h}{\lambda}$$

or

$$\lambda = \left(\frac{h^2}{2mE}\right)^{1/2}$$

Since the energy of an electron that has been accelerated through a potential difference ϕ is $E = e\phi$, the de Broglie wavelength is given by

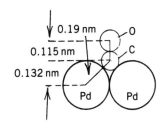

Figure 24.7 Structural model for CO adsorbed on palladium (100). The shaded circles in the top view are CO molecules. (From G. Ertl, *Pure Appl. Chem.* **52**:2051 (1980).)

$$\lambda = \left(\frac{h^2}{2me\phi}\right)^{1/2}$$

$$= \left(\frac{1.504 \times 10^{-18} \text{ V m}^2}{\phi}\right)^{1/2}$$

which is equation 24.12. For the crystal considered, the minimum accelerating potential is therefore

$$\phi = \frac{1.504 \times 10^{-18} \text{ V m}^2}{(0.2454 \times 10^{-9} \text{ m})^2} = 24.97 \text{ V}$$

24.5   Electron Emission from Surfaces

When photons are used to eject electrons the technique is called **photoelectron spectroscopy** (PES). When these photons are in the ultraviolet region, the abbreviation UPS is used, and when the photons are in the X-ray region, the abbreviation XPS is used. We have already discussed the use of ultraviolet photoelectron spectroscopy (UPS) of gas molecules in Section 15.11; this is the most direct method for determining the ionization potential I of a molecule and it provides additional information about molecular electronic structure. The kinetic energy of the ejected electron E_{el} is given by

$$E_{el} = h\nu - I - E_v - E_r \tag{24.13}$$

where I is the **adiabatic ionization potential,** the energy required to remove an electron from the molecule in its ground vibrational and rotational state to produce a molecular ion in its ground vibrational and rotational state. Since the ionization process may leave the molecular ion in an excited vibrational or rotational state, the vibrational energy E_v and rotational energy E_r above the ground state must also be subtracted from the energy of the incident photon.

Photons with energies in the ultraviolet range eject electrons from the valence orbitals of an adsorbed molecule or the valence bands of a solid (see Section 23.14), as shown in Fig. 24.8a. Electrons at these energies do not provide enough energy to eject electrons from atom cores.

UPS experiments are usually carried out with the He I line (21.2 eV) or the He II line (40.8 eV). Ultraviolet photon spectroscopy of a molecule physisorbed on a solid is a superposition of the ultraviolet photon spectrum of the valence band of the solid and the gas-phase molecular orbitals of the adsorbed molecule. Since the energy spread of the valence band of a solid is about 10 eV, the electrons ejected from the solid have energies spread over a similar range. The ultraviolet photon spectrum of the physisorbed molecule is quite similar to the corresponding gas molecule because the interaction with the surface is weak. But chemisorption affects valence orbitals of the molecule and the valence band of the solid, and so the sprectrum is complicated and reflects the type of bonding between the molecule and the surface. Important structural and chemical bonding information about chemisorbed species can be obtained from UPS.

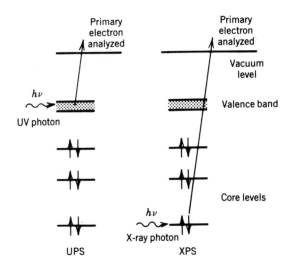

Figure 24.8 (a) In ultraviolet photoelectron spectroscopy (UPS) the energies of electrons ejected from the valence band are determined. (b) In X-ray photoelectron spectroscopy (XPS) the energies of electrons ejected from core levels are determined.

In **X-ray photoelectron spectroscopy** (XPS) core electrons are ejected from the metal and from adsorbed molecules, as illustrated in Fig. 24.8b. The usual sources of X-rays are Mg $K\alpha$ (1253.6 eV) or Al $K\alpha$ (1486.6 eV). Since the energies of atomic core levels are characteristic of each element, XPS can be used to obtain an elemental analysis of the surface. This type of spectroscopy is referred to as ESCA (electron spectroscopy for chemical analysis) because the relative amounts of elements in a sample can be determined. The ionization potential of an atomic core level depends to some extent on the chemical environment of the atom, so that information can also be obtained about the type of bonding between the adsorbate and the surface.

When a surface is irradiated with soft X-rays, the Auger effect may occur, and the energies of the secondary electrons can be determined. This effect, which can also be produced by bombarding with high-energy electrons, is the basis of another type of surface spectroscopy referred to as **Auger electron spectroscopy**, AES. In AES, an electron is ejected by an X-ray photon, as in XPS, but the emission of the secondary electrons is analyzed, rather than the primary electrons. When a core electron is ejected by absorption of an X-ray photon (or high-energy electron), as shown in Fig. 24.9a, an electron from a higher energy level may drop into the core vacancy. The energy liberated in this way can lead to the emission of a second, or Auger, electron, as illustrated in Fig. 24.8b. The energies of the Auger electrons are characteristic of the core levels, and so information about the atoms present and their bonding is obtained. Note that the energy of the Auger electron is independent of the energy of the exciting radiation. Auger electron spectroscopy is often used to measure the coverage of adsorbed species and in checking the cleanliness of a surface. AES can be studied using the same electron optics and collector as used in a LEED experiment.

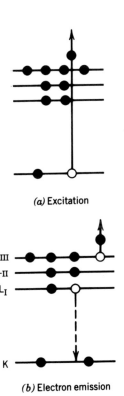

(a) Excitation

(b) Electron emission

Figure 24.9 Emission of an Auger electron with an energy equal to the difference in energy between levels L_I and L_{III}. (a) An electron in the K level is ejected by an X-ray photon. (b) An electron falls from the L_I level to the K level, and, as a consequence, an electron is ejected from the L_{III} level.

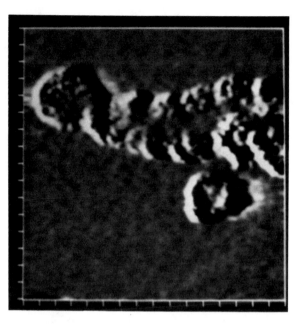

Figure 24.10 Direct observation of native DNA structures with the scanning tunneling microscope. (From T. P. Beebe et al., *Science* **243**:370 (1989). Copyright © 1989 by the AAAS.)

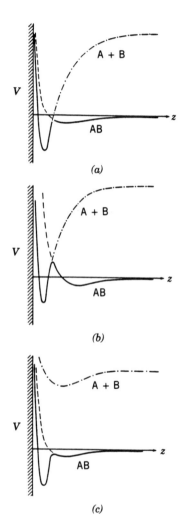

Figure 24.11 Schematic diagrams of an absorbate/substrate complex for three different ground-state configurations: (*a*) dissociative chemisorption, (*b*) molecular physisorption, and (*c*) molecular chemisorption. (From A. Zangwill, *Physics at Surfaces*. Cambridge, UK: Cambridge University Press, 1988.)

24.6 Scanning Tunneling Microscopy (STM)

The 1986 Nobel physics prize was given in part to G. Binnig and H. Rohrer for their development of this revolutionary microscope which makes it possible to ''see'' a single molecule adsorbed on a surface. A very sharp metal tip is moved over the surface of an electrical conductor at a height of about 500 pm. The tip is so close to the surface that there is an overlap of the wavefunctions of the atoms of the tip with those of the surface. If a potential difference is applied between the tip and the surface, quantum mechanical tunneling (Section 21.7) permits a current to flow through the vacuum gap. The potential of the tip with respect to the surface is held constant, and a piezoelectric feedback mechanism regulates the vertical motion of the tip so that the tunneling current is kept constant. Thus, the tip traces the surface topography as it is moved over the surface. These traces can be used to produce an image of the surface. Figure 24.10 is an image of native DNA obtained with a scanning tunneling microscope.

24.7 Theory of Surface Reactions

According to the Langmuir theory, a molecule is either bound to the surface or not. However, it is found experimentally that a molecule may initially be bound in a precursor state, which is much like a physisorbed state, and then later makes a transition to a chemisorbed state. Figure 24.11 gives schematic diagrams of the potential energy as a function of distance perpendicular to the surface for three cases. In Fig. 24.11a, an AB molecule is physisorbed on the surface but rapidly passes over the low potential barrier to dissociate to form a state in which atoms are chemisorbed; this is called **dissociative chemisorption.** If atoms existed in the gas phase, rather than AB molecules, the potential energy of the system would be given by the —·— line. The continuation of the potential energy curve for physisorbed AB molecules is shown as a dashed line. In Fig. 24.11b, the potential barrier between physisorbed AB molecules and chemisorbed atoms is much higher, so that only physisorption occurs at low temperatures. At higher temperatures, AB molecules pass over the barrier, spontaneously dissociate, and the atoms are chemisorbed. In this case, the physisorbed state is a **precursor state.** In Fig. 24.11c, a very low potential barrier separates the physisorbed state from a state of molecular chemisorption.

Since encounters of gaseous species with a solid surface can be considered to be a chemical reaction, the concepts developed earlier for the dynamics of chemical reactions in the gas phase can be used. Figure 24.12, which shows potentials for two types of surface reaction, is useful for thinking about the dynamics of surface reactions. If the incident particle has a very high kinetic energy, it will probably scatter back into the gas phase, as shown in Fig. 24.12a. If the incident particle has less energy, it can lose some energy to the surface and become trapped in an adsorption well, as also shown in Fig. 24.12a. If there is a barrier to adsorption, incident particles with low energy cannot be bound, but particles with high energies can be, as shown in Fig. 24.12b.

24.8 Heterogeneous Catalysis

Reactions that occur in a single phase are referred to as homogeneous reactions, and reactions that occur on a surface are referred to as heterogeneous reactions. The chemistry of a reaction on a surface is quite different from the chemistry in a gas or liquid phase, and many reactions can be carried out at lower temperatures in this way. Solid-state catalysts are used in many large-scale processes used in the chemical industry. The production of NH_3 from H_2 and N_2 is catalyzed by iron promoted with Al_2O_3 and K. The oxidation of NH_3 to NO in the production of HNO_3 is catalyzed by Pt-Rh. Enormous quantities of solid catalysts are used in the petroleum industry for "cracking," which is necessary to produce gasoline from petroleum, and for "reforming," which causes the rearrangement of the molecular structures and raises the octane rating of gasoline. To facilitate the regeneration of these catalysts, they may be circulated as fine particles in the gas stream.

The development of catalysts is especially important for slow exothermic reactions. A reaction of this type is $N_2 + 3H_2 = 2NH_3$, which is the basis for the Haber process for the manufacture of ammonia. Although the equilibrium constant may be favorable at room temperature, the rate may be so low that the reaction cannot be used practically. When the temperature is raised, the rate is increased, but at high temperatures the equilibrium constant may be unfavorable, as is readily surmised from Le Châtelier's principle.

The catalyst that is actually used in producing ammonia is a porous structure consisting of small Fe particles (doped with partially reduced K_2O adsorbed) interspersed with Al_2O_3. The purpose of the K_2O is to further enhance the catalytic activity of the Fe particles.

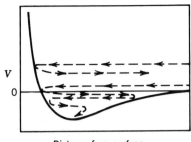

Distance from surface z

(a)

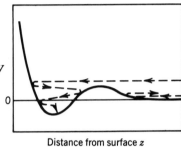

Distance from surface z

(b)

Figure 24.12 Schematic one-dimensional view of the approach of a gas particle to a solid surface: (a) a simple adsorption well; (b) a well with a barrier to adsorption. (From J. C. Tully, *Surf. Sci.* **111**:461 (1981).)

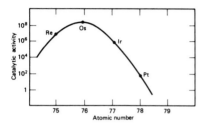

Figure 24.13 Relative specific activities for various metals as catalysts for the reaction $H_2 + C_2H_6 = 2CH_4$. (From J. H. Sinfelt, *Catal. Rev.* **9**:147 (1974), Dekker, New York.)

Substances that enhance the activity of catalysts are called **promoters,** and substances that inhibit catalytic activity are called **poisons.** Since only a fraction of the surface of the catalyst may be involved, it is easy to see how relatively small amounts of promoters and poisons may be effective. In the manufacture of sulfuric acid, the presence of a very minute amount of arsenic completely destroys the catalytic activity of the platinum catalyst by forming platinum arsenide at the surface.

The effectiveness of various metals as catalysts for a particular reaction often varies in a regular way with the position of the metal in the periodic table. For example, Fig. 24.13 shows the relative specific activities of rhenium, osmium, iridium, and platinum for the reaction $H_2 + C_2H_6 = 2CH_4$. These metals have atomic numbers 75, 76, 77, and 78. There are several possible explanations as to why catalytic activity might go through a maximum in this way. For example, significantly exothermic adsorption may be required for a significant amount of adsorption. On the other hand, if the metal binds the adsorbate too strongly it may not be an effective catalyst.

The catalysis of a reaction by a surface involves several steps: (1) diffusion of reactants to the surface, (2) adsorption, usually chemisorption, on the surface, (3) reaction on the surface, (4) desorption of products, and (5) diffusion of products away from the surface. If the third step is the slowest, the adsorption of various reactants and products will be in equilibrium. When this is the situation the Langmuir equation (equation 24.6) is often useful in deriving the rate equation for the surface catalyzed reaction, as shown in the following paragraph.

First, let us consider a reaction with a single reactant and a single product:

$$A \rightarrow P \qquad (24.14)$$

If the reaction on the surface is rate determining, the rate of the surface-catalyzed reaction is expected to be proportional to the surface area S_0 of the catalyst and the fraction θ_A of the surface covered by reactant:

$$-\frac{dn_A}{dt} = kS_0\theta_A \qquad (24.15)$$

where n_A is the number of moles of reactant in the system. If both reactant and product are adsorbed on the surface, the fraction of the surface occupied by A is given by

$$\theta_A = \frac{K_A P_A}{1 + K_A P_A + K_P P_P} \qquad (24.16)$$

according to the Langmuir adsorption isotherm, as shown in example 24.2. In this case the rate of the heterogeneous reaction is given by

$$-\frac{dn_A}{dt} = \frac{kS_0 K_A P_A}{1 + K_A P_A + K_P P_P} \qquad (24.17)$$

According to this rate equation the reaction will slow down as the product P accumulates because it competes with reactant A for the catalytic sites. When initial rates are studied for a reaction following this rate law, it is found that the rate increases linearly with P_A at low pressures and approaches a limiting value of kS_0 as P_A is increased.

Second, let us consider a reaction with two reactants and two products that are adsorbed on the same type of site:

$$A + B \rightarrow P + Q \qquad (24.18)$$

The reaction rate is assumed to be proportional to the surface area and the product of the fractions of the sites occupied by the two reactants:

$$-\frac{dn_A}{dt} = kS_0\Theta_A\Theta_B \qquad (24.19)$$

If all of the reactants and products are adsorbed, then

$$\Theta_A = \frac{K_A P_A}{1 + K_A P_A + K_B P_B + K_P P_P + K_Q P_Q} \qquad (24.20)$$

$$\Theta_B = \frac{K_B P_B}{1 + K_A P_A + K_B P_B + K_P P_P + K_Q P_Q} \qquad (24.21)$$

so that the reaction rate is

$$-\frac{dn_A}{dt} = \frac{kS_0 K_A K_B P_A P_B}{(1 + K_A P_A + K_B P_B + K_P P_P + K_Q P_Q)^2} \qquad (24.22)$$

According to this rate equation, if all partial pressures except P_A are held constant and P_A is increased, the rate will pass through a maximum.

Third, let us consider a reaction with two reactants that are adsorbed on different types of sites. Reactant A may be adsorbed on one type of site and reactant B may be adsorbed on another type of site. In the case of a metal oxide catalyst, one reactant might interact with the metal ions and the other reactant might interact with the oxide ions. If the reaction rate is proportional to the surface area and the product of the fraction Θ_{A1} of sites of type 1 occupied by A and the fraction Θ_{B2} of sites of type 2 occupied by B, then

$$-\frac{dn_A}{dt} = kS_0\Theta_{A1}\Theta_{B2}$$

$$= \frac{kS_0 K_{A1} K_{B2} P_A P_B}{(1 + K_{A1} P_A)(1 + K_{B2} P_B)} \qquad (24.23)$$

where the possible binding of product has been omitted. In contrast with rate equation 24.22, the rate in this case does not pass through a maximum when P_A is increased at constant P_B.

Many additional types of rate equations for heterogeneous reactions may be derived. The possibility that a reactant dissociates on the surface may have to be included. It is important to have the right rate equation in designing an industrial process.

24.9 Special Topic: Surface Reconstruction

When a crystalline solid is cleaved to expose a fresh surface, some "dangling" bonds are formed, and this may lead to a reconstruction of the surface. For example, when a diamond is cleaved, one hybrid orbital of each carbon atom

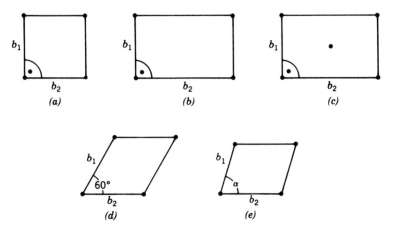

Figure 24.14 The five types of two-dimensional Bravais lattices: (*a*) square $b_1 = b_2$, $\alpha = 90°$; (*b*) primitive rectangular, $b_1 \neq b_2$, $\alpha = 90°$; (*c*) centered rectangular, $b_1 \neq b_2$, $\alpha = 90°$; (*d*) hexagonal, $b_1 = b_2$, $\alpha = 60°$; (*e*) oblique, $b_1 \neq b_2$, $\alpha \neq 90°$. (From G. Ertl and J. Küppers, *Low Energy Electrons and Surface Chemistry*, 2d ed. Weinheim: VCH, Verlag Chemie, 1985.)

dangles into the vacuum. Each such orbital is half occupied. These dangling bonds cause an adjustment of the positions of the atoms in the surface in which some atoms may rise above the surface plane and other atoms sink below the surface plane. Surface atoms may also be displaced horizontally. As the positions of surface atoms change to the geometrical configuration with the lowest Gibbs energy, energy gained by local bond formation is balanced by the work of elastic distortion. Furthermore, this type of surface reconstruction may also occur when atoms or molecules are adsorbed on the surface.

Arrangements of surface atoms can be described by the two-dimensional analog of equation 23.2, namely

$$T = h'\boldsymbol{b}_1 + k'\boldsymbol{b}_2 \tag{24.24}$$

where h' and k' are integers. The 14 Bravais lattices of three-dimensional structures are replaced by the five types of two-dimensional nets of equivalent lattice points that are shown in Fig. 24.14. The nomenclature for surface structures involves a comparison of the basis vectors ($\boldsymbol{b}_1$ and $\boldsymbol{b}_2$) of the surface lattice with the corresponding substrate lattice vectors ($\boldsymbol{a}_1$ and $\boldsymbol{a}_2$). The substrate lattice is defined as that plane parallel to the surface below which the three-dimensional (bulk) periodicity is found. The relation between the surface lattice and the substrate lattice is expressed by the ratios of the lengths of the basis vectors $|\boldsymbol{b}_1|/|\boldsymbol{a}_1|$ and $|\boldsymbol{b}_2|/|\boldsymbol{a}_2|$ and the angle of rotation between the two lattices. The nomenclature for surface lattices is illustrated in Fig. 24.15. The atoms in the surface lattice *a* are twice as far apart as in the substrate structure in two dimensions; the angle of rotation is omitted because it is zero. In the overlayer structure in *b* the designation $c(2 \times 2)$ indicates that the unit cell of the surface structure is centered. In Fig. 24.15*c* the ratios are $\sqrt{3}$, and the surface lattice is rotated 30° with respect to the substrate.

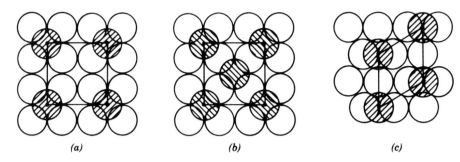

Figure 24.15 Examples for overlayer structures: (*a*) 2 × 2; (*b*) c(2 × 2); (*c*) √3 × √3/R30°. (From G. Ertl and J. Küppers, *Low Energy Electrons and Surface Chemistry,* 2d ed. Weinheim: VCH, Verlag Chemie, 1985.)

References

A. W. Adamson, *Physical Chemistry of Surfaces,* 4th ed. New York: Wiley-Interscience, 1982.

R. Aveyard and D. A. Haydon, *An Introduction to the Principles of Surface Chemistry.* Cambridge, UK: Cambridge University Press, 1973.

G. C. Bond, *Heterogeneous Catalysis: Principles and Applications.* Oxford, UK: Clarendon, 1987.

L. J. Clarke, *Surface Crystallography.* New York: Wiley-Interscience, 1985.

G. Ertl and J. Küppers, *Low Energy Electrons and Surface Chemistry,* 2d ed. Weinheim: VCH, Verlag Chemie, 1985.

R. P. H. Gasser, *An Introduction to Chemisorption and Catalysis by Metals.* Oxford, UK: Clarendon, 1985.

N. B. Hannay, *Solid-State Chemistry,* Chapter 10. Englewood Cliffs, NJ: Prentice-Hall, 1967.

G. A. Somorjai, *Principles of Surface Chemistry.* Englewood Cliffs, NJ: Prentice-Hall, 1972.

G. A. Somorjai, *Chemistry in Two Dimensions: Surfaces.* Ithaca, NY: Cornell University Press, 1981.

A. Zangwill, *Physics at Surfaces.* Cambridge, UK: Cambridge University Press, 1988.

Problems

24.1 In an ultrahigh vacuum chamber ($P = 5 \times 10^{-10}$ torr), how many molecules strike 1 cm^2 of surface in one second at 298 K, if (*a*) the gas is helium, and (*b*) the gas is mercury vapor?

24.2 A readily oxidized metal surface with 10^{15} metal atoms per square centimeter is exposed to molecular oxygen at 10^{-5} Pa at 298 K. How long will it take to completely oxidize the surface if the oxide formed is MO?

24.3   For the adsorption of nitrogen molecules on a certain sample of carbon, the pressure required to half saturate the surface at 298 K is 2×10^{-5} Pa. If the enthalpy of adsorption is -10 kJ mol^{-1} and the sticking coefficient s^* is unity, what is the rate constant of desorption k_d?

24.4   The pressures of nitrogen required for adsorption of 1.0 cm^3 g^{-1} (25 °C, 1.013 bar) of gas on graphitized carbon black are 24 Pa at 77.5 K and 290 Pa at 90.1 K. Calculate the enthalpy of adsorption at this fraction of surface coverage.

24.5   A mixture of A and B is adsorbed on a solid for which the adsorption isotherm follows the Langmuir equation. If the mole fractions in the gas at equilibrium are y_A and y_B, what is the equation for the adsorption isotherm in terms of total pressure? What is the expression for the mole fraction of A in the adsorbed gas in terms of K_A, K_B, y_A, and y_B?

24.6   The following table gives the volume of nitrogen (reduced to 0 °C and 1 bar) adsorbed per gram of active carbon at 0 °C at a series of pressures:

P/Pa	524	1731	3058	4534	7497
v/cm^{3}g^{-1}	0.987	3.04	5.08	7.04	10.31

Plot the data according to the Langmuir isotherm, and determine the constants.

24.7   According to problem 18.35, the rate with which oxygen molecules strike a surface at 1 bar and 25 °C is 2.69×10^{23} cm^{-2} s^{-1}. If the oxygen molecules are striking a platinum surface, what are the frequencies of collisions per atom on (100), (110) and (111) planes? (See problem 23.36.)

24.8   One gram of activated charcoal has a surface area of 1000 m^2. If complete surface coverage is assumed, as a limiting case, how much ammonia, at 25 °C and 1 bar, could be adsorbed on the surface of 45 g of activated charcoal? The diameter of the NH$_3$ molecule is 3×10^{-10} m, and it is assumed that the molecules just touch each other in a plane so that four adjacent spheres have their centers at the corners of a square.

24.9   Calculate the surface area of a catalyst that adsorbs 10^3 cm^3 of nitrogen (calculated at 1.013 bar and 0 °C) per gram in order to form a monolayer. The adsorption is measured at -195 °C, and the effective area occupied by a nitrogen molecule on the surface is 16.2×10^{-20} m^2 at this temperature.

24.10   What volume of oxygen, measured at 25 °C and 1 bar, is required to form an oxide film on 1 m^2 of a metal with

atoms in a square array 0.1 nm apart? Assume that one oxygen atom combines with each metal atom.

24.11   The following table gives data on the adsorption of benzene by graphitized carbon black (P-33) at two surface coverages and two temperatures:

v(cm^3 g^{-1} at 25 °C, 1 bar)	0.2	0.4
t/C°	Pressure (Pa)	
0	13	27
35.0	80	170

Calculate the enthalpy of adsorption $\Delta_{ads}H$ at each coverage. (S. Ross and J. P. Oliver, *On Physical Adsorption*, p. 238. New York: Wiley-Interscience, 1964.)

24.12   The adsorption of ammonia on charcoal is studied at 30 and 80 °C. It is found that the pressure required to adsorb a certain amount of NH$_3$ per gram of charcoal is 14.1 kPa at 30 °C and 74.6 kPa at 80 °C. Calculate the enthalpy of adsorption.

24.13   Hydrogen is dissociatively adsorbed on a metal and the pressure required to obtain half of the saturation coverage of the surface is 10 Pa. (*a*) What pressure will be required to reach $\Theta = 0.75$? (*b*) What pressure would have been required if the adsorption were not dissociative?

24.14   A fresh metal surface with 10^{15} atoms per square centimeter is prepared. This surface is exposed to oxygen at 10^{-2} Pa. If every oxygen molecule that strikes the surface reacts so that there is one oxygen atom per metal atom in the surface, how long will it take for half of the surface to become oxidized at 25 °C?

24.15   The volume of nitrogen gas v_m (measured at 1.013 bar and 0 °C) required to form a complete monolayer on a sample silica gel is 129 cm^3 g^{-1} of gel. Calculate the surface area per gram of the gel if each nitrogen molecule occupies 16.2×10^{-20} m^2.

24.16   The diameter of the hydrogen molecule is about 0.27 nm. If an adsorbent has a surface of 850 m^2 cm^{-3}, what volume of H$_2$ (measured at 25 °C and 1 bar) could be adsorbed by 100 mL of the adsorbent? It may be assumed that the adsorbed molecules just touch in a plane and are arranged so that four adjacent spheres have their centers at the corners of a square.

APPENDIX
A
Physical Quantities and Units

The measurement of any physical quantity consists of a comparison with a standard amount of that quantity, which is referred to as a unit. Thus, the statement of a physical measurement consists of two parts: (1) a number that represents the number of times the unit has to be used to give the physical quantity, and (2) the unit itself. The SI system of units is founded on the seven base units listed in the following table. (SI stands for Systeme International d'Unites.) This particular system was defined and given official status by the 11th Conference Generale des Poids et Mesures in 1960.*

Physical Quantity	Symbol for Quantity	Name of Unit	Symbol for SI Unit	Definition
Length	l	meter	m	The meter is the length of path traveled by light in vacuum during a time interval of 1/299 792 458 of a second.
Mass	m	kilogram	kg	The kilogram is equal to the mass of the international prototype of the kilogram.
Time	t	second	s	The second is the duration of 9 192 631 770 periods of the radiation corresponding to the transition between the two hyperfine levels of the ground state of the cesium-133 atom.
Electric current	I	ampere	A	The ampere is that constant current which, if maintained in two straight parallel conductors of infinite length, of negligible cross section, and placed 1 meter apart in a vacuum, would produce between these conductors a force equal to 2×10^7 newton per meter of length.

Note that symbols for quantities are always printed in italic type, and symbols for units are always printed in roman type.

* I. Mills, *Quantities, Units, and Symbols in Physical Chemistry*. Oxford, UK: Blackwell Scientific, 1988; *ISO Standards Handbook 2, Units of Measurement*. Geneva: ISO, 1982.

Physical Quantity	Symbol for Quantity	Name of Unit	Symbol for SI Unit	Definition
Thermodynamic temperature	T	kelvin	K	The kelvin is the fraction 1/273.16 of the thermodynamic temperature of the triple point of water.
Amount of substance	n	mole	mol	The mole is the amount of substance of a system that contains as many elementary entities as there are atoms in 0.012 kilograms of carbon-12. When the mole is used, the elementary entities must be specified and may be atoms, molecules, ions, electrons, other particles, or specified groups of such particles.
Luminous intensity	I_v	candela	cd	The candela is the luminous intensity, in a given direction, of a source that emits monochromatic radiation of frequency 540×10^{12} hertz and has a radiant intensity in that direction of $\frac{1}{683}$ watt per steradian.

In addition there are two supplementary units: for plane angle, the supplementary SI unit is the radian, with the symbol rad; for solid angle, the supplementary SI unit is the steradian, with the symbol sr. In 1980 the International Committee of Weights and Measures decided to interpret the class of supplemental units as dimensionless derived units. Although the coherent unit for both quantities is the number 1, it is convenient to use the special names radian and steradian instead of the number 1 in many practical cases.

The value of a physical quantity is equal to the product of a numerical value and a unit. Physical quantity = numerical value × unit.

All quantities may be expressed in SI units or in terms of derived units obtained algebraically by multiplication and division. Some derived units have their own special symbols. For example, the joule, which is the unit of work in SI, is defined in terms of base units by $kg\ m^2\ s^{-2}$ and is represented by the special symbol J. The principal derived units used in physical chemistry are given in the following table.

Quantity	Unit	Symbol	Definition
Force	newton	N	$kg\ m\ s^{-2}$
Work, energy, heat	joule	J	$N\ m\ (= kg\ m^2\ s^{-2})$
Power, radiant flux	watt	W	$J\ s^{-1}$
Pressure	pascal	Pa	$N\ m^{-2}$
Electric charge	coulomb	C	$A\ s$
Electric potential	volt	V	$kg\ m^2\ s^{-3}\ A^{-1}\ (= J\ A^{-1}s^{-1} = J\ C^{-1})$
Electric resistance	ohm	Ω	$kg\ m^2\ s^{-3}\ A^{-2}\ (= V\ A^{-1})$
Electric capacitance	farad	F	$A\ s\ V^{-1}\ (= m^{-2}\ kg^{-1}\ s^4\ A^2)$
Frequency	hertz	Hz	s^{-1} (cycle per second)
Magnetic flux density	tesla	T	$kg\ s^{-2}\ A^{-1}\ (= N\ A^{-1}\ m^{-1})$

It is important to be able to convert from other systems of units to SI. Some of the conversion factors are as follows:

Physical Quantity	Name of Unit	Symbol	Equivalent in SI Units
Length	angstrom	Å	10^{-10} m (10^{-1} nm)
Energy	electronvolt	eV	$1.602\ 177\ 33 \times 10^{-19}$ J
	wavenumber	cm^{-1}	$1.986\ 447 \times 10^{-23}$ J
	calorie (thermochemical)	cal	4.184 J
	erg		10^{-7} J
Force	dyne		10^{-5} N
Pressure	bar	bar	10^{5} N m^{-2}
	atmosphere	atm	101.325 kN m^{-2}
	torr		133.322 N m^{-2}
Electric charge	e.s.u.		3.334×10^{-10} C
Dipole moment	debye (10^{-18} e.s.u. cm)		3.334×10^{-30} C m
Magnetic flux density	gauss	G	10^{-4} T

In an equation relating physical quantities, the symbols represent numbers and associated units. If units are chosen arbitrarily, additional numerical factors may appear in equations relating different physical quantities. In practice it is more convenient to choose a system of units so that the equations between physical quantities have exactly the same form as the corresponding equations between pure numbers. A system of units that has this property is said to be coherent. The SI system is coherent. This means that if all quantities in a calculation are expressed in SI base units, the result will be expressed in SI base units without including any numerical factors. It is, however, a good habit to check a calculation to see that units cancel to yield the correct units for the final result.

A physical quantity may be converted from one unit to another by multiplying by a conversion factor. To find a conversion factor, it is necessary to express one unit in terms of another. For example, one calorie is equal to 4.184 joules; that is, 1 cal = 4.184 J. Dividing both sides of the equation by 1 cal yields 1 = 4.184 J cal^{-1}. To convert the change in enthalpy of a reaction from calories to joules, simply multiply the change in enthalpy in calories by 4.184 J cal^{-1}; note that this is equivalent to multiplying by one. A number of useful conversion factors are listed inside the back cover.

The following examples illustrate the procedure for calculating conversion factors in more complicated cases. The first step is to write down the equation relating the quantities of interest. The conversion factor for converting energies in J mol^{-1} to wavenumbers in cm^{-1} is based on the following equation:

$$E = N_A hc\tilde{\nu}$$

The ratio of wavenumbers in cm^{-1} to energy in J mol^{-1} is given by

$$\frac{\tilde{\nu}}{E} = (N_A hc)^{-1}$$

$$= [(6.022\ 136\ 7 \times 10^{23}\ \text{mol}^{-1})(6.626\ 075\ 5 \times 10^{-34}\ \text{J s})$$
$$\times (2.997\ 924\ 58 \times 10^{8}\ \text{m s}^{-1})]^{-1}$$

$$= 8.359\ 346\ \text{mol J}^{-1}\,\text{m}^{-1}$$
$$= (8.359\ 346\ \text{mol J}^{-1}\,\text{m}^{-1})(0.01\ \text{m cm}^{-1})$$
$$= 8.359\ 346 \times 10^{-2}\ (\text{J mol}^{-1})^{-1}\,\text{cm}^{-1}$$

The conversion factor for converting energies in J mol^{-1} to electronvolts is based on the following equation:

$$U = N_A E e$$

The potential in volts required to accelerate an electron to a given energy divided by that energy is

$$\frac{E}{U} = (N_A e)^{-1} = 1/F = [(6.022\ 136\ 7 \times 10^{23}\,\text{mol}^{-1})(1.602\ 177\ 33 \times 10^{-19}\,\text{C})]^{-1}$$
$$= 1.036\ 427 \times 10^{-5}\,\text{V}\,(\text{J mol}^{-1})^{-1}$$

since $V = J\ C^{-1}$.

Finally, since kT is an energy and the combination ϵ/kT is frequently found in statistical mechanical calculations, it is sometimes convenient to express energies in terms of temperature by dividing a molecular energy by the Boltzmann constant k. This makes it possible to write $\exp(-\epsilon/kT)$ as $\exp(-\theta/T)$, which is convenient both for making calculations and in thinking about the magnitude of an exponential term in an equation. If an energy E is known in J mol^{-1} the conversion to Kelvins is based on the following equation:

$$E = N_A kT = RT$$

The ratio of the temperature in kelvins to energy in J mol^{-1} is given by

$$\frac{T}{E} = (N_A k)^{-1} = \text{R}^{-1}$$
$$= [(6.022\ 136\ 7 \times 10^{23}\,\text{mol}^{-1})(1.380\ 658 \times 10^{-23}\,\text{J K}^{-1})]^{-1}$$
$$= 0.120\ 271\ 7\ \text{K}\,(\text{J mol}^{-1})^{-1}$$

B
Values of Physical Constants[a,b]

Quantity	Symbol	Value	Units
Speed of light in vacuum	c	299 792 458(exact)	m s^{-1}
Permeability of vacuum	μ_0	$4\pi \times 10^{-7}$ (exact)	N A^{-2}
		$= 12.566\ 370\ 614\ldots$	$10^{-7}\ \text{N A}^{-2}$
Permittivity of vacuum	ϵ_0	$1/\mu_0 c^2$ (exact)	$\text{C}^2\ \text{N}^{-1}\ \text{m}^{-2}$
		$= 8.854\ 187\ 817\ldots$	$10^{-12}\ \text{C}^2\ \text{N}^{-1}\ \text{m}^{-2}$
Planck constant	h	6.626 075 5(40)	$10^{-34}\ \text{J s}$
$h/2\pi$	$\hbar$	1.054 572 66(63)	$10^{-34}\ \text{J s}$
Elementary charge	e	1.602 177 33(49)	$10^{-19}\ \text{C}$
Bohr magneton, $e\hbar/2m_e$	μ_B	9.274 015 4(31)	$10^{-24}\ \text{J T}^{-1}$
Nuclear magneton, $e\hbar/2m_p$	μ_N	5.050 786 6(17)	$10^{-27}\ \text{J T}^{-1}$
Rydberg constant, $m_e e^4/8h^3 c\epsilon_0$	R_∞	10 973 731.534(13)	m^{-1}
Bohr radius, $h^2\epsilon_0/\pi m_e e^2$	a_0	0.529 177 249(24)	$10^{-10}\ \text{m}$
Hartree energy, $e^2/4\pi\epsilon_0 a_0$	E_h	4.359 748 2(26)	$10^{-18}\ \text{J}$
Electron mass	m_e	9.109 389 7(54)	$10^{-31}\ \text{kg}$
Proton mass	m_p	1.672 623 1(10)	$10^{-27}\ \text{kg}$
Neutron mass	m_n	1.674 928 6(10)	$10^{-27}\ \text{kg}$
Deuteron mass	m_d	3.343 586 0(20)	$10^{-27}\ \text{kg}$
Avogadro constant	N_A	6.022 136 7(36)	$10^{23}\ \text{mol}^{-1}$
Atomic mass constant			
$m_u = (1/12)m(^{12}\text{C})$	m_u	1.660 540 2(10)	$10^{-27}\ \text{kg}$
Faraday constant	F	96 485.309(29)	C mol^{-1}
Gas constant	R	8.314 510(70)	$\text{J K}^{-1}\ \text{mol}^{-1}$
		0.083 145 1	$\text{L bar K}^{-1}\ \text{mol}^{-1}$
		1.987 216	$\text{cal K}^{-1}\ \text{mol}^{-1}$
		0.082 057 8	$\text{L atm K}^{-1}\ \text{mol}^{-1}$
Boltzmann constant, R/N_A	k	1.380 658(12)	$10^{-23}\ \text{J K}^{-1}$

[a] E. R. Cohen and B. N. Taylor, The 1986 CODATA Recommended Values of the Fundamental Physical Constants. *J. Phys. Chem. Ref. Data* **17**:1795 (1988).

[b] The digits in parentheses are the one-standard-deviation uncertainty in the last digits of the given value.

C
Thermodynamic Tables

Table C.1   Chemical Thermodynamic Properties at 298.15 K and 1 bar[a]

Substance	$\Delta_f H°$ kJ mol^{-1}	$\Delta_f G°$ kJ mol^{-1}	$\overline{S}°$ J K^{-1} mol^{-1}	$\overline{C}_P°$ J K^{-1} mol^{-1}
$O(g)$	249.170	231.731	161.055	21.912
$O_2(g)$	0	0	205.138	29.355
$O_3(g)$	142.7	163.2	238.93	39.20
$H(g)$	217.965	203.247	114.713	20.784
$H^+(g)$	1536.202			
$H^+(ao)$	0	0	0	0
$H_2(g)$	0	0	130.684	28.824
$OH(g)$	38.95	34.23	183.745	29.886
$OH^-(ao)$	-229.994	-157.244	-10.75	-148.5
$H_2O(l)$	-285.830	-237.129	69.91	75.291
$H_2O(g)$	-241.818	-228.572	188.825	33.577
$H_2O_2(l)$	-187.78	-120.35	109.6	89.1
$He(g)$	0	0	126.150	20.786
$Ne(g)$	0	0	146.328	20.786
$Ar(g)$	0	0	154.843	20.786
$Kr(g)$	0	0	164.082	20.786
$Xe(g)$	0	0	169.683	20.786
$F(g)$	78.99	61.91	158.754	22.744
$F^-(ao)$	-332.63	-278.79	-13.8	-106.7
$F_2(g)$	0	0	202.78	31.30
$HF(g)$	-271.1	-273.2	173.779	29.133
$Cl(g)$	121.679	105.680	165.198	21.840
$Cl^-(ao)$	-167.159	-131.228	56.5	-136.4
$Cl_2(g)$	0	0	223.066	33.907
$ClO_4^-(ao)$	-129.33	-8.52	182.0	
$HCl(g)$	-92.307	-95.299	186.908	29.12
$HCl(ai)$	-167.159	-131.228	56.5	-136.4
HCl in 100 H_2O	-165.925			

[a] The values in Table C.1 are from The NBS Tables of Chemical Thermodynamic Properties (1982). The standard-state pressure is 1 bar (0.1 MPa). The compounds are in the order of elements used in these tables. For the elements represented in Table C.1, this order is O, H, He, F, Cl, Br, I, S, N, P, C, Pb, Al, Zn, Cd, Hg, Cu, Ag, Fe, Pt, Ti, Mg, Ca, Li, Na, K, Rb, and Cs. The standard state for a strong electrolyte in aqueous solution is the ideal solution at unit mean molality (unit activity). The thermodynamic properties of the completely dissociated electrolyte are designated by ai. The thermodynamic properties of undissociated molecules in water are designated by ao. The properties of organic substances with more than two carbon atoms are from D. R. Stull, E. F. Westrum, and G. C. Sinke, *The Chemical Thermodynamics of Organic Compounds*. New York: Wiley, 1969. The NBS Tables of Chemical Thermodynamic Properties have been published as a supplement to Volume II (1982) of the *Journal of Physical and Chemical Reference Data* and may be ordered from the American Chemical Society, 1155 Sixteenth St., NW, Washington, DC 20036. The conversion to the new standard-state pressure is described by R. D. Freeman, *J. Chem. Educ.* **62**:681 (1985).

Table C.1 Chemical Thermodynamic Properties at 298.15 K and 1 bar[a]

Substance	$\Delta_f H°$ kJ mol^{-1}	$\Delta_f G°$ kJ mol^{-1}	$\overline{S}°$ J K^{-1} mol^{-1}	$\overline{C}_P°$ J K^{-1} mol^{-1}
HCl in 200 H_2O	−166.272			
Br(g)	111.884	82.396	175.022	20.786
Br$^-$(ao)	−121.55	−103.96	82.4	−141.8
Br_2(l)	0	0	152.231	75.689
Br_2(g)	30.907	3.110	245.463	36.02
HBr(g)	−36.40	−53.45	198.695	29.142
I(g)	106.838	70.250	180.791	20.786
I$^-$(ao)	−55.19	−51.57	111.3	−142.3
I_2(sr)	0	0	116.135	54.438
I_2(g)	62.438	19.317	260.69	36.90
HI(g)	26.48	1.70	206.594	29.158
S(rhombic)	0	0	31.80	22.64
S(monoclinic)	0.33			
S(g)	278.805	238.250	167.821	23.673
S_2(g)	128.37	79.30	228.18	32.47
S^{2-}(ao)	33.1	85.8	−14.6	
SO_2(g)	−296.830	−300.194	248.22	39.87
SO_3(g)	−395.72	−371.06	256.76	50.67
SO_4^{-2}(ao)	−909.27	−744.53	2.01	−293
HS$^-$(ao)	−17.6	12.08	62.8	
H_2S(g)	−20.63	−33.56	205.79	34.23
H_2SO_4(l)	−813.989	−690.003	156.904	138.91
H_2SO_4(ao)	−909.27	−744.53	20.1	−293
N(g)	472.704	455.563	153.298	20.786
N_2(g)	0	0	191.61	29.125
NO(g)	90.25	86.57	210.761	29.844
NO_2(g)	33.18	51.31	240.06	37.20
NO_3^-(ao)	−205.0	−108.74	146.4	−86.6
N_2O(g)	82.05	104.20	219.85	38.45
N_2O_4(l)	−19.50	97.54	209.2	142.7
N_2O_4(g)	9.16	97.89	304.29	77.28
NH_3(g)	−46.11	−16.45	192.45	35.06
NH_3(ao)	−80.29	−26.50	111.3	
NH_4^+(ao)	−132.51	−79.31	113.4	79.9
HNO_3(l)	−174.10	−80.71	155.60	109.87
HNO_3(ai)	−207.36	−111.25	146.4	−86.6
NH_4OH(ao)	−366.121	−263.65	181.2	
P(s, white)	0	0	41.09	23.840
P(g)	314.64	278.25	163.193	20.786
P_2(g)	144.3	103.7	218.129	32.05
P_4(g)	58.91	24.44	279.98	67.15
PCl_3(g)	−287.0	−267.8	311.78	71.84
PCl_5(g)	−374.9	−305.0	364.58	112.8
C(graphite)	0	0	5.74	8.527
C(diamond)	1.895	2.900	2.377	6.113
C(g)	716.682	671.257	158.096	20.838
C_2(g)	0	−0.0330	144.960	29.196
CO(g)	−110.525	−137.168	197.674	29.116
CO_2(g)	−393.509	−394.359	213.74	37.11
CO_2(ao)	−413.80	−385.98	117.6	

Table C.1 (*continued*)

Substance	$\Delta_f H°$ kJ mol^{-1}	$\Delta_f G°$ kJ mol^{-1}	$\overline{S}°$ J K^{-1} mol^{-1}	$\overline{C}_P°$ J K^{-1} mol^{-1}
CO_3^{2-} (ao)	−677.14	−527.81	−56.9	
CH(g)	595.8			
CH_2(g)	392.0			
CH_3(g)	138.9			
CH_4(g)	−74.81	−50.72	186.264	35.309
C_2H_2(g)	226.73	209.20	200.94	43.93
C_2H_4(g)	52.26	68.15	219.56	43.56
C_2H_6(g)	−84.68	−32.82	229.60	52.63
HCO_3^- (ao)	−691.99	−586.77	91.2	
$HCHO$(g)	−117	−113	218.77	35.40
HCO_2H(l)	−424.72	−361.35	128.95	99.04
H_2CO_3(ao)	−699.65	−623.08	187.4	
CH_3OH(l)	−238.66	−166.27	126.8	81.6
CH_3OH(g)	−200.66	−161.96	239.81	43.89
$CH_3CO_2^-$ (ao)	486.01	−369.31	86.6	−6.3
C_2H_4O(ethylene oxide)(l)	−77.82	−11.76	153.85	87.95
CH_3CHO(l)	−192.30	−128.12	160.2	
CH_3CO_2H(l)	−484.5	−389.9	159.8	124.3
CH_3CO_2H(ao)	−485.76	−396.46	178.7	
C_2H_5OH(l)	−277.69	−174.78	160.7	111.46
C_2H_5OH(g)	−235.10	−168.49	282.70	65.44
$(CH_3)_2O$(g)	−184.05	−112.59	266.38	64.39
C_3H_6(propene)(g)	20.42	62.78	267.05	63.89
C_3H_6(cyclopropane)(g)	53.30	104.45	237.55	55.94
C_3H_8(propane)(g)	−103.89	−23.38	270.02	73.51
C_4H_8(1-butene)(g)	−0.13	71.39	305.71	85.65
C_4H_8(2-butene cis)(g)	−6.99	65.95	300.94	78.91
C_4H_8(2-butene trans)(g)	−11.17	63.06	296.59	87.82
C_4H_{10}(butane)(g)	−126.15	−17.03	310.23	97.45
C_4H_{10}(isobutane)(g)	−134.52	−20.76	294.75	96.82
C_6H_6(g)	82.93	129.72	269.31	81.67
C_6H_{12}(cyclohexane)(g)	−123.14	31.91	298.35	106.27
C_6H_{14}(hexane)(g)	−167.19	−0.07	388.51	143.09
C_7H_8(toluene)(g)	50.00	122.10	320.77	103.64
C_8H_8(styrene)(g)	147.22	213.89	345.21	122.09
C_8H_{10}(ethylbenzene)(g)	29.79	130.70	360.56	128.41
C_8H_{18}(octane)(g)	−208.45	16.64	466.84	188.87
Si(s)	0	0	18.83	20.00
SiO_2(s, alpha)	−910.94	−856.64	41.84	44.43
Sn(s, white)	0	0	51.55	26.99
Sn^{2+}(ao)	−8.8	−27.2	−17	
SnO(s)	−285.8	−256.9	56.5	44.31
SnO_2(s)	−580.7	−519.6	52.3	52.59
Pb(s)	0	0	64.81	26.44
Pb^{2+}(ao)	−1.7	−24.43	10.5	
PbO(s, yellow)	−217.32	−187.89	68.70	45.77
PbO_2(s)	−277.4	−217.33	68.6	64.64
Al(s)	0	0	28.33	24.35
Al(g)	326.4	285.7	164.54	21.38
Al_2O_3(s, alpha)	−1675.7	−1582.3	50.92	79.04

Table C.1   (*continued*)

Substance	$\Delta_f H°$ kJ mol^{-1}	$\Delta_f G°$ kJ mol^{-1}	$\overline{S}°$ J K^{-1} mol^{-1}	$\overline{C}_P°$ J K^{-1} mol^{-1}
$AlCl_3(s)$	−704.2	−628.8	110.67	91.84
$Zn(s)$	0	0	41.63	25.40
$Zn^{2+}(ao)$	−153.89	−147.06	−112.1	46
$ZnO(s)$	−348.28	−318.30	43.64	40.25
$Cd(s, gamma)$	0	0	51.76	25.98
$Cd^{2+}(ao)$	−75.90	−77.612	−73.2	
$CdO(s)$	−258.2	−228.4	54.8	43.43
$CdSO_4 \cdot \frac{8}{3}H_2O(s)$	−1729.4	−1465.141	229.630	213.26
$Hg(l)$	0	0	76.02	27.983
$Hg(g)$	61.317	31.820	174.96	20.786
$Hg^{2+}(ao)$	171.1	164.40	−32.2	
$HgO(s, red)$	−90.83	−58.539	70.29	44.06
$Hg_2Cl_2(s)$	−265.22	−210.745	192.5	
$Cu(s)$	0	0	33.150	244.35
$Cu^+(ao)$	71.67	49.98	40.6	
$Cu^{2+}(ao)$	64.77	65.49	−99.6	
$Ag(s)$	0	0	42.55	25.351
$Ag^+(ao)$	105.579	77.107	72.68	21.8
$Ag_2O(s)$	−31.05	−11.20	121.3	65.86
$AgCl(s)$	−127.068	−109.789	96.2	50.79
$Fe(s)$	0	0	27.28	25.10
$Fe^{2+}(ao)$	−89.1	−78.90	−137.7	
$Fe^{3+}(ao)$	48.5	−4.7	−315.9	
$Fe_2O_3(s, hematite)$	−824.2	−742.2	87.40	103.85
$Fe_3O_4(s, magnetite)$	−1118.4	−1015.4	146.4	143.43
$Ti(s)$	0	0	30.63	25.02
$TiO_2(s)$	−939.7	−884.5	49.92	55.48
$U(s)$	0	0	50.21	27.665
$UO_2(s)$	−1084.9	−1031.7	77.03	63.60
$UO_2^{2+}(ao)$	−1019.6	−953.5	−97.5	
$UO_3(s, gamma)$	−1223.8	−1145.9	96.11	81.67
$Mg(s)$	0	0	32.68	24.89
$Mg(g)$	147.70	113.10	148.650	20.786
$Mg^{2+}(ao)$	−466.85	−454.8	−138.1	
$MgO(s)$	−601.70	−569.43	26.94	37.15
$MgCl_2(ao)$	−801.15	−717.1	−25.1	
$Ca(s)$	0	0	41.42	25.31
$Ca(g)$	178.2	144.3	154.884	20.786
$Ca^{2+}(ao)$	−542.83	−553.58	−53.1	
$CaO(s)$	−635.09	−604.03	39.75	42.80
$CaCl_2(ai)$	−877.13	−816.01	59.8	
$CaCO_3(calcite)$	−1206.92	−1128.79	92.9	81.88
$CaCO_3(aragonite)$	−1207.13	−1127.75	88.7	81.25
$Li(s)$	0	0	29.12	24.77
$Li^+(ao)$	−278.49	−293.31	13.4	68.6
$Na(s)$	0	0	51.21	28.24
$Na^+(ao)$	−240.12	−261.905	59.0	46.4
$NaOH(s)$	−425.609	−379.494	64.455	59.54
$NaOH(ai)$	−470.114	−419.150	48.1	−102.1
$NaOH$ in 100 H_2O	−469.646			

Table C.1 (*continued*)

Substance	$\Delta_f H°$ kJ mol^{-1}	$\Delta_f G°$ kJ mol^{-1}	$\overline{S}°$ J K^{-1} mol^{-1}	$\overline{C}_P°$ J K^{-1} mol^{-1}
NaOH in 200 H$_2$O	-469.608			
NaCl(s)	-411.153	-384.138	72.13	50.50
NaCl(ai)	-407.27	-393.133	115.5	-90.0
NaCl in 100 H$_2$O	-407.066			
NaCl in 200 H$_2$O	-406.923			
K(s)	0	0	64.18	29.58
K$^+$(ao)	-252.38	-283.27	102.5	21.8
KOH(s)	-424.764	-379.08	78.9	64.9
KOH(ai)	-482.37	-440.50	91.6	-126.8
KOH in 100 H$_2$O	-481.637			
KOH in 200 H$_2$O	-481.742			
KCl(s)	-436.747	-409.14	82.59	51.30
KCl(ai)	-419.53	-414.49	159.0	-114.6
KCl in 100 H$_2$O	-419.320			
KCl in 200 H$_2$O	-419.191			
Rb(s)	0	0	76.78	10.148
Rb$^+$(ao)	-251.17	-283.98	121.50	
Cs(s)	0	0	85.23	32.17
Cs$^+$(ao)	-258.28	-292.02	133.05	-10.5

Table C.2 Chemical Thermodynamic Properties at Several Temperatures and 1 Bar from JANAF Thermochemical Tables[a]

T/K	$\overline{C}_P°$ J K^{-1} mol^{-1}	$\overline{S}°$ J K^{-1} mol^{-1}	$\overline{H}° - \overline{H}°_{298}$ kJ mol^{-1}	$\Delta_f H°$ kJ mol^{-1}	$\Delta_f G°$ kJ mol^{-1}
		C(graphite)			
0	0.000	0.000	-1.051	0.000	0.000
298	8.517	5.740	0.000	0.000	0.000
500	14.623	11.662	2.365	0.000	0.000
1000	21.610	24.457	11.795	0.000	0.000
2000	24.094	40.771	35.525	0.000	0.000
3000	26.611	51.253	61.427	0.000	0.000
		C(g)			
0	0.000	0.000	-6.536	711.185	711.185
298	20.838	158.100	0.000	716.670	671.244
500	20.804	168.863	4.202	718.507	639.906
1000	20.791	183.278	14.600	719.475	560.654
2000	20.952	197.713	35.433	716.577	402.694
3000	21.621	206.322	56.689	711.932	246.723

[a] M. Chase et al. JANAF Thermochemical Tables, 3d ed. *J. Phys. Chem. Ref. Data* **14**, Supplements 1 and 2 (1985). The values on hydrocarbons, other than CH$_4$, are from D. R. Stull, E. F. Westrum, and G. C. Sinke, *The Chemical Thermodynamics of Organic Compounds*. New York: Wiley, 1969.

Table C.2 (*continued*)

T/K	$\overline{C}_P^\circ$ J K^{-1} mol^{-1}	$\overline{S}^\circ$ J K^{-1} mol^{-1}	$\overline{H}^\circ - \overline{H}_{298}^\circ$ kJ mol^{-1}	$\Delta_f H^\circ$ kJ mol^{-1}	$\Delta_f G^\circ$ kJ mol^{-1}
			$CH_4(g)$		
0	0.000	0.000	−10.024	−66.911	−66.911
298	35.639	186.251	0.000	−74.873	−50.768
500	46.342	207.014	8.200	−80.802	−32.741
1000	71.795	247.549	38.179	−89.849	19.492
2000	94.399	305.853	123.592	−92.709	130.802
3000	101.389	345.690	222.076	−91.705	242.332
			$CO(g)$		
0	0.000	0.000	−8.671	−113.805	−113.805
298	29.142	197.653	0.000	−110.527	−137.163
500	29.794	212.831	5.931	−110.003	−155.414
1000	33.183	234.538	21.690	−111.983	−200.275
2000	36.250	258.714	56.744	−118.896	−286.034
3000	37.217	273.605	93.504	−127.457	−367.816
			$CO_2(g)$		
0	0.000	0.000	−9.364	−393.151	−393.151
298	37.129	213.795	0.000	−393.522	−394.389
500	44.627	234.901	8.305	−393.666	−394.939
1000	54.308	269.299	33.397	−394.623	−395.886
2000	60.350	309.293	91.439	−396.784	−396.333
3000	62.229	334.169	152.852	−400.111	−395.461
			$C_2H_4(g)$		
0	0.000	0.000	−10.518	60.986	60.986
298	43.886	219.330	0.000	52.467	68.421
500	63.477	246.215	10.668	46.641	80.933
1000	93.899	300.408	50.665	38.183	119.122
			$C_2H_6(g)$		
298	52.63	229.60	0.00	−84.68	−32.86
500	78.07	262.91	13.22	−93.89	4.96
1000	122.72	332.28	64.56	−105.77	109.55
			$C_4H_{10}(g)(n\text{-butane})$		
298	97.45	310.23	0.00	−126.15	−17.02
500	147.86	372.90	24.94	−140.21	61.10
1000	226.86	502.86	120.96	−155.85	270.31
			$C_6H_6(g)$		
298	81.67	269.31	0.00	82.93	129.73
500	137.24	325.42	22.43	73.39	164.29
1000	209.87	446.71	112.01	62.01	260.76
			$CH_3OH(g)$		
298	43.89	239.81	0.00	−201.17	−162.46
500	59.50	266.13	10.42	−207.94	−134.27
1000	89.45	317.59	48.41	−217.28	−56.16
			$Cl(g)$		
0	0.000	0.000	−6.272	119.621	119.621
298	21.838	165.189	0.000	121.302	105.306
500	22.744	176.752	4.522	122.272	94.203
1000	22.233	192.430	15.815	124.334	65.288
2000	21.341	207.505	37.512	127.058	5.081
3000	21.063	216.096	58.690	128.649	−56.297

(*continued*)

Table C.2 (*continued*)

T/K	$\overline{C}_P^\circ$ J K^{-1} mol^{-1}	$\overline{S}^\circ$ J K^{-1} mol^{-1}	$\overline{H}^\circ - \overline{H}_{298}^\circ$ kJ mol^{-1}	$\Delta_f H^\circ$ kJ mol^{-1}	$\Delta_f G^\circ$ kJ mol^{-1}
		HCl(g)			
0	0.000	0.000	−8.640	−92.127	−92.127
298	29.136	186.901	0.000	−92.312	−95.300
500	29.304	201.989	5.892	−92.913	−97.166
1000	31.628	222.903	21.046	−94.388	−100.799
2000	35.600	246.246	54.953	−95.590	−106.631
3000	37.243	261.033	91.478	−96.547	−111.968
		Cl$_2$(g)			
0	0.000	0.000	−9.180	0.000	0.000
298	33.949	223.079	0.000	0.000	0.000
500	36.064	241.228	7.104	0.000	0.000
1000	37.438	266.764	25.565	0.000	0.000
2000	38.428	293.033	63.512	0.000	0.000
3000	40.075	308.894	102.686	0.000	0.000
		H(g)			
0	0.000	0.000	−6.197	216.035	216.035
298	20.786	114.716	0.000	217.999	203.278
500	20.786	125.463	4.196	219.254	192.957
1000	20.786	139.871	14.589	222.248	165.485
2000	20.786	154.278	35.375	226.898	106.760
3000	20.786	162.706	56.161	229.790	46.007
		H$^+$(g)			
0	0.000	0.000	−6.197	1528.085	
298	20.786	108.946	0.000	1536.246	1516.990
500	20.786	119.693	4.196	1541.697	1502.422
1000	20.786	134.101	14.589	1555.084	1457.958
2000	20.786	148.509	35.375	1580.520	1350.840
3000	20.786	156.937	56.161	1604.198	1230.818
		H$^-$(g)			
0	0.000	0.000	−6.197	143.266	
298	20.786	108.960	0.000	139.032	132.282
500	20.786	119.707	4.196	136.091	128.535
1000	20.786	134.114	14.589	128.692	123.819
2000	20.786	148.522	32.375	112.557	125.012
3000	20.786	156.950	56.161	94.662	135.055
		HI(g)			
0	0.000	0.000	−8.656	28.535	28.535
298	29.156	206.589	0.000	26.359	1.560
500	29.736	221.760	5.928	−5.622	−10.088
1000	33.135	243.404	21.641	−6.754	−14.006
2000	36.623	267.680	56.863	−7.589	−21.009
3000	37.918	282.805	94.210	−10.489	−27.144
		H$_2$(g)			
0	0.000	0.000	−8.467	0.000	0.000
298	28.836	130.680	0.000	0.000	0.000
500	29.260	145.737	5.883	0.000	0.000
1000	30.205	166.216	20.680	0.000	0.000
2000	34.280	188.418	52.951	0.000	0.000
3000	37.087	202.891	88.740	0.000	0.000

(*continued*)

Table C.2   (*continued*)

T/K	$\overline{C}_P^\circ$ $J\ K^{-1}\ mol^{-1}$	$\overline{S}^\circ$ $J\ K^{-1}\ mol^{-1}$	$\overline{H}^\circ - \overline{H}_{298}^\circ$ $kJ\ mol^{-1}$	$\Delta_f H^\circ$ $kJ\ mol^{-1}$	$\Delta_f G^\circ$ $kJ\ mol^{-1}$
			$H_2O(g)$		
0	0.000	0.000	−9.904	−238.921	−238.921
298	33.590	188.834	0.000	−241.826	−228.582
500	35.226	206.534	6.925	−243.826	−219.051
1000	41.268	232.738	26.000	−247.857	−192.590
2000	51.180	264.769	72.790	−251.575	−135.528
3000	55.748	286.504	126.549	−253.024	−77.163
			$I(g)$		
0	0.000	0.000	−6.197	107.164	107.164
298	20.786	180.786	0.000	106.762	70.174
500	20.786	191.533	4.196	75.990	50.203
1000	20.795	205.942	14.589	76.937	24.039
2000	21.308	220.461	35.566	77.992	−29.410
3000	22.191	229.274	57.332	77.406	−82.995
			$I_2(g)$		
0	0.000	0.000	−10.116	65.504	65.504
298	36.887	260.685	0.000	62.421	19.325
500	37.464	279.920	7.515	0.000	0.000
1000	38.081	306.087	26.407	0.000	0.000
2000	42.748	332.521	66.250	0.000	0.000
3000	44.897	351.615	110.955	0.000	0.000
			$N(g)$		
0	0.000	0.000	−6.197	470.820	470.820
298	20.786	153.300	0.000	472.683	455.540
500	20.786	164.047	4.196	473.923	443.584
1000	20.786	178.454	14.589	476.540	412.171
2000	20.790	192.863	35.375	479.990	346.339
3000	20.963	201.311	56.218	482.543	278.946
			$NO(g)$		
0	0.000	0.000	−9.192	89.775	89.775
298	29.845	210.758	0.000	90.291	86.606
500	30.486	226.263	6.059	90.352	84.079
1000	33.987	248.536	22.229	90.437	77.775
2000	36.647	273.128	57.859	90.494	65.060
3000	37.466	288.165	94.973	89.899	52.439
			$NO_2(g)$		
0	0.000	0.000	−10.186	35.927	35.927
298	36.974	240.034	0.000	33.095	51.258
500	43.206	260.638	8.099	32.154	63.867
1000	52.166	293.889	32.344	32.005	95.779
2000	56.441	331.788	87.259	33.111	159.106
3000	57.394	354.889	144.267	32.992	222.058
			$N_2(g)$		
0	0.000	0.000	−8.670	0.000	0.000
298	29.124	191.609	0.000	0.000	0.000
500	29.580	206.739	5.911	0.000	0.000
1000	32.697	228.170	21.463	0.000	0.000
2000	35.971	252.074	56.137	0.000	0.000
3000	37.030	266.891	92.715	0.000	0.000

(*continued*)

Table C.2   (*continued*)

T/K	$\overline{C}_P^\circ$ $\text{J K}^{-1} \text{ mol}^{-1}$	$\overline{S}^\circ$ $\text{J K}^{-1} \text{ mol}^{-1}$	$\overline{H}^\circ - \overline{H}_{298}^\circ$ kJ mol^{-1}	$\Delta_f H^\circ$ kJ mol^{-1}	$\Delta_f G^\circ$ kJ mol^{-1}
		$N_2O_4(g)$			
0	0.000	0.000	-16.398	18.718	18.718
298	77.256	304.376	0.000	9.079	97.787
500	97.204	349.446	17.769	8.769	158.109
1000	119.208	425.106	72.978	15.189	305.410
2000	129.030	511.743	198.518	33.110	588.764
3000	131.200	564.555	328.840	49.178	862.983
		$NH_3(g)$			
0	0.000	0.000	-10.045	-38.907	-38.907
298	35.652	192.774	0.000	-45.898	-16.367
500	42.048	212.659	7.819	-49.857	4.800
1000	56.491	246.486	32.637	-55.013	61.910
2000	72.833	291.525	98.561	-54.833	179.447
3000	78.902	322.409	174.933	-50.433	295.689
		$O(g)$			
0	0.000	0.000	-6.725	246.790	246.790
298	21.911	161.058	0.000	249.173	231.736
500	21.257	172.197	4.343	250.474	219.549
1000	20.915	186.790	14.860	252.682	187.681
2000	20.826	201.247	35.713	255.299	121.552
3000	20.937	209.704	56.574	256.741	54.327
		$O^-(g)$			
0	0.000	0.000	-6.571	105.814	105.814
298	21.692	157.790	0.000	101.846	91.638
500	21.184	168.860	4.318	98.926	85.532
1000	20.899	183.426	14.817	90.723	75.219
2000	20.816	197.878	35.661	72.545	66.619
3000	20.800	206.314	56.467	53.146	67.810
		$O_2(g)$			
0	0.000	0.000	-8.683	0.000	0.000
298	29.376	205.147	0.000	0.000	0.000
500	31.091	220.693	6.084	0.000	0.000
1000	34.870	243.578	22.703	0.000	0.000
2000	37.741	268.748	59.175	0.000	0.000
3000	39.884	284.466	98.013	0.000	0.000
		$e^-(g)$			
0	0.000	0.000	-6.197	0.000	0.000
298	20.786	20.979	0.000	0.000	0.000
500	20.786	31.725	4.196	0.000	0.000
1000	20.786	46.133	14.584	0.000	0.000
2000	20.786	60.541	35.375	0.000	0.000
3000	20.786	68.969	56.161	0.000	0.000

D
Mathematical Relations

Series

$$e^x = 1 + x + \frac{x^2}{2!} + \frac{x^3}{3!} + \cdots \qquad\qquad [\text{all } x]$$

$$\ln(1 + x) = x - \frac{x^2}{2} + \frac{x^3}{3} - \frac{x^4}{4} + \cdots \qquad\qquad [x^2 < 1]$$

$$(1 + x)^{-1} = 1 - x + x^2 - x^3 + \cdots \qquad\qquad [x^2 < 1]$$

$$(1 - x)^{-1} = 1 + x + x^2 + x^3 + \cdots \qquad\qquad [x^2 < 1]$$

$$(1 - x)^{-2} = 1 + 2x + 3x^2 + 4x^3 + \cdots \qquad\qquad [x^2 < 1]$$

Vectors

A vector in three-dimensional space may be represented in terms of the unit vectors i, j, and k in the x, y, and z directions:

$$A = A_x i + A_y j + A_z k$$

where the components A_x, A_y, and A_z are the projections on the x, y, and z axes. From the Pythagorean theorem, it follows that the length of A is given by

$$|A| = A = (A_x^2 + A_y^2 + A_z^2)^{1/2}$$

The dot product (scalar product) $A \cdot B$ of two vectors is defined by

$$A \cdot B = |A||B| \cos \theta = B \cdot A$$

where θ is the angle between the vectors. The dot product is a scalar. In terms of components

$$A \cdot B = A_x B_x + A_y B_y + A_z B_z$$

The cross product (vector product) $A \times B$ is a vector with the magnitude

$$|A \times B| = |A||B| \sin \theta$$

pointed perpendicular to the plane defined by A and B in the direction such that A, B, and $A \times B$ form a right-handed system. The most convenient way of expressing the vector product in terms of the components of the individual vectors is the determinantal relation:

$$A \times B = \begin{vmatrix} i & j & k \\ A_x & A_y & A_z \\ B_x & B_y & B_z \end{vmatrix}$$

In terms of components

$$A \times B = (A_y B_z - A_z B_y)i + (A_z B_x - A_x B_z)j + (A_x B_y - A_y B_x)k$$

Matrices*

A matrix is an array of numbers. If a matrix has m rows and n columns it may be represented by

$$A = \begin{bmatrix} a_{11} & a_{12} & \cdots & a_{1n} \\ a_{21} & a_{22} & & \\ \vdots & & & \\ a_{m1} & a_{m2} & \cdots & a_{mn} \end{bmatrix}$$

The sum of two matrices is defined by

$$C = A + B$$

where $c_{ij} = a_{ij} + b_{ij}$ for every i and j.

The product of a scalar c and a matrix is defined by

$$B = cA$$

where $b_{ij} = ca_{ij}$ for every i and j.

The product of two matrices is similar to the scalar product of two vectors. If C is the product AB, then

$$c_{ij} = \sum_{k=1}^{n} a_{ik}b_{kj}$$

where n is the number of columns in A. If B is to be multiplied by A, it must have as many rows as A has columns. For example,

$$AB = \begin{bmatrix} a_{11} & a_{12} \\ a_{21} & a_{22} \\ a_{31} & a_{32} \end{bmatrix} \begin{bmatrix} b_{11} & b_{12} \\ b_{21} & b_{22} \end{bmatrix} = \begin{bmatrix} a_{11}b_{11} + a_{12}b_{21} & a_{11}b_{12} + a_{12}b_{22} \\ a_{21}b_{11} + a_{22}b_{21} & a_{21}b_{12} + a_{22}b_{22} \\ a_{31}b_{11} + a_{32}b_{21} & a_{31}b_{12} + a_{32}b_{22} \end{bmatrix}$$

Matrix multiplication is not commutative, as illustrated by

$$AB = \begin{bmatrix} 2 & 1 \\ 3 & -2 \end{bmatrix} \begin{bmatrix} 1 & 4 \\ 0 & 2 \end{bmatrix} = \begin{bmatrix} 2 \times 1 + 1 \times 0 & 2 \times 4 + 1 \times 2 \\ 3 \times 1 - 2 \times 0 & 3 \times 4 - 2 \times 2 \end{bmatrix} = \begin{bmatrix} 2 \\ 3 \end{bmatrix}$$

$$BA = \begin{bmatrix} 1 & 4 \\ 0 & 2 \end{bmatrix} \begin{bmatrix} 2 & 1 \\ 3 & -2 \end{bmatrix} = \begin{bmatrix} 1 \times 2 + 4 \times 3 & 1 \times 1 - 4 \times 2 \\ 0 \times 2 + 2 \times 3 & 0 \times 1 - 2 \times 2 \end{bmatrix} = \begin{bmatrix} 14 \\ 6 \end{bmatrix}$$

Simultaneous linear equations may be solved by use of matrices. For example, the set

$$a_{11}x_1 + a_{12}x_2 + a_{13}x_3 = c_1$$
$$a_{21}x_1 + a_{22}x_2 + a_{23}x_3 = c_2$$
$$a_{31}x_1 + a_{32}x_2 + a_{33}x_3 = c_3$$

may be written in matrix notation as

$$\begin{bmatrix} a_{11} & a_{12} & a_{13} \\ a_{21} & a_{22} & a_{23} \\ a_{31} & a_{32} & a_{33} \end{bmatrix} \begin{bmatrix} x_1 \\ x_2 \\ x_3 \end{bmatrix} = \begin{bmatrix} c_1 \\ c_2 \\ c_3 \end{bmatrix}$$

---

* G. Strang, *Linear Algebra and Its Applications*. New York: Academic, 1980; R. G. Mortimer, *Mathematics for Physical Chemistry*. New York: Macmillan, 1981.

or

$$AX = C$$

The inverse of a matrix A^{-1} has the property that

$$A^{-1}A = AA^{-1} = E$$

where E is the identity matrix:

$$E = \begin{bmatrix} 1 & 0 & \cdots & 0 \\ 0 & 1 & \cdots & 0 \\ \vdots & & & \\ 0 & 0 & & 1 \end{bmatrix}$$

If we multiply both sides of $AX = C$ by A^{-1}, we obtain

$$A^{-1}AX = X = A^{-1}C$$

Thus, the solution X of the simultaneous equations is obtained by multiplying C by the inverse of A. Small matrices may be inverted by hand by the Gauss elimination, and large matrices may be inverted with a computer to obtain the solution of the simultaneous linear equations.

Determinants

A determinant is a square array of numbers. Its value is defined as a certain sum of products of subsets of the elements. If the determinant has n rows and columns, each term in the sum will have n factors in it. For a determinant of order 2

$$\begin{vmatrix} a_1 & b_1 \\ a_2 & b_2 \end{vmatrix} = a_1b_2 - a_2b_1$$

The value of a large determinant may be obtained by expanding by minors. For a determinant of order 3

$$\begin{vmatrix} a_1 & b_1 & c_1 \\ a_2 & b_2 & c_2 \\ a_3 & b_3 & c_3 \end{vmatrix} = a_1 \begin{vmatrix} b_2 & c_2 \\ b_3 & c_3 \end{vmatrix} - b_1 \begin{vmatrix} a_2 & c_2 \\ a_3 & c_3 \end{vmatrix} + c_1 \begin{vmatrix} a_2 & b_2 \\ a_3 & b_3 \end{vmatrix}$$

$$= a_1b_2c_3 + a_2b_3c_1 + a_3b_1c_2 - a_3b_2c_1 - a_2b_1c_3 - a_1b_3c_2$$

Determinants of higher order are defined by an analogous row (or column) expansion.

Complex Numbers

A complex number z can be written as $z = x + iy$, where $i = \sqrt{-1}$ and x and y are real numbers. A complex number can be represented as a point in the complex plane, as shown in Fig. A.1. The real part of z is plotted on the horizontal axis and the imaginary part on the vertical axis. The distance r is called the absolute value or modulus of z:

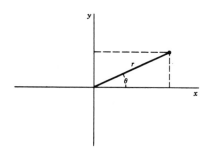

Figure A.1 Plot of a complex number.

$$|z| = r = (x^2 + y^2)^{1/2}$$

Since $x = r \cos \theta$ and $y = r \sin \theta$,

$$z = r \cos \theta + ir \sin \theta = re^{i\theta}$$

where the last relation is obtained using

$$e^{i\theta} = \cos \theta + i \sin \theta$$

The complex conjugate of z is defined by

$$z^* = x - iy = re^{-i\theta}$$

Forming the product of z and its complex conjugate we have

$$zz^* = x^2 + y^2 = r^2 = |z|^2$$

which is a real number.

Supplemental Reading

The following books deal with some of the mathematics used in physical chemistry. Since personal computers are increasingly used to carry out mathematical operations, these books include computer programs.

H. G. Hecht, *Mathematics in Chemistry*. Englewood Cliffs, NJ: Prentice-Hall, 1990.

K. J. Johnson, *Numerical Methods in Chemistry*. New York: Dekker, 1980.

D. W. Rogers, *Computational Chemistry Using the P. C.* New York: VCH, 1990.

E
Greek Alphabet

A	α	Alpha	N	ν	Nu
B	β	Beta	Ξ	ξ	Xi
Γ	γ	Gamma	O	o	Omicron
Δ	δ	Delta	Π	π	Pi
E	ϵ	Epsilon	P	ρ	Rho
Z	ζ	Zeta	Σ	σ	Sigma
H	η	Eta	T	τ	Tau
Θ	θ	Theta	Υ	υ	Upsilon
I	ι	Iota	Φ	ϕ	Phi
K	κ	Kappa	X	χ	Chi
Λ	λ	Lambda	Ψ	ψ	Psi
M	μ	Mu	Ω	ω	Omega

F
Symbols for Physical Quantities and Their SI Units[a,b]

Symbol	Name	Symbol for SI Unit
a	acceleration vector (2.1)	m s^{-2}
a	van der Waals' constant (1.9)	m^6 Pa mol^{-2}
a	activity (4.15)	1
a_0	Bohr radius (11.1)	m
A	area (2.7)	m^2
A	Helmholtz energy (4.3)	J
A	Debye–Hückel constant (8.5)	kg$^{1/2}$ mol$^{-1/2}$
A	pre-exponential factor in the Arrhenius equation (19.9)	(mol m^{-3})$^{1-n}$ s^{-1}
A	system formula matrix (5.13)	1
A_r	relative atomic mass (3.9)	1
A_{12}	Einstein probability of spontaneous emission (14.2)	s^{-1}
b	van der Waals' constant (1.9)	m^3 mol^{-1}
B	magnetic flux density (11.3)	T = kg s^{-2} A^{-1}
B	magnitude of the magnetic flux density (11.3)	T = kg s^{-2} A^{-1}
B	rotational constant (14.4)	m^{-1}
B_{12}	Einstein probability of stimulated emission (14.2)	s kg^{-1}
B_{21}	Einstein probability of stimulated absorption (14.2)	s kg^{-1}
$B(T)$	second virial coefficient (1.6)	m^3 mol^{-1}
c	(amount) concentration (8.4)	mol m^{-3}
c	number of components (6.6)	1
c	speed of light in a vacuum (10.1)	m s^{-1}
$c°$	standard concentration (5.10)	mol m^{-3}
C	capacitance (12.12)	F = C V^{-1} = A s V^{-1}

[a] Recommendations of the International Union of Pure and Applied Chemistry are given in I. Mills, *Quantities, Units and Symbols in Physical Chemistry,* Oxford, UK: Blackwell Scientific, 1988.

[b] The number in parentheses is the section number where the quantity is defined. For thermodynamic quantities a superscript degree sign indicates that the quantity has the value for the standard state, and a superscript E indicates the excess thermodynamic quantity (Section 7.15). A superscript * is used to designate the molar thermodynamic quantity for a pure substance. Molar thermodynamic quantities and partial molar thermodynamic quantities are indicated by an overbar. The overbar is not used in delta quantities such as $\Delta_{vap}H°$, $\Delta_f H°$, and $\Delta_r G°$. Concentrations are represented by use of square brackets. Quantum mechanical operators are indicated by a circumflex, as in $\hat{A}$ (Section 10.7). The SI unit for an operator is the same as for the physical quantity the operator represents. The expectation value of F is represented by $\langle F \rangle$. Vectors and matrices are represented by boldface italic symbols; the corresponding magnitudes are italicized and are not included here. A 1 in the symbol column indicates that the physical quantity does not have SI units.

Symbol	Name	Symbol for SI Unit
C_n	rotation operator (13.2)	1
$C(T)$	third virial coefficient (1.6)	$m^6 \, mol^{-2}$
C_P	heat capacity at constant pressure (2.12)	$J \, K^{-1}$
$\overline{C}_P$	molar heat capacity at constant pressure (2.13)	$J \, K^{-1} \, mol^{-1}$
C_V	heat capacity at constant volume (2.11)	$J \, K^{-1}$
$\overline{C}_V$	molar heat capacity at constant volume (2.11)	$J \, K^{-1} \, mol^{-1}$
$\Delta_r C_P^\circ$	reaction heat capacity at constant pressure (5.9)	$J \, K^{-1} \, mol^{-1}$
d	distance of closest approach of spherical molecules (17.14)	m
d_{hkl}	interplanar spacing (23.5)	m
D	centrifugal distortion constant (14.4)	m^{-1}
D	diffusion coefficient (18.10)	$m^2 \, s^{-1}$
D	electric displacement vector (12.12)	$C \, m^{-2}$
D_e	equilibrium dissociation energy (12.2)	J
D_0	spectroscopic dissociation energy (dissociation energy from ground state) (12.2)	J
e	charge of a proton (8.1)	C
E	electric field strength (8.1)	$V \, m^{-1}$
E	electromotive force (potential difference for a cell) (8.2)	V
E	energy (2.4)	J
E	energy of a particle (10.4)	J
E	identity operator (13.2)	1
E_a	activation energy in the Arrhenius equation (19.9)	$J \, mol^{-1}$
E_{ads}	activation energy for adsorption (24.2)	$J \, mol^{-1}$
E_{des}	activation energy for desorption (24.2)	$J \, mol^{-1}$
E_{ea}	electron affinity (11.13)	$J \, mol^{-1}$
E_{el}	electron kinetic energy (24.5)	J
E_h	Hartree energy (11.1)	J
$E(R)$	electronic energy as a function of internuclear distance R (12.1)	J
f	electrostatic factor (21.5)	1
f	force vector (2.1)	$kg \, m \, s^{-2}$
f	frictional coefficient (21.1)	$kg \, s^{-1}$
f	fugacity (4.9)	Pa
f	oscillator strength (15.5)	1
f_i	atomic scattering factor (23.6)	1
$f(v_x)$	probability density for velocity in the x direction (18.2)	$s \, m^{-1}$
$f(v_x, v_y, v_z)$	probability density for velocity (18.1)	$s^3 \, m^{-3}$
$f_p(p_x, p_y, p_z)$	probability density for momentum (18.1)	$kg^{-3} \, m^{-3} \, s^3$
F	Faraday constant (8.2)	$C \, mol^{-1}$
F	number of degrees of freedom (6.2)	1
$F(v)$	speed distribution function (Maxwell) (18.3)	$s \, m^{-1}$
$F(J)$	rotational term value for quantum number J (14.4)	m^{-1}
g	acceleration of gravity (1.12)	$m \, s^{-2}$

Symbol	Name	Symbol for SI Unit
g_e	electron g factor (11.4)	1
g_i	orbital in Hartree–Foch method (11.10)	
g_N	nuclear g factor (16.2)	1
G	Gibbs energy (4.3)	J
$G(v)$	vibrational term value for a diatomic molecule (14.6)	m^{-1}
$\overline{G}_i$	molar or partial molar Gibbs energy of i (4.12)	$J\ mol^{-1}$
$\Delta_{mix}G$	Gibbs energy of mixing (4.14)	$J\ mol^{-1}$
$\Delta_r G$	reaction Gibbs energy (5.1)	$J\ mol^{-1}$
$\Delta_r G^{\circ\prime}$	apparent reaction Gibbs energy (9.6)	$J\ mol^{-1}$
h	height (1.11)	m
h	Planck constant (10.1)	J s
$\hbar$	$h/2\pi$ (10.5)	J s
h_i	radial factor in orbital (11.10)	
H	enthalpy (2.10)	J
$\mathbf{H}$	magnetic field (16.1)	$A\ m^{-1}$
$\overline{H}$	molar enthalpy (2.10)	$J\ mol^{-1}$
$\overline{H}^{\circ}(T) - \overline{H}^{\circ}(298.15)$	molar enthalpy increment (2.15)	$J\ mol^{-1}$
$\hat{H}$	Hamiltonian operator (10.6)	J
H_{ij}	Hamiltonian matrix elements (12.7)	J
$H_v(x)$	Hermite polynomials (10.12)	1
H_{AA}	Coulomb integral (12.3)	J
H_{AB}	resonance integral (12.3)	J
$\Delta H^{\ddagger}$	enthalpy of activation (20.5)	$J\ mol^{-1}$
$\Delta_{ads}H$	enthalpy of adsorption (24.1)	$J\ mol^{-1}$
$\Delta_r H$	reaction enthalpy (2.17)	$J\ mol^{-1}$
$\Delta_f H^{\circ}$	standard enthalpy of formation (2.17)	$J\ mol^{-1}$
$\Delta_{vap}H$	enthalpy of vaporization (6.4)	$J\ mol^{-1}$
i	inversion operator (13.2)	1
I	adiabatic ionization potential (24.5)	V
I	intensity of transmitted light (15.4)	$J\ m^{-2}\ s^{-1}$
I	ionic strength (8.5)	$mol\ kg^{-1}$
I	moment of inertia (10.13)	$kg\ m^2$
I	nuclear spin quantum number (16.2)	1
I	total spin angular momentum of a nucleus (16.2)	J s
I_a	anodic partial current (21.12)	A
I_a	intensity of electromagnetic radiation absorbed (20.7)	$mol\ m^{-3}\ s^{-1}$
I_c	cathodic partial current (21.12)	A
I_z	z component of nuclear spin (16.2)	J s
J	photodissociation coefficient (20.12)	s^{-1}
J	quantum number of the total angular momentum of an atom (11.14)	1
J	rotational quantum number (10.14)	1
J	spin–spin coupling constant (16.6)	s^{-1}
J	total angular momentum vector for an atom (11.14)	J s
J_N	flux of particles (18.8)	$m^{-2}\ s^{-1}$
J_z	z component of the total angular momentum of an atom (11.14)	J s

Symbol	Name	Symbol for SI Unit
k	Boltzmann constant (3.12)	$J\ K^{-1}$
k	force constant (10.11)	$N\ m^{-1} = kg\ s^{-2}$
k	rate constant (19.2)	$(mol\ m^{-3})^{1-n}\ s^{-1}$
k_d	rate constant for desorption (24.2)	s^{-1}
K	constant in Langmuir adsorption isotherm (24.2)	Pa^{-1}
K	equilibrium constant (5.1)	1
K	quantum number of the component of angular momentum along the a axis of a symmetric molecule (14.5)	1
K_a	acid dissociation constant (9.1)	1
K_b	base dissociation constant (9.1)	1
K_c	equilibrium constant expressed in terms of $c/c°$ (5.10)	1
K_f	freezing point constant (7.9)	$K\ kg\ mol^{-1}$
K_P	equilibrium constant expressed in terms of $P_i/P°$, when needed to distinguish from K_c (5.10)	1
K_i	Henry law constant of species i (7.6)	Pa
K_w	ion product for water (9.1)	1
K_x, K_y	equilibrium constants in terms of mole fractions (5.6)	1
K'	apparent equilibrium constant at pH 7 (9.6)	1
$K^{\ddagger}$	equilibrium constant for the formation of a transition state (20.4)	1
K_I	inhibition constant for an enzymatic reaction (21.11)	$mol\ m^{-3}$
K_M	Michaelis constant (21.11)	$mol\ m^{-3}$
l	angular momentum quantum number (10.14)	1
l	length (2.1)	m
$\mathbf{l}$	vector length (2.1)	m
l_i	angular momentum of electron i in an atom (11.14)	J s
l_x	angular momentum in the x direction (10.7)	J s
l_{zi}	z component of the angular momentum of electron i in an atom (11.14)	J s
L	angular momentum (10.13)	J s
L	angular momentum vector for an atom (10.14)	J s
$\hat{L}_q$	operator for the angular momentum in the q direction (10.7)	J s
$\hat{L}^2$	operator for the square of the angular momentum (11.4)	$J^2\ s^2$
m	magnetic quantum number (10.14)	1
m	mass (2.1)	kg
m	number of initial conditions (6.6)	1
m	molality (8.4)	$mol\ kg^{-1}$
$m_{\pm}$	mean ionic molality (8.4)	$mol\ kg^{-1}$
m_e	rest mass of electron (11.1)	kg
m_i	quantum number for z component of the	1

Symbol	Name	Symbol for SI Unit
	orbital angular momentum of electron i in an atom (11.14)	
m_I	quantum number for z component of nuclear spin (16.2)	1
m_N	mass of nucleus (11.1)	kg
$m°$	standard molality (8.4)	$mol\ kg^{-1}$
m_s	spin quantum number for the z component of the spin angular momentum (11.4)	1
m_{si}	quantum number for the z component of the spin of electron i in an atom (11.14)	1
M	molar mass (1.3)	$kg\ mol^{-1}$
M	magnetization (magnetic dipole moment per unit volume) (16.1)	$A\ m^{-1}$
M^E	excess thermodynamic property where M is V, U, H, C_P, S, A, or G (7.15)	varies
M_J	quantum number for z component of total angular momentum for an atom (11.14)	1
M_L	quantum number for z component of the orbital angular momentum of an atom (11.14)	1
M_S	quantum number for z component of the spin angular momentum of an atom (11.14)	1
M_λ	exitance as a function of wavelength (10.1)	$J\ m^{-3}\ s^{-1}$
M_ν	exitance as a function of frequency (10.1)	$J\ m^{-2}\ s^{-1}$
$\overline{M}^E$	molar excess thermodynamic property (7.15)	varies
$\overline{M}_m$	mass average molar mass (7.10)	$kg\ mol^{-1}$
$\overline{M}_n$	number average molar mass (7.10)	$kg\ mol^{-1}$
n	amount of substance (1.1)	mol
n	number of electrons in a cell reaction (charge number) (8.3)	1
n	order of a reaction (19.2)	1
n	order of reflection in the Bragg equation (23.4)	1
n	principle quantum number (11.1)	1
n	refractive index (12.12)	1
N	number of particles (17.1)	1
N	number of species (2.16)	1
N_A	Avogadro constant (1.1)	mol^{-1}
p	momentum of a particle (10.4)	$kg\ m\ s^{-1}$
p	number of phases (6.6)	1
$p_{nl}(r)$	probability density (11.2)	m^{-3}
$\hat{p}_q$	linear momentum operator (10.7)	$kg\ m\ s^{-1}$
pH	$-\log a_{H^+}$ (8.11)	1
pK	$-\log K$ (9.2)	1
P	dielectric polarization (12.12)	$C\ m^{-2}$
P	pressure (1.2)	Pa

Symbol	Name	Symbol for SI Unit
P_i^*	equilibrium vapor pressure of i (7.1)	Pa
P°	standard state pressure (4.8)	Pa
P_c	critical pressure (1.8)	Pa
P_i	partial pressure of i (1.4)	Pa
P_r	reduced pressure (1.9)	1
P	molar polarization (12.12)	$m^3 \, mol^{-1}$
q	heat absorbed by a system (2.3)	J
q	molecular partition function (17.2)	1
Q	canonical ensemble partition function (17.16)	1
Q	electric charge (2.9)	C
Q	quotient of activities (5.1)	1
r	vector distance (8.1)	m
Q_i	charge on the ith ion (12.9)	C
$\langle r^2 \rangle^{1/2}$	root-mean-square end-to-end distance (22.4)	m
R	gas constant (1.3)	$J \, K^{-1} \, mol^{-1}$
R	number of independent reactions (5.13)	1
R	Rydberg constant (10.2)	m^{-1}
R_e	equilibrium internuclear distance in a diatomic molecule (12.1)	m
$R_{nl}(r)$	hydrogenlike radial wave function (11.1)	$m^{-3/2}$
s	electron spin quantum number (generally referred to as the spin) (11.4)	1
s	number of species (6.6)	1
s_i	quantum number of spin angular momentum of electron i in an atom (11.14)	1
s_i	spin angular momentum vector for electron i in an atom (11.14)	J s
s_i	stoichiometric number for elementary step i in a mechanism (19.10)	1
s_{zi}	z component of the spin angular momentum of electron i in an atom (11.14)	J s
s^*	sticking coefficient (24.2)	1
S	entropy (3.5)	$J \, K^{-1}$
S	quantum number of spin angular momentum of an atom (11.4)	1
S	reaction cross section (20.1)	m^2
S	sedimentation coefficient (22.11)	s
S	spin angular momentum vector (11.4)	J s
$\hat{S}$	spin angular momentum operator (11.4)	J s
S_n	improper rotation operator (13.2)	1
S_z	z component of the spin angular momentum (11.4)	J s
S_{AB}	overlap integral (12.3)	1
S_0	surface area of catalyst (24.8)	m^2
$\overline{S}_i$	molar or partial molar entropy of i (4.12)	$J \, K^{-1} \, mol^{-1}$
$\overline{S}_i^*$	molar entropy of pure substance i (4.14)	$J \, K^{-1} \, mol^{-1}$
$\hat{S}_z$	operator for the z component of the spin angular momentum (11.4)	J s
$\Delta S^{\ddagger}$	entropy of activation (20.5)	$J \, K^{-1} \, mol^{-1}$

Symbol	Name	Symbol for SI Unit
$\Delta_r S^\circ$	reaction entropy (3.15)	$J\ K^{-1}\ mol^{-1}$
$t_{1/2}$	half-life for a reaction (19.2)	s
T	kinetic energy (10.7)	J
T	thermodynamic temperature (1.3)	K
T_B	Boyle temperature (1.6)	K
T_c	critical temperature (1.8)	K
T_r	reduced temperature (1.9)	1
T_1	spin–lattice relaxation time (16.2)	s
T_2	transverse relaxation time (16.2)	s
$\hat{T}_x$	operator for the kinetic energy in the x direction (10.7)	J
u	electric mobility (21.3)	$m^2\ V^{-1}\ s^{-1}$
U	internal energy (2.2)	J
$\overline{U}_i$	molar or partial molar internal energy of i (2.2)	$J\ mol^{-1}$
v	rate of reaction (19.1)	$mol\ m^{-3}\ s^{-1}$
v	specific volume (22.12)	$m^3\ kg^{-1}$
v	velocity of a particle (10.4)	$m\ s^{-1}$
v	vibrational quantum number (10.12)	1
v_p	most probable speed (18.4)	$m\ s^{-1}$
v_s	speed of sound (18.4)	$m\ s^{-1}$
$\langle v_i \rangle$	average velocity in the ith direction (18.2)	$m\ s^{-1}$
$\langle v^2 \rangle^{1/2}$	root-mean-square speed (18.4)	$m\ s^{-1}$
V	potential energy (10.6)	J
V	volume (1.1)	m^3
$\overline{V}$	molar volume (1.1)	$m^3\ mol^{-1}$
$\overline{V}_i$	molar volume or partial molar volume of i (4.12)	$m^3\ mol^{-1}$
$\overline{V}_c$	critical molar volume (1.8)	$m^3\ mol^{-1}$
$\overline{V}_r$	reduced molar volume (1.9)	1
$\overline{V}_i^*$	molar volume of a pure substance (4.12)	$m^3\ mol^{-1}$
w	work done on a system (2.1)	J
W	number of ways distinguishable particles can be distributed between energy levels (17.1)	1
x_i	mole fraction of i in liquid phase (7.1)	1
$X(x)$	wavefunction in x direction (10.10)	$m^{-1/2}$
y_i	mole fraction of i in a gas (1.4)	1
$Y_l^m(\theta, \phi)$	spherical harmonic (10.13)	1
$Y(y)$	wavefunction in y direction (10.10)	$m^{-1/2}$
z_i	charge number of ion i (signed) (8.1)	1
Z	atomic number or proton number (11.1)	1
Z	compressibility factor (1.5)	1
$Z(z)$	wavefunction in z direction (10.10)	$m^{-1/2}$
Z_{11}	collision density for like molecules (18.6)	$m^{-3}\ s^{-1}$
Z_{12}	collision density for unlike molecules (18.6)	$m^{-3}\ s^{-1}$
$z_{1(1)}$	collision frequency for like molecules (18.6)	s^{-1}
$z_{1(2)}$	collision frequency for unlike molecules (18.6)	s^{-1}

Symbol	Name	Symbol for SI Unit
α	Coulomb integral (12.7)	
α	cubic expansion coefficient (1.10)	K^{-1}
α	polarizability (12.12)	$C^2\ m^2/V^{-1}$
α	spin function for an electron (spin up function) (11.4)	1
$[\alpha]$	specific rotation (15.11)	$deg\ m^2\ kg^{-1}$
α_a	anodic transfer coefficient (21.12)	1
α_c	cathodic transfer coefficient (21.12)	1
α_i	polarizability of ith molecule (12.11)	m^3
β	reciprocal temperature parameter ($1/kT$) (17.3)	J^{-1}
β	resonance integral (12.7)	
β	spin function for an electron (spin down function) (11.4)	1
γ	ratio of C_P to C_V (2.16)	1
γ	surface tension (2.9)	$N\ m^{-1}$
γ_e	gyromagnetic ratio of the electron (11.3)	$A\ m^2\ J^{-1}\ s^{-1}$
γ_i	activity coefficient of i based on deviations from Raoults' law (7.7)	1
γ_i'	activity coefficient of i based on deviations from Henry's law (7.8)	1
$\gamma_\pm$	mean ionic activity coefficient (8.4)	1
δ	chemical shift in NMR (16.5)	1
Δ_{AB}	electronegativity difference (12.8)	J
Δt	uncertainty in time (10.5)	s
Δx	extent of wave packet in space (10.5)	m
ΔE	uncertainty in energy (10.5)	J
$\Delta \nu$	range in frequency (10.5)	s^{-1}
ϵ	efficiency (3.3)	1
ϵ	energy of a photon (10.3)	J
ϵ	molar absorption coefficient (15.4)	$m^2\ mol^{-1}$
ϵ	parameter in Lennard–Jones equation (12.11)	J
ϵ_0	permittivity of vacuum (8.1)	$C^2\ N^{-1}\ m^{-2}$
ϵ_r	relative permittivity (dielectric constant) (8.1)	1
η	number of subsystems in a canonical ensemble (17.16)	1
η	overpotential (21.12)	V
η	viscosity (18.10)	$kg\ m^{-1}\ s^{-1} = Pa\ s$
$[\eta]$	intrinsic viscosity (22.10)	$m^2\ kg^{-1}$
Θ_e	characteristic electronic temperature (17.10)	K
Θ_r	characteristic rotational temperature (17.9)	K
Θ_v	characteristic vibrational temperature (17.8)	K
Θ_D	Debye temperature (2.14)	K
Θ_D	Debye temperature for a monatomic crystal (17.15)	K
Θ_E	Einstein temperature for a monatomic crystal (17.15)	K

Symbol	Name	Symbol for SI Unit
Θ	surface coverage (24.2)	1
κ	electric conductivity (21.3)	$\Omega^{-1}\,m^{-1}$
κ	isothermal compressibility (1.10)	Pa^{-1}
κ	naperian molar absorption coefficient (15.4)	$m^2\,mol^{-1}$
κ	thermal conductivity (18.10)	$J\,m^{-1}\,s^{-1}\,K^{-1}$
λ	mean free path (18.7)	m
λ	wave length (10.1)	m
λ_k	Lagrange multiplier (5.14)	1
Λ	quantum number for electronic angular momentum along molecular axis (12.4, 15.1)	1
Λ	thermal de Broglie wavelength (17.7)	m
μ	electric dipole moment (12.12)	C m
μ	magnitude of dipole moment (12.12)	C m
μ	reduced mass (10.11)	kg
$\hat{\mu}$	dipole moment operator (14.1)	C m
μ	magnetic dipole moment (11.3)	$J\,T^{-1} = A\,m^2$
μ_i	chemical potential of i (4.10)	$J\,mol^{-1}$
μ_i	magnitude of dipole moment of ith molecule (12.11)	C m
μ_i^*	chemical potential of pure component i at pressure P (4.14)	$J\,mol^{-1}$
μ_{nm}	transition dipole moment (14.1)	C m
μ_0	permeability of vacuum (16.1)	$N\,A^{-2}$
μ_s	magnetic dipole moment of an electron (11.4)	$A\,m^2$
μ_z	magnetic dipole moment in the z direction (11.3)	$J\,T^{-1} = A\,m^2$
μ_B	Bohr magneton (11.3)	$J\,T^{-1} = A\,m^2$
μ_{JT}	Joule–Thomson coefficient (2.13)	$K\,Pa^{-1}$
μ_N	nuclear magneton (16.2)	$J\,T^{-1}$
$\mu_i^*(1)$	reference chemical potential for solute i in a dilute real solution (7.6)	$J\,mol^{-1}$
ν	average kinetic chain length (22.9)	1
ν	frequency (10.1)	s^{-1}
ν	sum of stoichiometric numbers for a reaction (5.6)	1
ν_i	stoichiometric number for reactant i (positive for products, negative for reactants) (2.17)	1
$\boldsymbol{\nu}$	stoichiometric number matrix (5.13)	1
$\tilde{\nu}$	wavenumber (10.2)	m^{-1}
$\nu_\pm$	number of ions in an electrolyte molecule ($\nu_+ + \nu_-$) (8.4)	1
$\tilde{\nu}_R$	Raman frequency (14.9)	m^{-1}
ξ	extent of reaction (2.17)	mol
ξ'	dimensionless extent of reaction (5.4)	1
Π	osmotic pressure (7.10)	Pa
ρ	number density (18.6)	m^{-3}
$\rho(\nu, T)$	energy density in a cavity as a function of frequency (10.1)	$J\,m^{-3}\,s$

Symbol	Name	Symbol for SI Unit
$\rho(\lambda, T)$	energy density as a function of wavelength (10.1)	$J\ m^{-4}$
σ	absorption cross section (15.4)	m^2
σ	parameter in Lennard–Jones equation (12.11)	m
σ	reflection operator (13.2)	1
σ	shielding constant (16.5)	1
σ	standard deviation (21.2)	varies
σ	surface area (2.9)	m^2
σ	symmetry number (3.16)	1
$\sigma_g(l)$	molecular orbital (12.4)	
σ_x	standard deviation in x (10.9)	m
Σ	magnetic quantum number for spin angular momentum of a molecule (12.4, 15.1)	1
τ	radiative lifetime (14.2)	s
τ	turbidity (22.3)	m^{-1}
ϕ	eigenfunction of operator $\hat{A}$ (10.8)	varies
ϕ	electric potential difference (2.9)	V
ϕ	electric potential (8.1)	V
ϕ	fugacity coefficient (4.9)	1
ϕ	quantum yield (20.7)	1
ϕ	volume fraction (22.10)	1
χ	magnetic susceptibility (16.1)	1
ψ	wavefunction in three dimensions (10.6)	$m^{-3/2}$
ψ_{el}	electronic wavefunction (12.1)	$m^{-3/2}$
ψ_{nu}	nuclear motion wavefunction (12.1)	$m^{-3/2}$
$\psi(x)$	wavefunction in the x direction (10.6)	$m^{-1/2}$
$\psi(x, t)$	de Broglie wave in one dimension (10.5)	$m^{-1/2}$
$\Psi(r, t)$	time dependent wavefunction (10.15)	$m^{-3/2}$
ω	fundamental vibration frequency $(2\pi\nu)$ (10.11)	$rad\ s^{-1}$
ω_e	harmonic vibration wavenumber (12.2)	m^{-1}
$\omega_e x_e, \omega_e y_e$	anharmonicity constants (12.2)	m^{-1}
Ω	number of equally probable arrangements (3.12)	1
Ω	number of microstates in a macrostate (17.2)	1
Ω	quantum number for the total angular momentum along the internuclear axis of a diatomic molecule (15.1)	1

G
Answers to the First Set of Problems

Chapter 1   Zeroth Law of Thermodynamics and Equations of State

1.1 (a) 0.696, 0.304; (b) 0.522 bar, 0.228 bar; (c) 17.00 L.

1.2 21.0 g mol^{-1}; 0.643.

1.3 (a) 0.6884, 0.3116; (b) 18.49%.

1.4 0.014 L mol^{-1}.

1.5 276 cm^3 mol^{-1}.

1.6 $b = 2\pi d^3 N_A/3$, 21.7 cm^3 mol^{-1}.

1.7 (a) 0.603 L mol^{-1}; (b) 0.39 L mol^{-1}.

1.9 -0.048 L mol^{-1}, -0.026 L mol^{-1}; Fig. 1.9 yields -0.040 L mol^{-1} and -0.020 L mol^{-1}.

1.10 0.4719 L mol^{-1}.

1.11 $P_c = 8b/RT_c$.

1.12 112 K.

1.13 $V = K \exp(\beta T) \exp(-\kappa P)$.

1.14 $\alpha = 1/T$, $\kappa = 1/P$.

1.15 $V = V_0 \exp[\alpha(T - T_0)]$, $V = V_0[1 + \alpha(T - T_0)]$.

1.16 0.472 bar.

1.17 0.333 bar, $y(O_2) = 0.177$, $y(N_2) = 0.823$.

Chapter 2   First Law of Thermodynamics

2.1 914 m.

2.2 $w = -1.24$ kJ mol^{-1}. The work on the atmosphere is 1.24 kJ mol^{-1}.

2.3 (a) 4269 m; (b) 6.97 min; (c) 1.05 min; (d) 1.28 g.

2.4 The second function is exact because $(\partial M/\partial y)_x = (\partial N/\partial x)_y = -1/y^2$.

2.5 $(\partial/\partial P)(\partial V/\partial T)_P = (\partial/\partial T)(\partial V/\partial P)_T = K e^{\alpha T} e^{-\alpha T}(-\kappa\alpha)$.

2.6 (a) -5.03 kJ mol^{-1}; (b) -2.12 kJ mol^{-1}.

2.7 (a) $w = nRT \ln[(V_2 - nb)/(V_1 - nb)] + an^2[(1/V_1) - (1/V_2)]$; (b) (1) -9697 J; (2) -9575 J.

2.8 (a) -3.10 kJ mol^{-1}; (b) 40.69 kJ mol^{-1}; (c) 37.59 kJ mol^{-1}; (d) 40.69 kJ mol^{-1}.

2.9 (a) -5.70 kJ mol^{-1}; (b) 5.70 kJ mol^{-1}; (c) 0; (d) 0.

2.10 $T_{inv} = 2a/bR$; ΔT is negative.

2.11 41.572 kJ mol^{-1}.

2.12 21.468 kJ mol^{-1}.

2.13 (a) 13.07 kJ mol^{-1}; (b) 10.59 kJ mol^{-1}.

2.14 $\Delta T = 160.4$ °C, $P = 1.609$ bar.

2.15 $T = 118.70$ K, -2238 J mol^{-1}.

2.16 (a) -46.1 kJ; (b) -24.1 kJ.

2.17 0, -899.8, 899.8 J mol^{-1}.

2.18 0, -748, -748, -1247 J mol^{-1}.

2.19 (a) -542.2; (b) -184.62; (c) -103.71; (d) -9.48 kJ mol^{-1}.

2.20 (a) -13.4, -7.98, -13.6 MJ kg^{-1}; (b) -0.744, -0.299, -0.680.

2.21 -454.85 kJ mol^{-1}.

2.22 493.580, 498.346, 513.482 kJ mol^{-1}.

2.23 267.66, -39.84, -214.65, 13.17 kJ mol^{-1}; single reactor.

2.24 -5619 J mol^{-1}.

2.25 45,054 J mol^{-1}.

2.26 -89.849 kJ mol^{-1}.

2.27 493.486 kJ mol^{-1}, 5.115 eV.

2.28 41.4 kJ mol^{-1}.

2.29 -41.6 kJ mol^{-1}.

2.30 -81.21 kJ mol^{-1}.

2.31 -97.12 kJ mol^{-1}.

2.32 -73.965 kJ mol^{-1}.

2.33 (a) 0; (b) 13.05 kJ mol^{-1}.

2.34 Estimate $= 51.80$ J K^{-1} mol^{-1}, Table 2.2 yields 52.95 J K^{-1} mol^{-1}.

2.35 -126.62, -135.99 kJ mol^{-1}.

Chapter 3   Second and Third Laws of Thermodynamics

3.1 17 000 ft.

3.2 (a) 214 J; (b) 307 J, or 93 J more than (a).

3.3 0.079 J.

3.4 (a) 109.04; (b) 0 J K^{-1} mol^{-1}.

3.5 55.42 J K^{-1} mol^{-1}.

3.6 2.657 J K^{-1} mol^{-1}.

3.7 (a) 6.371; (b) 10.618 J K^{-1}mol^{-1}.

3.8 (a) 26.4 kJ mol^{-1}; (b) -4.96 kJ mol^{-1}; (c) 26.4 kJ mol^{-1}; (d) 21.4 kJ mol^{-1}; (e) 46.99 J K^{-1} mol^{-1}.

3.9 Since this ΔS is always positive, the change is always spontaneous.

3.10 (a) 15.56; (b) 0 J K^{-1} mol^{-1}.

3.11 (a) 0 J mol^{-1}, 10.01 J K^{-1} mol^{-1}, -3.003 kJ mol^{-1}, 3.003 kJ mol^{-1}; (b) 0, 10.01, 0, 0 J K^{-1} mol^{-1}.

3.12 (a) 19.14 J K^{-1} mol^{-1}, the same; (b) 0, 19.14 J K^{-1} mol^{-1}.

3.14 19.26 J K^{-1} mol^{-1}.

3.15 (a) -39.52; (b) 41.92; (c) 2.40 J K^{-1}, irreversible because $\Delta S_{\text{syst}} = 0$.

3.16 1.12 J K^{-1} mol^{-1}.

3.17 44.95 J K^{-1}.

3.19 14.92 J K^{-1} mol^{-1}.

3.20 -154.4 kJ mol^{-1}.

3.21 150.67 J K^{-1} mol^{-1}.

3.22 116.969, 130.002 J K^{-1} mol^{-1}.

3.23 (a) -163.34; (b) 20.066; (c) -162.0 J K^{-1} mol^{-1}.

3.24 -80.67 J K^{-1} mol^{-1}. Since the ions polarize the neighboring water molecules, the product state is more ordered than the reactant state.

3.25 (a) 309.3; (b) 303.8 J K^{-1} mol^{-1}.

3.13	Reversible	Irreversible	Isol. Rev.	Isol. Irrev.
w/kJ mol^{-1}	-5.71	0	0	0
q/kJ mol^{-1}	5.71	0	0	0
$\Delta \overline{U}$/kJ mol^{-1}	0	0	0	0
$\Delta \overline{H}$/kJ mol^{-1}	0	0	0	0
$\Delta \overline{S}$/J K^{-1}mol^{-1}	19.1	19.1	0	19.1

Chapter 4   Gibbs Energy and Helmholtz Energy

4.1 (a) 24.91; (b) 25.55 J K^{-1} mol^{-1}.

4.2 $\overline{C}_P - \overline{C}_V = R/\{1 - (2a/RT)[(\overline{V} - b)^2/\overline{V}^3]\}$.

4.3 $dS = (C_P/T)dT - \alpha V\, dP$.

4.4 -10.99 J K^{-1} mol^{-1}.

4.5 $\overline{C}_P - \overline{C}_V = R/(1 - 2a/RT\overline{V})$.

4.6 0.51 J K^{-1} mol^{-1}.

4.7 (a) $b - 2a/RT$; (b) -25 J bar^{-1} mol^{-1}.

4.8 (a) $\Delta G_2 = \Delta G_1 T_2/T_1 + \Delta H(1 - T_2/T_1)$; (b) $\Delta G_2 = \Delta G_1 T_2/T_1 + (\Delta H_1 + T_1 \Delta C_P)(1 - T_2/T_1) + T_2 \Delta C_P \ln(T_2/T_1)$.

4.9 1.8 kJ mol^{-1}.

4.11 1758 J mol^{-1}.

4.12 (a) -5708 J mol^{-1}; (b) -5708 J mol^{-1}.

4.13 166.848 J K^{-1} mol^{-1}, 11.41 kJ mol^{-1}.

4.14 (a) -4993 J mol^{-1}; (b) 4993 J mol^{-1}; (c) 4993 J mol^{-1}; (d) 4993 J mol^{-1}; (e) 0; (f) 0; (g) -13.38 J K^{-1} mol^{-1}.

4.15 (a) -3193 J mol^{-1}; (b) 33 340 J mol^{-1}; (c) 33 340 J mol^{-1}; (d) 30 147 J mol^{-1}; (e) 0; (f) 86.8 J K^{-1} mol^{-1}.

4.16 $S = -a + c/T^2$; $H = b + 2c/T$.

4.17 -65.1 J mol^{-1}.

4.18 -63.6 J mol^{-1}.

4.19 $\overline{G} = \overline{G}° + RT \ln(P/P°) + (b - a/RT)P$

$\overline{S} = \overline{S}° - RT \ln(P/P°) + aP/RT^2$

$\overline{A} = \overline{G}° + RT \ln(P/P°) - RT$

$\overline{U} = \overline{U}° - aP/RT$

$\overline{H} = \overline{H}° + (b - 2a/RT)P$

$\overline{V} = RT/P + b - a/RT$

4.20 106.9 bar.

4.21 51.7 bar.

4.22 29.3 cm^3 mol^{-1}.

4.23 -1239 J mol^{-1}; 4.159 J K^{-1} mol^{-1}.

4.24 3.4 kJ.

4.25 3.35×10^{-2} cm.

4.26 (a) 15.7 cm; (b) 7.86 mm.

4.27 (a) 200 Pa; (b) 20 Pa.

4.28

	$(\partial S/\partial l)_T/10^3$ N K^{-1} m^{-2}	$(\partial U/\partial l)_T/$N m^{-2}
200% elong.	-3.0	0.1
400% elong.	-7.8	-1.4

Chapter 5 Chemical Equilibrium

5.1 (a) 48.9 kJ mol^{-1}; (b) -2.6 kJ mol^{-1}; (c) Yes.

5.2 6.47×10^{-3}.

5.3 (a) 0.503; (b) 1.36; (c) 0.879.

5.4 1.83.

5.5 1.64×10^{-3}.

5.6 5.73 kJ mol^{-1}.

5.7 (a) 9.520 g; (b) 1.216, 0.042, 1.440 bar.

5.8 $K = \xi^2(2 - 2\xi)^2/(1 - \xi)(1 - 3\xi)^3(P/P°)$.

5.9 0.0819, 0.0337, 94.8%.

5.10 1.44, 0.545.

5.11 0.0055.

5.12 (a) 10.3; (b) 22.0 bar.

5.16 0.311.

5.17 231 bar.

5.18 (a) 12.7%; (b) No effect; (c) 54 bar.

5.19 (a) 5.97; (b) 0.29 bar.

5.20 (a) 2.03×10^{-3}; (b) Increase; (c) Increase; (d) No; (e) Decrease.

5.21 0.973.

5.22 (a) Decreases; (b) Increases; (c) No effect; (d) Decreases; (e) Increases.

5.23 1023 K.

5.24 75 kJ mol^{-1}.

5.25 0.0145, $K_c = [C_2H_4][H_2]/[C_2H_6]c°$.

5.26 (a) $y(N_2) = 0.1266$, $y(H_2) = 0.3798$, $y(NH_3) = 0.493$, $V = 22.27$ L; (b) $y(N_2) = 0.990$, $y(H_2) = 0.2970$, $y(NH_3) = 0.302$, $P = 0.698$ bar.

5.28 $P(H_2O) = 0.084$, $P(CO) = 0.458$, $P(H_2) = 0.458$ bar.

5.29 (a) 0.0204, 0.187; (b) 154 kJ mol^{-1}.

5.30 425 K.

5.32 3.2 bar.

5.33 1.82×10^{-13} bar.

5.34 20×10^{-5} bar.

5.35 (a) 1.4×10^{-5} bar; (b) 6.83×10^{-3} bar; (c) 0.0051, 0.72 bar.

5.36 $n(CuO) = 0.82$ μmol, $n(Cu_2O) = 0.14$ μmol, $n(O_2) = 0.02$ μmol.

5.37 -331.0 kJ mol^{-1}.

5.38 541.

5.39 (a) One, for example, $CO_2 + H_2 = CO + H_2O$;

(b) Two, for example, $CO_2 + H_2 = CO + H_2O$

$2CO \qquad = CO_2 + C$

5.40 $3C_2H_2 = C_6H_6$

$5C_2H_2 = C_{10}H_8 + H_2$

$$\begin{bmatrix} 2 & 0 & 6 & 10 \\ 2 & 2 & 6 & 8 \end{bmatrix} \begin{bmatrix} -3 & -5 \\ 0 & 1 \\ 1 & 0 \\ 0 & 1 \end{bmatrix} = \begin{bmatrix} 0 & 0 \\ 0 & 0 \end{bmatrix}$$

Chapter 6   Phase Equilibrium: One-Component Systems

6.2   The boiling point is reduced 0.4 °C to 68.3 °C.

6.3   90.6 °C.

6.4   (*a*) -38.81; (*b*) -16.6 °C.

6.5   0.258 bar.

6.6   11.548 kJ mol^{-1}, 6.190×10^{-3} bar.

6.7   3780 bar.

6.8   (*a*) 44.8 kJ mol^{-1}; (*b*) 97.03 °C; (*c*) 97.36 °C.

6.9   64.0 °C, 152.2 kPa.

6.10   166 Pa.

6.11   (*a*) 31.4 kJ mol^{-1}; (*b*) 13.9 kJ mol^{-1}; (*c*) 17.5 kJ mol^{-1}; (*d*) -88 °C.

6.12   (*a*) 6.85 °C; 5249 Pa; (*b*) 10.48 kJ mol^{-1}.

6.13   (*a*) 30.1 kJ mol^{-1}; (*b*) 0.755 bar.

6.14   (*a*) 3.17×10^3 Pa; (*b*) 373.5 K.

6.15   3, 4, and 5.

6.16   (*a*) 1, *T* or *P*; (*b*) 2, *T* and *P*; (*c*) 4, *T*, *P*, and two mole fractions; (*d*) 2, *T* and *P*.

6.17   4, *T*, *P*, H/C, O/C.

6.18   3, 2, 4.

6.19   25.552 mm of mercury or 3406 Pa.

Chapter 7   Phase Equilibrium: Two- and Three-Component Solutions

7.1   3988 J mol^{-1}.

7.3   (*a*) 0.590, 0.410; (*b*) 6980, 2430, 9410 Pa; (*c*) 0.742.

7.4   (*a*) 29.4 kPa; (*b*) 0.581.

7.5   (*a*)

t/°C	88	94	100
x_{benz}	0.633	0.422	0.244
y_{benz}	0.814	0.644	0.439

(*b*) Bubble point, 92 °C; $y_{benz} = 0.72$.

7.6   (*a*) 0.560, 0196; (*b*) 0.884, 0.577.

7.7   Three theoretical plates.

7.8   20.7%.

7.9 $\Delta_{mix}G = RT(x_1 \ln x_1 + x_2 \ln x_2) + wx_1x_2$

$\Delta_{mix}S = -R(x_1 \ln x_1 + x_2 \ln x_2)$

$\Delta_{mix}H = wx_1x_2$

$\Delta_{mix}V = 0$

7.10   (*a*) $y_{EtOH} = 0.69$; (*b*) $y_{EtOH} = 1$; (*c*) $y_{EtOH} = 0.69$, $y_{EtOH} = 0.90$.

7.11 -0.00233 °C.

7.13

y_{EtOH}	0	0.2	0.4	0.6	0.8	1.0
γ_{EtOH}	—	2.045	1.316	1.065	0.982	1.000
γ_{EtOH}	1	1.111	1.333	1.627	1.854	—

7.15 $\gamma_1 = \exp(wx_2^2/RT)$, $\gamma_2 = \exp(wx_1^2/RT)$.

7.16 $\dfrac{w}{RT} = \dfrac{P - (x_1P_1^* + x_2P_2^*)}{x_1x_2[P_1^* + x_1(P_2^* - P_1^*)]}$

7.17

x_2	0	0.2	0.4	0.6	0.8	1.0
γ_2	—	0.24	0.13	0.093	0.080	—
γ_1	1.00	1.15	1.52	2.09	2.82	—

7.18   (*a*) 2.3066 kPa; (*b*) -0.372 °C

7.19 (a) 719 Pa; (b) 73.4 mm of water.

7.20 255 kg mol^{-1}.

7.21 (a) 150 000; (b) 167 000 g mol^{-1}.

7.22 27.3 bar.

7.23 0.365, 0.516.

7.24 0.297.

7.25 0.0569, yes.

7.26 Sb_2Cd_3 is formed.

7.27 In the liquid region $v = 2$, in the two-phase regions $v = 1$, and at the eutectic points $v = 0$.

7.28 (a) Water layer: 5.23% isobutanol, 94.5% water. Benzene layer: 39.57% isobutanol, 57.09% benzene; (c) 10% water, 72% isobutanol; 18% benzene.

7.31 0.125, 0.186, 0.182, 0.166, 0.120.

Chapter 8 Electrochemical Equilibrium

8.1 14.398 J C^{-1}; 1389.3 kJ mol^{-1}.

8.2 277.8 kJ mol^{-1}.

8.3 5.04 miles.

8.4 (a) 6.34×10^{-3}; (b) 7.44×10^{-5}.

8.5 0.905.

8.6 (a) 0.24; (b) 0.80; (c) 0.06 mol kg^{-1}.

8.7 (a) 0.964; (b) 0.880; (c) 0.762.

8.8 (a) 0.710 V; (b) 0.804.

8.9 (a) Na/NaOH(m)/H$_2$/Pt
At 25 °C, $E = E° - 0.0591 \log[m^2\gamma_\pm^2 P(H_2)^{1/2}]$
(b) Pt/H$_2$/H$_2$SO$_4$(m)/Ag$_2$SO$_4$/Ag
At 25 °C, $E = E° - 0.0296 \log[4m^3\gamma_\pm^3 P(H_2)^{-1}]$

8.10 (a) Pb(s) + Hg$_2$SO$_4$(s) = PbSO$_4$(s) + 2Hg(l); (b) -186.16 kJ mol^{-1}, 33.58 J K^{-1} mol^{-1}, -176.15 kJ mol^{-1}.

8.11 (a) -131.260 kJ mol^{-1}, -167.127 kJ mol^{-1}, -120.3 kJ mol^{-1}; (b) -131.260 kJ mol^{-1}, -167.127 kJ mol^{-1}, 56.6 J K^{-1} mol^{-1}.

8.12 (a) 81.902 kJ mol^{-1}, 5.69×10^{-14}; (b) 81.673 kJ mol^{-1}, 6.28×10^{-14}.

8.13 0.828 V.

8.14 (a) 212.55; (b) -262.46; (c) -47.40; (d) -25.71; (e) -2.9 kJ mol^{-1}.

8.15 (a) R: Cu^{2+} (aq) + $2e^-$ = Cu(s), L: Zn^{2+} (aq) + $2e^-$ = Zn(s); (b) Zn(s) + Cu^{2+} (aq) = Zn^{2+} (aq) + Cu(s); (c) 1.100 V; (d) 1.62×10^{37}; (e) $K = m(ZnCl_2)^3\gamma_\pm(ZnCl_2)^3/m(CuCl_2)^3\gamma_\pm(CuCl_2)^3$.

8.16 -131.258, -157.26, -261.86 kJ mol^{-1}.

8.17 (a) 3.1×10^{51}; (b) 1.8×10^4; (c) 7.8×10^{-77}.

8.18 -0.152 V.

8.19 1.33×10^{-5} mol kg^{-1}.

8.20 (a) -3.040; (b) 2.889; (c) 1.458 V.

8.21 261.905 kJ mol^{-1}, 240.12 kJ mol^{-1}, -73.132 J K^{-1} mol^{-1}, -32.572 J K^{-1} mol^{-1}.

8.23 -261.92 kJ mol^{-1}.

8.24 (a) 6.49; (b) 3.24×10^{-7}.

8.25 1.30.

8.26 0.60.

8.27 (a) -817.90 kJ mol^{-1}, 1.0596 V; (b) -890.4 kJ mol^{-1}, 356 kJ mol^{-1}.

8.28 3.474 V.

8.29 0.018 V; The β phase is positive.

Chapter 9 Ionic Equilibria and Biochemical Reactions

9.2 4.70.

9.3 79.885 kJ mol^{-1}, 55.836 kJ mol^{-1}, -80.66 J K^{-1} mol^{-1}, 1.010×10^{-14}.

9.4 For the acetic acid dissociation, $\Delta_r S° = 91$ J K^{-1} mol^{-1}. The increase in order is due to the hydration of the ions that are formed. For aniline, $\Delta_r S° = -18$ J K^{-1} mol^{-1}.

This number is smaller because there is no change in the number of ions.

9.5 2.026, 6.831.

9.6 [His$^-$] = 6.3×10^{-4} mol L^{-1}, [HisH] = 9.03×10^{-2} mol L^{-1}, [HisH$_2^+$] = 9.03×10^{-3} mol L^{-1}, [HisH$_3^{2+}$] = 6×10^{-8} mol L^{-1}.

9.7 4.35.

9.8 No. 1.1×10^{-2} mol L^{-1}.

9.9 5.77 kJ mol^{-1}.

9.10 135 g.

9.11 -20.3 kJ mol^{-1}.

9.12 11.2.

9.13 -41.0 kJ mol^{-1}.

9.14 At pH 7, one mole of H$^+$ is produced for each mole of ethyl acetate hydrolyzed. Therefore, the enthalpy of dissociation of the buffer is needed.

9.15 $Y = ([A]^n/K)/(1 + [A]^n/K)$.

9.17 2.4, 4×10^7.

Chapter 10 Quantum Theory

10.1 (a) 9.07×10^5 J s^{-1} m^{-2}; (b) 24.9 s.

10.2 (a) 4830 K; (b) 3623 K.

10.3 (a) 3.61×10^{-19} J, 2.77×10^{20} s^{-1}; (b) 1.99×10^{-23} J, 5.03×10^{25} s^{-1}.

10.4 (a) $hc/4.965k$; (b) 5.9×10^{23} mol^{-1}.

10.5 1.8756, 1.2822, 1.0941 μm.

10.6 4.49 cm.

10.7 4.64 eV.

10.8 4.668×10^7, 4.634×10^7, 4.473×10^7 m s^{-1}.

10.9 (a) 0.0387 nm; (b) 2.24 nm.

10.10 0.012 J mol^{-1}.

10.11 0.0259 nm, 3.46 nm.

10.12 0.1452 nm.

10.13 (a) 2.21×10^{-33} m; (b) 6.626×10^{-19} m; (c) 6.626×10^{-10} m; (d) 3.1×10^{-9} m.

10.14 5.27×10^{-25} J, 0.318 J mol^{-1}.

10.15 (a) 1/2; (b) 0.609.

10.16 5.

10.17 (a) $-2axe^{-ax^2}$, $4ax^2e^{-ax^2} - 2ae^{ax^2}$; (b) $-b \sin bx$, $-b^2 \cos bx$ (Therefore, $\cos bx$ is an eigenfunction of d^2/dx^2 with eigenvalue $-b^2$.); (c) ike^{ikx}, $-k^2e^{ikx}$, (Therefore, e^{ikx} is an eigenfunction of both d/dx and d^2/dx^2 with eigenvalues ik and $(ik)^2 = -k^2$, respectively.).

10.18 $(\hat{x}\hat{p}_x - \hat{p}_x\hat{x}) = -\hbar/i$.

10.19 (a) 580.5, 2322, 5225 kJ mol^{-1}; (b) 68.7 nm.

10.20 (a) Approximately 27; (b) 2.7×10^4; (c) 2.7×10^8.

10.22 1, 3, 3.

10.23 (a) 323.3, 122.5, 575.1 N mol^{-1}; (b) 545 cm^{-1}.

10.25 $\frac{3}{2}h\nu_0$.

10.26 6.93×10^{13} s^{-1}, 4.52×10^{13} s^{-1}.

10.27 $\langle x \rangle = 0$, $\langle x^2 \rangle = \frac{3}{2}(\hbar^2/k\mu)^{1/2}$, $\Delta x = [\frac{3}{2}(\hbar^2/k\mu)^{1/2}]^{1/2}$.

10.28 8.67×10^{-24}, 2.60×10^{-23} J.

10.29 $i\hbar\hat{L}_z$, $-i\hbar\hat{L}_y$.

10.30 (a) 1.139×10^{-26} kg; (b) 1.449×10^{-46} kg m^2; (c) 2.58×10^{-3} m.

Chapter 11 Atomic Structure

11.1 13.598 42 eV.

11.2 (a) 145.9; (b) 82.05; (c) 82.05; (d) 36.47 kJ mol^{-1}.

11.3 $E/V = 54.44$, 122.49, 217.76, 340.25, 489.96.

11.4 (a) 13.598 48, 13.602 18 eV; (b) 656.470, 656.291 nm.

11.5 $1.646\ 397 \times 10^{-28}$ kg, $1.983\ 35 \times 10^9$ m^{-1}, $E = -(2461/n^2)$ eV, 0.2928 pm.

11.6 $Z = 5$, $n_1 = 2$, and $\lambda^{-1} = 25(0.010\ 973\ 732)(\frac{1}{4} - 1/n_2^2)$; $n_2 = 2 + m$, $m = 1, 2, \ldots$.

11.7 (a) 2; (b) 8; (c) 18. (Note degeneracy $= 2n^2$.)

11.8 105.8, 176.4 pm.

11.9 For 2s, $r_{node} = 2a_0/Z$. For 3s, $r_{node} = 7.098\ a_0/Z$, $1.902\ a_0/Z$.

11.10 (a) 317.5, 264.6 pm; (b) 105.8, 88.2 pm.

11.11 $\langle r \rangle = \frac{3}{2}a_0/Z$.

11.12 $\langle r \rangle = 5a_0/Z$.

11.14 For 2p, $|L| = 1.491 \times 10^{-34}$ J s, $L_z/\hbar = 1, 0, -1$. For 3d, $|L| = 2.583 \times 10^{-34}$ J s $L_z/\hbar = -2, -1, 0, +1, +2$.

11.15

	3s	3p	3d
$\lvert L \rvert = \sqrt{l(l+1)}\hbar$	0	$\sqrt{2}\hbar$	$\sqrt{6}\hbar$
Radial nodes	2	1	0
Angular nodes	0	1	2

11.16 1.8569×10^{-23} J, 1.0698×10^{-2} m, microwave.

11.18 $1s^2$, $1s^2$, $1s^2 2s^2\ 2p^6$, $1s^2\ 2s^2\ 2p^6$, $1s^2\ 2s^2\ 2p^6$.

11.19 2, 2, 6, 2, 6, 10.

Chapter 12   Molecular Electronic Structure

12.1 13.6 eV, 19.4 eV, -300 eV, -51.4 eV.

12.2 941.49 kJ mol^{-1}.

12.3 $[2(1+s)]^{-1/2}$, $[2(1-s)]^{-1/2}$.

12.4 See Fig. 12.5.

12.5 See Fig. 12.5.

12.6 Maximum at $R = 0$, minimum at $R = \infty$, $S = 0.766$.

12.7
$$\Psi_1 = \frac{1}{2^{1/2}}\begin{vmatrix} 1s_A(1)\alpha(1) & 1s_A(2)\alpha(2) \\ 1s_B(1)\beta(1) & 1s_B(2)\beta(2) \end{vmatrix}$$
$$-\frac{1}{2^{1/2}}\begin{vmatrix} 1s_A(1)\beta(1) & 1s_A(2)\beta(2) \\ 1s_B(1)\alpha(1) & 1s_B(2)\alpha(2) \end{vmatrix}$$
$$\Psi_2 = \frac{1}{2^{1/2}}\begin{vmatrix} 1s_A(1)\alpha(1) & 1s_A(2)\alpha(2) \\ 1s_B(1)\alpha(1) & 1s_B(2)\alpha(2) \end{vmatrix}$$

Ψ_3 is the same as Ψ_2 with β replacing α. Ψ_4 is obtained from Ψ_1 by replacing the minus sign with a plus sign.

12.9 For example, $\Psi_{sp^2(i)} = \dfrac{1}{4\sqrt{2\pi}}[(2-\sigma)+\sqrt{2}\sigma\cos\theta]$
$\times \dfrac{e^{-\sigma}}{\sqrt{3}}$

$\theta = 0°$: $\Psi = \dfrac{1}{4\sqrt{6\pi}}[2 + 0.4.14\sigma]e^{-\sigma}$

$\theta = 90°$: $\Psi = \dfrac{1}{4\sqrt{6\pi}}(2-\sigma)e^{-\sigma}$

$\theta = 180°$: $\Psi = \dfrac{1}{4\sqrt{6\pi}}(2-2.414\sigma)e^{-\sigma}$

12.10 The CH_3 radical is planar with the 3 two-electron bonds due to the overlapping of the sp^2 orbitals of C with the s orbitals of H. The odd electron is in the remaining unhybridized p orbital which is perpendicular to the plane.

12.11 (a) $10\sigma, 2\pi$; (b) $14\sigma, 0\pi$; (c) $18\sigma, 4\pi$; (d) $24\sigma, 6\pi$.

12.12 -153 kJ mol^{-1}.

12.13 Ψ_1, no nodes; Ψ_2, 1 node between atoms 2 and 3; Ψ_3, 2 nodes between 1 and 2 and between 3 and 4; Ψ_4, 3 nodes between 1 and 2, 2 and 3, and 3 and 4.

12.14 Butadiene, 0.472β; benzene, 2β; $\beta = -76.5$ kJ mol^{-1}; The extra stabilization of butadiene is -36.1 kJ mol^{-1}.

12.16 $r_m = 0.309$ nm.

12.17 5.70 eV, 34.7×10^{-30} C m.

12.18 3.78×10^{-29} C m. The ions polarize each other.

12.19 20.4×10^{-30} C m. The charges are not completely separated in HCl.

12.21 0.17, 0.12, 0.039. These are in accord with the electronegativities.

Chapter 13   Symmetry

13.1 C_{2v}

13.2 C_{3v}

13.3 D_{4h}

13.4 C_2

13.5 D_{3h}

13.6 C_1

13.7 C_{4v}

13.8 D_{3d}

13.9 D_{2h}

13.10 D_{2h}

13.11 D_{2d}

13.12 C_{2v}

13.13 D_{2h}

13.14 C_i

13.15

E	C_2^1	σ_h	i
E	E	C_2^1	σ_h	i
C_2^1	C_2^1	E	i	σ_h
σ_h	σ_h	i	E	C_2^1
i	i	σ_h	C_2^1	E

13.16 S_6^1 S_6^2 S_6^3 S_6^4 S_6^5 S_6^6
  C_3^1 i C_3^2 E

13.17 1,1-dichloroethylene, C_{2v}, dipole; *trans*-1,2-dichloroethylene, C_{2h}, no dipole; *cis*-1,2-dichloroethylene, C_{2v}, dipole.

13.18 D_{2d}, no dipole moment.

13.19 H_2S, PCl_3, $C_2H_4Cl_2$, CMeClBrH, IF_5, thiophene.

13.20 C(Me)ClBrH.

13.21 1,2-dichlorobenzene, C_{2v}, dipole; 1,3-dichlorobenzene, C_{2v}, dipole; 1,4-dichlorobenzene, D_{2h}, no dipole.

13.22

13.23 *cis*-bent, C_{2v}; *trans*-bent, C_{2h}.

Chapter 14 Molecular Spectroscopy

14.1 14.4, 1440, 144 000 K.

14.2 (*a*) 2991, 299.1 nm; (*b*) 3343, 33 430 cm^{-1}; (*c*) 0.415, 4.15 eV.

14.3 (*a*) 207 cm^{-1}; (*b*) 4.83×10^{-3} cm.

14.4 3.162×10^{-27} kg, 5.141×10^{-47} kg m^2.

14.5 0.091 83 cm.

14.6 112.8 pm, 11.589 cm^{-1}, 15.452 cm^{-1}.

14.7 3.693, 3.596, 3.766, 3.863 cm^{-1}.

14.8 (*a*) 4.37×10^{-47} kg m^2; (*b*) 163 pm.

14.10 (*a*) 3, 3.79; (*b*) 7, 8.90.

14.11 (*a*) 0.1162; (*b1*) 8.071×10^{-46} kg m^2; (*b2*) same.

14.12 $I = \frac{8}{3}mR^2$.

14.13 6.35 cm^{-1}, 9.95 cm^{-1}, $\lambda_{0\to1} = 1.01 \times 10^{-3}$ m, $\lambda_{1\to2}^{K=0} = 2.51 \times 10^{-4}$ m, $\lambda_{1\to2}^{K=\pm1} = 2.51 \times 10^{-4}$ m.

14.15 (*a*) 2.5 nm; (*b*) 3.5 nm.

14.16

	I_2	Br_2	Cl_2
D_0/eV	1.54	1.97	2.48
k/Nm^{-2}	172.2	246.3	323.0

14.17 -879 J mol^{-1}.

14.18 3112.1, 3811.6 cm^{-1}.

14.19 (*a*) 0.0133, 0.0018, 2×10^{-6}; (*b*) 0.0128, 0.0120, 0.0014.

14.20 (*a*) 6.2×10^{-10}; (*b*) 0.0165, 0.034, 0.007.

14.21 2886.30, 9.98, 10.19, 10.30, 0.21 cm^{-1}.

14.22 216.088 kJ mol^{-1}.

14.23 (*a*) 2632.72, 2666.61 cm^{-1}; (*b*) 3.798 34, 3.750 08 μm.

14.24 (*a*) 3; (*b*) 6; (*c*) 7; (*d*) 30.

14.25 (*a*) 3, 0, 0; (*b*) 3, 2, 1; (*c*) 3, 2, 4; (*d*) 3, 3, 6.

14.26 (*a*) $\bar{v}_4$ and $\bar{v}_5$; (*b*) $\bar{v}_3$ and $\bar{v}_5$; (*c*) $\bar{v}_1$, $\bar{v}_2$, and $\bar{v}_4$.

14.27

$\bar{v}_R$/cm^{-1}	214	312	454	759
λ/μm	46.8	32.0	22.0	13.2

14.28 110 pm.

14.29 2.343 6, 3.906 0, 5.468 4, 7.030 8 cm^{-1}.

Chapter 15 Electronic Spectroscopy of Molecules

15.1 13.93 eV; 1.012×10^4 m s^{-1}.

15.2 (a) 239.5 kJ mol^{-1}; (b) 90.53 kJ mol^{-1}, 0.938 eV.

15.3 5080 eV or 490.14 kJ mol^{-1}, $\Delta_f H°(0 \text{ K}) = 493.57$ kJ mol^{-1}.

15.4 (a) 804.35 nm; (b) 433.7 nm; (c) 94.32 pm.

15.5 (a) 1.2%; (b) 0.311 mol L^{-1}; (c) 32.8%; (d) 10.3 cm.

15.7 (a) Yes, 5.81×104 L mol^{-1} cm^{-1}; (b) 81%.

15.8 2.17×10^{-5}, 3.37×10^{-5} mol L^{-1}.

15.9 1.08, 0.152, and 0.002 2.

15.10

	Int. abs. coeff.		F
Strong	1.06×10^7 to 5.30×10^8	L mol^{-1} cm^{-2}	$0.046 - 2.29$
Weak	1.06×10^3	L mol^{-1} cm^{-2}	4.58×10^{-6}

15.11

	$|\mu_{12}|$/Cm	$|R_{12}|$/pm
Strong	9.72×10^{-30}	60.7
Moderate	3.08×10^{-31}	1.92
Weak	9.72×10^{-33}	0.061

15.12 Yes; There is no way to make the predictions agree for all four orbitals; $\beta = -5h^2/8(1.236)ma^2$.

15.13 5.3 cm^{-1}.

15.14 5.14, 4.09, 3.27 km s^{-1}.

15.15 553 eV.

15.16 (a) Since F is more electronegative than C and H, it pulls electrons toward itself, thereby decreasing the shielding at the carbon nucleus, and increasing the binding energy of the 1s electrons. (b) The binding energy should be greatest for the CF_3 carbon, next largest for the COO carbon, next for the OCH_2 carbon, and smallest for the CH_3 carbon.

15.17 67%.

Chapter 16 Magnetic Resonance Spectroscopy

16.1 1.409, 11.74 T.

16.2 2.394×10^{-5}, 1.995×10^{-2} kJ mol^{-1}; $E/RT = 0.960 \times 10^{-5}$, 0.800×10^{-2}.

16.3 $E/10^{-26}$ J $= -2.240, -0.7465, 0.7456, 2.240$; $v = 22.53$ MHz.

16.4 2.675×10^8 s^{-1} T^{-1}.

16.5 3.4485×10^{-6}.

16.6 0.500 003 43.

16.7 1.69×10^{-2}, 0.665.

16.8

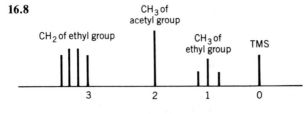

16.9 2.35, 11.7 T.

16.10 (a) 14.52; (b) 1210 Hz.

16.11 7.270×10^{-6}, which is usually given as 7.270.

16.12

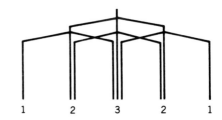

16.13 (a) 5.70 Hz; (b) 0.318 ppm; (c) 576.18, 581.88, 603.84, and 609.54 Hz.

16.14 At room temperature the rate of conversion of cyclohexane from boat to chair forms is so fast that the protons are at the average local magnetic field.

16.15 24.1 s^{-1}.

16.16 2.80 cm^{-1}.

16.17 2.80×10^4 MHz.

16.18 There will be five equally spaced lines with relative intensities 1, 4, 6, 4, 1.

16.19

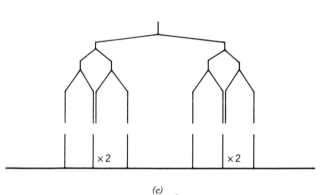

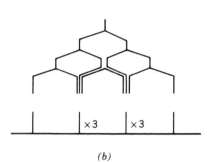

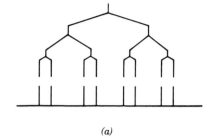

(a)

(b)

(c)

16.20

$a_1 = 9.60$ MHz

$a_2 = 1.40$ MHz

16.21

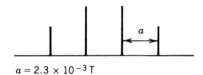

$a = 2.3 \times 10^{-3}$ T

Chapter 17   Statistical Mechanics

17.2   (a) 0.0080; (b) 0.0177.

17.3   (a) 1.253×10^{-17}; (b) 1.147×10^{-169}; (c) 1.100 $\times 10^{-4}$, 2.596×10^{-40}.

17.4   0.651, 0.274, 0.0649, 0.0086, 0.0006, . . .

17.5   0.472, 0.306, 0.149, 0.055, 0.015, 0.003, . . .

17.6   Same as problem 17.4.

17.8 $\mu = -kT \ln(q/N)$.

17.9   (a) 1.23; (b) 2.83×10^{-11}.

17.10 1.414×10^{30}.

17.11 8.84×10^{28}, 2.50×10^{29}.

17.12   (a) 139.86; (b) 82.4 J K^{-1} mol^{-1}.

17.13   146.3 J K^{-1} mol^{-1}.

17.14

	298 K	1000 K
P_0	0.933	0.553
P_1	0.063	0.247
P_2	0.004	0.111

17.17   12.472 J K^{-1} mol^{-1}, 3.718 kJ mol^{-1}, 47.822 J K^{-1} mol^{-1}, -10.54 kJ mol^{-1}.

17.18 1.43×10^{-6}.

17.19 5.00×10^{29}, 1.42×10^{30}, 2.60×10^{30}, 5.72, 36.8.

17.20 (a) 18; (b) 9; (c) 81; (d) 972.

17.21 For CH_4, $\sigma = 12$. For C_2H_4, $\sigma = 4$.

17.22 (a) 6; (b) 18.

17.23 11.53 J K^{-1} mol^{-1}; -3.44 kJ mol^{-1}.

17.24 114.718, 153.301 J K^{-1} mol^{-1}.

17.25

	298 K	1000 K
P_0	0.919	0.625
P_1	0.081	0.303
P_2	0.001	0.072

17.26 4.029, 4.568.

17.27 0.617, 0.294, 0.089.

17.28 0.076, 1.301 kJ mol^{-1}.

17.29 4×10^{-41}.

17.30

T/K	P_1	P_2
0	1	0
$\Delta\epsilon/k$	0.73	0.27
∞	$\frac{1}{2}$	$\frac{1}{2}$

17.31 55.67 J K^{-1} mol^{-1}.

17.32 191.52 J K^{-1} mol^{-1}.

17.33 62.36 J K^{-1} mol^{-1}.

17.34 7.18.

17.35 3.23.

17.37 $\overline{C}_V^\circ/R = 1.5, 3.5, 6.5, 12$
$\overline{C}_P^\circ/R = 2.5, 4.5, 7.5, 13$

17.38 33.26 J K^{-1} mol^{-1}, 58.20 J K^{-1} mol^{-1}.

17.39 0.395 nm, 146 K.

17.40 -61 cm^3 mol^{-1}, 6.9 cm^3 mol^{-1}.

Chapter 18 Kinetic Theory of Gases

18.1 (a) 8.6; (b) 390.

18.2

$v/10^2$ m s^{-1}	1	3	5	7	10
$f(v)/10^{-4}$ s m^{-1}	3.47	18.64	18.42	7.67	0.58

18.3 0.0877.

18.4 928, 982 m s^{-1}.

18.5 393, 444, 481 m s^{-1}.

18.6 $\frac{3}{2}RT$, $\frac{3}{2}R$.

18.7 352 m s^{-1} = 787 miles per hour.

18.8 (a) 7.21×10^9 s^{-1}; (b) 8.75×10^{28} cm^{-3} s^{-1}; (c) 0.354; (d) 4.

18.9 (a) 1.52×10^{-7}, 0.152 m; (b) 3.77×10^{-8}, 0.037 m.

18.10 (a) 0.724 cm^{-3}; (b) 5.92×10^{-10} s^{-1}; (c) 4.83×10^9 miles.

18.11 9.52×10^9 s^{-1}, 1.921×10^8 mol L^{-1} s^{-1}.

18.12 (a) 4.63×10^9 s^{-1}; (b) 9.258×10^9 s^{-1}; (c) 1.244×10^8 mol L^{-1} s^{-1}.

18.13 1.58×10^{-10} s, 7590.

18.14 (a) 1.075×10^{28}, 5.866×10^{27} m^{-2} s^{-1}; (b) 2.698×10^{27}, 1.473×10^{27} m^{-2} s^{-1}.

18.15 21.3 Pa.

18.16 326 kJ mol^{-1}.

18.17 347 s.

18.18 0.0711 g.

18.19 (a) 0.217 nm; (b) 1.28×10^{-4} m^2 s^{-1}.

18.20 1.26×10^{-5} m^2 s^{-1}.

Chapter 19 Experimental Kinetics and Gas Reactions

19.1 -1.5×10^{-4}, 3.0×10^{-4}, 0.75×10^{-4} mol L^{-1} s^{-1}.

19.2 (a) 0.75×10^{-2} mol L^{-1} s^{-1}; (b) 1.5×10^{-2} mol L^{-1} s^{-1}; (c) 2.25×10^{-2} mol L^{-1} s^{-1}.

19.3 2310 s.

19.4 First order; 6.27×10^{-5} s^{-1}.

19.5 1.32×10^{-4} s^{-1}.

19.7 (a) First; (b) 0.128 h^{-1}; (c) 0.325.

19.8 $k = 0.251$ min^{-1}; $t_{1/2} = 2.76$ min; $\tau = 3.98$ min.

19.9 15.7 s^{-1}.

19.10 (a) $t_{1/2} = 1/2k[HI]_0$; (b) 4.01×10^{-2} mol L^{-1} s^{-1}; (c) 6.17×10^{-4} bar^{-1} s^{-1}; (d) 6.66×10^{-23} cm^3 s^{-1}.

19.11 Second order; 0.675 L mol^{-1} min^{-1}.

19.12 (a) Second order; (b) 0.59 L mol^{-1} min^{-1}.

19.13 $t_{1/2} = (2^{n-1} - 1)/(n - 1)k[A]_0^{n-1}$.

19.14

t/h	P_A/bar	P_B/bar	P/bar
1	0.50	0.25	0.75
2	0.33	0.33	0.66
∞	0	0.50	0.50

19.15 1.70×10^{-3} s.

19.16 (a) 6.25; (b) 14.3; (c) 0%.

19.17 $a = 0$, $b = [R]_0 k$, $c = [R]_0 k^2/2$.

19.18 1.000 ± 0.005 mol L^{-1} h^{-1}.

19.19 $d[OI^-]/dt = (60\ \text{s}^{-1})[I^-][OCl]/[OH^-]$.

19.20 2.50×10^{-3} mol L^{-1}.

19.21

t/d	10	20	40	80
F	3.65×10^{-12}	6.39×10^{-12}	9.98×10^{-12}	13.14×10^{-12}

19.22 $-d[Br^-]/dt = k_b[Br^-][SO_4^{2-}]^3[H^+][SO_3^{2-}]^{-2}$.

19.23 $k_1 k_4 = k_2 k_3$.

19.24 (b) 90.3 kJ mol^{-1}; (c) 1.99×10^{12} s^{-1}.

19.25 (a) $76\ ^\circ$C; (b) $-3\ ^\circ$C.

19.26 1420 s.

19.27 2.76×10^{-39} cm^6 s^{-1}.

19.29 (a) $d[D]/dt = k_1 k_3[A][B]/(k_2 + k_3)$; (b) $E_{app} = E_{a1} + E_{a3} - E_{a2}$.

19.31 $d[D]/dt = k_1 k_3[A][B]/(k_2 + k_3[C])$.

19.32 (b) 3.4×10^3 cm^3 mol^{-1} s^{-1}.

19.34 4.5×10^8 mol L^{-1} s^{-1}.

19.35 $k' = (1.1 \times 10^{13}$ mol L^{-1} s$^{-1})$ exp$(-104\,000/8.314T)$.

19.37 $d[CH_3CHO]/dt = -k_2(k_1/2k_4)^{1/2}[CH_3CHO]^{3/2}$.

Chapter 20 Chemical Dynamics and Photochemistry

20.1 (a) $P(\epsilon > \epsilon_a) = \exp(-\epsilon_a/kT)$; (b) 3.12×10^{-4}; (c) 0.0902.

20.2 $E_a = N_A \epsilon_0 + RT/2$.

20.3 2.40×10^{-20} m^2, 87.4 pm.

20.4 (a) 8.00×10^{11} mol L^{-1} s^{-1}; (b) 8.1×10^{-11} s; (c) 6.2×10^{-5} s.

20.6 $A = (e^2 RT/c^\circ N_A h)\,e^{\Delta S^\ddagger/R} = 7.70 \times 10^{13}$ L mol^{-1} s^{-1}. The transition state has a lower entropy than the reactants.

20.7 (a) 124.3 kJ mol^{-1}; (b) -32.6 J K^{-1} mol^{-1}; (c) The transition state has a more organized structure than the reactant.

20.8 -108 J K^{-1} mol^{-1}. There is a decrease in entropy when two molecules of gas come together to make a transition complex.

20.9 (a) 43 J K^{-1} mol^{-1}; (b) -79 J K^{-1} mol^{-1}; (c) There is a loosening of the structure in forming the transition state in the forward reaction, and a loss of degrees of freedom in forming the transition state in the backward di-

rection; (d) 122 J K^{-1} mol^{-1} for a standard state of 1 mol L^{-1}.

20.10 (a) 3.14×10^{14} s^{-1}; (b) $10\,500$ cm^{-1}; (c) 925 nm; (d) 1.31 eV.

20.11 1.7×10^{-3} mol.

20.12 0.167.

20.13 1.3×10^6.

20.15 $v_{CH_3} = 5595$ m s^{-1}, $v_I = 662$ m s^{-1}.

20.16 6.4×10^{-8} s.

20.17 10^{-7} mol L^{-1}.

20.18 (a) 10.1; (b) A chain reaction is involved.

20.19 1.25×10^{-2} J s^{-1}.

20.20 83 kW L^{-1}.

20.21 $d[CCl_4]/dt = k_2 I_a^{1/2}[Cl_2]^{1/2}/k_3^{1/2} + 2I_a$.

20.22 1.36×10^{17} molecules cm^{-2}.

Chapter 21   Kinetics in the Liquid Phase

21.1 3.46 Pa s.

21.2 3.01×10^{-5} m s^{-1}.

21.3 2.36 kJ mol^{-1}, 1.41×10^{-3} Pa s.

21.4 (*a*) 5.23×10^{-10} m^2 s^{-1}; (*b*) 6.14×10^{-3} m.

21.5 (*a*) 1.75 h; (*b*) 56.5 h.

21.6 2.00×10^{-5} cm s^{-1}, 0.54 cm.

21.8 0.0426 Ω^{-1} m^{-1}.

21.9 (*a*) 2.16×10^{-3} Ω^{-1} m^{-1}; (*b*) 1.09×10^{-3} A.

21.10 9.26×10^4, 9.26×10^3, 9.26×10^2 Ω.

21.11 7.4×10^9 J K^{-1} mol^{-1}.

21.12 0.89 nm.

21.13 4.42×10^{10} mol L^{-1} s^{-1}, 7.65×10^5 s^{-1}.

21.14 See equation 21.45.

21.15 $1/\tau = k_1 + k_1' K_A + k_2 + k_2' K_B$.

21.16 7.9×10^5, 1.8×10^3, 25 s^{-1}.

21.17 d[Cl$^-$]/d$t = kK_1$[I$^-$][OCl]/[OH].

21.18 1.35×10^{-4} m^{-1}, 5.5×10^{-5} L mol^{-1} min^{-1}.

21.19 0.748, 3.20.

21.20 6.7 mol h^{-1}.

21.21 (*a*) $V_F = 6.5 \times 10^{-7}$ mol L^{-1} s^{-1},
$K_F = 3.9 \times 10^{-6}$ mol L^{-1}
$V_M = 4.0 \times 10^{-7}$ mol L^{-1} s^{-1},
$K_M = 1.03 \times 10^{-5}$ mol L^{-1}
$k_2 = 3.3 \times 10^2$ s^{-1}, $k_{-1} = 2.0 \times 10^2$ s^{-1}
(*b*) $k_1 = 1.4 \times 10^8$ mol L^{-1} s^{-1},
$k_{-2} = 5.1 \times 10^7$ mol L^{-1} s^{-1}
(*c*) 4.4.

21.22 See equation 21.81.

21.23 15.1, 0.38×10^{-3} mol L^{-1}, 8.6×10^{-6} mol L^{-1}.

21.24 When v/[E]$_0$[S] is plotted versus v/[E]$_0$, the intercept on the ordinate is k_{cat}/K_M, the slope is $-1/K_M$, and the intercept on the abscissa is k_{cat}.

21.25 234, 4.41×10^{-8} mol L^{-1}, 3.67×10^{-9} mol L^{-1}.

21.26 (*a*) 1.22; (*b*) 1.12.

21.27 0.0252 nm.

Chapter 22   Macromolecules

22.1 100 000 g mol^{-1}, 103 000 g mol^{-1}, 1.03.

22.2 0.9938.

22.3 $(\overline{r^2})^{1/2} = 31.6$.

22.4 12.7 nm, 17.8 nm.

22.5 0.13 kg^2 mol^{-2} min^{-1}, 0.36 kg^2 mol^{-2} min^{-1}, 63 kJ mol^{-1}.

22.6 258 min.

22.7 (*a*) 100; (*b*) 3.70×10^{-3}, (*c*) 3.70×10^{-3}.

22.8 (*a*) 0.9943; (*b*) 175; (*c*) 39 900 g mol^{-1}.

22.9 500 000 g mol^{-1}.

22.10 230 000 g mol^{-1}.

22.11 2.30×10^{-3} g cm^{-3}.

22.12 15 500 g mol^{-1}.

22.13 67 600 g mol^{-1}, 4.

22.14 0.236×10^{-13} s.

Chapter 23   Solid-State Chemistry

23.1 $d = ab/(a^2 + b^2)^{1/2}$.

23.2 5.74°, 11.54°, 17.46°.

23.3 $F_{hkl} = f_A + f_B(-1)^{h+k+l}$. If $h + k + l$ is even, the reflection will be strong.

23.4 3.62 g cm^{-3}.

23.5 21.32×10^3 kg m^{-3}.

23.6 (*a*) 316 pm; (*b*) $d_{200} = 158$, $d_{110} = 223$, $d_{222} = 21.2$ pm.

23.7 (*a*) Face-centered; (*b*) 361.6 pm; (*c*) 8.93×10^3 kg m^{-3}.

23.8 (*a*) Body-centered; (*b*) face-centered; (*c*) 286.8 pm; (*d*) 57.4 pm; (*e*) 7.84×10^3 kg m^{-3}.

23.9 (a) 3.995×10^3 kg m^{-3}; (b) 176 pm.

23.10 $6.022\ 093 \times 10^{23}$ mol^{-1}.

23.11 39.7 kg mol^{-1}.

23.12 273 pm.

23.13 2.33×10^3 kg m^{-3}.

23.14 12.

23.15 9.95°, 11.50°, 16.38°, 19.31°, 20.21°, 23.51°.

23.16 0.414.

23.17 282.8 pm.

23.18 4.517×10^3 kg m^{-3}.

23.19 0.0818 eV.

23.20 Interstitial, 0.51%.

23.21 5.61×10^{-14} m V^{-1} s^{-1}.

23.22 1.56×10^{-17}, 9.08×10^{-6}.

Chapter 24 Surface Dynamics

24.1 (a) 5.08×10^{-1} cm^{-2} s^{-1}; (b) 7.17×10^{10} cm^{-2} s^{-1}.

24.2 18.5 s.

24.3 3.26×10^{19} m^{-2} s^{-1}.

24.4 -11.6 kJ mol^{-1}.

24.5 $v = v_m(y_A K_A + y_B K_B)P/[1 + (y_A K_A + y_B K_B)P]$; $x_A = y_A K_A/(y_A K_A + y_B K_B)$

24.6 $v_m = 40$ cm^3 g^{-1}, $K = 4.8 \times 10^{-5}$ Pa^{-1}.

24.7 2.08×10^8, 2.94×10^8, 1.80×10^8 s^{-1}.

24.8 20.6 L.

24.9 449 m^2.

Index

RELATIVE ATOMIC MASSES AND ISOTOPIC ABUNDANCES[a,b]

Z	Symbol		Relative Atomic Mass, m_a/u	Isotopic Abundance, x/%
		A		
1	H	1	1.007 825 037(10)	99.985(1)
		2	2.014 101 787(21)	0.015(1)
		3*	3.016 049 286(32)	
2	He	3	3.016 029 297(33)	0.000 138(3)
		4	4.002 603 25(5)	99.999 862(3)
3	Li	6	6.015 123 2(8)	7.5(2)
		7	7.016 004 5(9)	92.5(2)
4	Be	9	9.012 182 5(4)	100
5	B	10	10.012 938 0(5)	19.9(2)
		11	11.009 305 3(5)	80.1(2)
6	C	12	12 (by definition)	98.90(3)
		13	13.003 354 839(17)	1.10(3)
		14*	14.003 241 993(24)	
7	N	14	14.003 074 008(23)	99.634(9)
		15	15.000 108 978(38)	0.366(9)
8	O	16	15.994 914 64(5)	99.762(15)
		17	16.999 130 6(8)	0.038(3)
		18	17.999 159 39(32)	0.200(12)
9	F	19	18.998 403 25(14)	100
10	Ne	20	19.992 439 1(5)	90.51(9)
		21	20.993 845 3(12)	0.27(2)
		22	21.991 383 7(6)	9.22(9)
11	Na	23	22.989 769 7(9)	100
12	Mg	24	23.985 045 0(8)	78.99(3)
		25	24.985 839 2(12)	10.00(1)
		26	25.982 595 4(10)	11.01(2)
13	Al	27	26.981 541 3(7)	100
14	Si	28	27.976 928 4(7)	92.23(1)
		29	28.976 496 4(9)	4.67(1)
14	Si	30	29.973 771 7(10)	3.10(1)
15	P	31	30.975 363 8(11)	100
16	S	32	31.972 071 8(6)	95.02(9)
		33	32.971 459 1(8)	0.75(1)
		34	33.967 867 74(29)	4.21(8)
		36	35.967 079 0(16)	0.02(1)
17	Cl	35	34.968 852 729(68)	75.77(5)
		37	36.965 902 62(11)	24.23(5)
35	Br	79	78.918 336 1(38)	50.69(5)
		81	80.916 290(6)	49.31(5)
53	I	127	126.904 477(5)	100

[a] IUPAC Commission on Atomic Weights, *Pure Appl. Chem.* **56**:653 (1984).

[b] An asterisk denotes an unstable nuclide. The standard error in parentheses is applicable to the last digits quoted.